KU-799-673

Modern Food Microbiology

Seventh Edition

FOOD SCIENCE TEXT SERIES

The Food Science Text Series provides faculty with the leading teaching tools. The Editorial Board has outlined the most appropriate and complete content for each food science course in a typical food science program, and has identified textbooks of the highest quality, written by leading food science educators.

EDITORIAL BOARD

Series Editor

Dennis R. Heldman, Ph.D., Heldman Associates, San Marcos, California.

Board Members

Richard W. Hartel, Professor of Food Engineering, Department of Food Science, University of Wisconsin.

Hildegarde Heymann, Professor of Sensory Science, Department of Food Science & Technology, University of California—Davis.

Joseph H. Hotchkiss, Professor, Institute of Food Science and Institute for Comparative and Environmental Toxicology, and Chair, Food Science Department, Cornell University.

James M. Jay, Professor Emeritus, Department of Biological Sciences, Wayne State University and Adjunct Professor, University of Nevada Las Vegas.

Kenneth Lee, Professor and Chair, Department of Food Science and Technology, The Ohio State University.

Norman G. Marriott, Emeritus Professor & Extension Food Scientist, Department of Food Science and Technology, Virginia Polytechnic and State University.

Joseph Montecalvo, Jr., Professor, Department of Food Science and Nutrition, California Polytechnic and State University—San Luis Obispo.

S. Suzanne Nielsen, Professor and Chair, Department of Food Science, Purdue University.

J. Antonio Torres, Associate Professor of Food Science, Department of Food Science and Technology, Oregon State University.

Titles

Elementary Food Science, Fourth Edition, Ernest R. Vieira (1996)
Essentials of Food Sanitation, Norman G. Marriott (1997)
Essentials of Food Science, Second Edition, Vickie A. Vaclavik and Elizabeth W. Christian (2003)
Food Analysis, Third Edition, S. Suzanne Nielsen (2003)
Food Analysis Laboratory Manual, S. Suzanne Nielsen (2003)
Food Science, Fifth Edition, Norman N. Potter and Joseph H. Hotchkiss (1995)
Fundamentals of Food Process Engineering, Second Edition, Romeo T. Toledo (1990)
Introduction to Food Process Engineering, P. G. Smith (2003)
Modern Food Microbiology, Seventh Edition, James M. Jay, Martin J. Loessner, and David A. Golden (2005)
Principles of Food Chemistry, Third Edition, John M. deMan (1999)
Principles of Food Processing, Dennis R. Heldman and Richard W. Hartel (1997)
Principles of Food Sanitation, Fourth Edition, Norman G. Marriott (1999)
Sensory Evaluation of Food: Principles and Practices, Harry T. Lawless and Hildegarde Heymann (1998)

Modern Food Microbiology
Seventh Edition

James M. Jay
University of Nevada Las Vegas
Las Vegas, Nevada

Martin J. Loessner
Institute of Food Science and Nutrition
Zürich, Switzerland

David A. Golden
University of Tennessee
Knoxville, Tennessee

 Springer

ISBN 0-387-23180-3
ISBN 0-387-23413-6 (e-Book)

©2005 Springer Science+Business Media, Inc.

All rights reserved. This work may not be translated or copied in whole or in part without the written permission of the publisher (Springer Science+Business Media, Inc., 233 Spring Street, New York, NY 10013, USA), except for brief excerpts in connection with reviews or scholarly analysis. Use in connection with any form of information storage and retrieval, electronic adaptation, computer software, or by similar or dissimilar methodology now known or hereafter developed is forbidden. The use in this publication of trade names, trademarks, service marks and similar terms, even if they are not identified as such, is not to be taken as an expression of opinion as to whether or not they are subject to proprietary rights.

Printed in the United States of America.   (BS/DH)

9 8 7 6 5 4 3 2 1

springeronline.com

Preface

The 7th edition of *Modern Food Microbiology*, like previous editions, focuses on the general biology of the microorganisms that are found in foods. All but one of the 31 chapters have been extensively revised and updated. The new material in this edition includes over 80 new bacterial and 10 new genera of fungi. This title is suitable for use in a second or subsequent course in a microbiology curriculum, or as a primary food microbiology course in a food science or food technology curriculum. Although organic chemistry is a desirable prerequisite, it is not necessary for one to get a good grasp of most of the topics covered.

When used as a microbiology text, the following sequence may be used. A synopsis of the information in Chapter 1 will provide students with a sense of the historical developments that have shaped this discipline and how it continues to evolve. Memorization of the many dates and events is not recommended since much of this information is presented again in the respective chapters. The material in Chapter 2 includes a synopsis of modern methods currently used to classify bacteria, taxonomic schemes for yeasts and molds, and brief information on the genera of bacteria and fungi encountered in foods. This material may be combined with the intrinsic and extrinsic parameters of growth in Chapter 3 as they exist in food products and as they affect the common foodborne organisms. Chapters 4 to 9 deal with specific food products, and they may be covered to the extent desired with appropriate reviews of the relevant topics in Chapter 3. Chapters 10 to 12 cover methods for culturing and identifying foodborne organisms and/or their products, and these topics may be dealt with in this sequence or just before foodborne pathogens. The food protection methods in Chapters 13 to 19 include some information that goes beyond the usual scope of a second course, but the principles that underlie each of these methods should be covered.

Chapters 20 and 21 deal with food sanitation, indicator organisms, HACCP, and FSO systems; and coverage of these topics is suggested before dealing with the pathogens. Chapters 22 to 31 deal with the known (and suspected) foodborne pathogens including their biology and methods of control. Chapter 22 is intended to provide an overview of the chapters that follow. Some of it includes ways in which foodborne pathogens differ from nonpathogens, their behavior in biofilms, and some information on the known roles of sigma factors and quorum sensing among foodborne organisms. The other material in this chapter that deals with the mechanisms of pathogenesis is probably best dealt with when the specific pathogens are covered in their respective chapters. The new Appendix section presents a simplified scheme for grouping foodborne and some general environmental bacterial genera by use of Gram, oxidase, and calalase reactions along with colony pigmentation.

For most semester courses with a 3-credit lecture and accompanying 2 or 3 credit laboratory, only about 65-70% of the material in this text is likely to be covered. The remainder is meant for reference puiposes.

The following individuals assisted us by critiquing various parts or sections of this edition, and we extend special thanks to each: B. P. Hedlund, K. E. Kesterson, J. Q. Shen, and H. H. Wang. Those who assisted with the previous six editions are acknowledged in the respective editions.

Contents

PART I

Historical Background

The material in this part provides a glimpse of some of the early events that ultimately led to the recognition of the significance and role of microorganisms in foods. Food microbiology as a defined subdiscipline does not have a precise beginning. Some of the early findings and observations are noted, along with dates. The selective lists of events noted for food preservation, food spoilage, food poisoning, and food legislation, are meant to be guideposts in the continuing evolution and development of food microbiology.

An excellent and more detailed review of the history of food microbiology has been presented by Hartman.

Hartman, P.A. 2001. The evolution of food microbiology. In *Food Microbiology—Fundamentals and Frontiers*, 2nd ed., eds. M.P. Doyle, L.R. Beuchat, and T.J. Montville, 3–12. Washington, D.C.: ASM Press.

CHAPTER 1

History of Microorganisms in Food

Although it is extremely difficult to pinpoint the precise beginning of human awareness of the presence and role of microorganisms in foods, the available evidence indicates that this knowledge preceded the establishment of bacteriology or microbiology as a science. The era prior to the establishment of bacteriology as a science may be designated the prescientific era. This era may be further divided into what has been called *the food-gathering period* and the *food-producing period*. The former covers the time from human origin over 1 million years ago up to 8,000 years ago. During this period, humans were presumably carnivorous, with plant foods coming into their diet later in this period. It is also during this period that foods were first cooked.

The food-producing period dates from about 8,000 to 10,000 years ago and, of course, includes the present time. It is presumed that the problems of spoilage and food poisoning were encountered early in this period. With the advent of prepared foods, the problems of disease transmission by foods and of faster spoilage caused by improper storage made their appearance. Spoilage of prepared foods apparently dates from around 6000 BC. The practice of making pottery was brought to Western Europe about 5000 BC from the Near East. The first boiler pots are thought to have originated in the Near East about 8,000 years ago.[11] The arts of cereal cookery, brewing, and food storage, were either started at about this time or stimulated by this new development.[10] The first evidence of beer manufacture has been traced to ancient Babylonia as far back as 7000 BC.[8] The Sumerians of about 3000 BC are believed to have been the first great livestock breeders and dairymen and were among the first to make butter. Salted meats, fish, fat, dried skins, wheat, and barley are also known to have been associated with this culture. Milk, butter, and cheese were used by the Egyptians as early as 3000 BC. Between 3000 BC and 1200 BC, the Jews used salt from the Dead Sea in the preservation of various foods.[2] The Chinese and Greeks used salted fish in their diet, and the Greeks are credited with passing this practice on to the Romans, whose diet included pickled meats. Mummification and preservation of foods were related technologies that seem to have influenced each other's development. Wines are known to have been prepared by the Assyrians by 3500 BC. Fermented sausages were prepared and consumed by the ancient Babylonians and the people of ancient China as far back as 1500 BC.[8]

Another method of food preservation that apparently arose during this time was the use of oils such as olive and sesame. Jensen[6] has pointed out that the use of oils leads to high incidences of staphylococcal food poisoning. The Romans excelled in the preservation of meats other than beef by around 1000 BC and are known to have used snow to pack prawns and other perishables, according to Seneca. The practice of smoking meats as a form of preservation is presumed to have emerged sometime during this period, as did the making of cheese and wines. It is doubtful whether people at this time understood the nature of these newly found preservation techniques. It is also doubtful whether the role of foods in the transmission of disease or the danger of eating meat from infected animals was recognized.

Few advances were apparently made toward understanding the nature of food poisoning and food spoilage between the time of the birth of Christ and AD 1100. *Ergot* poisoning (caused by *Claviceps purpurea*, a fungus that grows on rye and other grains) caused many deaths during the Middle Ages. Over 40,000 deaths due to ergot poisoning were recorded in France alone in AD 943, but it was not known that the toxin of this disease was produced by a fungus.[12] Meat butchers are mentioned for the first time in 1156, and by 1248 the Swiss were concerned with marketable and nonmarketable meats. In 1276, a compulsory slaughter and inspection order was issued for public abattoirs in Augsburg. Although people were aware of quality attributes in meats by the thirteenth century, it is doubtful whether there was any knowledge of the causal relationship between meat quality and microorganisms.

Perhaps the first person to suggest the role of microorganisms in spoiling foods was A. Kircher, a monk, who as early as 1658 examined decaying bodies, meat, milk, and other substances and saw what he referred to as "worms" invisible to the naked eye. Kircher's descriptions lacked precision, however, and his observations did not receive wide acceptance. In 1765, L. Spallanzani showed that beef broth that had been boiled for an hour and sealed remained sterile and did not spoil. Spallanzani performed this experiment to disprove the doctrine of the spontaneous generation of life. However, he did not convince the proponents of the theory because they believed that his treatment excluded oxygen, which they felt was vital to spontaneous generation. In 1837, Schwann showed that heated infusions remained sterile in the presence of air, which he supplied by passing it through heated coils into the infusion.[9] Although both of these men demonstrated the idea of the heat preservation of foods, neither took advantage of his findings with respect to application. The same may be said of D. Papin and G. Leibniz, who hinted at the heat preservation of foods at the turn of the eighteenth century.

The history of thermal canning necessitates a brief biography of Nicolas Appert (1749–1841). This Frenchman worked in his father's wine cellar early on, and he and two brothers established a brewery in 1778. In 1784, he opened a confectioner's store in Paris that was later transformed into a wholesale business. His discovery of a food preservation process occurred between 1789 and 1793. He established a cannery in 1802 and exported his products to other countries. The French navy began testing his preservation method in 1802, and in 1809 a French ministry official encouraged him to promote his invention. In 1810, he published his method and was awarded the sum of 12,000 francs.[7]

This, of course, was the beginning of canning as it is known and practiced today.[5] This event occurred some 50 years before L. Pasteur demonstrated the role of microorganisms in the spoilage of French wines, a development that gave rise to the rediscovery of bacteria. A. Leeuwenhoek in the Netherlands had examined bacteria through a microscope and described them in 1683, but it is unlikely that Appert was aware of this development and Leeuwenhoek's report was not available in French.

The first person to appreciate and understand the presence and role of microorganisms in food was Pasteur. In 1837, he showed that the souring of milk was caused by microorganisms, and in about 1860 he used heat for the first time to destroy undesirable organisms in wine and beer. This process is now known as pasteurization.

HISTORICAL DEVELOPMENTS

Some of the more significant dates and events in the history of food preservation, food spoilage, food poisoning, and food legislation are listed below. The latter pertains primarily to the United States.

Food Preservation

1782—Canning of vinegar was introduced by a Swedish chemist.

1810—Preservation of food by canning was patented by Appert in France.

—Peter Durand was issued a British patent to preserve food in "glass, pottery, tin, or other metals, or fit materials." The patent was later acquired by Hall, Gamble, and Donkin, possibly from Appert.[1,4]

1813—Donkin, Hall, and Gamble introduced the practice of postprocessing incubation of canned foods.

—Use of SO_2 as a meat preservative is thought to have originated around this time.

1825—T. Kensett and E. Daggett were granted a U.S. patent for preserving food in tin cans.

1835—A patent was granted to Newton in England for making condensed milk.

1837—Winslow was the first to can corn from the cob.

1839—Tin cans came into wide use in the United States.[3]

—L.A. Fastier was given a French patent for the use of brine bath to raise the boiling temperature of water.

1840—Fish and fruit were first canned.

1841—S. Goldner and J. Wertheimer were issued British patents for brine baths based on Fastier's method.

1842—A patent was issued to H. Benjamin in England for freezing foods by immersion in an ice and salt brine.

1843—Sterilization by steam was first attempted by I. Winslow in Maine.

1845—S. Elliott introduced canning to Australia.

1853—R. Chevallier-Appert obtained a patent for sterilization of food by autoclaving.

1854—Pasteur began wine investigations. Heating to remove undesirable organisms was introduced commercially in 1867–1868.

1855—Grimwade in England was the first to produce powdered milk.

1856—A patent for the manufacture of unsweetened condensed milk was granted to Gail Borden in the United States.

1861—I. Solomon introduced the use of brine baths to the United States.

1865—The artificial freezing of fish on a commercial scale was begun in the United States. Eggs followed in 1889.

1874—The first extensive use of ice in transporting meat at sea was begun.

—Steam pressure cookers or retorts were introduced.

1878—The first successful cargo of frozen meat went from Australia to England. The first from New Zealand to England was sent in 1882.

1880—The pasteurization of milk was begun in Germany.

1882—Krukowitsch was the first to note the destructive effects of ozone on spoilage bacteria.

1886—A mechanical process of drying fruits and vegetables was carried out by an American, A.F. Spawn.

1890—The commercial pasteurization of milk was begun in the United States.

—Mechanical refrigeration for fruit storage was begun in Chicago.

1893—The Certified Milk movement was begun by H.L. Coit in New Jersey.

1895—The first bacteriological study of canning was made by Russell.

1907—E. Metchnikoff and co-workers isolated and named one of the yogurt bacteria, *Lactobacillus delbrueckii* subsp. *bulgaricus*.

—The role of acetic acid bacteria in cider production was noted by B.T.P. Barker.

1908—Sodium benzoate was given official sanction by the United States as a preservative in certain foods.

1916—The quick freezing of foods was achieved in Germany by R. Plank, E. Ehrenbaum, and K. Reuter.

1917—Clarence Birdseye in the United States began work on the freezing of foods for the retail trade.

—Franks was issued a patent for preserving fruits and vegetables under CO_2.

1920—Bigelow and Esty published the first systematic study of spore heat resistance above 212°F. The "general method" for calculating thermal processes was published by Bigelow, Bohart, Richardson, and Ball; the method was simplified by C.O. Ball in 1923.

1922—Esty and Meyer established $z = 18$°F for *Clostridium botulinum* spores in phosphate buffer.

1928—The first commercial use of controlled-atmosphere storage of apples was made in Europe (first used in New York in 1940).

1929—A patent issued in France proposed the use of high-energy radiation for the processing of foods.

—Birdseye frozen foods were placed in retail markets.

1943—B.E. Proctor in the United States was the first to employ the use of ionizing radiation to preserve hamburger meat.

1950—The D value concept came into general use.

1954—The antibiotic nisin was patented in England for use in certain processed cheeses to control clostridial defects,

1955—Sorbic acid was approved for use as a food preservative.

—The antibiotic chlortetracycline was approved for use in fresh poultry (oxytetracycline followed a year later). Approval was rescinded in 1966.

1967—The first commercial facility designed to irradiate foods was planned and designed in the United States. The second became operational in 1992 in Florida.

1988—Nisin was accorded GRAS (generally regarded as safe) status in the United States.

1990—Irradiation of poultry was approved in the United States.

1997—The irradiation of fresh beef up to a maximum level of 4.5 kGy and frozen beef up to 7.0 kGy was approved in the United States.

1997—Ozone was declared GRAS by the U.S. Food and Drug Administration for food use.

Food Spoilage

1659—Kircher demonstrated the occurrence of bacteria in milk; Bondeau did the same in 1847.

1680—Leeuwenhoek was the first to observe yeast cells.

1780—Scheele identified lactic acid as the principal acid in sour milk.

1836—Latour discovered the existence of yeasts.

1839—Kircher examined slimy beet juice and found organisms that formed slime when grown in sucrose solutions.

1857—Pasteur showed that the souring of milk was caused by the growth of organisms in it.

1866—L. Pasteur's *Étude sur le Vin* was published.

1867—Martin advanced the theory that cheese ripening was similar to alcoholic, lactic, and butyric, fermentations.

1873—The first reported study on the microbial deterioration of eggs was carried out by Gayon.

—Lister was first to isolate *Lactococcus lactis* in pure culture.

1876—Tyndall observed that bacteria in decomposing substances were always traceable to air, substances, or containers.

1878—Cienkowski reported the first microbiological study of sugar slimes and isolated *Leuconostoc mesenteroides* from them.

1887—Forster was the first to demonstrate the ability of pure cultures of bacteria to grow at 0°C.

1888—Miquel was the first to study *thermophilic* bacteria.

1895—The first records on the determination of numbers of bacteria in milk were those of Von Geuns in Amsterdam.

—S.C. Prescott and W. Underwood traced the spoilage of canned corn to improper heat processing for the first time.

1902—The term *psychrophile* was first used by Schmidt-Nielsen for microorganisms that grow at 0°C.

1912—The term *osmophile* was coined by Richter to describe yeasts that grow well in an environment of high osmotic pressure.

1915—*Bacillus coagulans* was first isolated from coagulated milk by B.W. Hammer.

1917—*Geobacillus stearothermophilus* was first isolated from cream-style corn by P.J. Donk.

1933—Oliver and Smith in England observed spoilage by *Byssochlamys fulva*; first described in the United States in 1964 by D. Maunder.

Food Poisoning

1820—The German poet Justinus Kerner described "sausage poisoning" (which in all probability was botulism) and its high fatality rate.

1857—Milk was incriminated as a transmitter of typhoid fever by W. Taylor of Penrith, England.

1870—Francesco Selmi advanced his theory of ptomaine poisoning to explain illness contracted by eating certain foods.

1888—Gaertner first isolated *Salmonella enteritidis* from meat that had caused 57 cases of food poisoning.

1894—T. Denys was the first to associate staphylococci with food poisoning.

1896—Van Ermengem first discovered *Clostridium botulinum*.

1904—Type A strain of *C. botulinum* was identified by G. Landman.

1906—*Bacillus cereus* food poisoning was recognized. The first case of diphyllobothriasis was recognized.

1926—The first report of food poisoning by streptococci was made by Linden, Turner, and Thom.

1937—Type E strain of *C. botulinum* was identified by L. Bier and E. Hazen.

1937—Paralytic shellfish poisoning was recognized.

1938—Outbreaks of *Campylobacter* enteritis were traced to milk in Illinois.

1939—Gastroenteritis caused by *Yersinia enterocolitica* was first recognized by Schleifstein and Coleman.

1945—McClung was the first to prove the etiologic status of *Clostridium perfringens (welchii)* in food poisoning.

1951—*Vibrio parahaemolyticus* was shown to be an agent of food poisoning by T. Fujino of Japan.

1955—Similarities between cholera and *Escherichia coli* gastroenteritis in infants were noted by S. Thompson.

—Scombroid (histamine-associated) poisoning was recognized.

—The first documented case of anisakiasis occurred in the United States.

1960—Type F strain of *C. botulinum* identified by Moller and Scheibel.

—The production of aflatoxins by *Aspergillus flavus* was first reported.

1965—Foodborne giardiasis was recognized.

1969—*C. perfringens* enterotoxin was demonstrated by C.L. Duncan and D.H. Strong.

—*C. botulinum* type G was first isolated in Argentina by Gimenez and Ciccarelli.

1971—First U.S. foodborne outbreak of *Vibrio parahaemolyticus* gastroenteritis occurred in Maryland.

—First documented outbreak of *E. coli* foodborne gastroenteritis occurred in the United States.

1975—*Salmonella* enterotoxin was demonstrated by L.R. Koupal and R.H. Deibel.

1976—First U.S. foodborne outbreak of *Yersinia enterocolitica* gastroenteritis occurred in New York.

—Infant botulism was first recognized in California.

1977—The first documented outbreak of cyclosporiasis occurred in Papua, New Guinea; first in United States in 1990.

1978—Documented foodborne outbreak of gastroenteritis caused by the Norwalk virus occurred in Australia.

1979—Foodborne gastroenteritis caused by non-01 *Vibrio cholerae* occurred in Florida. Earlier outbreaks occurred in Czechoslovakia (1965) and Australia (1973).

1981—Foodborne listeriosis outbreak was recognized in the United States.

1982—The first outbreaks of foodborne hemorrhagic colitis occurred in the United States.

1983—*Campylobacter jejuni* enterotoxin was described by Ruiz-Palacios et al.

1985—The irradiation of pork to 0.3 to 1.0 kGy to control *Trichinella spiralis* was approved in the United States.

1986—Bovine spongiform encephalopathy (BSE) was first diagnosed in cattle in the United Kingdom.

Food Legislation

1890—The first national meat inspection law was enacted. It required the inspection of meats for export only.

1895—The previous meat inspection act was amended to strengthen its provisions.

1906—The U.S. Federal Food and Drug Act was passed by Congress.

1910—The New York City Board of Health issued an order requiring the pasteurization of milk.

1939—The new Food, Drug, and Cosmetic Act became law.

1954—The Miller Pesticide Chemicals Amendment to the Food, Drug, and Cosmetic Act was passed by Congress.

1957—The U.S. Compulsory Poultry and Poultry Products law was enacted.

1958—The Food Additives Amendment to the Food Drug, and Cosmetics Act was passed.

1962—The Talmadge-Aiken Act (allowing for federal meat inspection by states) was enacted into law.

1963—The U.S. Food and Drug Administration approved the use of irradiation for the preservation of bacon.

1967—The U.S. Wholesome Meat Act was passed by Congress and enacted into law on December 15.

1968—The Food and Drug Administration withdrew its 1963 approval of irradiated bacon.

—The Poultry Inspection Bill was signed into law.

1969—The U.S. Food and Drug Administration established an allowable level of 20 ppb of aflatoxin for edible grains and nuts.

1973—The state of Oregon adopted microbial standards for fresh and processed retail meat. They were repealed in 1977.

REFERENCES

1. Bishop, P.W. 1978. Who introduced the tin can? Nicolas Appert? Peter Durand? Bryan Donkin? *Food Technol.* 32:60–67.
2. Brandly, P.J., G. Migaki, and K.E. Taylor. 1966. *Meat Hygiene*, 3rd ed., chap. 1. Philadelphia: Lea & Febiger.
3. Cowell, N.D. 1995. Who introduced the tin can?—A new candidate. *Food Technol.* 49:61–64.
4. Farrer, K.T.H. 1979. Who invented the brine bath?—The Isaac Solomon myth. *Food Technol.* 33:75–77.
5. Goldblith, S.A. 1971. A condensed history of the science and technology of thermal processing. *Food Technol.* 25:44–50.
6. Jensen, L.B. 1953. *Man's Foods*, chaps. 1, 4, 12. Champaign, IL: Garrard Press.
7. Livingston, G.E., and J.P. Barbier. 1999. The life and work of Nicolas Appert, 1749–1841. Abstract # 7-1, p. 10, *Institute of Food Technol. Proceedings.*
8. Pederson, C.S. 1971. *Microbiology of Food Fermentations.* Westport, CT: AVI.
9. Schormüller, J. 1966. *Die Erhaltung der Lebensmittel.* Stuttgart: Ferdinand Enke Verlag.
10. Stewart, G.F., and M.A. Amerine. 1973. *Introduction to Food Science and Technology*, chap. 1. New York: Academic Press.
11. Tanner, F.W. 1944. *The Microbiology of Foods*, 2nd ed. Champaign, IL: Garrard Press.
12. Tanner, F.W., and L.P. Tanner. 1953. *Food-Borne Infections and Intoxications*, 2nd ed. Champaign, IL: Garrard Press.

PART II

Habitats, Taxonomy, and Growth Parameters

Many changes in the taxonomy of foodborne organisms have been made during the past two decades, and many are reflected in Chapter 2 along with the primary habitats of some organisms of concern in foods. The factors/parameters that affect the growth of microorganisms are treated in Chapter 3. See the following for more information:

Adams, M.R., and M.O. Moss. 2000. *Food Microbiology*, 2nd ed. New York: Springer-Verlag. A 494-page easy-to-read textbook.

Deak, T., and L.R. Beuchat. 1996. *Handbook of Food Spoilage Yeasts*. Boca Raton, FL: CRC Press. Detection, enumeration, and identification of foodborne yeasts.

Doyle, M.P., L.R. Beuchat, and T.J. Montville, eds. 2001. *Food Microbiology–Fundamentals and Frontiers*, 2nd ed. Washington, D.C.: ASM Press. Food spoilage as well as foodborne pathogens are covered in this 880-page work along with general growth parameters.

International Commission on Microbiological Specification of Foods (ICMSF). 1996. *Microorganisms in Foods*, 5th ed. New York: Kluwer Academic Publishers. All of the foodborne pathogens are covered in this 512-page work with details on growth parameters. Well referenced.

CHAPTER 2

Taxonomy, Role, and Significance of Microorganisms in Foods

Because human food sources are of plant and animal origin, it is important to understand the biological principles of the microbial biota associated with plants and animals in their natural habitats and respective roles. Although it sometimes appears that microorganisms are trying to ruin our food sources by infecting and destroying plants and animals, including humans, this is by no means their primary role in nature. In our present view of life on this planet, the primary function of microorganisms in nature is self-perpetuation. During this process, the heterotrophs and autotrophs carry out the following general reaction:

All organic matter
(carbohydrates, proteins, lipids, etc.)
↓
Energy + Inorganic compounds
(nitrates, sulfates, etc.)

This, of course, is essentially nothing more than the operation of the nitrogen cycle and the cycle of other elements. The microbial spoilage of foods may be viewed simply as an attempt by the food biota to carry out what appears to be their primary role in nature. This should not be taken in the teleological sense. In spite of their simplicity when compared to higher forms, microorganisms are capable of carrying out many complex chemical reactions essential to their perpetuation. To do this, they must obtain nutrients from organic matter, some of which constitutes our food supply.

If one considers the types of microorganisms associated with plant and animal foods in their natural states, one can then predict the general types of microorganisms to be expected on this particular food product at some later stage in its history. Results from many laboratories show that untreated foods may be expected to contain varying numbers of bacteria, molds, or yeasts, and the question often arises as to the safety of a given food product based on total microbial numbers. The question should be twofold: What is the total number of microorganisms present per gram or milliliter and what *types* of organisms are represented in this number? It is necessary to know which organisms are associated with a particular food in its natural state and which of the organisms present are not normal for that particular food. It is, therefore, of value to know the general distribution of bacteria in nature and the general types of organisms normally present under given conditions where foods are grown and handled.

BACTERIAL TAXONOMY

Many changes have taken place in the classification or taxonomy of bacteria in the past two decades. Many of the new taxa have been created as a result of the employment of molecular genetic methods, alone or in combination with some of the more traditional methods:

1. DNA homology and mol% G + C content of DNA
2. 23S, 16S, and 5S rRNA sequence similarities
3. Oligonucleotide cataloging
4. Numerical taxonomic analysis of total soluble proteins or of a battery of morphological and biochemical characteristics
5. Cell wall analysis
6. Serological profiles
7. Cellular fatty acid profiles

Although some of these have been employed for many years (e.g., cell wall analysis and serological profiles) others (e.g., ribosomal RNA [rRNA] sequence similarity) came into wide use only during the 1980s. The methods that are the most powerful as bacterial taxonomic tools are outlined and briefly discussed below.

rRNA Analyses

Taxonomic information can be obtained from RNA in the production of nucleotide catalogs and the determination of RNA sequence similarities. First, the prokaryotic ribosome is a 70S (Svedberg) unit, which is composed of two separate functional subunits: 50S and 30S. The 50S subunit is composed of 23S and 5S RNA in addition to about 34 proteins, whereas the 30S subunit is composed of 16S RNA plus about 21 proteins.

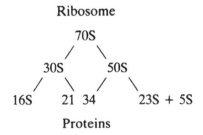

The 16S subunit is highly conserved and is considered to be an excellent chronometer of bacteria over time.[53] Using reverse transcriptase, 16S rRNA can be sequenced to produce long stretches (about 95% of the total sequence) to allow for the determination of precise phylogenetic relationships.[31] Alternatively, the 16S rDNA may be sequenced after amplification of specific regions by polymerase chain reaction (PCR)-based methods.

To sequence 16S rRNA, a single-stranded DNA copy is made by use of reverse transcriptase with the RNA as template. When the single-stranded DNA is made in the presence of dideoxynucleotides,

DNA fragments of various sizes result that can be sequenced by the Sanger method. From the DNA sequences, the template 16S rRNA sequence can be deduced. It was through studies of 16S rRNA sequences that led Woese and his associates to propose the establishment of three kingdoms of life forms: Eukaryotes, Archaebacteria, and Prokaryotes. The last include the cyanobacteria and the eubacteria, with the bacteria of importance in foods being eubacteria. Sequence similarities of 16S rRNA are widely employed, and some of the new foodborne taxa were created primarily by its use along with other information. It appears that the sequencing of 23S rDNA will become more widely used in bacterial taxonomy.

Nucleotide catalogs of 16S rRNA have been prepared for a number of organisms, and extensive libraries exist. By this method, 16S rRNA is subjected to digestion by RNase T1, which cleaves the molecule at G(uanine) residues. Sequences (-mers) of 6–20 bases are produced and separated, and similarities S_{AB} (Dice-type coefficient) between organisms can be compared. Although the relationship between S_{AB} and percentage similarity is not good below S_{AB} value of 0.40, the information derived is useful at the phylum level. The sequencing of 16S rRNA by reverse transcriptase is preferred to oligonucleotide cataloging, as longer stretches of rRNA can be sequenced.

Analysis of DNA

The mol% G + C of bacterial DNA has been employed in bacterial taxonomy for several decades, and its use in combination with 16S and 5S rRNA sequence data makes it even more meaningful. By 16S rRNA analysis, the Gram-positive eubacteria fall into two groups at the phylum level: one group with mol% G + C >55, and the other <50.[53] The former includes the genera *Streptomyces*, *Propionibacterium*, *Micrococcus*, *Bifidobacterium*, *Corynebacterium*, *Brevibacterium*, and others. The group with the lower G + C values include the genera *Clostridium*, *Bacillus*, *Staphylococcus*, *Lactobacillus*, *Pediococcus*, *Leuconostoc*, *Listeria*, *Erysipelothrix*, and others. The latter group is referred to as the *Clostridium* branch of the eubacterial tree. When two organisms differ in G + C content by more than 10%, they have few base sequences in common.

DNA–DNA or DNA–RNA hybridization has been employed for some time, and this technique continues to be of great value in bacterial systematics. It has been noted that the ideal reference system for bacterial taxonomy would be the complete DNA sequence of an organism.[49] It is generally accepted that bacterial species can be defined in phylogenetic terms by use of DNA–DNA hybridization results, where 70% or greater relatedness and 5°C or less T_m (melting point) defines a species.[50] When DNA–DNA hybridization is employed, phenotypic characteristics are not allowed to override except in exceptional cases.[50] Although a genus is more difficult to define phylogenetically, 20% sequence similarity is considered to be the minimum level of DNA–DNA homology.[50]

Even if there is not yet a satisfactory phylogenetic definition of a bacterial genus, the continued application of nucleic acid techniques, along with some of the other methods listed above, should lead ultimately to a phylogenetically based system of bacterial systematics. In the meantime, changes in the extant taxa may be expected to continue to occur.

The Proteobacteria

The Gram-negative bacteria of known importance in foods belong to the class *Proteobacteria*, which was established following extensive studies on the rRNA sequences of numerous genera of

Table 2–1 Subclasses of the *Proteobacteria* to Which Many Foodborne Genera Belong. *Campylobacter* and *Helicobacter* Belong to the δ-Subclass

Alpha	Beta	Gamma
Acetobacter	Acidovorax	Acinetobacter
Asaia	Alcaligenes	Aeromonas
Brevundimonas	Burkholderia	Alteromonas
Devosia	Chromobacterium	Azomonas
Gluconobacter	Comamonas	Bacteriodes
Paracoccus	Delftia	Carnimonas
Pseudoaminobacter	Hydrogenophaga	Enterobacteriaceae[a]
Sphingomonas	Janthinobacterium	Flavobacterium
Xanthobacter	Pandoraea	Halomonas
Zymomonas	Pseudomonas (plant pathogens)	Moraxella
	Ralstonia	Plesiomonas
	Telluria	Pseudoalteromonas
	Variovorax	Pseudomonas
	Vogesella	Psychrobacter
	Wautersia	Photobacterium
	Xylophilus	Shewanella
		Stenotrophomonas
		Vibrio
		Xanthomonas
		Xylella

[a]Include *Escherichia, Citrobacter, Salmonella, Shigella, Proteus, Raoultella, Proteus, Klebsiella, Edwardsiella*, etc.

Gram-negative bacteria.[43] The class is divided into five subclasses designated α, β, γ, etc. The subclasses are defined on the basis of their 16S rRNA sequences.[54–56] By extensive use of signature sequences (conserved inserts and deletions) of different proteins, an evolutionary relationship of the *Proteobacteria* has been proposed.[20] It has been suggested that the first eubacteria were low G + C Gram positives (e.g., *Clostridium, Bacillus, Lactobacillus*), followed by high G + C Gram positives (e.g., *Micrococcus, Propionibacterium, Rubrobacter*), and then by *Deinococcus-Thermus*. Next arose three groups that are not foodborne (not listed here), and then the *Proteobacteria* with ε and σ followed by α, β, and γ.[20] It has been stressed that these groups are related to each other in a linear rather than a tree-like manner.[20] It can be seen from Table 2–1 that most foodborne bacteria (especially foodborne pathogens) belong to the γ-subclass. The earliest prokaryotes are estimated to have arisen 3.5–3.8 billion years ago.[20]

Some of the important genera known to occur in foods are listed below in alphabetical order. Some are desirable in certain foods; others bring about spoilage or cause gastroenteritis. It should be noted that the bacterial genera in this list along with those in Table 2–2 are now somewhat problematic since most were defined largely on phenotypic data. They are placed here mainly on historical reports but the list may be expected to change as more phylogenetic data are employed.

Bacteria

Acinetobacter	*Erwinia*	*Proteus*
Aeromonas	*Escherichia*	*Pseudomonas*
Alcaligenes	*Flavobacterium*	*Psychrobacter*
Arcobacter	*Hafnia*	*Salmonella*
Bacillus	*Kocuria*	*Serratia*
Brevibacillus	*Lactococcus*	*Shewanella*
Brochothrix	*Lactobacillus*	*Shigella*
Burkholderia	*Leuconostoc*	*Sphingomonas*
Campylobacter	*Listeria*	*Stenotrophomonas*
Carnobacterium	*Micrococcus*	*Staphylococcus*
Citrobacter	*Moraxella*	*Vagococcus*
Clostridium	*Paenibacillus*	*Vibrio*
Corynebacterium	*Pandoraea*	*Weissella*
Enterobacter	*Pantoea*	*Yersinia*
Enterococcus	*Pediococcus*	

Molds

Alternaria	*Colletotrichum*	*Penicillium*
Aspergillus	*Fusarium*	*Rhizopus*
Aureobasidium	*Geotrichum*	*Trichothecium*
Botrytis	*Monilia*	*Wallemia*
Byssochlamys	*Mucor*	*Xeromyces*
Cladosporium		

Yeasts

Brettanomyces/Dekkera	*Issatchenkia*	*Schizosaccharomyces*
Candida	*Kluyveromyces*	*Torulaspora*
Cryptococcus	*Pichia*	*Trichosporon*
Debaryomyces	*Rhodotorula*	*Yarrowia*
Hanseniaspora	*Saccharomyces*	*Zygosaccharomyces*

Protozoa

Cryptosporidium parvum	*Entamoeba histolytica*	*Toxoplasma gondii*
Cyclospora cayetanensis	*Giardia lamblia*	

PRIMARY SOURCES OF MICROORGANISMS FOUND IN FOODS

The genera and species previously listed are among the most important normally found in food products. Each genus has its own particular nutritional requirements, and each is affected in predictable ways by the parameters of its environment. Eight environmental sources of organisms to foods are listed below, and these, along with the genera of bacteria and protozoa noted, are presented in Table 2–2 to reflect their primary food-source environments.

Soil and Water. These two environments are placed together because many of the bacteria and fungi that inhabit both have a lot in common. Soil organisms may enter the atmosphere by the action of wind and later enter water bodies when it rains. They also enter water when rainwater flows over soils into bodies of water. Aquatic organisms can be deposited onto soils through the actions of cloud formation and subsequent rainfall. This common cycling results in soil and aquatic organisms being one and the same to a large degree. Some aquatic organisms, however, are unable to persist in soils, especially those that are indigenous to marine waters. *Alteromonas* spp. are aquatic forms that require seawater salinity

Table 2–2 Relative Importance of Eight Sources of Bacteria and Protozoa to Foods

Organisms	Soil and Water	Plants/Products	Food Utensils	Gastrointestinal Tract	Food Handlers	Animal Feeds	Animal Hides	Air and Dust
Bacteria								
Acinetobacter	XX	X	X				X	X
Aeromonas	XX[a]	X						
Alcaligenes	X	X	X	X			X	
Alteromonas	XX[a]							
Arcobacter	X							
Bacillus	XX[b]	X	X		X	X	X	XX
Brochothrix		XX	X					
Brevibacillus	X	X						X
Burkholderia		XX						
Campylobacter		X	X	XX	X			
Carnobacterium	X	XX	X					
Citrobacter	X	X	X	XX	X	X	X	XX
Clostridium	XX[b]	X	X	X	X		X	X
Corynebacterium	XX[b]	X	X		X		X	X
Enterobacter	X	XX	X	X	X		X	
Enterococcus	X	X	X	XX	X	X	X	X
Erwinia	X	XX	X					
Escherichia	X	X		XX	X			
Flavobacterium	X	XX					X	
Hafnia	X	X		XX				
Kocuria	X	X	X		X		X	X
Lactococcus		XX	X	X			X	
Lactobacillus		XX	X	X			X	
Leuconostoc		XX	X	X			X	
Listeria	XX	XX			X	X	X	
Micrococcus	XX	X	X		X	X	X	XX
Mycobacterium[c]		X						
Moraxella	X	X					X	
Mycobacterium		X						

Organism							
Paenibacillus	XX				X		XX
Pandoraea	X		X				
Pectobacterium	XX						
Pantoea	X	X	X			X	
Pediococcus	XX	X	X			X	
Proteus	X	X	X	X		X	
Pseudomonas	XX	X	X	X	X	X	
Psychrobacter	XX	X	X	X		X	
Salmonella					XX	X	
Serratia	X	X	X	X		X	
Shewanella	X	X	X			X	
Sphingomonas	X	X					
Shigella			XX	XX			
Stenotrophomonas	X	XX				X	
Staphylococcus			X	XX			
Vagococcus	XX		XX	XX			
Vibrio	XX		X				
Weissella	XX	XX					
Yersinia	X	X	X	X			
Protozoa							
C. cayetanensis	X	X	X				
C. parvum	XX		X	X			
E. histolytica	XX		X	X			
G. lamblia	XX		X	XX			
T. gondii	XX	X	XX		X		

Note: XX indicates a very important source.

[a] Primarily water
[b] Primarily soil.
[c] Nontuberculous.

19

for growth and would not be expected to persist in soils. The bacterial biota of seawater is essentially Gram-negative, and Gram-positive bacteria exist there essentially only as transients. Contaminated water has been implicated in *Cyclospora* contamination of fresh raspberries.

Plants and Plant Products. It may be assumed that many or most soil and water organisms contaminate plants. However, only a relatively small number find the plant environment suitable to their overall well-being. Those that persist on plant products do so by virtue of a capacity to adhere to plant surfaces so that they are not easily washed away and because they are able to obtain their nutritional requirements. Notable among these are the lactic acid bacteria and some yeasts. Among others that are commonly associated with plants are bacterial plant pathogens in the genera *Corynebacterium, Curtobacterium, Pectobacterium, Pseudomonas,* and *Xanthomonas*; and fungal pathogens among several genera of molds.

Food Utensils. When vegetables are harvested in containers and utensils, one would expect to find some or all of the surface organisms on the products to contaminate contact surfaces. As more and more vegetables are placed in the same containers, a normalization of the microbiota would be expected to occur. In a similar way, the cutting block in a meat market along with cutting knives and grinders are contaminated from initial samples, and this process leads to a buildup of organisms, thus ensuring a fairly constant level of contamination of meat-borne organisms.

Gastrointestinal Tract. This biota becomes a water source when polluted water is used to wash raw food products. The intestinal biota consists of many organisms that do not persist as long in waters as do others, and notable among these are pathogens such as salmonellae. Any or all of the Enterobacteriaceae may be expected in fecal wastes, along with intestinal pathogens, including the five protozoal species already listed.

Food Handlers. The microbiota on the hands and outer garments of handlers generally reflect the environment and habits of individuals, and the organisms in question may be those from soil, water, dust, and other environmental sources. Additional important sources are those that are common in nasal cavities, the mouth, and on the skin, and those from the gastrointestinal tract that may enter foods through poor personal hygiene practices.

Animal Feeds. This is a source of salmonellae to poultry and other farm animals. In the case of some silage, it is a known source of *Listeria monocytogenes* to dairy and meat animals. The organisms in dry animal feed are spread throughout the animal environment and may be expected to occur on animal hides.

Animal Hides. In the case of milk cows, the types of organisms found in raw milk can be a reflection of the biota of the udder when proper procedures are not followed in milking and of the general environment of such animals. From both the udder and the hide, organisms can contaminate the general environment, milk containers, and the hands of handlers.

Air and Dust. Although most of the organisms listed in Table 2–2 may at times be found in air and dust in a food-processing operation, the ones that can persist include most of the Gram-positive organisms listed. Among fungi, a number of molds may be expected to occur in air and dust, along with some yeasts. In general, the types of organisms in air and dust would be those that are constantly reseeded to the environment. Air ducts are not unimportant sources.

SYNOPSIS OF COMMON FOODBORNE BACTERIA

These synopses are provided to give the reader glimpses of bacterial groups that are discussed throughout the textbook. They are not meant to be used for culture identifications. For the latter, one or more of the cited references should be consulted. Some of the phylogenetic features of these bacteria are presented in the Appendix.

Acinetobacter (A • ci • ne'to • bac • ter; Gr. *akinetos*, unable to move). These Gram-negative rods show some affinity to the family Neisseriaceae, and some that were formerly achromobacters and moraxellae are placed here. Also, some former acinetobacters are now in the genus *Psychrobacter*. They differ from the latter and the moraxellae in being oxidase negative. They are strict aerobes that do not reduce nitrates. Although rod-shaped cells are formed in young cultures, old cultures contain many coccoid-shaped cells. They are widely distributed in soil and water and may be found on many foods, especially refrigerated fresh products. The mol% G + C content of DNA for the genus is 39–47. (See Chapter 4 for a further discussion relative to meats.) It has been proposed, based on DNA–rRNA hybridization data, that the genera *Acinetobacter*, *Moraxella*, and *Psychrobacter* be placed in a new family (Moraxellaceae), but this proposal has not been approved.

Aeromonas (ae • ro • mo'nas; *gas producing*). These are typically aquatic Gram-negative rods formerly in the family Vibrionaceae but now in the family Aeromonadaceae.[32] As the generic name suggests, they produce copious quantities of gas from the fermented sugars. They are normal inhabitants of the intestines of fish, and some are fish pathogens. The mol% G + C content of DNA is 57–65. (The species that possesses pathogenic properties is discussed in Chapter 31.)

Alcaligenes (al • ca • li'ge • nes; *alkali producers*). Although Gram negative, these organisms sometimes stain Gram positive. They are rods that do not, as the generic name suggests, ferment sugars but instead produce alkaline reactions, especially in litmus milk. Nonpigmented, they are widely distributed in nature in decomposing matter of all types. Raw milk, poultry products, and fecal matter are common sources. The mol% G + C content of DNA is 58–70, suggesting that the genus is heterogeneous.

Alteromonas (al • te • ro • mo'nas; *another monad*). These are marine and coastal water inhabitants that are found in and on seafoods; all species require seawater salinity for growth. They are Gram-negative motile rods that are strict aerobe.[17]

Arcobacter (Ar'co • bac • ter; L. *arcus*, bow). This genus was created during revision of the genera *Campylobacter*, *Helicobacter*, and *Wolinella*,[45] and the three species were once classified as *Campylobacter*. They are Gram-negative curved or S-shaped rods that are quite similar to the campylobacters except they can grow at 15°C and are aerotolerant. They are found in poultry, raw milk, shellfish, and water; and in cattle and swine products.[51,52] These oxidase- and catalase-positive organisms cause abortion and enteritis in some animals, and the latter in humans is associated with *A. butzleri*.

Bacillus (ba • cil'lus). These are Gram-positive spore-forming rods that are aerobes in contrast to the clostridia, which are anaerobes. Although most are mesophiles, psychrotrophs and thermophiles exist. The genus contains only two pathogens: *B. anthracis* (cause of anthrax) and *B. cereus*. Although most strains of the latter are nonpathogens, some cause foodborne gastroenteritis (further discussed in Chapter 24). This genus has been delimited by the transfer of a number of its former species to eight new genera: *Alicyclobacillus*, *Aneurinibacillus*, *Brevibacillus*, *Gracilibacillus*, *Paenibacillus*, *Virgibacillus*, and *Salibacillus*.[5] Also, the former group 5 *Bacillus* species are now in the genus *Geobacillus*, and the former *B. stearothermophilus* is now *G. stearothermophilus*.[36]

Brevibacillus (Bre • vi • ba • cil'lus). Previously classified as *Bacillus* spp. as noted above, these organisms occur in soil and water, and are common on plants, and in air, and dust. At least nine species are recognized.

Brochothrix (bro • cho • thr'ix; Gr. *brochos*, loop; *thrix*, thread). These Gram-positive nonspore-forming rods are closely related to the genera *Lactobacillus* and *Listeria*,[40] and some of the common features are discussed in Chapter 25. Although they are not true coryneforms, they bear resemblance to this group. Typically, exponential-phase cells are rods, and older cells are coccoids, a feature typical of coryneforms. Their separate taxonomic status has been reaffirmed by rRNA data, although only two species are recognized: *B. thermosphacta* and *B. campestris*. They share some features with the genus *Microbacterium*. They are common on processed meats and on fresh and processed meats that are stored in gas-impermeable packages at refrigerator temperatures. In contrast to *B. thermosphacta*, *B. campestris* is rhamnose and hippurate positive.[44] The mol% G + C content of DNA is 36. They do not grow at 37°C.

Burkholderia (Burkholder • ia). Gram-negative rods that occur on plants (especially certain flowers), in raw milk, and cause vegetable spoilage. In a study of raw cow's milk in Northern Ireland, 14 out of 26 (54%) samples contained *B. cepacia*.[34] They are significant pathogens in cystic fibrosis patients. They were formerly classified in the genus *Pseudomonas*.

Campylobacter (cam • py' • lo • bac • ter; Gr. *campylo*, curved). Although most often pronounced "camp'lo • bac • ter," the technically correct pronunciation should be noted. These Gram-negative, spirally curved rods were formerly classified as vibrios. They are microaerophilic to anaerobic. The genus has been restructured since 1984. The once *C. nitrofigilis* and *C. cryaerophila* have been transferred to the new genus *Arcobacter*; the once *C. cinnaedi* and *C. fenneliae* are now in the genus *Helicobacter*; and the once *Wolinella carva* and *W. recta* are now *C. curvus* and *C. rectus*.[45] The mol% G + C content of DNA is 30–35. For more information, see reference 32 and Chapter 28.

Carnobacterium (car • no • bac • terium; L. *carnis*, of flesh-meat bacteria). This genus of Gram-positive, catalase-negative rods was formed to accommodate some organisms previously classified as lactobacilli. They are phylogenetically closer to the enterococci and vagococci than the lactobacilli.[6,12] They are heterofermentative, and most grow at 0°C and none at 45°C. Gas is produced from glucose by some species, and the mol% G + C for the genus is 33.0–37.2. They differ from the lactobacilli in being unable to grow on acetate medium and in their synthesis of oleic acid. They are found on vacuum-packaged meats and related products, as well as on fish and poultry meats.[11,23,48]

Citrobacter (cit • ro • bac'ter). These enteric bacteria are slow lactose-fermenting, Gram-negative rods that typically produce yellow colonies on plate count agar. All members can use citrate as the sole carbon source. *C. freundii* is the most prevalent species in foods, and it and the other species are not uncommon on vegetables and fresh meats. The mol% G + C content of DNA is 50–52.

Clostridium (clos • tri'di • um; Gr. *closter*, a spindle). These anaerobic spore-forming rods are widely distributed in nature, as are their aerobic counterparts, the bacilli. The genus contains many species, some of which cause disease in humans (see Chapter 24 for *C. perfringens* food poisoning and botulism). Mesotrophic, psychrotrophic, and thermophilic species/strains exist; their importance in the thermal canning of foods is discussed in Chapter 17. A reorganization of the genus created the following five new genera: *Caloramater*, *Filifactor*, *Moorella*, *Oxobacter*, and *Oxalophagus*.[8] The clostridial species of known importance in foods remain in the genus at this time. The five new genera appear to be unimportant in foods.

Corynebacterium (co • ry • ne • bac • ter' • i • um; Gr. *coryne*, club). This is one of the true coryneform genera of Gram-positive, rod-shaped bacteria that are sometimes involved in the spoilage of vegetable and meat products. Most are mesotrophs, although psychrotrophs are known, and one, *C. diphtheriae*, causes diphtheria in humans. The genus has been reduced in species with the transfer of some of the plant pathogens to the genus *Clavibacter* and others to the genus *Curtobacterium*. The mol% G + C content of DNA is 51–63.

Enterobacter (en • te • ro • bac'ter). These enteric Gram-negative bacteria are typical of other Enterobacteriaceae relative to growth requirements, although they are not generally adapted to the gastrointestinal tract. They are further characterized and discussed in Chapter 20. *E. agglomerans* has been transferred to the genus *Pantoea*. *E. sakazakii* is discussed in Chapter 31.

Enterococcus (en • te • ro • coc'cus). This genus was erected to accommodate some of the Lancefield serologic group D cocci. It has since been expanded to more than 16 species of Gram-positive ovoid cells that occur singly, in pairs, or in short chains. They were once in the genus *Streptococcus*. Some species do not react with group D antisera. The genus is characterized more thoroughly in Chapter 20, and its phylogenetic relationship to other lactic acid bacteria can be seen in Figure 25–1.

Erwinia (er • wi'ni • a). These Gram-negative enteric rods are especially associated with plants. At least three species have been transferred to the genus *Pantoea*,[33] and the former *E. carotovora* and *E. chrysanthemi* are now in the genus *Pectobacterium* as *P. carovovorum* and *P. chrysanthemi* (see Chapter 6).

Escherichia (esch • er • i'chi • a). This is clearly the most widely studied genus of all bacteria. Those strains that cause foodborne gastroenteritis are discussed in Chapter 27, and *E. coli* as an indicator of food safety is discussed in Chapter 20.

Flavobacterium (fla • vo • bac • te'ri • um). These Gram-negative rods are characterized by their production of yellow to red pigments on agar and by their association with plants. Some are mesotrophs, and others are psychrotrophs, where they participate in the spoilage of refrigerated meats and vegetables. Some of the former flavobacterial species have been placed in the following five new genera: *Empedobacter*, *Chryseobacterium*, *Myroides*, *Sphingomonas*, and *Sphingobacterium*.

Hafnia (haf'ni • a). These Gram-negative enteric rods are important in the spoilage of refrigerated meat and vegetable products; *H. alvei* is the only species at this time. It is motile and lysine and ornithine positive, and it has a mol% G + C content of DNA of 48–49.

Kocuria (Ko • cu'ri • a, after M. Kocur). A new genus split off from the genus *Micrococcus*.[42] The three species (*K. rosea*, *K. varians*, and *K. kristinae*) are oxidase negative and catalase positive, and the mol% G + C content of DNA is 66–75.

Lactobacillus (lac • to • ba • cil'lus). Taxonomic techniques that came into wide use during the 1980s have been applied to this genus, resulting in some of those in the ninth edition of *Bergey's Manual* being transferred to other genera. Based on 16S rRNA sequence data, three phylogenetically distinct clusters are revealed,[10] with one cluster encompassing *Weissella*. In all probability, this genus will undergo further reclassification. They are Gram-positive, catalase-negative rods that often occur in long chains. Although those in foods are typically microaerophilic, many true anaerobic strains exist, especially in the colon and the rumen. They typically occur on most, if not all, vegetables, along with some of the other lactic acid bacteria. Their occurrence in dairy products is common. One species, *L. suebicus*, was recovered from apple and pear mashes; it grows at pH 2.8 and in 12–16% ethanol.[28] Many fermented products are produced, and these are discussed in Chapter 7. Those that are common on refrigerator-stored, vacuum-packaged meats are discussed in Chapters 5 and 14.

Lactococcus (lac • to • coc'cus). The nonmotile Lancefield serologic group N cocci once classified in the genus *Streptococcus* have been elevated to generic status. They are Gram-positive, nonmotile, and catalase-negative, spherical, or ovoid cells, that occur singly, in pairs, or as chains. They grow at 10°C but not at 45°C, and most strains react with group N antisera. L-Lactic acid is the predominant end-product of fermentation.

Leuconostoc (leu • co • nos'toc; *colorless nostoc*). Along with the lactobacilli, this is another of the genera of lactic acid bacteria. They are Gram-positive, catalase-negative cocci that are heterofermentative. The genus has been reduced in number of species (see *Weissella* below). The former *L. oenos* has been transferred to a new genus, *Oenococcus* as *O. oeni*,[16] and the former *L. paramesenteroides* has been transferred to the new genus *Weissella*. These cocci are typically found in association with the lactobacilli.

Listeria (lis • te'ri • a). This genus of six species of Gram-positive, nonsporing rods is closely related to *Brochothrix*. The six species show 80% similarity by numerical taxonomic studies; they have identical cell walls, fatty acid, and cytochrome composition. They are more fully described and discussed in Chapter 25.

Micrococcus (mi • cro • coc'cus). These Gram-positive and catalase-positive cocci are inhabitants of mammalian skin and can grow in the presence of high levels of NaCl. This genus has been reduced by the creation of the following five new genera: *Dermacoccus*, *Kocuria*, *Kytococcus*, *Nesterenkonia*, and *Stomatococcus*. At the present time, *M. luteus* and *M. lylae* are the only two micrococcal species.

Moraxella (mo • rax • el'la). These short Gram-negative rods are sometimes classified as *Acinetobacter*. They differ from the latter in being sensitive to penicillin and oxidase positive and having a mol% G + C content of DNA of 40–46. The genus *Psychrobacter* includes some that were once placed in this genus. Their metabolism is oxidative, and they do not form acid from glucose.

Paenibacillus (pae • ba • cil'lus; *almost a bacillus*). This newly established genus comprises organisms formerly in the genera *Bacillus* and *Clostridium*, and it includes the following species: *P. alvei*, *P. amylolyticus*, *P. azotofixans*, *P. circulans*, *P. durum*, *P. larvae*, *P. macerans*, *P. macquariensis*, *P. pubuli*, *P. pulvifaciens*, and *P. validus*.[2,8] Recently, two new species were added (*P. lautus* and *P. peoriae*[22]). The paenibacilli are notable for their degradation of a number of macromolecules, their production of antibacterial and antifungal agents, and the capacity of some to fix N_2 in association with plants. A newly-named species was isolated from raw and UHT-treated milk.[38]

Pandoraea (Pan • do • rae'a). Although first isolated from sputa of cystic fibrosis patients,[7] these organisms are related to some of the pseudomonads. Although not demonstrated to be common in foods, one species, *P. norimbergenesis*, has been isolated from powdered milk.[35]

Pantoea (pan • toe'a). This genus consists of Gram-negative, noncapsulated, nonsporing straight rods, most of which are motile by peritrichous flagella. They are widely distributed and are found on plants and in seeds, in soil, water, and human specimens. Some are plant pathogens. The four recognized species were once classified as enterobacters or erwinias. *P. agglomerans* includes the former *Enterobacter agglomerans*, *Erwinia herbicola*, and *E. milletiae*; *P. ananas* includes the former *Erwinia ananas* and *E. uredovora*; *P. stewartii* was once *E. stewartii*; and *P. dispersa* is an original species.[18] The G + C content of DNA ranges from 49.7 to 60.6 mol%.[33]

Pediococcus (pe • di • o • coc'cus; *coccus growing in one plane*). These homofermentative cocci are lactic acid bacteria that exist in pairs and tetrads resulting from cell division in two planes. *P. acidilactici*, a common starter species, caused septicemia in a 53-year-old male.[19] Their mol% G + C content of DNA is 34–44; they are further discussed in Chapter 7. The once *P. halophilus* is now in the genus *Tetragenococcus* as *T. halophilus*. It can grow in 18% NaCl.

Proteus (pro'te • us). These enteric Gram-negative rods are aerobes that often display pleomorphism, hence the generic name. All are motile and typically produce swarming growth on the surface of moist agar plates. They are typical of enteric bacteria in being present in the intestinal tract of humans and animals. They may be isolated from a variety of vegetable and meat products, especially those that undergo spoilage at temperatures in the mesophilic range.

Pseudomonas (pseu • do'mo • nas; *false monad*). These are typical soil and water bacteria and they are widely distributed among fresh foods, especially vegetables, meats, poultry, and seafood products. Although once the largest genus of foodborne bacteria, the genus has been delimited by the transfer of many former species to at least 13 new genera: *Acidovorax, Aminobacter, Brevundimonas, Burkholderia, Comamonas, Delftia, Devosia, Herbaspirillium, Hydrogenophaga, Marinobacter, Ralstonia, Sphingomonas, Telluria*, and *Wautersia*. *P. fluorescens* and *P. aeruginosa* remain in the original genus (see reference 24).

Psychrobacter (psy • chro' • bac • ter). This genus was created primarily to accommodate some of the nonmotile Gram-negative rods that were once classified in the genera *Acinetobacter* and *Moraxella*. They are plump coccobacilli that often occur in pairs. Also, they are aerobic, nonmotile, and catalase and oxidase positive, and generally do not ferment glucose. Growth occurs in 6.5% NaCl and at 1°C, but generally not at 35°C or 37°C. They hydrolyze Tween 80, and most are egg-yolk positive (lecithinase). They are sensitive to penicillin and utilize γ-aminovalerate, whereas the acinetobacters do not. They are distinguished from the acinetobacters by being oxidase positive and aminovalerate users and from nonmotile pseudomonads by their inability to utilize glycerol or fructose. Because they closely resemble the moraxellae, they have been placed in the family Neisseriaceae. The genus contains some of the former achromobacters and moraxellae, as noted. They are common on meats, poultry, and fish, and in water.[26,39]

Salmonella (sal • mon • el'la). All members of this genus of Gram-negative enteric bacteria are considered to be human pathogens. It should be noted that the salmonellae have been placed in two species with those that affect humans placed in the species *Salmonella enterica*. The serotypes (serovars) of more than 2,400 are listed as follows: *Salmonella enterica* serotype Newport, or *Salmonella* Newport (note that the serotype is not italicized). (See Chapter 26 for a more detailed explanation.) The mol% G + C content of DNA is 50–53.

Serratia (ser • ra'ti • a). These Gram-negative rods that belong to the family Enterobacteriaceae are aerobic and proteolytic, and they generally produce red pigments on culture media and in certain foods, although nonpigmented strains are not uncommon. *S. liquefaciens* is the most prevalent of the foodborne species; it causes spoilage of refrigerated vegetables and meat products. The mol% G + C content of DNA is 53–59. Interestingly, a spore-forming *S. marcescens* isolate has been reported and named *S. marcescens* subsp. *sakuensis*.[1]

Shewanella (she • wa • nel'la). The bacterium once classified as *Pseudomonas putrefaciens* and later as *Alteromonas putrefaciens* has been placed in this genus as *S. putrefaciens*. They are Gram-negative, straight or curved rods, nonpigmented, and motile by polar flagella. They are oxidase positive and have a mol% G + C of 44–47. The other three species in this genus are *S. hanedai*, *S. benthica*, and *S. colwelliana*. All are associated with aquatic or marine habitats, and the growth of *S. benthica* is enhanced by hydrostatic pressure.[14,32]

Shigella (shi • gel'la). All members of this genus are presumed to be human enteropathogens; they are discussed further in Chapter 26.

Sphingomonas (Sphin • go • monas). There are at least 33 species of this genus of Gram-negative bacteria that typically produce yellow pigment, and which formerly were in the genus *Flavobacterium*. They are found in water, on certain vegetables, and cause human disease.[58]

Staphylococcus (staph • y • lo • coc'cus; *grape-like coccus*). These Gram-positive, catalase-positive cocci include *S. aureus*, which causes several disease syndromes in humans, including foodborne gastroenteritis. It and other members of the genus are discussed further in Chapter 23. The former *S. caseolyticus* has been transferred to the new genus *Macrococcus* as *M. caseolyticus*.[29]

Stenotrophomonas (Ste • no • tro • pho • mo'nas, a unit feeding on few substrates). These Gram-negative rods are common inhabitants of plants and they have been recovered from soil, water, and milk. They are growth-promoting or symbionts in the rhizosphere of several crop plants.[21] *S. maltophila* is regarded as the second most common nosocomial bacterium after *Pseudomonas aeruginosa* (see reference 21). One species, *S. rhizophila*, has been used to control fungal diseases in plants (see reference 57).

Vagococcus (va • go • coc'cus; *wandering coccus*). This genus was created to accommodate the group N lactococci based on 16S sequence data.[11] They are motile by peritrichous flagella, are Gram positive and catalase negative, and grow at 10°C but not at 45°C. They grow in 4% NaCl but not 6.5%, and no growth occurs at pH 9.6. The cell wall peptidoglycan is Lys-D-Asp, and the mol% G + C is 33.6. At least one species produces H_2S. They are found on fish, in feces, and in water and may be expected to occur on other foods.[11,47] Information on the phylogenetic relationship of the vagococci to other related genera is presented in Figure 25–1.

Vibrio (vib'ri • o). These Gram-negative straight or curved rods are members of the family Vibrionaceae. Several former species have been transferred to the genus *Listonella*.[32] Several species cause gastroenteritis and other human illness; they are discussed in Chapter 28. The mol% G + C content of DNA is 38–51. (See reference 13 for environmental distribution.)

Weissella (Weiss'ella, after N. Weiss). This genus of lactic acid bacteria was established in 1993 in part to accommodate the "leuconostoc branch" of the lactobacilli.[9] The seven species are closely related to the leuconostocs, and with the exception of *W. paramesenteroides* and *W. hellenica*, they produce DL-lactate from glucose. All produce gas from carbohydrates. *W. hellenica* is a new species associated with fermented Greek sausages.[9] The former *Leuconostoc paramesenteroides* is now *W. paramesenteroides*, and the following five species were formerly classified as *Lactobacillus* spp.: *W. confusa*, *W. halotolerans*, *W. kandleri*, *W. minor*, and *W. viridescens*. The G + C content of DNA is 37–47 mol%.

Yersinia (yer • si'ni • a). This genus includes the agent of human plague, *Y. pestis*, and at least one species that causes foodborne gastroenteritis, *Y. enterocolitica*. All foodborne species are discussed in Chapter 28. The mol% G + C content of DNA is 45.8–46.8. The sorbose-positive biogroup 3A strains have been elevated to species status as *Y. mollaretti* and the sorbose-negative strains as *Y. bercovieri*.[49]

SYNOPSIS OF COMMON GENERA OF FOODBORNE MOLDS

Molds are filamentous fungi that grow in the form of a tangled mass that spreads rapidly and may cover several inches of area in 2 to 3 days. The total of the mass or any large portion of it is referred to as *mycelium*. Mycelium is composed of branches or filaments referred to as *hyphae*. Those of greatest importance in foods multiply by ascospores, zygospores, or conidia. The *ascospores* of some genera are notable for their extreme degrees of heat resistance. One group forms pycnidia or acervuli (small, flask-shaped, fruiting bodies lined with conidiophores). *Arthrospores* result from the fragmentation of hyphae in some groups.

There were no radical changes in the systematics of foodborne fungi during the 1980s. The most notable changes involve the discovery of the sexual or perfect states of some well-known genera and species. In this regard, the *ascomycete* state is believed by mycologists to be the more important reproductive state of a fungus, and this state is referred to as the *teleomorph*. The species name given to a teleomorph takes precedence over that for the *anamorph*, the imperfect or conidial state. *Holomorph* indicates that both states are known, but the teleomorph name is used.

The taxonomic positions of the genera described are summarized below. (Consult references 3, 4, and 37 for identifications; see reference 25 for the types that exist in meats.)

Division: Zygomycota
   Class: Zygomycetes (nonseptate mycelium, reproduction by sporangiospores, rapid growth)
      Order: Mucorales
         Family: Mucoraceae
            Genus: *Mucor*
                   Rhizopus
                   Thamnidium

Division: Ascomycota
   Class: Plectomycetes (septate mycelium, ascospores produced in asci usually number 8)
      Order: Eurotiales
         Family: Trichocomaceae
            Genus: *Byssochlamys*
                   Emericella
                   Eupenicillium
                   Eurotium

Division: Deuteromycota (the "imperfects," anamorphs; perfect stages are unknown)
    Class: Coelomycetes
        Genus: *Colletotrichum*
    Class: Hypomycetes (hyphae give rise to conidia)
        Order: Hyphomycetales
            Family: Moniliaceae
                Genus: *Alternaria*
                        Aspergillus
                        Aureobasidium (*Pullularia*)
                        Botrytis
                        Cladosporium
                        Fusarium
                        Geotrichum
                        Helminthosporium
                        Monilia/Sclerotinium
                        Penicillium
                        Stachybotrys
                        Trichothecium

Some of the genera are listed below in alphabetical order.

Alternaria. Septate mycelia with conidiophores and large brown conidia are produced. The conidia have both cross and longitudinal septa and are variously shaped. They cause brown to black rots of stone fruits, apples, and figs. Stem-end rot and black rot of citrus fruits are also caused by species/strains of this genus. This is a field fungus that grows on wheat. Additionally, it is found on red meats. Some species produce mycotoxins (see Chapter 30).

Aspergillus. Chains of conidia are produced. Where cleistothecia with ascospores are developed, the perfect stage of those found in foods is *Emericella, Eurotium,* or *Neosartorya. Eurotium* (the former *A. glaucus* group) produces bright yellow cleistothecia, and all species are xerophilic. *E. herbariorum* has been found to cause spoilage of grape jams and jellies.[41] *Emericella* produces white cleistothecia, and *E. nidulans* is the teleomorph of *Aspergillus nidulans. Neosartorya* produces white cleistothecia and colorless ascospores. *N. fischeri* is heat resistant, and resistance of its spores is similar to those of *Byssochlamys.*[37] The aspergilli appear yellow to green to black on a large number of foods. Black rot of peaches, citrus fruits, and figs is one of the fruit spoilage conditions produced. They are found on country-cured hams and on bacon. Some species cause spoilage of oils, such as palm, peanut, and corn. *A. oryzae* and *A. soyae* are involved in the shogu fermentation and the former in koji. *A. glaucus* produces katsuobushi, a fermented fish product. The *A. glaucus–A. restrictus* group contains storage fungi that invade seeds, soybeans, and common beans. *A. niger* produces β-galactosidase, glucoamylase, invertase, lipase, and pectinase: and *A. oryzae* produces α-amylase. Several species produce aflatoxins, and others produce ochratoxin A and sterigmatocystin (see Chapter 30).

Aureobasidium (*Pullularia*). Yeast-like colonies are produced initially. They later spread and produce black patches. *A. pullulans* (*Pullularia pullulans*) is the most prevalent in foods. They are found in shrimp, are involved in the "black spot" condition of long-term stored beef, and are common on fruits and vegetables.

Botrytis. Long, slender, and often pigmented conidiophores are produced. Mycelium is septate; conidia are borne on apical cells and are gray in color, although black, irregular sclerotia are sometimes produced. *B. cinerea* is the most common in foods. They are notable as the cause of gray mold rot of apples, pears, raspberries, strawberries, grapes, blueberries, citrus, and some stone fruits (see Chapter 6).

Byssochlamys. This genus is the teleomorph of certain species of *Paecilomyces*, but the latter does not occur in foods.[37] The ascomycete *Byssochlamys* produces open clusters of asci, each of which contains eight ascospores. The latter are notable for their heat resistance, resulting in spoilage of some high-acid canned foods. In their growth, they can tolerate low oxidation–reduction potential (Eh) values. Some are pectinase producers, and *B. fulva* and *B. nivea* spoil canned and bottled fruits. These organisms are almost uniquely associated with food spoilage, and *B. fulva* possesses a thermal D value at $90°C$ between 1 and 12 minutes with a z value of $6–7°C$.[37]

Cladosporium. Septate hyphae with dark, tree-like, budding conidia variously branched, characterize this genus. In culture, growth is velvety and olive colored to black. Some conidia are lemon shaped. *C. herbarum* produces "black spot" on beef and frozen mutton. Some spoil butter and margarine, and some cause restricted rot of stone fruits and black rot of grapes. They are field fungi that grow on barley and wheat grains. *C. herbarum* and *C. cladosporiodes* are the two most prevalent on fruits and vegetables.

Colletotrichum. They belong to the class Coelomycetes and form conidia inside acervuli. Simple but elongated conidiophores and hyaline conidia that are one celled, ovoid, or oblong are produced. The acervuli are disc or cushion shaped, waxy, and generally dark in color. *C. gloeosporioides* is the species of concern in foods; it produces anthracnose (brown/black spots) on some fruits, especially tropical fruits such as mangos and papayas.

Fusarium. Extensive mycelium is produced that is cottony with tinges of pink, red, purple, or brown. Septate fusiform to sickle-shaped conidia (macroconidia) are produced. They cause brown rot of citrus fruits and pineapples and soft rot of figs. As field fungi, some grow on barley and wheat grains. Some species produce zearalenone, fumonisins, and trichothecenes (see Chapter 30).

Geotrichum (once known as *Oidium lactis* and *Oospora lactis*). These yeast-like fungi are usually white. The hyphae are septate, and reproduction occurs by formation of arthroconidia from vegetative hyphae. The arthroconidia have flattened ends. *G. candidum*, the anamorph of *Dipodascus geotrichum*, is the most important species in foods. It is variously referred to as "dairy mold" because it imparts flavor and aroma to many types of cheese, and as "machinery mold" because it builds up on food-contact equipment in food-processing plants, especially tomato canning plants. They cause sour rot of citrus fruits and peaches and the spoilage of dairy cream. They are widespread and have been found on meats and many vegetables. Some participate in the fermentation of gari.

Monilia/Sclerotinium. Pink, gray, or tan conidia are produced. *M. sitophila* is the conidial stage of *Neurospora intermedia*. *Monilia* is the conidial state of *Monilinia fructicola*. They produce brown rot of stone fruits such as peaches. *Monilina* sp. causes mummification of blueberries.

Mucor. Nonseptate hyphae are produced that give rise to sporangiophores that bear columella with a sporangium at the apex. No rhizoids or stolons are produced by members of this large genus. Cottony colonies are often produced. The conditions described as "whiskers" of beef and "black spot" of frozen mutton are caused by some species. At least one species, *M. miehei*, is a lipase producer. It is found in fermented foods, bacon, and many vegetables. One species ferments soybean whey curd.

Penicillium. When conidiophores and conidia are the only reproductive structures present, this genus is placed in the Deuteromycota. They are placed with the ascomycetes when cleistochecia with ascospores are formed as either *Talaromyces* or *Eupenicillium*. Of the two teleomorphic genera, *Talaromyces* is the most important in foods.[37] *T. flavus* is the teleomorph of *P. dangeardii*, and it has been involved in the spoilage of fruit juice concentrates.[27] It produces heat-resistant spores. When conidia are formed in the penicillus, they pinch off from *phialides*. Typical colors on foods are blue to blue-green. Blue and green mold rots of citrus fruits and blue mold rot of apples, grapes, pears, and stone fruits, are caused by some species. One species, *P. roqueforti*, produces blue cheese. Some species produce citrinin, yellow rice toxin, ochratoxin A, rubratoxin B, and other mycotoxins (see Chapter 30).

Rhizopus. Nonseptate hyphae are produced that give rise to stolons and rhizoids. Sporangiophores typically develop in clusters from ends of stolons at the point of origin of rhizoids. *R. stolonifer* is by far the most common species in foods. Sometimes referred to as "bread molds," they produce watery soft rot of apples, pears, stone fruits, grapes, figs, and others. Some cause "black spot" of beef and frozen mutton. They may be found on bacon and other processed meats. Some produce pectinases, and *R. oligosporus* is important in the production of oncom, bongkrek, and tempeh.

Thamnidium. These molds produce small sporangia borne on highly branched structures. *T. elegans* is the only species, and it is best known for its growth on refrigerated beef hindquarters where its characteristic growth is described as "whiskers." It is less often found in decaying eggs.

Trichothecium. Septate hyphae that bear long, slender, and simple conidiophores are produced. *T. roseum* is the only species, and it is pink in color and causes pink rot of fruits. It also causes soft rot of cucurbits and is common on barley, wheat, corn, and pecans. Some produce mycotoxins (see Chapter 30).

Other Molds. Two categories of organisms are presented here, the first being some miscellaneous genera that are found in some foods but are generally not regarded as significant. These are *Cephalosporium*, *Diplodia*, and *Neurospora*. *Cephalosporium* is a deuteromycete often found on frozen foods. The microspores of some *Fusarium* species are similar to those of this genus. *Diplodia* is another deuteromycete that causes stem-end rot of citrus fruits and water tan-rot of peaches. *Neurospora* is an ascomycete, and *N. intermedia* is referred to as the "red bread" mold. *Monilia sitophila* is the anamorph of *N. intermedia*. The latter is important in the oncom fermentation and has been found on meats. The "white spot" of beef is produced by *Sporotrichum* spp.; and rots of various fruits are caused by *Gloeosporium* spp. Some *Helminthosporium* spp. are plant pathogens and some are saprophytes.

Neosartorya fischeri (anamorph *Aspergillus fischerianus*) was first recognized in the early 1960s as the cause of spoilage of fruit products. Its ascospores are very heat resistant, being able to withstand boiling in distilled water for up to an hour. It has a D_{87} C of around 11 minutes in phosphate buffer. Interestingly, it produces several mycotoxins—fumitremorgin A, B, and C; terrein; verruculogen; and fischerin.

The second category consists of xerophilic molds, which are very important as spoilage organisms. In addition to *Aspergillus* and *Eurotium*, Pitt and Hocking[37] include six other genera among the xerophiles:

Basipetospora, Chrysosporium, Eremascus, Polypaecilum, Wallemia, and *Xeromyces.* These molds are characterized by the ability to grow below a_w (water activity) = 0.85. They are of significance in foods that owe their preservation to a low a_w. Only *Wallemia* and *Xeromyces* are discussed further below.

Wallemia produces deep-brown colonies on culture media and on foods. *W. sebi* (formerly *Sporendonema*), the most notable species, can grow at an a_w of 0.69. It produces the "dun" mold condition on dried and salted fish.

Xeromyces has only one species, *X. bisporus.* It produces colorless cleistothecia with evanescent asci that contain two ascospores. This organism has the lowest a_w growth of any other known organisms.[37] Its a_w high is ∼0.97, its optimum is 0.88, and its minimum is 0.61. Its thermal D at 82.2°C is 2.3 minutes. It causes problems in licorice, prunes, chocolate, syrup, and other similar products.

SYNOPSIS OF COMMON GENERA OF FOODBORNE YEASTS

Yeasts may be viewed as being unicellular fungi in contrast to the molds, which are multicellular; however, this is not a precise definition, as many of what are commonly regarded as yeasts actually produce mycelia to varying degrees.

Yeasts can be differentiated from bacteria by their larger cell size and their oval, elongate, elliptical, or spherical cell shapes. Typical yeast cells range from 5 to 8 μm in diameter, with some being even larger. Older yeast cultures tend to have smaller cells. Most of those of importance in foods divide by budding or fission.

Yeasts can grow over wide ranges of acid pH and in up to 18% ethanol. Many grow in the presence of 55–60% sucrose. Many colors are produced by yeasts, ranging from creamy, to pink, to red. The *asco-* and *arthrospores* of some are quite heat resistant. (Arthrospores are produced by some yeast-like fungi.)

Regarding the taxonomy of yeasts, newer methods have been employed in the past decade or so consisting of 5S rRNA, DNA base composition, and coenzyme Q profiles. Because of the larger genome size of yeasts, 5S rRNA sequence analyses are employed more than for larger RNA fractions. Many changes have occurred in yeast systematics, due in part of the use of newer methods but also to what appears to be a philosophy toward grouping rather than splitting taxa. One of the most authoritative works on yeast systematics is that edited by Kreger-van Rij and published in 1984.[30] In this volume, the former *Torulopsis* genus has been transferred to the genus *Candida,* and some of the former *Saccharomyces* have been transferred to *Torulaspora* and *Zygosaccharomyces.* The teleomorphic or perfect states of more yeasts are now known, and this makes references to the older literature more difficult.

The taxonomy of 15 or so foodborne genera is summarized below. For excellent discussions on foodborne yeasts, the publications by Deak and Beuchat,[15] Beneke and Stevenson,[3] and Pitt and Hocking[37] should be consulted. For identification, Deak and Beuchat[15] have presented an excellent simplified key to foodborne yeasts.

Division: Ascomycotina
  Family: Saccharomycetaceae (ascospores and arthrospores formed; vegetative
  reproduction by fission or budding)
    Subfamily: Nadsonioideae
      Genus: *Hanseniaspora*

    Subfamily: Saccharomycotoideae
      Genus: *Debaryomyces*
           Issatchenkia
           Kluyveromyces
           Pichia
           Saccharomyces
           Torulaspora
           Zygosaccharomyces
    Subfamily: Schizosaccharomycetoideae
      Genus: *Schizosaccharomyces*

Division: Deuteromycotina
    Family: Cryptococcaceae (the "imperfects"; reproduce by budding)
      Genus: *Brettanomyces*
           Candida
           Cryptococcus
           Rhodotorula
           Trichosporon

The above genera are listed below in alphabetical order.

Brettanomyces (The perfect stage is *Dekkera*). These asporogenous yeasts form ogival cells and terminal budding, and produce acetic acid from glucose only under aerobic conditions. *B. intermedius* is the most prevalent, and it can grow at a pH as low as 1.8. They cause spoilage of beer, wine, soft drinks, and pickles, and some are involved in afterfermentation of some beers and ales. *D. bruxellensis* is involved in some sourdough fermentations, and it contributes to biogenic amines in red wines.

Candida. This genus was erected in 1923 by Berkhout and has since undergone many changes in definition and composition.[46] It is regarded as being a heterogenous taxon that can be divided into 40 segments comprising 3 main groups, based mainly on fatty acid composition and electrophoretic karotyping.[47] The generic name means "shining white," and cells contain no carotenoid pigments. The ascomycetous imperfect species are placed here, including the former genus *Torulopsis*, as follows:

    Candida famata (Torulopsis candida; T. famata)
    Candida kefyr (Candida pseudotropicalis, T. kefyr; Torula cremoris)
    Candida stellata (Torulopsis stellata)
    Candida holmii (Torulopsis holmii)

Many of the *anamorphic* forms of *Candida* are now in the genera *Kluyveromyces* and *Pichia*.[15] *Candida lipolytica* is the anamorph of *Saccharomycopsis lipolytica*.

    Members of this genus are the most common yeasts in fresh ground beef and poultry, and *C. tropicalis* is the most prevalent in foods in general. Some members are involved in the fermentation of cacao beans, as a component of kefir grains, and in many other products, including beers, ales, and fruit juices.

Cryptococcus. This genus represents the anamorph of *Filobasidiella* and other *Basidiomycetes*. They are asporogenous, reproduce by multilateral budding, and are nonfermenters of sugars. They are hyaline and red or orange, and they may form arthrospores. They have been found on plants, in soils, on strawberries and other fruits, marine fish, shrimp, and fresh ground beef.

Debaryomyces. These ascosporogenous yeasts sometimes produce a pseudomycelium and reproduce by multilateral budding. They are one of the two most prevalent yeast genera in dairy products. *D. hansenii* represents what was once *D. subglobosus* and *Torulaspora hansenii*, and it is the most prevalent foodborne species. It can grow in 24% NaCl and at an a_w as low as 0.65. It forms slime on wieners, grows in brines and on cheeses, and causes spoilage of orange juice concentrate and yogurt.

Hanseniaspora. These are apiculate yeasts whose anamorphs are *Kloeckera* spp. They exhibit bipolar budding, and, consequently, lemonshaped cells are produced. The asci contain two to four hat-shaped spores. Sugars are fermented, and they can be found on a variety of foods, especially figs, tomatoes, strawberries, citrus fruits, and the cacao bean fermentation.

Issatchenkia. Members of this genus produce pseudomycelia and multiply by multilateral budding. Some species once in the genus *Pichia* have been placed here. The teleomorph of *Candida krusei* is *I. orientalis*. They typically form pellicles in liquid media. They contain coenzyme Q-7 and are prevalent on a wide variety of foods.

Kluyveromyces (*Fabospora*). These ascospore-forming yeasts reproduce by multilateral budding, and the spores are spherical. *K. marxianus* now includes the former *K. fragilis*, *K. lactis*, *K. bulgaricus*, *Saccharomyces lactis*, and *S. fragilis*. *K. marxianus* is one of the two most prevalent yeasts in dairy products. *Kluyveromyces* spp. produce β-galactosidase and are vigorous fermenters of sugars, including lactose. *K. marxianus* contains coenzyme Q-6 and is involved in the fermentation of kumiss. It is also used for lactase production from whey and as the organism of choice for producing yeast cells from whey. They are found on a wide variety of fruits, and *K. marxianus* causes cheese spoilage.

Pichia. This is the largest genus of true yeasts. They reproduce by multilateral budding, and the asci usually contain four spheroidal, hat- or saturn-shaped spores. Pseudomycelia and arthrospores may be formed. Some of the hat-shaped spore formers may be *Williopsis* spp., and some of the former species are now classified in the genus *Debaryomyces*. *P. guilliermondii* is the perfect state of *Candida guillermondii*. The anamorph of *P. membranaefaciens* is *Candida valida*. *Pichia* spp. typically form films on liquid media and are known to be important in producing indigenous foods in various parts of the world. Some have been found on fresh fish and shrimp, and they are known to grow in olive brines and to cause spoilage of pickles and sauerkraut.

Rhodotorula. These yeasts are anamorphs of Basidiomycetes. The teliospore producers are in the genus *Rhodosporidium*. They reproduce by multilateral budding and are nonfermenters. *R. glutinis* and *R. mucilaginosa* are the two most prevalent species in foods. They produce pink to red pigments, and most are orange or salmon pink in color. The genus contains many psychrotrophic species/strains that are found on fresh poultry, shrimp, fish, and beef. Some grow on the surface of butter.

Saccharomyces. These ascosporogenous yeasts multiply by multilateral budding and produce spherical spores in asci. They are diploid and do not ferment lactose. Those once classified as *S. bisporus* and *S. rouxii* are now in the genus *Zygosaccharomyces*, and the former *S. rosei* is now in the genus *Torulaspora*. All bakers', brewers', wine, and champagne yeasts are *S. cerevisiae*. They are found in kefir grains and can be isolated from a wide range of foods, such as dry-cured salami and numerous fruits. *S. cerevisiae* rarely causes spoilage.

Schizosaccharomyces. These ascosporogenous yeasts divide by lateral fission of cross-wall formation and may produce true hyphae and arthrospores. Asci contain from four to eight bean-shaped spores, and no buds are produced. They are regarded as being only distantly related to the true yeasts. *S. pombe* is the most prevalent species; it is osmophilic and resistant to some chemical preservatives.

Torulaspora. Multilateral budding is the method of reproduction with spherical spores in asci. Three haploid species formerly in the genus *Saccharomyces* are now in this genus. They are strong fermenters of sugars, and contain coenzyme Q-6. *T. delbrueckii* is the most prevalent species.

Trichosporon. These nonascospore-forming oxidative yeasts multiply by budding and by arthroconidia formation. They produce a true mycelium, and sugar fermentation is absent or weak. They are involved in cacao bean and idli fermentations and have been recovered from fresh shrimp, ground beef, poultry, frozen lamb, and other foods. *T. pullulans* is the most prevalent species, and it produces lipase.

Yarrowia. Formerly *Saccharomycopsis*, these yeasts belong to the order *Endomycetales* and they are common on fruits, vegetables, meats, and poultry. *Candida lipolytica* is the anamorph, and *Y. lipolytica* is the teleomorphic (perfect) stage.

Zygosaccharomyces. Multilateral budding is the method of reproduction, and the bean-shaped ascospores formed are generally free in asci. Most are haploid and they are strong fermenters of sugars. *Z. rouxii* is the most prevalent species, and it can grow at an a_w of 0.62, second only to *Xeromyces bisporus* in its ability to grow at a low a_w.[37] Some are involved in shoyu and miso fermentations, and some are common spoilers of mayonnaise and salad dressing, especially *Z. bailii*, which can grow at a pH of 1.8.[37]

REFERENCES

1. Ajithkumar, B., V.P. Ajithkumar, R. Iriye, Y. Doi, and T. Sakai. 2003. Spore-forming *Serratia marcescens* subsp. *sakuensis* subsp. nov., isolated from a domestic wastewater treatment tank. *Int. J. System. Evol. Microbiol.* 53:253–258.

2. Ash, C., F.G. Priest, and M.D. Collins. 1993. Molecular identification of rRNA group 3 bacilli (Ash, Farrow, Wallbanks and Collins) using a PCR probe test. *Antonie van Leeuwenhoek* 64:253–260.

3. Beneke, E.S., and K.E. Stevenson. 1987. Classification of food and beverage fungi. In *Food and Beverage Mycology*, 2d ed., ed. L.R. Beuchat, 1–50. New York: Kluwer Academic Publishing.

4. Beuchat, L.R., ed. 1987. *Food and Beverage Mycology*, 2d ed. New York: Van Nostrand Reinhold.

5. Berkeley, R.C.A., and N. Ali. 1994. Classification and identification of endospore-forming bacteria. *J. Appl. Bacteriol.* (Symp. Suppl.) 76:1S–8S.

6. Champomier, M.-C., M.-C. Montel, and R. Talon. 1989. Nucleic acid relatedness studies on the genus *Carnobacterium* and related taxa. *J. Gen. Microbiol.* 135:1391–1394.

7. Coenye, T., E. Falsen, B. Hoste, J. Ohlén, J. Goris, J.R.W. Govan, M. Gillis, and P. Vandamme. 2000. Description of *Pandoraea* gen. nov. with *Pandoraea apista* sp. nov., *Pandoraea pulmonicola* sp. nov., *Pandoraea pnomenusa* sp. nov., *Pandoraea sputorum* sp. nov. and *Pandoraea norimbergensis* comb. nov. *Int. J. System. Evol. Microbiol.* 50:887–899.

8. Collins, M.D., P.A. Lawson, A. Willems, J.J. Cordoba, J. Fernandez-Garayzabal, P. Garcia, J. Cai, H. Hippe, and J.A.E. Farrow. 1994. The phylogeny of the genus *Clostridium*: Proposal of five new genera and eleven new species combinations. *Int. J. Syst. Bacteriol.* 44:812–826.

9. Collins, M.D., J. Samelis, J. Metaxopoulos, and S. Wallbanks. 1993. Taxonomic studies on some leuconostoc-like organisms from fermented sausages: Description of a new genus *Weissella* for the *Leuconostoc paramesenteroides* group of species. *J. Appl. Bacteriol.* 75:595–603.

10. Collins M.D., U. Rodriguez, C. Ash, M. Aguirre, J.A.E. Farrow, A. Martinez-Murcia, B.A. Phillips, A.M. Williams, and S. Wallbanks. 1991. Phylogenetic analysis of the genus *Lactobacillus* and related lactic acid bacteria as determined by reverse transcriptase sequencing of 16S rRNA. *FEMS Microbiol. Lett.* 77:5–12.

11. Collins, M.D., C. Ash, J.A.E. Farrow, S. Wallbanks, and A.M. Williams. 1989. 16S ribosomal ribonucleic acid sequence analyses of lactococci and related taxa: Description of *Vagococcus fluvialis* gen. nov., sp. nov. *J. Appl. Bacteriol.* 67:453–460.

12. Collins, M.D., J.A.E. Farrow, B.A. Phillips, S. Ferusu, and D. Jones. 1987. Classification of *Lactobacillus divergens*, *Lactobacillus piscicola*, and some catalase-negative, asporogenous, rod-shaped bacteria from poultry in a new genus, *Carnobacterium. Int. J. Syst. Bacteriol.* 37:310–316

13. Colwell, R.R., ed. 1984. *Vibrios in the Environment.* New York: Wiley.

14. Coyne, V.E., C.J. Pillidge, D.D. Sledjeski, H. Hori, B.A. Ortiz-Conde, D.G. Muir, R.M. Weiner, and R.R. Colwell. 1989. Reclassification of *Alteromonas colwelliana* to the genus *Shewanella* by DNA–DNA hybridization, serology and 5S ribosomal RNA sequence data. *Syst. Appl. Microbiol.* 12:275–279.

15. Deak, T., and L.R. Beuchat. 1987. Identification of foodborne yeasts. *J. Food Protect.* 50:243–264.

16. Dicks, L.M.T., F. Dellaglio, and M.D. Collins. 1995. Proposal to reclassify *Leuconostoc oenos* as *Oenococcus oeni* [corrig.] gen. nov. comb. nov. *Int. J. Syst. Bacteriol.* 45:395–397.

17. Gauthier, G., M. Gauthier, and R. Christen. 1995. Phylogenetic analysis of the genera *Alteromonas, Shewanella*, and *Moritella* using genes coding for small-subunit rRNA sequences and division of the genus *Alternomas* into two genera, *Alteromonas* (amended) and *Pseudoalteromonas* gen. nov., and proposal of twelve new species combinations. *Int. J. Syst. Bacteriol.* 45:755–761.

18. Gavini, F., J. Mergaert, A. Beji, C. Mielcarek, D. Izard, K. Kersters, and J. de Ley. 1989. Transfer of *Enterobacter agglomerans* (Beijerinck 1888) (Ewing and Fife 1972) to *Pantoea* gen. nov. as *Pantoea agglomerans* comb. nov. and description of *Pantoea dispersa* sp. nov. *Int. J. Syst. Bacteriol.* 39:337–345.

19. Golledge, C.L., N. Stringmore, M. Aravena, and D. Joske. 1990. Septicemia caused by vancomycin-resistant *Pediococcus acidilactici. J. Clin. Microbiol.* 28:1678–1679.

20. Gupta, R.S. 2000. The phylogeny of proteobacteria: relationships to other eubacterial phyla and eukaryotes. *FEMS Microbiol. Rev.* 24:367–402.

21. Hauben, L., L. Vauterin, E.R.B. Moore, B. Hoste, and J. Swings. 1999. Genomic diversity of the genus *Stenotrophomonas. Int. J. System. Bacteriol.* 49:1749–1760.

22. Heyndrickx, M., K. Vandemeulebroecke, P. Scheldeman, K. Kersters, P. De Vos, N.A. Logan, A.M. Aziz, N. Ali, and R.C.W. Berkeley. 1996. A polyphasic reassessment of the genus *Paenibacillus*, reclassification of *Bacillus lautus* (Nakamura 1984) as *Paenibacillus lautus* comb. nov. and of *Bacillus peoriae* (Montefusco et al. 1993) as *Paenibacillus peoriae* comb. nov., and emended descriptions of *P. lautus* and of *P. peoriae. Int. J. Syst. Bacteriol.* 46:988–1003.

23. Holzapfel, W.H., and E.S. Gerber. 1983. *Lactobacillus divergens* sp. nov., a new heterofermentative *Lactobacillus* species producing L(+)-lactate. *Syst. Appl. Bacteriol.* 4:522–534.

24. Jay, J.M. 2003. A review of recent taxonomic changes in seven genera of bacteria commonly found in foods. *J. Food Protect.* 66:1304–1309.

25. Jay, J.M. 1987. Fungi in meats, poultry, and seafoods. In *Food and Beverage Mycology*, 2nd ed., ed. L.R. Beuchat, 155–173. New York: Kluwer Academic Publishers.

26. Juni, E., and G.A. Heym. 1986. *Psychrobacter immobilis* gen. nov., sp. nov.: Genospecies composed of Gram-negative, aerobic, oxidase-positive coccobacilli. *Int. J. Syst. Bacteriol.* 36:388–391.

27. King, A.D., Jr., and W.U. Halbrook. 1987. Ascospore heat resistance and control measures for *Talaromyces flavus* isolated from fruit juice concentrate. *J. Food Sci.* 52:1266, 1252–1254.

28. Kleynmans, U., H. Heinzl, and W.P. Hammes. 1989. *Lactobacillus suebicus* sp. nov., an obligately heterofermentative *Lactobacillus* species isolated from fruit mashes. *Syst. Appl. Bacteriol.* 11:267–271.

29. Kloos, W.E., D.N. Ballard, C.G. George, J.A. Webster, R.J. Hubner, W. Ludwig, K.H. Schleifer, F. Fiedler, and K. Schubert. 1998. Delimiting the genus *Staphylococcus* through description of *Macrococcus caseolyticus* gen nov., comb. nov. and *Macrococcus equipercicus* sp. nov., *Macrococcus bovicus* sp. nov. and *Macrococcus carouselicus* sp. nov. *Int. J. System. Bacteriol.* 48:859–877.

30. Kreger-van Rij, N.J.W., ed. 1984. *The Yeasts: A Taxonomic Study*. Amsterdam: Elsevier.

31. Lane, D.J., B. Pace, G.J. Olsen, D.A. Stahl, M.L. Sogin, and N.R. Pace. 1985. Rapid determination of 16S ribosomal RNA sequences for phylogenetic analysis. *Proc. Natl. Acad. Sci.* 82:6955–6959.

32. MacDonell, M.T., and R.R. Colwell. 1985. Phylogeny of the Vibrionaceae, and recommendation for two new genera, *Listonella* and *Shewanella*. *Syst. Appl. Microbiol.* 6:171–182.

33. Mergaert, J., L. Verdonck, and K. Kersters. 1993. Transfer of *Erwinia ananas* (synonym, *Erwinia uredovora*) and *Erwinia stewartii* to the genus *Pantoea* emend. as *Pantoea ananas* (Serrano 1928) comb. nov. and *Pantoea stewartii* (Smith 1898) comb. nov., respectively, and description of *Pantoea stewartii* subsp. *indologenes* subsp. nov. *Int. J. Syst. Bacteriol.* 43:162–173.

34. Moore, J.E., B. McIlhatton, A. Shaw, P.G. Murphy, and J.S. Elborn. 2001a. Occurrence of *Burkholderia cepacia* in foods and waters: Clinical implications for patients with cystic fibrosis. *J. Food Protect.* 64:1076–1078.

35. Moore, J.E., T. Coenye, P. Vandamme, and J.S. Elborn. 2001b. First report of *Pandoraea norimbergensis* isolated from food—Potential clinical significance. *Food Microbiol.* 18:113–114.

36. Nazina, T.N., T.P. Tourova, A.B. Poltaraus, E.V. Novikova, A.A. Grigoryan, A.E. Ivanova, A.M. Lysenko, V.V. Petrunyaka, G.A. Osipov, S.S. Belyaev, and M.V. Ivanov. 2001. Taxonomic study of aerobic thermophilic bacilli: Descriptions of *Geobacillus subterraneus* gen nov., sp. nov. and *Geobacillus uzenensis* sp. nov from petroleum reservoirs and transfer of *Bacillus stearothermophilus*, *Bacillus thermocatenulatus*, *Bacillus thermoleovorans*, *Bacillus kaustrophilus*, *Bacillus thermoglucosidasius* and *Bacillus thermodenitrificans* to *Geobacillus* as the new combinations *G. stearothermophilus*, *G. thermocatenulatus*, *G. thermoleovorans*, *G. kaustophilus*, *G. thermoglucosidasius*, and *G. thermodenitrificans*. *Int. J. System. Evol. Microbiol.* 51:433–446.

37. Pitt, J.I., and A.D. Hocking. 1985. *Fungi and Food Spoilage*. New York: Academic Press.

38. Scheldeman, P., K. Goossens, M. Rodriguez-Diaz, A. Pil, J. Goris, L. Herman, P. De Vos, N.A. Logan, and M. Heyndrickx. 2004. *Paenibacillus lactis* sp. nov., isolated from raw and heat-treated milk. *Int. J. Syst. Evol. Microbiol.* 54:885–891.

39. Shaw, B.G., and J.B. Latty. 1988. A numerical taxonomic study of non-motile non-fermentative Gram-negative bacteria from foods. *J. Appl. Bacteriol.* 65:7–21.

40. Sneath, P.H.A., and D. Jones. 1976. *Brochothrix*, a new genus tentatively placed in the family Lactobacteriaceae. *Int. J. Syst. Bacteriol.* 26:102–104.

41. Splittstoesser, D.F., J.M. Lammers, D.L. Downing, and J.J. Churney. 1989. Heat resistance of *Eurotium herbariorum*, a xerophilic mold. *J. Food Sci.* 54:683–685.

42. Stackebrandt, E., E.C. Koch, O. Gvozdiak, and P. Schumann. 1995. Taxonomic dissection of the genus *Micrococcus*: *Kocuria* gen. nov., *Nesterenkonia* gen. nov., *Kytococcus* gen. nov., *Dermacoccus* gen. nov., and *Micrococcus* Cohn 1872 gen. emend. *Int. J. Syst. Bacteriol.* 45:682–692.

43. Stackebrandt, E., R.G.E. Murray, and H.G. Trüper. 1988. *Proteobacteria* classis nov., a name for the phylogenetic taxon that includes the "purple bacteria and their relatives". *Int. J. Syst. Bacteriol.* 38:321–325.

44. Talon, R., P.A.D. Grimont, F. Grimont, and J.M. Boefgras. 1988. *Brochothrix campestris* sp. nov. *Int. J. Syst. Bacteriol.* 38:99–102.

45. Vandamme, P., P.E. Falsen, R. Rossau, B. Hoste, P. Segers, R. Tytgat, and J. deLey. 1991. Revision of *Campylobacter*, *Helicobacter*, and *Wolinella* taxonomy emendation of generic descriptions and proposal of *Arcobacter* gen. nov. *Int. J. Syst. Bacteriol.* 41:88–103.

46. Viljoen, B.C., and J.L.F. Kock. 1989a. The genus *Candida* Berkhout nom. con-serv.—A historical account of its delimitation. *Syst. Appl. Microbiol.* 12:183–190.

47. Viljoen, B.C., and J.L.F. Kock. 1989b. Taxonomic study of the yeast genus *Candida* Berkhout. *Syst. Appl. Microbiol.* 12:91–102.

48. Wallbanks, S., A.J. Martinez-Murcia, J.L. Fryer, B.A. Phillips, and M.D. Collins. 1990. 16S rRNA sequence determination for members of the genus *Carnobacterium* and related lactic acid bacteria and description of *Vagococcus salmoninarum* sp. nov. *Int. J. Syst. Bacteriol.* 40:224–230.

49. Wauters, G., M. Janssens, A.G. Steigerwalt, and D.J. Brenner. 1988. *Yersinia mollaretti* sp. nov. and *Yersinia bercovieri* sp. nov., formerly called *Yersinia entercolitica* biogroups 3A and 3B. *Int. J. Syst. Bacteriol.* 38:424–429.

50. Wayne, L.G., D.J. Brenner, R.R. Colwell, P.A.D. Grimont, O. Kandler, M.I. Krichovsky, L.H. Moore, W.E.C. Moore, R.G.E. Murray, E. Stackebrandt, M.P. Starr, and H.G. Truper. 1987. Report of the ad hoc committee on reconciliation of approaches to bacterial systematics. *Int. J. Syst. Bacteriol.* 37:463–464.

51. Wesley, I.V. 1997. *Helicobacter* and *Arcobacter*: Potential human foodborne pathogens? *Trends Food Sci. Technol.* 8:293–299.

52. Wesley, I.V. 1996. *Helicobacter* and *Arcobacter* species: Risks for foods and beverages. *J. Food Protect.* 59:1127–1132.

53. Woese, C.R. 1987. Bacterial evolution. *Microbiol. Rev.* 51:221–271.

54. Woese, C.R., W.G. Weisburg, C.M. Hahn, B.J. Paster, L.B. Zablen, B.J. Lewis, T.J. Macke, W. Ludwig, and E. Stackebrandt. 1985. The phylogeny of purple bacteria; The gamma subdivision. *System. Appl. Microbiol.* 6:25–33.

55. Woese, C.R., E. Stackebrandt, W.G. Weisburg, B.J. Paster, M.T. Madigan, V.J. Fowler, C.M. Hahn, P. Blanz, R. Gupta, K.H. Nealson, and G.E. Fox. 1984a. The phylogeny of purple bacteria: The alpha subdivision. *System. Appl. Microbiol.* 5:315–326.

56. Woese, C.R., W.G. Weisburg, B.J. Paster, C.M. Hahn, R.S. Tanner, N.R. Krieg, H.-P. Koops, H. Harms, and E. Stackebrandt. 1984b. The phylogeny of purple bacteria: The beta subdivision. *System. Appl. Microbiol.* 5:327–336.

57. Wolf, A., A. Fritze, M. Hagemann, and G. Berg. 2002. *Stenotrophomonas rhizophila* sp. nov., a novel plant-associated bacterium with antifungal properties. *Int. J. System. Evol. Microbiol.* 52:1937–1944.

58. Yabuuchi, E., Y. Kosako, N. Fujiwara, T. Naka, I. Matsunaga, H. Ogura, and K. Kobayashi. 2002. Emendation of the genus *Sphingomonas* Yabuuchi et al. 1990 and junior objective synonymy of the species of three genera, *Sphingobium*, *Novosphingobium*, and *Sphingopyxis*, in conjunction with *Blastomonas ursincola*. *Int. Syst. Evol. Microbiol.* 52:1485–1496.

CHAPTER 3

Intrinsic and Extrinsic Parameters of Foods That Affect Microbial Growth

As our foods are of plant and/or animal origin, it is worthwhile to consider those characteristics of plant and animal tissues that affect the growth of microorganisms. The plants and animals that serve as food sources have all evolved mechanisms of defense against the invasion and proliferation of microorganisms, and some of these remain in effect in fresh foods. By taking these natural phenomena into account, one can make effective use of each or all in preventing or retarding the growth of pathogenic and spoilage organisms in the products that are derived from them.

INTRINSIC PARAMETERS

The parameters of plant and animal tissues that are an inherent part of the tissues are referred to as *intrinsic parameter.*[33] These parameters are as follows:

1. pH
2. Moisture content
3. Oxidation–reduction potential (Eh)
4. Nutrient content
5. Antimicrobial constituents
6. Biological structures

Each of these substrate-limiting factors is discussed below, with emphasis placed on their effects on microorganisms in foods.

pH

It has been well established that most microorganisms grow best at pH values around 7.0 (6.6–7.5), whereas few grow below 4.0 (Figure 3–1). Bacteria tend to be more fastidious in their relationships

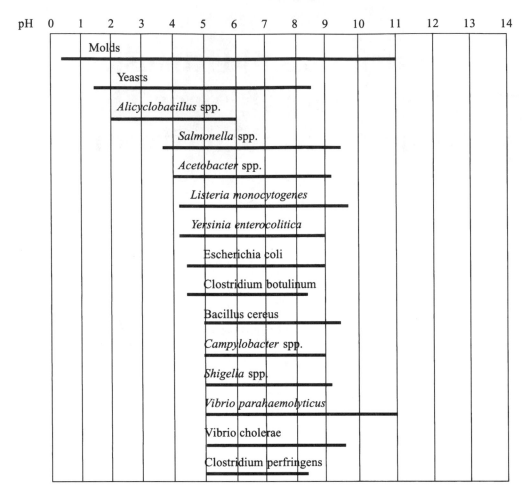

Figure 3–1 Approximate pH growth ranges for some foodborne organisms. The pH ranges for *L. monocytogenes* and *S. aureus* are similar.

to pH than molds and yeasts, with the pathogenic bacteria being the most fastidious. With respect to pH minima and maxima of microorganisms, those represented in Figure 3–1 should not be taken to be precise boundaries, as the actual values are known to be dependent on other growth parameters. For example, the pH minima of certain lactobacilli have been shown to be dependent on the type of acid used, with citric, hydrochloric, phosphoric, and tartaric acids permitting growth at a lower pH value than acetic or lactic acids. In the presence of 0.2 M NaCl, *Alcaligenes faecalis* has been shown to grow over a wider pH range than in the absence of NaCl or in the presence of 0.2 M sodium citrate (Figure 3–2). Of the foods presented in Table 3–1, it can be seen that fruits, soft drinks, vinegar, and wines all fall below the point at which bacteria normally grow. The excellent keeping quality of these products is due in great part to pH. It is a common observation that fruits generally undergo mold and yeast spoilage, and this is due to the capacity of these organisms to grow at pH values <3.5, which is considerably below the minima for most food-spoilage and all food-poisoning bacteria (see Table 3–2).

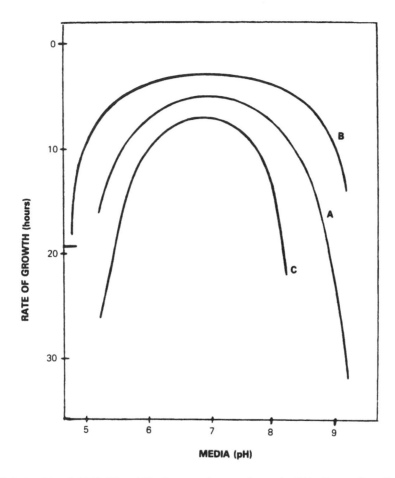

Figure 3–2   Relationship of pH, NaCl, and Na citrate on the rate of growth of *Alcaligenes faecalis* in 1% peptone: A = 1% peptone; B = 0.2 M NaCl; C = 1% peptone + 0.2 M Na citrate. *Source*: Redrawn from Sherman and Holm[48]; used with permission of the publisher.

It may be noted from Table 3–3 that most of the meats and seafoods have a final ultimate pH of about 5.6 and above. This makes these products susceptible to bacteria as well as to mold and yeast spoilage. Most vegetables have lower pH valucs than fruits, and, consequently, vegetables should be subject more to bacterial than fungal spoilage.

With respect to the keeping quality of meats, it is well established that meat from fatigued animals spoils faster than that from rested animals and that this is a direct consequence of final pH attained upon completion of rigor mortis. Upon the death of a well-rested meat animal, the usual 1% glycogen is converted to lactic acid, which directly causes a depression in pH values from about 7.4 to about 5.6, depending on the type of animal. Callow[11] found the lowest pH values for beef to be 5.1 and the highest 6.2 after rigor mortis. The usual pH value attained upon completion of rigor mortis of beef is around 5.6.[5] The lowest and highest values for lamb and pork were found by Callow to be 5.4 and 6.7, and 5.3 and 6.9, respectively. Briskey[8] reported that the ultimate pH of pork may be as low as

Table 3–1 Approximate pH Values of Some Fresh Fruits and Vegetables

Product	pH	Product	pH
Vegetables		**Fruits**	
Asparagus (buds and stalks)	5.7–6.1	Apples	2.9–3.3
Beans (string and Lima)	4.6–6.5	Apple cider	3.6–3.8
Beets (sugar)	4.2–4.4	Apple juice	3.3–4.1
Broccoli	6.5	Bananas	4.5–4.7
Brussels sprouts	6.3	Figs	4.6
Cabbage (green)	5.4–6.0	Grapefruit (juice)	3.0
Carrots	4.9–5.2; 6.0	Grapes	3.4–4.5
Cauliflower	5.6	Limes	1.8–2.0
Celery	5.7–6.0	Melons (honeydew)	6.3–6.7
Corn (sweet)	7.3	Oranges (juice)	3.6–4.3
Cucumbers	3.8	Plums	2.8–4.6
Eggplant	4.5	Watermelons	5.2–5.6
Lettuce	6.0		
Olives	3.6–3.8		
Onions (red)	5.3–5.8		
Parsley	5.7–6.0		
Parsnip	5.3		
Potatoes (tubers and sweet)	5.3–5.6		
Pumpkin	4.8–5.2		
Rhubarb	3.1–3.4		
Rutabaga	6.3		
Spinach	5.5–6.0		
Squash	5.0–5.4		
Tomatoes (whole)	4.2–4.3		
Turnips	5.2–5.5		

approximately 5.0 under certain conditions. The effect of pH of this magnitude on microorganisms, especially bacteria, is obvious. With respect to fish, it is known that halibut, which usually attains an ultimate pH of about 5.6, has better keeping qualities than most other fish, whose ultimate pH values range between 6.2 and 6.6.[42]

Some foods are characterized by inherent acidity; others owe their acidity or pH to the actions of certain microorganisms. The latter type is referred to as biological acidity and is displayed by products such as fermented milks, sauerkraut, and pickles. Regardless of the source of acidity, the effect on keeping quality appears to be the same.

Some foods are better able to resist changes in pH than others. Those that tend to resist changes in pH are said to be *buffered*. In general, meats are more highly buffered than vegetables. Contributing to the buffering capacity of meats are their various proteins. Vegetables are generally low in proteins and, consequently, lack the buffering capacity to resist changes in their pH during the growth of microorganisms (see Tables 6–4 and 6–5 for the general chemical composition of vegetables).

The capacity of *E. coli* to grow in three retail mustards was assessed, and with an inoculum of 10^6 cfu/g of this pathogen, its growth was inhibited in all three products.[31] The organism was not detected in dijon-style mustard (pH 3.55–3.60) beyond 3 h at room temperature, and after 2 days at 5°C. In yellow- and deli-style mustards (pH 3.30 and 3.38, respectively), the organism was not detectable beyond 1 h.[31]

Table 3–2   Reported Minimum pH Values for the Growth
of Some Foodborne Bacteria

Aeromonas hydrophila	ca. 6.0
Asaia siamensis	3.0
Alicyclobacillus acidocaldarius	2.0
Bacillus cereus	4.9
Botrytis cinerea	2.0
Clostridium botulinum, Group I	4.6
C. botulinum, Group II	5.0
C. perfringens	5.0
Escherichia coli 0157:H7	4.5
Gluconobacter spp.	3.6
Lactobacillus brevis	3.16
L. plantarum	3.34
L. sakei	3.0
Lactococcus lactis	4.3
Listeria monocytogenes	4.1
Penicillium roqueforti	3.0
Propioniibacterium cyclohexanicum	3.2
Plesiomonas shigelloides	4.5
Pseudomonas fragi	ca. 5.0
Salmonella spp.	4.05
Shewanella putrefaciens	ca. 5.4
Shigella flexneri	5.5–4.75
S. sonnei	5.0–4.5
Staphylococcus aureus	4.0
Vibrio parahaemolyticus	4.8
Yersinia enterocolitica	4.18
Zygosaccharomyces bailii	1.8

The natural or inherent acidity of foods, especially fruits, may have evolved as a way of protecting tissues from destruction by microorganisms. It is of interest that fruits should have pH values below those required by many spoilage organisms. The biological function of the fruit is the protection of the plant's reproductive body, the seed. This one fact alone has no doubt been quite important in the evolution of present-day fruits. Although the pH of a living animal favors the growth of most spoilage organisms, other intrinsic parameters come into play to permit the survival and growth of the animal organism.

Although acidic pH values are of greater use in inhibiting microorganisms, alkaline values in the range of pH 12–13 are known to be destructive, at least to some bacteria. For example, the use of $CaOH_2$ to produce pH values in this range has been shown to be destructive to *Listeria monocytogenes* and other foodborne pathogens on some fresh foods.

pH Effects

Adverse pH affects at least two aspects of a respiring microbial cell: the functioning of its enzymes and the transport of nutrients into the cell. The cytoplasmic membrane of microorganisms is relatively impermeable to H^+ and OH^- ions. Their concentration in the cytoplasm therefore probably remains

Table 3–3 Approximate pH Values of Dairy, Meat, Poultry, and Fish Products

Product	pH	Product	pH
Dairy products		**Fish and shellfish**	
Butter	6.1–6.4	Fish (most species)*	6.6–6.8
Buttermilk	4.5	Clams	6.5
Milk	6.3–6.5	Crabs	7.0
Cream	6.5	Oysters	4.8–6.3
Cheese (American mild	4.9; 5.9	Tuna fish	5.2–6.1
and cheddar)		Shrimp	6.8–7.0
Meat and poultry		Salmon	6.1–6.3
Beef (ground)	5.1–6.2	White fish	5.5
Ham	5.9–6.1		
Veal	6.0		
Chicken	6.2–6.4		
Liver	6.0–6.4		

*Just after death.

reasonably constant despite wide variations that may occur in the pH of the surrounding medium.[45] The intracellular pH of resting baker's yeast cells was found by Conway and Downey[16] to be 5.8. Although the outer region of the cells during glucose fermentation was found to be more acidic, the inner cell remained more alkaline. On the other hand, Peña et al.[37] did not support the notion that the pH of yeast cells remains constant with variations in pH of the medium. It appears that the internal pH of almost all cells is near neutrality. Bacteria such as *Sulfolobus* and *Methanococcus* may be exceptions, however. When microorganisms are placed in environments below or above neutrality, their ability to proliferate depends on their ability to bring the environmental pH to a more optimum value or range. When placed in acid environments, the cells must either keep H^+ from entering or expel H^+ ions as rapidly as they enter. Such key cellular compounds as DNA and ATP require neutrality. When most microorganisms grow in acid media, their metabolic activity results in the medium or substrate becoming less acidic, whereas those that grow in high pH environments tend to effect a lowering of pH. The amino acid decarboxylases that have optimum activity at around pH 4.0 and almost no activity at pH 5.5 cause a spontaneous adjustment of pH toward neutrality when cells are grown in the acid range. Bacteria such as *Clostridium acetobutylicum* raise the substrate pH by reducing butyric acid to butanol, whereas *Enterobacter aerogenes* produces acetoin from pyruvic acid to raise the pH of its growth environment. When amino acids are decarboxylated, the increase in pH occurs from the resulting amines. When grown in the alkaline range, a group of amino acid deaminases that have optimum activity at about pH 8.0 and cause the spontaneous adjustment of pH toward neutrality as a result of the organic acids that accumulate.

With respect to the transport of nutrients, the bacterial cell tends to have a residual negative charge. Therefore, nonionized compounds can enter cells, whereas ionized compounds cannot. At neutral or alkaline pH, organic acids do not enter, whereas at acid pH values, these compounds are nonionized and can enter the negatively charged cells. Also, the ionic character of side chain ionizable groups is affected on either side of neutrality, resulting in increasing denaturation of membrane and transport enzymes.

Among the other effects that are exerted on microorganisms by adverse pH is that of the interaction between H^+ and the enzymes in the cytoplasmic membrane. The morphology of some microorganisms can be affected by pH. The length of the hyphae of *Penicillium chrysogenum* has been reported to decrease when grown in continuous culture where pH values increased above 6.0. Pellets of mycelium rather than free hyphae were formed at about pH 6.7.[45] Extracellular H^+ and K^+ may be in competition where the latter stimulates fermentation, for example, while the former represses it. The metabolism of glucose by yeast cells in an acid medium was markedly stimulated by K^+.[46] Glucose was consumed 83% more rapidly in the presence of K^+ under anaerobic conditions and 69% more under aerobic conditions.

Other environmental factors interact with pH. With respect to temperature, the pH of the substrate becomes more acid as the temperature increases. Concentration of salt has a definite effect on pH growth rate curves, as illustrated in Figure 3–2, where it can be seen that the addition of 0.2 M NaCl broadened the pH growth range of *Alcaligenes faecalis*. A similar result was noted for *Escherichia coli* by these investigators. When the salt content exceeds this optimal level, the pH growth range is narrowed. An adverse pH makes cells much more sensitive to toxic agents of a wide variety, and young cells are more susceptible to pH changes than older or resting cells.

When microorganisms are grown on either side of their optimum pH range, an increased lag phase results. The increased lag would be expected to be of longer duration if the substrate is a highly buffered one in contrast to one that has poor buffering capacity. In other words, the length of the lag phase may be expected to reflect the time necessary for the organisms to bring the external environment within their optimum pH growth range. Analysis of the substances that are responsible for the adverse pH is of value in determining not only the speed of subsequent growth, but also the minimum pH at which salmonellae would initiate growth. Chung and Goepfert[14] found the minimum pH to be 4.05 when hydrochloric and citric acids were used, but 5.4 and 5.5 when acetic and propionic acids were used, respectively. This is undoubtedly a reflection of the ability of the organisms to alter their external environment to a more favorable range in the case of hydrochloric and citric acids as opposed to the other acids tested. It is also possible that factors other than pH come into play in the varying effects of organic acids as growth inhibitors. For more information on pH and acidity, see Corlett and Brown.[17]

Moisture Content

One of the oldest methods of preserving foods is drying or desiccation; precisely how this method came to be used is not known. The preservation of foods by drying is a direct consequence of removal or binding of moisture, without which microorganisms do not grow. It is now generally accepted that the water requirements of microorganisms should be described in terms of the *water activity* (a_w) in the environment. This parameter is defined by the ratio of the water vapor pressure of food substrate to the vapor pressure of pure water at the same temperature: $a_w = p/p_o$, where p is the vapor pressure of the solution and p_o is the vapor pressure of the solvent (usually water). This concept is related to relative humidity (RH) in the following way: $RH = 100 \times a_w$.[13] Pure water has an a_w of 1.00, a 22% NaCl solution (w/v) has an a_w of 0.86, and a saturated solution of NaCl has an a_w of 0.75 (Table 3–4).

The water activity (a_w) of most fresh foods is above 0.99. The minimum values reported for the growth of some microorganisms in foods are presented in Table 3–5 (see also Chapter 18). In general, bacteria require higher values of a_w for growth than fungi, with Gram-negative bacteria having higher requirements than Gram positives. Most spoilage bacteria do not grow below $a_w = 0.91$, whereas spoilage molds can grow as low as 0.80. With respect to food-poisoning bacteria, *Staphylococcus aureus* can grow as low as 0.86, whereas *Clostridium botulinum* does not grow below 0.94. Just as yeasts and molds grow over a wider pH range than bacteria, the same is true for a_w. The lowest reported value for foodborne bacteria is 0.75 for halophiles (literally, "salt-loving"), whereas xerophilic

Table 3–4 Relationship between Water Activity and
Concentration of Salt Solutions

Water Activity	Sodium Chloride Concentration	
	Molal	Percent, w/v
0.995	0.15	0.9
0.99	0.30	1.7
0.98	0.61	3.5
0.96	1.20	7
0.94	1.77	10
0.92	2.31	13
0.90	2.83	16
0.88	3.33	19
0.86	3.81	22

Source: From *The Science of Meat and Meat Products*, by the
American Meat Institute Foundation. W.H. Freeman and Company,
San Francisco; copyright © 1960.

Table 3–5 Approximate Minimum a_w Values for Growth of Microorganisms Important in Foods

Organisms	a_w	Organisms	a_w
Groups		**Groups**	
Most spoilage bacteria	0.9	Halophilic bacteria	0.75
Most spoilage yeasts	0.88	Xerophilic molds	0.61
Most spoilage molds	0.80	Osmophilic yeasts	0.61
Specific Organisms		**Specific Organisms**	
Clostridium botulinum, type E	0.97	*Candida scottii*	0.92
Pseudomonas spp.	0.97	*Trichosporon pullulans*	0.91
Acinetobacter spp.	0.96	*Candida zeylanoides*	0.90
Escherichia coli	0.96	*Geotrichum candidum*	ca. 0.9
Enterobacter aerogenes	0.95	*Trichothecium* spp.	ca. 0.90
Bacillus subtilis	0.95	*Byssochlamys nivea*	ca. 0.87
Clostridium botulinum, types A and B	0.94	*Staphylococcus aureus*	0.86
Candida utilis	0.94	*Alternaria citri*	0.84
Vibrio parahaemolyticus	0.94	*Penicillium patulum*	0.81
Botrytis cinerea	0.93	*Eurotium repens*	0.72
Rhizopus stolonifer	0.93	*Aspergillus glaucus**	0.70
Mucor spinosus	0.93	*Aspergillus conicus*	0.70
		Aspergillus echinulatus	0.64
		Zygosaccharomyces rouxii	0.62
		Xeromyces bisporus	0.61

*Perfect stages of the *A. glaucus* group are found in the genus *Eurotium*.

("dry-loving") molds and osmophilic (preferring high osmotic pressures) yeasts have been reported to grow at a_w values of 0.65 and 0.61, respectively (Table 3–5). When salt is employed to control a_w, an extremely high level is necessary to achieve a_w values below 0.80 (see Table 3–4).

Certain relationships have been shown to exist among a_w, temperature, and nutrition. First, at any temperature, the ability of microorganisms to grow is reduced as the a_w is lowered. Second, the range of a_w over which growth occurs is greatest at the optimum temperature for growth; and third, the presence of nutrients increases the range of a_w over which the organisms can survive.[32] The specific values given in Table 3–5, then, should be taken only as reference points, as a change in temperature or nutrient content might permit growth at lower values of a_w.

Effects of Low a_w

The general effect of lowering a_w below optimum is to increase the length of the lag phase of growth and to decrease the growth rate and size of final population. This effect may be expected to result from adverse influences of lowered water on all metabolic activities because all chemical reactions of cells require an aqueous environment. It must be kept in mind, however, that a_w is influenced by other environmental parameters such as pH, temperature of growth, and Eh. In their study of the effect of a_w on the growth of *Enterobacter aerogenes* in culture media, Wodzinski and Frazier[54] found that the lag phase and generation time were progressively lengthened until no growth occurred with a lowering of a_w. The minimum a_w was raised, however, when the incubation temperature was decreased. When both the pH and temperature of incubation were made unfavorable, the minimum a_w for growth was higher. The interaction of a_w, pH, and temperature on the growth of molds on jam was shown by Horner and Anagnostopoulos.[24] The interaction between a_w and temperature was the most significant.

In general, the strategy employed by microorganisms as protection against osmotic stress is the intracellular accumulation of compatible solutes. Halophiles (e.g., *Halobacterium* spp.) maintain osmotic equilibrium by maintaining the concentration of KCl in their cytoplasm equal to that of the suspending menstruum, and this is referred to as the "salt in cytoplasm" response. Nonhalophiles accumulate compatible solutes (osmolytes) in a biphasic manner. The first response is to increase K^+ (and endogenously synthesized glutamate), and the second is to increase, either by de novo synthesis or by uptake, compatible solutes. The latter are very soluble molecules that have no net charge at physiological pH, and they do not adhere to or react with intracellular macromolecules (see reference 49). The three most common compatible solutes in most bacteria are carnitine, glycine betaine, and proline. Carnitine may be synthesized de novo, but the other two are generally not. Proline is synthesized by some Gram-positive bacteria while it is transported by Gram negatives. The solubility of glycine betaine in 100 ml of water at $25°C$ is 160 g; it is 162 g for proline. Glycine betaine is employed by more living organisms that the other two osmolytes noted.

carnitine   glycine betaine   proline

The uptake of osmolytes is mediated by a transport system. In *L. monocytogenes*, glycine betaine is transported by BetL (it couples betaine accumulation to a Na^+-motive) and Gbu (transports betaine) whereas the transporter for carnitine is OpuC.[1,49] Although some Gram-positive bacteria accumulate proline, it is concentrated to higher levels by Gram-negative bacteria. The three transporter systems in *E. coli* and *S.* Typhimurium are PutP, ProP, and ProU, with ProP being the most effective. It has been shown that the overproduction of proline by mutants of *L. monocytogenes* did not lead to changes in mouse virulence.[49] Under salt stress, *L. monocytogenes* produces 12 proteins one of which is highly similar to the Ctc protein of *B. subtilis*, and it is involved in osmotic stress tolerance in the absence of osmoprotectants in the medium.[21] The sigma factor-B (δ^B; see Chapter 22) plays a major role in the regulation of carnitine utilization in *L. monocytogenes*, but it is not essential for betaine utilization.[20]

Because it can grow at $4°C$, evidence has been presented that low-temperature growth of *L. monocytogenes* is aided by the accumulation of glycine betaine.[29] The same is true for *Yersinia enterocolitica*, where osmotically stressed as well as cold-stressed cells accumulated osmolytes including glycine betaine.[36] Temperature downshock and osmotic upshock caused a 30-fold uptake of radiolabeled glycine betaine.[36] In at least one strain of *L. monocytogenes*, glycine betaine transport is mediated by Gbu and BetL; and to a lesser extent OpuC.[1]

With regard to specific compounds used to lower water activity, results akin to those seen with adsorption and desorption systems (see Chapter 18) have been reported. In a study on the minimum a_w for the growth and germination of *Clostridium perfringens*, Kang et al.[28] found the value to be between 0.97 and 0.95 in complex media when sucrose or NaCl was used to adjust a_w but 0.93 or below when glycerol was used. In another study, glycerol was found to be more inhibitory than NaCl to relatively salt-tolerant bacteria, but less inhibitory than NaCl to salt-sensitive species when compared at similar levels of a_w in complex media.[30] In their studies on the germination of *Bacillus* and *Clostridium* spores, Jakobsen and Murrell[25] observed strong inhibition of spore germination when a_w was controlled by NaCl or $CaCl_2$, but less inhibition when glucose or sorbitol was used, and very little inhibition when glycerol, ethylene, glycol, acetamide, or urea, were used. The germination of clostridial spores was completely inhibited at $a_w = 0.95$ with NaCl, but no inhibition occurred at the same a_w when urea, glycerol, or glucose was employed. In another study, the limiting a_w for the formation of mature spores by *B. cereus* strain T was shown to be about 0.95 for glucose, sorbitol, and NaCl, but about 0.91 for glycerol.[26] Both yeasts and molds have been found to be more tolerant to glycerol than to sucrose.[24] Using a glucose minimal medium and *Pseudomonas fluorescens*, Prior[39] found that glycerol permitted growth at lower a_w values than either sucrose or NaCl. It was further shown by this researcher that the catabolism of glucose, sodium lactate, and DL-arginine was completely inhibited by a_w values greater than the minimum for growth when a_w was controlled with NaCl. The control of a_w with glycerol allowed catabolism to continue at a_w values below that for growth on glucose. In all cases where NaCl was used by this investigator to adjust the a_w, substrate catabolism ceased at an a_w greater than the minimum for growth, whereas glycerol permitted catabolism at lower a_w values than the minimum for growth. In spite of some reports to the contrary, it appears that glycerol is less inhibitory to respiring organisms than agents such as sucrose and NaCl.

Osmophilic yeasts accumulate polyhydric alcohols to a concentration commensurate with their extracellular a_w. According to Pitt,[38] the xerophilic fungi accumulate compatible solutes or osmoregulators as a consequence of the need for high internal solutes if growth at a low a_w is to be possible. In a comparative study of xerotolerant and nonxerotolerant yeasts to water stress, Edgley and Brown[19] found that *Zygosaccharomyces rouxii* responded to a low a_w controlled by polyethylene glycol by retaining within the cells increasing levels of glycerol. However, the amount did not change greatly, nor did the level of arabitol change appreciably by a_w. On the other hand, a nontolerant *S. cerevisiae* responded to a lowering of a_w by synthesizing more glycerol but retaining less. The *Z. rouxii* response to a low a_w was at the level of glycerol permeation/transport, whereas that for *S. cerevisiae* was metabolic. It appears from this study that a low a_w forces *S. cerevisiae* to divert a greater proportion of its metabolic activity to glycerol production accompanied by an increase in the amount of glucose consumed during growth. In a later study, it was noted that up to 95% of the external osmotic pressure exerted on *S. cerevisiae, Z. rouxii,* and *Debaryomyces hansenii* may be counterbalanced by an increase in glycerol.[43] *Z. rouxii* accumulates more glycerol under stress, whereas ribitol remains constant.

It is known that the growth of at least some cells may occur in high numbers at reduced a_w values, while certain extracellular products are not produced. For example, reduced a_w results in the cessation of enterotoxin B production by *S. aureus* even though high numbers of cells are produced at the same time.[50,51] In the case of *Neurospora crassa*, a low a_w resulted in nonlethal alterations of permeability of the cell membrane, leading to a loss of several essential molecules.[12] Similar results were observed with electrolytes or nonelectrolytes.

Overall, the effect of a lowered a_w on the nutrition of microorganisms appears to be of a general nature where cell requirements that must be mediated through an aqueous milieu are progressively shut off. In addition to the effect on nutrients, a lowered a_w undoubtedly has adverse effects on the functioning of the cell membrane, which must be kept in a fluid state. The drying of internal parts of cells would be expected to occur upon placing cells in a medium of lowered a_w to a point where the equilibrium of water between cells and substrate occurs. Although the mechanisms are not entirely clear, all microbial cells may require the same effective internal a_w. Those that can grow under extreme conditions of a low a_w apparently do so by virtue of their ability to concentrate salts, polyols, and amino acids (and possibly other types of compounds) to internal levels sufficient not only to prevent the cells from losing water, but that it may allow the cell to extract water from the water-depressed external environment. For more information, see references 49, 51.

Oxidation–Reduction Potential

It has been known for decades that microorganisms display varying degrees of sensitivity to the oxidation–reduction potential (O/R, Eh) of their growth medium.[23] The O/R potential of a substrate may be defined generally as the ease with which the substrate loses or gains electrons. When an element or compound loses electrons, the substrate is oxidized, whereas a substrate that gains electrons becomes reduced:

$$Cu \underset{\text{reduction}}{\overset{\text{oxidation}}{\rightleftharpoons}} Cu + e$$

Oxidation may also be achieved by the addition of oxygen, as illustrated in the following reaction:

$$2Cu + O_2 \rightarrow 2CuO$$

Therefore, a substance that readily gives up electrons is a good reducing agent, and one that readily takes up electrons is a good oxidizing agent. When electrons are transferred from one compound to another, a potential difference is created between the two compounds. This difference may be measured by use of an appropriate instrument, and expressed as millivolts (mV). The more highly oxidized a substance, the more positive will be its electrical potential; the more highly reduced a substance, the more negative will be its electrical potential. When the concentration of oxidant and reductant is equal, a zero electrical potential exists. The O/R potential of a system is expressed by the symbol Eh. Aerobic microorganisms require positive Eh values (oxidized) for growth, whereas anaerobes require negative

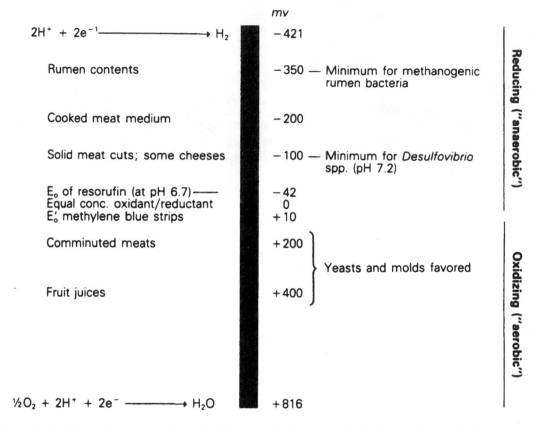

Figure 3–3 Schematic representation of oxidation–reduction potentials relative to the growth of certain microorganisms.

Eh values (reduced) (Figure 3–3). Among the substances in foods that help to maintain reducing conditions are –SH groups in meats and ascorbic acid, and reducing sugars in fruits and vegetables. In regard to the maximum positive and negative mV values in Fig. 3–3, not only are they not necessary for the growth of aerobes or anaerobes, but these extreme values can also be lethal to the respective group (see EO water section in Chapter 13).

The O/R potential of a food is determined by the following:

1. The characteristic O/R potential of the original food.
2. The *poising capacity;* that is, the resistance to change in potential of the food.
3. The oxygen tension of the atmosphere about the food.
4. The access that the atmosphere has to the food.

With respect to Eh requirements of microorganisms, some bacteria require reduced conditions for growth initiation (Eh of about −200 mV), whereas others require a positive Eh for growth. In the former category are the anaerobic bacteria such as the genus *Clostridium*; in the latter belong aerobic bacteria such as some members of the genus *Bacillus*. Some aerobic bacteria actually grow better under slightly reduced conditions, and these organisms are referred to as *microaerophiles*. Examples of microaerophilic bacteria are lactobacilli and campylobacters. Some bacteria have the capacity to grow under either aerobic or anaerobic conditions. Such types are referred to as *facultative anaerobes*. Most molds and yeasts encountered in and on foods are aerobic, although a few tend to be facultative anaerobes.

With regard to the Eh of foods, plant foods, especially plant juices, tend to have Eh values of from +300 to 400 mV. It is not surprising to find that aerobic bacteria and molds are the common cause of spoilage of products of this type. Solid meats have Eh values of around −200 mV; in minced meats, the Eh is generally around 200 mV. Cheeses of various types have been reported to have Eh values on the negative side, from −20 to around −200 mV.

With respect to the Eh of pre-rigor as opposed to post-rigor muscles, Barnes and Ingram[2,3] undertook a study of the measurement of Eh in muscle over periods of up to 30 hours postmortem and its effect on the growth of anaerobic bacteria. These authors found that the Eh of the sternocephalicus muscle of the horse immediately after death was +250 mV, at which time clostridia failed to multiply. At 30 hours postmortem, the Eh had fallen to about 30 mV in the absence of bacterial growth. When bacterial growth was allowed to occur, the Eh fell to about 250 mV. Growth of clostridia was observed at Eh values of 36 mV and below. These authors confirmed for horse meat the finding for whale meat: that anaerobic bacteria do not multiply until the onset of rigor mortis because of the high Eh in pre-rigor meat. The same is undoubtedly true for beef, pork, and other meats of this type.

Eh Effects

Microorganisms affect the Eh of their environments during growth just as they do pH. This is true especially of aerobes, which can lower the Eh of their environment while anaerobes cannot. As aerobes grow, O_2 in the medium is depleted, resulting in the lowering of Eh. Growth is not slowed, however, as much as might be expected due to the ability of cells to make use of O_2-donating or hydrogen-accepting substances in the medium. The result is that the medium becomes poorer in oxidizing and richer in reducing substances.[32] The Eh of a medium can be reduced by microorganisms by their production of certain metabolic byproducts such as H_2S, which has the capacity to lower Eh to −300 mV. Because H_2S reacts readily with O_2, it will accumulate only in anaerobic environments.

Eh is dependent on the pH of the substrate, and the direct relationship between these two factors is the rH value defined in the following way:

$$Eh = 2.303 \frac{RT}{F} (rH - 2pH)$$

where $R = 8.315$ joules, $F = 96,500$ coulombs, and T is the absolute temperature.[34] Therefore, the pH of a substrate should be stated when Eh is given. Normally Eh is taken at pH 7.0 (expressed Eh′). When taken at pH 7.0, 25°C, and with all concentrations at 1.0 M, Eh = Eh′$_0$ (simplified Nernst equation). In nature, Eh tends to be more negative under progressively alkaline conditions.

Among naturally occurring nutrients, ascorbic acid and reducing sugars in plants and fruits and −SH groups in meats are of primary importance. The presence or absence of appropriate quantities of oxidizing—reducing agents in a medium is of obvious value to the growth and activity of all microorganisms.

While the growth of anaerobes is normally believed to occur at reduced values of Eh, the exclusion of O_2 may be necessary for some anaerobes. When *Clostridium perfringens*, *Bacteroides fragilis*, and *Peptococcus magnus* were cultured in the presence of O_2, inhibition of growth occurred even when the medium was at a negative Eh of -50 mV.[52] These investigators found that growth occurred in media with an Eh as high as 325 mV when no O_2 was present.

With regard to the effect of Eh on lipid production by *Saccharomyces cerevisiae*, it has been shown that anaerobically grown cells produce a lower total level, a highly variable glyceride fraction, and decreased phospholipid and sterol components as compared to aerobically grown cells.[41] The lipid produced by anaerobically grown cells was characterized by a high content (up to 50% of total acid) of 8:0 to 14:0 acids and a low level of unsaturated fatty acid in the phospholipid fraction. In aerobically grown cells, 80–90% of the fatty acid component was associated with glyceride, and the phospholipid was found to be 16:1 and 18:1 acids. Unlike aerobically grown cells, anaerobically grown *S. cerevisiae* cells were found to have a lipid and sterol requirement.

Nutrient Content

In order to grow and function normally, the microorganisms of importance in foods require the following:

1. water
2. source of energy
3. source of nitrogen
4. vitamins and related growth factors
5. minerals

The importance of water to the growth and welfare of microorganisms is presented earlier in this chapter. With respect to the other four groups of substances, molds have the lowest requirement, followed by Gram-negative bacteria, yeasts, and Gram-positive bacteria.

As sources of energy, foodborne microorganisms may utilize sugars, alcohols, and amino acids. Some microorganisms are able to utilize complex carbohydrates such as starches and cellulose as sources of energy by first degrading these compounds to simple sugars. Fats are also used by microorganisms as sources of energy, but these compounds are attacked by a relatively small number of microbes in foods.

The primary nitrogen sources utilized by heterotrophic microorganisms are amino acids. A large number of other nitrogenous compounds may serve this function for various types of organisms. Some microbes, for example, are able to utilize nucleotides and free amino acids, whereas others are able to utilize peptides and proteins. In general, simple compounds such as amino acids will be utilized by almost all organisms before any attack is made on the more complex compounds such as high-molecular-weight proteins. The same is true of polysaccharides and fats.

Microorganisms may require B vitamins in low quantities, and almost all natural foods have an abundant quantity for those organisms that are unable to synthesize their essential requirements. In general, Gram-positive bacteria are the least synthetic and must therefore be supplied with one or more of these compounds before they will grow. The Gram-negative bacteria and molds are able to synthesize most or all of their requirements. Consequently, these two groups of organisms may be found growing on foods low in B vitamins. Fruits tend to be lower in B vitamins than meats, and this fact, along with the usual low pH and positive Eh of fruits, helps to explain the usual spoilage of these products by molds rather than bacteria.

Antimicrobial Constituents

The stability of some foods against attack by microorganisms is due to the presence of certain naturally occurring substances that possess and express antimicrobial activity. Some plant species are known to contain essential oils that possess antimicrobial activity. Among these are eugenol in cloves, allicin in garlic, cinnamic aldehyde and eugenol in cinnamon, allyl isothiocyanate in mustard, eugenol and thymol in sage, and carvacrol (isothymol) and thymol in oregano.[47] Cow's milk contains several antimicrobial substances, including lactoferrin (see below), conglutinin, and the lactoperoxidase system (see below). Raw milk has been reported to contain a rotavirus inhibitor that can inhibit up to 10^6 pfu (plaqueforming units)/ml. It is destroyed by pasteurization. Milk casein as well as some free fatty acids have been shown to be antimicrobial under certain conditions.

Eggs contain lysozyme, as does milk, and this enzyme, along with conalbumin, provides fresh eggs with a fairly efficient antimicrobial system. The hydroxycinnamic acid derivatives (*p*-coumaric, ferulic, caffeic, and chlorogenic acids) found in fruits, vegetables, tea, molasses, and other plant sources all show antibacterial and some antifungal activity. Lactoferrin is an iron-binding glycoprotein that is inhibitory to a number of foodborne bacteria and its use as a microbial blocking agent on beef carcasses is discussed in Chapter 13. Ovotransferrin appears to be the inhibitory substance in raw egg white that inhibits *Salmonella enteritidis*.[4]

Cell vacuoles of cruciferous plants (cabbage, Brussels sprouts, broccoli, turnips, etc.) contain *glucosinolates*, which upon injury or mechanical disruption, yield isothiocyanates. Some of the latter possess antifungal as well as antibacterial activity. More on antimicrobials in foods can be found in Chapter 13.

Lactoperoxidase System

This is an inhibitory system that occurs naturally in bovine milk, and it consists of three components: lactoperoxidase, thiocyanate, and H_2O_2. All three components are required for antimicrobial effects, and Gram-negative psychrotrophs such as the pseudomonads are quite sensitive. The quantity of lactoperoxidase needed is 0.5–1.0 ppm, whereas bovine milk normally contains about 30 ppm.[6] Although both thiocyanate and H_2O_2 occur normally in milk, the quantities vary. For H_2O_2, about 100 U/ml are required in the inhibitory system, whereas only 1–2 U/ml normally occur in milk. An effective level of thiocyanate is around 0.25 mM, whereas in milk the quantity varies between 0.02 and 0.25 mM.[6]

When the lactoperoxidase system in raw milk was activated by adding thiocyanate to 0.25 mM along with an equimolar amount of H_2O_2, the shelf life was extended to 5 days compared to 48 hours for controls.[6] The system was more effective at 30°C than at 4°C. The antibacterial effect increases with acidity, and the cytoplasmic membrane appears to be the cell target. In addition to the direct addition of H_2O_2, an exogenous source can be provided by the addition of glucose and glucose oxidase. To avoid the direct addition of glucose oxidase, this enzyme has been immobilized on glass beads so that glucose is generated only in the amounts needed by the use of immobilized β-galactosidase.[7] This system was effective in goat's milk against *P. fluorescens* and *E. coli* where the growth of the former was controlled for 3 days and the latter for 2 days at 8°C.[55]

The lactoperoxidase system can be used to preserve raw milk in countries where refrigeration is uncommon. The addition of about 12 ppm of SCN^- and 8 ppm of H_2O_2 should be harmless to the consumer.[44] An interesting aspect of this system is the effect it has on thermal properties. In one study, it was shown to reduce thermal D values at 57.8°C by around 80% for *L. monocytogenes* and by around 86% for *S. aureus* at 55.2°C.[27] Although the mechanism of this enhanced thermal destruction is unclear, some interesting implications can be envisioned.

Biological Structures

The natural covering of some foods provides excellent protection against the entry and subsequent damage by spoilage organisms. In this category are such structures as the testa of seeds, the outer covering of fruits, the shell of nuts, the hide of animals, and the shells of eggs. In the case of nuts such as pecans and walnuts, the shell or covering is sufficient to prevent the entry of all organisms. Once cracked, of course, nutmeats are subject to spoilage by molds. The outer shell and membranes of eggs, if intact, prevent the entry of nearly all microorganisms when stored under the proper conditions of humidity and temperature. Fruits and vegetables with damaged covering undergo spoilage much faster than those not damaged. The skin covering of fish and meats such as beef and pork prevents the contamination and spoilage of these foods, partly because it tends to dry out faster than freshly cut surfaces.

Taken together, these six intrinsic parameters represent nature's way of preserving plant and animal tissues from microorganisms. By determining the extent to which each exists in a given food, one can predict the general types of microorganisms that are likely to grow and, consequently, the overall stability of this particular food. Their determination may also aid one in determining age, and possibly the handling history of a given food.

EXTRINSIC PARAMETERS

The extrinsic parameters of foods are not substrate dependent. They are those properties of the storage environment that affect both the foods and their microorganisms. Those of greatest importance to the welfare of foodborne organisms are as follows:

1. temperature of storage
2. relative humidity of environment
3. presence and concentration of gases
4. presence and activities of other microorganisms

Temperature of Storage

Microorganisms, individually and as a group, grow over a very wide range of temperatures. Therefore, it is well to consider at this point the temperature growth ranges for organisms of importance in foods as an aid in selecting the proper temperature for the storage of different types of foods.

The lowest temperature at which a microorganism has been reported to grow is −34°C; the highest is somewhere in excess of 100°C. It is customary to place microorganisms into three groups based on their temperature requirements for growth. Those organisms that grow well at or below 7°C and have their optimum between 20°C and 30°C are referred to as *psychrotrophs* (see Chapter 16). Those that grow well between 20°C and 45°C with optima between 30°C and 40°C are referred to as *mesophiles*, whereas those that grow well at and above 45°C with optima between 55°C and 65°C are referred to as *thermophiles*. (Physiological properties of these groups are treated in Chapters 16 and 17.)

With regard to bacteria, psychrotrophic species and strains are found among the following genera of those presented in Chapter 2: *Alcaligenes, Shewanella, Brochothrix, Corynebacterium, Flavobacterium, Lactobacillus, Micrococcus, Pectobacterium, Pseudomonas, Psychrobacter, Enterococcus*, and others. The psychrotrophs found most commonly on foods are those that belong to the genera *Pseudomonas* and *Enterococcus* (see Chapter 16). These organisms grow well at refrigerator temperatures and cause spoilage at 5–7°C of meats, fish, poultry, eggs, and other foods normally held at this temperature. Standard plate counts of viable organisms on such foods are generally higher when the plates are incubated at about 7°C for at least 7 days than when incubated at 30°C and above. Mesophilic species and strains are known among all genera presented in Chapter 2 and may be found on foods held at refrigerator temperatures. They apparently do not grow at this temperature but do grow at temperatures within the mesophilic range if other conditions are suitable. It should be pointed out that some organisms can grow over a range from 0°C to >40°C. One such organism is *Enterococcus faecalis*.

Most thermophilic bacteria of importance in foods belong to the genera *Bacillus, Paenibacillus, Clostridium, Geobacillus, Alicyclobacillus*, and *Thermoanaerobacter*. Although not all species of these genera are thermophilic, they are of great interest to the food microbiologist and food technologist in the canning industry.

Just as molds are able to grow over wider ranges of pH, osmotic pressure, and nutrient content, they are also able to grow over wide ranges of temperature as do bacteria. Many molds are able to grow at refrigerator temperatures, notably some strains of *Aspergillus, Cladosporium*, and *Thamnidium*, which may be found growing on eggs, sides of beef, and fruits. Yeasts grow over the psychrotrophic and mesophilic temperature ranges but generally not within the thermophilic range.

The quality of the food product must also be taken into account in selecting a storage temperature. Although it would seem desirable to store all foods at refrigerator temperatures or below, this is not always best for the maintenance of desirable quality in some foods. For example, bananas keep better if stored at 13–17°C than at 5–7°C. A large number of vegetables are favored by temperatures of about 10°C, including potatoes, celery, cabbage, and many others. In every case, the success of storage temperature depends to a great extent upon the relative humidity (RH) of the storage environment and the presence or absence of gases such as CO_2 and O_3.

Temperature of storage is the most important parameter that affects the spoilage of highly perishable foods, and this fact has been emphasized by the work of Olley and Ratkowsky and their co-workers. According to these investigators, spoilage can be predicted by a spoilage rate curve.[34] The general spoilage curve has been incorporated into the circuitry of a temperature function integrator that reads out the equivalent days of storage at 0°C and thus makes it possible to predict the remaining shelf life at 0°C. It has been shown that the rate of spoilage of fresh poultry at 10°C is about twice that at 5°C, and that at 15°C is about three times that at 5°C.[18,22] Instead of using the Arrhenius law equation, the following was developed to describe the relationship between temperature and growth rate of microorganisms between the minimum and optimum temperatures.[40]

$$\sqrt{r} = B(T - T_0)$$

where r is the growth rate, B is the slope of the regression line, and T_0 is a conceptual temperature of no metabolic significance. The linear relationship has been shown to apply to spoilage bacteria and fungi when growing in foods or when utilizing amino acids.[40] The incorporation of growth data into mathematical equations to predict the behavior of microorganisms in food systems is discussed further in Chapter 20.

Relative Humidity of Environment

The RH of the storage environment is important both from the standpoint of a_w within foods and the growth of microorganisms at the surfaces. When the a_w of a food is set at 0.60, it is important that this food be stored under conditions of RH that do not allow the food to pick up moisture from the air and thereby increase its own surface and subsurface a_w to a point where microbial growth can occur. When foods with low a_w values are placed in environments of high RH, the foods pick up moisture until equilibrium has been established. Likewise, foods with a high a_w lose moisture when placed in an environment of low RH. There is a relationship between RH and temperature that should be borne in mind in selecting proper storage environments for foods. In general, the higher the temperature, the lower the RH, and vice versa.

Foods that undergo surface spoilage from molds, yeasts, and certain bacteria should be stored under conditions of low RH. Improperly wrapped meats such as whole chickens and beef cuts tend to suffer much surface spoilage in the refrigerator before deep spoilage occurs, due to the generally high RH of the refrigerator and the fact that the meat-spoilage biota is essentially aerobic in nature. Although it is possible to lessen the chances of surface spoilage in certain foods by storing under low conditions of RH, it should be remembered that the food itself will lose moisture to the atmosphere under such conditions and thereby become undesirable. In selecting the proper environmental conditions of RH, consideration must be given to both the possibility of surface growth and the desirable quality to be maintained in the foods in question. By altering the gaseous atmosphere, it is possible to retard surface spoilage without lowering the RH.

Presence and Concentration of Gases in the Environment

Carbon dioxide (CO_2) is the single most important atmospheric gas that is used to control microorganisms in foods.[15,35] It along with O_2 are the two most important gases in modified atmosphere packaged (MAP) foods, and this is discussed in Chapter 14.

Ozone (O_3) is the other atmospheric gas that has antimicrobial properties, and it has been tried over a number of decades as an agent to extend the shelf life of certain foods. It has been shown to be effective against a variety of microorganisms,[9] but because it is a strong oxidizing agent, it should not be used on high-lipid-content foods since it would cause an increase in rancidity. Ozone was tested against *Escherichia coli* 0157:H7 in culture media, and at 3 to 18 ppm the bacterium was destroyed in 20 to 50 minutes.[10] The gas was administered from an ozone generator and on tryptic soy agar, the D value for 18 ppm was 1.18 minutes, but in phosphate buffer, the D value was 3.18 minutes. To achieve a 99% inactivation of about 10,000 cysts of *Giardia lamblia* per milliliter, the average concentration time was found to be 0.17 and 0.53 mg-min/L at 25°C and 5°C, respectively.[53] The protozoan was about three times more sensitive to O_3 at 25°C than at 5°C. It is allowed in foods in Australia, France, and Japan; and in 1997 it was accorded GRAS (generally regarded as safe) status in the United States for food use. Overall, O_3 levels of 0.15 to 5.00 ppm in air have been shown to inhibit the growth of some spoilage bacteria as well as yeasts. The use of ozone as a food sanitizing agent is presented in Chapter 13.

Presence and Activities of Other Microorganisms

Some foodborne organisms produce substances that are either inhibitory or lethal to others; these include antibiotics, bacteriocins, hydrogen peroxide, and organic acids. The bacteriocins and some antibiotics are discussed in Chapter 13. The inhibitory effect of some members of the food biota on others is well established, and this is discussed under the section Biocontrol in Chapter 13.

REFERENCES

1. Angelidis, A.S., and G.M. Smith. 2003. Three transportors mediate uptake of glycine betaine and carnitine in *Listeria monocytogenes* in response to hyperosmotic stress. *Appl. Environ. Microbiol.* 69:1013–1022.

2. Barnes, E.M., and M. Ingram. 1955. Changes in the oxidation–reduction potential of the sterno-cephalicus muscle of the horse after death in relation to the development of bacteria. *J. Sci. Food Agric.* 6:448–455.

3. Barnes, E.M., and M. Ingram. 1956. The effect of redox potential on the growth of *Clostridium welchii* strains isolated from horse muscle. *J. Appl. Bacteriol.* 19:117–128.

4. Baron, F., M. Gautier, and G. Brule. 1997. Factors involved in the inhibition of *Salmonella enteritidis* in liquid egg white. *J. Food Protect.* 60:1318–1323.

5. Bate-Smith, E.C. 1948. The physiology and chemistry of rigor mortis, with special reference to the aging of beef. *Adv. Food Res.* 1:1–38.

6. Björck, L. 1978. Antibacterial effect of the lactoperoxidase system on psychrotrophic bacteria in milk. *J. Dairy Res.* 45:109–118.

7. Björck, L., and C.-G. Rosen. 1976. An immobilized two-enzyme system for the activation of the lactoperoxidase antibacterial system in milk. *Biotechnol. Bioeng.* 18:1463–1472.

8. Briskey, E.J. 1964. Etiological status and associated studies of pale, soft, exudative porcine musculature. *Adv. Food Res.* 13:89–178.

9. Burleson, G.R., T.M. Murray, and M. Pollard. 1975. Inactivation of viruses and bacteria by ozone, with and without sonication. *Appl. Microbiol.* 29:340–344.

10. Byun, M.-W., L.-J. Kwon, H.-S. Yook, and K.-S. Kim. 1998. Gamma irradiation and ozone treatment for inactivation of *Escherichia coli* 0157:H7 in culture media. *J. Food Protect.* 61:728–730.

11. Callow, E.H. 1949. Science in the imported meat industry. *J. R. Sanitary Inst.* 69:35–39.

12. Charlang, G., and N.H. Horowitz. 1974. Membrane permeability and the loss of germination factor from *Neurospora crassa* at low water activities. *J. Bacteriol.* 117:261–264.

13. Christian, J.H.B. 1963. Water activity and the growth of microorganisms. In *Recent Advances in Food Science*, ed. J.M. Leitch and D.N. Rhodes, vol. 3, 248–255. London: Butterworths.

14. Chung, K.C., and J.M. Goepfert. 1970. Growth of *Salmonella* at low pH. *J. Food Sci.* 35:326–328.

15. Clark, D.S., and C.P. Lentz. 1973. Use of mixtures of carbon dioxide and oxygen for extending shelf-life of prepackaged fresh beef. *Can. Inst. Food Sci. Technol. J.* 6:194–196.

16. Conway, E.J., and M. Downey. 1950. pH values of the yeast cell. *Biochem. J.* 47:355–360.

17. Corlett, D.A., Jr., and M.H. Brown. 1980. pH and acidity. In *Microbial Ecology of Foods*, 92–111. New York: Academic Press.

18. Daud, H.B., T.A. McMeekin, and J. Olley. 1978. Temperature function integration and the development and metabolism of poultry spoilage bacteria. *Appl. Environ. Microbiol.* 36:650–654.

19. Edgley, M., and A.D. Brown. 1978. Response of xerotolerant and nontolerant yeasts to water stress. *J. Gen. Microbiol.* 104:343–345.

20. Fraser, K.R., D. Sue, M. Wiedmann, K. Boor, and C.P. O'Bryne. 2003. Role of σ^B in regulating the compatible solute uptake systems of *Listeria monocytogenes*: Osmotic induction of *opuC* is σ^B dependent. *Appl. Environ. Microbiol.* 69:2015–2022.

21. Gardan, R., O. Duché, S. Leroy-Sétrin, European *Listeria* genome consortium, and J. Labadie. 2003. Role of *ctc* from *Listeria monocytogenes* in osmotolerance. *Appl. Environ. Microbiol.* 69:154–161.

22. Goepfert, J.M., and H.U. Kim. 1975. Behavior of selected foodborne pathogens in raw ground beef. *J. Milk Food Technol.* 38:449–452.

23. Hewitt, L.F. 1950. *Oxidation–Reduction Potentials in Bacteriology and Biochemistry*, 6th ed. Edinburgh: Livingston.

24. Horner, K.J., and G.D. Anagnostopoulos. 1973. Combined effects of water activity, pH and temperature on the growth and spoilage potential of fungi. *J. Appl. Bacteriol.* 36:427–436.

25. Jakobsen, M., and W.G. Murrell. 1977. The effect of water activity and the a_w-controlling solute on germination of bacterial spores. *Spore Res.* 2:819–834.

26. Jakobsen, M., and W.G. Murrell. 1977. The effect of water activity and a_w-controlling solute on sporulation of *Bacillus cereus* T. *J. Appl. Bacteriol.* 43:239–245.

27. Kamau, D.N., S. Doores, and K.M. Pruitt. 1990. Enhanced thermal destruction of *Listeria monocytogenes* and *Staphylococcus aureus* by the lactoperoxidase system. *Appl. Environ. Microbiol.* 56:2711–2716.

28. Kang, C.K., M. Woodburn, A. Pagenkopf, and R. Cheney. 1969. Growth, sporulation, and germination of *Clostridium perfringens* in media of controlled water activity. *Appl. Microbiol.* 18:798–805.

29. Ko, R., L.T. Smith, and G.M. Smith. 1994. Glycine betaine confers enhanced osmotolerance and cryotolerance on *Listeria monocytogenes*. *J. Bacteriol.* 176:426–431.

30. Marshall, B.J., F. Ohye, and J.H.B. Christian. 1971. Tolerance of bacteria to high concentrations of NaCl and glycerol in the growth medium. *Appl. Microbiol.* 21:363–364.

31. Mayerhauser, C.M. 2001. Survival of enterohemorrhagic *Escherichia coli* 0157:H7 in retail mustard. *J. Food Protect.* 64:783–787.

32. Morris, E.O. 1962. Effect of environment on microorganisms. In *Recent Advances in Food Science*, ed. J. Hawthorn and J.M. Leitch, vol. 1, 24–36. London: Butterworths.

33. Mossel, D.A.A., and M. Ingram. 1955. The physiology of the microbial spoilage of foods. *J. Appl. Bacteriol.* 18:232–268.

34. Olley, J., and D.A. Ratkowsky. 1973. The role of temperature function integration in monitoring fish spoilage. *Food Technol. N.Z.* 8:13–17.

35. Parekh, K.G., and M. Solberg. 1970. Comparative growth of *Clostridium perfringens* in carbon dioxide and nitrogen atmospheres. *J. Food Sci.* 35:156–159.

36. Park, S., L.T. Smith, and G.M. Smith. 1995. Role of glycine betaine and related osmolytes in osmotic stress adaptation in *Yersinia entercolitica* ATCC 9610. *Appl. Environ. Microbiol.* 61:4378–4381.

37. Peña, A., G. Cinco, A. Gomez-Puyou, and M. Tuena. 1972. Effect of pH of the incubation medium on glycolysis and respiration in *Saccharomyces cerevisiae*. *Arch. Biochem. Biophys.* 153:413–425.

38. Pitt, J.I. 1975. Xerophilic fungi and the spoilage of foods of plant origin. In *Water Relations of Foods*, ed. R.B. Duckworth, 273–307. London: Academic Press.

39. Prior, B.A. 1978. The effect of water activity on the growth and respiration of *Pseudomonas fluorescens*. *J. Appl. Bacteriol.* 44:97–106.

40. Ratkowsky, D.A., J. Olley, T.A. McMeekin, and A. Ball. 1982. Relationship between temperature and growth rate of bacterial cultures. *J. Bacteriol.* 149:1–5.

41. Rattray, J.B.M., A. Schibeci, and D.K. Kidby. 1975. Lipids of yeasts. *Bacteriol. Rev.* 39:197–231.

42. Reay, G.A., and J.M. Shewan. 1949. The spoilage of fish and its preservation by chilling. *Adv. Food Res.* 2:343–398.

43. Reed, R.K., J.A. Chudek, K. Foster, and G.M. Gadd. 1987. Osmotic significance of glycerol accumulation in exponentially growing yeasts. *Appl. Environ. Microbiol.* 53:2119–2123.

44. Reiter, B., and G. Harnulv. 1984. Lactoperoxidase antibacterial system: Natural occurrence, biological functions and practical applications. *J. Food Protect.* 47:724–732.

45. Rose, A.H. 1965. *Chemical Microbiology*, chap. 3. London: Butterworths.

46. Rothstein, A., and G. Demis. 1953. The relationship of the cell surface to metabolism: The stimulation of fermentation by extracellular potassium. *Arch. Biochem. Biophys.* 44:18–29.

47. Shelef, L.A. 1983. Antimicrobial effects of spices. *J. Food Safety.* 6:29–44.

48. Sherman, J.M., and G.E. Holm. 1922. Salt effects in bacterial growth. II. The growth of *Bacterium coli* in relation to H-ion concentration. *J. Bacteriol.* 7:465–470.

49. Sleator, R.D., and C. Hill. 2001. Bacterial osmoadaptation: The role of osmolytes in bacterial stress and virulence. *FEMS Microbiol. Rev.* 26:49–71.

50. Stier, R.F., L. Bell, and K.A. Ito, B.D. Shafer, L.A. Brown, M.L. Seeger, B.H. Allen, M.N. Porcuna, and P.A. Lerke. 1981. Effect of modified atmosphere storage on *C. botulinum* toxigenesis and the spoilage microflora of salmon fillets. *J. Food Sci.* 46:1639–1642.

51. Troller, J.A. 1986. Water relations of foodborne bacterial pathogens: an updated review. *J. Food Protect.* 49:656–670.

52. Walden, W.C., and D.J. Hentges. 1975. Differential effects of oxygen and oxidation–reduction potential on the multiplication of three species of anaerobic intestinal bacteria. *Appl. Microbiol.* 30:781–785.

53. Wickramanayake, G.B., A.J. Rubin, and O.J. Sproul. 1984. Inactivation of *Giardia lamblia* cysts with ozone. *Appl. Environ. Microbiol.* 48:671–672.

54. Wodzinski, R.J., and W.C. Frazier. 1961. Moisture requirements of bacteria. II. Influence of temperature, pH, and maleate concentration on requirements of *Aerobacter aerogenes. J. Bacteriol.* 81:353–358.

55. Zapico, P., P. Gaya, and M. Nuñez, and M. Medina. 1994. Activity of goats' milk lactoperoxidase system on *Pseudomonas fluorescens* and *Escherichia coli* at refrigeration temperatures. *J. Food Protect.* 58:1136–1138.

PART III

Microorganisms in Foods

Chapters 4 to 9 address the numbers and types of microorganisms found in various food products, and the roles they play in microbial spoilage. Fermentation is discussed in Chapter 7, along with fermented dairy products. Nondairy fermented products are presented in Chapter 8. More details are provided in the references below.

Davies, A., and R. Board, eds. 1998. *The Microbiology of Meat and Poultry*. Gaithersburg, MD: Aspen Publishers, Inc. Excellent coverage of meat and poultry microbiology.

International Commission on Microbiological Specifications of Foods (ICMSF). 1996. *Microorganisms in Foods*, 5th ed. New York: Kluwer Academic Publishers. Detailed coverage of the parameters that affect the titled organisms that include viruses and animal parasites.

Kraft, A.A. 1992. *Psychrotrophic Bacteria in Foods: Disease and Spoilage*. Boca Raton, FL: CRC Press. Contains historical information on the titled organisms.

Lamikanra, O. 2002. *Fresh-Cut Fruits and Vegetables: Science, Technology and Market*. General microbiology of fresh-cut produce along with MAP and some others, 480 pp.

Mossel, D.A.A., J. Corry, C. Struijk, et al., eds. 1995. *Essentials of the Microbiology of Foods*. New York: John Wiley & Sons. This 736-page work is an excellent reference on most areas of food microbiology.

Robinson, R.K., ed. 2002. *Dairy Microbiology Handbook*, 3rd ed. New York: Wiley & Sons. Thorough coverage of many dairy products.

UNIVERSITY OF HERTFORDSHIRE LRC

CHAPTER 4

Fresh Meats and Poultry

It is generally agreed that the internal tissues of healthy slaughter animals are free of bacteria at the time of slaughter, assuming that the animals are not in a state of exhaustion. When one examines fresh meat and poultry at the retail level, varying numbers and types of microorganisms are found. The following are the primary sources and routes of microorganisms to fresh meats with particular emphasis on red meats:

1. *The stick knife.* After being stunned and hoisted by the hind legs, animals such as steers are exsanguinated by slitting the jugular vein with what is referred to as a "stick knife." If the knife is not sterile, organisms are swept into the bloodstream, where they may be deposited throughout the carcass.
2. *Animal hide.* Organisms from the hide are among those that enter the carcass via the stick knife. Others from the hide may be deposited onto the dehaired carcass or onto freshly cut surfaces. Some hide biota becomes airborne and can contaminate dressed out carcasses as noted below. See the section on carcass sanitizing and washing towards the end of this chapter.
3. *Gastrointestinal tract.* By way of punctures, intestinal contents along with the usual heavy load of microorganisms may be deposited onto the surface of freshly dressed carcasses. Especially important in this regard is the paunch or rumen of ruminant animals, which typically contains $\sim 10^{10}$ bacteria per gram.
4. *Hands of handlers.* As noted in Chapter 2, this is a source of human pathogens to freshly slaughtered meats. Even when gloves are worn, organisms from one carcass can be passed on to other carcasses.
5. *Containers.* Meat cuts that are placed in nonsterile containers may be expected to become contaminated with the organisms in the container. This tends to be a primary source of microorganisms to ground or minced meats.
6. *Handling and storage environment.* Circulating air is not an insignificant source of organisms to the surfaces of all slaughtered animals; this is noted in Chapter 2.
7. *Lymph nodes.* In the case of red meats, lymph nodes that are usually embedded in fat often contain large numbers of organisms, especially bacteria. If they are cut through or added to portions that are ground, one may expect this biota to become prominent.

In general, the most significant of the above are nonsterile containers. When several thousand animals are slaughtered and handled in a single day in the same abattoir, there is a tendency for the external carcass biota to become normalized among carcasses, although a few days may be required. The practical effect of this is the predictability of the biota of such products at the retail level.

BIOCHEMICAL EVENTS THAT LEAD TO RIGOR MORTIS

Upon the slaughter of a well-rested beef animal, a series of events takes place that leads to the production of meat. Lawrie[106] discussed these events in great detail, and they are presented here only in outline form. The following are stages of an animal's slaughter:

1. Its circulation ceases: the ability to resynthesize ATP (adenosine triphosphate) is lost; lack of ATP causes actin and myosin to combine to form actomyosin, which leads to a stiffening of muscles.
2. The oxygen supply falls, resulting in a reduction of the O/R (oxidation–reduction) potential.
3. The supply of vitamins and antioxidants ceases, resulting in a slow development of rancidity.
4. Nervous and hormonal regulations cease, thereby causing the temperature of the animal to fall and fat to solidify.
5. Respiration ceases, which stops ATP synthesis.
6. Glycolysis begins, resulting in the conversion of most glycogen to lactic acid, which depresses pH from about 7.4 to its ultimate level of about 5.6. This pH depression also initiates protein denaturation, liberates and activates cathepsins, and completes rigor mortis. Protein denaturation is accompanied by an exchange of divalent and monovalent cations on the muscle proteins.
7. The reticuloendothelial system ceases to scavenge, thus allowing microorganisms to grow unchecked.
8. Various metabolites accumulate that also aid protein denaturation.

These events require between 24 and 36 hours at the usual temperatures of holding freshly slaughtered beef (2–5°C). Meanwhile, part of the normal biota of this meat has come from the animal's own lymph nodes,[109] the stick knife used for exsanguination, the hide of the animal, intestinal tract, dust, hands of handlers, cutting knives, storage bins, and the like. Upon prolonged storage at refrigerator temperatures, microbial spoilage begins. In the event that the internal temperatures are not reduced to the refrigerator range, the spoilage that is likely to occur is caused by bacteria of internal sources. Chief among these are *Clostridium perfringens* and genera in the Enterobacteriaceae family.[90] On the other hand, bacterial spoilage of refrigerator-stored meats is, by and large, a surface phenomenon reflective of external sources of the spoilage biota.[90]

THE BIOTA OF MEATS AND POULTRY

The term "biota" is used throughout this text in lieu of "flora" as a general reference to bacteria. Flora refers to plant life. "Bacterial flora" dates back to the time when it was believed that bacteria were primitive plants. Since bacteria are not plants, "bacterial biota or microbiota" is preferred to flora. The major genera of bacteria, yeasts, and molds that are found in these products before spoilage are listed in Tables 4–1 and 4–2. In general, the biota is reflective of the slaughtering and processing environments as noted above, with Gram-negative bacteria being predominant. Among Gram-positives, the enterococci are the biota most often found along with lactobacilli. Because of their ubiquity in meat-processing environments, a rather large number of mold genera may be expected, including *Penicillium, Mucor,*

Table 4–1   Genera of Bacteria Most Frequently Found on Meats and Poultry

Genus	Gram Reaction	Fresh Meats	Fresh Livers	Poultry
Acinetobacter	−	XX	X	XX
Aeromonas	−	XX		X
Alcaligenes	−	X	X	X
Arcobacter	−	X		
Bacillus	+	X		X
Brochothrix	+	X	X	X
Campylobacter	−			XX
Carnobacterium	+	X		
Caseobacter	+	X		
Citrobacter	−	X		X
Clostridium	+	X		X
Corynebacterium	+	X	X	XX
Enterobacter	−	X		X
Enterococcus	+	XX	X	X
Erysipelothrix	+	X		X
Escherichia	−	X	X	
Flavobacterium	−	X	X	X
Hafnia	−	X		
Kocuria	+	X	X	X
Kurthia	+	X		
Lactobacillus	+	X		
Lactococcus	+	X		
Leuconostoc	+	X	X	
Listeria	+	X		XX
Microbacterium	+	X		X
Micrococcus	+	X	XX	XX
Moraxella	−	XX	X	X
Paenibacillus	+	X		X
Pantoea	−	X		X
Pediococcus	+	X		
Proteus	−	X		X
Pseudomonas	−	XX		XX
Psychrobacter	−	XX		X
Salmonella	−	X		X
Serratia	−	X		X
Shewanella	−	X		
Staphylococcus	+	X	X	X
Vagococcus	+			XX
Weissella	+	X	X	
Yersinia	−	X		

Note: X = known to occur; XX = most frequently reported.

Table 4–2   Genera of Fungi Most Often Found on Meats and Poultry

Genus	Fresh and Refrigerated Meats	Poultry
Molds		
Alternaria	X	X
Aspergillus	X	X
Aureobasidium	X	
Cladosporium	XX	X
Eurotium	X	
Fusarium	X	
Geotrichum	XX	X
Monascus	X	
Monilia	X	
Mucor	XX	X
Neurospora	X	
Penicillium	X	X
Rhizopus	XX	X
Sporotrichum	XX	
Thamnidium	XX	
Yeasts		
Candida	XX	XX
Cryptococcus	X	X
Debaryomyces	X	XX
Hansenula	X	
Pichia	X	X
Rhodotorula	X	XX
Saccharomyces		X
Torulopsis	XX	X
Trichosporon	X	X
Yarrowia		XX

Note: X = known to occur; XX = most frequently found.

Source: Taken from the literature and from references 34, 35, and 94.

and *Cladosporium*. The most ubiquitous yeasts found in meats and poultry are members of the genera *Candida* and *Rhodotorula* (Table 4–2). For an extensive review, see Dillon.[35]

INCIDENCE/PREVALENCE OF MICROORGANISMS IN FRESH RED MEATS

The incidence and prevalence of microorganisms in some red meats are presented in Table 4–3. The aerobic plate counts (APCs) of the fresh ground beef in this table are considerably higher than those reported by the U.S. Department of Agriculture (USDA[176]). In that survey of 563 raw ground beef samples from throughout the United States, the $\log_{10}$ mean number for APC was only 3.90; and 1.98, 1.83, and 1.49 for coliforms, *Clostridium perfringens*, and *Staphylococcus aureus*, respectively. To what extent these lower numbers are reflective of a trending-down of bacteria in fresh ground beef

Table 4–3 Relative Percentage of Organisms in Red Meats That Meet Specified Target Numbers (Numbers Reported are $\log_{10}$ cfu/g or ml)

Products	Number of Samples	Microbial Group/Target (All Numbers Are $\log_{10}$)	% Samples Meeting Target	Reference
Raw beef patties	735	APC: $\log_{10}$ 6.00 or less/g	76	170
	735	Coliforms: log 2.00 or less/g	84	170
	735	*E. coli*: log 2.00 or less/g	92	170
	735	*S. aureus*: 2.00 or less/g	85	170
	735	Presence of salmonellae	0.4	170
Fresh ground beef*	1,830	APC: 6.70 or less/g	89	21
	1,830	*S. aureus*: 3.00 or less/g	92	21
	1,830	*E. coli*: 1.70 or less/g	84	21
	1,830	Presence of salmonellae	2	21
	1,830	Presence of *C. perfringens*	20	21
Fresh ground beef	1,090	APC: $\geq$7.00 or less/g at 35°C	88	142
	1,090	Fecal coliforms: $\leq$2.00/g	76	142
	1,090	*S. aureus*: <2.00/g	91	142
Frozen ground beef patties	605	APC: 6.00 or less/g	67	74
	604	*E. coli*: <2.70/g	85	74
	604	*E. coli*: >3.00/g MPN	9	74
Fried hamburger	107	APC at 21°C; 72 h, <3.00/g	76	43
		Absence of enterococci, coliforms, *S. aureus*,	100	43
	107	Salmonellae		
Comminuted big game meats	113	Coliforms: 2.00 or less/g	42	163
	113	*E. coli*: 2.00 or less/g	75	163
	113	*S. aureus*: 2.00 or less/g	96	163

*Under Oregon law that was in effect at the time.

Note: APC = Aerobic plate count; MPN = most probable number.

or of laboratory methodology is unclear. For many decades, comminuted meats have been shown to contain higher numbers of microorganisms than noncomminuted meats such as steaks, and there are reasons for this:

1. Commercial ground meats consisting of trimmings from various cuts that are handled excessively generally contain high levels of microbial contamination. Ground meats that are produced from large cuts tend to have lower microbial numbers.
2. Ground meat provides a greater surface area, which itself accounts in part for the increased biota. It should be recalled that as particle size is reduced, the total surface area increases with a consequent increase in surface energy.
3. This greater surface area of ground meat favors the growth of aerobic bacteria, the usual low-temperature spoilage biota.
4. In some commercial establishments, the meat grinders, cutting knives, and storage utensils are rarely cleaned as often and as thoroughly as is necessary to prevent the successive buildup of

microbial numbers. This may be illustrated by data obtained from a study of the bacteriology of several areas in the meat department of a large grocery store. The blade of the meat saw and the cutting block were swabbed immediately after they were cleaned on three different occasions with the following mean results: the saw blade had a total $\log_{10}/in.^2$ count of 5.28, with 2.3 coliforms, 3.64 enterococci, 1.60 staphylococci, and 3.69 micrococci; the cutting block had a mean $\log_{10}/in.^2$ count of 5.69, with 2.04 coliforms, 3.77 enterococci, <1.00 staphylococci, and 3.79 micrococci. These are among the sources of the high total bacterial count to comminuted meats.

5. One heavily contaminated piece of meat is sufficient to contaminate others, as well as the entire lot, as they pass through the grinder. This heavily contaminated portion is often in the form of lymph nodes, which are generally embedded in fat. These organs have been shown to contain high numbers of microorganisms and account in part for hamburger meat having a generally higher total count than ground beef. In some states, the former may contain up to 30% beef fat, whereas the latter should not contain more than 20% fat.

Bacteria

The high prevalence of enterococci in meats is illustrated by a study conducted in 2001–2002 on retail meats in the state if Iowa. Of 255 pork samples, 247 (97%) were positive for these organisms with 54% of isolates being *Enterococcus faecalis* and 38% *E. faecium*.[84] Of 262 beef samples, all contained enterococci with 65% of isolates identified as *E. faecium*, 17% *E. faecalis*, and 14% *E. hirae*.[84]

Members of the genera *Paenibacillus*, *Bacillus*, and *Clostridium*, are found in meats of all types. In a study of the incidence of putrefactive anaerobe (PA) spores in fresh and cured pork trimmings and canned pork luncheon meat, Steinkraus and Ayres[165] found these organisms to occur at very low levels, generally less than 1/g. In a study of the incidence of clostridial spores in meats, Greenberg et al.[76] found a mean PA spore count per gram of 2.8 from 2,358 meat samples. Of the 19,727 PA spores isolated, only one was a *Clostridium botulinum* spore, and it was recovered from chicken. The large number of meat samples studied by these investigators consisted of beef, pork, and chicken, obtained from all parts of the United States and Canada. The significance of PA spores in meats is due to the problems encountered in the heat destruction of these forms in the canning industry (see Chapter 17).

Erysipelothrix rhusiopathiae was isolated from about 34% of retail pork samples in Japan and from 4% to 54% of pork loins in Sweden. A variety of serovars has been found in pork, and nine were found among chicken isolates in Japan.[133] The latter investigators suggested chickens as a possible reservoir of *Erysipelothrix* spp. for human infections (see Chapter 31 for more on this bacterium).

The incidence of *Clostridium perfringens* in a variety of American foods was studied by Strong et al.[169] They recovered the organism from 16.4% of raw meats, poultry, and fish tested; from 5% of spices; from 3.8% of fruits and vegetables; from 2.7% of commercially prepared frozen foods; and from 1.8% of home-prepared foods. Others have found low numbers of this organism in both fresh and processed meats. In ground beef, *C. perfringens* at 100 or less per gram was found in 87% of 95 samples, whereas 45 of the 95 (47%) samples contained this organism at levels <1,000/g.[103] One group was unable to recover *C. perfringens* from pork carcasses, hearts, and spleens, but 21.4% of livers were positives.[13] Commercial pork sausage was found to have a prevalence of 38.9%. A study in the United States in 2001–2002 of 445 whole muscle, ground, and emulsified samples of raw pork, beef, and chicken products found that *C. perfringens* spores did not exceed 2.0 $\log_{10}$ and averaged 1.56 $\log_{10}$ cfu/g.[173] When several products were inoculated with ca. 3.0 $\log_{10}/g$ of three *C. perfringens* strains then cooked and stored for up to 14 days under vacuum at 4°C, the inoculated cells showed a slight decline and remained essentially unchanged as product temperature decreased from 54.5 to 7.2°C.[173] The significance of this organism in foods is discussed in Chapter 24.

Some members of the family Enterobacteriaceae have been found to be common in fresh and frozen beef, pork, and related meats. Of 442 meat samples examined by Stiles and Ng,[169] 86% yielded enteric bacteria, with all 127 ground beef samples being positive. The most frequently found were *Escherichia coli* biotype I (29%), *Serratia liquefaciens* (17%), and *Pantoea agglomerans* (12%). A total of 721 isolates (32%) were represented by *Citrobacter freundii, Klebsiella pneumoniae, Enterobacter cloacae,* and *E. hafniae.* In an examination of 702 foods for fecal coliforms by the most-probable-numbers (MPN) method representing 10 food categories, the highest number was found in the 119 ground beef samples, with the geometric mean by the AOAC (Association of Official Analytical Chemists) procedure being 59/g.[3] The mean number for 94 pork sausage samples was 7.9/g. From 32 samples of minced goat meat, the mean coliform, Enterobacteriaceae, and APC counts were, respectively, 2.88, 3.07, and 6.57 $\log_{10}$.[131] More information on the incidence/prevalence of coliforms, enterococci, and other indicator organisms can be found in Chapter 20.

From the 563 samples of ground beef examined in the United States as noted above, 53% contained *C. perfringens* and 30% *S. aureus.*[176] Using a nested polymerase chain reaction (PCR) assay, enterotoxigenic *Clostridium perfringens* was found in 2%, 12%, and 0% of 50 beef, chicken, and pork samples, respectively, in Japan.[128]

A study of 470 fresh sheep carcasses in Australia found the mean APC (determined at 25°C after 72 hours) to be 3.92 $\log_{10}/cm^2$ and 3.48 $\log_{10}/cm^2$ when determined at 5°C after a 14-day incubation.[179] For a more extensive coverage of Gram-positive bacteria in meats, see reference 87.

Escherichia coli (Biotype I)

This bacterium is the most widely used as an indicator of the sanitary state of fresh foods, and it along with other indicator organisms is defined and discussed in Chapter 20. An international committee has stressed the desirability of testing for indicator organisms rather than specific pathogens in assessing the safety of beef.[17] Some findings of this organism in fresh meats are summarized below.

In a study of frozen beef patties in the United States, the mean aerobic plate count (APC) was <3.0 $\log_{10}$ cfu/g, and coliforms and *E. coli* biotype I were <1.0 $\log_{10}$ cfu/g.[144] These investigators noted a lack of correlation between low numbers of *E. coli* biotype I and *E. coli* 0157:H7. A Canadian study found that coliforms and *E. coli* recovered from the table top and conveyor belt in a meat processing facility were comparable to those recovered from beef cuts and sides, which emphasizes the importance of conveying equipment as sources of these organisms to beef cuts.[70]

The incidence and prevalence of biotype I strains of *E. coli* vary widely among retail or finished red meats. From 470 sheep carcasses studied in Australia, 75% contained this organism[181] while from 812 Australian beef carcasses processed for export only, 11% were positive.[141] In the United States, *E. coli* was recovered from 25% of 404 ground beef samples;[186] from 30% of 100 postexsanginated pork carcasses; and from 30% of the chilled carcasses tested.[172]

Arcobacter and Campylobacter spp.

These genera are closely related phylogenetically, and it is not surprising that they share common habitats. Summaries of their incidence and prevalence in a variety of meats and poultry are presented in Table 4–4. In general, *Arcobacter* spp. appear to be more common among poultry than red meat products, and this is true for *Campylobacter* spp. *A. butzleri* is common, and it was found on all 25

Table 4–4 Incidence/Prevalence of *Arcobacter, Campylobacter*, and *Helicobacter* spp. in Fresh and Frozen Meats and Poultry

Product	Genus	% Positive/ Total Tested	Country	Reference
Pork	*Arcobacter*	32/200	United States	13
Beef cattle	*Arcobacter butzleri*	9/200	United States	75
Turkey meat	*Arcobacter*	77/391	United States	117
Broilers	*Arcobacter*	95/480	Belgium	88
Broilers, chickens	*Arcobacter*	60/25	Denmark	4
Chicken	*Arcobacter*	40/45	Mexico	181
Pork	*Arcobacter*	64/200	United States	134
Beef	*Arcobacter*	29/45	Mexico	181
Pork	*Arcobacter*	5/45	Mexico	181
Fresh chicken	*Campylobacter*	94/63	North Ireland	130
Frozen chicken	*Campylobacter*	77/44	North Ireland	130
Fresh chicken	*Campylobacter*	85/35	The Netherlands	42
Frozen chicken	*Campylobacter*	87/38	The Netherlands	42
Chicken meats	*Campylobacter*	83/90	United Kingdom	102
Lamb liver	*Campylobacter*	73/96	United Kingdom	102
Pork liver	*Campylobacter*	72/99	United Kingdom	102
Pork liver	*Campylobacter*	ca. 6/400	North Ireland	129
Ox liver	*Campylobacter*	54/96	United Kingdom	129
Retail pork	*Campylobacter*	1.3/384	United States	41
Broilers	*Campylobacter*	88/1,297	United States	177
Sheep carcasses	*Campylobacter*	1.3/470	Australia	179
Ground beef	*Campylobacter*	<1/563	United States	176
Swine samples	*Campylobacter*	0.99/202	United States	138
Turkey carcasses, pre-chill	*Campylobacter*	41.3/1,198	United States	115
Broilers	*Campylobacter*	27/12,233	United Kingdom	139
Fresh meats	*Campylobacter*	12/405	United States	167
Frozen meats	*Campylobacter*	2.3/396	United States	166
Chicken	*Campylobacter*	30/360	United States	166
Red meats	*Campylobacter*	5/1,800	United States	178
Beef cuts	*Helicobacter pylori*	0/20	United States	168
Rumen, mucosal samples[a]	*Helicobacter*	0/105	United States	168

[a]Rumen and abomasum mucosal cattle samples.

chicken carcasses examined in Denmark.[4] *A. cryaerophilus* was recovered from 13 of the 25 carcasses, and *A. skirrowii* from only two.

In their study of 200 fresh pork samples in the United States using different recovery methods, Ohlendort and Murano[136] found that 20% of low-fat but only 4% of high-fat samples contained *Acrobacter* spp.; and that these organisms were more frequently isolated from younger than older hogs.

Wild and migratory birds also carry *Campylobacter* spp. Among 1,794 birds representing 107 species in Europe, 22.2% harbored *Campylobacter* spp. consisting of 5.6, 4.9, and 0.95% of *C. lari*, *C. jejuni*, and *C. coli*, respectively.[182] The highest percentage of *Campylobacter* spp. was 76.8 among

Table 4–5 Prevalence of *Salmonella* in Some Fresh and Frozen Meats and Poultry Products

Product	% Positive/Total Tested	Country	Reference
Broilers	20/1,297	United States	177
Broilers	25.9/27	Korea	26
Egg yolks	0/1,620	Korea	26
Frozen ground turkey	38/50	United States	78
Turkey carcasses[a]	12/208	United States	18
Turkey carcasses	69/230	Canada	105
Turkey raw rolls[b]	27/336	United States	18
Chicken carcasses	61/670	Canada	105
Chicken carcasses	34.8/69	Canada	44
Chicken carcasses	91/45	Venezuela	150
Chicken carcasses	60/192	Spain	22
Ground beef	20/55	Botswana	65
Ground beef	7.5/563	United States	176
Ground beef	11/88	Mexico	85
Butcher shop beef	9.9/354	Botswana	65
Beef carcasses	0/62	Belgium	99
Beef carcasses	2.6/666	Canada	105
Beef carcasses[c]	0/812	Australia	141
Steer/heifer carcasses	1/2089	United States	178
Sheep carcasses	5.7/470	Australia	179
Pork carcasses	27/49	Belgium	99
Pork carcasses	17.5/596	Canada	105
Swine carcasses[d]	73/100	United States	172
Swine carcasses, chilled	0.7/122	United States	172
Hogs	1/8,066	United States	10

[a]Preprocessed; [b] Turkey carcasses processed into raw rolls; [c]Export samples only; [d]Post exsanguinated.

the 383 shoreline-foraging invertebrate feeders. Of the 464 arboreal insectivore feeders, only 0.6% were positive for *Campylobacter*.[182]

Salmonellae

Summaries of the occurrence of *Salmonella* spp. on meat and poultry are presented in Table 4–5. As is the case for *Arcobacter* and *Campylobacter* spp., meat and poultry meats continue to be common sources of these organisms.

Salmonellae were found in 9.1% of 109 packs of chilled and 7.5% of 53 frozen packs of sausages or 8.6% overall in the United Kingdom in 2000.[119] Some were isolated from fried, grilled, and barbecued samples. Samples that were grilled for 12 min. or more reached internal temperatures >75°C all of which were salmonellae-negative. None of 51 packages contained *Campylobacter* spp.

In regards to the source of salmonellae in preharvest pork production, a study in Brazil found that the holding pens are significant sources of *Salmonella enterica*.[152] These findings are based on the study of a larger number of animals. Another study in the United States on salmonellae in the ecosystem of slaughter hogs examined 8,066 samples and found salmonellae (percentage occurrence) in the

Table 4–6 Prevalence of *Listeria monocytogenes* in Some Fresh Meat and Poultry Products

Product	% Positive/No. tested	Country	Reference
Broilers	15/1,297	United States	177
Chicken	30.2/86	Korea	7
Broiler parts (raw)	62/61	Finland	127
Poultry parts	13/160	United States	67
Turkey meat	5/180	United States	66
Ground beef	12/563	United States	176
Ground beef	16/88	Mexico	85
Beef	4.3/70	Korea	7
Beef carcasses	22/62	Belgium	99
Steer/heifer carcasses	4/2,089	United States	178
Lamb carcasses	4.3/69	Brazil	3
Pork carcasses	2/49	Belgium	99
Pork	19.1/84	Korea	7

following places: 83 swine, 54 pen floor, 32 boots, 16 flies, 9 mice, 3 cats, and 3 birds.[10] These investigators noted that cats and worker boots were the two most salmonellae-abundant ecological niches in their study. The most common serotypes found were *S*. Derby, Agona, Worthington, and Uganda.[10] In contrast to the above studies, five Swedish pig slaughterhouses were studied for the incidence of salmonellae, and of 3,388 samples cultured, all were negative.[174] In regards to *S*. Typhimurium, 3.5% of 404 samples of ground beef collected throughout the United States in 1998 were positive with five of the 14 isolates being strain DT-104A (*S*. Typhimurium var. Copenhagen), and they were all isolated from samples obtained in the San Francisco area.[186] Of the 404 samples, 25% contained type I strains of *E. coli*. In a Canadian study of the feces contents of 1,420 healthy 5-month-old pigs, 5.2% were positive for 12 serovars, with *S*. Brandenburg accounting for 42%.[111] Of 112 strains of salmonellae recovered from a poultry slaughterhouse in Spain in 1992, 77% were *S*. Enteritidis.[22]

To better understand how salmonellae are distributed throughout a broiler operation, samples were collected and tested from the following points (along with percent positive for salmonellae): Breeder farm (6%), hatchery (98%), previous grow-out flock (24%), flock during grow-out (60%), and carcasses after processing (7%).[8] This study pointed to the hatchery as the primary site that requires disinfection. Along lines similar to the above study, several sites in 60 small poultry slaughterhouses (<200 birds/day) in Brazil were examined with the following results and percent positive for salmonellae: Carcasses (42%), utensils (23%), water (71%), and freezer and refrigerator (71%, 62). Overall, 41% of samples contained salmonellae, which included 17 serotypes with *S*. Enteritidis being the most predominant at 30%; and *S*. Albany and Hadar at 12% each being the next most predominant.[62]

Listeria and Yersinia spp.

The prevalence of *L. monocytogenes* varies widely among raw red meats and poultry with the four poultry products listed in Table 4–6 having contamination rates from 5 to 62%. The latter consisted of raw broiler pieces, and the serotypes found were 1/2a, 1/2c, and 4b. The isolates represented 14 different PFGE (pulsed field gel electrophoresis—see Chapter 11) types.[127]

Raw pork and chicken products were examined for the presence of *Yersinia* spp. in Mexico, and 27% were positive for this genus.[147] Of 706 yersiniae-like isolates, 24% were confirmed with 49%

of these being *Y. enterocolitica*, 25% *Y. kristensenii*, 15% *Y. intermedia*, and 9% *Y. frederiksenii*. In a study of 43 pork samples from a slaughterhouse, eight contained the following yersiniae species: *Y. enterocolitica, Y. intermedia, Y. kristensenii*, and *Y. frederiksenii*.[82] From a study of hogs in the United States, 95 of 103 (92%) lots carried at least one *Y. enterocolitica* isolate, and 99% of the pathogenic isolates were serotype 0:5, and 3.7% were 0:3.[61] In Finland, 92% of 51 tongue and 25% of 255 ground meat samples contained *Y. enterocolitica*.[57] By using PCR and a culture method, >98% of the pork tongues were positive with biotype IV being the most common. From 31 pork tongues from freshly slaughtered animals, 21 strains were isolated with 0:8 being the most common, and 0:6, 30 the next most common.[38] A study of yersiniae in raw beef and chicken in Brazil revealed that 80% contained these organisms with ground beef and liver accounting for 60% and pork for 20%.[183]

In a study of the fate of *Yersinia enterocolitica* in a Turkish dry sausage (sucuk), the pathogen decreased from around 5.0 to 1.8 $\log_{10}$ cfu after 4 days of fermentation and to 0.5 $\log_{10}$ cfu/g after 12 days of drying without the direct addition of lactic acid bacteria.[25] With the addition of about 7.0 $\log_{10}$ cfu/g each of *Lactobacillus sakei* and *Pediococcus acidilactici*, the pathogen was reduced to 0.5 $\log_{10}$ cfu/g after 3 days of fermentation, and none was detectable after 4 days.[25]

Pathogenic Escherichia coli Strains

E. coli 0157:H7 on red meats and poultry varies widely with none being found on 990 samples of boneless beef to 17% for two red meats (Table 4–7). The 296 ground beef and cattle fecal samples were taken from the Seattle, WA area (where the 1993 ground beef outbreak occurred). Of non-0157:H7 Stx-producing strains isolated by Brooks et al.,[16] the most common serovar was 0128:H2, which produced both Stx1 and 2 toxins, along with three others. No 0157:H7 strains were found in this study of 218 samples.

In large and medium diameter Lebanon-style bologna, a 5-log reduction of *E. coli* 0157:H7 was achieved during the smoking process.[69] The fermentation process consisted of 8 hr at 26.7°C, then 24 h at 37.8°C, and finally 24 h at 43.3H°C, all internal temperatures. The starting material had a pH of 4.4 with 4.0% NaCl and low fat (10–13%), and it was inoculated with *E. coli* 0157:H7 at a level of 7.5 to 7.9 $\log_{10}$ cfu/g.[69]

Soy-Extended Ground Meats

The addition of soy protein (soybean flour, soy flakes, texturized soy protein) at levels of 10–30% to ground meat patties is fairly widespread in the fast-food industry, at least in the United States, and the microbiology of these soy blends has been investigated. The earliest, most detailed study is that of Craven and Mercuri[30] who found that when ground beef or chicken was extended with 10% or 30% soy, APCs of these products increased over unextended controls when both were stored at 4°C for up to 8–10 days. Whereas coliforms were also higher in beef-soy mixtures than in controls, this was not true for the chicken-soy blends. In general, APCs were higher at the 30% level of soy than at 10%. In one study in which 25% soy was used with ground beef, the mean time to spoilage at 4°C for the beef-soy blend was 5.3 days compared to 7.5 days for the unextended ground beef.[14] In another study using 10%, 20%, and 30% soy, the APC increased significantly with both time and concentration of soy in the blend.[97]

With regard to the microbiological quality of soy products, the geometric mean APC of 1,226 sample units of seasoned product was found to be 1,500/g, with fungi, coliforms, *E. coli*, and *Staphylococcus aureus*, counts of 25, 3, 3, and 10/g, respectively.[171]

Table 4–7 Prevalence of *Escherichia coli* 0157:H7 and Related Pathogenic Serotypes in Some Fresh and Frozen Meat and Poultry Products, and Some Slaughter Animals and/or Their Products

Product	% Positive/No. Tested	Country	Reference
Chicken	0/36[a]	New Zealand	16
Ground beef	17/296	United States	154
Boneless beef	0/990	Australia	141
Beef carcasses	0.1/1,275	Australia	141
Beef carcasses	1.4/1,500	United Kingdom	27
Healthy cattle	1.5/201	United States	19
Downer cattle	4.9/203	United States	19
Steer/heifer carcasses	0.2/2,081	United States	178
Lamb/mutton	17/37	New Zealand	16
Lamb carcasses	0.7/1,500	United Kingdom	27
Lamb products	7.4/7,200	United Kingdom	27
Sheep carcasses (frozen)	0.3/343	Australia	179
Pork	4/35[a]	New Zealand	16
Retail meats	12/91[a]	New Zealand	16
Raw meats	0.44/4,983	United Kingdom	27
Cattle feces	18/296	United States	154
Beef carcasses, export only	0.1/812	Australia	141
Cattle feed	14.9/504	United States	36
Calves[b], <1 month	31.4/35	Japan	98
Calves, 1 to 3 months	8.4/107	Japan	98
Heifers, >3 to 6 months	26.1/88	Japan	98
Heifers, >6 months	14.5/214	Japan	98
Beef products	12.9/4,800	United Kingdom	27

[a]Non-0157:H7 Stx-producing strains; [b]Rectal stool samples.

Why bacteria grow faster in the meat-soy blends than in nonsoy controls is not clear. The soy itself does not alter the initial biota, and the general spoilage pattern of meat-soy blends is not unlike that of all-meat controls. One notable difference is a slightly higher pH (0.3–0.4 unit) in soy-extended products, and this alone could account for the faster growth rate. This was assessed by Harrison et al.[83] by using organic acids to lower the pH of soy blends to that of beef. By adding small amounts of a 5% solution of acetic acid to 20% blends, spoilage was delayed about 2 days over controls, but not all of the inhibitory activity was due to pH depression alone. With 25% fat in the ground meat, bacterial counts did not increase proportionally to those of soy-extended beef.[97] It is possible that soy protein increases the surface area of soy–meat mixtures so that aerobic bacteria of the type that predominate on meats at refrigerator temperatures are favored, but data along these lines are wanting. The spoilage of soy-meat blends is discussed below. For more information, see reference 39.

Mechanically Deboned Meats

When meat animals are slaughtered for human consumption, meat from the carcasses is removed typically by meat cutters. However, the most economical way to salvage the small bits and pieces of lean meat left on carcass bones is by mechanical means (mechanical deboning). Mechanically deboned meat (MDM) is removed from bones by machines. The production of MDM began in the 1970s, preceded by chicken meat in the late 1950s, and fish in the late 1940s.[53,58] During the deboning process, small quantities of bone powder become part of the finished product, and the 1978 U.S. Department of Agriculture (USDA) regulation limits the amount of bone (based on calcium content) to no more than 0.75% (the calcium content of meat is 0.01%). MDM must contain a minimum of 14% protein and no more than 30% fat. The most significant parametrical difference between MDM and conventionally processed meat relative to microbial growth is the higher pH of the former, typically 6.0–7.0.[53,54] The increased pH is due to the incorporation of marrow in MDM.

Although most studies on the microbiology of MDM have shown these products to be not unlike those produced by conventional methods, some have found higher counts. The microbiological quality of deboned poultry was compared to other raw poultry products, and although the counts were comparable, MPN coliform counts of the commercial MDM products ranged from 460 to >1,100/g. Six of 54 samples contained salmonellae, four contained *C. perfringens*, but none contained *S. aureus*.[137] The APC of handboned lamb breasts was found to be 680,000, whereas for mechanically deboned lamb allowed to age for 1 week, the APC was 650,000/g.[55] Commercial samples of mechanically deboned fish were found to contain tenfold higher numbers of organisms than conventionally processed fish, but different methods were used to perform the counts on fish frames and the mechanically deboned flesh (MDF[146]). These investigators did not find *S. aureus* and concluded that the spoilage of MDF was similar to that for the traditionally processed products. In a later study, MDM was found to support the more rapid growth of psychrotrophic bacteria than lean ground beef.[149]

Several studies have revealed the absence of *S. aureus* in MDM, reflecting perhaps the fact that these products are less handled by meat cutters. In general, the mesophilic biota count is a bit higher than that for psychrotrophs, and fewer Gram negatives tend to be found. Field[53] concluded that with good manufacturing practices, MDM should present no microbiological problems, and a similar conclusion was reached by Froning[58] relative to deboned poultry and fish.

Hot-Boned Meats

In the conventional processing of meats (cold boning), carcasses are chilled after slaughter for 24 hours or more and processed in the chilled state (postrigor). Hot boning (hot processing) involves the processing of meats generally within 1–2 hours after slaughter (prerigor) while the carcass is still "hot."

In general, the microbiology of hot-boned meats is comparable to that of cold-boned meats, but some differences have been reported. One of the earliest studies on hot-boned hams evaluated the microbiological quality of cured hams made from hot-boned meat (hot-processed hams). These hams were found to contain a significantly higher APC (at 37°C) than cold-boned hams, and 67% of the former yielded staphylococci to 47% of the latter.[145] Mesophiles counted at 35°C were significantly higher on hot-boned prime cuts than comparable cold-boned cuts, both before and after vacuum-packaged storage at 2°C for 20 days.[101] Coliforms, however, were apparently not affected by hot boning. Another early study is that of Barbe et al.,[9] who evaluated 19 paired hams (hot and cold boned) and found that the former contained 200 bacteria per gram, whereas 220 per gram were found in the latter. In a study of hot-boned carcasses held at 16°C and cold-boned bovine carcasses held at 2°C for up to 16 hours postmortem, no significant differences in mesophilic and psychrotrophic counts were found.[96] Both hot-boned and cold-boned beef initially contained low bacterial counts, but after a 14-day storage period, the hot-boned meats contained higher numbers than the cold boned.[59] These investigators found that the temperature control of hot-boned meat during the early hours of chilling is critical and in a later study found that chilling to 21°C within 3–9 hours was satisfactory.[60]

In a study of sausage made from hot-boned pork, significantly higher counts of mesophiles and lipolytics were found in the product made from hot-boned pork than in the cold-boned product, but no significant differences in psychrotrophs were found.[114]

The effect that delayed chilling might have on the biota of hot-boned beef taken about 1 hour after slaughter was examined by McMillin et al.[125] Portions were chilled for 1, 2, 4, and 8 hours after slaughter and subsequently ground, formed into patties, frozen, and examined. No significant differences were found between this product and a cold-boned product relative to coliforms, staphylococci, psychrotrophs, and mesophiles. A numerical taxonomy study of the biota from hot-boned and cold-boned beef at both the time of processing and after 14 days of vacuum storage at 2°C revealed no statistically significant differences in the biota.[108] The predominant organisms, after storage, for both products were "streptococci" (most likely enterococci) and lactobacilli, whereas in the freshly prepared hot-boned product (before storage), more staphylococci and bacilli were found. Overall, though, the two products were comparable.

Restructured lamb roast made from 10% and 30% MDM and hot-boned meat was examined for microorganisms; overall, the two uncooked products were of good quality.[148] The uncooked products had counts $<3.0 \times 10^4$/g, with generally higher numbers in products containing the higher amounts of MDM. Coliforms and fecal coliforms especially were higher in products with 30% MDM, and this was thought to be caused by contamination of shanks and pelvic regions during slaughtering and evisceration. Not detected in either uncooked product (in 0.1 g) were *S. aureus* and *C. perfringens*; no salmonellae, *Yersinia enterocolitica*, or *Campylobacter jejuni*, were found in 25-g samples. Cooking reduced cell counts in all products to <30/g.

A summary of the work of 10 groups of investigators made by Kotula[100] on the effect of hot boning on the microbiology of meats revealed that six found no effect, three found only limited effects, and only one found higher counts. Kotula concluded that hot boning per se has no effect on microbial counts. Hot boning is often accompanied by prerigor pressurization consisting of the application of around 15,000 psi for 2 minutes. This process improves muscle color and overall shelf appearance and increases tenderization. It appears not to have any effect on the microbiota.

Effect of Electrical Stimulation

If the temperature of a beef carcass falls to <10°C before carcass pH is <5.9 or so, the meat will "cold shorten" and thus become tough. Electrical stimulation increases the rate of pH drop by stimulating the speed of conversion of glycogen to lactic acid and thus eliminating the toughening. By this method, an electric stunner is attached to a carcass, and repeated pulses of 0.5–1.0 or more seconds are administered to the product at 400+V potential differences between the electrodes. A summary of the findings of 10 groups of researchers on what effect, if any, electrical stimulation had on the microbiota revealed that 6 found no effect, two found a slight effect, and two found some effect.[100] The meats studied included beef, lamb, and pork.

Among investigators who found a reduction of APC by electrical stimulation were Ockerman and Szczawinski[135] who found that the process significantly reduced the APC of samples of beef inoculated before electrical stimulation, but when samples were inoculated immediately after the treatment, no significant reductions occurred. The latter finding suggests that the disruption of lysosomal membranes and the consequent release of catheptic enzymes, which has been shown to accompany electrical stimulation,[45] should not affect microorganisms. The tenderization associated with electrical stimulation of meats is presumed to be, at least in part, the result of lysosomal destruction.[45] In one study, no significant reduction in surface organisms was observed, whereas significant reduction was found to occur on the muscle above the aitch bone of beef carcasses.[113] These workers exposed meat-borne bacteria to electrical stimulation on culture media and found that Gram-positive bacteria were the most sensitive to electrical stimulation, followed by Gram negatives and sporeformers. When exposed to a 30-V, 5-minute treatment in saline or phosphate-buffered saline, a five log-cycle reduction occurred with *E. coli*, *Shewanella putrefaciens*, and *Pseudomonas fragi*, whereas in 0.1% peptone or 2.5 M sucrose solutions, essentially no changes occurred. It appears that electrical stimulation per se does not exert measurable effects on the microbial biota of hot-boned meats.

Prerigor meat can be tenderized by high-pressure treatments such as the application of about 15,000 lb/in.^{2} for several minutes, or by a process called Hydrodyne. The latter tenderizes beef by employing a small amount of explosive to generate a hydrodynamic shock wave in water.[164] It is not clear if this treatment affects the bacterial biota, but when applied at 55 to 60 megaPascal (MPa), it did not destroy the infectivity of *Trichinella spiralis* in pork.[63]

Organ and Variety Meats

The meats discussed in this section are livers, kidneys, hearts, and tongues (and others) of bovine, porcine, and ovine origins. They differ from the skeletal muscle parts of the respective animals in having both higher pH and glycogen levels, especially in the case of liver. The pH of fresh beef and pork liver ranges from 6.1 to 6.5 and that of kidneys from 6.5 to 7.0. Most investigators have found generally low numbers of microorganisms on these products, with surface numbers ranging from $\log_{10}$ 1.69 to 4.20/cm^2 for fresh livers, kidneys, hearts, and tongues. The initial biota has been reported to consist largely of Gram-positive cocci, coryneforms, aerobic sporeformers, *Moraxella-Acinetobacter*, and *Pseudomonas* spp. In detailed studies by Hanna et al.,[79] micrococci, "streptococci", and coryneforms were clearly the three most dominant groups on fresh livers, kidneys, and hearts. In one study, coagulase-positive staphylococci, coliforms, and *C. perfringens* counts ranged from $\log_{10}$ 0.9 to $\log_{10}$ 1.37/cm^2, but no salmonellae were found.[153] APC and coliform counts on some other organ and variety meats are presented in Table 4–8.

Table 4–8 Summary of Mean APC and Coliform Counts on Beef Variety Meats Ready for Shipment (Summarized from Delmore et al.[33])

Products	Mean $\log_{10}$ cfu/g = APC	Mean $\log_{10}$ cfu/g = coliforms
Abomasum	5.1	3.0
Heart	4.2	2.2
Large intestine	4.9	3.3
Liver	4.5	2.6
Mountain chain tripe	5.0	2.5
Omasum	6.0	2.9
Oxtail	4.7	2.7
Sweetbread	4.3	2.1
Tongue	5.6	2.0

MICROBIAL SPOILAGE OF FRESH RED MEATS

Most studies dealing with the spoilage of meats have been done with beef, and most of the discussion in this section is based on beef studies. Pork, lamb, veal, and similar meats, are presumed to spoil in a similar way.

Meats are the most perishable of all major foods, and some reasons are shown in Table 4–9, which lists the chemical composition of a typical adult mammalian muscle postmortem. Meats contain

Table 4–9 Chemical Composition of Typical Adult Mammalian Muscle after Rigor Mortis but Before Degradative Changes Postmortem (%, w/w)

Water	75.5%
Protein	18.0
Myofibrillar	
Myosin, tropomyosin, X protein	7.5
Actin	2.5
Sarcoplasmic	
Myogen, globulins	5.6
Myoglobin	0.36
Hemoglobin	0.04
Mitochondrial—cytochrome C	ca. 0.002
Sarcoplasmic reticulum, collagen, elastin,	
"reticulin," insoluble enzymes, connective tissue	2.0
Fat	3.0
Soluble nonprotein substances	3.5
Nitrogenous	
Creatine	0.55
Inosine monophosphate	0.30
Di- and triphosphopyridine nucleotides	0.07
Amino acids	0.35
Carnosine, anserine	0.30
Carbohydrate	
Lactic acid	0.90
Glucose-6-phosphate	0.17
Glycogen	0.10
Glucose	0.01
Inorganic	
Total soluble phosphorus	0.20
Potassium	0.35
Sodium	0.05
Magnesium	0.02
Calcium	0.007
Zinc	0.005
Traces of glycolytic intermediates,	
trace metals, vitamins, etc.	ca. 0.10

Source: Reprinted with permission from R.A. Lawrie[106]: Meat Science, copyright © 1966, Pergamon Press.

Table 4–10 Meat and Meat Products: Approximate Percentage Chemical Composition

Meats	Water	Carbohydrates	Proteins	Fat	Ash
Beef, hamburger	55.0	0	16.0	28.0	0.8
Beef, round	69.0	0	19.5	11.0	1.0
Bologna	62.4	3.6	14.8	15.9	3.3
Chicken (broiler)	71.2	0	20.2	7.2	1.1
Frankfurters	60.0	2.7	14.2	20.5	2.7
Lamb	66.3	0	17.1	14.8	0.9
Liver (beef)	69.7	6.0	19.7	3.2	1.4
Pork, medium	42.0	0	11.9	45.0	0.6
Turkey, medium fat	58.3	0	20.1	20.2	1.0

Source: Watt and Merrill.[184]

an abundance of all nutrients required for the growth of bacteria, yeasts, and molds, and an adequate quantity of these constituents exists in fresh meats in available form. The general chemical composition of a variety of meats is presented in Table 4–10.

The genera of bacteria often found on fresh and spoiled meats and poultry are listed in Table 4–1. Not all of the genera indicated for a given product are found at all times, of course. Those that are more often found during spoilage are indicated under the various products. In Table 4–2 are listed the genera of yeasts and molds most often identified from meats and related products. When spoiled meat products are examined, only a few of the many genera of bacteria, molds, or yeasts are found, and in almost all cases, one or more genera are found to be characteristic of the spoilage of a given type of meat product The presence of the more-varied biota on nonspoiled meats, then, may be taken to represent the organisms that exist in the original environment of the product in question or contaminants picked up during processing, handling, packaging, and storage.

The question arises, then, as to why only a few types predominate in spoiled meats. It is helpful here to return to the intrinsic and extrinsic parameters that affect the growth of spoilage microorganisms. Fresh meats such as beef, pork, and lamb, as well as fresh poultry, seafood, and processed meats, have pH values within the growth range of most of the organisms listed in Table 4–1. Nutrient and moisture contents are adequate to support the growth of all organisms listed. Although the O/R potential of whole meats is low, O/R conditions at the surfaces tend to be higher so that strict aerobes and facultative anaerobes, as well as strict anaerobes, generally find conditions suitable for growth. Effective levels of antimicrobial constituents are not known to occur in products of the type in question. Of the extrinsic parameters, temperature of storage stands out as being of utmost importance in controlling the types of microorganisms that develop on meats, as these products are normally held at refrigerator temperatures. Essentially all studies on the spoilage of meats, poultry, and seafood carried out over the past 50 years or so have dealt with low-temperature-stored products.

With respect to fungal spoilage of fresh meats, especially beef, the following genera of molds have been recovered from various spoilage conditions of whole beef: *Thamnidium*, *Mucor*, and *Rhizopus*, all of which produce "whiskers" on beef; *Cladosporium*, a common cause of "black spot"; *Penicillium*, which produces green patches; and *Sporotrichum* and *Chrysosporium*, which produce "white spot." Molds generally do not grow on meats if the storage temperature is below 5°C.[116] Among genera of yeasts recovered from refrigerator-spoiled beef with any consistency are *Candida* and *Rhodotorula*, with *C. lipolytica* and *C. zeylanoides* being the two most abundant species in spoiled ground beef.[89]

Unlike the spoilage of fresh beef carcasses, ground beef or hamburger meat is spoiled exclusively by bacteria, with the following genera being the most important: *Pseudomonas, Alcaligenes, Acinetobacter, Moraxella,* and *Aeromonas*. Those generally agreed to be the primary cause of spoilage are *Pseudomonas* and *Acinetobacter-Moraxella* spp., with others playing relatively minor roles in the process. Findings from two studies suggest that *Acinetobacter* and *Moraxella* spp. may not be as abundant in spoiled beef as once reported.[50,51] In another study, *Psychrobacter* and *Moraxella* were relatively abundant on fresh lamb carcasses but few were detected after carcass spoilage.[143]

It should be noted that over the past 70 years, the *Pseudomonas* spp. have been shown to be the most dominant spoilage bacteria of refrigerated fresh meats, poultry, and seafoods, but the true position of the newly defined genus has been called into question by the transfer of at least 40 species to >10 new genera (see Chapter 2). As presently defined, the genus *Pseudomonas* belongs to the gamma-subclass of the *Proteobacteria*, and it includes *P. fluorescence, P. fragi*; the type species *P. aeruginosa*, and others. Some of the species formerly in this genus have been transferred to the alpha-subclass of the *Proteobacteria* (*Brevundimonas, Devosia, Sphingomonas*), while others (*Acidovorax, Comamonas,* and *Telluria*) are now in the beta-subclass. It remains to be determined where these genera belong among the organisms previously classified as pseudomonads. Until this is sorted out, references to pseudomonads in this and subsequent chapters include all species of the genus *Pseudomonas* as defined in the 1986 edition of *Bergey's Manual*.

A study of the aerobic Gram-negative bacteria recovered from beef, lamb, pork, and fresh sausage, revealed that all 231 polarly flagellated rods were pseudomonads and that of 110 nonmotile organisms, 61 were *Moraxella* and 49 were *Acinetobacter*.[32] The pseudomonads that cause meat spoilage at low temperatures generally do not match the named species in *Bergey's Manual*.

Numerical taxonomic studies by Shaw and Latty[156,157] led them to group most of their isolates into four clusters based on carbon source utilization tests. Of 787 *Pseudomonas* strains isolated from meats, 89.7% were identified, with 49.6% belonging to their cluster 2, 24.9% to cluster 1, and 11.1% to cluster 3.[157] The organisms in clusters 1 and 2 were nonfluorescent and eggyolk negative and resembled *P. fragi*; those in cluster 3 were fluorescent and gelatinase positive. *P. fluorescens* biotype I strains were represented by 3.9%, biotype III by 0.9%, and *P. putida* by only one strain. The relative incidence of the clusters on beef, pork, and lamb, and on fresh and spoiled meats, was similar.[157]

Beef rounds and quarters are known to undergo deep spoilage, usually near the bone, especially the "aitch" bone. This type of spoilage is often referred to as "bone taint" or "sours." Only bacteria have been implicated, with the genera *Clostridium* and *Enterococcus* being the primary causative agents.

Temperature of incubation is the primary reason that only a few genera of bacteria are found in spoiled meats as opposed to fresh. In one study, only four of the nine genera present in fresh ground beef could be found after the meat underwent frank spoilage at refrigerator temperatures.[93] It was noted by Ayres[5] that after processing, more than 80% of the total population of freshly ground beef may be composed of chromogenic bacteria, molds, yeasts, and spore-forming bacteria, but after spoilage, only nonchromogenic, short Gram-negative rods are found. Although some of the bacteria found in fresh meats can be shown to grow at refrigerator temperatures on culture media, they apparently lack the capacity to compete successfully with the *Pseudomonas* and *Acinetobacter-Moraxella* types.

Beef cuts, such as steaks or roasts, tend to undergo surface spoilage; whether the spoilage organisms are bacteria or molds depends on available moisture. Freshly cut meats stored in a refrigerator with high humidity invariably undergo bacterial spoilage preferential to mold spoilage. The essential feature

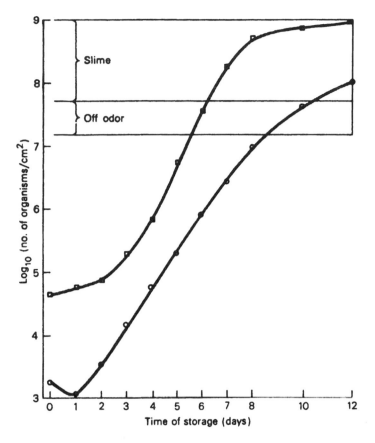

Figure 4–1 The development of off-odor and slime on dressed chicken (squares) and packaged beef (circles) during storage at 5°C. *Source*: From Ayres.[5]

of this spoilage is surface sliminess in which the causative organisms can nearly always be found. The relatively high O/R, availability of moisture, and low temperature favor the pseudomonads. It is sometimes possible to note discrete bacterial colonies on the surface of beef cuts, especially when the level of contamination is low. The slime layer results from the coalescence of surface colonies and is largely responsible for the tacky consistency of spoiled meats. Ayres[5] presented evidence that odors can be detected when the surface bacterial count is between log 7.0 and log 7.5/cm^2, followed by detectable slime with surface counts usually about $\log_{10}$ 7.5 to log 8.0/cm^2 (Figure 4–1). This is further depicted in Figure 4–2, which relates numbers of bacteria not only to surface spoilage of fresh poultry but to red meats and seafoods as well.

Molds tend to predominate in the spoilage of beef cuts when the surface is too dry for bacterial growth or when beef has been treated with antibiotics such as the tetracyclines. Molds virtually never develop on meats when bacteria are allowed to grow freely. The reason appears to be that bacteria grow faster than molds, thus consuming available surface oxygen, which molds also require for their activities.

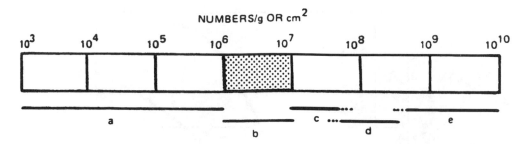

Figure 4–2 Significance of total viable microbial numbers in food products relative to their use as indicators of spoilage. (a) Microbial spoilage generally not recognized with the possible exception of raw milk, which may sour in the 10^5–10^6 range. (b) Some food products show incipiency in this range. Vacuum-packaged meats often display objectionable odors and may be spoiled. (c) Off-odors generally associated with aerobically stored meats and some vegetables. (d) Almost all food products display obvious signs of spoilage. Slime is common on aerobically stored meats. (e) Definite structural changes in product occur at this stage.

Unlike the case of beef cuts or beef quarters, visible mold growth is nonexistent on ground beef except when antibacterial agents have been used as preservatives or when the normal bacterial load has been reduced by long-term freezing. Among the early signs of spoilage of ground beef is the development of off-odors followed by tackiness, which indicate the presence of bacterial slime. The slime layer that develops on fresh meat, poultry, and seafood products, as they undergo microbial spoilage at refrigerator temperatures is a *biofilm*, which is further described in Chapter 22.

In the spoilage of soy-extended ground meats, nothing indicates that the pattern differs from that of unextended ground meats, although their rate of spoilage is faster.

The precise roles played by spoilage microorganisms that result in the spoilage of meats are not fully understood at this time, but significant progress has been made. Some of the earlier views on the mechanism of meat spoilage are embodied in the many techniques proposed for its detection (Table 4–11).

Mechanism

It is reasonable to assume that reliable methods of determining meat spoilage should be based on the cause and mechanism of spoilage. The chemical methods in Table 4–11 embody the assumption that as meats undergo spoilage, some utilizable substrate is consumed, or some new product or products are created by the spoilage biota. It is well established that the spoilage of meats at low temperature is accompanied by the production of off-color compounds such as ammonia, H_2S, indole, and amines. The drawbacks to the use of these methods are that not all spoilage organisms are equally capable of producing them. Inherent in some of these methods is the incorrect belief that low-temperature spoilage is accompanied by a breakdown of primary proteins.[91] The physical and direct bacteriological methods all tend to show what is obvious: Meat that is clearly spoiled from the standpoint of organoleptic characteristics (odor, touch, appearance, and taste) is, indeed, spoiled. They do not allow one to predict spoilage or shelf life, which a meat freshness test should ideally do.

Among the metabolic byproducts of meat spoilage, the diamines, cadaverine, and putrescine, have been studied as spoilage indicators of meats. The production of these diamines occurs in the

Table 4–11 Some Methods Proposed for Detecting Microbial Spoilage in Meats, Poultry, and Seafood

Chemical Methods
a. Measurement of H_2S production
b. Measurement of mercaptans produced
c. Determination of noncoagulable nitrogen
d. Determination of di- and trimethylamines
e. Determination of tyrosine complexes
f. Determination of indole and skatol
g. Determination of amino acids
h. Determination of volatile reducing substances
i. Determination of amino nitrogen
j. Determination of biochemical oxygen demand
k. Determination of nitrate reduction
l. Measurement of total nitrogen
m. Measurement of catalase
n. Determination of creatinine content
o. Determination of dye-reducing capacity
p. Measurement of hypoxanthine
q. Measurement of ATP from microorganisms
r. Radiometric measurement of CO_2
s. Ethanol production (fish spoilage)
t. Measurement of lactic acid
u. Change in color
v. CO_2 evolution rate measurement

Physical methods
a. Measurement of pH changes
b. Measurement of refractive index of muscle juices
c. Determination of alteration in electrical conductivity
d. Determination of surface tension
e. Measurement of ultraviolet illumination (fluorescence)
f. Determination of surface charges
g. Determination of cryoscopic properties
h. Impedance changes
i. Micro-calorimetry
j. Measurement of proton efflux from and influx into bacterial cells

Direct bacteriological methods
a. Determination of total aerobes
b. Determination of total anaerobes
c. Determination of ratio of total aerobes to anaerobes
d. Determination of one or more of above at different temperatures
e. Determination of Gram-negative endotoxins

Physicochemical methods
a. Determination of extract-release volume
b. Determination of water-holding capacity
c. Determination of viscosity
d. Determination of meat swelling capacity
e. Volatile organic acid detection with proton transfer reaction-mass spectrophometry (PTR-MS)

following manner:

$$\text{Lysine} \xrightarrow{\text{decarboxylase}} \underset{\text{Cadaverine}}{H_2N(CH_2)_5NH_2}$$

$$\text{Ornithine or arginine} \xrightarrow{\text{decarboxylase}} \underset{\text{Putrescine}}{H_2N(CH_2)_5NH_2}$$

Their use as quality indicators of vacuum-packaged beef that was stored at 1°C for up to 8 weeks has been investigated.[46] Cadaverine increased more than putrescine in vacuum-packaged meats, the reverse of findings for aerobically stored samples. Cadaverine levels attained over the incubation period were tenfold higher than the initial levels at total viable counts of $10^6/cm^2$, whereas there was little change in putrescine at this level. Overall, the findings suggested that these diamines could be of value for vacuum-packaged meats. In fresh beef, pork, and lamb, putrescine occurred at levels from 0.4 to 2.3 ppm and cadaverine from 0.1 to 1.3 ppm.[47,132,185] Putrescine is the major diamine produced by pseudomonads, whereas cadaverine is produced more by Enterobacteriaceae.[162] It may be noted from Table 4–12 that putrescine increased from 1.2 to 26.1 ppm in one sample of naturally contaminated beef stored at 5°C for 4 days; cadaverine levels were much lower. In another sample, the two diamines increased to higher levels under the same conditions. Cadaverine was the only amine that correlated with coliforms in ground beef in one study.[155] That significant changes in putrescine and cadaverine do not occur in beef until the APC exceeds about 4×10^7 raises questions about their utility to predict meat spoilage.[47] This is a common problem with most, if not all, single metabolites because their production and concentration tend to be related to specific organisms. In one study, the changes in histamine and tyramine in beef and pork were found to be too small to be of value

Table 4–12 Development of Microbial Number and Diamine Concentrations on Naturally Contaminated Minced Beef Stored at 5°C

Sample[a]	Storage Time (days)	Putrescine[b] ($\mu g/g$)	Cadaverine[b] ($\mu g/g$)	Enterobacteriaceae (log_{10}/g)	Aerobic Plate Count (log_{10}/g)
E	0	1.2	0.1	3.81	6.29
	1	1.8	0.1	3.56	7.66
	2	4.2	0.5	4.57	8.49
	3	10.0	0.5	5.86	9.48
	4	26.1	0.6	7.54	9.97
F	0	2.3	1.3	6.18	7.49
	1	3.9	4.5	6.23	7.85
	2	12.4	17.9	6.69	8.73
	3	29.9	35.2	7.94	9.69
	4	59.2	40.8	9.00	9.91

[a]Samples E and F were obtained from two different retail outlets.
[b]Diamine values are the mean of two determinations.

Source: Edwards et al.,[47] copyright © 1983, Blackwell Scientific Publications, Ltd.

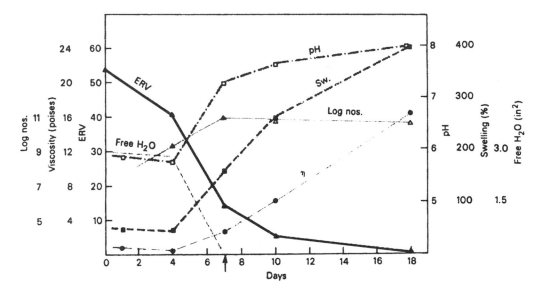

Figure 4–3 The response of several physicochemical meat spoilage tests as fresh ground beef was held at 7°C until definite spoilage had occurred. The arrow indicates the first day off-odors were detected. ERV = extract-release volume; free H_2O = measurement of water-holding capacity (inversely related); Sw = meat swelling; η = viscosity; and log nos. = total aerobic bacteria/g. *Source*: From Shelef and Jay,[159] copyright © 1969, Institute of Food Technologists.

---

as spoilage predictors, but putrescine for beef and cadaverine for pork showed greater changes as spoilage occurred.[20]

The extract-release volume (ERV) technique, first described in 1964, has been shown to be of value in determining incipient spoilage in meats as well as in predicting refrigerator shelf life.[92] The technique is based on the volume of aqueous extract released by a homogenate of beef when allowed to pass through filter paper for a given period of time. By this method, beef of good organoleptic and microbial quality releases large volumes of extract, whereas beef of poor microbial quality releases smaller volumes or none (Figure 4–3). One of the more important aspects of this method is the information that it has provided concerning the mechanism of low-temperature beef spoilage.

The ERV method reveals two aspects of the spoilage mechanism. First, low-temperature meat spoilage occurs in the absence of any significant breakdown of primary proteins—at least not complete breakdown. Although this fact has been verified by total protein analyses on fresh and spoiled meats, it is also implicit in the operation of the method; that is, as meats undergo microbial spoilage, ERV is decreased rather than increased, which would be the case if complete hydrolysis of proteins occurred. The second aspect of meat spoilage revealed by ERV is the increase in hydration capacity of meat proteins by some as-yet unknown mechanism, although amino sugar complexes produced by the spoilage biota have been shown to play a role.[160] In the absence of complete protein breakdown, the question arises as to how the spoilage organisms obtain their nutritional needs for growth.

When fresh meats are placed in storage at refrigerator temperatures, those organisms capable of growth at the particular temperature begin their growth. In the case of fresh meats that have an ultimate pH of around 5.6, enough glucose and other simple carbohydrates are present to support about 10^8

organisms/cm.[71,73] Among the heterogeneous fresh-meat biota, the organisms that grow the fastest and utilize glucose at refrigerator temperatures are the pseudomonads, and available surface O_2 has a definite effect on their ultimate growth.[73] *Brochothrix thermosphacta* also utilizes glucose and glutamate, but because of its slower growth rate, it is a poor competitor of the pseudomonads. Upon reaching a surface population of about 10^8/cm^2, the supply of simple carbohydrates is exhausted, and off-odors may or may not be evident at this point, depending on the extent to which free amino acid utilization has occurred. Once simple carbohydrates have been exhausted, pseudomonads along with other Gram-negative psychrotrophs such as *Moraxella, Alcaligenes, Aeromonas, Serratia,* and *Pantoea* utilize free amino acids and related simple nitrogenous compounds as sources of energy. *Acinetobacter* spp. utilize amino acids first, lactate next, and their growth is reduced at and below pH 5.7.[73] With regard to poultry, the conversion of glucose to glucsonate appears to give pseudomonads the competitive advantage.[95]

Employing lamb juice at pH 6.0 and 4°C, one group of investigators suggested that the dominance of *P. fragi* was due to its ability to utilize creatine and creatinine.[40] It has been observed by a number of investigators that *P. fluorescens* is more abundant on fresh meats than *P. fragi* but that the latter becomes dominant over time.[107]

The foul odors generally associated with spoiling meats owe their origin to free amino acids and related compounds (H_2S from sulfur-containing amino acids, NH_3 from many amino acids, and indole from tryptophan). Off-odors and off-flavors appear only when amino acids begin to be utilized (see below). In the case of dark, firm, and dry (DFD) meats, which have ultimate pH > 6.0 and a considerably lower supply of simple carbohydrates, spoilage is more rapid and off-odors are detectable with cell numbers around 10^6/cm^2.[134] With normal or DFD meats, the primary proteins are not attacked until the supply of the simpler constituents has been exhausted. It has been shown, for example, that the antigenicity of salt-soluble beef proteins is not destroyed under the usual conditions of low-temperature spoilage.[118]

In the case of fish spoilage, it has been shown that raw fish press juice displays all the apparent aspects of fish spoilage as may be determined by use of the whole fish.[110] This can be taken to indicate a general lack of attack on soluble proteins by the fish-spoilage organisms because these proteins were absent from the filtered press juice. The same is apparently true for beef and related meats. Incipient spoilage is accompanied by a rise in pH, an increase in bacterial numbers, and an increase in the hydration capacity of meat proteins, along with other changes. In ground beef, pH may rise as high as 8.5 in putrid meats, although at the time of incipient spoilage, mean pH values of about 6.5 have been found.[161] By plotting the growth curve of the spoilage biota, the usual phases of growth can be observed and the phase of decline may be ascribed to the exhaustion of utilizable nutrients by most of the biota and the accumulation of toxic byproducts of bacterial metabolism. Precisely how the primary proteins of meat are destroyed at low temperatures is not well understood. For a review, see reference 91.

Dainty et al.[31] inoculated beef slime onto slices of raw beef and incubated them at 5°C. Off-odors and slime were noted after 7 days with counts at 2×10^9/cm^2. Proteolysis was not detected in either sarcoplasmic or myofibrillar fractions of the beef slices. No changes in the sarcoplasmic fractions could be detected even 2 days later, when bacterial numbers reached 10^{10}/cm^2. The first indication of breakdown of myofibrillar proteins occurred at this time with the appearance of a new band and the weakening of another. All myofibrillar bands disappeared after 11 days, with the weakening of several bands of the sarcoplasmic fraction. With naturally contaminated beef, odors and slime were first noted after 12 days when the numbers were 4×10^8/cm^2. Changes in myofibrillar proteins were not noted until 18 days of holding. By the use of pure culture studies, these workers showed that pseudomonads were active against myofibrillar proteins, whereas others were more active against

Table 4–13   pH and Concentrations in 10 Fresh Livers of Glycogen, Glucose, Lactic Acid, and Ammonia

Component	Average Concentration and Range
Glucose	2.73 (0.68–6.33) mg/g
Glycogen	2.98 (0.70–5.43) mg/g
Lactic acid	4.14 (3.42–5.87) mg/g
Ammonia	7.52 (6.44–8.30) μmol/g
pH	6.41 (6.26–6.63)

Source: Gill and DeLacy,[72] copyright © 1982, American Society for Microbiology.

sarcoplasmics. *Aeromonas* spp. were active on both myofibrillar and sarcoplasmic proteins. With pure cultures, protein changes were not detected until counts were above 3.2×10^9/cm^2. Earlier, Borton et al.[15] showed that *P. fragi* effected the loss of protein bands from inoculated pork muscle, but no indication was given as to the minimum numbers that were necessary. For more on the microbial spoilage of muscle foods, see the review by Ellis and Goodacre.[49]

SPOILAGE OF FRESH LIVERS

The events that occur in the spoilage of beef, pork, and lamb livers are not as well defined as for meats. The mean content of carbohydrate, NH$_3$, and pH of 10 fresh lamb livers is presented in Table 4–13.[72] Based on the relatively high content of carbohydrates and mean pH of 6.41, these may be expected to undergo a fermentative spoilage, with the pH decreasing below 6.0. This would undoubtedly occur if livers were comminuted or finely diced and stored at refrigerator temperatures, but most studies have been conducted on whole livers, where growth was assessed at the surface, from drip, or from deep tissue. In a study of the spoilage of diced beef livers the initial pH of 6.3 decreased to about 5.9 after 7–10 days at 5°C and the predominant biota at spoilage consisted of lactic acid bacteria.[158] In most other studies, the predominant biota at spoilage was found to consist essentially of the same types of organisms that are dominant in the spoilage of muscle meats. In pork livers held at 5°C for 7 days, the predominant organisms found in one study were *Pseudomonas*, *Alcaligenes*, *Escherichia*, lactic streptococci, and *B. thermosphacta*.[64] In five beef livers stored at 2°C for 14 days, *Pseudomonas* constituted from 7% to 100% of the spoilage biota, while the mean initial pH of 6.49 decreased to 5.93 over the 14-day period.[82] In another study of beef, pork, and lamb livers, the predominant biota after 5 days at 2°C differed for the three products, with beef livers being dominated by streptococci, yeasts, coryneforms, and pseudomonads; lamb by coryneforms, micrococci, and "streptococci"; and pork livers by staphylococci, *Moraxella-Acinetobacter*, and "streptococci".[80] The mean initial pH of each of the three livers declined upon storage, although only slightly. In a study of spoilage of lamb livers by Gill and DeLacy,[72] the spoiled surface biota was dominated by *Pseudomonas*, *Acinetobacter*, and *Enterobacter*; drip from the whole livers was dominated by *Pseudomonas* and *Enterobacter*; whereas *Enterobacter* and lactobacilli were dominant in the deep tissues. It was shown in this study that the initial pH of around 6.4 decreased to around 5.7 in antibiotic-treated samples, indicating that liver glycolytic events can lead to a decrease in pH in the absence of organisms even though these samples did contain $<10^4$ organisms/cm^2. The high glucose level was sufficient to allow visible surface colony growth before off-odors developed, and herein may lie the explanation for the dominance of the spoilage biota of livers by nonlactic types.

Because most psychrotrophic oxidative, Gram-negative bacteria grow at a faster rate and are more favored by the higher surface O/R than the lactic fermentative Gram positives, their dominance in whole liver spoilage may not be unexpected. The higher concentration of carbohydrates would delay the onset of amino acid utilizers and explain in part why pH does not increase with whole liver spoilage as it does for meats. In this regard, comminuted livers would be expected to support the growth of lactic acid bacteria because of the redistribution of the surface biota throughout the sample where the lactics would be favored by the high carbohydrate content and reduced O/R away from the surface. This would be somewhat analogous to the surface spoilage of meat carcasses, where the slower-growing yeasts and molds develop when conditions are not favorable for bacterial growth. Fungi never dominate the spoilage of fresh comminuted meats unless special steps are taken to inhibit bacteria. By this analogy, lactic acid bacteria are inconspicuous in the spoilage of whole livers because conditions favor the faster-growing, psychrotrophic Gram-negative bacteria.

INCIDENCE/PREVALENCE OF MICROORGANISMS IN FRESH POULTRY

Whole poultry tends to have a lower microbial count than cut-up poultry. Most of the organisms on such products are at the surface, so surface counts/cm^2 are generally more valid than counts on surface and deep tissues. May[120] showed how the surface counts of chickens build up through successive stages of processing. In a study of whole chickens from six commercial processing plants, the initial mean total surface count was $\log_{10}$ 3.30/cm^2. After the chickens were cut up, the mean total count increased to $\log_{10}$ 3.81 and further increased to $\log_{10}$ 4.08 after packaging. The conveyor over which these birds moved showed a count of $\log_{10}$ 4.76/cm^2. When the procedures were repeated for five retail grocery stores, May found that the mean count before cutting was $\log_{10}$ 3.18, which increased to $\log_{10}$ 4.06 after cutting and packaging. The cutting block was shown to have a total count of $\log_{10}$ 4.68/cm^2.

Campylobacter jejuni is found less often on turkey products than salmonellae. Fertile turkey eggs and newly hatched turkey poults were free of this organism in one study.[2] However, fecal samples were positive about 2 weeks after hatching in up to 76% of those in one brooder house. The organism could not be recovered from either the surface or the drip of frozen, thawed turkey carcasses at the wholesale or retail level, and the scalding and carcass washing steps appear to have been responsible.[1]

Of the various cooked poultry products, precooked turkey rolls have been found to have considerably lower microbial numbers of all types (Table 4–14). In an examination of 118 samples of cooked broiler products, *C. perfringens* was found in 2.6%.[112] In a study of chicken carcasses in Argentina, 7 of 70 contained *Yersinia* spp. including *Y. enterocolitica* and *Y. frederiksenii* (4.3% for each); and *Y. intermedia* (1.4%). All *Y. enterocolitica* isolates belonged to biogroup 1A, serotype 0:5, and phagotype X$_2$.[56] Enterococci are common on poultry products. Of 227 turkey samples examined in the state of Iowa in 2001–2002, 226 were positive for these organisms with 60% of isolates identified as *E. faecium* and 31% of *E. faecalis*.[84] Of 234 chicken samples, 236 were positive with 79% of isolates being *E. faecium* and 16% *E. faecalis*.

The changes in enteric bacteria during various stages of poultry chilling were studied by Cox et al.[29] who found that carcass counts before chilling were 3.17 $\log_{10}$ cfu/cm^2 for APC and 2.27 $\log_{10}$ cfu/cm^2 for Enterobacteriaceae. After chilling, the latter organisms were reduced more than the APC. On day 0, *E. coli* constituted 85% of enterics but after 10 days at 4°C, they were reduced to 14% whereas *Enterobacter* spp. increased from 6 to 88% during the same time. In another study, *Micrococcus*

Table 4–14 General Microbiological Quality of Some Turkey Meat Products (All log Numbers are log$_{10}$)

Products	No. of Samples	Microbial Group/Target	% Samples Meeting Target	Reference
Precooked turkey rolls	6	APC: log 3.00/g	100	126
	6	Coliforms: log 2.00 or less/g	67	126
	6	Enterococci: log 2.00 or less/g	83	126
	48	Presence of salmonellae	4	126
	48	Presence of *C. perfringens*	0	126
Precooked turkey rolls/sliced turkey meat	30	APC: <log 2.00/g	20	187
	29	Presence of coliforms	21	187
	29	Presence of *E. coli* or salmonellae	0	187
Ground fresh turkey meat	74	APC: log 7.00 or less/g	51	77
	75	Presence of coliforms	99	77
	75	Presence of *E. coli*	41	77
	75	Presence of "fecal streptococci"	95	77
	75	Presence of *S. aureus*	69	77
	75	Presence of salmonellae	28	77
Frozen ground turkey meat	50	APC 32°C: <10^6/g	54	78
	50	Psychrotrophs: <10^6/g	32	78
	50	MPN *E. coli*: <10/g	80	78
	50	MPN *S. aureus*: <10/g	94	78
	50	MPN "fecal streptococci": <10/g	54	78

Note: APC = Aerobic plate count; MPN = most probable number.

spp. were found to be the single most abundant genus of bacteria on poultry during processing, with more organisms on neck-skin samples than feather-associated samples for both pre- and postscalded carcasses.[68] *Corynebacterium* spp. were abundant in air samples in the same study. In a 1994–1995 study of 1,297 broiler carcasses thoughout the United States, *Clostridium perfringens* was found in 43% and *Staphylococcus aureus* in 64%.[177] The prevalence of *Arcobacter* and *Campylobacter* spp. on some poultry products is summarized in Table 4–4; salmonellae in Table 4–5; *L. monocytogenes* in Table 4–6; and *E. coli* 0157:H7 in Table 4–7.

The yeasts that developed on broiler carcasses stored at 4°C for up to 14 days were isolated and identified, and at least 7 genera were identified with a *Candida* being the most predominant followed by *Cryptococcus* and *Yarrowia*.[86]

MICROBIAL SPOILAGE OF POULTRY

Studies on the bacterial biota of fresh poultry by many investigators have revealed over 25 genera (Table 4–1). However, when these meats undergo low-temperature spoilage, almost all workers agree that the primary spoilage organisms belong to the genus *Pseudomonas*. In a study of 5,920 isolates from chicken carcasses,[104] pseudomonads were found to constitute 30.5%, *Acinetobacter* 22.7%, *Flavobacterium* 13.9%, and *Corynebacterium* 12.7%, with yeasts, Enterobacteriaceae, and others in lower numbers. Of the pseudomonads, these investigators found that 61.8% were fluorescent on King's medium and that 95.2% of all pseudomonads oxidized glucose. A previous characterization of pseudomonads on poultry undergoing spoilage was made by Barnes and Impey,[11] who showed that the pigmented pseudomonads decreased from 34% to 16% from initial storage to the development of strong off-odors, whereas the nonpigmented actually increased from 11% to 58%. *Acinetobacter* and other species of bacteria decreased along with the pseudomonads. A similar process occurs in spoiling fish.

Fungi are of considerably less importance in poultry spoilage except when antibiotics are employed to suppress bacterial growth. When antibiotics are employed, however, molds become the primary agents of spoilage. The genera *Candida, Rhodotorula, Debaryomyces,* and *Yarrowia* are the most important yeasts found on poultry (Table 4–2). The essential feature of poultry spoilage is sliminess at the outer surfaces of the carcass or cuts. The visceral cavity often displays sour odors or what is commonly called visceral taint. This is especially true of the spoilage of New York-dressed poultry, where the viscera are left inside. The causative organisms here are also bacteria of the type noted above in addition to enterococci.

In a study of yeasts on fresh and spoiling poultry carcasses in South Africa, *Candida* and *Debaryomyces* spp. were the two most dominant genera on both fresh and spoiled carcasses, while *Rhodotorula* was not found on spoiled carcasses.[180] *Trichosporon* spp. were not found on fresh poultry but they were on 5% of spoiled while 3% of fresh and 11% of spoiled contained *Yarrowia.* The two most abundant species found on fresh and spoiled were *Candida zeylanoides* and *Debaryomyces hansenii.*[180]

S. putrefaciens grows well at 5°C and produces potent off-odors in 7 days when growing on chicken muscle.[124] Among odor producers in general, there is a selection of types that produce strong odors among the varied biota that exists on fresh poultry.[122] The study noted was conducted with chicken breast muscle, which spoils differently than leg muscles because the latter have a higher pH. With chicken leg muscle stored at 2°C for 16 days, 79% of the biota consisted of pseudomonads, 17% of *Acinetobacter-Moraxella,* and 4% of *S. putrefaciens.*[123] All isolates of the latter produced sulfide-like odors and this organism produces H_2S, methyl mercaptan, and dimethyl sulfide. It was not of significance in the spoilage of chicken breast muscles.

When New York-dressed poultry undergoes microbial spoilage, the organisms make their way through the gut walls and invade inner tissues of the intestinal cavity. The characteristic sharpness associated with the spoilage of this type of poultry is referred to as "visceral taint."

As poultry undergoes spoilage, off-odors are generally noted before sliminess, with the former being first detected when $\log_{10}$ numbers/cm^2 are about 7.2 to 8.0. Sliminess generally occurs shortly after the appearance of off-odors, with the $\log_{10}$ counts/cm^2 about 8.[6] Total aerobic plate counts/cm^2 of slimy surface rarely go higher than $\log_{10}$ 9.5. With the initial growth first confined to poultry surfaces, the tissue below the skin remains essentially free of bacteria for some time. Gradually, however, bacteria begin to enter the deep tissues, bringing about increased hydration of muscle proteins, much as occurs with beef. Whether autolysis plays an important role in the spoilage of inner poultry tissues is not clear.

The primary reasons that poultry spoilage is mainly restricted to the surfaces are as follows. The inner portions of poultry tissue are generally sterile, or contain relatively few organisms, which generally do not grow at low temperatures. The spoilage biota, therefore, is restricted to the surfaces and hide where it is deposited from water, processing, and handling. The surfaces of fresh poultry stored in an environment of high humidity are susceptible to the growth of aerobic bacteria such as pseudomonads. These organisms grow well on the surfaces, where they form minute colonies that later coalesce to produce the sliminess (biofilm) characteristic of spoiled poultry. May et al.[121] showed that poultry skin supports the growth of the poultry spoilage biota better than even the muscle tissue. In the advanced stages of poultry spoilage, the surfaces will often fluoresce when illuminated with ultraviolet light because of the presence of large numbers of fluorescent pseudomonads. Surface spoilage organisms can be recovered directly from the slime for plating, or one can prepare slides for viewing by smearing with portions of slime. Upon Gram staining, one may note the uniform appearance of organisms indistinguishable from those listed. Tetrazolium (2,3,5-triphenyltetrazolium chloride) can be used also to assess microbial activity on poultry surfaces. When an eviscerated carcass is sprayed with this compound, a red pigment develops in areas of high microbial activity. These areas generally consist of cut muscle surfaces and other damaged areas such as feather follicles.[140] When the skin is removed from a fresh chicken carcass, leg muscles are more likely to spoil faster than breast muscles since the pH of the former is typically in the pH range of 6.3–6.6 while the latter is lower, i.e., 5.7–5.9.

Pseudomonads are favored at the lowest growth temperature. When poultry was spoiled at 1°C, these organisms dominated while at 10°C and 15°C, enteric and other bacteria became significant.[12] More information on poultry spoilage can be found in reference 28. The spoilage of poultry and other meats under vacuum and modified atmosphere packaging is covered in Chapter 14.

CARCASS SANITIZING/WASHING

Just prior to slaughter, the outer surfaces of meat animals are laden with dust, dirt, and fecal matter. It is inevitable that some of the microorganisms from these sources will be found on the carcasses of slaughtered animals, and although most are nonpathogens, pathogens may be present. In an effort to reduce the number and types of pathogens on dressed carcasses and finished products, a number of methods have emerged:

1. Trimming—the excising of skin or outer tissue
2. Washing—the use of plain water at varying temperatures and hose pressures
3. Organic acids—the addition to wash water of acetic, citric, or lactic acid at concentrations of 2% to 5%
4. Other chemicals—the addition to wash water of hydrogen peroxide, chlorine dioxide, or chlorhexidine
5. Steam vacuum treatments—the application of steam for 5 to 10 seconds at 80°C or higher as the final carcass preparation step
6. Combinations—the use of two or more of the above

In the USDA's pathogen reduction program for beef carcasses, 1 out of every 300 carcasses is to be examined by sponging 100-cm^2 sections from three carcass areas (rump, flank, and brisket) for *E. coli*, which should be <5 cfu/cm^2.[175] The sponging method is one of six that were compared for beef carcasses.[37]

Overall, a large number of studies have been conducted on most of the methods noted above for removing microorganisms from slaughter carcasses, and reduction of APCs on the order of 1 to 3 log cycles is common. Many studies have employed laboratory and genetically modified strains of certain pathogens that were mixed with fresh animal feces and then rubbed onto meat cuts. The removal of biota applied in this way may be expected to be different from that acquired naturally, but comparative studies are wanting. The long-term effect of acid and steam treatments on meat biota is unknown because these procedures are relatively new for commercial use. The emergence of acid-resistant organisms after prolonged use is a likely outcome based on the long-term and widespread use of antimicrobials in general. It has been noted that multiple treatments are better than any one method alone[23] and this approach could reduce the emergence of resistant organisms. For catfish, the shelf life of fillets was extended by spraying with 4% lactic or 2–4% propionic acid.[52]

The combined use of a hot water wash followed by organic acid rinse was more effective for hog carcasses than either alone, and they effected about a 2-log cycle reduction.[48] These investigators suggested using water at 80°C. Using cold beef carcass sufaces spiked with *E. coli* 0157:H7 and *S.* Typhimurium, a 30 s 4% lactic acid spray effected a 5.2-log reduction of the two pathogens.[24] Using a postchill 30 s lactic acid (4%) spray at 55°C effected an additional reduction of *E. coli* 0157:H7 of 2 to 2.4 log cycles and of 1.6–1.9 for *S.* Typhimurium.

The stages of microbial contamination of pig carcasses was investigated in an Iberian slaughterhouse. Microbial numbers of *E. coli* were decreased by scalding and singing but increased by dehairing.[151] The *E. coli* count was decreased significantly by closure of the anus and evisceration. A final carcass wash with high pressure potable water failed to decrease microbial surface number.[151]

REFERENCES

1. Acuff, G.R., C. Vanderzant, M.O. Hanna, J.G. Ehlers, F.A. Golan, and F.A. Gardner. 1986. Prevalence of *Campylobacter jejuni* in turkey carcass processing and further processing of turkey products. *J. Food Protect.* 49:712–717.

2. Acuff, G.R., C. Vanderzant, F.A. Gardner, and F.A. Golan. 1982. Examination of turkey eggs, poults, and brooder house facilities for *Campylobacter jejuni. J. Food Protect.* 45:1279–1281.

3. Antoniollo, P.G., F. da Silva Bandeira, M.M. Jantzen, E.H. Duval, and W.P. da Silva. 2003. Prevalence of *Listeria* spp. in feces and carcasses of a lamb packing plant in Brazil. *J. Food Protect.* 66:328–330.

4. Atabay, H.I., J.E.L. Corry, and S.L.W. On. 1998. Diversity and prevalence of *Arcobacter* spp. in broiler chickens. *J. Appl. Microbiol.* 84:1007–1016.

5. Ayres, J.C. 1960. The relationship of organisms of the genus *Pseudomonas* to the spoilage of meat, poultry and eggs. *J. Appl. Bacteriol.* 23:471–486.

6. Ayres, J.C., W.S. Ogilvy, and G.F. Stewart. 1950. Post mortem changes in stored meats. I. Microorganisms associated with development of slime on eviscerated cut-up poultry. *Food Technol.* 4:199–205.

7. Baek, S.-Y., S.-Y. Lim, D.-H. Lee, K.-H. Min, and C.-M. Kim. 2000. Incidence and characterization of *Listeria monocytogenes* from domestic and imported foods in Korea. *J. Food Protect.* 63:186–189.

8. Bailey, J.S., N.A. Cox, S.E. Craven, and D.E. Cosby 2002. Serotype tracking of *Salmonella* through integrated broiler chicken operations. *J. Food Protect.* 65:742–745.

9. Barbe, C.D., R.W. Mandigo, and R.L. Henrickson. 1966. Bacterial flora associated with rapid-processed ham. *J. Food Sci.* 31:988–993.

10. Barber, D.A., P.B. Bahnson, R. Isaacson, C.J. Jones, and R.M. Weigei. 2002. Distributioon of *Salmonella* in swine production ecosystems. *J. Food Protect.* 65:1861–1868.

11. Barnes, E.M., and C.S. Impey. 1968. Psychrophilic spoilage bacteria of poultry. *J. Appl. Bacteriol.* 31:97–107.

12. Barnes, E.M., and M.J. Thornley. 1966. The spoilage flora of eviscerated chickens stored at different temperatures. *J. Food Technol.* 1:113–119.

13. Bauer, F.T., J.A. Carpenter, and J.O. Reagan. 1981. Prevalence of *Clostridium perfringens* in pork during processing. *J. Food Protect.* 44:279–283.

14. Bell, W.N., and L.A. Shelef. 1978. Availability and microbial stability of retail beef-soy blends. *J. Food Sci.* 43:315–318, 333.

15. Borton, R.J., J. Bratzler, and J.F. Price. 1970. Effects of four species of bacteria on porcine muscle. 2. Electrophoretic patterns of extracts of salt-soluble protein. *J. Food Sci.* 35:783–786.

16. Brooks, H.J.L., B.D. Mollison, K.A. Bettelheim, K. Matejka, K.A. Patterson, and V.K. Ward. 2001. Occurrence and virulence factors of non-0157 Shiga toxin-producing *Escherichia coli* in retail meat in Dunedin, New Zealand. *Lett. Appl. Microbiol.* 32:118–122.

17. Brown, M.H., C.O. Gill, J. Hollingsworth, R. Nickelson, II, S. Seward, J.J. Sheridan, T. Stevenson, J.L. Sumner, D.M. Theno, W.R. Usborne, and D. Zink. 2000. The role of microbiological testing in systems for assuring the safety of beef. *Int. J. Food Microbiol.* 62:7–16.

18. Bryan, F.L., J.C. Ayres, and A.A. Kraft. 1968. Salmonellae associated with further-processed turkey products. *Appl. Microbiol.* 16:1–9.

19. Byrne, C.M., I. Erol, J.E. Call, C.W. Kasper, D.R. Buege, C.J. Hiemke, P.J. Fedorka-Cray, A.K. Benson, F.M. Wallace, and J.B. Luchansky. 2003. Characterization of *Escherichia coli* 0157:H7 from downer and healthy dairy cattle in the upper midwest region of the United States. *Appl. Environ. Microbiol.* 69:4683–4688.

20. Byun, J.-S., J.S. Min, I.S. Kim, J.-W. Kim, M.-S. Chung, and M. Lee. 2003. Comparison of indicators of microbial quality of meat during aerobic cold storage. *J. Food Protect.* 66:1733–1737.

21. Carl, K.E. 1975. Oregon's experience with microbial standards for meat. *J. Milk Food Technol.* 38:483–486.

22. Carramiñana, J.J., J. Vangüella, D. Blanco, C. Rota, A.I. Agustin, A. Ariño, and A. Herrera. 1997. *Salmonella* incidence and distribution of serotypes throughout processing in a Spanish poultry slaughterhouse. *J. Food Protect.* 60:1312–1317.

23. Castillo, A.L., M. Lucia, K.J. Goodson, J.W. Savell, and G.R. Acuff. 1998. Comparison of water wash, trimming, and combined hot water and lactic acid treatments for reducing bacteria of fecal origin on beef carcasses. *J. Food Protect.* 61:823–828.

24. Castillo, A., L.M. Lucia, D.B. Roberson, T.H. Stevenson, L. Mercado, and G.R. Acuff. 2001. Lactic acid sprays reduce bacterial pathogens on cold beef carcass surfaces and in subsequently produced ground beef. *J. Food Protect.* 64:58–62.

25. Ceylan, E., and D.Y.C. Fung. 2000. Destruction of *Yersinia enterocolitica* by *Lactobacillus sake* and *Pediococcus acidi-lactici* duing low-temperature fermentation of Turkish dry sausage (sucuk). *J. Food Sci.* 65:876–879.

26. Chang, Y.H. 2000. Prevalence of *Salmonella* spp. in poultry broilers and shell eggs in Korea. *J. Food Protect.* 63:655–658.

27. Chapman, P.A., A.T.C. Malo, M. Ellin, R. Ashton, and M.A. Harkin. 2001. *Escherichia coli* 0157:H7 in cattle and sheep at slaughter, on beef and lamb carcasses and in raw beef and lamb products in South Yorkshire, UK. *Int. J. Food Microbiol.* 64:139–150.

28. Cox, N.A., S.M. Russell, and J.S. Bailey. 1998. The microbiology of stored poultry. In *The Microbiology of Meat and Poultry*, ed. A. Davies and R. Board, 266–287. New York: Kluwer Academic Publishers, Inc.

29. Cox, N.A., A.J. Mercuri, and J.E. Thompson. 1975. *Enterobacteriaceae* at various stages of poultry chilling. *J. Food Sci.* 40:44–46.

30. Craven, S.E., and A.J. Mercuri. 1977. Total aerobic and coliform counts in beef-soy and chicken-soy patties during refrigerated storage. *J. Food Protect.* 40:112–115.

31. Dainty, R.H., B.G. Shaw, K.A. DeBoer, and E.S.J. Scheps. 1975. Protein changes caused by bacterial growth on beef. *J. Appl. Bacteriol.* 39:73–81.

32. Davidson, C.M., M.J. Dowdell, and R.G. Board. 1973, Properties of Gram negative aerobes isolated from meats. *J. Food Sci.* 38:303–305.

33. Delmore, R.J., J.N. Sofos, K.E. Belk, W.R. Lloyd, G.L. Bellinger, G.R. Schmidt, and G.C. Smith. 1999. Good manufacturing practices for improving the microbiological quality of beef variety meats. *Dairy Fd. Environ. Sanit.* 19:742–752.

34. Dillon, V.M., and R.G. Board. 1991. Yeasts associated with red meats. *J. Appl. Bacteriol.* 71:93–108.

35. Dillon, V.M. 1998. Yeasts and moulds associated with meat and meat products. In *The Microbiology of Meat and Poultry*, ed. A. Davies and R. Board, 85–117. New York: Kluwer Academic Publishers, Inc.

36. Dodd, C.C., M.W. Sanderson, J.M. Sargeant, T.G. Nagaraja, R.D. Oberst, R.A. Smith, and D.D. Griffin. 2003. Prevalence of *Escherichia coli* 0157 in cattle feeds in midwestern feedlots. *Appl. Environ. Microbiol.* 69:5243–5247.

37. Dorsa, W.J., E.N. Cutter, and G.R. Siragusa. 1996. Evaluation of six sampling methods for recovery of bacteria from beef carcass surfaces. *Lett. Appl. Microbiol.* 22:39–41.

38. Doyle, M.P., M.B. Hugdahl, and S.L. Taylor. 1981. Isolation of virulent *Yersinia enterocolitica* from porcine tongues. *Appl. Environ. Microbiol.* 42:661–666.

39. Draughon, F.A. 1980. Effect of plant-derived extenders on microbiological stability of foods. *Food Technol.* 34:69–74.

40. Drosinos, E.H., and R.G. Board. 1994. Metabolic activities of pseudomonads in batch cultures in extract of minced lamb. *J. Appl. Bacteriol.* 77:613–620.

41. Duffy, E.A., K.E. Belk, J.N. Sofos, G.R. Bellinger, A. Pape, and G.C. Smith. 2001. Extent of microbial contamination in United States pork retail products. *J. Food. Protect.* 64:172–178.

42. Defrenne, J., W. Ritmeester, E. Delfgou-van Asch, F. van Leusden, and R. de Jonge. 2001. Quantification of the contamination of chicken and chicken products in the Netherlands with *Salmonella* and *Campylobacter*. *J. Food Protect.* 64:538–541.

43. Duitschaever, C.L., D.H. Bullock, and D.R. Arnott. 1977. Bacteriological evaluation of retail ground beef, frozen beef patties, and cooked hamburger. *J. Food Protect.* 40:378–381.

44. Duitschaever, C.L. 1977. Incidence of *Salmonella* in retailed raws cut-up chicken. *J. Food Protect.* 40:191–192.

45. Dutson, T.R., G.C. Smith, and Z.L. Carpenter. 1980. Lysosomal enzyme distribution in electrically stimulated ovine muscle. *J. Food Sci.* 45:1097–1098.

46. Edwards, R.A., R.H. Dainty, and C.M. Hibbard. 1985. Putrescine and cadaverine formation in vacuum packed beef. *J. Appl. Bacteriol.* 58:13–19.

47. Edwards, R.A., R.H. Dainty, and C.M. Hibbard. 1983. The relationship of bacterial numbers, and types of diamine concentration in fresh and aerobically stored beef, pork and lamb. *J. Food Technol.* 18:777–788.

48. Eggenberger-Solarzano, L.S.E. Niebuhr, G.R. Acuff, and J.S. Dickson. 2002. Hot water and organic acid interventions to control microbiological contamination on hog carcasses during processing. *J. Food Protect.* 65:1248–1252.

49. Ellis, D.I., and R. Goodacre. 2001. Rapid and quantitative detection of the microbial spoilage of muscle foods: current status and future trends. *Trends Fd. Sci. Technol.* 12:414–424.

50. Eribo, B.E., and J.M. Jay. 1985. Incidence of *Acinetobacter* spp. and other Gram-negative oxidase-negative bacteria in fresh and spoiled ground beef. *Appl. Environ. Microbiol.* 49:256–257.

51. Eribo, B.E., S.D. Lall, and J.M. Jay. 1985. Incidence of *Moraxella* and other Gram-negative, oxidase-positive bacteria in fresh and spoiled ground beef. *Food Microbiol.* 2:237–240.

52. Fernandes, C.F., G.J. Flick, J. Cohen, and T.B. Thomas. 1998. Role of organic acids during processing to improve quality of channel catfish fillets. *J. Food Protect.* 61:495–498.

53. Field, R.A. 1976. Mechanically deboned red meat. *Food Technol.* 30:38–48.

54. Field, R.A. 1981. Mechanically deboned red meat. *Adv. Food Res.* 27:23–107.

55. Field, R.A., and M.L. Riley. 1974. Characteristics of meat from mechanically deboned lamb breasts. *J. Food Sci.* 39:851–852.

56. Floccari, M.E., M.M. Carranza, and J.L. Parada. 2000. *Yersinia enterocolitica* biogroup 1A, serotype 0:5 in chicken carcasses. *J. Food Protect.* 63:1591–1593.

57. Frediksson-Ahomaa, M., S. Hielm, and H. Korkeala. 1999. High prevalence of *yadA*-positive *Yersinia enterocolitica* on pig tongues and minced meat at the retail level in Finland. *J. Food Protect.* 62:123–127.

58. Froning, G.W. 1981. Mechanical deboning of poultry and fish. *Adv. Food Res.* 27:109–147.

59. Fung, D.Y.C., C.L. Kastner, M.C. Hunt, M.E. Dikeman, and D.H. Kropf. 1980. Mesophilic and psychro-trophic bacterial populations on hot-boned and conventionally processed beef. *J. Food Protect.* 43:547–550.

60. Fung, D.Y.C., C.L. Kastner, C.-Y. Lee, M.C. Hunt, M.E. Dikeman, and D.H. Kropf. 1981. Initial chilling rate effects of bacterial growth on hot-boned beef. *J. Food Protect.* 44:539–544.

61. Funk, J.A., H.F. Troutt, R.E. Isaacson, and C.P. Fossler. 1998. Prevalence of pathogenic *Yersinia enterocolitica* in groups of swine at slaughter. *J. Food Protect.* 61:677–682.

62. Fuzihara, T.O., S.A. Fermandes, and B.D.G.M. Franco. 2000. Prevalence and dissemination of *Salmonella* serotypes along the slaughtering process in Brazilian small poultry slaughterhouses. *J. Food Protect.* 63:1749–1753.

63. Gamble, H.R., M.B. Solomon, and J.B. Long. 1998. Effects of hydrodynamic pressure on the viability of *Trichinella spiralis* in pork. *J. Food Protect.* 61:637–639.

64. Gardner, G.A. 1971. A note on the aerobic microflora of fresh and frozen porcine liver stored at 5°C. *J. Food Technol.* 6:225–231.

65. Gashe, B.A., and S. Mpuchane. 2000. Prevalence of salmonellae on beef products at butacheries and their antibiotic resistance profiles. *J. Food Sci.* 65:880–883.

66. Genigeorgis, C.A., P. Oanca, and D. Dutulescu. 1990. Prevalence of *Listeria* spp. in turkey meat at the supermarket and slaughterhouse level. *J. Food Protect.* 53:288.

67. Genigeorgis, C.A.D. Dutulescu, and J.F. Garayzabal. 1989. Prevalence of *Listeria* spp. in poultry meat ta the supermarket and slaughter level. *J. Food Protect.* 53:282–288.

68. Geornaras, I., A.E. de Jesus, and A. von Holy. 1998. Bacterial populations associated with the dirty area of a South African poultry abattoir. *J. Food Protect.* 61:700–703.

69. Getty, K.J.K., R.K. Phebus, J.L. Marsden, J.R. Schwenke, and C.J. Kastner. 1999. Control of *Escherichia coli* 0157:H7 in large (115 mm) and intermediate (90 mm) diameter Lebanon-style bologna. *J. Food Sci.* 64:1100–1107.

70. Gill, C.O., J.C. McGinnis, and J. Bryant. 2001. Contamination of beef chucks with *Escherichia coli* during carcass breaking. *J. Food Protect.* 64:1824–1827.

71. Gill, C.O. 1976. Substrate limitation of bacterial growth at meat surfaces. *J. Appl. Bacteriol.* 41:401–410.

72. Gill, C.O., and K.M. DeLacy. 1982. Microbial spoilage of whole sheep livers. *Appl. Environ. Microbiol.* 43:1262–1266.

73. Gill, C.O., and K.G. Newton. 1977. The development of aerobic spoilage flora on meat stored at chill temperatures. *J. Appl. Bacteriol.* 43:189–195.

74. Goepfert, J.M. 1977. Aerobic plate count and *Escherichia coli* determination on frozen ground-beef patties. *Appl. Environ. Microbiol.* 34:458–460.

75. Golla, S.C., E.A. Murano, L.G. Johnson, N.C. Tipton, E.A. Currengton, and J.W. Savell. 2002. Determination of the occurrence of *Arcobacter butzleri* in beef and dairy cattle from Texas by various isolation methods. *J. Food Protect.* 65:1849–1853.

76. Greenberg, R.A., R.B. Tompkin, B.O. Bladel, R.S. Kittaka, and A. Anellis. 1966. Incidence of mesophilic *Clostridium* spores in raw pork, beef, and chicken in processing plants in the United States and Canada. *Appl. Microbiol.* 14:789–793.

77. Guthertz, L.S., J.T. Fruin, R.L. Okoluk, and J.L. Fowler. 1977. Microbial quality of frozen comminuted turkey meat. *J. Food Sci.* 42:1344–1347.

78. Guthertz, L.S., J.T. Fruin, D. Spicer, and J.L. Fowler. 1976. Microbiology of fresh comminuted turkey meat. *J. Milk Food Technol.* 39:823–829.

79. Hanna, M.O., G.C. Smith, J.W. Savell, F.K. McKeith, and C. Vanderzant. 1982. Microbial flora of livers, kidneys and hearts from beef, pork and lamb: Effects of refrigeration, freezing and thawing. *J. Food Protect.* 45:63–73.

80. Hanna, M.O., G.C. Smith, J.W. Savell, F.K. McKeith, and C. Vanderzant. 1982. Effects of packaging methods on the microbial flora of livers and kidneys from beef or pork. *J. Food Protect.* 45:74–81.

81. Hanna, M.O., G.C. Smith, J.W. Savell, F.K. McKeith, and C. Vanderzant. 1982. Effects of packaging methods on the microbial flora of livers and kidneys from beef or pork. *J. Food Protect.* 45:74–81.

82. Harmon, M.C., B. Swaminathan, and J.C. Forrest. 1984. Isolation of *Yersinia enterocolitica* and related species from porcine samples obtained from an abattoir. *J. Appl. Bacteriol.* 56:421–427.

83. Harrison, M.A., F.A. Draughton, and C.C. Melton. 1983. Inhibition of spoilage bacteria by acidification of soy extended ground beef. *J. Food Sci.* 48:825–828.

84. Hayes, J.R., L.L. English, P.J. Carter, T. Proescholdt, K.Y. Lee, D.D. Wagner, and D.G. White. 2003. Prevalence and antimicrobial resistance of *Enterococcus* species isolated from retail meats. *Appl. Environ. Microbiol.* 69:7153–7160.

85. Heredia, N., S. Garcia, G. Rojas, and L. Salazar. 2001. Microbiological condition of ground meat retailed in Monterrey, Mexico. *J. Food Protect.* 64:1249–1251.

86. Hinton, A. Jr., J.A. Cason, and K.D. Ingram. 2002. Enumeration and identification of yeasts associated with commercial poultry processing and spoilage of refrigerated broiler carcasses. *J. Food. Protect.* 65:993–998.

87. Holzapfel, W.H. 1998. The Gram-positive bacteria associated with meat and meat products. In *The Microbiology of Meat and Poultry*, ed. A. Davies and R. Board, 35–84. New York: Kluwer Academic Publishers.

88. Houf, K., L. De Zutter, J. Van Hoof, and P. Vandamme. 2002. Occurrence and distribution of *Arcobacter* species in poultry processing. *J. Food Protect.* 65:1233–1239.

89. Hsieh, D.Y., and J.M. Jay. 1984. Characterization and identification of yeasts from fresh and spoiled ground beef. *Int. J. Food Microbiol.* 1:141–147.

90. Ingram, M., and R.H. Dainty. 1971. Changes caused by microbes in spoilage of meats. *J. Appl. Bacteriol.* 34:21–39.

91. Jay, J.M., and L.A. Shelef. 1991. The effect of psychrotrophic bacteria on refrigerated meats. In *Biodeterioration and Biodegradation*, 8th ed., ed. H.W. Rossmoore, 147–159. London: Elsevier Applied Sciences.

92. Jay, J.M. 1964. Beef microbial quality determined by extract-release volume (ERV). *Food Technol.* 18:1637–1641.

93. Jay, J.M. 1967. Nature, characteristics, and proteolytic properties of beef spoilage bacteria at low and high temperatures. *Appl. Microbiol.* 15:943–944.

94. Jay, J.M. 1987. Meats, poultry, and seafoods. In *Food and Beverage Mycology*, 2nd ed., ed. L.R. Beuchat, chap. 5. New York: Kluwer Academic Publishers.

95. Kakouri, A., and G.J.E. Nychas. 1994. Storage of poultry meat under modified atmospheres or vacuum packs: Possible role of microbial metabolites as indicator of spoilage. *J. Appl. Bacteriol.* 76:163–172.

96. Kastner, C.L., L.O. Leudecke, and T.S. Russell. 1976. A comparison of microbial counts on conventionally and hot-boned carcasses. *J. Milk Food Technol.* 39:684–685.

97. Keeton, J.T., and C.C. Melton. 1978. Factors associated with microbial growth in ground beef extended with varying levels of textured soy protein. *J. Food Sci.* 43:1125–1129.

98. Kobayashi, H., A. Miura, H. Hayashi, T. Ogawa, T. Endo, E. Hata, M. Eguchi, and K. Yamamoto. 2003. Prevalence and characteristics of *eae*-positive *Escherichia coli* from healthy cattle in Japan. *Appl. Environ. Microbiol.* 69:5690–5692.

99. Korsak, N., G. Daube, Y. Ghafir, A. Chahed, S. Jolly, and H. Vindevogel. 1998. An efficient sampling technique used to detect four foodborne pathogens on pork and beef carcasses in nine Belgian abattoirs. *J. Food Protect.* 61:535–541.

100. Kotula, A.W. 1981. Microbiology of hot-boned and electrostimulated meat. *J. Food Protect.* 44:545–549.

101. Kotula, A.W., and B.S. Emswiler-Rose. 1981. Bacteriological quality of hot-boned primal cuts from electrically stimulated beef carcasses. *J. Food Sci.* 46:471–474.

102. Kramer, J.M., J.A. Frost, F.J. Bolton, and D.R.A. Wareing. 2000. *Campylobacter* contamination of raw meat and poultry at retail sale: Identification of multiple types and comparison with isolates from human infection. *J. Food Protect.* 63:1654–1659.

103. Ladiges, W.C., J.F. Foster, and W.M. Ganz. 1974. Incidence and viability of *Clostridium perfringens* in ground beef. *J. Milk Food Technol.* 37:622–623.

104. Lahellec, C., C. Meurier, and G. Benjamin. 1975. A study of 5,920 strains of psychrotrophic bacteria isolated from chickens. *J. Appl. Bacteriol.* 38:89–97.

105. Lammerding, A.M., M.W. Garcia, E.D. Mann, Y. Robinson, W.J. Dorward, R.B. Truscott, and F. Tittiger. 1988. Prevalence of *Salmonella* and thermophilic campylobacters in fresh pork, beef, veal and poultry in Canada. *J. Food Protect.* 51:47–52.

106. Lawrie, R.A. 1966. *Meat Science*. New York: Pergamon Press.

107. Lebert, I., C. Begot, and A. Lebert. 1998. Growth of *Pseudomonas fluorescens* and *Pseudomonas fragi* in a meat medium as affected by pH (5.8–7.0), water activity (0.97–1.00) and temperature (7–25°C). *Int. J. Food Microbiol.* 39:53–60.

108. Lee, C.Y., D.Y.C. Fung, and E.L. Kastner. 1982. Computer-assisted identification of bacteria on hot-boned and conventionally processed beef. *J. Food Sci.* 47:363–367, 373.

109. Lepovetsky, B.C., H.H. Weiser, and F.E. Deatherage. 1953. A microbiological study of lymph nodes, bone marrow and muscle tissue obtained from slaughtered cattle. *Appl. Microbiol.* 1:57–59.

110. Lerke, P., R. Adams, and L. Farber. 1963. Bacteriology of spoilage of fish muscle. I. Sterile press juice as a suitable experimental medium. *Appl. Microbiol.* 11:458–462.

111. Letellier, A., S. Messier, and S. Quessay. 1999. Prevalence of *Salmonella* spp. and *Yersinia enterocolitica* in finishing swine at Canadian abattoirs. *J. Food Protect.* 62:22–25.

112. Lillard, H.S. 1971. Occurrence of *Clostridium perfringens* in broiler processing and further processing operations. *J. Food Sci.* 36:1008–1010.

113. Lin, C.K., W.H. Kennick, W.E. Sandine, and M. Koohmaraie. 1984. Effect of electrical stimulation on meat microflora: Observations on agar media, in suspensions and on beef carcasses. *J. Food Protect.* 47:279–283.

114. Lin, H.-S., D.G. Topel, and H.W. Walker. 1979. Influence of prerigor and postrigor muscle on the bacteriological and quality characteristics of pork sausage. *J. Food Sci.* 44:1055–1057.

115. Logue, C.M., J.S. Sherwood, L.M. Elijah, P.A. Olah, and M.R. Dockter. 2003. The incidence of *Campylobacter* spp. on processed turkey from processing plants in the midwestern United States. *J. Appl. Microbiol.* 95:234–241.

116. Lowry, P.D., and C.O. Gill. 1984. Temperature and water activity minima for growth of spoilage moulds from meat. *J. Appl. Bacteriol.* 56:193–199.

117. Manke, T.R., I.V. Wesley, J.S. Dickson, and K.M. Harmon. 1998. Prevalence and genetic variability of *Arcobacter* species in mechanically separated turkey. *J. Food Protect.* 61:1623–1628.

118. Margitic, S., and J.M. Jay. 1970. Antigenicity of salt-soluble beef muscle proteins held from freshness to spoilage at low temperatures. *J. Food Sci.* 35:252–255.

119. Mattick, K.L., R.A. Bailey, F. Jorgensen, and T.J. Humphrey. 2002. The prevalence and number of *Salmonella* in sausages and their destruction by frying, grilling or barbecuing. *J. Appl. Microbiol.* 93:541–547.

120. May, K.N. 1962. Bacterial contamination during cutting and packaging chicken in processing plants and retail stores. *Food Technol.* 16:89–91.

121. May, K.N., J.D. Irby, and J.L. Carmon. 1961. Shelf life and bacterial counts of excised poultry tissue. *Food Technol.* 16:66–68.

122. McMeekin, T.A. 1975. Spoilage association of chicken breast muscle. *Appl. Microbiol.* 29:44–47.

123. McMeekin, T.A. 1977. Spoilage association of chicken leg muscle. *Appl. Microbiol.* 33:1244–1246.

124. McMeekin, T.A., and J.T. Patterson. 1975. Characterization of hydrogen sulfide-producing bacteria isolated from meat and poultry plants. *Appl. Microbiol.* 29:165–169.

125. McMillin, D.J., J.G. Sebranek, and A.A. Kraft. 1981. Microbial quality of hot-processed frozen ground beef patties processed after various holding times. *J. Food Sci.* 46:488–490.

126. Mercuri, A.J., G.J. Banwart, J.A. Kinner, and A.R. Sessoms. 1970. Bacteriological examination of commercial precooked Eastern-type turkey rolls. *Appl. Microbiol.* 19:768–771.

127. Miettinen, M.K., l. Palmu, K.J. Björkroth, and H. Korkeala. 2001. Prevalence of *Listeria monocytogenes* in broilers at the abattoir, processing plant, and retail level. *J. Food Protect.* 64:994–999.

128. Miwa, N., T. Nishina, S. Kubo, M. Atsumi, and H. Honda. 1998. Amount of enterotoxogenic *Clostridium perfringens* in meat detected by nested PCR. *Int. J. Food Microbiol.* 42:195–200.

129. Moore, J.E., and R.H. Madden. 1998. Occurrence of thermophilic *Campylobacter* spp. in porcine liver in Northern Ireland. *J. Food Protect.* 61:409–413.

130. Moore, J.E., T.S. Wilson, D.R.A. Wareing, T.J. Humphrey, and P.G. Murphy. 2002. Prevalence of thermophilic *Campylobacter* spp. in ready-to-eat foods and raw poultry in Northern Ireland. *J. Food Protect.* 65:1326–1328.

131. Murthy, T.R.K. 1984. Relative numbers of coliforms, *Enterobacteriaceae* (by two methods), and total aerobic bacteria counts as determined from minced goat meat. *J. Food Protect.* 47:142–144.

132. Nakamura, M., Y. Wada, H. Sawaya, and T. Kawabata. 1979. Polyamine content in fresh and processed pork. *J. Food Sci.* 44:515–517.

133. Nakazawa, H., H. Hayashidani, J. Higashi, K.-I. Kaneko, T. Takahashi, and M. Ogawa. 1998. Occurrence of *Erysipelothrix* spp. in broiler chickens at an abattoir. *J. Food Protect.* 61:907–909.

134. Newton, K.G., and C.O. Gill. 1978. Storage quality of dark, firm, dry meat. *Appl. Environ. Microbiol.* 36:375–376.

135. Ockerman, H.W., and J. Szczawinski. 1983. Effect of electrical stimulation on the microflora of meat. *J. Food Sci.* 48:1004–1005, 1007.

136. Ohlendorf, D.S., and E.A. Murano. 2002. Prevalence of *Arcobacter* spp. in raw ground pork from several geographical regions according to various isolation methods. *J. Food Protect.* 65:1700–1705.

137. Ostovar, K., J.H. MacNeil, and K. O'Donnell. 1971. Poultry product quality. 5. Microbiological evaluation of mechanically deboned poultry meat. *J. Food Sci.* 36:1005–1007.

138. Pearce, R.A., F.M. Wallace, J.E. Call, R.L. Dudley, A. Oser, L. Yoder, J.J. Sheridan, and J.B. Luchansky. 2003. Prevalence of *Campylobacter* within a swine slaughter and processing facility. *J. Food Protect.* 66:1550–1556.

139. Pearson, A.D., M.H. Greenwood, R.K.A. Feltham, T.D. Healing, J. Donaldson, D.M. Jones, and R.R. Colwell. 1996. Microbial ecology of *Campylobacter jejuni* in a United Kingdom chicken supply chain: Intermittent common source, vertical transmission, and amplification by flock propagation. *Appl. Environ. Microbiol.* 62:4614–4620.

140. Peel, J.L., and J.M. Gee. 1976. The role of micro-organisms in poultry taints. In *Microbiology in Agriculture, Fisheries and Food*, ed. F.A. Skinner and J.G. Carr, 151–160. New York: Academic Press.

141. Phillips, D.J. Sumner, J.F. Alexander, and K.M. Dutton. 2001. Microbiological quality of Australian beef. *J. Food Protect.* 64:692–696.

142. Pivnick, H., I.E. Erdman, D. Collins-Thompson, G. Roberts, M.A. Johnston, D.R. Conley, G. Lachapelle, U.T. Purvis, R. Foster, and M. Milling. 1976. Proposed microbiological standards for ground beef based on a Canadian survey. *J. Milk Food Technol.* 39:408–412.

143. Prieto, M., M.R. Garcia-Armesto, M.L. Garcia-López, A. Otero, and B. Moreno. 1992. Numerical taxonomy of Gram-negative nonmotile, nonfermentative bacteria isolated during chilled storage of lamb carcasses. *Appl. Environ. Microbiol.* 58:2245–2249.

144. Pruett, W.P. Jr., T. Biela, C.P. Lattuada, P.M. Mrozinski, W.M. Barbour, R.S. Flowers, W. Osborne, J.O. Reagan, D. Theno, V. Cook, A.M. McNamara, and B. Rose. 2002. Incidence of *Escherichia coli* 0157:H7 in frozen beef patties produced over an 8-hour shift. *J. Food Protect.* 1363–1370.

145. Pulliam, J.D., and D.C. Kelley. 1965. Bacteriological comparisons of hot processed and normally processed hams. *J. Milk Food Technol.* 28:285–286.

146. Raccach, M., and R.C. Baker. 1978. Microbial properties of mechanically deboned fish flesh. *J. Food Sci.* 43:1675–1677.

147. Ramírez, E.I.Q, C. Vázquez-Salinas, O.R. Rodas-Suárez, and F.F. Pedroche. 2000. Isolation of *Yersinia* from raw meat (pork and chicken) and precooked meat (porcine tongues and sausage) collected from commercial establishments in Mexico City. *J. Food Protect.* 63:542–544.

148. Ray, B., and R.A. Field. 1983. Bacteriology of restructured lamb roasts made with mechanically deboned meat. *J. Food Protect.* 46:26–28.

149. Ray, B., C. Johnson, and R.A. Field. 1984. Growth of indicator, pathogenic and psychrotrophic bacteria in mechanically separated beef, lean ground beef and beef bone marrow. *J. Food Protect.* 47:672–677.

150. Rengel, A., and S. Medoza. 1984. Isolation of *Salmonella* from raw chicken in Venezuela. *J. Food Protect.* 47:213–216.

151. Rivas, T., J.A. Vizcaino, and F.J. Herrera. 2000. Microbial contamination of carcasses and equipment from an Iberian pig slaughterhouse. *J. Food Protect.* 63:1670–1675.

152. Rostagno, M.H., H.S. Hurd, J.D. McKean, C.J. Ziemer, J.K. Gailey, and R.C. Leite. 2003. Preslaughter holding environment in pork plants is highly contaminated with *Salmonella enterica. Appl. Environ. Microbiol.* 69:4489–4494.

153. Rothenberg, C.A., B.W. Berry, and J.L. Oblinger. 1982. Microbiological characteristics of beef tongues and livers as affected by temperature-abuse and packaging systems. *J. Food Protect.* 45:527–532.

154. Samadpour, M., M. Kubler, F.C. Buck, G.A. Dapavia, E. Mazengia, J. Stewart, P. Yang, and D. Alfi. 2002. Prevalence of Shiga toxin-producing *Escherichia coli* in ground beef and cattle feces from King County, Washington. *J. Food Protect.* 65:1322–1325.

155. Sayem-El-Daher, N., and R.E. Simard. 1985. Putrefactive amine changes in relation to microbial counts of ground beef during storage. *J. Food Protect.* 48:54–58.

156. Shaw, B.G., and J.B. Latty. 1982. A numerical taxonomic study of *Pseudomonas* strains from spoiled meat. *J. Appl. Bacteriol.* 52:219–228.

157. Shaw, B.G., and J.B. Latty. 1984. A study of the relative incidence of different *Pseudomonas* groups on meat using a computer-assisted identification technique employing only carbon source tests. *J. Appl. Bacteriol.* 57:59–67.

158. Shelef, L.A. 1975. Microbial spoilage of fresh refrigerated beef liver. *J. Appl. Bacteriol.* 39:273–280.

159. Shelef, L.A., and J.M. Jay. 1969. Relationship between meat-swelling, viscosity, extract-release volume, and water-holding capacity in evaluating beef microbial quality. *J. Food Sci.* 34:532–535.

160. Shelef, L.A., and J.M. Jay. 1969. Relationship between amino sugars and meat microbial quality. *Appl. Microbiol.* 17:931–932.

161. Shelef, L.A., and J.M. Jay. 1970. Use of a titrimetric method to assess the bacterial spoilage of fresh beef. *Appl. Microbiol.* 19:902–905.

162. Slemr, J. 1981. Biogene Amine als potentieller chemischer Qualitätsindikator für Fleisch. *Fleischwirt.* 61:921–925.

163. Smith, F.C., R.A. Field, and J.C. Adams. 1974. Microbiology of Wyoming big game meat. *J. Milk Food Technol.* 37:129–131.

164. Solomon, M.B., J.B. Long, and J.S. Eastridge. 1997. The Hydrodyne: A new process to improve beef tenderness. *J. Anim. Sci.* 75:1534–1537.

165. Steinkraus, K.H., and J.C. Ayres. 1964. Incidence of putrefactive anaerobic spores in meat. *J. Food Sci.* 29:87–93.

166. Stern, N.J., M.P. Hernandez, L. Blankenship, K.E. Deibel, S. Doores, M.P. Doyle, H. Ng, M.D. Pierson, J.N. Sofos, W.H. Sveum, and D.C. Westhoff. 1985. Prevalence and distribution of *Campylobacter jejuni* and *Campylobacter coli* in retail meats. *J. Food Protect.* 48:595–599.

167. Stern, N.J., S.S. Green, N. Thaker, D.J. Krout, and J. Chiu. 1984. Recovery of *Campylobacter jejuni* from fresh and frozen meat and poultry collected at slaughter. *J. Food Protect.* 47:372–374.

168. Stevenson, T.H., N. Bauer, L.M. Lucia, and G.R. Acuff. 2000. Attempts to isolate *Helicobacter* from cattle and survival of *Helicobacter pylori* in beef products. *J. Food Protect.* 63:174–178.

169. Stiles, M.E., and L.-K. Ng. 1981. Biochemical characteristics and identification of *Enterobacteriaceae* isolated from meats. *Appl. Environ. Microbiol.* 41:639–645.

170. Surkiewicz, B.F., M.E. Harris, R.P. Elliott, J.F. Macaluso, and M.M. Strand. 1975. Bacteriological survey of raw beef patties produced at establishments under federal inspection. *Appl. Microbiol.* 29:331–334.

171. Swartzentruber, A., A.H. Schwab, B.A. Wentz, A.P. Duran, and R.B. Read, Jr. 1984. Microbiological quality of biscuit dough, snack cakes and soy protein meat extender. *J. Food Protect.* 47:467–470.

172. Tamplin, M.L., I. Feder, S.A. Palumbo, A. Oser, L. Yoder, and J.B. Luchansky. 2001. *Salmonella* spp. and *Escherichia coli* biotype I on swine carcasses processed under the hazard analysis and critical control point-based inspection models project. *J. Food Protect.* 64:1305–1308.

173. Taormina, P.J., G.W. Bartholomew, and W.J. Dorsa. 2003. Incidence of *Clostridium perfringens* in commercially produced cured raw meat product mixtures and behavior in cooked products during chilling and refrigerated storage. *J. Food Protect.* 66:72–81.

174. Thorberg, B.-M., and A. Engvall. 2001. Incidence of *Salmonella* in five Swedish slaughterhouses. *J. Food Protect.* 64:542–545.

175. United States Department of Agriculture (USDA). 1996. Pathogen reduction; hazard analysis and critical control point (HACCP) systems; final rule. *Federal Register* 61:38806.

176. USDA. 1996. *Nationwide Federal Plant Raw Ground Beef Microbiological Survey.* Washington, D.C.: USDA.

177. USDA. 1996. *Nationwide Broiler Chicken Microbiological Baseline Data Collection Program.* Washington, D.C.: USDA.

178. USDA. 1994. *Nationwide Beef Microbiological Baseline Data Collection Program: Steers and Heifers.* Washington, D.C.: USDA.

179. Vanderlinde, P.B., B. Shay, and J. Murray. 1999. Microbiological status of Australian sheep meat. *J. Food Protect.* 62:380–385.

180. Viljoen, B.C., I. Geornaras, A. Lamprecht, and A. von Holy. 1998. Yeast populations associated with processed poultry. *Food Microbiol.* 15:113–117.

181. Villarruel-López, A.M. Marquess-González, L.E. Garay-Martinez, H. Zepeda, A. Castillo, L. mota de la Garza, E.A. Murano, and R. Torres-Vitela. 2003. Isolation of *Arcobacter* spp. from retail meats and cytotoxic effects of isolates against Vero cells. *J. Food Protect.* 66:1374–1378.

182. Waldenström, J., T. Broman, I. Carlsson, D. Hasselquist, R.P. Achterberg, J.A. Wagenaar, and B. Olsen. 2002. Prevalence of *Campylobacter jejuni, Campylobacter lari*, and *Campylobacter coli* in different ecological guilds and taxa of migrating birds. *Appl. Environ. Microbiol.* 68:5911–5917.

183. Warnken, M.B., M.P. Nunez, and A.L.S. Noleto. 1987. Incidence of *Yersinia* species in meat samples purchased in Rio de Janeiro, Brazil. *J. Food Protect.* 50:578–579, 583.

184. Watt, B.K., and A.L. Merrill. 1950. Composition of foods—Raw, processed, prepared. *Agricultural Handbook No. 8.* Washington, D.C.: USDA.

185. Yamamoto, S., H. Itano, H. Kataoka, and M. Makita. 1982. Gas–liquid chromatographic method for analysis of di- and polyamines in foods. *J. Agric. Food Chem.* 30:435–439.

186. Zhao, T., M.P. Doyle, P.J. Fedorka-Cray, P. Zhao, and S. Ladely. 2002. Occurrence of *Salmonella enterica* serotype Typhimurium DT 104A in retail ground beef. *J. Food Protect.* 65:403–407.

187. Zottola, E.A., and F.F. Busta. 1971. Microbiological quality of further-processed turkey products. *J. Food Sci.* 36:1001–1004.

CHAPTER 5

Processed Meats and Seafoods

PROCESSED MEATS

Processed meats are those meat products that are cured, smoked, or cooked. The microbiota most often associated with these products is listed in Table 5–1. The behavior of processed meats stored under vacuum or modified atmospheres is discussed in Chapter 14.

Curing

Although curing was used in ancient times as a means of meat preservation, it is employed now more for flavor and color development. The classic meat cure ingredients are NaCl, nitrite or nitrate, and sugar (sucrose or glucose), with NaCl being the major ingredient. In addition to these, some products may contain curing adjuncts such as phosphates, sodium ascorbate or erythorbate, potassium sorbate, monosodium glutamate, hydrolyzed vegetable proteins, lactates, or spices.

In dry curing, no water is added to the NaCl, nitrite or nitrate, and sugar mixtures. In pickle curing, these ingredients are added to water to form a brine.

Salt serves to prevent microbial growth during and after curing, and up to 2.5% may be found in finished products. Nitrite or nitrate serves to stabilize red meat color, contribute to cured meat flavor, retard rancidity, and prevent the germination of clostridial spores. The isomers sodium ascorbate and erythorbate are used to stabilize color, to speed curing, and to make the cure more uniform. Since erythorbate is more stable than its isomer, its use is preferred, and it increases the yield of nitric oxide from nitrite and nitrous acid. At a level of 550 ppm, ascorbate or erythorbate reduces nitrosamine formation. Sugar is involved in at least three curing functions: color stabilization, flavoring, and substrate for lactic fermentation. Also, it moderates the harsh flavor of NaCl. Corn syrups, molasses, or honey may be substituted for flavor.

Phosphates are used in most pumped meats (bacon, ham, roast beef, pastrami, etc.) to increase water binding. In curing brines, sodium tripolyphosphate is most commonly used, but a mixture of tripolyphosphate and sodium hexametaphosphate is used widely. See Chapter 13 for information on polyphosphates as antimicrobial agents.

Table 5–1   Genera of Bacteria and Fungi Most Frequently Found on Processed Meats

Bacteria			Fungi	
Genus	Gram Reaction	Relative Prevalence	Genus	Relative Prevalence
Acinetobacter	−	X	**Yeasts**	
Aeromonas	−	X	Candida	X
Alcaligenes	−	X	Debaryomyces	XX
Bacillus	+	X	Saccharomyces	X
Brochothrix	+	X	Trichosporon	X
Carnobacterium	+	X	Yarrowia	X
Corynebacterium	+	X	**Molds**	
Enterobacter	−	X	Alternaria	X
Enterococcus	+	X	Aspergillus	XX
Hafnia	+	X	Botrytis	X
Kocuria	+	X	Cladosporium	X
Kurthia	+	X	Fusarium	X
Lactobacillus	+	XX	Geotrichum	X
Lactococcus	+	X	Monilia	X
Leuconostoc	+	X	Mucor	X
Listeria	+	X	Penicillium	XX
Microbacterium	+	X	Rhizopus	X
Micrococcus	+	X	Scopulariopsis	X
Moraxella	−	X	Thamnidium	X
Paenibacillus	+	X		
Pediococcus	+	X		
Pseudomonas	−	XX		
Serratia	−	X		
Staphylococcus	+	X		
Vibrio	−	X		
Weissella	+	X		
Yersinia	−	X		
Carnimonas	−	X		
Clostridium	+	XX		
Macrococcus	+	X		
Shewanella	−	X		

Note: X = known to occur; XX = most frequently reported.

Sausages (*L. salsus*, salted or preserved) constitute one of the major groups of cured meat products and they may be classified as follows:

1. fresh (patties, links)
2. uncooked smoked (mettwurst and Polish sausage)
3. cooked smoked (bologna and wieners)
4. cooked (liver sausage)
5. dry (Genoa salami, pepperoni)
6. semidry (Lebanon bologna, cervelat)

Semidry sausages have a final pH around 4.7–5.0, and refrigeration is required. The final pH of dry sausages is about the same as that for semidry but these products are shelf stable because of their lower moisture content. The relative safety of these products is discussed below.

The common cured bacon in the United States is either dry or pickle cured, with the latter being the more common. Following cure, it may be smoked. Canadian bacon is characterized by being quite lean, since it comes from the large muscle of pork loins. Wiltshire bacon is prepared from the sides of selected hogs, followed by pumping of cure ingredients and subsequent storage in pickle brines.

Most hams in commerce are of the pickle-cure variety and they are cured following injection of the pickle cure by artery pumping, single-needle stitch, or multiple-needle stitch. For dry-cured or country-cured hams, the dry curing salts are applied by rubbing followed by storage at refrigerator temperatures for 28 to 50 days, depending on size and thickness.

All curing ingredients may be expected to contain microorganisms, and care should be taken to ensure that undesirable ones are not introduced to products during ingredient application.

Smoking

This process is applied to many cured meats, and the primary purposes of smoking meat are (1) development of aroma and flavor, (2) preservation, (3) creation of new products, (4) development of color, (5) formation of a protective skin on emulsion-type sausages, and (6) protection from oxidation.[73] Smoke, whether directly from wood or in liquid form, contains phenols, alcohol, organic acids, carbonyls, hydrocarbons, and gases. The antimicrobial properties of smoking result from the activities of some of the smoke ingredients and the heat that is associated with wood smoking. Liquid smoke contains all of the essential ingredients of wood smoke, but it is free of the carcinogen benzopyrene.

SAUSAGE, BACON, BOLOGNA, AND RELATED PRODUCTS

In addition to the meat components, sausages and frankfurters have additional sources of organisms in the seasoning and formulation ingredients that are usually added in their production. Many spices and condiments have high microbial counts. The lactic acid bacteria and yeasts in some composition products are usually contributed by milk solids. In the case of pork sausage, natural casings have been shown to contain large numbers of bacteria. In their study of salt-packaged casings, Riha and Solberg[76] found counts to range from $\log_{10}$ 4.48 to $\log_{10}$ 7.77 cfu/g and from $\log_{10}$ 5.26 to $\log_{10}$ 7.36 cfu/g for wet-packaged casings. Over 60% of the isolates from these natural casings consisted of *Bacillus* spp., followed by clostridia and pseudomonads. Of the individual ingredients of fresh pork sausage, casings have been shown to contribute the largest number of bacteria.[76,88]

Processed meats such as bologna and salami may be expected to reflect the sum of their ingredient makeup with regard to microbial numbers and types. The biota of frankfurters has been shown to consist largely of Gram-positive organisms with micrococci, bacilli, lactobacilli, microbacteria, enterococci, and leuconostocs along with yeast.[24] In a study of slime from frankfurters, these investigators found that 275 of 353 isolates were bacteria, and 78 were yeast. *B. thermosphacta* was the most conspicuous single isolate. With regard to the incidence of *C. botulinum* spores in liver sausage, 3 of 276 heated (75°C for 20 minutes) and 2 of 276 unheated commercial preparations contained type A botulinal toxin.[43] The most probable number (MPN) of botulinal spores in this product was estimated to be 0.15/kg.

Wiltshire bacon has been reported to have a total count generally in the range of $\log_{10}$ 5–6/g,[53] whereas high-salt vacuum-packaged bacon has been reported to have a generally lower count—about

$\log_{10}$ 4/g. The biota of vacuum-packaged sliced bacon consists largely of catalase-positive cocci, such as micrococci and coagulase-negative staphylococci, as well as catalase-negative bacteria of the lactic acid types, such as lactobacilli, leuconostocs, pediococci, and enterococci.[3,13,59] The biota in cooked salami has been found to consist mostly of lactobacilli.

So-called soul foods may be expected to contain large numbers of organisms, as they consist of offal parts that are in direct contact with the intestinal-tract biota, as well as other parts, such as pig feet and pig ears, that do not receive much care during slaughtering and processing. This was confirmed in a study by Stewart[86] who found a geometric mean aerobic plate count (APC) of $\log_{10}$ 7.92/g for chitterlings (pig intestines), $\log_{10}$ 7.51/g for maws, and $\log_{10}$ 7.32/g for liver pudding. For *S. aureus*, $\log_{10}$ numbers of 5.18, 5.70, and 5.15/g, respectively, were found for chitterlings, maws, and liver pudding.

Jerky is a dried shelf-stable product made from lightly salted and spiced slices of meat or fish—most often beef. When drying to reduce a_w to or below 0.86 is carried out within 3 hours, no problems are likely to result from pathogens, but when drying is not rapid and extends over a long period of time at temperatures <60°C, *S. aureus* can survive.[49] In 1993, beef jerky was the vehicle food in New Mexico for 93 cases of salmonellosis caused by three serovars—*S*. Montevideo, Kentucky, and Typhimurium.[14] The product was produced in a commercial establishment, but it is unclear how it became contaminated. To reduce a_w to 0.86 during jerky processing, it has been found that a period of 2.5–3.0 hours of drying at 52.9°C is needed.[50] This is not lethal to foodborne pathogens, but it would make the product stable to the growth of *Staphylococcus aureus* in the case of post-processing contamination. For beef jerky, a period of 10 hours of drying at 60°C has been shown capable of reducing *E. coli* 0157:H7, *L. monocytogenes*, and *Salmonella* serovar Typhimurium by $\log_{10}$ 5.5–6.0 units.[42] In an evaluation of homestyle dehydrators, it was found that in order to achieve a $\log_{10}$ 5 reduction of *E. coli* 0157:H7, the following relation needed to be observed: about 20 hours of drying at 125°F, about 12 hours at 135°F, about 8 hours at 145°F, or 4 hours at 155°F.[11] This organism was more sensitive in meat with 5% fat than with 20%. For example, for jerky with 5% fat, a log 5 reduction could be achieved in about 8 hours at 125°F (51.7°C).

From 32,800 packages of frankfurters examined by the FDA in the United States, *L. monocytogenes* was recovered from 532 (1.6%) and about 90% of all isolates were serotype 1/2a.[96] In an extensive study of luncheon meats for *L. monocytogenes* in the states of Maryland and California, the organism was found in 82 of 9,199 or 0.89%.[38]

Spoilage

Spoilage of these products is generally of three types: sliminess, souring, and greening. *Slimy spoilage* occurs on the outside of casings, especially of frankfurters, and may be seen in its early stages as discrete colonies, which may later coalesce to form a uniform layer of gray slime. Yeasts, lactic acid bacteria of the genera *Lactobacillus*, *Enterococcus*, *Weissella*, and *B. thermosphacta*, may be isolated from the slimy material. *W. viridescens* produces both sliminess and greening. Slime formation is favored by a moist surface and is usually confined to the outer casing. Removal of this material with hot or warm water leaves the product essentially unchanged.

Souring generally takes place underneath the casing of these meats and results from the growth of lactobacilli, enterococci, and related organisms. The usual sources of these organisms to processed meats are milk solids. The souring results from the utilization of lactose and other sugars by the organisms and the production of acids. Sausage usually contains a more varied biota than most other processed meats due to the different seasoning agents employed, almost all of which contribute their own biota. *B. thermosphacta* has been found by many investigators to be the most predominant spoilage organism for sausage.

The changes that fresh meat bacteria bring about as processed meats undergo souring are unlikely to occur in the latter since many members of the fresh meat Gram-negative biota are unable to proliferate at the lower a_w and pH of processed meats. Even in fresh meats, definitive spoilage changes affecting structural proteins do not occur until APC is in the 10^9 to 10^{10} range[54].

Although mold spoilage of these meats is not common, it can and does occur under favorable conditions. When the products are moist and stored under conditions of high humidity, they tend to undergo bacterial and yeast spoilage. Mold spoilage is likely to occur only when the surfaces become dry or when the products are stored under other conditions that do not favor bacteria or yeasts.

Two types of *greening* occur on stored and processed red meats: one caused by H_2O_2 and the other by H_2S. The former occurs commonly on frankfurters as well as on other cured and vacuum-packaged meats. It generally appears after an anaerobically stored meat product is exposed to air. Upon exposure to air, H_2O_2 forms and reacts with nitrosohemochrome to produce a greenish oxidized porphyrin.[73] H_2O_2 may accumulate when heating if nitrite destroys catalase, and the peroxide reacts with meat pigments to form choleglobin, which is green. Greening also occurs from growth of causative organisms in the interior core, where the low oxidation–reduction (Eh) potential allows H_2O_2 to accumulate. *Weissella viridescens* is the most common organism in this type of greening, but leuconostocs, *Enterococcus faecium*, and *Enterococcus faecalis* are capable of producing greening of products. Greening can also be produced by H_2O_2 producers such as *Lactobacillus fructivorans* and *Lactobacillus jensenii*. *W. viridescens* is resistant to >200 ppm $NaNO_2$, and it can grow in the presence of 2–4% NaCl but not in 7%.[73] *W. viridescens* has been recovered from anaerobically spoiled frankfurters and from both smoked pork loins and frankfurter sausage stored in atmospheres of CO_2 and N_2.[8] In spite of the discoloration, the green product is not known to be harmful if eaten.

The second type of greening occurs generally on fresh red meats that are held at 1–5°C and stored in gas-impermeable or vacuum-packaging containers; it is caused by H_2S production. H_2S reacts with myoglobin to form sulphmyoglobin (Table 5–2). This type of greening does not usually occur when meat pH is below 6.0. The organism responsible in one study was thought to be *Pseudomonas mephitica*[71] but in another study of DFD meats, *S. putrefaciens* was the H_2 producer.[37] In the latter, greening occurred even with glucose present, and it could be prevented by lowering pH to below 6.0. H_2S-producing lactobacilli were recovered from vacuum-packaged fresh beef and found to produce H_2S in the pH range of 5.4–6.5.[81] Only slight greening was produced, and the H_2S was from cysteine, a system that was plasmid borne. The organism reached $3 \times 10^7/cm^2$ after 7 days, and ultimately reached about $10^8/cm^2$ at 50°C. No offness of vacuum-packaged sliced luncheon meat was observed when another lactobacillus attained $10^8/cm^2$.

At least one strain of *Lactobacillus sakei* has been shown to produce H_2S on vacuum-packaged beef; and the effect of pH and glucose on production is presented in Table 5–3.[26] The investigators found that greening by *L. sakei* was not as intense as that caused by *S. putrefaciens* and that it occurred only after about 6 weeks at 0°C. Further, the lactobacillus produced H_2S only in the absence of O_2 and utilizable sugars. No greening was observed when films with an O_2 transmission rate of 1 ml O_2/m^2 or 300 ml O_2/m^2 were used, but it did occur with films that had O_2 transmission rates between 25 and 200 ml/m²/24 hours.[26] Visible greening was seen only on samples packaged in films with O_2 transmission rates of 100 and 200 ml/m²/24 hours and only after 75 days' storage. With meat in the pH 6.4–6.6 range, H_2S was detected when cell numbers reached $10^8/g$.

A *yellow* discoloration of vacuum-packaged luncheon-style meat was caused apparently by *Enterococcus casseliflavus*. The discoloration appeared as small spots on products stored at 4.4°C, and

Table 5–2 Pigments Found in Fresh, Cured, or Cooked Meat

Pigment	Mode of Formation	State of Iron	State of Haematin Nucleus	State of Globin	Color
1. Myoglobin	Reduction of metmyoglobin; deoxygenation of oxymyoglobin	Fe^{2+}	Intact	Native	Purplish red
2. Oxymyoglobin	Oxygenation of myoglobin	Fe^{2+}	Intact	Native	Bright red
3. Metmyoglobin	Oxidation of myoglobin, oxymyoglobin	Fe^{3+}	Intact	Native	Brown
4. Nitric oxide myoglobin	Combination of myoglobin with nitric oxide	Fe^{3+}	Intact	Native	Bright red
5. Metmyoglobin nitrite	Combination of metmyoglobin with excess nitrite	Fe^{3+}	Intact	Native	Red
6. Globin haemochromogen	Effect of heat, denaturing agents on myoglobin, oxymyoglobin; irradiation of globin haemichromogen	Fe^{2+}	Intact	Denatured	Dull red
7. Globin haemichromogen	Effect of heat, denaturing agents on myoglobin, oxymyoglobin, metmyoglobin, haemochromogen	Fe^{3+}	Intact	Denatured	Brown
8. Nitric oxide haemochromogen	Effect of heat, salts on nitric oxide myoglobin	Fe^{2+}	Intact	Denatured	Bright red
9. Sulphmyoglobin	Effect of H_2S and oxygen on myoglobin	Fe^{3+}	Intact but reduced	Denatured	Green
10. Choleglobin	Effect of hydrogen peroxide on myoglobin or oxymyoglobin; effect of ascorbic or other reducing agent on oxymyoglobin	Fe^{2+} or Fe^{3+}	Intact but reduced	Denatured	Green
11. Verdohaem	Effect of reagents as in 9 in excess	Fe^{3+}	Porphyrin ring opened	Denatured	Green
12. Bile pigments	Effects of reagents as in 9 in large excess	Fe absent	Porphyrin ring destroyed; chain of porphyrins	Absent	Yellow or colorless

Source: Reprinted with permission from R.A. Lawrie, *Meat Science*, copyright © 1966, Pergamon Press.

Table 5–3 Effect of Meat pH and Glucose on Hydrogen Sulfide Production by a Pure Culture of *Lactobacillus sakei* L13 Growing under Anaerobic Conditions at 5°C on Beef

	Hydrogen Sulfide Production		
Days	*pH 5.6–5.7*	*pH 6.4–6.6*	*pH 6.4–6.6 with 250 µg Glucose per Gram of Meat*
8	−*	−	−
9	−	+‡	−
11	−	+	−
15	−	+	−
18	+†	+	+‡
21	+	+	+

*Each treatment done in triplicate. − = All three tubes negative; + = all three positive.
†One tube out of three positive.
‡Two tubes out of three positive.

Source: Egan et al.[26]

it was fluorescent under long-wave ultraviolet light.[10] Between 3 and 4 weeks were required for the condition to develop, and the responsible organism survived 71.1°C for 20 minutes but not 30 minutes. In addition to 4.4°C, it occurred also at 10°C but not at 20°C or above. Although tentatively identified as *E. casseliflavus*, the causative organism did not react with Group D antisera. The other yellow-pigmented enterococcal species is *E. mundtii*; and both are discussed further in Chapter 20. A summary of several spoilage conditions of processed meats is presented in Table 5–4.

Table 5–4 Summary of Some Microbial Spoilage Conditions of Processed Meats

Condition	*Products Affected*	*Etiology*	*Reference*
Greening	Reference vacuum-packaged bologna	*C. viridans*	48
Greening	Vacuum-packaged beef	*L. sakei*	26
Greening	Fresh red meats	*P. mephitica, S. putrefaciens*	41, 71
Greening	DFD meats	*S. purefaciens*	37
Greening/sliminess	Wieners, bologna	*W. viridescens*	Many
Yellowing	Vacuum-packaged luncheon meats	*E. casseliflavus*	101
Black spot	Cured meats	*C. nigrificans*	36
Souring	Sausage	*B. thermosphacta*	66
"Blown pack"	Vacuum-packaged meats	*C. frigidicarnis, C. gasigenes*	9, 10
General spoilage	Vacuum-packaged meats	*L. algidus; L. fuchuensis*	57, 79

BACON AND CURED HAMS

The nature of these products and the procedures employed in preparing certain ones, such as smoking and brining, make them relatively insusceptible to spoilage by most bacteria. The most common form of bacon spoilage is moldiness, which may be due to *Aspergillus, Alternaria, Fusarium, Mucor, Rhizopus, Botrytis, Penicillium,* and other molds (Table 5–1). The high fat content and low a_w make it somewhat ideal for this type of spoilage. Bacteria of the genera *Enterococcus, Lactobacillus,* and *Micrococcus* are capable of growing well on certain types of bacon such as Wiltshire, and *E. faecalis* is often present on several types. Vacuum-packaged bacon tends to undergo souring due primarily to micrococci and lactobacilli. Vacuum-packed, low-salt bacon stored above 20°C may be spoiled by staphylococci.[92]

Cured hams undergo a type of spoilage different from that of fresh or smoked hams. This is due primarily to the fact that curing solutions pumped into the hams contain sugars that are fermented by the natural biota of the ham and also by those organisms pumped into the product in the curing solution, such as lactobacilli. The sugars are fermented to produce conditions referred to as "*sours*" of various types, depending on their location within the ham. A large number of genera of bacteria have been implicated as the cause of ham sours, among which are *Acinetobacter, Bacillus, Pseudomonas, Lactobacillus, Proteus, Micrococcus,* and *Clostridium.* Gassiness is not unknown to occur in cured hams where members of the genus *Clostridium* have been found.

In their study of vacuum-packed sliced bacon, Cavett[13] and Tonge et al.[92] found that when high-salt bacon was held at 20°C for 22 days, the catalase-positive cocci dominated the biota, whereas at 30°C the coagulase-negative staphylococci became dominant. In the case of low-salt bacon (5–7% NaCl versus 8–12% in high-salt bacon) held at 20°C, the micrococci as well as *E. faecalis* became dominant; at 30°C the coagulase-negative staphylococci as well as *E. faecalis* and micrococci became dominant. In a study of Iberian dry-cured hams, over 97% of the isolates were staphylococci with the four predominant species being *S. equorum, S. xylosus, S. saprophyticus,* and *S. cohnii.*[77] Interestingly, one *S. xylosus* isolate hybridized with a DNA probe for staphylococcal enterotoxins C and D, but the investigators noted that probe-positive isolates do not always produce enterotoxins.

In a study of lean Wiltshire bacon stored aerobically at 5°C for 35 days or 10°C for 21 days, Gardner[35] found that nitrates were reduced to nitrites when the microbial load reached about $10^9/g$. The predominant organisms at this stage were micrococci, vibrios, and the yeast genera *Candida* and *Torulopsis.* Upon longer storage, microbial counts reached about $10^{10}/g$ with the disappearance of nitrites. At this stage, *Acinetobacter, Alcaligenes,* and *Arthrobacter-Corynebacterium* spp. became more important. Micrococci were always found, whereas vibrios were found in all bacons with salt contents >4%. In a study of Italian dry fermented sausages, the most frequently isolated staphylococci were *S. xylosus* followed by *S. saprophyticus, S. aureus,* and *S. sciuri.*[34] *S. xylosus* appears to be the most frequently isolated from several Italian dry sausages. In Iberian dry cured hams, the two predominant organisms in the ripening process are *Staphylococcus equorum* and *S. xylosus,* and both are believed to contribute to product flavor.

Safety

Overall, fermented meat products have a long history of safety throughout the world. This is not to imply that they are never the vehicles of foodborne illness outbreaks, but when such incidents have occurred they have been sporadic. Several outbreaks of illness occurred in the United States in the 1990s involving fermented meat products as vehicles. As a consequence, the USDA mandated a $\log_{10}$

5 reduction in the number of pathogens, especially *E. coli* 0157:H7, in the manufacture of dry and semidry fermented sausages. As a result, a number of studies have been conducted on the efficacy of domestic and commercial processing to achieve the pathogen reduction goal.

An outbreak of *E. coli* 0157:H7 from dry-cured salami occurred in the states of California and Washington in 1994, and there were 23 victims.[15] Following this outbreak, a series of studies were conducted on the conditions of pepperoni manufacture that are needed to effect a log 5 reduction in numbers of specific pathogens. Using a 5-strain cocktail of *E. coli* 0157:H7 at a level of $\geq 2 \times 10^7$/g, it was found that the traditional nonthermal process destroyed only about log 2 units/g and that in order to effect a log 5–6 reduction, postfermentation heating to an internal temperature of 63°C instantaneous or 53°C for 60 minutes was necessary.[46] In a more extensive study, pepperoni sticks were fermented at 36°C and 85% relative humidity (RH) to a pH $\leq$4.8 and then dried at 13°C and 65% RH to a moisture/protein ratio of $\leq$1.6:1.[29] The five-strain pathogen mixture was reduced only about log 2 units. To achieve a log 5 reduction, storage for at least 2 weeks at ambient temperature in air was necessary for sliced pepperoni. In another study, a >log 5 decrease in *E. coli* 0157:H7 could be achieved by fermentation at 41°C to a pH of 4.6 or 5.0 and postfermentative heating of summer sausage chubs to an internal temperature of 54°C for 30 minutes.[12] In a similar study of pepperoni manufacture and storage with *S.* Typhimurium DT104, it was found that this pathogen is more sensitive to destruction than *E. coli* 0157:H7 and thus methods that will reduce the latter by log 5 are more than adequate for the former.[52] The fate of *S. aureus* in country-cured hams was investigated by the spraying of fresh hams with four strains at levels of $\log_{10}$ 8.57 and $\log_{10}$ 8.12 followed by curing, cold smoking, and aging. After 4 months of aging, *S. aureus* was below detectable levels by plating, although some cells were recovered by enrichment.[74] Sodium nitrite was employed in some cures and a_w was controlled with 4.45 or 3.37% NaCl. Forty percent of the inoculated and 50% of controls contained enterotoxin following the aging period.

SEAFOODS

Fish and Shellfish

As used in this chapter, the term *seafood* covers fish, shellfish, and mollusks from all waters–fresh, marine, warm, or cold. In general, the biota of a fresh seafood animal is reflective of the waters from which it is taken. As is the case for meat animals, the inner tissues of a healthy fish are sterile. With fish, the microbial biota is found generally in three places: the outer slime, gills, and the intestines of feeding fish. Fresh or warm-water fish tend to have a biota that is composed of more mesophilic Gram-positive bacteria than cold-water fish, which tends to be largely Gram negative (the indigenous bacterial biota of marine water is Gram negative).

The organisms that make up the biota of seafoods are listed in Table 5–5, and what is known about their interplay in bringing about the spoilage of these products is discussed in the section on spoilage of fish and shellfish.

Microorganisms

As noted above, the overall sanitary quality of the waters from which these animals are taken is key to the overall microbial quality of finished products. Beyond the water source, microbes are picked up at various processing steps such as peeling, shucking, evisceration, breading, and the like.

Table 5-5 Genera of Bacteria, Yeasts, and Molds, Most Often Found on Fresh and Spoiled Fish and Other Seafoods

Bacteria	Gram	Prevalence	Yeasts	Prevalence	Molds	Prevalence
Acinetobacter	–	X	Candida	XX	Aspergillus	X
Aeromonas	–	XX	Cryptococcus	XX	Aureobasidium (Pullularia)	XX
Alcaligenes	–	X	Debaryomyces	X	Penicillium	X
Bacillus	+	X	Hansenula	X	Scopulariopsis	X
Corynebacterium	+	X	Pichia	X		
Enterobacter	–	X	Rhodotorula	XX		
Enterococcus	+	X	Sporobolomyces	X		
Escherichia	–	X	Trichosporon	X		
Flavobacterium	–	X				
Lactobacillus	+	X				
Listeria	+	X				
Microbacterium	+	X				
Moraxella	–	X				
Photobacterium	–	X				
Pseudomonas	–	XX				
Psychrobacter	–	X				
Shewanella	–	XX				
Vibrio	–	XX				
Weissella	+	X				
Pseudoalteromonas	–	X				

Note: X = known to occur; XX = most frequently reported.

In their study of 91 samples of shrimp of various types, Silverman et al.[84] found that all precooked samples except one had total counts of $<\log_{10}$ 4.00/g. Of the raw samples, 59% had total counts below $\log_{10}$ 5.88, whereas 31% had below $\log_{10}$ 5.69/g. In a study of 204 samples of frozen, cooked, and peeled shrimp, 52% had total counts $<\log_{10}$ 4.70/g, and 71% had counts of $\log_{10}$ 5.30/g or less/g.[62] The general microbiological quality of a variety of seafood is presented in Table 5–6.

In a study of haddock fillets, most microbial contamination was found to occur during filleting and subsequent handling prior to packaging.[70] These investigators showed that the total count increased from $\log_{10}$ 5.61/g in the morning to $\log_{10}$ 5.65/g at noon and to $\log_{10}$ 5.94/g in the evening for one particular processor. According to their study, results obtained in other companies were generally similar if the night time cleanup was good. In the case of shucked, soft-shell clams, the same general pattern of buildup was demonstrated from morning to evening. The mean clostridial count for both haddock fillets and soft-shell clams was less than 2/g, with clams having slightly higher counts than haddock fillets for these organisms, although both were low. Total counts on fresh perch fillets produced under commercial conditions were found to average $\log_{10}$ 5.54/g with yeast and mold counts of about $\log_{10}$ 2.69/g.[58]

Clams may be expected to contain the organisms that inhabit the waters from which they are obtained. Of 60 clam samples from the coast of Florida, 43% contained salmonellae, which were also found in oysters at a level of 2.2/100 g oyster meats.[33] Hard-shell clams have been shown to retain *S. Typhimurium* more efficiently than *E. coli*.[33]

The initial biota of herring fillets has been found to be dominated by *S. putrefaciens* and *Pseudomonas* spp., with the latter dominating at 2°C and *S. putrefaciens* more predominant at 2–15°C.[69] In general, frozen seafood and other frozen products have lower microbial counts than the comparable fresh products. In a study of 597 fresh and frozen seafood from retail stores, the APC $\log_{10}$ geometric means ranged from 3.54 to 4.97/g for the 240 frozen products and from 4.89 to 8.43/g for the 357 fresh products.[32] For coliforms, geometric mean MPN counts ranged from 1 to 7.7 cells/g for frozen and from 7.78 to 4,800/g for fresh products. By MPN, only 4.7% of the 597 were positive for *E. coli*, 7.9% were positive for *S. aureus*, and 2% were positive for *C. perfringens*. All were negative for salmonellae and *Vibrio parahaemolyticus* (Table 5–6).

Plate counts are generally higher on seafood when incubated at 30°C than at 35°C, and this is reflected in results from fresh crabmeat, clams, and oysters, evaluated by Wentz et al.[100] The APC geometric means were: for 896 crabmeat samples, $\log_{10}$ 5.15 at 35°C and $\log_{10}$ 5.72 at 30°C; for 1,337 shucked oysters, $\log_{10}$ 5.59/g at 35°C and $\log_{10}$ 5.95 at 30°C; and for 358 soft-shell clams, $\log_{10}$ 2.83/g at 35°C and $\log_{10}$ 4.43 at 30°C. This was also seen in raw in-shell shrimp and frozen raw lobster tails, where the geometric mean APC for shrimp was $\log_{10}$ 5.48 at 35°C and $\log_{10}$ 5.90/g at 30°C, whereas for lobster tail, it was $\log_{10}$ 4.62 at 35°C and $\log_{10}$ 5.15/g at 30°C.[89]

In a study of the prevalence of aeromonads on catfish, 228 channel catfish fillets from three processing plants in the Mississippi Delta were examined, and both *A. hydrophila* and *A. sobria* were found on 36% while *A. caviae* was found on 11%.[97] Most of the two predominant species produced alpha hemolysis on sheep red blood cells. In a study of the biota of catfish processing equipment in two plants, *Aeromonas* and *Pseudomonas* spp. were the most abundant genera found.[22]

Between 1995 and 1998, a 3-year monthly study of viral contamination of 108 oysters and 73 mussels for four groups of viruses was conducted in southern France. Results showed that virus concentrations generally were highest during the cold months (November to March) but variations were seen with different viruses.[61] Rotaviruses showed no seasonal distribution but varied from month to month with July being the month with the lowest prevalence as was the case for enteroviruses and astroviruses. The peak for noroviruses was November; and December and January were the peak months for the other three groups with at least 70% of samples being positive during the peaks.

Table 5–6 Genera of Bacteria, Yeasts, and Molds, Most Often Found on Fresh and Spoiled Fish and Other Seafoods

Products	Number of Samples	Microbial Group/Target	% Samples Meeting Target	Reference
Frozen catfish fillet	41	APC 32°C: 10^5/g or less	100	32
	41	MPN coliforms: <3/g	100	32
	41	MPN *S. aureus*: <3/g	100	32
Frozen salmon steaks	43	APC 32°C: 10^5/g or less	98	32
	43	MPN coliforms: <3/g	93	32
	43	MPN *S. aureus*: <3/g	98	32
Fresh clams	53	APC 32°C: 10^5/g or less	53	32
	53	MPN coliforms: <3/g	51	32
	53	MPN *S. aureus*: <3/g	91	32
Fresh oysters	59	APC 32°C: 10^7/g or less	49	32
	59	MPN coliforms: 1,100 or less/g	22	32
	59	MPN *S. aureus*: <3/g	90	32
Shucked oysters (retail)	1,337	APC 30°C: 10^6/g or less	51	100
	1,337	MPN coliforms: 460 or less/g	94	100
	1,337	MPN fecal coliforms: 460 or less/g	96	100
Blue crabmeat (retail)	896	APC 30°C: 10^6/g or less	61	100
	896	MPN coliforms: 1,100/g or less	93	100
	896	MPN *E. coli*: <3/g	97	100
	896	MPN *S. aureus*: 1,100/g or less	94	100
Hard-shell clams (wholesale)	1,124	APC 30°C: 10^6/g or less	99.8	100
	1,130	MPN coliforms: 460/g or less	96	100
	161	MPN fecal coliforms: <3/g	91	100
Soft-shell clams (wholesale)	351	APC 30°C: 10^6/g or less	96	100
	363	MPN coliforms: 460/g or less	98	100
	75	MPN fecal coliforms: <3/g	72	100

Food	n	Test	%	Ref
Peeled shrimp (raw)	1,468	APC 30°C: 10^7/g or less	94	89
	1,468	MPN coliforms: 64/g or less	97	89
	1,468	MPN E. coli: <3/g	97	89
	1,468	MPN S. aureus: 64/g or less	97	89
Peeled shrimp (cooked)	1,464	APC 30°C: 10^5/g or less	81	89
	1,464	MPN coliforms: <3/g	86	89
	1,464	MPN S. aureus: <3/g	99	89
	1,464	MPN S. aureus: <3/g	99	89
Lobster tail (frozen, raw)	1,315	APC 30°C: 10^6/g or less	74	89
	1,315	MPN coliforms: 64/g or less/g	91	89
	1,315	MPN E. coli: <3/g	95	89
	1,315	MPN S. aureus: <3/g	76	89
Retail frozen, breaded, raw shrimp	27	APC: 6.00 or less/g	52	95
	27	Coliforms: 3.00 or less/g	100	95
	27	Presence of E. coli	4	95
	27	Presence of S. aureus	59	95
Fresh channel catfish	335	APC: ≤7.00/g	93	4
	335	Fecal coliforms: 2.60/g	70.7	4
	335	Presence of salmonellae	4.5	4
Frozen channel catfish	342	APC: ≤7.00/g	94.5	4
	342	Fecal coliforms: 2.60/g	92.4	4
	342	Presence of salmonellae	1.5	4
Frozen, cooked, peeled shrimp	204	APC: <4.70/g	52	62
	204	APC: 5.30 or less/g	71	62
	204	Coliforms: none or <0.3/g	52.4	62
	204	Coliforms: <3/g	75.2	62
Fresh rainbow trout[a]	74	APC range: $\log_{10}$ 2.4–8.6; mean APC $\log_{10}$ 6.2 cfu/g	–	25
Seafoods, various	82	2.4% pos. for salmonellae	–	72

Note: APC = aerobic plate count; MPN = most probable number.

[a]51% contained *L. monocytogenes.*

113

In 1998–1999, 370 lots of oysters from coastal waters and 29 states in the United States were examined at 275 establishments. Gulf Coast oysters contained the highest number of *Vibrio vulnificus* and *V. parahaemolyticus* with numbers exceeding 10^5 MPN/g while *V. vulnificus* was <0.2 MPN/g and none exceeded 100 MPN/g in oysters from the north Atlantic, Pacific, and Canadian coasts.[21] Thermostable direct hemolysin was detected in 9 of 3,429 (0.3%) *V. parahaemolyticus* cultures and in 4% of oyster lots. Among 345 samples of retail oysters from ca. 20 states and two Canadian provinces, the highest number of *V. vulnificus* was 25×10^5 in oysters in Florida. The highest number of *V. vulnificus* was found in oysters from the state of Mississippi, $>8.8 \times 10^5$ MPN/g.[21] In an earlier study of oysters from 39 sites in the state of Washington in 1997, 9 contained <3/g.[23] In the same study, 34 oysters from Galveston Bay, Texas contained a mean of $\log_{10}$ 2.36–2.73 cfu/g. The level of alarm for *V. parahaemolyticus* in oysters in the United States is an MPN of 10^4/g.[27] In a study of 671 samples of molluscan shellfish harvested in 1999 and 2000 from 14 sites in waters off the American Gulf and Atlantic coast states, 6.0% were positive for *V. parahaemolyticus* by gene probe and culture methods.[20] The numbers of positives correlated with water temperature, with numbers being higher in the warmer waters. This organism was found in north Atlantic waters only during the summer but they were found during all seasons in Gulf waters. Although they are generally associated with pork, 21 *Erysipelothrix* spp. were isolated from seafoods in Australia.[30]

In regard to noroviruses in oysters, 87 samples (of 435 oysters) tested in Switzerland that were imported from three European countries between November 2001 and February 2002 showed that 11.5% were positive (all of which belonged to serogroup II—see Chapter 31) and 2.3% contained enteroviruses (coxsackie and ECHO). No hepatitis A viruses were found.[7] In a study of 147 roes from three fish species in Finland, *Listeria* spp. were found in 17% and *L. monocytogenes* in 4.7%, with trout containing the highest percentage of the latter.[68] The mean APC was $\log_{10}$ 6.6 cfu/g, and coliforms were $\log_{10}$ 3.2 cfu/g, placing these products in the "moderate" class based on APC, but in the "unacceptable" class based on coliform numbers.

Employing a PCR-ELISA detection technique, a survey of *C. botulinum* spores was conducted in northern France and this organism was found in 31 of 214 environmental samples. Most positive samples contained <10 spores/25 g fish, and 16.6% of seawater fish and 4% of sediment samples were positive with 70% being type B; 22.5% type A; and 9.6% type E.[28] No type F spores were found.

A study of 106 Gram-negative nonmotile rods isolated from ice-stored freshwater fish in Spain found that 64 were *Psychrobacter* spp. followed by 24 *Acinetobacter*, 6 *Moraxella*, 5 *Chryseobacterium*, 2 *Myroides*, 1 each of *Flavobacterium* and *Empedobacter*; and 3 unknown.[39] Chryseobacteria, empedobacteria, and myroides were previously classified in the genus *Flavobacterium*.

In a study of *Listeria* spp. in raw and processed crawfish in the United States in 2001, they were found in 31 of 337 (9.2%), but only 4 were positive for *L. monocytogenes*.[91] When a total of 2,446 ready-to-eat seafood salads from two states in the United States were examined for *L. monocytogenes* in 2000 and 2001, 4.7% were positive.[38] In the same survey, *L. monocytogenes* was found in 4.3% of 2,644 ready-to-eat smoked seafoods.

For the 9-year period 1990–1999, the FDA in the United States examined imported and domestic fish and seafoods for salmonellae. Of the 11,312 imported samples, 7.2% were positive while only 1.3% of the 768 domestic samples were positive.[44] The most common serovar found in this survey was *S.* Weltvreden.

From a study of 50 frozen fish samples in Spain for the prevalence of nontuberculous mycobacteria, 20% were found to contain these bacteria with *M. fortuitum* and *M. nonchromogenicum* being the two most common species.[67] Of the approximate one-half of isolates that could be identified, they consisted of six species. Little, if anything, is known about the significance of these organisms in seafoods. It may be assumed that they play no role in seafood spoilage because of their slow growth.

SPOILAGE OF FISH AND SHELLFISH

Fish

Both saltwater and freshwater fish contain comparatively high levels of proteins and other nitroge-nous constituents (Table 5–7). The carbohydrate content of these fish is nil, whereas the fat content varies from very low to rather high values depending on the species. Of particular importance in fish flesh is the nature of the nitrogenous compounds. The relative percentages of total N and protein N are presented in Table 5–8, from which it can be seen that not all nitrogenous compounds in fish are in the form of proteins. Among the nonprotein nitrogen compounds are the free amino acids, volatile nitrogen bases such as ammonia and trimethylamine; creatine, taurine, the betaines, uric acid, anserine, carnosine, and histamine.

The microorganisms known to cause fish spoilage are indicated in Table 5–5. Fresh iced fish are invariably spoiled by bacteria, whereas salted and dried fish are more likely to undergo fungal spoilage. The bacterial biota of spoiling fish is found to consist of asporogenous, Gram-negative rods of the *Pseudomonas* and *Acinetobacter-Moraxella* types. Many fish-spoilage bacteria are capable of good growth between 0°C and 1°C. Shaw and Shewan[80] found that a large number of *Pseudomonas* spp. are capable of causing fish spoilage at 3°C, although at a slow rate.

The spoilage of saltwater and freshwater fish appears to occur in essentially the same manner, with the chief differences being the requirement of the saltwater biota for a seawater-type environment and the differences in chemical composition between various fish with respect to nonprotein nitrogenous constituents. The most susceptible part of fish is the gill region, including the gills. The earliest signs of organoleptic spoilage may be noted by examining the gills for the presence of off-odors. If feeding fish are not eviscerated immediately, intestinal bacteria soon make their way through the intestinal walls and

Table 5–7   Fish and Shellfish: Approximate Percentage Chemical Composition

	Water	Carbohydrates	Proteins	Fat	Ash
Bony fish					
Bluefish	74.6	0	20.5	4.0	1.2
Cod	82.6	0	16.5	0.4	1.2
Haddock	80.7	0	18.2	0.1	1.4
Halibut	75.4	0	18.6	5.2	1.0
Herring (Atlantic)	67.2	0	18.3	12.5	2.7
Mackerel (Atlantic)	68.1	0	18.7	12.0	1.2
Salmon (Pacific)	63.4	0	17.4	16.5	1.0
Swordfish	75.8	0	19.2	4.0	1.3
Crustaceans					
Crab	80.0	0.6	16.1	1.6	1.7
Lobster	79.2	0.5	16.2	1.9	2.2
Mollusks					
Clams, meat	80.3	3.4	12.8	1.4	2.1
Oysters	80.5	5.6	9.8	2.1	2.0
Scallops	80.3	3.4	14.8	0.1	1.4

Source: Watt and Merrill.[99]

Table 5–8   Distribution of Nitrogen in Fish and Shellfish Flesh

Species	Percentage Total N	Percentage Protein N	Ratio of Protein N: Total N
Cod (Atlantic)	2.83	2.47	0.87
Herring (Atlantic)	2.90	2.53	0.87
Sardine	3.46	2.97	0.86
Haddock	2.85	2.48	0.87
Lobster	2.72	2.04	0.75

Source: Jacquot.[55] © 1961, Academic Press.

into the flesh of the intestinal cavity. This process is believed to be aided by the action of proteolytic enzymes, which are from the intestines and may be natural enzymes inherent in the intestines of the fish, or enzymes of bacterial origin from the inside of the intestinal canal, or both. Fish-spoilage bacteria apparently have little difficulty in growing in the slime, and on the outer integument of fish. Slime is composed of mucopolysaccharide components, free amino acids, trimethylamine oxide, piperidine derivatives, and other related compounds. As in the case with poultry spoilage, plate counts are best done on the surface of fish, with the number of organisms expressed per square centimeter of examined surface.

It appears that the spoilage organisms first utilize the simpler compounds and in the process release various volatile off-odor components. According to Shewan[83] trimethylamine oxide, creatine, taurine, anserine, and related compounds along with certain amino acids decrease during fish spoilage with the production of trimethylamine, ammonia, histamine, hydrogen sulfide, indole, and other compounds. Fish flesh differs from mammalian flesh with regard to autolysis. Flesh of the former type seems to undergo autolysis at more rapid rates. Although the occurrence of this process along with microbial spoilage is presumed by some investigators to aid either the spoilage biota or the spoilage process,[45] attempts to separate and isolate the two events have proved difficult. In a detailed study of fish isolates with respect to the capacity to cause typical fish spoilage by use of sterile fish muscle press juice, Lerke et al.[64] found that the spoilers belonged to the genera *Pseudomonas* and *Acinetobacter-Moraxella*, with none of the coryneforms, micrococci, or flavobacteria, being spoilers. In characterizing the spoilers with respect to their ability to utilize certain compounds, these workers found that most spoilers were unable to degrade gelatin or digest egg albumin. This suggests that fish spoilage proceeds much as does that of beef–in the general absence of complete proteolysis by the spoilage biota. Pure culture inoculations of cod and haddock muscle blocks failed to effect tissue softening.[45] In fish that contain high levels of lipids (herrings, mackerel, salmon, and others), these compounds undergo rancidity as microbial spoilage occurs. It should be noted that the skin of fish is rich in collagen. The scales of most fish are composed of a scleroprotein belonging to the keratin group, and it is quite probable that these are among the last parts of fish to be decomposed.

In a study of 159 Gram-negative isolates from spoiled freshwater fish with total aerobic biota of about 10^8 cfu/g, about 46% were pseudomonads and 38% were *Shewanella* spp.[85] Because the latter produce H_2S and reduce trimethylamine-N-oxide (TMAO), they are believed by some to be the most significant fish spoilage bacteria.

Studies on the skin biota of four different fish revealed the following as the most common organisms: *Pseudomonas-Alteromonas*, 32–60%, and *Moraxella-Acinetobacter*, 18–37%.[47] The initial biota of herring fillets was dominated by *S. putrefaciens* and pseudomonads, and after spoilage in air, these organisms constituted 62–95% of the biota.[69] When allowed to spoil in 100% CO_2 at 4°C, the herring fillets were dominated almost completely by lactobacilli.[69] In the case of rock cod fillets stored in 80% CO_2 + 20% air at 4°C for 21 days, the biota consisted of 71–87% lactobacilli and some tan-colored pseudomonads.[56] In a study of psychrotrophic Enterobacteriaceae isolated from vacuum- and CO_2-packaged cold-smoked salmon, *Pantoea agglomerans* and *Serratia liquefaciens* were the most common species.[40] In this study, numbers of Entcrobacteriaceae found in spoiled products ranged from 10^3 to 1.2×10^7/g, but their role in the spoilage process is unclear. A study of *Pseudomonas* spp. from Mediterranean sea bream stored aerobically and under MAP conditions found *Pseudomonas lundensis* and *P. fluorescens* to be the two dominant species during aerobic spoilage.[93] In some spoiled fish from marine waters, *Photobacterium phosphoreum* is often found.

Phenethyl alcohol has been shown to be produced consistently in fish by a specific organism designated "Achromobacter" by Chen et al.[17] and Chen and Levin.[16] The compound, along with phenol, was recovered from a high-boiling fraction of the haddock fillets held at 2°C. None of the ten known *Acinetobacter* and only one of the nine known *Moraxella* produced phenethyl alcohol under similar conditions. Ethanol, propanol, and iso-propanol are produced by fish spoilers, and of 244 bacteria isolated from king salmon and trout and tested in fish extracts, all produced ethanol, 241 (98.8%) produced isopropanol, and 227 (93%) produced propanol.[2]

In detecting microbial fish spoilage, the reduction of TMAO to trimethylamine (TMA) has been used with some success:

$$\text{Trimethylamine-}N\text{-oxide} \rightarrow \quad \begin{array}{c} H_3C \\ \diagdown \\ \qquad N-CH_3 \\ \diagup \\ H_3C \end{array}$$

$$\text{Trimethylaminc}$$

TMAO is a normal constituent of seafish, whereas little or no TMA is found in freshly caught fish. The presence of TMA is generally regarded to be of microbial origin, although some fish contain muscle enzymes that reduce TMAO. Also, some TMAO may be reduced to dimethylamine. Not all bacteria are equal in their capacity to reduce TMAO to TMA, and its reduction is pH dependent. Methods employed to detect TMA include its extraction from fish with toluene and potassium hydroxide followed by reaction with picric acid or its flushing from extracts and subsequent extraction by use of alkaline permanganate solutions.[31,90] Gas chromatography was used to detect headspace TMA, and sampling and analysis could be completed in less than 5 minutes, with results being consistent with sensory tests.[60]

The detection of CO_2 using an infrared CO_2 analyzer was suggested as a rapid method for spoilage detection in refrigerated catfish.[18] Results by this CO_2 rate evolution method could be obtained in less than 4 hours and they correlated well with APC data. Histamine, diamines, and total volatile substances are used also as fish-spoilage indicators. Histamine is produced from the amino acid histidine by microbially produced histidine decarboxylase, as noted:

$$\text{Histidine} \xrightarrow{\text{Decarboxylase}} \text{Histamine}$$

Histamine is associated with scombroid poisoning (discussed further in Chapter 31). Cadaverine and putrescine are the most important diamines evaluated as spoilage indicators, and they have been used for fish as well as for meats and poultry.

Tyramine is produced by some fish-spoilage organisms, and its production by *Carnobacterium piscicola* and *Weissella viridescens* from isolates of vacuum-packaged sugar-salted fish has been reported.[63] This product, normally kept under refrigeration for 2–4 weeks, may contain $\log_{10}$ 7–10 cfu/g of lactics, and tyramine production was reduced by lowering storage temperatures from 9°C to 4°C.[63] Tyramine results from decarboxylation of the amino acid tyrosine:

$$\underset{\text{OH}}{\overset{\overset{\displaystyle NH_2}{\vert}}{\text{C}-\text{C}-\text{COOH}}} \quad\longrightarrow\quad \underset{\text{OH}}{\text{C}-\text{C}-NH_2} \quad + CO_2$$

Total volatile compounds include total volatile bases (TVB), total volatile acids (TVA), total volatile substances (TVS), and total volatile nitrogen (TVN). TVB includes ammonia, dimethylamine, and trimethylamine; TVN includes TVB and other nitrogen compounds that are obtained by steam distillation of samples; and TVS are those that can be aerated from a product and reduce alkaline permanganate solutions. Because of the reducing capacity of these products, it is sometimes referred to as the volatile reducing substances (VRS) methods. TVA includes acetic, propionic, and related organic acids. TVN has been employed in Australia and Japan for shrimp, where a maximum level for acceptable quality products is 30 mg of TVN/100 g along with a maximum of 5 mg of trimethylamine nitrogen/100 g. Clear-cut offness of shrimp has been noted when TVN is more than 30 mg N/100 g.[19] TVB values of approximately 45 mg of TVB-N/100 g of fish were found to correspond to about 10,000 ng of lipopolysaccharide in one study and to be reflective of lean fish of marginal quality.[87] Among the advantages of these methods for fish freshness is their lack of reliance on a single metabolite. Among the drawbacks is their inability to measure spoilage incipiency.

Shellfish

Crustaceans

The most widely consumed shellfish within this group are shrimp, lobsters, crabs, and crayfish. Unless otherwise specified, spoilage of each is presumed or known to be essentially the same. The chief differences in spoilage of these various foods are referable, generally, to the way in which they are handled and their specific chemical composition.

Crustaceans differ from fish in having about 0.5% carbohydrate as opposed to none for the fish presented (Table 5–7). Shrimp has been reported to have a higher content of free amino acids than fish and to contain cathepticlike enzymes that rapidly break down proteins.

The bacterial biota of freshly caught crustaceans should be expected to reflect the waters from which these foods are caught, and contaminants from the deck, handlers, and washing waters. Many of the organisms reported for fresh fish have been reported on these foods, with pseudomonads, *Acinetobacter-Moraxella*, and yeast spp. being predominant on microbially spoiled crustacean meats. When shrimp was allowed to spoil at 0°C for 13 days, *Pseudomonas* spp. were the dominant spoilers, with only 2% of the spoilage biota being Gram positive, in contrast to 38% for the fresh product.[65] *Moraxella*-dominated spoilage at 5.6°C and 11.1°C, whereas at 16.7°C and 22.2°C *Proteus* was dominant (Table 5–9).

Table 5–9 Most Predominant Bacteria in Shrimp
Held to Spoilage

Temperature (°C)	Days Held	Organisms
0	13	*Pseudomonas*
5.6	9	*Moraxella*
11.1	7	*Moraxella*
16.7	5	*Proteus*
22.2	3	*Proteus*

Source: Matches.[65]

The spoilage of crustacean meats appears to be quite similar to that of fish flesh. Spoilage would be expected to begin at the outer surfaces of these foods due to the anatomy of the organisms. It has been reported that the crustacean muscle contains over 300 mg of nitrogen/100 g of meat, which is considerably higher than that for fish.[94] The presence of higher quantities of free amino acids in particular, and of higher quantities of nitrogenous extractives in crustacean meats in general, makes them quite susceptible to rapid attack by the spoilage biota. Initial spoilage of crustacean meats is accompanied by the production of large amounts of volatile base nitrogen, much as is the case with fish. Some of the volatile base nitrogen arises from the reduction of trimethylamine oxide present in crustacean shellfish (lacking in most mollusks). Creatine is lacking among shellfish, both crustacean and molluscan, and arginine is prevalent. Shrimp microbial spoilage is accompanied by increased hydration capacity in a manner similar to that for meats or poultry.[82]

Mollusks

The molluscan shellfish considered in this section are oysters, clams, squid, and scallops. These animals differ in their chemical composition from both teleost fish and crustacean shellfish in having a significant content of carbohydrate material and a lower total quantity of nitrogen in their flesh. The carbohydrate is largely in the form of glycogen, and with levels of the type that exist in molluscan meats, fermentative activities may be expected to occur as part of the microbial spoilage. Molluscan meats contain high levels of nitrogen bases, much as do other shellfish. Of particular interest in molluscan muscle tissue is a higher content of free arginine, aspartic, and glutamic acids than is found in fish. The most important difference in chemical composition between crustacean shellfish and molluscan shellfish is the higher content of carbohydrate in the latter. For example, clam meat and scallops have been reported to contain 3.4% and oysters 5.6% carbohydrate, mostly as glycogen. The higher content of carbohydrate materials in molluscan shellfish is responsible for the different spoilage pattern of these foods over other seafood.

The microbiota of molluscan shellfish may be expected to vary considerably, depending on the quality of the water from which these fish are taken and the quality of wash water and other factors. The following genera of bacteria have been recovered from spoiled oysters: *Serratia*, *Pseudomonas*, *Proteus*, *Clostridium*, *Bacillus*, *Escherichia*, *Enterobacter*, *Pseudoalteromonas*, *Shewanella*, *Lactobacillus*, *Flavobacterium*, and *Micrococcus*. As spoilage sets in and progresses, *Pseudomonas* and *Acinetobacter-Moraxella* spp. predominate, with enterococci, lactobacilli, and yeasts dominating the late stage of spoilage. A *Pseudoalteromonas* sp. from the marine environment around Chile was found to be the most abundant bacterium in oysters spoiled at 18°C.[78]

Due to the relatively high level of glycogen, the spoilage of molluscan shellfish is basically fermentative. Several investigators, including Hunter and Linden[51] and Pottinger[75] proposed the following pH scale as a basis for determining microbial quality in oysters:

pH 6.2–5.9	Good
pH 5.8	"off"
pH 5.7–5.5	Musty
pH 5.2 and below	sour or putrid

A measure of pH decrease is apparently a better test of spoilage in oysters and other molluscan shellfish than volatile nitrogen bases. A measure of volatile acids was attempted by Beacham[6] and found to be unreliable as a test of oyster freshness. Although pH is regarded by many investigators as being the best objective technique for examining the microbial quality of oysters, Abbey et al.[1] found that organoleptic evaluations and microbial counts were more desirable indexes of microbial quality in this product.

Clams and scallops appear to display essentially the same patterns of spoilage as do oysters, but squid meat does not. In squid meat, volatile base nitrogen increases as spoilage occurs much in the same manner as for the crustacean shellfish. An extensive review of fish and shellfish spoilage has been presented by Ashie et al.[5]

REFERENCES

1. Abbey, A., R.A. Kohler, and S.D. Upham. 1957. Effect of aureomycin chlortetracycline in the processing and storage of freshly shucked oysters. *Food Technol.* 11:265–271.

2. Ahmed, A., and J.R. Matches. 1983. Alcohol production by fish spoilage bacteria. *J. Food Protect.* 46:1055–1059.

3. Allen, J.R., and E.M. Foster. 1960. Spoilage of vacuum-packed sliced processed meats during refrigerated storage. *Food Res.* 25:1–7.

4. Andrews, W.H., C.R. Wilson, P.L. Poelma, and A. Romero. 1977. Bacteriological survey of the channel catfish (*Ictalurus punctalus*) at the retail level. *J. Food Sci.* 42:359–363.

5. Ashie, I.N.A., J.P. Smith, and B.K. Simpson. 1996. Spoilage and shelf-life extension of fresh fish and shellfish. *Crit. Rev. Food Sci. Nutr.* 36:87–121.

6. Beacham, L.M. 1946. A study of decomposition in canned oysters and clams. *J. Assoc. Off. Anal. Chem.* 29:89–92.

7. Beuret, C., A. Baumgardner, and J. Schluep. 2003. Virus-contaminated oysters: A three-month monitoring of oysters imported to Switzerland. *Appl. Environ. Microbiol.* 69:2292–2297.

8. Blickstad, E., and G. Molin. 1983. The microbial flora of smoked pork loin and frankfurter sausage stored in different gas atmospheres at 4°C. *J. Appl. Bacteriol.* 54:45–56.

9. Broda, D.M., P.A. Lawson, R.G. Bell, and D.R. Musgrave. 1999. *Clostridium frigidicarnis* sp. nov., a psychrotolerant bacterium associated with "blown pack" spoilage of vacuum-packed meats. *Int. J. Syst. Bacteriol.* 49:1539–1550.

10. Broda, D.M., G.J. Saul, P.A. Lawson, R.G. Bell, and D.R. Musgrave. 2000. *Clostridium gasigenes* sp. nov., a psychrophile causing spoilage of vacuum-packaged meat. *Int. J. Syst. Enol. Microbiol.* 50:107–118.

11. Buege, D., and J. Luchansky. 1999. Ensuring the safety of home-prepared jerky. *Meat Poultry.* 45(2):56, 59.

12. Calicioglu, M., N.G. Faith, D.R. Buege, and J.B. Luchansky. 1997. Viability of *Escherichia coli* 0157:H7 in fermented semidry low-temperature-cooked beef summer sausage. *J. Food Protect.* 60:1158–1162.

13. Cavett, J.J. 1962. The microbiology of vacuum packed sliced bacon. *J. Appl. Bacteriol.* 25:282–289.

14. Centers for Disease Control and Prevention. 1995. Outbreak of salmonellosis associated with beef jerky—New Mexico, 1995. *Morb. Mort. Wkly. Rept.* 44:785–788.

15. Centers for Disease Control and Prevention. 1995. *Escherichia coli* 0157:H7 outbreak linked to commercially distributed dry-cured salami—Washington and California, 1994. *Morb. Mort. Wkly. Rept.* 44:157–160.

16. Chen, T.C., and R.E. Levin. 1974. Taxonomic significance of phenethyl alcohol production by *Achromobacter* isolates from fishery sources. *Appl. Microbiol.* 28:681–687.

17. Chen, T.C., W.W. Nawar, and R.E. Levin. 1974. Identification of major high-boiling volatile compounds produced during refrigerated storage of haddock fillets. *Appl. Microbiol.* 28:679–680.

18. Chew, S.-Y., and Y.-H.P. Hsieh. 1988. Rapid CO_2 evolution method for determining shelf life of refrigerated catfish. *J. Food Sci.* 63:768–771.

19. Cobb, B.F., III, and C. Vanderzant. 1985. Development of a chemical test for shrimp quality. *J. Food Sci.* 40:121–124.

20. Cook, D.W., J.C. Bowers, and A. DePaola. 2000. Density of total and pathogenic (tdh+) *Vibrio parahaemolyticus* in Atlantic and Gulf Coast molluscan shellfish at harvest. *J. Food Protect.* 65:1873–1880.

21. Cook, D.W., P.O. Leary, J.C. Hunsucker, E.M. Sloan, J.C. Bowsers, R.J. Blodgett, and A. DePaola. 2002. *Vibrio vulnificus* and *Vibrio parahaemolyticus* in U.S. retail shell oysters: A national survey from June 1998 to July 1999. *J. Food Protect.* 65:79–87.

22. Cotton, L.N., and D.L. Marshall. 1998. Predominant microflora on catfish processing equipment. *Dairy Food Environ. Sanit.* 18:650–654.

23. DePaola, A., C.A. Kaysner, J. Bowers, and D.W. Cook. 2000. Environmental investigation of *Vibrio parahaemolyticus* in oysters after outbreaks in Washington, Texas, and New York (1997 and 1998). *Appl. Environ. Microbiol.* 66:4649–4654.

24. Drake, S.D., J.B. Evans, and C.F. Niven. 1958. Microbial flora of packaged frankfurters and their radiation resistance. *Food Res.* 23:291–296.

25. Draughon, F.A., B.A. Anthony, and M.E. Denton. 1999. *Listeria* species in fresh rainbow trout purchased from retail markets. *Dairy Fd. Environ. Sanit.* 19:90–94.

26. Egan, A.F., B.J. Shaw, and P.J. Rogers. 1989. Factors affecting the production of hydrogen sulphide by *Lactobacillus sake* L13 growing on vacuum-packaged beef. *J. Appl. Bacteriol.* 67:255–262.

27. Ellison, R.K., E. Malnari, A. DePaola, J. Bowers, and G.E. Rodrick. 2001. Populations of *Vibrio parahaemolyticus* in retail oysters from Florida using two methods. *J. Food Protect.* 64:682–686.

28. Fach, P., S. Perelle, F. Dilasser, J. Grout, C. Dargaignaratz, L. Botella, J.-M. Gourreau, F. Carlin, M.R. Popoff, and V. Broussolc. 2002. Detection by PCR-enzyme-linked immunosorbent assay of *Clostridium botulinum* in fish and environmental samples from a coastal area in northern France. *Appl. Environ. Microbiol.* 68:5870–5876.

29. Faith, N.G., N. Parniere, T. Larson, T.D. Lorang, and J.B. Luchansky. 1998. Viability of *Escherichia coli* 0157:H7 in pepperoni during the manufacture of sticks and the subsequent storage of slices at 21, 4, and -20°C under air, vacuum, and CO_2. *Int. J. Food Microbiol.* 37:47–54.

30. Fidalgo, S.G., Q. Wang, and T.V. Riley. 2000. Comparison of methods for detection of *Erysipelothrix* spp. and their distribution in some Australasian seafoods. *Appl. Environ. Microbiol.* 66:2066–2070.

31. Fields, M.L., B.S. Richmond, and R.E. Baldwin. 1968. Food quality as determined by metabolic by-products of microorganisms. *Adv. Food Res.* 16:161–229.

32. Foster, J.F., J.L. Fowler, and J. Dacey. 1977. A microbial survey of various fresh and frozen seafood products. *J. Food Protect.* 40:300–303.

33. Fraiser, M.B., and J.A. Koburger. 1984. Incidence of salmonellae in clams, oysters, crabs and mullet. *J. Food Protect.* 47:343–345.

34. Gardini, E., R. Tofalo, and G. Suzzi. 2003. A survey of antibiotic resistance in *Micrococcaceae* isolated from Italian dry fermented sausages. *J. Food Protect.* 66:937–945.

35. Gardner, G.A. 1971. Microbiological and chemical changes in lean Wiltshire bacon during aerobic storage. *J. Appl. Bacteriol.* 34:645–654.

36. Garriga, M., M.A. Ehrmann, J. Arnau, M. Hugas, and R.F. Vogel. 1998. *Carnimonas nigrificans* gen. nov., sp. nov., a bacterial causative agent for black spot formation on cured meat products. *Int. J. Syst. Bacteriol.* 48:677–686.

37. Gill, C.O., and K.G. Newton. 1979. Spoilage of vacuum-packaged dark, firm, dry meat at chill temperatures. *Appl. Environ. Microbiol.* 37:362–364.

38. Gombas, E.E., Y. Chen, R.S. Clavero, and V.N. Scott. 2003. Survey of *Listeria monocytogenes* in ready-to-eat foods. *J. Food Protect.* 66:559–569.

39. González, C.J., J.A. Santos, M.-L. Carcía-López, and A. Otero. 2000. Psychrobacters and related bacteria in freshwater fish. *J. Food Protect.* 63:315–321.

40. Gram, L., A.B. Christensen, L. Ravn, S. Molin, and M. Givskov. 1999. Production of acylated homoserine lactones by psychrotrophic members of the *Enterobacteriaceae* isolated from foods. *Appl. Environ. Microbiol.* 65:3458–3463.

41. Grant, G.F., A.R. McCurdy, and A.D. Osborne. 1988. Bacterial greening in cured meats: A review. *Can. Inst. Food Sci. Technol. J.* 21:50–56.

42. Harrison, J.A., and M.A. Harrison. 1996. Fate of *Escherichia coli* 0157:H7, *Listeria monocytogenes,* and *Salmonella* Typhimurium during preparation and storage of beef jerky. *J. Food Protect.* 59:1336–1338.

43. Hauschild, A.H.W., and R. Hilsheimer. 1983. Prevalence of *Clostridium botulinum* in commercial liver sausage. *J. Food Protect.* 46:243–244.

44. Heinitz, M.L., R.D. Ruble, D.E. Wagner, and S.R. Tatini. 2000. Incidence of *Salmonella* in fish and seafood. *J. Food Protect.* 63:579–592.

45. Herbert, R.A., M.S. Hendrie, D.M. Gibson, and J.M. Shewan. 1971. Bacteria active in the spoilage of certain sea foods. *J. Appl. Bacteriol.* 34:41–50.

46. Hinkens, J.C., N.G. Faith, T.D. Lorang, P. Bailey, D. Buege, C.W. Kaspar, and J.B. Luchansky. 1996. Validation of pepperoni processes for control of *Escherichia coli* 0157:H7. *J. Food Protect.* 59:1260–1266.

47. Hobbs, G. 1983. Microbial spoilage of fish. In *Food Microbiology: Advances and Prospects*, ed. T.A. Roberts and F.A. Skinner, 217–229. London: Academic Press.

48. Holley, R.A., T.Y. Guan, M. Peirson, and C.K. Yost. 2002. *Carnobacterium viridans* sp. nov., an alkaliphilic, facultative anaerobe isolated from refrigerated, vacuum-packed bologna sausage. *Int. J. Syst. Evol. Microbiol.* 52:1881–1885.

49. Holley, R.A. 1985. Beef jerky: Fate of *Staphylococcus aureus* in marinated and corned beef during jerky manufacture and 2.5°C storage. *J. Food Protect.* 48:107–111.

50. Holley, R.A. 1985. Beef jerky: Viability of food-poisoning microorganisms on jerky during its manufacture and storage. *J. Food Protect.* 48:100–106.

51. Hunter, A.C., and B.A. Linden. 1923. An investigation of oyster spoilage. *Amer. Food J.* 18:538–540.

52. Ihnot, A.M., A.M. Roering, R.K. Wierzba, N.G. Faith, and J.B. Luchansky. 1998. Behavior of *Salmonella* Typhimurium DT104 during the manufacture and storage of pepperoni. *Int. J. Food Microbiol.* 40:117–121.

53. Ingham, M. 1960. Bacterial multiplication in packed Wiltshire bacon. *J. Appl. Bacteriol.* 23:206–215.

54. Ingram, M., and R.H. Dainty. 1971. Changes caused by microbes in spoilage of meats. *J. Appl. Bacteriol.* 34:21–39.

55. Jacquot, R. 1961. Organic constituents of fish and other aquatic animal foods. In *Fish as Food*, ed. G. Borgstrom, vol. 1, 145–209. New York: Academic Press.

56. Johnson, A.R., and D.M. Ogrydziak. 1984. Genetic adaptation to elevated carbon dioxide atmospheres by *Pseudomonas-*like bacteria isolated from rock cod (*Sebastes* spp.). *Appl. Environ. Microbiol.* 48:486–490.

57. Kato, Y., R.M. Sakala, H. Hayashidani, A. Kiuchi, C. Kaneuchi, and M. Ogawa. 2000. *Lactobacillus algidus* sp. nov., a psychrophilic lactic acid bacterium isolated from vacuum-packaged refrigerated beef. *Int. J. Syst. Evol. Microbiol.* 50:1143–1149.

58. Kazanas, N., J.A. Emerson, H.L. Seagram, and L.L. Kempe. 1966. Effect of γ-irradiation on the microflora of fresh-water fish. I. Microbial load, lag period, and rate of growth on yellow perch (*Perca flavescens*) fillets. *Appl. Microbiol.* 14:261–266.

59. Kitchell, A.G. 1962. Micrococci and coagulase negative staphylococci in cured meats and meat products. *J. Appl. Bacteriol.* 25:416–431.

60. Krzymien, M.E., and L. Elias. 1990. Feasibility study on the determination of fish freshness by trimethylamine headspace analysis. *J. Food Sci.* 55:1228–1232.

61. LeGuyader, F., L. Haugarreau, L. Miossec, E. Dubois, and M. Pommepuy. 2000. Three-year study to assess human enteric viruses in shellfish. *Appl. Environ Microbiol.* 66:3241–3248.

62. Leininger, H.V., L.R. Shelton, and K.H. Lewis. 1971. Microbiology of frozen cream-type pies, frozen cooked-peeled shrimp, and dry food-grade gelatin. *Food Technol.* 25:224–229.

63. Leisner, J.J., J.C. Millan, H.H. Huss, and L.M. Larsen. 1994. Production of histamine and tyramine by lactic acid bacteria isolated from vacuum-packaged sugar-salted fish. *J. Appl. Bacteriol.* 76:417–423.

64. Lerke, P., R. Adams, and L. Farber. 1965. Bacteriology of spoilage of fish muscle. III. Characteristics of spoilers. *Appl. Microbiol.* 13:625–630.

65. Matches, J.R. 1982. Effects of temperature on the decomposition of Pacific coast shrimp (*Pandalus jordani*). *J. Food Sci.* 47:1044–1047, 1069.

66. McLean, R.A., and W.L. Sulzbacher. 1953. *Microbacterium thermosphactum* spec. nov., a non-heat resistant bacterium from fresh pork sausage. *J. Bacteriol.* 65:428–432.

67. Mediel, M.J., V. Rodriguez, G. Codina, and N. Martin-Casabona. 2000. Isolation of mycobacteria from frozen fish destined for human consumption. *Appl. Environ. Microbiol.* 66:3637–3638.

68. Miettinen, H., A. Arvola, T. Luoma, and G. Wirtanen. 2003. Prevalence of *Listeria monocytogenes* in, and microbiological sensory quality of, rainbow trout, whitefish, and vendace roes from Finnish retail markets. *J. Food Protect.* 66:1832–1839.

69. Molin, G., and I.-M. Stenstrom. 1984. Effect of temperature on the microbial flora of herring fillets stored in air or carbon dioxide. *J. Appl. Bacteriol.* 56:275–282.

70. Nickerson, J.T.R., and S.A. Goldblith. 1964. A study of the microbiological quality of haddock fillets and shucked, soft-shelled clams processed and marketed in the greater Boston area. *J. Milk Food Technol.* 27:7–12.

71. Nicol, D.J., M.K. Shaw, and D.A. Ledward. 1970. Hydrogen sulfide production by bacteria and sulfmyoglobin formation in prepacked chilled beef. *Appl. Microbiol.* 19:937–939.

72. Ng, D.L.K., B.B. Koh, L. Tay, and M. Yeo. 1999. The presence of *Salmonella* in local food and beverage items in Singapore. *Dairy Fd. Environ. Sanit.* 19:848–852.

73. Pearson, A.M., and T.A. Gillett. 1999. *Processed Meats.* New York: Kluwer Academic Publishers.

74. Portocarrero, S.M., M. Newman, and B. Mikel. 2002. *Staphylococcus aureus* survival, staphylococcal enterotoxin production and shelf stability of country-cured hams manufactured under different processing procedures. *Meat Sci.* 62:267–273.

75. Pottinger, S.R. 1948. Some data on pH and the freshness of shucked eastern oysters. *Comm. Fisheries Rev.* 10(9):1–3.

76. Riha, W.E., and M. Solberg. 1970. Microflora of fresh pork sausage casings. 2. Natural casings. *J. Food Sci.* 35:860–863.

77. Rodríguez, M., F. Núñez, J.J. Córdoba, E. Bermúdez, and M.A. Asensio. 1996. Gram-positive, catalase-positive cocci from dry cured Iberian ham and their enterotoxigenic potential. *Appl. Environ. Microbiol.* 62:1897–1902.

78. Romero, J., N. González, and R.T. Espero. 2002. Marine *Pseudoalteromonas* sp. composes most of the bacterial population developed in oysters (*Tiostrea chilensis*) spoiled during storage. *J. Food Sci.* 67:2300–2303.

79. Sakala, R.M., Y. Kato, H. Hayashidanik, M. Murakami, C. Kaneuchi, and M. Ogawa. 2002. *Lactobacillus fuchuensis* sp. nov., isolated from vacuum-packaged refrigerated beef. *Int. J. Syst. Evol. Microbiol.* 52:1151–1154.

80. Shaw, B.G., and J.M. Shewan. 1968. Psychrophilic spoilage bacteria of fish. *J. Appl. Bacteriol.* 31:89–96.

81. Shay, B.J., and A.F. Egan. 1981. Hydrogen sulphide production and spoilage of vacuum-packaged beef by a *Lactobacillus*. In *Micro-Organisms in Spoilage and Pathogenicity*, ed. T.A. Roberts, G. Hobbs, J.H.B. Christian, et al., 241–251. London: Academic Press.

82. Shelef, L.A., and J.M. Jay. 1971. Hydration capacity as an index of shrimp microbial quality. *J. Food Sci.* 36:994–997.

83. Shewan, J.M. 1961. The microbiology of sea-water fish. In *Fish as Food*, ed. G. Borgstrom, vol. 1, 487–560. New York: Academic Press.

84. Silverman, G.J., J.T.R. Nickerson, D.W. Duncan, N.S. Davis, J.S. Schachter, and M.M. Joselow. 1961. Microbial analysis of frozen raw and cooked shrimp. I. General results. *Food Technol.* 15:455–458.

85. Stenstrom, I.-M., and G. Molin. 1990. Classification of the spoilage flora of fish, with special reference to *Shewanella putrefaciens*. *J. Appl. Bacteriol.* 68:601–618.

86. Stewart, A.W. 1983. Effect of cooking on bacteriological population of "soul foods". *J. Food Protect.* 46:19–20.

87. Sullivan, J.D., Jr., P.C. Ellis, R.G. Lee, W.S. Combs, Jr., and S.W. Watson. 1983. Comparison of the *Limulus amoebocyte* lysate test with plate counts and chemical analyses for assessment of the quality of lean fish. *Appl. Environ. Microbiol.* 45:720–722.

88. Surkiewicz, B.F., M.E. Harris, R.P. Elliott, J.F. Macaluso, and M.M. Strand. 1975. Bacteriological survey of raw beef patties produced at establishments under federal inspection. *Appl. Microbiol.* 29:331–334.

89. Swartzentruber, A., A.H. Schwab, A.P. Duran, B.A. Wentz, and R.B. Read, Jr. 1980. Microbiological quality of frozen shrimp and lobster tail in the retail market. *Appl. Environ. Microbiol.* 40:765–769.

90. Tarr, H.L.A. 1954. Microbiological deterioration of fish post mortem, its detection and control. *Bacteriol. Rev.* 18:1–15.

91. Thimothe, J., J. Walker, V. Suvanich, K.L. Call, M.W. Moody, and M. Wiedmann. 2002. Detection of *Listeria* in crawfish processing plants and in raw, whole crawfish and processed crawfish (*Procambarus* spp.). *J. Food Protect.* 65:1735–1739.

92. Tonge, R.J., A.C. Baird-Parker, and J.J. Cavett. 1964. Chemical and microbiological changes during storage of vacuum packed sliced bacon. *J. Appl. Bacteriol.* 27:252–264.

93. Tryfinopoulou, P., E. Tsakalidou, and G.-J.E. Nychas. 2002. Characterization of *Pseudomonas* spp. associated with spoilage of gilt-head sea bream stored under various conditions. *Appl. Environ. Microbiol.* 68:65–72.

94. Velankar, N.K., and T.K. Govindan. 1958. A preliminary study of the distribution of nonprotein nitrogen in some marine fishes and investebrates. *Proc. Indian Acad. Sci. Secft. B* 47:202–209.

95. Vanderzant, C., A.W. Matthys, and B.F. Cobb, III. 1973. Microbiological, chemical, and organoleptic characteristics of frozen breaded raw shrimp. *J. Milk Food Technol.* 36:253–261.

96. Wallace, R.M., J.E. Call, A.C.S. Porto, G.J. Cocoma, the ERRC Sepc. Proj. Team, and J.B. Luchansky. 2003. Recovery rate of *Listeria monocytogenes* from commercially prepared frankfurters during extended refrigerated storage. *J. Food Protect.* 66:584–591.

97. Wang, C., and J.L. Silva. 1999. Prevalence and characteristics of *Aeromonas* species isolated from processed channel catfish. *J. Food Protect.* 62:30–34.

98. Wardlaw, F.B., G.C. Skelley, M.G. Johnson, and J.C. Ayres. 1973. Changes in meat components during fermentation, heat processing and drying of a summer sausage. *J. Food Sci.* 38:1228–1231.

99. Watt, B.K., and A.L. Merrill. 1950. Composition of foods—Raw, processed, prepared. *Agricultural Handbook No. 8.* Washington, D.C.: U.S. Department of Agriculture.

100. Wentz, B.A., A.P. Duran, A. Swartzentruber, A.H. Schwab, and R.B. Read, Jr. 1983. Microbiological quality of fresh blue crabmeat, clams and oysters. *J. Food Protect.* 46:978–981.

101. Whiteley, A.M., and M.D. D'Souza. 1989. A yellow discoloration of cooked cured meat products—Isolation and characterization of the causative organisms. *J. Food Protect.* 52:392–395.

CHAPTER 6

Vegetable and Fruit Products

The microbial biota of land-grown vegetables may be expected to reflect that of the soils in which they are grown, although exceptions occur. The actinomycetes (Gram-positive branching forms) are the most abundant bacteria in stable soils, yet they are rarely reported on vegetable products. On the other hand, the lactic acid bacteria are rarely found in soil per se, but they are significant parts of the bacterial biota of plants and plant products. The overall exposure of plant products to the environment provides many opportunities for contamination by microorganisms. The protective cover of fruits and vegetables and the possession by some of pH values below which many organisms cannot grow are important factors in the microbiology of these products.

An attempt is made in this chapter to treat fruits and vegetables separately even though this is difficult. In common usage, products such as tomatoes and cucumbers are referred to as vegetables and yet from the botanical standpoint they are fruits. Lemons, oranges, and limes are fruits botanically as well as in common usage. By and large, the distinctions between fruits and vegetables are based on pH, irrespective of the lack of scientific merit.

FRESH AND FROZEN VEGETABLES

The incidence of microorganisms in vegetables may be expected to reflect the sanitary quality of the processing steps and the microbiological condition of the raw product at the time of processing. In a study of green beans before blanching, Splittstoesser et al.[50] showed that the total counts ranged from $5.60 \log_{10}$ to over $6.00 \log_{10}$ in two production plants. After blanching, the total numbers were reduced to $3.00–3.60 \log_{10}/g$. After passing through the various processing stages and packaging, the counts ranged from 4.72 to $5.94 \log_{10}/g$. In the case of French-style beans, one of the greatest buildups in numbers of organisms occurred immediately after slicing. This same general pattern was shown for peas and corn. Preblanched green peas from three factories showed total counts per gram between 4.94 and $5.95 \log_{10}$. These numbers were reduced by blanching but again increased successively with each processing step. In the case of whole-kernel corn, the postblanch counts rose both after cutting and at the end of the conveyor belt to the washer. Whereas the immediate postblanch count was about $3.48 \log_{10}$, the product had total counts of about $5.94 \log_{10}/g$ after packaging. Between 40% and 75% of the bacterial biota of peas, snap beans, and corn was shown to consist of leuconostocs and "streptococci," whereas many of the Gram-positive, catalase-positive rods resembled corynebacteria.[48,49]

125

Table 6–1　Microbial Numbers in Some Fresh Vegetables

Products	$Log_{10}cfu/g$	Reference
Red and green chicory and carrot mixture	APC, 7.94	57
	Coliforms, 7.03	57
	Fecal coliforms, 6.74	57
	Lactic acid bacteria, 6.18	57
Red chicory, endive, and carrot mixture	APC, 6.14	57
	Coliforms, 4.68	57
	Fecal coliforms, 4.51	57
	Lactic acid bacteria, 5.86	57
Bean sprouts	APC, 7.26, 7.99	23
	Coliforms, 7.49, 6.99	
Broccoli	APC, 3.97	37
Carrots	APC, 4.20	37
Cauliflower	APC, 6.97	37
Celery	APC, 10.0	37
Coleslaw	APC, 7.00	37
Radish	APC, 6.04	37
Sprouts	APC 8.7; Coliforms 7.2	55
Lettuce	APC 8.6; Coliforms 5.6	55
Celery	APC 7.5; Coliforms	55
Cauliflower	APC 7.4; Coliforms 2.9	55
Broccoli	APC 6.3; Coliforms 4.8	55
Fresh retail lettuce[a]	APC 6.94; Coliforms 3.25	30

Note: APC = aerobic plate count; cfu = colony-forming unit.

[a]Mean of 10 samples, which also contained 1.64 *E. coli* and 5.62 fungi.

Numbers of bacteria that have been reported for a number of different fresh vegetables are summarized in Table 6–1. It may be noted that APCs of around 6 to 7 log_{10} cfu/g are common among the vegetables listed, and that coliform numbers of around 5–6 log_{10} are not uncommon.

Lactic acid cocci are associated with many raw and processed vegetables.[29] These cocci have been shown to constitute from 41% to 75% of the aerobic plate count (APC) biota of frozen peas, snap beans, and corn.[46] It has been shown that fresh peas, green beans, and corn all contained coagulase-positive staphylococci after processing.[48] Peas were found to have the highest count (0.86 log_{10}/g), whereas 64% of corn samples contained this organism. These authors found that a general buildup of staphylococci occurred as the vegetables underwent successive stages of processing, with the main source of organisms coming from the hands of employees. Although staphylococci may be found on vegetables during processing, they are generally unable to proliferate in the presence of the more normal lactic biota. Both coliforms (but not *Escherichia coli*) and enterococci have been found at most stages during raw vegetable processing, but they appear to present no public health hazard.[47]

In a study of the incidence of *Clostridium botulinum* in 100 commercially available frozen vacuum pouch-pack vegetables, the organism was not found in 50 samples of string beans, but types A and B spores were found in 6 of 50 samples of spinach.[20]

The total numbers of bacteria on frozen vegetables tend to be lower than on comparable nonfrozen products. This is due primarily to, (1) blanching of products prior to freezing, (2) selection of higher quality products for freezing, and (3) the die-off of some bacteria while in the frozen state (see

Table 6–2 General Microbiological Quality of Frozen Vegetables

Products	No. of Samples	Microbial Group/Target	% Samples Meeting Target	Reference
Cauliflower	1,556	APC at 35°C: 10^5/g or less	75	5
	1,556	MPN coliforms: <20/g	79	5
	1,556	MPN *E. coliforms*: <3/g	98	5
Corn	1,542	APC at 35°C: 10^5/g or less	94	5
	1,542	MPN coliforms: <20/g	71	5
	1,542	MPN *E. coli*: <3/g	99	5
Peas	1,564	APC at 35°C: 10^5/g or less	95	5
	1,564	MPN coliforms: <20/g	78	5
	1,564	MPN *E. coli*: <3/g	99	5
Blanched vegetables (17 different)	575	Absence of fecal coliforms	63	49
	575	$n = 5, c = 3, m = 10, M = 10^3$	33	49
	575	$n = 5, c = 3, m = 50, M = 10^3$	70	49
Cut green beans, leaf spinach, peas	144	Mean APC range for group: 4.73–4.93 $\log_{10}$/g	–	47
Lima beans, corn, broccoli spears, brussels sprouts	170	Mean APC range for group: 5.30–5.36 $\log_{10}$/g	–	47
French-style green beans, chopped greens, squash	135	Mean APC range: 5.48–5.51 $\log_{10}$/g	–	47
Chopped spinach, cauliflower	80	Mean APC range: 5.54–5.65 $\log_{10}$/g	–	47
Chopped broccoli	45	Mean APC: 6.26 $\log_{10}$/g	–	47

Note: APC = Aerobic plate count; MPN = most probable number.

Chapter 16). It can be seen from Table 6–2 that APCs of the frozen green vegetables listed is around 5.0 $\log_{10}$ cfu/g in contrast to fresh green vegetables in Table 6–1, which are higher. A summary of the $\log_{10}$ APC for some frozen processed potato products is presented in Table 6–3. Since the inside of a fresh and undamaged potato is free of bacteria, the numbers summarized in this table may be taken to reflect post-cooking contamination for those that received heat treatment.

Table 6–3 Summary of $\log_{10}$ APC for Some Frozen Processed Potato Products[12]

Products	APC Range
Potato salad	2.6–5.1
Dried potatoes	2.5–5.5
Baked potatoes	4.0–6.4
French fries	2.0–6.7
Hash browns	3.2–8.7

Table 6–4 General Chemical Composition of Higher Plant Materials

---

Carbohydrates and related compounds

1. Polysaccharides—pentosan (araban), hexosans (cellulose, starch, xylans, fructans, mannans, galactans, levans)
2. Oligosaccharides—tetrasaccharide (stachyose), trisaccharides (robinose, mannotriose, raffinose), disaccharides (maltose, sucrose, cellobiose, melibiose, trehalose)
3. Monosaccharides—hexoses (mannose, glucose, galactose, fructose, sorbose), pentoses (arabinose, xylose, ribose, L-rhamnose, L-fucose)
4. Sugar alcohols—glycerol, ribitol, mannitol, sorbitol, inositols
5. Sugar acids—uronic acids, ascorbic acid
6. Esters—tannins
7. Organic acids—citric, shikimic, D-tartaric, oxalic, lactic, glycolic, malonic, etc.

Proteins—albumins, globulins, glutelins, prolamines, peptides, and amino acids
Lipids—fatty acids, fatty acid esters, phospholipids, glycolipids, etc.
Nucleic acids and derivatives—purine and pyrimidine bases, nucleotides, etc.
Vitamins—fat soluble (A, D, E), water soluble (thiamine, niacin, riboflavin, etc.)
Minerals—Na, K, Ca, Mg, Mn, Fe, etc.
Water
Others—alkaloids, porphyrins, aromatics, etc.

Spoilage

The general composition of higher plants is presented in Table 6–4, and the composition of 21 common vegetables is presented in Table 6–5. The average water content of vegetables is about 88%, with an average content of 8.6% carbohydrates, 1.9% proteins, 0.3% fat, and 0.84% ash. The total percentage composition of vitamins, nucleic acids, and other plant constituents is generally less than 1%. From the standpoint of nutrient content, vegetables are capable of supporting the growth of molds, yeasts, and bacteria and, consequently, of being spoiled by any or all of these organisms. The higher water content of vegetables favors the growth of spoilage bacteria, and the relatively low carbohydrate and fat contents suggest that much of this water is in available form. The pH range of most vegetables is within the growth range of a large number of bacteria, and it is not surprising, therefore, that bacteria are common agents of vegetable spoilage. The relatively high oxidation–reduction (O/R) potential of vegetables and their lack of high poising capacity suggest that the aerobic and facultative anaerobic types would be more important than the anaerobes. This is precisely the case; some of the most ubiquitous etiologic agents in the bacterial spoilage of vegetables are species of the genera *Erwinia* and *Pectobacterium*, and they are associated with plants and vegetables in their natural growth environment. The common spoilage pattern displayed by these organisms is referred to as *bacterial soft rot*.

Bacterial Agents

The genus *Erwinia* has been delimited by the transfer of ten species and 5 subspecies to two new genera, and they are listed in Table 6–6. It should be noted that some taxonomists are not in full accord with the transfer of some of the erwiniae to the genus *Pectobacterium*, suggesting that further changes are likely to occur (see reference 63). Many of the plant-associated pseudomonads have

Table 6–5 Vegetable Foods: Approximate Percentage Chemical Composition

Vegetable	Water	Carbohydrates	Proteins	Fat	Ash
Beans, green	89.9	7.7	2.4	0.2	0.8
Beets	87.6	9.6	1.6	0.1	1.1
Broccoli	89.9	5.5	3.3	0.2	1.1
Brussels sprouts	84.9	8.9	4.4	0.5	1.3
Cabbage	92.4	5.3	1.4	0.2	0.8
Cantaloupe	94.0	4.6	0.2	0.2	0.6
Cauliflower	91.7	4.9	2.4	0.2	0.8
Celery	93.7	3.7	1.3	0.2	1.1
Corn	73.9	20.5	3.7	1.2	0.7
Cucumbers	96.1	2.7	0.7	0.1	0.4
Lettuce	94.8	2.9	1.2	0.2	0.9
Onions	87.5	10.3	1.4	0.2	0.6
Peas	74.3	17.7	6.7	0.4	0.9
Potatoes	77.8	19.1	2.0	0.1	1.0
Pumpkin	90.5	7.3	1.2	0.2	0.8
Radish	93.6	4.2	1.2	0.1	1.0
Spinach	92.7	3.2	2.3	0.3	1.5
Squash, summer	95.0	3.9	0.6	0.1	0.4
Sweet potatoes	68.5	27.9	1.8	0.7	1.1
Tomatoes	94.1	4.0	1.0	0.3	0.6
Watermelon	92.1	6.9	0.5	0.2	0.3
Mean	88.3	8.6	2.0	0.3	0.8

Source: Watt and Merrill.[60]

Table 6–6 The Transfer of 10 Former *Erwinia* spp. and Five Subspecies to the New Genera *Pectobacterium* and *Brenneria* (Summarized from Reference 18)

Pectobacterium	Formerly Erwinia spp.
P. carotovorum subsp. atrosepticum	E. carotovora subsp. atroseptica
P. carotovorum subsp. carotovorum	E. carotovora subsp. carotovora
P. carotovorum subsp. betavasculorum	E. carotovora subsp. betavasculorum
P. carotovorum subsp. odoriferum	E. carotovora subsp. odorifera
P. carotovorum subsp. wasabiae	E. carotovorum sp. wasabiae
P. cacticidum	E. cacticida
P. chrysanthemi (formerly Erwinia)	
P. cypripedii (formerly Erwinia)	
Brenneria (all former Erwinia spp.)	
B. alni, B. nigrifluens, B. paradisiaca, B. quercina, B. rubrifaciens, and B. salicis	

Table 6–7 Some Bacteria that Cause Field and Storage Spoilage of Vegetables and Fruits (See Text for Others)

Organisms	Spoilage Condition/Products
Corynebacterium michiganenese	Vascular wilt, canker; leaf and fruit spot on tomatoes, others
C. nebraskense	Leaf spot, leaf blight, and wilt of corn
C. sepedonicum	Tuber rot of white potatoes
Curtobacterium flaccumfaciens (formerly Corynebacterium)	Bacterial wilt of beans
Pseudomonas agarici and P. tolaasii	Drippy gill of mushrooms
P. corrupata	Tomato pith necrosis
Pseudomonas cichorii group	Bacterial zonate spot of cabbage and lettuce
Pseudomonas marginalis group	Soft rot of vegetables, side slime of lettuce
P. morsprunorum group (formerly P. phaseolicola)	Halo blight of beans
P. syringae pv. syringae	Bacterial canker of stone fruit trees
Formerly P. glycinea	Disease of soybeans
Formerly P. lachrymans	Angular leaf spot of cucumbers
Formerly P. pisi	Bacterial blight of pears
P. tomato group	Bacterial speck of tomatoes
Xanthomonas campestris	
pv. campestris	Black rot of cabbage and cauliflower
X. oryzae pv. oryzae	Bacterial blight of rice
pv. oryzicola	Bacterial leaf streak of rice
Rathyibacter spp.	Gumming diseases of plants
Janthinobacterium agaricidamnosum	Soft rot of mushrooms
Streptomyces spp.	Potato scab
Xanthomonas axonopodis pv. citri	Citrus canker
Xylella fastidiosa	Pierce's disease of grapes
Ralstonia spp.	Wilt of tomatoes
Erwinia amylovora	Fire blight of apple/pear trees
Acidovorax valerianellae	Black water-soaked spots of lamb's lettuce

been transferred to new genera including *Acidovorax*, *Burkholderia*, and *Hydrogenophaga*. Changes have been made to the genus *Xanthomonas*, and it appears that this process will continue with the emphasis more on molecular genetic methods than the traditional phenotypic approaches. The genus *Pantoea* is closely related to *Erwinia*, and it along with *Citrobacter* and *Klebsiella* are probably more important in the storage spoilage of vegetables than is now apparent. The bacterial genera most often associated with field and storage spoilage of vegetables are *Pseudomonas*, *Pectobacterium*, *Erwinia*, *Xanthomonas*, and some of the specific spoilage conditions are listed in Table 6–7.

Soft rots occur in plants of a number of species and those of carrots are well known. A "soft" rot refers to the mushy consistency of the plant or vegetable in contrast to some other spoilage conditions where the product remains firm. The bacteria most commonly associated with the soft rotting of carrots are *Pectobacterium* spp., especially *P. carotovorum* subsp. *carotovorum* and *P. carotovorum* subsp. *odoriferum*. After the spoilage process is initiated, a number of soil-dwelling bacteria are involved, and they include *Pseudomonas* spp. as well as *Bacillus*, *Paenibacillus*, and *Clostridium*. For fruits and vegetables like tomatoes and potatoes, *P. carotovorum* subsp. *carotovorum* causes soft rot by entering surface wounds. For root vegetables such as carrots, the roots are protected for some. Plant roots are protected from invading microorganisms by their possession of hydrogen peroxide and superoxide, and invading microorganisms produce catalase and superoxide dismutase to overcome this defense. The *Pseudomonas syringae* group as well as erwiniae produce these enzymes.

The cementing substance of the vegetable body induces the formation of pectinases, which act by hydrolyzing pectin, thereby producing the mushy consistency. In potatoes, tissue maceration has been shown to be caused by an endopolygalacturonate transeliminase of *Pectobacterium* origin.[28] *P. chrysanthemi* produces two pectin methyesterases, at least seven pectate lyases, a polygalacturonase, and a pectin lyase.[19] Once pectin is destroyed, oligogalacturonidases are formed and utilized by the varied microbiota. Because of the early and relatively rapid growth of bacteria, molds, which tend to be crowded out, are of less consequence in the spoilage of vegetables that are susceptible to bacterial agents.

Once the outer plant barrier has been destroyed by these pectinase producers, nonpectinase producers no doubt enter the plant tissues and help bring about fermentation of the simple carbohydrates that are present. The quantities of simple nitrogenous compounds present, the vitamins (especially the B-complex group), and minerals are adequate to sustain the growth of the invading organisms until the vegetables have been essentially consumed or destroyed. The malodors that are produced are the direct result of volatile compounds (such as NH_3, volatile acids, and the like) produced by the biota. When growing in acid media, microorganisms tend to decarboxylate amino acids, leaving amines that cause an elevation of pH toward the neutral range and beyond. Complex carbohydrates such as cellulose are generally the last to be degraded, and a varied biota consisting of molds and other soil organisms is usually responsible, as cellulose degradation by *Erwinia* spp. is doubtful. Aromatic constituents and porphyrins are probably not attacked until late in the spoilage process, and again by a varied biota of soil types. The genus *Brenneria* causes diseases of trees such as bark cankers, necrosis on walnut trees, oozing of sap from acorns, etc.[18]

The genes of *P. carotovorum* subsp. *carotovorum* that are involved in potato tuber maceration have been cloned. Plasmids containing cloned DNA mediated the production of endopectate lyases, exopectate lyase, endopolygalacturonase, and cellulases.[39] The *Escherichia coli* strains that contained cloned plasmids showed that endopectate lyases with endopolygalacturonase or exopectate lyase caused maceration of potato tuber slices. These enzymes, along with phosphatidase C and phospholipase A, are involved in soft rot by this organism. Carrots infected with *Agrobacterium tumefaciens* undergo senescence at a faster rate because of increased ethylene synthesis. In normal uninfected plants, ethylene synthesis is regulated by auxins, but *A. tumefaciens* increases the synthesis of indoleacetic acid, which results in increased levels of ethylene.

The citrus canker pathogen, *Xanthomonas anonopodis* pv. (pathovar) *citri* wreaked havoc on the Florida citrus industry in the late 1990s. The canker consists of external scabs and cork-like lesions that occur on oranges, limes, grapefruits, and other citrus fruits. The organism enters fruit trees through leaf stomata. Once inside, they employ a type III secretion system (see Chapter 22) to induce increased cell division that leads to the brownish cankers. The infected fruits produce more ethylene, which leads to senescence, increased ripening, and premature falling of fruits from trees. This infection does not destroy trees. The bacterium is spread from tree to tree by wind, rain, insects, and other nonspecific means.

The genus *Xanthomonas* is undergoing reclassification, thus some of the species and pathovars noted in Table 6–6 are likely to be changed. Most form yellow mucoid and smooth colonies and

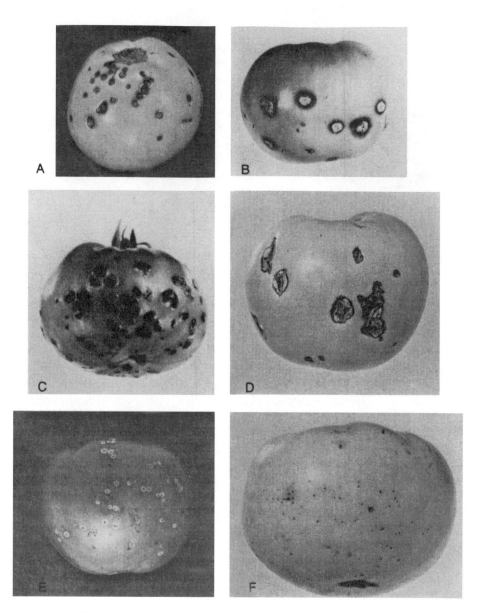

Figure 6–1   Tomato diseases—(A and B) nailhead spot; (C and D) bacterial spot; (E) bacterial canker; (F) bacterial spot. *Source*: From *Agriculture Handbook 28*, USDA, 1968, "Fungus and Bacterial Diseases of Fresh Tomatoes."

produce the yellow-pigmented *xanthomonadins*. The mucoid colonies are due to *xanthans*, which are typical of the genus.[56] Bacterial canker of stone fruits is caused by *P. syringae* pv. *syringae*, and this pathovar has been reported to cause disease in over 180 species of plants.[25]

Some of the more important bacteria that cause field and storage spoilage of vegetables are presented in Table 6–7. Some genera and species listed are undergoing taxonomic changes. The plant

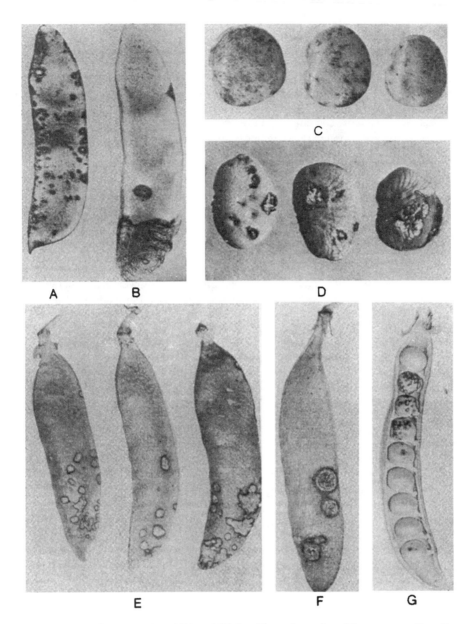

Figure 6–2 Lima bean diseases—(A and B) pod blight; (C) seed spotting; (D) yeast spot. Pea diseases—(E) pod spot; (F) anthracnose; (G) scab. *Source*: From *Agriculture Handbook 303*, USDA, 1966, Chapter 5.

corynebacteria represent a diverse collection, many of which do not belong to this genus. Some have been transferred to the genus *Curtobacter*. The plant pathogenic and field spoilage pseudomonads and xanthomonads are also diverse.

The appearance of some market vegetables undergoing bacterial and fungal spoilage is shown in Figures 6–1 and 6–2.

Fungal Agents

A synopsis of some of the common spoilage conditions of vegetables and fruits is presented in Table 6–8. Some of these spoilage conditions are initiated preharvest and others postharvest. Among the former, *Botrytis* invades the flower of strawberries to cause gray mold rot, *Colletotrichum* invades the epidermis of bananas to initiate banana *anthracnose*, and *Gloeosporium* invades the lenticels of apples to initiate lenticel rot.[13] The largest number of market fruit and vegetable spoilage conditions occur after harvesting, and although the fungi most often invade bruised and damaged products, some enter specific areas. For example, *Thielaviopsis* invades the fruit stem of pineapples to cause black rot of this fruit, and *Colletotrichum* invades the crown cushion of bananas to cause banana crown rot.[13] Black rot of sweet potatoes is caused by *Ceratocystis*, neck rot of onions by *Botrytis allii*, and downey mildew of lettuce by *Bremia* spp.[9] Some of the spoilage conditions listed in Table 6–8 are discussed below.

Gray mold rot. This condition is caused by *Botrytis cinerea*, which produces a gray mycelium. This type of spoilage is favored by high humidity and warm temperatures. Among the vegetables affected are asparagus, onions, garlic, beans (green, lima, and wax), carrots, parsnips, celery, tomatoes, endives, globe artichokes, lettuce, rhubarb, cabbage, Brussels sprouts, cauliflower, broccoli, radish, rutabagas, turnips, cucumbers, pumpkin, squash, peppers, and sweet potatoes. In this disease, the causal fungus grows on decayed areas in the form of a prominent gray mold. It can enter fruits and vegetables through the unbroken skin or through cuts and cracks.

Sour rot (oospora rot, watery soft rot). This condition of vegetables is caused by *Geotrichum candidum* and other organisms. Among the vegetables affected are asparagus, onions, garlic, beans (green, lima, and wax), carrots, parsnips, parsley, endives, globe artichokes, lettuce, cabbage, Brussels sprouts, cauliflower, broccoli, radishes, rutabagas, turnips, and tomatoes. The causal fungus is widely distributed in soils and on decaying fruits and vegetables. *Drosophila melanogaster* (fruit fly) carries spores and mycelial fragments on its body from decaying fruits and vegetables to growth cracks and wounds in healthy fruits and vegetables. Because the fungus cannot enter through the unbroken skin, infections usually start in openings of one type or another.[27]

Rhizopus soft rot. This condition is caused by *Rhizopus stolonifer* and other species that make vegetables soft and mushy. Cottony growth of the mold with small black dots of sporangia often covers the vegetables. Among those affected are beans (green, lima, and wax), carrots, sweet potatoes, potatoes, cabbage, Brussels sprouts, cauliflower, broccoli, radish, rutabagas, turnips, cucumbers, cantaloupes, pumpkins, squash, watermelons, and tomatoes. This fungus is spread by *D. melanogaster*, which lays its eggs in the growth cracks on various fruits and vegetables. The fungus is widespread and is disseminated by other means also. Entry usually occurs through wounds and other skin breaks.

Phytophthora rot. This market condition, caused by *Phytophthora* spp., occurs largely in the field as a blight and fruit rot of market vegetables. It appears to be more variable than some other market "diseases" and affects different plants in different ways. Among the vegetables affected are asparagus, onions, garlic, cantaloupes, watermelons, tomatoes, eggplants, and peppers.

Anthracnose. This plant disease is characterized by spotting of leaves, fruit, or seed pods. It is caused by *Colletotrichum coccodes* and other species. These fungi are considered weak plant pathogens. They live from season to season on plant debris in the soil and on the seed of various plants such as the tomato. Their spread is favored by warm, wet, weather. Among the vegetables affected are beans, cucumbers, watermelons, pumpkins, squash, tomatoes, and peppers.

Blue mold rot is a post-harvest disease of apples and pears that is caused by *Penicillium expansum* while **gray mold** rot is caused by *Botrytis cinerea*. In an attempt to control these fungi on apples and pears, two biocontrol agents have been tested—*Candida sake* and *Pantoea agglomerans*.[32] When

Table 6–8 Common Fungal Fruit and Vegetable Spoilage Conditions, Etiologic Agents, and Typical Products Affected

Spoilage Condition	Etiologic Agent	Typical Products Affected
Alternaria rot	*A. tenuis*	Citrus fruits
Anthracnose (bitter rot)	*Colletotrichum musae*	Bananas
Anthracnose	*C. lindemuthianum*	Beans
	C. lagenarium	Watermelons
Black rot	*Aspergillus niger, Alternaria*	Onions, cabbage
Black rot	*Ceratocystis fimbriata*	Sweet potatoes
Blue mold rot	*Penicillium digitatum*	Citrus fruits
Brown rot	*Monilinia fructicola* (= *Sclerotinia fructicola*)	Peaches, cherries
Brown rot	*Phytophthora* spp.	Citrus fruits
Cladosporium rot	*C. herbarum*	Cherries, peaches
Crown rot	*Colletotrichum musae* (= *Gloeosporium musarum*), *Fusarium roseum*, *Verticillium theobromae*, *Ceratocystis paradoxa*	Bananas
Downy mildew	*Plasmapara viticola, Phytophthora* spp., *Bremia* spp.	Grapes
Dry rot	*Fusarium* spp.	Potatoes
Gray mold rot	*Botrytis cinerea*	Grapes, many others
Green mold rot	*Penicillium digitatum*	Citrus fruits
Lenticel rot	*Cryptosporiopsis malicorticis* (= *Gloeosporium perennans*), *Phylctaena vagabunda*	Apples, pears
Pineapple black rot	*Ceratocystis paradoxa* (= *Thielaviopsis paradoxa*)	Pineapples
Phytophthora rot	*Colletotrichum coccodes*	Vegetables
Pink mold rot	*Trichothecium roseum*	
Rhizopus soft rot	*Rhizopus stolonifer*	Sweet potatoes, tomatoes
Slimy brown rot	*Rhizoctonia* spp.	Vegetables
"Smut" (black mold rot)	*Aspergillus niger*	Peaches, apricots
Sour rot	*Geotrichum candidum*	Tomatoes, citrus fruits
Stem-end rot	*Phomopsis citri, Diplodia natalensis, Alternaria citri*	Citrus fruits
Watery soft rot	*Sclerotinia sclerotiorum*	Carrots
Wheat scab	*Fusarium graminearum*	Wheat, barley
Corn smut	*Ustilago maydia*	Corn
Rice blast	*Magnaporaathe grisea*	Rice
Potato blight	*Phytophthora infestans*	Potatoes
Blight	*cinnamomi*	Chestnut trees
Root rot	*sojae*	Soybeans
Sudden oak tree death	*ramorum*	oak trees

Figure 6–3   Onion diseases—(A) white rot; (B) black mold rot; (C) diplodia stain. *Source*: From *Agriculture Handbook 303*, USDA, 1966, Chapter 5.

applied to fruits in a 50:50 ratio at levels up to 2×10^7 and 8×10^7 cfu/ml and held at room temperature, no blue mold rot was observed, and gray mold lesion size was reduced by >95%.

Several spoilage conditions of onions are shown in Figure 6–3.

SPOILAGE OF FRUITS

The general composition of 18 common fruits is presented in Table 6–9, which shows that the average water content is about 85%, and the average carbohydrate content is about 13%. The fruits differ from vegetables in having somewhat less water but more carbohydrate. The mean protein, fat, and ash content of fruits are, respectively, 0.9%, 0.5%, and 0.5%—somewhat lower than vegetables except for ash content. Fruits contain vitamins and other organic compounds, just as vegetables do. On the basis of nutrient content, these products would appear to be capable of supporting the growth of bacteria, yeasts, and molds. However, the pH of fruits is below the level that generally favors bacterial growth. This one fact alone would seem to be sufficient to explain the general absence of bacteria in the incipient spoilage of fruits. The wider pH growth range of molds and yeasts suits them as spoilage agents of fruits. With the exception of pears, which sometimes undergo *Erwinia* rot, bacteria are of no known importance in the initiation of fruit spoilage. Just why pears with a reported pH range of 3.8 to 4.6 should undergo bacterial spoilage is not clear. It is conceivable that *Erwinia* and *Pectobacterium* spp. initiate their growth on the surface of this fruit where the pH is presumably higher than on the inside.

A variety of yeast genera can usually be found on fruits, and these organisms often bring about the spoilage of fruit products, especially in the field. Many yeasts are capable of attacking the sugars

Table 6–9 Common Fruits: Approximate Percentage Composition

Fruit	Water	Carbohydrate	Protein	Fat	Ash
Apples	84.1	14.9	0.3	0.3	0.4
Apricots	85.4	12.9	1.0	0.6	0.1
Bananas	74.8	23.0	1.2	0.8	0.2
Blackberries	84.8	12.5	1.2	0.5	1.0
Cherries, sweet and sour	83.0	14.8	1.1	0.6	0.5
Figs	78.0	19.6	1.4	0.6	0.4
Grapefruit	88.8	10.1	0.5	0.4	0.2
Grapes, American type	81.9	14.9	1.4	0.4	1.4
Lemons	89.3	8.7	0.9	0.5	0.6
Limes	86.0	12.3	0.8	0.8	0.1
Oranges	87.2	11.2	0.9	0.5	0.2
Peaches	86.9	12.0	0.5	0.5	0.1
Pears	82.7	15.8	0.7	0.4	0.4
Pineapples	85.3	13.7	0.4	0.4	0.2
Plums	85.7	12.9	0.7	0.5	0.2
Raspberries	80.6	15.7	1.5	0.6	1.6
Rhubarb	94.9	3.8	0.5	0.7	0.1
Strawberries	89.9	8.3	0.8	0.5	0.5
Mean	84.9	13.2	0.88	0.53	0.46

Source: Watt and Merrill.[60]

found in fruits and bringing about fermentation with the production of alcohol and carbon dioxide. Due to their generally faster growth rate than molds, they generally precede the latter organisms in the spoilage process of fruits in certain circumstances. It is not clear whether some molds are dependent on the initial action of yeasts in the process of fruit and vegetable spoilage. The utilization or destruction of the high-molecular-weight constituents of fruits is brought about more by molds than yeasts. Many molds are capable of utilizing alcohols as sources of energy, and when these and other simple compounds have been depleted, these organisms proceed to destroy the remaining parts of fruits, such as the structural polysaccharides and rinds. **Fire blight** of apple and pear trees is caused by *Erwinia amylovora*. One method that has been studied to control the fire blight organism involves the use of *Pantoea agglomerans* and *P. dispersa* as biological control agents since they do not harm apple or pear trees, yet they prevent the pathogen's invasion. In vitro studies have revealed the *P. agglomerans* inhibition to be an antibiotic complex designated pantocin A and pantocin B.[61] These inhibitors appear to be effective against other Gram-negative bacteria.

FRESH-CUT PRODUCE

The production of precut packaged fruit and vegetable salads (minimally processed) has led to an explosion in the sale and consumption of these commodities during the past decade, and this trend shows signs of continuing. In essence, salad vegetables such as lettuce and carrots, and fruits such as cantaloupes and watermelons are cut, sliced, and packaged in see-through containers that are stored at chill temperatures, such that they are ready-to-use (RTU) upon purchase. If packaged in high-oxygen permeable films, the primary concerns are product quality and enzymatic browning in the case of light-colored products. However, when low-O_2 permeable packaging is used with long-term storage, the possibility exists for the growth of microbial pathogens such as *C. botulinum* and *L. monocytogenes*. This concern has led to numerous studies on the safety of the final RTU produce, and some of these are summarized below. Since modified atmosphere/vacuum packaging is often used for these products, some relevant information can be found in Chapter 14. More extensive information can be found in reference 14.

Microbial Load

Overall, RTU (ready-to-use, ready-to-eat) produce is by no means microbe-free. In their preparation, intact vegetables are washed, typically with water that contains chlorine from 50 to 200 ppm, followed by cutting and packaging. While washing reduces microbial numbers, the cutting operation has the potential to recontaminate. Also, the freshcut vegetables provide a higher level of moisture, more simple nutrients, and a higher surface area, all of which make the RTU product more susceptible to microbial growth than the original.

The APCs of eight RTU vegetables in Ontario, Canada, recorded on day 0 and day 4 after storage at $4°C$ are presented in Table 6–10.[33] It can be seen that the initial numbers ranged from 4.82 $\log_{10}/g$ to near 6.0 $\log_{10}/g$ on day 0, but after a 4-day storage, they ranged from 5.45 to >7.0 $\log_{10}/g$. In an earlier study, the APC of RTU vegetables at harvest was around $10^5 - 10^8/g$, and after storage at $7°C$, the APC at time of sell-by date +1 day for 12 vegetables ranged between 7.7 and 9.0 $\log_{10}/g$, a time when all products were organoleptically acceptable.[10] In the first study,[33] coliforms ranged from 5.1 to 7.2 $\log_{10}/g$, but no type 1 *E. coli* strains were found. The most predominant organisms were *Pseudomonas* and *Pantoea*. In a study[4] of the types of organisms on RTU, spinach that was stored at $10°C$ for 12 days, mesophiles ranged between 10^7 and $10^{10}/g$, psychrotrophs and pseudomonads between 10^6 and

Table 6–10 Log_{10} Aerobic Plate Counts (per Gram) of RTU Vegetables Held at $4°C^a$

Product	Day 0	Day 4
Chopped lettuce	4.85	5.63
Salad mlx	5.35	6.05
Cauliflower florets	4.82	5.45
Sliced celery	5.67	6.59
Coleslaw mix	5.14	6.95
Carrot sticks	5.13	6.27
Broccoli florets	5.58	6.59
Green peppers	5.99	7.22

^{a}The products had a 7-day recommended shelf life.

Source: Data from Odumeru et al.[33]

10^{10}/g, and enteric bacteria between 10^4 and 10^7/g. The APC of some lettuce and fennel tested in Italy in the 1970s is presented in Chapter 20.

Seed Sprouts

These products are produced by the germination of certain plant seeds (e.g., alfalfa, radish, clover, mung beans). Their popularity in the United States has increased significantly during the past 20 years, and the current interest in their microbiology is the result of these products being the vehicle for a number of foodborne illness outbreaks

The production of sprouts in the home or on a small-scale is carried out by using about 25–30 g of germinatable seeds, which are placed in jars and soaked upon addition of 400–500 ml of tap water. Ideally, sterile or at least boiled water should be used. The initial soaking is done for at least 3 h followed by drainage of the soaking water and incubation of seeds in trays in the dark at ca. 25°C. The swollen and germinating seeds are rinsed daily until the sprouts are ready for collection or harvesting. On a commercial scale, large rotary drums are used that can hold 10 kg or more of seeds. These containers are automatically rotated at specified intervals with periodic additions of irrigation water. Sprouts may be harvested after 3 to 7 days.

Since germinating seeds are rich in simple nutrients (to nourish the young plant until it can make its own food by photosynthesis), it is not surprising that seed sprouts may contain high numbers of microorganisms, especially bacteria. In one study, the sprout bacterial biota was ca. 2–3 logs on day one but ca. 10^8 cfu/g on day two.[15] In an earlier study of the APC of mung beans, the average initial number of 10^6/g increased to 10^8 after 2 days of germination;[45] and in another study, the initial 10^4 cfu/g on mung bean seeds increased to 7.7×10^8 after two days of sprouting.[1] Numbers of seven to nine log_{10} cfu/g are not uncommon, and some investigators have found as many as 10^{11} cfu/g (see reference 26).

Among the bacterial genera found on alfalfa sprouts, *Pseudomonas* spp. were the most predominant followed by *Pantoea* and *Acinetobacter* spp. with one species each of the genera *Escherichia, Erwinia, Enterobacter,* and *Stenotrophomonas.*[26] The numbers of bacteria in spent irrigation water have been found to be reflective of numbers on sprouts.[15] In another study, 69% of the sprout biota were Gram-negative bacteria followed by 17% Gram-positive rods.[45] A number of molds have been found on mung seeds with 98% of 750 nondisinfected and only 1.8% of surface-disinfected seeds containing a number of mold genera including aflatoxigenic species although no aflatoxins were found.[1] On alfalfa seeds, only 21% of 500 contained yeasts and molds.

Since sprouts are eaten without being heated or cooked, problems arise when the sprout seed stock contains foodborne pathogens such as *Salmonella* or *E. coli* 0157:H7. Because of the excellent nutrient state of germinating seeds as noted above, the pathogens grow very well. In an effort to control pathogens, sprout seeds are treated in various ways to destroy these forms. The United States Food and Drug Administration (FDA) recommends the use of 20,000 ppm calcium hypochlorite for 15 min. to reduce pathogens, and this treatment does not significantly reduce germination efficiency or sprout length.

Pathogens

The pathogen of greatest concern in RTU vegetables is *C. botulinum* and reasons for this concern are pointed out by several studies. In one, five RTU vegetables (butternut squash, mixed salad, rutabagas, romaine lettuce, and a stir-fry mix) were inoculated with a 10-strain cocktail—five each of proteolytic and nonproteolytic spores.[3] The products were sealed in polystyrene trays with an oxygen transmission rate (OTR) (see Chapter 14) of 2,100 ml and incubated at 5, 10, or 25°C. All five vegetables became toxic at some point during their storage. The time to toxin detection for nonproteolytic strains in butternut squash was 7 days at 10°C with CO_2 at 27.8%; and for proteolytics in this product, 3 days at 25°C with 64.7% CO_2.[3] In butternut squash at 5°C with an inoculum of nonproteolytic strains of 10^3/g, toxin was detectable in 21 days. At the time of toxin detection in all samples, O_2 was <1%. Although the packaging material was by no means of "zero" barrier quality, respiration of the products decreased O_2 and increased CO_2 to the levels noted. It was the opinion of these investigators that the temperature of storage of RTU vegetables of the type noted was of critical importance to their safety. Most products were in states of detectable spoilage at the time of toxin detection.

Another study employed cabbage and lettuce inoculated with about 10^2 spores/g of a 10-strain cocktail as above and packaged in film with an OTR of either 3,000 low OTR (LOTR) or 7,000 high OTR (HOTR) and stored at 4, 13, or 21°C for 21 or 28 days.[17] Toxin was not detected under any conditions, and both vegetables were organoleptically spoiled before toxin could be produced. In the cabbage stored at 21°C for 10 days, the LOTR contained 69.4% and the HOTR 41.9% CO_2, while in lettuce at 21°C after 8 days, CO_2 was 41.9% and 9.0% in LOTR and HOTR, respectively. In contrast to the study by Austin et al.[3] where O_2 was <1%, both packaging materials allowed O_2 ranging from 1.0% to 7.9%. In the former study, the packaging material had an OTR of 2,100, while in the latter OTRs were 3,000 and 7,000. The more permeable film may have allowed the growth of other organisms that interfered with *C. botulinum*.

A 10-strain cocktail of 7 proteolytics and 3 nonproteolytics was used in the study by Larson et al.[24] in which five vegetables (broccoli, cabbage, carrots, lettuce, and green beans) were inoculated. Botulinal toxin was found in all grossly spoiled broccoli stored at 21°C, in one-half of those grossly spoiled at 12°C, and in one-third of the grossly spoiled lettuce stored at 21°C. No toxin was detected prior to spoilage, and no toxin was found in the other three vegetables. In contrast to the two studies noted above, the packaging material used in this study had varying OTRs ranging from 3,000 to 16,544, and the vegetables were sealed under vacuum with vacuum pulled. Interestingly, broccoli was packaged in material with OTRs of 13,013 to 16,544, while cabbage (which did not become toxic) was packaged in 3,000 to 8,000 OTR materials. The broccoli packs stored at 21°C for seven days contained <2% O_2 and about 12% CO_2, while lettuce at 21°C for six days contained up to 40% CO_2.[24] The APC of spoiled products was in the 10^8 to $> 10^9$ range.

In a fourth study, romaine lettuce and shredded cabbage were each inoculated with a nine strain cocktail of proteolytic and nonproteolytic spores at a level of about 100 spores per gram, and the samples were stored in vented and nonvented plastic bags.[35] The latter were vacuum packed but vacuum was not pulled. After seven days at 21°C, the cabbage packaged in nonvented bags became toxic, but not when stored at 4.4 or 12.7°C for up to 28 days. Romaine lettuce became toxic after 14 days at 21°C in nonvented packs and in 21 days in vented packs. The toxic samples were organoleptically spoiled prior to toxin detection.

A potential health hazard for RTU vegetables is pointed out by the above studies relative to botulinal toxin. However, these studies as well as others point to the importance of storage temperature in controlling not only this pathogen but others, including those below. Temperature and time of storage of RTU products are obviously critical to their safety. More on *C. botulinum* and other pathogens in vacuum/modified atmosphere packaged foods is presented in Chapter 14; and on *C. botulinum* in fresh foods in Table 4–7.

L. monocytogenes has been demonstrated to grow on refrigerated vegetables, including lettuce, broccoli, cauliflower, and asparagus. Although it grew on raw tomatoes at 21°C, it did not at 10°C.[7] Not only did this organism not grow on raw carrots, the numbers were actually reduced, with as little as 1% added to a broth base being effective (Figure 6–4). The antilisterial effect was destroyed when carrots were cooked.[8]

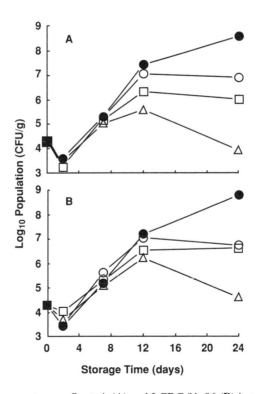

Figure 6–4 Growth of *L. monocytogenes* Scott A (**A**) and LCDC 81–86 (**B**) in tryptic phosphate broth (TPB) containing 0% (●), 1% (○), 10% (□), and 50% (□) (v/v) carrot juice in place of water. *Source*[8]: Reprinted with permission from L.R. Beuchat and R.E. Brackett, Inhibitory Effects of Raw Carrots on *Listeria Monocytogenes*, *Applied Environmental Microbiology*, vol. 56, p. 1741; copyright © 1990, American Society for Microbiology.

A study on the survival of *Shigella sonnei* in shredded cabbage revealed that numbers of this organism remained essentially unchanged for 1 to 3 days under three conditions of packaging—aerobic, vacuum, and in 30% N_2 + 70% CO_2.[40] After 3 days, however, numbers decreased concomitant with decreasing pH. Thus, the organism could survive under refrigerator or room temperature conditions, but it did not grow.

Internalization of Pathogens

A number of studies have demonstrated the capacity of certain foodborne pathogens to enter vegetable plants and their fruits from the time of seed germination or flowering, and several are summarized below. For a review, see Burnett and Beuchat.[11]

An increase in the numbers of salmonellae during the sprouting of alfalfa seeds has been demonstrated. In one study, the naturally contaminated seeds contained <1 organism/g by an MPN procedure and the numbers increased to $10^2 - 10^3$/g in one seed lot, and to $10^2 - 10^4$ in another lot during the germination process.[51] The maximum numbers developed after 48 h. When added to mung bean seeds, *S*. Anatum and *S*. Montevideo increased from ca. 10^2 to ca. 10^7 after 2 days of germination.[1]

From the time that *Salmonella* cells were inoculated on tomato plants during the flowering stage, they persisted through fruit ripening and led to 37% of 30 tomatoes from the inoculated plants being positive.[16] Surface and stem-scar samples contained the organisms at rates of 82 and 73% respectively. *S*. Montevideo was isolated 49 days after its inoculation, and *S*. Poona was present in 5 of 11 *Salmonella*-positive tomatoes.[16] In regard to *L. monocytogenes*, a study of 425 heads of cabbage harvested in 1999 from farms in the southern part of the state of Texas revealed that 20 or 4.7% contained this bacterium.[36] Six additional isolates were obtained from water and environmental samples from the same farms. The cabbage was not washed between harvesting and testing. The growth of *S*. Stanley on alfalfa seeds over a period of 342 h is depicted in Figure 6–5.[22]

To study the internalization of *E. coli* in apples, a strain was applied to top soil followed by the placing of three apple varieties on the contaminated soil to simulate their fall from trees. *E. coli* was found in the inner core and flesh samples of all three apple varieties when tested up to 10 days after

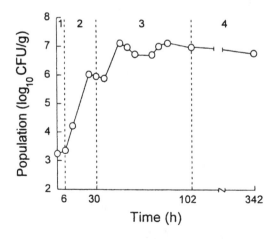

Figure 6–5 Growth of *S*. Stanley on alfalfa seeds during soaking,[1] germination,[2] sprouting,[3] and survival[4] during refrigerated storage.[22]

soil exposure.[41] From a study of 14 apple and pear orchards throughout the United States in 1999, coliforms were found in 74% of the fruits sampled and they were internalized in the cores of 40% of fruits tested.[38] *E. coli* 0157:H7 was not recovered from any fruit samples. No *E. coli* 0157:H7 or *Salmonella* spp. were found in 202 mushroom and 206 alfalfa sprout samples in the United States, but one sprout sample was positive for *L. monocytogenes*.[52] In another study, *E. coli* 0157:H7 containing a mutant of GFP (EGFP) which produced a longer wavelength than GFP was used as a marker to determine attachment to produce, and the results revealed preferential attachment to damaged tissues of green leaf lettuce and tomatoes, in contrast to intact tissue and stem surfaces.[53] The *E. coli* cells were tracked by confocal scanning laser microscopy. The internalization of a green fluorescent protein-containing strain of *E. coli* 0157:H7 from manure-contaminated soil and irrigation water in lettuce has been demonstrated. In this study, the investigators used confocal and epifluorescence microscopy to demonstrate the distribution of the bacterial cells throughout the root system of the edible parts of the lettuce.[44] With seed inocula of 10^4 cfu/ml of the pathogen in soil, lettuce seedlings were not invaded, but they were invaded by the pathogen at levels of 10^6 and 10^8 cfu/g.

A bioluminescent-labeled *E. coli* was assessed on growing spinach plants and the organism became internalized in root tissue and to a lesser extent within the hypocotyls.[58] When added to soil with seeds and cultivated for 42 days, the bacterium could be cultivated from roots and leaves. Under hydroponic conditions, the pathogen at a level of 10^2 or 10^3 cfu/ml of solution became internalized, and this process appeared to occur more under hydroponic conditions than in soils.[59] Using bioluminescent-labeled strains of *E. coli* and *S.* Montevideo, both became internalized on mung bean sprouts after the initial 24 h germinating period.[59] The use of 20,000 ppm sodium hypochlorite was not successful in freeing sprouts of the pathogens. The inner tissues and stomata of cotyledons of radish sprouts grown from seeds inoculated with *E. coli* 0157:H7 were internalized by this bacterium, which was not removed by the surface treatment of hypocotyl with $HgCl_2$.[21]

The incidence and prevalence of nontuberculous *Mycobacterium* spp. in a number of fresh produce and juice products in the United States were studied, and from a total of 121 products, 25 or 20.7% yielded seven different mycobacterial species.[2] The produce included mushrooms, sprouts, broccoli, lettuce, leeks, parsley, and apple juice. The most frequently isolated species was *M. avium*, which accounted for 12 of the 29 identified isolates. Three isolates each of *M. simian* and *gordonae*; and two of *M. flavescens* were identified. The natural occurrence of the so-called fast growing mycobacteria on fresh produce is not surprising since these organisms are part of the stable microbiota of farm soils. Due to their slow growth rate relative to that of the soil zymogenous biota, their internalization of produce or seed sprouts seems unlikely.

The internalization of *S.* Enteritidis in immature and ripened mangoes has been demonstrated. The immersion of these fruits in 21°C water containing the pathogen (which contained a green fluorescent protein) revealed that it was internalized in 80% of the immature and 87% of the ripened mangoes.[34] The stem-end segment was more affected than other mango parts, and the pathogen was detected in fruit pulp one week after incubation at temperatures of 10–30°C.

A list of pathogens isolated from vegetables has been produced by Beuchat,[6] as well as a synopsis of the model HACCP system for fresh-cut produce issued by the International Fresh-Cut Produce Association.

Disease Outbreaks

A thorough review of outbreaks in Canada for the years 1981–1999 revealed that nine produce products were responsible for 18 outbreaks. The produce products were alfalfa sprouts ($n = 5$), raspberries

($n = 4$); two each for cantaloupes and potato salad; and one each for coleslaw, lettuce, parsley, blackberries, and vegetable salad.[42] Seven outbreaks were caused by *Salmonella* spp. followed by *Cyclospora cayetanensis*. Two outbreaks each were caused by *E. coli* 0157:H7 and a calicivirus; and one each by *L. monocytogenes*, *Shigella sonnei*, and *Staphylococcus aureus*. The four protozoan outbreaks were traced to imported blackberries and raspberries, and each of the seven salmonellae outbreaks was caused by a different serovar.[42]

Among fruit juices that have served as vehicles of foodborne illness, apple juice is the most common (single outbreaks contaminated with salmonellae, *E. coli* 0157:H7, and *Cryptosporidium parvum*). Coconut milk has been the vehicle for a *Vibrio cholerae* outbreak.[11]

Because coleslaw has been incriminated in an outbreak of *E. coli* 0157:H7 food poisoning, a study was undertaken to determine the residence time of this pathogen in the product. Two coleslaw preparations of pH 4.53 and 4.5 were inoculated with 5.31 $\log_{10}$ cfu/g and held for 3 days at 4, 11, and 21°C.[62] No growth of the pathogen was detected at either temperature, rather there was a decrease of 0.4–0.5 $\log_{10}$ cfu/g in cell numbers at 21°C. The investigators speculated that the decrease may have been due, at least in part, to competition by the normal biota.[62]

Intestinal viruses are not uncommon on fresh produce, and they often originate from the wash water used. Among the most common are the noroviruses (see Chapter 31). Hepatitis A and E along with rotavirus and astrorvirus may be expected when polluted water is used as is the case with any gastrointestinal illness agent. More information on intestinal viruses can be found in Chapter 31, and an extensive review of viruses and fresh produce is that of Seymour and Appleton.[43]

For a review of infections from the consumption of seed sprouts, see Taormina et al.[54] and NACM.[31] The latter report was produced by the National Advisory Committee on Microbiological Criteria for Foods (in the United States), and it covers not only sprout-associated diseases, but many other aspects associated with these products from farm to table.

REFERENCES

1. Andrews, W.H., P.B. Mislivec, C.R. Wilson, V.R. Bruce, P.L. Poelma, R. Gibson, M.W. Trucksess, and K. Young. 1982. Microbial hazards associated with bean sprouting. *J. Assoc. Off. Anal. Chem.* 65:241–248.

2. Argueta, C., S. Yoder, A.E. Holtzman, T.W. Aronson, N. Glover, G.G.W. Berlin, G.N. Stelma Jr., S. Froman, and P. Tomasek. 2000. Isolation and identification of nontuberculous mycobacteria from foods as possible exposure sources. *J. Food Protect.* 63:930–933.

3. Austin, J.W., K.L. Doss, B. Blanchfield, and J.M. Farber. 1998. Growth and toxin production by *Clostridium botulinum* on inoculated fresh-cut packaged vegetables. *J. Food Protect.* 61:324–328.

4. Babic, I.,S. Roy, A.E. Watada, and W.P. Wergin. 1996. Changes in microbial populations on fresh cut spinach. *Int. J. Food Microbiol.* 31:107–119.

5. Bernard, R.J., A.P. Duran, A. Swartzentruber, A.H. Schwab, B.A. Wentz, and R.B. Read, Jr. 1982. Microbiological quality of frozen cauliflower, corn, and peas obtained at retail markets. *Appl. Environ. Microbiol.* 44:54–58.

6. Beuchat, L.R. 1996. Pathogenic microorganisms associated with fresh produce. *J. Food Protect.* 59:204–216.

7. Beuchat, L.R., and R.E. Brackett. 1991. Behavior of *Listeria monocytogenes* inoculated into raw tomatoes and processed tomato products. *Appl. Environ. Microbiol.* 57:1367–1371.

8. Beuchat, L.R., and R.E. Brackett. 1990. Inhibitory effects of raw carrots on *Listeria monocytogenes*. *Appl. Environ. Microbiol.* 56:1734–1742.

9. Brackett, R.E. 1987. [Fungal spoilage of] vegetables and related products. In *Food and Beverage Mycology*, 2nd ed., 129–154, ed. L.R. Beuchat. New York: Kluwer Academic Publishers.

10. Brocklehurst, T.F., C.M. Zaman-Wong, and B.M. Lund. 1987. A note on the microbiology of retail packs of prepared salad vegetables. *J. Appl. Microbiol.* 63:409–415.

11. Burnett, S.L., and L.R. Beuchat. 2000. Human pathogens associated with raw produce and unpasteurized juices, and difficulties in decontamination. *J. Ind. Microbiol. Biotechnol.* 25:281–287.

12. Doan, C.H., and P.M. Davidson. 2000. Microbiology of potatoes and potato products: A review. *J. Food Protect.* 63:668–683.

13. Eckert, J.W. 1979. Fungicidal and fungistatic agents: Control of pathogenic microorganisms on fresh fruits and vegetables after harvest. In *Food Mycology*, 164–199, ed. M.E. Rhodes. Boston: Hall.

14. Francis, G.A., C. Thomas, and D. O'Beirne. 1999. Review paper: The microbiological safety of minimally processed vegetables. *Int. J. Food Sci. Technol.* 34:1–22.

15. Fu, T., D. Stewart, K. Reineke, J. Ulaszek, J. Schlesser, and M. Tortorello. 2001. Use of spent irrigation water for microbiological analysis of alfalfa sprouts. *J. Food Protect.* 64:802–806.

16. Guo, X., J. Chen, R.E. Brackett, and L.R. Beuchat. 2001. Survival of salmonellae on and in tomato plants from the time of inoculation at flowering and early stages of fruit development through fruit ripening. *Appl. Environ. Microbiol.* 67:4760–4764.

17. Hao, Y.Y., R.E. Brackett, L.R. Beuchat, and M.P. Doyle. 1998. Microbiological quality and the inability of proteolytic *Clostridium botulinum* to produce toxin in film-packaged fresh-cut cabbage and lettuce. *J. Food Protect.* 61:1148–1153.

18. Hauben, L., E.R.B. Moore, L. Vauterin, M. Steenackers, J. Mergaert, L. Verdonck, and J. Swings. 1998. Phylogenetic position of phytopathogens within the Enterobacteriaceae. *System. Appl. Microbiol.* 21:384–397.

19. Hugouvieux-Cotte-Pattat, N., G. Condemine, W. Nasser, and S. Reverchon. 1996. Regulation of pectinolysis in *Erwinia chrysanthemi. Annu. Rev. Microbiol.* 50:213–257.

20. Insalata, N.F., J.S. Witzeman, J.H. Berman, and E. Berker. 1968. A study of the incidence of the spores of *Clostridium botulinum* in frozen vacuum pouch-pack vegetables. *Proc., 96th. Ann. Meet., Amer. Pub. Hlth. Assoc.*, 124.

21. Ito, Y., Y. Sugita-Konishi, F. Kasuga, M. Iwaki, Y. Hara-Kudo, N. Saito, Y. Noguchi, H. Konuma, and S. Kumagai. 1998. Enterohemorrhagic *Escherichia coli* 0157:H7 present in radish sprouts. *Appl. Environ. Microbiol.* 64:1532–1535.

22. Jacquette, C.B., L.R. Beuchat, and B.E. Mahon. 1996. Efficacy of chlorine and heat treatment in killing *Salmonella* Stanley inoculated onto alfalfa seeds and growth and survival of the pathogen during sprouting and storage. *Appl. Environ. Microbiol.* 62:2212–2215.

23. Jinneman, K.C., P.A. Trost, W.E. Hill, S.D. Weagant, J.L. Bryant, C.A. Kaysner, and M.M. Wekell. 1995. Comparison of template-preparation methods from foods for amplification of *Escherichia coli* 0157 Shiga-like toxins type I and II DNA by multiplex polymerase chain reaction. *J. Food Protect.* 58:722–726.

24. Larson, A.E., E.A. Johnson, C.R. Barmore, and M.D. Hughes. 1997. Evaluation of the botulism hazard from vegetables in modified atmosphere packaging. *J. Food Protect.* 60:1208–1214.

25. Little, E.L., R.M. Bostock, and B.C. Kirkpatrick. 1998. Genetic characterization of *Pseudomonas syringae* pv. *syringae* strains from some fruits in California. *Appl. Environ. Microbiol.* 64:3818–3823.

26. Matsos, A., J.L. Garland, and W.F. Fett. 2002. Composition and physiological profiling of sprout-associated microbial communities. *J. Food Protect.* 65:1903–1908.

27. McColloch, L.P., H.T. Cook, and W.R. Wright. 1968. Market diseases of tomatoes, peppers, and eggplants. *Agricultural Handbook No. 28.* Washington, D.C.: Agricultural Research Service.

28. Mount, M.S., D.F. Bateman, and H.G. Basham. 1970. Induction of electrolyte loss, tissue maceration, and cellular death of potato tissue by an endopolygalacturonate trans-eliminase. *Phytopathology* 60:924–1000.

29. Mundt, J.O., W.F. Graham, and I.E. McCarty. 1967. Spherical lactic acid-producing bacteria of southern-grown raw and processed vegetables. *Appl. Microbiol.* 15:1303–1308.

30. Nascimento, M.S., N. Silva, M.P.L. Catanozi, and K.C. Silva. 2003. Effects of different disinfection treatments on the natural microbiota of lettuce. *J. Food Protect.* 66:1697–1700.

31. National Advisory Committee on Microbiological Criteria for Foods. 1999. Microbiological safety evaluations and recommendations on sprouted seeds. *Int. J. Food Microbiol.* 52:123–153.

32. Nunes, C., J. Usall, N. Teixidó, R. Torres, and I. Viñas. 2002. Control of *Penicillium expansum* and *Botrytis cinerea* on apples and pears with the combination of *Candida sake* and *Pantoea agglomerans. J. Food Protect.* 65:178–184.

33. Odumeru, J.A., S.J. Mitchell, D.M. Alves, J.A. Lynch, A.J. Yee, S.L. Wang, S. Styliadis, and J. Farber. 1997. Assessment of the microbiological quality of ready-to-eat vegetables for health-care food services. *J. Food Protect.* 60:954–960.

34. Penteado, A.L., B.S. Eblen, and A.J. Miller. 2004. Evidence of *Salmonella* internalization into fresh mangos during simulated postharvest insect disinfestation procedures. *J. Food Protect.* 67:181–184.

35. Petran, R.L., W.H. Sperber, and A.B. Davis. 1995. *Clostridium botulinum* toxin formation in romaine lettuce and shredded cabbage: Effect of storage and packaging conditions. *J. Food Protect.* 58:624–627.

36. Prazak, A.M., E.A. Murano, I. Mercado, and G.R. Acuff. 2002. Prevalence of *Listeria monocytogenes* during production and postharvest processing of cabbage. *J. Food Protect.* 65:1728–1734.

37. Rafil, F., M.A. Holland, W.E. Hill, and C.E. Cerniglia. 1995. Survival of *Shigella flexneri* on vegetables and detection by polymerase chain reaction. *J. Food Protect.* 58:727–732.

38. Riordan, D.C.R., G.M. Sapers, T.R. Hankinson, M. Magee, A.M. Mattrazzo, and B.A. Annous. 2001. A study of U.S. orchards to identify potential sources of *Escherichia coli* 0157:H7. *J. Food Protect.* 64:1320–1327.

39. Roberts, D.P., P.M. Berman, C. Allen, V.K. Stromberg, G.H. Lacy, and M.S. Mound. 1986. Requirement for two or more *Erwinia carotovora* subsp. *carotovora* pectolytic gene products for maceration of potato tuber tissue by *Escherichia coli*. *J. Bacteriol.* 167:279–284.

40. Satchell, F.B., P. Stephenson, W.H. Andrews, L. Estela, and G. Allen. 1990. The survival of *Shigella sonnei* in shredded cabbage. *J. Food Protect.* 53:558–562.

41. Seeman, B.K., S.S. Sumner, R. Marini, and K.E. Kniel. 2002. Internalization of *Escherichia coli* in apples under natural conditions. *Dairy Fd. Environ. Sanit.* 22:667–673.

42. Sewell, A.M., and J.M. Farber. 2001. Foodborne outbreaks in Canada linked to produce. *J. Food Protect.* 64:1863–1877.

43. Seymour, I.J., and H. Appleton. 2001. Foodborne viruses and fresh produce. *J. Appl Microbiol.* 91:759–773.

44. Solomon, E.B., S. Yaron, and K.R. Matthews. 2002. Transmission of *Escherichia coli* 0157:H7 from contaminated manure and irrigation water to lettuce plant tissue and its subsequent internalization. *Appl. Environ. Microbiol.* 68:397–400.

45. Splittstoesser, D.F., D.T. Queale, and B.W. Andaloro. 1983. The microbiology of vegetable sprouts during commercial production. *J. Food Safety* 5:79–86.

46. Splittstoesser, D.F. 1973. The microbiology of frozen vegetables: How they get contaminated and which organisms predominate. *Food Technol.* 27:54–56.

47. Splittstoesser, D.F., and D.A. Corlett, Jr. 1980. Aerobic plate counts of frozen blanched vegetables processed in the United States. *J. Food Protect.* 43:717–719.

48. Splittstoesser, D.F., G.E.R. Hervey II, and W.P. Wettergreen. 1965. Contamination of frozen vegetables by coagulase-positive staphylococci. *J. Milk Food Technol.* 28:149–151.

49. Splittstoesser, D.F., D.T. Queale, J.L. Bowers, and M. Wilkison. 1980. Coliform content of frozen blanched vegetables packed in the United States. *J. Fd. Safety* 2:1–11.

50. Splittstoesser, D.F., W.P. Wettergreen, and C.S. Pederson. 1961. Control of microorganisms during preparation of vegetables for freezing. I. Green beans. *Food Technol.* 15:329–331.

51. Stewart, D.S., K.F. Reineke, J.M. Ulaszek, and M.L. Tortorello. 2001. Growth of *Salmonella* during sprouting of alfalfa seeds associated with salmonellosis outbreaks. *J. Food Protect.* 64:618–622.

52. Strapp, C.M., A.E.H. Shearer, and R.D. Joerger. 2003. Survey of retail alfalfa sprouts and mushrooms for the presence of *Escherichia coli* 0157:H7, *Salmonella*, and *Listeria* with BAX, and evaluation of this polymerase chain reaction-based system with experimentally contaminated samples. *J. Food Protect.* 66:182–187.

53. Takeuchi, K., and J.F. Frank. 2001. Expression of red-shifted green fluorescent protein by *Escherichia coli* 0157:H7 as a marker for the detection of cells on fresh produce. *J. Food Protect.* 64:298–304.

54. Taormina, P.J., L.R. Beuchat, and L. Slutsker. 1999. Infections associated with eating seed sprouts: An international concern. *Emerg. Inf. Dis.* 5:626–634.

55. Thunberg, R.L., T.T. Tran, R.W. Bennett, R.N. Matthews, and N. Belay. 2002. Microbial evaluation of selected fresh produce obtained in retail markets. *J. Food Protect.* 65:677–682.

56. Vauterin, L., B. Hoste, K. Kersters, and J. Swings. 1995. Reclassification of *Xanthomonas*. *Int. J. Syst. Bacteriol.* 45:472–489.

57. Vescova, M., C. Orsi, G. Scolari, and S. Torriani. 1995. Inhibitory effect of selected lactic acid bacteria on microflora associated with ready-to-eat vegetables. *Lett. Appl. Microbiol.* 21:121–125.

58. Warriner, K., F. Ibrahim, M. Dickinson, C. Wright, and W.M. Waites. 2003a. Interaction of *Escherichia coli* with growing salad spinach plants. *J. Food Protect.* 66:1790–1797.

59. Warriner, K., S. Spaniolas, M. Dickinson, C. Wright, and W.M. Waites. 2003b. Internalization of bioluminescent *Escherichia coli* and *Salmonella* Montevideo in growing bean sprouts. *J. Appl. Microbiol.* 95:719–727.

60. Watt, B.K., and A.L. Merrill. 1950. Composition of foods—Raw, processed, prepared. *Agric. Handbook No. 8*. Washington, D.C.: U.S. Department of Agriculture.

61. Wright, S.A., C.H. Zumoff, L. Schneider, and S.V. Beer. 2001. *Pantoea agglomerans* strain EH318 produces two antibiotics that inhibit *Erwinia amylovora* in vitro. *Appl. Environ. Microbiol.* 67:284–292.

62. Wu, F.M., L.R. Beuchat, M.P. Doyle, V. Garrett, J.G. Wells, and B. Swaminathan. 2002. Fate of *Escherichia coli* 0157:H7 in coleslaw during storage. *J. Food Protect.* 65:845–847.

63. Yap, M.N., J.D. Barak, and A.O. Charkowski. 2004. Genomic diversity of *Erwinia carotovora* subsp. *carotovora* and its correlation with virulence. *Appl. Environ. Microbiol.* 70:3013–3023.

CHAPTER 7

Milk, Fermentation, and Fermented and Nonfermented Dairy Products

Although this chapter is devoted to milk and other dairy products, it begins with a discussion of fermentation because of the importance of this process to dairy products.

FERMENTATION

Background

Numerous food products owe their production and characteristics to the fermentative activities of microorganisms. Many foods such as ripened cheeses, pickles, sauerkraut, and fermented sausages are preserved products in that their shelf life is extended considerably over that of the raw materials from which they are made. In addition to being made more shelf stable, all fermented foods have aroma and flavor characteristics that result directly or indirectly from the fermenting organisms. In some instances, the vitamin content of the fermented food is increased along with an increased digestibility of the raw materials. The fermentation process reduces the toxicity of some foods (for example, gari and peujeum), whereas others may become extremely toxic during fermentation (as in the case of bongkrek). From all indications, no other single group or category of foods or food products is as important as these are and have been relative to nutritional well-being throughout the world.

The microbial ecology of food and related fermentations has been studied for many years in the case of ripened cheeses, sauerkraut, wines, and so on, and the activities of the fermenting organisms are dependent on the intrinsic and extrinsic parameters of growth discussed in Chapter 3. For example, when the natural raw materials are acidic and contain free sugars, yeasts grow readily, and the alcohol they produce restricts the activities of most other naturally contaminating organisms. If, on the other hand, the acidity of a plant product permits good bacterial growth and at the same time the product is high in simple sugars, lactic acid bacteria may be expected to grow, and the addition of low levels of NaCl will ensure their growth preferential to yeasts (as in sauerkraut fermentation).

Products that contain polysaccharides but no significant levels of simple sugars are normally stable to the activities of yeasts and lactic acid bacteria due to the lack of amylase in most of these organisms. To effect fermentation, an exogenous source of saccharifying enzymes must be supplied. The use of barley malt in the brewing and distilling industries is an example of this. The fermentation of sugars to ethanol that results from malting is then carried out by yeasts. The use of *koji* in the fermentation of soybean products is another example of the way in which alcoholic and lactic acid fermentations may be carried out on products that have low levels of sugars but high levels of starches and proteins. Whereas the saccharifying enzymes of barley malt arise from germinating barley, the enzymes of koji are produced by *Aspergillus oryzae* growing on soaked or steamed rice or other cereals (the commercial product takadiastase is prepared by growing *A. oryzae* on wheat bran). The koji hydrolysates may be fermented by lactic acid bacteria and yeasts, as is the case for soy sauce, or the koji enzymes may act directly on soybeans in the production of products such as Japanese miso.

Defined and Characterized

The word *fermentation* has had many shades of meaning in the past. According to one dictionary definition, it is "a process of chemical change with effervescence . . . a state of agitation or unrest . . . any of various transformations of organic substances." The word came into use before Pasteur's studies on wines. Prescott and Dunn[56] and Doelle[12] have discussed the history of the concept of fermentation, and the former authors note that in the broad sense in which the term is commonly used, it is "a process in which chemical changes are brought about in an organic substrate through the action of enzymes elaborated by microorganisms." It is in this broad context that the term is used in this chapter. In the brewing industry, a *top fermentation* refers to the use of a yeast strain that carries out its activity at the upper parts of a large vat, such as in the production of ale; a *bottom fermentation* requires the use of a yeast strain that will act in lower parts of the vat, such as in the production of lager beer.

Biochemically, fermentation is the metabolic process in which carbohydrates and related compounds are partially oxidized with the release of energy in the absence of any external electron acceptors. The final electron acceptors are organic compounds produced directly from the breakdown of the carbohydrates. Consequently, incomplete oxidation of the parent compound occurs, and only a small amount of energy is released during the process. The products of fermentation consist of some organic compounds that are more reduced than others.

The Lactic Acid Bacteria

This group is composed of 13 genera of Gram-positive bacteria at this time:

Carnobacterium	*Oenococcus*
Enterococcus	*Pediococcus*
Lactococcus	*Paralactobacillus*
Lactobacillus	*Streptococcus*
Lactosphaera	*Tetragenococcus*
Leuconostoc	*Vagococcus*
	Weissella

With the enterococci and lactococci having been removed from the genus *Streptococcus*, the member of this genus of most importance in foods is *S. salivarius* subsp. *thermophilus*. *S. diacetilactis* has been reclassified as a citrate-utilizing strain of *Lactococcus lactis* subsp. lactis.

Related to the lactic acid bacteria but not considered to fit the group are genera such as *Aerococcus*, *Microbacterium*, and *Propionibacterium*, among others. The last genus has been reduced by the transfer of some of its species to the new genus *Propioniferax*, which produces propionic acid as its principal carboxylic acid from glucose.[80]

The history of our knowledge of the lactic streptococci and their ecology has been reviewed by Sandine et al.[63] These authors believe that plant matter is the natural habitat of this group, but they note the lack of proof of a plant origin for *Lactococcus cremoris*. It has been suggested that plant streptococci may be the ancestral pool from which other species and strains developed.[47]

Although the lactic acid group is loosely defined with no precise boundaries, all members share the property of producing lactic acid from hexoses. As fermenting organisms, they lack functional heme-linked electron transport systems or cytochromes, and they obtain their energy by substrate-level phosphorylation while oxidizing carbohydrates; they do not have a functional Krebs cycle.

Kluyver divided the lactic acid bacteria into two groups based on end products of glucose metabolism. Those that produce lactic acid as the major or sole product of glucose fermentation are designated *homofermentative* (Figure 7–1(A)). The homolactics are able to extract about twice as much energy from a given quantity of glucose as are the heterolactics. The homofermentative pattern is observed when glucose is metabolized but not necessarily when pentoses are metabolized, for some homolactics produce acetic and lactic acids when utilizing pentoses. Also the homofermentative

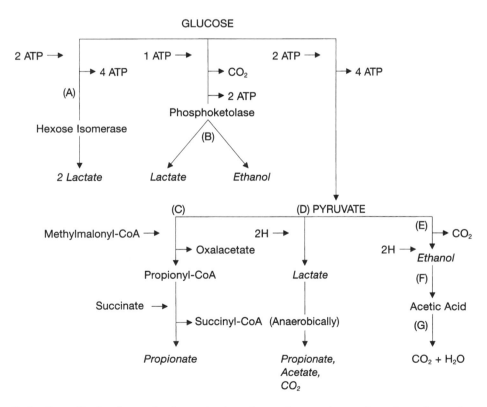

Figure 7–1 Generalized pathways for the production of some fermentation products from glucose by various organisms. (A) Homofermentative lactics; (B) heterofermentative lactics; (C) and (D) *Propionibacterium* (see Figure 7–3); (E) *Saccharomyces* spp.; (F) *Acetobacter* spp.; and (G) *Acetobacter* "overoxidizers."

Table 7–1 Some of the Homo- and Heterofermentative Lactic
Acid Bacteria

Homofermentative	Heterofermentative
Lactobacillus	*Lactobacillus*
L. acetotolerans	L. brevis
L. acidipiscis	L. buchneri
L. acidophilus	L. cellobiosus
L. alimentarius	L. coprophilus
L. casei	L. fermentum
	L. hilgardii
L. coryniformis	L. sanfranciscensis
L. curvatus	L. trichoides
subsp. curvatus	L. pontis
subsp. melibiosus	L. fructivorans
L. delbrueckii	L. kimchii
subsp. bulgaricus	L. paralimentarius
subsp. delbrueckii	L. panis
subsp. lactis	L. sakei
L. fuchuensis	subsp. sakei
L. helveticus	subsp. carnosus
L. jugurti	*Leuconostoc*
L. jensenii	L. argentinum
L. kefiranofaciens	L. citreus
subsp. kefiranofaciens	L. fallax
subsp. kefirgranum	L. carnosum
L. leichmannii	L. gelidum
L. mindensis	L. inhae
L. plantarum	L. kimchii
L. salivarius	L. lactis
Lactococcus	L. mesenteroides
L. lactis	subsp. cremoris
subsp. lactis	subsp. dextranicum
subsp. cremoris	subsp. mesenteroides
subsp. diacetylactis	*Carnobacterium*
subsp. hordniae	C. divergens
L. garvieae	C. gallinarum
L. plantarum	C. mobile
L. raffinolactis	C. piscicola
Paralactobacillus	C. viridans
P. selangorensis	*Oenococcus*
Pediococcus	O. oeni
P. acidilactici	*Weissella*
P. claussenii	W. cibaria
P. pentosaceus	W. confusa
P. damnosus	W. hellenica
P. dextrinicus	W. halotolerans
P. inopinatus	W. kandleri
P. parvulus	W. kimchii

(continued)

Table 7–1   (continued)

Streptococcus	W. minor
S. bovis	W. thialandensis
S. salivarius	W. paramesenteroides
subsp. salivarius	W. viridescens
subsp. thermophilus	W. koreensis
Tetragenococcus	
T. halophilus	
T. muriaticus	
Vagococcus	
V. fluvialis	
V. salmoninarum	

character of homolactics may be shifted for some strains by altering growth conditions such as glucose concentration, pH, and nutrient limitation.[8,42]

Those lactics that produce equal molar amounts of lactate, carbon dioxide, and ethanol from hexoses are designated *heterofermentative* (Figure 7–1(B)). All members of the genera *Pediococcus, Streptococcus, Lactococcus,* and *Vagococcus* are homofermenters, along with some of the lactobacilli. Heterofermenters consist of *Leuconostoc, Oenococcus, Weissella, Carnobacterium, Lactosphaera,* and some lactobacilli (Table 7–1). The heterolactics are more important than the homolactics in producing flavor and aroma components such as acetylaldehyde and diacetyl (Figure 7–2).

The genus *Lactobacillus was* subdivided historically into three subgenera: *Betabacterium, Streptobacterium,* and *Thermobacterium.* All of the heterolactic lactobacilli in Table 7–1 are betabacteria. The streptobacteria (for example, *L. casei* and. *plantarum*) produce up to 1.5% lactic acid with an optimal growth temperature of 30°C, whereas the thermobacteria (such as *L. acidophilus* and *L. delbrueckii* subsp. *bulgaricus*) can produce up to 3% lactic acid and have an optimal temperature of 40°C.[43]

More recently, the genus *Lactobacillus* has been arranged into three groups based primarily on fermentative features.[70] Group 1 includes *obligate homofermentative* species (*L. acidophilus, L. delbrueckii* subsp. *bulgaricus,* etc.). These are the thermobacteria, and they do not ferment pentoses. Group 2 consists of *facultative heterofermentative* species (*L. casei, L. plantarum, L. sakei;* etc.). Members of this group ferment pentoses. Group 3 consists of the obligate heterofermentative species, and it includes *L. fermentum, L. brevis, L. reuteri, L. sanfranciscensis,* and others. They produce CO_2 from glucose. The lactobacilli can produce a pH of 4.0 in foods that contain a fermentable carbohydrate, and they can grow up to a pH of about 7.1.[70]

In terms of their growth requirements, the lactic acid bacteria require preformed amino acids, B vitamins, and purine and pyrimidine bases—hence their use in microbiological assays for these compounds. Although they are mesophilic, some can grow below 5°C and some as high as 45°C. With respect to growth pH, some can grow as low as 3.2, some as high as 9.6, and most grow in the pH range 4.0–4.5. The lactic acid bacteria are only weakly proteolytic and lipolytic.[69]

The cell mucopeptides of lactics and other bacteria have been reviewed by Schleifer and Kandler.[64] Although there appear to be wide variations within most of the lactic acid genera, the homofermentative lactobacilli of the subgenus *Thermobacterium* appear to be the most homogeneous in this regard in having L-lysine in the peptidoglycan peptide chain and D-aspartic acid as the interbridge peptide. The lactococci have similar wall mucopeptides.

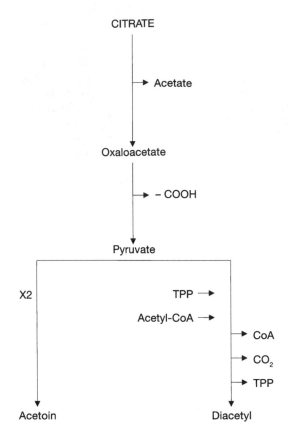

Figure 7–2   The general pathway by which acetoin and diacetyl are produced from citrate by group N lactococci and *Leuconostoc* spp. Pyruvate may be produced from lactate, and acetyl coenzyme A (CoA) from acetate.

Molecular genetics have been employed by McKay and co-workers to stabilize lactose fermentation by *L. lactis*. The genes responsible for lactose fermentation by some lactic cocci are plasmidborne, and loss of the plasmid results in the loss of lactose fermentation. In an effort to make lactose fermentation more stable, *lac*[+] genes from *L. lactis* were cloned into a cloning vector, which was incorporated into a *Streptococcus sanguis* strain.[28] Thus, the *lac* genes from *L. lactis* were transformed into *S. sanguis* via a vector plasmid, or transformation could be effected by use of appropriate fragments of DNA through which the genes were integrated into the chromosome of the host cells.[29] In the latter state, lactose fermentation would be a more stable property than when the *lac* genes are plasmidborne.

Metabolic Pathways and Molar Growth Yields

The end-product differences between homo- and heterofermenters when glucose is attacked are a result of basic genetic and physiological differences (Figure 7–1). The homolactics possess the enzymes aldolase and hexose isomerase but lack phosphoketolase (Figure 7–1(A)). They use the Embden–Meyerhof–Parnas (EMP) pathway toward their production of two lactates/glucose molecule. The heterolactics, on the other hand, have phosphoketolase but do not possess aldolase and hexose iso-merase, and instead of the EMP pathway for glucose degradation, these organisms use the hexose monophosphate or pentose pathway (Figure 7–1(B)).

The measurement of molar growth yields provides information on fermenting organisms relative to their fermentation substrates and pathways. By this concept, the microgram dry weight of cells produced per micromole of substrate fermented is determined as the *molar yield constant*, indicated by Y. It is tacitly assumed that essentially none of the substrate carbon is used for cell biosynthesis, that oxygen does not serve as an electron or hydrogen acceptor, and that all of the energy derived from the metabolism of the substrate is coupled to cell biosynthesis.[25] When the substrate is glucose, for example, the molar yield constant for glucose, Y_G, is determined by

$$Y_G = \frac{\text{g dry weight of cells}}{\text{moles glucose fermented}}$$

If the adenosine triphosphate (ATP) yield or moles of ATP produced per mole of substrate used is known for a given substrate, the amount of dry weight of cells produced per mole of ATP formed can be determined by

$$Y_{ATP} = \frac{\text{g dry weight of cells/moles ATP formed}}{\text{moles substrate fermented}}$$

A large number of fermenting organisms has been examined during growth and found to have $Y_{ATP} = 10.5$ or close thereto. This value is assumed to be a constant, so that an organism that ferments glucose by the EMP pathway to produce 2 ATP/mole of glucose fermented should have $Y_G = 21$ (i.e., it should produce 21 g of cells dry weight/mole of glucose). This has been verified for *E. faecalis*, *Saccharomyces cerevisiae*, *Saccharomyces rosei*, and *L. plantarum* on glucose (all $Y_G = 21$, $Y_{ATP} = 10.5$, within experimental error). A study by Brown and Collins[8] indicates that Y_G and Y_{ATP} values for *Lactococcus lactis* subsp. *lactis* biovar *diacetylactis* and *Lactococcus lactis* subsp. *cremoris* differ when cells are grown aerobically on a partially defined medium with low and higher levels of glucose, and further when grown on a complex medium. On a partially defined medium with low glucose levels (1–7 μmol/ml), values for *L. lactis* subsp. *lactis* biovar *diacetylactis* were $Y_G = 35.3$ and $Y_{ATP} = 15.6$, whereas for *L. lactis* subsp. *cremoris*, $Y_G = 31.4$ and $Y_{ATP} = 13.9$. On the same medium with higher glucose levels (1–15 μmol/ml), Y_G for *L. lactis* subsp. *lactis* biovar *diacetylactis* was 21, Y_{ATP} values for these two organisms on the complex medium with glucose 2 μmol/ml were 21.5 and 18.9 for *L. lactis* subsp. *lactis* biovar *diacetylactis* and *L. lactis* subsp. *cremoris*, respectively. Anaerobic molar growth yields for enterococcal species on low levels of glucose have been studied by Johnson and Collins.[36] *Zymomonas mobilis* utilizes the Entner–Doudoroff pathway to produce only 1 ATP/mole of glucose fermented ($Y_G = 8.3$, $Y_{ATP} = 8.3$). If and when the produced lactate is metabolized further, the molar growth yield would be higher. *Bifidobacterium bifidum* produces 2.5–3 ATP/mole of glucose fermented resulting in $Y_G = $ and $Y_{ATP} = 13$.[71]

ACETIC ACID BACTERIA

These Gram-negative bacteria belong to the family Acetobacteriaceae, and to the alpha-subclass of *Proteobacteria*. The recognized genera are: *Acetobacter*, *Asaia*, *Acidomonas*, *Gluconobacter*, *Gluconacetobacter*, and *Kozakia*.[79] With the exception of *Asaia*, they produce large quantities of acetic acid from ethanol, and can grow in the presence of 0.35% acetic acid. The metabolic pathway employed by the acetic acid producing strains is shown in Figure 7–1(F) and (G). *Asaia*, on the other hand, produces little or no acetic acid from ethanol, and its species do not grow in the presence of 0.35% acetic acid.[79] The three recognized species oxidize acetate and lactate to CO_2 and water.

DAIRY PRODUCTS

Milk

Milk is used throughout the world as a human food in at least one form, and from at least one of a number of different mammals. Bovine milk is typical of other milk types and it is the basis of the discussion that follows. Many of the aspects of milk microbiology not covered below have been presented or reviewed by Frank[17] and Murphy and Boor.[50]

Composition

From the general chemical composition of cow's milk in Table 7–2, some differences between this product and red meats in Table 4–9 are readily evident. The protein content of milk is considerably lower (3.5 vs. 18.0%) while the carbohydrate content is considerably higher (14.9 vs. ca. 1.0%). The higher structural protein content of red meats enables these products to exist as solids. Although the average water content near the surface of fresh meats of ca. 75.5% is lower than the average of 87% for milk, the a_w of both products is near 1.0. The milk of goats and sheep is similar in composition to that of cows.

Milk protein consists mainly of casein, and it exists in several classes: α, β, etc. If milk pH falls below 4.6, the casein precipitates. Although casein represents 80–85% of total milk protein, when precipitation occurs, the liquid portion is referred to as *whey*. The remaining proteins are found in whey and they include serum albumin, immunoglobulins, α-lactalbumin, etc. Milk carbohydrate is principally lactose and its content is fairly consistent among breeds of milk cows at around 5.0%. Although lactose is the main sugar, smaller quantities of glucose and citric acid exist. The fat content varies between ca. 3.5 and 5.0% depending upon cattle breed, and it consists mainly of triglycerides composed of C_{14}, C_{16}, C_{18}, and $C_{18:1}$ fatty acids. Smaller quantities of diglycerides and phospholipids occur. Milk lipids exist largely in the form of fat globules that are surrounded by a phospholipid layer. The ash content of around 0.7% consists of a relatively high level of Ca^{2+} and a lower level of Fe^{2+}. Overall, the nonfat solids in cow's milk average ca. 9.0% while total solids range between 12.5 and 14.5%, and average ca. 12.9% depending upon breed.

Table 7–2 Average Chemical Composition (%) of Whole Bovine Milk (Summarized from the Literature)

Water	87.0
Protein	3.5
Fat	3.9
Carbohydrate	4.9
Ash	0.7

The pH of fresh whole milk is around 6.6 but it may reach ca. 6.8 from a cow that has mastitis. Mastitis is an infection of the udder that is most often caused by *Streptococcus agalactiae* and *S. uberis* but sometimes by *Staphylococcus aureus* or *Streptococcus dysgalactiae*. Fresh milk from a mastitic cow typically contains leucocytes (white blood cells) $>10^6$/ml in contrast to nonmastitic milk that contains leucocytes around 70,000/ml.

Milk contains a very adequate supply of B vitamins with pantothenic acid and riboflavin being the two most abundant. Vitamins A and D are added for human consumption, and their presence has no known effect on the activity of microorganisms.

Overall, the chemical composition of whole cow's milk makes it an ideal growth medium for heterotrophic microorganisms, including the nutritionally fastidious Gram-positive lactic acid bacteria. How the milk microbiota utilize these constituents and bring about its spoilage is covered below under spoilage.

Processing

Milk is processed in a number of ways to produce a variety of products such as cream, cheese, and butter. Whole fresh milk is processed to produce a number of fluid products. Skim milk (0.5% fat) or reduced fat milk (up to 2.0% fat) is produced by high-speed centrifugation following heating to ca. 100°F to remove butter fat as cream, or by use of skim milk to which the desired fat content is added. The latter is pasteurized either at 150–155°F (65.5–68.3°C) for 30 minutes or at 166–175°F (74.4–79.4°C) for 15 sec prior to cooling to around 4°C.[78]

Evaporated milk is produced by the removal of about 60% water from whole milk which results in the lactose content being about 11.5%. *Sweetened condensed* milk is produced by the addition of sucrose or glucose before evaporation. This leads to a product with a sugar content of about 54% or $>64\%$ in solution.

In the United States, grade A raw milk that is to be pasteurized should not have an APC that exceeds 300,000 cfu/ml for commingled or blended milk, or should not exceed 100,000/ml for milk from an individual producer. After pasteurization, the APC should not exceed 20,000 cfu/ml, and the coliform count should not exceed 10/ml.[15] Raw milk should not be held longer than 5 days at 40°F (4.4°C) prior to pasteurization.

Chocolate milk is processed at a slightly higher temperature than unflavored milk (75°C for 15 sec rather than 72°C). A study of chocolate milk from four plants revealed that the APC was higher at 14 days post-processing than unflavored milk even though the initial numbers for both types were essentially the same.[13] On day 14, 76.1% of unflavored and 91.6% of chocolate milk had APCs $>20,000$ cfu/ml with 26.1% of the former and 53.7% of the latter products having APCs $>10^6$ cfu/ml. These investigators suggested that the chocolate flavor powder contributed to increased growth.[13] The higher numbers were not due to higher numbers in the chocolate powder per se.

Pasteurization

The objective of milk pasteurization is the destruction of all disease-causing microorganisms. Endospores of pathogens such as *Clostridium botulinum* and spoilage organisms such as *Clostridium tyrobutyricum*, *C. sporogenes*, or *Bacillus cereus* are not destroyed. Although pathogens can be destroyed by nonthermal means, milk pasteurization is achieved solely by heating.

The low temperature-long time (LTLT) method consists of heating the coolest part to 145°F (63°C) for 30 minutes. This is referred to as the batch method. The other more widely used method is the high temperature-short time (HTST) method, and it consists of heating to 161°F (72°C) for 15 sec. This is the flash method, and it is inherently less destructive than the batch method. The basis for the heating time and temperature is the thermal death time (TDT) of the most heat-resistant non-sporeforming milk-borne pathogens. Prior to 1950, the LTLT method involved heating at 143°F for 30 minutes, which was the TDT of *Mycobacterium tuberculosis*. However, after the discovery of the Q fever agent (*Coxiella burnetti*) and the determination of its presence in bovine, goat, and sheep milk, the LTLT method was changed so that it involved heating at 145°F for 30 minutes to correspond to the TDT of this pathogen. In properly pasteurized milk, the naturally occurring enzyme alkaline phosphatase is destroyed.[76]

UHT (ultra-high temperature) is another thermal treatment that destroys non-sporeforming pathogens in milk, but in addition some sporeformers are severally reduced in numbers. The UHT treatment is achieved by heating at temperatures of 275–284°F (135–140°C) for a few sec (the minimum treatment is 130°C for 1 sec). UHT-treated milk is commercially sterile with a shelf life of 40–45 days at 40°F when aseptically packaged in sterile containers.[7] UHT-treated whole milk is said to be more flavorful, due apparently to formation of some Maillard products.

Although pasteurized milk is free of non-sporeforming pathogens, it is not sterile. The efficacy of either LTLT or HTST to destroy the mycobacterial subspecies that is associated with Crohn's disease in humans has been called in to question, and this is discussed further below under milk-borne diseases. Most if not all Gram-negative bacteria (especially psychrotrophs) are destroyed along with many Gram positives. Thermoduric Gram positives belonging to the genera *Enterococcus*, *Streptococcus* (especially *Streptococcus salivarius* subsp. *thermophilus*), *Microbacterium*, *Lactobacillus*, *Mycobacterium*, *Corynebacterium*, and most if not all sporeformers survive. Among the survivors are a number of psychrotrophic species of the genus *Bacillus*.[45]

General Microbiota of Milk

Theoretically, milk that is secreted to the udder of a healthy cow should be free of microorganisms. However, freshly drawn milk is generally not free of microorganisms. Numbers of several hundred to several thousand cfu/ml are often found in freshly drawn milk, and they represent the movement up the teat canal of some and the presence of others at the lower ends of teats. Although the APC of milk from healthy cows is generally $<10^3$ cfu/ml, numbers of 10^4/ml are not uncommon.[50]

Milk-Borne Pathogens

Since it is such an excellent nutrient source and because milk-producing animals may harbor organisms that cause human diseases, it is not surprising that raw milk can be a source of diseases. Some of the most obvious are the animal diseases below to which humans are susceptible and which may occur in milk of cows:

Brucellosis	Anthrax
Tuberculosis	Listeriosis
Salmonellosis	Q fever
Campylobacteriosis	Crohn's disease (?)
Enterohemorrhagic colitis	Staph./Strep. Mastitis

Prior to the general use of mechanical milking devices, raw milk was the source of both human respiratory diseases (e.g., diphtheria) as well as enteric infections (e.g., typhoid fever). When milking a cow by hand where the milk is collected in an open pail, an infected person (or carrier) may contaminate the milk by coughing or by hand contact. It was largely the connection between raw milk and cases of human scarlet fever, diphtheria, and typhoid fever in the early 1900s that led the New York City Board of Health to require milk pasteurization in 1910. The adoption of this law followed a large outbreak of typhoid fever in New York City in the previous year, and it was traced to a common milk supply where a chronic typhoid carrier worked.[58] The etiologic agents of all of the above-listed diseases are destroyed by the milk pasteurization process.

In spite of the widespread use of pasteurization, milk continues to be a vehicle for some diseases. In the case of campylobacteriosis, it is not surprising that its etiologic agent can be found in milk since the organism exists in cow feces. In a survey of 108 samples from bulk tanks of raw milk in Wisconsin, only 1 was positive for *C. jejuni*, whereas the feces of 64% of the cows in a grade A herd were *C. jejuni* positive.[14] In the Netherlands, 22% of 904 cow fecal and 4.5% of 904 raw milk samples contained *C. jejuni*.[6] In regard to *Helicobacter pylori*, it was not detected in 120 raw bovine milk samples, and when it was added to sterile milk and refrigerated at $4°C$, it could not be detected after 6 days.[35] An outbreak of 75 cases of campylobacteriosis traced to raw milk occurred in Wisconsin in 2001.[10] The milk was from a grade A organic farm.

For the years 1973–1992, 46 outbreaks and 1,733 cases of human gastroenteritis were traced to raw milk and reported to the Centers for Disease Control and Prevention in the United States. Of the 1,733 cases, 57% were caused by *Campylobacter*, 26% by *Salmonella*, and 2% by *E. coli* 0157:H7.[31] Of the 50 states in the United States, 28 allow the sale of raw milk but <1% of the milk-borne outbreaks during the period noted were traced to raw milk. The interstate shipment/sale of raw milk was banned in the United States in 1987. Some of the earliest outbreaks of human listeriosis were traced to milk (see Chapter 25). Although the outbreaks traced to dairy products may be presumed to result from the shedding of virulent strains into milk, this is not always confirmed. In one study, *L. monocytogenes* was found to be shed in milk from the left forequarter of a mastitic cow, but milk from the other quarters was uninfected.[20] When bulk tank milk from 474 dairy herds in the Pacific northwest of the United States were examined in 2000, 4.9% were positive for *L. monocytogenes* and in 2001, 7.0% were positive with serotype 1/2a being the most common each year.[48]

Yersinia enterocolitica has been found in several studies on raw milk samples. The first documented outbreak of this organism in the United States was traced to chocolate milk, and this outbreak is further described in Chapter 28. Twelve of 100 raw milk samples in the state of Wisconsin were positive for *Y. enterocolitica* but only 1 pasteurized sample was positive.[46] Of 219 raw milk samples tested in Brazil, 37 (16.9%) contained *Listeria* spp. and 32.4% were *Y. enterocolitica*.[74] Of 280 pasteurized milk samples, 13.7% were positive for *Yersinia* spp. with 41.5% being *Y. enterocolitica*. The latter species was the most common in raw milk while *Y. frederiksenii* at 56.1% was the most prevalent in pasteurized milk.[74] The two largest human outbreaks of salmonellosis to occur in the United States involved milk and ice cream, and they are described in Chapter 26.

In regard to aflatoxin M_1 in milk, a study of 290 2-liter samples of pasteurized and ultrapasteurized milk was carried out in Mexico on the seven most widely used brands with varying fat content. Forty percent contained AFM_1 at levels ≥ 0.05 ppm and 9.7% contained ≥ 0.5 ppm, and the range was 0–8.35 ppm in 40% of those with the lower concentration and in 10% of those with the higher concentration.[9] Milk with the highest fat content had a slightly higher probability of containing AFM_1.[9]

A study of the prevalence of *E. coli* 0157:H7 in fecal samples from cull cows and bulk tank milk in east Tennessee found 8 of 415 (2%) fecal samples and 2 of 268 (0.7%) bulk tank milk samples positive.[49] At least one large outbreak of *E. coli* 0157:H7 was traced to raw milk (see Chapter 27).

A bacterium of continuing concern in milk is the etiologic agent of Johne's disease of cattle, which appears to play some role in Crohn's disease of humans. The organisms in question are classified as follows: *Mycobacterium avium* subsp. *avium* causes tuberculosis in birds and it is infectious for AIDS patients. *M. avium* subsp. *paratuberculosis* is an obligate pathogen of ruminants and it is thought to be involved in the etiology of Crohn's disease. *M. avium* subsp. *silvaticum* is an obligate pathogen of animals where it causes paratuberculosis in mammals and tuberculosis in birds.[73] *M. avium* subsp. *paratuberculosis* is the strain of interest in cow's milk because of its possible role in Crohn's disease. Of primary concern is whether the pasteurization methods in use are adequate to destroy this organism. In one study, neither the HTST nor the LTLT method destroyed 10^3–10^4 cfu/ml in all milk samples,[24] but in another study, up to 10^6 cfu/ml were destroyed by HTST carried out at 72°C for 15 sec.[68] Crohn's disease is an inflammatory bowel disease (regional ileitis), a condition wherein the terminal ileum and sometimes the cecum and ascending colon are thickened and ulcerated. The lumen of the affected region is much narrowed, resulting in intestinal obstruction.

In a survey of 814 bovine milk samples in the United Kingdom over a 17-month period in 1999–2000, a mean of 7.8% of raw and 11.8% of pasteurized samples were positive for *M. avium* subsp. *paratuberculosis* DNA.[22] Culture confirmation was achieved in 1.6% of the raw and 1.8% of the pasteurized samples. The pasteurized milk samples in this study were phosphatase negative. In another study of raw and pasteurized cows' milk in the United Kingdom, *M. avium* subsp. *paratuberculosis* was found in 4 of 40 (6.7%) raw and 10 of 144 (6.9%) pasteurized milk samples.[23] The investigators found viable organisms in milk samples processed by four different treatments including heating at 73°C for 25 sec.

The fate of this organism in the ripening of cheese has been investigated.[72] According to the FDA, two options are available for producing safe cheese: (1) use only pasteurized milk, or (2) hold finished cheese for at least 60 days at 2°C. In the study noted above, a soft white cheese using pasteurized milk that was spiked with ca. 10^6 *M. avium* subsp. *paratuberculosis* was prepared by varying pH and NaCl content and tested for viable cells after storage (ripening). It was found that cheese made from HTST-pasteurized milk, pH 6.0, 2% NaCl, and cured for 60 days resulted in ca. log-3 reduction of *M. avium* subsp. paratuberculosis cells.[72] In a review of Crohn's disease and the role of *M. avium* subsp. *paratuberculosis*, Harris and Lammerding[30] concluded that the evidence for cause and effect is not conclusive.

Spoilage

As the only natural source of the disaccharide lactose, milk undergoes microbial spoilage in a way that is unique. Only a relatively small number of milk-borne bacteria can obtain energy from this sugar (especially at refrigerator temperatures) in contrast to the disaccharides sucrose and maltose, and the lactic acid bacteria are well suited to this task. The coliform bacteria are the most conspicuous utilizers of lactose among Gram-negative bacteria. Thus, the bacterial spoilage of either raw or pasteurized milk is conspicuous by the production of lactic acid by lactose users, with the normal pH of around 6.6 being reduced to 4.5 or so that leads to the precipitation of casein (curdling). The thermoduric *Streptococcus salivarius* subsp. *thermophilus* strains preferentially use the glucose moiety of lactose and excrete galactose, which is a ready substrate for nonlactose users.

The spoilage of UHT milk is caused by *Bacillus* spp. that survive the UHT process. Anaerobic spores appear not to be a problem because of the relatively high Eh of milk. Among the *Bacillus* species that have been recovered from spoiled products are *B. cereus*, *B. licheniformis*, *B. badius*, and *B. sporothermodurans*.[55] *Paenibacillus* spp. have been isolated also from UHT-treated products.

During cold storage of pasteurized milk, psychrotrophic *Bacillus weihenstephanensis* causes "sweet curdling" due to its production of proteases and peptidases. The spoilage of UHT-treated milk can result from the actions of heat-resistant proteases and lipases that are produced by some psychrotrophs in raw milk, and more on this can be found in reference 17.

Ropiness is a condition sometimes seen in raw milk that is caused by *Alcaligenes viscolactis*. Its growth is favored by low-temperature maintenance of raw milk for several days. The "rope" consists of a slime-layer material produced by the bacterial cells, and it gives the product a stringy consistency.

PROBIOTICS AND PREBIOTICS

The definition of a probiotic that dates back to the mid-1960s has been modified during the past decade. In general, it is a consumable product that contains live organisms that are or are believed to be beneficial to the consumer. The ingestion of live organisms such as those in yogurt or fermented milk is critical to the original concept. The term has been applied to consumable products that produce a number of clinical benefits, but which do not contain viable cells. One example of this is the easing of symptoms of lactose intolerance by the ingestion of heated yogurt (for a review, see reference 53). A definition of probiotic that encompasses the above has been proposed by Salminen et al.[61] In the absence of viable organisms, the term "abiotics" has been suggested.[37,67] To make the definition of a probiotic even wider, the term has been applied to bacteria that act as control agents in an aquaculture environment.[77] For a more detailed review of probiotics, see reference 37.

Yogurt appears to be the most widely consumed of probiotic products (especially in the United States) and while it is consumed by some because it is a fermented dairy product, it is consumed by others because of its real or presumed health benefits. Although the typical starter cultures for yogurt are *Streptococcus salivarius* subsp. *thermophilus* and *Lactobacillus delbrueckii* subsp. *bulgaricus*, some preparations are made by the addition of bifidobacteria. Regarding the number of viable cells that should exist in probiotic and probiotic-like products, the Swiss Food Regulation and the International Standard of FIL/IDF require that such products contain at least 10^6 cfu/ml while the Fermented Milks and Lactic Acid Beverages Association in Japan requires at least 10^7 cfu/ml of viable bifidobacteria (see reference 66). It is unclear whether viable cell numbers can be of both lactic acid bacteria and bifidobacteria, or only one of these groups. When bifidobacteria are involved in the yogurt fermentation, they tend to die out during storage. These organisms are anaerobes that require negative Eh conditions and a pH near neutrality.[66] The effect of a whey protein hydrolysate on some probiotic bacteria in milk was investigated and while the hydrolysate initially increased the growth of *Bifidobacterium longum* along with two lactobacilli, after 28 days at refrigerator temperatures, the probiotic organisms were at about the same level as in the control milk.[44]

Live *Bacillus* spp. spores are used as probiotics and the three species most often used are *B. clausii*, *B. pumilus*, and *B. cereus* at levels of ca. 10^9/g. Positive effects have been demonstrated and they appear to be due to the immunogenic properties of the spores, not the vegetative cells.

The effect of three starter cultures on the survival of *Yersinia enterocolitica* in yogurt is shown in Figure 7–3 where the rapid acid-producing starter effected a log-5.0 reduction in 72 hours, and a slow acid-producing strain effected a log-5.6 reduction in 96 hours.[7] Inhibitory efforts of this type by probiotic-type products against foodborne pathogens have been demonstrated for a number of pathogens, and while pH reduction is one factor in the inhibition, factors such as bacteriocins and organic acid toxicity are undoubtedly involved, and this is discussed further in Chapter 13. The cause of lactose intolerance and its detection are described below.

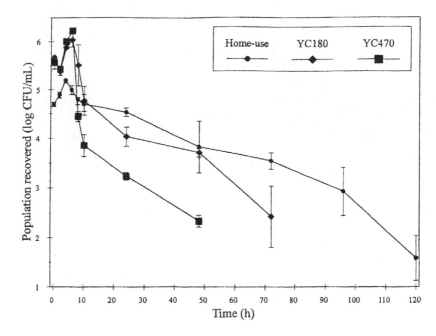

Figure 7–3   Survival of *Yersinia enterocolitica* in yogurt during fermentation at 44°C and storage at 4°C. Yogurt was prepared using slow (YC 180) and rapid (YC 470) acid-producing starter cultures and a "home-use" starter culture. Error bars represent standard errors of the mean,[7] copyright © 1998, used with permission from Institute of Food Technologists

Prebiotics are not microorganisms; they are substrates for the indigenous probiotic-type bacteria that reside in the colon. These substrates are nondigestible as they pass through the small intestine, and they consist of oligosaccharides such as fructooligosaccharides of which inulin is an example. They are metabolized by the bidifobacteria and the anaerobic lactobacilli (both of which are indigenous to the colon) where the Eh favors their growth and activity, which results in an environment that is antagonistic to aerobic pathogens. Unlike probiotics, these substrates can be added to a number of food types that do not support cell viability over long periods of time. Their use obviates the need for cultures that can persist in the small intestines.

Lactose Intolerance

Lactose intolerance (lactose malabsorption, intestinal hypolactemia) is the normal state for adult mammals, including most adult humans, and many more groups are intolerant to lactose than are tolerant.[40] Among the relatively few groups that have a majority of adults who tolerate lactose are northern Europeans, white Americans, and members of two nomadic pastoral tribes in Africa.[40] When lactose malabsorbers consume certain quantities of milk or ice cream, they immediately experience flatulence and diarrhea. The condition is due to the absence or reduced amounts of intestinal lactase, and this allows the bacteria in the colon to utilize lactose with the production of gases. The breath

Table 7–3 Summary of Some of the Many Claimed and Presumed Human Benefits from Probiotics (Summarized from references 37, 53, 61, and 62)

Benefits	Comments
Lactose intolerance	Well established benefits (see text)
Acute gastroenteritis	Incidence/duration often reduced
Serum cholestrol reduction	Positive in vitro results; mixed in vivo results
Tumor size reduction/survival rate	Limited studies
Reducing secondary tumor growth	Positive results from limited studies
Production of IFNα,β	Positive results from limited studies
Vaginal candidiases	Mixed results
Antihypertensive effects	Positive results shown
Anticolon cancer	Some positive effects shown
Helicobacter pylori infection	Inhibitors are produced
Enteric pathogen resistance	Some positive effects demonstrated
Travelers'diarrhea	Positive effects reported
Attenuation of colonic hyperplasia	Positive results in mice
Increase in human longevity	Not yet proven

hydrogen test for lactose intolerance is based on the increased levels of H_2 produced by anaerobic and facultatively anaerobic bacteria utilizing the nonabsorbed lactose.

"Sweet" acidophilus milk has been reported by some to prevent symptoms of lactose intolerance, whereas others have found this product to be ineffective. Developed by M.L. Speck and co-workers, it consists of normal pasteurized milk to which is added large numbers of viable *L. acidophilus* cells as frozen concentrates. As long as the milk remains under refrigeration, the organisms do not grow, but when it is drunk, the consumer gets the benefit of viable *L. acidophilus* cells. It is "sweet" because it lacks the tartness of traditional acidophilus milk. When 18 lactase-deficient patients ingested unaltered milk for 1 week, followed by "sweet" acidophilus milk for an additional week, they were as intolerant to the latter product as to the unaltered milk.[51] Lactose-free milk is available in Germany that is pretreated with β-galactosidase. In one study with rats, the yogurt bacteria had little effect in preventing the malabsorption of lactose. The indigenous lactics in the gut tended to be suppressed by yogurt, and the rat lactobacillus biota changed from one that was predominantly heterofermentative to one that was predominantly homofermentative. A number of other possible health benefits from both viable and nonviable probiotic bacteria have been studied, and an extensive review has been provided.[62] Summaries of some health benefits are presented in Table 7–3.

STARTER CULTURES, FERMENTED PRODUCTS

The products discussed in this subsection require the use of an appropriate *starter* culture. A lactic starter is a basic starter culture with widespread use in the dairy industry. For cheese making of all kinds, lactic acid production is essential, and the lactic starter is employed for this purpose. Lactic starters are also used for preparing butter, cultured buttermilk, cottage cheese, and cultured sour cream and are often referred to by product (butter starter, buttermilk starter, and so on). Lactic starters always

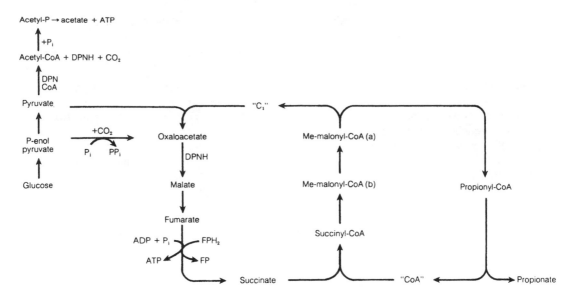

Figure 7–4   Reactions of the propionic acid fermentation and the formation of acetate, CO_2, propionate, and ATP. Me-malonyl-CoA is methylmalonyl-CoA and (a) and (b) are the two isomers. FP is flavoprotein, and FPH_2 is reduced flavoprotein. Summary: 1.5 glucose + 6 Pi + 6 ADF → 6 ATP + $2H_2O$ + CO_2 + acetate + 2 propionate. *Source*: Allen et al.,[3] copyright © 1964 by American Society for Microbiology.

include bacteria that convert lactose to lactic acid, usually *L. lactis* subsp. *lactis*, *L. lactis* subsp. *cremoris*, or *L. lactis* subsp. *lactis* biovar *diacetylactis*. Where flavor and aroma compounds such as diacetyl are desired, the lactic starter will include a heterolactic such as *Leuconostoc mesenteroides* subsp. *cremoris*, *L. lactis* subsp. *lactis* biovar *diacetylactis*, or *Leuconostoc mesenteroides* subsp. *dextranicum* (for biosynthetic pathways, see Figure 7–4). Starter cultures may consist of single or mixed strains. They may be produced in bulk and preserved by freezing in liquid nitrogen[19] or by freeze drying. The lactococci generally make up around 90% of a mixed dairy starter population, and a good starter culture can convert most of the lactose to lactic acid. The titratable acidity may increase to 0.8–1.0%, calculated as lactic acid, and the pH usually drops to 4.3–4.5.[16]

Fermented Products

Butter, *buttermilk*, and *sour cream* are produced generally by inoculating pasteurized cream or milk with a lactic starter culture and holding until the desired amount of acidity is attained. In the case of butter, where cream is inoculated, the acidified cream is then churned to yield butter, which is washed, salted, and packaged.[54] Buttermilk, as the name suggests, is the milk that remains after cream is churned for the production of butter. The commercial product is usually prepared by inoculating skim milk with a lactic or buttermilk starter culture and holding until souring occurs. The resulting curd is broken up into fine particles by agitation, and this product is termed *cultured buttermilk*. Cultured sour cream is produced generally by fermenting pasteurized and homogenized light cream with a lactic starter. These products owe their tart flavor to lactic acid and their buttery aroma and taste to diacetyl.

An outbreak of *Campylobacter* enteritis in the state of Louisiana in 1995 where garlic butter was incriminated led to studies on the fate of foodborne pathogens in butter with or without garlic. In one study, garlic butter was inoculated with ca. 10^4 and 10^6 cfu/g of *C. jejuni* and held at 5 or 21°C. At 5°C, the pathogen decreased to <10 cfu/g within 3 hours for two batches and within 24 hours for a third batch.[81] In butter with no garlic, *C. jejuni* survived for 13 days at 5°C. At 21°C, the pathogen decreased to <10 cfu/g within 5 hours for two preparations and to 50 cfu/g in 5 hours for the third.[81] Overall, up to 10^5 cells were killed within a few hours in butter with garlic, but the organism could survive for days in refrigerated butter. In another study, *Salmonella, E. coli* 0157:H7, and *L. monocytogenes* were found not to grow in unsalted butter with or without 20% garlic when held at 4.4, 21, or 37°C for up to 48 hours.[1]

Yogurt (yoghurt) is produced with a yogurt starter, which is a mixed culture of *S. salivarius* subsp. *thermophilus* and *Lactobacillus delbrueckii* subsp. *bulgaricus* in a 1:1 ratio. The coccus grows faster than the rod and is primarily responsible for initial acid production at a higher rate than that produced by either when growing alone, and more acetaldehyde (the chief volatile flavor component of yogurt) is produced by *L. delbrueckii* subsp. *bulgaricus* when growing in association with *S. salivarius* subsp. *thermophilus* (see reference 57). The coccus can produce about 0.5% lactic acid and the rod about 0.6–0.8% (pH of 4.2–4.5). However, if incubation is extended, pH can decrease to about 3.5 with lactic acid increasing to about 2%.[32]

The product is prepared either by reducing the water content of either whole or skim milk by at least one-fourth (may be done in a vacuum pan following sterilization of milk), or by adding about 5% milk solids followed by water reduction (condensing). The concentrated milk is then heated to 82–93°C for 30–60 minutes and cooled to around 45°C.[54] The yogurt starter is now added at a level of around 2% by volume and incubated at 45°C for 3–5 hours followed by cooling to 5°C. The titratable acidity of a good finished product is around 0.85–0.90%, and to get this amount of acidity the fermenting product should be removed from 45°C when the titratable acidity is around 0.65–0.70%.[11] Good yogurt keeps well at 5°C for 1–2 weeks. The coccus grows first during the fermentation followed by the rod, so that after around 3 hours, the numbers of the two organisms should be approximately equal. Higher amounts of acidity, such as 4%, can be achieved by allowing the product to ferment longer, with the effect that the rods will exceed the cocci in number. The streptococci tend to be inhibited at yogurt pH values of 4.2–4.4, whereas the lactobacilli can tolerate pH values in the 3.5–3.8 range. The lactic acid of yogurt is produced more from the glucose moiety of lactose than the galactose moiety. Goodenough and Kleyn[21] found only a trace of glucose throughout yogurt fermentation, whereas galactose increased from an initial trace to 1.2%. Samples of commercial yogurts showed only traces of glucose, but galactose varied from around 1.5% to 2.5%.

Freshly produced yogurt typically contains around 10^9 organisms/g, but during storage, numbers may decrease to 10^6/g, especially when stored at 5°C for up to 60 days.[27] The rod generally decreases more rapidly than the coccus. The addition of fruits to yogurt does not appear to affect the numbers of fermenting organisms.[27] The International Dairy Federation norm for yogurt is 10^7/g or above. In one study, *E. coli* 0157:H7 did not survive in skim milk at pH 3.8, and the organism was inactivated in yogurt, sour cream, and buttermilk similarly.[26]

The antimicrobial qualities of yogurt, buttermilk, sour cream, and cottage cheese have been examined by inoculating *Enterobacter aerogenes* and *Escherichia coli* separately into commercial products and studying the fate of these organisms when the products were stored at 7.2°C. A sharp decline of both coliforms was noted in yogurt and buttermilk after 24 hours. Neither could be found in yogurt generally beyond 3 days. Although the numbers of coliforms were reduced also in sour cream, they were not reduced as rapidly as in yogurt. Some cottage cheese samples actually supported an increase in coliform numbers, probably because the products had higher pH values. The initial pH ranges for the products studied by these workers were as follows: 3.65–4.40 for yogurts, 4.1–4.9 for buttermilk, 4.18–4.70 for sour creams, and 4.80–5.10 for cottage cheese samples. In another study, commercially produced yogurts in Ontario were found to contain the desired 1:1 ratio of coccus to rod in only 15% of 152 products examined.[5] Staphylococci were found in 27.6% and coliforms in around 14% of these yogurts. Twenty-six percent of the samples had yeast counts more than 1,000/g and almost 12% had psychrotroph counts more than 1,000/g. In his study of commercial unflavored yogurt in Great Britain, Davis[11] found counts of the two starters to range from a low of around 82 million to a high of over 1 billion/g, and the final pH to range from 3.75 to 4.20. The antimicrobial activities of lactic acid bacteria are discussed further in Chapters 3 and 13.

Kefir is prepared by the use of kefir grains, which contain one or more bacterial species of the genera *Acetobacter*, *Lactobacillus*, *Lactococcus*, *Leuconostoc*, and one or more yeast species of the genera *Candida*, *Kluyveromyces*, and *Saccharomyces*. These symbionts are held together by coagulated protein.[18] The important *Lactobacillus* spp. in kefir are: *L. kefiri*, *L. parakefiri*, *L. kefiranofaciens* subsp. *kefiranofaciens*, and *L. kefiranofaciens* subsp. *kefirgranum*.[75] The last two are responsible for the production of kefiran (a water-soluble polysaccharide), which accounts for about 24% of kefir grains.[75] *Kumiss* is similar to kefir except that mare's milk is used, the culture organisms do not form grains, and the alcohol may reach 2%.

Acidophilus milk is produced by the inoculation of an intestinal implantable strain of *L. acidophilus* into sterile skim milk. The inoculum of 1–2% is added, followed by holding the product at 37°C until a smooth curd develops. A popular variant of this product that is produced commercially in the United States consists of adding a concentrated implantable strain culture of *L. acidophilus* to a pasteurized and cold vat of whole milk (or skim or 2% milk), and it is bottled immediately. It has the pH of normal milk and is more palatable than the more acidic product. The numbers of *L. acidophilus* should be in the 10^7–10^8/ml range.[32] *Bulgarian buttermilk* is produced in a similar manner by the use of *L. bulgaricus* as the inoculum or starter, but unlike *L. acidophilus*, *L. bulgaricus* is not implantable in the human intestines. A summary of fermented milk is presented in Table 7–4.

Butter contains around 15% water, 81% fat, and generally less than 0.5% carbohydrate and protein. Although it is not a highly perishable product, it does undergo spoilage by bacteria and molds. The main source of microorganisms for butter is cream, whether sweet or sour, pasteurized or nonpasteurized. The biota of whole milk may be expected to be found in cream because as the fat droplets rise to the surface of milk, they carry up microorganisms. The processing of both raw and pasteurized creams to yield butter brings about a reduction in the numbers of all microorganisms, with values for finished cream ranging from several hundred to over 100,000/g having been reported for finished salted butter. *Salted butter* may contain up to 2% salt, and this means that water droplets throughout may contain an effective level of about 10%, thus making this product even more inhibitory to bacterial spoilage.[32]

Bacteria cause two principal types of spoilage in butter. The first is a condition known as *"surface taint"* or putridity. This condition is caused by *Pseudomonas putrefaciens* as a result of its growth on the surface of finished butter. It develops at temperatures within the range 4–7°C and may become apparent within 7–10 days. The odor of this condition is apparently due to certain organic acids, especially isovaleric acid. Surface taint along with an apple odor is caused also by *Chryseobacterium joostei*.[34] The second most common bacterial spoilage condition of butter is *rancidity*. This condition is caused by the hydrolysis of butterfat with the liberation of free fatty acids. Lipase from sources other than microorganisms can cause the effect. The causative organism is *Pseudomonas fragi*, although *P. fluorescens* is sometimes found. Bacteria may cause three other less common spoilage conditions in butter. *Malty flavor* is reported to be due to the growth of *Lactococcus lactis* var. *maltigenes*. *Skunklike*

Table 7–4   Some Fermented Milk Products

Foods and Products	Raw Ingredients	Fermenting Organisms	Where Produced
Acidophilus milk	Milk	*Lactobacillus acidophilus*	Many countries
Bulgarian buttermilk		*L. delbrueckii* subsp. *bulgaricus*	Balkans, other areas
Cheeses (ripened)	Milk curd	Lactic starters	Worldwide
Kefir	Milk	*Lactococcus lactis, L. delbrueckii* subsp. *bulgaricus*, "Torula" spp.	Southwestern Asia
Kumiss	Raw mare's milk	*Lactobacillus leichmannii, L. delbrueckii* subsp. *bulgaricus*, "Torula" spp.	Russia
Taette	Milk	*S. lactis* var. *taette*	Scandinavian peninsula
Tarhana*	Wheat meal and yogurt	Lactics	Turkey
Yogurt†	Milk, milk solids	*L. delbrueckii* subsp. *bulgaricus, S. salivarius* subsp. *thermophilus*	Worldwide
Bioghurt	Milk, milk solids	*L. acidophilus, Lactococcus lactis*	Worldwide

*Similar to Kishk in Syria and Kushuk in Iran.
†Also yoghurt (matzoon in Armenia; leben in Egypt; naja in Bulgaria; gioddu in Italy; dadhi in India).

odor is reported to be caused by *Pseudomonas mephitica*; black discolorations of butter have been reported to be caused by *P. nigrifaciens*.

Butter undergoes fungal spoilage rather commonly by species of *Cladosporium, Alternaria, Aspergillus, Mucor, Rhizopus, Penicillium,* and *Geotrichum,* especially *G. candidum (Oospora lactis)*. These organisms can be seen growing on the surface of butter, where they produce colorations referable to their particular spore colors. Black yeasts of the genus *Torula* also have been reported to cause discolorations on butter. The microscopic examination of moldy butter reveals the presence of mold mycelia some distances from the visible growth. The generally high lipid content and low water content make butter more susceptible to spoilage by molds than by bacteria.

Cottage cheese undergoes spoilage by bacteria, yeasts, and molds. The most common spoilage pattern displayed by bacteria is a condition known as *slimy curd. Alcaligenes* spp. have been reported to be among the most frequent causative organisms, although *Pseudomonas, Proteus, Enterobacter,* and *Acinetobacter* spp. have been implicated. *Penicillium, Mucor, Alternaria,* and *Geotrichum* all grow well on cottage cheese, to which they impart stale, musty, moldy, and yeasty flavors. The shelf life of commercially produced cottage cheese in Alberta, Canada was found to be limited by yeasts and molds.[59] Although 48% of fresh samples contained coliforms, these organisms did not increase upon storage in cottage cheese at 40°F for 16 days. For more on fermented dairy products, see references 52, 54.

Cheeses

Most but not all cheeses result from a lactic fermentation of milk. In general, the process of manufacture consists of two important steps:

1. Milk is prepared and inoculated with an appropriate lactic starter. The starter produces lactic acid, which, with added rennin, gives rise to curd formation. The starter for cheese production may differ depending on the amount of heat applied to the curds. *S. salivarius* subsp. *thermophilus* is employed for acid production in cooked curds (up to 60°C) because it is more heat tolerant than either of the other more commonly used lactic starters; or a combination of *S. salivarius* subsp. *thermophilus* and *L. lactis* subsp. *lactis* is employed for curds that receive an intermediate cook.
2. The curd is shrunk and pressed, followed by salting, and, in the case of ripened cheeses, allowed to ripen under conditions appropriate to the cheese in question.

Although most ripened cheeses are the product of metabolic activities of the lactic acid bacteria, several well-known cheeses owe their particular character to other related organisms. In the case of Swiss cheese, a mixed culture of *L. delbrueckii* subsp. *bulgaricus* and *S. salivarius* subsp. *thermophilus* is usually employed along with a culture of *Propionibacterium shermanii* or *P. freundenreichii* added to function during the ripening process in flavor development and eye formation. (See Figure 7–1(C) and (D)) for a summary of propionibacteria pathways and Figure 7–4 for pathway in detail.) These organisms have been reviewed extensively by Hettinga and Reinbold.[33] For blue cheeses such as Roquefort, the curd is inoculated with spores of *Penicillium roqueforti*, which effect ripening and impart the blue-veined appearance characteristic of this type of cheese. In a similar fashion, either the milk or the surface of Camembert cheese is inoculated with spores of *Penicillium camemberti*.

Two coryneform bacteria of the genus *Brachybacterium* have been recovered from the surfaces of French Gruyère and Beaufort cheeses[65] but the role these organisms play in the ripening process is unclear. In a study of *L. monocytogenes* in European red smear cheese (soft, semisoft, and hard), 5.8% of 329 test samples contained *Listeria* spp. with 6.4% being *L. monocytogenes* and 10.6% *L. innocua.*[60] Eight samples contained >100 *L. monocytogenes*/cm^2; and two samples contained 10^4 cfu/cm^2.

There are over 400 varieties of cheeses representing fewer than 20 distinct types, and these are grouped or classified according to texture or moisture content, whether ripened or unripened, and if ripened, whether by bacteria or molds. The three textural classes of cheeses are hard, semihard, and soft. Examples of hard cheeses are all cheddar, Provolone, Romano, Parmesan, Gruyère, Emmental, and Edam. All hard cheeses are ripened by bacteria over periods ranging from 2 to 16 months. Semihard cheeses include Muenster, Roquefort, Limburger, and Gouda and are ripened by bacteria over periods of 1–8 months. Blue and Roquefort are two examples of semihard cheeses that are mold ripened for 2–12 months. Limburger is an example of a soft bacteria-ripened cheese, and Brie and Camembert are examples of soft mold-ripened cheeses. Among unripened cheeses are cottage, cream, Mozzarella, and Neufchatel.

The low moisture content of hard and semihard ripened cheeses makes them insusceptible to spoilage by most organisms, although molds can and do grow on these products as would be expected. Some ripened cheeses have sufficiently low oxidation–reduction potentials to support the growth of anaerobes. It is not surprising to find that anaerobic bacteria sometimes cause the spoilage of these products when a_w (water activity) permits growth to occur. *Clostridium* spp., especially *C. pasteurianum*, *C. butyricum*, *C. sporogenes*, and *C. tyrobutyricum*, have been reported to cause late gassiness of cheeses. One of these (*C. tyrobutyricum*) is well established as the cause of a butyric acid

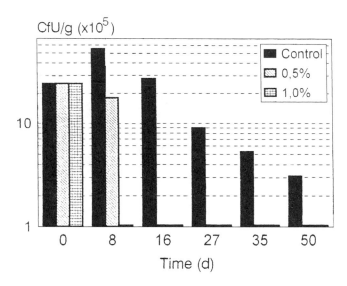

Figure 7–5   Inhibition of *Clostridium tyrobutyricum* in processed cheese spread (cheese blend B) by 0.5% and 1.0% of HBS polyphosphate. Reprinted with permission from *J. Food Protect.*, (c) held by the Int. Assoc. Food Protect., Des Moines, IA, USA. Source: Loessner et al.[41]

fermentation or the late-blowing defect in cheeses such as Gouda and Emmentaler.[39] With 0.5% of a long-chain polyphosphate mixture, growth of *C. tyrobutyricum* was inhibited for at least 8 days and could not be detected after 16–50 days (see Figure 7–5). Growth was completely inhibited by 1.0% polyphosphate,[41] and it was due to the sequestration of Ca^{2+}/Mg^{2+} by poly-P, which led to filamentous cells and lysis. An aerobic sporeformer, *Paenibacillus polymyxa*, has been reported to cause gassiness. This condition is the result of CO_2 being produced from lactic acid.

For the years 1973–1992, there were 32 cheese-associated disease outbreaks in the United States with 1,700 cases and 58 deaths with 52 of the latter caused by *L. monocytogenes* in the 1985 California outbreak.[4] The most common vehicle was soft cheeses, and improper pasteurization was common.

DISEASES CAUSED BY LACTIC ACID BACTERIA

Although the beneficial aspects of the lactic acid bacteria to human and animal health are unquestioned, some of these bacteria are associated with human illness. This subject has been reviewed by Aguirre and Collins,[2] who noted that around 68 reports of involvement of lactobacilli in human clinical illness were made over about a 50-year period. Several species of leuconostocs were implicated in about 27 reports in 7 years, the pediococci in 18 reports over 3 years, and the enterococci in numerous reports. The enterococci are the third leading cause of nosocomial (hospital acquired) infections, with *E. faecalis* and *E. faecium* being the two most common species. It appears that lactic acid bacteria are opportunists that are not capable of initiating infection in normal healthy individuals. To determine whether vancomycin-resistant enterococci (VRE) existed in ground beef and pork in Germany, 555 samples were examined for VRE, and overall their incidence in ground beef was too low to be a significant source of nosocomial infections.[38]

REFERENCES

1. Adler, B.B., and L.R. Beuchat. 2002. Death of *Salmonella*, *Escherichia coli* 0157:H7, and *Listeria monocytogenes* in garlic butter as affected by storage temperature. *J. Food Protect.* 65:1976–1980.

2. Aguirre, M., and M.D. Collins. 1993. Lactic acid bacteria and human clinical infection. *J. Appl. Bacteriol.* 75:95–107.

3. Allen, S.H.G., R.W. Killermeyer, R.L. Stjernholm, and H.G. Wood. 1964. Purification and properties of enzymes involved in the propionic acid fermentation. *J. Bacteriol.* 87:171–187.

4. Altekruse, S.F., B.B. Timbo, J.C. Mobray, N.H. Bean, and M.E. Potter. 1998. Cheese-associated outbreaks of human illness in the United States, 1973 to 1992: Sanitary manufacturing practices protect consumers. *J. Food Protect.* 61:1405–1407.

5. Arnott, D.R., C.L. Duitschaever, and D.H. Bullock. 1974. Microbiological evaluation of yogurt produced commercially in Ontario. *J. Milk Food Technol.* 37:11–13.

6. Beumer, R.R., J.J.M. Cruysen, and I.R.K. Birtantie. 1988. The occurrence of *Campylobacter jejuni* in raw cows' milk. *J. Appl. Bacteriol.* 65:93–96.

7. Bodnaruk, P.W., R.G. Williams, and D.A. Golden. 1998. Survival of *Yersinia enterocolitica* during fermentation and storage of yogurt. *J. Food Sci.* 63:535–537.

8. Brown, W.V., and E.B. Collins. 1977. End products and fermentation balances for lactic streptococci grown aerobically on low concentrations of glucose. *Appl. Environ. Microbiol.* 33:38–42.

9. Carvajal, M., A. Bolanos, F. Rojo, and I. Méndez. 2003. Aflatoxin M_1 in pasteurized and ultrapasteurized milk with different fat content in Mexico. *J. Food Protect.* 66:1885–1892.

10. Centers for Disease Control and Prevention. 2002. Outbreak of *Campylobacter jejuni* infections associated with drinking unpasteurized milk procured through a cow-leasing program—Wisconsin, 2001. *Morb. Mort. Wkly. Rept.* 51:548–549.

11. Davis, J.G. 1975. The microbiology of yoghurt. In *Lactic Acid Bacteria in Beverages and Food*, ed. J.G. Carr, C.V. Cutting, and G.C. Whiting, 245–263. New York: Academic Press.

12. Doelle, H.A. 1975. *Bacterial Metabolism*. New York: Academic Press.

13. Douglas, S.A., M.J. Gray, A.D. Crandall, and K.J. Boor. 2000. Characterization of chocolate milk spoilage patterns. *J. Food Protect.* 63:516–521.

14. Doyle, M.P., and D.J. Roman. 1982. Prevalence and survival of *Campylobacter jejuni* in unpasteurized milk. *Appl. Environ. Microbiol.* 44:1154–1158

15. Food and Drug Administration, United States. 1995. Grade A pasteurized milk ordinance. Washington, D.C.: U.S. Department of Health and Human Services, Public Health Service.

16. Foster, E.M., F.E. Nelson, M.L. Speck, R.N. Doetsch, and J.C. Olson. 1957. *Dairy Microbiology*. Englewood Cliffs, N.J.: Prentice-Hall.

17. Frank, J.F. 2001. Milk and dairy products, In *Food Microbiology: Fundamentals and Frontiers*, 2nd ed., ed. M.P. Doyle, L.R. Beuchat, and T.J. Montville, 111–126. Washington, DC: ASM Press.

18. Garrote, G.L., A.G. Abraham, and G.L. de Antoni. 2000. Inhibitory power of kefir: The role of organic acids. *J. Food Protect.* 63:364–369.

19. Gilliland, S.E., and M.L. Speck. 1974. Frozen concentrated cultures of lactic starter bacteria: A review. *J. Milk Food Technol.* 37:107–111.

20. Gitter, M., R. Bradley, and P.H. Blampied. 1980. *Listeria monocytogenes* infection in bovine mastitis. *Vet. Rec.* 107:390–393.

21. Goodenough, E.R., and D.H. Kleyn. 1976. Qualitative and quantitative changes in carbohydrates during the manufacture of yoghurt. *J. Dairy Sci.* 59:45–47.

22. Grant, I.R., H.J. Ball, and M.T. Rowe. 2002a. Incidence of *Mycobacterium paratuberculosis* in bulk raw and commercially pasteurized cows' milk from approved dairy processing establishments in the United Kingdom. *Appl. Environ. Microbiol* 68:2428–2435.

23. Grant, I.R., E.I. Hitchings, A. McCartney, F. Ferguson, and M.T. Rowe. 2002b. Effect of commercial-scale high-temperature, short-time pasteurization on the viability of *Mycobacterium paratuberculosis* in naturally infected cows' milk. *Appl. Environ. Microbiol.* 68:602–607.

24. Grant, I.R., H.J. Ball, S.D. Neill, and M.T. Rowe. 1996. Inactivation of *Mycobacterium paratuberculosis* in cows' milk at pasteurization temperatures. *Appl. Environ. Microbiol.* 62:631–636.

25. Gunsalus, I.C., and C.W. Shuster. 1961. Energy yielding metabolism in bacteria. In *The Bacteria*, ed. I.C. Gunsalus and R.Y. Stanier, vol. 2, 1–58. New York: Academic Press.

26. Guraya, R., J.F. Frank, and A.N. Hassan. 1998. Effectiveness of salt, pH, and diacetyl as inhibitors of *Escherichia coli* 0157:H7 in dairy foods stored at refrigeration temperatures. *J. Food Protect.* 61:1098–1102.

27. Hamann, W.T., and E.H. Marth. 1984. Survival of *Streptococcus thermophilus* and *Lactobacillus bulgaricus* in commercial and experimental yogurts. *J. Food Protect.* 47:781–786.

28. Harlander, S.K., and L.L. McKay. 1984. Transformation of *Streptococcus sanguis Challis* with *Streptococcus lactis* plasmid DNA. *Appl. Environ. Microbiol.* 48:342–346.

29. Harlander, S.K., L.L. McKay, and C.F. Schachtels. 1984. Molecular cloning of the lactose-metabolizing genes from *Streptococcus lactis. Appl. Environ. Microbiol.* 48:347–351.

30. Harris, J.E., and A.M. Lammerding. 2001. Crohn's disease and *Mycobacterium avium* subsp. *paratuberculosis*: Current issues. *J. Food Protect.* 64:2103–2110.

31. Headrick, M.L., S. Korangy, N.H. Bean, F.J. Angulo, S.F. Altekruse, M.E. Potter, and K.C. Klontz. 1998. The epidemiology of raw milk-associated foodborne disease outbreaks reported in the United States, 1973 through 1992. *J. Amer. Public Health* 88:1219–1221.

32. Henning, D.R. 1999. Personal communication.

33. Hettinga, D.H., and G.W. Reinbold. 1972. The propionic-acid bacteria—A review. *J. Milk Food Technol.* 35:295–301, 358–372, 436–447.

34. Hugo, C.J., P. Segers, B. Hoste, M. Vancanneyt, and K. Kersters. 2003. *Chryseobacterium joostei* sp. nov., isolated from the dairy environment. *Int. J. Syst. Evol. Microbiol.* 53:771–777.

35. Jiang, X., and M.P. Doyle. 2002. Optimizing enrichment culture conditions for detecting *Helicobacter pylori* in foods. *J. Food Protect.* 65:1949–1954.

36. Johnson, M.G., and E.B. Collins. 1973. Synthesis of lipoic acid by *Streptococcus faecalis* 10C1 and end-products produced anaerobically from low concentrations of glucose. *J. Gen. Microbiol.* 78:47–55.

37. Klaenhammer, T.R. 2001. Probiotics and prebiotics. In *Food Microbiology: Fundamentals and Frontiers*, 2nd ed., ed. M.P. Doyle, L.R. Beuchat, and T.J. Montville, 97–811. Washington, D.C.: ASM Press.

38. Klein, G., A. Pack, and G. Reuter. 1998. Antibiotic resistance patterns of enterococci and occurrence of vancomycin-resistant enterococci in raw minced beef and pork in Germany. *Appl. Environ. Microbiol.* 64:1825–1830.

39. Klijn, N., F.F.J. Nieuwenhof, J.D. Hoolwerf, C.B. van der Waals, and A.H. Weerkamp. 1995. Identification of *Clostridium tyrobutyricum* as the causative agent of late blowing in cheese by species-specific PCR amplification. *Appl. Environ. Microbiol.* 61:2919–2924.

40. Kretchmer, N. 1972. Lactose and lactase. *Sci. Am.* 227(10):71–78.

41. Loessner, M.J., S.K. Maier, P. Schiwek, and S. Scherer. 1997. Long-chain polyphosphates inhibit growth of *Clostridium tyrobutyricum* in processed cheese spreads. *J. Food Protect.* 60:493–498.

42. London, J. 1976. The ecology and taxonomic status of the lactobacilli. *Ann. Rev. Microbiol.* 30:279–301.

43. Marth, E.H. 1974. Fermentations. In *Fundamentals of Dairy Chemistry*, ed. B.H. Webb, A.H. Johnson, and J.A. Alford, chap. 13. Westport, C.T.: AVI.

44. McComas, K.A., Jr., and S.E. Gilliland. 2003. Growth of probiotic and traditional yogurt cultures in milk supplemented with whey protein hydrolysate. *J. Food Sci.* 68:2090–2095.

45. Meer, R.R., J. Baker, F.W. Bodyfelt, and M.W. Griffiths. 1991. Psychrotrophic *Bacillus* spp. in fluid milk products: A review. *J. Food Protect.* 54:969–979.

46. Moustafa, M.K., A.A.-H. Admed, and E.H. Marth. 1983. Occurrence of *Yersinia enterocolitica* in raw and pasteurized milk. *J. Food Protect.* 46:276–278.

47. Mundt, J.O. 1975. Unidentified streptococci from plants. *Int. J. Syst. Bacteriol.* 25:281–285.

48. Muraoka, W., C. Gay, D. Knowles, and M. Borucki. 2003. Prevalence of *Listeria monocytogenes* subtypes in bulk milk of the Pacific Northwest. *J. Food Protect.* 66:1413–1419.

49. Murinda, S.E., K.T. Nguyen, S.J. Ivey, B.E. Gillespie, R.A., Almeida, F.A. Draughon, and S.P. Oliver. 2002. Prevalence and molecular characterization of *Escherichia coli* 0157:H7 in bulk tank milk and fecal samples from cull cows: A 12-month survey of dairy farms in east Tennessee. *J. Food Protect.* 65:752–759.

50. Murphy, S.C., and K.J. Boor. 2000. Trouble-shooting sources and causes of high bacteria counts in raw milk. *Dairy Fd. Environ. Sanit.* 20:606–611.

51. Newcomer, A.D., H.S. Park, P.C. O'Brien, and D.B. McGill. 1983. Response of patients with irritable bowel syndrome and lactase deficiency using unfermented acidophilus milk. *Am. J. Clin. Nutr.* 38:257–263.

52. National Academy of Science, USA. 1992. *Applications of Biotechnology to Traditional Fermented Foods.* Washington, D.C.: National Academy Press.

53. Ouwehand, A.C., and S.J. Salminen. 1998. The health effects of cultured milk products with viable and non-viable bacteria. *Int. Dairy J.* 8:749–758.

54. Pederson, C.S. 1979. *Microbiology of Food Fermentations*, 2nd ed. Westport, C.T.: AVI.

55. Pettersson, B., F. Lembke, P. Hammer, E. Stackebrandt, and F.G. Priest. 1996. *Bacillus sporothermodurans*, a new species producing highly heat-resistant endospores. *Int. J. System. Bacteriol.* 46:759–764.

56. Prescott, S.C., and C.G. Dunn. 1957. *Industrial Microbiology.* New York: McGraw-Hill.

57. Radke-Mitchell, L., and W.E. Sandine. 1984. Associative growth and differential enumeration of *Streptococcus thermophilus* and *Lactobacillus bulgaricus*: A review. *J. Food Protect.* 47:245–248.

58. Rosen, G. 1958. *A History of Public Health*, 358–360. New York: MD Publications.

59. Roth, L.A., L.F.L. Clegg, and M.E. Stiles. 1971. Coliforms and shelf life of commercially produced cottage cheese. *Can. Inst. Food Technol. J.* 4:107–111.

60. Rudolf, M., and S. Scherer. 2001. High incidence of *Listeria monocytogenes* in European red smear cheese. *Int. J. Food Microbiol.* 63:91–98.

61. Salminen, S., A. Ouwehand, Y.H. Benno, and Y.K. Lee. 1999. Probiotics: How should they be defined? *Trends Food. Sci. Technol.* 10:107–110.

62. Sanders, M.E. 1999. Probiotics. *Food Technol.* 53(11):67–77.

63. Sandine, W.E., P.C. Radich, and P.R. Elliker. 1972. Ecology of the lactic streptococci: A review. *J. Milk Food Technol.* 35:176–185.

64. Schleifer, K.H., and O. Kandler. 1972. Peptidoglycan types of bacterial cell walls and their taxonomic implications. *Bacteriol. Rev.* 36:407–477.

65. Schubert, K., W. Ludwig, N. Springer, R.M. Kroppenstedt, J.-P. Accolas, and F. Fiedler. 1996. Two coryneform bacteria isolated from the surface of French Gruyère and Beaufort cheeses are new species of the genus *Brachybacterium*: *Brachybacterium alimentarium* sp. nov. and *Brachybacterium tyrofermentans* sp. nov. *Int. J. Syst. Bacteriol.* 46:81–87.

66. Shin, M.-S., J.-H. Lee, J.J. Pestka, and Z. Ustunol. 2000. Viability of bifidobacteria in commercial dairy products during refrigerated storage. *J. Food Protect.* 63:327–331.

67. Shortt, C. 1998. The probiotic century: Historical and current perspectives. *Trends Food Sci. Technol.* 10:411–417.

68. Stabel, J.R., E.M. Steadham, and C.A. Bolin. 1997. Heat inactivation of *Mycobacterium paratuberculosis* in raw milk: Are current pasteurization conditions effective? *Appl. Environ. Microbiol.* 63:4975–4977.

69. Stamer, J.R. 1976. Lactic acid bacteria. In *Food Microbiology: Public Health and Spoilage Aspects*, ed. M.P. deFigueiredo and D.F. Splittstoesser, 404–426. New York: Kluwer Academic Publishers.

70. Stiles, M.E., and W.H. Holzapfel. 1997. Lactic acid bacteria of foods and their current taxonomy. *Int. J. Food Microbiol.* 36:1–29.

71. Stouthamer, A.H. 1969. Determination and significance of molar growth yields. *Methods Microbiol.* 1:629–663.

72. Sung, N., and M.T. Collins. 2000. Effect of three factors in cheese production (pH, salt, and heat) on *Mycobacterium avium* subsp. *paratuberculosis* viability. *Appl. Environ. Microbiol.* 66:1334–1339.

73. Thorel, M.-F., M. Krichevsky, and V.V. Levy-Frébault. 1990. Numerical taxonomy of mycobactin-dependent mycobacteria, emended description of *Mycobacterium avium*, and description of *Mycobacterium avium* subsp. *avium* subsp. nov., and *Mycobacterium avium* subsp. *silvaticum* subsp. nov. *Int. J. Syst. Bacteriol.* 40:254–260.

74. Tibana, A., M.B. Warnken, M.P. Nunes, I.D. Ricciardi, and A.L.S. Noleto. 1987. Occurrence of *Yersinia* species in raw and pasteurized milk in Rio de Janeiro, Brazil. *J. Food Protect.* 50:580–583.

75. Vancanneyt, M., J. Mengaud, I. Cleenwerck, K. Vanhonacker, H. Hoste, P. Dawyndt, M.C. Degivry, D. Ringuet, D. Janssens, and J. Swings. 2004. Reclassification of *Lactobacillus kefirgranum* Takizawa et al. 1994 *Lactobacillus kefiranofaciens* subsp. *kefirgranum* subsp. nov. and emended description of *L. kefiranofaciens* Fujisawa et al. 1988. *Int. J. Syst. Evol. Microbiol.* 54:551–556.

76. Vaclavik, V.A., and E.W. Christian. 2003. *Essentials of Food Science*, 2nd ed. New York: Springer.

77. Verschuere, L., G. Rombaut, P. Sorgeloos, and W. Verstraete. 2000. Probiotic bacteria as biological control agents in aquaculture. *Microbiol. Mol. Biol. Rev.* 64:655–671.

78. Vieira, E.R. 1996. *Elementary Food Science*, 4th ed., chap. 15. New York: Kluwer/Plenum Publishing, Inc.

79. Yukphan, P., W. Potacharoen, S. Tanasupawat, M. Tanticharoen, and Y. Yamada. 2004. *Asaia krungthepensis* sp. nov., an acetic acid bacterium in the alpha-*Proteobacteria*. *Int. J. Syst. Evol. Microbiol.* 54:313–316.

80. Yokota, A., T. Tamura, M. Takeuchi, N. Weiss, and E. Stackebrandt. 1994. Transfer of *Propionibacterium innocuum* Pitcher and Collins 1991 to *Propioniferax* gen. nov. as *Propioniferax innocua* comb. nov. *Int. J. Syst. Bacteriol.* 44:579–582.

81. Zhao, T., M.P. Doyle, and D.E. Berg. 2000. Fate of *Campylobacter jejuni* in butter. *J. Food Protect.* 63:120–122.

CHAPTER 8

Nondairy Fermented Foods and Products

MEAT PRODUCTS

Fermented sausages are produced generally as dry or semidry products, although some are intermediate. Dry or Italian-type sausages contain 30–40% moisture, are generally not smoked or heat processed, and are eaten usually without cooking.[58] In their preparation, curing and seasonings are added to ground meat, followed by its stuffing into casings and incubation for varying periods of time at 80–95°F. Incubation times are shorter when starter cultures are employed. The curing mixtures include glucose as substrate for the fermenters and nitrates and/or nitrites as color stabilizers. When only nitrates are used, it is necessary for the sausage to contain bacteria that reduce nitrates to nitrites, usually micrococci present in the sausage biota or added to the mix. Following incubation, during which fermentation occurs, the products are placed in drying rooms with a relative humidity of 55–65% for periods ranging from 10 to 100 days, or, in the case of Hungarian salami, up to 6 months.[47] Genoa and Milano salamis are other examples of dry sausages.

In one study of dry sausages, the pH was found to decrease from 5.8 to 4.8 during the first 15 days of ripening and remained constant thereafter.[36] Nine different brands of commercially produced dry sausages were found by these investigators to have pH values ranging from 4.5 to 5.2, with a mean of 4.87. With respect to the changes that occur in the biota of fermenting dry sausage when starters are not used, Urbaniak and Pezacki[82] found the homofermenters to predominate overall, with *Lactobacillus. plantarum* being the most commonly isolated species. Heterofermenters such as *L. brevis* and *L. buchneri* increased during the six-day incubation period as a result of changes in pH and Eh brought about by the homofermenters.

Semidry sausages are prepared in essentially the same way as dry sausages but are subjected to less drying. They contain about 50% moisture and are finished by heating to an internal temperature of 140–154°F (60–68°C) during smoking. Cervelat, summer sausage, and Lebanon bologna are some examples of semidry sausages. "Summer sausage" refers to those traditionally of northern European origin, made during colder months, stored, aged, and then eaten during summer months. They may be dry or semidry.

Lebanon bologna is typical of a semidry sausage. This product, originally produced in the Lebanon, Pennsylvania area, is an all-beef, heavily smoked, spiced product that may be prepared by the use of a *Pediococcus cerevisiae* starter.[19] The product is made by the addition of approximately 3% NaCl along with sugar, seasoning, and either nitrate, nitrite, or both, to raw cubed beef. The salted beef is allowed to age at refrigerator temperature for about 10 days during which time the growth of naturally occurring lactic acid bacteria or the starter organisms are encouraged and Gram negatives are inhibited. A higher level of microbial activity occurs along with some drying during the smoking step at higher temperatures. A controlled production process for this product has been studied,[52] and it consists of aging salted beef at 5°C for 10 days and smoking at 35°C with high relative humidity (RH) for 4 days. Fermentation may be carried out either by the natural biota of the meat or by the use of a commercial starter of *P. cerevisiae* or *P. acidilactici*. The amount of acidity produced in Lebanon bologna may reach 0.8–1.2%.[8,57]

The hazard of eating improperly prepared, homemade, fermented sausage was indicated by an outbreak of trichinosis: of the 50 persons who actually consumed the raw summer sausage, 23 fell ill with trichinosis.[62] The sausage was made on two different days in three batches according to a family recipe that called for smoking at cooler smoking temperatures, believed to produce a better-flavored product. All three batches of sausages contained home-raised beef. In addition, two batches eaten by victims contained pork inspected by the U.S. Department of Agriculture (USDA) in one case and home-raised pork in the other, but *Trichinella spiralis* larvae were found only in the USDA-inspected pork. This organism can be destroyed by a heat treatment that results in internal temperatures of at least 60°C or 140°F (see Chapter 29).

In the production of dry sausages, lactobacilli produce aminopeptidases that aid in the generation of amino acids from sausage proteins. The amino acids contribute to the overall flavor of dry sausages. In the case of *Lactobacillus sakei*, it produces decarboxylases that give rise to biogenic amines, and these compounds can inhibit aminopeptidases and thus reduce flavor enhancement in dry-fermented sausages (see reference 71).

Fermented sausages produced without the use of starters have been found to contain large numbers of lactobacilli such as *L. plantarum*.[20] The use of a *P. cerevisiae* starter leads to the production of a more desirable product.[19,36] In their study of commercially produced fermented sausages, Smith and Palumbo[77] found total aerobic plate counts to be in the 10^7–10^8/g range, with a predominance of lactic acid types. When starter cultures were used, the final pH of the products ranged from 4.0 to 4.5, whereas those produced without starters ranged between 4.6 and 5.0. For summer-type sausages, pH values of 4.5–4.7 have been reported for a 72-hour fermentation.[2] These investigators found that fermentation at 30°C and 37°C led to a lower final pH than at 22°C and that the final pH was directly related to the amount of lactic acid produced. The pH of fermented sausage may actually increase by 0.1 or 0.2 unit during long periods of drying due to uneven buffering produced by increases in the amounts of basic compounds.[90] The ultimate pH attained following fermentation depends on the type of sugar added. Although glucose is most widely used, sucrose has been found to be an equally effective fermentable sugar for low pH production.[1] The effect of a commercial frozen concentrate starter (*P. acidilactici*) in fermenting various sugars added to a sausage preparation is illustrated in Figure 8–1. *Lactobacillus gasseri*, when employed in a meat fermentation was shown to prevent enterotoxin formation by *Staphylococcus aureus* in a model sausage preparation.[6] This species was the most effective of five other *Lactobacillus* species.

Prior to the late 1950s, the production of fermented sausages was facilitated by either back inoculations, or a producer took the chance of the desired organisms being present in the raw materials. Until recently, the manufacture of these, as well as of many other fermented foods, has been more of an art than a science. With the advent of pure culture starters, not only has production time been shortened, but more uniform and safer products can be produced.[25] Although the use of starter cultures has been

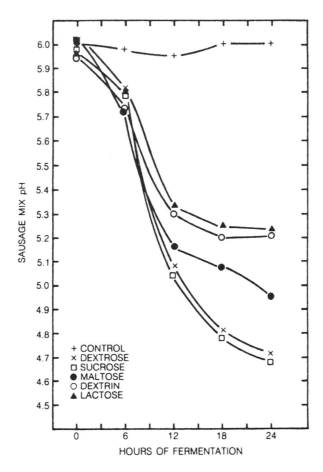

Figure 8–1   Rate of pH reduction in fermenting sausage containing 0% or 1% of various carbohydrates. *Source*: From Acton et al.[1] copyright © 1977 by Institute of Food Technologists.

in effect for many years in the dairy industry, their use in many nondairy products worldwide is a recent development with great promise. "*Micrococcus aurantiacus*" has been employed along with starters in the production of some European sausages.[47] The addition of a *Micrococcus* or a *Staphylococcus*, especially *S. carnosus*, to a lactic culture is a common practice in Europe. The nonlactic member reduces nitrates to nitrites and produces catalase that benefits the lactic culture.

Molds are known to contribute to the quality of dry European-type sausages such as Italian salami. In an extensive study of the fungi of ripened meat products, Ayres et al.[7] found nine species of penicillia and seven of aspergilli on fermented sausages and concluded that the organisms play a role in the preservation of products of this type. Fewer species of other mold genera were found. A study of the fungal biota of naturally fermented sausages in northern Italy revealed that penicillia made up 96% and aspergilli 4%.[5] The initial biota of the sausage was made up of >95% yeasts. After 2 weeks, yeasts and molds were about 50:50, but after 4–8 weeks, molds constituted >95% of the biota.[5] Fifty percent of the mold biota was *P. nalgiovensis*. The addition of *Penicillium camemberti* and *P. nalgiovensis* during the curing of raw, dry, sausages was used in an effort to prevent the growth of mycotoxigenic house molds, and it was more successful than potassium sorbate.[11]

Country-cured hams are dry-cured hams produced in the southern United States. During the curing and ripening period of 6 months to 2 years, heavy mold growth occurs on the surfaces. Although Ayres et al.[7] noted that the presence of molds is incidental and that a satisfactory cure does not depend on their presence, it seems likely that some aspects of flavor development of these products derive from the heavy growth of such organisms, and to a lesser extent from yeasts. Heavy mold growth obviates the activities of food-poisoning and food-spoilage bacteria, and in this sense the mold biota aids in preservation. Ayres et al. found aspergilli and penicillia to be the predominant types of molds on country-cured hams.[7]

The processing of country-cured hams takes place during the early winter and consists of rubbing sugar cure into the flesh side and onto the hock end. This is followed some time later by rubbing NaCl into all parts of the ham not covered by skin. The hams are then wrapped in paper and individually placed in cotton fabric bags and left lying flat for several days between 32°C and 40°C. The hams are hung shank end down in ham houses for 6 weeks or longer and may be given a hickory smoke during this time, although smoking is not essential for a desirable product.

Italian-type country-cured hams are produced with NaCl as the only cure. Curing is carried out for about a month, followed by washing, drying, and ripening for 6–12 months or longer.[33] Although halophilic and halotolerant bacteria increase as Italian hams ripen, the biota, in general, is thought to play only a minor role.[56] In Europe, molds are critical in the production of safe and high quality products such as salami and hams (Parma from Italy and Solano from Spain). For more detailed information on meat starter cultures and formulations for fermented sausages, along with cure ingredients for country-style hams, see references 6 and 57.

FISH PRODUCTS

Fermented fishery products are rather widespread in parts of Asia where marine sources contribute more protein to the human diet than is the case in the Western world. More on fermentation can be found in Chapter 7. Only two classes of fermented seafood products are noted below—sauces and pastes.

Fish sauces are popular products in Southeast Asia, where they are known by various names such as *ngapi* (Burma), *nuoc-mam* (Cambodia and Vietnam), *nam-pla* (Laos and Thailand), *ketjap-ikan* (Indonesia), and so on. The production of some of these sauces begins with the addition of salt to uneviscerated fish at a ratio of approximately 1:3, salt to fish. The salted fish are then transferred to fermentation tanks generally constructed of concrete and built into the ground or placed in earthenware pots and buried in the ground. The tanks or pots are filled and sealed off for at least 6 months to allow the fish to liquefy. The liquid is collected, filtered, and transferred to earthenware containers and ripened in the sun for 1–3 months. The finished product is described as being clear, dark-brown in color with a distinct aroma and flavor.[70] In a study of fermenting Thai fish sauce by the latter investigators, the pH from start to finish ranged from 6.2 to 6.6 with the NaCl content around 30% over the 12-month fermentation period.[70] These parameters, along with the relatively high fermentation temperature, result in the growth of halophilic aerobic spore formers as the predominant microorganisms of these products. Lower numbers of streptococci, micrococci, and staphylococci were found, and they, along with the *Bacillus* spp., were apparently involved in the development of flavor and aroma. Some part of the liquefaction that occurs is undoubtedly due to the activities of fish proteases. Although the temperature and pH of the fermentation are well within the growth range of a large number of undesirable organisms, the safety of products of this type is due to the 30–33% NaCl.

Fish pastes are also common in Southern Asia, but the role of fermenting microorganisms in these products appears to be minimal. Among the many other fermented fish, fish-paste, and fish-sauce products, are the following: *mam-tom* of China; *mam-ruoc* of Cambodia; *bladchan* of Indonesia; *shiokara* of Japan; *belachan* of Malaya; *bagoong* of the Philippines; *kapi, hoi-dong*, and *pla-mam* of Thailand; *fessik* of Africa; *nam-pla*, pla-ra, *pla-chom*, and *pla-com* of Thailand. A fermented shrimp product of Thailand is *kung-chom*.

Soy sauces are fermented condiments of various plant materials. Typically, the plant material first undergoes a fungal fermentation followed by a brine fermentation in which *Tetragenococcus* spp. are active. In Chinese soy sauce only soy beans are used, whereas in Japanese both wheat and soy beans are used. *T. halophilus*, which can tolerate 18% NaCl, is active in the brine of the soy sauces noted.[68] Another *Tetragenococcus* sp., *T. muriaticus*, has been isolated from fermented squid liver sauce.[72] This species can grow in 1–25% NaCl, and it produces histamine. Some soy sauces are made by acid hydrolysis of soy beans.

BREADS

San Francisco *sourdough* bread is similar to sourdough breads produced in various countries. Historically, the starter for sourdough breads consists of the natural biota of baker's barm (sour ferment or mother sponge, with a portion of each inoculated dough saved as starter for the next batch). The barm generally contains a mixture of yeasts and lactic acid bacteria. In the case of San Francisco sourdough bread, the yeast has been identified as *Saccharomyces exiguus* (*Candida holmii*[80] and the responsible bacteria are *Lactobacillus sanfranciscensis, L. fermentum, L. fructivorans*, some *L. brevis* strains, and *L. pontis*.[89] The key bacterium is *L. sanfranciscensis*, and it preferentially ferments maltose rather than glucose and it requires fresh yeast extractives and unsaturated fatty acids.[34] The souring is caused by acids produced by these bacteria, and the yeast is responsible for the leavening action, although some CO_2 is produced by the bacterial biota. The pH of these sourdoughs ranges from 3.8 to 4.5. Both acetic and lactic acids are produced, with the former accounting for 20–30% of the total acidity.[40] *Lactobacillus paralimentarius* is another of the sourdough bacteria.[15]

Sourdoughs are placed into three groups and each has its unique fermentation consortium. Type I sourdoughs are fermented at 20–30°C and the two primary organisms are *L. sanfranciscensis* and *L. pontis*. Type II doughs employ baker's yeast as a leavening agent, and the dominant lactics are *L. pontis, L. panis*, and from one to nine other lactobacilli. Type III doughs are dried products of traditional fermentations (see reference 21). Greek wheat sourdoughs belong to Type I, and the fermentation consortium in the traditional wheat product consists of *L. sanfranciscensis, L. brevis, L. paralimentarius*, and *Weissella cibaria*.[21] Among other organisms found in some sourdough fermentations are *Candida humilis, Dekkera bruxellensis, Saccharomyces cerevisiae*, and *Saccharomyces uvarum*.

Idli is a fermented bread-type product common in southern India. It is made from rice and black gram mung (urd beans). These two ingredients are soaked in water separately for 3–10 hours and then ground in varying proportions, mixed, and allowed to ferment overnight. The fermented and raised product is cooked by steaming and served hot. It is said to resemble a steamed, sourdough bread.[78] During the fermentation, the initial pH of around 6.0 falls to values of 4.3–5.3. In a particular study, a batter pH of 4.70 after a 20-hour fermentation was associated with 2.5% lactic acid, based on dry grain weight.[46] In their studies of idli, Steinkraus et al.[78] found total bacterial counts of 10^8–10^9/g after 20–22 hours of fermentation. Most of the organisms consisted of Gram-positive cocci or short rods, with *L. mesenteroides* being the single most abundant species, followed by *E. faecalis*. The leavening action of idli is produced by *L. mesenteroides*. This is the only known instance of a lactic acid bacterium having this role in a naturally fermented bread.[46] The latter authors confirmed the work of others in finding the urd beans to be a more important source of lactic acid bacteria than rice. *L. mesenteroides* reaches its peak at around 24 hours, with *E. faecalis* becoming active only after about 20 hours. Other probable fermenters include *L. delbrueckii* subsp. *delbrueckii*, *L. fermentum*, and *Bacillus* spp.[69] Only after idli has fermented for more than 30 hours does *P. cerevisiae* become active. The product is not fermented generally beyond 24 hours because maximum leavening action occurs at this time and decreases with longer incubations. When idli is allowed to ferment longer, more acidity is produced. It has been found that total acidity (expressed as grams of lactic acid per gram of dry grains) increased from 2.71% after 24 hours to 3.70% after 71 hours, whereas the pH decreased from 4.55 to 4.10 over the same period.[65] (A review of idli fermentation has been made by Reddy et al.[65])

PLANT PRODUCTS

Sauerkraut

Sauerkraut is a fermentation product of fresh cabbage. The starter for sauerkraut production is usually the normal mixed biota of cabbage. The addition of 2.25–2.5% salt restricts the activities of Gram-negative bacteria, while the lactic acid rods and cocci are favored. *Leuconostoc mesenteroides*, *Lactobacillus plantarum*, and *Leuconostoc fallax* are the three most dominant lactics in sauerkraut production, with the two *Leuconostoc* spp. having the shorter generation time and the shorter life span. The activities of the cocci usually cease when acid content increases to 0.7–1.0%. The final stages of kraut production are effected by *L. plantarum* and *L. brevis*. *P. cerevisiae* and *E. faecalis* may also contribute to product development. The final total acidity is generally 1.6–1.8%, with lactic acid at 1.0–1.3% and pH in the range of 3.1 to 3.7.

The microbial spoilage of sauerkraut generally falls into the following categories: soft kraut, slimy kraut, rotted kraut, and pink kraut. Soft kraut results when bacteria that normally do not initiate growth until the late stages of kraut production actually grow earlier. Slimy kraut is caused by the rapid growth of *Lactobacillus cucumeris* and *L. plantarum*, especially at elevated temperatures. Rotted sauerkraut may be caused by bacteria, molds, and/or yeasts, whereas pink kraut is caused by the surface growth of *Torula* spp., especially *T. glutinis*. Due to the high acidity, finished kraut is generally spoiled by molds growing on the surface. The growth of these organisms effects an increase in pH to levels where a large number of bacteria can grow which were previously inhibited by conditions of high acidity.

Olives

Olives to be fermented (Spanish, Greek, or Sicilian) are done so by the natural biota of green olives, which consists of a variety of bacteria, yeasts, and molds. The olive fermentation is quite similar to that of sauerkraut except that it is slower, involves a lye treatment, and may require the addition of starters. The lactic acid bacteria become prominent during the intermediate stage of fermentation. *L. mesenteroides* and *P. cerevisiae* are the first lactics to become prominent, and these are followed by lactobacilli, with *L. plantarum* and *L. brevis* being the most important.[87]

The olive fermentation is preceded by a treatment of green olives with 1.6 to 2.0% lye, depending on the type of olive, at 21–25°C for 4–7 hours for the purpose of removing some of the bitter principal. Following the complete removal of lye by soaking and washing, the green olives are placed in oak barrels and brined so as to maintain a constant 28°–30° salinometer level. Inoculation with *L. plantarum* may be necessary because of destruction of organisms during the lye treatment. The fermentation may take as long as 6–10 months, and the final product has a pH of 3.8–4.0 following up to 1% lactic acid production.

Among the types of microbial spoilage that olives undergo, one of the most characteristic is *zapatera spoilage*. This condition, which sometimes occurs in brined olives, is characterized by a malodorous fermentation. The odor is due apparently to propionic acid, which is produced by certain species of *Propionibacterium*.[61] In addition to propionic acid, formic, butyric, succinic, isobutyric, *n*-valeric, and cyclohexacarbolic acids, as well as methanol, ethanol, 2-butanol, and *n*-butanol may be produced (see reference 31).

A softening condition of Spanish-type green olives has been found to be caused by the yeasts *Rhodotorula glutinis* var. *glutinis*, *R. minuta* var. *minuta*, and *R. rubra*.[88] All of these organisms produce polygalacturonases, which effect olive tissue softening. Under appropriate cultural contions, the organisms were shown to produce pectin methyl esterase as well as polygalacturonase. A sloughing type of spoilage of California ripe olives was shown by Patel and Vaughn[55] to be caused by *Cellulomonas flavigena*. This organism showed high cellulolytic activity, which was enhanced by the growth of other organisms such as *Xanthomonas*, *Enterobacter*, and *Escherichia* spp.

The production of some biogenic amines has been shown to occur primarily in Spanish-style green olives during the brining process.[32] The amines found were cadaverine, histamine, tyramine, tryptamine, and putrescine with the latter being in highest concentration after 3 months of brining. The others were found in samples taken after 12 months.[32]

Pickles

Pickles are fermentation products of fresh cucumbers, and as is the case for sauerkraut production, the starter culture generally consists of the normal mixed biota of cucumbers. In the natural production of pickles, the following lactic acid bacteria are involved in the process in order of increasing prevalence: *L. mesenteroides*, *E. faecalis*, *P. cerevisiae*, *L. brevis*, and *L. plantarum*. Of these, the pediococci and *L. plantarum* are the most involved, with *L. brevis* being undesirable because of its capacity to produce gas. *L. plantarum* is the most essential species in pickle production, as it is for sauerkraut.

In the production of pickles, selected cucumbers are placed in wooden brine tanks with initial brine strengths as low as 5% NaCl (20° salinometer). Brine strength is increased gradually during the course of the 6- to 9-week fermentation, until it reaches around 60° salinometer (15.9% NaCl). In addition to exerting an inhibitory effect on the undesirable Gram-negative bacteria, the salt extracts water and water-soluble constituents from the cucumbers, such as sugars, which are converted by the lactic acid bacteria to lactic acid. The product that results is a salt-stock pickle from which pickles such as sour, mixed sour, chowchow, and so forth may be made.

The general technique of producing brine-cured pickles has been in use for many years, but it often leads to serious economic loss because of pickle spoilage from such conditions as bloaters, softness, off-colors, and so on. The controlled fermentation of cucumbers brined in bulk has been achieved, and this process not only reduces economic losses of the type noted, but leads to a more uniform product over a shorter period of time. The controlled fermentation method employs a chlorinated brine of 25° salinometer, acidification with acetic acid, the addition of sodium acetate, and inoculation with

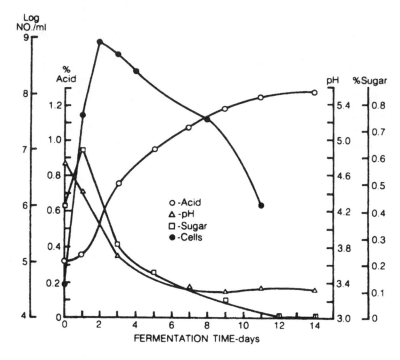

Figure 8–2   Controlled fermentation of cucumbers brined in bulk. Equilibrated brine strength during fermentation, 6.4% NaCl; incubation temperature = 27°C. *Source*: From Etchells et al.[24]; copyright © 1975 by Academic Press.

P. cerevisiae and *L. plantarum*, or the latter alone. The course of the 10- to 14-day fermentation is represented in Figure 8–2.

With a final pH of ∼4.0, pickles undergo spoilage by bacteria and molds. Pickle blackening may be caused by *Bacillus nigrificans*, which produces a dark water-soluble pigment. *Enterobacter* spp., lactobacilli, and pediococci have been implicated as causes of a condition known as "bloaters," produced by gas formation within the individual pickles. Pickle softening is caused by pectolytic organisms of the genera *Bacillus, Fusarium, Penicillium, Phoma, Cladosporium, Alternaria, Mucor, Aspergillus,* and others. The actual softening of pickles may be caused by any one or several of these or related organisms. Pickle softening results from the production of pectinases, which break down the cementlike substance in the wall of the product.

BEER, ALE, WINES, CIDER, AND DISTILLED SPIRITS

Beer and Ale

Beer and ale are malt beverages produced by brewing. An essential step in the brewing process is the fermentation of carbohydrates to ethanol. Because most of the carbohydrates in grains used for brewing exist as starches, and because the fermenting yeasts do not produce amylases to degrade the starch, a necessary part of beer brewing includes a step whereby malt or other exogenous sources of amylase are provided for the hydrolysis of starches to sugars. The malt is first prepared by allowing barley grains to germinate. This serves as a source of amylases (fungal amylases may be used also). Both β- and α-amylases are involved, with the latter acting to liquefy starch and the former to increase sugar formation. In brief, the brewing process begins with the mixing of malt, malt adjuncts, hops, and water. Malt adjuncts include certain grains, grain products, sugars, and other carbohydrate products to serve as fermentable substances. Hops are added as sources of pyrogallol and catechol tannins, resins, essential oils, and other constituents for the purpose of precipitating unstable proteins during the boiling of wort and to provide for biological stability, bitterness, and aroma. The process by which the malt and malt adjuncts are dissolved and heated and the starches digested is called mashing. The soluble part of the mashed materials is called *wort* (compare with *koji*). In some breweries, lactobacilli are introduced into the mash to lower the pH of wort through lactic acid production. The species generally used for this purpose is *L. delbrueckii* subsp. *delbrueckii*.[39]

Wort and hops are mixed and boiled for 1.5–2.5 hours for the purpose of enzyme inactivation, extraction of soluble hop substances, precipitation of coagulable proteins, concentration, and sterilization. Following the boiling of wort and hops, the wort is separated, cooled, and fermented. The fermentation of the sugar-laden wort is carried out by the inoculation of *S. cerevisiae*. Ale results from the activities of top-fermenting yeasts, which depress the pH to around 3.8, whereas bottom-fermenting yeasts (*S. "carlsbergensis"* strains) give rise to lager and other beers with pH values of 4.1–4.2. A top fermentation is complete in 5–7 days; a bottom fermentation requires 7–12 days. The freshly fermented product is aged and finished by the addition of CO_2 to a final content of 0.45–0.52% before it is ready for commerce. The pasteurization of beer at 140°F (60°C) or higher, may be carried out for the purpose of destroying spoilage organisms. When lactic acid bacteria are present in beers, the lactobacilli are found more commonly in top fermentations, whereas pediococci are found in bottom fermentations.[39]

The industrial spoilage of beers and ales is commonly referred to as beer infections. This condition is caused by yeasts and bacteria. The spoilage patterns of beers and ales may be classified into four groups: ropiness, sarcinae sickness, sourness, and turbidity. *Ropiness* is a condition in which the liquid becomes characteristically viscous and pours as an "oily" stream. It is caused by *Acetobacter*, *Lactobacillus*, *Pediococcus cerevisiae*, and *Gluconobacter oxydans* (formerly *Acetomonas*).[26,64,91] *Sarcinae sickness* is caused by *P. cerevisiae*, which produces a honeylike odor. This characteristic odor is the result of diacetyl production by the spoilage organism in combination with the normal odor of beer. *Sourness* in beers is caused by *Acetobacter* spp. These organisms are capable of oxidizing ethanol to acetic acid, and the sourness that results is referable to increased levels of acetic acid. *Turbidity* and off-odors in beers are caused by *Zymomonas anaerobia* (formerly *Achromobacter anaerobium*) and several yeasts such as *Saccharomyces* spp. The growth of bacteria is possible in beers because of a normal pH range of 4–5 and a good content of utilizable nutrients.

Some Gram-negative obligately anaerobic bacteria have been isolated from spoiled beers and pitching yeasts, and the six species are represented by four genera:

Megasphaera cerevisiae	*Selenomonas lacticifex*
Pectinatus cerevisiiphilus	*Zymophilus paucivorans*
P. frisingensis	*Z. raffinosivorans*

All but *M. cerevisiae* produce acetic and propionic acids, and *S. lacticifex* also produces lactate.[73] Although *M. cerevisiae* produces negligible to minor amounts of acetic and propionic acids, it produces large quantities of isovaleric acid in addition to H_2S.[23] *P. cerevisiiphilus* was the first of these to be associated with spoiled beer when it was isolated from turbid and off-flavor beer in 1978.[43] It has since been found in breweries not only in the United States, but also in several European countries and Japan. Among the unusual features of these organisms as beer spoilers is their Gram reaction and obligately anaerobic status. Previously, the typical beer spoilers were regarded as being either lactic acid bacteria or yeasts. *Megasphaera* and *Selenomonas* are best known as members of the rumen biota. In addition to the organic acids noted above, *Pectinatus* spp. also produce H_2S and acetoin. The beers most susceptible to their growth are those that contain <4.4% alcohol.

With respect to spoiled packaged beer, one of the major contaminants found is *Saccharomyces diastaticus*, which is able to utilize dextrins that normal brewers' yeasts (*S. "carlsbergensis"* and *S. cerevisiae*) cannot.[39] Pediococci, *Flavobacterium proteus* (formerly *Obesumbacterium*), and *Brettanomyces* are sometimes found in spoiled beer.

Wines

Wines are normal alcoholic fermentations of sound grapes followed by aging. A large number of other fruits such as peaches, pears, and so forth may be fermented for wines, but in these instances the wine is named by the fruit, such as peach wine, pear wine, and the like. Because fruits already contain fermentable sugars, the use of exogenous sources of amylases is not necessary, as it is when grains are used for beers or whiskeys. Wine making begins with the selection of suitable grapes, which are crushed and then treated with a sulfite such as potassium metabisulfite to retard the growth of acetic acid bacteria, wild yeasts, and molds. The pressed juice, called *must*, is inoculated with a suitable wine strain of *S. "ellipsoideus."* The fermentation is allowed to continue for 3–5 days at temperatures between 70°F and 90°F (21°C and 32°C), and good yeast strains may produce up to 14–18% ethanol.[58] Following fermentation, the wine is racked–that is, drawn off from the lees or sediment, which contains potassium bitartrate (cream of tartar). The clearing and development of flavor occur during the storage and aging process. Red wines are made by initially fermenting the crushed grape must "on the skins" during which pigment is extracted into the juice; white wines are prepared generally from the juice of white grapes. Champagne, a sparkling wine made by a secondary fermentation of wine, is produced by adding sugar, citric acid, and a champagne yeast starter to bottles of a previously prepared, selected table wine. The bottles are corked, clamped, and stored horizontally at suitable temperatures for about 6 months. They are then removed, agitated, and aged for an additional period of up to 4 years. The final sedimentation of yeast cells and tartrates is accelerated by reducing the temperature of the wine to around 25°C and holding for 1–2 weeks. Clarification of the champagne is brought about by working the sediment down the bottle onto the cork over a period of 2–6 weeks by frequent rotation of the bottle. Finally, the sediment is frozen and disgorged upon removal of the cork.

Table wines undergo spoilage by bacteria and yeasts with *Candida valida* being the most important yeast. Growth of this organism occurs at the surface of wines, where a thin film is formed. The organisms attack alcohol and other constituents from this layer and create an appearance that is sometimes referred to as *wine flowers*. Among the bacteria that cause wine spoilage are members of the genus *Acetobacter*, which oxidize alcohol to acetic acid (produce vinegar). The most serious and the most common disease of table wines is referred to as *tourne disease*. Tourne disease is caused by a facultative anaerobe or an anaerobe that utilizes sugars and seems to prefer conditions of low alcohol content. This type of spoilage is characterized by an increased volatile acidity, a silky type of cloudiness, and later in the course of spoilage, a "mousy" odor and taste caused by *Brettanomyces* spp. commonly found in Bordeaux wines.

Malo-lactic fermentation is a spoilage condition of great importance in wines. Malic and tartaric acids are two of the predominant organic acids in grape must and wine, and in the malolactic fermentation, contaminating bacteria degrade malic acid to lactic acid and CO_2:

$$L(-)\text{-Malic acid} \xrightarrow{\text{``malo-lactic enzyme''}} L(+)\text{-Lactic acid} + CO_2$$

L-Malic acid may be decarboxylated also to yield pyruvic acid.[41] The effect of these conversions is to reduce the acid content and affect flavor. The malo-lactic fermentation (which may also occur in cider) can be carried out by many lactic acid bacteria, including leuconostocs, pediococci, and lactobacilli.[63] Although the function of the malolactic fermentation to the fermenting organism is not well understood, it has been shown that *Oenococcus oeni* is actually stimulated by the process.[60] The decomposition in wines of tartaric acid is undesirable also, and this process can be achieved by some strains of *Lactobacillus plantarum* in the following general manner:

$$\text{Tartaric acid} \longrightarrow \text{Lactic acid} + \text{Acetic acid} + CO_2$$

The effect is to reduce the acidity of wine. Unlike the malo-lactic fermentation, few lactic acid bacteria break down tartaric acid. Sluggish or stuck alcoholic fermentations of wines is caused by *Lactobacillus kunkeei* and *L. nagelii*.

The bacterium *Oenococcus oeni* is an acidophile that can grow in grape must and wine at pH 3.5–3.8, and actually prefers an initial growth pH of 4.8.[22] It can grow in the presence of 10% ethanol but requires special growth factors found in grape or tomato juice. For a review, see reference 44.

Cider

Cider, in the United States, is a product that represents a mild fermentation of apple juice by naturally occurring yeasts. In making apple cider, the fruits are selected, washed, and ground into a pulp. The pulp "cheeses" are pressed to release the juice. The juice is strained and placed in a storage tank, where sedimentation of particulate matter occurs, usually for 12–36 hours or several days if the temperature is kept at 40°F (4.4°C) or below. The clarified juice is cider. If pasteurization is desired, this is accomplished by heating at 170°F (76.7°C) for 10 minutes. The chemical preservative most often used is sodium sorbate at a level of 0.10%. Preservation may also be effected by chilling or freezing. The finished product contains small amounts of ethanol in addition to acetaldehyde. The holding of nonpasteurized or unpreserved cider at suitable temperatures invariably leads to the development of cider vinegar, which indicates the presence of acetic acid bacteria in these products. The pathway employed by acetic acid bacteria is summarized in Chapter 7, Figure 7–1F, G.

In their study of the ecology of the acetic acid bacteria in cider manufacture, Passmore and Carr[53] found six species of *Acetobacter* and noted that those that display a preference for sugars tend to be found early in the cider process, whereas those that are more acid tolerant and capable of oxidizing alcohols appear after the yeasts have converted most of the sugars to ethanol. *Zymomonas* spp., Gram-negative bacteria that ferment glucose to ethanol, have been isolated from ciders, but they are presumed to be present in low numbers. *Saccharobacter fermentatus* is similar to *Zymomonas* in that it ferments glucose to ethanol and CO_2.[92] It was isolated from agave leaf juice, but its presence and possible role in spoiled ciders have yet to be determined. *Zymobacter palmae* is an ethanol fermentor isolated from palm sap.[51] It produces ethanol from mannitol.[37]

Following several illness outbreaks traced to apple cider, the microbial load of finished ciders produced in the state of Iowa was investigated, and in apples from 21 producers, APC ranged from 15 to > 1.1×10^5/ml; coliforms from <1 to 2.1×10^3; and *E. coli* was <10/ml.[18] The fate of *E. coli* 0157:H7 in fermenting apple cider was studied by Semanchek and Golden[75] who found that $6.4 \log_{10}$ cfu/ml of this organism were reduced to <0.5 $\log_{10}$ cfu/ml after 3 days at 20°C, while in nonfermenting ciders, the initial number was reduced only to 2.9 $\log_{10}$ cfu/ml after 10 days at 20°C. Although the pH of the two products was not significantly different, ethanol concentration in the fermenting products reached 6.1% after 10 days at 20°C. The investigators believed the combination of low pH and the ethanol were important in the reduction of the pathogen. In another study of *E. coli* 0157:H7 survival in four apple varieties (Golden Delicious, Red Delicious, Rome, and Winesap), the organism behaved essentially the same in each variety.[28]

Distilled Spirits

Distilled spirits are alcoholic products that result from the distillation of yeast fermentations of grain, grain products, molasses, or fruit or fruit products. Whiskeys, gin, vodka, rum, cordials, and liqueurs are examples of distilled spirits. Although the process for producing most products of these types is quite similar to that for beers, the content of alcohol in the final products is considerably higher than for beers. *Rye* and *bourbon* are examples of whiskeys. In the former, rye and rye malt, or rye and barley malt, are used in different ratios, but at least 51% rye is required by law. Bourbon is made from corn, barley malt, or wheat malt, and usually another grain in different proportions, but at least 51% corn is required by law. A sour wort is maintained to keep down undesirable organisms, the souring occurring naturally or by the addition of acid. The mash is generally soured by inoculating with a homolactic such as *L. delbrueckii* subsp. *delbrueckii*, which is capable of lowering the pH to around 3.8 in 6–10 hours.[58] The malt enzymes (diastases) convert the starches of the cooked grains to dextrins and sugars, and upon completion of diastatic action and lactic acid production, the mash is heated to destroy all microorganisms. It is then cooled to 75–80°F (24–27°C) and pitched (inoculated) with a suitable strain of *S. cerevisiae* for the production of ethanol. Upon completion of fermentation, the liquid is distilled to recover the alcohol and other volatiles, and these are handled and stored under special conditions relative to the type of product being made. *Scotch whisky* is made primarily from barley and is produced from barley malt dried in kilns over peat fires. *Rum* is produced from the distillate of fermented sugar cane or molasses. *Brandy* is a product prepared by distilling grape or other fruit wines.

Palm wine or Nigerian palm wine is an alcoholic beverage consumed throughout the tropics and is produced by a natural fermentation of palm sap. The sap is sweet and dirty brown in color, and it contains 10–12% sugar, mainly sucrose. The fermentation process results in the sap becoming milky-white in appearance due to the presence of large numbers of fermenting bacteria and yeasts. This product is unique in that the microorganisms are alive when the wine is consumed. The fermentation has been reviewed and studied by Faparusi and Bassir[26] and Okafor[49] who found the following genera of bacteria to be the most predominant in finished products: *Micrococcus*, *Leuconostoc*, "*Streptococcus*," *Lactobacillus*, and *Acetobacter*. The predominant yeasts found were *Saccharomyces* and *Candida* spp., with the former being the more common.[48] The fermentation occurs over a 36- to 48-hour period, during which the pH of sap falls from 7.0 or 7.2 to <4.5. Fermentation products consist of organic acids in addition to ethanol. During the early phases of fermentation, *Serratia* and *Enterobacter* spp. increase in numbers, followed by lactobacilli and leuconostocs. After a 48-hour fermentation, *Acetobacter* spp. begin to appear.[26,50]

Sake is an alcoholic beverage commonly produced in Japan. The substrate is the starch from steamed rice, and its hydrolysis to sugars is carried out by *A. oryzae* to yield the koji. Fermentation is carried out by *Saccharomyces sake* over periods of 30–40 days, resulting in a product containing 12–15% alcohol and around 0.3% lactic acid.[58] The latter is produced by hetero- and homolactic lactobacilli. Other fermented products of this type are further summarized in Table 8–1.

Table 8–1 Summary of a Variety of Fermented Products

Products	Substrate	Fermenters	Where Found
Nonbeverage plant products			
Bongkrek	Coconut presscake	*Rhizopus oligosporus*	Indonesia
Cocoa beans	Cacao fruit (pods)	*Candida krusei* (*Issatchenkia orientalis*), *Geotrichum* spp.	Africa, South America
Coffee beans	Coffee cherries	*Erwinia dissolvens*, *Saccharomyces* spp.	Brazil, Congo, Hawaii, India
Gari	Cassava	"*Corynebacterium manihot*," *Geotrichum* spp.	West Africa
Kenkey	Corn	*Aspergillus* spp., *Penicillium* spp., lactobacilli, yeasts	Ghana, Nigeria
Kimchi	Cabbage and other vegetables	Lactic acid bacteria	Korea
Miso	Soybeans	*Aspergillus oryzae*, *Zygosaccharomyces rouxii*	Japan
Ogi	Corn	*L. plantarum*, *L. lactis*, *Zygosaccharomyces rouxii*	Nigeria
Olives	Green olives	*L. mesenteroides*, *L. plantarum*	Worldwide
Ontjom*	Peanut presscake	*Neurospora sitophila*	Indonesia
Peujeum	Cassava	Molds	Indonesia
Pickles	Cucumbers	*P. cerevisiae*, *L. plantarum*	Worldwide
Poi	Taro roots	Lactics	Hawaii
Sauerkraut	Cabbage	*L. mesenteroides*, *L. plantarum*	Worldwide
Soy sauce (shoyu)	Soybeans	*A. oryzae*; or *A. soyae*; *Z. rouxii, L. delbrueckii*	Japan
Sufu	Soybeans	*Mucor* spp.	China and Taiwan
Tao-si	Soybeans	*A. oryzae*	Philippines
Tempeh	Soybeans	*Rhizopus oligosporus*; *R. oryzae*	Indonesia, New Guinea, Surinam
Beverages and related products			
Arrack	Rice	Yeasts, bacteria	Far East
Beer and ale	Cereal wort	*Saccharomyces cerevisiae*	Worldwide
Binuburan	Rice	Yeasts	Philippines
Bourbon whiskey	Corn, rye	*S. cerevisiae*	United States
Bouza beer	Wheat grains	Yeasts	Egypt
Cider	Apples; others	*Saccharomyces* spp.	Worldwide
Kaffir beer	Kaffircorn	Yeasts, molds, lactics	Nyasaland (Malawi)
Magon	Corn	*Lactobacillus* spp.	Bantus of South Africa
Mezcal	Century plant	Yeasts	Mexico
Oo	Rice	Yeasts	Thailand
Pulque†	Agave juice	Yeasts and lactics	Mexico, U.S. Southwest
Sake	Rice	*Saccharomyces sake* (*S. cerevisiae*)	Japan

continues

Table 8–1   continued

Products	Substrate	Fermenters	Where Found
Scotch whiskey	Barley	*S. cerevisiae*	Scotland
Teekwass	Tea leaves	*Acetobacter xylinum,* *Schizosaccharomyces* *pombe*	
Thumba	Millet	*Endomycopsis fibuliges*	West Bengal, India
Tibi	Dried figs; raisins	*Betabacterium vermiforme,* *Saccharomyces* *intermedium*	
Vodka	Potatoes	Yeasts	Russia, others
Wines	Grapes, other fruits	*Saccharomyces "ellipsoideus"* strains	Worldwide
Vinegar	Cider, wine	*Acetobacter* spp.	Worldwide
Palm wine	Palm sap	*Acetobacter* spp., lactics, yeasts	Nigeria
Breads			
Idli	Rice and bean flour	*Leuconostoc mesenteroides*	Southern India
Rolls, cakes, etc.	Wheat flours	*S. cerevisiae*	Worldwide
San Francisco sourdough bread	Wheat flour	*S. exiguus,* *L. sanfranciscensis*	Northern California
Sour pumpernickel	Wheat flour	*L. mesenteroides*	Switzerland, other areas
Vodka	Potatoes	Yeasts	Russia, Scandinavia

*N. sitophila is used to make red ontjom; *R. oligosporus* for white ontjom.
†Distilled to produce tequila.

Kombucha is a home-prepared tea that is produced by fermenting sweetened black tea with a mixed culture of bacteria and yeasts. It is consumed mainly in China, Russia, and Germany. It has been consumed for over 2000 years, and is believed (at least by some) to provide a number of health benefits. The tea fermentation occurs at room temperature for 7 to 10 days, and the finished product contains organic acids, tea components, vitamins, minerals, and is slightly carbonated.[35] The most predominant organism in the fermentaion mat is *Acetobacter xylinum*. Among the large number of yeasts found are species of *Brettanomyces, Candida, Pichia, Saccharomyces,* and *Zygosaccharomyces*.

MISCELLANEOUS PRODUCTS

Coffee beans, which develop as berries or cherries in their natural state, have an outer pulpy and mucilaginous envelope that must be removed before the beans can be dried and roasted. The wet method of removal of this layer seems to produce the most desirable product, and it consists of depulping and demucilaging followed by drying. Whereas depulping is done mechanically, demucilaging is accomplished by natural fermentation. The mucilage layer is composed largely of pectic substances,[29] and pectinolytic microorganisms are important in their removal. *Erwinia dissolvens* has been found to be the most important bacterium during the demucilaging fermentation in Hawaiian[30] and Congo coffee cherries,[83] although Pederson and Breed[59] indicated that the fermentation of coffee berries from Mexico and Colombia was carried out by typical lactic acid bacteria (leuconostocs and lactobacilli). Agate and Bhat[3] in their study of coffee cherries from the Mysore state of India found that the following pectinolytic yeasts predominated and played important roles in the loosening and removal of the mucilaginous layers: *Saccharomyces marxianus, S. bayanus, S. "ellipsoideus,"* and S*chizosaccharomyces* spp. Molds are common on green coffee beans, and in one study, 99.1% of products from 31 countries contained these organisms, generally on the surface.[45] Seven species of aspergilli dominated the biota, with *A. ochraceus* being the most frequently recovered from beans before surface disinfection, followed by *A. niger* and species of the *A. glaucus* group. The toxigenic molds, *A. flavus* and *A. versicolor*, were found, as were *P. cyclopium, P. citrinum,* and *P. expansum,* but the penicillia were less frequently found than the aspergilli.[45] Microorganisms do not contribute to the development of flavor and aroma in coffee beans as they do in cocoa beans.

Cocoa beans (actually cacao beans–cocoa is the powder and chocolate is the manufactured product), from which chocolate is derived, are obtained from the fruits or pods of the cacao plant in parts of Africa, Asia, and South America. The beans are extracted from the fruits and fermented in piles, boxes, or tanks for 2–12 days, depending on the type and size of beans. During the fermentation, high temperatures (45–50°C) and large quantities of liquid develop. Following sun or air drying, during which the water content is reduced to less than 7.5%, the beans are roasted to develop the characteristic flavor and aroma of chocolate. The fermentation occurs in two phases. In the first, sugars from the acidic pulp (about pH 3.6) are converted to alcohol. The second phase consists of the alcohol being oxidized to acetic acid. In a study of Brazilian cocoa beans by Camargo et al.,[16] the biota on the first day of fermentation at 21°C consisted of yeasts. On the third day, the temperature had risen to 49°C, and the yeast count had decreased to no more than 10% of the total biota. Over the seven-day fermentation, the pH increased from 3.9 to 7.1. The cessation of yeast and bacterial activity around the third day is due in part to the unfavorable temperature, lack of fermentable sugars, and increase in alcohol. Although some decrease in acetic acid bacteria occurs because of high temperature, not all of these organisms are destroyed. The importance of lactic acid in the overall process was shown earlier.[54,66]

In one study, the cocoa fermentation was carried out with a defined microbial cocktail consisting of only five organisms rather than the 50 or so that have been isolated from natural fermentations.[74] The five consisted of *Saccharomyces cerevisiae* var. *chevalieri, Lactobacillus plantarum, L. lactis, Acetobacter aceti,* and *Gluconobacter oxydans* subsp. *suboxydans.* The defined inoculum led to a product highly similar to that produced by natural fermentation. The key roles for the yeasts involved elevating pH from about 3.5 to 4.2, breaking down citric acid in pulp, producing ethanol, producing organic acids (oxalic, succinic, malic, etc.) that destroy bean cotyledons, producing volatile substances that may play a role in chocolate flavor, and reducing viscosity of pulp. *S. cerevisiae* was the most important organism in the above activities.

Although yeasts play important roles in producing alcohol in cocoa bean fermentation, their presence appears even more essential to the development of the final, desirable, chocolate flavor of roasted beans. Levanon and Rossetini[42] found that the endoenzymes released by autolyzing yeasts are responsible for the development of chocolate precursor compounds. The acetic acid apparently makes the bean tegument permeable to the yeast enzymes. It has been shown that chocolate aroma occurs only after cocoa beans are roasted and that the roasting of unfermented beans does not produce the characteristic aroma. Reducing sugars and free amino acids are in some way involved in the final chocolate aroma development.[67] For an extensive review, see reference 81.

Soy sauce or shoyu is produced in a two-stage manner. The first stage, the koji (analogous to malting in the brewing industry), consists of inoculating either soybeans or a mixture of beans and wheat flour with *A. oryzae* or *A. soyae* and allowing them to stand for 3 days. This results in the production of large amounts of fermentable sugars, peptides, and amino acids. The second stage, the moromi, consists of adding the fungal-covered product to around 18% NaCl and incubating at room temperatures for at least a year. The liquid obtained at this time is soy sauce. During the incubation of the moromi, lactic acid bacteria, *L. delbrueckii* subsp. *delbruckeii* in particular, and yeasts such as *Zygosaccharomyces rouxii* carry out an anaerobic fermentation of the koji hydrolysate. Pure cultures of *A. oryzae* for the koji and *L. delbrueckii* subsp. *delbrueckii* and *Z. rouxii* for the moromi stages have been shown to produce good quality soy sauce.[93]

Tempeh is a fermented soybean product. Although there are many variations in its production, the general principle of the Indonesian method for tempeh consists of soaking soybeans overnight in order to remove the seed coats or hulls. Once seed coats are removed, the beans are cooked in boiling water for about 30 minutes and spread on a bamboo tray to cool and surface dry. Small pieces of tempeh from a previous fermentation are incorporated as starter, followed by wrapping with banana leaves. The wrapped packages are kept at room temperature for 1 or 2 days during which mold growth occurs and binds the beans together as a cake—the tempeh. An excellent product can be made by storing in perforated plastic bags and tubes with fermentations completed in 24 hours at 31°C.[27] The desirable organism in the fermentation is *Rhizopus oligosporus*, especially for wheat tempeh. Good soybean tempeh can be made with *R. oryzae* or *R. arrhizus*. During the fermentation, the pH of soybeans rises from around 5.0 to values as high as 7.5.

Miso, a fermented soybean product common in Japan, is prepared by mixing or grinding steamed or cooked soybeans with koji and salt and allowing fermentation to take place usually over a 4- to 12-month period. White or sweet miso may be fermented for only a week, whereas the higher-quality dark brown product (*mame*) may ferment for 2 years. In Israel, Ilany-Feigenbaum et al.[38] prepared miso-type products by using defatted soybean flakes instead of whole soybeans and fermenting for around 3 months. The koji for these products was made by growing *A. oryzae* on corn, wheat, barley, millet or oats, potatoes, sugar beets, or bananas, and the investigators found that the miso-type products compared favorably to Japanese-prepared miso. Because of the possibility that *A. oryzae* may produce toxic substances, koji was prepared by fermenting rice with *Rhizopus oligosporus* at 25°C for 90 days; the product was found to be an acceptable alternative to *A. oryzae* as a koji fungus.[76]

Ogi is a staple cereal of the Yorubas of Nigeria, and it is often the first food given to babies at weaning. It is produced generally by soaking corn grains in warm water for 2–3 days followed by wet-milling and sieving through a screen mesh. The sieved material is allowed to sediment and ferment and is marketed as wet cakes wrapped in leaves. Various food dishes are made from the fermented cakes or the ogi.[10] During the steeping of corn, *Corynebacterium* spp. become prominent and appear to be responsible for the diastatic action necessary for the growth of yeasts and lactic acid bacteria.[4] Along with the corynebacteria, *S. cerevisiae* and *L. plantarum* have been found to be prominent in the traditional ogi fermentation, as are *Cephalosporium*, *Fusarium*, *Aspergillus*, and *Penicillium*. Most of the acid produced is lactic, which depresses the pH of desirable products to around 3.8. The corynebacteria develop early, and their activities cease after the first day; those of the lactobacilli and yeasts continue beyond the first day of fermentation. A newer process for making ogi has been developed, tested, and found to produce a product of better quality than the traditional process.[9] By this newer method, corn is dry-milled into whole corn and dehulled corn flour. Upon the addition of water, the mixture is cooked, cooled, and then inoculated with a mixed culture (starter) of *L. plantarum*, *L. lactis*, and *Z. rouxii*. The inoculated preparation is incubated at 32°C for 28 hours, during which time the pH of the corn drops from 6.1 to 3.8. This process eliminates the need for starch-hydrolyzing bacteria. In addition to the shorter fermentation time, there is also less chance for faulty fermentations.

Gari is a staple food of West Africa prepared from the root of the cassava plant. Cassava roots contain cyanogenic glucosides, *linamarin* and *lotaustralin*, which make them poisonous if eaten fresh or raw. The roots can be detoxified by the addition of linamarase, which acts on both.[13] In practice the roots are rendered safe by a fermentation during which the toxic glucoside decomposes with the liberation of gaseous hydrocyanic acid. In the home preparation of gari, the outer peel and the thick cortex of the cassava roots are removed, followed by grinding or grating the remainder. The pulp is pressed to remove the remaining juice and placed in bags for 3 or 4 days to allow fermentation to occur.[17] The organisms most responsible for the product include *L. plantarum*, *E. faecium*, and *Leuconostoc mesenteroides*.[13] The fermented product is cooked by frying.

Bongkrek is an example of a fermented food product that in the past has led to a large number of deaths. Bongkrek or semaji is a coconut presscake product of central Indonesia, and it is the homemade product that may become toxic. The safe products fermented by *R. oligosporus* are finished cakes covered with and penetrated by the white fungus. In order to obtain the desirable fungal growth, it appears to be essential that conditions permit good growth within the first 1 or 2 days of incubation. If, however, bacterial growth is favored during this time and if the bacterium *Burkholderia cocovenenans* is present, it grows and produces two toxic substances–toxoflavin and bongkrekic acid.[84,85,86,94] Both of these compounds show antifungal and antibacterial activity, are toxic for humans and animals, and are heat stable. Production of both is favored by growth of the organisms on coconut (toxoflavin can be produced in complex culture media). The structural formulas of the two antibiotics–toxoflavin, which acts as an electron carrier, and bongkrekic acid, which inhibits oxidative phosphorylation in mitochondria–follow:

Toxoflavin

Bongkrekic acid

Bongkrekic acid has been shown to be cidal to all 17 molds studied by Subik and Behun[79] by preventing spore germination and mycelial outgrowth. The growth of *B. cocovenenans* in the preparation of bongkrek is not favored if the acidity of starting materials is kept at or below pH 5.5.[84] It has been shown that 2% NaCl in combination with acetic acid to produce a pH of 4.5 will prevent the formation of the bongkrek toxin in tempeh.[14]

A fermented cornmeal product that is prepared in parts of China has been the cause of food poisoning by strains *B. cocovenenans*. The product is prepared by soaking corn in water at room temperature for 2–4 weeks, washing in water, and grinding the wet corn into flour for various uses. The toxic organisms apparently grow in the moist product during its storage at room temperature. The responsible organism produced both bongkrekic acid and toxoflavin, as do the strains of *B. cocovenenans* in bongkrek.

Ontjom (oncom) is a somewhat similar but more popular fermented product of Indonesia made from peanut presscake, the material that remains after oil has been extracted from peanuts. The presscake is soaked in water for about 24 hours, steamed, and pressed into molds. The molds are covered with banana leaves and inoculated with *Neurospora sitophila* or *R. oligosporus*. The product is ready for consumption 1 or 2 days later. A more detailed description of ontjom fermentation and the nutritive value of this product has been provided by Beuchat.[12]

REFERENCES

1. Acton, J.C., R.L. Dick, and E.L. Norris. 1977. Utilization of various carbohydrates in fermented sausage. *J. Food Sci.* 42:174–178.

2. Acton, J.C., J.G. Williams, and M.G. Johnson. 1972. Effect of fermentation temperature on changes in meat properties and flavor of summer sausage. *J. Milk Food Technol.* 35:264–268

3. Agate, A.D., and J.V. Bhat. 1966. Role of pectinolytic yeasts in the degradation of mucilage layer of *Coffea robusta* cherries. *Appl. Microbiol.* 14:256–260.

4. Akinrele, I.A. 1970. Fermentation studies on maize during the preparation of a traditional African starch-cake food. *J Sci. Food Agric.* 21:619–625.

5. Andersen, S.J. 1995. Compositional changes in surface mycoflora during ripening of naturally fermented sausages. *J. Food Protect.* 58:426–429.

6. Arihara, K., H.Ota, M.Itoh, Y. Kondo, T. Sameshima, M. Akimoto, S. Kanai, and T. Miki. 1998. *Lactobacillus acidophilus* group lactic acid bacteria applied to meat fermentation. *J. Food Sci.* 63:544–547.

7. Ayres, J.C., D.A. Lillard, and L. Leistner. 1967. Mold ripened meat products. In *Proceedings of the 20th Annual Reciprocal Meat Conference*, 156–168. Chicago: National Live Stock and Meat Board.

8. Bacus, J. 1984. *Utilization of Microorganisms in Meat Processing: A Handbook for Meat Plant Operators.* New York: John Wiley & Sons

9. Banigo, E.O.I., J.M. deMan, and C.L. Duitschaever. 1974. Utilization of high-lysine corn for the manufacture of ogi using a new, improved processing system. *Cereal Chem.* 51:559–572.

10. Banigo, E.O.I., and H.G. Muller. 1972. Manufacture of ogi (a Nigerian fermented cereal porridge): Comparative evaluation of corn, sorghum and millet. *Can. Inst. Food Sci. Technol. J.* 5:217–221.

11. Berwal, J.S., and D.Dincho. 1995. Molds as protective cultures for raw dry sausages. *J. Food Protect.* 58:817–819.

12. Beuchat, L.R. 1976. Fungal fermentation of peanut press cake. *Econ. Bot.* 30:227–234.

13. Bokanga, M. 1995. Biotechnology and cassava processing in Africa. *Food Technol.* 49:86–90.

14. Buckle, K.A., and E. Kartadarma. 1990. Inhibition of bongkrek acid and toxoflavin production in tempe bongkrek containing *Pseudomonas cocovenenans. J. Appl. Bacteriol.* 68:571–576.

15. Cai, Y., H. Okada, H. Mori, Y. Benno, and T. Nakase. 1999. *Lactobacillus paralimentarius* sp. nov., isolated from sough-dough. *Int. J. Syst. Bacteriol.* 49:1451–1455.

16. Camargo, R.de, J.Leme, Jr., and A.M. Filho. 1963. General observations on the microflora of fermenting cocoa beans (*Theobroma cacao*) in Bahia (Brazil). *Food Technol.* 17:1328–1330.

17. Collard, P., and S. Levi. 1959. A two-stage fermentation of cassava. *Nature* 183:620–621.

18. Cummings, A., C. Reitmeier, L. Wilson, and B. Glatz. 2002. A survey of apple cider production practices and microbial loads in cider in the state of Iowa. *Dairy Fd. Environ. Sanit.* 22:745–751.

19. Deibel, R.H., and C.F. Niven, Jr. 1957. *Pediococcus cerevisiae*: A starter culture for summer sausage. *Bacteriol. Proc.* 14–15.

20. Deibel, R.H.,C.F. Niven, Jr., and G.D. Wilson. 1961. Miccrobiology of meat curing. III. Some microbiological and related technological aspects in the manufacture of fermented sausages. *Appl. Microbiol.* 9:156–161.

21. De Vuyst,L.V.Schrijvers, S.Paramithiotis, B. Hoste, M. Vancanneyt, J. Swings, G. Kalantzopoulos, E. Tsakalidou, and W. Messens. 2002. The biodiversity of lactic acid bacteria in Greek traditional wheat sourdoughs is reflected in both composition and metabolite formation. *Appl. Environ. Microbiol.* 68:6059–6069.

22. Dicks, L.M.T., F. Dellaglio, and M.D. Collins. 1995. Proposal to reclassify *Leuconostoc oenos* to *Oenococcus oeni* [corrig.]. gen. nov., comb. nov. *Int. J. Syst. Bacteriol.* 45:395–397.

23. Engelmann, U., and N. Weiss. 1985. *Megasphaera cerevisiae* sp. nov.: A new Gram-negative obligately anaerobic coccus isolated from spoiled beer. *Syst. Appl. Microbiol.* 6:287–290.

24. Etchells, J.L., H.P. Fleming, and T.A. Bell. 1975. Factors influencing the growth of lactic acid bacteria during the fermentation of brined cucumbers. In *Lactic Acid Bacteria in Beverages and Food*, ed. J.G. Carr, C.V. Cutting, and G.C. Whiting, 281–305. New York: Academic Press.

25. Everson, C.W., W.E. Danner, and P.A. Hammes. 1970. Improved starter culture for semidry sausage. *Food Technol.* 24:42–44.

26. Faparusi, S.I., and O. Bassir. 1971. Microflora of fermenting palm-wine. *J. Food Sci. Technol.* 8:206–210.

27. Filho, A.M., and C.W. Hesseltine. 1964. Tempeh fermentation: Package and tray fermentations. *Food Technol.* 18:761–765.

28. Fisher, T.L., and D.A. Golden. 1998. Fate of *Escherichia coli* 0157:H7 in ground apples used in cider production. *J. Food Protect.* 61:1372–1374.

29. Frank, H.A., N.A. Lum, and A.S. Dela Cruz. 1965. Bacteria responsible for mucilage-layered composition in Kona coffee cherries. *Appl. Microbiol.* 13:201–207.

30. Frank, H.A., and A.S. Dela Cruz. 1964. Role of incidental microflora in natural decomposition of mucilage layer in Kona coffee cherries. *J. Food Sci.* 29:850–853.

31. García-García, P., R. Barranco, M.C. Durán Quintana, and A. Garrido-Fernández. 2004. Biogenic amine formation and "zapatera" spoilage of fermented green olives: Effect of storage temperature and debittering process. *J. Food Protect.* 67:117–123.

32. García-García, P., M.Brenes-Balbuena, D. Hornero-Méndez, A. García-Borrego, and A. Garrido-Fernández. 2000. Content of biogenic amines in table olives. *J. Food Protect.* 63:111–116.

33. Giolitti, G., C.A. Cantoni, M.A. Bianchi, and P. Renon. 1971. Microbiology and chemical changes in raw hams of Italian type. *J. Appl. Bacteriol.* 34:51–61.

34. Gobbetti, M., and A. Corsetti. 1997. *Lactobacillus sanfrancisco* a key sourdough lactic acid bacterium: A review. *Food Microbiol.* 14:175–187.

35. Greenwalt, C.J., K.H. Steinkraus, and R.A. Ledford. 2000. Kombucha, the fermented tea: Mcrobiology, composition, and claimed health effects. *J. Food Protect.* 63:976–981.

36. Harris, D.A., L.Chaiet, R.P. Dudley, and P. Ebert. 1957. The development of commercial starter culture for summer sausages. *Bacteriol. Proc. Amer. Soc. Microbiol.*, 15.

37. Horn, S.J., I.M. Aasen, and K. Østgaard. 2000. Production of ethanol from mannitol by *Zymobacter palmae*. *J. Ind. Microbiol. Biotechnol.* 24:51–57.

38. Ilany-Feigenbaum, J.J. Diamant, S. Laxer, and A. Pinsky. 1969. Japanese miso-type products prepared by using defatted soybean flakes and various carbohydrate-containing foods. *Food Technol.* 23:554–556.

39. Kleyn, J., and J.Hough. 1971. The microbiology of brewing. *Annu. Rev. Microbiol.* 25:583–608.

40. Kline, L., and T.F. Sugihara. 1971. Microorganisms of the San Francisco sour dough bread process. II. Isolation and characterization of undescribed bacterial species responsible for the souring activity. *Appl. Microbiol.* 21:459–465.

41. Kunkee, R.E. 1975. A second enzymatic activity for decomposition of malic acid by malo-lactic bacteria. In *Lactic Acid Bacteria in Beverages and Food*, ed. J.G. Carr, C.V. Cutting, and G.C. Whiting, 29–42. New York: Academic Press.

42. Levanon, Y., and S.M.O. Rossetini. 1965. A laboratory study of farm processing of cocoa beans for industrial use. *J. Food Sci.* 30:719–722.

43. Lee, S.Y., M.S. Mabee, and N.O. Jangaard. 1978. *Pectinatus*, a new genus of the family *Bacteroidaceae*. *Int. J. Syst. Bacteriol.* 28:582–594.

44. Liu, S.-Q. 2002. Malolactic fermentation in wine beyond deacidification. *J. Appl. Microbiol.* 92:589–601

45. Mislivec, P.B., V.R. Bruce, and R. Gibson. 1983. Incidence of toxigenic and other molds in green coffee beans. *J. Food Protect.* 46:969–973.

46. Mukherjee, S.K., M.N. Albury, C.S. Pederson, A.G. van Veen, and K.H. Steinkraus. 1965. Role of *Leuconostoc mesenteroides* in leavening the batter of idli, a fermented food of India. *Appl. Microbiol.* 13:227–231.

47. Niinivaara, F.P., M.S. Pohja, and S.E. Komulainen. 1964. Some aspects about using bacterial pure cultures in the manufacture of fermented sausages. *Food Technol.* 18:147–153.

48. Okafor, N. 1972. Palm-wine yeasts from parts of Nigeria. *J. Sci. Food Agric.* 23:1399–1407.

49. Okafor, N. 1975. Microbiology of Nigerian palm wine with particular reference to bacteria. *J. Appl. Bacteriol.* 38:81–88.

50. Okafor, N. 1975. Preliminary microbiological studies on the preservation of palm wine. *J. Appl. Bacteriol.* 38:1–7.

51. Okamoto, T., H. Taguchi, K. Nakamura, H. Ikenaga, H. Kuraishi, and K. Yamasato. 1993. *Zymobacter palmae* gen. nov., sp. nov., a new ethanol-fermenting peritrichous bacterium isolated from palm sap. *Arch. Microbiol.* 160:333–337.

52. Palumbo, S.A., J.L. Smith, and S.A. Kerman. 1973. Lebanon bologna. I. Manufacture and processing. *J. Milk Food Technol.* 36:497–503.

53. Passmore, S.M., and J.G. Carr. 1975. The ecology of the acetic acid bacteria with particular reference to cider manufacture. *J. Appl. Bacteriol.* 38:151–158.

54. Passos, F.M.L., D.O. Silva, A. Lopez, C.L.L.F. Ferreira, and W.V. Guimaraes. 1984. Characterization and distribution of lactic acid bacteria from traditional cocoa bean fermentations in Bahia. *J. Food Sci.* 49:205–208.

55. Patel, I.B., and R.H. Vaughn. 1973. Cellulolytic bacteria associated with sloughing spoilage of California ripe olives. *Appl. Microbiol.* 25:62–69.

56. Pearson, A.M., and T.A. Gillett. 1999. *Processed Meats.* New York: Kluwer Academic Publishers.

57. Pearson, A.M., and F.W. Tauber. 1984. *Processed Meats,* 2nd ed. New York: Kluwer Academic Publishers.

58. Pederson, C.S. 1979. *Microbiology of Food Fermentations,* 2nd ed. New York: Kluwer Academic Publishers.

59. Pederson, C.S., and R.S. Breed. 1946. Fermentation of coffee. *Food Res.* 11:99–106.

60. Pilone, G.J., and R.E. Kunkee. 1976. Stimulatory effect of malo-lactic fermentation on the growth rate of *Leuconostoc oenos. Appl. Environ. Microbiol.* 32:405–408.

61. Plastourgos, S., and R.H. Vaughn. 1957. Species of *Propionibacterium* associated with zapatera spoilage of olives. *Appl. Microbiol.* 5:267–271.

62. Potter, M.E., M.B. Kruse, M.A. Matthews, R.O. Hill, and R.J. Martin. 1976. A sausage-associated outbreak of trichinosis in Illinois. *Amer. J. Pub. Hlth.* 66:1194–1196.

63. Radler, F. 1975. The metabolism of organic acids by lactic acid bacteria. In *Lactic Acid Bacteria in Beverages and Food,* ed. J.G. Carr, C.V. Cutting, and G.C. Whiting, 17–27. New York: Academic Press.

64. Rainbow, C. 1975. Beer spoilage lactic acid bacteria. In *Lactic Acid Bacteria in Beverages and Food,* ed. J.G. Carr, C.V. Cutting, and G.C. Whiting, 149–158. New York: Academic Press.

65. Reddy, N.R., S.K. Sathe, M.D. Pierson, and D.K. Salunkha. 1981. Idli, an Indian fermented food: A review. *J. Food Qual.* 5:89–101.

66. Roelofsen, P.A. 1958. Fermentation, drying, and storage of cacao beans. *Adv. Food Res.* 8:225–296.

67. Rohan, T.A., and T. Stewart. 1966. The precursors of chocolate aroma: Changes in the sugars during the roasting of cocoa beans. *J. Food Sci.* 31:206–209.

68. Röling, W.F.M., and H.W. van Verseveld. 1996. Charcteristics of *Tetragenococcus helophila* populations of Indonesian soy mash (kecap) fermentation. *Appl. Environ. Microbiol.* 62:1203–1207.

69. Rose, A.H. 1982. *Fermented Foods.* Economic Microbiology Series, 7. New York: Academic Press.

70. Saisithi, P., B.-O. Kasemsarn, J. Liston, and A.M. Dollar. 1966. Microbiology and chemistry of fermented fish. *J. Food Sci.* 31:105–110.

71. Sanz, Y., and F. Toldra. 1998. Aminopeptidases from *Lactobacillus sake* affected by amines in dry sausages. *J. Food Sci.* 63:894–896.

72. Satomi, M., B. Kimura, M. Mizoi, T. Sato, and T. Fujii. 1997. *Tetragenococcus muriaticus* sp. nov., a new moderately halophilic lactic acid bacterium isolated from fermented squid liver sauce. *Int. J. Syst. Bacteriol.* 47:832–836.

73. Schleifer, K.H., M. Leuteritz, N. Weiss, W. Ludwig, G. Kirchhof, and H. Seidel-Rufer. 1990. Taxonomic study of anaerobic, Gram-negative, rodshaped bacteria from breweries: Emended description of *Pectinatus cerevisiiphilus* and description of *Pectinatus frisingensis* sp. nov., *Selenomonas lacticifex* sp. nov., *Zymophilus paucivorans* sp. nov. *Int. J. System. Bacteriol.* 49:19–27.

74. Schwan, R.F. 1998. Cocoa fermentations conducted with a defined microbial cocktail inoculum. *Appl. Environ. Microbiol.* 64:1477–1483.

75. Semanchek, J.J., and D.A. Golden. 1996. Survival of *Escherichia coli* 0157:H7 during fermentation of apple cider. *J. Food Protect.* 59:1256–1259.

76. Shieh, Y.-S.G., and L.R. Beuchat. 1982. Microbial changes in fermented peanut and soybean pastes containing kojis prepared using *Aspergillus oryzae* and *Rhizopus oligosporus. J. Food Sci.* 47:518–522.

77. Smith, J.L., and S.A. Palumbo. 1973. Microbiology of Lebanon bologna. *Appl. Microbiol.* 26:489–496.

78. Steinkraus, K.H., A.G. van Veen, and D.B. Thiebeau. 1967. Studies on idli—An Indian fermented black gram-rice food. *Food Technol.* 21:916–919.

79. Subik, J., and M. Behun. 1974. Effect of bongkrekic acid on growth and metabolism of filamentous fungi. *Arch. Microbiol.* 97:81–88.

80. Sugihara, T.F., L.Kline, and M.W. Miller. 1971. Microorganisms of the San Francisco sour dough bread process. I. Yeasts responsible for the leavening action. *Appl. Microbiol.* 21:456–458.

81. Thompson, S.S., K.B. Miller, and A.S. Lopez. 2001. Cocoa and coffee. In *Food Microbiology: Fundamentals and Frontiers*, 2nd ed., ed. M.P. Doyle, L.R. Beuchat, and T.J. Montville, 721–733. Washington, DC: ASM Press.

82. Urbaniak, L., and W. Pezacki. 1975. Die Milchsäure bildende Rohwurst-Mikroflora und ihre technologisch bedingte Veränderung. *Fleischwirts.* 55:229–237.

83. Van Pee, W., and J.M. Castelein. 1972. Study of the pectinolytic microflora, particularly the *Enterobacteriaceae*, from fermenting coffee in the Congo. *J. Food Sci.* 37:171–174.

84. van Veen, A.G. 1967. The bongkrek toxins. In *Biochemistry of Some Foodborne Microbial Toxins*, ed. R.I. Mateles and G.N. Wogan, 43–50. Cambridge, MA: MIT Press.

85. van Veen, A.G., and W.K. Mertens. 1934. Die Gifstoffe der sogenannten Bongkrek-vergiftungen auf Java. *Rec. Trav. Chim.* 53:257–268.

86. van Veen, A.G., and W.K. Mertens. 1934. Das Toxoflavin, der gelbe Gifstoff der Bongkrek. *Rec. Trav. Chim.* 53:398–404.

87. Vaughn, R.H. 1975. Lactic acid fermentation of olives with special reference to California conditions. In *Lactic Acid Bacteria in Beverages and Food*, ed. J.G. Carr, C.V. Cutting, and G.C. Whiting, 307–323. New York: Academic Press.

88. Vaughn, R.H., T.Jakubczyk, J.D. MacMillan, T.E. Higgins, B.A. Dave, and V.M. Crampton. 1969. Some pink yeasts associated with softening of olives. *Appl. Microbiol.* 18:771–775.

89. Vogel, R.F., G. Böcker, P. Stolz, M. Ehrmann, D. Fanta, W. Ludwig, B. Pot, K. Kersters, K.H. Schleifer, and W.P. Hammes. 1994. Identification of lactobacilli from sourdough and description of *Lactobacillus pontis* sp. nov. *Int. J. Syst. Bacteriol.* 44:223–229.

90. Wardlaw, F.B., G.C. Skelley, M.G. Johnson, and J.C. Acton. 1973. Changes in meat components during fementation, heat processing and drying of a summer sausage. *J. Food Sci.* 38:1228–1231.

91. Williamson, D.H. 1959. Studies on lactobacilli causing ropiness in beer. *J. Appl. Bacteriol.* 22:392–402.

92. Yaping, J., L. Xiaoyang, and Y. Jiaqi. 1990. *Saccharobacter fermentatus* gen. nov., sp. nov., a new ethanol-producing bacterium. *Int. J. Syst. Bacteriol.* 40:412–414.

93. Yong, F.M., and B.J.B. Wood. 1974. Microbiology and biochemistry of the soy sauce fermentation. *Adv. Appl. Microbiol.* 17:157–194.

94. Zhao, N., C. Qu, E. Wang, and W. Chen. 1995. Phylogenetic evidence for the transfer of *Pseudomonas cocovenenans* (van Damme et al. 1960) to the genus *Burkholderia* as *Burkholderia cocovenenans* (van Damme et al. 1960) comb. nov. *Int. J. Syst. Bacteriol.* 45:600–603.

CHAPTER 9

Miscellaneous Food Products

This chapter contains brief descriptions of a wide variety of food products along with the microbial biota of both fresh and spoiled products.

DELICATESSEN AND RELATED FOODS

Delicatessen foods, such as salads and sandwiches, are sometimes involved in food-poisoning outbreaks. These foods are often prepared by hand, and this direct contact may lead to an increased incidence of food-poisoning agents such as *Staphylococcus*. Once organisms such as these enter meat salads or sandwiches, they may grow well because of the reduction in numbers of the normal food biota by the prior cooking of salad ingredients.

In a study of retail salads and sandwiches, 36% of 53 salads were found to have total counts $\log_{10}$/g >6.00, but only 16% of the 60 sandwiches had counts as high.[6] With respect to coliforms, 57% of sandwiches were found to harbor $\log_{10}$/g <2.00. *S. aureus* was present in 60% of sandwiches and 39% of salads. Yeasts and molds were found in high numbers, with six samples containing $\log_{10}$/g >6.00.

In a study of 517 salads from around 170 establishments, 71–96% were found to have aerobic plate counts (APCs) $\log_{10}$/g <5.00.[45] Almost all (96–100%) salads contained coagulase positive *S. aureus* at levels $\log_{10}$/g <2.00. Salads included chicken, egg, macaroni, and shrimp. *S. aureus* was recovered in low numbers from 6 to 64 salads in another study.[12] The 12 different salads examined by these investigators had total counts between $\log_{10}$ 2.08 and 6.76, with egg, shrimp, and some of the macaroni salads having the highest counts. Neither salmonellae nor *C. perfringens* were found in any product. A study of 42 salads by Harris et al.[18] revealed the products to be of generally good microbial quality. The mean APC was $\log_{10}$/g 5.54, and the mean coliform count was $\log_{10}$ 2.66/g for the six different products. Staphylococci were found in some products, especially ham salad.

Fresh green salads (greens, mixed greens, and coleslaw) were found to contain mean total counts of $\log_{10}$ 6.67 for coleslaw to $\log_{10}$ 7.28 for green salads.[13] Fecal coliforms were found in 26% of mixed, 28% of green, and 29% of coleslaw, whereas the respective percentage findings for *S. aureus* were 8, 14, and 3. With respect to parsley, *E. coli* was found on 11 of 64 samples of fresh and unwashed products and on over 50% of frozen samples.[24] The mean APC of fresh washed parsley was 7.28/g log. Neither salmonellae nor *S. aureus* was found in any sample.

In a study of the microbiological quality of imitation-cream pies from plants operated under poor sanitary conditions, Surkiewicz[56] found that the microbial load increased successively as the products were carried through the various processing steps. For example, in one instance, the final mixture of the synthetic pie base contained fewer than $\log_{10}$ 2.00 bacteria per gram after final heating to 160°F (71.1°C). After overnight storage, however, the count rose to $\log_{10}$ 4.15. The pie topping ingredients to be mixed with the pie base had a rather low count: log 2.78/g. After being deposited on the pies, the pie topping showed a total count of $\log_{10}$ 7.00/g. In a study of the microbiological quality of french fries, Surkiewicz et al.[57] demonstrated the same pattern—that is, the successive buildup of microorganisms as the fries underwent processing. Because these products are cooked late in their processing, the incidence of organisms in the finished state does not properly reflect the actual state of sanitation during processing.

The geometric mean APC of 1,187 sample units of refrigerated biscuit dough was found to be 34,000/g, whereas for fungi, coliforms, *E. coli*, and *S. aureus*, the mean counts were 46, 11, <3, and <3/g, respectively.[58] In the same study, the geometric mean APC of 1,396 units of snack cake was 910/g, with <3/g of coliforms, *E. coli*, and *S. aureus* (see Table 9–1).

A bacteriological study of 580 frozen cream-type pies (lemon, coconut, chocolate, and banana) showed them to be of excellent quality, with 98% having an APC of log 4.70 or less/g.[33] The overall microbiological quality of other related products is presented in Table 9–1. A summary of mean numbers of *L. monocytogenes* and *Salmonella* spp. in 11 ready-to-eat products in the United States for the 10-year period 1990–1999 is presented in Table 9–2,[35] and the prevalence of *Aeromonas* spp. in four ready-to-eat foods in Italy is presented in Table 9–3.[62]

EGGS

The hen's egg is an excellent example of a product that normally is well protected by its intrinsic parameters. Externally, a fresh egg has three structures, each effective to some degree in retarding the entry of microorganisms: the outer waxy shell membrane; the shell; and the inner shell membrane (Figure 9–1). Internally, lysozyme is present in egg white. This enzyme is quite effective against Gram-positive bacteria. Egg white also contains avidin, which forms a complex with biotin, thereby making this vitamin unavailable to microorganisms. In addition, egg white has a high pH (about 9.3) and contains conalbumin, which forms a complex with iron, thus rendering it unavailable to microorganisms. On the other hand, the nutrient content of the yolk material and its pH in fresh eggs (about 6.8) make it an excellent source of growth for most microorganisms.

Freshly laid eggs are generally sterile. However, in a relatively short period of time after laying, numerous microorganisms may be found on the outside and, under the proper conditions, may enter eggs, grow, and cause spoilage. The speed at which microbes enter eggs is related to temperature of storage, age of eggs, and level of contamination. The use of cryogenic gas (CO_2) to effect the rapid cooling of eggs led to fewer bacteria in the interior compared to conventional cooling, even though the differences were less significant after 30-day storage at 7°C.[8] A study of the migration of artificially contaminated *S*.Enteritidis from the albumen into the egg yolk using 860 eggs revealed that this bacterium could be detected in yolk within a day, depending on storage temperature and contamination level. Migration occurred in 1 day at 30°C but not until 14 days at 7°C.[2] Also, 1-day-old eggs were more resistant than 4-week-old eggs, and the speed of migration was positively correlated with the level of contamination. The deposing of *S*.Enteritidis in eggs by the hen before they are laid has been demonstrated (see reference 22). Among the bacteria found were members of the following genera: *Pseudomonas, Acinetobacter, Proteus, Aeromonas, Alcaligenes, Escherichia, Micrococcus, Salmonella, Serratia, Enterobacter, Flavobacterium,* and *Staphylococcus*. Among the molds generally found are members of the genera *Mucor, Penicillium, Hormodendron, Cladosporium,* and others; "*Torula*" is the only yeast found with any degree of consistency. The most common form of bacterial spoilage of eggs is a condition known as *rotting*. Green rots are caused by *Pseudomonas* spp., especially

Table 9–1 General Microbiological Quality of Miscellaneous Food Products

Products	No. of Samples	Microbial Group/Target	% Samples Meeting Target	Reference
Frozen cream-type pies	465	APC: $\leq 10^4$/g	96	60
	465	Fungi: 10^3/g or less	98	60
	465	Coliforms: <10/g	89	60
	465	*E. coli*: 10/g or less	99	60
	465	*S. aureus*: <25/g	99	60
	465	0 salmonellae	100	60
Frozen breaded onion rings (pre- or partially cooked)	1,590	APC 30°C: 10^5/g or less	99	66
	1,590	MPN coliforms: <3/g	89	66
	1,590	MPN *E. coli*: <3/g	99	66
	1,590	MPN *S. aureus*: <10/g	99.6	66
Frozen tuna pot pies	1,290	APC 30°C: 10^5/g or less	97.6	66
	1,290	MPN coliforms: 64/g or less	93	66
	1,290	MPN *E. coli*: <3/g	97	66
	1,290	MPN *S. aureus*: <10/g	98	66
Tofu (commercial)	60	APC: $>10^6$/g	83	50
	60	Psychrotrophs: $<10^4$/g	83	50
	60	Coliforms: $<10^3$/g	67	50
	60	*S. aureus*: <10/g	100	50
Dry food-grade gelatin	185	APC: 3.00* or less/g	74	33
Delicatessen salads	764	Within Army and Air Force Exchange Service microbial limits	44	12
	764	APC: 5.00* or less/g	84	12
	764	Coliforms: 1.00* or less/g	78	12
	764	Yeasts and molds: 1.30* or less/g	55	12
	764	"Fecal streptococci": 1.00*/g	77	12
	764	Presence of *S. aureus*	9	12
	764	Pres. of *C. perfringens*; salmonellae	0	12
	517	APC: 5.00* or less/g	26–85	44
	517	Coliforms: 2.00* or less/g	36–79	44
	517	*S. aureus*: 2.00* or less/g	96–100	44
Retail trade salads	53	APC: >6.00*/g	36	6
	53	Coliforms: 2.00* or less/g	57	6
	53	Presence of *S. aureus*	39	6
Retail trade sandwiches	62	APC: >6.00*/g	16	6
	62	Coliforms: >3.00*/g	12	6
	62	Presence of *S. aureus*	60	6
Imported spices and herbs	113	APC: 6.00* or less/g	73	23
	114	Spores: 6.00* or less/g	75	23
	113	Yeasts/molds: 5.00* or less/g	97	23

(continues)

Table 9–1   (continued)

Products	No. of Samples	Microbial Group/Target	% Samples Meeting Target	Reference
	114	TA spores: 3.00* or less/g	70	23
	114	Pres. of E. coli, S. aureus, salmonellae	0	23
Processed spices	114	APC: 5.00* or less/g	70	49
	114	APC: 6.00 or less/g	91	49
	114	Coliforms: 2.00* or less/g	97	49
	114	Yeasts/ molds: 4.00* or less/g	96	49
	114	C. perfringens: <2.00*/g	89	49
	110	Presence of B. cereus	53	48
Dehydrated space foods	129	APC: <4.00*/g	93	47
	129	Coliforms: <1/g	98	47
	129	E. coli: negative in 1 g	99	47
	102	"Fecal streptococci": 1.30*/g	88	47
	104	S. aureus: negative in 5 g	100	47
	104	Salmonellae: negative in 10 g	98	47
Ready-to-eat salad vegetables[a]	2,950	Listeria spp. , 4%; L. monocytogenes, 3% (10^2 in only one)		51
RTE foods, U.K.	4,469	Campylobacter = 0		39
Various restaurant foods—Spain	103	L. monocytogenes = 2.9%		55

Note: APC = aerobic plate count; MPN = most probable number.

[a]Negative for Campylobacter spp., E. coli 0157:H7, and Salmonella spp.
*$\log_{10}$ numbers.

Table 9–2   Mean Prevalence of *Listeria monocytogenes* and *Salmonella* Species in Seven Ready-to-Eat Meat/Poultry Products in the United States for the 10-Year Period 1990–1999.[35] Number of Foods Positive/Number Tested/Percent Positive.

Products	L. monocytogenes	Salmonella
Cooked, roast, corned beef	163/5,272/3.1	12/5,444/0.2
Sliced ham, luncheon meats	118/2,287/5.2	5/2,293/0.2
Small cooked sausages	243/6,820/3.6	14/6,996/0.2
Large cooked sausages	56/4,262/1.3	3/4,328/0.07
Jerky	4/770/0.5	2/648/0.3
Cooked poultry products	145/6,836/2.1	7/7,020/0.1
Salads, spreads, patés	119/3,932/3.0	2/4,204/0.05

Table 9–3 Summary of *Aeromonas* spp. in Ready-to-Eat Foods in Naples, Italy[62]

Products	Number	Percent Positive	A. hydrophila Isolates	A. caviae Isolates	A. sobria Isolates
Vegetables[a]	100	25	11	15	0
Cheeses[a]	100	10	6	5	1
Meat products[a]	100	11	10	3	0
Ice cream	20	0	0	0	0

[a]Five different products included.

P. fluorescens; colorless rots by *Pseudomonas, Acinetobacter*, and other species; black rots by *Proteus, Pseudomonas*, and *Aeromonas*; pink rots by *Pseudomonas*; red rots by *Serratia* spp., and "custard" rots by *Proteus vulgaris* and *P. intermedium*. Mold spoilage of eggs is generally referred to as pinspots, from the appearance of mycelial growth on the inside upon candling. *Penicillium* and *Cladosporium* spp. are among the most common causes of pinspots and fungal rotting in eggs. Bacteria also cause a condition in eggs known as mustiness. *Pseudomonas graveolens* and *Proteus* spp. have been implicated in this condition, with *P. graveolens* producing the most characteristic spoilage pattern.

The entry of microorganisms into whole eggs is favored by high humidity. Under such conditions, growth of microorganisms on the surface of eggs is favored, followed by penetration through the shell and inner membrane. The latter structure is the most important barrier to the penetration of bacteria into eggs, followed by the shell and the outer membrane.[36] More bacteria are found in egg yolk than in egg white, and the reason for a general lack of microorganisms in egg white is quite possibly its content of antimicrobial substances. In addition, upon storage, the thick white loses water to the yolk, resulting in a thinning of yolk and a shrinking of the thick white. This phenomenon makes it possible for the yolk to come into direct contact with the inner membrane, where it may be infected directly by

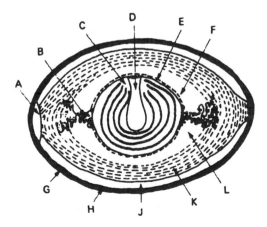

Figure 9–1   Structure of the hen's egg as shown by a section through the long axis. *Source*: From Brooks and Hale,[3] reproduced with permission from Elsevier Publishing Co.

microorganisms. Once inside the yolk, bacteria apparently grow in this nutritious medium, producing byproducts of protein and amino acid metabolism such as H_2S and other foul-smelling compounds. The effect of significant growth is to cause the yolk to become "runny" and discolored. Molds generally multiply first in the region of the air sac, where oxygen favors growth of these forms. Under conditions of high humidity, molds may be seen growing over the outer surface of eggs. Under conditions of low humidity and low temperatures, surface growth is not favored, but eggs lose water at a faster rate and thereby become undesirable as products of commerce.

The antimicrobial systems of eggs are noted in Chapter 3. In addition, hen egg albumen contains ovotransferrin, which chelates metal ions, particularly Fe^{3+}, and ovoflavoprotein, which binds riboflavin. At its normal pH of 9.0–10.0, egg albumen is cidal to Gram-positive bacteria and yeasts at both 30°C and 39.5°C.[61] The addition of iron reduces the antimicrobial properties of egg albumen.

In regards to the destruction of salmonellae in boiled shell eggs, it is generally recognized that cooking until the entire yolk is solidified is sufficient to destroy *S.* enterica serotype Enteritidis. In one study, boiling for 7 minutes destroyed 10^8 cfu of artifically inoculated *S.* Typhimurium.[1] In another study, the American Egg Board method of heating to 100°C, removing from heat, and holding for 15 minutes was found to be better than two other methods tested.[5] The recommended handling of fresh eggs to prevent salmonellosis is presented in Chapter 26.

MAYONNAISE AND SALAD DRESSING

Mayonnaise can be defined as a semisolid emulsion of edible vegetable oil, egg yolk or whole egg, vinegar, and/or lemon juice, and other ingredients such as salt and other seasonings and glucose, in a finished product containing not less than 50% edible oil. The pH of this product ranges from 3.6 to 4.0, with acetic acid as the predominant acid, representing 0.29–0.5% of total product (aqueous phase up to 2%) with a water activity (a_w) of 0.925. The aqueous phase contains 9–11% salt and 7–10% sugar.[53] Salad dressings are quite similar in composition to mayonnaise, but the finished product contains at least 30% edible vegetable oil and has an a_w of 0.929, a pH of 3.2 to 3.9, with acetic acid usually the predominant acid accounting for 0.9 to 1.2% of total product. The aqueous phase contains 3.0 to 4.0% salt and 20 to 30% sugar.[53] Although the nutrient content of these products is suitable as food sources for many spoilage organisms, the pH, organic acids, and low a_w restrict spoilers to yeasts, a few bacteria, and molds. The yeast *Zygosaccharomyces bailii* is known to cause the spoilage of salad dressings, tomato catsup, carbonated beverages, and some wines. Yeasts of the genus *Saccharomyces* have been implicated in the spoilage of mayonnaise, salad dressing, and French dressing. The two main spoilers for these products are *Lactobacillus fructivorans* and *Z. bailii*. In the spoilage of mayonnaise, *Z. bailii* produces product separation and a "yeasty" odor. In one study, the shelf-life of mayonnaise was extended by the addition of encapsulated cells of *Bifidobacterium bifidum* and *B. infantis*.[27] With the addition of nearly 10^7 cfu/g of the bifidobacteria, yeasts and molds were delayed for about 12 weeks compared to uninoculated controls; and the sensory quality of the preparation was improved by the bifidobacteria. Another spoilage organism is *Lactobacillus brevis* subsp. *lindneri*. Growth of the latter in buttermilk ranch dressing at pH 3.8–4.2 was inhibited by 200-ppm nisin over a 90-day incubation period.[41]

Bacillus vulgatus has been recovered from spoiled Thousand Island dressing, where it caused darkening and separation of the emulsion. In one study of the spoilage of Thousand Island dressing, pepper and paprika were shown to be the sources of *B. vulgatus*.[46] Mold spoilage of products of this type occurs only at the surfaces when sufficient oxygen is available. Separation of the emulsion is generally one of the first signs of spoilage of these products, although bubbles of gas and the rancid odor of butyric acid may precede emulsion separation. The spoilage organisms apparently attack the sugars fermentatively. It appears that the pH remains low, thereby preventing the activities of most proteolytic and lipolytic bacteria. It is not surprising to find yeasts and lactic acid bacteria under these conditions. In a study of 17 samples of spoiled mayonnaise, mayonnaise-like, and blue cheese dressings, Kurtzman et al.[31] found high yeast counts in most samples and high lactobacilli counts in two. The pH of samples ranged from 3.6 to 4.1. Two-thirds of the spoiled samples yielded *Z. bailii*. Common in some samples was *L. fructivorans*, with aerobic spore formers being found in only two samples. Of ten unspoiled samples tested, microorganisms were in low numbers or not detectable at all.

It is well established that foodborne pathogens will not grow in commercially produced mayonnaise or dressings that have a pH of 4.4 and water-phase titratable acidity of at least 0.43 for acetic acid.[53] Foodborne pathogens typically die-off in these products, but some studies have found that *E. coli* 0157:H7 can persist for several weeks (see reference 53). Home-made mayonnaise has been the source of several food poisoning outbreaks typically associated with the use of contaminated raw eggs and the lack of sufficient organic acid to produce a safe pH level to destroy *S.*Enteritidis. It has been recommended that with pure lemon juice [citric acid concentration $\geq 5\%$ (w/v)], the pH should be 3.30 or below, or at least 20 ml of pure lemon juice/egg yolk.[67] It is recommended that a product of this composition should be held at least 72 hours at 22°C or above. Pathogenic strains of *E. coli* from mayonnaise are covered in Chapter 27.

CEREALS, FLOUR, AND DOUGH PRODUCTS

The microbial biota of wheat, rye, corn, and related products may be expected to be that of soil, storage environments, and those picked up during the processing of these commodities. Although these products are high in proteins and carbohydrates, their low a_w is such as to restrict the growth of all microorganisms if stored properly. The microbial biota of flour is relatively low, as some of the bleaching agents reduce the load. When conditions of a_w favor growth, bacteria of the genus *Bacillus* and molds of several genera are usually the only ones that develop. Many aerobic spore formers are capable of producing amylase, which enables them to utilize flour and related products as sources of energy, provided that sufficient moisture is present to allow growth to occur. With less moisture, mold growth occurs and may be seen as typical mycelial growth and spore formation. Members of the genus *Rhizopus* are common and may be recognized by their black spores.

The spoilage of fresh refrigerated dough products, including buttermilk biscuits, dinner and sweet rolls, and pizza dough, is caused mainly by lactic acid bacteria. In a study by Hesseltine et al.,[19] 92% of isolates were Lactobacillaceae, with more than half belonging to the genus *Lactobacillus*, 35% to the genus *Leuconostoc*, and 3% to "*Streptococcus*." Molds were found generally in low numbers in spoiled products. The fresh products showed lactic acid bacterial numbers as high as 8.38 $\log_{10}$/g.

BAKERY PRODUCTS

Commercially produced and properly handled bread generally lacks sufficient amounts of moisture to allow for the growth of any organisms except molds. One of the most common is *Rhizopus stolonifer*, often referred to as the "bread mold." The "red bread mold," *Neurospora sitophila*, may also be seen from time to time. Storage of bread under conditions of low humidity retards mold growth, and this type of spoilage is generally seen only when bread is stored at high humidity or when wrapped while still warm. Homemade breads may undergo a type of spoilage known as ropiness, which is caused by the growth and amylase production of certain strains of *Bacillus subtilis* (*B. mesentericus*). The ropiness may be seen as stringiness by carefully breaking a batch of dough into two parts. The source of the organisms is flour, and their growth is favored by holding the dough for sufficient periods of time at suitable temperatures. In a recent study, part-baked soda bread (pH 7–9) stored at room temperature developed ropiness after 2 days, and three species were isolated from the ropy product: *B. subtilis, B. pumilus*, and *B. licheniformis*.[34]

Cakes of all types rarely undergo bacterial spoilage due to their unusually high concentrations of sugars, which restrict the availability of water. The most common form of spoilage displayed by these products is moldiness. Common sources of spoilage molds are any and all cake ingredients, especially sugar, nuts, and spices. Although the baking process is generally sufficient to destroy these organisms, many are added in icings, meringues, toppings, and so forth. Also, molds may enter baked cakes from handling and from the air. Growth of molds on the surface of cakes is favored by conditions of high humidity. On some fruitcakes, growth often originates underneath nuts and fruits if they are placed on the surface of such products after baking. Continued growth of molds on breads and cakes results in a hardening of the products.

FROZEN MEAT PIES

The microbiological quality of frozen meat pies has steadily improved since these products were first marketed. Any and all of the ingredients added may increase the total number of organisms, and the total count of the finished product may be taken to reflect the overall quality of ingredients, handling, and storage. Many investigators have suggested that these products should be produced with total counts not to exceed $\log_{10}$/g 5.00. In a study of 48 meat pies, 84% had an APC $\log_{10} < 5$,[38] whereas in another study of 188 meat pies, 93% had counts $\log_{10} < 5.00$.[25] Accordingly, a microbiological criterion of $\log_{10}$ 5.00 seems attainable for such products (see Chapter 21 for further information on microbiological standards and criteria). In a study of 1,290 frozen tuna pot pies, the geometric mean APC at 35°C was $\log_{10}$ 3.20, whereas at 30°C it was $\log_{10}$/g 3.38.[66] Coliforms averaged 5/g, *E. coli* <3/g, and *S. aureus* <10/g (Table 9–1).

SUGARS, CANDIES, AND SPICES

These products rarely undergo microbial spoilage if properly prepared, processed, and stored, primarily because of the lack of sufficient moisture for growth. Both cane and beet sugars may be expected to contain microorganisms. The important bacterial contaminants are members of the genera *Bacillus, Paenibacillus*, and *Clostridium*, which sometimes cause trouble in the canning industry (see Chapter 17). If sugars are stored under conditions of extremely high humidity, growth of some of these organisms is possible, usually at the exposed surfaces. The successful growth of these organisms depends, of course, on their getting an adequate supply of moisture and essential nutrients other than carbohydrates. "*Torula*" and osmophilic strains of *Saccharomyces* (*Zygosaccharomyces* spp.) have been reported to cause trouble in high-moisture sugars. These organisms have been reported to cause inversion of sugar. One of the most troublesome organisms in sugar refineries is *Leuconostoc mesenteroides*. This organism hydrolyzes sucrose and synthesizes a glucose polymer referred to as dextran. This gummy and slimy polymer sometimes clogs the lines and pipes through which sucrose solutions pass.

Among candies that have been reported to undergo microbial spoilage are chocolate creams, which sometimes undergo explosions. The causative organisms have been reported to be *Clostridium* spp., especially *C. sporogenes*, which finds its way into these products through sugars, starch, and possibly other ingredients.

Although spices do not undergo microbial spoilage in the usual sense of the word, molds and a few bacteria do grow in those that do not contain antimicrobial principals, provided sufficient moisture is available. Prepared mustard has been reported to undergo spoilage by yeasts and by *Proteus* and *Bacillus* spp. usually with a gassy fermentation. The usual treatment of spices with propylene oxide reduces their content of microorganisms, and those that remain are essentially spore formers and molds. No trouble should be encountered from microorganisms as long as the moisture level is kept low.

The microbial profile of some spices is presented in Table 9–1. In a study of products on the Austrian market, no confirmed *S. aureus* could be found in the 160 samples, and only one sample was positive for a *Salmonella—S.* Arizonae.[30] The single highest numbers of organisms found were 2.6×10^7/g in China spice and 2.2×10^7/g in black pepper. Over half of the 160 samples were positive for enteric bacteria, and over half had an APC of 10^4–10^6 cfu/g.[30] Only 3 (all paprika) of the 160 samples contained potentially aflatoxigenic fungi.

NUTMEATS

Due to the extremely high fat and low water content of products such as pecans and walnuts (Table 9–4), these products are quite refractory to spoilage bacteria. Molds can and do grow on them if they are stored under conditions that permit sufficient moisture to be picked up. Examination of nutmeats will reveal molds of many genera that are picked up by the products during collecting, cracking, sorting, and packaging. (See Chapter 30 for a discussion of aflatoxins as related to nutmeats.)

Table 9–4 Percentage Composition of Miscellaneous Foods

Food	Water	Carbohydrates	Proteins	Fat	Ash
Beer (4% alcohol)	90.2	4.4	0.6	0.0	0.2
Bread, enriched white	34.5	52.3	8.2	3.3	1.7
Butter	15.5	0.4	0.6	81.0	2.5
Cake (pound)	19.3	49.3	7.1	23.5	0.8
Figbars	13.8	75.8	4.2	4.8	1.4
Jellies	34.5	65.0	0.2	0.0	0.3
Margarine	15.5	0.4	0.6	81.0	2.5
Mayonnaise	1.7	21.0	26.1	47.8	3.4
Peanut butter	16.0	3.0	1.5	78.0	1.5
Almonds (dried)	4.7	19.6	18.6	34.1	3.0
Brazil nuts	5.3	11.0	14.4	65.9	3.4
Cashews	3.6	27.0	18.5	48.2	2.7
Peanuts	2.6	23.6	26.9	44.2	2.7
Pecans	3.0	13.0	9.4	73.0	1.6
Mean	3.8	18.8	17.6	57.1	2.7

Source: Watt and Merrill.[65]

DEHYDRATED FOODS

In a detailed study of the microbiology of dehydrated soups, Fanelli et al.[10,11] showed that approximately 17 different kinds of dried soups from nine different processors had total counts of less than $\log_{10}/g$ 5.00. These soups included chicken noodle, chicken rice, beef noodle, vegetable, mushroom, pea, onion, tomato, and others. Some of these products had total counts as high as $\log_{10}/g$ 7.30, and some had counts as low as around $\log_{10}$ 2.00. These investigators further found that reconstituted dehydrated onion soup showed a mean total count of $\log_{10}/ml$ 5.11, with $\log_{10}$ 3.00 coliforms, $\log_{10}$ 4.00 aerobic spore formers, and $\log_{10}/ml$ 1.08 of yeast and molds. Upon cooking, the total counts were reduced to a mean of $\log_{10}$ 2.15, whereas coliforms were reduced to <0.26, spore formers to $\log_{10}$ 1.64, and yeasts and molds to $\log_{10}/ml$ <1.00. In a study of dehydrated sauce and gravy mixes, soup mixes, spaghetti sauce mixes, and cheese sauce mixes, *C. perfringens* was isolated from 10 of 55 samples.[42] The facultative anaerobe counts ranged from $\log_{10}/g$ 3.00 to >6.00. In a study of 185 samples of food-grade dry gelatin, no samples exceeded an APC of $\log_{10}/g$ 3.70.[33] Of 129 dehydrated space food samples examined, 93% contained total counts $\log_{10}/g$ <4.00.[47]

Powdered eggs and milk often contain high numbers of microorganisms—on the order of $\log_{10}/g$ 6–8. One reason for the generally high numbers in dried products is that the organisms have been concentrated on a per gram basis along with product concentration. The same is generally true for fruit juice concentrates, which tend to have higher numbers of microorganisms than the fresh, nonconcentrated products.

ENTERAL NUTRIENT SOLUTIONS (MEDICAL FOODS)

Enteral nutrient solutions (ENS), also known as medical foods, are liquid foods administered by tube. They are available as powdered products requiring reconstitution or as liquids. They are generally administered to certain patients in hospitals or other patient care facilities, but may also be administered in the home. Administration is by continuous drip from enteral feeding bags, and the process may go on for 8 hours or longer, with the ENS at room temperature. Enteral foods are made by several commercial companies as complete diets that only require reconstituting with water before use, or as incomplete meals that require supplementation with milk, eggs, or the like prior to use. ENS-use preparations are nutritionally complete, with varying concentrations of proteins, peptides, carbohydrates, and so forth, depending on patient need.

The microbiology of ENS has been addressed by some hospital researchers, who have found the products to contain varying numbers and types of bacteria and to be the source of patient infections. Numbers as high as $10^8/ml$ have been found in some ENS at time of infusion.[14] In a study of one reconstituted commercial ENS, the initial count of $9 \times 10^3/ml$ increased to $7 \times 10^4/ml$ after 8 hours at room temperature.[20] Numbers as high as $1.2 \times 10^5/ml$ were found in another sample of the same preparation. The most frequently isolated organism was *Staphylococcus epidermidis*, with *Corynebacterium, Citrobacter*, and *Acinetobacter* spp. among the other isolates. From a British study, enteral feeds yielded $10^4 - 10^6$ organisms/ml, with coliforms and *Pseudomonas aeruginosa* as the predominant types.[15]

The capacity of five different commercial ENS to support the growth of *Enterobacter cloacae* under use conditions has been demonstrated,[9] and the addition of 0.2% potassium sorbate was shown to reduce numbers of this organism by three log cycles over controls. Patients are known to have contracted *E. cloacae* and *Salmonella* Enteritidis infections from ENS.[4,14] Procedures that should be employed in the preparation/handling of ENS to minimize microbial problems have been noted.[17] For a review of the history and other nonmicrobial aspects of ENS or medical foods (see reference 52).

SINGLE-CELL PROTEIN (SCP)

The cultivation of unicellular microorganisms as a direct source of human food was suggested in the early 1900s. The expression *single-cell protein* (SCP) was coined at the Massachusetts Institute of Technology around 1966 to depict the idea of microorganisms as food sources.[54] Although SCP is a misnomer in that proteins are not the only food constituent represented by microbial cells, it obviates the need to refer to each product generically as in "algal protein," "yeast cell protein," and so on. Although SCP as a potential and real source of food for humans differs from the other products covered in this chapter, with the exception of that from algal cells, it is produced in a similar manner.

Rationale for SCP Production

It is imperative that new food sources should be found in order that future generations are adequately fed. A food source that is nutritionally complete and requires a minimum of land, time, and cost to produce is highly desirable. In addition to meeting these criteria, SCP can be produced on a variety of waste materials. Among the overall advantages of SCP over plant and animal sources of proteins are the following:[28]

1) Microorganisms have a very short generation time and can thus provide a rapid mass increase.
2) Microorganisms can be easily modified genetically—to produce cells that bring about desirable results.
3) The protein content is high.
4) The production of SCP can be based on raw materials readily available in large quantities.
5) SCP production can be carried out in continuous culture and thus be independent of climatic changes.

The greater speed and efficiency of microbial protein production compared to plant and animal sources may be illustrated as follows: a 1,000-lb steer produces about 1 lb of new protein per day; soybeans (prorated over a growing season) produce about 80 lb, and yeasts produce about 50 tons.

Organisms and Fermentation Substrates

A large number of algae, yeasts, molds, and bacteria have been studied as SCP sources. Among the most promising genera and species studied are the following:

1) Algae: *Chlorella* spp., and *Scenedesmus* spp.
2) Yeasts: *Candida guilliermondii, C. utilis, C. lipolytica*, and *C. tropicalis; Debaryomyces kloeckeri; Candida famata, C. methanosorbosa; Pichia* spp.; *Kluyveromyces fragilis; Hansenula polymorpha; Rhodotorula* spp.; and *Saccharomyces* spp.
3) Filamentous fungi: *Agaricus* spp.; *Aspergillus* spp.; *Fusarium* spp.; *Penicillium* spp.; *Saccharomycopsis fibuligera*; and *Trichosporon cutaneum*.
4) Bacteria: *Bacillus* spp.; *Acinetobacter calcoaceticus; Cellulomonas* spp.; *Nocardia* spp.; *Methylomonas* spp.; *Aeromonas hydrophila; Alcaligenes eutrophus* (*Hydrogenomonas eutropha*), *Mycobacterium* sp.; *Spirulina maxima*, and *Rhodopseudomonas* sp.

Of these groups, yeasts have received, by far, the most attention.

The choice of a given organism is dictated in large part by the type of substrate or waste material in question. The cyanobacterium *Spirulina maxima* grows in shallow waters high in bicarbonate at a temperature of 30°C and a pH of 8.5–11.0. It can be harvested from pond waters and dried for food use. This cell has been eaten by the people of the Chad Republic for many years.[54] Other cyanobacteria require sunlight, CO_2, minerals, water, and proper growth temperatures. However, the large-scale use of such cells as SCP sources is said to be practical only in areas below 35° latitude, where sunlight is available most of the year.[37]

Bacteria, yeasts, and molds can be grown on a wide variety of materials, including food-processing wastes (such as cheese whey and brewery, potato processing, cannery, and coffee wastes), industrial wastes (such as sulfite liquor in the paper industry and combustion gases), and cellulosic wastes (including bagasse, newsprint mill, and barley straw). In the case of cellulosic wastes, it is necessary to use organisms that can utilize cellulose, such as a *Cellulomonas* sp. or *Trichoderma viride*. A mixed culture of *Cellulomonas* and *Alcaligenes* has been employed. For starchy materials, a combination of *Saccharomycopsis fibuligera* and a *Candida* sp. such as *C. utilis* has been employed, in which the former effects hydrolysis of starches and the latter subsists on the hydrolyzed products to produce biomass. Some other representative substrates and organisms are listed in Table 9–5.

Table 9–5 Substrate Materials that Support the Growth of Microorganisms in the Production of SCP

Substrates	Microorganisms
CO_2 and sunlight	*Chlorella pyrenoidosa*
	Scenedesmus quadricauda
	Spirulina maxima
n-Alkanes, kerosene	*Candida intermedia, C. lipolytica, C. tropicalis*
	Nocardia spp.
Methane	*Methylomonas* sp. (*Methanomonas*)
	Methylococcus capsulatus
	Trichoderma spp.
H_2 and CO_2	*Alcaligenes eutrophus* (*Hydrogenomonas eutropha*)
Gas oil	*Acinetobacter calcoaceticus* (*Micrococcus cerificans*)
	Candida lipolytica
Methanol	*Methylomonas methanica* (*Methanomonas methanica*)
Ethanol	*Candida utilis*
	Acinetobacter calcoaceticus
Sulfite liquor wastes	*Candida utilis*
Cellulose	*Cellulomonas* spp.
	Trichoderma viride
Starches	*Saccharomycopsis fibuligera*
Sugars	*Saccharomyces cerevisiae*
	Candida utilis
	Kluyveromyces fragilis

SCP Products

The cells may be used directly as a protein source in animal feed formulations, thereby freeing animal feed, such as corn, for human consumption, or they may be used as a protein source or food ingredient for human food. In the case of animal feed or feed supplements, the dried cells may be used without further processing. Whole cells of *Spirulina maxima* are consumed by humans in at least one part of Africa as noted above.

For human use, the most likely products are SCP concentrates or isolates that can be further processed into textured or functional SCP products. To produce functional protein fibers, cells are mechanically disrupted, cell walls are removed by centrifugation, proteins are precipitated from disrupted cells, and the resulting protein is extruded from syringelike orifices into suitable menstra such as acetate buffer, $HCIO_4$, acetic acid, and the like. The SCP fibers may now be used to form textured protein products. Baker's yeast protein is one product of this type approved for human food ingredient use in the United States.

Nutrition and Safety of SCP

Chemical analyses of the microorganisms evaluated for SCP reveal that they are comparable in amino acid content and type to plant and animal sources with the possible exception of methionine, which is lower in some SCP sources. All are relatively high in nitrogen. For example, the approximate percentage composition of nitrogen on a dry weight basis is as follows: bacteria 12–13, yeast 8–9, algae 8–10, and filamentous fungi 5–8.[28] In addition to proteins, microorganisms contain adequate levels of carbohydrates, lipids, and minerals and are excellent sources of B vitamins. The fat content varies among these sources, with algal cells containing the highest levels and bacteria the lowest. On a dry weight basis, nucleic acids average 3–8% for algae, 6–12% for yeasts, and 8–16% for bacteria.[28] B vitamins are high in all SCP sources. The digestibility of SCP in experimental animals has been found to be lower than for animal proteins such as casein. A thorough review of the chemical composition of SCP from a large variety of microorganisms has been made.[7,64]

Success has been achieved in rat-feeding studies with a variety of SCP products, but human-feeding studies have been less successful, except in the case of certain yeast cell products. Gastrointestinal disturbances are common complaints following the consumption of algal and bacterial SCP, and these and other problems associated with the consumption of SCP have been reviewed elsewhere.[64] When Gram-negative bacteria are used as SCP sources for human use, the endotoxins must be removed or detoxified.

The high nucleic acid content of SCP leads to kidney stone formation and/or gout. The nucleic acid content of bacterial SCP may be as high as 16%, whereas the recommended daily intake is about 2 g. The problems are caused by an accumulation of uric acid, which is sparingly soluble in plasma. Upon the breakdown of nucleic acids, purine and pyrimidine bases are released. Adenine and guanine (purines) are metabolized to uric acid. Lower animals can degrade uric acid to the soluble compound allantoin (they possess the enzyme uricase), and, consequently, the consumption of high levels of nucleic acids does not present metabolic problems to these animals as it does to humans. Although high nucleic acid contents presented problems in the early development and use of SCP, these compounds can be reduced to levels below 2% by techniques such as acid precipitation, acid or alkaline hydrolysis, or use of endogenous and bovine pancreatic RNases.[37]

BOTTLED WATER

"Water that is vended in a bottle" is a very simple yet incomplete definition of bottled water as it exists in commerce. It is defined by the Codex Alimentarius Commission, the European Economic Community (Council Directive for Potable Waters), and regulated in most countries by the same body that regulates foods. The most common bottled waters fall into one of the following groups (extracted from references 21, 29):

Natural mineral water (*eaux plates* or "flat waters" in Europe)
Naturally carbonated natural mineral water
Non-carbonated natural mineral water
Carbonated natural mineral waters
Ground water
Spring water

However classified, bottled water must be potable, which means it must be free of pathogens, toxic substances, objectionable odor, color, turbidity, or taste (see reference 21). The pH of non-carbonated water should be around neutrality, whereas that of carbonated water is typically between 3 and 4.0— ideally at or below pH 3.5.

In general, the microbiology of drinking water is complex, and this subject has been reviewed by Szewzyk et al.[59] Although bottled waters are generally free of intestinal pathogens that persist in the drinking water supply, *Legionella pneumophila* (causes legionellosis) has been shown to survive for months in sterile drinking water (see reference 59). The safety of bottled water is determined by testing for coliforms—not specific fecal coliforms or *E. coli*. However, if a fecal coliform test is run, the WHO recommendation of 1993 is for <1 cfu/100 ml. When tested by the membrane filter method, not more than 3 coliforms/100 ml should be present. APC determinations on bottled water are of little value in determining safety since the numbers should not exceed 10^2/100 ml. Because of the destructive effects of the low pH of carbonated bottled water, testing for coliforms is not necessary.

The types of organisms that are found in potable bottled water consist of Gram negatives since they are so much more synthetic than Gram positives, and thus, can survive on traces of organic matter. Pseudomonads are not unknown among this group, and the presence of *P. aeruginosa* or *Burkholderia cepacia* is of special concern in bottled water because of the infectious nature of these organisms for debilitated individuals. The presence of organisms of this type is minimized by the use of ozone in the bottled water industry. The FDA in the United States approved in 1997 ozonation at a maximum level at the time of bottling of 0.4 ppm.[29]

Results from various investigators on the presence of *E. coli* and other coliforms; and the fate of inoculated *E. coli* 0157:H7 in bottled waters are varied. An analysis of 104 brands of bottled waters from 10 countries failed to detect either *E. coli* or coliforms.[16] In another study on noroviruses (see chapter 31), none could be found in 1,436 analyses of bottled water in different sized containers although by RT-PCR, 34 (2.4%) were presumptively positive but by DNA sequence analysis, they did not confirm.[32] *E. coli* 0157:H7 was shown to survive for >300 days when a 10-strain mixture was added at levels of $\log_{10}$ 3.24– 6.54 cfu/ml to bottled spring and mineral waters.[63] These investigators presented evidence for biofilm formation by the inoculated organisms on the sides of containers. Previous investigators found that *E. coli* numbers were reduced by 3 to 5 logs in mineral water held at 20 to 25°C.

In a detailed study with a nontoxigenic strain of *E. coli* 0157:H7 added to three types of bottled natural and noncarbonated mineral waters at levels of 10^3 and 10^6 cfu/ml and stored at 15°C for 10 weeks, no differences were found between cell survival in natural noncarbonated and sterile natural mineral and natural noncarbonated waters up to 35 days.[26] Overall, the inoculated cells survived the longest in the natural noncarbonated water. Evidence was presented to indicate that the cells lysed during storage with cell contents providing a nutrient source for the autochthonous bacterial biota.

In regards to ice used to cool drinks, 9% of 3,528 samples contained coliforms and 1% contained either *E. coli* or enterococci with both being in excess of 10^2 cfu/100 ml.[43] The APC (at 37°C) of 11% of these samples was > 10^3 cfu/ml. Ice used in food displays was of poorer quality with 23% containing coliforms and 5% *E. coli*.[43]

Fruit-flavored bottle water was the source of a relatively new acetic acid bacterium, *Asaia* sp.[40] The bottled water containing this organism was noted prior to its distribution because of its turbidity. When tested, the acetic acid bacterium was found at a level equal to or above 10^6 cfu/ml.[40]

REFERENCES

1. Baker, R.C., S. Hogarty, and W. Poon. 1983. Survival of *Salmonella* Typhimurium and *Staphylococcus aureus* in egg products cooked by different methods. *Poult. Sci.* 62:1211–1216.

2. Braun, P., and K. Fehlhaber. 1995. Migration of *Salmonella* Enteritidis from the albumen into the egg yolk. *Int. J. Food Microbiol.* 25:95–99.

3. Brooks, J., and H.P. Hale. 1959. The mechanical properties of the thick white of the hen's egg. *Biochem. Biophys. Acta* 32:237–250.

4. Casewell, M.W., J.E. Cooper, and M. Webster. 1981. Enteral feeds contaminated with *Enterobacter cloacae* as a cause of septicaemia. *Br. Med. J.* 282:973.

5. Chantarapanont, W., L. Slutsker, R.V. Tauxe, and L.R. Beuchat. 2000. Factors influencing inactivation of *Salmonella* Enteritidis in hard-cooked eggs. *J. Food Protect.* 63:36–43.

6. Christiansen, L.N., and N.S. King. 1971. The microbial content of some salads and sandwiches at retail outlets. *J. Milk Food Technol.* 34:289–293.

7. Cooney, C.L., C. Rha, and S.R. Tannenbaum. 1980. Single-cell protein: Engineering, economics, and utilization in foods. *Adv. Food Res.* 26:1–52.

8. Curtis, P.A., K.E., anderson, and F.T. Jones. 1995. Cryogenic gas for rapid cooling of commercially processed shell eggs before packaging. *J. Food Protect.* 58:389–394.

9. Fagerman, K.E., J.D. Paauw, M.A. McCamish, and R.E Dean. 1984. Effects of time, temperature, and preservative on bacterial growth in enteral nutrient solutions. *Am. J. Hosp. Pharm.* 41:1122–1126.

10. Fanelli, M.J., A.C. Peterson, and M.F. Gunderson. 1965. Microbiology of dehydrated soups. I. A survey. *Food Technol.* 19:83–86.

11. Fanelli, M.J., A.C. Peterson, and M.F. Gunderson. 1965. Microbiology of dehydrated soups. III. Bacteriological examination of rehydrated dry soup mixes. *Food Technol.* 19:90–94.

12. Fowler, J.L., and W.S. Clark, Jr. 1975. Microbiology of delicatessen salads. *J. Milk Food Technol.* 38:146–149.

13. Fowler, J.L., and J.F. Foster. 1976. A microbiological survey of three fresh green salads: Can guidelines be recommended for these foods? *J. Milk Food Technol.* 39:111–113.

14. Furtado, D., A. Parrish, and P. Beyer. 1980. Enteral nutrient solutions (ENS): In vitro growth supporting properties of ENS for bacteria. *J. Paren. Ent. Nutr.* 4:594.

15. Gill, K.J., and P. Gill. 1981. Contaminated enteral feeds. *Br. Med. J.* 282:1971.

16. Grant, M.A. 1998. Analysis of bottled water for *Escherichia coli* and total coliforms. *J. Food Protect.* 61:334–338.

17. Gröschel, D.H.M. 1983. Infection control considerations in enteral feeding. *Nutr. Suppl. Serv.* 3:48–49.

18. Harris, N.D., S.R. Martin, and L. Ellias. 1975. Bacteriological quality of selected delicatessen foods. *J. Milk Food Technol.* 38:759–761.

19. Hesseltine, C.W., R.R. Graves, R. Rogers, and H.R. Burmeister. 1969. Aerobic and facultative microflora of fresh and spoiled refrigerated dough products. *Appl. Microbiol.* 18:848–853.

20. Hostetler, C., T.O. Lipman, M. Geraghty, and H.R. Parker. 1982. Bacterial safety of reconstituted continuous drip tube feeding. *J. Paren. Ent. Nutr.* 6:232–235.

21. ICMSF. 1986. *Microorganisms in Foods. 2. Sampling for Microbiological Analysis: Principles and Specific Applications*, 2nd ed., 234–243. Toronto: University of Toronto Press.

22. Jones, D.R., J.K. Northcutt, M.T. Musgrove, P.A. Curtis, K.E. Anderson, D.L. Fletcher, and N.A. Cox. 2003. Survey of shell egg processing plant sanitation programs: Effects of egg contact surfaces. *J. Food Protect.* 66:1486–1489.

23. Julseth, R.M., and R.H. Deibel. 1974. Microbial profile of selected spices and herbs at import. *J. Milk Food Technol.* 37:414–419.

24. Käferstein, F.K. 1976. The microflora of parsley. *J. Milk Food Technol.* 39:837–840.

25. Kereluk, K., and M.F. Gunderson. 1959. Studies on the bacteriological quality of frozen meat pies. I. Bacteriological survey of some commercially frozen meat pies. *Appl. Microbiol.* 7:320–323.

26. Kerr, M., M. Fitzgerald, J.J. Sheridan, D.A. McDowell, and I.S. Blair. 1999. Survival of *Escherichia coli* 0157:H7 in bottled natural mineral water. *J. Appl. Microbiol.* 87:833–841.

27. Khalil, A.H., and E.H. Mansour. 1998. Alginate encapsulated bifidobacteria survival in mayonnaise. *J. Food Sci.* 63:702–705.

28. Kihlberg, R. 1972. The microbe as a source of food. *Annu. Rev. Microbiol.* 26:427–466.

29. Kim, H., and P. Feng. 2001. Bottled water. In *Compendium of Methods for the Microbiological Examinaation of Foods*, 4th ed., 573–576. ed. F.P. Downes and K. Ito. Washington, DC: American Public Health Association.

30. Kneifel, W., and E. Berger. 1994. Microbiological criteria of random samples of spices and herbs retailed on the Austrian market. *J. Food Protect.* 57:893–901.

31. Kurtzman, C.P., R.Rogers, and C.W. Hesseltine. 1971. Microbiological spoilage of mayonnaise and salad dressings. *Appl. Microbiol.* 21:870–874.

32. Lamothe, G.T., T. Putallaz, H. Joosten, and J.D. Marugg. 2003. Reverse transcription-PCR analysis of bottled and natural mineral waters for the presence of noroviruses. *Appl. Environ. Microbiol.* 69:6541–6549.

33. Leininger, H.V., L.R. Shelton, and K.H. Lewis. 1971. Microbiology of frozen cream-type pies, frozen cooked-peeled shrimp, and dry food-grade gelatin. *Food Technol.* 25:224–229.

34. Leuschner, R.G.K., M.J.A. O'Callaghan, and E.K. Arendt. 1998. Bacilli spoilage in part-baked and rebaked brown soda bread. *J. Food Sci.* 63:915–918.

35. Levine, P., B. Rose, S. Green, G. Ransom, and W. Hill. 2001. Pathogen testing of ready-to-eat meat and poultry products collected at federally inspected establishments in the United States, 1990 to 1999. *J. Food Protect.* 64:1188–1193.

36. Lifshitz, A., R.G. Baker, and H.B. Naylor. 1964. The relative importance of chicken egg exterior structures in resisting bacterial penetration. *J. Food Sci.* 29:94–99.

37. Litchfield, J.H. 1977. Single-cell proteins. *Food Technol.* 31:175–179.

38. Litsky, W., I.S. Fagerson, and C.R. Fellers. 1957. A bacteriological survey of commercially frozen beef, poultry and tuna pies. *J. Milk Food Technol.* 20:216–219.

39. Meldrum, R.J., and C.D. Ribeiro. 2003. *Campylobacter* in ready-to-eat foods: The result of a 15-month survey. *J. Food Protect.* 66:2135–2137.

40. Moore, J.E., M. McCalmont, J. Xu, B.C. Miller, and N. Heaney. 2002. *Asaia* sp., an unusual spoilage organism of fruit-flavored bottled water. *Appl. Environ. Microbiol.* 68:4130–4131.

41. Muriana, P.M., and L. Kanach. 1995. Use of Nisaplin™ to inhibit spoilage bacteria in buttermilk ranch dressing. *J. Food Protect.* 58:1109–1113.

42. Nakamura, M., and K.D. Kelly. 1968. *Clostridium perfringens* in dehydrated soups and sauces. *J. Food Sci.* 33:424–426.

43. Nichols, G., I. Gillespie, and J. de Louvois. 2000. The microbiological quality of ice used to cool drinks and ready-to-eat food from retail and catering premises in the United Kingdom. *J. Food Protect.* 63:78–82.

44. Pace, P.J. 1975. Bacteriological quality of delicatessen foods: Are standards needed? *J. Milk Food Technol.* 38:347–353.

45. Paradis, D.C., and M.E. Stiles. 1978. A study of microbial quality of vacuum packaged, sliced bologna. *J. Food Protect.* 41:811–815.

46. Pederson, C.S. 1930. Bacterial spoilage of a thousand island dressing. *J. Bacteriol.* 20:99–106.

47. Powers, E.M., C. Ay, H.M. El-Bisi, and D.B. Rowley. 1971. Bacteriology of dehydrated space foods. *Appl. Microbiol.* 22:441–445.

48. Powers, E.M., T.G. Latt, and T. Brown. 1976. Incidence and levels of *Bacillus cereus* in processed spices. *J. Milk Food Technol.* 39:668–670.

49. Powers, E.M., R. Lawyer, and Y. Masuoka. 1975. Microbiology of processed spices. *J. Milk Food Technol.* 38:683–687.

50. Rehberger, T.G., L.A. Wilson, and B.A. Glatz. 1984. Microbiological quality of commercial tofu. *J. Food Protect.* 47:177–181.

51. Sagoo, S.K., C.L. Little, and R.T. Mitchell. 2003. Microbiological quality of open ready-to-eat salad vegetables: Effectiveness of food hygiene training of management.*J. Food Protect.* 66:1581–1586.

52. Schmidl, M.K., and T.P. Labuza. 1992. Medical foods. *Food Technol.* 46:87–96.

53. Smittle, R.B. 2000. Microbiological safety of mayonnaise, salad dressings, and sauces produced in the United States: A review. *J. Food Protect.* 63:1144–1153.

54. Snyder, H.E. 1970. Microbial sources of protein. *Adv. Food Res.* 18:85–140.

55. Soriano, J.M., H. Rico, J.C. Molto, and J. Mañes. 2001. Incidence of microbial flora in lettuce, meat and Spanish potato omelette from restaurants. *Food Microbiol.* 18:159–163.

56. Surkiewicz, B.F. 1966. Bacteriological survey of the frozen prepared foods industry. *Appl. Microbiol.* 14:21–26.

57. Surkiewicz, B.F., R.J. Groomes, and A.P. Padron. 1967. Bacteriological survey of the frozen prepared foods industry. III. Potato products. *Appl. Microbiol.* 15:1324–1331.

58. Swartzentruber, A., A.H. Schwab, B.A. Wentz, A.P. Duran, and R.B Read, Jr. 1984. Microbiological quality of biscuit dough, snack cakes and soy protein meat extender. *J. Food Protect.* 47:467–470.

59. Szewzyk, U., R. Szewzyk, W. Manx, and K.-H. Schleifer. 2000. Microbiological safety of drinking water. *Ann. Rev. Microbiol.* 54:81–127.

60. Todd, E.C.D., G.A. Jarvis, K.F. Weiss, and S. Charbonneau. 1983. Microbiological quality of frozen cream-type pies sold in Canada. *J. Food Protect.* 46:34–40.

61. Tranter, H.S., and R.G. Board. 1984. The influence of incubation temperature and pH on the antimicrobial properties of hen egg albumen. *J. Appl. Bacteriol.* 56:53–61.

62. Villari, P., M. Crispino, P. Montuori, and S. Stanzione. 2000. Prevalence and molecular characterization of *Aeromonas* spp. in ready-to-eat foods in Italy. *J. Food Protect.* 63:1734–1757.

63. Warburton, D.W., J.W. Austin, B.H. Harrison, and G. Sanders. 1998. Survival and recovery of *Escherichia coli* 0157:H7 in inoculated bottled water.*J. Food Protect.* 61:948–952.

64. Waslien, C.I. 1976. Unusual sources of proteins for man. *CRC Crit. Rev. Food Sci. Nutr.* 6:77–151.

65. Watt, B.K., and A.L. Merrill. 1950. Composition of foods—Raw, processed, prepared. *Agricultural Handbook No. 8.* Washington, DC: USDA.

66. Wentz, B.A., A.P. Duran, A. Swartzentruber, A.B Schwab, and R.B. Read, Jr. 1984. Microbiological quality of frozen breaded onion rings and tuna pot pies. *J. Food Protect.* 47:58–60.

67. Xiong, R., G. Xie, and A.S. Edmondson. 1999. The fate of *Salmonella* Enteritidis PT4 in home-made mayonnaise prepared with citric acid. *Lett. Appl. Microbiol.* 28:36–40.

PART IV

Determining Microorganisms and/or Their Products in Foods

If one assumes that a given science discipline is no better than its methodology, the material presented in these three chapters is critical to food microbiology. Traditional methods are presented in Chapters 10 and 11, along with some newer developments that are designed to be more precise, accurate, and rapid than the former. The areas covered are being pursued actively in research laboratories, and complete treatments go beyond the scope of this work. The references noted below should be consulted for in-depth coverage. The animal and tissue culture assay methods covered in Chapter 12 are designed to provide the basic principles of these bioassay methods. For most foodborne pathogens, additional information is provided in the chapters in Part VII.

1. Feng, P. 1997. Impact of molecular biology on the detection of foodborne pathogens. *Mol. Biotechnol.* 7:267–278. Contains lists of commercially available sources of molecular and immunological test kits along with descriptions of the most widely used methods.
2. Hill, W.E. 1996. The polymerase chain reaction: Applications for the detection of foodborne pathogens. *Crit. Rev. Food Sci. Nutr.* 36:123–173. Contains an extensive review of PCR basic methods of operation and its application to a number of foodborne organisms. An extensive list of primers for foodborne pathogens is provided.
3. McKillip, J.L., and M. Drake. 2004. Real-time nucleic acid-based detection methods for pathogenic bacteria in food. *J. Food Protect.* 67:823–832. An up-to-date review of the titled methodology along with a list of commercially available test systems.
4. McMeekin, T.A., ed. 2003. *Detecting Pathogens in Foods.* Boca Raton, FL: CRC Press. A 370-page coverage of a number of the methods presented in Chapters 10 and 11.
5. Scheu, P.M., K. Berghof, and U. Stahl. 1998. Detection of pathogenic and spoilage microorganisms in food with the polymerase chain reaction. *Food Microbiol.* 15:13–31. An extensive compilation of PCR systems for foodborne organisms is provided.
6. Tortorello, M.L., and S.M. Gendel, eds. 2002. *Food Microbiology and Analytical Methods.* New York: Marcel Dekker.

CHAPTER 10

Culture, Microscopic, and Sampling Methods

The examination of foods for the presence, types, and numbers of microorganisms and/or their products is basic to food microbiology. In spite of the importance of this, none of the methods in common use permits the determination of exact numbers of microorganisms in a food product. Although some methods of analysis are better than others, every method has certain inherent limitations associated with its use.

The four basic methods employed for "total" numbers are as follows:

1. Standard plate counts (SPC) or aerobic plate counts (APC) for viable cells or colony forming units (cfu).
2. The most probable numbers (MPN) method as a statistical determination of viable cells.
3. Dye reduction techniques to estimate numbers of viable cells that possess reducing capacities.
4. Direct microscopic counts (DMC) for both viable and nonviable cells.

All of these are discussed in this chapter, along with their uses in determining microorganisms from various sources. Detailed procedures for their use can be obtained from references in Table 10–1. In addition, variations of these basic methods for examining the microbiology of surfaces are presented along with a summary of methods and attempts to improve their overall efficiency.

CONVENTIONAL STANDARD PLATE COUNT

By the conventional SPC method, portions of food samples are blended or homogenized, serially diluted in an appropriate diluent, plated in or onto a suitable agar medium, and incubated at an appropriate temperature for a given time, after which all visible colonies are counted by use of a Quebec or electronic counter. The SPC is by far the most widely used method for determining the numbers of viable cells or colony-forming units (cfu) in a food product. When total viable counts are reported for a product, the counts/numbers should be viewed as a function of at least some of the following factors:

1. Sampling methods employed
2. Distribution of the organisms in the food sample

Table 10–1 Some Standard References for Methods of Microbiological Analysis of Foods

	Reference							
	72	*12*	*79*	*73*	*31*	*80*	*36*	*89*
Direct microscopic counts				X	X	X	X	X
Standard plate counts			X	X	X		X	X
Most probable numbers			X	X	X		X	X
Dye reductions				X				
Coliforms			X	X	X		X	X
Fungi		X			X		X	X
Fluorescent antibodies					X		X	X
Sampling plans			X		X	X	X	
Parasites	X							

3. Nature of the food biota
4. Nature of the food material
5. The preexamination history of the food product
6. Nutritional adequacy of the plating medium employed
7. Incubation temperature and time used
8. pH, water activity (a_w), and oxidation–reduction potential (Eh) of the plating medium
9. Type of diluent used
10. Relative number of organisms in food sample
11. Existence of other competing or antagonistic organisms.

In addition to the limitations noted, plating procedures for selected groups are further limited by the degree of inhibition and effectiveness of the selective and/or differential agents employed.

Although the SPC is often determined by pour plating, comparable results can be obtained by surface plating. By the latter method, prepoured and hardened agar plates with dry surfaces are employed. The diluted specimens are planted onto the surface of replicate plates, and, with the aid of bent glass rods ("hockey sticks"), the 0.1-mm inoculum per plate is carefully and evenly distributed over the entire surface. Surface plating offers advantages in determining the numbers of heat-sensitive psychrotrophs in a food product because the organisms do not come in contact with melted agar. It is the method of choice when the colonial features of a colony are important to its presumptive identification and for most selective media. Strict aerobes are obviously favored by surface plating, but microaerophilic organisms tend to grow slower. Among the disadvantages of surface plating are the problem of spreaders (especially when the agar surface is not adequately dry prior to plating) and the crowding of colonies, which makes enumeration more difficult. See Spiral Plater section below.

Homogenization of Food Samples

Prior to the mid- to late 1970s, microorganisms were extracted from food specimens for plating almost universally by use of mechanical blenders (Waring type). Around 1971, the Colwell Stomacher was developed in England by Sharpe and Jackson[114] and this device is now the method of choice in many laboratories for homogenizing foods for counts. The Stomacher, a relatively simple device, homogenizes specimens in a special plastic bag by the vigorous pounding of two paddles. The pounding effects the shearing of food specimens, and microorganisms are released into the diluent. Several models of the instrument are available, but model 400 is most widely used in food microbiology laboratories. It can handle samples (diluent and specimen) of 40–400 ml.

The Stomacher has been compared to a high-speed blender for food analysis by a large number of investigators. Plate counts from Stomacher-treated samples are similar to those treated by blender. The instrument is generally preferred over blending for the following reasons:

1. The need to clean and store blender containers is obviated.
2. Heat buildup does not occur during normal operational times (usually 2 minutes).
3. The homogenates can be stored in the Stomacher bags in a freezer for further use.
4. The noise level is not as unpleasant as that of mechanical blenders.

In a study by Sharpe and Harshman[113] the Stomacher was shown to be less lethal than a blender to *Staphylococcus aureus*, *Enterococcus faecalis*, and *Escherichia coli*. One investigator reported that counts using a Stomacher were significantly higher than when a blender was used[129] whereas other investigators obtained higher overall counts by blender than by Stomacher.[5] The latter investigators showed that the Stomacher is food specific; it is better than high-speed blending for some types of foods but not for others. In another study, SPC determinations made by Stomacher, blender, and shaking were not significantly different, although significantly higher counts of Gram-negative bacteria were obtained by Stomacher than by either of the other two methods.[63] Another advantage of the Stomacher over blending is the homogenization of meats for dye reduction tests. Holley et al.[54] showed that the extraction of bacteria from meat by using a Stomacher does not cause extensive disruption of meat tissue, and, consequently, fewer reductive compounds were present to interfere with resazurin reduction; whereas with blending, the level of reductive compounds released made resazurin reduction results meaningless.

Another device, the Pulsifier®, is somewhat similar to the Stomacher. It creates a high level of turbulence on food samples resulting in the release of microorganisms from the sample.

The Spiral Plater

The spiral plater is a mechanical device that distributes the liquid inoculum on the surface of a rotating plate containing a suitable poured and hardened agar medium. The dispensing arm moves from the near center of the plate toward the outside, depositing the sample in an Archimedes spiral. The attached special syringe dispenses a continuously decreasing volume of sample so that a concentration range of up to 10,000:1 is effected on a single plate. Following incubation at an appropriate temperature, colony development reveals a higher density of deposited cells near the center of the plate, with progressively fewer toward the edge.

The enumeration of colonies on plates prepared with a spiral plater is achieved by use of a special counting grid (Figure 10–1A). Depending on the relative density of colonies, colonies that appear in one or more specific areas of the superimposed grid are counted. An agar plate prepared by a spiral plater is shown in Figure 10–1B, and the corresponding grid area counted is shown in Figure 10–1C. In this example, a total sample volume of 0.0018 ml was deposited, and the two grid areas counted contained 44 and 63 colonies, respectively, resulting in a total count of 6.1×10^4 bacteria per milliliter.

The spiral plating device described here was devised by Gilchrist et al.[44] although some of its principles were presented by earlier investigators, among whom were Reyniers[105] and Trotman.[128] The method has been studied by a rather large number of investigators and compared to other methods of enumerating viable organisms. It was compared to the SPC method by using 201 samples of raw and pasteurized milk; overall good agreement was obtained.[30] A collaborative study from six analysts

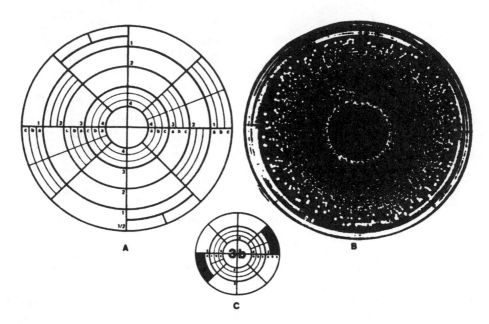

Figure 10–1   Special counting grid for spiral plater (A); growth of organisms on an inoculated spiral plate (B); and areas of plate enumerated (C). In this example, the inoculum volume was 0.0018 ml counts for the two areas shown were 44 and 63, and the averaged count was 6.1×10^4 bacteria per milliliter. Courtesy of Spiral System Instruments, Bethesda, Maryland.

on milk samples showed that the spiral plater compared favorably with the SPC. A standard deviation of 0.109 was obtained by using the spiral plater compared to 0.110 for the SPC.[94] In another study, the spiral plater was compared with three other methods (pour, surface plating, and drop count), and no difference was found among the methods at the 5% level of significance.[62] In yet another study, the spiral plate maker yielded counts as good as those by the droplet method.[51] Spiral plating is an official Association of Official Analytical Chemists (AOAC) method.

Among the advantages of the spiral plater over standard plating are the following: less agar is used; fewer plates, dilution blanks, and pipettes are required; and three to four times more samples per hour can be examined.[69] Also, 50–60 plates per hour can be prepared, and little training is required for its operation.[62] Among the disadvantages is the problem that food particles may cause blocking in the dispensing stylus. It is more suited for use with liquid foods such as milk. A laser-beam counter has been developed for use with the plater. Because of the expense of the device, it is not likely to be available in laboratories that do not analyze large numbers of plates. The method is further described in reference 36.

MEMBRANE FILTERS

Membranes with a pore size that will retain bacteria (generally 0.45 μm) but allow water or diluent to pass are used. Following the collection of bacteria upon filtering a given volume, the membrane is placed on an agar plate or an absorbent pad saturated with the culture medium of choice and incubated appropriately. Following growth, the colonies are enumerated. Alternatively, a DMC can be made. In this case, the organisms collected on the membrane are viewed and counted microscopically following appropriate staining, washing, and treatment of the membrane to render it transparent. These methods are especially suited for samples that contain low numbers of bacteria. Although relatively large volumes of water can be passed through a membrane without clogging it, only small samples of dilute homogenates from certain foods can be used for a single membrane.

The overall efficiency of membrane filter methods for determining microbial numbers by the DMC has been improved by the introduction of fluorescent dyes. The use of fluorescent dyes and epifluorescent microscopes to enumerate bacteria in waters has been employed rather widely since the early 1970s. Cellulose filters were among the earliest used; however, polycarbonate Nucleopore filters offer the advantage of retaining all bacteria on top of the filter. When lake and ocean waters were examined using the two kinds of membranes, counts were twice as high with Nucleopore membranes as with cellulose membranes.[52]

Direct Epifluorescent Filter Technique

This membrane filter technique may be viewed as an improved modification of the basic method. The direct epifluorescent filter technique (DEFT) employs fluorescent dyes and fluorescent microscopy,[52] and it has been evaluated by a number of investigators as a rapid method for microorganisms in foods. Typically, a diluted food homogenate is filtered through a 5-μm nylon filter, and the filtrate is collected and treated with 2 ml of Triton X-100 and 0.5 ml of trypsin. The latter reagents are used to lyse somatic cells and to prevent clogging of filters. After incubation, the treated filtrate is passed through a 0.6-μm Nucleopore polycarbonate membrane, and the filter is stained with acridine orange. After drying, the stained cells are enumerated by epifluorescence microscopy, and the number of cells per gram is calculated by multiplying the average number per field by the microscope factor. Results can be obtained in 25–30 minutes, and numbers as low as around 6,000 cfu/g can be obtained from meats and milk products.

DEFT has been employed on milk[97] and found to compare favorably with results obtained by aerobic plate count (APC), and standard Breed DMC on raw milk that contained between 5×10^3 and 5×10^8 bacteria per milliliter. It has been adapted to the enumeration of viable Gram-negative and all Gram-positive bacteria in milk in about 10 minutes.[108] As few as 5,700 bacteria per milliliter could be detected in heat-treated milk and milk products in about 20 minutes.[98] In a collaborative study by six laboratories that compared DEFT and APC, the correlation coefficient was generally above 0.9, but the repeatability of DEFT was 1.5 times worse than APC, and reproducibility was only three times that for APC.[96] Solid foods can be examined by DEFT after proper filtrations, and <60,000 organisms per gram could be detected in one study.[99] DEFT has been employed successfully to estimate numbers of microorganisms on meat and poultry[120] and on food contact surfaces.[53] For more information, see reference 95.

Microcolony-DEFT

DEFT allows for the direct microscopic determination of cells; microcolony-DEFT is a variation that allows one to determine viable cells only. Typically, food homogenates are filtered through DEFT membranes, and the latter are then placed on the surface of appropriate culture media and incubated for microcolony development. A 3-hour incubation can be used for Gram-negative bacteria and a 6-hour incubation for Gram positives.[107] The microcolonies that develop must be viewed with a microscope. For coliforms, pseudomonads, and staphylococci, as few as 10^3/g could be detected within 8 hours.[107]

In another variation, a microcolony epifluorescence microscopy method that combined DEFT with hydrophobic grid membrane filter (HGMF) was devised.[106] By this method, nonenzyme detergent-treated samples are filtered through Nucleopore polycarbonate membranes, which are transferred to the surface of a selective agar medium and incubated for 3 or 6 hours for Gram-negative or Gram-positive bacteria as for microcolony-DEFT. The membranes are then stained with acridine orange, and the microcolonies are enumerated by epifluorescence microscopy. The method allows results to be obtained in <6 hours without a repair step for injured organisms, and in about 12 hours when a repair step was employed.[106]

Hydrophobic Grid Membrane Filter (HGMF)

The hydrophobic grid membrane filter (HGMF) technique was advanced by Sharpe and Michaud,[118,119] and it has since been further developed and used to enumerate microorganisms from a variety of food products. The method employs a specially constructed filter that consists of 1600 wax grids on a single membrane filter that restricts growth and colony size to individual grids. On one filter, from 10 to 9×10^4 cells can be enumerated by an MPN procedure, and enumeration can be automated.[19] The method can detect as few as 10 cells per gram, and results can be achieved in 24 hours or so.[116] It can be used to enumerate all cfus or specific groups such as indicator organisms,[8,15,33] fungi,[17] salmonellae,[32] and pseudomonads.[66] It has been given AOAC approval for total coliforms, fecal coliforms, salmonellae, and yeasts and molds. The ISO-GRID method for fungi employs a special plating medium that contains two antibacterial antibiotics and trypan blue. The latter gives fungal colonies a blue color, and as few as 10 cfu can be detected in 48 hours.

In a typical application, 1 ml of a 1:10 homogenate is filtered through a filter membrane, followed by the placing of the membrane on a suitable agar medium for incubation overnight to allow colonies to develop. The grids that contain colonies are enumerated, and the MPN is calculated. The method allows the filtering of up to 1 g of food per membrane.[117] The ISOGRID method employing SD-39 agar has been shown to be more versatile than ISO-GRID with lactose monensin glucuronate (LMG) agar in conjunction with buffered MUG (4-methylumbelliferyl-β-D-glucuronide) agar for the detection of E. coli in foods since it enables the simultaneous detection of E. coli O157:H7 and β-glucuronidase-positive E. coli.[34] The SD-39 agar method provides results in about 24 hours with a sensitivity of <10, while LMG requires about 30 hours.

When compared to a five-tube MPN for coliforms, the HGMF method, employing a resuscitative step, produced statistically equivalent results for coliforms and fecal coliforms.[19] In the latter application, HGMF filters were placed first on trypticase soy agar for 4–5 hours at 35°C (for resuscitation of injured cells) followed by removal to m-FC agar for additional incubation. An HGMF-based enzyme-labeled antibody (ELA) procedure has been developed for the recovery of E. coli O157:H7 (hemorrhagic colitis, HC) strains from foods.[126] The method employs the use of a special plating medium that permits HC strains to grow at 44.5°C. The special medium, HC agar, contains only 0.113% bile salt #3 in contrast to 0.15%. With its use, about 90% of HC strains could be recovered from ground beef.[124] The HGMF-ELA method employs the use of HC agar incubated at 43°C for 16 hours, washing of colony growth from membranes, exposure of membranes to a blocking solution, and immersion in a horseradish peroxidase-protein A-monoclonal antibody complex. By the method, ELA-positive colonies stain purple, and 95% of HC strains could be recovered within 24 hours with a detection limit of 10 HC strains per gram of meat.

MICROSCOPE COLONY COUNTS

Microscope colony count methods involve the counting of microcolonies that develop in agar layered over microscope slides. The first was that of Frost, which consisted of spreading 0.1 ml of milk-agar mixture over a 4-cm^2 area on a glass slide. Following incubation, drying, and staining, microcolonies are counted with the aid of a microscope. In another method, 2 ml of melted agar are mixed with 2 ml of warmed milk and, after mixing, 0.1 ml of the inoculated agar is spread over a 4-cm^2 area. Following staining with thionin blue, the slide is viewed with the 16-mm objective of a wide-field microscope.[65]

AGAR DROPLETS

In the agar droplet method of Sharpe and Kilsby,[115] the food homogenate is diluted in tubes of melted agar (at 45°C). For each food sample, three tubes of agar are used, the first tube being inoculated with 1 ml of food homogenate. After mixing, a sterile capillary pipette (ideally delivering 0.033 ml/drop) is used to transfer a line of 5 × 0.1-ml droplets to the bottom of an empty Petri dish. With the same capillary pipette, three drops (0.1 ml) from the first 9-ml tube are transferred to the second tube, and, after mixing, another line of 5 × 0.1-ml droplets is placed next to the first. This step is repeated for the third tube of agar. Petri plates containing the agar droplets are incubated for 24 hours, and colonies are enumerated with the aid of a 10× viewer. Results using this method from pure cultures, meats, and vegetables compared favorably to those obtained by conventional plate counts; droplet counts from ground meat were slightly higher than plate counts. The method was about three times faster, and 24-hour incubations gave counts equal to those obtained after 48 hours by the conventional plate count. Dilution blanks are not required, and only one Petri dish per sample is needed.

DRY FILM AND RELATED METHODS

A rehydratable dry film method consisting of two plastic films attached together on one side and coated with culture medium ingredients and a cold-water-soluble jelling agent was developed by the 3M Company and designated Petrifilm. The method can be used with nonselective ingredients to make aerobic plate counts (APCs), and, with selective ingredients, certain specific groups can be detected. Use of this method to date indicates that it is an acceptable alternative to SPC methods that employ Petri dishes, and it has been approved by AOAC.

For use, 1 ml of diluent is placed between the two films and spread over the nutrient area by pressing with a special flat-surface device. Following incubations, microcolonies appear red on the nonselective film because of the presence of a tetrazolium dye in the nutrient phase. In addition to its use for APC, Petrifilm methods exist for the detection and enumeration of specific groups, such as coliforms and *E. coli*. For APC determination on 108 milk samples, this dry film method correlated highly with the conventional plate count method and was shown to be a suitable alternative.[45] When compared to violet red bile agar (VRBA) and MPN for coliform enumeration on 120 samples of raw milk, Petrifilm-VRB compared favorably to VRBA counts, and both were comparable to MPN results.[84] A dry medium EC (*E. coli*) count method has been developed; it employs the substrate for β-glucuronidase so that *E. coli* is distinguished from other coliforms by the formation of a blue halo around colonies. When compared to the classical confirmed MPN and VRBA on 319 food samples, the EC dry medium gave comparable results.[75]

Redigel is a plating medium that does not use agar as a solidifying agent. It is employed by inoculating presterilized ingredients with food homogenates or diluents followed by mixing and holding to allow for solidification, which occurs in about 30 minutes. It is attractive for enumerating psychrotrophic organisms because there is no exposure to hot molten agar, which can lower numbers of psychrotrophs since some are extremely heat sensitive. On the other hand, colonies on Redigel tend to be rather small in size. In a comparison of this method with Petrifilm, ISO-GRID, and the spiral plater using seven different foods, all were statistically comparable.[21]

SimPlate® is a culture method that is based on the activity of several enzymes common to many foodborne organisms. The growth medium contains substrates that are hydrolyzed by enzymes to release MUG (see Chapter 11), and this fluorescent compound is visible under long-wave ultraviolet light. The special plates have holes or wells, and they come in two sizes—84 or 198 incubation wells. The technique is in essence an MPN method. Unlike conventional plating methods, it does not allow for the characterization of colony features. In a comparative study employing seafoods, no significant differences were found among aerobic plate counts by Petrifilm, Redigel, ISO-GRID, and SimPlate.[26] In a study employing 751 food samples, SimPlate was found to be a suitable alternative to the conventional plate method, Petrifilm, and Redigel.[16] However, some foods (raw liver, wheat flour, and nuts) gave false-positive results. A comparison of SimPlate to the standard plate count by six laboratories on the enumeration of heterotrophic bacteria in water found that the two methods produced comparable results.[61]

MOST PROBABLE NUMBERS

In this method, dilutions of food samples are prepared as for the SPC. Three serial aliquots or dilutions are then planted into 9 or 15 tubes of appropriate medium for the three- or five-tube method, respectively. Numbers of organisms in the original sample are determined by use of standard MPN tables. The method is statistical in nature, and MPN results are generally higher than SPC results.

This method was introduced by McCrady in 1915. It is not a precise method of analysis; the 95% confidence intervals for a three-tube test range from 21 to 395. When the three-tube test is used, 20 of the 62 possible test combinations account for 99% of all results, whereas with the five-tube test, 49 of the possible 214 combinations account for 99% of all results.[131] In a collaborative study on coliform densities in foods, a three-tube MPN value of 10 was found to be as high as 34, whereas in another phase of the study, the upper limit could be as high as 60.[121] Although Woodward[131] concluded that many MPN values are improbable, this method of analysis has gained popularity. Among the advantages it offers are the following:

1. It is relatively simple.
2. Results from one laboratory are more likely than SPC results to agree with those from another laboratory.
3. Specific groups of organisms can be determined by use of appropriate selective and differential media.
4. It is the method of choice for determining fecal coliform densities.

Among the drawbacks to its use are the large volume of glassware required (especially for the five-tube method), the lack of opportunity to observe the colonial morphology of the organisms, and its lack of precision.

TEMPO® is an MPN-based method that employs an enumerating card with a specific medium that allows rapid fluorescent detection of target organisms, and results may be obtained within 24 h. It obviates the need for serial dilutions.

DYE REDUCTION

Two dyes are commonly employed in this procedure to estimate the number of viable organisms in suitable products: methylene blue and resazurin. To conduct a dye-reduction test, properly prepared supernatants of foods are added to standard solutions of either dye for reduction from blue to white for methylene blue; and from slate blue to pink or white for resazurin. The time for dye reduction to occur is inversely proportional to the number of organisms in the sample.

Methylene blue and resazurin reduction by 100 cultures was studied in milk; with two exceptions, a good agreement was found between numbers of bacteria and time needed for reduction of the two dyes.[43] In a study of resazurin reduction as a rapid method for assessing ground beef spoilage, reduction to the colorless state, odor scores, and SPC correlated significantly.[111] One of the problems of using dye reduction for some foods is the existence of inherent reductive substances. This is true of raw meats, and Austin and Thomas[9] reported that resazurin reduction was less useful than with cooked meats. For the latter, approximately 600 samples were successfully evaluated by resazurin reduction by adding 20 ml of a 0.0001% resazurin solution to 100 g of sliced meat in a plastic pouch. Another way of getting around the reductive compounds in fresh meats is to homogenize samples by Stomacher rather than by Waring blender. By using Stomacher homogenates, raw meat was successfully evaluated by resazurin reduction when Stomacher homogenates were added to a solution of resazurin in 10% skim milk.[54] Stomacher homogenates contained less disrupted tissue and, consequently, lower concentrations of reductive compounds. The method of Holley et al.[54] was evaluated further by Dodsworth and Kempton,[29] who found that raw meat with an SPC $>10^7$ bacteria per gram could be detected within 2 hours. When compared to nitroblue tetrazolium (NT) and indophenyl nitrophenyl tetrazolium (INT), resazurin produced faster results.[104] With surface samples from sheep carcasses, resazurin was reduced in 30 minutes by 18,000 cfu/m^2, NT in 600 minutes by 21,000 cfu/m^2, and INT in 660 minutes by 18,000 cfu/m^2.[108] Methylene blue reduction was compared to APC on 389 samples of frozen peas, and the results were linear over the APC range of log 2–6 cfus. Average decolorization times were 8 and 11 hours for 10^5 and 10^4 cfu/g, respectively.

Dye-reduction tests have a long history of use in the dairy industry for assessing the overall microbial quality of raw milk. Among their advantages are that they are simple, rapid, and inexpensive; and only viable cells actively reduce the dyes. Disadvantages are that not all organisms reduce the dyes equally, and they are not applicable to food specimens that contain reductive enzymes unless special steps are employed. The use of fluorogenic and chromogenic substrates in food microbiology is discussed in Chapter 11.

ROLL TUBES

Screw-capped tubes or bottles of varying sizes are used in this method. Predetermined amounts of the melted and inoculated agar are added to the tube and the agar is made to solidify as a thin layer on the inside of the vessel. Following appropriate incubation, colonies are counted by rotating the vessel. It has been found to be an excellent method for enumerating fastidious anaerobes. For a review of the method, see Anderson and Fung.[3]

DIRECT MICROSCOPIC COUNT (DMC)

In its simplest form, the DMC consists of making smears of food specimens or cultures onto a microscope slide, staining with an appropriate dye, and viewing and counting cells with the aid of a microscope (oil immersion objective). DMCs are most widely used in the dairy industry for assessing the microbial quality of raw milk and other dairy products, and the specific method employed is that originally developed by R.S. Breed (Breed count). Briefly, the method consists of adding 0.01 ml of a sample to a 1-cm^2 area on a microscope slide, and following fixing, defatting of sample, and staining, the organisms or clumps of organisms are enumerated. The latter involves the use of a calibrated microscope (for further details, see reference 73). The method lends itself to the rapid microbiological examination of other food products, such as dried and frozen foods.

Among the advantages of DMC are that it is rapid and simple, cell morphology can be assessed, and it lends itself to fluorescent probes for improved efficiency. Among its disadvantages are that it is a microscopic method and therefore fatiguing to the analyst, both viable and nonviable cells are enumerated, food particles are not always distinguishable from microorganisms, microbial cells are not uniformly distributed relative to single cells and clumps, some cells do not take the stain well and may not be counted, and DMC counts are invariably higher than counts by SPC. In spite of its drawbacks, it remains the fastest way to make an assessment of microbial cells in a food product.

A slide method to detect and enumerate viable cells has been developed.[11] The method employs the use of the tetrazolium salt (p-iodophenyl-3-p-nitrophenyl)-5-phenyl tetrazolium chloride (INT). Cells are exposed to filter-sterilized INT for 10 minutes at 37°C in a water bath followed by filtration on 0.45-μm membranes. Following drying of membranes for 10 minutes at 50°C, the special membranes are mounted in cottonseed oil and viewed with coverslip in place. The method was found to be workable for pure cultures of bacteria and yeasts, but it underestimated APC by 1–1.5 log cycles when compared using milk. By use of fluorescence microscopy and Viablue (modified aniline blue fluorochrome), viable yeast cells could be differentiated from nonviable cells.[60,67] Viable cells can be determined by staining with acridine orange (0.01%) followed by epifluorescence microscopy and enumeration of those that fluoresce orange. This is the gist of the acridine orange direct count (AODC) method.

Howard Mold Counts

This is a microscope slide method developed by B.J. Howard in 1911 primarily for the purpose of monitoring tomato products. The method requires the use of a special chamber (slide) designed to enumerate mold mycelia. It is not valid on tomato products that have been comminuted. Similar to the Howard mold count is a method for quantifying *Geotrichum candidum* in canned beverages and fruits, and this method, as well as the Howard mold count method, is fully described by AOAC.[89] The DEFT method has been shown to correlate well with the Howard mold count method on autoclaved and unautoclaved tomato concentrate, and it could be used as an alternative to the Howard mold count.[100]

MICROBIOLOGICAL EXAMINATION OF SURFACES

The need to maintain food contact surfaces in a hygienic state is of obvious importance. The primary problem that has to be overcome when examining surfaces or utensils for microorganisms is the removal of a significant percentage of the resident biota. Although a given method may not recover all organisms, its consistent use in specified areas of a food-processing plant can still provide valuable information as long as it is realized that not all organisms are being recovered. The most commonly used methods for surface assessment in food operations are presented below.

Swab/Swab-Rinse Methods

Swabbing is the oldest and most widely used method for the microbiological examination of surfaces not only in the food and dairy industries but also in hospitals and restaurants. The swab-rinse method was developed by W.A. Manheimer and T. Ybanez. Either cotton or calcium alginate swabs are used. If one wishes to examine given areas of a surface, templates may be prepared with openings corresponding to the size of the area to be swabbed, for example, 1 in^2 or 1 cm^2. The sterile template is placed over the surface, and the exposed area is rubbed thoroughly with a moistened swab. The exposed swab is returned to its holder (test tube) containing a suitable diluent and stored at refrigerator temperatures until plated. The diluent should contain a neutralizer, if necessary. When cotton swabs are used, the organisms must be dislodged from the fibers. When calcium alginate swabs are used, the organisms are released into the diluent upon dissolution of the alginate by sodium hexametaphosphate. The organisms in the diluent are enumerated by a suitable method such as SPC, but any of the culture media may be used to test specifically for given groups of organisms. In an innovation in the swab-rinse method presented by Koller,[68] 1.5 ml of fluid is added to a flat surface, swabbed for 15 seconds over a 3-cm^2 area, and volumes of 0.1 and 0.5 ml collected in microliter pipettes. The fluid may be surface or pour plated using plate count agar or selective media.

Concerning the relative efficacy of cotton and calcium alginate swabs, most investigators agree that higher numbers of organisms are obtained by use of the latter. Using swabs, some researchers recovered as few as 10% of organisms from bovine carcasses,[87] 47% of *Bacillus subtilis* spores from stainless-steel surfaces,[7] and up to 79% from meat surfaces.[22,93] Swab results from bovine carcasses were on the average 100 times higher than by contact plate method, and the deviation was considerably lower.[87] The latter investigators found the swab method to be best suited for flexible, uneven, and heavily contaminated surfaces. The ease of removal of organisms depends on the texture of the surface and the nature and types of biota. Even with its limitations, the swab-rinse method remains a rapid, simple, and inexpensive way to assess the microbiological biota of food surfaces and utensils.

The use of the ATP assay system to detect the presence of cells within 2–5 minutes after swabbing allows it to be used on-line. Although the ATP assay as used in this regard is not specific for bacteria, it provides valuable information on the level of cell contamination of a surface and can be used to make quick assessments of the relative efficacy of surface cleaning methods. The basis of the ATP assay is described in Chapter 11.

Contact Plate

The replicate organism direct agar contact (RODAC) method employs special Petri plates, which are poured with 15.5–16.5 ml of an appropriate plating medium, resulting in a raised agar surface. When the plate is inverted, the hardened agar makes direct contact with the surface. Originated by Gunderson and Gunderson in 1945, it was further developed in 1964 by Hall and Hartnett. When surfaces are examined that have been cleaned with certain detergents, it is necessary to include a neutralizer (lecithin, Tween 80, and so on) in the medium. Once exposed, plates are covered and incubated, and the colonies enumerated.

Perhaps the most serious drawbacks to this method are the covering of the agar surface by spreading colonies, and its ineffectiveness for heavily contaminated surfaces. These can be minimized by using plates with dried agar surfaces and by using selective media.[28] The RODAC plate has been shown to be the method of choice when the surfaces to be examined are smooth, firm, and nonporous.[7,87] Although it is not suitable for heavily contaminated surfaces, it has been estimated that a solution that contaminates a surface needs to contain at least 10 cells per milliliter before results can be achieved either by contact or by swabs.[87] The latter investigators found that the contact plate removed only about 0.1% of surface biota. This suggests that 10 cfu/cm^2 detected by this method are referable to a surface that actually contains about 10^4 cfu/cm^2. When stainless-steel surfaces were contaminated by *B. subtilis* endospores, 41% were recovered by the RODAC plate compared to 47% by the swab method.[7] In another study, swabs were better than contact plates when the contamination level was 100 or more organisms per 21–25 cm^2.[112] On the other hand, contact plates give better results where low numbers exist. In terms of ranking of surface contamination, the two methods correlated well.

Agar Syringe/"Agar Sausage" Methods

The agar syringe method was proposed by W. Litsky in 1955 and subsequently modified.[6] By this method, a 100-ml syringe is modified by removing the needle end to create a hollow cylinder that is filled with agar. A layer of agar is pushed beyond the end of the barrel by means of the plunger and pressed against the surface to be examined. The exposed layer is cut off and placed in a Petri dish, followed by incubation and colony enumeration. The "agar sausage" method proposed by ten Cate[125] is similar but employs plastic tubing rather than a modified syringe. The latter method has been used largely by European workers for assessing the surfaces of meat carcasses, as well as for food plant surfaces. Both methods can be viewed as variations of the RODAC plate, and both have the same disadvantages: spreading colonies and applicability limited to low levels of surface contaminants. Because clumps or chains of organisms on surfaces may yield single colonies, the counts obtained by these methods are lower than those obtained by methods that allow for the breaking up of chains or clumps.

For the examination of meat carcasses, Nortje et al.[88] compared three methods: a double swab, excision, and agar sausage. Although the excision method was found to be the most reliable of the three, the modified agar sausage method correlated more closely with it than the double swab, and the investigators recommended the agar sausage method because of its simplicity, speed, and accuracy.

Other Surface Methods

Direct Surface

A number of workers have employed direct surface agar plating methods, in which melted agar is poured onto the surface or utensil to be assessed. Upon hardening, the agar mold is placed in a Petri dish and incubated. Angelotti and Foter[6] proposed this as a reference for assessing surface contamination, and it is excellent for enumerating particulates containing viable microorganisms.[35] It was used successfully to determine the survival of *Clostridium sporogenes* endospores on stainless-steel surfaces.[83] Although effective as a research tool, the method does not lend itself to routine use for food plant surfaces.

Sticky Film

The sticky film method of Thomas has been used with some success by Mossel et al.[82] The method consists of pressing sticky film or tape against the surface to be examined and pressing the exposed side on an agar plate. It was shown to be less effective than swabs in recovering bacteria from wooden surfaces.[82] An adhesive tape method has been employed successfully to assess microorganisms on meat surfaces.[41] In a recent study, the swab, RODAC, and adhesive tape (Mylar) methods were compared for the examination of pork carcasses, and the correlation between adhesive tape and RODAC was better than that between adhesive tape and swab or between RODAC and swab.[25] Plastic strips attached to pads containing culture media have been used to monitor microorganisms on bottles.[27]

Swab/Agar Slant

The swab/agar slant method described in 1962 by N.-H. Hansen has been used with success by some European workers. The method involves sampling with cotton swabs that are transferred directly to slants. Following incubation, slants are grouped into one-half $\log_{10}$ units based on estimated numbers of developed colonies. The average number of colonies is determined by plotting the distribution on probability paper. A somewhat similar method, the swab/agar plate, was proposed by Ølgaard.[90] It requires a template, a comparator disc, and a reference table, making it a bit more complicated than the other methods noted.

Ultrasonic Devices

Ultrasonic devices have been used to assess the microbiological contamination of surfaces, but the surfaces to be examined must be small in size and removable so that they can be placed inside a container immersed in diluent. Once the container is placed in an ultrasonic apparatus, the energy generated effects the release of microorganisms into the diluent. A more practical use of ultrasonic energy may be the removal of bacteria from cotton swabs in the swab-rinse method.[102]

Spray Gun

A spray gun method was devised by Clark[22,23] based on the impingement of a spray of washing solution against a circumscribed area of surface and the subsequent plating of the washing solution. Although the device is portable, a source of air pressure is necessary. It was shown to be much more effective than the swab method in removing bacteria from meat surfaces.

METABOLICALLY INJURED ORGANISMS

When microorganisms are subjected to environmental stresses such as sublethal heat and freezing, many of the individual cells undergo metabolic injury, resulting in their inability to form colonies on selective media that uninjured cells can tolerate. Whether a culture has suffered metabolic injury can be determined by plating aliquots separately on a nonselective and a selective medium and enumerating the colonies that develop after suitable incubation. The colonies that develop on the nonselective medium represent both injured and uninjured cells, whereas only the uninjured cells develop on the selective medium. The difference between the number of colonies on the two media is a measure of the number of injured cells in the original culture or population. This principle is illustrated in Figure 10–2 by data from Tomlins et al.[127] on sublethal heat injury of *S. aureus*. These investigators subjected the organism to 52°C for 15 minutes in a phosphate buffer at pH 7.2 to inflict cell injury. The plating of cells at zero time and up to 15 minutes of heating on nonselective trypticase soy agar (TSA) and selective TSA + 7.0% NaCl (stress medium; TSAS) revealed only a slight reduction in numbers on TSA, whereas the numbers on TSAS were reduced considerably, indicating a high degree of injury

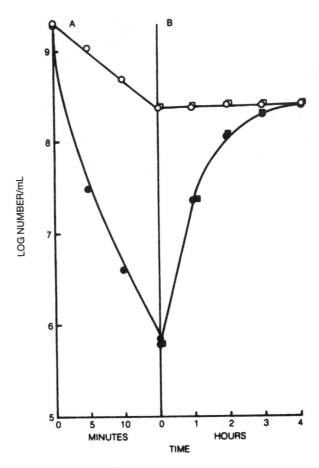

Figure 10–2   Survival and recovery curve for *S. aureus* MF = 31. (A) Heat injury at 52°C for 15 minutes in 100 mM potassium phosphate buffer. (B) Recovery from heat injury in nutrient broth (NB) at 37°C. Symbols: ○, samples plated on TSA to give a total viable count; ●, samples plated on TSAS to give an estimate of the uninjured population—cells recovered in NB containing 100 μg/ml of chloramphenicol; ■, samples plated on TSAS. *Source*: Tomlins et al.,[127] reproduced by permission of National Research Council of Canada from *Canadian Journal of Microbiology* 17:759–765, 1971.

relative to a level of salt that uninjured *S. aureus* can withstand. To allow the heat-injured cells to repair, the cells were placed in nutrient broth (recovery medium) followed by incubation at 37°C for 4 hours. With hourly plating of aliquots from the recovery medium onto TSAS, it can be seen that the injured cells regained their capacity to withstand the 7.0% NaCl in TSAS after the 4-hour incubation.

The existence of metabolically injured cells in foods and their recovery during culturing procedures is of great importance not only from the standpoint of pathogenic organisms but also for spoilage organisms. The data cited suggest that if a high-salt medium had been employed to examine a heat-pasteurized product for *S. aureus*, the number of viable cells found would have been lower than the actual number by a factor of 3 log cycles. Injury of foodborne microorganisms has been shown by a large number of investigators to be induced not only by sublethal heat and freezing but also by freeze drying, drying, irradiation, aerosolization, dyes, sodium azide, salts, heavy metals, antibiotics, essential oils, and other chemicals, such as ethylenediaminetetraacetic acid (EDTA) and sanitizing compounds.

The recognition of sublethal stresses on foodborne microorganisms and their effect on growth under varying conditions dates back to around 1900. However, a full appreciation of this phenomenon did not come until the late 1960s. During the early 1960s, it was observed that an initial rapid decrease in numbers of a metabolically injured organism was followed by only a limited recovery during the resuscitation process ("Phoenix phenomenon"). The increased nutritional requirement of bacteria that had undergone heat treatment was noted by Nelson[85] in 1943. (Nelson also reviewed the work of others up to that time.) Gunderson and Rose[46] noted the progressive decrease in numbers of coliforms from frozen chicken products that grew on VRBA with increasing storage time of products. Hartsell[50] inoculated foods with salmonellae, froze the inoculated foods, and then studied the fate of the organisms during freezer storage. More organisms could be recovered on highly nutritive nonselective media than on selective media such as MacConkey, deoxycholate, or VRBA. The importance of the isolation medium in recovering stressed cells was also noted by Postgate and Hunter[101] and by Harris.[48] In addition to the more exacting nutritional requirements of foodborne organisms that undergo environmental stresses, these organisms may be expected to manifest their injury via increased lag phases of growth, increased sensitivity to a variety of selective media agents, damage to cell membranes and tricarboxylic acid (TCA)-cycle enzymes, breakdown of ribosomes, and DNA damage. Although damage to ribosomes and cell membranes appears to be a common consequence of sublethal heat injury, not all harmful agents produce identifiable injuries.

Recovery/Repair

Metabolically injured cells can recover, at least in *S. aureus*, in no-growth media[59] and at a temperature of 15°C but not 10°C.[42] In some instances at least, the recovery process is not instantaneous, for it has been shown that not all stressed coliforms recover to the same degree but that the process takes place in a stepwise manner.[76] Not all cells in a population suffer the same degree of injury. Hurst et al.[56] found dry-injured *S. aureus* cells that failed to develop on the nonselective recovery medium (TSA), but did recover when pyruvate was added to this medium. These cells were said to be severely injured in contrast to injured and uninjured cells. It has been found that sublethally heated *S. aureus* cells may recover their NaCl tolerance before certain membrane functions are restored.[58] It is well established that injury repair occurs in the general absence of cell wall and protein synthesis. It can be seen from Figure 10–2 that the presence of chloramphenicol in the recovery medium had no effect on the recovery of *S. aureus* from sublethal heat injury. The repair of cell ribosomes and membrane appears to be essential for recovery, at least from sublethal heat, freezing, drying, and irradiation injuries.

The protection of cells from heat and freeze injury is favored by complex media and menstra or certain specific components thereof. Milk provides more protection than saline or mixtures of amino acids,[81] and the milk components that are most influential appear to be phosphate, lactose, and casein. Sucrose appears to be protective against heat injury[2,70] whereas glucose has been reported to decrease heat protection for *S. aureus*.[81] Nonmetabolizable sugars and polyols such as arabinose, xylose, and sorbitol have been found to protect *S. aureus* against sublethal heat injury, but the mechanism of this action is unclear.[122]

The consequences of not employing a recovery step have been reviewed by Busta.[20] The use of trypticase soy broth (TSB) with incubations ranging from 1 to 24 hours at temperatures from 20°C to

37°C is widely used for various organisms. The enumeration of sublethally heated *S. aureus* strains on various media has been studied.[14,37,56] In one of these studies, seven staphylococcal media were compared on their capacity to recover 19 strains of sublethally heated *S. aureus*, and the Baird–Parker medium was found to be clearly the best of those studied, including nonselective TSA. Similar findings by others led to the adoption of this medium in the official methods of AOAC for the direct determination of *S. aureus* in foods that contain ≥ 10 cells per gram. The greater efficacy of the Baird–Parker medium has been shown to result from its content of pyruvate. The use of this medium following recovery in an antibiotic-containing, nonselective medium has been suggested.[56] Although this approach may be suitable for *S. aureus* recovery, some problems may be expected to occur with the widespread use of antibiotics in recovery media to prevent cell growth. It has been shown that heat-injured spores of *C. perfringens* are actually sensitized to polymyxin and neomycin,[10] and it is well established that the antibiotics that affect cell wall synthesis are known to induce L-phase variations in many bacteria.

Pyruvate is well established as an injury repair agent not only for injured *S. aureus* cells but also for other organisms such as *E. coli*. Higher counts are obtained on media containing this compound when injured by a variety of agents. When added to TSB containing 10% NaCl, higher numbers of both stressed and nonstressed *S. aureus* were achieved,[14] and the repair–detection of freeze- or heat-injured *E. coli* was significantly improved by pyruvate.[77]

Catalase is another agent that increases recovery of injured aerobic organisms. First reported by Martin et al.,[74] it has been found effective by many other investigators. It is effective for sublethally heated *S. aureus*, *Pseudomonas fluorescens*, *Salmonella* Typhimurium, and *E. coli*.[74] It is also effective for *S. aureus* in the presence of 10% NaCl,[14] and for water-stressed *S. aureus*.[37] Another compound, shown to be as effective as pyruvate for heat-injured *E. coli*, is 3,3'-thiodipropionic acid.[77]

Radiation injury of *Clostridium botulinum* type E spores by 4 kGy resulted in the inability to grow at 10°C in the presence of polymyxin and neomycin.[110] The injured cells had a damaged postgermination system and formed aseptate filaments during outgrowth, but the germination lytic system was not damaged. The radiation injury was repaired at 30°C in about 15 hours on tellurite polymyxin egg yolk (TPEY) agar without antibiotics. When *C. botulinum* spores are injured with hypochlorite, the L-alanine germination sites are modified, resulting in the need for higher concentrations of alanine for repair.[39] The L-alanine germination sites could be activated by lactate, and hypochlorite-treated spores could be germinated by lysozyme, indicating that the chloride removed spore coat proteins.[40] More detailed information on spore injury has been provided by Foegeding and Busta.[38]

Sublethally heat-stressed yeasts are inhibited by some essential oils (spices at concentrations as low as 25 ppm).[24] The spice oils affect colony size and pigment production.

Special plating procedures have been found by Speck et al.[123] and Hartman et al.[49] to allow for recovery from injury and subsequent enumeration in essentially one step. The procedures consist of using the agar overlay plating technique with one layer consisting of TSA, onto which are plated the stressed organisms. Following a 1- to 2-hour incubation at 25°C for recovery, the TSA layer is overlaid with VRBA and incubated at 35°C for 24 hours. The overlay method of Hartman et al. involved the use of a modified VRBA. The principle involved in the overlay technique could be extended to other selective media, of course. An overlay technique has been recommended for the recovery of coliforms. By this method, coliforms are plated with TSA and incubated at 35°C for 2 hours followed by an overlay of VRBA.

In their comparison of 18 plating media and seven enrichment broths to recover heat-stressed *Vibrio parahaemolyticus*, Beuchat and Lechowich[13] found that the two most efficient plating media were water blue-alizarin, yellow agar and arabinose–ammonium–sulfate–cholate agar; arabinose-ethyl violet broth was the most suitable enrichment broth.

Mechanism of Repair

Pyruvate and catalase both act to degrade peroxides, suggesting that metabolically injured cells lack this capacity. The inability of heat-damaged *E. coli* cells to grow as well when surface plated as when pour plated with the same medium[47] may be explained by the loss of peroxides.

A large number of investigators have found that metabolic injury is accompanied by damage to cell membranes, ribosomes, DNA, or enzymes. The cell membrane appears to be the most universally affected.[55] The lipid components of the membrane are the most likely targets, especially for sublethal heat injuries. Ribosomal damage is believed to result from the loss of Mg^{2+} and not to heat effects per se.[57] On the other hand, ribosome-free areas have been observed by electron microscopy in heat-injured *S. aureus* cells.[64] Following prolonged heating at 50°C, virtually no ribosomes were detected, and, in addition, the cells were characterized by the appearance of surface blebs and exaggerated internal membranes.[64] When *S. aureus* was subjected to acid injury by exposure to acetic, hydrochloric, and lactic acids at 37°C, coagulase and thermostable nuclease activities were reduced in injured cells.[132] Although acid injury did not affect cell membranes, RNA synthesis was affected. For more information on cell injury and on methods of recovery, see reference 4).

VIABLE BUT NONCULTURABLE ORGANISMS

Under certain conditions and in some environments, standard plate count results suggest either an absence of colony-forming units or numbers that may be considerably lower than the actual viable population. Although this might appear to be the result of metabolic injury as outlined above, the viable but nonculturable cells (VBNC) are in a state that sets them apart from injured cells. For example, metabolically injured cells will repair when plated onto a nonselective medium that does not contain inhibitors, but cells in the VBNC state will not.

The VBNC state was first noted with marine vibrios, which were difficult to culture from marine waters during winter months. A downshift in temperature to around 5°C is known to induce this state. In an early study with *Campylobacter jejuni*, log phase cells were predominantly spiral shaped, whereas late stationary phase cells were mainly coccoids.[109] The VBNC state was maintained at 4°C for >4 months. The cells in the VBNC state yielded low numbers by standard plate count, but by direct viable count (DVC) and acridine orange direct count methods, viable cell numbers were found to be about 7 logs higher; this phenomenon is illustrated in Figure 10–3.

Cells in the VBNC state are coccoid in shape, and in one study with *V. vulnificus*, this state was induced in nutrient-limited artificial seawater after 27 days at 5°C.[86] In another study, the VBNC state was induced in *V. vulnificus* within 7 days following temperature downshift to 5°C.[91] Resuscitation normally occurs within 24 hours of return to temperatures around 21°C.[92] Among internal cellular changes known to occur as organisms enter the VBNC state are changes in cellular lipids and protein synthesis. When the temperature was decreased from 23°C to 13°C for *V. vulnificus*, the generation time increased from 3.0 to 13.1 hours and 40 new proteins were synthesized.[78] While in the VBNC state, *V. vulnificus* has been shown to retain its virulence, although at reduced levels.[91] The VBNC state has been demonstrated for *Salmonella* Enteritidis, *Shigella*, *Vibrio cholerae*, and enteropathogenic *E. coli*, as well as those noted above. Although in one study evidence suggested that *E. coli* O157:H7 could enter the VBNC state in water,[130] investigators in another study were unable to induce the VBNC state in a number of enteric bacteria, including *E. coli*.[18]

Using a green fluorescent protein-tagged *Pseudomonas fluorescens* culture, cells that were stressed at 37.5°C and became VBNC fluoresced at an intensity of about 50% of nonstressed cells and those that

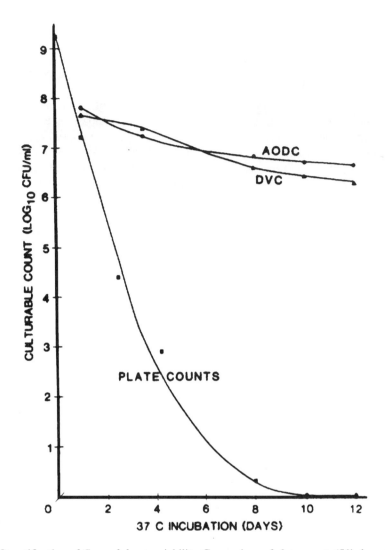

Figure 10–3   Quantification of *Campylobacter* viability. Comparison of plate counts (5% sheep blood agar). (■): DVC assaying protein synthesis in the absence of DNA replication (▲); and AODC (●) as indices of viability for stream-water stationary microcosms. *Source*: Rollins and Colwell,[109] Copyright © 1986 American Society for Microbiology.

---

became VBNC after starvation had fluorescence intensities that were 90–120% of nonstarved cells.[71] Since dead cells did not fluoresce, these findings indicate that VBNC cells retain their viability. When *Vibrio harveyi* and *V. fischeri* were induced into the VBNC state by nutrient limitation, both lost the capacity of luminesce but luminescence was restored when nutrient limitation was relieved by the addition of nutrients (*103*).

REFERENCES

1. Alcock, S.J., L.P. Hall, and J.H. Blanchard. 1987. Methylene blue test to assess the microbial contamination of frozen peas. *Food Microbiol.* 4:3–10.

2. Allwood, M.C., and A.D. Russell. 1967. Mechanism of thermal injury in *Staphylococcus aureus.* I. Relationship between viability and leakage. *Appl. Microbiol.* 15:1266–1269.

3. Anderson, K.L., and D.Y.C. Fung. 1983. Anaerobic methods, techniques and principles for food bacteriology: A review. *J. Food Protect.* 46:811–822.

4. Andrew, M.H.E., and A.D. Russell. 1984. *The Revival of Injured Microbes.* London: Academic Press.

5. Andrews, W.H., C.R. Wilson, P.L. Poelma, A. Romero, R.A. Rude, A.P. Duran, F.D. McClure, and D.E. Gentile. 1978. Usefulness of the Stomacher in a microbiological regulatory laboratory. *Appl. Environ. Microbiol.* 35:89–93.

6. Angelotti, R., and M.J. Foter. 1958. A direct surface agar plate laboratory method for quantitatively detecting bacterial contamination on nonporous surfaces. *Food Res.* 23:170–174.

7. Angelotti, R., J.L. Wilson, W.Litsky, and W.G. Walter. 1964. Comparative evaluation of the cotton swab and rodac methods for the recovery of *Bacillus subtilis* spore contamination from stainless steel surfaces. *Health Lab. Sci.* 1:289–296.

8. Association of Official Analytical Chemists. 1983. Enumeration of coliforms in selected foods. Hydrophobic grid membrane filter method, official first action. *J. Assoc. Off. Anal. Chem.* 66:547–548.

9. Austin, B.L., and B. Thomas. 1972. Dye reduction tests on meat products. *J. Sci. Food Agric.* 23:542.

10. Barach, J.T., R.S. Flowers, and D.M. Adams. 1975. Repair of heat-injured *Clostridium perfringens* spores during outgrowth. *Appl. Microbiol.* 30:873–875.

11. Betts, R.P., P. Bankes, and J.G. Board. 1989. Rapid enumeration of viable micro-organisms by staining and direct microscopy. *Lett. Appl. Microbiol.* 9:199–202.

12. Beuchat, L.R., ed. 1987. *Food and Beverage Mycology*, 2nd ed. New York: Kluwer Academic Publishers.

13. Beuchat, L.R., and R.V. Lechowich. 1968. Effect of salt concentration in the recovery medium on heat-injured *Streptococcus faecalis. Appl. Microbiol.* 16:772–776.

14. Brewer, D.G., S.E. Martin, and Z.J. Ordal. 1977. Beneficial effects of catalase or pyruvate in a most-probable-number technique for the detection of *Staphylococcus aureus. Appl. Environ. Microbiol.* 34:797–800.

15. Brodsky, M.H., P. Entis, A.N. Sharpe, and G.A. Jarvis. 1982. Enumeration of indicator organisms in foods using the automated hydrophobic grid membrane filter technique. *J. Food Protect.* 45:292–296.

16. Beuchat, L.R., F. Copeland, M.S. Curiale, D. Danisavich, V. Ganger, B.W. King, T.L. Lawlis, R.O. Likin, J. Owkusoa, C.E. Smith, and D.E. Townsend. 1998. Comparison of SimPlate total plate count method with Petrifilm, Redigel, and conventional pour-plate methods for enumerating aerobic microorganisms in foods. *J. Food Protect.* 61:14–18.

17. Brodsky, M.H., P. Entis, M.P. Entis, A.N. Sharpe, and G.A. Jarvis. 1982. Determination of aerobic plate and yeast and mold counts in foods using an automated hydrophobic grid membrane filter technique. *J. Food Protect.* 45:301–304.

18. Bogosian, G., P.J.L. Morris, and J.P. O'Neil. 1998. A mixed culture recovery method indicates that enteric bacteria do not enter the viable but nonculturable state. *Appl. Environ. Microbiol.* 64:1736–1742.

19. Brodsky, M.H., P. Boleszczuk, and P. Entis. 1982. Effect of stress and resuscitation on recovery of indicator bacteria from foods using hydrophobic grid-membrane filtration. *J. Food Protect.* 45:1326–1331.

20. Busta, F.F. 1976. Practical implications of injured microorganisms in food. *J. Milk Food Technol.* 39:138–145.

21. Chain, V.S., and D.Y.C. Fung. 1991. Comparison of Redigel, Petrifilm, Spiral plate system, Isogrid, and aerobic plate count for determining the numbers of aerobic bacteria in selected foods. *J. Food Protect.* 54:208–211.

22. Clark, D.S. 1965. Method of estimating the bacterial population of surfaces. *Can. J. Microbiol.* 11:407–413.

23. Clark, D.S. 1965. Improvement of spray gun method of estimating bacterial populations on surfaces. *Can. J. Microbiol.* 11:1021–1022.

24. Conner, D.E., and L.R. Beuchat. 1984. Sensitivity of heat-stressed yeasts to essential oils of plants. *Appl. Environ. Microbiol.* 47:229–233.

25. Cordray, J.C., and D.L. Huffman. 1985. Comparison of three methods for estimating surface bacteria on pork carcasses. *J. Food Protect.* 48:582–584.

26. Cormier, A., S. Chiasson, and A. Léger. 1993. Comparison of maceration and enumeration procedures for aerobic count in selected seafoods by standard method, Petrifilm, Redigel, and Isogrid. *J. Food Protect.* 56:249–255.

27. Cousin, M.A. 1982. Evaluation of a test strip used to monitor food processing sanitation. *J. Food Protect.* 45:615–619, 623.

28. deFigueiredo, M.P., and J.M. Jay. 1976. Coliforms, enterococci, and other microbial indicators. In *Food Microbiology: Public Health and Spoilage Aspects*, ed. M.P. deFigueiredo and D.F. Splittstoesser, 271–297. New York: Kluwer Academic Publishers.

29. Dodsworth, P.J., and A.G. Kempton. 1977. Rapid measurement of meat quality by resazurin reduction. II. Industrial application. *Can. Inst. Food Sci. Technol. J.* 10:158–160.

30. Donnelly, C.B., J.E. Gilchrist, J.T. Peeler, and J.E. Campbell. 1976. Spiral plate count method for the examination of raw and pasteurized milk. *Appl. Environ. Microbiol.* 32:21–27.

31. Downes, F.P., and K. Ito, eds. 2001. *Compendium of Methods for the Microbiological Examination of Foods.* Washington, DC: American Public Health Association.

32. Entis, P. 1985. Rapid hydrophobic grid membrane filter method for *Salmonella* detection in selected foods. *J. Assoc. Off. Anal. Chem.* 68:555–564.

33. Entis, P. 1983. Enumeration of coliforms in non-fat dry milk and canned custard by hydrophobic grid membrane filter method: Collaborative study. *J. Assoc. Off. Anal. Chem.* 66:897–904.

34. Entis, P., and I. Lerner. 1998. Enumeration of β-glucuronidase-positive *Escherichia coli* in foods by using the ISO-GRID method with SD-39 agar. *J. Food Protect.* 61:913–916.

35. Favero, M.S., J.J. McDade, J.A. Robertsen, R.K. Hoffman, and R.W. Edwards. 1968. Microbiological sampling of surfaces. *J. Appl. Bacteriol.* 31:336–343.

36. *FDA Bacteriological Analytical Manual*, 8th ed. 1995. McLean, VA: Association of Official Analytical Chemists Int. (Also, http://www.cfsan.fda.gov/ ebam/bam-4a.html) www.cfsan.fda.gov/~ebam/bam-4a.html)

37. Flowers, R.S., S.E. Martin, D.G. Brewer, and Z.J. Ordal. 1977. Catalase and enumeration of stressed *Staphylococcus aureus* cells. *Appl. Environ. Microbiol.* 33:1112–1117.

38. Foegeding, P.M., and F.F. Busta. 1981. Bacterial spore injury—An update. *J. Food Protect.* 44:776–786.

39. Foegeding, P.M., and F.F. Busta. 1983. Proposed role of lactate in germination of hypochlorite-treated *Clostridium botulinum* spores. *Appl. Environ. Microbiol.* 45:1369–1373.

40. Foegeding, P.M., and F.F. Busta. 1983. Proposed mechanism for sensitization by hypochlorite treatment of *Clostridium botulinum* spores. *Appl. Environ. Microbiol.* 45:1374–1379.

41. Fung, D.Y.C., C.-Y. Lee, and C.L. Kastner. 1980. Adhesive tape method for estimating microbial load on meat surfaces. *J. Food Protect.* 43:295–297.

42. Fung, D.Y., and L.L. VandenBosch. 1975. Repair, growth, and enterotoxigenesis of *Staphylococcus aureus* S-6 injured by freeze-drying. *J. Milk Food Technol.* 38:212–218.

43. Garvie, E.I., and A. Rowlands. 1952. The role of micro-organisms in dye-reduction and keeping-quality tests. II. The effect of micro-organisms when added to milk in pure and mixed culture. *J. Dairy Res.* 19:263–274.

44. Gilchrist, J.E., J.E. Campbell, C.B. Donnelly, J.T. Peeler, and J.M. Delany. 1973. Spiral plate method for bacterial determination. *Appl. Microbiol.* 25:244–252.

45. Ginn, R.E., V.S. Packard, and T.L. Fox. 1984. Evaluation of the 3M dry medium culture plate (Petrifilm SM) method for determining numbers of bacteria in raw milk. *J. Food Protect.* 47:753–755.

46. Gunderson, M.F., and K.D. Rose. 1948. Survival of bacteria in a precooked, fresh-frozen food. *Food Res.* 13:254–263.

47. Harries, D., and A.D. Russell. 1966. Revival of heat-damaged *Escherichia coli. Experientia.* 22:803–804.

48. Harris, N.D. 1963. The influence of the recovery medium and the incubation temperature on the survival of damaged bacteria. *J. Appl. Bacteriol.* 26:387–397.

49. Hartman, P.A., P.S. Hartman, and W.W. Lanz. 1975. Violet red bile 2 agar for stressed coliforms. *Appl. Microbiol.* 29:537–539.

50. Hartsell, S.E. 1951. The longevity and behavior of pathogenic bacteria in frozen foods: The influence of plating media. *Am. J. Public Health.* 41:1072–1077.

51. Hedges, A.J., R. Shannon, and R.P. Hobbs. 1978. Comparison of the precision obtained in counting viable bacteria by the spiral plate maker, the droplette and the Miles and Misra methods. *J. Appl. Bacteriol.* 45:57–65.

52. Hobbie, J.E., R.J. Daley, and S. Jasper. 1977. Use of nucleopore filters for counting bacteria by fluorescence microscopy. *Appl. Environ. Microbiol.* 33:1225–1228.

53. Holah, J.T., R.P. Betts, and R.H. Thorpe. 1988. The use of direct epifluorescent microscopy (DEM) and the direct epifluorescent filter technique (DEFT) to assess microbial populations on food contact surfaces. *J. Appl. Bacteriol.* 65:215–221.

54. Holley, R.A., S.M. Smith, and A.G. Kempton. 1977. Rapid measurement of meat quality by resazurin reduction. I. Factors affecting test validity. *Can. Inst. Food Sci. Technol. J.* 10:153–157.

55. Hurst, A. 1977. Bacterial injury: A review. *Can. J. Microbiol.* 23:935–944.

56. Hurst, A., G.S. Hendry, A. Hughes, and B. Paley. 1976. Enumeration of sublethally heated staphylococci in some dried foods. *Can. J. Microbiol.* 22:677–683.

57. Hurst, A., and A. Hughes. 1978. Stability of ribosomes of *Staphylococcus aureus* S-6 sublethally heated in different buffers. *J. Bacteriol.* 133:564–568.

58. Hurst, A., A. Hughes, J.L. Beare-Rogers, and D.L. Collins-Thompson. 1973. Physiological studies on the recovery of salt tolerance by *Staphylococcus aureus* after sublethal heating. *J. Bacteriol.* 116:901–907.

59. Hurst, A., A. Hughes, D.L. Collins-Thompson, and B.G. Shah. 1974. Relationship between loss of magnesium and loss of salt tolerance after sublethal heating of *Staphylococcus aureus*. *Can. J. Microbiol.* 20:1153–1158.

60. Hutcheson, T.C., T. McKay, L. Farr, and B. Seddon. 1988. Evaluation of the stain Viablue for the rapid estimation of viable yeast cells. *Lett. Appl. Microbiol.* 6:85–88.

61. Jackson, R.W., K. Osborne, G. Barnes, C. Jolliff, D. Zamani, B. Roll, A. Stillings, D. Herzog, S. Cannon, and S. Loveland. 2000. Multiregional evaluation of the SimPlate heterotrophic plate count method compared to the standard plate count agar pour plate method in water. *Appl. Environ. Microbiol.* 66:453–454.

62. Jarvis, B., V.H. Lach, and J.M. Wood. 1977. Evaluation of the spiral plate maker for the enumeration of micro-organisms in foods. *J. Appl. Bacteriol.* 43:149–157.

63. Jay, J.M., and S. Margitic. 1979. Comparison of homogenizing, shaking, and blending of the recovery of microorganisms and endotoxins from fresh and frozen ground beef as assessed by plate counts and the *Limulus* amoebocyte lysate test. *Appl. Environ. Microbiol.* 38:879–884.

64. Jones, S.B., S.A. Palumbo, and J.L. Smith. 1983. Electron microscopy of heat-injured and repaired *Staphylococcus aureus*. *J. Food Safety* 5:145–157.

65. Juffs, H.S., and F.J. Babel. 1975. Rapid enumeration of psychrotrophic bacteria in raw milk by the microscopic colony count. *J. Milk Food Technol.* 38:333–336.

66. Knabel, S.J., H.W. Walker, and A.A. Kraft. 1987. Enumeration of fluorescent pseudomonads on poultry by using the hydrophobic-grid membrane filter method. *J. Food Sci.* 52:837–841, 845.

67. Koch, H.A., R. Bandler, and R.R. Gibson. 1986. Fluorescence microscopy procedure for quantification of yeasts in beverages. *Appl. Environ. Microbiol.* 52:599–601.

68. Koller, W. 1984. Recovery of test bacteria from surfaces with a simple new swab-rinse technique: A contribution to methods for evaluation of surface disinfectants. *Zent. Bakteriol. Hyg. I. Orig. B.* 179:112–124.

69. Konuma, H., A. Suzuki, and H. Kurata. 1982. Improved Stomacher 400 bag applicable to the spiral plate system for counting bacteria. *Appl. Environ. Microbiol.* 44:765–769.

70. Lee, A.C., and J.M. Goepfert. 1975. Influence of selected solutes on thermally induced death and injury of *Salmonella* Typhimurium. *J. Milk Food Technol.* 38:195–200.

71. Lowder, M., A. Unge, N. Maraha, J.K. Jansson, J. Swiggett, and J.D. Oliver. 2000. Effect of starvation and the viable-but-nonculturable state on green fluorescent protein (GFP) fluorescence in GFP-tagged *Pseudomonas fluorescens* A506. *Appl. Environ. Microbiol.* 66:3160–3165.

72. P. Murray, E. Baron, J. Jorgensen, M. Pfaller, and M. Yolken, eds. 2003. *Manual of Clinical Microbiology*, 8th ed. Washington, DC: ASM Press.

73. Marshall, R.T., ed. 1993. *Standard Methods for the Examination of Dairy Products*, 16th ed. Washington, DC: American Public Health Association.

74. Martin, S.E., R.S. Flowers, and Z.J. Ordal. 1976. Catalase: Its effect on microbial enumeration. *Appl. Environ. Microbiol.* 32:731–734.

75. Matner, R.R., T.L. Fox, D.E. McIver, and M.S. Curiale. 1990. Efficacy of Petrifilm™ count plates for *E. coli* and coliform enumeration. *J. Food Protect.* 53:145–150.

76. Maxcy, R.B. 1973. Condition of coliform organisms influencing recovery of subcultures on selective media. *J. Milk Food Technol.* 36:414–416.

77. McDonald, L.C., C.R. Hackney, and B. Ray. 1983. Enhanced recovery of injured *Escherichia coli* by compounds that degrade hydrogen peroxide or block its formation. *Appl. Environ. Microbiol.* 45:360–365.

78. McGovern, V.P., and J.D. Oliver. 1995. Induction of cold-responsive proteins in *Vibrio vulnificus. J. Bacteriol.* 177:4131–4133.

79. *Microorganisms in Foods.* 1982. Vol. 1, *Their Significance and Methods of Enumeration,* 2nd ed. ICMSF, Toronto: University of Toronto Press.

80. *Microorganisms in Foods.* 1986. Vol. 2, *Sampling for Microbiological Analysis: Principles and Specific Applications,* 2nd ed. ICMSF, Toronto: University of Toronto Press.

81. Moats, W.A., R. Dabbah, and V.M. Edwards. 1971. Survival of *Salmonella anatum* heated in various media. *Appl. Microbiol.* 21:476–481.

82. Mossel, D.A.A., E.H. Kampelmacher, and L.M. Van Noorle Jansen. 1966. Verification of adequate sanitation of wooden surfaces used in meat and poultry processing. *Zent. Bakteriol. Parasiten., Infek. Hyg. Abt. I.* 201:91–104.

83. Neal, N.D., and H.W. Walker. 1977. Recovery of bacterial endospores from a metal surface after treatment with hydrogen peroxide. *J. Food Sci.* 42:1600–1602.

84. Nelson, C.L., T.L. Fox, and F.F. Busta. 1984. Evaluation of dry medium film (Petrifilm VRB) for coliform enumeration. *J. Food Protect.* 47:520–525.

85. Nelson, F.E. 1943. Factors which influence the growth of heat-treated bacteria. I. A comparison of four agar media. *J. Bacteriol.* 45:395–403.

86. Nilsson, L., J.D. Oliver, and S. Kjelleberg. 1991. Resuscitation of *Vibrio vulnificus* from the viable but nonculturable state. *J. Bacteriol.* 173:5054–5059.

87. Niskanen, A., and M.S. Pohja. 1977. Comparative studies on the sampling and investigation of microbial contamination of surfaces by the contact plate and swab methods. *J. Appl. Bacteriol.* 42:53–63.

88. Nortje, G.L., E. Swanepoel, R.T. Naude, W.H. Holzapfel, and P.L. Steyn. 1982. Evaluation of three carcass surface microbial sampling techniques. *J. Food Protect.* 45:1016–1017, 1021.

89. *Official Methods of Analysis,* 16th ed., Vol. I. 1998. McLean, VA: Association of Official Analytical Chemists Int.

90. Ølgaard, K. 1977. Determination of relative bacterial levels on carcasses and meats—A new quick method. *J. Appl. Bacteriol.* 42:321–329.

91. Oliver, J.D., and R. Bockian. 1995. In vivo resuscitation, and virulence towards mice, of viable but nonculturable cells of *Vibrio vulnificus. Appl. Environ. Microbiol.* 61:2620–2623.

92. Oliver, J.D., F. Hite, D. McDougald, N.L. Andon, and L.M. Simpson. 1995. Entry into, and resuscitation from, the viable but nonculturable state by *Vibrio vulnificus* in an estuarine environment. *Appl. Environ. Microbiol.* 61:2624–2630.

93. Patterson, J.T. 1971. Microbiological assessment of surfaces. *J. Food Technol.* 6:63–72.

94. Peeler, J.T., J.E. Gilchrist, C.B. Donnelly, and J.E. Campbell. 1977. A collaborative study of the spiral plate method for examining milk samples. *J. Food Protect.* 40:462–464.

95. Pettipher, G.L. 1983. *The Direct Epifluorescent Filter Technique for the Rapid Enumeration of Microorganisms.* New York: Wiley.

96. Pettipher, G.L., R.J. Fulford, and L.A. Mabbitt. 1983. Collaborative trial of the direct epifluorescent filter technique (DEFT), a rapid method for counting bacteria in milk. *J. Appl. Bacteriol.* 54:177–182.

97. Pettipher, G.L., R. Mansell, C.H. McKinnon, and C.M. Cousins. 1980. Rapid membrane filtration-epifluorescent microscopy technique for direct enumeration of bacteria in raw milk. *Appl. Environ. Microbiol.* 39:423–429.

98. Pettipher, G.L., and U.M. Rodrigues. 1981. Rapid enumeration of bacteria in heat-treated milk and milk products using a membrane filtration-epifluorescent microscopy technique. *J. Appl. Bacteriol.* 50:157–166.

99. Pettipher, G.L., and U.M. Rodrigues. 1982. Rapid enumeration of microorganisms in foods by the direct epifluorescent filter technique. *Appl. Environ. Microbiol.* 44:809–813.

100. Pettipher, G.L., R.A. Williams, and C.S. Gutteridge. 1985. An evaluation of possible alternative methods to the Howard mould count. *Lett. Appl. Microbiol.* 1:49–51.

101. Postgate, J.R., and J.R. Hunter. 1963. Metabolic injury in frozen bacteria. *J. Appl. Bacteriol.* 26:405–414.

102. Puleo, J.R., M.S. Favero, and N.J. Petersen. 1967. Use of ultrasonic energy in assessing microbial contamination on surfaces. *Appl. Microbiol.* 15:1345–1351.

103. Ramaiah, N., J. Ravel, W.L. Straube, R.T. Hill, and R.R. Colwell. 2002. Entry of *Vibrio harveyi* and *Vibrio fischeri* into the viable but nonculturable state. *J. Appl. Microbiol.* 93:108–116.

104. Rao, D.N., and V.S. Murthy. 1986. Rapid dye reduction tests for the determination of microbiological quality of meat. *J. Food Technol.* 21:151–157.

105. Reyniers, J.A. 1935. Mechanising the viable count. *J. Pathol. Bacteriol.* 40:437–454.

106. Rodrigues, U.M., and R.G. Kroll. 1989. Microcolony epifluorescence microscopy for selective enumeration of injured bacteria in frozen and heat-treated foods. *Appl. Environ. Microbiol.* 55:778–787.

107. Rodrigues, U.M., and R.G. Kroll. 1988. Rapid selective enumeration of bacteria in foods using a microcolony epifluorescence microscopy technique. *J. Appl. Bacteriol.* 64:65–78.

108. Rodrigues, U.M., and R.G. Kroll. 1985. The direct epifluorescent filter technique (DEFT): Increased selectivity, sensitivity and rapidity. *J. Appl. Bacteriol.* 59:493–499.

109. Rollins, D.M., and R.R. Colwell. 1986. Viable but nonculturable stage of *Campylobacter jejuni* and its role in survival in the natural aquatic environment. *Appl. Environ. Microbiol.* 52:531–538.

110. Rowley, D.B., R. Firstenberg-Eden, and G.E. Shattuck. 1983. Radiation-injured *Clostridium botulinum* type E spores: Outgrowth and repair. *J. Food Sci.* 48:1829–1831, 1848.

111. Saffle, R.L., K.N. May, H.A. Hamid, and J.D. Irby. 1961. Comparing three rapid methods of detecting spoilage in meat. *Food Technol.* 15:465–467.

112. Scott, E., S.F. Bloomfield, and C.G. Barlow. 1984. A comparison of contact plate and calcium alginate swab techniques of environmental surfaces. *J. Appl. Bacteriol.* 56:317–320.

113. Sharpe, A.N., and G.C. Harshman. 1976. Recovery of *Clostridium perfringens*, *Staphylococcus aureus*, and molds from foods by the Stomacher: Effect of fat content, surfactant concentration, and blending time. *Can. Inst. Food Sci. Technol. J.* 9:30–34.

114. Sharpe, A.N., and A.K. Jackson. 1972. Stomaching: A new concept in bacteriological sample preparation. *Appl. Microbiol.* 24:175–178.

115. Sharpe, A.N., and D.C. Kilsby. 1971. A rapid, inexpensive bacterial count technique using agar droplets. *J. Appl. Bacteriol.* 34:435–440.

116. Sharpe, A.N., M.P. Diotte, I. Dudas, S. Malcolm, and P.I. Peterkin. 1983. Colony counting on hydrophobic grid-membrane filters. *Can. J. Microbiol.* 29:797–802.

117. Sharpe, A.N., P.I. Peterkin, and N. Malik. 1979. Improved detection of coliforms and *Escherichia coli* in foods by a membrane filter method. *Appl. Environ. Microbiol.* 38:431–435.

118. Sharpe, A.N., and G.L. Michaud. 1974. Hydrophobic grid-membrane filters: New approach to microbiological enumeration. *Appl. Microbiol.* 28:223–225.

119. Sharpe, A.N., and G.L. Michaud. 1975. Enumeration of high numbers of bacteria using hydrophobic grid membrane filters. *Appl. Microbiol.* 30:519–524.

120. Shaw, B.G., C.D. Harding, W.H. Hudson, and L. Farr. 1987. Rapid estimation of microbial numbers on meat and poultry by direct epifluorescent filter technique. *J. Food Protect.* 50:652–657.

121. Silliker, J.H., D.A. Gabis, and A. May. 1979. ICMSF methods studies. XI. Collaborative/comparative studies on determination of coliforms using the most probable number procedure. *J. Food Protect.* 42:638–644.

122. Smith, J.L., R.C. Benedict, M. Haas, and S.A. Palumbo. 1983. Heat injury in *Staphylococcus aureus* 196E: Protection by metabolizable and non-metabolizable sugars and polyols. *Appl. Environ. Microbiol.* 46:1417–1419.

123. Speck, M.L., B. Ray, and R.B. Read, Jr. 1975. Repair and enumeration of injured coliforms by a plating procedure. *Appl. Microbiol.* 29:549–550.

124. Szabo, R.A., E.C.D. Todd, and A. Jean. 1986. Method to isolate *Escherichia coli* O157:H7 from food. *J. Food Protect.* 49:768–772.

125. ten Cate, L. 1963. An easy and rapid bacteriological control method in meat processing industries using agar sausage techniques in Rilsan artificial casing. *Fleischwarts.* 15:483–486.

126. Todd, E.C.D., R.A. Szabo, P. Peterkin, A. N. Sharpe, L. Parrington, D. Bundle, M.A.J. Gidney, and M.B. Perry. 1988. Rapid hydrophobic grid membrane filter-enzyme-labeled antibody procedure for identification and enumeration of *Escherichia coli* O157 in foods. *Appl. Environ. Microbiol.* 54:2526–2540.

127. Tomlins, R.I., M.D. Pierson, and Z.J. Ordal. 1971. Effect of thermal injury on the TCA cycle enzymes of *Staphylococcus aureus* MF 31 and *Salmonella* Typhimurium 7136. *Can. J. Microbiol.* 17:759–765.

128. Trotman, R.E. 1971. The automatic spreading of bacterial culture over a solid agar plate. *J. Appl. Bacteriol.* 34:615–616.

129. Tuttlebee, J.W. 1975. The Stomacher—Its use for homogenization in food microbiology. *J. Food Technol.* 10:113–122.

130. Wang, G., and M.P. Doyle. 1998. Survival of enterohemorrhagic *Escherichia coli* O157:H7 in water. *J. Food Protect.* 61:662–667.

131. Woodward, R.L. 1957. How probable is the most probable number? *J. Am. Water Works Assoc.* 49:1060–1068.

132. Zayaitz, A.E.K., and R.A. Ledford. 1985. Characteristics of acid-injury and recovery of *Staphylococcus aureus* in a model system. *J. Food Protect.* 48:616–620.

CHAPTER 11

Chemical, Biological, and Physical Methods

Most of the methods for detecting and characterizing microorganisms covered in this chapter have been developed since 1960. Many can be used to estimate numbers of cells or quantity of cellular byproducts. Unlike direct microscopic counts, most of those that follow are based on metabolic activity of microorganisms on given substrates, measurements of growth response, measurements of some part of cells including nucleic acids, or combinations of these.

CHEMICAL METHODS

The methods covered in this section are used primarily to detect, enumerate, or identify foodborne organisms or their products:

> Thermostable nuclease (for *Staphylococcus aureus*)
> *Limulus* amoebocyte lysate (LAL) assay (for Gram-negative bacteria)
> ATP assay (for live cells)
> Radiometry
> Fluorogenic/chromogenic substrates (to identify/differentiate microbial species or strains).

The relative sensitivity of these methods compared to others in this chapter can be seen from Table 11–1.

Thermostable Nuclease

The presence of *S. aureus* in significant numbers in a food can be determined by examining the food for the presence of thermostable nuclease (DNase). This is possible because of the high correlation between the production of coagulase and thermostable nuclease by *S. aureus* strains, especially enterotoxin producers. For example, in one study, 232 of 250 (93%) enterotoxigenic strains produced coagulase, and 242 or 95% produced thermostable nuclease.[118] Non-*S. aureus* species that produce DNAse are discussed in Chapter 23.

The examination of foods for this enzyme was first carried out by Chesbro and Auborn[32] employing a spectrophotometric method for nuclease determination. They showed that as the numbers of cells

Table 11–1 Reported Minimum Detectable Levels of Toxins or Organisms by Biological, Chemical, and Physical Methods of Analysis

Methods	Toxin or Organism	Sensitivity
Flow cytometry	S. Typhimurium in milk	10^3/ml within 40 minutes, 10/ml after 6 hours nonselective enrichment
Impedance	Coliforms in meats	10^3/g in 6.5 hours
	Coliforms in culture media	10 in 3.8 hours
Microcalorimetry	S. aureus cells	2 cells in 12–13 hours
	S. aureus	Minimum HPR* ~10^4 cells/ml
ATP measurement	Beef carcass	10^2/cm^2 in ~5 minutes
Radiometry	Frozen orange juice flora	10^4 cells/g in 6–10 hours
	Coliforms in water	1–10 cells in 6 hours
Fluorescent antibody	Salmonellae	10^6 cells/ml
	Staph. enterotoxin B	~50ng/ml
Thermostable nuclease	From S. aureus	10 ng/g
	From S. aureus	2.5–5 ng
Limulus lysate test	Gram-negative endotoxins	2–6 pg of E. coli LPS
Radioimmunoassay	Staph. enterotoxins A, B, C, D, and E in foods	0.5–1.0 ng/g
	Staph. enterotoxin B in nonfat dry milk	2.2 ng/ml
	Staph. enterotoxins A and B	0.1 ng/ml for A; 0.5 ng/ml for B
	Staph. enterotoxin C_2	100 pg
	E. coli ST_a enterotoxin	50–500 pg/tube
	Aflatoxin M_1 in milk	0.5 ng/ml
	Ochratoxin A	20 ppb
	Bacterial cells	500–1,000 cells in 8–10 minutes
	Aflatoxin B_1 in corn, wheat, peanut butter	6 ng/g
	Deoxynivalenol in corn, wheat	20 ng/g
Electroimmunodiffusion	C. perfringens enterotoxin	10 ng
	Botulinal toxins	3.7–5.6 mouse LD_{50}/0.1 ml
Micro-Ouchterlony	S. aureus enterotoxins A and B	10–100 ng/ml
	C. perfringens type A toxin	500 ng/ml
Lux-phage	Listeria monocytogenes	<1 cell/g of food
Passive immune hemolysis	E. coli LT enterotoxin	<100 ng
Aggregate-hemagglutination	B. cereus enterotoxin	4 ng/ml
Latex agglutination	E. coli LT enterotoxin	32 ng/ml
Single radial immunodiffusion	S. aureus enterotoxins	0.3 μg/ml
Hemagglutination-inhibition	Staph. enterotoxin B	1.3 ng/ml
Reverse passive hemagglutination	Staph. enterotoxin B	1.5 ng/ml
	C. perfringens type A toxin	1 ng/ml

(continued)

Table 11–1   (continued)

Methods	Toxin or Organism	Sensitivity
ELISA	Staph. Enterotoxin A in wieners	0.4 ng
	Staph. enterotoxins A, B, and C in foods	0.1 ng/ml
	Staph. enterotoxins A, B, C, D, and E in foods	≥ 1 ng/g
	Botulinal toxin type A	About 9 mouse LD_{50}/ml with monoclonal antibody
	Botulinal toxin type A	50–100 mouse i.p. LD_{50}
	Botulinal toxin type E	100 mouse LD_{50}
	Aflatoxin B_1	25 pg/assay
	Aflatoxin M_1 in milk	0.25 ng/ml
	Aflatoxin B_1	<1 pg/assay
	Salmonellae	10^4–10^5 cells/ml
	AFB_1	0.2 ng/ml (monoclonal)
	AFB_1	0.4 ng/ml (polyclonal)
	AFM_1	1.0 ng/ml (monoclonal)
	Fumonisins in feed	250 ng/g
	Zearalenone in corn	1 ng/g
	E. coli O157:H7 in ground beef	<1 cell/g
	C. perfringens enterotoxin	1 pg/ml
	AFB_1 in peanut butter	2.5 ng/g
	Ochratoxin in barley	1 ng/ml
Polymerase chain reaction (PCR)	E. coli	1–5 cells/100 ml H_2O
	E. coli	1 cell
	L. monocytogenes	1–10 cells
	V. vulnificus	10^2 cfu/g (oysters)
	C. perfringens	<1 cfu, 2–6 hours
	Stx1 and Stx2 of E. coli	1 cell/g in 12 hours
	C. botulinum toxins A to E	10 fg ($\sim$3 cells)
	Y. enterocolitica	10–30 cfu/g meat
Fluorogenic PCR-based assay	E. coli	0.5 cfu/g
Real-time PCR	Listeria monocytogenes	1 cell
Real-time PCR	Giardia lamblia	1 oocyst
Probelia system	Salmonellae in foods	10^2 threshold sensitivity
RT-PCR	Cryptosporidium parvum	1 oocyte/l, 3 hours
Multiplex PCR	Listeria	1–5 cfu/25 g
Multiplex PCR	E. coli 0157:H7	<1 cfu/g
Fluorescent PCR assay	E. coli	3 cfu/25 g
Immunomagnetic separation	E. coli O157:H7	<10^3 cfu/g
Ice nucleation	Salmonellae	ca. 25/g

*HPR, exothermic heat production rate.

increased in ham sandwiches, there was an increase in the amount of extractable thermostable nuclease of staphylococcal origin. They suggested that the presence of 0.34 unit of nuclease indicated certain staphylococcal growth and that at this level, it was unlikely that enough enterotoxin was present to cause food poisoning. The 0.34 unit was shown to correspond to $9.5 \times 10^{-3} \mu g$ of enterotoxin by *S. aureus* strain 234. The reliability of the thermostable nuclease assay as an indicator of *S. aureus* growth has been shown by others.[50] It has been found to be as good as coagulase in testing for entero-toxigenic strains,[146] and in another study, all foods that contained enterotoxin contained thermostable nuclease, which was present in most foods with 1×10^6 *S. aureus* cells per gram.[156] On the other hand, thermostable nuclease is produced by some enterococci. Of 728 enterococci from milk and milk products, about 30% produced nuclease, with 4.3% of the latter (31 of the 728) being positive for thermostable nuclease.[11]

The mean quantity of thermostable nuclease produced by enterotoxigenic strains is less than that for nonenterotoxigenic strains, with 19.4 and 25.5 $\mu g/ml$, respectively, as determined in one study.[146] For detectable levels of nuclease, 10^5–10^6 cells are needed, whereas for detectable enterotoxin, $>10^6$ cells per milliliter are needed.[147] During the recovery of heat-injured cells in trypticase soy broth (TSB), nuclease was found to increase during recovery but later decreased.[228] The reason for the decrease was found to be proteolytic enzymes, and the decrease was reversed by the addition of protease inhibitors.

Among the advantages of testing for heat-stable nuclease as an indicator of *S. aureus* growth and activity are the following:

1. Because of its heat-stable nature, the enzyme will persist even if the bacterial cells are destroyed by heat, chemicals, or bacteriophage or if they are induced to L-forms.
2. The heat-stable nuclease can be detected faster than enterotoxin (about 3 hours versus several days).[115]
3. The nuclease appears to be produced by enterotoxigenic cells before enterotoxins appear (Figure 11–1).
4. The nuclease is detectable in unconcentrated cultures of food specimen, whereas enterotoxin detection requires concentrated samples.
5. The nuclease of concern is stable to heat, as are the enterotoxins.

Although *S. epidermidis* and some micrococci produce nuclease, it is not as stable to heating as that produced by *S. aureus*.[118] Thermostable nuclease will withstand boiling for 15 minutes. It has been found to have a D value (D_{130}) of 16.6 minutes in brain–heart infusion (BHI) broth at pH 8.2, and a z value of 51.[50]

Limulus Lysate for Endotoxins

Gram-negative bacteria are characterized by their production of endotoxins, which consist of a lipopolysaccharide (LPS) layer (outer membrane) of the cell envelope and lipid A, which is buried in the outer membrane. The LPS is pyrogenic and responsible for some of the symptoms that accompany infections caused by Gram-negative bacteria.

The *Limulus* amoebocyte lysate (LAL) test employs a lysate protein obtained from the blood (actually hemolymph) cells (amoebocytes) of the horseshoe crab (*Limulus polyphemous*). The lysate protein is the most sensitive substance known for endotoxins. Of six different LAL preparations tested from five commercial companies, they were found to be 3–300 times more sensitive to endotoxins than the U.S. Pharmacopeia rabbit pyrogen test.[214] The LAL test is performed by adding aliquots

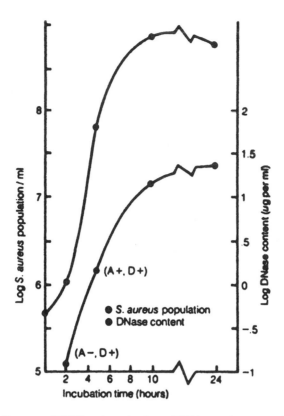

Figure 11–1 Growth of *S. aureus* (196E) and production of DNAse and enterotoxins in brain–heart infusion broth at 37°C. DNAse and enterotoxin D were detectable within 4 hours at a population of 2×10^6, whereas enterotoxin A was detected after 4 hours at higher cell populations. DNAse was detectable in unconcentrated cultures, and enterotoxins at 50-fold concentrates. *Source*: Reprinted from *Journal of Food Science*, Vol. 40, p. 353, 1975, Copyright © by Institute of Food Technologies.

of food suspensions or other test materials to small quantities of a lysate preparation, followed by incubation at 37°C for 1 hour. The presence of endotoxins causes gel formation of the lysate material. LAL reagent is available that can detect 1.0 pg of LPS. Because the *E. coli* cell contains about 3.0 fg of LPS, it is possible to detect <300 Gram-negative cells. From studies with meatborne pseudomonads, as few as 10^2 cfu/ml were detected.[54] The various LAL methods employed to detect microorganisms in foods have been reviewed.[36,95]

The first food application was the use of LAL to detect the microbial spoilage of ground beef.[90,91] Endotoxin titers increase in proportion to viable counts of Gram-negative bacteria.[94] Since the normal spoilage of refrigerated fresh meats is caused by Gram-negative bacteria, the LAL test is a good, rapid indicator of the total numbers of Gram-negative bacteria. The method has been found to be suitable for the rapid evaluation of the hygienic quality of milk relative to the detection of coliforms before and after pasteurization.[206] For raw and pasteurized milk, it represents a method that can be used to determine the history of a milk product relative to its content of Gram-negative bacteria. Because both viable and nonviable Gram-negative bacteria are detected by LAL, a simultaneous plating is necessary to determine the numbers of cfus. The method has been applied successfully to monitor milk and milk products,[88,227] microbial quality of raw fish,[199] and cooked turkey rolls. In the last case, LAL titers and numbers of Enterobacteriaceae in vacuum-packaged rolls were found to have a statistically significant linear relationship.[45]

LAL titers for foods can be determined either by direct serial dilutions or by MPN, with results by the two methods being essentially similar.[185] To extract endotoxins from foods, the Stomacher has been found to be generally better than the Waring blenders or the shaking of dilution bottles.[93]

In this test, the proclotting enzyme of the *Limulus* reagent has been purified. It is a serine protease with a molecular weight of about 150,000 daltons. When activated with Ca^{2+} and endotoxin, gelation of the natural clottable protein occurs. The *Limulus* coagulogen has a molecular weight of 24,500. When it is acted upon by the *Limulus* clotting enzyme, the coagulogen releases a soluble peptide of about 45 amino acid residues and an insoluble coagulin of about 170 amino acids. The latter interacts with itself to form the clot, which involves the cleavage of –arg–lys– or –arg–gly– linkages.[204] The process, as summarized from reference 144, may be viewed as noted below.

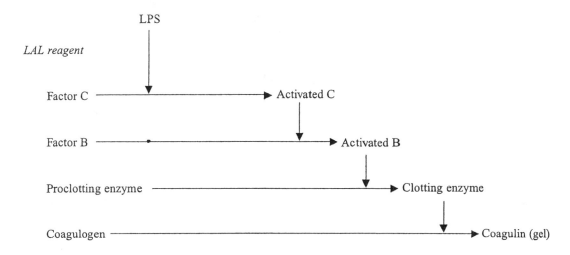

Commercial substrates are available that contain amino acid sequences similar to coagulogen. The chromogenic substrates used for endotoxin consist of these linked to *p*-nitroaniline. When the endotoxin-activated enzyme attacks the chromogenic substrate, free *p*-nitroaniline results and can be read at 405 nm. The amount of the chromogenic compound liberated is proportional to the quantity of endotoxin in the sample. Employing a chromogenic substrate, Tsuji et al.[210] devised an automated method for endotoxin assay, and the method was shown to be sensitive to as little as 30 pg of endotoxin per milliliter.

Assuming that the quantity of endotoxin per Gram-negative bacterial cell is fairly constant, and assuming further that cells of all genera contain the same given quantity, it is possible to calculate the number of cells (viable and nonviable) from which the experimentally determined endotoxin was derived. With a further assumption that the ratio of Gram-negative to Gram-positive bacteria is more or less constant for given products, one can make a 1-hour estimate of the total numbers of bacteria in food products such as fresh ground beef.[92] Low values by this procedure are more meaningful than high values, and the latter need to be confirmed by other methods.

Overall, the value of the LAL test lies in the speed at which results can be obtained. Foods that have high LAL titers can be candidates for further testing by other methods; those that have low titers may be placed immediately into categories of lower risk relative to numbers of Gram-negative bacteria.

Adenosine Triphosphate Measurement

Adenosine triphosphate (ATP) is the primary source of energy in all living cells. It disappears within 2 hours after cell death, and the amount per cell is generally constant,[208] with values of 10^{-18} to 10^{-17} mole per bacterial cell, which corresponds to around 4×10^4 M ATP/10^5 cfu of bacteria.[208] Among procaryotes, ATP in exponentially growing cells is regularly around 2–6 nmole ATP/mg dry weight regardless of mode of nutrition.[105] In the case of rumen bacteria, the average cellular content was found to be 0.3 fg per cell, with higher levels found in rumen protozoal cells.[151] The complete extraction and accurate measurement of cellular ATP can be equated to individual groups of microorganisms in the same general way as endotoxins for Gram-negative bacteria.

One of the simplest ways to measure ATP is by use of the firefly luciferin–luciferase system. In the presence of ATP, luciferase emits light, which is measured with a luminometer. The amount of light produced by firefly luciferase is directly proportional to the amount of ATP added.[157]

The application of ATP measurement as a rapid method for estimating microbial numbers has been used in clinical microbiology. In the clinical laboratory, it has been employed to screen urine specimens. The successful use of the method for bacteriuria and for assessing biomass in activated sludge.[157] suggested that it should be of value for foods. It lends itself to automation and represents an excellent potential method for the rapid estimation of microorganisms in foods. The major problem that has to be overcome for food use is the removal of nonmicrobial ATP. The method was suggested for food use by Sharpe et al.[187] Thore et al.[208] used Triton X-100 and apyrase selectively to destroy nonbacterial ATP in urine specimens and found that the resultant ATP levels were close to values observed in laboratory cultures with detection at 10^5 bacteria per milliliter. In meats, the problem of nonmicrobal ATP was addressed by Stannard and Wood[194] by use of a three-stage process consisting of centrifugation, use of cation exchange resin, and filtration to get rid of food particles and collect bacteria on 0.22-μm filters. ATP analyses were carried out on bacteria eluted from the filter membranes, and 70–80% of most microorganisms were recovered on the filters. A linear relationship was shown between microbial ATP and bacterial numbers over the range 10^6–10^9 cfu/g. By the methods employed, results on ground beef were obtained in 20–25 minutes. In another study, 75 samples of ground beef were evaluated, and a high correlation was found between $\log_{10}$ APC and $\log_{10}$ ATP when samples were incubated at 20°C.[108] In this study, the amount of ATP/cfu ranged from 0.6 to 17.1 fg, with 51 of the 75 samples containing $\leq$5.0 fg of ATP. The ATP assay has been employed successfully for seafoods and for the determination of yeasts in beverages.

The ATP assay has been adapted for the determination of microbial load on chicken carcasses[12] as well as pork and beef.[190] Chicken carcasses were rinsed and results were obtained within 10 minutes, but the method could not reliably detect $<1 \times 10^4$/ml due to carcass ATP.[12] A 500-cm^2 area for beef carcasses, and a 50-cm^2 area for pork were surface-wiped with an ATP-free sponge. The entire test could be completed in about 5 minutes with the minimum detectable number being log 2.0 cfu/cm^2 for beef and log 3.2/cm^2 for pork.[190]

The ATP assay is widely used as a rapid and on-the-spot method for monitoring food handling surfaces by swabbing designated areas and reading the relative light units (RLU) from a luminometer. Since nonmicrobial ATP can contribute to RLU readings, these methods, while valuable for monitoring purposes, should not be used to indicate numbers of microorganisms.

Radiometry

The radiometric detection of microorganisms is based on the incorporation of a ^{14}C-labeled metabolite in a growth medium so that when the organisms utilize this metabolite, ^{14}CO$_2$ is released and measured by use of a radioactivity counter. For organisms that utilize glucose, ^{14}C-glucose is usually employed. For those that cannot utilize this compound, others such as ^{14}C-formate or ^{14}C-glutamate are used. The overall procedure consists of using capped 15-ml serum vials to which are added anywhere from 12 to 36 ml of medium containing the labeled metabolite. The vials are made either aerobic or anaerobic by sparging with appropriate gases and are then inoculated. Following incubation, the headspace is tested periodically for the presence of ^{14}CO$_2$. The time required to detect the labeled CO$_2$ is inversely related to the number of organisms in a product. The Bactec is a commercially available detection system.

The use of radiometry to detect the presence of microorganisms was first suggested by Levin et al.[124] It is confined largely to clinical microbiology, but some applications have been made to foods and water. The experimental detection of *S. aureus*, *Salmonella* Typhimurium, and spores of putrefactive anaerobe 3679 and *Clostridium botulinum* in beef loaf was studied by Previte.[165] The inocula employed ranged from about 10^4 to 10^6/ml of medium, and the detection time ranged from 2 hours for *S*.Typhimurium to 5–6 hours for *C. botulinum* spores. For these studies, 0.0139 μCi of ^{14}C-glucose per milliliter of tryptic soy broth was employed. In another study, Lampi et al.[119] found that one cell per milliliter of *S*.Typhimurium or *S. aureus* could be detected by a radiometric method in 9 hours. For 10^4 cells, 3–4 hours were required. With respect to spores, a level of 90 of putrefactive anaerobe (PA) 3679 was detected in 11 hours, whereas 10^4 were detectable within 7 hours. These and other investigators have shown that spores required 3–4 hours longer for detection than vegetable cells. From the findings of Lampi et al., the radiometric detection procedure could be employed as a screening procedure for foods containing high numbers of organisms, for such foods produced results by this method within 5–6 hours, whereas those with lower numbers required longer times.

The detection of nonfermenters of glucose by this method is possible when metabolites such as labeled formate and/or glutamate are used. It has been shown that a large number of foodborne organisms can be detected by this method in 1–6 hours. The radiometric detection of 1–10 coliforms in water within 6 hours was achieved by Bachrach and Bachrach[9] by employing ^{14}C-lactose with incubation at 37°C in a liquid medium. It is conceivable that a differentiation can be made between fecal *E. coli* and total coliforms by employing 45.5°C incubation along with 37°C incubation.

Radiometry has been used to detect organisms in frozen orange juice concentrate.[78] The investigators used ^{14}C-glucose, four yeasts, and four lactic acid bacteria, and at an organism concentration of 10^4 cells, detection was achieved in 6–10 hours. Of 600 juice samples examined, 44 with counts of 10^4/ml were detected in 12 hours and 41 of these in 8 hours. No false negatives occurred, and only two false positives were noted. The method was used for cooked foods to determine if counts were <10^5 cfu/ml, and the results were compared to APC. Of 404 samples consisting of seven types of foods, around 75% were correctly classified as acceptable or unacceptable within 6 hours.[174] No more than five were incorrectly classified. The study employed ^{14}C-glucose, glutamic acid, and sodium formate.

Fluorogenic and Chromogenic Substrates

Some of the fluorogenic and chromogenic substrates employed in culture media in food microbiology are

1. 4-methylumbelliferyl-β-D-glucuronide (MUG),
2. 4-methylumbelliferyl-β-D-galactoside (MUGal),
3. 4-methylumbelliferyl phosphate (MUP),
4. *o*-nitrophenyl-β-D-galactopyranoside (ONPG),

5. L-alanine-p-nitroanilide (LAPN),
6. 5-bromo-4-chloro-3-indolyl-β-D-glucuronic acid (sodium or cyclohexylammonium salt— variously BCIG, X-Gluc, X-GlcA),
7. 5-bromo-4-chloro-3-indolyl-β-D-galactopyranoside (X-Gal), and
8. indoxyl-β-D-glucuronide (IBDG).

These substrates are employed in various ways in plating media, MPN broths, and for membrane filtration methods. MUG is the most widely used of the fluorogenic substrates, and it is hydrolyzed by β-D-glucuronidase (GUD) to release the fluorescent 4-methylumbelliferyl moiety, which is detected with long-wave ultraviolet light. The value of MUG is that *E. coli* is the primary producer of GUD, and this substrate has found wide use as a differential agent in media and methods for this organism. A few salmonellae and shigellae are also GUD positive, as are some corynebacteria.

The first to employ MUG for *E. coli* detection were Feng and Hartman,[57] who incorporated it into lauryl tryptose broth (LTB) and other coliform-selective media and found that in LTB-MUG, one *E. coli* cell could be detected in 20 hours. Because about 10^7 *E. coli* cells are needed to produce enough GUD to provide detectable MUG results, the time for results depends on the initial number of cells. Although most positive reactions occurred in 4 hours, some weak GUD-positive strains required up to 16 hours for reaction. An important aspect of this method is the occurrence of fluorescence before gas production from lactose. Employing the Feng–Hartman method, another group examined 1,020 specimens by a three-tube MPN and were able to detect more *E. coli*-positive samples than with a conventional MPN.[213] The greater effectiveness of LTB-MUG resulted because some *E. coli* strains are anaerogenic. No false-negative results were obtained. In an evaluation of MUG added to lauryl sulfate broth (LSB), a 94.8% agreement was obtained on 270 product samples with 4.8% false positives but no false negatives with LSB-MUG.[172] Oysters contain endogenous glucuronidase, but an *E. coli* (EC) broth–MUG method was employed successfully in one study where 102 of 103 fluorescing tubes were positive for *E. coli*.[113] A 20-minute tube procedure employing MUG was applied to 682 *E. coli* cultures, and 630 (92.4%) were positive.[207] Of 188 0157 strains of *E. coli*, 166 were MUG negative and all were positive for the vero toxin. By use of this 20-minute method, MUG-negative *E. coli* are very likely to be verotoxigenic.[207] In a study of molluscan shellfish, EC-MUG broth with MUG employed at 50 ppm, 95% of *E. coli*, were positive with 11% being false negative.[169] When compared to the Association of Official Analytical Chemists (AOAC) method for *E. coli*, LST-MUG (lauryl sulfate tryptose) was found to be equivalent for one product and better than AOAC for some others tested,[163] whereas in another, LST-MUG was found to be comparable to the AOAC MPN method.[164]

MUGal has received limited study as a fluorogenic substrate, but in one study it was used to detect fecal coliforms in water by use of a membrane filter method where as few as 1 cfu/100 ml of water could be detected in 6 hours.[16] It has also been used to differentiate species of enterococci.[128] The method employed dyed starch along with the substrate, both of which were added to a medium selective for enterococci. By observing for starch hydrolysis and fluorescence, 86% of enterococci from environmental samples were correctly differentiated. ONPG is a colorithmetric substrate that is specific for coliforms. The substrate is hydrolyzed by β-galactosidase to produce a yellow color that can be quantitated at 420 nm. To determine *E. coli* in water, the organisms are collected on a 0.45-μm membrane and incubated in EC medium for 1 hour followed by the addition of filter-sterilized ONPG. Incubation is continued at 45.5°C until color develops that can be read at 420 nm.[218] The sensitivity of ONPG is similar to that of MUG, with about 10^7 cells required to produce measurable hydrolysis. ONPG is employed in a modification of the classical presence–absence method for coliforms in water.[49] By this modified method, tubes that contain coliforms become yellow. To detect *E. coli*, each yellow tube is viewed with a hand-held fluorescent lamp (366 nm); those that contain *E. coli* fluoresce brightly. The Colilert and ColiQuik systems employ both ONPG and MUG as sole nutrient substrates where total coliforms are indicated by a yellow color; *E. coli* is indicated by MUG fluorescence.

BCIG or X-Gluc is employed in plating media for the detection of *E. coli*. When added at 500 ppm to a peptone–Tergitol agar, *E. coli* produced a blue color in 24 hours that did not diffuse from colonies, and did not require fluorescent light.[64] In another study, no differences were observed between results from a three-tube standard MPN on 50 ground beef samples.[170] When used in lauryl tryptose agar at a final concentration of 100 ppm, only 1% of 1,025 presumptively positive *E. coli* cultures did not produce the blue color, whereas 5% of 583 non-*E. coli* colonies were false positive.[219] The plating medium was incubated at 35°C for 2 hours and then at 44.5°C for 22–24 hours.

The LAPN substrate is specific for Gram-negative bacteria on the premise that aminopeptidase is restricted to this group. The enzyme cleaves L-alanine-*p*-nitroanilide to yield *p*-nitroaniline, a yellow compound that is read spectrophotometrically at 390 nm.[29] When used to determine Gram-negative bacteria in meats, 10^4–5×10^5 cfu was the minimum detectable number.[41] Numbers of 10^6–10^7 cfu/cm^2 could be detected in 3 hours. The *Limulus* test has received more study for Gram-negative bacteria, and because it gives results within 1 hour, the LAPN method cannot be considered to be comparable.

The combination of MUP and ONPG has been used in HEPES buffer as a test for *Clostridium perfringens*, and 164 of 333 presumptive isolates from TSC agar were positive compared to 153 by standard identification methods, and results were obtained in 4 hours.[1] To determine the contamination level of beef carcass swab samples, a luminescence-based phosphatase test kit was employed on 70 beef carcass swab samples, and the results were highly correlated with APC and obtained in 10 minutes.[104]

A chromogenic substrate (BCM; Biosynth, Inc.) available in a *Listeria monocytogenes* plating agar is specific for this species and *L. ivanovii* based on its capacity to respond to phosphatidylinositol-specific phospholipase C (PI-PLC). When using a combination of BCM with either Oxford or Palcam agars, one group of investigators found Oxford-BCM to be 99.3; Palcam-BCM 99.2; and Oxford-Palcam 90.2% sensitive when tested on 2,000 food and environmental samples.[98]

Since the botulinal neurotoxins (BoNTs) are metalloproteases that exhibit stringent substrate requirements relative to amino acid sequences, Schmidt and Stafford[180] developed fluorogenic protease assays for types A, B, and F toxins. The synthetic peptide substrates consisted of having the P_1 and P_3' residues substituted with 2,4-dinitrophenyl-lysine (for P_1) and *S*-(*N*-[4-methyl-7-dimethylamino-coumarin-3-yl]-carboxamidomethyl)-cystine (for P_3'). When the BoNTs were added to these synthetic substrates, fluorescence increased over time and results were obtained in 1 or 2 minutes with BoNT concentrations of 60 ng/ml. The three BoNTs cleaved the substrates at the same locations, and the substrates were selective.

IMMUNOLOGICAL METHODS

Serotyping

Serotyping is most widely applied to Gram-negative enteric bacterial pathogens such as *Salmonella* and *Escherichia*. Among Gram positives, serotyping is important for the genus *Listeria*. The gist of a typical serotyping scheme is the use of specific antibodies (antiserum) to identify homologous antigens. In the case of many foodborne pathogens, the antigens are particulate, and agglutination methods are employed. For soluble antigens such as toxins, methods such as gel diffusion may be used. The O and H antigens of enteric bacteria are illustrated in Figure 11–2. The serologic classification of salmonellae was begun by Kauffmann in the early 1940s.[106] He defined and numbered the first 20 O groups. This

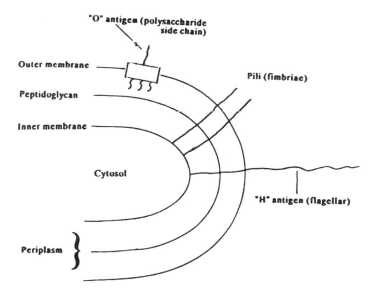

Figure 11–2   Exaggerated section of a Gram-negative bacterial cell showing the relative locations of O and H antigens.

typing scheme results in the recognition of three antigenic sites—somatic (O, Ger. *ohne*), capsular (K, Ger. *kapsel*), and flagellar (H, Ger. *hauch*). The O antigens consist of the O polysaccharide side chains that are exposed on the surface (see Figure 11–2). These are heterogeneous structures, and antigenic specificity is determined by the composition and linkage of the O group sugars. Mutations that affect sugars and/or their linkages lead to new O antigens. About 2,400 O serovars have been recognized for salmonellae, and over 200 for *E. coli*. The O antigens are quite stable to heat (can withstand boiling), whereas the K and H antigens are heat labile. Because flagellar proteins are less heterogeneous than the carbohydrate side chains, considerably fewer H antigenic types exist—around 30 for *E. coli*.

Fluorescent Antibody

This technique has had extensive use in both clinical and food microbiology since its development in 1942. An antibody to a given antigen is made fluorescent by coupling it to a fluorescent compound and when the antibody reacts with its antigen, the antigen–antibody complex emits fluorescence and can be detected by the use of a fluorescence microscope. The fluorescent markers used are rhodamine B, fluorescein isocyanate, and fluorescein isothiocyanate with the last being the most widely used. The fluorescent antibody (FA) technique can be carried out by use of either of two basic methods. The direct method employs antigen and specific antibody to which is coupled the fluorescent compound (antigen coated by specific antibody with fluorescent label). With the indirect method, the homologous antibody is not coupled with the fluorescent label, but instead an antibody to this antibody is prepared and coupled (antigen coated by homologous antibody, which is, in turn, coated by antibody to the homologous antibody bearing the fluorescent label). In the indirect method, the labeled compound detects the presence of the homologous antibody; in the direct method, it detects the presence of the antigen. The use of the indirect method eliminates the need to prepare FA for each organism of interest. The FA technique obviates the necessity of pure culture isolations of salmonellae if H antisera are employed. A commonly employed conjugate is polyvalent salmonellae OH globulin labeled with fluorescein isothiocyanate with somatic groups A to Z represented. Because of the cross-reactivity of salmonellae antisera with other closely related organisms (e.g., *Arizona*, *Citrobacter*, *E. coli*), false-positive results are to be expected when naturally contaminated foods are examined. The early history and development of the FA technique for clinical microbiology has been reviewed by Cherry and Moody[31] and for food applications by Ayres[7] and Goepfert and Insalata.[69] The popularity of this method for foodborne pathogens has waned with the advent of molecular and other detection methods.

Enrichment Serology

The use of enrichment serology (ES) is a more rapid method for recovering salmonellae from foods than the conventional culture method (CCM). Originally developed by Sperber and Deibel,[193] it is carried out in four steps: pre-enrichment in a nonselective medium for 18 hours; selective enrichment in selenite–cystine and/or tetrathionate broth for 24 hours; elective enrichment in M broth for either 6–8 hours or 24 hours; and agglutination with polyvalent H antisera at 50°C for 1 hour. Results can be obtained in 50 hours (depending on elective enrichment time used) compared to 96–120 hours by CCM. A modified ES method has been proposed involving a 6-hour pre-enrichment, thus making it possible to obtain results in 32 hours.[200]

Overall, the ES method provides results in 32–50 hours compared to 92–120 hours for CCM, results are comparable to both CCM and FA, and no specialized equipment or training is needed. Possible disadvantages to its use are the need for a minimum of about 10^7 cells per milliliter and its lack of response to nonmotile salmonellae. The latter can be overcome by use of a slide agglutination test from the elective enrichment broth employing polyvalent O antiserum.[193]

The Oxoid *Salmonella* rapid test (OSRT) is a variation of ES. It consists of a culture vessel containing two tubes, each of which contains dehydrated enrichment media in the lower compartments and dehydrated selective media in the upper compartments. The media are hydrated with sterile distilled water, and a special salmonellae elective medium is added to the culture vessel along with a novobiocin disk, followed by 1 ml of pre-enrichment culture of sample. Following incubation at 41°C for 24 hours, media in the upper compartment (selective media) of each tube are examined for color change, indicating the presence of salmonellae. Positive tubes are further tested with the Oxoid *Salmonella* latex test (2 minutes). Final confirmation of salmonellae is made by use of traditional biochemical and serologic tests.

Salmonella 1–2 Test

This method is similar to ES and OSRT. ES relies on antibody reaction with flagellated salmonellae strains. Unlike ES, the 1–2 Test employs the use of a semisolid phase. The test is conducted in a specially designed plastic device that has two chambers, one for selective broth and the other for a nonselective motility medium. In addition to selective ingredients, the latter contains the amino acid L-serine, which is elective for salmonellae. Following inoculation of the selective medium chamber, the device is incubated, during which time motile salmonellae move into the nonselective medium chamber. The latter contains flagellar antibodies, and when the motile organisms enter the antibody area, an immunoband develops, indicating antigen–antibody reaction. Following nonselective enrichment, test results can be obtained in 8–14 hours.[37]

When compared to a culture method on 196 food and feed samples, the 1–2 test detected 34 positive samples and the culture method detected 26 positive samples.[145] With the addition of a tetrathionate brilliant green broth enrichment step for the 1–2 test, 84 of 314 samples were salmonellae positive—3 more than the culture method—and results could be obtained 1 day before the culture method.[145] That this method produces better results with a pre-enrichment step was shown by others on foods that contained large numbers of nonsalmonellae.

Radioimmunoassay

This technique consists of adding a radioactive label to an antigen, allowing the labeled antigen to react with its specific antibody, and measuring the amount of antigen that combined with the antibody by the use of a counter to measure radioactivity. Solid-phase radioimmunoassay (RIA) refers to methods that employ solid materials or surfaces onto which a monolayer of antibody molecules binds electrostatically. The solid materials used include polypropylene, polystyrene, and bromacetylcellulose. The ability of antibody-coated polymers to bind specifically with radioactive tracer antigens is essential to the basic principle of solid-phase RIA. When the free-labeled antigen is washed out, the radioactivity measurements are quantitative. The label used by many workers is ^{125}I.

Johnson et al.[100] developed a solid-phase RIA procedure for the determination of *S. aureus* enterotoxin B and found the procedure to be 5–20 times more sensitive than the immunodiffusion technique. Employing polystyrene and counting radioactivity with an integral counter, these investigators found the sensitivity of the test to be in the 1–5-ng range. Collins et al.[34] employed RIA for enterotoxin B with the concentrated antibody coupled to bromacetylcellulose. Their findings indicated the procedure to be 100-fold more sensitive than immunodiffusion and to be reliable at an enterotoxin level of 0.01 μg/ml. Staphylococcal enterotoxin A was extracted from a variety of foods, including ham, milk products, and crabmeat, by Collins et al.[33] and measured by RIA, all within 3–4 hours. They agreed with earlier workers that the method was highly sensitive and useful up to 0.001 μg/ml and quantitatively reliable up to 0.01 μg/ml of enterotoxin A.

By iodination of enterotoxins, solid-phase RIA can be used to detect as little as 1 ng of toxin per gram.[17] When protein A was used as immunoabsorbent to separate the antigen–antibody complex from unreacted toxin, a sensitivity of <1.0 ng/g for staphylococcal enterotoxin A (SEA), SEB, SEC, SED, and SEE was achieved within 1 working day.[17,142] In another study, 0.1 ng/ml of SEA and 0.5 ng/ml of SEA and 0.5 ng/ml of SEB could be detected when protein A was used.[3] A sensitivity of 100 pg for SEC$_2$ was achieved by use of a double-antibody RIA.[171]

The RIA technique lends itself to the examination of foods for other biological hazards such as endotoxins, paralytic shellfish toxins, and the like. The detection and identification of bacterial cells within 8–10 minutes have been achieved[198] by use of ^{125}I-labeled homologous antibody filtered and washed on a Millipore membrane. Because of its requirement for an isotope and its lack of portability, the RIA method is rarely used now in the context of food microbiology.

ELISA

The enzyme-linked immunosorbent assay (ELISA, enzyme immunoassay, or EIA) is an immunological method similar to RIA but employing an enzyme coupled to either an antigen or an antibody rather than a radioactive isotope. Essentially synonymous with ELISA are the enzyme-multiplied immunoassay technique (EMIT) and the indirect enzyme-linked antibody technique (ELAT). A typical ELISA is performed with a solid-phase (polystyrene) coated with antigen and incubated with antiserum. Following incubation and washing, an enzyme-labeled preparation of anti-immunoglobulin is added. After gentle washing, the enzyme remaining in the tube or microtiter well is assayed to determine the amount of specific antibodies in the initial serum. A commonly used enzyme is horseradish peroxidase and its presence is measured by the addition of peroxidase substrate. The amount of enzyme present is ascertained by the colorimetric determination of enzyme substrate. Variations of this basic ELISA consist of a "sandwich" ELISA in which the antigen is required to have at least two binding sites. The antigen reacts first with excess solid-phase antibody, and following incubation and washing, the bound antigen is treated with excess labeled antibody. The "double sandwich" ELISA is a variation of the latter method, and it employs a third antibody.

The ELISA technique is used widely to detect and quantitate organisms and/or their products in foods, and synopses of some of these applications are presented below.

Salmonellae

1. Employing a polyclonal EIA with the immunoglobulin G (IgG) fraction of polyvalent flagellar antibodies and horseradish peroxidase, 92.2% agreement was found with the classic culture method on 142 food samples. False positives by the EIA were 6.4%, and a 95.8% agreement with FA was achieved.[201]
2. A polyclonal EIA was used with polystyrene microtiter plates, a capture antibody technique, and a MUG assay. The sensitivity threshold was 10^7 cells/ml, and results could be obtained in 3 working days.[143]
3. The Salmonella-TEK micro-ELISA method with monoclonal antibodies could detect 1–5 cfu/25 g with results in 31 hours. The sensitivity threshold was 10^4–10^5 cells.[213]
4. Monoclonal antibodies were used in a microtiter plate antibody capture method, and 10 cells/25 g could be detected in 19 hours with no cross-reactions with other organisms.[123]

S. aureus and its enterotoxins

1. A double-antibody EIA was developed for staphylococcal enterotoxin A (SEA) that detected 0.4 ng in 20 hours in wieners, 3.2 ng/ml in 1–3 hours in milk, and 1.6 ng/ml from mayonnaise.[178]
2. With polystyrene balls coated individually with respective antibodies, SEA, SEB, and SEC were detected at 0.1 ng or less per milliliter.[196]
3. A standard ELISA for SEA, SEB, SEC, and SEE was used for ground meat, and the method detected <0.5 μg/100 g.[149]

Molds and mycotoxins

1. Both viable and nonviable molds could be detected with an ELISA method, which produced comparable or better results than the Howard mold count method.[127]
2. For the detection of aflatoxin B_1 (AFB$_1$), an ELISA method employing monoclonal antibodies could detect 0.1 ng/ml,[166] 0.2 ng/ml,[26] and 0.5 ng/ml.[44] A commercially available kit could detect 5 ppb (Environmental Diagnostics); a tube ELISA could detect <10 pg/ml; a polystyrene microtiter plate method could detect 25 pg per assay,[161] and a nylon bead or Terasaki plate method could detect 0.1 ng/ml.[160]

3. A commercially available field kit could detect 5 ppb of AFB_2 and AFG_1; T-2 toxin has been detected at a level as low as 0.05 ng/ml; and ochratoxin A at a level of 25 pg per assay (162).

Botulinal toxins

1. For type A toxin a "double-sandwich" ELISA detected 50–100 mouse LD_{50} of type A and <100 mouse i.p. LD_{50} of type F; and a "double-sandwich" ELISA with alkaline phosphatase and polystyrene plates was shown capable of detecting 1 mouse i.p. median lethal dose of type G toxin.[125]

E. coli enterotoxins

1. A monoclonal antibody specific for enterohemorrhagic strains of *E. coli* (EHEC) was shown to be highly specific when used in an ELISA to detect EHEC strains.[154]
2. Two "sandwich" ELISAs were developed based on toxin-specific murine monoclonal capture antibodies and rabbit polyclonal second antibodies specific for the Stx1 and Stx2 genes of *E. coli*. The Stx1 ELISA could detect 200 pg of purified Stx1 toxin, whereas the Stx2 could detect 75 pg of Stx2 toxin.[47]

Gel Diffusion

Gel diffusion methods have been widely used for the detection and quantitation of bacterial toxins and enterotoxins. The four most widely used methods are the single-diffusion tube (Oudin), microslide double diffusion, micro-Ouchterlony slide, and electroimmunodiffusion. They have been employed to measure enterotoxins of staphylococci and *C. perfringens*; and the toxins of *C. botulinum*. The relative sensitivity of the various methods is presented in Table 11–1. Although they should be usable for any soluble protein of which an antibody can be made, they require that the antigen be in precipitable form. Perhaps the most widely used is the Crowle modification of the Ouchterlony slide test as modified by Casman and Bennett[28] and Bennett and McClure.[15] The micro-Ouchterlony method can detect 0.1–0.01 μg of staphylococcal enterotoxin, which is the same limit for the Oudin test. The double-diffusion tube test can detect levels as low as 0.1 μg/ml, but the incubation period required for such low levels is 3–6 days. This immunodiffusion method requires that extracts from a 100-g sample be concentrated to 0.2 ml. Although other methods such as RIA and RPH (reverse passive hemagglutination) are more sensitive and rapid than the gel diffusion methods, the latter has been more widely used. Their reliability within their range of sensitivity is unquestioned. Recent studies suggest that results can be obtained in <8 hours when slides are incubated at 45°C.

Immunomagnetic Separation

This method employs paramagnetic beads (about 2–3 μm in size, about 10^6–10^8/ml) that are surface activated and can be coated with antibody by incubating in the refrigerator for varying periods of time up to 24 hours. The unabsorbed antibody is removed by washing. When properly treated, the coated beads are added to a food slurry that contains the homologous antigen (toxin or whole cells in the case of Gram-negative bacteria), thoroughly mixed, and allowed to incubate from a few minutes to several hours to allow for reaction of antigen with antibody-coated beads. The latter complex is collected by a magnet followed by elution of antigen or measurement on beads. The concentrated antigen is assayed by other methods. In one study, immunomagnetic separation was combined with flow cytometry for the detection of *E. coli* 0157:H7. The antigens were labeled with fluorescent antibody, which was measured by flow cytometry, and the combined method could detect $<10^3$ cfu/g of pure culture or 10^3–10^4 cfu/g in ground beef.[186] This method may be used for a number of other organisms including viruses and protozoa.

Hemagglutination

Whereas gel diffusion methods generally require at least 24 hours for results, two comparable serologic methods yield results in 2–4 hours: hemagglutination–inhibition (HI) and reverse passive hemagglutination (RPH). Unlike the gel diffusion methods, antigens are not required to be in precipitable form for these two tests.

In the HI test, specific antibody is kept constant and enterotoxin (antigen) is diluted out. Following incubation for about 20 minutes, treated sheep red blood cells (SRBCs) are added. Hemagglutination (HA) occurs only when antibody is not bound by antigen. HA is prevented (inhibited) where toxin is present in optimal proportions with antibody. The sensitivity of HI in detecting enterotoxins is noted in Table 11–1.

In contrast to HI, antitoxin globulin in RPH is attached directly to SRBCs and used to detect toxin. When diluted toxin preparations are added, the test is read for HA after incubation for 2 hours. HA occurs only where optimal antigen antibody levels occur. No HA occurs if no toxin or enterotoxin is present. The levels of two enterotoxins detected by RPH are indicated in Table 11–1.

MOLECULAR GENETIC METHODS

Although phenotypic methods for the identification and characterization of foodborne bacteria continue to be widely used, the trend is clearly in the direction of making more use of genotypic methods, especially for foodborne pathogens. This trend is aided by the availability of test kits produced by a number of companies, and some are listed below under Polymerase Chain Reaction. The importance of 16S rRNA in classifying bacteria along genotypic lines is noted in Chapter 2, and a glimpse of how this is achieved is presented below.

As noted in Chapter 2, the 70S bacterial ribosome contains three definable rRNA fractions: 5S, 16S, and 23S. The 5S contains ca. 120 nucleotides while 16S and 23S contain ca. 1,500 and ca. 3,000, respectively. The 16S rRNA has been shown to be an excellent chronometer of life forms, especially bacteria. It was by the use of 16S rRNA sequence and hybridization data that the class *Proteobacteria* was established, as noted in Chapter 2. By use of these sequences, new bacterial genera can be defined on a genotypic rather than a phenotypic basis. As a result of this and other molecular genetic information, the genus *Pseudomonas* has been reduced by the transfer of over 50 species to 10 new genera (see Chapter 2). The existence of a number of gene banks for many other bacterial taxa of importance in foods suggests that generic and species realignments will continue.

The extraction and amplification of 16S rRNA from bacterial cells is relatively simple. With ca. 1 g of wet cells, they are broken open in the presence of DNase (to destroy all DNA), and RNA may be extracted by use of commercial kits. With the addition of nucleotide primers, reverse transcriptase (RT), and deoxyribonucleotides, RT reads the rRNA template and makes DNA (cDNA) copies. The original 16S rRNA sequence can be deduced after a series of manipulations.

Nucleic Acid (DNA) Probes

A DNA probe consists of the DNA sequence from an organism of interest that can be used to detect homologous DNA or RNA sequences. In effect, the probe DNA must hybridize with that of the strain to be sought. Ideally, the probe contains sequences that code for a specific product. The probe DNA must be labeled in some way in order to be able to assess whether hybridization has occurred. Radioisotopes are the most widely used labels, and they include ^{32}P, ^{3}H, ^{125}I, and ^{14}C, with ^{32}P being the most widely used. Reporter groups such as alkaline phosphatase, peroxidase, fluorescein, haptens, digoxigenin, and biotin have been developed and used.[46] When biotin is used, its presence is detected with avidin-enzyme conjugates, antibiotin, or photobiotin. Chromosomal DNA is often the source of target nucleic acid, but it contains only one copy per cell in most cases. Multiple targets are provided by mRNA, rRNA, and plasmid DNA, thus making for a more sensitive detection system. Synthetic oligonucleotide probes may be constructed of 20–50 bases, and under proper conditions, hybridization times of 30–60 minutes are possible.

In a typical probe application, DNA fragments of unknown organisms are prepared by the use of restriction endonucleases. After separating fragment strands by electrophoresis, they are transferred to cellulose nitrate filters and hybridized to the labeled probe. After gentle washing to remove unreacted probe DNA, the presence of hybridized product is assessed by autoradiography.

The minimum number of bacterial cells that can be detected with a standard probe is in the 10^6–10^7 range, although some investigators report the detection of 10^4 cells. When probes are used on foods where perhaps only 1 cfu/ml of the target organism exists, enrichment procedures must be employed to allow this cell to attain a level that will provide enough DNA for detection. With an initial cell number of 10^8 cells, probe results may be obtained in around 10–12 hours when radiolabels are employed. When enrichments are necessary, the time required for results would be enrichment time plus probe assay time, generally 44 hours or more. Some examples of probe results for foodborne organisms are summarized below. For a review, see reference 202.

1. For enterotoxigenic *C. perfringens* in raw beef, a digoxigenin-labeled probe was used to detect ≤10 cfu/g in 48 hours.[10]
2. A colorimetric DNA probe was compared to the Food and Drug Administration (FDA) culture method for *L. monocytogenes* in dairy, meat, and seafood samples.[21] Of 660 dairy and seafood samples examined, 354 were found positive by the FDA and 393 were found positive by the probe. Of 540 meat samples examined, 261 were found positive by the FDA and 378 by the probe. Probe results could be obtained in 48 hours, whereas 3–4 days were required for the FDA/USDA method.
3. For *Salmonella* spp., a colorimetric probe correctly identified all of 110 serovars and gave no false positives from 61 nonsalmonellae.[38]
4. Radiolabeled probe for salmonellae was tested on 269 poultry carcass and water samples, and as few as 0.03 cell/ml could be detected when two enrichments were used.[86] The probe could detect as few as ~10^4 cells per milliliter and it is AOAC approved.

A probe for *S. aureus* enterotoxins was constructed from DNA encoding amino acids 207–219 of SEB and SEC reacted with the genes for SEB and 3 SEC enterotoxins.[150]

DNA probes are used in colony hybridization methods where microcolonies of macrocolonies of the target organism are allowed to develop directly on a membrane following incubation on a suitable agar medium. A replica plate is produced as a duplicate of the master plate or membrane. Colonies that have grown on the duplicate plate are lysed directly on the membrane to release nucleic acid and to convert DNA into single strands. Some of the DNA is transferred to nitrocellulose filters, where hybridization is carried out by applying a labeled DNA or RNA probe. A modification of the traditional DNA colony hybridization technique has been made such that 60 filters with up to 48 organisms per filter can be used.[107]

The colony hybridization method developed by Grunstein and Hogness[71]) has been employed successfully to detect *Listeria monocytogenes*, enterotoxigenic *E. coli*, and *Yersinia enterocolitica*. In one study, synthetic polynucleotide probes were constructed that were homologous to a region of the ST enterotoxin gene of *E. coli* and applied for the detection of strains produced by DNA colony hybridization.[80] For the latter, colonies were placed on paper filters to free and denature cellular DNA, hybridized overnight at 40°C, and exposed to autoradiograms. By this procedure, as few as 10^5 ST-producing cells could be detected. In an earlier study from the same laboratory, colony hybridization was used to detect *E. coli* in artificially contaminated food without enrichment, and the method could detect 100–1,000 cells per gram, or about 1–10 cells per filter.[79] More information on nucleic acid probes can be obtained from the review by Wolcott.[225]

Polymerase Chain Reaction

This method is fast becoming the most widely used of all molecular genetic methods for detecting and identifying bacteria and viruses in foods. Its increasing use is due to its high sensitivity, specificity, its availability in a number of formats, and the commercial availability of PCR-based methods in kit-like formats.

This technique, first outlined in 1971 by Kleppe et al.[112] is applicable more to the identification of foodborne organisms than to their enumeration. The currently used methodology is that further developed by scientists at the Perkin Elmer-Cetus Corp.[177,197] among others. For his efforts in the development of PCR, K.B. Mullis was co-winner of the Nobel Prize in chemistry in 1993.

A general outline of a PCR test is as follows. When the starting genomic material is dsDNA, it is heated to ca. 95°C to separate the strands. When the starting material is RNA (e.g. RNA viruses), it is converted to dsDNA by use of reverse transcriptase (RT-PCR). After heating to separate the DNA strands, they are cooled to ca. 55°C in the presence of oligonucleotide primers, which anneal to the single DNA strands. DNA polymerase plus dATP, dCTP, dTTP, and dGTP are added resulting in the synthesis of complementary strands (the "d" is deoxynucleotide; A = adenine; TP = triphosphate, etc. for the other bases. These are precursors of DNA synthesis). When this process is repeated, the two strands become four, the four become eight, and so on for each additional cycle, resulting in several million copies of the original if enough cycles are run. Among commercially available kits are the following:

BAX system (Qualicon, Dupont Corp.)
Probelia (Sanofi Diagnostics Pasteur)
Foodproof (Biotecon Diagnostics)

AG-9600 Amplisensor Analyzer is an automated fluorescence-based system for detecting PCR products. BAX is the oldest of these detection systems and its application, along with that of other PCR-based methods, is among the synopses presented below.

1. Multiplex PCR
   a. For *Escherichia coli* 0157:H7, the primers used were: hly_{933}k, $fliC_{h7}$, *stx* 1, *stx* 2, *eaeA*. Could detect ≤1 cfu/g with results obtained in 24 hours. Food and bovine fecal specimen used.[65]

b. For *Staphylococcus aureus*, the primers used were: *entC* (SEC gene) and *huc* (TNAse gene). Amplification by RFLP could detect 10 cfu/ml in skim milk and 20 cfu/20 g in Cheddar cheese with results in <6 hours.[205]

c. For *Listeria*, used primer pairs within the 16S or rRNA that are specific for *Listeria* and *L. monocytogenes* to simultaneously detect each. Could detect 1–5 cfu/25 g of food. Only live cells detected.[192]

d. *Salmonella* and *Campylobacter* capture probes used: *invA* (*Salmonella*) and *ceuE* (*Campylobacter*). ELISA used to detect the PCR products. Could detect 2×10^2 cfu/ml of salmonellae, and 4×10^1 *Campylobacter*.[82]

2. RT-PCR

a. For *Cryptosporidium parvum*, dsRNA extracted and hybridized to the Xtra Bind Capture System and amplified while on the Xtra Bind material. Could detect as few as 1 oocyte/l in ca. 2 hours.[116]

b. Norwalk virus detected by replacing RT and *Taq* polymerase with *rTth* polymerase in single tube. The generated amplicons were detected using biotinylated oligoprobes in an ELISA-based format. Results from oysters and clams obtained in 1 day.[181]

c. Polio and hepatitis A viruses added to oysters at 10^1–10^5 pfu, concentrated with polyethylene glycol. Could detect 10 pfu by RT-PCR.[96]

d. One commercially available system (Genevision™, Warnex Diagnostics, Laval, Canada) employs RT-PCR to identify *Salmonella* spp. in 1 day, and *Listeria* spp. in about 2 days following enrichments.

3. Probelia system

a. For *Salmonella* spp. in foods, method had threshold sensitivity of 10^2 cfu/ml. As few as 3 cfu/25 g could be detected after 18 hours pre-enrichment. With 285 contaminated food samples, a 99.6% agreement was found in a collaborative study. Findings agreed with those by ISO 6579 method.[53]

b. *Listeria monocytogenes*, when compared to ISO culture method 11290 for recovery from salmon, the two methods produced comparable results but Probelia detected 20 cfu/ml in 48–50 hours compared to 5+ days for ISO method.[216]

4. BAX for screening *E. coli*

a. When compared to other methods for detecting levels of <3 cfu/g in ground beef, BAX detected 96.5% of positives compared to 71.5% by immunoassay methods, and 39% for the best culture method.[101]

5. Molecular beacon PCR

a. For *Escherichia coli* 0157:H7, a fluorescently labeled oligonucleotide probe designed to hybridize with a region of the *slt*-II gene such that it fluoresces when the hairpin-stem conformation is linear to the target sequence. In PCR reactions containing DNA from contaminated skim milk, fluorescence increased with increasing numbers of organisms, and method produced results without need for electrophoresis or Southern blotting.[138]

6. PCR-DGGE

a. For *Listeria* spp., used PCR amplification of a fragment of the *iap* gene from 5 *Listeria* spp. PCR products analyzed by denaturing gradient gel electrophoresis (DGGE). Species specific DGGE migrations allowed identifications of all listerial species. Can be used to detect the listeriae and *L. monocytogenes* in foods.[35]

A number of other PCR-based methods have been developed and tested, and this is an area of active pursuit with those below being among some that have been used for foodborne organisms.

LAMP—loop-mediated isothermal amplification

Rep-PCR—surveys repetitive sequences only

ERIC—enterobacterial repetitive intergenic concensus

Molecular Beacon—uses fluorescently labeled oligonucleotides in a qPCR format

QC-PCR—quantitative competitive

VNTR—variable number of tandem repeats (minisatellites)

Real-time PCR (qPCR, RTi-PCR)—PCR results in real time. *TaqMan* is a fluorogenic 5′ nuclease assay that produces results in real time.

A few applications of some of the above are summarized below.

FISH (fluorescent in situ hybridization) is an RNA probe method that targets 23S rRNA. To detect *Salmonella* spp. in foods, two established oligonucleotide probes (Sal-1 and Sal-3) and a newly constructed one were used. The probes were labeled with Cy3 (indocarboxyanine) and applied to cultures of whole cells on slides.[55] Fluorescence occurred when the probes hybridized with the 23S rRNA of target organisms (16S RNA sequences are more often used for bacteria). After slides were prepared, they were counter-stained with DAPI ([4′, 6′-diamidino-2-phenylindole; a non-intercalating, DNA-specific stain that fluoresces blue or bluish-white when bound to DNA and excited with light of wavelength 365 nm),[109] and fluorescence was detected by epifluorescence microscopy. When FISH was compared to a culture method for detecting *Salmonella* spp. in 18 different foods, up to 56 were found positive by at least one of the three FISH probes but only 30 of the 225 samples were found positive by the culture method.[55] Overall, FISH detected salmonellae in 64% of the naturally contaminated samples while culture methods detected in 13%. Of the 52 tested *Salmonella* serovars, FISH hybridized with all 52. FISH was negative when tested on 46 non-salmonellae strains of 22 Enterobacteriaceae species, and on 14 species of 12 non-Enterobacteriaceae. Living, dead, and VBNC (viable but nonculturable cells) could be detected. Among viable cells, those exposed to unfavorable environmental conditions were detectable even though they may have been nonculturable. In this study, investigators could detect salmonellae in food specimens 2–3 days sooner than by culturing. Although it is unclear as to the minimum number of cells needed for a positive FISH response, this number is probably $>10^4$ cfu/ml since it is a microscopic method.

LAMP is an in situ DNA amplification technique for the microscopic detection of bacteria. It is similar to FISH in that they are microscope slide methods that emit fluorescence from Cy3- or DAPI-stained cells. LAMP employs a low molecular weight DNA polymerase that enters whole cells; a reaction temperature of 63°C that makes it possible to use fluorescent antibody (FA) for simultaneous cell detection; and generates large tandem repeats that prevent amplicons from leaking out of cells.[134] After DNA amplification using primers for the *E. coli* 0157:H7 *stx* 2 gene, the identification of these cells was possible with FA labeling.

qPCR is a method whereby amplified gene products are detected by using fluorescence probes during PCR cycling, and results can be obtained in 30–90 minutes. A number of fluorescent probes are used such as SYBR Green I and fluorescence resonance energy transfer (FRET). These probes allow the monitoring of the complete process, and they obviate the need for electrophoresis of end products. FRET is a short oligonucleotide that is complementary to one DNA strand.[215] When this method was employed to detect *Giardia lamblia* (the β-giardin gene was targeted) and *Cryptosporidium parvum* (COWP gene was targeted), one cyst of the former and 100 cysts of the latter could be detected in water and sewage samples.[74]

Using SYBR Green I, melting curve analyses were made at the end of PCR cycles to identify the PCR products, which have specific melting temperatures. The method was used to simultaneously detect *Salmonella* and *L. monocytogenes* employing specific genes for each group. After an overnight enrichment, the method could detect 2.5 cells of *Salmonella* serovars and 1 *L. monocytogenes* cell.[103] Twenty-nine salmonellae and 18 *L. monocytogenes* strains were employed. qPCR has been also used to simultaneously detect *stx*1 and *stx*2 genes of *E. coli*.[99] A qPCR in which the SYBR Green I fluorescent dye was used was developed for the identification of *Vibrio vulnificus* in oyster tissue homogenates and Gulf waters.[155] After a 5-hour enrichment, the method detected one cell. Without enrichment, 10^2 cells could be detected in 1 g of oyster homogenate or 10 ml of Gulf water. The method targeted the hemolysin specific gene, *vvh*. The entire assay could be completed within 8 hours.

Lux Gene Luminescence

Luminescence in marine bacteria such as *Vibrio fischeri* and *V. harveyi* is controlled by genes, and the capacity to produce luminescence can be transferred to other organisms by effecting the transfer of some of these genes. The primary genes (designated *lux*) for luciferase are *lux* A and *lux* B. The former encodes the synthesis of the luciferase α-subunit and the latter the β-subunit. The other eight genes in the bioluminescence operon of the organisms noted do not need to be transferred. In the food microbiology application of *lux* phages, one starts with bacteriophages that are specific for the bacterium of interest and thus takes advantage of the highly specific relationship that exists between phages and their hosts (see the section on Bacteriophage Typing later in this chapter and the section on bacteriophages in Chapter 20). If *Y. enterocolitica* is the bacterium of interest, one selects a phage that will infect the widest range of strains and yet not infect closely related species. To this phage, the *lux* genes are inserted by recombination methods, which amounts to about 2 kb of DNA. By themselves, these transduced phages are not luminous because they lack all components necessary to produce light. When added to their specific host bacteria, the *lux* gene-bearing phages enter and multiply, and thus cause the host cells to luminesce by the increased production of more *lux* genes. The light-emitting reaction requires the components in the following equation:

$$FMNH_2 + RCHO + O_2 \xrightarrow{\text{luciferase}} FMN + RCOOH + H_2O + light$$

where $FMNH_2$ is reduced flavin mononucleotide and RCHO is a long-chain aliphatic aldehyde such as dodecanal. The emitted light can be measured by luminometry as in the ATP assay. Time for results depends on the time required for the phage to enter host cells and begin their multiplication phase; this is typically 30–50 minutes.

The addition of *lux* genes to a phage genome was first described by Ulitzur and Kuhn[212] who showed that as few as ten *E. coli* cells could be detected within 10 minutes. The on-line method for the enteric bacteria in swabs from a meat-processing plant could detect 10^4 cfu/g of cm^2.[114] A number of studies have shown that around 100 salmonellae can be detected in about 1 hour. As few as one *S.* Typhimurium cell/100 ml of water could be detected within 24 hours in one study using an MPN method.[211] The *lux* gene methodology can be adapted to the detection of a wide range of bacteria in foods by the direct addition of phage constructs. Where the initial numbers are low, enrichments are necessary. The method does not lend itself well to Gram-positive bacteria, as light emission is typically 100-fold less than that for Gram negatives.[195]

A broad host-range reporter bacteriophage for *Listeria monocytogenes* has been constructed that caries the *Vibrio harveyi* LuxAB protein.[130] After a 2-hour incubation, as few as 5×10^2 to 10^3 cells/ml could be detected with a single-tube luminometer following an enrichment step. Less than one cell of *L. monocytogenes*/g of artificially contaminated salad could be identified.[130] In meat and soft cheese, as few as 10 cells/g could be detected. Of 348 natural food and environmental samples tested, 55 were found positive by the *lux*-phage method compared to 57 by a plating method.[129] The *lux*-phage method

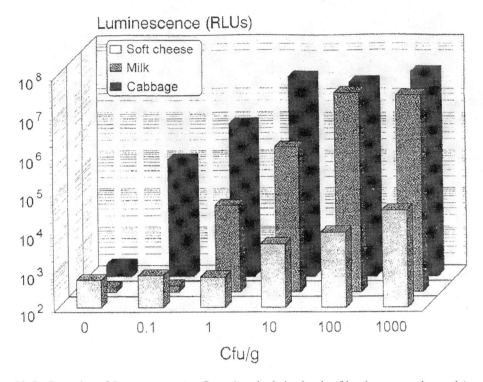

Figure 11–3   Detection of *L. monocytogenes* Scott A and relative levels of luminescence observed (*y* axis) in different artificially contaminated foods, spiked with various levels of cells (*x* axis). Samples were taken after 3 days of storage at 4°C, enriched for 44 hours in selective culture, and subsequently assayed by A511::*lux*AB. Detection limits in this experiment were 0.1 (cabbage), 1 (milk), and 10 (Camembert soft cheese) cfu/g,[129] Copyright © 1997, American Society for Microbiology. Used with permission.

produced results in 24 hours compared to 4 days for the plating method. The relative effectiveness of this system to detect *L. monocytogenes* strain Scott A in three different foods is illustrated in Figure 11–3.[129]

Along the lines of the *lux*-phage assay noted above a phage that is specific for *E. coli* 0157:H7 was labeled with the green fluorescent protein (GFP), and it has been used to detect the bacterium not only in its normal viable state but also heat-killed and viable but nonculturable cells.[152] After a 1-hour incubation of the phage with the bacterium at a multiplicity of infection of 1,000, culture fluorescence increased (because of replication of the GFP phages) and it plateaued at 3 hours of incubation. Fluorescence was measured under fluorescence microscopy. Although these phage constructs attached to non-dividing cells, fluorescence intensity did not increase after incubation.[152]

Ice Nucleation Assay

This technique is quite similar to *lux* gene luminescence above in that a specific gene is carried from one bacterium to another via a bacteriophage. A number of genera of Gram-negative, plant-inhabiting bacteria carry a gene (*ina*) that encodes the synthesis of a protein that acts as an ice nucleator. One of the most common is *Pseudomonas syringae* whose *ina* gene consists of about 3,600 base pairs (bp) of DNA that will yield a single ice nucleation *ina* protein. These proteins facilitate the freezing of water at warmer temperatures than would otherwise be the case, and they lead to frost damage to many field plants since they lead to supercooling at temperatures of $-6°C$ or lower before nucleation becomes active. This is an example of heterogeneous ice nucleation where supercooled water is bound to a non-water material, and it can occur at temperatures as low as $-2°C$.[224] The application of these ice nucleators to food products such as egg white, salmon muscle, and others can lead to a reduction in freezing time and energy savings.[126]

The bacterial ice nucleation diagnostic (BIND) test, developed by scientists at the DNA Plant Technology Corporation, was developed for the detection of salmonellae. In a nutshell, the *ina* gene from *P. syringae* is cloned into genetically engineered bacteriophages that are specific for salmonellae. If salmonellae are present, the phages infect and lead to the synthesis of the ice nucleation protein as part of the outer cell membrane. This is evidenced by the formation of ice crystals at a temperature around $-9°C$. By coupling a fluorescent freeze indicator dye, a green color indicates freezing and thus the presence of salmonellae while an orange color indicates no freezing. With salmonellae phage P22, as few as 25 cells per gram can be detected within 24 hours.

FINGERPRINTING METHODS

A number of methods are in use for the purpose of characterizing (fingerprinting, differentiating, typing) species and strains of organisms of interest in foods, including some of the molecular genetic methods above, and most are listed below and briefly described.

Phage typing
Amplified-fragment length polymorphism (AFLP)
Multilocus enzyme electrophoresis (MEE)
Restriction enzyme analysis (REA)
Random amplification of polymorphic DNA (RAPD)
Pulsed field gel electrophoresis (PFGE)
Restriction fragment length polymorphism (RFLP)
Ribotyping
Microarrays

Bacteriophage Typing

Phage typing is based on the specificity of a given phage for its host bacterium, and this relationship allows one to use known phages to identify their specific hosts. All foodborne pathogens can be phage typed, but the practice is applied more to some than others. More on phage typing as it relates to specific foodborne pathogens may be found in the respective chapters.

One of the earliest and perhaps most elaborate of phage typing schemes was that developed for *S. aureus* in the 1950s. Although the routine use of staphylococcal phage typing has waned, it has emerged as an important tool in studying the epidemiology of *L. monocytogenes*.

Since they were first described in 1945, bacteriophages specific for *Listeria* have been studied by a number of investigators relative to their uses for species and strain differentiation, and their epidemiologic value. *Listeria* phages contain dsDNA and belong to two groups: Siphoviridae (non-contractile tails) and Myoviridae (contractile tails). The phage receptors on *Listeria monocytogenes* cells are N-acetylglucosamine and rhamnose substituents of teichoic acids, or the peptidoglycan itself.[58,221] A phage-resistant strain of *L. monocytogenes* lacked glucosamine in its wall structure.[221] In a study of 823 strains of *L. monocytogenes* collected in France over the period 1958–1978, 69.4% were serotype 4, and a phage typing system was defined using 12 principal and 3 secondary phages.[5] Six phages could be used to differentiate serotype 1, nine phages for serotype 4, and only two phages for serotype 2 strains. By employing a set of 20 phages, these investigators were able to type 78.4% of the 823 strains, with 88% of serotype 4 and 57% of serotype 1 being typable. For 552 of the 645 typable strains, 8 principal phage patterns could be established.[5] When a set of 29 phages was employed in a multicenter study, 77% and 54% of serotypes 4 and 1/2, respectively, were typable.[173] The typing set of Audurier and co-workers is divided into three groups: 12 phages for 1/2 strains, 16 for 4b strains, and 7 for other strains.[4] Typability of 826 serovar 4b strains isolated in France in the period 1985–1987 was 84% compared with 49% of 1,644 serovar 1 strains employing the 35 phages. By use of this scheme, *L. monocytogenes* isolates involved in three outbreaks of human listeriosis were shown to be of the same phage type, whether recovered from victims or foods.

The phage typing of 80 cultures of *L. monocytogenes* was compared by six laboratories in Europe using an international phage set in five laboratories and phage sets unique to two laboratories. A high level of agreement among the laboratories was found, and suggestions for the possible improvement of the international set were presented.[110,139]

In a study that employed 127 isolates of *L. monocytogenes*,[203] lytic patterns allowed eight phage groups to be established, but findings suggested that the lytic agents were *monocins*—defective phages with tails that lack head region.[153] Although monocin susceptibility appeared to be associated with serotypes, no relationship was found relative to animal source or geographic origin of *L. monocytogenes* strains.[203] Monocins of the listeriae are cryptic prophages that are closely related to intact phages, suggesting that they are incompletely assembled phage particles.[231] Monocin proteins are highly similar to the major protein of phage tails, and 75% of listerial phage DNA hybridize to DNA of monocin-producing strains.[231] The tails contain a lytic principle on their base plate region, which can lead to cell lysis if/when sufficient numbers of tails attach to cells. *L. monocytogenes* monocins do not attach to other bacteria.[231]

In a study of 807 *L. monocytogenes* cultures collected in Britain from human cases over the period 1967–1984, phage typing was shown to be an effective tool for common source cases of listeriosis involving more than one patient or for recurrent episodes in the same patient.[140] The 807 cultures belonged to serotypes 1/2, 3, and 4. In another study that employed a set of 16 phages recovered from lysogenic and environmental sources, 464 strains representing five species were placed in four groups.[132] Although the results were highly reproducible, species and serovar specificities did not conform to any lytic patterns. The phage susceptibility of *L. monocytogenes* was highest for serotype 4 (98%) followed by serotype 1 (90%) and serotype 3 (10%). No phages were restricted to either one species or serovar in lytic patterns. *L. grayi* was not lysed by any of the selected phages.[132] A reversed typing procedure that employs ready-to-use plates containing phage suspensions on tryptose agar plates was developed by Loessner.[131] With a set of 21 genus-specific phages, the overall typability rate was 89.5% on 1,087 listerial strains.

The phage typing of 105 strains of toxigenic *E. coli* 0157:H7 recovered from humans in Finland between 1990 and 1999 revealed that 56% belonged to PT2 and 11% to PT54.[176] Seventy percent of the PT2 strains contained the *stx* 2 gene alone and in combinations. In a study of 166 strains of *Bacillus cereus* from food-poisoning cases, 97% were phage typable using a set of 12 phages.[2] Interestingly, most *B. thuringiensis* strains were also typable by the same set of phages.

Amplified Fragment Length Polymorphism

This is a PCR-based fingerprinting technique that requires small amounts of pure dsDNA—10–100 ng from 1–3 bacterial colonies.[89] The DNA is digested with two restriction enzymes (such as *Eco*RI and *Mse*I). Double-stranded oligonucleotide adaptors are applied, and an aliquot is subjected to 1- to 2-PCR amplifications under highly stringent conditions with adapter-specific primers. The primer that spans the average frequency restriction site is fluorescently labeled. After subjecting to PAGE (polyacrylamide gel electrophoresis), from 40 to 200 bands are obtained, and it has been found to be more reproducible than RAPD.[19]

In a study of 147 strains of bacteria from nine species, AFLP analysis clustered all strains within their respective species.[89] The method was shown to produce results consistent with DNA homology data when the genomic DNA of 98 *Aeromonas* strains were digested with *Apa*I and *Taq*I.[85] Using an automated method with two restriction endonucleases, 25 *Campylobacter* isolates from poultry farms in the Netherlands grouped in three *C. jejuni* clusters that were different from a *C. coli* cluster.[48] A fluorescent-AFLP (FAFLP) employs fluorophore-labeled primers, and it has been employed to discriminate between 30 phage type 4 (PT 4) strains of *S.* Enteritidis from different sources. When AFLP was used with *Xba*I, 73% of the PT 4 strains were classified as a single type.[43] However, when FAFLP was employed using *Eco*RI + O and *Mse*I + C, 23 profiles (with 1–61 amplified-fragment differences) were obtained with a discriminatory power of 0.98 compared to 0.47 for PFGE.[43] Genetic diversity of clinical and environmental isolates of *Vibrio cholerae* was investigated by using *Apa*I and *Taq*I, and 20–30 bands each were generated from the genomic DNA of *V. cholerae* serotypes 01 and 0139 strains.[97] Of the 79 strains tested, 26 of 01 showed identical patterns and were represented by the 01 El Tor strain of the seventh pandemic. For more information on this technique, see reference 179.

Multilocus Enzyme Electrophoresis Typing

This technique may be employed to estimate the overall genomic relationships among strains of organisms within species by determining the relative electrophoretic mobilities of a set of water-soluble cellular enzymes. The variation in electrophoretic mobility can be related to allelic variation and to levels of genetic variation within populations of a species. Typically, 15–25 enzymes are tested for on starch gels. Because some of the enzymes may have different mobilities (be polymorphic) among strains of the same species, multilocus enzyme electrophoresis (MEE) typing can be used to characterize strains for epidemiologic purposes in the same general way as serotyping or phage type. The basic technique has been described in detail and reviewed.[183] Some applications are summarized as follows:

1. After examining 175 isolates of *L. monocytogenes* for allelic variation among 16 enzymes, 45 allelic profiles or MEE types were distinguished. The above could be divided into two primary phylogenetic divisions with all 4a, 4b, and 1/2b serovars in the same division.
2. A study of 245 strains of *L. monocytogenes* from a variety of sources in Denmark revealed 33 MEE types with 73% of strains belonging to one of two MEE types.[148] One MEE type was found most frequently among food isolates. In a related study, 47 clinical isolates and 72 fish isolates were found by MEE not to constitute two distinct lineages but that they belonged to two separate populations.[20]
3. When compared to restriction fragment length polymorphism (RFLP; see later in this chapter) on 141 strains of *L. monocytogenes*, the two methods were in substantial agreement on recurrent strains in certain food products.[77]

Restriction Enzyme Analysis

By this method, chromosomal DNA of test strains is digested by use of an appropriate restriction endonuclease. The latter class of enzymes makes double-stranded breaks in DNA at specific nucleotide sequences. One of the most widely used restriction endonucleases is *Eco*RI (obtained from *Escherichia coli*), which recognizes the DNA base sequence GAATTC and cleaves between GA. Another endonuclease is *Hha*I (obtained from *Haemophilus influenzae*), and it recognizes the sequence GTPyPuAC (Py = any pyrimidine, Pu = any purine base). The cleavage site for *Hha*I is between PyPu, and it has been found to be of value in studying the epidemiology of *L. monocytogenes*.[223]

After some *L. monocytogenes* serovar 4b strains associated with three food-associated outbreaks were subjected to restriction enzyme analysis (REA) using *Hha*I, the method was found to be valuable as both a taxonomic tool and an epidemiologic tracer.[222] Of 32 isolates associated with the 1981 outbreak in Nova Scotia, Canada, 29 showed restriction enzyme patterns identical to the strain recovered from coleslaw. Also, the patterns of nine clinical isolates from the 1983 Boston cases were identical to each other. Some of the isolates from the 1985 California outbreak were subjected to REA, and those examined from patients, suspect cheese samples, and cheese factory environmental samples were found to be identical.[222]

The combined use of REA and PCR for subtyping of *L. monocytogenes* has been presented.[51] Employing 133 strains of serovar 4b from a variety of sources along with 22 other serovars, PCR-REA divided the strains into two groups, I and II, with the former containing 37 and the latter 96 strains. Seventy-four of the 4b serovars belonged to phagovar 2389:2425: 3274:2671:47:108:340, and all fell in the same group, II, when cleaved with the nuclease *Alu*I.

Random Amplification of Polymorphic DNA

In essence, random amplification of polymorphic DNA (RAPD) employs the use of PCR to obtain randomly amplified polymorphic DNA electrophoretic profiles. Briefly, cells are harvested, suspended in water, and lysed for their DNA. The DNA, along with a specific primer such as a 10-mer (10-bp section), is mixed with DNA and *Taq* polymerase. PCR is carried out at varying temperatures for 40 or more cycles, followed by electrophoresis of the products on an agarose gel. Following staining of gel (typically with ethidium bromide), the bands are photographed and analyzed. Purified genomic DNA is not needed for RAPD, nor is there a need for prior sequence data. By using a capillary air thermal cycler, which was able to complete 30 cycles in <1.0 hour, RAPD results could be obtained in 3 hours starting with colony growth.[18]

RAPD analysis has been used to fingerprint outbreak strains of *L. monocytogenes* by a number of investigators. When 289 strains from a poultry-processing environment were subjected to RAPD using a 10-mer primer, 18 profiles were identified with 64% of strains displaying a single profile.[121] Using the same 10-mer primer, 29 strains of *L. monocytogenes* from raw milk yielded seven profiles, which were specific for milk isolates.[122] In the latter study, RAPD in combination with serotyping allowed for a higher level of differentiation than either alone. RAPD was found to be more rapid and less labor-intensive than restriction fragment length polymorphism, and pure DNA was not needed.[122]

When RAPD analysis was compared to phage typing on 104 strains of *L. monocytogenes* from six different outbreaks, a 98% agreement was found and RAPD was suggested as an alternative to phage typing.[135] RAPD was found to be far better than 16S rRNA sequence data in discriminating between strains of *L. monocytogenes*, and it showed differentiation even in strains with the same 16S rRNA sequence.[39] Employing three 10-mer primers, 34 banding profiles were obtained with one of the primers on 52 strains of *L. monocytogenes* representing five species.[56]

Although RAPD amplification does not occur with starved and viable but nonculturable cells (VBNC), both cell types can be detected by supplying starved cells with nutrients and resuscitation of VBNC cells by temperature upshift.[217]

In an epidemiologic study of *L. monocytogenes*, RAPD was one of five methods compared, and all 4b strains were distributed into two RAPD and four pulsed field gel electrophoresis (PFGE) types.[122] RAPD was one of the top three discriminating methods along with PFGE and ribotyping.

Pulsed Field Gel Electrophoresis

This method entails the digestion of genomic DNA by one or more restriction enzymes, separation of the restriction fragments by field inversion electrophoresis, and resolution of fragments in agarose gels. In contrast to conventional electrophoresis where a gel is run in one direction, PFGE is carried out with pulse times ramped from 1 to 100 sec over varying periods of time, which is determined by the sizes of molecules. The alternating electrical fields force molecules to change directions, and the electrophoretic profiles are designated *pulsovars*. It has been used to fingerprint foodborne outbreak strains of several pathogens.

Using two restriction enzymes (*Asc*I and *Apa*I), 176 strains of *L. monocytogenes* and 22 other listerial species/strains generated 87 genomically distinct groups, with *Apa*I generating the largest number of bands.[23] In another study, 42 serovar 4b strains of *L. monocytogenes* were divided into at least 24 different genomic varieties using one of three restriction enzymes.[24] Although all 42 cultures could be typed using PFGE, only 89% were phage typable.[24] When serovar 4b strains of *L. monocytogenes* from 279 human listeriosis cases were subjected to PFGE (along with other methods), 34 pulsovars were obtained, with 89% being pulsovar 2/1/3, the human epidemic strain.[87] Using three restriction enzymes, the strain of *L. monocytogenes* that caused the 1992 human outbreak in France was shown to be genomically closely related to those that caused outbreaks in California, Denmark, and Switzerland.[87]

In addition to *L. monocytogenes*, PFGE has been employed on a number of other bacteria of importance in foods. Outbreak and sporadic strains of *E. coli* O157:H7 involved in the 1994 Pacific Northwest outbreak of hemorrhagic colitis were differentiated,[102] and the close relationship of the 0139 serogroup *V. cholerae* to the 01 E1 Tor biotype has been substantiated by PFGE.[75] Using *Sma*I digests and PFGE, the genome sizes of three staphylococcal species were extrapolated.[67] In order to trace the source of this organism on processed cold-smoked rainbow trout, Autio et al.[6] did PFGE analyses on 303 isolates and found that those on the final product were associated with brining and slicing, and that those associated with raw fish were not detected on the final product.

Restriction Fragment Length Polymorphism

DNA restriction fragment length polymorphisms (RFLP) are heritable differences in the lengths of DNA fragments that arise when DNA is digested by a restriction endonuclease. In brief, cellular DNA is digested with a restriction enzyme, separated by electrophoresis, followed by Southern blot hybridization with a DNA probe from a given gene library of the organism in question. Along with MEE, it was used to demonstrate the recurrence of strains of *L. monocytogenes* in raw milk and nondairy foods.[77]

Table 11–2 Summaries of Some Ribotyping Applications to Foodborne Organisms

Organisms	Ribotype applications	Reference
Clostridium botulinum	With Qualicon riboprinter system, used EcoRI and tested 31 strains of the four major types. Found 15 ribogroups	191
Escherichia coli strains	Using HindIII, unable to differentiate between strains from different animal species, but may be used to differentiate human and animal derived strains	182
E. coli from humans and animals	Used 40 human and 247 animal isolates (seven different animals). Probe was BamHI fragment from the plasmid PKK containing E. coli 16S and 23S rRNA genes. Concluded that ribotyping is a valuable means of determining sources of fecal pollution	27
Bacillus sporothermodurans	Used automated ribotyping of repetitive extragenic palindromic (REP)-PCR fingerprinting; and PvuII and EcoRI. Found two main clusters among the 38 strains from different sources	72

Ribotyping

DNA is extracted from cells and digested with an endonuclease such as *Eco*RI, and the fragments are separated by agarose gel electrophoresis. Separated fragments are transferred to a nylon membrane and hybridized with an appropriately labeled copy DNA (cDNA) probe derived from ribosomal RNA (rRNA) by reverse transcriptase. The chemiluminescent pattern that is created is recorded. An automated ribotyping system can process eight samples simultaneously. The automated device creates *riboprints* that are matched or compared to those of known strains stored on computer software.

When ribotyping and MEE were performed on 305 strains of *L. monocytogenes*, 28 ribotypes and 78 electrophoretic types (ET) by MEE resulted. The strains were divided into two subgroups by both methods but neither was satisfactory for serogroups 1/2b and 4b. Overall, MEE was more discriminating than ribotyping. When compared to PFGE employing 73 isolates of *Acinetobacter*, ribotyping distinguished 39 patterns using two endonucleases, but 49 were distinguished by PFGE.[184] In a study of *Salmonella* serotype Enteritidis, ribotyping was the most discriminating and accurate of the genetic methods used to distinguish among food, water, and pathogenic strains, with phage typing being best for further differentiation of the ribo groups.[120] For more ribotype application, see Table 11–2.

Microarrays

A very simple microarray may consist of a solid surface (such as a nylon membrane, glass slide, or silicon chips) onto which are attached small quantities of single-stranded DNA (ssDNA) from different known bacterial species. When ssDNA from as many unknown species is exposed to this array (*DNA chip*), complementary strains will bind to their respective sites on the chip. If a reporter molecule is used, the identity of the unknown species can be confirmed. A DNA microarray is, in essence, a dot blot set-up with the capacity to obtain and process large amounts of data. A DNA microarray for microorganisms begins with the construction of oligonucleotides (primers), probes, hybridization, and data analysis. The overall process has been presented and reviewed by Ye et al.[230]

As currently used, several hundred to several thousand specimens or probes may be applied to a solid surface. In a study of *Xanthomonas* pathovars, a 47-probe microarray was employed to fingerprint 14 closely related strains, and the fingerprints showed clear differences between the test strains.[111] To determine the relative numbers and genera of bacteria in the microbial consortium of ready-to-eat vegetable salads, a study was undertaken when the salads were fresh and after storage at 4 and 10°C for up to 12 days. While stored in modified-atmosphere packages, specific probes from the 16S sequences were used to identify bacterial genera in the salads without cultural isolations.[175] The investigators concluded that the DNA array-based method gave an accurate picture of the heterogeneous bacterial community that was dominated by pseudomonads after 4°C storage and by enteric bacteria after 10°C storage. A fiber optic DNA microarray has been developed for the detection of some foodborne pathogens, with as few as 100 cfu being detectable in <1 hour.

It appears that microarrays represent great potential for identifying microbial species and strains in foods and for the fingerprinting of biotypes. Array platforms are available from a number of commercial companies. In addition to DNA, RNA and protein microarrays are in use, and protein chips are available for proteomics research.

PHYSICAL METHODS

Biosensors

In a broad sense, a biosensor is a device, method, or procedure that can be used to detect the presence or activity of an organism—living or dead. A more concise definition is "... a device containing a biological sensing element connected to a transducer." In this definition, the transducer is the unit that converts the change into a measurable signal. Not included are biochemical or immunological methods that are used primarily to measure enzyme–substrate or antigen–antibody reactions although these reactions may be components of a biosensor. Some biosensors are based on principles of physics (e.g., fiber optics) while others are based on biological principles (e.g., *lux* gene luminescence). Those that have been demonstrated to be of value for foodborne microorganisms are listed and briefly described below.

Piezoelectric Crystals (Accoustical Biosensors)

Piezoelectric is electricity or electric polarity due to pressure in a crystalline substance such as quartz. A vibrating quartz is an extremely sensitive weight indicator. If a crystal is coated with an antibody, a flow injection analysis (FIA) system can be used to detect the addition of its homologous antigen. The working principle of a piezoelectric biosensor is depicted in Figure 11–4. With the quartz coated with an antibody, the target analyte is the homologous antigen which, when it binds to the antibody, changes the mass with a resulting decrease in frequency.

In one study, gold-coated quartz crystal surfaces were used to develop a FIA system to detect *S.* Typhimurium.[229] To mobilize the antibody, crystals were first coated with DSP (dithiobis-succinimidly-propionate), then the *S.* Typhimurium antibody, and finally with *S.* Typhimurium cells. It can be seen from Figure 11–5 that the ΔF (Hz) response for DSP alone was 9.0; 123 for DSP+antibody;

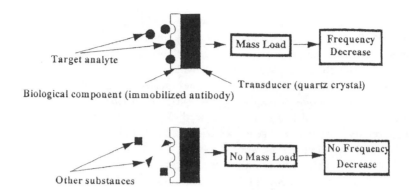

Figure 11–4   Working principle of a piezoelectric biosensor (Babacan et al.[8]).

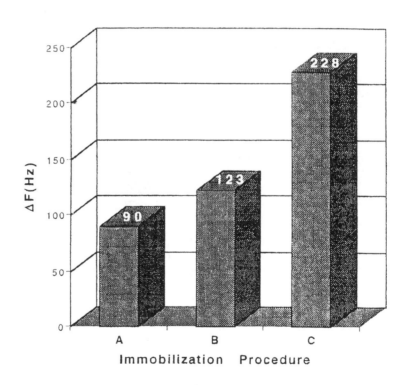

A: DSP
B: DSP+Antibody
C: DSP+Antibody+Antigen

Figure 11–5   *Salmonella* piezoelectric biosensor dry measurement of the frequency changes (ΔF) at each step of DSP coupling for anti-*Salmonella* spp. antibody attachment and antigen binding to quartz crystals (Ye et al.[229]).

and 228 for DSP+antibody+antigen The hertz changes represent the effect of attached material on the quartz surface. The method was linear with *S*. Typhimurium from 10^5 to 10^9 cfu/ml with ΔF changing from 90 to 170 Hz.

The minimum detectable level was 10^4 cfu/ml, and the biosensor had a response time of ca. 25 minutes.[229] In another piezoelectric FIA system, protein A was used as an antibody binding site rather than DSP.[8] The latter system could detect *S*. Typhimurium in 30–40 minutes but the minimum detectable level was 2.1×10^6 cfu/ml.

Fiber optics

An optical fiber is a "light wire" (optical waveguide) made of glass or polymeric material, and the light waves are propagated along the fiber by total internal reflection. A fiber optic biosensor uses electronic or optical transduction to monitor a biological reaction, and reports it as an optical signal.

The typical format of a fiber optic system consists of a tapered fiber optic probe coated with an antibody of interest. Light from a diode laser travels through an all-fiber system to the fiber top and then penetrates as an *evanescent wave* in the area outside the tips. When a fluorescently-labeled homologous antigen binds to the antibody on the fiber tip, it interacts with the evanescent wave of the fiber optic waveguide and the fluorescent signal radiates in all directions with some traveling back up the fiber tip to the detection system.[167] Fluorescent dyes (such as Cy5) appear to be the light sources of choice.

With a *surface plasmon resonance* (SPR) fiber optic system, antibodies bind to the surface of a thin film on a precious metal that is on the reflecting surface of an optically transparent glass waveguide (see reference 167). When visible or near-infrared light is passed through the waveguide, a reflection occurs from the waveguide. The reflected light interacts with a plasma of electrons on the metal surface and a resonance effect causes a strong absorbance which is a consequence of the concentration of the antibody–antigen complex on the reflecting surface of the waveguide. The greater the antibody–antigen reaction, the longer the wavelengths.

Among the commercially available biosensors are BIAcore, an SPR system produced in Sweden; the Raptor, developed and produced in the state of Washington and an immunomagnetic system developed at the University of Rhode Island and produced in Massachusetts by Pierson Scientific.[167]

A portable evanescent wave fiber optic biosensor (Analyte 2000, developed at the U.S. Naval Research Laboratories) was evaluated for its capacity to detect *E. coli* 0157:H7 in ground beef.[42] Using two waveguides, the system detected 9×10^3 and 5.2×10^2 cfu/g. No false-positive reactions occurred, and results were obtained within 25 minutes of sample preparation. The above test system employed a standard sandwich immunoassay and Cy5 to illuminate the captured antigens. The Analyte 2000 was used to detect *L. monocytogenes*, and with an inoculum of <10 cfu/ml followed by a 20-hour enrichment, biosensor results were obtained in 20–45 minutes.[209] The BIAcore 3000 (another SPR system primarily for research use) was used to detect staphylococcal enterotoxin B (SEB) in milk and meat, and results were obtained in 5 minutes using one antibody, or 8 minutes using two antibodies.[168] The system could detect ca. 10 ng/ml of SEB. Another BIAcore instrument is available for food analyses. A 20-minute assay for the detection of a minimum of 5×10^5 cfu/ml of *S*. Typhimurium in spent alfalfa sprout irrigation water was developed by Kramer and Lim.[117] The method employs a portable RAPTOR automated fiber optic-based biosensor system. The system used a 635-nm laser diode for excitation of light. A *S*. Typhimurium antibody was used to capture the pathogen, and a Cy5-labeled monoclonal antibody was used. By this method, *S*. Typhimurium colonies could be recovered from the waveguides used, and background microbiota was not detected.

Impedance

Although the concept of electrical impedance measurement of microbial growth was advanced by G.N. Stewart in 1899, it was not until the 1970s that the method was employed for this purpose. Impedance is the apparent resistance in an electric circuit to the flow of alternating current, corresponding to the actual electrical resistance to a direct current. When microorganisms grow in cultured media, they metabolize substrates of low conductivity into products of higher conductivity and thereby decrease the impedance of the media. When the impedance of broth cultures is measured, the curves are reproducible for species and strains, and mixed cultures can be identified by use of specific growth inhibitors. The technique has been shown capable of detecting as few as 10–100 cells (Table 11–1). Cell populations of 10^5–10^6/ml can be detected in 3–5 hours and 10^4–10^5/ml in 5–7 hours.[226] The times noted are required for the organisms in question to attain a threshold of 10^6–10^7 cells per milliliter. Some applications of impedance to foods are summarized below.

1. In one study, 200 samples of puréed vegetables were assessed, and a 90–95% agreement was found between impedance measurements and plate count results relative to unacceptable levels of bacteria.[76] Impedance analyses required 5 hours, and the method was found to be applicable to cream pies, ground meat, and other foods.

2. The microbiological quality of pasteurized milk was assessed by using the impedance detection time (IDT) of 7 hours or less, which was equivalent to an aerobic plate count (APC) of 10^4/ml or more bacteria.[25] Of 380 samples evaluated, 323 (85%) were correctly assessed by impedance. Using the same criterion for 27 samples of raw milk, 10 hours were required for assessment. In a collaborative study of raw milk involving six laboratories, impedance results varied less than standard plate count (SPC) results among laboratories.[59] In yet another study with raw milk, impedance was found useful when a 7-hour cutoff time (10^5 cfu/ml) was used to screen samples.[68] A scattergram relating IDT to APC on 132 raw milk samples is presented in Figure 11–6.

3. The brewing industry test for detecting spoilage organisms in beer was shortened from 3 weeks or more to only 2–4 days by use of impedance.[52] Yeasts growing in wort caused an increase in impedance, whereas bacteria caused a decrease.

4. For raw beef, IDTs for 48 samples were plotted against log bacterial numbers and a regression coefficient of 0.97 was found.[59] The IDT for meats was found to be about 9 hours. In another study, the relative level of contamination of meat surfaces by impedance was assessed.[22] With 10^7 cells/cm^2 and above,[22] detection could be made accurately within 2 hours.

5. For frozen orange juice concentrate, an impedance method was used to classify acceptable ($<10^4$ cfu/ml) or unacceptable ($>10^4$ cfu/ml).[220] By using cutoff times of 10.2 hours for bacteria, and 15.8 hours for yeasts, 96% of 468 retail samples could be correctly classified.

6. For coliforms in ground beef, a new selective medium was used, and impedance was assessed on 70 samples.[133] In this study, 79% of impedimetric results fell within the 95% confidence limits of the three-tube most probable number (MPN) procedure for coliforms, and fewer than 100–21,000 cells per gram could be detected with results obtained within 24 hours. In another study, a new selective medium was developed that yielded impedance signals that were consistent with cfu results.[60] From an inoculum of 10 coliforms into the new medium, the average IDT was 3.8 hours and of 96 meat samples, a correlation coefficient of 0.90 was found between impedance and corrected coliform counts on violet red bile agar (VRBA). Further, an IDT of 6.5 hours was required for meat samples with 10^3 coliforms, and it was suggested that an impedance signal in 5.5 hours or less denoted meat with coliforms $>10^3$/g, whereas the inability to detect in 7.6 hours

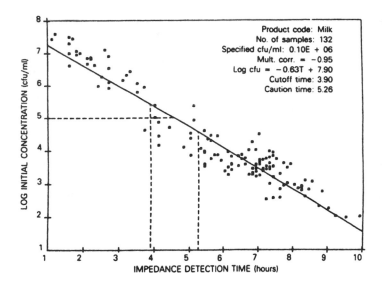

Figure 11–6 Scattergram relating IDT to APC on 132 samples of raw milk. Samples containing $> 10^5$ mesophiles per milliliter were detected within 4 hours. Courtesy of Ruth Firstenberg-Eden.

denoted coliform levels $< 10^3/g$.[60] In yet another study on coliforms in raw and pasteurized milk and two other dairy products, an IDT of < 9 hours indicated coliforms were $> 10/ml$, whereas an IDT of > 12 hours indicated < 10 cells/ml.[61]

7. Impedance measurement was used to assess growth (negative test) or no growth (positive test) of *E. coli* 0157:H7 strains in a sorbitol-containing medium to which a specific bacteriophage (AR1) was added. Readings were obtained at 6 minutes intervals for up to 20 hours. At 30 minutes, the absence of conductance changes in tubes that contained the AR1 phage was recorded as positive.[30] Only 1 of 155 *E. coli* 0157:H7 strains grew in the presence of the phage.

Microcalorimetry

This is the study of small heat changes: the measurement of the enthalpy change involved in the breakdown of growth substrates. The heat production that is measured is closely related to the cell's catabolic activities.[63]

There are two types of calorimeters: batch and flow. Most of the early work was done with batch-type instruments. The thermal events measured by microcalorimetry are those from catabolic activities, as already noted. One of the most widely used microcalorimeters for microbiological work is the Calvet instrument, which is sensitive to a heat flow of 0.01 cal/hour from a 10-ml sample.[63] With respect to its use as a rapid method, most attention has been devoted to the identification and characterization of foodborne organisms. Microcalorimetric results vary according to the history of the organism, inoculum size, fermentable substrates, and the like. One group of investigators found the variations such that the identification of microorganisms by this method was questioned, but in a later study in which a synthetic medium was used, Perry et al.[159] successfully characterized commercial yeast strains. The utility of the method to identify yeasts has been questioned,[14] but by the use of flow microcalorimetry, yeasts could be characterized. The latter method is one in which a microcalorimeter is filled with a flow-through calorimetric vessel. By the use of a chemically defined medium containing seven sugars, thermograms were produced by nine lactic acid bacteria (belonging to the genera *Enterococcus*, *Leuconostoc*, and *Lactobacillus*) distinctive enough to recommend the method for their identification.[66] All cultures were run at 37°C except "*S. cremoris*," which was run at 30°C, and results were obtained within 24 hours.

This method has been used to study spoilage in canned foods, to differentiate between the Enterobacteriaceae, to detect the presence of *S. aureus*, and to estimate bacteria in ground meat. In detecting *S. aureus*, results were achieved in 2 hours using an initial number of 10^7–10^8 cells per milliliter and in 12–13 hours when only 2 cells per milliliter were used.[119] As a monitoring device, flow microcalorimetry was used to determine the viability of recovered frozen cells of *S. cerevisiae* within 3 hours after thawing.[13] When applied to comminuted meat, the peak exothermic heat production rate (HPR) could be recorded within 24 hours for meats that contained 10^5–10^8 cfu/g, and results correlated well with plate count results.[70] With 10^2 cfu/ml, a measurable HPR was produced after 6 hours, with a peak HPR at 10 hours.

Flow Cytometry

Flow cytometry is the science of measuring components (cells) and the properties of individual cells in liquid suspension. In essence, suspended cells, one by one, are brought to a detector by means of a flow channel. Fluidic devices under laminar flow define the trajectories and velocities that cells traverse the detector, and among the cell properties that can be detected are fluorescence, absorbance, and light scatter. By use of flow sorting, individual cells can be sorted on the basis of their measured properties, and 1–3 or more global properties of the cell can be measured.[141] Flow cytometers and cell sorters make use of one or more excitation sources such as argon, krypton, or helium–neon ion lasers and one or two fluorescent dyes to measure and characterize several thousand cells per second. When a dye is used, its excitation spectrum must match the light wavelengths of the excitation source.[40] Two dyes may be used in combination to measure, for example, total protein and DNA content. In these instances, both dyes must excite at the same wavelength and emit at different wavelengths so that the light emitted by each dye is measured separately. The early history of flow cytometry has been reviewed by Horan and Wheeless.[83]

Although most studies have been conducted on mammalian cells, both DNA and protein have been measured in yeast cells. Typically, yeast cells are grown, fixed, and incubated in an RNase solution for 1 hour. Cell protein may be stained with fluorescein isothiocyanate and DNA with propidium iodide. Following necessary washing, the stained cells are suspended in a suitable buffer and are now ready for application to a flow cytometer. The one used by Hutter et al.[84] was equipped with a 50-mW argon laser. Yeast cells were excited at different wavelengths with the aid of special optical filters. By this method, baker's yeast was found to contain 4.6×10^{-14} g of DNA per cell, and the protein content per cell was found to be 1.1×10^{-11} g.

Flow cytometry when combined with fluorescently labeled monoclonal antibodies detected *S.* Typhimurium in eggs and milk within 40 minutes with a sensitivity of 10^3/ml.[136] When a 6-hour nonselective enrichment was used, the detection limit was ten cells per milliliter for milk and one cell per milliliter for eggs.

A flow cytometric method for the detection and enumeration of bacteria in milk was developed by enzymatically clearing the milk of lipid particles and proteins. When bacteria were added to UHT-treated milk and the numbers determined by PCA and flow cytometry, the numbers recovered by the two methods were highly comparable, but those by flow cytometry were obtained in ca. 1 hour.[73] A flow cytometric method for assessing the Gram reaction of bacteria in raw milk was developed by using two fluorescent stains one of which stains only Gram-positive cells while the other stains both Gram positives and Gram negatives. With seven Gram positives and five Gram negatives, the method correctly recognized the Gram reactions of 99%.[81]

BioSys Instrument

The BioSys-32 instrument makes automatic and computer-analyzed changes in the color of reaction vials as organisms grow in specified culture media that contain a substrate that changes color in proportion to increases in numbers of cells. It was used to simultaneously detect *Salmonella* and *Listeria* spp. in 70 naturally contaminated foods by the decrease in light transmission upon the production of H_2S by *Salmonella* spp.[158] By this method, 10–50 salmonellae or listeriae/25 g could be detected in 24 hours with an additional 6 hours needed for PCR confirmation of results. The BioSys instrument has been used successfully to measure amino acid decarboxylation in enteric bacteria,[188] *Listeria* spp. in environmental swabs and sponges,[62] the effects of citrate and lactate on raw meat biota,[189] and the antilisterial activity of lactate and diacetate in ready-to-eat meats.[137]

REFERENCES

1. Adcock, P.W., and C.P. Saint. 2001. Rapid confirmation of *Clostridium perfringens* by using chromogenic and fluorogenic substrates. *Appl. Environ. Microbiol.* 67:4382–4384.

2. Ahmed, R., P. Sankar-Mistry, S. Jackson, H.W. Ackermann, and S.S. Kasatiya. 1995. *Bacillus cereus* phage typing as an epidemiological tool in outbreaks of food poisoning. *J. Clin. Microbiol.* 33:636–640.

3. Areson, P.W.D., S.E. Charm, and B.L. Wong. 1980. Determination of staphylococcal enterotoxins A and B in various food extracts, using staphylococcal cells containing protein. *J. Food Sci.* 45:400–401.

4. Audurier, A., and C. Martin. 1989. Phage typing of *Listeria monocytogenes. Int. J. Food Microbiol.* 8:251–257.

5. Audurier, A., R. Chatelain, F. Chalons, and M. Piéchaud. 1979. Lysotypie de 823 souches de *Listeria monocytogenes* isolées en France de 1958 à 1978. *Ann. Microbiol.* (*Inst. Pasteur*). 130B:179–189.

6. Autio, T., S. Hielm, M. Miettinen, A.-M. Sjöberg, K. Aarnisalo, J. Björkroth, T. Mattila-Sandholm, and H. Korkeala. 1999. Sources of *Listeria monocytogenes* contamination in a cold-smoked rainbow trout processing plant detected by pulsed-field gel electrophoresis typing. *Appl. Environ. Microbiol.* 65:150–155.

7. Ayres, J.C. 1967. Use of fluorescent antibody for the rapid detection of enteric organisms in egg, poultry and meat products. *Food Technol.* 21:631–640.

8. Babacan, S., P. Pivarnik, S. Letcher, and A. Rand. 2002. Piezoelectric flow injection analysis biosensor for the detection of *Salmonella* Typhimurium. *J. Food Sci.* 67:314–320.

9. Bachrach, U., and Z. Bachrach. 1974. Radiometric method for the detection of coliform organisms in water. *Appl. Microbiol.* 28:169–171.

10. Baez, L.A., and V.K. Juneja. 1995. Nonradioactive colony hybridization assay for detection and enumeration of entero-toxigenic *Clostridium perfringens* in raw beef. *Appl. Environ. Microbiol.* 61:807–810.

11. Bastish, V.K., H. Chander, and G. Ranganathan. 1984. Incidence of enterococcal thermonuclease in milk and milk products. *J. Food Sci.* 49:1610–1611, 1615.

12. Bautista, D.A., J.P. Vaillancourt, R.A. Clarke, S. Renwick, and M.W. Griffiths. 1995. Rapid assessment of the microbiological quality of poultry carcasses using ATP bioluminescence. *J. Food Protect.* 58:551–554.

13. Beezer, A.E., D. Newell, and H.J.V. Tyrrell. 1976. Application of flow microcalorimetry to analytical problems: The preparation, storage and assay of frozen inocula of *Saccharomyces cerevisiae. J. Appl. Bacteriol.* 41:197–207.

14. Beezer, A.E., D. Newell, and H.J.V. Tyrrell. 1978. Characterisation and metabolic studies of *Saccharomyces cerevisiae* and *Kluyveromyces fragilis* by flow microcalorimetry. *Antonie Van Leeuwenhoek* 45:55–63.

15. Bennett, R.W., and F. McClure. 1976. Collaborative study of the serological identification of staphylococcal enterotoxins by the microslide gel double diffusion test. *J. Assoc. Off. Anal. Chem.* 59:594–600.

16. Berg, J.D., and L. Fiksdal. 1988. Rapid detection of total and fecal coliforms in water by enzymatic hydrolysis of 4-methylumbelliferone-β-D-galactoside. *Appl. Environ. Microbiol.* 54:2118–2122.

17. Bergdoll, M.S., and R. Reiser. 1980. Application of radioimmunoassay of detection of staphylococcal enterotoxins in foods. *J. Food Protect.* 43:68–72.

18. Black, S.F., D.I. Gray, D.B. Fenton, and R.G. Kroll. 1995. Rapid RAPD analysis for distinguishing *Listeria* species and *Listeria monocytogenes* serotypes using a capillary air thermal cycler. *Lett. Appl. Microbiol.* 20:188–189.

19. Blears, M.J, S.A. De Grandis, H. Lee, and J.T. Trevors. 1999. Amplified fragment length polymorphism (AFLP): Review of the procedure and its applications. *J. Ind. Microbiol. Biotechnol.* 21:99–114.

20. Boerlin, P., F. Boerlin-Petzold, E. Bannerman, J. Bille, T. Jemmi. 1999. Typing *Listeria monocytogenes* isolates from fish products and human listeriosis cases. *Appl. Environ. Microbiol.* 63:1338–1343.

21. Bottari, D.A., C.D. Emmett, C.E. Nichols, K.D. Whippie, D. Rodriguez, G.W. Durbin, K.M. Keough, E.P. Groody, M.A. Mazola, and G.N. Reynolds. 1995. Comparative study of a colorimetric DNA hybridization method and conventional culture procedures for the detection of *Listeria* spp. in foods. *J. Food Protect.* 58:1083–1090.

22. Bulte, M., and G. Reuter. 1984. Impedance measurement as a rapid method for the determination of the microbial contamination of meat surfaces, testing two different instruments. *Int. J. Food Microbiol.* 1:113–125.

23. Brosch, R., H. Chen, and J.B. Luchansky. 1994. Pulsed-field fingerprinting of listeriae: Identification of genomic divisions for *Listeria monocytogenes* and their correlation with serovar. *Appl. Environ. Microbiol.* 60:2584–2592.

24. Brosch, R., C. Buchrieser, and J. Rocourt. 1991. Subtyping of *Listeria monocytogenes* serovar 4b by use of low-frequency-cleavage restriction endonucleases and pulsed-field gel electrophoresis. *Res. Microbiol.* 142:667–675.

25. Cady, P., D. Hardy, S. Martins, S.W. Duforu, and S.J. Kraeger. 1978. Automated impedance measurements for rapid screening of milk microbiol content. *J. Food Protect.* 41:277–283.

26. Candlish, A.A.G., W.H. Stimson, and J.E. Smith. 1985. A monoclonal antibody to aflatoxin B_1: Detection of the mycotoxin by enzyme immunoassay. *Lett. Appl. Microbiol.* 1:57–61.

27. Carson, C.A, B.L. Shear, M.R. Ellersieck, and A. Asfaw. 2001. Identification of fecal *Escherichia coli* from humans and animals by ribotyping. *Appl. Environ. Microbiol.* 67:1503–1507.

28. Casman, E.P., and R.W. Bennett. 1965. Detection of staphylococcal enterotoxin in food. *Appl. Microbiol.* 13:181–189.

29. Cernic, G. 1976. Method for the distinction of Gram-negative from Gram-positive bacteria. *Eur. J. Appl. Microbiol.* 3:223–225.

30. Chang, T.C., H.C. Ding, and S. Chen. 2002. A conductance method for the identification of *Escherichia coli* 0157:H7 using bacteriophage AR1. *J. Food Protect.* 65:12–17.

31. Cherry, W.B., and M.D. Moody. 1965. Fluorescent-antibody techniques in diagnostic bacteriology. *Bacteriol. Rev.* 29:222–250.

32. Chesbro, W.R., and K. Auborn. 1967. Enzymatic detection of the growth of *Staphylococcus aureus* in foods. *Appl. Microbiol.* 15:1150–1159.

33. Collins, W.S., II, A.D. Johnson, J.F. Metzger, and R.W. Bennett. 1973. Rapid solid-phase radioimmunoassay for staphylococcal enterotoxin A. *Appl. Microbiol.* 25:774–777.

34. Collins, W.S., II, J.F. Metzger, and A.D. Johnson. 1972. A rapid solid phase radioimmunoassay for staphylococcal B enterotoxin. *J. Immunol.* 108:852–856.

35. Cocolin, L., K. Rantsiou, L. Iacumin, C. Cantoni, and G. Comi. 2002. Direct identification in food samples of *Listeria* spp., and *Listeria monocytogenes* by molecular methods. *Appl. Environ. Microbiol.* 68:6273–6282.

36. Cousin, M.A., J.M. Jay, and P.C. Vasavada. 2001. Psychrotrophic microorganisms. In *Compendium of Methods for the Microbiological Examination of Foods*, 2nd ed., eds. F.P. Downes, and K. Ito. Washington, DC: American Public Health Association.

37. D'Aoust, J.-Y., and A.M. Sewell. 1988. Reliability of the immunodiffusion 1–2 Test system for detection of *Salmonella* in foods. *J. Food Protect.* 51:853–856.

38. D'Aoust, J.-Y., A.M. Sewell, P. Greco, M.A. Mazola, and R.E. Colvin. 1995. Performance assessment of the GENE-TRAK^R colorimetric probe assay for the detection of foodborne *Salmonella* spp. *J. Food Protect.* 58:1069–1076.

39. Czajka, J., N. Bsat, M. Piani, W. Russ, K. Sultana, M. Wiedmann, R. Whitaker, and C.A. Batt. 1993. Differentiation of *Listeria monocytogenes* and *Listeria innocua* by 16S rRNA genes and intraspecies discrimination of *Listeria monocytogenes* strains by random amplified polymorphic DNA polymorphisms. *Appl. Environ. Microbiol.* 59:304–308.

40. Dean, P.N., and D. Pinkel. 1978. High resolution dual laser flow cytometry. *J. Histochem. Cytochem.* 26:622–627.

41. de Castro, B.P., M.A. Asenio, B. Sanz, and J.A. Ordonez. 1988. A method to assess the bacterial content of refrigerated meat. *Appl. Environ. Microbiol.* 54:1462–1465.

42. DeMarco, D.M., and D.V. Lim. 2002. Detection of *Escherichia coli* 0157:H7 in 10- and 25-gram ground beef samples with an evanescent-wave biosensor with silica and polystyrene waveguides. *J. Food Protect.* 65:596–602.

43. Desai, M., E.J. Threlfall, and J. Stanley. 2001. Fluorescent amplified-fragment length polymorphism subtyping of the *Salmonella enterica* serovar Enteritidis phage type 4 clone complex. *J. Clin. Microbiol.* 39:201–206.

44. Dixon-Holland, D.E., J.J. Pestka, B.A. Bidigare, W.L. Casale, R.L. Warner, B.P. Ram, and L.R. Hart. 1988. Production of sensitive monoclonal antibodies to aflatoxin B_1 and aflatoxin M_1 and their application to ELISA of naturally contaminated foods. *J. Food Protect.* 51:201–204.

45. Dodds, K.L., R.A. Holley, and A.G. Kempton. 1983. Evaluation of the catalase and *Limulus* ameobocyte lysate tests for rapid determination of the microbiol quality of vacuum-packed cooked turkey. *Can. Inst. Food Sci. Technol. J.* 16:167–172.

46. Dovey, S., and K.J. Towner. 1989. A biotinylated DNA probe to detect bacterial cells in artificially contaminated foodstuffs. *J. Appl. Bacteriol.* 66:43–47.

47. Downes, F.P., J.H. Green, K. Greene, N. Stockbine, J.G. Wells, and I.K. Wachsmuth. 1989. Development and evaluation of enzyme-linked immunosorbent assays for detection of Shiga-like toxin I and Shiga-like toxin II. *J. Clin. Microbiol.* 27:1292–1297.

48. Duim, B., T.M. Wassenaar, A. Rigter, and J. Wagenaar. 1999. High-resolution genotyping of *Campylobacter* strains isolated from poultry and humans with amplified fragment length polymorphism fingerprinting. *Appl. Environ. Microbiol.* 65:2369–2375.

49. Edberg, S.C., M.J. Allen, and D.B. Smith, and the National Collaboratiave study. 1989. National field evaluation of a defined substrate method for the simultaneous detection of total coliforms and *Escherichia coli* from drinking water: Comparison with presence–absence techniques. *Appl. Environ. Microbiol.* 55:1003–1008.

50. Erickson, A., and R.H. Deibel. 1973. Turbidimetric assay of staphylococcal nuclease. *Appl. Microbiol.* 25:337–341.

51. Ericsson, H., P. Stalhandske, M.-L. Danielsson-Tham, E. Bannerman, J. Bille, C. Jacquet, J. Rocourt, and W. Tham. 1995. Division of *Listeria monocytogenes* serovar 4b strains into two groups by PCR and restriction enzyme analysis. *Appl. Environ. Microbiol.* 61:3872–3874.

52. Evans, H.A.V. 1982. A note on two uses for impedimetry in brewing microbiology. *J. Appl. Bacteriol.* 53:423–426.

53. Fach, P., F. Dilasser, J. Grout, and J. Tache. 1999. Evaluation of a polymerase chain reaction-based test for detecting *Salmonella* spp. in food samples: Probelia *Salmonella* spp. *J. Food Protect.* 62:1387–1393.

54. Fallowfield, H.J., and J.T. Patterson. 1985. Potential value of the *Limulus* lysate assay for the measurement of meat spoilage. *J. Food Technol.* 20:467–479.

55. Fang, Q., S. Brockmann, K. Botzenhart, and A. Wiedenmann. 2003. Improved detection of *Salmonella* spp. in foods by fluorescent in situ hybridization with 23S rRNA probes: A comparison with conventional culture methods. *J. Food Protect.* 66:723–731.

56. Farber, J.M., and C.J. Addison. 1994. RAPD typing for distinguishing species and strains in the genus *Listeria*. *J. Appl. Bacteriol.* 77:242–250.

57. Feng, P.C.S., and P.A. Hartman. 1982. Fluorogenic assays for immediate confirmation of *Escherichia coli*. *Appl Environ. Microbiol.* 43:1320–1329.

58. Fiedler, F., J. Seger, A. Schrettenbrunner, and H.P.R. Seeliger. 1984. The biochemistry of murein and cell wall teichoic acids in the genus *Listeria*. *Syst. Appl. Microbiol.* 5:360–376.

59. Firstenberg-Eden, R. 1984. Collaborative study of the impedance method for examining raw milk samples. *J. Food Protect.* 47:707–712.

60. Firstenberg-Eden, R., and C.S. Klein. 1983. Evaluation of a rapid impedimetric procedure for the quantitative estimation of coliforms. *J. Food Sci.* 48:1307–1311.

61. Firstenberg-Eden, R., M.L. Van Sise, J. Zindulis, and P. Kahn. 1984. Impedimetric estimation of coliforms in dairy products. *J. Food Sci.* 49:1449–1452.

62. Firstenberg-Eden, R., and L.A. Shelef. 2000. A new rapid automated method for the detection of *Listeria* from environmental swabs and sponges. *Int. J. Food Microbiol.* 56:231–237.

63. Forrest, W.W. 1972. Microcalorimetry. *Methods Microbiol.* 6B:385–318.

64. Frampton, E.W., L. Restaino, and N. Blaszko. 1988. Evaluation of the B-glucuronidase substrate 5-bromo-4-chloro-3-indolyl-β-D-glucuronide (X-GLUC) in a 24-hour direct plating method for *Escherichia coli*. *J. Food Protect.* 51:402–404.

65. Fratamico, P.M, L.K. Bagi, and T. Pepe. 2000. A multiplex polymerase chain reaction assay for rapid detection and identification of *Escherichia coli* 0157:H7 in foods and bovine feces. *J. Food Protect.* 63: 1032–1037.

66. Fujita, T., P.R. Monk, and I. Wadso. 1978. Calorimetric identification of several strains of lactic acid bacteria. *J. Dairy Res.* 45:457–463.

67. George, C.G., and W.B. Kloos. 1994. Comparison of the *Sma*I-digested chromosomes of *Staphylococcus epidermidis* and the closely related species *Staphylococcus capitis* and *Staphylococcus caprae. Int. J. Syst. Bacteriol.* 44:404–409.

68. Gnan, S., and L.O. Luedecke. 1982. Impedance measurements in raw milk as an alternative to the standard plate count. *J. Food Protect.* 45:4–7.

69. Goepfert, J.M., and N.F. Insalata. 1969. Salmonellae and the fluorescent antibody technique: A current evaluation. *J. Milk Food Technol.* 32:465–473.

70. Gram, L., and H. Sogaard. 1985. Microcalorimetry as a rapid method for estimation of bacterial levels in ground meat. *J. Food Protect.* 48:341–345.

71. Grunstein, M., and D.S. Hogness. 1975. Colony hybridization: A method for the isolation of cloned DNAs that contain a specific gene. *Proc. Nat. Acad. Sci. USA* 72:3961–3965.

72. Guillaume-Gentil, G., P. Scheldeman, J. Marugg, L. Herman, M. Joosten, and M. Hendrickx. 2002. Genetic heterogeneity in *Bacillus sporothermodurans* as demonstrated by ribotyping and repetitive extragenic palindromic-PCR fingerprinting. *Appl. Environ. Microbiol.* 68:4216–4224.

73. Gunasekera, T.S, P.V. Attfield, and D.A. Veal. 2000. A flow cytometry method for rapid detection and enumeration of total bacteria in milk. *Appl. Environ. Microbiol.* 66:1228–1232.

74. Guy, R.A., P. Payment, U.J. Krull, and P.A. Horgen. 2003. Real-time PCR for quantification of *Giardia* and *Cryptosporidium* in environmental water samples and sewage. *Appl. Environ. Microbiol.* 69:5178–5185.

75. Hall, R.H., F.M. Khambaty, M.H. Kothary, S.P. Keasler, and B.D. Tall. 1994. *Vibrio cholerae* non-01 serogroup associated with cholera gravis genetically and physiologically resembles 01 E1 Tor cholera strains. *Infect. Immun.* 62:3859–3863.

76. Hardy, D., S.W. Dufour, and S.J. Kraeger. 1975. Rapid detection of frozen food bacteria by automated impedance measurements. *Proc. Inst. Food Technol.*

77. Harvey, J., and A. Gilmour. 1994. Application of multilocus enzyme electrophoresis and restriction fragment length polymorphism analysis to the typing of *Listeria monocytogenes* strains isolated from raw milk, nondairy foods, and clinical and veterinary sources. *Appl. Environ. Microbiol.* 60:1547–1553.

78. Hatcher, W.S., S. DiBenedetto, L.E. Taylor, and D.L. Murdock. 1977. Radiometric analysis of frozen concentrated orange juice for total viable microorganisms. *J. Food Sci.* 42:636–639.

79. Hill, W.E., J.M. Madden, B.A. McCardell, D.B. Shah, J.A. Jagow, and B.K. Boutin. 1983. Foodborne enterotoxigenic *Escherichia coli*. Detection and enumeration by DNA colony hybridization. *Appl. Environ. Microbiol.* 45:1324–1330.

80. Hill, W.E., W.L. Payne, G. Zon, and S.L. Moseley. 1985. Synthetic oligodeoxyribonucleotide probes for detecting heat-stable enterotoxin-producing *Escherichia coli* by DNA colony hybridization. *Appl. Environ. Microbiol.* 50:1187–1191.

81. Holm, C., and L. Jespersen. 2003. A flow-cytometric Gram-staining technique for milk-associated bacteria. *Appl. Environ. Microbiol.* 69:2857–2863.

82. Hong, Y., M.E. Berrang, T. Liu, C.L. Hofacre, S. Sanchez, L. Wang, and J.J. Maurer. 2003. Rapid detection of *Campylobacter coli*, *C. jejuni*, and *Salmonella enterica* on poultry carcasses by using PCR-enzyme-linked immunosorbent assay. *Appl. Environ. Microbiol.* 69:3492–3499.

83. Horan, P.K., and L.L. Wheeless, Jr. 1977. Quantitative single cell analysis and sorting. *Science* 198:149–157.

84. Hutter, K.-J., M. Stöhr, and H.E. Eipel. 1980. Simultaneous DNA and protein measurements of microorganisms. In *Flow Cytometry*, ed. O.D. Lacrum, T. Lindmo, and E. Thorud, Vol. 4, 100–102. Bergen: Universitetsforlaget.

85. Huys, G., R. Coopman, P. Janssen, and K. Kersters. 1996. High-resolution genotypic analysis of the genus *Aeromonas* by AFLP fingerprinting. *Int. J. Syst. Bacteriol.* 46:572–580.

86. Izat, A.L., C.D. Driggers, M. Colberg, M.A. Reiber, and M.H. Adams. 1989. Comparison of the DNA probe to culture methods for the detection of *Salmonella* on poultry carcasses and processing waters. *J. Food Protect.* 52:564–570.

87. Jacquet, C., B. Catimel, R. Brosch, C. Buchrieser, P. Dehaumont, V. Goulet, A. Lepoutre, P. Veit, and J. Rocourt. 1995. Investigations related to the epidemic strain involved in the French listeriosis outbreak in 1992. *Appl. Environ. Microbiol.* 61:2242–2246.

88. Jaksch, V.P., K.-J. Zaadhof, and G. Terplan. 1982. Zur Bewertung der hygienischen Qualität von Milchprodukten mit dem *Limulus*-Test. *Molkerei-Zeitung Welt der Milch* 36:5–8.

89. Janssen, P., R. Coopman, G. Huys, J. Swings, N. Blecker, P. Vos, M. Zabeau, and K. Kersters. 1996. Evaluation of the DNA fingerprinting method AFLP as a new tool in bacterial taxonomy. *Microbiology* 152:1881–1893.

90. Jay, J.M. 1974. Use of the *Limulus* lysate endotoxin test to assess the microbiol quality of ground beef. *Bacteriol. Proc.* 13.

91. Jay, J.M. 1977. The *Limulus* lysate endotoxin assay as a test of microbial quality of ground beef. *J. Appl. Bacteriol.* 43:99–109.

92. Jay, J.M. 1981. Rapid estimation of microbial numbers in fresh ground beef by use of the *Limulus* test. *J. Food Protect.* 44:275–278.

93. Jay, J.M., and S. Margitic. 1979. Comparison of homogenizing, shaking, and blending on the recovery of microorganisms and endotoxins from fresh and frozen ground beef as assessed by plate counts and the *Limulus* amoebocyte lysate test. *Appl. Environ. Microbiol.* 38:879–884.

94. Jay, J.M., S. Margitic, A.L. Shereda, and H.V. Covington. 1979. Determining endotoxin content of ground beef by the *Limulus* amoebocyte lysate test as a rapid indicator of microbial quality. *Appl. Environ. Microbiol.* 38:885–890.

95. Jay, J.M. 1989. The *Limulus* amoebocyte lysate (LAL) test. In *Progress in Industrial Microbiology: Rapid Methods in Food Microbiology*, ed. M.R. Adams and C.F.A. Hope, 101–119. Amsterdam: Elsevier.

96. Jaykus, L.-A., R. De Leon, and M.D. Sobsey. 1996. A virion concentration method for detection of human enteric viruses in oysters by PCR and oligoprobe hybridization. *Appl. Environ. Microbiol.* 62:2074–2080.

97. Jiang, S.C., M. Matte, G. Matte, A. Huo, and R.R. Colwell. 2000. Genetic diversity of clinical and environmental isolates of *Vibrio cholerae* determined by amplified fragment length polymorphism fingerprinting. *Appl. Environ. Microbiol.* 66:148–153.

98. Jinneman, K.C., J.M. Hunt, C.A. Eklund, J.S. Wernberg, P.N. Sado, J.M. Johnson, R.S. Richter, S.T. Torres, E. Ayotte, S.J. Eliasberg, P. Istafanos, D. Bass, N. Kexel-Calabresa, W. Lin, and C.N. Barton. 2003a. Evaluation and interlaboratory validation of a selective agar for phosphatidylinositol-specific phospholipase C activity using a chromogenic substrate to detect *Listeria monocytogenes* from foods. *J. Food Protect.* 66:441–445.

99. Jinneman, K.C., K.J. Yoshitomi, and S.D. Weagant. 2003b. Multiplex real-time PCR method to identify Shiga toxin genes *stx*1 and *stx*2 and *Escherichia coli* O157:H7 serotype. *Appl. Environ. Microbiol.* 69:6327–6333.

100. Johnson, H.M., J.A. Bukovic, P.E. Kauffman, and J.T. Peeler. 1971. Staphylococcal enterotoxin B: Solid-phase radioimmunoassay. *Appl. Microbiol.* 22:837–841.

101. Johnson, J.L., C.L. Brooke, and S.J. Fritschel. 1998. Comparison of the BAX for screening/*E. coli* O157:H7 method with conventional methods for detection of extremely low levels of *Escherichia coli* O157:H7 in ground beef. *Appl. Environ. Microbiol.* 64:4390–4395.

102. Johnson, J.M., S.D. Weagant, K.C. Jinneman, and J.L. Bryant. 1995. Use of pulsed-field gel electrophoresis for epidemiological study of *Escherichia coli* O157:H7 during a food-borne outbreak. *Appl. Environ. Microbiol.* 61:2806–2808.

103. Jothikumar, N., X. Wang, and M.W. Griffiths. 2003. Real-time multiplex SYBR Green I-based PCR assay for simultaneous detection of *Salmonella* serovars and *Listeria monocytogenes*. *J. Food Protect.* 66:2141–2145.

104. Kang, D.-H., and G.R. Siragusa. 2002. Monitoring beef carcass surface microbial contamination with a luminescence-based bacterial phosphatase assay. *J. Food Protect.* 65:50–52.

105. Karl, D.M. 1980. Cellular nucleotide measurements and applications in microbial ecology. *Microbiol. Rev.* 44:739–796.

106. Kauffmann, F. 1944. Zur Serologie der Coli-Gruppe. *Acta Path. Microbiol. Scand.* 21:20–45.

107. Kaysner, C.A., S.D. Weagant, and W.E. Hill. 1988. Modification of the DNA colony hybridization technique for multiple filter analysis. *Molec. Cell. Probes* 2:255–260.

108. Kennedy, J.E., Jr., and J.L. Oblinger. 1985. Application of bioluminescence to rapid determination of microbial levels in ground beef. *J. Food Protect.* 48:334–340.

109. Kepner, R.L. Jr., and J.R. Pratt. 1994. Use of fluorochromes for direct enumeration of total bacteria in environmental samples: Past and present. *Microbiol. Rev.* 58:603–615.

110. Kerouanton, A., A. Brisabois, E. Denoyer, F. Dilasser, J. Grout, G. Salvat, and B. Picard. 1998. Comparison of five typing methods for the epidemiological study of *Listeria monocytogenes*. *Int. J. Food Microbiol.* 43:61–71

111. Kingsley, M.T, T.M. Straub, D.R. Call, D.S. Daly, S.C. Wunschel, and D.P. Chandler. 2002. Fingerprinting closely related *Xanthomonas* pathovars with random monamer oligonucleotide microarrays. *Appl. Environ. Microbiol.* 68:6361–6370.

112. Kleppe, K., E. Ohtsuka, R. Kleppe, I. Molineux, and H.G. Khorana. 1971. Studies on polynucleotides. XCVI. Rapid replication of short synthetic DNA's as catalyzed by DNA polymerases. *J. Mol. Biol.* 56:341–361.

113. Koburger, J.A., and M.L. Miller. 1985. Evaluation of a fluorogenic MPN procedure for determining *Escherichia coli* in oysters. *J. Food Protect.* 48:244–245.

114. Kodikara, C.P., H.H. Crew, and G.S.A.B. Stewart. 1991. Near on-line detection of enteric bacteria using *lux* recombinant bacteriophage. *FEMS Microbiol. Lett.* 83:261–266.

115. Koupal, A., and R.H. Deibel. 1978. Rapid qualitative method for detecting staphylococcal nuclease in foods. *Appl. Environ. Microbiol.* 35:1193–1197.

116. Kozwich, D., K.A. Johansen, K. Landau, C.A. Roehl, S. Woronoff, and P.A. Roehl. 2000. Development of a novel, rapid integrated *Cryptosporidium parvum* detection assay. *Appl. Environ. Microbiol.* 66:2711–2717.

117. Kramer, M.F., and D.V. Lim. 2004. A rapid and automated fiber optic-based biosensor assay for the detection of *Salmonella* in spent irrigation water used in the sprouting of sprout seeds. *J. Food Protect.* 67:46–52.

118. Lachica, B.V., K.F. Weiss, and R.H. Deibel. 1969. Relationships among coagulase, enterotoxin, and heat-stable deoxyribonuclease production by *Staphylococcus aureus*. *Appl. Microbiol.* 18:126–127.

119. Lampi, R.A., D.A. Mikelson, D.B. Rowley, J.J. Previte, and R.E. Wells. 1974. Radiometry and microcalorimetry—techniques for the rapid detection of foodborne microorganisms. *Food Technol.* 28(10):52–55.

120. Landeras, E., M.A. González-Hevia, and M.C. Mendoza. 1998. Molecular epidemiology of *Salmonella* serotype Enteritidis. Relationships between food, water and pathogenic strains. *J. Food Microbiol.* 43:81–90.

121. Lawrence, L.M., and A. Gilmour. 1995. Characterization of *Listeria monocytogenes* isolated from poultry products and from the poultry-processing environment by random amplification of polymorphic DNA and multilocus enzyme electrophoresis. *Appl. Environ. Microbiol.* 61:2139–2144.

122. Lawrence, L.M., J. Harvey, and A. Gilmour. 1993. Development of a random amplification of polymorphic DNA typing method for *Listeria monocytogenes*. *Appl. Environ. Microbiol.* 59:3117–3119.

123. Lee, H.A., G.M. Wyatt, S. Bramham, and M.R.A. Morgan. 1990. Enzyme-linked immunosorbent assay for *Salmonella typhimurium* in food: Feasibility of 1-day *Salmonella* detection. *Appl. Environ. Microbiol.* 56:1541–1546.

124. Levin, G.V., V.B. Harrison, and W.C. Hess. 1956. Preliminary report on a one-hour presumptive test for coliform organisms. *J. Am. Water Works Assoc.* 18:75–80.

125. Lewis, G.E., Jr., S.S. Kulinski, D.W. Reichard, and J.F. Metzger. 1981. Detection of *Clostridium botulinum* Type G toxin by enzyme-linked immunosorbent assay. *Appl. Environ. Microbiol.* 42:1018–1022.

126. Li, J., and T.-C. Lee. 1995. Bacterial ice nucleation and its potential application in the food industry. *Trends Food Sci. Technol.* 6:259–265.

127. Lin, H.H., and M.A. Cousin. 1987. Evaluation of enzyme-linked immunosorbent assay for detection of molds in food. *J. Food Sci.* 52:1089–1094, 1096.

128. Littel, K.J., and P.A. Hartman. 1983. Fluorogenic selective and differential medium for isolation of fecal streptococci. *Appl. Environ. Microbiol.* 45:622–627.

129. Loessner, M.J., M. Rudolf, and S. Scherer. 1997. Evaluation of luciferase reporter bacteriophage A511::*lux*AB for detection of *Listeria monocytogenes* in contaminated foods. *Appl. Environ. Microbiol.* 63:2961–2965.

130. Loessner, M.J., C.E.D. Rees, G.S.A.B. Stewart, and S. Scherer. 1996. Construction of luciferase reporter bacteriophage A511::*lux*AB for rapid and sensitive detection of viable Listeria cells. *Appl. Environ. Microbiol.* 62:1133–1140.

131. Loessner, M.J. 1991. Improved procedure for bacteriophage typing of *Listeria* strains and evaluation of new phages. *Appl. Environ. Microbiol.* 57:882–884.

132. Loessner, M.J., and M. Busse. 1990. Bacteriophage typing of *Listeria* species. *Appl. Environ. Microbiol.* 56:1912–1918.

133. Martins, S.B., and M.J. Selby. 1980. Evaluation of a rapid method for the quantitative estimation of coliforms in meat by impedimetric procedures. *Appl. Environ. Microbiol.* 39:518–524.

134. Maruyama, F., T. Kenzaka, N. Yamaguchi, K. Tani, and M. Nasu. 2003. Detection of bacteria carrying the stx_2 gene by in site loop-mediated isothermal amplification. *Appl. Environ. Microbiol.* 69:5023–5028.

135. Mazurier, S.-I., A. Audurier, N. Marquet-Van der Mee, S. Notermans, and K. Wernars. 1992. A comparative study of randomly amplified polymorphic DNA analysis and conventional phage typing for epidemiological studies of *Listeria monocytogenes* isolates. *Res. Microbiol.* 143:507–512.

136. McClelland, R.G., and A.C. Pinder. 1994. Detection of *Salmonella typhimurium* in dairy products with flow cytometry and monoclonal antibodies. *Appl. Environ. Microbiol.* 60:4255–4262.

137. Mbandi, E., and L.A. Shelef. 1998. Automated measurements of antilisterial activities of lactate and diacetate in ready-to-eat meat. *Microbiol. Meth.* 49:307–314.

138. McKillip, J.L., and M. Drake. 2000. Molecular beacon polymerase chain reaction detection of *Escherichia coli* 0157:H7 in milk. *J. Food Protect.* 63:855–859.

139. McLauchlin, J., A. Audurier, A. Frommett, P. Gerner-Smidt, Ch. Jacquet, M.J. Loessner, N. van der Mee-Marquet, J. Rocourt, S. Shah, and D. Wilhelms. 1996. WHO study on subtyping *Listeria monocytogenes*: results of phage-typing. *Int. J. Food Microbiol.* 32:289–299.

140. McLauchlin, J., A. Audurier, and A.G. Taylor. 1986. Aspects of the epidemiology of human *Listeria monocytogenes* infections in Britain 1967–1984: The use of serotyping and phage typing. *J. Med. Microbiol.* 22:367–377.

141. Mendelsohn. M.L. 1980. The attributes and applications of flow cytometry. In *Flow Cytometry*, ed. O.D. Laerum, T. Lindmo, and E. Thorud, Vol. 4, 15–27. Bergen: Universitetsforlaget.

142. Miller, B.A., R.F. Reiser, and M.S. Bergdoll. 1978. Detection of staphylococcal enterotoxins A, B, C, D, and E in foods by radioimmunoassay, using staphylococcal cells containing protein A as immunoadsorbent. *Appl. Environ. Microbiol.* 36:421–426.

143. Minnich, S.A., P.A. Hartman, and R.C. Heimsch. 1982. Enzyme immunoassay for detection of salmonellae in foods. *Appl. Environ. Microbiol.* 43:877–883.

144. Nakamura, T., T. Morita, and S. Iwanga. 1986. Lipopolysaccharide-sensitive serine-protease zymogen (factor C) found in *Limulus* hemocytes. Isolation and characterization. *Eur. J. Biochem.* 154:511–521.

145. Nath, E.J., E. Neidert, and C.J. Randall. 1989. Evaluation of enrichment protocols for the 1–2 Test™ for *Salmonella* detection in naturally contaminated foods and feeds. *J. Food Protect.* 52:498–499.

146. Niskanen, A., and L. Koiranen. 1977. Correlation of enterotoxin and thermonuclease production with some physiological and biochemical properties of staphylococcal strains isolated from different sources. *J. Food Protect.* 40:543–548.

147. Niskanen, A., and E. Nurmi. 1976. Effect of starter culture on staphylococcal enterotoxin and thermonuclease production in dry sausage. *Appl. Environ. Microbiol.* 31:11–20.

148. Norrung, B., and N. Skovgaard. 1993. Application of multilocus enzyme electrophoresis in studies of the epidemiology of *Listeria monocytogenes* in Denmark. *Appl. Environ. Microbiol.* 59:2817–2822.

149. Notermans, S., J. Dufrenne, and M. van Schothorst. 1978. Enzyme-linked immunosorbent assay for detection of *Clostridium botulinum* toxin type A. *Jpn. J. Med. Sci. Biol.* 31:81–85.

150. Notermans, S., K.J. Heuvelman, and K. Wernars. 1988. Synthetic enterotoxin B DNA probes for detection of enterotoxigenic *Staphylococcus aureus* strains. *Appl. Environ. Microbiol.* 54:531–533.

151. Nuzback, D.E., E.E. Bartley, S.M. Dennis, T.G. Nagaraja, S.J. Galitzer, and A.D. Dayton. 1983. Relation of rumen ATP concentration to bacterial and protozoal numbers. *Appl. Environ. Microbiol.* 46:533–538.

152. Oda, M., M. Morita, H. Unno, and Y. Tanji. 2004. Rapid detection of *Escherichia coli* 0157:H7 by using green fluorescent protein-labeled PPO1 bacteriophage. *Appl. Environ. Microbiol.* 70:527–534.

153. Ortel, S. 1989. Listeriocins (monocins). *Int. J. Food Microbiol.* 8:249–250.

154. Padhye, N.V., and M.P. Doyle. 1991. Production and characterization of a monoclonal antibody specific for enterohemorrhagic *Escherichia coli* of serotypes 0157:H7 and 026:H11. *J. Clin. Microbiol.* 29:99–103.

155. Panicker, G., M.L. Myers, and A.K. Bej. 2004. Rapid detection of *Vibrio vulnificus* in shellfish and Gulf of Mexico water by real-time PCR. *Appl. Environ. Microbiol.* 70:498–507.

156. Park, C.E., H.B. El Derea, and M.K. Rayman. 1978. Evaluation of staphylococcal thermonuclease (TNase) assay as a means of screening foods for growth of staphylococci and possible enterotoxin production. *Can. J. Microbiol.* 24:1135–1139.

157. Patterson, J.W., P.L. Brezonik, and H.D. Putnam. 1970. Measurement and significance of adenosine triphosphate in activated sludge. *Environ. Sci. Technol.* 4:569–575.

158. Peng, M., and L.A. Shelef. 1998. Automated simultaneous detection of low levels of listeriae and salmonellae in foods. *Int. J. Food Microbiol.* 63:225–235.

159. Perry, B.F., A.E. Beezer, and R.J. Miles. 1983. Characterization of commercial yeast strains by flow microcalorimetry. *J. Appl. Bacteriol.* 54:183–189.

160. Pestka, J.J., and F.S. Chu. 1984. Enzyme-linked immunosorbent assay of mycotoxins using nylon bead and Terasaki plate solid phases. *J. Food Protect.* 47:305–308.

161. Pestka, J.J., P.K. Gaur, and F.S. Chu. 1980. Quantitation of aflatoxin B_1 and aflatoxin B_1 antibody by an enzyme-linked immunosorbent microassay. *Appl. Environ. Microbiol.* 40:1027–1031.

162. Pestka, J.J., V. Li, W.O. Harder, and F.S. Chu. 1981. Comparison of radioimmunoassay and enzyme-linked immunosorbent assay for determining aflatoxin M₁ in milk. *J. Assoc. Off. Anal. Chem.* 64:294–301.

163. Peterson, E.H., M.L. Nierman, R.A. Rude, and J.T. Peeler. 1987. Comparison of AOAC method and fluorogenic (MUG) assay for enumerating *Escherichia coli* in foods. *J. Food Sci.* 52:409–410.

164. Poelma, P.L., C.R. Wilson, and W.H. Andrews. 1987. Rapid fluorogenic enumeration of *Escherichia coli* in selected, naturally contaminated high moisture foods. *J. Assoc. Off. Anal. Chem.* 70:991–993.

165. Previte, J.J. 1972. Radiometric detection of some food-borne bacteria. *Appl. Microbiol.* 24:535–539.

166. Ramakrishna, N., J. Lacey, A.A.G. Candish, J.E. Smith, and I.A. Goodbrand. 1990. Monoclonal antibody-based enzyme linked immunosorbent assay of aflatoxin B₁, T-2 toxin, and ochratoxin A in barley. *J. Assoc. Off. Anal. Chem.* 73:71–76.

167. Rand, A.G., J. Ye, C.W. Brown, and S.V. Letcher. 2002. Optical biosensors for food pathogen detection. *Food Technol.* 56(3):32–39.

168. Rasooly, A. 2001. Surface plasmon reasonance analysis of staphylococcal enterotoxin B in food. *J. Food Protect.* 64:37–43.

169. Rippey, S.R., L.A. Chandler, and W.D. Watkins. 1987. Fluorometric method for enumeration of *Escherichia coli* in molluscan shellfish. *J. Food Protect.* 50:685–690.

170. Restaino, L., E.W. Frampton, and R.H. Lyon. 1990. Use of the chromogenic substrate 5-bromo-4-chloro-3-indolyl-β-glucuronide (X-GLUC) for enumerating *Escherichia coli* in 24 h from ground beef. *J. Food Protect.* 53:508–510.

171. Robern, H., M. Dighton, Y. Yano, and N. Dickie. 1975. Double-antibody radioimmunoassay for staphylococcal enterotoxin C₂. *Appl. Microbiol.* 30:525–529.

172. Robison, B.J. 1984. Evaluation of a fluorogenic assay for detection of *Escherichia coli* in foods. *Appl. Environ. Microbiol.* 48:285–288.

173. Rocourt, J., A. Audurier, A.L. Courtieu, J. Durst, S. Ortel, A. Schrettenbrunner, and A.G. Taylor. 1985. A multi-centre study on the phage typing of *Listeria monocytogenes. Zentralbl. Bakteriol. Mikrobiol. Hyg. [A].* 259:489–497.

174. Rowley, D.B., J.J. Previte, and H.P. Srinivasa. 1978. A radiometric method for rapid screening of cooked foods for microbial acceptability. *J. Food Sci.* 43:1720–1722.

175. Rudi, K., S.L. Flateland, J.F. Hanssen, G. Bengtsson, and H. Nissen. 2002. Development and evolution of a 16S ribosomal DNA array-based approach for describing complex microbial communities in ready-to-eat vegetable salads packed in a modified atmosphere. *Appl. Environ. Microbiol.* 68:1146–1156.

176. Saari, M., T. Cheasty, K. Leino, and A. Siitonen. 2001. Phage types and genotypes of Shiga toxin-producing *Escherichia coli* O157:H7 in Finland. *J. Clin. Microbiol.* 39:1140–1143.

177. Saiki, R.K., D.H. Gelfand, S. Stoffel, S.J. Sharf, R. Higuchi, G.T. Horn, K.B. Mullis, and H.A. Erlich. 1988. Primer-directed enzymatic amplification of DNA with a thermostable DNA polymerase. *Science* 239:487–491.

178. Saunders, G.C., and M.L. Bartlett. 1977. Double-antibody solid-phase enzyme immunoassay for the detection of staphylococcal enterotoxin A. *Appl. Environ. Microbiol.* 34:518–522.

179. Savelkoul, P.H.N., H.J.M. Aarts, J. de Haas, L. Dijkshoorn, B. Duim, M. Otsen, J.L.W. Rademaker, L. Schouls, and J.A. Lenstra. 1999. Amplified-fragment length polymorphism analysis: The state of the art. *J. Clin. Microbiol.* 37:3083–3091.

180. Schmidt, J.J., and R.G. Stafford. 2003. Fluorogenic substrates for the protease activities of botulinum neurotoxins, serotypes A, B, and F. *Appl. Environ. Microbiol.* 69:297–303.

181. Schwab, K.J., F.H. Neill, F. le Guyader, M.K. Estes, and R.L. Atmar. 2001. Deveopmment of a reverse transcription-PCR-DNA enzyme immunoassay for detection of "Norwalk-like" viruses and hepatitis A virus in stool and shellfish. *Appl. Environ. Microbiol.* 67:742–749.

182. Scott, T.M., S. Parveen, K.M. Portier, J.B. Rose, M.L. Tamplin, S.R. Farrah, A. Koo, and J. Lukasik. 2003. Geographical variation in ribotype profiles of *Escherichia coli* isolates from humans, swine, poultry, beef, and dairy cattle in Florida. *Appl. Environ. Microbiol.* 69:1089–1092.

183. Selander, R.K., D.A. Caugant, H. Ochman, J.M. Musser, M.N. Gilmour, and T.S. Whittam. 1986. Methods of multilocus enzyme electrophoresis for bacterial population genetics and systematics. *Appl. Environ. Microbiol.* 51:873–884.

184. Seifert, H., and P. Gerner-Smidt. 1995. Comparison of ribotyping and pulsed-field gel electrophoresis for molecular typing of *Acinetobacter* isolates. *J. Clin. Microbiol.* 33:1402–1407.

185. Seiter, J.A., and J.M. Jay. 1980. Comparison of direct serial dilution and most-probable-number methods for determining endotoxins in meats by the *Limulus* amoebocyte lysate test. *Appl. Environ. Microbiol.* 40:177–178.

186. Seo, K.H., R.E. Brackett, J.F. Frank, nd S. Hilliard. 1998. Immunomagnetic separation and flow cytometry for rapid detection of *Escherichia coli* O157:H7. *J. Food Protect.* 61:812–816.

187. Sharpe, A.N., M.N. Woodrow, and A.K. Jackson. 1970. Adenosine-triphosphate (ATP) levels in foods contaminated by bacteria. *J. Appl. Bacteriol.* 33:758–767.

188. Shelef, L.A., A. Surtani, K. Kanagapandian, and W. Tan. 1998. Automated detection of amino acid decarboxylation in salmonellae and other enterobacteriaceae. *Food Microbiol.* 15:199–205.

189. Shelef, L.A., S. Mohammed, W. Tan, and M.L. Webber. 1997. Rapid optical measurements of microbial contamination in raw ground beef and effects of citrate and lactate. *J. Food Protect.* 60:673–676.

190. Siragusa, G.R., and C.N. Cutter. 1995. Microbial ATP bioluminescence as a means to detect contamination on artificially contaminated beef carcass tissue. *J. Food Protect.* 58:764–769.

191. Skinner, G.E., S.M. Gendel, G.A. Fingerhut, H.A. Solomon, and J. Ulaszek. 2000. Differentiation between types and strains of *Clostridium botulinum* by riboprinting. *J. Food Protect.* 63:1347–1352.

192. Somer, L., and Y. Kashi. 2003. A PCR method based on 16S rRNA sequence for simultaneous detection of the genus *Listeria* and the species *Listeria monocytogenes* in food products. *J. Food Protect.* 66:1658–1665.

193. Sperber, W.H., and R.H. Deibel. 1969. Accelerated procedure for *Salmonella* detection in dried foods and feeds involving only broth cultures and serological reactions. *Appl. Microbiol.* 17:533–539.

194. Stannard, C.J., and J.M. Wood. 1983. The rapid estimation of microbial contamination of raw meat by measurement of adenosine triphosphate (ATP). *J. Appl. Bacteriol.* 55:429–438.

195. Stewart, G.S.A.B., and P. Williams. 1992. *Lux* genes and the applications of bacterial bioluminescence. *J. Gen. Microbiol.* 138:1289–1300.

196. Stiffler-Rosenberg, G., and H. Fey. 1978. Simple assay for staphylococcal enterotoxins A, B, and D: Modification of enzyme-linked immunosorbent assay. *J. Clin. Microbiol.* 8:473–479.

197. Stoflet, E.S., D.D. Koeberi, G. Sarkar, and S.S. Summer 1988. Genomic amplification with transcript sequencing. *Science* 239:491–494.

198. Strange, R.E., E.O. Powell, and T.W. Pearce. 1971. The rapid detection and determination of sparse bacterial populations with radioactively labelled homologous antibodies. *J. Gen. Microbiol.* 67:349–357.

199. Sullivan, J.D., Jr., P.C. Ellis, R.G. Lee, W.S. Combs, Jr., and S.W. Watson. 1983. Comparison of the *Limulus* amoebocyte lysate test with plate counts and chemical analyses for assessment of the quality of lean fish. *Appl. Environ. Microbiol.* 45:720–722.

200. Surdy, T.E., and S.G. Haas. 1981. Modified enrichment-serology procedure for detection of salmonellae in soy products. *Appl. Environ. Microbiol.* 42:704–707.

201. Swaminathan, B., and J.C. Ayres. 1980. A direct immunoenzyme method for the detection of salmonellae in foods. *J. Food Sci.* 45:352–355, 361.

202. Swaminathan, B., and P. Feng. 1994. Rapid detection of food-borne pathogenic bacteria. *Annu. Rev. Microbiol.* 48:401–426.

203. Sword, C.P., and M.J. Pickett. 1961. The isolation and characterization of bacteriophages from *Listeria monocytogenes*. *J. Gen. Microbiol.* 25:241–248.

204. Tai, J.Y., R.C. Seid, Jr., R.D. Hurn, and T.-Y. Liu. 1977. Studies on *Limulus* amoebocyte lysate. II. Purification of the coagulogen and the mechanism of clotting. *J. Biol. Chem.* 252:4773–4776.

205. Tamarapu, S., J.L. McKillip, and M. Drake. 2001. Development of a multiplex polymerase chain reaction assay for detection and differentiation of *Staphylococcus aureus* in dairy products. *J. Food Protect.* 64:664–668.

206. Terplan, V.G., K.-J. Zaadhof, and S. Buchholz-Berchtold. 1975. Zum nachweis von Endotoxinen gramnegativer Keime in Milch mit dem *Limulus*-test. *Arch Lebensmittelhyg.* 26:217–221.

207. Thompson, J.S., D.S. Hodge, and A.A. Borczyk. 1990. Rapid biochemical test to identify verocytotoxin-positive strains of *Escherichia coli* serotype 0157. *J. Clin. Microbiol.* 28:2165–2168.

208. Thore, A., S. Ånséhn, A. Lundin, and S. Bergman. 1975. Detection of bacteriuria by luciferase assay of adenosine triphosphate. *J. Clin. Microbiol.* 1:1–8.

209. Tims, T.B., S.S. Dickey, D.R. DeMarco, and D.V. Lim. 2001. Detection of low levels of *Listeria monocytogenes* within 20 hours using an evanescent wave biosensor. *Amer. Clin. Lab.* 20(8):28–29.

210. Tsuji, K., P.A. Martin, and D.M. Bussey. 1984. Automation of chromogenic substrate *Limulus* amebocyte lysate assay method for endotoxin by robotic system. *Appl. Environ. Microbiol.* 48:550–555.

211. Turpin, P.E., K.A. Maycroft, J. Bedford, C.L. Rowlands, and E.M.H. Wellington. 1993. A rapid luminescent-phage based MPN method for the enumeration of *Salmonella typhimurium* in environmental samples. *Lett. Appl. Microbiol.* 16:24–27.

212. Ulitzur, S., and J. Kuhn. 1987. Introduction of *lux* genes into bacteria, a new approach for specific determination of bacteria and their antibiotic susceptibility. In *Bioluminescence and Chemiluminescence: New Perspectives*, ed. J. Schlomerich, R. Andreesen, A. Knapp, et al., 463–472. New York: Wiley.

213. Van Wart, M., and L.J. Moberg. 1984. Evaluation of a novel fluorogenic-based method for detection of *Escherichia coli*. *Bacteriol. Proc.* 201.

214. Wachtel, R.E., and K. Tsuji. 1977. Comparison of *Limulus* amebocyte lysates and correlation with the United States Pharmacopeial pyrogen test. *Appl. Environ. Microbiol.* 33:1265–1269.

215. Walker, N.J. 2002. A technique whose time has come. *Science* 296:557–559.

216. Wan, J., K. King, S. Forsyth, and M.J. Coventry. 2003. Detection of *Listeria monocytogenes* in salmon using the Probelia polymerase chain reaction system. *J. Food Protect.* 66:436–440.

217. Warner, J.M., and J.D. Oliver. 1998. Randomly amplified polymorphic DNA analysis of starved and viable but nonculturable *Vibrio vulnificus* cells. *Appl. Environ. Microbiol.* 64:3025–3028.

218. Warren, L.S., R.E. Benoit, and J.A. Jessee. 1978. Rapid enumeration of fecal coliforms in water by a colorimetric β-galactosidase assay. *Appl. Environ. Microbiol.* 35:136–141.

219. Watkins, W.D., S.B. Rippey, C.S. Clavet, D.J. Kelley-Reitz, and W. Burkhardt III. 1988. Novel compound for identifying *Escherichia coli*. *Appl. Environ. Microbiol.* 54:1874–1875.

220. Weihe, J.L., S.L. Seist, and W.S. Hatcher, Jr. 1984. Estimation of microbial populations in frozen concentrated orange juice using automated impedance measurements. *J. Food Sci.* 49:243–245.

221. Wendlinger, G., M.J. Loessner, and S. Scherer. 1996. Bacteriophage receptors on *Listeria monocytogenes* cells are the N-acetylglucosamine and rhamnose substituents of teichoic acids or the peptidoglycan itself. *Microbiology* 142:985–992.

222. Wesley, I.V., and F. Ashton. 1991. Restriction enzyme analysis of *Listeria monocytogenes* strains associated with foodborne epidemics. *Appl. Environ. Microbiol.* 57:969–975.

223. Wesley, I.V., R.D. Wesley, J. Heisick, F. Harrel, and D. Wagner. 1990. Restriction enzyme analysis in the epidemiology of *Listeria monocytogenes*. In *Symposium on Cellular and Molecular Modes of Action of Selected Microbial Toxins in Foods and Feeds*, ed. J.L. Richard, 225–238. New York: Plenum.

224. Wolber, P.K. 1993. Bacterial ice nucleation. *Adv. Microbiol. Physiol.* 34:203–237.

225. Wolcott, M.J. 1991. DNA-based rapid methods for the detection of foodborne pathogens. *J. Food Protect.* 54:387–401.

226. Wood, J.M., V. Lach, and B. Jarvis. 1977. Detection of food-associated microbes using electrical impedance measurements. *J. Appl. Bacteriol.* 43:14–15.

227. Zaadhof, K.-J., and G. Terplan. 1981. Der *Limulus*-Test—ein Verfahren zur Beurteilung der mikrobiologischen Qualität von Milch und Milch produkten. *Deutsch Molkereizeitung.* 34:1094–1098.

228. Zayaitz, A.E.K., and R.A. Ledford. 1982. Proteolytic inactivation of thermonuclease activity of *Staphylococcus aureus* during recovery from thermal injury. *J. Food Protect.* 45:624–626.

229. Ye, J., S.V. Letcher, and A.G. Rand. 1997. Piezoelectric biosensor for detection of *Salmonella typhimurium*. *J. Food Sci.* 62:1067–1071, 1086.

230. Ye, R.W, T. Wang, L. Bedzyk, and K.M. Croker. 2001. Applications of DNA microarrays in microbial systems. *J. Microbiol. Meth.* 47:257–272.

231. Zink, R., M.J. Loessner, and S. Scherer. 1995. Characterization of cryptic prophages (monocins) in *Listeria* and sequence analysis of a holin/endolysin gene. *Microbiology* 141:2577–2584.

CHAPTER 12

Bioassay and Related Methods

After establishing the presence of pathogens or toxins in foods or food products, the next important concern is whether the organisms/toxins are biologically active. For this purpose, experimental animals are employed where feasible. When it is not feasible to use whole animals or animal systems, a variety of tissue culture systems have been developed that, by a variety of responses, provide information on the biological activity of pathogens or their toxic products. These bioassay and related tests are the methods of choice for some foodborne pathogens, and some of the principal ones are listed in Table 12–1.

WHOLE-ANIMAL ASSAYS

Mouse Lethality

This method was first employed for foodborne pathogens around 1920 and continues to be an important bioassay method. To test for botulinal toxins in foods, appropriate extracts are made and portions are treated with trypsin (for toxins of nonproteolytic *Clostridium botulinum* strains). Pairs of mice are injected intraperitoneally (IP) with 0.5 ml of trypsin-treated and untreated preparations. Untreated preparations that have been heated for 10 minutes at 100°C are injected into a pair of mice. All injected mice are observed for 72 hours for symptoms of botulism or death. Mice injected with the heat preparations should not die because the botulinal toxins are heat labile. Specificity in this test can be achieved by protecting mice with known botulinal antitoxin, and in a similar manner, the specific serologic type of botulinal toxin can be determined (see Chapter 24 for toxin types).

Mouse lethality may be employed for other toxins. Stark and Duncan[45] used the method for *Clostridium perfringens* enterotoxin. Mice were injected IP with enterotoxin preparations and observed for up to 72 hours for lethality. The mouse-lethal dose was expressed as the reciprocal of the highest dilution that was lethal to the mice within 72 hours. Genigeorgis et al.[18] employed the method by use of intravenous (IV) injections. *C. perfringens* enterotoxin preparations were diluted in phosphate buffer, pH 6.7, to achieve a concentration of 5–12 μg/ml. From each dilution prepared, 0.25 ml was injected IV into six male mice weighing 12–20 g, the number of deaths were recorded, and the LD_{50} was calculated. The mouse is the most widely used animal for virulence assessment of *Listeria* spp. The LD_{50} for *L. monocytogenes* in normal adult mice is 10^5–10^6, and for 15-g infant mice, as few as 50 cells may be lethal (see Chapter 25).

Table 12–1   Some Bioassay Models Used to Assess the Biological Activity of Various Foodborne Pathogens and/or Their Products (Taken from the Literature)

Organism	Toxin/Product	Bioassay Method	Sensitivity
A. hydrophila	Cytotoxic enterotoxin	Infant mouse intestines	~30 ng
B. cereus	Diarrheagenic toxin	Monkey feeding	
	Diarrheagenic toxin	Rabbit ileal loop	
	Diarrheagenic toxin	Rabbit skin	
	Diarrheagenic toxin	Guinea pig skin	
	Diarrheagenic toxin	Mouse lethality	
	Emetic toxin	Rhesus monkey emesis	
	Emetic toxin	*Suncus murinus*	ED_{50} 12.9 μg/kg
C. jejuni	Viable cells	Adult mice	10^4 cells
	Viable cells	Chickens	90 cells
	Viable cells	Neonatal mice	
	Culture supernatants	Adult rat jejunal loops	
	Enterotoxin	Rat ileal loop	
C. botulinum	A, B, E, F, G, toxins	Mouse lethality	
C. perfringens A	Enterotoxin	Mouse lethality, LD_{50}	1.8 μg
	Enterotoxin	Mouse ileal loop, 90-min test	1.0 μg
	Enterotoxin	Rabbit ileal loop, 90-min test	6.25 μg
	Enterotoxin	Guinea pig skin (erythemal activity)	0.06–0.125 mg/ml
Infant botulism	Endospores	7- to 12-day-old rats	1,500 spores
	Endospores	9-day-old mice	700 spores
	Endospores	Adult germ-free mice	10 spores
E. coli	LT	Rabbit ileal loop, 18-hour test	
	ST	Suckling mouse (fluid accumulation)	
	ST	Rabbit ileal loop, 6-hour test	
	ST_a	Suckling mouse	
	ST_a	1- to 3-day-old piglets	
	ST_b	Jejunal loop of pig	
	ST_b	Weaned piglets, 7–9 weeks old	
E. coli O157:H7	ETEC	Mouse colonization	
Salmonella enterica	Gastroenteritis	N.Z. White rabbit	$>10^5$ cells
Salmonella spp.	Heat-labile cytotoxin	Rabbit ileal loop (protein synthesis inhibition)	
S. aureus	SEB	Skin of specially sensitized guinea pigs	0.1–1.0 pg
	All enterotoxins	Emesis in rhesus monkeys	5 μg/2–3 kg body wt
	SEA, SEB	Emesis in suckling kittens	0.1, 0.5 μg/kg body wt
V. parahaemolyticus	Broth cultures	Rabbit ileal loop; response in 50% animals	10^2 cells
	Viable cells	Adult rabbit ileal loop, invasiveness	

continues

Table 12–1   continued

Organism	Toxin/Product	Bioassay Method	Sensitivity
	Thermostable direct toxin	Mouse lethality, death in 1 minute	5 μg/mouse
	Thermostable direct toxin	Mouse lethality, LD$_{50}$ by IP route	1.5 μg
	Thermostable direct toxin	Rabbit ileal loop	250 μg
	Thermostable direct toxin	Guinea pig skin	2.5 μg/g
V. vulnificus	Culture filtrates	Rabbit skin permeability	
V. cholerae (non-01)	Enterotoxin	Suckling mice	
Y. enterocolitica	Heat-stable toxin	Sereny test	
	Heat-stable toxin	Suckling mouse (oral)	110 ng
	Enterotoxin	Rabbit ileal loop, 6- and 18-hour tests	
	Viable cells	Rabbit diarrhea	50% infectious dose = 2.9 $\times$ 10^{8}
	Viable cells	Lethality in suckling mice by IP injection	14 cells
	Viable cells	Lethality of gerbils by IP injection	100 cells

Note: LT = heat-labile toxin; ST = heat-stable toxin; SEA = staphylococcal enterotoxin A; ED$_{50}$ = see text; ETEC = enterotoxigenic *E. coli*.

Suncus murinus

This small animal has been used in Japan as an experimental model for emesis research using a variety of drugs,[48] and it has been shown to respond to *cereulide*, the emetic toxin of *Bacillus cereus*.[1] *Suncus murinus* is referred to as the Japanese house shrew, and adults do not exceed 100 g in weight. For experimental use, those weighing 50–80 g are used. In their study of the emetic toxin of *B. cereus*, Agata et al.[1] found the ED$_{50}$ (quantity of toxin required for emesis in one half of the exposed animals) in *Suncus* by oral administration to be 12.9 μg/kg. The ED$_{50}$ by the intraperitoneal route was 9.8 μg/kg. Whether *Suncus* is a suitable animal model for the staphylococcal or other enterotoxins is unclear.

Ferrets

Adult female ferrets (mean body weight of 735 g) have been evaluated as sensitive animals for staphylococcal enterotoxin B (SEB). SEB was administered to the stomach via tube to animals deprived of food for 24 hours but with free access to water. Doses of SEB of 1, 2, or 5 mg were added to sterile saline, and animals were observed for body temperature, pressure, incidence of retching, vomiting, and defecation for a period of 3 hours with observations every 5 minutes.[52] Those given 5 mg SEB showed significant increase in subcutaneous temperature after 75 minutes. compared to controls that received only saline. All animals given 5 mg SEB retched (after ca. 105 minutes, total of 18 times) and vomited (after ca. 106 minutes, total of 2 times). The ferret was shown previously to respond to SEB when injected intravenously.

Suckling (Infant) Mouse

This animal model was introduced by Dean et al.[12] primarily for *Escherichia coli* enterotoxins and is now used for some other foodborne pathogens. Typically, mice are separated from their mothers and given oral doses of the test material consisting of 0.05–0.1 ml with the aid of a blunt 23-gauge hypodermic needle. A drop of 5% Evans blue dye per milliliter of test material may be used to determine the presence of the test material in the small intestine. The animals are usually held at 25°C for 2 hours and then killed. The entire small intestine is removed, and the relative activity of test material is determined by the ratio of gut weight to body weight (GW/BW). Giannella[19] found the following GW/BW ratios for *E. coli* enterotoxins: <0.074 = negative test; 0.075–0.082 = intermediate (should be retested); and >0.083 = positive test. The investigator found the day-to-day variability among various *E. coli* strains to range from 10.5% to 15.7% and about 9% for replicate tests with the same strain. A GW/BW of 0.060 was considered negative for *E. coli* ST_a by Mullan et al.[32] In studies with *E. coli* ST, Wood et al.[51] treated as positive GW/BW ratios that were >0.087, whereas Boyce et al.[6] held mice at room temperature for 4 hours for *Yersinia enterocolitica* heat-stable enterotoxin and considered a GW/BW of 0.083 or greater to be positive. In studies with *Y. enterocolitica*, Okamoto et al.,[36] keeping mice for 3 hours at 25°C, considered a GW/BW of 0.083 to be positive.

In using the suckling mouse model, test material may also be injected percutaneously directly in the stomach through the mouse's translucent skin or by administration orogastrically or intraperitoneally. For the screening of large numbers of cultures, the intestines may be examined visually for dilation and fluid accumulation.[38] Infant mice along with 1- to 3-day-old piglets are the animals of choice for *E. coli* enterotoxin ST_a; ST_b is inactive in the suckling mouse but active in piglets and weaned pigs.[7,27] The infant mouse assay does not respond to choleragen or to the heat-labile toxin (LT) of *E. coli*. It correlates well with the 6-hour rabbit ileal loop assay for the ST_a of *E. coli*.

Suckling mice have been used for lethality studies by employing IP injections. Aulisio et al.[3] used 1- to 3-day-old Swiss mice and injected 0.1 ml of diluted culture. The mice were observed for 7 days; deaths that occurred within 24 hours were considered nonspecific, whereas deaths occurring between days 2 and 7 were considered specific for *Y. enterocolitica*. By this method, an LD_{50} can be calculated relative to numbers of cells per inoculum. In the case of *Y. enterocolitica*, Aulisio et al. found the LD_{50} to be 14 cells, and the average time for death of mice to be 3 days.

Rabbit and Mouse Diarrhea

Rabbits and mice have been employed to test for diarrheagenic activity of some foodborne pathogens. Employing young rabbits weighing 500–800 g, Pai et al.[37] inoculated orogastrically with approximately 10^{10} cells of *Y. enterocolitica* suspended in 10% sodium bicarbonate. Diarrhea developed in 87% of 47 rabbits after a mean time of 5.4 days. Bacterial colonization occurred in all animals regardless of dose of cells.

Mice deprived of water for 24 hours were used by Schiemann[41] to test for the diarrheogenic activity of *Y. enterocolitica*. The animals were given inocula of 10^9 cells/ml in peptone water, and fresh drinking water was allowed 24 hours later. After 2 days, feces of mice were examined for signs of diarrhea.

Infant rabbits have been used by Smith[42] to assay enterotoxins *E. coli* and *Vibrio cholerae*. Infant rabbits 6–9 days old are administered 1–5 ml of culture filtrate via stomach tube. Following return to their mothers, they are observed for diarrhea. Diarrhea after 6–8 hours is a positive response. If death of animals occurs, a large volume of yellow fluid is found in the small and large intestines. The quantitation of enterotoxin is achieved by ascertaining the ratio of intestinal weight to total body weight. Young pigs have been used in a similar way to assay porcine strains of *E. coli* for enterotoxin activity. Infant rabbits have been employed to detect Shiga-like toxins of *E. coli*.[34]

Monkey Feeding

The use of rhesus monkeys (*Macaca mulatta*) to assay staphylococcal enterotoxins was developed in 1931 by Jordan and McBroom.[26] Next to humans, this is perhaps the animal most sensitive to staphylococcal enterotoxins. When enterotoxins are to be assayed by this method, young rhesus monkeys weighing 2–3 kg are selected. The food homogenate, usually in solution in 50-ml quantities, is administered via stomach tube. The animals are then observed continuously for 5 hours. Vomiting in at least two of six animals denotes a positive response. Rhesus monkeys have been shown to respond to levels of enterotoxins A and B as low as approximately 5 μg per 2–3 kg of body weight.[31]

Kitten (Cat) Test

This method was developed by Dolman et al.[15] as an assay for staphylococcal enterotoxins. The original test employed the injection of filtrates into the abdominal cavity of very young kittens (250–500 g). This procedure leads to false-positive results. The most commonly used method consists of administering the filtrates IV and observing the animals continuously for emesis. When cats weighing 2–4 kg are used, positive responses occur in 2–6 hours.[10] Emesis has been reported to occur with 0.1 and 0.5 μg of staphylococcal enterotoxin A (SEA) and SEB per kilogram of body weight.[4] The test tends to lack the specificity of the monkey-feeding test because staphylococcal culture filtrates containing other byproducts may also induce emesis. Kittens are much easier to obtain and maintain than rhesus monkeys, and in this regard the test has value.

Rabbit and Guinea Pig Skin Tests

The skin of these two animals is used to assay toxins for at least two properties. The vascular permeability test is generally done by use of albino rabbits weighing 1.5–2.0 kg. Typically, 0.05–0.1 ml of culture filtrate is inoculated intradermally (ID) in a shaved area of the rabbit's back and sides. From 2 to 18 hours later, a solution of Evans blue dye is administered IV, and 1–2 hours are allowed for permeation by the dye. The diameters of two blue zones are measured and the area approximated by squaring the average of the two values. Areas of 25 cm^2 are considered positive. *E. coli* LT gives a positive response in this assay.[16] Employing this assay, permeability has been shown to be a function of the *E. coli* diarrheagenic enterotoxin.

Similar to the permeability factor test is a test of erythemal activity that employs guinea pigs. The method has been employed by Stark and Duncan[45] to test for erythemal activity of *C. perfringens* enterotoxin. Guinea pigs weighing 300–400 g are depilated (back and sides) and marked in 2.5-cm squares, and duplicate 0.05-ml samples of toxic preparations are injected ID in the center of the squares. Animals are observed after 18–24 hours for erythema at the injection site. In the case of *C. perfringens* enterotoxin, a concentric area of erythema is produced without necrosis. A unit of erythemal activity is defined as the amount of enterotoxin producing an area of erythema 0.8 cm in diameter. The enterotoxin preparation used by Stark and Duncan contained 1,000 erythemal units/ml. To enhance readings, 1 ml of 0.5% Evans blue can be injected intracardially (IC) 10 minutes following the skin injections and the diameters read 80 minutes later.[18] The specificity of the skin reactions can be determined by neutralizing the enterotoxin with specific antisera prior to injections. The erythema test was found to be 1,000 times more sensitive than the rabbit ileal loop technique for assaying the enterotoxin of *C. perfringens*.[23]

Sereny and Anton Tests

The Sereny method is used to test for virulence of viable bacterial cultures. It was proposed by Sereny in 1955, and the guinea pig is the animal most often used. The test consists of administering, with the aid of a loop, a drop of cell suspension, containing 1.5×10^{10} to 2.3×10^{10}/ml in phosphate-buffered saline, into the conjunctivae of guinea pigs weighing about 400 g each. The animal's eyes are examined daily for 5 days for evidence of keratoconjunctivitis. When strains of unknown virulence are evaluated, it is important that known positive and negative strains are tested also.

A mouse Sereny test has been developed using Swiss mice and administering half of the dose noted above. A Sereny test for shigellae and enteroinvasive *E. coli* (EIEC) strains has been developed and found useful.[33]

The Anton test is similar to the Sereny; it is used to assess the virulence of *Listeria* spp. Conjunctivitis is produced when about 10^6 cells of *L. monocytogenes* are administered into the eye of a rabbit or guinea pig.[2]

ANIMAL MODELS REQUIRING SURGICAL PROCEDURES

Ligated Loop Techniques

These techniques are based on the fact that certain enterotoxins elicit fluid accumulation in the small intestines of susceptible animals. Although they may be performed with a variety of animals, rabbits are most often employed. Young rabbits 7–20 weeks old and weighing 1.2–2.0 kg are kept off food and water for a period of 24 hours or off food for 48–72 hours with water ad libitum prior to surgery. Under local anesthesia, a midline incision about 2 inches long is made just below the middle of the abdomen through the muscles and peritoneum in order to expose the small intestines.[11] A section of the intestine midway between its upper and lower ends or just above the appendix is tied with silk or other suitable ligatures in 8 to 12-cm segments with intervening sections of at least 1 cm. Up to six sections may be prepared by single or double ties.

Meanwhile, the specimen or culture to be tested is prepared, suspended in sterile saline, and injected intraluminally into the ligated segments. A common inoculum size is 1 ml, although smaller or larger doses may be used. Different doses of test material may be injected into adjacent loops or into loops separated by a blank loop or by a sham (inoculated with saline). Following injection, the abdomen is closed with surgical thread, and the animal is allowed to recover from anesthesia. The recovered animal may be kept off food and water for an additional 18–24 hours, or water or feed or both may be allowed. With ligatures intact, the animals may not survive beyond 30–36 hours.[8]

To assess the effect of the materials previously injected into ligated loops, the animal is killed, and the loops are examined and measured for fluid accumulation. The fluid may be aspirated and measured. The reaction can be quantitated by measuring loop fluid volume to loop length ratios,[8] or by determining the ratio of fluid volume secreted per milligram of dry weight intestine.[30] The minimum amount of *C. perfringens* enterotoxin necessary to produce a loop reaction has been reported variously to be 28–40 μg and as high as 125 μg of toxin by the standard loop technique. The 90-minute loop technique has been found to respond to as little as 6.25 μg and the standard technique to 29 μg of toxin.[18]

This technique was developed to study the mode of action of the cholera organism in producing the disease.[11] It has been employed widely in studies on the virulence and pathogenesis of foodborne pathogens, including *Bacillus cereus*, *C. perfringens*, *E. coli*, and *Vibrio parahaemolyticus*.

Although the rabbit loop is the most widely used of ligated loop methods, other animal models are used. The mouse intestinal loop may be used for *E. coli* enterotoxins. As used by Punyashthiti and Finkelstein.[40] Swiss mice (18–22 g) are deprived of food 8 hours before use. The abdomen is opened under light anesthesia, and two 6-cm loops separated by 1-cm interloops are prepared. The loops are inoculated with 0.2 ml of test material, followed by closing of the abdomen. Animals are deprived of food and water and killed 8 hours later. Fluid is measured and the length of the loops determined. Results are considered positive when the ratio of fluid to length is 50 or more mg/cm. In this study, positive loops generally had ratios between 50 and 100, but occasionally approached 200 or more. Alternatively, the net increase in weight of loops in milligrams can be used to measure the intensity of a toxic reaction.[53] With the mouse loop, 1 μg of enterotoxin can be detected.[53] A rat jejunal loop assay has been presented for detecting the ST$_b$ enterotoxin of *E. coli*. A linear dose response was found using 250- to 350-g rats but only after the endogenous protease activity was blocked with soybean trypsin inhibitor.[49]

The RITARD Model

The removable intestinal tie–adult rabbit diarrhea (RITARD) method was developed by Spira et al.[44] Rabbits weighing 1.6–2.7 kg are kept off food for 24 hours but allowed water. Under local anesthesia, the cecum is brought out and ligated close to the ileocecal junction. The small intestine is now brought out and a slip knot tied to close it in the area of the mesoappendix. Test material in 10 ml of phosphate-buffered saline is injected into the lumen of the anterior jejunum. After injection, intestine and cecum are returned to the peritoneal cavity and the incision closed. With the animal kept in a box, the temporary tie is removed 2–4 hours after administration of the test dose, and the slip knot in the intestine is released. Sutures are applied as needed. The animal is now returned to its cage and provided with food and water. Animals are observed for diarrhea or death at 2-hour intervals up to 124 hours. At autopsy, small intestine and adjacent sections are tied and removed for fluid measurement. Enterotoxigenic strains of *E. coli* produce severe and watery diarrhea, and the susceptibility of animals to *V. cholerae* infections is similar in this system to that in the infant rabbit model.

The gist of the RITARD model is that the animals are not altered except that the cecum is ligated to prevent it from taking up fluid from the small intestine, and a temporary reversible obstruction is placed on the ileum long enough to allow the inoculated organism to initiate colonization of the small intestine. The method has been successfully used as an animal model for *Campylobacter jejuni* infection[9] and to test virulence of *Aeromonas* strains.[39]

CELL CULTURE SYSTEMS

A variety of cell culture systems are employed to assess certain pathogenic properties of viable cells. The properties often assessed are invasiveness, permeability, cytotoxicity, adherence/adhesion/binding, and other more general biological activities. Some cell cultures are used to assess various properties of toxins and enterotoxins. Some examples of these models are summarized in Table 12–2, and brief descriptions are presented below.

Human Mucosal Cells

As employed by Ofek and Beachey,[35] human buccal mucosa cells (about 2×10^5 in phosphate-buffered saline) are mixed with 0.5 ml of washed *E. coli* cells—2×10^8/ml. The mixture is rotated for 30 minutes at room temperature. Epithelial cells are separated from the bacteria by differential centrifugation, followed by drying and staining with gentian violet. Adherence is determined by the microscopic counting of bacteria per epithelial cell. As employed by Thorne et al.,[47] *E. coli* cells are labeled with ^{3}H-amino acids (alanine and leucine) or fluorescein isothiocyanate. In another use of this method, *V. parahaemolyticus* cells were mixed with mucosal epithelial cells and incubated at 37°C for 5 minutes followed by filtering. The unbound cells were washed off, and the culture dried, fixed, and stained with Giemsa. Adherence was quantitated by counting the total number of *V. parahaemolyticus* adhering to 50 buccal cells as compared to controls. Best results were obtained when approximately 10^9 bacterial cells and 10^5 buccal cells were suspended together in phosphate-buffered saline at pH 7.2 for 5 minutes. All 12 strains tested adhered. Adherence apparently bears no relationship to pathogenicity for *V. parahaemolyticus*.

Human Fetal Intestine

By this adherence model, human fetal intestine (HFI) cells are employed in monolayers. The monolayers are thoroughly washed, inoculated with a suspension of *V. parahaemolyticus*, and incubated at 37°C for up to 30 minutes. Adherence is determined by the microscopic examination of stained cells after washing away unattached bacteria. All strains of *V. parahaemolyticus* tested adhered, but those from food-poisoning cases had a higher adherence ability than those from foods.[22] By use of this method, the adherence of an enteropathogenic strain of *E. coli* of human origin was found to be plasmid mediated.[50]

Human Ileal and Intestinal Cells

To study adherence of enterotoxigenic *E. coli* (ETEC), Deneke et al.[13] used ileal cells from adult humans in a filtration–binding assay. The cells were mixed with bacteria grown in ^{3}H-alanine and leucine. The amount of binding was determined with a scintillation counter. ETEC strains of human origin bound to a greater extent than controls. Binding to human ileal cells was 10- to 100-fold greater than to human buccal cells.

Monolayers of human intestine cells were used by Gingras and Howard[20] to study adherence of *V. parahaemolyticus*. The bacterium was grown in the presence of ^{14}C-labeled valine, and the labeled cells were added to monolayers and incubated for up to 60 minutes. Following incubation, unattached cells were removed, and those adhering were counted by radioactive counts of monolayers. The adhered cells were also enumerated microscopically. The Kanagawa-positive and -negative organisms adhered similarly. No correlation was found between hemolysis production and adherence.

Guinea Pig Intestinal Cells

To study adherence of *V. parahaemolyticus*, Iijima et al.[25] employed adult guinea pigs weighing about 300 g and fasted them for 2 days before use. Under anesthesia, the abdomen was opened and the

Table 12–2 Some Tissue and Cell Culture Systems Employed to Study Biological Activity of Gastroenteritis-Causing Organisms or Their Products (Taken from the Literature)

Culture System	Pathogen/Toxin	Demonstration/Use
CHO monolayer	E. coli LT; V. cholerae toxin	Biological activity
	V. parahaemolyticus	Biological activity
	Salmonella toxin	Biological activity
	C. jejuni enterotoxin	Biological activity
CHO floating cell assay	Salmonella toxin	Biological activity
HeLa cells	E. coli	Invasiveness
	Y. enterocolitica	Invasiveness
	V. parahaemolyticus	Adherence
	C. jejuni	Invasiveness
Vero	E. coli O157: H7	Shiga-like toxin receptors
Vero cells	C. perfringens enterotoxin	Mode of action
	E. coli LT	Biological activity, assay
	A. hydrophila toxin	Cytotoxicity
	C. perfringens enterotoxin	Binding
	C. perfringens enterotoxin	Biological activity
	Salmonella cytotoxin	Protein synthesis inhib.
	V. vulnificus	Cytotoxicity
Y-1 adrenal cells	E. coli LT	Biological activity, assay
	V. cholerae toxin	Biological activity, assay
	V. mimicus	Biological activity
Rabbit intestine epithelial cells	C. perfringens enterotoxin	Binding
	Salmonella cytotoxin	Protein synthesis inhib.
Murine spleen cells	Staph. enterotoxins A, B, and E	Binding
Macrophages	Y. enterocolitica	Phagocytosis
Human peripheral lymphocytes	Staph. enterotoxin A	Biological effects
Human laryngeal carcinoma	E. coli, Shigella	Invasiveness
Henle 407 human intestine	E. coli, Shigella	Invasiveness
Henle 407	L. monocytogenes	Invasiveness
	E. coli O157: H7	Adherence
Caco-2	V. cholerae non-01	Adherence
	ETEC	Adhesions
	L. monocytogenes	Invasion
HT29.74	C. parvum	Infection model
	C. perfringens	Cell lethality
Peritoneal macrophages	L. monocytogenes	Intracellular survival
Murine embryo primary fibroblasts	L. monocytogenes	Interleukin production
Human fetal intestinal cells	V. parahaemolyticus	Adherence
	Enteropathogenic E. coli	Adherence
	B. cereus toxins	Biological activity
Human intestinal cells	V. parahaemolyticus	Adherence
Human ileal cells	Enterotoxigenic E. coli	Adherence
Human mucosal cells	E. coli	Adherence
	V. parahaemolyticus	Adherence
Human uroepithelial cells	E. coli	Adhesion
Viable human duodenal biopsies	E. coli	Adherence
Rat hepatocytes	C. perfringens enterotoxin	Amino acid transport
	C. perfringens enterotoxin	Membrane permeability
Guinea pig intestinal cells	V. parahaemolyticus	Adherence
HEp-2 cells	B. cereus cereulide	Vacuole formation (mitochondrial swelling)

Note: LT = heat-labile toxin; ETEC = enterotoxigenic E. coli.

small intestine tied approximately 3 cm distal from the stomach. The intestine was injected with 1.0 ml of a suspension of 2×10^8 cells of adherence-positive and adherence-negative strains, followed by closing of the abdomen. Six hours later, the animals were killed, and the small intestine was removed and cut into four sections. Following homogenization with 3% NaCl, the number of cells in the homogenate was determined by plating. With adherence-positive cells, larger numbers were found in the homogenates, especially in the upper section of the intestine.

Another adhesion model consists of immobilizing soluble mucosal glycoproteins from mouse intestines on polystyrene.[28] Using this model, it was shown that two plasmid-bearing strains of *E. coli* (K88 and K99) adhered readily, as do other adhering strains of this organism.

HeLa Cells

This cell line has had wide use in testing for the invasive potential of intestinal pathogens as well as for adherence. Although HeLa cells seem to be preferred, other cell lines such as human laryngeal carcinoma and Henle 407 human intestine may be employed. In general, monolayers of cells are prepared by standard culture techniques on a chamber slide and inoculated with 0.2 ml of a properly prepared test culture suspension. Following incubation for 3 hours at 35°C to allow for bacterial growth, monolayer cells are washed, fixed, and stained for viewing under the light microscope. In the case of invasive *E. coli*, cells will be present in the cytoplasm of monolayer cells but not in the nucleus. In addition, invasive strains are phagocytized to a greater extent than noninvasives and the number of bacteria/cell is >5. According to the *Bacteriological Analytical Manual* (BAM),[17] at least 0.5% of the HeLa cells should contain no less than five bacteria. Positive responses to this test are generally confirmed by the Sereny test (see reference 17).

A modification is used for invasive *Yersinia*. By this method, 0.2 ml of a properly prepared bacterial suspension is inoculated into chamber slides containing the HeLa cell monolayer. Following incubation for 1.5 hours at 35°C, the cells are washed, fixed, and stained for microscopic examination. Invasive *Y. enterocolitica* are present in the cytoplasm—usually in the phagolysosome. Infectivity rates are generally greater than 10%. Although invasive *E. coli* are confirmed by the Sereny test, this is not done with *Y. enterocolitica*, even though invasive, because this organism may not yield a positive Sereny test.

HeLa cells have been used to test for adherence of *V. parahaemolyticus* and to study the penetration of *Y. enterocolitica*. Strains of the latter that gave an index of 3.7–5.0 were considered penetrating.[41] The infectivity of HeLa cells by *Y. enterocolitica* has been studied by use of cell monolayers in roller tubes. The number of infecting bacterial cells is counted at random in 100 stained HeLa cells for up to 24 hours.[14]

Chinese Hamster Ovary Cells

The Chinese hamster ovary (CHO) assay was developed by Guerrant et al.[21] for *E. coli* enterotoxins and employs CHO cells grown in a medium containing fetal calf serum. Upon establishment of a culture of cells, enterotoxin is added. Microscopic examinations are made 24–30 hours later to determine whether cells have become bipolar and elongated to at least three times their width and whether their knoblike projections have been lost. The morphological changes in CHO cells caused by both cholera toxin and *E. coli* enterotoxin have been shown to parallel the elevation of cyclic AMP. It has been found to be 100–10,000 times more sensitive than skin permeability and ileal loop assays for *E. coli* enterotoxins. For the LT of *E. coli*, CHO has been found to be 5–100 times more sensitive than skin permeability and rabbit ileal loop assays.[21]

Vero Cells

This monolayer consists of a continuous cell line derived from African green monkcy kidneys; it was employed by Speirs et al.[43] to assay for *E. coli* LT. Vero cell results compare favorably with Y-1 adrenal cells (see below), and the test was found by these investigators to be the simple and more economical of the two to maintain in the laboratory. Toxigenic strains produce a morphological response to Vero cells similar to Y-1 cells. The use of Vero cells to study *E. coli* toxins is discussed in Chapter 27.

A highly sensitive and reproducible biological assay for *C. perfringens* enterotoxin employing Vero cells was developed by McDonel and McClane.[29] The assay is based on the observation that the enterotoxin inhibits plating efficiency of Vero cells grown in culture. The inhibition of plating efficiency detected as little as 0.1 ng of enterotoxin, and a linear dose-response curve was obtained with 0.5–5 ng (5–50 ng/ml). The investigators proposed a new unit of biological activity—the plating efficiency unit (PEU)—as that amount of enterotoxin that causes a 25% inhibition of the plating of 200 cells inoculated into 100 μl of medium.

Y-1 Adrenal Cell Assay

In this widely used assay, mouse adrenal cells (Y-1) are grown in a monolayer using standard cell culture techniques. With monolayer cells in microtiter plate wells, test extracts or filtrates are added to the microtiter wells followed by incubation at 37°C. In testing *E. coli* LT, heated and unheated culture filtrates of known positive and negative LT-producing strains are added to monolayers in microtiter plates and results are determined by microscopic examinations. The presence of 50% or more rounded cells in monolayers of unheated filtrates and 10% or less for heated filtrates denotes a positive response. The specificity of the response can be determined by the use of specific antibodies in toxin-containing filtrates. Details of this method for foodborne pathogens are presented in *BAM*.[17]

Other Assays

An immunofluorescence method was employed by Boutin et al.[5] using 6-week-old rabbit ileal loops inoculated with *V. parahaemolyticus*. The loops were removed 12–18 hours after infection and placed in trays, cut into tissue sections, and cleaned by agitation. Tissue sections were fixed and stained with fluorescein isothiocyanate-stained agglutinins to *V. parahaemolyticus*. The reaction of the tagged antibody with *V. parahaemolyticus* cells in thc tissue was assessed microscopically. By use of immunofluorescence, it was possible to demonstrate the penetration by this organism into the lamina propria of the ileum and thus the tissue invasiveness of the pathogen. Both Kanagawa-positive and -negative cells penetrated the lamina according by this method.

The chorioallantoic membrane of 10-day-old chick embryos was used to assess the pathogenicity of *Listeria* spp. *L. monocytogenes* causes death within 2–5 days with as few as 100 cells, and *L. ivanovii* cells are lethal at levels of 100–30,000 per egg with death within 72 hours.[46] Culture filtrates of *L. monocytogenes* and *L. ivanovii* release lactate dehydrogenase (LDH) in rat hepatocyte monolayers following 3-hour exposures, but other listerial species had no effect.[24]

REFERENCES

1. Agata, N., M. Ohta, M. Mori, and M. Isobe. 1995. A novel dodecadepsipeptide, cereulide, is an emetic toxin of *Bacillus cereus*. *FEMS Microbiol. Lett.* 129:17–20.

2. Anton, W. 1934. Kritisch-experimenteller Beitrag zur Biologie des Bacterium monocytogenes. Mit besonderer Berucksichtigung seiner Beziehung zue infektiosen Mononucleose des Menschen. *Zbt. Bakteriol. Abt. I. Orig.* 131:89–103.

3. Aulisio, C.C.G., W.E. Hill, J.T. Stanfield, and J.A. Morris. 1983. Pathogenicity of *Yersinia enterocolitica* demonstrated in the suckling mouse. *J. Food Protect.* 46:856–860.

4. Bergdoll, M.S. 1982. The enterotoxins. In *The Staphylococci*, ed. J.O. Cohen, 301–331. New York: Wiley.

5. Boutin, B.K., S.F. Townsend, P.V. Scarpino, and R.M. Twedt. 1979. Demonstration of invasiveness of *Vibrio parahaemolyticus* in adult rabbits by immunofluorescence. *Appl. Environ. Microbiol.* 37:647–653.

6. Boyce, J.M., D.J. Evans, Jr., D.G. Evans, and H.L. DePont. 1979. Production of heat-stable, methanol-soluble enterotoxin by *Yersinia enterocolitica*. *Infect. Immun.* 25:532–537.

7. Burgess, M.N., R.J. Bywater, C.M. Cowley, N.A. Mullan, and P.M. Newsome. 1978. Biological evaluation of a methanol-soluble, heat-stable *Escherichia coli* enterotoxin in infant mice, pigs, rabbits, and calves. *Infect. Immun.* 21:526–531.

8. Burrows, W., and G.M. Musteikis. 1966. Cholera infection and toxin in the rabbit ileal loop. *J. Infect. Dis.* 116:183–190.

9. Caldwell, M.B., R.I. Walker, S.D. Stewart, and J.E. Rogers 1983. Simple adult rabbit model for *Campylobacter jejuni* enteritis. *Infect. Immun.* 42:1176–1182.

10. Clark, W.G., and J.S. Page. 1968. Pyrogenic responses to staphylococcal enterotoxins A and B in cats. *J. Bacteriol.* 96:1940–1946.

11. De, S.N., and D.N. Chatterje. 1953. An experimental study of the mechanism of action of *Vibrio cholerae* on the intestinal mucous membrane. *J. Path. Bacteriol.* 66:559–562.

12. Dean, A.G., Y.C. Ching, R.G. Williams, and L.B. Harden. 1972. Test for *Escherichia coli* enterotoxin using infant mice: Application in study of diarrhea in children in Honolulu. *J. Infect. Dis.* 125:407–411.

13. Deneke, C.F., K. McGowan, G.M. Thorne, and S.L. Gerbach. 1983. Attachment of enterotoxigenic *Escherichia coli* to human intestinal cells. *Infect. Immun.* 39:1102–1106.

14. Devenish, J.A., and D.A. Schiemann. 1981. HeLa cell infection by *Yersinia enterocolitica*: Evidence for lack of intracellular multiplication and development of a new procedure for quantitative expression of infectivity. *Infect. Immun.* 32:48–55.

15. Dolman, C.E., R.J. Wilson, and W.H. Cockroft. 1936. A new method of detecting *Staphylococcus* enterotoxin. *Can. J. Public Health.* 27:489–493.

16. Evans, D.J., Jr., D.G. Evans, and S.L. Gorbach. 1973. Production of vascular permeability factor by enterotoxigenic *Escherichia coli* isolated from man. *Infect. Immun.* 8:725–730.

17. *FDA Bacteriological Analytical Manual*, 8th ed. 1995. McLean, VA: Association of Official Analytical Chemists Int.

18. Genigeorgis, C., G. Sakaguchi, and H. Riemann. 1973. Assay methods for *Clostridium perfringens* type A enterotoxin. *Appl. Microbiol.* 26:111–115.

19. Giannella, R.A. 1976. Suckling mouse model for detection of heat-stable *Escherichia coli* enterotoxin: Characteristics of the model. *Infect. Immun.* 14:95–99.

20. Gingras, S.P., and L.V. Howard. 1980. Adherence of *Vibrio parahaemolyticus* to human epithelial cell lines. *Appl. Environ. Microbiol.* 39:369–371.

21. Guerrant, R.L., L.L. Brunton, T.C. Schaitman, L.L. Rebhun, and A.G. Gilman. 1974. Cyclic adenosine monophosphate and alteration of Chinese hamster ovary cell morphology: A rapid, sensitive in vitro assay for the enterotoxins of *Vibrio cholerae* and *Escherichia coli*. *Infect. Immun.* 10:320–327.

22. Hackney, C.R., E.G. Kleeman, B. Ray, and M.L. Speck. 1980. Adherence as a method for differentiating virulent and avirulent strains of *Vibrio parahaemolyticus*. *Appl. Environ. Microbiol.* 40:652–658.

23. Hauschild, A.H.W. 1970. Erythemal activity of the cellular enteropathogenic factor of *Clostridium perfringens* type A. *Can. J. Microbiol.* 16:651–654.

24. Huang, J.C., H.S. Huang, M. Jurima-Romet, F. ashton, and B.H. Thomas. 1990. Hepatocidal toxicity of *Listeria* species. *FEMS Microbiol. Lett.* 72:249–252.

25. Iijima, Y., H. Yamada, and S. Shinoda. 1981. Adherence of *Vibrio parahaemolyticus* and its relation to pathogenicity. *Can. J. Microbiol.* 27:1252–1259.

26. Jordan, E.O., and J. McBroom. 1931. Results of feeding *Staphylococcus* filtrates to monkeys. *Proc. Soc. Exp. Biol. Med.* 29:161–162.

27. Kennedy, D.J., R.N. Greenberg, J.A. Dunn, R. Abernathy, J.S. Ryerse, and R.L. Guerrant. 1984. Effects of *Escherichia coli* heat-stable enterotoxin ST_b on intestines of mice, rats, rabbits, and piglets. *Infect. Immun.* 46:639–641.

28. Laux, D.C., E.F. McSweegan, and P.S. Cohen. 1984. Adhesion of enterotoxigenic *Escherichia coli* to immobilized intestinal mucosal preparations: A model for adhesion to mucosal surface components. *J. Microbiol. Meth.* 2:27–39.

29. McDonel, J.L., and B.A. McClane. 1981. Highly sensitive assay for *Clostridium perfringens* enterotoxin that uses inhibition of plating efficiency of Vero cells grown in culture. *J. Clin. Microbiol.* 13:940–946.

30. Mehlman, J.J., M. Fishbein, S.L. Gorbach, A.C. Sandes, E.L. Eide, and J.C. Olson, Jr. 1976. Pathogenicity of *Escherichia coli* recovered from food. *J. Assoc. Off. Anal. Chem.* 59:67–80.

31. Minor, T.E., and E.H. Marth. 1976. *Staphylococci and Their Significance in Foods*. New York: Elsevier.

32. Mullan, N.A., M.N. Burgess, and P.M. Newsome. 1978. Characterization of a partially purified, methanol-soluble heat-stable *Escherichia coli* enterotoxin in infant mice. *Infect. Immun.* 19:779–784.

33. Murayama, S.Y., T. Sakai, S. Makino, T. Kurata, C. Sasakawa, and M. Yoshikawa. 1986. The use of mice in the Sereny test as a virulence assay of shigellae and enteroinvasive *Escherichia coli*. *Infect. Immun.* 51:696–698.

34. O'Brien, A.D., and R.K. Holmes. 1987. Shiga and Shiga-like toxins. *Microbiol. Rev.* 51:206–220.

35. Ofek, I., and E.H. Beachey. 1978. Mannose binding and epithelial cell adherence of *Escherichia coli*. *Infect. Immun.* 22:247–254.

36. Okamoto, K., T. Inoue, K. Shimizu, S. Hara, and A. Miyama. 1982. Further purification and characterization of heat-stable enterotoxin produced by *Yersinia enterocolitica*. *Infect. Immun.* 35:958–964.

37. Pai, C.H., V. Mors, and T.A. Seemayer. 1980. Experimental *Yersinia enterocolitica* enteritis in rabbit. *Infect. Immun.* 28:238–244.

38. Pai, C.H., V. Mors, and S. Toma. 1978. Prevalence of enterotoxigenicity in human and nonhuman isolates of *Yersinia enterocolitica*. *Infect. Immun.* 22:334–338.

39. Pazzaglia, G., R.B. Sack, A.L. Bourgeois, J. Froehlich, and J. Eckstein. 1990. Diarrhea and intestinal invasiveness of *Aeromonas* strains in the removable intestinal tie rabbit model. *Infect. Immun.* 58:1924–1931.

40. Punyashthiti, K., and R.A. Finkelstein. 1971. Enteropathogenicity of *Escherichia coli*. I. Evaluation of mouse intestinal loops. *Infect. Immun.* 39:721–725.

41. Schiemann, D.A. 1981. An enterotoxin-negative strain of *Yersinia enterocolitica* serotype 0:3 is capable of producing diarrhea in mice. *Infect. Immun.* 32:571–574.

42. Smith, H.W. 1972. The production of diarrhea in baby rabbits by the oral administration of cell-free preparations of enteropathogenic *Escherichia coli* and *Vibrio cholerae*: The effect of antisera. *J. Med. Microbiol.* 5:299–303.

43. Speirs, J.I., S. Stavric, and J. Konowalchuk. 1977. Assay of *Escherichia coli* heat-labile enterotoxin with Vero cells. *Infect. Immun.* 16:617–622.

44. Spira, W.M., R.B. Sack, and J.L. Froehlich. 1981. Simple adult rabbit model for *Vibrio cholerae* and enterotoxigenic *Escherichia coli* diarrhea. *Infect. Immun.* 32:739–747.

45. Stark, R.L., and C.L. Duncan. 1971. Biological characteristics of *Clostridium perfringens* type A enterotoxin. *Infect. Immun.* 4:89–96.

46. Terplan, G., and S. Steinmeyer. 1989. Investigations on the pathogenicity of *Listeria* spp. by experimental infection of the chick embryo. *Int. J. Food Microbiol.* 8:277–280.

47. Thorne, G.M., C.F. Deneke, and S.L. Gorbach. 1979. Hemagglutination and adhesiveness of toxigenic *Escherichia coli* isolated from humans. *Infect. Immun.* 23:690–699.

48. Ueno, S., N. Matsuki, and H. Saito, 1987. *Suncus murinus*: A new experimental model in emesis research. *Life Sci.* 41:513–516.

49. Whipp, S.C. 1990. Assay for enterotoxigenic *Escherichia coli* heat-stable toxin b in rats and mice. *Infect. Immun.* 58:930–934.

50. Williams, P.H., M.I. Sedgwick, N. Evans, P.J. Turner, R.H. George, and A.S. McNeish. 1978. Adherence of an enteropathogenic strain of *Escherichia coli* to human intestinal mucosa is mediated by a colicinogenic conjugative plasmid. *Infect. Immun.* 22:393–402.

51. Wood, L.V., W.H. Wolfe, G. Ruiz-Palacios, W.S. Foshee, L.I. Corman, F. McCleskey, J.A. Wright, and H.L. DuPont. 1983. An outbreak of gastroenteritis due to a heat-labile enterotoxin-producing strain of *Escherichia coli. Infect. Immun.* 41:931–934.

52. Wright, A., P.L.R. Andrews, and R.W. Titball. 2000. Induction of emetic, pyrexic, and hehavioral effects of *Staphylococcus aureus* enterotoxin B in the ferret. *Infect. Immun.* 68:2386–2389.

53. Yamamoto, K., I. Ohishi, and G. Sakaguchi. 1979. Fluid accumulation in mouse-ligated intestine inoculated with *Clostridium perfringens* enterotoxin. *Appl. Environ. Microbiol.* 37:181–186.

PART V

Food Protection and Some Properties of Psychrotrophs, Thermophiles, and Radiation-Resistant Bacteria

The microbiology of a variety of food protection methods is covered in Chapters 13 through 19. In previous editions of this work, the words "food preservation" were applied to these methods but with this edition, we opted to use "protection" rather than preservation. The latter word is more traditional and its usage is more restrictive. For example, fresh meat from an infected animal can be preserved by covering it with NaCl but if the animal tissues contained the pathogen, they would likely remain when preserved along with the spoilage biota. Food protection is meant to imply the control of both spoilage and disease-causing organisms. In Chapter 13, a more extensive coverage of individual food sanitizers is presented along with methods that are based on activities of protective cultures under the heading "Biocontrol". More information may be obtained from the listed sources.

Barbosa-Canovas, G.V., and Q.H. Zhang. 2001. *Pulsed Electric Fields in Food Processing*. Boca Raton, FL: CRC Press.

Davidson, P.M., and A.L. Branen, ed. 1995. *Antimicrobials in Foods*, 2nd ed. New York: Marcel Dekker, Inc.

Farber, J.M., and K. Dodds, ed. 1995. *Principles of Modified-Atmosphere and Sous-Vide Product Packaging*. Lancaster, PA: Technomic Publishing.

Hendrickx, M.E.G., and D. Knorr, ed. 2002. *Ultra High Pressure Treatments of Foods*. New York: Kluwer Academic Publishers.

Leistner, L., and G.W. Gould. 2002. *Hurdle Technologies: Combination Treatments for Food Safety, Stability, and Quality*. New York: Kluwer Academic Publishers.

Naidu, A.S., ed. 2000. *Natural Food Antimicrobial Systems*. Boca Raton, FL: CRC Press.

Novak, J.S., G.M. Sapers, and V.K. Juneja, ed. 2003. *Microbial Safety of Minimally Processed Foods*. Boca Raton, FL: CRC Press

Russell N.J., and G.W. Gould, ed. 2003. *Food Preservatives*. New York: Kluwer Academic Publishers.

Sofos, J.N. 1989. *Sorbate Food Preservatives*. Boca Raton, FL: CRC Press.

CHAPTER 13

Food Protection with Chemicals, and by Biocontrol

The use of chemicals to prevent or delay the spoilage of foods derives in part from the fact that such compounds are used with great success in the treatment of diseases of humans, animals, and plants. This is not to imply that any and all chemotherapeutic compounds can or should be used as food preservatives. On the other hand, there are some chemicals of value as food preservatives that would be ineffective or too toxic as chemotherapeutic compounds. With the exception of certain antibiotics, none of the food preservatives now used find any real use as chemotherapeutic compounds in people and animals. Although a large number of chemicals have been described that show potential as food preservatives, only a relatively small number are allowed in food products, due in large part to the strict rules of safety adhered to by the Food and Drug Administration (FDA) and to a lesser extent to the fact that not all compounds that show antimicrobial activity in vitro do so when added to certain foods. Described below are those compounds most widely used, their modes of action where known, and the types of foods in which they are used. Those chemical preservatives generally recognized as safe (GRAS) are summarized in Table 13–1.

BENZOIC ACID AND THE PARABENS

Benzoic acid (C_6H_5COOH) and its sodium salt ($C_7H_5NaO_2$), along with the esters of *p*-hydroxybenzoic acid (parabens), are considered together in this section. Sodium benzoate was the first chemical preservative permitted in foods by the FDA, and it continues in wide use today in a large number of foods. Its approved derivatives have structural formulas as noted:

Methylparaben
Methyl *p*-Hydroxybenzoate

HO— (benzene ring) —$COOCH_2$

Propylparaben
Propyl *p*-Hydroxybenzoate

HO— (benzene ring) —$COO(CH_2)_2CH_3$

Heptylparaben
n-Heptyl-*p*-Hydroxybenzoate

HO— (benzene ring) —$COO(CH_2)_6CH_3$

Table 13-1   Summary of Some GRAS Chemical Food Preservatives

Preservatives	Maximum Tolerance	Organisms Affected	Foods
Propionic acid/ propionates	0.32%	Molds	Bread, cakes, some cheeses, rope inhibitor in bread dough
Sorbic acid/sorbates	0.2%	Molds	Hard cheeses, figs, syrups, salad dressings, jellies, cakes
Benzoic acid/ benzoates	0.1%	Yeasts and molds	Margarine, pickle relishes, apple cider, soft drinks, tomato catsup, salad dressings
Parabens*	0.1%[†]	Yeasts and molds	Bakery products, soft drinks, pickles, salad dressings
SO_2/sulfites	200–300 ppm	Insects, microorganisms	Molasses, dried fruits, wine making, lemon juice (not to be used in meats or other foods recognized as sources of thiamine)
Ethylene/propylene oxides[‡]	700 ppm	Yeasts, molds, vermin	Fumigant for spices, nuts
Sodium diacetate	0.32%	Molds	Bread
Nisin	1%	Lactics, clostridia	Certain pasteurized cheese spreads
Dehydroacetic acid	65 ppm	Insects	Pesticide on strawberries, squash
Sodium nitrite[‡]	120 ppm	Clostridia	Meat-curing preparations
Caprylic acid	–	Molds	Cheese wraps
Sodium lactate	Up to 4.8%	Bacteria	Pre-cooked meats
Ethyl formate	15–220 ppm[§]	Yeasts and molds	Dried fruits, nuts

Note: GRAS (generally recognized as safe) per Section 201[32] (s) of the U.S. Food, Drug, and Cosmetic Act as amended.

*Methyl-, propyl-, and heptyl-esters of *p*-hydroxybenzoic acid.
[†]Heptyl-ester—12 ppm in beers; 20 ppm in noncarbonated and fruit-based beverages.
[‡]May be involved in mutagenesis and/or carcinogenesis.
[§]As formic acid.

The antimicrobial activity of benzoate is related to pH, the greatest activity being at low pH values. The antimicrobial activity resides in the undissociated molecule (see below). These compounds are most active at the lowest pH values of foods and essentially ineffective at neutral values. The pK of benzoate is 4.20 and at a pH of 4.00, 60% of the compound is undissociated, whereas at a pH of 6.0, only 1.5% is undissociated. This results in the restriction of benzoic acid and its sodium salts to high-acid products such as apple cider, soft drinks, tomato catsup, and salad dressings. High acidity alone is generally sufficient to prevent growth of bacteria in these foods but not that of certain molds and yeasts. As used in acidic foods, benzoate acts essentially as a mold and yeast inhibitor, although it is effective against some bacteria in the 50–500-ppm range. Against yeasts and molds at around pH 5.0–6.0, 100–500 ppm are effective in inhibiting the former, whereas for the latter, 30–300 ppm are inhibitory. In foods such as fruit juices, benzoates may impart disagreeable tastes at the maximum level of 0.1%. The taste has been described as being "peppery" or burning.

The three parabens that are permissible in foods in the United States are heptyl-, methyl-, and propylparaben; butyl- and ethylparabens are permitted in food in certain other countries. As esters of *p*-hydroxybenzoic acid, they differ from benzoate in their antimicrobial activity in being less sensitive to pH. Although not as much data have been presented on heptylparaben, it appears to be quite effective against microorganisms, with 10–100 ppm effecting complete inhibition of some Gram-positive and Gram-negative bacteria. Propylparaben is more effective than methylparaben on a parts per million basis, with up to 1,000 ppm of the former and 1,000–4,000 ppm of the latter needed for bacterial inhibition, with Gram-positive bacteria being more susceptible than Gram negatives to the parabens in general.[38] Heptylparaben is reported to be effective against the malo-lactic bacteria. In a reduced-broth medium, 100 ppm propylparaben delayed germination and toxin production by *Clostridium botulinum* type A; and 200 ppm effected inhibition up to 120 hours at 37°C.[157] In the case of methylparaben, 1,200 ppm were required for inhibition similar to that for the propyl analog.

The parabens appear to be more effective against molds than yeasts. As with bacteria, the propyl derivative appears to be the most effective where 100 ppm or less are capable of inhibiting some yeasts and molds, whereas for heptyl- and methylparabens, 50–200 and 500–1,000 ppm, respectively, are required.

Like benzoic acid and its sodium salt, the methyl- and propylparabens are permissible in foods up to 0.1%, and heptylparaben is permitted in beers to a maximum of 12 ppm and up to 20 ppm in fruit drinks and beverages. The pK for these compounds is around 8.47, and their antimicrobial activity is not increased to the same degree as for benzoate with the lowering of pH as noted. They have been reported to be effective at pH values up to 8.0. For a more thorough review of these preservatives, see reference 38.

Similarities between the modes of action of benzoic and salicylic acids have been noted.[17] Both compounds, when taken up by respiring microbial cells, blocked the oxidation of glucose and pyruvate at the acetate level in *Proteus vulgaris*. With *P. vulgaris*, benzoic acid caused an increase in the rate of O_2 consumption during the first part of glucose oxidation.[17] The benzoates, like propionate and sorbate, act against microorganisms by inhibiting the cellular uptake of substrate molecules.[62] The stage of endospore germination most sensitive to benzoate is noted in Figure 13–1.

The undissociated form is essential to the antimicrobial activity of benzoate as well as for other lipophilics such as sorbate and propionate. In this state, these compounds are soluble in the cell membrane and act apparently as proton ionophores.[73] As such, they facilitate proton leakage into cells and thereby increase energy output of cells to maintain their usual internal pH. With the disruption in membrane activity, amino acid transport is adversely affected.[73] The mechanism of action is further described under section "Sorbic Acid" below.

SORBIC ACID

Sorbic acid ($CH_3CH–CHCH–CHCOOH$) is employed as a food preservative, usually as the calcium, sodium, or potassium salt. These compounds are permissible in foods at levels not to exceed 0.2%. Like sodium benzoate, they are more effective in acid foods than neutral foods and tend to

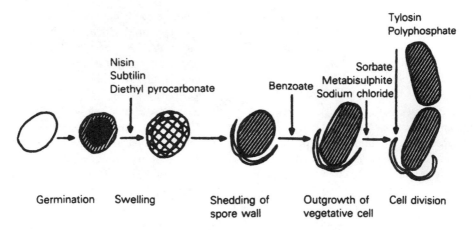

Figure 13–1 Diagrammatic representation of growth of an endospore into vegetable cells showing stages arrested by minimum inhibitory concentrations of some food preservatives. *Source*: From Gould.[72]

be on par with the benzoates as fungal inhibitors. Sorbic acid works best below a pH of 6.0 and is generally ineffective above pH 6.5. These compounds are more effective than sodium benzoate between pH 4.0 and 6.0. At pH values of 3.0 and below, the sorbates are slightly more effective than the propionates but about the same as sodium benzoate. The pK of sorbate is 4.80, and at a pH of 4.0, 86% of the compound is undissociated, whereas at a pH of 6.0, only 6% is undissociated. Sorbic acid can be employed in cakes at higher levels than propionates without imparting flavor to the product.

The sorbates are primarily effective against molds and yeasts, but research has shown them to be effective against a wide range of bacteria. In general, the catalase-positive cocci are more sensitive than the catalase negatives, and aerobes are more sensitive than anaerobes. The resistance of the lactic acid bacteria to sorbate, especially at pH 4.5 or above, permits its use as a fungistat in products that undergo lactic fermentations. Its effectiveness has been shown against *Staphylococcus aureus*, salmonellae, coliforms, psychrotrophic spoilage bacteria (especially the pseudomonads), and *Vibrio parahaemolyticus*. Against the last organism, concentrations as low as 30 ppm have been shown to be effective. Shelf-life extensions have been obtained by use of sorbates on fresh poultry meat, vacuum-packaged poultry products, fresh fish, and perishable fruits. For further information, see nitrite–sorbate combinations later in this chapter and the review by Sofos.[182]

The sorbates have been studied by a large number of groups for use in meat products in combination with nitrites. Bacon formulations that contain 120 ppm $NaNO_2$ without sorbate yield products that maintain their desirable organoleptic qualities in addition to being protected from *C. botulinum* growth. When 0.26% (2,600 ppm) potassium sorbate is added along with 40 ppm nitrite, no significant differences are found in the organoleptic qualities or in botulinal protection.[90,145] The combination of 40 ppm of $NaNO_2$ and 0.26% potassium sorbate (along with 550 ppm of sodium ascorbate or sodium erythrobate) was proposed by the U.S. Department of Agriculture (USDA) in 1978 but postponed in 1979. The later action was prompted not by the failure of the reduced nitrite level in combination with sorbate but because of taste panel results that characterized finished bacon as having "chemical"-like flavors and producing prickly mouth sensations.[14] The combination of sorbate plus reduced nitrite is effective in a variety of cured meat products against not only *C. botulinum* but other bacteria such as

S. aureus and a spoilage *Clostridium* (putrefactive anaerobe [P.A.] 3679). With the latter, a noninhibitory concentration of nitrite and sorbate was bactericidal.[162]

The widest use of sorbates is as a fungistat in products such as cheeses, bakery products, fruit juices, beverages, salad dressings, and the like. In the case of molds, inhibition may be due to inhibition of the dehydrogenase enzyme system. Against germinating endospores, sorbate prevents the outgrowth of vegetative cells (Figure 13–1).

As lipophilic acids, sorbate, benzoate, and propionate appear to inhibit microbial cells by the same general mechanism. The mechanism involves the proton motive force (PMF). Briefly, hydrogen ions (protons) and hydroxyl ions are separated by the cytoplasmic membrane, with the former, outside the cell, giving rise to acidic pH and the latter, inside the cell, giving rise to pH near neutrality. The membrane gradient thus created represents electrochemical potential that the cell employs in the active transport of some compounds such as amino acids. Weak lipophilic acids act as protonophores. After diffusing across the membrane, the undissociated molecule ionizes inside the cell and lowers intracellular pH. This results in a weakening of the transmembrane gradient such that amino acid transport is affected adversely. This hypothesis has been supported by research on P.A. 3679 where sorbate inhibited phenylalanine uptake, decreased protein synthesis, and altered phosphorylated nucleotide accumulation.[162,163] Although alteration of the PMF by lipophilic acids has wide support, other factors may be involved in their mode of action.[52] For example, a H^+-ATPase in the plasma membrane of *S. cerevisiae* aids in maintenance of cell homeostasis by exporting protons. The efficacy of this plasma membrane appears to be responsible, at least in part, for the adaptation of *S. cerevisiae* cells to sorbic acid.[85] With respect to safety, sorbic acid is metabolized in the body to CO_2 and H_2O in the same manner as fatty acids normally found in foods.[44]

THE PROPIONATES

Propionic acid is a three-carbon organic acid with the structure CH_3CH_2COOH. This acid and its calcium and sodium salts are permitted in breads, cakes, certain cheese, and other foods, primarily as a mold inhibitor. Propionic acid is employed also as a "rope" inhibitor in bread dough. The tendency toward dissociation is low with this compound and its salts, and they are consequently active in low-acid foods. They tend to be highly specific against molds, with the inhibitory action being primarily fungistatic rather than fungicidal.

With respect to the antimicrobial mode of action of propionates, they act in a manner similar to that of benzoate and sorbate. The pK of propionate is 4.87 and at a pH of 4.00, 88% of the compound is undissociated, whereas at a pH of 6.0, only 6.7% remains undissociated. The undissociated molecule of this lipophilic acid is necessary for its antimicrobial activity. The mode of action of propionic acid is noted above with benzoic acid. See also the section on medium-chain fatty acids and esters in this chapter, and the review by Doores[47] for further information.

SULFUR DIOXIDE AND SULFITES

Sulfur dioxide (SO_2) and the sodium and potassium salts of sulfite ($=SO_3$), bisulfite ($-HSO_3$), and metabisulfite ($=S_2O_5$) all appear to act similarly and are treated together here. Sulfur dioxide is used in its gaseous or liquid form or in the form of one or more of its neutral or acid salts on dried fruits, in lemon juice, molasses, wines, fruit juices, and others. The parent compound has been used as a food preservative since ancient times. Its use as a meat preservative in the United States dates back to at least 1813; however, it is not permitted in meats or other foods recognizable as sources of thiamine. Although SO_2 possesses antimicrobial activity, it is also used in certain foods as an antioxidant.

The predominant ionic species of sulfurous acid depends on pH of milieu, with SO_2 being favored by pH <3.0, HSO_3^- by pH between 3.0 and 5.0, and SO_3^{2-} by pH $\geq$6.0.[144] SO_2 has pKs of 1.76 and 7.2. The sulfites react with various food constituents including nucleotides, sugars, disulfide bonds, and others.

With regard to its effect on microorganisms, SO_2 is bacteriostatic against *Acetobacter* spp. and the lactic acid bacteria at low pH, with concentrations of 100–200 ppm being effective in fruit juices and beverages. It is bactericidal at higher concentrations. When added to temperature-abused comminuted pork, 100 ppm of SO_2 or higher were required to effect significant inhibition of spores of *C. botulinum* at target levels of 100 spores per gram.[198] The source of SO_2 was sodium metabisulfite. Employing the same salt to achieve an SO_2 concentration of 600 ppm, Banks and Board[9] found that growth of salmonellae and other Enterobacteriaceae were inhibited in British fresh sausage. The most sensitive bacteria were eight salmonellae serovars, which were inhibited by 15–109 ppm at a pH of 7.0; *Serratia liquefaciens*, *S. marcescens*, and *Hafnia alvei* were the most resistant, requiring 185–270 ppm free SO_2 in broth.

Yeasts are intermediate to acetic and lactic acid bacteria and molds in their sensitivity to SO_2, and the more strongly aerobic species are generally more sensitive than the more fermentative species. Sulfurous acid at levels of 0.2–20 ppm was effective against some yeasts, including *Saccharomyces*, *Pichia*, and *Candida*, whereas *Zygosaccharomyces bailii* required up to 230 ppm for inhibition in certain fruit drinks at pH 3.1.[119] Yeasts can actually form SO_2 during juice fermentation; some "*S. carlsbergensis*" and *S. bayanus* strains produce up to 1,000 and 500 ppm, respectively.[144] Molds such as *Botrytis* can be controlled on grapes by periodic gassing with SO_2, and bisulfite can be used to destroy aflatoxins.[48] Both aflatoxins B_1 and B_2 can be reduced in corn.[76] Sodium bisulfite was found to be comparable to propionic acid in its antimicrobial activity in corn containing up to 40% moisture.[76] (Aflatoxin degradation is discussed further in Chapter 30.)

Although the actual mechanism of action of SO_2 is not known, several possibilities have been suggested, each supported by some experimental evidence. One suggestion is that the undissociated sulfurous acid or molecular SO_2 is responsible for the antimicrobial activity. Its greater effectiveness at low pH tends to support this. Vas and Ingram[202] suggested the lowering of pH of certain foods by addition of acid as a means of obtaining greater preservation with SO_2. It has been suggested that the antimicrobial action is due to the strong reducing power that allows these compounds to reduce oxygen tension to a point below which aerobic organisms can grow or by direct action on some enzyme system. SO_2 is also thought to be an enzyme poison, inhibiting growth of microorganisms by inhibiting essential enzymes. Its use in the drying of foods to inhibit enzymatic browning is based on this assumption. Because the sulfites are known to act on disulfide bonds, it may be presumed that certain essential enzymes are affected and that inhibition ensues. The sulfites do not inhibit cellular transport. It may be noted from Figure 13–1 that metabisulfite acts on germinating endospores during the outgrowth of vegetative cells.

NITRITES AND NITRATES

Sodium nitrate ($NaNO_3$) and sodium nitrite ($NaNO_2$) are used in curing formulas for meats because they stabilize red meat color, inhibit some spoilage and food poisoning organisms, and contribute to flavor development. The role of NO_2 in cured meat flavor has been reviewed.[29,74] NO_2 has been shown to disappear on both heating and storage. It should be recalled that many bacteria are capable of utilizing nitrate as an electron acceptor and in the process effect its reduction to nitrite. The nitrite ion is by far the more important of the two in preserved meats. This ion is highly reactive and is capable of serving as a reducing and an oxidizing agent. In an acid environment, it ionizes to yield nitrous acid (3HONO), which further decomposes to yield nitric oxide (NO), the important product from the standpoint of color fixation in cured meats. Ascorbate or erythrobate acts also to reduce NO_2 to NO. Nitric oxide reacts with myoglobin under reducing conditions to produce the desirable red pigment nitrosomyoglobin (see also Table 5–2, Chapter 5).

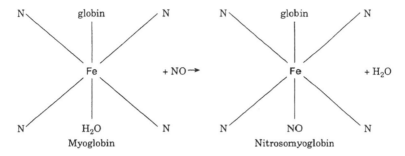

When the meat pigment exists in the form of oxymyoglobin, as would be the case for fresh comminuted meats, this compound is first oxidized to metmyoglobin (brown color). Upon the reduction of the latter, nitric oxide reacts to yield nitrosomyoglobin. Because nitric oxide is known to be capable of reacting with other porphyrin-containing compounds such as catalase, peroxidases, cytochromes, and others, it is conceivable that some of the antibacterial effects of nitrites against aerobes may be due to this action (the mechanism is discussed below). It has been shown that the antibacterial effect of NO_2 increases as pH is lowered within the acid range, and this effect is accompanied by an overall increase in the undissociated HNO_2.

The cooked cured meat pigment is dinitrosyl ferrohemochrome (DNFH). It forms when globin in nitrosomyoglobin is replaced with a second NO group.[176] A nitrite-free curing formula for wieners has been developed by adding 35 ppm of encapsulated DNFH prepared from bovine erythrocytes, *t*-butylhydroxyquinoline (TBHQ) as an antioxidant,[220] and 3,000 ppm of sodium hypophosphite as antibotulinal agent.[142,210] Essentially no microbial growth occurred through 4 weeks of storage with this formula, similar to the control nitrite-containing formulation. Sodium hypophosphite was the best of a variety of compounds tested as NO_2 replacements.[210] In spite of all efforts, no alternatives to nitrite are in sight.[29] As far as cured meat color is concerned, it was found in a recent study that metmyoglobin could be converted to a red myoglobin in salami by avoiding the use of nitrite or nitrate and by inoculating with a culture of *Staphylococcus xylosus*.[131]

Organisms Affected

Although the single microorganism of greatest concern relative to nitrite inhibition is *C. botulinum*, the compound has been evaluated as an antimicrobial for other organisms. During the late 1940s it was evaluated as a fish preservative and found to be somewhat effective but generally only at low pH. It is effective against *S. aureus* at high concentrations and, again, the effectiveness increases as the pH is lowered. The compound is generally ineffective against Enterobacteriaceae, including the salmonellae, and against the lactic acid bacteria, although some effects are noted in cured and in vacuum-packaged meats and are probably due to the interaction of nitrite with other environmental parameters rather than to nitrite alone. Nitrite is added to cheeses in some countries to control gassiness caused by *Clostridium butyricum* and *C. tyrobutyricum*. It is effective against other clostridia, including *C. sporogenes* and *C. perfringens*, which are often employed in laboratory studies to assess potential antibotulinal effects not only of nitrites but of other inhibitors that might have value as nitrite adjuncts or sparing agents.[161]

The Perigo Factor

The almost total absence of botulism in cured, canned, and vacuum-packed meats and fish products led some investigators in the mid-1960s to seek reasons as to why meat products that contained viable endospores did not become toxic. Employing culture medium, it was shown in 1967 that about ten times more nitrite was needed to inhibit clostridia if it were added after instead of before the medium was autoclaved. It was concluded that the heating of the medium with nitrite produced a substance or agent about ten times more inhibitory than nitrite alone.[147,148] This agent is referred to as the Perigo factor. The existence of this factor or effect has been confirmed by some and questioned by others. Although the Perigo factor may be questionable in cured and perishable cured meats, the evidence for an inhibitory factor in culture media involving nitrite, iron, and SH groups is more conclusive.[194]

This inhibitory or antibotulinal effect that results from the heat processing or smoking of certain meat and fish products containing nitrite warrants the continued use of nitrite in such products. The antibotulinal activity of nitrite in cured meats is of greater public health importance than the facts of color and flavor development. For the latter, initial nitrite levels as low as 15–50 ppm have been found to be adequate for various meat products, including Thuringer sausage.[43] Nitrite levels of 100 ppm or more have been found to make for maximum flavor and appearance in fermented sausages.[111] The antibotulinal effect requires at least 120 ppm for bacon[18,35] comminuted cured ham[34] and canned, shelf-stable luncheon meat.[32] Many of these canned products are given a low heat process (F_0 of 0.1–0.6).

Interaction with Cure Ingredients and Other Factors

The interplay of all ingredients and factors involved in heat-processed, cured meats on antibotulinal activity was noted about 40 years ago and several investigators have pointed out that curing salts in semipreserved meats are more effective in inhibiting heat-injured spores than noninjured.[49,160] With brine and pH alone, higher concentrations of the former are required for inhibition as pH increases, and Chang et al.[32] suggested that the inhibitory effect of salt in shelf-stable canned meats against heat-injured spores may be more important than the Perigo-type factor. With smoked salmon inoculated with 10^2 spores per gram of *C. botulinum* types A and E and stored in O_2-impermeable film, 3.8% and 6.1% water-phase NaCl alone inhibited toxin production in 7 days by types E and A, respectively.[145] With 100 ppm or more of NO_2, only 2.5% NaCl was required for inhibition of toxin production by type E, and for type A 3.5% NaCl + 150 ppm of $NaNO_2$ was inhibitory. With longer incubations or larger spore inocula, more NaCl or $NaNO_2$ is needed.

The interplay of NaCl, $NaNO_2$, $NaNO_3$, isoascorbate, polyphosphate, thermal process temperatures, and temperature/time of storage on spore outgrowth and germination in pork slurries has been studied extensively by Roberts et al.,[158] who found that significant reductions in toxin production could be achieved by increasing the individual factors noted. It is well known that low pH is antagonistic to growth and toxin production by *C. botulinum*, whether the acidity results from added acids or the growth of lactic acid bacteria. When 0.9% sucrose was added to bacon along with *Lactobacillus plantarum*, only 1 of 49 samples became toxic after 4 weeks, whereas with sucrose and no lactobacilli, 50 of 52 samples became toxic in 2 weeks.[189] When 40 ppm nitrite was used alone, 47 of 50 samples became toxic after 2 weeks, but when 40 ppm of nitrite was accompanied by 0.9% sucrose and an inoculum of *L. plantarum*, none of 30 became toxic. Although this was most likely a direct pH effect, other factors may have been involved. In later studies, bacon was prepared with 40 or 80 ppm of $NaNO_2$ + 0.7% sucrose followed by inoculation with *Pediococcus acidilactici*. When inoculated with *C. botulinum* types A and B spores, vacuum packaged, and incubated up to 56 days at 27°C, the bacon was found to have greater antibotulinal properties than control bacon prepared with 120 ppm of $NaNO_2$ but not sucrose or lactic inoculum.[188] Bacon prepared by the above formulation, called the Wisconsin process, was preferred by a sensory panel to that prepared by the conventional method.[187] The Wisconsin process employs 550 ppm of sodium ascorbate or sodium erythrobate, as does the conventional process.

Nitrosamines

When nitrite reacts with secondary amines, nitrosamines are formed, and many are known to be carcinogenic. The generalized way in which nitrosamines may form is as follows:

$$R_2NH_2 + HONO \xrightarrow{H^+} R_2N - NO + H_2O$$

The amine dimethylamine reacts with nitrite to form *N*-nitrosodimethylamine:

$$\begin{array}{ccc} H_3C & & H_3C \\ \diagdown & & \diagdown \\ N-H + NO_2 \rightarrow & & N-N=O \\ \diagup & & \diagup \\ H_3C & & H_3C \end{array}$$

In addition to secondary amines, tertiary amines and quaternary ammonium compounds also yield nitrosamines with nitrite under acidic conditions. Nitrosamines have been found in cured meat and fish products at low levels. Isoascorbate has an inhibitory effect on nitrosamine formation.

It has been shown that lactobacilli, enterococci, clostridia, and other bacteria will nitrosate secondary amines with nitrite at neutral pH values.[79] The fact that nitrosation occurred at near-neutral pH values was taken to indicate that the process was enzymatic, although no cell-free enzyme was obtained.[80] Several species of catalase-negative cocci, including *E. faecalis*, *E. faecium*, and *L. lactis*, have been shown to be capable of forming nitrosamines, but the other lactic acid bacteria and pseudomonads tested did not.[36] These investigators found no evidence for an enzymatic reaction. *S. aureus* and halobacteria obtained from Chinese salted marine fish (previously shown to contain nitrosamines) produced nitrosamines when inoculated into salted fish homogenates containing 40 ppm of nitrate and 5 ppm of nitrite.[58]

Nitrite–Sorbate and Other Nitrite Combinations

In an effort to reduce the potential hazard of *N*-nitrosamine formation in bacon, the USDA in 1978 reduced the input NO_2 level for bacon to 120 ppm and set a 10-ppb maximum level for nitrosamines. Although 120 ppm of nitrite along with 550 ppm of sodium ascorbate or sodium erythrobate are adequate to reduce the botulism hazard, it is desirable to reduce nitrite levels even further if protection against botulinal toxin production can be achieved. To this end, a proposal to allow the use of 40 ppm of nitrite in combination with 0.26% potassium sorbate for bacon was made in 1978 but rescinded a year later when taste panel studies revealed undesirable effects. Meanwhile, many groups of researchers

Table 13–2 Effect of Nitrite and Sorbate on Toxin Production in Bacon Inoculated with *C. botulinum* Types A and B Spores and Held up to 60 Days at 27°C

Treatment	Percentage Toxigenic
Control (no NO_2, no sorbate)	90.0
0.26% sorbate, no $NaNO_2$	58.8
0.26% sorbate + 40 ppm $NaNO_2$	22.0
0.26% sorbate + 80 ppm $NaNO_2$	0.0
No sorbate, 120 ppm $NaNO_2$	0.4

Source: Sofos et al.[184]

have shown that 0.26% sorbate in combination with 40 or 80 ppm of nitrite is effective in preventing botulinal toxin production.

In an early study of the efficacy of 40 ppm of nitrite + sorbate to prevent or delay botulinal toxin production in commercial-type bacon, Ivey et al.[90] used an inoculum of 1,100 types A and B spores per gram and incubated the product at 27°C for up to 110 days. The time for the appearance of toxic samples when neither nitrite nor sorbate was used was 19 days. With 40 ppm of nitrite and no sorbate, toxic samples appeared in 27 days, and for samples containing 40 ppm of nitrite plus 0.26% sorbate or no nitrite and 0.26% sorbate, more than 110 days were required for toxic samples. This reduced nitrite level resulted in lower levels of nitrosopyrrolidine in cooked bacon. Somewhat different findings were reported by Sofos et al.[184] (Table 13–2), with 80 ppm of nitrite being required for the absence of toxigenic samples after 60 days. In addition to its inhibitory effect on *C. botulinum*, sorbate slows the depletion of nitrite during storage.[183]

The action of isoascorbate is to enhance nitrite inhibition by sequestering iron, although under some conditions it may reduce nitrite efficiency by causing a more rapid depletion of residual nitrite.[195,197] Ethylenediaminetetraacetic acid (EDTA) at 500 ppm appears to be even more effective than erythrobate in potentiating the nitrite effect, but only limited studies have been reported. Another chelate, 8-hydroxyquinoline, has been evaluated as a nitrite-sparing agent. When 200 ppm were combined with 40 ppm of nitrite, a *C. botulinum* spore mixture of types A and B strains was inhibited for 60 days at 27°C in comminuted pork.[150]

In an evaluation of the interaction of nitrite and sorbate, the relative effectiveness of the combination has been shown to be dependent on other cure ingredients and product parameters. Employing a liver–veal agar medium at a pH of 5.8–6.0, the germination rate of *C. botulinum* type E spores decreased to nearly zero with 1.0%, 1.5%, or 2.0% sorbate, but with the same concentrations at a pH of 7.0–7.2, germination and outgrowth of abnormally shaped cells occurred.[175] When 500 ppm of nitrite was added to the higher-pH medium along with sorbate, cell lysis was enhanced. These investigators also found that 500 ppm of linoleic acid alone at the higher pH prevented emergence and elongation of spores. Potassium sorbate significantly decreased toxin production by types A and B spores in pork slurries when NaCl was increased or pH and storage temperature were reduced.[159] For chicken frankfurters, a sorbate–betalains mixture was found to be as effective as a conventional nitrite system for inhibiting *C. perfringens* growth.[201]

Mode of Action

It appears that nitrite inhibits *C. botulinum* by interfering with iron–sulfur enzymes such as ferredoxin and thus preventing the synthesis of adenosine triphosphate (ATP) from pyruvate. The first direct finding in this regard was that of Woods et al.,[212] who showed that the phosphoroclastic system of *C. sporogenes* is inhibited by nitric oxide and later that the same occurs in *C. botulinum*, resulting in the accumulation of pyruvic acid in the medium.[211]

The phosphoroclastic reaction involves the breakdown of pyruvate with inorganic phosphate and coenzyme A to yield acetyl phosphate. In the presence of adenosine diphosphate (ADP), ATP is synthesized from acetyl phosphate with acetate as the other product. In the breakdown of pyruvate, electrons are transferred first to ferredoxin and from ferredoxin to H^+ to form H_2 in a reaction catalyzed by hydrogenase. Ferredoxin and hydrogenase are iron–sulfur (nonheme) proteins or enzymes.

Following the work of Woods and Wood,[211] the next most significant finding was that of Reddy et al.,[155] who subjected extracts of nitrite–ascorbate-treated *C. botulinum* to electron spin resonance and found that nitric oxide reacted with iron–sulfur complexes to form iron–nitrosyl complexes. The presence of the latter results in the destruction of iron–sulfur enzymes such as ferredoxin.

The resistance of the lactic acid bacteria to nitrite inhibition is well known, but the basis is just now clear: these organisms lack ferredoxin. The clostridia contain both ferredoxin and hydrogenase, which function in electron transport in the anaerobic breakdown of pyruvate to yield ATP, H_2, and CO_2. The ferredoxin in clostridia has a molecular weight of 6,000 and contains eight Fe atoms/mole and eight-labile sulfide atoms/mole.

Although the first definitive experimental finding was reported in 1981, earlier work pointed to iron–sulfur enzymes as the probable nitrite targets. Among the first were O'Leary and Solberg,[143] who showed that a 91% decrease occurred in the concentration of free–SH groups of soluble cellular compounds of *C. perfringens* inhibited by nitrite. Two years later, Tompkin et al.[196] offered the hypothesis that nitric oxide reacted with iron in the vegetative cells of *C. botulinum*, perhaps the iron in ferredoxin. The inhibition by nitrite of active transport and electron transport was noted by several investigators, and these effects are consistent with nitrite inhibition of nonheme enzymes such as ferredoxin and hydrogenase.[159,217] The enhancement of inhibition in the presence of sequestering agents may be due to the reaction of sequestrants to substrate iron: more nitrite becomes available for nitric oxide production and reaction with microorganisms.

Summary of Nitrite Effects

When added to processed meats such as wieners, bacon, smoked fish, and canned cured meats followed by substerilizing heat treatments, nitrite has definite antibotulinal effects. It also forms desirable product color and enhances flavor in cured meat products. The antibotulinal effect consists of inhibition of vegetative cell growth and the prevention of germination and growth of spores that survive heat processing or smoking during postprocessing storage. Clostridia other than *C. botulinum* are affected in a similar manner. Whereas low initial levels of nitrite are adequate for color and flavor development, considerably higher levels are necessary for the antimicrobial effects.

When nitrite is heated in certain laboratory media, an antibotulinal factor or inhibitor is formed, the exact identity of which is not yet known. The inhibitory factor is the Perigo effect/factor or Perigo inhibitor. It does not form in filter-sterilized media. It develops in canned meats only when nitrite is present during heating. The initial level of nitrite is more important to antibotulinal activity than the residual level. Once formed, the Perigo factor is not affected greatly by pH changes. Measurable preheating levels of nitrite decrease considerably during heating in meats and during postprocessing storage—more at higher storage temperatures than at lower.

The antibotulinal activity of nitrite is interdependent with pH, salt content, temperature of incubation, and numbers of botulinal spores. Heat-injured spores are more susceptible to inhibition than uninjured. Nitrite is more effective under oxidation–reduction potential minus (E̅h) than under Eh+ conditions.

Nitrite does not decrease the heat resistance of spores. It is not affected by ascorbate in its antibotulinal actions but does act synergistically with ascorbate in pigment formation. Lactic acid bacteria are relatively resistant to nitrite (see above). Endospores remain viable in the presence of the antibotulinal effect and will germinate when transferred to nitrite-free media.

Nitrite has a pK of 3.29 and, consequently, exists as undissociated nitrous acid at low pH values. The maximum undissociated state and consequent greatest antibacterial activity of nitrous acid are between pH 4.5 and 5.5.

With respect to its depletion or disappearance in ham, Nordin[141] found the rate to be proportional to its concentration and to be exponentially related to both temperature and pH. The depletion rate doubled for every 12.2°C increase in temperature or a 0.86 pH unit decrease and was not affected by heat denaturation of the ham. These relationships did not apply at room temperature unless the product was first heat treated, suggesting that viable organisms aided in its depletion.

It appears that the antibotulinal activity of nitrite is due to its inhibition of nonheme, iron–sulfur enzymes.

FOOD SANITIZERS

A large number of chemicals have been assessed for their efficacy in destroying pathogens on fruits, vegetables, and meat surfaces. The pathogens of primary concern are the enterohemorrhagic *E. coli* strains, *L. monocytogenes*, and salmonellae. One of the desirable objectives of a food sanitizer is the capacity to effect a 5-log reduction of the pathogen of concern. In addition to their application to the surfaces of foods, some of these chemicals are applied directly to the surfaces of food handling and storage equipment. The chemicals in this section are not food additives as are those in other sections of this chapter. Brief summaries of those that have received the most attention are presented in the sections that follow. For a review, see reference 60.

Acidified Sodium Chlorite

Produced by the Alcide Corp., acidified sodium chlorite (ASC) is a product of citric or phosphoric acid and NaCl, and it is used either as a spray (ca. 5 sec) or dip (ca. 5 minutes exposure) at concentrations of 1,000 to 1,200 ppm. The antimicrobial species is a product of the dissociation of chlorite that breaks or disrupts oxidative bonds on the surface of cell membranes in a nonspecific manner.[103] It has been approved by the U.S. Food and Drug Administration as a sanitizer for poultry, red meat surfaces, seafoods, some fruits and vegetables, and some processed meats. For poultry, it is used pre-chill and post-chill on whole or cut-up birds. It may be used in water or ice. Acidic calcium sulfate is a closely related preparation.

Electrolized oxidizing water

Electrolized oxidizing (EO) water was first demonstrated in Russia in the 1970s, and further shown in Japan to possess antimicrobial properties. It is prepared in a special device following the addition

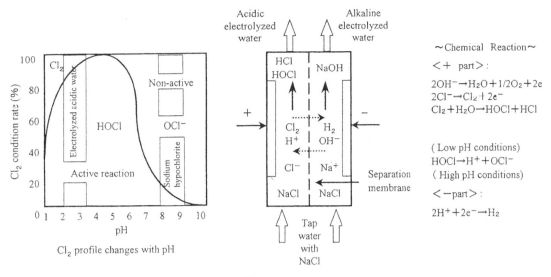

Figure 13–2 Principle of EO water production.[10] See text for explanations, copyright © 2003, Int. Assoc. for Food Protection

of tap water and NaCl. A diagram of the overall process is presented in Figure 13–2.[10] The water and salt (ca. 12%) are separated by a membrane. When voltage is applied, the product that is produced by the cathode has a pH of ca. 11.4 and an oxidation-reduction potential (ORP) of −795 mV while that from the anode has a pH of ca. 2.4–2.6 and an ORP of ca. +1,150 mV. The acidic water contains some free Cl_2 (10–80 ppm) and hypochlorous acid and it is more antimicrobial than the cathodic water (see below). The lethal effect of EO water appears to be due more to the extreme ORP than to other factors although some evidence points to other constituents such as hypochlorous acid. Gram-negative bacteria appear to be more sensitive than Gram positives.

From a number of studies, EO water has been shown to effect a 2- to 5-log reduction of the pathogens of primary concern on fresh produce, bean sprouts, etc.[55] When compared against *S.* Typhimurium and *L. monocytogenes* for 5 and 15 minute exposures at 4°C, *S.* Typhimurium was reduced >5 logs by acidic water after 15 days while *L. monocytogenes* was reduced by >4 logs (Table 13–3). The acidic EO water was even more effective at 25°C.

Table 13–3 The Relative Effectiveness of Acidic and Basic EO Waters on Cultures of *L. monocytogenes* and *S.* Typhimurium Stored at 4°C for 5 and 15 Days (Summarized from Fabrizio et al.[55]). The Numbers are log_{10} APC/ml.

	Water	Water	Basic	Basic	Acidic	Acidic
Days →	5	15	5	15	5	15
Salmonella Typhimurium	8.47	8.39	7.98	7.87	5.13	3.32
Listeria monocytogenes	8.73	8.74	8.69	8.77	5.36	4.60

Using culture media, 7–10 log reductions of a 5-strain mixture of *E. coli* 0157:H7, *S.* Enteritidis, and *L. monocytogenes* were achieved after exposure at 4 or 25°C in acidic EO water.[203] After a 10-minute exposure, complete inactivation occurred. The water in this study had a pH in the range of 2.4–2.6, ORP around 1,150, and free Cl_2 of 43–86 ppm.

Activated Lactoferrin (ALF, Activin)

As noted in Chapter 3, lactoferrin is a normal component of fresh milk that has been known for many years to possess antimicrobial properties. It is a glyoprotein with a molecular weight of around 80,000 da. It occurs also in saliva, tears, and some other body fluids. Activin is a more potent antimicrobial than plain lactoferrin, and it was developed by A.S. Naidu. In the activated form, lactoferrin is immobilized on food-grade polysaccharides such as carrageenan and solubilized in a citric/bicarbonate buffer with NaCl.[134] It has been accorded GRAS status by the U.S. Food and Drug Administration.

Contributing to its antimicrobial activity is its capacity to chelate Fe^{2+} along with HCO_3^-. It binds to cell surfaces and has a high affinity for the outer membrane proteins (OMP) of Gram-negative bacteria. Naidu characterizes AFL as a blocking agent that interferes with the adhesion of bacterial cells to animal tissues.[135] It also inhibits growth and neutralizes endotoxins. Its activity against both RNA and DNA viruses suggests that it also interacts with nucleic acids. It has been approved at a level of 65.2 ppm for beef carcasses, and may be applied either as a mist or by spraying. It is not a cidal agent but acts primarily by preventing pathogens from establishing a niche on meat surfaces. More information can be obtained from reference 134.

A feline kidney cell culture was used to propagate a norovirus (feline calicivirus, FCV), and a monkey kidney cell line was used to propagate poliovirus (PV). When feline kidney cells were incubated with PCV either prior to or together with bovine lactoferrin, substantial reductions of FCV infections were noted.[122] Lactoferrin bound to both cell lines and apparently prevented virus attachment. Lactoferrin B (cationic peptide from the N-terminal domain of bovine lactoferrin) reduced FCV attachment but not PV.[122]

Ozone (O_3)

This gaseous compound has been known for over 120 years to possess antimicrobial activity. Like chlorine, it is a powerful oxidant, and is ca. 1.5 times more potent than chlorine. It is effective in solution and in its gaseous form. Because it is more effective in killing *Cryptosporidium parvum* than chlorine, its use in water treatment systems is increasing. It is normally supplied from ozone generators. It leaves no residue after it reacts, but its antimicrobial activity is antagonized by organic matter.

The odor of ozone can be detected at around 0.01 ppm. The threshold limit for long-term human exposure (set by the U.S. Office for Safety and Health Administration, OSHA) is 0.1 ppm/day/work week but for short-term exposure, it is 0.3 ppm for 15 minutes.[214] The cell target for O_3 is the membrane where it disrupts permeability functions. Ozone is GRAS for bottled water use, and for use on a variety of fresh foods, but its strong oxidizing power does not recommend its use for red meats. A typical concentration used is 0.1–0.5 ppm, which is effective against Gram-positive and Gram-negative bacteria as well as viruses and protozoa.[108] Ozone is allowed in foods in Australia, France, and Japan; and in 1997 it was accorded GRAS status in the United Stated as noted above.

Ozone has been shown to be effective in reducing pathogens on a number of food products, and its effect on apple surfaces and the stem and calyx is shown in Figure 13–3 where *E. coli* 0157:H7 was killed much more efficiently on apple surfaces than on stem/calyx.[2] These investigators found

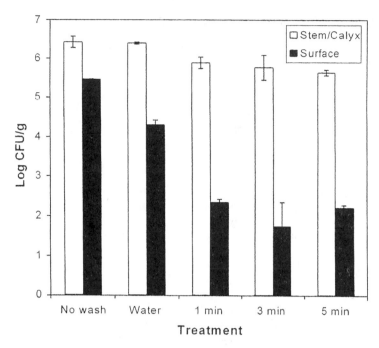

Figure 13–3   Counts of *Escherichia coli* 0157:H7 ($\log_{10}$ cfu/g) on inoculated apples that were unwashed, water-washed, or treated with bubbling ozone in water at 22–25°C; residual ozone concentrations at 1 minute: 20.8 mg/l, 3 minutes: 24.5 mg/l, and 5 minutes: 27.7 mg/l. Error bars represent the standard deviation of the mean of six apples (Achen and Yousef[2]).

that bubbling O_3 was more effective for apples than dipping in ozonated water. Using cultures and 40 sec exposures, 2.5 ppm reduced four bacterial species (including *E. coli* 0157:H7) by 5–6 logs.[107] On the other hand, 0.3–1.0 ppm exposed for 40 sec in maple sap was essentially ineffective as a sanitizer apparently due to the antagonism of sucrose to ozone's action.[112] When 95 ppm/inch^2 of O_3 was applied to beef carcasses at 28°C, the reduction of *E. coli* 0157:H7 and *S.* Typhimurium was much better than a water wash.[30] For a review of ozone as a food sanitizer, see reference 108.

Ozone was tested against *Escherichia coli* 0157:H7 in culture media, and at 3–18 ppm, the bacterium was destroyed in 20–50 minutes (Byun, M.-W. et al. 1998. *J. Food Protect.* 61:728–730). The gas was administered from an ozone generator and on tryptic soy agar, the *D* value for 18 ppm was 1.18 minutes, but in phosphate buffer, the *D* value was 3.18 minutes. To achieve a 99% inactivation of about 10,000 cysts of *Giardia lamblia*/ml, the average concentration time was 0.17 and 0.53 mg-minute/l at 25°C and 5°C, respectively (Wickramanayake, G. B. et al. 1984. *Appl. Environ. Microbiol.* 48:671–672). The protozoan was ca. three times more sensitive to O_3 at 25 than at 5°C.

Hydrogen Peroxide (H_2O_2)

This compound is a weak acid that is formed to some extent by all aerobic organisms, and it is enzymatically degraded by the enzyme *catalase*:

$$2H_2O_2 \longrightarrow 2H_2O + O_2$$

Table 13–4   *D* Values for Four Chemical Sterilants of Some Foodborne Microorganisms

Organisms	*D**	Concentration	Temperature[†]	Condition	Reference
			Hydrogen peroxide		
C. botulinum 169B	0.03	35%	88		192
B. coagulans	1.8	26%	25		193
G. stearothermophilus	1.5	26%	25		193
C. sporogenes	0.8	378 ppm	25	Vapor	123
G. stearothermophilus	1.5	370 ppm	25	Vapor	123
B. subtilis ATCC 95244	1.5	20%	25		186
B. subtilis A	7.3	26%	25		192
			Ethylene oxide		
C. botulinum 62A	11.5	700 mg/l	40	47% RH	167
C. botulinum 62A 7.4		700 mg/l	40	23% RH	208
C. sporogenes ATCC 7955	3.25	500 mg/l	54.4	40% RH	105
B. coagulans	7.0	700 mg/l	40	33% RH	16
B. coagulans	3.07	700 mg/l	60	33% RH	16
G. stearothermophilus ATCC 7953	2.63	500 mg/l	54.4	40% RH	105
L. brevis	5.88	700 mg/l	30	33% RH	16
D. radiodurans	3.00	500 mg/l	54.4	40% RH	105
			Sodium hypochlorite		
A. niger conidiospores	0.61	20 ppm[‡]	20	pH 3.0	33
A. niger conidiospores	1.04	20 ppm[‡]	20	pH 5.0	33
A. niger conidiospores	1.31	20 ppm[‡]	20	pH 7.0	33
			Iodine ($\frac{1}{2}I_2$)		
A. niger conidiospores	0.86	20 ppm[‡]	20	pH 3.0	33
A. niger conidiospores	2.04	20 ppm	20	PH 7.0	33
A. niger conidiospores	1.15	20 ppm	20	PH 5.0	33

* = In minutes
† = °C
‡ = As chlorine

As noted above, it is a strong oxidizing agent. In combination with heat, it has had limited use in milk pasteurization and sugar processing. It is used as a sterilant for food-contact surfaces of olefin polymers and polyethylene in aseptic packaging systems. It is used also at minimum levels of 0.08 or 0.05% in the flow-injection pasteurization of egg white by at least two commercial processors. Hydrogen peroxide vapors are quite microbiocidal, and *D* values of some foodborne microorganisms are presented in Table 13–4. H_2O_2 prevented spores of *Bacillus cereus* from swelling properly during the germination process, but it did not effect the release of dipicolinic acid.[126]

Interest in this compound as a food sanitizer has increased during the past decade or so, and it has been employed in combination with other agents. Now it is not permitted by the United States' Food and Drug Administration as a food sanitizer except when used with acetic acid to form peroxyacetic acid. However, a 1% wash is permissible by the U.S. Environmental Protection Agency for some postharvest uses.[168]

When 5% H_2O_2 was employed on apples inoculated with a nonpathogenic *E. coli* strain, it was found to be only marginally effective due to attachment of the organism to inaccessible sites, and its survival and growth in punctures.[169] The life of this compound on fresh produce may be expected to be shortened by the presence of catalase in the produce. For a review of its use on fruits and vegetables, see reference 168.

In regard to *Cryptosporidium parvum* in apple cider, orange juice, and grape juice, 0.025% reduced infectivity by >5 logs as assessed by using a human ileocecal cell line.[110] Malic, citric, and tartaric acids at 1–5% inhibited the protozoan by up to 88%. The treatment of lettuce leaves with 2% H_2O_2 at 50°C for 60 sec effected a $\leq$4-log reduction of *E. coli* 0157:H7 and a 3-log reduction of *S. Enteritidis* without affecting sensory quality.[117]

Chlorine and Other Agents

Chlorine (Cl_2) does not exist in this state in nature but rather in one of many combined forms such as calcium chloride ($CaCl_2$). However, Cl_2 is the form that is needed for sanitizing purposes, and it is supplied from compounds such as sodium chlorite ($ClNaO_2$), sodium hypochlorite ($ClNaO$), chlorine dioxide (ClO_2), and others. It is similar in some ways to O_3 and H_2O_2 since they, too, are strong oxidants but not as strong as ozone. However, in contrast to Cl_2 and O_3, H_2O_2 affects DNA. Chlorine is widely used to sanitize drinking and swimming pool waters, and it has a long history of use as a sanitizing agent on food contact surfaces, work surfaces, and floor drains.

Among the other agents tested as potential food sanitizers are cetylpyridinium chloride; hydrogen peroxide; quarternary ammonium compounds (various); peracetic (peroxyacetic) acid; Fit (an alkaline produce wash product that contains seven GRAS ingredients); Tsunami™(solution of peroxyacetic acid); Avgard™(contains phosphate); and calcinated calcium (natural product of oyster shells that are ground, and the pearl layer treated electrically with ohmic heating at 220 V for 10–50 minutes). Shells are crushed into a powder and used as a food supplement in Japan.[11] Summaries of the comparative effects of some of these agents as food sanitizers are presented in the next three tables.

The relative efficacy of Cl_2 and calcinated calcium to achieve log number reductions of *E. coli* 0157:H7, *S.* Typhimurium, and *L. monocytogenes* on the surface of tomatoes is presented in Table 13–5 where calcinated calcium effected reductions of >7 logs compared to 2–3 logs for chlorine.[11]

Chlorine, ASC, H_2O_2, and Tsunami are compared in Table 13–6 on the reduction of APCs on cantaloupes, honeydew melons, and asparagus.[146] It can be seen that 2,000 ppm Cl_2 and 1,200 ppm ASC were the most effective of the three sanitizers. Asparagus was the most difficult of these products for the four agents.

Table 13–5 Effectiveness of Calcinated Calcium and Chlorine in Effecting log Reductions of *E. coli* 0157:H7, *S.* Enteritidis, and *L. monocytogenes* on the Surface of Tomatoes (Summarized from Reference 11). Numbers Represent logs Reduced/tomato.

Organisms	Chlorine	Calcinated Calcium
Escherichia coli 0157:H7	3.4	7.85
Salmonella Enteritidis	2.1	7.36
Listeria monocytogenes	2.3	7.59

Table 13–6 The Relative Effectiveness of a 3-minute Exposure Time for Four Sanitizers on the APC of Cantaloupes, Honeydew Melons, and Asparagus (Summarized from Park and Beuchat[146]). Numbers Represent $\log_{10}$ cfu APC.

Treatments	Cantaloupes	Honeydew Melons	Asparagus
Water (control)	5.49	3.81	6.71
Chlorine, 200 ppm	4.73	3.48	6.35
Chlorine, 2,000 ppm	2.86	1.48	6.05
Ascid. Na chlorite (ASC), 850 ppm	3.48	2.14	6.14
ASC, 1,200 ppm	3.35	1.32	6.19
H_2O_2, 0.2%	4.53	3.40	6.58
H_2O_2, 1.0%	5.15	2.89	6.31
Tsunami™, 40 ppm	4.61	2.44	6.51
Tsunami™, 80 ppm	4.87	3.13	6.49

A number of studies have shown how biofilm formation on food-handling surfaces and on certain foods antagonizes the action of most if not all sanitizers in removing or destroying foodborne pathogens. In a detailed study, Frank et al.[61] compared the efficacy of a number of sanitizers to inactivate *L. monocytogenes*, and the mean log reductions of this organism in biofilms coated with protein and fat are summarized in Table 13–7. ASC was the most effective of the four sanitizers/combinations tested followed by peracetic acid + octanoic acid, and a quarternary ammonium compound. Biofilms are discussed in Chapter 22.

To reduce the numbers of *E. coli* 0157:H7 on whole fresh Bracburn apples, three commercial sanitizing preparations were tested: Tsunami 100; AgClor 300/Decco buffer (Cl_2 phosphate buffer solution); and Oxine (ClO_2). Apples were exposed for up to 15 minutes. A water wash reduced the pathogen by ca. 2 logs, but no sanitizer effected a 5-log reduction when used at its recommended concentration. However, when used at ca. 2–16 times recommended concentrations, Tsunami (at 1,280 ppm) achieved a 5.5-log reduction; AgClor (at 3,200 ppm) ca. 4.5 log reduction; and Oxine (80 ppm) ca. 4.0. logs.[209] Chlorine dioxide did not achieve a 5-log reduction at any concentration

Table 13–7 Comparison of Sanitizer Treatments for Inactivation of *Listeria monocytogenes* Biofilm Coated with Protein (4 mg) and Fat (365 mg) Statistical Analysis on Combined Data for All Treatment Times and at Recommended Usage Level,[61] copyright © 2003, Int. Assoc. for Food Protection.

Sanitizer Treatments	Mean $\log_{10}$ Reduction*
Acidified sodium chlorite	6.02[A]
Peracetic + octanoic acid	5.50[B]
Quarternary ammonium compound	5.47[B]
Peracetic acid	4.78[C]
Sodium hypochlorite	4.75[C]
Water	<2.0[D]

*Mean separation by Duncan's multiple range test; means with different superscripts (A, B, C, D) significantly different at $P = 0.05$.

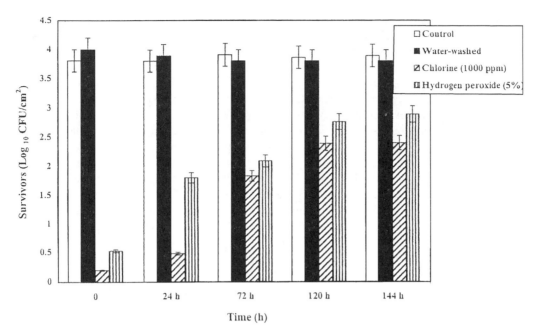

Figure 13–4   Effect of sanitizer treatments on viability of attached *Salmonella* Stanley HO558 on the surface of cantaloupes stored at 20°C for up to 6 days after inoculation. Values represent mean +SE of values from three separate experiments,[200] copyright © 2001, Int. Assoc. for Food Protection.

tested. The peroxyacetic acid-containing product was the most effective in this study. The effect of 1,000 ppm Cl_2 and 5% H_2O_2 on *S*. Stanley on cantaloupe surfaces is presented in Figure 13–4.[200] It can be seen that both agents were more effective than a water wash after 6 days at 20°C.

Nine volatile chemicals were tested for the capacity to destroy *Salmonella* spp. on alfalfa seeds and sprouts, and the three most effective were acetic acid, cinnimic aldehyde, and thymol, and they effected >3 $\log_{10}$ cfu/g reductions compared to ca. 1.9 $\log_{10}$ cfu/g for the control.[206] The compounds (1,000 mg/l of air) were exposed for up to 7 hours at 50°C. Allylisothiocyanate (AIT) was lethal to the pathogen but it adversely affected sensory quality of products. To destroy *Salmonella* and *E. coli* 0157:H7 on alfalfa seeds, FIT and chlorine were tested. At 20,000 ppm, chlorine reduced pathogens by up to 2.5 logs after a 30-minute exposure while the reduction by Fit was up to 2.3 logs.[15] Although numbers were reduced, neither agent eliminated the pathogens from the seeds.

To sanitize retail-store lettuce, four agents were compared to 200 ppm sodium hypochlorite and after a 15-minute exposure for each, the following were found to be equivalent to or more effective than NaOCl: 4% acetic acid; 80 ppm peracetic acid; 200 ppm sodium dichloroisocyanurate; and 50% vinegar.[136] When Iceberg lettuce was inoculated with ca. 8.0 log cfu of lettuce spoilage biota and treated with Cl_2 (up to 200 ppm), ozone (up to 7.5 ppm), in addition to these sanitizers combined; greatest reduction was achieved with 7.5 ppm ozone +150 ppm chlorine, which effected a 1.4-log reduction.[66]

The difficulty of destroying pathogens on sprout seeds, such as alfalfa, without destroying their capacity to germinate, can be seen from a study that evaluated seven sanitizers against a cocktail of six *Salmonella* serotypes previously associated with foodborne outbreaks (*Salmonella* Montevideo,

Gaminara, Infantis, Anatum, Cubana, and Stanley). After a 10-minute exposure, the reductions ranged between 2 and 3.2 logs for the following sanitizers: 20,000 ppm $Ca(OCl)_2$; 5% Na_3PO_4; 8% H_2O_2; 1% $Ca(OH)_2$; 1% calcinated calcium, 5% lactic acid, and 5% citric acid.[207]

The effect of isothiocyanate vapors against antibiotic-resistant *E. coli* 0157:H7 and *S.* Montevideo; and *L. monocytogenes* inoculated onto lettuce was examined.[118] Iceberg lettuce was inoculated with up to 10^7-10^8 cfu/g and held at 4°C for up to 4 days. A reduction of up to 8 logs of *E. coli* 0157:H7 (in 2 days) and *S.* Montevideo (in 4 days) was achieved with vapor generated from 400 μl of allyl isothiocyanate. Overall, allyl isothiocyanate was more effective than methyl isothiocyanate against the two Gram-negative pathogens but the methyl form was more effective against *L. monocytogenes*.[118]

The control of *L. monocytogenes* on turkey frankfurters was achieved by the use of any one of four GRAS chemicals employed at a level of 0.3%/frank. With an inoculum of a five-strain mixture of 10^6 cfu/ml, the franks were dipped for 1 minute each and held at 4, 13, and 22°C.[89] An immediate decrease of 1–2 logs was achieved by all agents tested alone, and after 14 days at 4°C, the decrease was 3–4 logs while untreated franks spoiled within 7 days at 22°C. When chopped parsley was inoculated with ca. 10^3 or 10^6 cfu/g of *Shigella sonnei* and held at 21°C for up to 14 days, the pathogen increased about 3 logs after 1 day while that held at 4°C decreased by 2.5–3.0 $\log_{10}$ cfu/g over the 14-day period.[213] With vinegar (5.2% acetic acid) or 200 ppm free Cl_2 for 5 minutes at 21°C, a >6-log reduction was achieved, and with 7.6% acetic acid or 250 ppm free Cl_2, a reduction of 7 to 7.3 logs was achieved.[213] The effect of hypochlorite and ClO_2 on bacterial spores has been studied, and in one study using *Bacillus subtilis* spores, neither hypochlorite nor chlorine dioxide caused the release of the spore core's depot of dipicolinic acid (DPA) but spores so treated more readily released DPA upon a subsequent normally sub-lethal heat treatment than did untreated spores.[219])

NaCl AND SUGARS

These compounds are grouped together because of the similarity in their modes of action in preserving foods. NaCl has been employed as a food preservative since ancient times. The early food uses of salt were for the purpose of preserving meats. This use is based on the fact that at high concentrations, salt exerts a drying effect on both food and microorganisms. Salt (saline) in water at concentrations of 0.85–0.90% produces an isotonic condition for nonmarine microorganisms. Because the amounts of NaCl and water are equal on both sides of the cell membrane, water moves across the cell membranes equally in both directions. When microbial cells are suspended in, say, a 5% saline solution, the concentration of water is greater inside the cells than outside (concentration of H_2O is highest where solute concentration is lowest). In diffusion, water moves from its area of high concentration to its area of low concentration. In this case, water passes out of the cells at a greater rate than it enters. The result to the cell is plasmolysis, which results in growth inhibition and possibly death. This is essentially what is achieved when high concentrations of salt are added to fresh meats for the purpose of preservation. Both the microbial cells and those of the meat undergo plasmolysis (shrinkage), resulting in the drying of the meat, as well as inhibition or death of microbial cells. Enough salt must be used to effect hypertonic conditions. The higher the concentration, the greater are the preservative and drying effects. In the absence of refrigeration, fish and other meats may be effectively preserved by salting. The inhibitory effects of salt are not dependent on pH, as are some other chemical preservatives. Most nonmarine bacteria can be inhibited by 20% or less NaCl, whereas some molds generally tolerate higher levels. Organisms that can grow in the presence of and require high concentrations of salt are referred to as halophiles; those that can withstand but not grow in high concentrations are referred to as halodurics. (The interaction of salt with nitrite and other agents in the inhibition of *C. botulinum* has been discussed earlier under Nitrites and Nitrates.

Sugars, such as sucrose, exert their preserving effect in essentially the same manner as salt. One of the main differences lies in relative concentrations. It generally requires about six times more sucrose than NaCl to effect the same degree of inhibition. The most common uses of sugars as preserving agents are in the making of fruit preserves, candies, condensed milk, and the like. The shelf stability of certain pies, cakes, and other such products is due in large part to the preserving effect of high concentrations of sugar, which, like salt, makes water unavailable to microorganisms. Bacterial pathogens inoculated into liquid sweeteners (such as high-fructose corn syrup) at levels of about 10^5/g could not be detected after 3 days at normal storage temperatures[139] and these investigators suggested that the incidental contamination of such products by pathogens should be of no public health concern.

Microorganisms differ in their response to hypertonic concentrations of sugars, with yeasts and molds being less susceptible than bacteria. Some yeasts and molds can grow in the presence of as much as 60% sucrose, whereas most bacteria are inhibited by much lower levels. Organisms that are able to grow in high concentrations of sugars are designated osmophiles; osmoduric microorganisms are those that are unable to grow but are able to withstand high levels of sugars. Some osmophilic yeasts such as *Zygosaccharomyces rouxii* can grow in the presence of extremely high concentrations of sugars.

INDIRECT ANTIMICROBIALS

The compounds and products in this section are added to foods primarily for effects other than antimicrobial and are thus multifunctional food additives.

Antioxidants

Although used in foods primarily to prevent the auto-oxidation of lipids, the phenolic antioxidants listed in Table 13–8 have been shown to possess antimicrobial activity against a wide range of

Table 13–8 Some GRAS Indirectly Antimicrobial Chemicals Used in Foods

Compound	Primary Use	Most Susceptible Organisms
Butylated hydroxyanisole (BHA)	Antioxidant	Bacteria, some fungi
Butylated hydroxytoluene (BHT)	Antioxidant	Bacteria, viruses, fungi
t-Butylhydroxyquinoline (TBHQ)	Antioxidant	Bacteria, fungi
Propyl gallate (PG)	Antioxidant	Bacteria
Nordihydroguaiaretic acid	Antioxidant	Bacteria
Ethylenediaminetetraacetic acid (EDTA)	Sequestrant/stabilizer	Bacteria
Sodium citrate	Buffer/sequestrant	Bacteria
Lauric acid	Defoaming agent	Gram-positive bacteria
Monolaurin	Emulsifier	Gram-positive bacteria, yeasts
Diacetyl	Flavoring	Gram-negative bacteria, fungi
d- and *l*-Carvone	Flavoring	Fungi, Gram-positive bacteria
Phenylacetaldehyde	Flavoring	Fungi, Gram-positive bacteria
Menthol	Flavoring	Bacteria, fungi
Vanillin, ethyl vanillin	Flavoring	Fungi
Phosphates	H_2O binding, flavoring	Bacteria
Spices/spice oils	Flavoring	Bacteria, fungi

microorganisms, including some viruses, mycoplasmas, and protozoa. These compounds have been evaluated extensively as nitrite-sparing agents in processed meats and in combination with other inhibitors, and several excellent reviews have been made.[19,63]

Butylated hydroxyanisole (BHA), butylated hydroxytoluene (BHT), and TBHQ are inhibitory to Gram-positive and Gram-negative bacteria, as well as to yeasts and molds at concentrations ranging from about 10–1,000 ppm, depending on substrate. In general, higher concentrations are required to inhibit in foods than in culture media, especially in high-fat foods. BHA was about 50 times less effective against *Bacillus* spp. in strained chicken than in nutrient broth.[180] BHA, BHT, TBHQ, and propyl gallate (PG) were all less effective in ground pork than in culture media.[65] Although strains of the same bacterial species may show wide variation in sensitivity to either of these antioxidants, it appears that BHA and TBHQ are more inhibitory than BHT to bacteria and fungi, whereas the latter is more viristatic. To prevent growth of *C. botulinum* in a prereduced medium, 50 ppm of BHA and 200 ppm of BHT were required; 200 ppm of PG were ineffective.[156] Employing 16 Gram-negative and 8 Gram-positive bacteria in culture media, Gailani and Fung[65] found the Gram positives to be more susceptible than Gram negatives to BHA, BHT, TBHQ, and PG, with each being more effective in nutrient agar than in brain heart infusion (BHI) broth. In nutrient agar, the relative effectiveness was BHA > PG > TBHQ > BHT, whereas in BHI, TBHQ > PG > BHA > BHT. Conidial germination of four *Fusarium* spp. was inhibited by 200 ppm BHA or propyl paraben (PP) over the pH range 4–10, but overall, PP was more inhibitory than BHA.[191]

Foodborne pathogens such as *Bacillus cereus*, *V. parahaemolyticus*, salmonellae, and *S. aureus* are effectively inhibited at concentrations <500 ppm, whereas some are sensitive to as little as 10 ppm. The pseudomonads, especially *P. aeruginosa*, are among the most resistant bacteria. Three toxin-producing penicillia were inhibited significantly in salami by BHA, TBHQ, and a combination of these two at 100 ppm, whereas BHT and PG were ineffective.[116] Combinations of BHA/sorbate and BHT/monolaurin have been shown to be synergistic against *S. aureus*[19,39] and BHA/sorbate against *S.typhimurium*.[39] BHT/TBHQ has been shown to be synergistic against aflatoxin-producing aspergilli.[116]

Flavoring Agents

Of the many agents used to impart aromas and flavors to foods, some possess definite antimicrobial effects. In general, flavor compounds tend to be more antifungal than antibacterial. The nonlactic, Gram-positive bacteria are the most sensitive, and the lactic acid bacteria are rather resistant. The essential oils and spices have received the most attention by food microbiologists, and the aroma compounds have been studied more for their use in cosmetics and soaps.

Of 21 flavoring compounds examined in one study, about half had minimal inhibitory concentrations (MIC) of 1,000 ppm or less against either bacteria or fungi.[97] All were pH sensitive, with inhibition increasing as pH and temperature of incubation decreased. Some of these compounds are noted in Table 13–8.

One of the most effective flavoring agents is diacetyl, which imparts the aroma of butter.[94] It is somewhat unique in being more effective against Gram-negative bacteria and fungi than against Gram-positive bacteria. In plate count agar at pH 6.0 and incubation at 30°C, all but 1 of 25 Gram-negative bacteria and 15 of 16 yeasts and molds were inhibited by 300 ppm.[91] At pH 6.0 and incubation at 5°C in nutrient broth, <10 ppm inhibited *Pseudomonas fluorescens*, *P. geniculata*, and *E. faecalis*; under the same conditions except with incubation at 30°C, about 240 ppm were required to inhibit these and other organisms.[97] It appears that diacetyl antagonizes arginine utilization by reacting with arginine-binding proteins of Gram-negative bacteria. The greater resistance of Gram-positive bacteria appears to be due to their lack of similar periplasmic binding proteins and their possession of larger amino acid pools. Another flavor compound that imparts the aroma of butter is 2,3-pentanedione, and it has been found to be inhibitory to a limited number of Gram-positive bacteria and fungi at 500 ppm or less.[95,97]

The agent *l*-carvone imparts spearmintlike aromas and the agent *d*-carvone imparts carawaylike aromas, and both are antimicrobial, with the *l*-isomer being more effective than the *d*-isomer; both are more effective against fungi than bacteria at 1,000 ppm or less.[97] Phenylacetaldehyde imparts a hyacinthlike aroma and has been shown to be inhibitory to *S. aureus* at 100 ppm; and *Candida albicans* at 500 ppm.[97,132] Menthol, which imparts a peppermintlike aroma, was found to inhibit *S. aureus* at 32 ppm, and *E. coli* and *C. albicans* at 500 ppm.[97,132] Vanillin and ethyl vanillin are inhibitory, especially to fungi at levels <1,000 ppm.

Spices and Essential Oils

Although used primarily as flavoring and seasoning agents in foods, many spices possess significant antimicrobial activity. In all instances, antimicrobial activity is due to specific chemicals or essential oils (some are noted in Chapter 3). The search for nitrite-sparing agents generated new interest in spices and spice extracts in the late 1970s.[179]

It would be difficult to predict what antimicrobial effects, if any, are derived from spices as they are used in foods; the quantities employed differ widely depending on taste, and the relative effectiveness varies depending on product composition. Because of the varying concentrations of the antimicrobial constituents in different spices and because many studies have been conducted employing them on a dry weight basis, it is difficult to ascertain the MIC of given spices against specific organisms. Another reason for conflicting results by different investigators is the assay method employed. In general, higher MIC values are obtained when highly volatile compounds are evaluated on the surface of plating media than when they are tested in pour plates or broth. When eugenol was evaluated by surface plating onto plate count agar (PCA) at pH 6, only 9 of 14 Gram-negative and 12 of 20 Gram-positive bacteria (including 8 lactics) were inhibited by 493 ppm, whereas in nutrient broth at the same pH, MICs of 32 and 63 were obtained for *Torulopsis candida* and *Aspergillus niger*, and *S. aureus* and *Escherichia coli*, respectively.[95] Spice extracts are less inhibitory in media than spices, probably due to a slower release of volatiles by the latter.[181] In spite of the difficulties of comparing results from study to study, the antimicrobial activity of spices is unquestioned, and a large number of investigators have shown the effectiveness of at least 20 spices or their extracts against most food-poisoning organisms, including mycotoxigenic fungi.[179]

In general, spices are less effective in foods than in culture media, and Gram-positive bacteria are more sensitive than Gram negatives, with the lactic acid bacteria being the most resistant among Gram positives.[221] Although results concerning them are debatable, the fungi appear in general to be more sensitive than Gram-negative bacteria. Some Gram negatives, however, are highly sensitive. Antimicrobial substances vary in content from the allicin of garlic (with a range of 0.3–0.5%) to eugenol in cloves (16–18%).[179] When whole spices are employed, MIC values range from 1% to 5% for sensitive organisms. Sage and rosemary are among the most antimicrobial as reported by various researchers, and it has been reported that 0.3% in culture media inhibited 21 of 24 Gram-positive bacteria and were more effective than allspice.[181] White mustard essential oil has been found to be effective in a number of foods at levels of ca. 25–100 ppm against Gram-negative and Gram-positive bacteria as well as against some yeasts.

With respect to specific inhibitory levels of extracts and essential oils, Huhtanen[87] made ethanol extracts of 33 spices, tested them in broth against *C. botulinum*, and found that achiote and mace extracts produced an MIC of 31 ppm and were the most effective of the 33. Next most effective were nutmeg, bay leaf, and white and black peppers, with MICs of 125 ppm. Employing the essential oils of oregano, thyme, and sassafras, Beuchat (in 1976) found that 100 ppm were cidal to *V. parahaemolyticus* in broth. Growth and aflatoxin production by *Aspergillus parasiticus* in broth were inhibited by 200–300 ppm of cinnamon and clove oils, by 150 ppm cinnamic aldehyde, and by 125 ppm eugenol.[23]

The mechanisms by which spices inhibit microorganisms are unclear and may be presumed to be different for unrelated groups of spices. That the mechanism for oregano, rosemary, sage, and thyme may be similar is suggested by the finding that resistance development by some lactic acid bacteria to one was accompanied by resistance to the other three.[221] Of nine spice oils tested for activity against mycotoxin production by *Aspergillus parasiticus* and *Fusarium moniliforme*, the most effective was eugenol at 0.25 ppm followed by cinnamic aldehyde; thymol and carvacol; oregano; and mace oils—myristin.[98]

Phosphates

These salts are commonly added to certain processed meats to increase their water holding capacity. They also contribute to flavor, and they are antioxidative.

Food-grade phosphates range from one P (e.g. trisodium phosphate) to at least 13 (sodium polyphosphate). In the 1970s and 1980s, they were shown to possess antibotulinal activity, especially when combined with nitrites. In one study, a combination of 40 ppm $NaNO_2$, 0.26% potassium sorbate, and 0.4% sodium acid pyrophosphate (SAPP) delayed *C. botulinum* neurotoxin production in frankfurter emulsions to 12–18 days compared to 6–12 days for controls at the same pH.[204] Using 13 Gram-positive and 12 Gram-negative bacteria in culture media, 0.5% of 3 polyphosphates at pH 7 and 25°C, tripoly- and hexametaphosphate were the most effective, and the Gram-positive bacteria were more susceptible than Gram negatives.[223] Filter-sterilized phosphate preparations were more inhibitory than autoclaved in the above study. As noted in Chapter 7, sodium phosphate glassy at 0.5% delayed the growth of *Clostridium tyrobutyricum* in cheese spreads for 3 weeks, and a 1% concentration effected complete inhibition in the product that was inoculated with 5×10^5 spores.[120]

Trisodium phosphate (TSP) was tested on whole chicken legs and excised chicken leg fragments for its action against *L. monocytogenes*. Using an inoculum of ca. 10^8 cfu/ml, chicken parts were dipped in 10 or 12% TSP, and sterile water as control.[27] After 5 days at 2°C, TSP was found to be more effective on excised skin than on whole legs for 9 of the 12 tested for a decrease of 4.28 logs for excised products compared to a decrease of only 3.3 logs for whole legs.[27]

As to the mechanism by which phosphates inhibit bacteria, *Bacillus cereus* cells and spores were subjected to 0.1% or more of a long-chain sodium polyphosphate glassy (polyP) and the compound was bactericidal to log-phase cells where it caused cell lysis (see Figure 13–5). Active growth of cells was necessary for this effect. The inhibition of *B. cereus* spore germination and outgrowth were effected by 0.1%, and a 1.0% concentration was sporicidal.[121] The antispore activity was antagonized by divalent cations. While using *L. monocytogenes* in a culture medium, its inhibition by a long-chain sodium polyphosphate was shown to be reversed by cations.[222] With the elongation of vegetative cells by a polyP concentration of 0.05% after 4 hours, Maier et al.[121] concluded that polyP may affect the cell division protein FtsZ and thus block polymerization of the Z ring. The FtsZ protein exhibits Mg^{2+}-dependent GTPase activity.

Medium-Chain Fatty Acids and Esters

Acetic, propionic, and sorbic acids are short-chain fatty acids used primarily as preservatives. Medium-chain fatty acids are employed primarily as surface-active or emulsifying agents. The

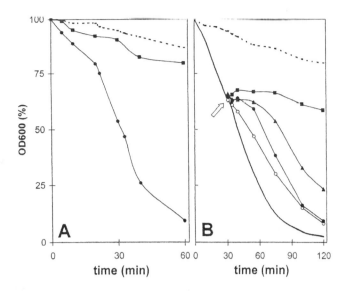

Figure 13–5 Lytic effect of the polyP JOHA HBS on exponential-phase cultures of *B. cereus* WSBC 10030 and influence of divalent metal cations. (A) Lytic effects of 0.1% polyP. Addition of 70 μg of chloramphenicol/ml to an exponential-phase culture for 20 minutes decreased the lytic effect of polyP. Dashed line = control cells in water. Symbols: ●, polyP; ■, polyP and chloramphenicol. (B) Influence of cations on polyP-induced lysis. Thirty minutes after 0.1% polyP was added to cell suspensions, 10 mM (■), 5 mM (▲), or 1 mM (●) Mg^{2+} or 1 mM (○) Ca^{2+} was added. Dashed line, control cells in water without polyP but with 10 mM Mg^{2+}; solid line, control cells with polyP and no cations,[121] (c) 1999, American Society for Microbiology. Used with permission.

antimicrobial activity of the medium-chain fatty acids is best known from soaps, which are salts of fatty acids. Those most commonly employed are composed of 12–16 carbons. For saturated fatty acids, the most antimicrobial chain length is C_{12} for monounsaturated (containing one double bond), $C_{16:1}$; and for polyunsaturated (containing more than one double bond), $C_{18:2}$ is the most antimicrobial.[99] In general, fatty acids are effective primarily against Gram-positive bacteria and yeasts. Although the C_{12} to C_{16} chain lengths are the most active against bacteria, the C_{10} to C_{12} are most active against yeasts.[99] Fatty acids and esters and the structure–function relationships among them have been reviewed and discussed by Kabara.[99] Saturated aliphatic acids effective against *C. botulinum* have been evaluated by Dymicky and Trenchard.[50]

The monoesters of glycerol and the diesters of sucrose are more antimicrobial than the corresponding free fatty acids and compare favorably with sorbic acid and the parabens as antimicrobials. Monolaurin is the most effective of the glycerol monoesters, and sucrose dicaprylate is the most effective of the sucrose diesters. Monolaurin (lauricidin) has been evaluated by a large number of investigators and found to be inhibitory to a variety of Gram-positive bacteria and some yeasts at 5–100 ppm.[19] Unlike the short-chain fatty acids, which are most effective at low pH, monolaurin is effective over the range 5.0–8.0.[100]

Because the fatty acids and esters have a narrow range of effectiveness, and GRAS substances such as EDTA, citrate, and phenolic antioxidants also have limitations as antimicrobial agents when used alone, Kabara[99] has stressed the "preservative system" approach for the control of microorganisms in foods by using combinations of chemicals to fit given food systems and preservation needs. By this approach, a preservative system might consist of three compounds, for example monolaurin/EDTA/BHA. Although EDTA possesses little antimicrobial activity by itself, it renders Gram-negative bacteria more susceptible by rupturing the outer membrane and thus potentiating the effect of fatty acids or fatty acid esters. An antioxidant such as BHA would exert effects against bacteria and molds and serve as an antioxidant at the same time. By use of such a system, the development of resistant strains could be minimized and the pH of a food could become less important relative to the effectiveness of the inhibitory system.

ACETIC AND LACTIC ACIDS

These two organic acids are among the most widely employed as preservatives. In most instances, their presence in the subject foods is due to their production within the food by lactic acid bacteria. Products such as pickles, sauerkraut, and fermented milks, among others, are created by the fermentative activities by various lactic acid bacteria, which produce acetic, lactic, and other acids.

The antimicrobial effects of organic acids such as propionic and lactic acid is due to both the depression of pH below the growth range and metabolic inhibition by the undissociated acid molecules. In determining the quantity of organic acids in foods, titratable acidity is of more value than pH alone, because the latter is a measure of hydrogen-ion concentration and organic acids do not ionize completely. In measuring titratable acidity, the amount of acid that is capable of reacting with a known amount of base is determined. The titratable acidity of products such as sauerkraut is a better indicator of the amount of acidity present than pH. When *E. coli* 0157:H7 and five other genera of foodborne pathogens were exposed to 10% acetic acid at 30°C for 4 days, none grew.[54] The same concentration of acetic acid reduced *E. coli* 0157:H7 by 6 log cycles in 1 minute. Lactic acid has been shown to function as a permeabilizer of the outer membrane of Gram-negative bacteria and thus possibly acts as a potentiator of other antimicrobials.[3]

The bactericidal effect of acetic acid can be demonstrated by its action on certain pathogens. When two species of *Salmonella* were added to an oil-and-vinegar-based salad dressing, the initial inoculum of 5×10^6 S. Enteritidis could not be detected after 5 minutes nor could S. Typhimurium be detected after 10 minutes.[129]

Organic acids are employed to wash and sanitize animal carcasses after slaughter to reduce their carriage of pathogens and to increase product shelf life, and this topic is discussed in Chapter 4.

Salts of Acetic and Lactic Acids

The sodium and potassium salts of acetic and lactic acid are widely used in foods, and they have a long history of use. For example, sodium diacetate ($CH_2COONa \cdot CH_3COOH \cdot xH_2O$) is used widely in the baking industry to prevent moldiness of bread and cakes. Interest in these multifunctional compounds has increased during the past 2–3 decades in large part because of their potential to extend shelf life of processed meats. The targeted shelf life for refrigerated cooked meat products in the United States is 75–90 days,[13] and some of these compounds are important in that regard. Another reason for the increased interest is because of their activity against psychrotrophic pathogens such as *L. monocytogenes*. The salts in this section are bacteriostatic rather than cidal. Some of the more recent findings are summarized below. For a review of lactates, see Shelef[177] and Shelef and Seiter.[178]

To control *L. monocytogenes* in frankfurters, sodium lactate (3 or 6%), sodium acetate (0.25 or 0.5%), and sodium diacetate (0.25 or 0.5%) were tested on the surface of peeled frankfurters inoculated with $\log_{10}$ 304 cfu/cm^2 of the pathogen and stored at 4°C vacuum packaged, and the pathogen was inhibited for 20–70 days.[13] The most effective was 3% sodium lactate (0.25% sodium acetate was the least effective). Growth of the pathogen was completely inhibited for more than 90 days with sodium lactate at 6% or sodium diacetate at 0.5%. In another study, wiener surfaces were inoculated with 10^5 cfu of *L. monocytogenes*, vacuum packaged, and stored for 60 days at 4.5°C; and bratwurst for 84 days at 3 and 7°C. Sodium lactate at ≥3% and combinations of ≥1% lactate and 0.1% diacetate prevented growth of the pathogen on wieners for 60 days at 4.5°C.[68] Inclusion of the agents was more effective than dipping. Potassium lactate at 2 or 3% was found to be effective in preventing the growth of a 5-strain mixture of *L. monocytogenes* in frankfurters.[152] With a pathogen inoculum of up to 500 cfu/package, storage in nylon-polyethylene bags, and incubation at 4° or 10°C for 60 or 90 days, a 4 to 5-long reduction of pathogen was found in lactate-treated meats compared to uninoculated controls.

To control growth of *Clostridium perfringens* in cooked vacuum-packaged restructured roast beef, the following compounds were assessed: Sodium citrate (2 or 4.8%, pH 4.4–5.6); sodium lactate (60% solution, 2 or 4.8%); sodium acetate (0.25%); and sodium diacetate (0.25%). The meat was inoculated with a three-strain mixture of *C. perfringens* spores, vacuum packaged, cooked at 75°C for 20 minutes, and slow cooled for 18 hours.[165] Spores were not destroyed by cooking, but growth was reduced to <1 log over the 18-hour cooling period by the citrate, lactate, and diacetate additives.[165]

In a study to determine the effect of sodium diacetate (up to 0.5%) and sodium lactate (up to 2.5%) on cook-in-bag turkey breast, the product was inoculated with 9–30 spores of a psychrotrophic *Clostridium* sp., vacuum packaged, cooked to an internal temperature of 71.1°C, chilled, and incubated at 4°C for up to 22 weeks.[127] Spoilage occurred within 6 weeks with no inhibitors, but products that contained either 0.25% diacetate or 1.25% lactate did not spoil until 13 weeks.

ANTIBIOTICS

Historically, antibiotics are secondary metabolites produced by microorganisms that inhibit or kill a wide spectrum of other microorganisms. Most of the useful ones are produced by molds and bacteria of the genus *Streptomyces*, and a few by *Bacillus and Paenibacillus* spp. Many of the clinically useful agents now in use are synthetic products.

Two antibiotics have been investigated extensively as heat adjuncts for canned foods: subtilin and tylosin. Chlortetracycline and oxytetracycline were once widely studied for their application to fresh foods, whereas natamycin is employed as a food fungistat. According to the Animal Health Institute, 22 million pounds of antibiotics were sold in the United States in 2002 for use in both farm and companion animals, a slight increase over 2001. Ninety percent of these agents were used to treat, control, and prevent diseases.

In general, the use of chemical preservatives in foods is not popular among many consumers; the idea of employing antibiotics is even less popular. Some risks may be anticipated from the use of any food additive, but the risks should not outweigh the benefits overall. The general view in the United States is that the benefits to be gained by using antibiotics in foods do not outweigh the risks, some of which are known and some of which are presumed. Some 15 considerations on the use of antibiotics as food preservatives were noted by Ingram et al., and several of the key ones are summarized as follows:

1. The antibiotic agent should kill, not inhibit, the flora and should ideally decompose into innocuous products or be destroyed on cooking for products that require cooking.
2. The antibiotic should not be inactivated by food components or products of microbial metabolism.
3. The antibiotic should not readily stimulate the appearance of resistant strains.
4. The antibiotic should not be used in foods if used therapeutically or as an animal feed additive.

The tetracyclines are used both clinically and as feed additives, and tylosin is used in animal feeds and only in the treatment of some poultry diseases (Table 13–9). Neither nisin nor subtilin is used

Table 13–9   Properties of Some Antibiotics

Property	Tetracyclines	Subtilin	Tylosin	Nisin	Natamycin
Widely used in foods	No	No	No	Yes	Yes
First food use	1950	1950	1961	1951	1956
Chemical nature	Tetracycline	Polypeptide	Macrolide	Polypeptide	Polyene
Used as heat adjunct	No	Yes	Yes	Yes	No
Heat stability	Sensitive	Stable	Stable	Stable	Stable
Microbial spectrum	G+, G−	G+	G+	G+	Fungi
Used medically	Yes	No	Yes*	No	Yes†
Used in feeds	Yes	No	Yes	No	No

*In treating poultry diseases.
†Limited.

Source: reference 96.

medically or in animal feeds, and although nisin is used in many countries, subtilin is not. The structural similarities of these two antibiotics may be noted from Figure 13–6.

Monensin

This antibiotic was approved by the FDA as a cattle feed additive in the 1970s, and it is used primarily to improve feed efficiency in ruminants. Its amino acid-sparing action has been demonstrated in fistulated cows.[113] It inhibits Gram-positive bacteria, and thus its long-term use has the potential of shifting the gastrointestinal tract bacterial biota from one that is normally Gram positive to one that is more Gram negative. Like nisin, monensin is an ionophore (destroys selective permeability of cell membranes), and the two agents compare favorably as feed additives.[26] See Chapter 27 for possible effect on *E. coli* 0157:H7 in animal feces.

Natamycin

This antibiotic (also known as pimaricin, tennecetin, and myprozine) is a polyene that is quite effective against yeasts and molds but not bacteria. Natamycin is the international nonproprietary name, as it was isolated from *Streptomyces natalensis*. Its structural formula is presented in Figure 13–6.

In granting the acceptance of natamycin as a food preservative, the joint Food and Agriculture Organization/the World Health Organization (FAO/WHO) Expert Committee[59] took the following into consideration: it does not affect bacteria, it stimulates an unusually low level of resistance among fungi, it is rarely involved in cross-resistance among other antifungal polyenes, and DNA transfer between fungi does not occur to the extent that it does with some bacteria. Also, from Table 13–9, it may be noted that its use is limited as a clinical agent, and it is not used as a feed additive. Natamycin has been shown by numerous investigators to be effective against both yeasts and molds, and many of these reports have been summarized.[96]

The relative effectiveness of natamycin was compared to that of sorbic acid and four other antifungal antibiotics by Klis et al.[109] for the inhibition of 16 different fungi (mostly molds), and although 100–1,000 ppm of sorbic acid were required for inhibition, 1–25 ppm of natamycin were effective against the same strains in the same media. To control fungi on strawberries and raspberries, natamycin was

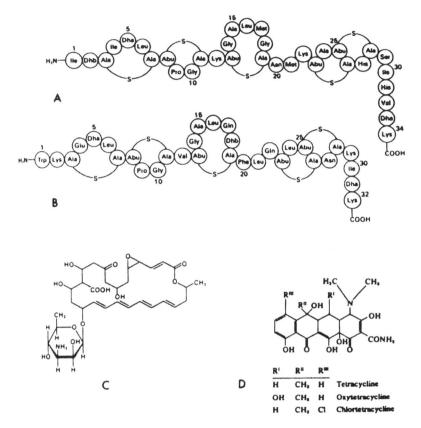

Figure 13–6   Structural formulas of nisin (*A*), subtilin (*B*), natamycin (*C*), and the tetracyclines (*D*).

compared with rimocidin and nystatin, and it, along with rimocidin, was effective at levels of 10–20 ppm, whereas 50 ppm of nystatin were required for effectiveness. In controlling fungi on salami, the spraying of fresh salami with a 0.25% solution was found to be effective by one group of investigators,[81] but another researcher was unsuccessful in his attempts to prevent surface-mold growth on Italian dry sausages when they were dipped in a 2,000-ppm solution.[84] Natamycin spray (2 × 1,000 ppm) was as good as or slightly better than 2.5% potassium sorbate.

Natamycin appears to act in the same manner as other polyene antibiotics—by binding to membrane sterols and inducing distortion of selective membrane permeability.[77] Because bacteria do not possess membrane sterols, their lack of sensitivity to this agent is thus explained.

Tetracyclines

Chlortetracycline (CTC) and oxytetracycline (OTC) were approved by the FDA in 1955 and 1956, respectively, at a level of 7 ppm to control bacterial spoilage in uncooked refrigerated poultry, but these approvals were subsequently rescinded. The efficacy of this group of antibiotics in extending the shelf life of refrigerated foods was first established by Tarr and associates working with fish in Canada.[190] Subsequent research by a large number of workers in many countries established the effectiveness of CTC and OTC in delaying bacterial spoilage of not only fish and seafoods but poultry, red meats, vegetables, raw milk, and other foods (for a review of food applications, see reference 96). CTC is generally more effective than OTC. The surface treatment of refrigerated meats with 7–10 ppm typically results in shelf-life extensions of at least 3–5 days and a shift in ultimate spoilage biota from Gram-negative bacteria to yeasts and molds. When CTC is combined with sorbate to delay the spoilage of fish, the combination has been shown to be effective for up to 14 days. Rockfish fillets dipped in a solution of 5 ppm of CTC and 1% sorbate had significantly lower aerobic plate counts (APCs) after vacuum-package storage at 2°C after 14 days than controls.[128]

The tetracyclines are both heat sensitive and storage labile in foods, and these factors were important in their initial acceptance for food use. They are used to treat diseases in humans and animals and are used also in feed supplements in the United States. The risks associated with their use as food preservatives in developed countries seem clearly to outweigh the benefits.

Subtilin

This antibiotic was discovered and developed by scientists at the Western Regional Laboratory of the USDA, and its properties were described by Dimick et al.[46] It is structurally similar to nisin (Figure 13–6), although it is produced by some strains of *Bacillus subtilis*. Like nisin, it is effective against Gram-positive bacteria, is stable to acid, and possesses enough heat resistance to withstand destruction at 121°C for 30–60 minutes. Subtilin is effective in canned foods at levels of 5–20 ppm in preventing the outgrowth of germinating endospores, and its site of action is the same as for nisin (Figure 13–1). Like nisin, it is used neither in the treatment of human or animal infections nor as a feed additive. This antibiotic may be just as effective as nisin, although it has received little attention since the late 1950s. Its mode of action is discussed above along with that of nisin, and its development and evaluation have been reviewed.[96]

Tylosin

This antibiotic is a nonpolyene macrolide, as are the clinically useful antibiotics erythromycin, oleandomycin, and others. It is more inhibitory than nisin or subtilin. Denny et al.[41] were apparently the first to study its possible use in canned foods. When 1 ppm was added to cream-style corn containing flat-sour spores and given a "botulinal" cook, no spoilage of product occurred after 30 days with incubation at 54°C.[41] Similar findings were made by others in the 1960s, and these have been summarized.[41]

Unlike nisin, subtilin, and natamycin, tylosin is used in animal feeds and also to treat some diseases of poultry. As a macrolide, it is most effective against Gram-positive bacteria. It inhibits protein synthesis by associating with the 50S ribosomal subunit and shows at least partial cross-resistance with erythromycin.

ANTIFUNGAL AGENTS FOR FRUITS

Listed in Table 13–10 are some compounds applied to fruits after harvest to control fungi, primarily molds. Benomyl is applied uniformly over the entire surface of fruits (examples are noted in Table 13–10). It is applied at concentrations of 0.5–1.0 g/l. It can penetrate the surface of some vegetables and is used worldwide to control crown rot and anthracnose of bananas, and stem-end rots of citrus fruits. It is more effective than thiabendazole and penetrates with greater ease. Both benomyl and thiabendazole are effective in controlling dry rot caused by *Fusarium* spp. To prevent the spread of *Botrytis* from

Table 13–10 Some Chemical Agents Employed to Control Fungal Spoilage of Fresh Fruits

Compound	Fruits
Thiabendazole	Apples, pears, citrus fruits, pineapples
Benomyl	Apples, pears, bananas, citrus fruits, mangoes, papayas, peaches, cherries, pineapples
Biphenyl	Citrus fruits
SO$_2$ fumigation	Grapes
Sodium-α-phenylphenate	Apples, pears, citrus fruits, pineapples

Source: Eckert.[51]

grape to grape, SO$_2$ is employed for long-term storage. It is applied shortly after harvest and about once a week thereafter. A typical initial treatment consists of a 20-minute application of a 1% preparation and about 0.25% in subsequent treatments (the use of SO$_2$ in other foods is discussed above).

An extract of a *Trichoderma* sp. (6-pentyl-α-pyrone, 6-PAP) is an effective inhibitor of *Botrytis* and *Armillaria* strains that destroy kiwi fruit in New Zealand. The effectiveness of 6-PAP on other fungi is unclear.

ETHYLENE AND PROPYLENE OXIDES

Ethylene and propylene oxides, along with ethyl and methyl formate (HCOOC$_2$H$_5$ and HCOOCH$_3$, respectively), are treated together in this section because of their similar actions. The structures of the oxide compounds are as follows:

$$\begin{array}{ccc} H_2C & & CH_2 \\ | \quad \diagdown O & O \diagup & | \\ H_2C & & CH \cdot CH_3 \end{array}$$

The oxides exist as gases and are employed as fumigants in the food industry. The oxides are applied to dried fruits, nuts, spices, and so forth, primarily as antifungal compounds.

Ethylene oxide is an alkylating agent. Its antimicrobial activity is presumed to be related to this action in the following manner. In the presence of labile H atoms, the unstable three-membered ring of ethylene oxide splits. The H atom attaches itself to the oxygen, forming a hydroxyl ethyl radical, CH$_2$CH$_2$OH, which attaches itself to the position in the organic molecule left vacant by the H atom. The hydroxyl ethyl group blocks reactive groups within microbial proteins, thus resulting in inhibition. Among the groups capable of supplying a labile H atom are —COOH, —NH$_2$, —SH, and O—OH. Ethylene oxide appears to affect endospores of *C. botulinum* by alkylation of guanine and adenine components of spore DNA.[167,208]

Ethylene oxide is used as a gaseous sterilant for flexible and semirigid containers for packaging aseptically processed foods. All of the gas dissipates from the containers following their removal from treatment chambers. With respect to its action on microorganisms, it is not much more effective against vegetative cells than against endospores, as can be seen from the *D* values given in Table 13–4.

MISCELLANEOUS CHEMICAL PRESERVATIVES

Chitosans

These are cationic polysaccharides made from chitin by acid or enzymatic hydrolysis; they are deacetylated derivatives of chitin. The latter may be achieved by use of chitosanase (see reference 106). O-carbomethylated (O-CM) chitosan is water soluble and has a broader antimicrobial spectrum than some other preparations. Chitosans vary widely in molecular size with some having molecular weights as low as 30 to over 1,000. These compounds appear to be more effective against Gram-positive than Gram-negative bacteria, and they are being investigated as antimicrobial compounds for use in packaging films (bioactive packaging). The polycationic chitosans bind to negatively charged bacterial cells and interfere with membrane and transport functions.

Using an agar plate disk assay and six chitosan oligomers and six chitosans at 0.1% concentration, they were tested against some tofu spoilage bacteria. The six chitosans completely inhibited *Bacillus* spp. after 24 hours at 37°C, and a 3- to 4-log reduction of *Bacillus megaterium* and *B. cereus* was observed with chitosans.[140] A concentration of 0.04% of chitosans completely eliminated *Enterobacter sakazakii*. From a study of chitosans in oil-in-water emulsions against levels of 10^7 cfu/ml of *L. monocytogenes* and *S.* Typhimurium, the investigators found that 0.1% chitosan polysaccharide should be sufficient to inhibit these two pathogens at either 10°C or 25°C with *L. monocytogenes* being more sensitive than *S.* Typhimurium.[225]

Dimethyl Dicarbonate

This compound is used as a yeast inhibitor in wine and in some fruit drinks at a level of 0.025%. Upon hydrolysis, it yields methanol and CO_2. With a concentration of 0.25% in apple cider stored at 4°C, a significant reduction (to <1 cfu/ml) of *E. coli* 0157:H7 occurred after 3 days compared to controls with no DMDC or cider preserved with sodium bisulfite or sodium benzoate where this pathogen survived for up to 15 days.[56]

Ethanol

This alcohol is present in flavoring extracts and effects preservation by virtue of its desiccant and denaturant properties. Ethanol vapors, produced by a vapor generator, can be produced within the headspace of a package, and the vapors have been shown to be effective against some bacteria and fungi. Ethanol has been shown to sensitize *L. monocytogenes* to low pH, organic acids, and osmotic stress, and these effects can be demonstrated with 5% concentrations. At a pH of 3.0, >3-log killing of the pathogen occurred after a 40-minute exposure to 5% but with a 10% concentration, the same reduction occurred after 10 minutes.[12] The most potent combination was pH 3.0+50 mM formate +5% ethanol, which produced a 5-log kill in 4 minutes. It appears that ethanol alters membrane permeability and thus makes cells more susceptible to certain other agents. Spores of *Bacillus subtilis* treated with either alkali or ethanol release their DPA. Spores treated with ethanol did not germinate in nutrient and nonnutrient germinants, and they did not recover after lysozyme treatment.[174]

Dehydroacetic acid (below) is used to preserve squash.

Diethylpyrocarbonate has been used in bottled wines and soft drinks as a yeast inhibitor. It decomposes to form ethanol and CO_2 by either hydrolysis or alcoholysis. Hydrolysis (reaction with water):

$$
\begin{array}{c}
C_2H_5O - CO \\
\diagdown \\
\qquad\qquad O \xrightarrow{H_2O} 2C_2H_5OH + 2CO_2 \\
\diagup \\
C_2H_5O - CO
\end{array}
$$

Alcoholysis (reaction with ethyl alcohol):

$$
\begin{array}{cc}
C_2H_5O - CO & C_2H_5O \\
\diagdown & \diagdown \\
\quad O \xrightarrow{C_2H_5OH} & \quad C = O + CO_2 + C_2H_5OH \\
\diagup & \diagup \\
C_2H_5O - CO & C_2H_5O
\end{array}
$$

Saccharomyces cerevisiae and conidia of *A. niger* and *Byssochlamys fulva* have been shown to be destroyed by this compound during the first half hour of exposure, whereas the ascospores of *B. fulva* required 4–6 hours for maximal destruction.[185] Cidal concentrations for yeasts range from about 20 to 1,000 ppm, depending on species or strain. *L. plantarum* and *Leuconostoc mesenteroides* required 24 hours or longer for destruction. Spore-forming bacteria are quite resistant to this compound. Sometimes urethane is formed when this compound is used, and because it is a carcinogen, the use of diethylpyrocarbonate is no longer permissible in the United States.

Glucose Oxidase

This enzyme catalyzes the oxidation of glucose, in the presence of O_2, to gluconic acid and H_2O_2. The enzyme is produced by some molds, and the products of the reaction have been shown to suppress the growth of at least some Gram-negative bacteria in culture media.

Polyamino Acids

At least two amino acid cationic polymers have been shown to be inhibitory to a number of foodborne bacteria and fungi, and to possess minimal toxicity to humans. They apparently act by combining and interfering with the activity of microbial cell membranes. The two noted below are reported to have GRAS status.

Epsilon-polylysine is produced in Japan, and levels as low as 5 ppm have been found to be inhibitory, especially of some Gram-positive bacteria. It is water soluble, effective over a wide pH range, and is degraded to lysine by proteases. The other compound is ethyl-*N*-dodecanoyl-L-arginine HCl (produced in Spain). Its effectiveness against meat and poultry microbiota has been demonstrated at levels of ca. 200 ppm. In the body, it is metabolized to arginine.

BIOCONTROL

Simply stated, biocontrol is the use of one or more organisms to inhibit or control other organisms. The control may require a living organism (such as phages) or it may be effected by indirect actions or agents (such as the production of bacteriocins). Antibiotics, covered in the preceding section of this chapter, are not included here. Biocontrol as related to food protection encompasses the activities of the lactic acid bacteria, bacteriocins, endolysins, bacteriophages, and "protective cultures" in general.

Microbial Interference

The food products covered in this section are favorably affected by selective members of their microbiota, which do not alter the identity of the food products. In contrast, the fermented products covered in Chapters 7 and 8 are essentially new and finished products with an identity of their own, e.g., yogurt from milk; pickles from cucumbers; wine from fruit juice, etc. Although fermented foods are not included under microbial interference, it should be noted that many of the effective organisms are also involved in fermentations. The interference biota consists essentially of biocontrol agents.

Microbial interference refers to the general nonspecific inhibition or destruction of one microorganism by other members of the same habitat or environment. Whereas lactic antagonism is a specific example of microbial interference, there are other less well-defined ways in which inhibition occurs and some of these are outlined below. The expression "bacterial interference" was suggested by R. Dubos to describe the early work in this area, which dealt primarily with the antagonism of certain human pathogens by the normal background biota of the skin. More specifically, a number of clinical researchers showed in the 1960s and 1970s that the normal harmless staphylococcal biota of the nares prevented colonization by more virulent staphylococcal strains. This was demonstrated by spraying or inoculating the nares of newborn infants with live avirulent strains, which prevented subsequent colonization by virulent strains. Examples of bacterial interference dating back to around 1877 have been noted and reviewed by Florey.[57] Among the earliest published studies of general microbial interference in foods were those of Dack and Lippitz,[37] Peterson et al.,[149] and Goepfert and Kim.[69] Dack and Lippitz observed that the natural biota of frozen pot pies inhibited inoculated cells of *Staphylococcus aureus*, *E. coli*, and *S.* Typhimurium. The repression of *S. aureus* in pot pies by around 10^5/g of the normal biota was shown by Peterson et al. The inability of foodborne pathogens to grow in fresh ground beef with a background biota of ca. 10^5/g was demonstrated by Goepfert and Kim. Since these early studies, the antagonism of the normal food biota against *L. monocytogenes* and against pathogenic strains of *E. coli* has been demonstrated. The suppressive effects of a sufficiently large aerobic bacterial biota against the growth of *Clostridium botulinum* in fresh meats is well established, as is the suppression of yeasts and molds by the bacterial biota of comminuted fresh meats.[93]

The mechanisms of general microbial interference are not clear, but some observations are worthy of note. First, the background biota needs to be larger in a number of viable cells than the organism to be inhibited. Second, the interfering biota is generally not homogeneous, and the specific roles that individual species play are unclear. Among the explanations offered over the years are (1) competition for nutrients, (2) competition for attachment/adhesion sites, (3) unfavorable alteration of the environment, and (4) combinations of these. Since interference typically occurs when the APC is at least 10^6 cells/g, it is not inconceivable that biofilm formation and the occurrence of quorum sensing play some as-yet unknown role(s) in this phenomenon. Some specific examples of interference and lactic antagonism are presented below.

A somewhat unusual example of what might be called "biotic interference" has been demonstrated with a soil nematode. The free-living nematode, *Caenorhabditis elegans*, was shown to disperse bacteria with an apparent preference for Gram-negative cells over Gram positives although members of both groups were ingested in a laboratory study.[5] *Salmonella* Poona was ingested by this nematode, which resulted in the protection of this bacterium from the effect of sanitizers.[25] The dispersal capacity of soil-inhabiting nematodes could prove to be significant to the dispersal and persistence of some foodborne bacterial pathogens in soils, especially among seed sprouts.

Lactic antagonism

The phenomenon of a lactic acid bacterium inhibiting or killing closely related and food-poisoning and food spoilage organisms when in mixed culture has been observed for more than 80 years. Commonly referred to as lactic antagonism, the precise mechanisms are yet unclear. Among factors identified are the production of antibiotics, H_2O_2, diacetyl, and bacteriocins in addition to pH depression and nutrient depletion. Lactic antagonism, thus, is an example of microbial interference.

In one study, a pediocin-producing strain of *Lactococcus lactis* was genetically enhanced to produce enough pediocin to control growth of *L. monocytogenes* in ripening Cheddar cheese. In control cheese, the pathogen increased to about 10^7/g after 2 weeks and then decreased to about 10^3 after 6 months but in the experimental cheese, the pathogen decreased to 10^2/g within 1 week and then to only 10/g within 3 months.[24] *Propionibacterium freudenreichii* subsp. *shermanii* produces an ill-defined, multicomponent inhibitory system when cultured in pasteurized skim milk that is effective against Gram-negative bacteria and molds in cottage cheese. One such product is *Microgard. Reuterin* (3-hydroxypropionalaldehyde) is produced by *Lactobacillus reuteri* from glycerol. At a concentration of 100 AU/g, a 5-log reduction of *E. coli* 0157:H7 was achieved in raw ground pork after 1 day at 7°C.[53] When used alone, Reuterin at 4 AU/ml inhibited the growth of *E. coli* and 8 AU/ml inhibited *L. monocytogenes*. This inhibitor was even more effective in combination with lactic acid.[53]

Lactobacillus reuteri and allyl isothiocyanate (AITC, at ca. 1,300 ppm) alone and in combination were tested for their effectiveness against a five-strain mixture of *E. coli* 0157:H7 in ground beef at 4°C for 25 days. The *E. coli* inoculum levels were 3- and 6-$\log_{10}$ cfu/g, and 250 mM glycerol/kg of meat was added. *E. coli* was killed in meat with the two levels of glycerol before day 20 by *L. reuteri*.[133] AITC completely eliminated the 3-$\log_{10}$ inoculum of *E. coli* 0157:H7 and at the higher concentration, it effected a >4.5-$\log_{10}$ cfu/g reduction after 25 days.

To study the protective effect of a lactic organism, *Lactobacillus casei* and its culture permeate were tested on ready-to-use salad vegetables at 8°C.[199] After 6 days of storage, 3% culture permeate reduced APC from 6 to 1 $\log_{10}$ cfu/g, and suppressed coliforms, enterococci, and *Aeromonas hydrophila*. Coliforms were reduced by about 2 logs and fecal coliforms by about 1 log by the use of 1% lactic acid.[199] The inhibitory principal produced by a *Staphylococcus equorum* strain has been identified as a macrocyclic peptide antibiotic, *micrococcin* P_1, and the bacterium was effective in inhibiting *L. monocytogenes* on the surface of a soft cheese.[28] Another antilisterial compound, *coagulin*, was isolated from a strain of *Bacillus subtilis* and placed in the pediocin family of bacteriocins.[115]

The efficacy of five lactic acid bacteria to control *L. monocytogenes* in pasteurized milk was assessed by inoculating the pathogen at ca. 10^4 cfu/ml with incubation at 30°C. The lactics employed were *Lactococcus lactis*, *Leuconostoc cremoris*, *Lactobacillus plantarum*, *L. delbrueckii* subsp. *bulgaricus*, and *Streptococcus salivarious* subsp. *thermophilus*.[151] An inhibition of 89–100% was achieved with final pH of 4.17–4.21 at 30°C or 37°C. Complete inhibition was achieved by the two lactobacilli after 20 hours at 37°C and 64 hours at 30°C. In another study, 49 isolates of lactic acid bacteria from ready-to-eat products were screened on agar spot plates against *L. monocytogenes*.[4] The three most effective isolates were identified as *Lactobacillus casei*, *Lactobacillus paracasei*, and *Pediococcus acidilactici*. Using MRS broth incubated at 5°C for 28 days, a 3.5-log reduction of *L. monocytogenes* was found while in frankfurters, a 4.2–4.7 log reduction occurred. In cooked ham stored at 5°C for 28 days under vacuum packaging, a 2.6-log reduction was noted.[4]

A number of other investigators have demonstrated the efficacy of various lactic acid bacteria to inhibit foodborne pathogens in meats and meat products. A meat starter culture of *Pediococcus acidilactici* was shown to effect a 2.3-log reduction of *E. coli* 0157:H7, *L. monocytogenes*, and *Staphylococcus aureus* during a salami fermentation after 24 hours compared to a 1.3-log reduction in controls.[101] When added to the surfaces of beef steaks inoculated with *E. coli* or *S.*Typhimurium and stored at 5°C, *Lactobacillus delbrueckii* subsp. *lactis* effected a significant reduction of psychrotrophs and coliforms, and a slight reduction in *E. coli*.[173] Significant reductions of both pathogens as well as psychrotrophs occurred when the lactic culture was applied to the surfaces of freshly slaughtered beef and pork carcass samples. The effectiveness of *Leuconostoc carnosum* against *L. monocytogenes* on cooked, sliced, and modified atmosphere packaged cooked meat (saveloy) has been demonstrated.[92] The most effective method used was spraying the lactic organism on meat surfaces, which kept *L. monocytogenes* to 10 cfu/g for 4 weeks at 10°C compared to controls where the pathogen increased to ca. 10^7 cfu/g.[92]

Regarding further reports on the protective effects of lactics in meat products, a study in Norway using cooked, sliced, modified-atmosphere-packaged ham and sausage meats that were inoculated with 10^4–10^5/g of a five-strain mixture of *Lactobacillus sakei*, it prevented the growth of added *L. monocytogenes* and *E. coli* 0157:H7 when held at 8°C for 21 days although a serotype 0:3 strain of *Yersinia enterocolitica* was unaffected.[20] All meats were acceptable at the end of the storage period.

When 1,180 psychrotrophic isolates from salad vegetables were tested by agar plate assay against *S. aureus*, *E. coli* 0157:H7, *L. monocytogenes*, and *S.* Montevideo, 37 (3.2%) of the isolates displayed varying degrees of inhibition against at least one of the four pathogens.[171] Thirty-four of the 37 inhibitory cultures were Gram negative. A hydrogen-peroxide-producing strain of *Lactobacillus delbrueckii* subsp. *lactis* was added to several fresh-cut vegetables along with *E. coli* 0157:H7 and *L. monocytogenes* and incubated at 7°C for up to 6 days. There was no reduction of pathogens, apparently because the catalase in the cut vegetables destroyed the hydrogen peroxide by the lactic culture.[78] When a foodborne strain of *Staphylococcus equorum* was tested against 95 *Listeria* strains (all species included), all were inhibited.[28] All but one of 131 other species and strains of Gram-positive bacteria were inhibited while 37 strains of 6 Gram-negative species were not.[28]

Protective cultures refer to the microorganisms that can be found in or added to a food product to effect preservation/protection, and this concept was advanced by Holzapfel et al.[83] The organisms noted above under lactic antagonism fit the definition of protective cultures. Among the properties that the latter should possess are: (1) they should present no health risks, (2) provide beneficial effects on the product, (3) have no negative impact on sensory properties, and (4) serve as "indicators" under abuse conditions.[83] Again, the lactic acid bacteria constitute the largest and most important group that falls under this category.

Nisin and Other Bacteriocins

Nisin

Nisin is produced by some strains of *Lactococcus lactis*, and it is a lantibiotic (contains the rare amino acids, *meso*-lanthionine and 3-methyl-lanthione). It is the prototype of foodborne bacteriocins, and its polypeptide structure is shown in Figure 13–6. The C-terminal amino acids are similar; the N-terminals are not. The first use of nisin as food was shown by Hurst[88] to prevent the spoilage of Swiss cheese by *Clostridium butyricum*. It is clearly the most widely used of these compounds for food preservation, with around 50 countries permitting its use in foods to varying degrees.[40] It was approved in 1988 for use in food in the United States, its use being limited to pasteurized processed cheese spreads. It is a hydrophobic compound, and it can be degraded by metabisulfite, titanium oxide, and certain proteolytic enzymes. The compound is effective against Gram-positive bacteria, primarily spore formers, and is ineffective against fungi and Gram-negative bacteria. *Enterococcus faecalis* is one of the most resistant Gram positives.

Among some of its desirable properties as a food preservative are the following:

- It is nontoxic.
- It is produced naturally by *Lactococcus lactis* strains.
- It is heat stable and has excellent storage stability.
- It is destroyed by digestive enzymes.
- It does not contribute to off-flavors or off-odors.
- It has a narrow spectrum of antimicrobial activity.

A large amount of research has been carried out with nisin as a heat adjunct in canned foods or as an inhibitor of heat-shocked spores of *Bacillus* and *Clostridium* strains, and the MIC for preventing outgrowth of germinating spores ranges widely from 3 to >5,000 IU/ml or <1 to >125 ppm (1 μg of pure nisin is about 40 IU or RU—Reading unit).[88] Depending on the country and the food product, typical usable levels are in the range of about 2.5–100 ppm, although some countries do not impose concentration limits. Nisin has been combined with low heat to destroy *L. monocytogenes* in cold-pack lobster meat. When using a brine at about pH 8.0 and nisin at 25 mg/kg of can contents at 60°C for 5 minutes using two can sizes, a 3- to 5-log reduction of inoculated cells was achieved, whereas with nisin alone the reduction was only 1–3 logs.[22]

A conventional heat process for low-acid canned foods requires an F_0 treatment of 6–8 (see Chapter 17) to inactivate the endospores of both *C. botulinum* and spoilage organisms. By adding nisin, the heat process can be reduced to an F_o of 3 (to inactivate *C. botulinum* spores), resulting in increased product quality of low-acid canned foods. Whereas the low-heat treatment will not destroy the endospores of spoilage organisms, nisin prevents their germination by acting early in the endospore germination cycle (Figure 13–1). In addition to its use in certain canned foods, nisin is most often employed in dairy products—processed cheeses, condensed milk, pasteurized milk, and so on. Some countries permit its use in processed tomato products and canned fruits and vegetables.[88] It is most stable in acidic foods.

Because of the effectiveness of nisin in preventing the outgrowth of germinating endospores of *C. botulinum* and the search to find safe substances that might replace nitrites in processed meats, this agent has been studied as a possible replacement for nitrite. Although some studies showed encouraging results employing *C. sporogenes* and other nonpathogenic organisms, a study employing *C. botulinum* types A and B spores in pork slurries indicated the inability of nisin at concentrations up to 550 ppm in combination with 60 ppm of nitrite to inhibit spore outgrowth.[154] Employed in culture media without added nitrite, the quantity of nisin required for 50% inhibition of *C. botulinum* type E spores was 1–2 ppm, 10–20 ppm for type B, and 20–40 ppm for type A.[172] The latter investigators found that higher levels were required for inhibition in cooked meat medium than in TPYG medium and suggested that nisin was approximately equivalent to nitrite in preventing the outgrowth of *C. botulinum* spores.

A system of classifying bacteriocins that places them into one of four classes has been presented. The Klaenhammer system is based primarily on the genetics and biochemistry of these compounds. Class I includes the lantibiotics such as nisin; class II are small heat-stable peptides such as lactacin F; class III are large heat-labile proteins such as helveticin J; and class IV are proteins that form a complex with other factors.

Unlike antibiotics, bacteriocins generally inhibit only closely related species and strains of Gram-positive bacteria. They consist of small proteins, and most are plasmid mediated. It appears that some species and strains of all genera of lactic acid bacteria possess the capacity to produce bacteriocins or bacteriocin-like compounds. Although early attention was focused on the lactics associated with dairy products, producing species and strains have been recovered from meats and other nondairy

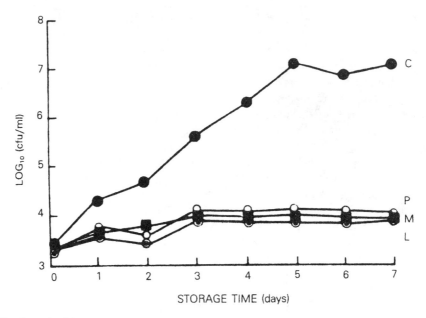

Figure 13–7   Growth of *S. aureus* in pure culture (*C*) and in association with *L. plantarum* (*L*), *P. cerevisiae* (*P*), and the mixture (*M*) in cooked mechanically deboned poultry meat (MDPM) at 15°C. Lactic acid bacteria were added at a concentration of 10^9 cells/g. *Source:* From Raccach and Baker,[153] copyright © 1978, International Association for Food Protection. Used with permission.

---

fermented products. The repression of growth of *S. aureus* by *Pediococcus cerevisiae* and *L. plantarum* is illustrated in Figure 13–7.

With respect to mode of action, nisin and subtilin appear to be identical. The structural genes appear to be the same for nisin, subtilin, and other antibiotics. The cell target for these agents is the cytoplasmic membrane, where they depolarize energized bacterial membranes (reduce transmembrane potential) and form voltage-dependent multistate pores.[1,166] The result of a pore formation is the loss of accumulated amino acids and the inhibition of amino acid transport. A nisin-resistant mutant of *L. monocytogenes* has been shown to contain significantly less phospholipids in its membrane.[130] On the assumption that the membrane targets for nisin are phospholipids, fewer would make membranes less susceptible to pore formation.[139] Unlike nisin (a class I bacteriocin), class II bacteriocins, such as lactococcin B, possess narrow host ranges and their membrane activity leads to the leakage of ions, ATP depletion, and proton motive force depletion. A vast research literature on bacteriocins has accumulated in the past decade, and it is beyond the scope of this text to provide adequate coverage of this field. For more detailed information, see references 8, 86, 91, and 170.

The ineffectiveness of nisin on fresh beef is due apparently to its inactivation by glutathione S-transferase.[164] It was shown that three glutathione molecules conjugated with one nisin molecule. In another study, synergism between nisin and CO_2 was shown by the leakage of carboxylfluorescein from liposomes in wild-type strains of *L. monocytogenes* (this same effect is caused by *enterocin P* that is produced by *Enterococcus faecium* P13 (see reference 82) after exposure to 2.5 ppm nisin in an atmosphere of 100% CO_2.[137] With a nisin-resistant strain, 2.5 ppm caused no reduction in numbers but a 2-log reduction was seen with a wild-type strain in air and a 4.1-log reduction in 100% CO_2.[137] Synergy was observed also against *L. monocytogenes* when zinc and aluminum lactates or zinc and aluminum chlorides were used with 100 IU/ml nisin, and the results showed that pretreatment with zinc lactate sensitized the organism to nisin.[124] These findings point to the bacterial cell membrane as the target for nisin, and this is further supported by the finding that nisin and vancomycin use the same target, specifically, the membrane anchored cell wall precursor lipid II for which nisin has a high affinity.[21]

Nisin was combined with lysozyme and EDTA in a gelatin coating to assess its effect on the spoilage biota. The products tested were cooked ham and bologna sausage, and they were coated with 0.2 g of 7% gelatin + 25.5 g/l lysozyme-nisin (1:3) + 25.5 g/l of EDTA. Each treatment was inoculated (with 4–5 $\log_{10}$ cfu) with six bacterial species, vacuum packaged, and stored at 8°C for 4 weeks.[67] An immediate reduction in numbers occurred in the antimicrobial gels and up to 4 log cfu/cm^2 for the four Gram positives (*Brochothrix thermosphacta*, *Lactobacillus sakei*, *Leuconostoc mesenteroides*, and *Listeria monocytogenes*) and further growth was inhibited during the 4 weeks. *E. coli* 0157:H7 was reduced by 2 logs on ham, but the antimicrobials were ineffective against this species on bologna.[67]

Using a Doehlert design, it was found that nisin and a_w were synergistic and that cell numbers could be reduced by 4–5 log cycles with 1,000 to 1,400 IU of nisin/ml at pH 5.5–6.5 and a_w of 0.97 and 0.98.[31] The noted effect was not solute specific relative to a_w control.

Other Bacteriocins

Two strains of *Carnobacterium piscicola* were added to cold-smoked salmon stored at 5°C and one was effective in reducing *L. monocytogenes* from 10^3 to <10 cfu/ml after 32 days.[138] This strain was antilisterial by agar diffusion assay, and the nonbacteriocin producer prevented the pathogen from growing on salmon. Another strain of *C. piscicola* was tested in cold-smoked salmon against L. monocytogenes, and it was also found to be bactericidal to the pathogen within 21 and 12 days at 4 and 12°C, respectively.[215] Cell-free extracts of the lactic were inhibitory to the pathogen by plate assay. Two hundred food isolates and food industry cultures of *L. monocytogenes* were tested for their susceptibility to the class IIa bacteriocins sakacin P_1, sakacin A, and pediocin PA-1 along with nisin. The 50% inhibitory concentrations (IC_{50}) were determined by plate assay with compounds at the following ng concentrations: Pediocin PA-1 (0.10–7.34); sakacin A (0.15–44.2); and nisin (2.2–781). None of the listerial strains were resistant to the class IIa bacteriocins,[102] and sakacin P_1 divided the strains into two distinct groups.

ENDOLYSINS

Upon the maturation of newly formed bacteriophages inside their host bacterial cells, they effect their own release by the consecutive use of two small hydrophobic proteins. *Holins* disrupt the cell membrane and form holes through which endolysins can pass.[218] Endolysins target bonds in the peptidoglycan, and upon the destruction of this cell barrier, the phage progeny is released. For a review, see reference 205. In addition to their lysis of bacterial cells from within, endolysins from Gram-positive bacteria also lyse bacteria exogenously (see reference 224). The production and use of phage endolysins to control some foodborne bacterial pathogens have been demonstrated, and three examples are outlined below.

An examination of the cell wall lysis system of *Clostridium perfringens* phage Φ3626 revealed that it produces a holin and an endolysin. Holin function was demonstrated by its ability to substitute for the deleted holin of phage lambda in a modified phage vector, and the endolysin gene (*ply*3626) was cloned and expressed in *E. coli*. When tested against 48 strains of *C. perfringens*, the phage endolysin destroyed all by its lytic activity when applied exogenously.[224] *Clostridium fallax* was the only other species that was lysed, and it has a peptidoglycan structure similar to that of *C. perfringens*.

Endolysins from *L. monocytogenes* phages have been introduced into a lactic starter culture, enabling the phage enzyme to reduce or eliminate the pathogen during cheese ripening. In order to optimize the release of the intracellularly synthesized endolysin from the bacterial cells onto the cheese surface, the endolysin encoding gene was modified to carry a signal peptide. When this construct was introduced into a dairy starter culture of *Lactococcus lactis*, a clone was identified, which expressed a strong lytic activity that was quantitatively exported from the lactococcal cells into the surrounding medium where it caused rapid lysis of *L. monocytogenes* cells.[64] The vector was also introduced into a lactose-utilizing strain of *L. lactis*, where a functional enzyme was produced and the vector was shown to be compatible with native lactococcal plasmids.[64] These recombinants were also used in preliminary dairy fermentation experiments to control *L. monocytogenes*, and a 95% reduction of the pathogen at the end of a camembert cheese ripening period was demonstrated (unpublished results: M.J. Loessner et al.).

A broad-spectrum endolysin from a *Lactobacillus helveticus* phage was shown to lyse different species of lactobacilli and also some lactococci, pediococci, *Enterococcus faecium*, and some other Gram-positive bacteria.[45] *Listeria innocua*, *Streptococcus salivarius* subsp. *thermophilis*, three species/strains of propionibacteria as well as *Escherichia coli*, *Pseudomonas fluorescens*, and *Salmonella* Abortus-ovis were not affected by the construct.

BACTERIOPHAGES AS BIOCONTROL AGENTS

Lytic phages specific for given bacterial species and strains are known to be effective in destroying their host cells, and this is the basis of phage typing, which is described in Chapter 11. To control pathogens and spoilage bacteria on foods, the question is whether phages can destroy their specific host cells in this environment. In other words, do food substrates prevent phage attachment to host cells and if not, do other factors come into play to prevent cell lysis? Dating back to the 1960s, research has been published on the efficacy of bacteriophages to destroy their host cells in a number of settings including meats, poultry, and certain human illnesses; and some of these have been reviewed.[6,70,75]

Phages were shown in the late 1960s to lyse their host cells recovered from fish, meats, and skim milk but these early studies employed phages and their host bacteria in broth and meat extract cultures. It appears that the first study of phages directly added to meat to control meat spoilage bacteria was that of Greer.[75] In this study, rib–eye steaks, a *Pseudomonas* sp. previously isolated from spoiled beef, and a homologous phage at a level of 10^8 pfu/ml were used. Four days after adding the phage to the steaks (held at 7°C) that were surface inoculated with the host bacterium, a 1–2-log reduction of the bacterium and a 2-log increase in phage numbers were noted. Adding 10^8 pfu/ml of phage caused steak case-life to increase from 1.6 to 2.9 days.[75] Overall, surface discoloration and retail acceptance were improved by the phage treatments.

The reduction of *S.* Enteritidis and *Campylobacter jejuni* by specific phages on excised chicken skin was investigated.[70] On *S.* Enteritidis, phages increased from an initial of 1.0 to 3.49 pfu/cm^2 after 48 hours while those on untreated skin decreased. These investigators employed a multiplicity of infection (MOI) factor, which relates the relative numbers of phages to their host bacteria. At an MOI of 1, phages increased in numbers and reduced the two pathogens by <1 log/cm^2. At an MOI of 100–1,000, the numbers of bacteria were reduced rapidly by up to 2 logs over 48 hours. No salmonellae were recovered with MOI of 10^7.[70]

A study on the incidence and prevalence of *C. jejuni* phages on retail poultry found phages in 34 of 300 fresh chicken parts but none were found in 150 frozen samples.[7] The efficacy of host-specific phages to reduce numbers of *C. jejuni* on artificially contaminated chicken skin has been demonstrated.[7] The *Campylobacter* phages are dsDNA and belong to the families Myxoviridae and Siphoviridae. In order to effect cell lysis, phages need host cells that are dividing and since *C. jejuni* does not grow below ca. 30°C, it appears not to be a good candidate for phage control on refrigerated products. However, if phages attach to the refrigerated cells, they may become active during later host cell growth.

Phages have been shown to reduce numbers of foodborne pathogens such as *L. monocytogenes* on surface-ripened cheeses as well as *E. coli* 0157:H7 and salmonellae on fresh poultry. They were used openly in the former Soviet Union to treat certain human bacterial infections. Coliphages are very common on fresh poultry where they apparently reduce the numbers of viable *E. coli*.[104] Phages present problems for dairy and meat starter culture users if one or more starter strains is lysed leading to faulty fermentations. The true potential of bacterial viruses to reduce spoilage and food poisoning bacteria in foods requires more research.

Vibrio vulnificus phages have been isolated from oysters taken from estaurine waters at several locations around the United States, and numbers/g of oyster tissue ranged from 10^1 to 10^5. All but one isolate was specific for this pathogen with one causing lysis of *V. parahaemolyticus*.[42] Two coliphages have been demonstrated to possess a wide host range that allowed them to lyse many strains of *E. coli* in addition to *Proteus mirabilis*, *Shigella dysenteriae*, and two salmonellae strains.[71]

THE HURDLE CONCEPT

Under intrinsic and extrinsic growth parameters presented in Chapter 3, the effect of single factors on the welfare of microorganisms is presented. In the hurdle concept, multiple factors or techniques are employed to effect the control of microorganisms in foods. Barrier technology, combination preservation, and combined methods are among some of the other descriptions of this concept. Referred to as "hurdle technology" since the mid-1980s by L. Leistner in Germany, the practice has been applied to some foods for over a century.

A simple example of the hurdle concept or barrier technology is demonstrated by preventing the germination of spores of proteolytic or group I strains of *Clostridium botulinum*. Among the intrinsic and extrinsic parameters that are known to prevent their germination and growth are: pH <4.6; a_w <0.94; NaCl of 10% or more; $NaNO_2$ ca. 120 ppm; incubation temperature <10°C; and a large aerobic bacterial biota. Foods that employ the hurdle concept in their formulation would embody a series of the above, thus making for a multitargeted approach to preventing germination and growth of these spores. In order for *C. botulinum* to grow, it must "hurdle" a series of the barriers noted. Note that the hurdles listed above include "aerobic bacterial biota", which is microbial interference. The important parameters of pH and a_w may be controlled by the growing food biota, especially the lactic acid bacteria. The hurdle technology concept is much broader than the above sketch presents, and more details can be found in Leistner and Gould.[114]

The concept of growth/no growth (G/NG) has been advanced to better quantify the hurdle concept by employing the synergy that exists between two or more parameters[125] Implicit in this concept is the interaction between two or more parameters to a point where growth ceases—the G/NG interface. Precise definitions and determinations of those factors/parameters that permit and prevent growth of a given organism should make it possible to devise models for the hurdle concept. For more information, see reference 125.

REFERENCES

1. Abee, T. 1995. Pore-forming bacteriocins of Gram-positive bacteria and self-protection mechanisms of producer organisms. *FEMS Microbiol. Lett.* 129:1–10.

2. Achen, M., and A.E. Yousef. 2001. Efficacy of ozone against *Escherichia coli* 0157:H7 on apples. *J. Food Sci.* 66:1380–1384.

3. Alakomi, H.-L., E. Skytta, M. Saarela, T. Mattila-Sandholm, K. Latva-Kala, and I.M. Helander. 2000. Lactic acid permeabilizes Gram-negative bacteria by disrupting the outer membrane. *Appl. Environ. Microbiol.* 66:2001–2005.

4. Amézquita, A., and M.M. Brashears. 2002. Competitive inhibition of *Listeria monocytogenes* in ready-to-eat meat products by lactic acid bacteria. *J. Food Protect.* 65:316–325.

5. Anderson, G.L., K.C. Caldwell, L.R. Beuchat, and P.L. Williams. 2003. Interaction of a free-living soil nematode, *Caenorhabditis elgans*, with surrogates of foodborne pathogenic bacteria. *J. Food Protect.* 66:1543–1549.

6. Atterbury, R.J., P.L. Connerton, C.E.R. Dodd, C.E.D. Rees, and I.F. Connerton. 2003. Application of host-specific bacteriophages to the surface of chicken skin leads to a reduction in recovery of *Campylobacter jejuni. Appl. Environ. Microbiol.* 69:6302–6306.

7. Atterbury, R.J., P.L. Connerton, C.E.R. Dodd, C.E.D. Rees, and I.F. Connerton. 2003. Isolation and characterization of *Campylobacter* bacteriophages from retail poultry. *Appl. Environ. Microbiol.* 69:4511–4518.

8. Aymerich, M.T., M. Hugas, and J.M. Monfort. 1998. Review: Bacteriocinogenic lactic acid bacteria associated with meat products. *Food Sci. Technol. Int.* 4:141–158.

9. Banks, J.G., and R.G. Board. 1982. Sulfite inhibition of *Enterobacteriacae* including *Salmonella* in British fresh sausage and in culture systems. *J. Food Protect.* 45:1292–1297, 1301.

10. Bari, M.L., Y. Sabina, S. Isobe, T. Uemura, and K. Isshiki. 2003. Effectiveness of electrolyzed acidic water in killing *Escherichia coli* 0157:H7, *Salmonella* Enteritidis, and *Listeria monocytogenes* on the surfaces of tomatoes. *J. Food Protect.* 66:542–548.

11. Bari, M.L., Y. Inatsu, S. Kawasaki, E. Nazuka, and K. Isshiki. 2002. Calcinated calcium killing of *Escherichia coli* 0157:H7, *Salmonella*, and *Listeria monocytogenes* on the surface of tomatoes. *J. Food Protect.* 65:1706–1711.

12. Barker, C., and S.F. Park. 2001. Sensitization of *Listeria monocytogenes* to low pH, organic acids, and osmotic stress by ethanol. *Appl. Environ. Microbiol.* 67:1594–1600.

13. Bedie, G.K., J. Samelis, J.N. Sofos, K.E. Belk, J.A. Scanga, and G.C. Smith. 2001. Antimicrobials in the formulation to control *Listeria monocytogenes* postprocessing contamination on frankfurters stored at $4°C$ in vacuum packages. *J. Food Protect.* 64:1949–1955.

14. Berry, B.W., and T.N. Blumer. 1981. Sensory, physical, and cooking characteristics of bacon processed with varying levels of sodium nitrite and potassium sorbate. *J. Food Sci.* 46:321–327.

15. Beuchat, L.R., T.E. Ward, and C.A. Pettigrew. 2001. Comparison of chlorine and a prototype produce wash product for effectiveness in killing *Salmonella* and *Escherichia coli* 0157:H7 on alfalfa seeds. *J. Food Protect.* 64:152–158.

16. Blake, D.F., and C.R. Stumbo. 1970. Ethylene oxide resistance of microorganisms important in spoilage of acid and high-acid foods. *J. Food Sci.* 35:26–29.

17. Bosund, I. 1962. The action of benzoic and salicylic acids on the metabolism of microorganisms. *Adv. Food Res.* 11:331–353.

18. Bowen, V.G., and R.H. Deibel. 1974. Effects of nitrite and ascorbate on botulinal toxin formation in wieners and bacon. In *Proceedings of the Meat Industry Research Conference*, 63–68. Chicago: American Meat Institute Foundation.

19. Branen, A.L., P.M. Davidson, and B. Katz. 1980. Antimicrobial properties of phenolic, antioxidants and lipids. *Food Technol.* 34(5):42–53, 63.

20. Bredholt, S., T. Nasbakken, and A. Holck. 1999. Protective cultures inhibit growth of *Listeria monocytogenes* and *Escherichia coli* 0157:H7 in cooked, sliced, vacuum- and gas-packaged meat. *Int. J. Food Microbiol.* 53:43–52.

21. Breukink, E., I. Wiedemann, C. van Kraaij, O.P. Kulpers, H.-G. Sahl, and B. de Kruijff. 1999. Use of the cell wall precursor lipid II by a pore-forming peptide antibiotic. *Science* 286:2361–2364.

22. Budu-Amoako, E., R.F. Ablett, J. Harris, and J. Delves-Broughton. 1999. Combined effect of nisin and moderate heat on destruction of *Listeria monocytogenes* in cold-pack lobster meat. *J. Food Protect.* 62:46–50.

23. Bullerman, L.B., F.Y. Lieu, and S.A. Seier. 1977. Inhibition of growth and aflatoxin production by cinnamon and clove oils, cinnamic aldehyde and eugenol. *J. Food Sci.* 42:1107–1109, 1116.

24. Buyong, N., J. Kok, and J.B. Luchansky. 1998. Use of a genetically enhanced, pedio-producing starter culture, *Lactococcus lactis* subsp. *lactis* MM217, to control *Listeria monocytogenes* in Cheddar cheese. *Appl. Environ. Microbiol.* 64:4842–4845.

25. Caldwell, K.N., B.B. Adler, G.L. Anderson, P.I. Williams, and L.R. Beuchat. 2003. Ingestion of *Salmonella enterica* serotype Poona by a free-living nematode, *Caenorhabditis elegans*, and protection against inactivation by produce sanitizers. *Appl. Environ. Microbiol.* 69:4103–4110.

26. Callaway, T.R., A.M.S. Carneiro de Melo, and J.B. Russell. 1997. The effect of nisin and monensin on ruminal fermentations in vitro. *Curr. Microbiol.* 35:90–96.

27. Capita, R., C. Alonso-Calleja, M. Prieto, M. del Camino Garcia-Fernández, and B. Moreno. 2003. Effectiveness of trisodium phosphate against *Listeria monocytogenes* on excised and nonexcised chicken skin. *J. Food Protect.* 66:61–64.

28. Carnio, M.C., A. Höltzel, M. Rudolf, T. Henle, G. Jung, and S. Scherer. 2000. The macrocyclic peptide antibiotic microccoccin P₁ is secreted by the food-borne bacterium *Staphylococcus equorum* WS 2733 and inhibits *Listeria monocytogenes* on soft cheese. *Appl. Environ. Microbiol.* 66:2378–2384.

29. Cassens, R.G. 1995. Use of sodium nitrite in cured meats today. *Food Technol.* 49(7):72–80, 115.

30. Castillo, A., K.S. McKenzie, L.M. Lucia, and G.R. Acuff. 2003. Ozone treatment for reduction of *Escherichia coli* 0157:H7 and *Salmonella* serotype Typhimurium on beef carcass surfaces. *J. Food Protect.* 66:775–779.

31. Cerrutti, P., M.R. Terebiznik, M.S. de Huergo, R. Jagus, and A.M.R. Pilosof. 2001. Combined effect of water activity and pH on the inhibition of *Escherichia coli* by nisin. *J. Food Protect.* 64:1510–1514.

32. Chang, P.-C., S.M. Akhtar, T. Burke, and H. Pivnick. 1974. Effect of sodium nitrite on *Clostridium botulinum* in canned luncheon meat: Evidence for a Perigo-type factor in the absence of nitrite. *Can Inst. Food Sci. Technol. J.* 7:209–212.

33. Cheng, M.K.C., and R.E. Levin. 1970. Chemical destruction of *Aspergillus niger* conidiospores. *J. Food Sci.* 35:62–66.

34. Christiansen, L.N., R.W. Johnston, D.A. Kautter, J.W. Howard, and W.J. Aunan. 1973. Effect of nitrite and nitrate on toxin production by *Clostridium botulinum* and on nitrosamine formation in perishable canned comminuted cured meat. *Appl. Microbiol.* 25:357–362.

35. Christiansen, L.N., R.B. Tompkin, A.B. Shaparis, T.V. Kueper, R.W. Johnston, D.A. Kautter, and O.J. Kolari. 1974. Effect of sodium nitrite on toxin production by *Clostridium botulinum* in bacon. *Appl. Microbiol.* 27:733–737.

36. Collins-Thompson, D.L., N.P. Sen, B. Aris, and L. Schwinghamer. 1972. Nonenzymic in vitro formation of nitrosamines by bacteria isolated from meat products. *Can. J. Microbiol.* 18:1968–1971.

37. Dack, G.M., and G. Lippitz. 1962. Fate of staphylococci and enteric microorganisms introduced into slurry of frozen pot pies. *Appl. Microbiol.* 10:472–479.

38. Davidson, P.M. 1983. Phenolic compounds. In *Antimicrobials in Foods*, ed. A.L. Branen and P.M. Davidson, 37–73. New York: Marcel Dekker.

39. Davidson, P.M., C.J. Brekke, and A.L. Branen. 1981. Antimicrobial activity of butylated hydroxyanisole, tertiary butylhydroquinone, and potassium sorbate in combination. *J. Food Sci.* 46:314–316.

40. Delves-Broughton, J. 1990. Nisin and its uses as a food preservative. *Food Technol.* 44(11):100, 102, 104, 106, 108, 111–112, 117.

41. Denny, C.B., L.E. Sharpe, and C.W. Bohrer. 1961. Effects of tylosin and nisin on canned food spoilage bacteria. *Appl. Microbiol.* 9:108–110.

42. DePaola, A., M.L. Motes, A.M. Chan, and C.A. Suttle. 1998. Phages infecting *Vibrio vulnificus* are abundant and diverse in oysters (*Crassostrea virginica*) collected from the Gulf of Mexico. *Appl. Environ. Microbiol.* 64:346–351.

43. Dethmers, A.E., H. Rock, T. Fazio, and R.W. Johnston. 1975. Effect of added sodium nitrite and sodium nitrate on sensory quality and nitrosamine formation in Thuringer sausage. *J. Food Sci.* 40:491–495.

44. Deuel, H.J., Jr., C.E. Calbert, L. Anisfeld, H. McKeechan, and H.D. Blunden. 1954. Sorbic acid as a fungistatic agent for foods. II. Metabolism of α,β-unsaturated fatty acids with emphasis on sorbic acid. *Food Res.* 19:13–19.

45. Deutsch, S.-M., S. Guezenec, M. Piot, S. Foster, and S. Lortal. 2004. Mur-LH, the broad-spectrum endolysin of *Lactobacillus helveticus* temperate-bacteriophage (Φt0303). *Appl. Environ. Microbiol.* 70:96–103.

46. Dimick, K.P., G. Alderton, J.C. Lewis, H.D. Lightbody, and H.L. Fevold. 1947. Purification and properties of subtilin. *Arch. Biochem.* 15:1–11.

47. Doores, S. 1983. Organic acids. In *Antimicrobials in Foods*, ed. A.L. Branen and P.M. Davidson, 75–107. New York: Marcel Dekker.

48. Doyle, M.P., and E.H. Marth. 1978. Bisulfite degrades aflatoxins. Effect of temperature and concentration of bisulfite. *J. Food Protect.* 41:774–780.

49. Duncan, C.L., and E.M. Foster. 1968. Role of curing agents in the preservation of shelf-stable canned meat products. *Appl. Microbiol.* 16:401–405.

50. Dymicky, M., and H. Trenchard. 1982. Inhibition of *Clostridium botulinum* 62A by saturated *n*-aliphatic acids, *n*-alkyl formates, acetates, propionates and butyrates. *J. Food Protect.* 45:1117–1119.

51. Eckert, J.W. 1979. Fungicidal and fungistatic agents: Control of pathogenic microorganisms on fresh fruits and vegetables after harvest. In *Food Mycology*, ed. M.E. Rhodes, 164–199. Boston: Hall.

52. Eklund, T. 1985. The effect of sorbic acid and esters of *p*-hydroxybenzoic acid on the protonmotive force in *Escherichia coli* membrane vesicles. *J. Gen. Microbiol.* 131:73–76.

53. El-Ziney, M.G., M.G.T. van den Tempel, and J. Debevere. 1999. Application of reuterin produced by *Lactobacillus reuteri* 12002 for meat decontamination and preservation. *J. Food Protect.* 62:257–261.

54. Entani, E., M. Asai, S. Tsujihata, Y. Tsukamoto, and M. Ohta. 1998. Antibacterial action of vinegar against food-borne pathogenic bacteria including *Escherichia coli* O157:H7. *J. Food Protect.* 61:953–959.

55. Fabrizio, K.A., and C.N. Cutter. 2003. Stability of electrolyzed oxidizing water and its efficacy against cell suspensions of *Salmonella* Typhimurium and *Listeria monocytogenes*. *J. Food Protect.* 66:1379–1384.

56. Fisher, T.L., and D.A. Golden. 1998. Survival of *Escherichia coli* O157:H7 in apple cider as affected by dimethyl dicarbonate, sodium bisulfite, and sodium benzoate. *J. Food Sci.* 63:904–906.

57. Florey, H.W. 1946. The use of micro-organisms for therapeutic purposes. *Yale J. Biol. Med.* 19:101–118.

58. Fong, Y.Y., and W.C. Chan. 1973. Bacterial production of di-methyl nitrosamine in salted fish. *Nature* 243:421–422.

59. Food and Agriculture Organization/World Health Organization (FAO/WHO). 1976. *Evaluation of Certain Food Additives*. WHO Technical Report Series 599.

60. Francis, G.A., C. Thomas, and D. O'Beirne. 1999. Review paper: The microbiological safety of minimally processed vegetables. *Int. J. Food Sci. Technol.* 34:1–22.

61. Frank, J.F., J. Ehlers, and L. Wicker. 2003. Removal of *Listeria monocytogenes* and poultry soil-containing biofilms using chemical cleaning and sanitizing agents under static conditions. *Food Protect. Trends* 23:654–663.

62. Freese, E., C.W. Sheu, and E. Galliers. 1973. Function of lipophilic acids as antimicrobial food additives. *Nature* 241:321–325.

63. Fung, D.Y.C., C.C.S. Lin, and M.B. Gailani. 1985. Effect of phenolic antioxidants on microbial growth. *CRC Crit. Rev. Microbiol.* 12:153–183.

64. Gaeng, S., S. Scherer, H. Neve, and M.J. Loessner. 2002. Gene cloning and expression and secretion of *Listeria monocytogenes* bacteriophage-lytic enzymes in *Lactococcus lactis*. *Appl. Environ. Microbiol.* 66:2951–2958.

65. Gailani, M.B., and D.Y.C. Fung. 1984. Antimicrobial effects of selected antioxidants in laboratory media and in ground pork. *J. Food Protect.* 47:428–433.

66. Garcia, A., J.R. Mount, and P.M. Davidson. 2003. Ozone and chlorine treatment of minimally processed lettuce. *J. Food Sci.* 68:2747–2751.

67. Gill, A.O., and R.A. Holley. 2000. Surface application of lysozyme, nisin, and EDTA to inhibit spoilage and pathogenic bacteria on ham and bologna. *J. Food Protect.* 63:1338–1346.

68. Glass, K.A., D.A. Granberg, A.L. Smith, A.M. McNamara, M. Hardin, J. Mattias, K. Ladwig, and E.A. Johnson. 2002. Inhibition of *Listeria monocytogenes* by sodium diacetate and sodium lactate on wieners and cooked bratwurst. *J. Food Protect.* 65:116–123.

69. Goepfert, J.M., and H.U. Kim. 1975. Behavior of selected foodborne pathogens in raw ground beef. *J. Milk Food Technol.* 35:449–452.

70. Goode, D., V.M. Allen, and P.A. Barrow. 2003. Reduction of experimental *Salmonella* and *Campylobacter* contamination of chicken skin by application of lytic bacteriophages. *Appl. Environ. Microbiol.* 69:5032–5036.

71. Goodridge, L., A. Gallaccio, and M.W. Griffiths. 2003. Morphological, host range, and genetic characterization of two coliphages. *Appl. Environ. Microbiol.* 69:5364–5371.

72. Gould, G.W. 1964. Effect of food preservatives on the growth of bacteria from spores. In *Microbial Inhibitors in Foods*, ed. G. Molin, 17–24. Stockholm: Almquist & Wiksell.

73. Gould, G.W., M.H. Brown, and B.C. Fletcher. 1983. Mechanisms of action of food preservation procedures. In *Food Microbiology: Advances and Prospects*, ed. T.A. Roberts and F.A. Skinner, 67–84. New York: Academic Press.

74. Gray, J.I., and A.M. Pearson. 1984. Cured meat flavor. *Adv. Food Res.* 29:1–86.

75. Greer, G.G. 1986. Homologous bacteriophage control of *Pseudomonas* growth and beef spoilage. *J. Food Protect.* 49:104–109.

76. Hagler, W.M., Jr., J.E. Hutchins, and P.B. Hamilton. 1982. Destruction of aflatoxin in corn with sodium bisulfite. *J. Food Protect.* 45:1287–1291.

77. Hamilton-Miller, J.M.T. 1974. Fungal sterols and the mode of action of the polyene antibiotics. *Adv. Appl. Microbiol.* 17:109–134.

78. Harp, E., and S.E. Gilliland. 2003. Evaluation of a select strain of *Lactobacillus delbrueckii* subsp. *lactis* as a biological control agent for pathogens on fresh-cut vegetables stored at 7°C. *J. Food Protect.* 66:1013–1018.

79. Hawksworth, G., and M.J. Hill. 1971. The formation of nitrosamines by human intestinal bacteria. *Biochem. J.* 122:28–29P.

80. Hawksworth, G., and M.J. Hill. 1971. Bacteria and the *N*-nitrosation of secondary amines. *Brit. J. Cancer* 25:520–526.

81. Hechelman, H., and L. Leistner. 1969. Hemmung von unerwunschtem Schimmelpilzwachstum auf Rohwursten durch Delvocid (Pimaricin). *Fleischwirtschaft* 49:1639–1641.

82. Herranz, C., V. Chen, H.-J. Chung, L.M. Cintas, P.E. Hernández, T.J. Montville, and M.L. Chikindas. 2001. Enterocin P selectively dissipates the membrane potential of *Enterococcus faecium* T136. *Appl. Environ. Microbiol.* 67:1689–1692.

83. Holzapfel, W.H., R. Geisen, and U. Schillinger. 1995. Biological preservation of foods with reference to protective cultures, bacteriocins and food-grade enzymes. *Int. J. Food Microbiol.* 24:343–362.

84. Holley, R.A. 1981. Prevention of surface mold growth on Italian dry sausage by natamycin and potassium sorbate. *Appl. Environ. Microbiol.* 41:422–429.

85. Holyoak, C.D., M. Stratford, A. McMullin, M.B. Cole, K. Crimmins, A.J.P. Brown, and P.J. Coote. 1996. Activity of the plasma membrane H-ATPase and optimal glycolytic flux are required for rapid adaptation and growth of *Saccharomyces cerevisiae* in the presence of the weak-acid preservative sorbic acid. *Appl. Environ. Microbiol.* 62:3158–3164.

86. Hoover, D.G., and L.R. Steenson, ed. 1993. *Bacteriocins of Lactic Acid Bacteria*. New York: Academic Press.

87. Huhtanen, C.N. 1980. Inhibition of *Clostridium botulinum* by spice extracts and aliphatic alcohols. *J. Food Protect.* 43:195–196, 200.

88. Hurst, A. 1981. Nisin. *Adv. Appl. Microbiol.* 27:85–123.

89. Islam, M., J. Chen, M.P. Doyle, and M. Chinnan. 2002. Control of *Listeria monocytogenes* on turkey frankfurters by generally-recognized-as-safe preservatives. *J. Food Protect.* 65:1411–1416.

90. Ivey, F.J., K.J. Shaver, L.N. Christiansen, and R.B. Tompkin. 1978. Effect of potassium sorbate on toxinogenesis by *Clostridium botulinum* in bacon. *J. Food Protect.* 41:621–625.

91. Jack, R.W., J.R. Tagg, and B. Ray. 1995. Bacteriocins of Gram-positive bacteria. *Microbiol. Rev.* 59:171–200.

92. Jacobsen, T., B.B. Budde, and A.G. Koch. 2003. Application of *Leuconostoc carnosum* for biopreservation of cooked meat products. *J. Appl. Microbiol.* 95:242–249.

93. Jay, J.M. 1997. Do background microorganisms play a role in the safety of fresh foods? *Trends Food Sci. Technol.* 8:421–424.

94. Jay, J.M. 1982. Antimicrobial properties of diacetyl. *Appl. Environ. Microbiol.* 44:525–532.

95. Jay, J.M. 1982. Effect of diacetyl on foodborne microorganisms. *J. Food Sci.* 47:1829–1831.

96. Jay, J.M. 1983. Antibiotics as food preservatives. In *Food Microbiology*, ed. A.H. Rose, 117–143. New York: Academic Press.

97. Jay, J.M., and G.M. Rivers. 1984. Antimicrobial activity of some food flavoring compounds. *J. Food Safety* 6:129–139.

98. Juglal, S., R. Govinden, and B. Odhav. 2002. Spice oils for the control of co-occurring mycotoxin-producing fungi. *J. Food Protect.* 65:683–687.

99. Kabara, J.J. 1983. Medium-chain fatty acids and esters. In *Antimicrobials in Foods*, ed. A.L. Branen and P.M. Davidson, 109–139. New York: Marcel Dekker.

100. Kabara, J.J., H. Vrable, and M.S.F. Lie Ken Jie. 1977. Antimicrobial lipids: Natural and synthetic fatty acids and monoglycerides. *Lipids* 12:753–759.

101. Kang, D.-H., and D.Y.C. Fung. 2000. Stimulation of starter culture for further reduction of foodborne pathogens during salami fermentation. *J. Food Protect.* 63:1492–1495.

102. Katla, T., K. Naterstad, M. Vancanneyt, J. Swings, and L. Axelsson. 2003. Differences in susceptibility of *Listeria monocytogenes* strains to sakacin P, sakacin A, pediocin PA-1, and nisin. *Appl. Environ. Microbiol.* 69:4431–4437.

103. Kemp, G.K., M.L. Aldrich, and A.L. Waldroup. 2000. Acidified sodium chlorite antimicrobial treatment of broiler carcasses. *J. Food Protect.* 63:1087–1092.

104. Kennedy, J.E., Jr., J.L. Oblinger, and B. Bitton. 1984. Recovery of coliphages from chicken, pork sausage and delicatessen meats. *J. Food Protect.* 47:623–626.

105. Kereluk, K., H.A. Gammon, and R.S. Lloyd. 1970. Microbiological aspects of ethylene oxide sterilization. II. Microbial resistance to ethylene oxide. *Appl. Microbiol.* 19:152–156.

106. Kim, K.W., R.L. Thomas, C. Lee, and H.J. Park. 2003. Antimicrobial activity of native chitosan, degraded chitosan, and o-carboxymethylated chitosan. *J. Food Protect.* 66:1495–1498.

107. Kim, J.-G., and A.E. Yousef. 2000. Inactivation kinetics of foodborne spoilage and pathogenic bacteria by ozone. *J. Food Sci.* 65:521–528.

108. Kim, J.-G., A.E. Yousef, and S.A. Dave. 1999. Application of ozone for enhancing the microbiological safety and quality of foods: A review. *J. Food Protect.* 62:1071–1087.

109. Klis, J.B., L.D. Witter, and Z.J. Ordal. 1964. The effect of several antifungal antibiotics on the growth of common food spoilage fungi. *Food Technol.* 13:124–128.

110. Kniel, K.E., S.S. Sumner, D.S. Lindsay, C.R. Hackney, M.D. Pierson, A.M. Zajac, D.A. Golden, and D. Fayer. 2003. Effect of organic acids and hydrogen peroxide on *Cryptosporidium parvum* viability in fruit juices. *J. Food Protect.* 66:1650–1657.

111. Kueper, T.V., and R.D. Trelease. 1974. Variables affecting botulinum toxin development and nitrosamine formation in fermented sausages. In *Proceedings of the Meat Industry Research Conference*, 69–74. Chicago: American Meat Institute Foundation.

112. Labbe, R.G., M. Kinsley, and J. Wu. 2001. Limitations in the use of ozone to disinfect maple sap. *J. Food Protect.* 64:104–107.

113. Lana, R.P., and J.B. Russell. 1997. Effect of forage quality and monensin on the ruminal fermentation of fistulated cows fed continuously at a constant intake. *J. Anim. Sci.* 75:224–229.

114. Leistner, L., and G. Gould. 2002. *Hurdle Technologies—Combination Treatments for Food Stability, Safety and Quality.* New York: Kluwer Academic Publishers.

115. LeMarrec, C., B. Hyronimus, P. Bressollier, B. Verneuil, and M.C. Urdaci. 2000. Biochemical and genetic characterization of coagulin, a new antilisterial bacteriocin in the pediocin family of bacteriocins, produced by *Bacillus coagulans* I_4. *Appl. Environ. Microbiol.* 66:5213–5220.

116. Lin, C.C.S., and D.Y.C. Fung. 1983. Effect of BHA, BHT, TBHQ, and PG on growth and toxigenesis of selected aspergilli. *J. Food Sci.* 48:576–580.

117. Lin, C.-M., S.S. Moon, M.P. Doyle, and K.H. McWatters. 2002. Inactivation of *Escherichia coli* 0157:H7, *Salmonella enterica* serotype Enteritidis, and *Listeria monocytogenes* on lettuce by hydrogen peroxide and lactic acid and by hydrogen peroxide with mild heat. *J. Food Protect.* 65:1215–1220.

118. Lin, C.-M., J. Kim, W.-X. Du, and C.I. Wei. 2000. Bactericidal activity of isothiocyanate against pathogens on fresh produce. *J. Food Protect.* 63:25–30.

119. Lloyd, A.C. 1975. Preservation of comminuted orange products. *J. Food Technol.* 10:565–567.

120. Loessner, M.J., S.K. Maier, P. Schiwek, and S. Scherer. 1997. Long-chain polyphosphates inhibit growth of *Clostridium tyrobutyricum* in processed cheese spreads. *J. Food Protect.* 60:493–498.

121. Maier, S.K., S. Scherer, and M.J. Loessner. 1999. Long-chain polyphosphate causes cell lysis and inhibits *Bacillus cereus* septum formation, which is dependent on divalent cations. *Appl. Environ. Microbiol.* 65:3942–3949.

122. McCann, K.B., A. Lee, J. Wan, H. Roginski, and M. J. Coventry. 2003. The effect of bovine lactoferrin and lactoferricin B on the ability of feline calicivirus (a norovirus surrogate) and poliovirus to infect cell cultures. *J. Appl. Microbiol.* 95:1026–1033.

123. McDonnell, G., G. Grignol, and K. Ankloga. 2002. Vapor phase hydrogen peroxide decontamination of food contact surfaces. *Dairy Food Environ. Sanit.* 22:868–873.

124. McEntire, J.C., T.J. Montville, and M.L. Chikindas. 2003. Synergy between nisin and select lactates against *Listeria monocytogenes* is due to the metal cations. *J. Food Protect.* 66:1631–1636.

125. McMeekin, T.A., K. Presser, D. Ratkowsky, T. Ross, M. Salter, and S. Tienungoon. 2000. Quantifying the hurdle concept by modeling the bacterial growth/no growth interface. *Int. J. Food Microbiol.* 55:93–98.

126. Melly, E., A.E. Cowan, and P. Setlow. 2002. Studies on the mechanism of killing of *Bacillus subtilis* spores by hydrogen peroxide. *J. Appl. Microbiol.* 93:316–325.

127. Meyer, J.D., J.G. Cerveny, and J.B. Luchansky. 2003. Inhibition of nonproteolytic, psychrotrophic clostridia and anaerobic sporeformers by sodium diacetate and sodium lactate in cook-in-bag turkey breast. *J. Food Protect.* 66:1474–1478.

128. Miller, S.A., and W.D. Brown. 1984. Effectiveness of chlortetracycline in combination with potassium sorbate or tetra-sodium ethylene-diaminetetraacetate for preservation of vacuum packed rockfish fillets. *J. Food Sci.* 49:188–191.

129. Miller, M.L., and E.D. Martin. 1990. Fate of *Salmonella* Enteritidis and *Salmonella* Typhimurium into an Italian salad dressing with added eggs. *Dairy Food Environ. Sanit.* 10(1):12–14.

130. Ming, X., and M.A. Daeschel. 1995. Correlation of cellular phospholipid content with nisin resistance of *Listeria monocytogenes* Scott A. *J. Food Protect.* 58:416–420.

131. Morita, H., R. Sakata, and Y. Nagata. 1998. Nitric oxide complex of iron (II) myoglobin converted from metmyoglobin by *Staphylococcus xylosus. J. Food Sci.* 63:352–355.

132. Morris, J.A., A. Khettry, and E.W. Seitz. 1979. Antimicrobial activity of aroma chemicals and essential oils. *J. Am. Oil Chem. Soc.* 56:595–603.

133. Muthukumarasamy, P.J., H. Han, and R.A. Holley. 2003. Bactericidal effects of *Lactobacillus reuteri* and allyl isohiocyanate on *Escherichia coli* 0157:H7 in refrigerated ground beef. *J. Food Protect.* 66:2038–2044.

134. Naidu, A.S. 2002. Activated lactoferrin—A new approach to meat safety. *Food Technol.* 56(3):40–45.

135. Naidu, A.S. 2000. Microbial blocking agents: A new approach to meat safety. *Food Technol.* 54(2):112.

136. Nascimento, H.S., N. Silva, M.P.L.M. Catanozi, and K.C. Silva. 2003. Effects of different disinfection treatments on the natural microbiota of lettuce. *J. Food Protect.* 66:1697–1700.

137. Nilsson, L., V. Chen, M.L. Chikindas, H.H. Huss, L. Gram, and T.J. Montville. 2000. Carbon dioxide and nisin act synergistically on *Listeria monocytogenes. Appl. Microbiol. Environ.* 66:769–774.

138. Nilsson, L., L. Gram, and H.H. Huss. 1999. Growth control of *Listeria monocytogenes* on cold-smoked salmon using a competitive lactic acid bacteria flora. *J. Food Protect.* 62:336–342.

139. Niroomand, F., W.H. Sperber, V.J. Lewandowski, and L.J. Hobbs. 1998. Fate of bacterial pathogens and indicator organisms in liquid sweeteners. *J. Food Protect.* 61:295–299.

140. No, H.K., N.Y. Park, S.H. Lee, H.J. Hwang, and S.P. Meyers. 2002. Antibacterial activities of chitosans and chitosan oligomers with different molecular weights on spoilage bacteria isolated from tofu. *J. Food Sci.* 67:1511–1514.

141. Nordin, H.R. 1969. The depletion of added sodium nitrite in ham. *Can. Inst. Food Sci. Technol. J.* 2:79–85.

142. O'Boyle, A.R., L.J. Rubin, L.L. Diosady, N. Aladin-Kassam, F. Comer, and W. Brightwell. 1990. A nitrite-free curing system and its application to the production of wieners. *Food Technol.* 44(5):88, 90–91, 93, 95–96, 98, 100, 102–104.

143. O'Leary, V., and M. Solberg. 1976. Effect of sodium nitrite inhibition on intracellular thiol groups and on the activity of certain glycolytic enzymes in *Clostridium perfringens. Appl. Environ. Microbiol.* 31:208–212.

144. Ough, C.S. 1983. Sulfur dioxide and sulfites. In *Antimicrobials in Foods*, ed. A.L. Branen and P.M. Davidson, 177–203. New York: Marcel Dekker.

145. Paquette, M.W., M.C. Robach, J.N. Sofos, and F. Busta. 1980. Effects of various concentrations of sodium nitrite and potassium sorbate on color and sensory qualities of commercially prepared bacon. *J. Food Sci.* 45:1293–1296.

146. Park, C.M., and L.R. Beuchat. 1999. Evaluation of sanitizers for killing *Escherichia coli* 0157:H7, *Salmonella*, and naturally occurring microorganisms on cantaloupes, honeydew melons, and asparagus. *Dairy Food Environ. Sanit.* 19:842–847.

147. Perigo, J.A., and T.A. Roberts. 1968. Inhibition of clostridia by nitrite. *J. Food Technol.* 3:91–94.

148. Perigo, J.A., E. Whiting, and T.E. Bashford. 1967. Observations on the inhibition of vegetative cells of *Clostridium sporogenes* by nitrite which has been autoclaved in a laboratory medium, discussed in the context of sublethally processed meats. *J. Food Technol.* 2:377–397.

149. Peterson, A.C., J.J. Black, and M.F. Gunderson. 1962. Staphylococci in competition. I. Growth of naturally occurring mixed populations in precooked frozen foods during defrost. *Appl. Microbiol.* 10:16–22.

150. Pierson, M.D., and N.R. Reddy. 1982. Inhibition of *Clostridium botulinum* by antioxidants and related phenolic compounds in comminuted pork. *J. Food Sci.* 47:1926–1929, 1935.

151. Pitt, W.M., T.J. Harden, and R.R. Hull. 2000. Behavior of *Listeria monocytogenes* in pasteurized milk during fermentation with lactic acid bacteria. *J. Food Protect.* 63:916–920.

152. Porto, A.C.S., B.D.G.M. Franco, E.S. Sant'Anna, J.K. Call, A. Piva, and J.B. Luchansky. 2002. Viability of a five-strain mixture of *Listeria monocytogenes* in vacuum-sealed packages of frankfurters, commercially prepared with and without 2.0 or 3.0% added potassium lactate, during extended storage at 4 and 10°C. *J. Food Protect.* 65:308–315.

153. Raccach, M., and R.C. Baker. 1978. Lactic acid bacteria as an antispoilage and safety factor in cooked, mechanically deboned poultry meat. *J. Food Protect.* 41:703–705.

154. Rayman, K., N. Malik, and A. Hurst. 1983. Failure of nisin to inhibit outgrowth of *Clostridium botulinum* in a model cured meat system. *Appl. Environ. Microbiol.* 46:1450–1452.

155. Reddy, D., J.R. Lancaster, Jr., and D.P. Cornforth. 1983. Nitrite inhibition of *Clostridium botulinum*: Electron spin resonance detection of iron–nitric oxide complexes. *Science* 221:769–770.

156. Robach, M.C., and M.D. Pierson. 1979. Inhibition of *Clostridium botulinum* types A and B by phenolic antioxidants. *J. Food Protect.* 42:858–861.

157. Robach, M.C., and M.D. Pierson. 1978. Influence of *para*-hydroxybenzoic acid esters on the growth and toxin production of *Clostridium botulinum* 10755A. *J. Food Sci.* 43:787–789, 792.

158. Roberts, T.A., A.M. Gibson, and A. Robinson. 1981. Factors controlling the growth of *Clostridium botulinum* types A and B in pasteurized, cured meats. II. Growth in pork slurries prepared from "high" pH meat (range 6.3–6.8). *J. Food Technol.* 16:267–281.

159. Roberts, T.A., A.M. Gibson, and A. Robinson. 1982. Factors controlling the growth of *Clostridium botulinum* types A and B in pasteurized, cured meats. III. The effect of potassium sorbate. *J. Food Technol.* 17:307–326.

160. Roberts, T.A., and M. Ingram. 1966. The effect of sodium chloride, potassium nitrate and sodium nitrite on the recovery of heated bacterial spores. *J. Food Technol.* 1:147–163.

161. Roberts, T.A., and J.L. Smart. 1974. Inhibition of spores of *Clostridium* spp. by sodium nitrite. *J. Appl. Bacteriol.* 37:261–264.

162. Ronning, I.E., and H.A. Frank. 1988. Growth response of putrefactive anaerobe 3679 to combinations of potassium sorbate and some common curing ingredients (sucrose, salt, and nitrite), and to noninhibitory levels of sorbic acid. *J. Food Protect.* 51:651–654.

163. Ronning, I.E., and H.A. Frank. 1987. Growth inhibition of putrefactive anaerobe 3679 caused by stringent-type response induced by protonophoric activity of sorbic acid. *Appl. Environ. Microbiol.* 53:1020–1027.

164. Rose, N.L., M.M. Palcic, P. Sporns, and L.M. McMullen. 2002. Nisin: A novel substrate for glutathione S-transferase isolated from fresh beef. *J. Food Sci.* 67:2288–2293.

165. Sabah, J R., H. Thippareddi, J.L. Marsden, and D.Y.C. Fung. 2003. Use of organic acids for the control of *Clostridium perfringens* in cooked vacuum-packaged restructured roast beef during an alternative cooling procedure. *J. Food Protect.* 66:1408–1412.

166. Sahl, H.-G., M. Kordel, and R. Benz. 1987. Voltage-dependent depolarization of bacterial membranes and artificial lipid bilayers by the peptide antibiotic nisin. *Arch. Microbiol.* 149:120–124.

167. Savage, R.A., and C.R. Stumbo. 1971. Characteristics of progeny of ethylene oxide treated *Clostridium botulinum* type 62A spores. *J. Food Sci.* 36:182–184.

168. Sapers, G.M., and J.E. Sites. 2003. Efficacy of 1% hydrogen peroxide wash in decontaminating apples and cantaloupe melons. *J. Food. Sci.* 68:1793–1797.

169. Sapers, G.M., R.L. Miller, M. Jantschke, and A.M. Mattrazzo. 2000. Factors limiting the efficacy of hydrogen peroxide washes for decontamination of apples containing *Escherichia coli*. *J. Food Sci.* 65:529–532.

170. Schillinger, U., R. Geisen, and W.H. Holzapfel. 1996. Potential of antagonistic microorganisms and bacteriocins for the biological preservation of foods. *Trends Food Sci. Technol.* 7:158–164.

171. Schuenzel, K.M., and M.A. Harrison. 2002. Microbial antagonists of foodborne pathogens on fresh, minimally processed vegetables. *J. Food Protect.* 65:1909–1915.

172. Scott, V.N., and S.L. Taylor. 1981. Effect of nisin on the outgrowth of *Clostridium botulinum* spores. *J. Food Sci.* 46:117–120, 126.

173. Senne, M.M., and S.E. Gilliland. 2003. Antagonistic action of cells of *Lactobacillus delbrueckii* subsp. *lactis* against pathogenic and spoilage microorganisms in fresh meat systems. *J. Food Protect.* 66:418–425.

174. Setlow, B, C.A. Loshon, P.C. Genest, A.W. Cowan, C. Setlow, and P. Setlow. 2002. Mechanisms of killing spores of *Bacillus subtilis* by acid, alkali and ethanol. *J. Appl. Microbiol.* 92:362–375.

175. Seward, R.A., R.H. Deibel, and R.C. Lindsay. 1982. Effects of potassium sorbate and other antibotulinal agents on germination and outgrowth of *Clostridium botulinum* type E spores in microcultures. *Appl. Environ. Microbiol.* 44:1212–1221.

176. Shahidi, F., L.J. Rubin, L.L. Diosady, V. Chew, and D.F. Wood. 1984. Preparation of dinitrosyl ferrohemochrome from hemin and sodium nitrite. *Can. Inst. Food Sci. Technol. J.* 17:33–37.

177. Shelef, L.A. 1994. Antimicrobial effects of lactates: A review. *J. Food Protect.* 57:445–450.

178. Shelef, L.A., and J.A. Seiter. 1993. Indirect antimicrobials. In *Antimicrobials in Foods*, 2nd ed., ed. P.M. Davidson, 539–569. New York: Marcel Dekker.

179. Shelef, L.A. 1983. Antimicrobial effects of spices. *J. Food Safety* 6:29–44.

180. Shelef, L.A., and P. Liang. 1982. Antibacterial effects of butylated hydroxyanisole (BHA) against *Bacillus* species. *J. Food Sci.* 47:796–799.

181. Shelef, L.A., O.A. Naglik, and D.W. Bogen. 1980. Sensitivity of some common food-borne bacteria to the spices sage, rosemary, and allspice. *J. Food Sci.* 45:1042–1044.

182. Sofos, J.N. 1989. *Sorbate Food Preservatives*. Boca Raton, FL: CRC Press.

183. Sofos, J.N., F.F. Busta, and C.E. Allen. 1980. Influence of pH on *Clostridium botulinum* control by sodium nitrite and sorbic acid in chicken emulsions. *J. Food Sci.* 45:7–12.

184. Sofos, J.N., F.F. Busta, K. Bhothipaksa, C.E. Allen, M.C. Robach, and M.W. Paquette. 1980. Effects of various concentrations of sodium nitrite and potassium sorbate on *Clostridium botulinum* toxin production in commercially prepared bacon. *J. Food Sci.* 45:1285–1292.

185. Splittstoesser, D.F., and M. Wilkison. 1973. Some factors affecting the activity of diethylpyrocarbonate as a sterilant. *Appl. Microbiol.* 25:853–857.

186. Swartling, P., and B. Lindgren. 1968. The sterilizing effect against *Bacillus subtilis* spores of hydrogen peroxide at different temperatures and concentrations. *J. Dairy Res.* 35:423–428.

187. Tanaka, N., N.M. Gordon, R.C. Lindsay, L.M. Meske, M.P. Doyle, and E. Traisman. 1985. Sensory characteristics of reduced nitrite bacon manufactured by the Wisconsin process. *J. Food Protect.* 48:687–692.

188. Tanaka, N., L. Meske, M.P. Doyle, E. Traisman, D.W. Thayer, and R.W. Johnston. 1985. Plant trials of bacon made with lactic acid bacteria, sucrose and lowered sodium nitrite. *J. Food Protect.* 48:679–686.

189. Tanaka, N., E. Traisman, M.H. Lee, R.G. Cassens, and E.M. Foster. 1980. Inhibition of botulinum toxin formation in bacon by acid development. *J. Food Protect.* 43:450–457.

190. Tarr, H.L.A., B.A. Southcott, and H.M. Bissett. 1952. Experimental preservation of flesh foods with antibiotics. *Food Technol.* 6:363–368.

191. Thompson, D.P., L. Metevia, and T. Vessel. 1993. Influence of pH alone and in combination with phenolic antioxidants on growth and germination of mycotoxigenic species of *Fusarium* and *Penicillium*. *J. Food Protect.* 56:134–138.

192. Toledo, R.T. 1975. Chemical sterilants for aseptic packaging. *Food Technol.* 29(5):102–107.

193. Toledo, R.T., F.E. Escher, and J.C. Ayres. 1973. Sporicidal properties of hydrogen peroxide against food spoilage organisms. *Appl. Microbiol.* 26:592–597.

194. Tompkin, R.B. 1983. Nitrite. In *Antimicrobials in Foods*, ed. A.L. Branen and P.M. Davidson, 205–206. New York: Marcel Dekker.

195. Tompkin, R.B., L.N. Christiansen, and A.B. Shaparis. 1978. Enhancing nitrite inhibition of *Clostridium botulinum* with isoascorbate in perishable canned cured meat. *Appl. Environ. Microbiol.* 35:59–61.

196. Tompkin, R.B., L.N. Christiansen, and A.B. Shaparis. 1978. Causes of variation in botulinal inhibition in perishable canned cured meat. *Appl. Environ. Microbiol.* 35:886–889.

197. Tompkin, R.B., L.N. Christiansen, and A.B. Shaparis. 1979. Iron and the antibotulinal efficacy of nitrite. *Appl. Environ. Microbiol.* 37:351–353.

198. Tompkin, R.B., L.N. Christiansen, and A.B. Shaparis. 1980. Antibotulinal efficacy of sulfur dioxide in meat. *Appl. Environ. Microbiol.* 39:1096–1099.

199. Torriani, S., C. Orsi, and M. Vescova. 1997. Potential of *Lactobacillus casei* culture permeate, and lactic acid to control microorganisms in ready-to-use vegetables. *J. Food Protect.* 60:1564–1567.

200. Ukuku, D.O., and G.M. Sapers. 2001. Effect of sanitizer treatments on *Salmonella* Stanley attached to the surface of cantaloupe and cell transfer to fresh-cut tissues during cutting practices. *J. Food Protect.* 64:1286–1292.

201. Vareltzis, K., E.M. Buck, and R.G. Labbe. 1984. Effectiveness of a betalains/potassium sorbate system versus sodium nitrite for color development and control of total aerobes, *Clostridium perfringens* and *Clostridium sporogenes* in chicken frankfurters. *J. Food Protect.* 47:532–536.

202. Vas, K., and M. Ingram. 1949. Preservation of fruit juices with less SO_2. *Food Manuf.* 24:414–416.

203. Venkitanarayanan, K.S., G.O. Ezeike, Y.-C Hung, and M.P. Doyle. 1999. Efficacy of electrolyzed oxiding water for inactivating *Escherichia coli* 0157:H7, *Salmonella* Enteritidis, and *Listeria monocytogenes*. *Appl. Environ. Microbiol.* 65:4276–4279.

204. Wagner, M.K., and F.F. Busta. 1983. Effect of sodium acid pyrophosphate in combination with sodium nitrite or sodium nitrite/potassium sorbate on *Clostridium botulinum* growth and toxin production in beef/pork frankfurter emulsions. *J. Food Sci.* 48:990–991, 993.

205. Wang, I.-N., D.L. Smith, and R. Young. 2000. Holins: The protein clocks of bacteriophage infections. *Ann. Rev. Microbiol.* 54:799–825.

206. Weissinger, W.R., K.H. Watters, and L.R. Beuchat. 2001. Evaluation of volatile chemical treatments for lethality to *Salmonella* on alfalfa seeds and sprouts. *J. Food Protect.* 64:442–450.

207. Weissinger, W.R., and L.R. Beuchat. 2000. Comparison of aqueous chemical treatments to eliminate *Salmonella* on alfalfa seeds. *J. Food Protect.* 63:1475–1482.

208. Winarno, F.G., and C.R. Stumbo. 1971. Mode of action of ethylene oxide on spores of *Clostridium botulinum* 62A. *J. Food Sci.* 36:892–895.

209. Wisniewsky, M.A., B.A. Glatz, M.L. Gleason, and C.A. Reitmeier. 2000. Reduction of *Escherichia coli* 0157:H7 counts on whole fresh apples by treatment with sanitizers. *J. Food Protect.* 63:703–708.

210. Wood, D.S., D.L. Collins-Thompson, W.R. Usborne, and B. Pickard. 1986. An evaluation of antibotulinal activity in nitrite-free curing systems containing dinitrosyl ferrohemochrome. *J. Food Protect.* 49:691–695.

211. Woods, L.F.J., and J.M. Wood. 1982. A note on the effect of nitrite inhibition on the metabolism of *Clostridium botulinum*. *J. Appl. Bacteriol.* 52:109–110.

212. Woods, L.F.J., J.M. Wood, and P.A. Gibbs. 1981. The involvement of nitric oxide in the inhibition of the phosphoroclastic system in *Clostridium sporogenes* by sodium nitrite. *J. Gen. Microbiol.* 125:399–406.

213. Wu, F.M., M.P. Doyle, L.R. Beuchat, J.G. Wells, E.D. Mintz, and B. Swaminathan. 2000. Fate of *Shigella sonnei* on parsley and methods of disinfection. *J. Food Protect.* 63:568–572.

214. Xu, L. 1999. Use of ozone to improve the safety of fresh fruits and vegetables. *Food Technol.* 53(10):58–61, 63.

215. Yamazaki, K., M. Suzuki, Y. Kawai, N. Inoue, and T.J. Montville. 2003. Inhibition of *Listeria monocytogenes* in cold-smoked salmon by *Carnobacterium piscicola* CS526 isolated from frozen surimi. *J. Food Protect.* 66:1420–1425.

216. Yang, H., B.L. Svem, and Y. Li. 2003. The effect of pH on inactivation of pathogenic bacteria on fresh-cut lettuce by dipping treatment with electrolyzed water. *J. Food Sci.* 68:1013–1017.

217. Yarbrough, J.M., J.B. Rake, and R.G. Egon. 1980. Bacterial inhibitory effects of nitrite: Inhibition of active transport, but not of group translocation, and of intracellular enzymes. *Appl. Environ. Microbiol.* 39:831–834.

218. Young, R., and U. Bläsi. 1995. Holins: Form and function in bacteriophage lysis. *FEMS Microbiol. Rev.* 17:191–205.

219. Young, S.B., and P. Setlow. 2003. Mechanisms of killing of *Bacillus subtilis* spores by hypochlorite and chlorine dioxide. *J. Appl. Microbiol.* 95:54–67.

220. Yun, J., F. Shahidi, L.J. Rubin, and L.L. Diosady. 1987. Oxidative stability and flavour acceptability of nitrite-free curing systems. *Can. Inst. Food. Sci. Technol. J.* 20:246–251.

221. Zaika, L.L., J.C. Kissinger, and A.E. Wasserman. 1983. Inhibition of lactic acid bacteria by herbs. *J. Food Sci.* 48:1455–1459.

222. Zaika, L.L., O.J. Scullen, and J.S. Fanelli. 1997. Growth inhibition of *Listeria monocytogenes* by sodium polyphosphate as affected by polyvalent metal ions. *J. Food Sci.* 62:867–869, 872.

223. Zessin, K.G., and L.A. Shelef. 1988. Sensitivity of *Pseudomonas* strains to polyphosphates in media systems. *J. Food Sci.* 53:669–670.

224. Zimmer, M., N. Vukov, S. Scherer, and M.J. Loessner. 2002. The murein hydrolase of the bacteriophage Φ3626 dual lysis system is active against all tested *Clostridium perfringens* strains. *Appl. Environ. Microbiol.* 68:5311–5317.

225. Zivanovic, S., C.C. Basurto, S.Chi, P.M. Davidson, and J. Weiss. 2004. Molecular weight of chitosan influences—antimicrobial activity in oil-in-water emulsions. *J. Food Protect.* 67:952–959.

CHAPTER 14

Food Protection with Modified Atmospheres

This chapter addresses the various methods of modified-atmosphere packaging (MAP) that are used to alter the gaseous environment on and around foods for the purpose of extending self-life. By and large, it consists of the various ways in which carbon dioxide (CO_2) is used as a food preservative. It has been known since 1882 that increased concentrations of CO_2 will extend the self-life of fresh meats, and the actual application of this gas to extend the self-life of red meats has been practiced for many decades (Table 14–1). The effect of CO_2 on some plant products was recorded as early as 1821.[84] About 90% of boxed beef in the United States is packed under vacuum/MAP, and 90–95% of fresh pasta sold in the United Kingdom is MAP.[76]

DEFINITIONS

There has been a lack of consensus on the terminology that is used to describe the various ways in which increased levels of CO_2 and decreased levels of O_2 are achieved. The most widely used terminology is defined and briefly described below.

Hypobaric (Low Pressure) Storage

Foods are stored in air under low pressure, low temperature, and high humidity, all of which are precisely controlled along with ventilation. The hypobaric state results in reduced concentrations of O_2, which also results in reduced fat oxidation. Atmospheres of about 10 mm Hg have been found to be effective for meats and seafoods; 10–80 mm Hg for fruits and vegetables; and 10–50 mm Hg for cut flowers (1 atm = 760 mm Hg). In one study using pork loins, a pressure of 10 mm Hg along with a temperature of 0°F and 95% humidity was up to six times more effective on self-life than air storage.[49] This method was first outlined around 1960 by Stanley Burg, and a commercial hypobaric container was developed in 1976 (Table 14–1). The use of hypobaric storage is limited, and is not discussed further in this text.

351

Table 14–1 Synopsis of the Early History of Use of Modified-Atmospheres and Related Technologies

Year	Event
1882	Elevated levels of CO_2 were shown to extend the storage life of meats for 4 to 5 weeks.
1889	The antibacterial activity of CO_2 was established
1895	Lopriore observed that 100% CO_2 inhibited the germination of mold spores
1910	Modified-atmosphere packaging was quite widely used to preserve certain foods
1938	About 26% of New Zealand and 60% of Australian beef was shipped under CO_2 atmospheres
1960	The hypobaric system was outlined by S. Burg
1972	The Tectrol process was introduced in the United States for long-distance transportation of meats, poultry, and seafoods
1972	A cryogenic O_2–N_2 atmosphere (liquid O_2–N_2) system was patented by the Union Carbide Corporation
1976	The Grumman Corporation built the Dormavac, a hypobaric highway storage container based on Burg's hypothesis

Vacuum Packaging

By this method, air is evacuated from gas-impermeable pouches followed by sealing. This has the effect of reducing the residual air pressure from the usual 1 bar to 0.3–0.4 bar, hence some O_2 is removed (1 bar = 0.9869 atm). Upon storage of a vacuum-packaged food product, an increase in CO_2 occurs as a result of both tissue and microbial respiration where O_2 is consumed and CO_2 is released in equal volumes. In the case of meats, up to 10–20% CO_2 may develop within four hours and the concentration may ultimately reach 30% from respiratory activities of the aerobic biota (Table 14–2).

A vacuum pack can be achieved by placing foods in high-barrier plastic pouches followed by the evacuation of pouches under vacuum (10–745 mm Hg) and heat sealing or heat shrinking by dipping in

Table 14–2 Percentage of CO_2 and O_2 in Gas-Impermeable Packages of Fresh Pork Stored between 3 Hours and 14 Days at 2°C and 16°C

Storage Time	2°C		16°C	
	CO_2	O_2	CO_2	O_2
After 3 hours	3–5	20	3–5	—
After 4 days	13	20	30	1
After 5 days	—	—	30	1
After 10 days	15	1	—	—
After 14 days	15	1	—	—

Source: Adapted with permission from G.A. Gardner et al.,[28] Bacteriology of Prepacked Pork with Reference to the Gas Composition within the Pack, *Journal of Applied Bacteriology*, Vol. 30, pp. 321–333, copyright © 1967, Blackwell Scientific Publishers, Ltd.

Table 14–3 Some Examples of Gas and Water Vapor Transmission Properties of Film Used for Food Packaging*

Transmission Properties	Comment
1. OTR 7.8–9.3 ml/m^2/24 hours/37.8°C/70% RH	Extremely high barrier
2. a. OTR 8 ml/m^2/24 hours/4°C/100% RH	Extremely high barrier
b. CO$_2$TR 124 ml/m^2/24 hours/100% RH	
c. WVTR 18.6 g/m^2/24 hours/37°C/100% RH	
3. OTR 10 ml/m^2/24 hours/22.8°C/0% RH	
4. a. OTR 32 ml/m^2/24 hours/23.9°C/50% RH	High barrier
b. CO$_2$TR 47 ml/m^2/24 hours/23.9°C/70% RH	
c. WVTR 0.8–1.8 g/m^2/24 hours/23.9°C/70% RH	
5. OTR 52 ml/m^2/24 hours/1 atm/25°C/75% RH	High barrier
6. OTR 154 ml/m^2/24 hours	Whirl-Pak bags
7. OTR 300 ml/m^2/24 hours/25°C/1 atm/100% RH	Commonly used for vacuum packaging
8. OTR 1,000 ml/m^2/24 hours/25°C/1 atm/90% RH	Essentially aerobic
9. OTR 6,500 ml/m^2/24 hours/23°C/0% RH	Highly permeable
10. OTR 7,800–13,900 ml/m^2/24 hours	PVC film
WVTR 240–419 g/m^2/24 hours	
11. OTR 6,500 ml/m^2/24 hours/23°C/0% RH	Stretch-wrapped film

*OTR = oxygen transmission rate; RH = relative humidity; WVTR = water vapor transmission; PVC = polyvinyl chloride.

80–90°C water. A method that is suitable for raw meats consists of simply squeezing out the excess air from a pouch followed by heat sealing. In addition to retarding aerobic spoilage organisms, vacuum packaging minimizes product shrinkage, and retards both fat oxidation and discoloration. Gas and water vapor transmission properties of some plastics used to vacuum-pack foods are listed in Table 14–3. In general, CO$_2$ permeability of such films is always higher than for O$_2$ by a factor of around 2 to 5.

Modified Atmosphere Packaging

Overall, MAP is a hyperbaric process that consists of altering the chamber or package atmosphere by flushing with varying mixtures of CO$_2$, N$_2$, and/or O$_2$. The initial gas concentration cannot be readjusted during storage. Two types of MAP are recognized.[29]

1. In high-O$_2$ MAP, up to 70% of O$_2$ along with about 20–30% CO$_2$ and 0–20% N$_2$ may be used. Growth of aerobes is slowed but not suppressed by the moderate concentration of CO$_2$. This method is suitable for the packaging of red meats, as the high level of O$_2$ will aid in maintenance of the red color. With time, the gas composition may be expected to change.
2. In a low-O$_2$ MAP system, O$_2$ levels may be as high as 10% while CO$_2$ is maintained in the 20–30% range with N$_2$ added as necessary.

Equilibrium-Modified Atmosphere

Equilibrium-modified atmosphere (EMA) packaging is achieved by flushing a gas-permeable pack with gas, or sealing the pack without alteration. EMA is used for fresh fruits and vegetables.[76]

Controlled-Atmosphere Packaging or Storage

Although controlled-atmosphere packaging or storage (CAP, CAS) is regarded by some as being different from MAP, it may be considered a form of MAP. While in a typical MAP the compositions may change upon storage, in CAP the gas compositions remain unchanged for the duration of the storage period. While low- and high-O_2 MAP systems may be prepared with high-barrier plastic films, CAP requires aluminum foil laminates, metal or glass containers, since single plastic film is not entirely impervious to gases.

Since vacuum, MAP, or CAP methods alter the concentrations of O_2 and CO_2, albeit in different ways, the distinction between them is often obscured in studies on the effectiveness and mode of action of CO_2. In the remaining sections of this chapter, the inhibitory effects of increased levels of CO_2 on foodborne microorganisms and food quality are examined without regard to methodology.

PRIMARY EFFECTS OF CO_2 ON MICROORGANISMS

The following facts are well established following prolonged exposure of microorganisms to around 10% and above of CO_2.

1. The inhibitory activity increases as incubation or storage temperatures decrease. This is due in part to the greater solubility of CO_2 in water at the lower temperatures, and in part to the additive effect of a less than optimal growth temperature. At 1 atm, 100 ml of water will absorb 88 ml of CO_2 at 20°C, but only 36 ml at 60°C.
2. Although concentrations from about 5–100% have been used, 20–30% seems optimal, with no additional benefits derived from higher levels. This is especially true for fresh meats, where 20% is about ideal.[30,31] Higher levels can be used for seafoods. To maintain red meat color, they can first be exposed to carbon monoxide (CO) before CO_2, or be stored in a 20:80 mixture of $CO_2 + O_2$.
3. Inhibition increases as pH is decreased into the acid range. One practical effect of this is that CO_2 is more effective for fresh red meats than for seafoods. The vacuum packaging of red meats with pH >6.0 is not effective. The self-life of vacuum-packaged fish is shortened by the growth of *Photobacterium phosphoreum*[16] and *Shewanella putrefaciens*.[1]
4. In general, the Gram-negative bacteria are more sensitive to CO_2 inhibition than Gram positives, with pseudomonads being among the most sensitive and clostridia the most resistant (Table 14–4). Upon prolonged storage of meats, CO_2 affects a rather dramatic shift in biota from one that is largely Gram negative in fresh products to one that is largely or exclusively Gram positive. This can be seen in Table 14–5 for smoked pork loins and frankfurter sausage.[6]
5. Both lag and logrithmic phases of growth are retarded.
6. CO_2 under pressure is considerably more antimicrobial than when not under pressure, and pressures of 6–30 megapascal (MPa) can destroy bacteria and fungi under varying conditions (see High Hydrostatic Pressure in Chapter 19). The destructive action is believed to occur when pressure is released suddenly.

Mode of Action

As to the mechanism of CO_2 inhibition of microorganisms, two explanations have been offered. King and Nagel[57] found that CO_2 blocked the metabolism of *Pseudomonas aeruginosa* and appeared to affect a mass action on enzymatic decarboxylations. Sears and Eisenberg[79] found that CO_2 affected

Table 14–4 Relative Sensitivity of Microorganisms to CO_2 Relative to Vacuum and Modified-Atmosphere Packaging

Pseudomonas spp.	(Most sensitive)
Aeromonas spp.	
Bacillus spp.	
Molds	
Enterobacteriaceae	
Enterococcus spp.	
Brochothrix spp.	
Lactobacillus spp.	
Clostridium spp.	(Most resistant)

Source: Adapted with permission from G. Molin,[68] The Resistance to Carbon Dioxide of Some Food Related Bacteria, *European Journal of Applied Microbiology and Biotechnology*, Vol. 18, pp. 214–217, Copyright © 1983, Springer-Verlag, New York.

Table 14–5 Effect of Storage on the Microbiota of Two Meats Held from 48 to 140 Days at 4°C

	Smoked Pork Loins			
	0 Day	*Vacuum 48 Days*	*CO₂ 48 Days*	*N₂ 48 Days*
Log APC/g	2.5	7.6	6.9	7.2
pH	5.8	5.8	5.9	5.9
Dominant biota (%)	Flavo (20)	Lactos (52)[a]		
	Arthro (20)		Lactos (74)[b]	
	Yeasts (20)			Lactos (67)[c]
	Pseudo (11)			
	Coryne (10)			

	Frankfurter Sausage			
	0 Day	*Vacuum 98 Days*	*CO₂ 140 Days*	*N₂ 140 Days*
Log APC/g	1.7	9.0	2.4	4.8
pH	5.9	5.4	5.6	5.9
Dominant biota (%)	Bac (34)	Lactos (38)	Lactos (88)[d]	Lactos (88)[e]
	Coryne (34)			
	Flavo (8)			
	Broch (8)			

Note: Percentage biota represented by *Weissella viridescens*: [a]40; [b]72; [c]50; [d]22; [e]35. APC = aerobic plate count; Flavo = *Flavobacterium*; Arthro = *Arthrobacter*; Pseudo = *Pseudomonas*; Coryne = *Corynebacterium*; Bac = *Bacillus*; Broch = *Brochothrix*.

Source: Adapted from Blickstad and Molin.[6]

the permeability of cell membranes, and Enfors and Molin[22] found support for the latter hypothesis in their studies on the germination of *Clostridium sporogenes* and *C. perfringens* endospores. At 1 atm CO_2, spore germination of these two species was stimulated, whereas *B. cereus* spore germination was inhibited. As was shown by others, CO_2 is more stimulatory at low pH than at high pH. With 55 atm CO_2, only 4% germination of *C. sporogenes* spores occurred, whereas with *C. perfringens*, 50 atm reduced germination to 4%.[22] These authors suggested that CO_2 inhibition was due to its accumulation in the membrane lipid bilayer such that increased fluidity results. An adverse effect on cell permeability has been suggested by others. If CO_2 is dissolved in the form of carbonic acid, HCO_3^- would be present as a dissociation product, and thus can cause changes in cell permeability.[17] When dissolved in water, CO_2 products are as follows:

$$CO_2 + H_2O \rightleftharpoons H_2CO_3 \rightleftharpoons H^+ + HCO_3^- \rightleftharpoons 2H + CO_3^{2-}$$

The antimicrobial spectrum of CO_2 and diacetyl is quite similar, and while this per se does not mean they possess identical modes of action, the striking similarities seem worthy of note. Diacetyl is an arginine antagonist and its mode of action along with some other α-dicarbonyl compounds has been discussed.[51] The greater sensitivity of Gram-negative bacteria to α-dicarbonyl inhibitors appears to be due to their capacity to inactivate amino acid-binding proteins of the cell's periplasm, especially the arginine-binding proteins. Thus, it is not inconceivable that the site of action of CO_2 is the periplasm, where it interferes with the normal functioning of amino acid-binding proteins.

Food Products

The successful use of vacuum packaging, MAP, and CAS to extend the self-life of a wide variety of food products is well documented, and some of the specific antimicrobial aspects are outlined below.

Fresh and Processed Meats

Among the first to demonstrate the effectiveness of high levels of CO_2 in preserving cut-up meats was J. Brooks in England, who in 1933 studied the effect of CO_2 on lean meat; E. Callow in England, who studied pork and bacon; R.B. Haines, also in England, who was among the first to show the effect of CO_2 on spoilage organisms; and W.A. Empey in Australia, who in 1933 applied CO_2 to beef.[73]

In general, the self-life of red meats can be extended for up to 2 months if packaged in 75% O_2 + 25% CO_2 and stored at $-1°C$. The high level of oxygen ensures that the red-meat color is maintained. It has been shown that at least 15% CO_2 is necessary to retard microbial growth on beef steaks, and that the mixture of 15% CO_2 + 75% O_2 + 10% N_2 was more effective than vacuum both for red-meat color and microbial quality.[4] The importance of the temperature of storage of MAP meats was shown early by Jaye et al.[52] who found striking differences in quality when ground beef was stored at 30°C versus 38°. They compared the use of the more gas-impermeable Saran to the gas-permeable cellophane packs. Earlier, Halleck et al.[38] showed the dramatic inhibitory effect of vacuum packaging and storage at 1.1–3.3°C. The importance of temperature of storage was demonstrated in another study using the packaging system known as the Captech process, which combines hygienic processing, storage at $-1.5°C$, high CO_2, low O_2, and gas-impermeable packaging.[36] The process was applied to pork loins, with the temperature of holding for simulated retail display being raised to 8°C. Lactic acid bacteria grew without perceptible decrease in lag phase, and reached $10^7/cm^2$ within 9 weeks. The behavior of the biota of smoked pork loins and frankfurter sausage stored under vacuum and CO_2 is presented in

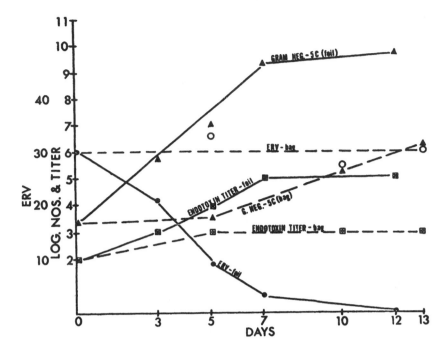

Figure 14–1 Lack of increase in hydration capacity of fresh ground beef stored in high-barrier bags at 7°C for 13 days as measured by extract-release volume (ERV). The foil-wrapped samples underwent aerobic spoilage as evidenced by increased hydration and endotoxin titers.

Table 14–5. As is typical of MAP meats, the initial heterogeneous biota became homogeneous upon long-term storage under vacuum or MAP with pH being decreased due to predominance of lactic acid bacteria.[6]

The relative effectiveness of MAP/vacuum packaging of red meats can be assessed by determining changes that occur in hydration capacity. When fresh ground beef was stored in high-barrier bags and held at 7°C for up to 13 days, the hydration capacity was essentially unchanged in comparison to the samples that were loosely wrapped in foil to allow for aerobic conditions (Figure 14–1). This is reflected by extract-release volume (ERV; see Chapter 4). Over the holding period, Gram-negative bacteria increased by about 6 $\log_{10}$ but by only 3 $\log_{10}$ for the aerobically stored foil-wrapped and high-barrier bag-stored meats, respectively. Similar results can be obtained by using the filter-paper press method to measure hydration capacity.[50] The increased hydration is brought about by the preferential growth of Gram-negative bacteria accompanied by increases in pH into the alkaline range.

In their study of beef and pork livers, and beef kidneys, packaged in high-barrier bags, Hanna et al.[43] found that pH decreased in each product when held at 2°C for up to 28 days. ERV has been used to assess the spoilage of vacuum-packaged meats.[75]

In a study of normal and DFD ground beef stored at 3°C in 100% CO_2 for up to 11 days, self-life increased by ca. 3—4 days with the predominant bacterial biota consisting of lactic acid bacteria and *Brochothrix thermosphacta* in contrast to samples stored in air where the pseudomonads predominated.[72] In the normal low-pH meat under CO_2, lactate appeared to increase, while in air-stored samples it decreased over the 11-day holding period. When Greek taverna sausage was stored under vacuum and

100% CO_2 at 4 and 10°C, it was dominated by *Lactobacillus sakei/curvatus* (92–96% of the biota) after 30 days.[77] D-lactate increased more under CO_2 and vacuum than it did in aerobically stored products. The authors concluded that CO_2 had no significant effect on extending the self-life of taverna sausage.

Overall, the low-temperature storage of fresh meats under vacuum or MAP is very satisfactory. This is in large part a reflection of the existence of lactic acid and related bacteria on fresh meats, and when these products are stored under low O_2 and high CO_2 conditions at low temperatures, the normal biota prevents the growth of pathogens by virtue of depressed pH, competition for O_2, possible production of antimicrobial substances, and other factors.

Poultry

The effectiveness of MAP for the storage of fresh poultry was demonstrated in the early 1950s[73] and since that time a number of studies have been reported. Hotchkiss[48] used from 60% to 80% CO_2 on raw poultry in glass jars and found an increase in self-life to at least 35 days at 2°C. In another study, when high-barrier film (oxygen transmission rate [OTR] ca. 18 ml) was used to pack cut-up or whole chicken that was held at 5°C, the chicken had lower numbers of bacteria and kept longer than that which was stretch-wrapped with a film that had an OTR of 6,500 ml, and this is illustrated in Figure 14–2.[58] With poultry stored in air, the aerobic plate count (APC) of drip after 16 days at 10°C was 9.40 log_{10}, whereas in 20% CO_2 the APC was 6.14 log_{10}.[92]

Overall, the generally higher initial pH of fresh poultry meat is primarily responsible for this product's not having the MAP self-life of products such as fresh beef.

Seafoods

MAP/vacuum packaging has been shown to extend the self-life of cod fillets, red snapper, rainbow trout, herrings, mackerel, sardines, catfish, and others. In 1933, F.P. Coyne of England was apparently the first person to show the preservative effects of CO_2 on fish (see reference 73).

For fish, using 80% CO_2+ air, log numbers after 14 days at 35°C were approximately 6.00/cm^2 compared to air controls with log_{10} numbers >10.5 cm^2. The pH of CO_2-stored products after 14 days decreased from around 6.75 to around 6.30, whereas controls increased to around 7.45.[74] The self-life of rockfish and salmon at 4.5°C was extended by 20–80% CO_2.[8] At least 1 log difference in bacterial counts over controls was obtained when trout and croaker were stored in CO_2 environments at 4°C.[39] When fresh shrimp or prawns were packed in ice with an atmosphere of 100% CO_2, they were edible for up to 2 weeks, and bacterial counts after 14 days were lower than air-packed controls after 7 days.[64] When cod fillets were stored at 2°C, air-stored samples spoiled in 6 days, with APC of log_{10} 7.7, whereas samples stored in 50% CO_2 + 50% O_2 or 50% CO_2 + 50% N_2 or 100% CO_2 did not show bacterial spoilage until, respectively, 26, 34, and 34 days, with respective APCs of 7.2, 6.6, and 5.5 log_{10}/g.[88] It was suggested that the use of 50% CO_2 + 50% O_2 is technically more feasible than the use of 100% CO_2. Whereas the practical upper limit of CO_2 for red meats is around 20%, higher concentrations can be used with fish because they contain lower levels of myoglobin.

When salmon fillets were packaged in 60% CO_2:40% N_2 and stored at −2°C and 4°C for 24 days, the product stored at −2°C was good for 24 days based on sensory and microbial analyses while that stored in air had a 21-day sensory shelf-life.[83] The MAP fillets stored at 4°C spoiled after 10 days while those stored in air spoiled in 7 days. The overall sensory quality of the fillets was not adversely affected by storage under MAP.[83]

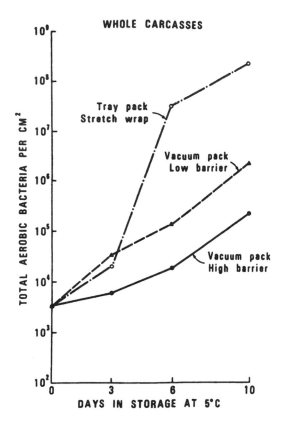

Figure 14–2 Numbers of total aerobic mesophilic bacteria from packaged whole chickens under aerobic (tray pack) and vacuum-pack storage. *Source*: Reprinted with permission from J. Kraft et al.,[58] Microbiological Quality of Vacuum Packaged Poultry with or without Chlorine Treatment, *Journal of Food Science*, Vol. 47, p. 381, Copyright © 1982, Institute of Food Technologists.

The concern over the use of MAP for fishery products has to do with the fact that nonproteolytic botulism strains are found in waters and they can grow at temperatures <4°C coupled with the fact that the pH of seafoods in general is higher and more favorable to growth by pathogens. For more information on the MAP of seafoods and their relative safety, see below.

THE SAFETY OF MAP FOODS

Clostridium botulinum

As a general rule, foods that are to be subjected to MAP should possess one or more of the following antibotulinal hurdles:

1. have a water activity (a_w) < 0.93,
2. have a pH of 4.6 or less,

3. cured with NaCl or NO_2,
4. contain high levels of nonpathogens (for raw meat, poultry, and the like),
5. maintained in frozen state,
6. maintained at 40°F (4.4°C) or below, and
7. have a definitive self-life (e.g., not to exceed 10 days).

Since this organism is of greatest concern in such products, a number of studies have been conducted relative to its behavior under MAP conditions.

The question of the organoleptic state of MAP fish products at the time of botulinal toxin production has been addressed by a number of researchers, among whom are Garcia et al.[27] With an inoculum of 13 nonproteolytic types B, E, and F spores at levels of 10^1–10^4, 50 g-samples of salmon fillets were stored at varying temperatures and tested for toxin under vacuum, 100% CO_2, and 70% CO_2 + 30% air. Overall, toxin detection coincided with spoilage at 30°C but preceded spoilage at 8°C and 12°C, and followed spoilage at 4°C.[27] Regarding start-up time to toxin detection, at 30°C the fillets were toxic after 1 day, after 2 days at 16°C, 6 days at 12°C, 6–12 days at 8°C, and no toxin at 4°C in 60 days. Only type B toxin was detected. In another study, channel catfish was inoculated with a mixture of four strains of type E at a level of three or four spores per gram and stored in 80:20 CO_2:N_2 in O_2-barrier bags at 4°C and 10°C.[9] Those stored at 10°C all contained toxin by day 6. Those at 4°C contained toxin on day 9 in overwrapped packages (O_2 permeable) but not until day 18 in those stored under MAP. These investigators found that toxin detection and spoilage coincided at 10°C, while at 4°C spoilage preceded toxin detection.[9] When raw beef was inoculated with types A and B spores and stored for up to 15 days at 25°C, toxin was first detected after 6 days, always accompanied by significant organoleptic changes, indicating that the vacuum-packaged toxic samples would be rejected before consumption.[45] In a later study, sliced raw potatoes were stored in O_2-impermeable bags with 30% N_2 + 70% CO_2 and incubated at 22°C.[85] The untreated potatoes became toxic in 4 to 5 days while those treated with $NaHSO_3$ were toxic after 4 days, but appeared to be acceptable through day 7. Type A toxin appeared earlier than type B. Potato salad was the vehicle food for two cases of botulism in Colorado.[7] This product was temperature-abused in the home, and type A toxin was found in leftovers.

Five vegetables (lettuce, cabbage, broccoli, carrots, and green beans) were used in a study assessing the botulism hazard of MAP. They were inoculated with a spore cocktail of ten types A, B, and E (7 proteolytics, 3 nonproteolytics) and stored in bags with OTRs that varied between 3,000 and 16,544 ml at 4°C, 12°C, or 21°C.[59] No toxin was found in any at 4°C for up to 50 days, and no toxin was found in cabbage, carrots, or green beans. Toxin was detected in all broccoli at 21°C, in one-half of the broccoli held at 12°C, and in one-third of the lettuce held at 21°C.[59] Both toxigenic vegetables were grossly spoiled. These investigators concluded that toxin production does not precede gross spoilage. In a study of MAP cabbage, an inoculum of 7 type A and 7 proteolytic type B strains was added to shredded cabbage stored in high-barrier bags containing 70% CO_2 + 30% N_2 and incubated at 22–25°C.[86] The inoculum size was 100–200 spores per gram. Only type A strains grew and produced toxin as early as day 4 while the cabbage was still organoleptically acceptable. No toxin was detected on day 3, and the product was organoleptically unacceptable by day 7.[86] The type B strains did not produce toxin even with an inoculum of 14,000 spores per gram. In a survey of 1,118 commercially available one-pound packages of precut MAP vegetables in the United States, only 4 contained type A spores—1 each of shredded cabbage, chopped green pepper, and an Italian salad mix; another salad mix contained types A and B.[62]

Fresh Italian pasta was inoculated with types A, B, and F spores at a level of 1.2×10^2 spores/g and stored at 12°C and 20°C for up to 50 days in an atmosphere of 15% CO_2 + 83% N_2 + 2% O_2.[18] No toxin was detected in any tortelli that was stored at 12°C but at 20°C, toxin was detected in the salmon-filled tortelli at day 30 (pH was 6.1, a_w 0.95) and in the meat and ricotta–spinach tortelli at day 50 but not in the artichoke-filled tortelli at day 50.[18]

In an assessment of *C. botulinum* toxin production in cooked, uncured, turkey meat inoculated with nonproteolytic type B, packaged in O_2-impermeable bags stored under MAP, toxin was produced in 7 days at 15°C, in 14 days at 10°C, and in 28 days at 4°C.[60] At the latter temperature, toxin appeared in 14 days under 100% N_2 preceded or coincided with sensory spoilage.

Using cooked turkey breast that was inoculated with a nonproteolytic type B strain of *C. botulinum*, samples were stored in O_2-impermeable bags with either 100% N_2 or 70% N_2 + 30% CO_2 and observed for toxin production. The turkey meat was positive for toxin at all temperatures and MAP combinations used after sufficient incubation times. At 15°C, toxin appeared after 7 days; after 4 days at 10°C; and after 14 days at 4°C (in 100% N_2 and 28 days in CO_2 + N_2 combination).[60]

Listeria monocytogenes

The fact that this bacterium can grow in the refrigerator temperature range raises concerns about its presence and potential for growth in MAP foods. In a study using ground fresh top beef rounds with a pH of 5.47 that were vacuum packaged and held at 4°C for up to 56 days, one strain increased in numbers by 2.3 logs (from 4.25 to 6.53) after 35 days, another increased by 1.8 logs after 35 days, and a third was unchanged after 56 days.[3] With high-pH (6.14) beef, three strains of *L. monocytogenes* increased significantly in 28 days, but strain Scott A did not.

In a study of the growth and survival of *L. monocytogenes* in vacuum-packaged ground beef (initial pH 5.4) inoculated with *Lactobacillus alimentarius* (the FloraCarn L-2 strain, which is a homofermentative psychrotroph that can grow at 2°C), *L. monocytogenes* numbers were reduced by the antilisterial effects of the lactobacillus due apparently to its production of lactic acid, and this is illustrated in Figure 14–3.[54]

Regarding the behavior of this organism on vacuum-packaged beef, it has been shown that critical factors are storage temperature, pH, and type of tissue, and whether lean or fat.[35] The organism grew more extensively on fat than lean beef, and the background biota had no effect on its growth. At 5.3°C, it grew from about 10^3 to about 10^7 cfu/cm^2 in 16 days on fat, and in 20 days to 10^6 cfu/cm^2 on lean tissue.[35] The organism grew faster on sirloins with pH 6.0–6.1 than those with pH 5.5–5.7. After 76 days at 0°C, the organism reached 10^2/cm^2 on fat and 10^4 cfu/cm^2 on lean.[32] In a prevalence study of *Listeria* spp. on vacuum-packaged processed meats in the retail trade in Australia, listeriae were found in 93 of 175 samples and *L. monocytogenes* was found on 78 of 93 samples.[34] It was found mainly on corned beef and ham, and on two corned beef samples the numbers were >10^4 cfu/g. Strain Scott A, when inoculated onto turkey roll slices at a level of about 10^3/g and packaged in high-barrier bags under 70% CO_2 + 30% N_2, did not grow after 30 days at either 4°C or 10°C.[25] A 50:50 mixture of CO_2:N_2 was less inhibitory.

In a study of the growth of *L. monocytogenes* and *Yersinia enterocolitica* on cooked MAP poultry, the product was stored in a CO_2:N_2 mixture of 44:56 and stored at 3.5°C, 6.5°C, or 10°C for up to 5 weeks.[2] Both organisms grew under all test conditions, and the naturally occurring microbiota did not influence the growth of either. In a study of the effect of MAP and nisin on this organism in cooked pork tenderloins, 100% CO_2 or 80% CO_2 + 20% air with 10^3 or 10^4 IU of nisin decreased the growth not only of *L. monocytogenes* but also that of *Pseudomonas fragi*.[23]

In a study of the changes in populations of *L. monocytogenes* on seven fresh-cut vegetables inoculated with a five-strain cocktail at about 10^3/g and stored in bags with an OTR of 2,100 cm^3, the organism remained constant on all vegetables that were stored at 4°C for 9 days, but numbers declined on carrots and increased on butternut squash.[24] At 10°C, growth occurred on all except chopped carrots. The antilisterial properties of carrots is well established (see Chapter 6). When prepeeled potatoes were

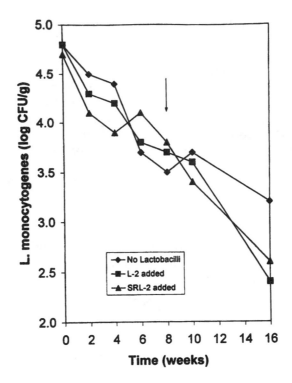

Figure 14–3 Survival of *L. monocytogenes* (five-strain compositive) in vacuum-packaged ground beef during refrigerated storage without or with added *L. alimentarius* FloraCarn L-2 or the antibiotic-resistant mutant SRL-2. The arrow indicates shift of the beef samples from 4°C to 7°C. Experiments were repeated twice, and the results were averaged and reported in log_{10} cfu/g. *Source*: Reprinted with permission from B.J. Juven (Virginia Polytechnic Institute and State University, Blacksburg, Virginia) et al.,[54] Growth and Survival of *Listeria Monocytogenes* in Vacuum-packaged Ground Beef Inoculated with *Lactobacillus alimentarius* FloraCarn L-2, *Journal of Food Protection*, Vol. 61, p. 553. Copyright © 1998, held by the International Association of Milk, Food and Environmental Sanitarians, Inc.

---

inoculated with *L. monocytogenes* and vacuum packaged in high-barrier bags, no growth occurred when stored at 4°C for 21 days but at 15°C, the organisms grew to 7 log_{10}/g within 12 days.[53] Two browning inhibitors were used in this study but they were not inhibitory to growth of the organisms at 15°C.

Other Pathogens

When cooked bologna-type sausage was vacuum packaged, the growth of *Yersinia enterocolitica* and salmonellae was restricted but not that of *Staphylococcus aureus*.[70] *Clostridium perfringens* was also inhibited, and growth inhibition was attributed to the normal biota.

In a study of the relative survival of *Campylobacter jejuni* on ground beef under four atmospheres, no striking differences in numbers were noted except that 100% N_2 allowed the highest number of survivors.[89] The meat was inoculated at a level of about 10^5 cfu/g and stored in 5% O_2 + 10%

CO_2 + 85% N_2, 80% CO_2 + 20% N_2, under vacuum, and in pure nitrogen at 4°C for up to 2 weeks. In another study of *C. jejuni* in beef, two of three strains decreased slightly when stored at 20°C and 4°C for 48 hours but increased when stored at 37°C under all gas mixtures used.[44] The storage conditions consisted of vacuum packaging, 20% CO_2 + 80% N_2, 5% O_2 + 10% CO_2 + 85% N_2, and the background biota apparently had no effect on *C. jejuni*. When heat-injured and uninjured cells of *Aeromonas hydrophila* were stored at 5°C for 22 days under 100% N_2 or 100% CO_2, both types steadily declined in numbers, but growth of both was enhanced when incubated under N_2.[33] The heat-injured cells apparently were not particularly disadvantaged under the conditions noted.

In a study of raw crayfish stored at 4–10°C for up to 30 days, *Aeromonas hydrophila*, *S. aureus*, *C. perfringens*, and *C. botulinum* type E were not found.[63] When cooked crawfish was inoculated with 10^3 cfu of *C. botulinum* type E spores/g of crawfish tail, vacuum packaged with both high-barrier film and air-impermeable bags followed by storage at 4 and 10°C for 30 days, type E toxin was not detected in any packages.[63]

SPOILAGE OF MAP AND VACUUM-PACKAGED MEATS

From the research of many groups, it is clear that when vacuum-packaged meats undergo long-term refrigerator spoilage, often the predominant organisms are lactobacilli, other lactics, *B. thermosphacta*, or all of these. Other organisms can be found and, indeed, others may predominate. Among the determining factors are the following:

1. whether the product is raw or cooked;
2. concentration of nitrites present;
3. relative load of psychrotrophic bacteria;
4. the degree to which the vacuum-package film excludes O_2;
5. product pH.

Cooked or partially cooked meats, along with dark, firm, and dry (DFD) and dark-cutting meats, have a higher pH than raw and light-cutting meats, and the organisms that dominate these products during vacuum storage are generally different from those found in vacuum-packaged normal meats. In vacuum-packaged DFD meats held at 2°C for 6 weeks, the dominant biota consisted of *Yersinia enterocolitica*, *Serratia liquefaciens*, *Shewanella putrefaciens*, and a *Lactobacillus* sp.[32] *S. putrefaciens* caused greening of product, but a pH <6.0 was inhibitory to its growth. When dark-cutting beef of pH 6.6 was vacuum packaged and stored at 0–2°C, lactobacilli were dominant after 6 weeks, but after 8 weeks psychrotrophic Enterobacteriaceae became dominant.[75] Most of the Enterobacteriaceae resembled *S. liquefaciens* and the remainder resembled *Hafnia alvei*. In vacuum-packaged beef with pH 6.0, *Y. enterocolitica*-like organisms were found at levels of 10^7/g after 6 weeks at 0–2°C, but on meats with pH <6.0, their numbers did not exceed 10^5/g even after 10 weeks.[80] The high-pH meat also yielded *S. putrefaciens* with counts as high as log 6.58/g after 10 weeks.

When normal raw beef with an ultimate pH of about 5.6 is vacuum packaged, lactobacilli and other lactic acid bacteria predominate. When the beef was allowed to spoil aerobically, acidic/sour odors were noted when the APC was about 10^7–10^8/cm^2 with approximately 15% of the biota being *Pseudomonas* spp.; but when vacuum-packaged samples spoiled, the product was accompanied by a slight increase in pH with a general increase in extract-release volume (ERV).[42] After a 9-week storage at 0–1°C, Hitchener et al.[46] found that 75% of the biota of vacuum-packaged raw beef consisted of catalase-negative organisms. Upon further characterization of 177 isolates, 18 were found to be *Leuconostoc mesenteroides*, 115 were heterofermentative, and 44 were homofermentative lactobacilli. Using high-barrier oxygen film, the dominant biota of vacuum-packaged beef loin steaks after 12 and 24 days consisted of heterofermentative lactobacilli, with *Lactobacillus cellobiosus* being isolated from 92% of the steaks.[90] In 59% of the samples, *L. cellobiosus* constituted 50% or more of the biota. The latter investigators found that when medium-oxygen-barrier film was used, high percentages of organisms such as *Aeromonas*, *Enterobacter*, *Hafnia*, *B. thermosphacta*, pseudomonads, and *Morganella morganii* were usually found.

When high concentrations of nitrites are present, they generally inhibit *B. thermosphacta* and psychrotrophic Enterobacteriaceae, and the lactic acid bacteria become dominant because they are relatively insensitive to nitrites.[69] However, low concentrations of nitrites appear not to affect *B. thermosphacta* growth, especially in cooked, vacuum-packaged products. When Egan et al.[19] inoculated this organism and a homo- and a heterofermentative lactobacillus into corned beef and sliced ham containing 240 ppm nitrate and 20 ppm nitrite, *B. thermosphacta* grew with no detectable lag phase. It had a generation time of 12–16 hours at 5°C, whereas the generation time for the heterofermentative lactobacillus was 13–16 hours and 18–22 hours for the homofermentative. Times to reach 10^8 cells/g were 9, 9–12, and 12–20 days, respectively. Although off-flavors developed 2–3 days after the numbers attained 10^8/g for *B. thermosphacta*, the same did not occur for the homo- and heterofermentative lactobacilli until 11 and 21 days, respectively. The lactic acid bacteria are less significant than *B. thermosphacta* in the spoilage of vacuum-packaged luncheon meats.[20] On the other hand, this organism has a longer lag phase and a slower growth rate than the lactobacilli.[31] When the two groups are present in equal numbers, the lactobacilli generally dominate. In a study of spoiled vacuum-packed smoked Vienna sausage, 540 isolates were examined and 58% were homofermentative lactics and 36.3% were leuconostocs, and they attained numbers of 10^7–10^8/g of spoiled product.[91] No carnobacters were found in these products.

It appears that at least two *Leuconostoc* spp. are uniquely adapted to vacuum-packaged and MAP meats: *L. carnosum* and *L. gelidum*. Unnamed *Leuconostoc* spp. were found in one study of loins packaged under high O_2 and CO_2 to constitute from 88% to 100% of the biota[78] and to be the dominant members of the biota in another similar study.[40] Following an extensive study of lactic acid bacteria isolated from vacuum-packaged meats, *L. carnosum* and *L. geldium* were established.[81] Both species grow at 1°C but not at 37°C, and both produce gas from glucose. In spoiled vacuum-packaged, sliced, cooked ham, *Leuconostoc carnosum* was found to be the specific spoilage organism.[5]

The genus *Carnobacterium* is important in the spoilage of MAP and vacuum-stored meats. These catalase-negative bacteria are heterofermentative, produce only L(+)-lactic acid, and produce gas from glucose (the typical heterofermentative betabacteria produce both D- and L-lactate). Prior to 1987, the carnobacteria were regarded as being lactobacilli. Of 159 isolates of lactobacilli from vacuum-packaged beef, 115 could not be identified to species level.[46] Similar strains were isolated from vacuum-packaged beef, pork, lamb, and bacon.[81] Another group of investigators isolated similar organisms from vacuum-packaged beef, and upon further study named these unique organisms *Lactobacillus divergens*.[47] In the latter study, this organism constituted 6.7% of 120 psychrotrophic isolates, none of which grew either at pH 3.9 or at 4°C in MRS broth. Following DNA–DNA hybridization and other studies, the genus *Carnobacterium* was erected, and *L. divergens* and two other species were placed in the new genus.[12] *C. divergens* is associated with vacuum-packaged meats, and *C. piscicola* and *C. mobile* are associated with fish and irradiated chicken, respectively. Because it does not produce H_2S or other foul-odor compounds, *C. divergens* may not be a spoilage bacterium. Indeed, it, along with the two leuconostocs noted, has the potential of being beneficial in gas-impermeable packages, where they may produce enough CO_2 to inhibit undesirable organisms. In spoiled packaged beef, *Carnobacterium* was favored by 100% N_2 at −1°C, whereas vacuum and 100% CO_2 favored leuconostocs.[71]

B. thermosphacta grows on beef at pH 5.4 when incubated aerobically but does not grow anaerobically at pH <5.8.[10] Under the latter conditions, the apparent minimum growth pH is 6.0. *S. putrefaciens* is also pH sensitive and does not grow on beef of normal pH but grows on DFD meats.

The effect of MAP on the growth of molds on bakery products was assessed by the use of a sponge cake analog. No fungal growth occurred up to 28 days at 25°C with 100% CO_2 regardless of a_w.[37] At a_w values of 0.80–0.90, there was no significant influence on fungal growth. Lag phase doubled when a_w was 0.85 and CO_2 increased to 70% in the headspace.

The spoilage of fresh and thawed MAP-stored salmon was studied at 2°C and the dominant organism was found to be *Photobacterium phosphoreum*, which reduced self-life to ca. 14 days for fresh and 21 days for thawed products.[21] This organism was eliminated by freezing and storage at 2°C and thus product self-life was extended by 1–2 weeks. The spoilage biota of thawed MAP salmon was dominated by *Carnobacterium piscicola*, which was apparently responsible for the tyramine found at the end of the storage period. Numbers of *P. phosphoreum* exceeded 10^6 cfu/g and it was considered to be the specific spoilage organism.[21]

Volatile Components of Vacuum-Packaged Meats and Poultry

The off-odors and off-flavors produced in vacuum-packaged meat products by the spoilage biota are summarized in Table 14–6. In general, short-chain fatty acids are produced by both lactobacilli and *B. thermosphacta*, and spoiled products may be expected to contain these compounds, which confer sharp off-odors. In vacuum-packaged luncheon meats, acetoin and diacetyl have been found to be the most significant relative to spoiled meat odors.[87] Using a culture medium (all purpose Tween—APT) containing glucose and other simple carbohydrates, the formation by *B. thermosphacta* of isobutyric and isovaleric acids was favored by low glucose and near-neutral pH, whereas acetoin, acetic acid, 2,3-butanediol, 3-methylbutanol, and 3-methyl-propanol production were favored by high glucose and low pH.[13,14] According to these investigators, acetoin is the major volatile compound produced on raw and cooked meats in O_2-containing atmospheres. This suggests that the volatile compounds produced by *B. thermosphacta* may be expected to vary between products with high and low glucose concentrations. The addition of 2% glucose to raw ground beef has been shown to decrease pH and delay off-odor and slime development without affecting the general spoilage flora,[82] and although the studies noted were not conducted with vacuum-packaged meats, it would seem to be a way to shift the volatile components from short-chain fatty acids to acetoin and other compounds that derive from glucose. Because vacuum-packaged, high-pH meats have a much shorter self-life, the addition of glucose could be of benefit in this regard.

In a study of spoiled vacuum-packaged steaks, a sulfide odor was evident with numbers of 10^7–10^8/cm^2.[41] The predominant organisms isolated were *H. alvei*, lactobacilli, and *Pseudomonas. H. alvei* was the likely cause of the sulfide odor. Within 1 week after processing, vacuum-packaged refrigerated raw beef underwent spoilage, which was characterized by large amounts of H_2-smelling gas along with extensive proteolysis.[56] The causative agent was *Clostridium laramie*, a species that is psychrotrophic[55] Another psychrotrophic *Clostridium* recovered from vacuum-packaged refrigerated pork is *C. algidicarnis*. From spoiled refrigerated vacuum-packaged meat was isolated a bacterium that produced large amounts of H_2, CO_2, butanol, and butanoic acid, along with esters and volatile sulfur-containing compounds.[15] The isolate was a psychrophile that grows between 1°C and 15°C but not at 22°C, and it has been classified as *Clostridium estertheticum*.[11]

From the summary of volatiles in Table 14–6, it is evident that all organisms produced either dimethyl di- or trisulfide, or methyl mercaptan, except *B. thermosphacta*. Dimethyl disulfide was produced in

Table 14–6   Volatile Compounds Produced by the Spoilage Biota or Spoilage Organisms in Meats, Poultry, Seafood, or Culture Media

Organism/Inoculum	Substrate/Conditions	Principal Volatiles	Reference
Shewanella putrefaciens	Sterile fish muscle, 1–2°C, 15 days	Dimethyl sulfide, dimethyl trisulfide, methyl mercaptan, trimethylamine, propionaldehyde, 1-penten-3-ol, H_2S, etc.	66
"Achromobacter" sp.	As above	Same as above except no dimethyl trisulfide or H_2S	66
P. fluorescens	As above	Methyl sulfide, dimethyl disulfide	66
P. perolens	As above	Dimethyl trisulfide, dimethyl disulfide, methyl mercaptan, 2-methoxy-3-isopropylpyrazine (potatolike odor)	67
Moraxella sp.	TSY agar, 2–4°C, 14 days	16 compounds including dimethyl disulfide, dimethyl trisulfide, methyl isobutyrate, and methyl-2-methyl butyrate	61
P. fluorescens	As above	15 compounds including all the above except methyl isobutyrate	61
P. putida	As above	14 compounds including the same for Moraxella sp. above except methyl isobutyrate and methyl-2-methyl butyrate	61
B. thermosphacta	Inoculated vacuum-packaged corned beef, 5°C	7 compounds including diacetyl, acetoin, nonane, 3-methyl-butanal, and 2-methyl-butanol	87
	Aerobically stored, inoculated beef slices, 1°C, 14 days, pH 5.5–5.8	Acetoin, acetic acid, isobutyric/isovaleric acids; acetic acid increased fourfold after 28 days	13
	As above; pH 6.2–6.6	Acetic acid, isobutyric, isovaleric, and n-butyric acids	13
	APT broth, pH 6.5, 0.2% glucose	Acetoin, acetic acid, isobutyric and isovaleric acids	13
	APT broth, pH 6.5, no glucose	Same as above but no acetoin	13
B. thermosphacta (15 strains)	APT broth, pH 6.5, 0.2% glucose	Acetoin, acetic acid, isobutyric and isovaleric acids, traces of 3-methylbutanol	14
S. putrefaciens	Grown in radappertized chicken, 5 days, 10°C	H_2S, methyl mercaptan, dimethyl disulfide, methanol, ethanol	26
P. fragi	As above	Methanol, ethanol, methyl and ethyl acetate, dimethyl sulfide, methanol, ethanol	26
B. thermosphacta	As above	Methanol, ethanol	26
Flora	Spoiled chicken	H compounds including H_2S, methanol, ethanol, methyl mercaptan, dimethyl sulfide, dimethyl disulfide	26

chicken by 8 of 11 cultures evaluated by Freeman et al.,[26] ethanol by 7, and methanol and ethyl acetate by 6 each. *S. putrefaciens* consistently produces H_2 in vacuum-packaged meats on which it grows. From chicken breast muscle inoculated with *Pseudomonas* strains and held at 2°C for 14 days, odors detected from chromatograph peaks were described by McMeekin[65] as being "sulfide-like," "evaporated milk," and "fruity."

REFERENCES

1. Adams, M.R., and M.O. Moss. 1995. *Food Microbiology*, Chap. 4. Cambridge, England: Royal Society of Chemistry.

2. Barakat, R.K., and L.J. Harris. 1999. Growth of *Listeria monocytogenes* and *Yersinia enterocolitica* on cooked modified-atmosphere-packaged poultry in the presence and absence of a naturally occurring microbiota. *Appl. Environ. Microbiol.* 65:342–345.

3. Barbosa, W.B., J.N. Sofos, G.R. Schmidt, and G.C. Smith. 1995. Growth potential of individual strains of *Listeria monocytogenes* in fresh vacuum-packaged refrigerated ground top round of beef. *J. Food Protect.* 58:398–403.

4. Bartkowski, L., F.D. Dryden, and J.A. Marchello. 1982. Quality changes of beef steaks stored in controlled gas atmospheres containing high or low levels of oxygen. *J. Food Protect.* 45:42–45.

5. Björkroth, K.J., P. Vandamme, and H.J. Korkeala. 1998. Identification and characterization of *Leuconostoc carnosum*, associated with production and spoilage of vacuum-packaged sliced, cooked ham. *Appl. Environ. Microbiol.* 64:3313–3319.

6. Blickstad, E., and G. Molin. 1983. The microbial flora of smoked pork loin and frankfurter sausage stored in different gas atmospheres at 4°C. *J. Appl. Bacteriol.* 54:45–56.

7. Brent, J., H. Gomez, F. Judson, K. Miller, A. Rossi-Davis, P. Shillam, C. Hatheway, L. McCroskey, E. Mintz, K. Kallander, C. McKee, J. Romer, E. Singleton, J. Yager, and J. Sofos. 1995. Botulism from potato salad. *Dairy Food Environ. Sanit.* 15:420–422.

8. Brown, W.D., M. Albright, D.A. Watts, B. Heyer, B. Spruce, and R.J. Price. 1980. Modified atmosphere storage of rockfish (*Sebastes miniatus*) and silver salmon (*Oncorhynchus kisutch*). *J. Food Sci.* 45:93–96.

9. Cai, P., M.A. Harrison, Y.-W. Huang, and J.L. Silva. 1997. Toxin production by *Clostridium botulinum* type E in packaged channel catfish. *J. Food Protect.* 60:1358–1363.

10. Campbell, R.J., A.F. Egan, F.H. Grau, and B.J. Shay. 1979. The growth of *Microbacterium thermosphactum* on beef. *J. Appl. Bacteriol.* 47:505–509.

11. Collins, M.D., U.M. Rodrigues, R.H. Dainty, R.A. Edwards, and T.A. Roberts. 1992. Taxonomic studies on a psychrophilic *Clostridium* from vacuum-packed beef: Description of *Clostridium estertheticum* sp. nov. *FEMS Microbiol. Lett.* 96:235–240.

12. Collins, M.D., J.A.E. Farrow, B.A. Phillips, S. Ferusu, and D. Jones. 1987. Classification of *Lactobacillus divergens*, *Lactobacillus piscicola*, and some catalase-negative, asporogenous, rod-shaped bacteria from poultry in a new genus, *Carnobacterium*. *Int. J. System. Bacteriol.* 37:310–316.

13. Dainty, R.H., and C.M. Hibbard. 1980. Aerobic metabolism of *Brochothrix thermosphacta* growing on meat surfaces and in laboratory media. *J. Appl. Bacteriol.* 48:387–396.

14. Dainty, R.H., and F.J.K. Hofman. 1983. The influence of glucose concentration and culture incubation time on end-product formation during aerobic growth of *Brochothrix thermosphacta*. *J. Appl. Bacteriol.* 55:233–239.

15. Dainty, R.H., R.A. Edwards, and C.M. Hibbard. 1989. Spoilage of vacuum-packed beef by a *Clostridium* sp. *J. Sci. Food Agric.* 49:473–486.

16. Dalgaard, P., L.G. Munoz, and O. Mejlholm. 1998. Specific inhibition of *Photobacterium phosphoreum* extends the self-life of modified-atmosphere-packaged cod fillets. *J. Food Protect.* 61:1191–1194.

17. Daniels, J.A., R. Krishnamurthi, and S.S.H. Rizvi. 1985. A review of effects of carbon dioxide on microbial growth and food quality. *J. Food Protect.* 48:532–537.

18. Del Torre, M., M.L. Stecchini, and M.W. Peck. 1998. Investigation of the ability of proteolytic *Clostridium botulinum* to multiply and produce toxin in fresh Italian pasta. *J. Food Protect.* 61:988–993.

19. Egan, A.F., A.L. Ford, and B.J. Shay. 1980. A comparison of *Microbacterium thermosphactum* and lactobacilli as spoilage organisms of vacuum-packaged sliced luncheon meats. *J. Food Sci.* 45:1745–1748.

20. Egan, A.F., and B.J. Shay. 1982. Significance of lactobacilli and film permeability in the spoilage of vacuum-packaged beef. *J. Food Sci.* 47:1119–1122, 1126.

21. Emborg, J., B.G. Laursen, T. Rathjen, and P. Dalgaard. 2002. Microbial spoilage and formation of biogenic amines in fresh and thawed modified atmosphere-packaged salmon (*Salmo solar*) at 2°C. *J .Appl. Microbiol.* 92:790–799.

22. Enfors, S.-O., and G. Molin. 1978. The influence of high concentrations of carbon dioxide on the germination of bacterial spores. *J. Appl. Bacteriol.* 45:279–285.

23. Fang, T.J., and L.-W. Lin. 1994. Growth of *Listeria monocytogenes* and *Pseudomonas fragi* on cooked pork in a modified atmosphere packaging/nisin combination system. *J. Food Protect.* 57:479–485.

24. Farber, J.M., S.L. Wang, Y. Cai, and S. Zhang. 1998. Changes in populations of *Listeria monocytogenes* inoculated on packaged fresh-cut vegetables. *J. Food Protect.* 61:192–195.

25. Farber, J.M., and E. Daley. 1994. Fate of *Listeria monocytogenes* on modified-atmosphere packaged turkey roll slices. *J. Food Protect.* 57:1098–1100.

26. Freeman, L.R., G.J. Silverman, P. Angelini, C. Merritt, Jr., and W.B. Esselen. 1976. Volatiles produced by microorganisms isolated from refrigerated chicken at spoilage. *Appl. Environ. Microbiol.* 32:222–231.

27. Garcia, G.W., C. Genigeorgis, and S. Lindroth. 1987. Risk of growth and toxin production by *Clostridium botulinum* nonproteolytic types B, E, and F in salmon fillets stored under modified atmospheres at low and abused temperatures. *J. Food Protect.* 50:330–336.

28. Gardner, G.A., A.W. Carson, and J. Patton. 1967. Bacteriology of prepacked pork with reference to the gas composition within the pack. *J. Appl. Bacteriol.* 30:321–333.

29. Gill, C.O., and G. Molin. 1998. Modified atmospheres and vacuum packaging. In *Food Preservatives*, ed. N.J. Russell, and G.W. Gould, 172–199. New York: Kluwer Academic Publishers.

30. Gill, C.O., and K.H. Tan. 1980. Effect of carbon dioxide on growth of meat spoilage bacteria. *Appl. Environ. Microbiol.* 39:317–319.

31. Gill, C.O., 1983. Meat spoilage and evaluation of the potential storage life of fresh meat. *J. Food Protect.* 46:444–452.

32. Gill, C.O., and K.G. Newton. 1979. Spoilage of vacuum-packaged dark, firm, dry meat at chill temperatures. *Appl. Environ. Microbiol.* 37:362–364.

33. Golden, D.A., M.J. Eyles, and L.R. Beuchat. 1989. Influence of modified-atmosphere storage on the growth of uninjured and heat-injured *Aeromonas hydrophilia*. *Appl. Environ. Microbiol.* 55:3012–3015.

34. Grau, F.H., and P.B. Vanderlinde. 1992. Occurrence, numbers, and growth of *Listeria monocytogenes* on some vacuum-packaged processed meats. *J. Food Protect.* 55:4–7.

35. Grau, F.H., and P.B. Vanderlinde. 1990. Growth of *Listeria monocytogenes* on vacuum-packaged beef. *J. Food Protect.* 53:739–741.

36. Greer, G.G., B.D. Stilts, and L.E. Jeremiah. 1993. Bacteriology and retail case life of pork after storage in carbon dioxide. *J. Food Protect.* 56:689–693.

37. Guynot, M.E., S. Marin, Y. Sanchis, and A.J. Ramos. 2003. Modified atmosphere packaging for prevention of mold spoilage of bakery products with different pH and water activity levels. *J. Food Protect.* 66:1864–1872.

38. Halleck, F.E., C.O. Ball, and E.F. Stier. 1958. Factors affecting the quality of prepackaged meat. IV. Microbiological studies. B. Effect of package characteristics and of atmospheric pressure in package upon bacterial flora of meat. *Food Technol.* 12:301–306.

39. Hanks, H., R. Nickelson, II, and G. Finne. 1980. Shelf-life studies on carbon dioxide packaged finfish from the Gulf of Mexico. *J. Food Sci.* 45:157–162.

40. Hanna, M.O., C. Vanderzant, G.C. Smith, and J.W. Savell. 1981. Packaging of beef loin steaks in 75% O_2 + 25% CO_2. II. Microbiological properties. *J. Food Protect.* 44:928–933.

41. Hanna, M.O., G.C. Smith, L.C. Hall, and C. Vanderzant. 1979. Role of *Hafnia alvei* and a *Lactobacillus* species in the spoilage of vacuum-packaged strip loin steaks. *J. Food Protect.* 42:569–571.

42. Hanna, M.O., C. Vanderzant, Z.L. Carpenter, and G.C. Smith. 1977. Microbial flora of vacuum-packaged lamb with special reference to psychrotrophic, Gram-positive, catalase-positive pleomorphic rods. *J. Food Protect.* 40:98–100.

43. Hanna, M.O., G.C. Smith, J.W. Savell, F.K. McKeith, and C. Vanderzant. 1982. Effects of packaging methods on the microbial flora of livers and kidneys from beef or pork. *J. Food Protect.* 45:74–81.

44. Hänninen, M.-L., H. Korkeala, and P. Pakkala. 1984. Effect of various gas atmospheres on the growth and survival of *Campylobacter jejuni* on beef. *J. Appl. Bacteriol.* 57:89–94.

45. Hauschild, A.H.W., L.M. Poste, and R. Hilsheimer. 1985. Toxin production by *Clostridium botulinum* and organoleptic changes in vacuum-packaged raw beef. *J. Food Protect.* 48:712–716.

46. Hitchener, B.J., A.F. Egan, and P.J. Rogers. 1982. Characteristics of lactic acid bacteria isolated from vacuum-packaged beef. *J. Appl. Bacteriol.* 52:31–37.

47. Holzapfel, W.H., and E.S. Gerber. 1983. *Lactobacillus divergens* sp. nov., a new heterofermentative *Lactobacillus* species producing L(+)-lactate. *Syst Appl. Bacteriol.* 4:522 534.

48. Hotchkiss, J.H., R.C. Baker, and R.A. Qureshi. 1985. Elevated carbon dioxide atmospheres for packaging poultry. II. Effects of chicken quarters and bulk packages. *Poultry Sci.* 64:333–340.

49. Jamieson, W. 1980. Use of hypobaric conditions for refrigerated storage of meats, fruits, and vegetables. *Food Technol.* (3):64–71.

50. Jay, J.M. 1966. Response of the extract-release volume and water-holding capacity phenomena to microbiologically spoiled beef and aged beef. *Appl. Microbiol.* 14:492–496.

51. Jay, J.M. 1983. Antimicrobial properties of α-dicarbonyl and related compounds. *J. Food Protect.* 46:325–329.

52. Jaye, M., R.S. Kittaka, and Z.J. Ordal. 1962. The effect of temperature and packaging material on the storage life and bacterial flora of ground beef. *Food Technol.* 16(4):95–98.

53. Juneja, V.K., S.T. Martin, and G.M. Sapers. 1998. Control of *Listeria monocytogenes* in vacuum-packaged pre-peeled potatoes. *J. Food Sci.* 63:911–914.

54. Juven, B.J., S.F. Barefoot, M.D. Pierson, L.H. McCaskill, and B. Smith 1998. Growth and survival of *Listeria monocytogenes* in vacuum-packaged ground beef inoculated with *Lactobacillus alimentarius* FloraCarn L-2. *J. Food Protect.* 61:551–556.

55. Kalchayanand, N., B. Ray, and R.A. Field. 1993. Characteristics of psychrotrophic *Clostridium laramie* causing spoilage of vacuum-packaged refrigerated fresh and roasted beef. *J. Food Protect.* 56:13–17.

56. Kalchayanand, N., B. Ray, R.A. Field, and M.C. Johnson. 1989. Spoilage of vacuum-packaged refrigerated beef by *Clostridium. J. Food Protect.* 52:424–426.

57. King, A.D., Jr., and C.W. Nagel. 1975. Influence of carbon dioxide upon the metabolism of *Pseudomonas aeruginosa. J. Food Sci.* 40:362–366.

58. Kraft, A.A., K.V. Reddy, R.J. Hasiak, K.D. Lind, and D.E. Galloway. 1982. Microbiological quality of vacuum packaged poultry with or without chlorine treatment. *J. Food Sci.* 47:380–385.

59. Larson, A.E., E.A. Johnson, C.R. Barmore, and M.D. Hughes. 1997. Evaluation of the botulism hazard from vegetables in modified atmosphere packaging. *J. Food Protect.* 60:1208–1214.

60. Lawlor, K.A., M.D. Pierson, C.R. Hackney, J.R. Claus, and J.E. Marcy. 2000. Nonproteolytic *Clostridium botulinum* toxigenesis in cooked turkey stored under modified atmospheres. *J. Food Protect.* 63:1511–1516.

61. Lee, M.L., D.L. Smith, and L.R. Freeman. 1979. High-resolution gas chromatographic profiles of volatile organic compounds produced by microorganisms at refrigerated temperatures. *Appl. Environ. Microbiol.* 37:85–90.

62. Lilly, T., Jr., H.M. Solomon, and E.J. Rhodehamel. 1996. Incidence of *Clostridium botulinum* in vegetables packaged under vacuum or modified atmosphere. *J. Food Protect.* 59:59–61.

63. Lyon, W.J., and C.S. Reddmann. 2000. Bacteria associated with processed crawfish and potential toxin production by *Clostridium bottulinum* type E in vacuum-packaged and aerobically packaged crawfish tails. *J. Food Protect.* 63:1687–1696.

64. Matches, J.R., and M.E. Lavrisse. 1985. Controlled atmosphere storage of spotted shrimp (*Pandalus platyceros*). *J. Food Protect.* 48:709–711.

65. McMeekin, T.A. 1975. Spoilage association of chicken breast muscle. *Appl. Microbiol.* 29:44–47.

66. Miller, A., III, R.A. Scanlan, J.S. Lee, and L.M. Libbey. 1973. Volatile compounds produced in sterile fish muscle (*Sebastes melanops*) by *Pseudomonas putrefaciens*, *Pseudomonas fluorescens*, and an *Achromobacter* species. *Appl. Microbiol.* 26:18–21.

67. Miller, A., III, R.A. Scanlan, J.S. Lee, L.M. Libbey, and M.E. Morgan. 1973. Volatile compounds produced in sterile fish muscle (*Sebastes melanops*) by *Pseudomonas perolens. Appl. Microbiol.* 25:257–261.

68. Molin, G. 1983. The resistance to carbon dioxide of some food related bacteria. *Eur. J. Appl. Microbiol. Biotechnol.* 18:214–217.

69. Nielsen, H.-J.S. 1983. Influence of nitrite addition and gas permeability of packaging film on the microflora in a sliced vacuum-packed whole meat product under refrigerated storage. *J. Food Technol.* 18:573–585.

70. Nielsen, H.-J.S., and P. Zeuthen. 1985. Influence of lactic acid bacteria and the overall flora on development of pathogenic bacteria in vacuum-packed, cooked, emulsion-style sausage. *J. Food Protect.* 48:28–34.

71. Nissen, H., O. Serheim, and R. Dainty. 1996. Effects of vacuum, modified atmospheres and storage temperature on the microbial flora of packaged beef. *Food Microbiol.* 13:183–191.

72. Nychas, G.J., and J.S. Arkoudelos. 1990. Microbiological and physicochemical changes in minced meats under carbon dioxide, nitrogen or air at 3°C. *Int. J. Food Sci. Technol.* 25:389–398.

73. Ogilvy, W.S., and J.C. Ayres. 1951. Post-mortem changes in stored meats. II. The effect of atmospheres containing carbon dioxide in prolonging the storage life of cut-up chicken. *Food Technol.* 5:97–102.

74. Parkin, K.L., M.J. Wells, and W.D. Brown. 1981. Modified atmosphere storage of rockfish fillets. *J. Food Sci.* 47:181–184.

75. Patterson, J.T., and P.A. Gibbs. 1977. Incidence and spoilage potential of isolates from vacuum-packaged meat of high pH value. *J. Appl. Bacteriol.* 43:25–38.

76. Phillips, C.A. 1996. Review: Modified atmosphere packaging and its effects on the microbiological quality and safety of produce. *Int. J. Food Sci. Technol.* 31:463–479.

77. Samelis, J., and K.G. Georgiadou. 2000. The microbial association of Greek taverna sausage stored at 4 and 10°C in air, vacuum or 100% carbon dioxide, and its spoilage potential. *J. Appl. Microbiol.* 88:58–86.

78. Savell, J.W., M.O. Hanna, C. Vanderzant, and G.C. Smith. 1981. An incident of predominance of *Leuconostoc* sp. in vacuum-packaged beef strip loins—Sensory and microbial profile of steaks stored in O_2–CO_2–N_2 atmospheres. *J. Food Protect.* 44:742–745.

79. Sears, D.F., and R.M. Eisenberg. 1961. A model representing a physiological role of CO_2 at the cell membrane. *J. Gen. Physiol.* 44:869–887.

80. Seelye, R.J., and B.J. Yearbury. 1979. Isolation of *Yersinia enterocolitica*-resembling organisms and *Alteromonas putrefaciens* from vacuum-packaged chilled beef cuts. *J. Appl. Bacteriol.* 46:493–499.

81. Shaw, B.G., and C.D. Harding. 1984. A numerical taxonomic study of lactic acid bacteria from vacuum-packed beef, pork, lamb and bacon. *J. Appl. Bacteriol.* 56:25–40.

82. Shelef, L.A. 1977. Effect of glucose on the bacterial spoilage of beef. *J. Food Sci.* 42:1172–1175.

83. Silvertsvik, M., J.T. Rosnes, and G.H. Kleiber. 2003. Effect of modified atmosphere packaging and superchilled storage on the microbial and sensory quality of Atlantic salmon (*Salmo solar*) fillets. *J. Food Sci.* 68:1467–1472.

84. Smith, W.H. 1964. The use of carbon dioxide in the transport and storage of fruits and vegetables. *Adv. Food Res.* 12:95–146.

85. Solomon, H.M., E.J. Rhodehamel, and D.A. Kautter. 1998. Growth and toxin production by *Clostridium botulinum* on sliced raw potatoes in a modified atmosphere with and without sulfite. *J. Food Protect.* 61:126–128.

86. Solomon, H.M., D.A. Kautter, T. Lilly, and E.J. Rhodehamel. 1990. Outgrowth of *Clostridium botulinum* in shredded cabbage at room temperature under a modified atmosphere. *J. Food Protect.* 53:831–833.

87. Stanley, G., K.J. Shaw, and A.F. Egan. 1981. Volatile compounds associated with spoilage of vacuum-packaged sliced luncheon meat by *Brochothrix thermosphacta*. *Appl. Environ. Microbiol.* 41:816–818.

88. Stenstrom, I.-M. 1985. Microbial flora of cod fillets (*Gadus morhua*) stored at 2°C in different mixtures of carbon dioxide and nitrogen/oxygen. *J. Food Protect.* 48:585–589.

89. Stern, N.J., M.D. Greenberg, and D.M. Kinsman. 1986. Survival of *Campylobacter jejuni* in selected gaseous environments. *J. Food Sci.* 51:652–654.

90. Vanderzant, C., M.O. Hanna, J.G. Ehlers, J.W. Savell, G.C. Smith, B. Griffin, R.N. Terrell, K.D. Lind, and D.E. Galloway. 1982. Centralized packaging of beef loin steaks with different oxygen-barrier films: Microbiological characteristics. *J. Food Sci.* 47:1070–1079.

91. von Holy, A., T.E. Cloete, and W.H. Holzapfel. 1991. Quantification and characterization of microbial populations associated with spoiled, vacuum-packed Vienna sausages. *Food Microbiol.* 8:95–104.

92. Wabeck, C.J., C.E. Parmalee, and W.J. Stadelman. 1968. Carbon dioxide preservation of fresh poultry. *Poultry Sci.* 47:468–474.

CHAPTER 15

Radiation Protection of Foods, and Nature of Microbial Radiation Resistance

Although a patent was issued in 1929 for the use of radiation as a means of preserving or protecting foods, it was not until shortly after World War II that this method of food protection received any serious consideration. While the application of radiation has been somewhat slow in reaching its maximum potential use, the full application of this method presents some interesting challenges to food microbiologists and other food scientists.

Radiation may be defined as the emission and propagation of energy through space or through a material medium. The type of radiation of primary interest in food preservation is electromagnetic. The electromagnetic spectrum is presented in Figure 15–1. The various radiations are separated on the basis of their wavelengths, with the shorter wavelengths being the most damaging to microorganisms. The electromagnetic spectrum may be further divided as follows with respect to the radiations of interest in food preservation: microwaves, ultraviolet rays, X-rays, and gamma rays. The radiations of primary interest in food preservation are ionizing radiations, defined as those radiations that have wavelengths of 2000 Å or less—for example, alpha particles, beta rays, gamma rays, and X-rays. Their quanta contain enough energy to ionize molecules in their paths. Because they destroy microorganisms without appreciably raising the temperature, the process is termed cold sterilization.

In considering the application of radiation to foods, there are several useful concepts that should be clarified. A Roentgen is a unit of measure used for expressing an exposure dose of X-ray or gamma radiation. A milliroentgen is equal to 1/1,000 of a roentgen. A Curie is a quantity of radioactive substance in which 3.7×10^{10} radioactive disintegrations occur per second. For practical purposes, 1 g of pure radium possesses the radioactivity of 1 curie of radium. The unit for a curie is the Becquerel (Bq). A rad is a unit equivalent to the absorption of 100 ergs/g of matter. A kilorad (krad) is equal to 1,000 rads, and a megarad (Mrad) is equal to 1 million rads. The newer unit of absorbed dose is the Gray (1 Gy = 100 rads = 11 joule/kg; 1 kGy = 10^5 rads). The energy gained by an electron in moving through 1 V is designated eV (electron volt). An meV is equal to 1 million electron volts. Both the rad and eV are measurements of the intensity of irradiation.

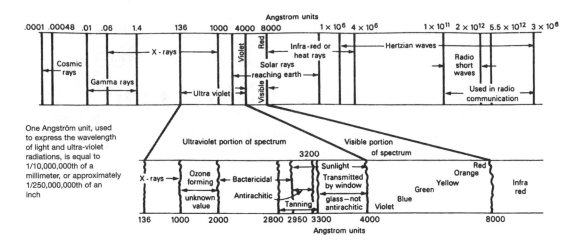

Figure 15–1 Spectrum charts. *Source*: From the *Westinghouse Sterilamp and the Rentschler-James Process of Sterilization*, courtesy of the Westinghouse Electric and Manufacturing Co., Inc.

CHARACTERISTICS OF RADIATIONS OF INTEREST IN FOOD PRESERVATION

Ultraviolet Light

Ultraviolet (UV) light is a powerful bactericidal agent, with the most effective wavelength being about 2,600 Å. It is nonionizing and is absorbed by proteins and nucleic acids, in which photochemical changes are produced that may lead to cell death. The mechanism of UV death in the bacterial cell is the production of lethal mutations as a result of action on cell nucleic acids. The poor penetrative capacities of UV light limit its food use to surface applications, where it may catalyze oxidative changes that lead to rancidity, discolorations, and other reactions. Small quantities of ozone may also be produced when UV light is used for the surface treatment of certain foods. UV light is sometimes used to treat the surfaces of baked fruitcakes and related products before wrapping.

Beta Rays

Beta rays may be defined as streams of electrons emitted from radioactive substances. Cathode rays are the same except that they are emitted from the cathode of an evacuated tube. These rays possess poor penetration power. Among the commercial sources of cathode rays are Van de Graaff generators and linear accelerators. The latter seem better suited for food protection uses. There is some concern over the upper limit of energy level of cathode rays that can be employed without inducing radioactivity in certain constituents of foods.

Gamma Rays

These are electromagnetic radiations emitted from the excited nucleus of elements such as ^{60}Co and ^{137}Cs. This is the cheapest form of radiation for food preservation, because the source elements are either byproducts of atomic fission or atomic waste products. Gamma rays have excellent penetration power, as opposed to beta rays. [60]Co has a half-life of about 5 years; and the half-life for ^{137}Cs is about 30 years.

X-Rays

These rays are produced by the bombardment of heavy-metal targets with high-velocity electrons (cathode rays) within an evacuated tube. They are essentially the same as gamma rays in other respects.

Microwaves

Microwave energy may be illustrated in the following way.[24] When electrically neutral foods are placed in an electromagnetic field, the charged asymmetric molecules are driven first one-way and then another. During this process, each asymmetric molecule attempts to align itself with the rapidly changing alternating-current field. As the molecules oscillate about their axes while attempting to go to the proper positive and negative poles, intermolecular friction is created and manifested as a heating effect. This is microwave energy. Most food research has been carried out at two frequencies: 915 and 2,450 megacycles. At the microwave frequency of 915 megacycles, the molecules oscillate back and forth 915 million times per second.[24] Microwaves lie between the infrared and radio frequency portions of the electromagnetic spectrum (Figure 15–1). The problem associated with the microwave destruction of trichina larvae in pork products is discussed in Chapter 29.

PRINCIPLES UNDERLYING THE DESTRUCTION OF MICROORGANISMS BY IRRADIATION

Several factors should be considered when the effects of radiation on microorganisms are considered, and these are discussed in the following subsections.

Types of Organisms

Gram-positive bacteria are more resistant to irradiation than Gram negatives. In general, spore formers are more resistant than non-spore-formers (with the exception of species discussed later in this chapter). Among spore formers, *Paenibacillus larvae* seem to possess a higher degree of resistance than most other aerobic spore formers. Spores of *Clostridium botulinum* type A appear to be the most resistant of all clostridial spores. Apart from the extremely resistant species, *Enterococcus faecium* R53, micrococci, and the homofermentative lactobacilli are among the most resistant of non-spore-forming bacteria. Most sensitive to radiations are the pseudomonads and flavobacters (including the new genera created by the delimiting of these genera as noted in Chapter 2), with other Gram-negative bacteria being intermediate. A general spectrum of radiation sensitivity from enzymes to higher animals is illustrated in Figure 15–2. Possible mechanisms of radioresistance are discussed below.

With the exception of endospores and the extremely resistant species already noted, radioresistance generally parallels heat resistance among bacteria.

With respect to the radiosensitivity of molds and yeasts, the latter have been reported to be more resistant than the former, with both groups in general being less sensitive than Gram-positive bacteria.

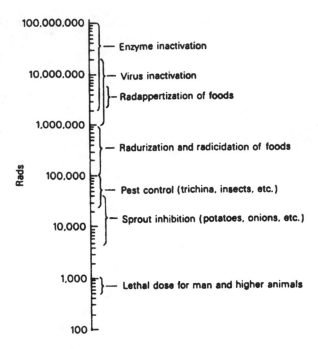

Figure 15–2　Dose ranges of irradiation for various applications. *Source*: Adapted from Grünewald.[28]

Some *Candida* strains have been reported to possess resistance comparable to that of some bacterial endospores.

Numbers of Organisms

The numbers of organisms have the same effect on the efficacy of radiations as is the case for heat, chemical disinfection, and certain other phenomena: The larger the number of cells, the less effective is a given dose.

Composition of Suspending Menstruum (Food)

Microorganisms in general are more sensitive to radiation when suspended in buffer solutions than in protein-containing media. For example, Midura et al.[52] found radiation D values for a strain of *Clostridium perfringens* to be 0.23 in phosphate buffer, whereas in cooked-meat broth, the D value was 3 kGy. Proteins exert a protective effect against radiations, as well as against certain antimicrobial chemicals and heat. Several investigators have reported that the presence of nitrites tends to make bacterial endospores more sensitive to radiation.

Presence or Absence of Oxygen

The radiation resistance of microorganisms is greater in the absence of oxygen than in its presence. Complete removal of oxygen from the cell suspension of *Escherichia coli* has been reported to increase its radiation resistance up to threefold.[58] The addition of reducing substances such as sulfhydryl compounds generally has the same effect in increasing radiation resistance as an anaerobic environment has.

Physical State of Food

The radiation resistance of dried cells is, in general, considerably higher than that for moist cells. This is most likely a direct consequence of the radiolysis of water by ionizing radiations, which is discussed later in this chapter. Radiation resistance of frozen cells is greater than that of nonfrozen cells. Grecz et al.[26] found that the lethal effects of gamma radiation decreased by 47% when ground beef was irradiated at $-196°C$ as compared to $0°C$.

Age of Organisms

Bacteria tend to be most resistant to radiation in the lag phase just prior to active cell division. The cells become more radiation sensitive as they enter and progress through the log phase and reach their minimum at the end of this phase.

PROCESSING OF FOODS FOR IRRADIATION

Prior to being exposed to ionizing radiations, several processing steps must be carried out in much the same manner as for the freezing or canning of foods.

Selection of Foods

Foods to be irradiated should be carefully selected for freshness and overall desirable quality. Especially to be avoided are foods that are already in incipient spoilage.

Cleaning of Foods

All visible debris and dirt should be removed. This will reduce the numbers of microorganisms to be destroyed by the radiation treatment.

Packaging

Foods to be irradiated should be packed in containers that will afford protection against post irradiation contamination. Clear glass containers undergo color changes when exposed to doses of radiation of around 10 kGy, and the subsequent color may be undesirable.

Blanching or Heat Treatment

Sterilizing doses of radiation are insufficient to destroy the natural enzymes of foods (Figure 15–2). In order to avoid undesirable postirradiation changes, it is necessary to destroy these enzymes. The best method is a heat treatment—that is, the blanching of vegetables and mild heat treatment of meats prior to irradiation.

APPLICATION OF RADIATION

The two most widely used techniques of irradiating foods are gamma radiation from either ^{60}Co or ^{137}Cs, and the use of electron beams from linear accelerators.

Gamma Radiation

The advantage of gamma radiation is that ^{60}Co and ^{137}Cs are relatively inexpensive byproducts of atomic fission. In a common experimental radiation chamber employing these elements, the radioactive material is placed on the top of an elevator that can be moved up for use and down under water when not in use. Materials to be irradiated are placed around the radioactive material (the source) at a suitable distance for the desired dosage. Once the chamber has been vacated by all personnel, the source is raised into position, and the gamma rays irradiate the food. Irradiation at desired temperatures may be achieved either by placing the samples in temperature-controlled containers or by controlling the temperature of the entire concrete-leadwalled chamber. Among the drawbacks to the use of radioactive material is that the isotope source emits rays in all directions and cannot be turned "on" or "off" as may be desired (Figure 15–3). Also, the half-life of ^{60}Co (5.27 years) requires that the source be changed periodically in order to maintain a given level of radioactive potential. This drawback is overcome by the use of ^{137}Cs, which has a half-life of around 30 years.

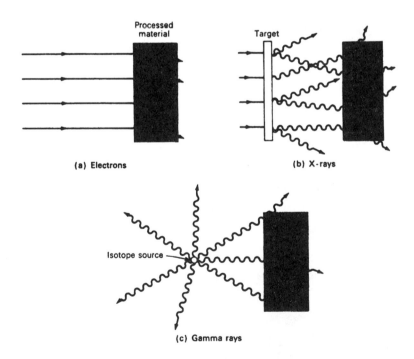

(a) Electrons

(b) X-rays

(c) Gamma rays

Figure 15–3   The three basic techniques for radiation processing—interactions of electrons, X-rays, and gamma rays in the medium. *Source*: From Koch and Eisenhower,[39] 1965, *Radiation Preservation of Foods*, Publication 1273, Advisory Board on Military Personnel Supplies, National Academy of Sciences, National Research Council.

Electron Beams/Accelerated Electrons

The use of electron accelerators offers certain advantages over radioactive elements that make this form of radiation somewhat more attractive to potential commercial users. Koch and Eisenhower[39] listed the following:

1. High efficiency for the direct deposition of energy of the primary electron beams means high plant-product capacity.
2. The efficient convertibility of electron power to X-ray power means the capability of handling very thick products that cannot be processed by electron or gamma-ray beams.
3. The easy variability of electron-beam current and energy means flexibility in the choice of surface and depth treatments for a variety of food items, conditions, and seasons.
4. The mono directional characteristic of the primary and secondary electrons and X-rays at the higher energies permits a great flexibility in the food package design.
5. The ability to program and to regulate automatically from one instant to the next with simple electronic detectors and circuits and various beam parameters means the capability of efficiently processing small, intricate, or nonuniform shapes.
6. The ease with which an electron accelerator can be turned off or on means the ability to shut down during off-shifts or off-seasons without a maintenance problem, and the ability to transport the radiation source without a massive radiation shield.

Two differences between gamma rays and accelerated electrons are worthy of note. First, with regard to penetration capacity, gamma is higher than accelerated electrons, but the penetration capacity of the latter increases with their energy. For example, electrons at 10 MeV are more penetrating than those at 4 MeV. The second difference is dose rate. The gamma rate from ^{60}Co is 1–100 Gy/min, whereas electron beams from an electron accelerator are 10^3–10^6 Gy/sec.

RADAPPERTIZATION, RADICIDATION, AND RADURIZATION OF FOODS

Definitions

Initially, the destruction of microorganisms in foods by ionizing radiation was referred to by terminology carried over from heat and chemical destruction of microorganisms. Although microorganisms can indeed be destroyed by chemicals, heat, and radiation, there is, nevertheless, a lack of precision in the use of this terminology for radiation-treated foods. Consequently, in 1964 an international group of microbiologists suggested the following terminology for radiation treatment of foods.[25]

Radappertization is equivalent to radiation sterilization or "commercial sterility," as it is -understood in the canning industry. Typical levels of irradiation are 30–40 kGy. The term was coined to honor N. Appert.

Radicidation is equivalent to pasteurization—of milk, for example. Specifically, it refers to the reduction of the number of viable specific non-spore-forming pathogens, other than viruses, so that none is detectable by any standard method. Typical levels to achieve this process are 2.5–10 kGy.

Radurization may be considered equivalent to pasteurization. It refers to the enhancement of the keeping quality of a food by causing substantial reduction in the numbers of viable specific spoilage microbes by radiation. Common dose levels are 0.75–2.5 kGy for fresh meats, poultry, seafood, fruits, vegetables, and cereal grains.

Table 15–1   Effect of Irradiation Temperature on D Values of Two Load Levels of *C. botulinum* 33A in Precooked Ground Beef

Temperature (°C)	D (kGy)	
	~5 × 10⁶ *Spores/Can*	~2 × 10⁸ *Spores/Can*
−196	5.77	5.95
−150	5.32	5.43
−100	4.83	4.86
−50	4.34	4.30
0	3.85	3.73
25	3.60	3.45
65	3.21	2.99

Note: Data are based on linear spore destruction.

Source: Grecz et al.[27] reproduced by permission of National Research Council of Canada from *Canadian Journal of Microbiology* 17:135–142, 1971.

Radappertization

Radappertization of any foods may be achieved by application of the proper dose of radiation under the proper conditions. The effect of this treatment on endospores and exotoxins of *C. botulinum* is of obvious interest. Type E spores have been reported to possess radiation D values on the order of 1.2–1.7 kGy.[71] Types A and B spores were found by Kempe[37] to have D values of 2.79 and 2.38 kGy, respectively. Type E spores are the most radiation sensitive of these three types.

The effect of temperature of irradiation on D values of *C. botulinum* spores is presented in Table 15–1: resistance increases at the colder temperatures and decreases at warmer temperatures.[27] Different inoculum levels had no significant effect on D values whose calculations were based on a linear destruction rate. D values of four *C. botulinum* strains in three food products are presented in Table 15–2, from which it can be seen that each strain displayed different degrees of radiation resistance in each product. Also, irradiation of cured meat products produced the lowest D values. (The possible significance of this is discussed in Chapter 13 under nitrates and nitrites.). The minimum radiation doses (MRD) in kGy for the radappertization of nine meat and fish products are indicated below.[3,13,35] With the exception of bacon (irradiated at ambient temperatures), each was treated at −30°C +10:

Bacon	23
Beef	47
Chicken	45
Ham	37
Pork	51
Shrimp	37
Codfish cakes	32
Corned beef	25
Pork sausage	24–27

Table 15–2 Variations in Radiation *D* Values of Strains of *C. botulinum* at 30°C in Three Meat Products

Strain Number	D (kGy)		
	Codfish Cake	Corned Beef	Pork Sausage
33A	2.03	1.29	1.09
77A	2.38	2.62	0.98
41B	2.45	1.92	1.84
53B	3.31	1.83	0.76

Note: Computed by the Schmidt equation.

Source: Anellis et al.,[3] copyright © 1972, American Society for Microbiology.

To achieve 12*D* treatments of meat products at about 30°C, the following kGy values are necessary:[74] beef and chicken, 41.2–42.7; ham and codfish cake, 31.4–31.7; pork, 43.7; and corned beef and pork sausage, 25.5–26.9. Irradiation treatments of the types noted do not make the foods radioactive.[74]

The radiation resistance of *C. botulinum* spores in aqueous media was studied by Roberts and Ingram,[72] and these values are considerably lower than those obtained in meat products. On three type A strains, *D* ranged from 1.0 to 1.4; on two strains of type B, 1.0–1.1; on two strains of type E, 0.8–1.6; and the one type F strain examined by these authors showed a *D* value of 2.5 kGy. All strains were irradiated at 18–23°C and an exponential death rate was assumed in the *D* calculations.

With respect to the effect of radiation on *C. perfringens*, each of five different strains (types A, B, C, E, and F) was found to have *D* values between 1.5 and 2.5 kGy in an aqueous environment.[72] The 12*D* values for 8 strains of this organism were found to range between 30.4 and 41.4 kGy, depending upon the strain and method of computing 12*D* doses.[8]

Radiation D_{10} values for *Listeria monocytogenes* in mozzarella cheese and ice cream were found to be 1.4 and 2.0 kGy, respectively, with strain Scott A irradiated at 78°C.[29] The respective calculated 12*D* values were 16.8 and 24.4 kGy. To effect radappertization of ice cream and frozen yogurt, 40 kGy was sufficient but not for mozzarella or cheddar cheeses.[30] The radappertization dose for *Bacillus cereus* in cheese and ice cream was 40–50 kGy.

As indicated in Figure 15–2, viruses are considerably more resistant to radiation than bacteria. Sullivan et al. found radiation *D* values of 30 viruses[78] to range between 3.9 and 5.3 kGy in Eagle's minimal essential medium supplemented with 2% serum. The 30 viruses included coxsackie-, echo-, and poliovirus. Of five selected viruses subjected to ^{60}Co rays in distilled water, the *D* values ranged from 1.0 to 1.4 kGy. *D* values of coxsackievirus B-2 in various menstra at −30 and −90°C are presented in Table 15–3. The use of a radiation 12*D* process for *C. botulinum* in meat products would result in the survival of virus particles unless previously destroyed by other methods such as heating.

Enzymes are also highly resistant to radiation, and a dose of 20–60 kGy has been found to destroy only up to 75% of the proteolytic activity of ground beef.[48] When blanching at 64 or 70°C was combined with radiation doses of 45–52 kGy, however, at least 95% of the beef proteolytic activity was destroyed. Radiation *D* values for a variety of organisms are presented in Table 15–4.

The main drawbacks to the application of radiation to some foods are color changes and/or the production of off-flavors. Consequently, those food products that undergo relatively minor changes in color and flavor have received the greatest amount of attention for commercial radappertization.

Table 15–3　*D* Values of Coxsackievirus B-2

Suspending Menstrum	D (kGy)	
	−30° C	−90° C
Eagle's minimal essential medium + 2% serum	6.9	6.4
Distilled water	—	5.3
Cooked ground beef	6.8	8.1
Raw ground beef	7.5	6.8

Note: A linear model was assumed in *D* calculations.

Source: Sullivan et al.,[78] copyright © 1973, American Society for Microbiology.

Bacon is one product that undergoes only slight changes in color and flavor development following radappertization. Mean preference scores on radappertized versus control bacon were found to be rather close, with control bacon being scored just slightly higher.[95] Acceptance scores on a larger variety of irradiated products were in the favorable range.[35]

Radappertization of bacon is one way to reduce nitrosamines. When bacon containing 20 ppm $NaNO_2$ +550 ppm sodium ascorbate was irradiated with 30 kGy, the resulting nitrosamine levels were similar to those in nitrite-free bacon.[18]

From a review of 539 D values obtained from 39 published papers, the most radiation resistant spore-formers noted were *Geobacillus stearothermophilus* and *Clostridium sporogenes*, while the most resistant nonsporeformers were *Enterococcus faecium*, *Alcaligenes* spp., and the *Moraxella-Acinetobacter* group.[91] Overall, Gram-negative bacteria were more sensitive than Gram positives from published reports.

Table 15–4　Some Radiation *D* Values Reported

Organism/Substance	D (kGy)	Reference
Bacteria		
Acinetobacter calcoaceticus	0.26	87
Aeromonas hydrophila	0.14	60
Bacillus pumilus spores, ATCC 27142	1.40	87
Arcobacter butzleri	0.27	10
Bacillus cereus	1.485	42
Campylobacter jejuni (5 strains)	0.175–0.235	7
C. jejuni	0.19	10
Clostridium botulinum, type E spores	1.1–1.7	19,46
C. botulinum, type E Beluga	0.8	48
C. botulinum, 62A spores	1.0	48
C. botulinum, type A spores	2.79	27
C. botulinum, type B spores	2.38	27
C. botulinum, type F spores	2.5	48

continues

Table 15–4   continued

Organism/Substance	D (kGy)	Reference
C. botulinum A toxin in meat slurry	36.08	73
C. bifermentans spores	1.4	48
C. butyricum spores	1.5	48
C. perfringens, type A spores	1.2	48
C. sporogenes spores (PA 3679/S$_2$)	2.2	48
C. sordellii spores	1.5	48
Enterobacter cloacae	0.18	87
Escherichia coli	0.20	87
E. coli 0157:H7 (gr. beef, -20°C)	0.98	80
E. coli 0157:H7 (gr. beef, 4°C)	0.39	80
E. coli O157:H7 (5 strains)	0.241–0.307	7
Klebsiella pneumoniae	0.183	42
Listeria monocytogenes	0.42–0.55	61
L. monocytogenes (mean of 7 strains)	0.35	32
L. monocytogenes	0.42–0.43	1
on beef at 5°C	~ 0.44	83
on beef at 0°C	0.45	83
on beef at −20°C	1.21	83
Moraxella phenylpyruvica	0.86	62
M. osloensis	0.191	42
Pseudomonas putida	0.08	62
P. aeruginosa	0.13	87
Salmonella Typhimurium	0.50	61
S. Enteritidis in poultry meat at 22°C	0.37	54
in egg white at 15°C	0.33	54
Salmonella sp.	0.13	87
Salmonellae spp.*	0.621–0.800	7
S. Mbandaka (alfalfa seeds, 20°C)	0.98	81
Staphylococcus aureus (gr. beef, 0°C)	0.51	80
S. aureus (gr. beef, −20°C)	0.88	80
Staphylococcus aureus	0.16	87
S. aureus ent. toxin A in meat slurry	61.18; 208.49	73
Yersinia enterocolitica, beef, 25°C	0.195	16
Y. enterocolitica, ground beef at 30°C	0.388	16
Fungi		
Aspergillus flavus spores (mean)	0.66	70
A. flavus	0.055–0.06	75
A. niger	0.042	75
Penicillium citrinum, NRRL 5452 (mean)	0.88	70
Penicillium sp.	0.42	87
Viruses		
Adenovirus (4 strains)	4.1-4.9	50
Coxsackievirus (7 strains)	4.1–5.0	50
Echovirus (8 strains)	4.4–5.1	50
Herpes simplex	4.3	50
Poliovirus (6 strains)	4.1–5.4	50

*Five strains including serotypes Dublin, Enteritidis, and Typhimurium.

Radicidation

Irradiation at levels of 2–5 kGy has been found by many to be effective in destroying non-spore-forming and nonviral pathogens and to present no health hazard. Kampelmacher[36] notes that raw poultry meats should be given the highest priority because they are often contaminated with salmonellae and because radicidation is effective on prepackaged products, thus eliminating the possibilities of cross-contamination. The treatment of refrigerated and frozen chicken carcasses with 2.5 kGy was highly effective in destroying salmonellae.[53,54] A radiation dosage up to 7 kGy (0.7 Mrad) has been approved by the World Health Organization (WHO) as being "unconditionally safe for human consumption".[19] When whole cacao beans were treated with 5 kGy, 99.9% of the bacterial biota was destroyed, and *Penicillium citrinum* spores were reduced by about 5 logs/g, and at a level of 4 kGy, *Aspergillus flavus* spores were reduced by about 7 logs/g.[70] Fresh poultry, cod and red fish, and spices and condiments have been approved for radicidation in some countries (Table 15–5).

Table 15–5   Some Food and Food Products Approved for Irradiation by Various Countries and by WHO

Products	Objective	Dose Range (kGy)	Number of Countries*
Potatoes	Sprout inhibition	0.1–0.15	17
Onions	Sprout inhibition	0.1–0.15	10
Garlic	Sprout inhibition	0.1–0.15	2
Mushrooms	Growth inhibition	2.5 max	
Wheat, wheat flour	Insect disinfestation	0.2–0.75	4
Dried fruits	Insect disinfestation	1.0	2
Cocoa beans	Insect disinfestation	0.7	1
Dry food concentrates	Insect disinfestation	0.7–1.0	1
Poultry, fresh	Radicidation[†]	7.0 max	2
Cod and redfish	Radicidation	2.0–2.2	1
Spices/condiments	Radicidation	8.0–10.0	1
Semipreserved meats	Radurization	6.0–8.0	1
Fresh fruits[‡]	Radurization	2.5	6
Asparagus	Radurization	2.0	1
Raw meats	Radurization	6.0–8.0	1
Cod and haddock fillets	Radurization	1.5 max	1
Poultry (eviscerated)	Radurization	3.0–6.0	2
Shrimp	Radurization	0.5–1.0	1
Culinary prepared meat products	Radurization	8.0	1
Deep-frozen meals	Radappertization	25.0 min	2
Papaya	Radurization	250 Gy	
Shell eggs	Radurization	3.0	
Fresh, tinned/liquid foodstuffs	Radappertization	25.0 min	1

*Including WHO recommendations.
[†]For salmonellae.
[‡]Includes tomatoes, peaches, apricots, strawberries, cherries, grapes, and so forth.

Source: Urbain[89] and the literature.

The irradiation of steaks at 1.5 kGy inoculated with $\sim 10^5$/g of *Escherichia coli* 0157:H7 resulted in complete elimination of cells.[21] *Yersinia enterocolitica* was reduced to undetectable levels under the same conditions. Irradiation of mechanically deboned chicken meat that was inoculated with ~ 400 spores of 20 strains of C. *botulinum* types A and B at 1.5 or 3.0 kGy resulted in no samples becoming demonstrably toxic after refrigerated storage for four weeks, but samples that were temperature abused at 28°C became toxic within 18 hours.[79] In a similar product, an initial level of *Salmonella* Enteritidis 3.86 $\log_{10}$/g was reduced to <10 cfu/g after four weeks at 5°C.

Live oysters were placed in a tank of water to which were added cultures of *S*. Enteritidis, *S*. Infantis, and *Vibrio parahaemolyticus*, and after 13 hours they contained ca. 4–6 log cfu/g.[33] Irradiation at 3 kGy reduced the two salmonellae by 5 to 6 logs, and 1.0 kGy reduced *V. parahaemolyticus* by 6 logs. The oysters were not killed by up to 3 kGy.

SEED SPROUTS AND OTHER VEGETABLES

The *D* values of meat and vegetable isolates of *E. coli* 0157:H7 on alfalfa and broccoli were found to be 0.34 and 0.30 kGy, respectively, while a salmonellae cocktail of meat and vegetable isolates was found to be 0.54 and 0.46 kGy, respectively.[68] No salmonellae were detected on sprouts given a 0.5 kGy or higher dose. The *D* value for *S*. Mbandaka on alfalfa seeds was found to be 0.81 ± 0.02 kGy.[81] In this study, an absorbed dose of 4 kGy but not 3 kGy eliminated *S*. Mbandaka from naturally contaminated seeds. In another study, an irradiation dose of 2 kGy reduced the APC from 10^{5-8} to 10^{3-5} cfu/g while coliforms were reduced even more.[67] This treatment extended keeping quality by 10 days over controls with little adverse effect on seed germination and sprout quality. Although the U.S. Food and Drug Administration has approved the use of 8 kGy for alfalfa seeds for sprouts, irradiation doses above 3 kGy lead to a reduction of germination. From a later study of *E. coli* 0157:H7 and *Salmonella* on alfalfa seeds, a 2-kGy dose effected a 3.3-log reduction of *E. coli* and *Salmonella*, respectively, with no significant loss of sprouting.[82] These authors found these organisms to be more radiation resistant on alfalfa seeds than in meats or poultry. In another study, a nonvegetable isolate of *E. coli* 0157:H7 had a *D* value of 1.43 kGy while a vegetable isolate was 1.11.[69] The background biota of broccoli sprouts was reduced from 10^6–10^7 to 10^4–10^5 cfu/g resulting in an increased shelf life of 10 days.

A combination of irradiation and MAP was evaluated for the preservation of fresh-cut lettuce in two studies. In one, the initial APC of 10^5–10^6 cfu/g was reduced by ca. 1.5 $\log_{10}$ cfu/g after 14 days at 4°C with a dose of 0.35 kGy.[64] This treatment resulted in a 10% loss in product firmness. The Romaine lettuce was packaged in laminate polyethylene bags with an initial concentration of ca. 1.5% O_2, 4% CO_2, and the remainder N_2.[64] In the other study, Iceberg lettuce was best with doses of 1–2 kGy, which reduced the amount of leakage seen in samples exposed to doses >2 kGy.[17] The O_2 content of the 3°C-stored samples ranged from 1.5 to 3.7% after day 5, and the CO_2 content was higher than in non-irradiated controls. In a study of the effect of irradiation on the destruction of a 5-strain cocktail of *L. monocytogenes* in six different ready-to-eat meats at doses of 2.0 and 4.0 kGy followed by storage at 4 and 10°C, no cells could be detected when 4.0 kGy was used.[20] With 2.0 kGy and 10°C, survivors were recovered after the second week but with 2.0 kGy-treated products stored at 4°C, survivors were recovered after 5 weeks.

Radurization

Irradiation treatments to extend the shelf life of seafoods, vegetables, and fruits have been verified in many studies. The shelf life of shrimp, crab, haddock, scallops, and clams may be extended from twofold to sixfold by radurization with doses from 1 to 4 kGy. Similar results can be achieved for fish and shellfish under various conditions of packaging.[59] In one study, scallops stored at 0°C had a shelf life of 13 days, but after irradiation doses of 0.5, 1.5, and 3.0 kGy, shelf life was 18, 23, and 42 days, respectively.[63] The Gram-negative non-spore-forming rods are among the most radiosensitive of all bacteria, and they are the principal spoilage organisms for these foods. Following the irradiation of vacuum-packaged ground pork at 1.0 kGy and storage at 5°C for 9 days, 97% of the irradiated biota consisted of Gram-positive bacteria, with most being coryneforms.[15] The Gram-negative coccobacillary rods belonging to the genera *Moraxella* and *Acinetobacter* have been found to possess degrees of radiation resistance higher than for all other Gram negatives. In studies on ground beef subjected to doses of 272 krad, Tiwari and Maxcy[86] found that 73–75% of the surviving biota consisted of these related genera. In unirradiated meat, they constituted only around 8% of the biota. Of the two genera, the *Moraxella* spp. appeared to be more resistant than *Acinetobacter* spp., with D_{10} values of 0.539 and 0.583 kGy, whereas the D_{10} for *M. osloensis* strains was 477 up to 1,000 krad (0.477–1.0 kGy).

In comparing the radiosensitivity of some non–spore-forming bacteria in phosphate buffer at −80°C, Anellis et al.[2] found that *Deinococcus radiodurans* survived 18 kGy, *Enterococcus faecium* strains survived 9–15 kGy, *E. faecalis* survived 6–9 kGy, and *Lactococcus lactis* did not survive 6 kGy. *Staphylococcus aureus*, *Lactobacillus casei*, and *Lactobacillus arabinosus* did not survive 3-kGy exposures. It was shown that radiation sensitivity decreased as the temperature of irradiation was lowered, as is the case for endospores.

The ultimate spoilage of radurized, low-temperature-stored foods is invariably caused by one or more of the *Acinetobacter-Moraxella* or lactic acid types noted above. The application of 2.5 kGy to ground beef destroyed all pseudomonads, Enterobacteriaceae, and *Brochothrix thermosphacta*; and reduced aerobic plate counts (APCs) from 6.18/g to 1.78 $\log_{10}$/g, but reduced lactic acid bacteria only by 3.4 log/g.[57]

Radurization of fruits with doses of 2–3 kGy brings about an extension of shelf life of at least 14 days. Radurization of fresh fruits is permitted by at least six countries, with some meats, poultry, and seafood permitted by several others (Table 15–5). In general, shelf-life extension is not as great for radurized fruits as for meats and seafood because molds are generally more resistant to irradiation than the Gram-negative bacteria that cause spoilage of the latter products. When ground beef patties were subjected to 2.0 kGy under vacuum, they remained unspoiled after 60 days in the refrigerator.[55] In another study, unirradiated ground beef patties that originally contained 10^6 APC/g contained 10^8/g after 8 days at 4°C, but the samples that were irradiated at 2 kGy (range 1.9–2.4) reached only 10^6/g after 55 days at 4°C.[56] In regards to pathogens in ground beef, it was concluded that an applied dose of 2.5 kGy would be sufficient to destroy $10^{8.1}$ *E. coli*. 0157:H7, $10^{3.1}$ salmonellae, and $10^{10.6}$ *Campylobacter jejuni*.[7]

Insect eggs and larvae can be destroyed by 1 kGy, and cysticerci of the pork tapeworm (*Taenia solium*) and the beef tapeworm (*T. saginata*) can be destroyed with even lower doses, with cysticercosis-infested meat being rendered free of parasites by exposure to 0.2–0.5 kGy.[92]

LEGAL STATUS OF FOOD IRRADIATION

At least 40 countries had approved the irradiation of some foods as of mid-1989.[47] At least 20 different food–packaging materials have been approved by the U.S. Food and Drug Administration (FDA) at levels of 10 or 60 kGy. In 1983, the FDA permitted spices and vegetable seasonings to be irradiated up to 10 kGy (*U.S. Federal Register*, July 15, 1983). In 1985 the FDA granted permission for the irradiation of pork at up to 1 kGy to control *Trichinella spiralis* (U.S. Federal Register, July 22,

1985). In 1986, fermented pork sausage (Nham) was irradiated in Thailand at a minimum of 2.0 kGy, and the product was sold in Bangkok.[47] Puerto Rican mangoes were irradiated in 1986 at up to 1.0 kGy, flown to Miami, Florida, and sold. Hawaiian papayas were treated at doses of 0.41–0.51 kGy to control pests in 1987 and later sold to the public. USDA approval was granted for Hawaiian papayas in 1989 for insect control. In May 1990, the USDA approved the irradiation of poultry up to 3.0 kGy, and on September 2, 1993, irradiated poultry was sold in a retail grocery store in Illinois for the first time.[65] Strawberries were irradiated at 2.0 kGy and sold in Lyon, France, in 1987, and in the United States on January 25, 1992, in the state of Florida. In 1995, the states of Maine and New York repealed their bans on the sale of irradiated foods. Sprout inhibition and insect disinfestations continue to be the most widely used direct applications of food irradiation.

In 1981, a joint Food and Agriculture Organization (FAO)/International Atomic Energy Agency (IAEA)/WHO Expert Committee on food irradiation found that foods given an overall average of up to 10.0 kGy were unconditionally safe. At least 40 countries have approved irradiation of one or more food products, and 29 are using food irradiation commercially. For the control of salmonellae in animal feed and pet foods in the United States, 2–25 kGy was approved in 1995; and, in 1997, 4.5 kGy was approved for refrigerated raw, and 7.5 kGy for frozen, raw ground beef.

In the early 1970s, Canada approved for test marketing a maximum dose of 1.5 kGy for fresh cod and haddock fillets. In 1983, the Codex Alimentarius Commission suggested 1.5 or 2.2 kGy for teleost fish and fish products.[19] One of the obstacles to getting food irradiation approved on a wider scale in the United States is the way irradiation is defined. It is considered an additive rather than a process, which it is. This means that irradiated foods must be labeled as such. Another area of concern is the fate of *C. botulinum* spores (see below), and yet another is the concern that nonpathogens may become pathogens or that the virulence of pathogens may be increased after exposure to subradappertization doses. There is no evidence that the latter occurs.[74]

As of the first quarter of 2003, over 7,000 supermarkets and other retail stores in the United States were selling irradiated ground beef, according to the American Council on Science and Health. Food irradiation has been approved by the American Medical Association, the American Dietetic Association, the Institute of Food Technologists, and the United Nations. The U.S. Centers for Disease Control and Prevention has estimated that if only one-half of the ground beef, pork, poultry, and processed luncheon meats in the U.S. were irradiated, there would be over 880,000 fewer cases of foodborne illness (Food Protection Trends, July 2003). An agency of the U.S. government has approved the use of irradiated ground beef for school disricts, and this product is to be made available by the fall of 2004.

When low-acid foods are irradiated at doses that do not effect the destruction of *C. botulinum* spores, legitimate questions about the safety of such foods are raised, especially when they are held under conditions that allow for growth and toxin production. Because these organisms would be destroyed by radappertization, only products subjected to radicidation and radurization are of concern here. In regard to the radurization of fish, Giddings[22] has pointed out that the lean whitefish species are the best candidates for irradiation, whereas high-fat fishes such as herring arc not, because they are more botulogenic. This investigator noted that when botulinal spores are found on edible lean whitefish, they occur at less than 1/g.

EFFECT OF IRRADIATION ON FOOD QUALITY

The undesirable changes that occur in certain irradiated foods may be caused directly by irradiation or indirectly as a result of post irradiation reactions. Water undergoes radiolysis when irradiated in the

Table 15–6   Methods for Reducing Side Effects in Foodstuffs Exposed to Ionizing Radiations

Method	Reasoning
Reducing temperature	Immobilization of free radicals
Reducing oxygen tension	Reduction of numbers of oxidative free radicals to activated molecules
Addition of free-radical scavengers	Competition for free radicals by scavengers
Concurrent radiation distillation	Removal of volatile off-flavor, off-odor precursors
Reduction of dose	Obvious

Source: Goldblith.[23]

following manner:

$$3H_2O \xrightarrow{\text{radiolysis}} H + OH + H_2O_2 + H_2$$

In addition, free radicals are formed along the path of the primary electron and react with each other as diffusion occurs.[13] Some of the products formed along the track escape and can then react with solute molecules. By irradiating under anaerobic conditions, off-flavors and off-odors are somewhat minimized due to the lack of oxygen to form peroxides. One of the best ways to minimize off-flavors is to irradiate at subfreezing temperatures.[88] The effect of subfreezing temperatures is to reduce or halt radiolysis and its consequent reactants. Other ways to reduce side effects in foodstuffs are presented in Table 15–6.

Other than water, proteins and other nitrogenous compounds appear to be the most sensitive to irradiation effects in foods. The products of irradiation of amino acids, peptides, and proteins depend on the radiation dose, temperature, amount of oxygen, amount of moisture present, and other factors. The following are among the products reported: NH_3, hydrogen, CO_2, H_2S, amides, and carbonyls. With respect to amino acids, the aromatics tend to be more sensitive than the others and undergo changes in ring structure. Among the most sensitive to irradiation are methionine, cysteine, histidine, arginine, and tyrosine. The amino acid most susceptible to electron-beam irradiation is cystine. Johnson and Moser[34] reported that about 50% of this amino acid was lost when ground beef was irradiated. Tryptophan suffered a 10% loss, whereas little or no destruction of the other amino acids occurred. Amino acids have been reported to be more stable to gamma irradiation than to electron-beam irradiation.

Several investigators have reported that the irradiation of lipids and fats results in the production of carbonyls and other oxidation products such as peroxides, especially if irradiation and/or subsequent storage take(s) place in the presence of oxygen. The most noticeable organoleptic effect of lipid irradiation in air is the development of rancidity.

It has been observed that high levels of irradiation lead to the production of "irradiation odors" in certain foods, especially meats. Wick et al.[94] investigated the volatile components of raw ground beef irradiated with 20–60 kGy at room temperature and reported finding a large number of odorous compounds. Of the 45 or more constituents identified by these investigators, there were 17 containing sulfur-, 14 hydrocarbons, and 9 carbonyls, and 5 or more were basic and alcoholic in nature. The higher the level of irradiation, the greater is the quantity of volatile constituents produced. Many of these constituents have been identified in various extracts of nonirradiated, cooked ground beef.

With regard to B vitamins, Liuzzo et al.[46] found that levels of ^{60}Co irradiation between 2 and 6 kGy effected partial destruction of the following B vitamins in oysters: thiamine, niacin, pyridoxine, biotin, and B_{12}. Riboflavin, pantothenic acid, and folic acid reportedly increased by irradiation, probably owing to the release of bound vitamins. Overall, the reported effects on water-soluble vitamins are not striking.[84]

In addition to flavor and odor changes produced in certain foods by irradiation, certain detrimental effects have been reported for irradiated fruits and vegetables. One of the most serious is the softening of these products caused by the irradiation–degradation of pectin and cellulose, the structural polysaccharides of plants. This effect was shown by Massey and Bourke[49] to be caused by radappertization doses of irradiation. Ethylene synthesis in apples is affected by irradiation so that this fruit fails to mature as rapidly as nonirradiated controls.[49] In green lemons, however, ethylene synthesis is stimulated upon irradiation, resulting in a faster ripening than in controls.[50]

Among radiolytic products that develop upon irradiation are some that are antibacterial when exposed in culture media. When 15 kGy were applied to meats, however, no antimicrobial activity was found in the meats.[12] The overall wholesomeness and toxicology of irradiated foods have been reviewed.[76,85]

STORAGE STABILITY OF IRRADIATED FOODS

Foods subjected to radappertization doses of ionizing radiation may be expected to be as shelf stable as commercially heat-sterilized foods. There are, however, two differences between foods processed by these two methods that affect storage stability: Radappertization does not destroy inherent enzymes, which may continue to act, and some postirradiation changes may be expected to occur. Employing 45 kGy and enzyme-inactivated chicken, bacon, and fresh and barbecued pork, Heiligman[31] found the products to be acceptable after storage for up to 24 months. Those stored at 70°F were more acceptable than those stored at 100°F. Licciardello et al. reported the effect of irradiation on beefsteak, ground beef, and pork sausage held at refrigerator temperatures for 12 years.[45] These foods were packed with flavor preservatives and treated with 10.8 kGy. The investigators described the appearance of the meats as excellent after 12 years of storage. A slight irradiation odor was perceptible, but was not considered objectionable. The meats were reported to have a sharp, bitter taste, which was presumed to be caused by the crystallization of the amino acid tyrosine. The free amino nitrogen content of the beefsteak was 75 and 175 mg%, respectively, before and after irradiation storage; and 67 and 160 mg% before and after storage, respectively, for hamburger.

Foods subjected to radurization ultimately undergo spoilage from the surviving biota if stored at temperatures suitable for growth of the organisms in question. The normal spoilage biota of seafoods is so sensitive to ionizing radiations that 99% of the total biota of these products is generally destroyed by doses on the order of 2.5 kGy. Ultimate spoilage of radurized products is the property of the few microorganisms that survive the radiation treatment. For further information on all aspects of food irradiation, see references 66 and 89.

NATURE OF RADIATION RESISTANCE OF MICROORGANISMS

The most sensitive bacteria to ionizing radiation are Gram-negative rods such as the pseudomonads; the coccobacillary-shaped Gram-negative cells of moraxellae and acinetobacters are among the

Table 15–7 Effects of Oxidizing and Reducing
Conditions on Resistance to Radiation of *Deinococcus
radiodurans* (Table of Means)

Condition	Log of Surviving Fraction*
Buffer, unmodified	−3.11542
Oxygen flushed	−3.89762
Nitrogen flushed	−2.29335
H_2O_2 (100 ppm)	−3.47710
Thioglycolate (0.01 M)	−1.98455
Cysteine (0.1 M)	−0.81880
Ascorbate (0.1 M)	−5.36050

Note: Determined by count reduction after exposure to 10 kGy
of gamma radiation in 0.05 M phosphate buffer. LSD: $P = 0.05$
(1.98116); $P = 0.01$ (2.61533).
*Averages of four replicates.

Source: Giddings.[22]

most resistant of Gram negatives. Gram-positive cocci are the most resistant of nonsporing bacteria, including micrococci, staphylococci, and enterococci. What makes one organism more sensitive or resistant than another is not only a matter of fundamental biological interest but also is of interest in the application of irradiation to the preservation/protection of foods. A better understanding of resistance mechanisms can lead to ways of increasing radiation sensitivity and, consequently, to the use of lower doses for food use.

The effect of oxidizing and reducing conditions on the resistance of *Deinococcus radiodurans* in phosphate buffer has been studied; and the findings are presented in Table 15–7. The flushing of buffer suspensions with nitrogen or O_2 had no significant effect on radiation sensitivity when compared to the control, nor did the presence of 100 ppm H_2O_2. Treatment with cysteine rendered the cells less sensitive, and ascorbate increased their sensitivity. A study of *N*-ethylmaleimide (NEM) and indoleacetic acid (IAA) on resistance showed that IAA reduced resistance but NEM did not when tested at nontoxic levels.[41] The presence or absence of O_2 had no effect on these two compounds.

Biology of Extremely Resistant Species

The most resistant of all known non–sporeforming bacteria consist of four species of the genus *Deinococcus* and one each of *Deinobacter*, *Rubrobacter*, and *Acinetobacter*. Some characteristics of these species are presented in Table 15–8. The deinococci were originally assigned to the genus *Micrococcus*, but they, along with *Deinobacter* and the archaebacterial genus, *Thermus*, constitute one of the ten major phyla based on 16S ribosomal RNA (rRNA).[90,93,96] The deinococci occur in pairs or tetrads, contain red water-insoluble pigments, have optimum growth at 30°C, contain L-ornithine as the basic amino acid in their murein (unlike the micrococci, which contain lysine), and are characterized by mol% G + C content between 62 and 70. They do not contain teichoic acids. One of the most

Table 15–8 Some of the Extremely Radiation-Resistant Non-Spore-Forming Bacteria

Organisms	Gram Rx	Morph.	Pigment	Outer Membr
Deinococcus radiodurans	+	C	Red	+
D. radiophilus	+	C	Red	+
D. proteolyticus	+	C	Red	+
D. radiopugnans	+	C	Red	+
D. murrayi	+	C	Orange	+
Deinobacter grandis	−	R	Red/pink	+
D. geothermalis	+	C	Orange	+
Hymenobacter actinosclerus	−	R	Red	+
Kineococcus radiotolerans	+	C	Orange	−
Kocuria erythromyxa	+	C	Red	+
Methylobacterium radiotolerans	−	R	Orange	+
Rubrobacter xylanophilus	+	R	Pink	+

Morph.: C = coccus; R = rod; + = positive; − = negative.

unusual features of this genus is the possession of an outer membrane, unlike other Gram-positive bacteria. They have been characterized as being Gram-negative clones of ancient lineage.[11]

Among other unusual features of deinococci is their possession of palmitoleate (16:1), which makes up about 60% of the fatty acids in their envelope and about 25% of the total cellular fatty acid. The high content of fatty acids is another feature characteristic of Gram-negative bacteria. The predominant isoprenoid quinone in their plasma membrane is a menaquinone. The menaquinones represent one of the two groups of naphthoquinones that are involved in electron transport, oxidative phosphorylation, and perhaps active transport.[9] The length of the C-3 isoprenyl side chains ranges from 1 to 14 isoprene units (MK), and the deinococci are characterized by the possession of MK-8, as are some micrococci, planococci, staphylococci, and enterococci.[9] The deinococci do not contain phosphatidylgylcerol or diphosphatidylglycerol in their phospholipids but contain, instead, phosphoglycolipids as the major component.

The genus *Deinobacter* shares many of the deinococcal features except that its members are Gram-negative rods. *Rubrobacter radiotolerans* is a Gram-positive rod that is highly similar to the deinococci, but the basic amino acid in its murein is L-lysine rather than L-ornithine. *Acinetobacter radioresistens* is a Gram-negative coccobacillary rod that differs in several ways from deinococci. Its mol% G + C content of DNA is in the range 44.1–44.8, and its predominant isoprenoid quinone is Q-9, not MK-8.

Deinococci have been isolated from ground beef, pork sausage, haddock, the hides of animals, and from creek water.[40] They have been reported to occur in fcccs, in sawdust, and in air. *Deinobacter* was isolated from animal feces and freshwater fish, *Rubrobacter* from a radioactive hot spring in Japan, and *A. radioresistens* from cotton and soils.

The species noted in Table 15–8 are aerobic, catalase positive, and generally inactive on substrates for biochemical tests. The deinococci possess a variety of carotenoids, and their isolated plasma membrane is bright red.

Radiation *D* values of the nondeinococcal species are 1.0–2.2 kGy, whereas many strains of the deinococci can survive 15 kGy. *D. radiophilus* is the most radioresistant species.

Apparent Mechanisms of Resistance

Why these organisms are so resistant to radiation is unclear. The extreme resistance of deinococci to desiccation has been observed and presumed to be related in some way to radioresistance. The complicated cell envelope of these organisms may be a factor, but precise data are wanting. All are highly pigmented and contain various carotenoids, a fact that suggests some relationship to radiation resistance. However, these pigments have been found to play no role in the resistance of *D. radiophilus*.[38,44] Some of the chemical events that occur in organic matter after irradiation are outlined in Figure 15–4. The radiolysis of water leads to the formation of free radicals and peroxides, and radiation-sensitive organisms appear to be unable to overcome their deleterious effects. Chemicals

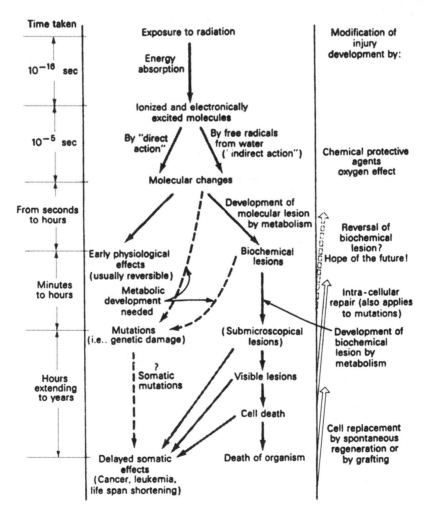

Figure 15–4   Summary of radiation and postirradiation effects in organic matter. *Source*: From Bacq and Alexander[5] reprinted with permission of the authors, *Fundamentals of Radiobiology*, copyright © 1961, Pergamon Press.

that contain –SH groups tend to be radioprotective,[14] but what role these play, if any, in the extreme resistance of bacteria is still unclear.

One of the unusual things about *D. radiodurans* is that each cell in the stationary phase carries about four genomes (i.e., about four copies of its chromosome). Actively dividing cells may contain four to ten copies. Although this abundance of DNA may not be necessary for extreme radiation resistance, it is conceivable that upon exposure to radiation the extra DNA makes it possible for the damaged cell to synthesize a new genome. It has been found that upon exposure to radiation, these organisms undergo an immediate and extensive breakdown of chromosomal DNA, and this appears to be a part of the DNA repair process. For more on irradiation resistance in *D. radiodurans*, see reference 6.

Among the other unusual features of *D. radiodurans* is its possession of an outer membrane, which is characteristic of Gram-negative bacteria. It may be noted from Table 15–8 that a number of other Gram-positive radioresistant bacteria possess an outer membrane. Although it possesses an outer membrane, it does not possess lipopolysaccharide or lipid A, as do the typical Gram-negative bacteria. Its outer envelope consists of five layers outside the plasma membrane.[51] The genome of *D. radiodurans* is unusual in that it has a tight and ordered packaging of its DNA that may facilitate repair.43

REFERENCES

1. Andrews, L.S., D.L. Marshall, and R.M. Grodner. 1995. Radiosensitivity of *Listeria monocytogenes* at various temperatures and cell concentrations. *J. Food Protect.* 58:748–751.

2. Anellis, A., D. Berkowitz, and D. Kemper. 1973. Comparative resistance of nonsporogenic bacteria to low-temperature gamma irradiation. *Appl. Microbiol.* 25:517–523.

3. Anellis, A., D. Berkowitz, W. Swantak, and C. Strojan. 1972. Radiation sterilization of prototype military foods: Low-temperature irradiation of codfish cake, corned beef, and pork sausage. *Appl. Microbiol.* 24:453–462.

4. Anellis, A., E. Shattuck, D.B. Rowley, E.W. Ross, Jr., D.N. Whaley, and V.R. Dowell, Jr. 1975. Low-temperature irradiation of beef and methods for evaluation of a radappertization process. *Appl. Microbiol.* 30:811–820.

5. Bacq, Z.M., and P. Alexander. 1961. *Fundamentals of Radiobiology*, 2nd ed. Oxford: Pergamon.

6. Battista, J.R. 1997. Against all odds: The survival strategies of *Deinococcus radiodurans*. *Ann. Rev. Microbiol.* 51:203–224.

7. Clavero, M.R.S., J.D. Monk, L.R. Beuchat, M.P. Doyle, and R.E. Brackett. 1994. Inactivation of *Escherichia coli* 0157:H7, *salmonellae*, and *Campylobacter jejuni* in raw ground beef by gamma irradiation. *Appl. Environ. Microbiol.* 60:2069–2075.

8. Clifford, W.J., and A. Anellis. 1975. Radiation resistance of spores of some *Clostridium perfringens* strains. *Appl. Microbiol.* 29:861–863.

9. Collins, M.D., and D. Jones. 1981. Distribution of isoprenoid quinone structural types in bacteria and their taxonomic implications. *Microbiol. Rev.* 45:316–354.

10. Collins, C.I., E.A. Murano, and I.V. Wesley. 1996. Survival of *Arcobacter butzleri* and *Campylobacter jejuni* after irradiation treatment in vacuum-packaged ground pork. *J. Food Protect.* 59:1164–1166.

11. Counsell, T.J., and R.G.E. Murray. 1986. Polar lipid profiles of the genus *Deinococcus*. *Int. J. Syst. Bacteriol.* 36:202–206.

12. Dickson, J.S., and R.B. Maxcy. 1984. Effect of radiolytic products on bacteria in a food system. *J. Food Sci.* 49:577–580.

13. Doty, D.M. 1965. Chemical changes in irradiated meats. In *Radiation Preservation of Foods*, 121–125. Washington, DC: National Research Council, National Academy of Sciences.

14. Duggan, D.E., A.W. Anderson, and P.R. Elliker. 1963. Inactivation of the radiation-resistant spoilage bacterium *Micrococcus radiodurans*. II. Radiation inactivation rates as influenced by menstruum temperature, preirradiation heat treatment, and certain reducing agents. *Appl. Microbiol.* 11:413–417.

15. Ehioba, R.M., A.A. Kraft, R.A. Molins, H.W. Walker, D.G. Olson, G. Subbaraman, and R.P. Skowronski. 1988. Identification of microbial isolates from vacuum-packaged ground pork irradiated at 1 kGy. *J. Food Sci.* 53:278–279, 281.

16. EI-Zawahry, Y.A., and D.B. Rowley. 1979. Radiation resistance and injury of *Yersinia enterocolitica*. *Appl. Environ. Microbiol.* 37:50–54.

17. Fan, X., and K.J.B. Sokorai. 2002. Sensorial and chemical quality of gamma-irradiated fresh-cut Iceberg lettuce in modified atmosphere packages. *J. Food Protect.* 65:1760–1765.

18. Fiddler, W., R.A. Gates, J.W. Pensabene, J.G. Phillips, and E. Wierbicki. 1981. Investigations on nitrosamines in irradiation-sterilized bacon. *J Agric. Food Chem.* 29:551–554.

19. Food and Agriculture Organization/IAEA/World Health Organization. 1977. *Wholesomeness of Irradiated Food.* Report of joint FAO/IAEA/WHO Expert Committee, WHO Technical Report Series 604.

20. Foong, S.C.C., G.L. Gonzalez, and J.S. Dickson. 2004. Reduction and survival of *Listeria monocytogenes* in ready-to-eat meats after irradiation. *J. Food Protect.* 67:77–82.

21. Fu, A.H., J.G. Sebranek, and E.A. Murano. 1995. Survival of *Listeria monocytogenes*, *Yersinia enterocolitica*, and *Escherichia coli* 0157:H7 and quality changes after irradiation of beef steaks and ground beef. *J. Food Sci.* 60:972–977.

22. Giddings, G.G. 1984. Radiation processing of fishery products. *Food Technol.* 38(4):61–65, 94–97.

23. Goldblith, S.A. 1963. Radiation preservation of foods—Two decades of research and development. In *Radiation Research*, 155–167. Washington, DC: U.S. Department of Commerce, Office of Technical Services.

24. Goldblith, S.A. 1966. Basic principles of microwaves and recent developments. *Adv. Food Res.* 15:277–301.

25. Goresline, H.E., M. Ingram, P. Macuch, G. Mocquot, D.A.A. Mossell, C.F. Niven, and F.S. Thatcher. 1964. Tentative classification of food irradiation processes with microbiological objectives. *Nature* 204:237–238.

26. Grecz, N., O.P. Snyder, A.A. Walker, and A. Anellis. 1965. Effect of temperature of liquid nitrogen on radiation resistance of spores of *Clostridium botulinum.* *Appl. Microbiol.* 13:527–536.

27. Grecz, N., A.A. Walker, A. Anellis, and D. Berkowitz. 1971. Effects of irradiation temperature in the range −196 to 95°C on the resistance of spores of *Clostridium botulinum* 33A in cooked beef. *Can J. Microbiol.* 17:135–142.

28. Grünewald, T. 1961. Behandlung von Lebensmitteln mit energiereichen Strahlen. *Ernährungs-Umschau* 8:239–244.

29. Hashisaka, A.E., S.D. Weagant, and F.M. Dong. 1989. Survival of *Listeria monocytogenes* in mozzarella cheese and ice cream exposed to gamma irradiation. *J. Food Protect.* 52:490–492.

30. Hashisaka, A.E., J.R. Matches, Y. Batters, F.P. Hungate, and F.M. Dong. 1990. Effects of gamma irradiation at −78°C on microbial populations in dairy products. *J. Food Sci.* 55:1284–1289.

31. Heiligman, F. 1965. Storage stability of irradiated meats. *Food Technol.* 19:114–116.

32. Huhtanen, C.N., R.K. Jenkins, and D.W. Thayer. 1989. Gamma radiation sensitivity of *Listeria monocytogenes*. *J. Food Protect.* 52:610–613.

33. Jakabi, M., D.S. Gelli, J.C.M.D. Torre, M.A.B. Rodas, B.D.G.M. Franco, M.T. Destro, and M. Landgraf. 2003. Inactivation by ionizing radiation of *Salmonella* Enteritidis, *Salmonella* Infantis, and *Vibrio parahaemolyticus* in oysters (*Crassostrea brasiliana*). *J. Food Protect.* 66:1025–1029.

34. Johnson, B., and K. Moser. 1967. Amino acid destruction in beef by high energy electron beam irradiation. In *Radiation Preservation of Foods*, 171–179. Washington, DC: American Chemical Society.

35. Josephson, E.S., A. Brynjolfsson, and E. Wierbicki. 1975. The use of ionizing radiation for preservation of food and feed products. In *Radiation Research—Biomedical, Chemical, and Physical Perspectives*, ed. O.F. Nygaard, H.I. Adler, and W.K. Sinclair, 96–117. New York: Academic Press.

36. Kampelmacher, E.H. 1983. Irradiation for control of *Salmonella* and other pathogens in poultry and fresh meats. *Food Technol.* 37(4):117–119, 169.

37. Kempe, L.L. 1965. The potential problems of type E botulism in radiation-preserved seafoods. In *Radiation Preservation of Foods*, 211–215. Washington, DC: National Research Council, National Academy of Science.

38. Kilburn, R.E., W.D. Bellamy, and S.A. Terni. 1958. Studies on a radiation-resistant pigmented *Sarcina* sp. *Radiat. Res.* 9:207–215.

39. Koch, H.W., and E.H. Eisenhower. 1965. Electron accelerators for food processing. In *Radiation Preservation of Foods*, 149–180. Washington, DC: National Research Council, National Academy of Science.

40. Krabbenhoft, K.L., A.W. Anderson, and P.R. Elliker. 1965. Ecology of *Micrococcus radiodurans.* *Appl. Microbiol.* 13:1030–1037.

41. Lee, J.S., A.W. Anderson, and P.R. Elliker. 1963. The radiation-sensitizing effects of N-ethylmaleimide and iodoacetic acid on a radiation-resistant *Micrococcus.* *Radiat. Res.* 19:593–598.

42. Lefebvre, N., C. Thibault, and R. Charbonneau. 1992. Improvement of shelf-life and wholesomeness of ground beef by irradiation. 1. Microbial aspects. *Meat Sci.* 32:203–213.

43. Levin-Zaidman, S., J. Englander, E. Shimoni, A.K. Sharma, K.W. Minton, and A. Minsky. 2003. Ringlike sstructure of the *Deinococcus radiodurans* genome: A key to radioresistance? *Science* 299:254–256.

44. Lewis, N.F., D.A. Madhavesh, and U.S. Kumta. 1974. Role of carotenoid pigments in radio-resistant micrococci. *Can. J. Microbiol.* 20:455–459.

45. Licciardello, J.J., J.T.R. Nickerson, and S.A. Goldblith. 1966. Observations on radio-pasteurized meats after 12 years of storage at refrigerator temperatures above freezing. *Food Technol.* 20:1232.

46. Liuzzo, J.S., W.B. Barone, and A.F. Novak. 1966. Stability of B-vitamins in Gulf oysters preserved by gamma radiation. *Fed. Proc.* 25:722.

47. Loaharanu, P. 1989. International trade in irradiated foods: Regional status and outlook. *Food Technol.* 43(7):77–80.

48. Losty, T., J.S. Roth, and G. Shults. 1973. Effect of irradiation and heating on proteolytic activity of meat samples. *J. Agric. Food Chem.* 21:275–277.

49. Massey, L.M., Jr., and J.B. Bourke. 1967. Some radiation-induced changes in fresh fruits and vegetables. In *Radiation Preservation of Foods*, 1–11. Washington, DC: American Chemical Society.

50. Maxie, E., and N. Sommer. 1965. Irradiation of fruits and vegetables. In *Radiation Preservation of Foods*, 39–52. Washington, DC: National Research Council, National Academy of Science.

51. Makarova, K.S., L. Aravind, Y.I. Wolf, R.L. Tatusov, K.W. Minton, E.V. Koonin, and M.J. Daly. 2001. Genome of the extremely radiation-resistant bacterium *Deinococcus radiodurans* viewed from the perspective of comparative genomics. *Microbiol. Mol. Biol. Rev.* 65:44–79

52. Midura, T.F., L.L. Kempe, J.T. Graikoski, and N.A. Milone. 1965. Resistance of *Clostridium perfringens* type A spores to gamma-radiation. *Appl. Microbiol.* 13:244–247.

53. Mulder, R.W. 1984. Ionizing energy treatment of poultry. *Food Technol. Aust.* 36:418–420.

54. Mulder, R.W., S. Notermans, and E.H. Kampelmacher. 1977. Inactivation of salmonellae on chilled and deep frozen broiler carcasses by irradiation. *J. Appl. Bacteriol.* 42:179–185.

55. Murano, E.A. 1995. Irradiation of fresh meats. *Food Technol.* 49(12):52–54.

56. Murano, P.S., E.A. Murano, and D.G. Olson. 1998. Irradiated ground beef: Sensory and quality changes during storage under various packaging conditions. *J. Food Sci.* 63:548–551.

57. Niemand, J.G., H.J. van derLinde, and W.H. Holzapfel. 1983. Shelf-life extension of minced beef through combined treatments involving radurization. *J. Food Protect.* 46:791–796.

58. Niven, C.F., Jr. 1958. Microbiological aspects of radiation preservation of food. *Annu. Rev. Microbiol.* 12:507–524.

59. Novak, A.F., R.M. Grodner, and M.R.R. Rao. 1967. Radiation pasteurization of fish and shellfish. In *Radiation Preservation of Foods*, 142–151. Washington, DC: American Chemical Society.

60. Palumbo, S.A., R.K. Jenkins, R.L. Buchanan, and D.W. Thayer. 1986. Determination of irradiation D-values for *Aeromonas hydrophila*. *J. Food Protect.* 49:189–191.

61. Patterson, M. 1989. Sensitivity of *Listeria monocytogenes* to irradiation on poultry meat and in phosphate buffered saline. *Lett. Appl. Microbiol.* 8:181–184.

62. Patterson, M.F. 1988. Sensitivity of bacteria to irradiation on poultry meat under various atmospheres. *Lett. Appl. Microbiol.* 7:55–58.

63. Poole, S.E., P. Wilson, G.E. Mitchell, and P.A. Wills. 1990. Storage life of chilled scallops treated with low dose irradiation. *J. Food Protect.* 53:763–766.

64. Prakash, A., A.R. Guner, E. Caporado, and D.M. Foley. 2000. Effects of low-dose gamma irradiation on the shelf life and quality characteristics of cut Romaine lettuce packaged under modified atmosphere. *J. Food Sci.* 65:549–553.

65. Pszczola, D. 1993. Irradiated poultry makes U.S. debut in midwest and Florida markets. *Food Technol.* 47(11):89–96.

66. Radomyski, T., E.A. Murano, D.G. Olson, and P.S. Murano. 1994. Elimination of pathogens of significance in food by low-dose irradiation: A review. *J. Food Protect.* 57:73–86.

67. Rajkowski, K.T., and D.W. Thayer. 2001. Alfalfa seed germination and yield ratio and alfalfa sprout microbial keeping quality following irradiation of seeds and sprouts. *J. Food Protect.* 64:1988–1995.

68. Rajkowski, K.T., and D.W. Thayer. 2000. Raduction of *Salmonella* spp. and strains of *Escherichia coli* 0157:H7 by gamma radiation of inoculated sprouts. *J. Food Protect.* 63:871–875.

69. Rajkowski, K.T., G. Boyd, and D.W. Thayer. 2003. Irradiation *D*-values for *Escherichia coli* 0157:H7 and *Salmonella* sp. on inoculated broccoli seeds and effects of irradiation on broccoli sprout keeping quality and seed viability. *J. Food Protect.* 66:760–766.

70. Restaino, L., J.J.J. Myron, L.M. Lenovich, S. Bills, and K. Tschernoff. 1984. Antimicrobial effects of ionizing radiation on artificially and naturally contaminated cacao beans. *Appl. Environ. Microbiol.* 47:886–887.

71. Roberts, T.A., and M. Ingram. 1965. The resistance of spores of *Clostridium botulinum* Type E to heat and radiation. *J. Appl. Bacteriol.* 28:125–141.

72. Roberts, T.A., and M. Ingram. 1965. Radiation resistance of spores of *Clostridium* species in aqueous suspension. *J. Food Sci.* 30:879–885.

73. Rose, S.A., N.K. Modi, H.S. Tranter, N.E. Bailey, M.F. Stringer, and P. Hambleton. 1988. Studies on the irradiation of toxins of *Clostridium botulinum* and *Staphylococcus aureus*. *J. Appl. Bacteriol.* 65:223–229.

74. Rowley, D.B., and A. Brynjolfsson. 1980. Potential uses of irradiation in the processing of food. *Food Technol.* 34(10):75–77.

75. Saleh, Y.G., M.S. Mayo, and D.G. Ahearn. 1988. Resistance of some common fungi to gamma irradiation. *Appl. Environ. Microbiol.* 54:2134–2135.

76. Skala, J.H., E.L. McGown, and P.P. Waring. 1987. Wholesomeness of irradiated foods. *J. Food Protect.* 50:150–160.

77. Sullivan, R., A.C. Fassolitis, E.P. Larkin, R.B. Read, Jr., and J.T. Peeler. 1971. Inactivation of thirty viruses by gamma radiation. *Appl. Microbiol.* 22:61–65.

78. Sullivan, R., P.V. Scarpino, A.C. Fassolitis, E.P. Larkin, and J.T. Peeler 1973. Gamma radiation inactivation of coxsackievirus B-2. *Appl. Microbiol.* 26:14–17.

79. Thayer, D.W., G. Boyd, and C.N. Huhtanen. 1995. Effects of ionizing radiation and anaerobic refrigerated storage on indigenous microfiora, *Salmonella*, and *Clostridium botulinum* types A and B in vacuum canned, mechanically deboned chicken meat. *J. Food Protect.* 58:752–757.

80. Thayer, D.W., and G. Boyd. 2001. Effect of irradiation temperature on inactivation of *Escherichia coli* 0157:H7 and *Staphylococcus aureus*. *J. Food Protect.* 64:1624–1626.

81. Thayer, D.W., G. Boyd, and W.F. Fett. 2003. Gamma-radiation decontamination of alfalfa seeds naturally contaminated with *Salmonella* Mbandaka. *J. Food Sci.* 68:1777–1781.

82. Thayer, D.W., K.T. Rajkowski, G. Boyd, P.H. Cooke, and D.S. Soroka. 2003b. Inactivation of *Escherichia coli* 0157:H7 and *Salmonella* by gamma irradiation of alfalfa seed intended for production of food sprouts. *J. Food Protect.* 66:175–181.

83. Thayer, D.W., and G. Boyd. 1995. Radiation sensitivity of *Listeria monocytogenes* on beef as affected by temperature. *J. Food Sci.* 60:237–240.

84. Thayer, D.W., J.B. Fox, Jr., and L. Lakritz. 1991. Effects of ionizing radiation on vitamins. In *Food Irradiation*, ed. S. Thorne, 285–325. New York: Elsevier Applied Science.

85. Thayer, D.W., J.P. Christopher, L.A. Campbell, D.C. Ronning, R.R. Dahlgren, G.M. Thomson, and E. Wierbicki. 1987. Toxicology studies of irradiation-sterilized chicken. *J. Food Protect.* 50:278–288.

86. Tiwari, N.P., and R.B. Maxcy. 1972. *Moraxella-Acinetobacter* as contaminants of beef and occurrence in radurized product. *J. Food Sci.* 37:901–903.

87. Tsuji, K. 1983. Low-dose cobalt 60 irradiation for reduction of microbial contamination in raw materials for animal health products. *Food Technol.* 37(2):48–54.

88. Urbain, W.M. 1965. Radiation preservation of fresh meat and poultry. In *Radiation Preservation of Foods*, 87–98. Washington, DC: National Research Council, National Academy of Science.

89. Urbain, W.M. 1978. Food irradiation. *Adv. Food Res.* 24:155–227.

90. Van den Eynde, H., Y. Van de Peer, H. Vandenabeele, M. van Bogaert, and R. de Wachter. 1990. 5S rRNA sequences of myxobacteria and radioresistant bacteria and implications for eubacterial evolution. *Int. J. Syst. Bacteriol.* 40:399–404.

91. Van Gerwen, S.J.C., F.M. Rombouts, K. van't Riet, and M.H. Zwietering. 1999. A data analysis of the irradiation parameter D_{10} for bacteria and spores under various conditions. *J. Food Protect.* 62:1024–1032.

92. Verster, A., T.A. du Plessis, and L.W. van den Heever. 1977. The eradication of tapeworms in pork and beef carcasses by irradiation. *Radiat. Phys. Chem.* 9:769–771.

93. Weisburg, W.G., S.J. Giovannoni, and C.R. Woese. 1989. The *Deinococcus-Thermus* phylum and the effect of rRNA composition on phylogenetic tree construction. *Syst. Appl. Microbiol.* 11:128–134.

94. Wick, E., E. Murray, J. Mizutani, and M. Koshika. 1967. Irradiation flavor and the volatile components of beef. In *Radiation Preservation of Foods*, 12–25. Washington, DC: American Chemical Society.

95. Wierbicki, E., M. Simon, and E.S. Josephson. 1965. Preservation of meats by sterilizing doses of ionizing radiation. In *Radiation Preservation of Foods*, 383–409. Washington, DC: National Research Council, National Academy of Science.

96. Woese, C.R. 1987. Bacterial evolution. *Microbiol. Rev.* 51:221–271.

CHAPTER 16

Protection of Foods with Low-Temperatures, and Characteristics of Psychrotrophic Microorganisms

The use of low temperatures to preserve foods is based on the fact that the activities of microorganisms can be slowed at temperatures above freezing and generally stopped at subfreezing temperatures. The reason is that all metabolic reactions of microorganisms are enzyme catalyzed and that the rate of enzyme-catalyzed reactions is dependent on temperature. With a rise in temperature, there is an increase in reaction rate. The temperature coefficient (Q_{10}) may be generally defined as follows:

$$Q_{10} = \frac{(\text{Velocity at a given temp.} +10°\text{C})}{\text{Velocity at T}}$$

The Q_{10} for most biological systems is 1.5–2.5, so that for each 10°C rise in temperature within the suitable range, there is a twofold increase in the rate of reaction. For every 10°C decrease in temperature, the reverse is true. Because the basic feature of low-temperature food preservation consists of its effect on pathogens and spoilage organisms, most of the discussion that follows will be devoted to the effect of low temperatures on these microorganisms. It should be remembered, however, that temperature is related to relative humidity (RH) and that subfreezing temperatures affect RH as well as pH, and possibly other parameters of microbial growth as well.

DEFINITIONS

The term psychrophile was coined by Schmidt-Nielsen in 1902 for microorganisms that grow at 0°C.[31] This term is now applied to organisms that grow over the range of subzero to 20°C, with an optimum range of 10–15°C.[47] Around 1960, the term psychrotroph (*psychros*, cold, and *trephein*, to nourish or to develop) was suggested for organisms able to grow at 5°C or below.[13,50] It is now widely accepted among food microbiologists that a psychrotroph is an organism that can grow at temperatures between 0°C and 7°C and produce visible colonies (or turbidity) within 7–10 days in this temperature range. Because some psychrotrophs can grow at temperatures at least as high as 43°C, they are, in fact, mesophiles. By these definitions, psychrophiles would be expected to occur only on products from oceanic waters or from extremely cold climes. The organisms that cause the spoilage of meats, poultry, and vegetables in the 0–5°C range are psychrotrophs.

Because all psychrotrophs do not grow at the same rate over the 0–7°C range, the terms *eurypsychrotroph* (*eurys*, wide or broad) and *stenopsychrotroph* (*stenos*, narrow, little, or close) have been suggested. Eurypsychrotrophs typically do not form visible colonies until sometime between 6 and 10 days; stenopsychrotrophs typically form visible colonies in about 5 days.[34] It has been suggested that psychrotrophs can be distinguished from nonpsychrotrophs by their inability to grow on a nonselective medium at 43°C in 24 hours, whereas the latter do grow.[50] It has been shown that some bacteria that grow well at 7°C within 10 days also grow well at 43°C, and among these are *Enterobacter cloacae*, *Hafnia alvei*, and *Yersinia enterocolitica* (ATCC 27739).[34] These are eurypsychrotrophs, although there are others that grow well at 43°C but only poorly at 7°C in 10 days. Typical of stenopsychrotrophs are *Pseudomonas fragi* (ATCC 4973) and *Aeromonas hydrophila* (ATCC 7965), which grow well at 7°C in 3–5 days but do not grow at 40°C.[34]

There are three distinct temperature ranges for low-temperature stored foods. Chilling temperatures are those between the usual refrigerator (5–7°C) and ambient temperatures, usually about 10–15°C. These temperatures are suitable for the storage of certain vegetables and fruits such as cucumbers, potatoes, and limes. Refrigerator temperatures are those between 0°C and 7°C (ideally no higher than 40°F or 4.4°C). Freezer temperatures are those at or below −18°C. Under normal circumstances, growth of all microorganisms is prevented at freezer temperatures; nevertheless, some can and do grow within the freezer range but at an extremely slow rate.

TEMPERATURE GROWTH MINIMA

Bacterial species and strains that can grow at or below 7°C are rather widely distributed among the Gram-negative and less so among Gram-positive genera (Tables 16–1 and 16–2). The lowest recorded temperature of growth for a microorganism of concern in foods is −34°C, in this case a pink yeast. Growth at temperatures below 0°C is more likely to be that of yeasts and molds, than bacteria. This is consistent with the growth of fungi under lower water activity (a_w) conditions. Bacteria have been reported to grow at −20°C and around −12°C.[45] Foods that are likely to support microbial growth at subzero temperatures include fruit juice concentrates, bacon, ice cream, and certain fruits. These products contain cryoprotectants that depress the freezing point of water.

PREPARATION OF FOODS FOR FREEZING

The preparation of vegetables for freezing includes selecting, sorting, washing, blanching, and packaging prior to actual freezing. Foods in any state of detectable spoilage should be rejected for freezing. Meats, poultry, seafoods, eggs, and other foods should be as fresh as possible.

Blanching is achieved either by a brief immersion of foods into hot water or by the use of steam. Its primary functions are as follows:

1. inactivation of enzymes that might cause undesirable changes during freezing storage;
2. enhancement or fixing of the green color of certain vegetables;
3. reduction in the numbers of microorganisms on the foods;

Table 16–1   Some Bacterial Genera that Contain Species/Strains Known to Grow at or Below 7°C

Gram Negatives	Relative Numbers	Gram Positives	Relative Numbers
Acinetobacter	XX	Bacillus	XX
Aeromonas	XX	Brevibacterium	X
Alcaligenes	X	Brochothrix	XXX
Alteromonas	XX	Carnobacterium	XXX
Cedecea	X	Clostridium	XX
Chromobacterium	X	Corynebacterium	X
Citrobacter	X	Deinococcus	X
Enterobacter	XX	Enterococcus	XXX
Erwinia	XX	Kurthia	X
Escherichia	X	Lactobacillus	XX
Flavobacterium	XX	Lactococcus	XX
Halobacterium	X	Leuconostoc	X
Hafnia	XX	Listeria	XX
Klebsiella	X	Micrococcus	XX
Moraxella	XX	Pediococcus	X
Morganella	X	Propionibacterium	X
Photobacterium	X	Vagococcus	XX
Pantoea	XX	Macrococcus	X
Proteus	X	Paenibacillus	X
Providencia	X	Staphylococcus	X
Pseudomonas	XXX		
Psychrobacter	XX		
Salmonella	X		
Serratia	XX		
Shewanella	XXX		
Vibrio	XXX		
Burkholderia	X		
Chryseobacterium	X		
Frigoribacterium	XX		
Janthinobacterium	XX		
Acetobacterium	XX		
Yersinia	XX		

Note: Relative importance and dominance as psychrotrophs: X = minor, XX = intermediate, XXX = very significant.

---

4. facilitating the packing of leafy vegetables by inducing wilting;
5. displacement of entrapped air in the plant tissues.

The method of blanching employed depends on the products in question, the size of the packs, and other related information. When water is used, it is important that bacterial spores not be allowed to build up sufficiently to contaminate foods. Reductions of initial microbial loads as high as 99% have been claimed upon blanching. Remember that most vegetative bacterial cells can be destroyed at milk pasteurization temperatures (145°F or 62.3°C for 30 minutes). This is especially true of most bacteria of importance in the spoilage of vegetables. Although it is not the primary function of blanching to

Table 16–2 Minimum Reported Growth Temperatures of Some Foodborne Microbial Species and Strains that Grow at or Below 7°C

Species/Strains	Temperature (°C)	Comments
Pink yeast	−34	
Pink yeasts (2)	−18	
Unspecified molds	−12	
Vibrio spp.	−5	True psychrophiles
Cladosporium cladosporiodes	−5	
Yersinia enterocolitica	−2	
Bacillus psychrotolerans	−2 to 1	
Acetobacterium bakii	1.0	
Carnobacterium viridans	2.0	
Clostridium algidixylanolyticum	2.5	
Janthinobacterium agaricidamnosum 2.0	2.0	
Frigoribacterium faeni	2.0	
Lactobacillus algidus	0	
Bacillus psychrodurans	−2 to 0	
Unspecified coliforms	−2	
Brochothrix thermosphacta	−0.8	Within 7 days; 4°C in 10 days
Aeromonas hydrophila	−0.5	
Enterococcus spp.	0	Various species/strains
Leuconostoc carnosum	1.0	
L. gelidum	1.0	
Listeria monocytogenes	1.0	
Thamnidium elegans	∼ 1	
Leuconostoc sp.	2.0	Within 12 days
L. sakei/curvatus	2.0	Within 12 days; 4°C in 10 days
Lactobacillus alimentarius	2.0	
C. botulinum B, E, F	3.3	
Pantoea agglomerans	4.0	
Salmonella Panama	4.0	In 4 weeks
Bacillus weihenstephanensis	4.0	
Serratia liquefaciens	4.0	
Vibrio parahaemolyticus	5.0	
Vagococcus salmonirarum	5.0	
Salmonella Heidelberg	5.3	
Pediococcus sp.	6.0	Weak growth in 8 days
Lactobacillus brevis	6.0	In 8 days
Weissella viridescens	6.0	In 8 days
Salmonella Typhimurium	6.2	
Staphylococcus aureus	6.7	
Klebsiella pneumoniae	7.0	
Bacillus spp.	7.0	165 of 520 species/strains
Salmonella spp.	7.0	65 of 109, within 4 weeks

Source: Data from Bonde,[9] Mossel et al.,[49] Reuter,[57] and the literature.

Exhibit 16–1 Comparison of Freezing Methods

Quick Freezing	Slow Freezing
• Small ice crystals formed	• Large ice crystals formed
• Blocks or suppresses metabolism	• Breakdown of metabolic rapport
• Brief exposure to concentration of adverse constituents	• Longer exposure to adverse or injurious factors
• No adaptation to low temperatures	• Gradual adaptation
• Thermal shock (too brutal a transition)	• No shock effect
• No protective effect	• Accumulation of concentrated solutes with beneficial effects
• Microorganisms frozen into crystals?	
• Avoid internal metabolic imbalance	

destroy microorganisms, the amount of heat necessary to effect destruction of most food enzymes is also sufficient to reduce vegetative cells significantly.

FREEZING OF FOODS AND FREEZING EFFECTS

The two basic ways to achieve the freezing of foods are quick and slow freezing. *Quick* or *fast freezing* is the process by which the temperature of foods is lowered to about $-20°C$ within 30 minutes. This treatment may be achieved by direct immersion or indirect contact of foods with the refrigerant and the use of air blasts of frigid air blown across the foods being frozen. *Slow freezing* refers to the process whereby the desired temperature is achieved within 3–72 hours. This is essentially the type of freezing utilized in the home freezer. Quick freezing possesses more advantages than slow freezing, from the standpoint of overall product quality. The two methods are compared in Exhibit 16–1.

With respect to crystal formation upon freezing, slow freezing favors large extracellular crystals, and quick freezing favors the formation of small intracellular ice crystals. Crystal growth is one of the factors that limit the freezer life of certain foods, because ice crystals grow in size and cause cell damage by disrupting membranes, cell walls, and internal structures to the point where the thawed product is quite unlike the original in texture and flavor. Upon thawing, foods frozen by the slow freezing method tend to lose more drip (drip for meats; leakage in the case of vegetables) than quick-frozen foods held for comparable periods of time. The overall advantages of small crystal formation to frozen food quality may also be viewed from the standpoint of what takes place when a food is frozen. During the freezing of foods, water is removed from the solution and transformed into ice crystals of a variable but high degree of purity.[17] In addition, the freezing of foods is accompanied by changes in properties such as pH, titratable acidity, ionic strength, viscosity, osmotic pressure, vapor pressure, freezing point, surface and interfacial tension, and oxidation–reduction (O/R) potential (see reference below).

STORAGE STABILITY OF FROZEN FOODS

A large number of microorganisms have been reported by many investigators to grow at and below $0°C$. In addition to factors inherent within these organisms, their growth at and below freezing temperatures is dependent on nutrient content, pH, and the availability of liquid water. The a_w of foods may be expected to decrease as temperatures fall below the freezing point. The relationship between

Table 16–3   Vapor Pressures of Water and Ice at Various Temperatures

Temperature (°C)	Liquid Water (mm Hg)	Ice (mm Hg)	$a_w = \frac{P_{ice}}{P_{water}}$
0	4.579	4.579	1.00
−5	3.163	3.013	0.953
−10	2.149	1.950	0.907
−15	1.436	1.241	0.864
−20	0.943	0.776	0.823
−25	0.607	0.476	0.784
−30	0.383	0.286	0.75
−40	0.142	0.097	0.68
−50	0.048	0.030	0.62

Source: Scott.[60]

temperature and the a_w of water and ice is presented in Table 16–3. For water at 0°C, a_w is 1.0 but falls to about 0.8 at −20°C and to 0.62 at about −50°C. Organisms that grow at subfreezing temperatures, then, must be able to grow at the reduced a_w levels, unless a_w is favorably affected by food constituents with respect to microbial growth. In fruit juice concentrates, which contain comparatively high levels of sugars, these compounds tend to maintain a_w at levels higher than would be expected in pure water, thereby making microbial growth possible even at subfreezing temperatures. The same type of effect can be achieved by the addition of glycerol to culture media. Not all foods freeze at the same initial point (Table 16–4). The initial freezing point of a given food is due in large part to the nature of its solute constituents and the relative concentration of those that have freezing-point depressing properties.

Table 16–4   Approximate Freezing Temperatures of Selected Foods (Modified from Desrosier[11])

Food Products	°F	°C
Peanuts	17	−8.3
Walnuts	20	−6.7
Coconut	24.5	−4.4
Bananas	25	−3.9
Garlic	25.5	−3.6
Lamb, veal	27	−2.8
Potatoes	28	−2.2
Beef, fish	28.5	−2.0
Carrots	29	−1.7
Raspberries	29.5	−1.4
Asparagus	30	−1.1
Peas	30.5	−0.9
Cauliflower	31	−0.6
Lettuce, cabbage	31.5	−0.3
Water	32	0.0

Although the metabolic activities of all microorganisms can be stopped at freezer temperatures, frozen foods may not be kept indefinitely if the thawed product is to retain the original flavor and texture. Most frozen foods are assigned a freezer life. The suggested maximum holding time for frozen foods is not based on the microbiology of such foods but on such factors as texture, flavor, tenderness, color, and overall nutritional quality upon thawing, and subsequent cooking.

Some foods that are improperly wrapped during freezer storage undergo freezer burn, characterized by a browning of light-colored foods such as the skin of chicken meat.The browning results from the loss of moisture at the surface, leaving the product more porous than the original at the affected site. The condition is irreversible and is known to affect certain fruits, poultry, meats, and fish, both raw and cooked.

EFFECT OF FREEZING ON MICROORGANISMS

In considering the effect of freezing on those microorganisms that are unable to grow at freezing temperatures, it is well known that freezing is one means of preserving microbial cultures, with freeze drying being perhaps the best method known. However, freezing temperatures have been shown to effect the killing of certain microorganisms of importance in foods. Ingram[32] summarized the salient facts of what happens to certain microorganisms upon freezing:

1. There is a sudden mortality immediately on freezing, varying with species.
2. The proportion of cells surviving immediately after freezing die gradually when stored in the frozen state.
3. This decline in numbers is relatively rapid at temperatures just below the freezing point, especially about −2°C, but less so at lower temperatures, and it is usually slow below −20°C.

Bacteria differ in their capacity to survive during freezing, with cocci being generally more resistant than Gram-negative rods. Of the food-poisoning bacteria, salmonellae are less resistant than *Staphylococcus aureus* or vegetative cells of clostridia, whereas endospores and food-poisoning toxins are apparently unaffected by low temperatures.[21] The effect of freezing several species of *Salmonella* to −25.5°C and holding up to 270 days is presented in Table 16–5. Although a significant reduction in viable numbers occurred over the 270-day storage period with most species, in no instance did all the cells die off.

Table 16–5 Survival of Pure Cultures of Enteric Organisms in Chicken Chow Mein at −25.5°C

Organism	*Bacterial Count ($\times 10^5$/g) after Storage for (Days)*								
	0	*2*	*5*	*9*	*14*	*28*	*50*	*92*	*270*
Salmonella Newington	7.5	56.0	27.0	21.7	11.1	11.1	3.2	5.0	2.2
S. Typhimurium	167.0	245.0	134.0	118.0	11.0	95.5	31.0	90.0	34.0
S. Typhi	128.5	45.5	21.8	17.3	10.6	4.5	2.6	2.3	0.86
S. Gallinarium	68.5	87.0	45.0	36.5	29.0	17.9	14.9	8.3	4.8
S. Anatum	100.0	79.0	55.0	52.5	33.5	29.4	22.6	16.2	4.2
S. Paratyphi B	23.0	205.0	118.0	93.0	92.0	42.8	24.3	38.8	19.0

Source: From Gunderson and Rose,[25] copyright © 1948 by Institute of Food Technologists.

From the strict standpoint of food preservation, freezing should not be regarded as a means of destroying foodborne microorganisms. The type of organisms that lose their viability in this state differ from strain to strain and depend on the type of freezing employed, the nature and composition of the food in question, the length of time of freezer storage, and other factors, such as temperature of freezing. Low freezing temperatures of about $-20°C$ are less harmful to microorganisms than the median range of temperatures, such as $-10°C$. For example, more microorganisms are destroyed at $-4°C$ than at $-15°C$ or below. Temperatures below $-24°C$ seem to have no additional effect. Food constituents such as egg white, sucrose, corn syrup, fish, glycerol, and undenatured meat extracts have all been found to increase freezing viability, especially of food-poisoning bacteria, whereas acid conditions have been found to decrease cell viability.[21] Consider some of the events that are known to occur when cells freeze:

1. The water that freezes is the so-called free water. Upon freezing, the free water forms ice crystals. The growth of ice crystals occurs by accretion, so that all of the free water of a cell might be represented by a relatively small number of ice crystals. In slow freezing, ice crystals are extracellular; in fast freezing, they are intracellular. Bound water remains unfrozen. The freezing of cells depletes them of usable liquid water and thus dehydrates them.

2. Freezing results in an increase in the viscosity of cellular matter, a direct consequence of water being concentrated in the form of ice crystals.

3. Freezing results in a loss of cytoplasmic gases such as O_2 and CO_2. A loss of O_2 to aerobic cells suppresses respiratory reactions. Also, the more diffuse state of O_2 may make for greater oxidative activities within the cell.

4. Freezing causes changes in pH of cellular matter. Various investigators have reported changes ranging from 0.3 to 2.0 pH units. Increases and decreases of pH upon freezing and thawing have been reported.

5. Freezing effects concentration of cellular electrolytes. This effect is also a consequence of the concentration of water in the form of ice crystals.

6. Freezing causes a general alteration of the colloidal state of cellular protoplasm. Many of the constituents of cellular protoplasm such as proteins exist in a dynamic colloidal state in living cells. A proper amount of water is necessary to the well-being of this state.

7. Freezing causes some denaturation of cellular proteins. Precisely how this effect is achieved is not clear, but it is known that upon freezing, some –SH groups disappear and such groups as lipoproteins break apart from others. The lowered water content, along with the concentration of electrolytes, no doubt affects this change in state of cellular proteins.

8. Freezing induces temperature shock in some microorganisms. This is true more for thermophiles and mesophiles than for psychrophiles. More cells die when the temperature decline above freezing is sudden than when it is slow.

9. Freezing causes metabolic injury to some microbial cells such as certain *Pseudomonas* spp. Some bacteria have increased nutritional requirements upon thawing from the frozen state and as much as 40% of a culture may be affected in this way.

Clearly, the effects of the freezing process on living cells such as bacteria and other microorganisms, as well as on foods, are complex. According to Mazur[41] the response of microorganisms to subzero temperatures appears to be largely determined by solute concentration and intracellular freezing, although there are only a few cases of clear demonstration of this conclusion.

Why are some bacteria killed by freezing but not all cells? Some small and microscopic organisms are unable to survive freezing as can most bacteria. Examples include the foot-and-mouth disease virus and the causative agent of trichinosis (*Trichinella spiralis*). Protozoa are generally killed when frozen below $-5°C$ or $-10°C$, if protective compounds are not present.[41]

Effect of Thawing

Of great importance in the freezing survival of microorganisms is the process of thawing. Repeated freezing and thawing will destroy bacteria by disrupting cell membranes. Also, the faster the thaw, the greater the number of bacterial survivors. Why this is so is not entirely clear. From the changes listed above that occur during freezing, it can be seen that the thawing process becomes complicated if it is to lead to the restoration of viable activity. It has been pointed out that thawing is inherently slower than freezing and follows a pattern that is potentially more detrimental. Among the problems attendant on the thawing of specimens and products that transmit heat energy primarily by conduction, are the following:[18]

1. Thawing is inherently slower than freezing when conducted under comparable temperature differentials.
2. In practice, the maximum temperature differential permissible during thawing is much less than that which is feasible during freezing.
3. The time–temperature pattern characteristic of thawing is potentially more detrimental than that of freezing. During thawing, the temperature rises rapidly to near the melting point and remains there throughout the long course of thawing, thus affording considerable opportunity for chemical reactions, recrystallization, and even microbial growth, if thawing is extremely slow.

It has been stated that microorganisms die not upon freezing but, rather, during the thawing process. As to why some organisms are able to survive freezing while others do not, Luyet[39] suggested that it is a question of the ability of an organism to survive dehydration and to undergo dehydration when the medium freezes. With respect to survival after freeze-drying, Luyet has stated that it may be due to the fact that bacteria do not freeze at all but merely dry up. (See Chapter 18 for further discussion of the effect of freeze-drying on microorganisms.)

It is fairly well established that the freeze-thaw cycle leads to: (1) ice nucleation, (2) dehydration, and (3) oxidative damage. During thawing, an oxidative burst has been shown to occur and superoxide dismutase (SOD) provides resistance to the deleterious oxidative effects. In a study of *Campylobacter coli*, it was found that SOD, but not catalase, formed during the freeze–thaw.[62] SOD is also important in the resistance of *Campylobacter* cells to the oxidative stress that occurs during their survival in foods, during intestinal colonization of poultry, and during their survival in macrophages (see reference 62).

Most frozen-foods processors advise against the refreezing of foods once they have been thawed. Although the reasons are more related to the texture, flavor, and other nutritional qualities of the frozen product, the microbiology of thawed frozen foods is pertinent. Some investigators have pointed out that foods from the frozen state spoil faster than similar fresh products. There are textural changes associated with freezing that would seem to aid the invasion of surface organisms into deeper parts of the produce and, consequently, facilitate the spoilage process. Upon thawing, surface condensation of water is known to occur. There is also, at the surface, a general concentration of water-soluble substances such as amino acids, minerals, B vitamins, and, possibly, other nutrients. Freezing has the effect of destroying many thermophilic and some mesophilic organisms, making for less competition among the survivors upon thawing. It is conceivable that a greater relative number of psychrotrophs on thawed foods might increase the spoilage rate. Some psychrotrophic bacteria have been reported to have Q_{10} values in excess of 4.0 at refrigerator temperatures. For example, *P. fragi* has been reported to possess a Q_{10} of 4.3 at 0°C. Organisms of this type are capable of doubling their growth rate with only a 4–5°C rise in temperature. Whether frozen thawed foods do, in fact, spoil faster than fresh foods would depend on a large number of factors, such as the type of freezing, the relative numbers and types of organisms on the product prior to freezing, and the temperature at which the product is held to thaw. Although there are no known toxic effects associated with the refreezing of frozen and thawed foods, this act should be minimized in the interest of the overall nutritional quality of the products. One effect of freezing and thawing animal tissues is the release of lysosomal enzymes consisting of cathepsins, nucleases, phosphatases, glycosidases, and others. Once released, these enzymes may act to degrade macromolecules and thus make available simpler compounds that are more readily utilized by the spoilage biota.

SOME CHARACTERISTICS OF PSYCHROTROPHS AND PSYCHROPHILES

There is an increase in unsaturated fatty acid residues. The usual lipid content of most bacteria is between 2% and 5%, most or all of which is in the cell membrane. Bacterial fats are glycerol esters of two types: neutral lipids, in which all three or only one or two of the –OH groups of glycerol are esterified with long-chain fatty acids; and phospholipids, in which one of the –OH groups is linked through a phosphodiester bond to choline, ethanolamine, glycerol, inositol, or serine. The other two –OH groups are esterified with long-chain fatty acids.[58]

Many psychrotrophs synthesize neutral lipids and phospholipids containing an increased proportion of unsaturated fatty acids when grown at low temperatures compared with growth at higher temperatures. As much as a 50% increase in the content of unsaturated bonds of fatty acids from mesophilic and psychrotrophic *Candida* spp. was found in cells grown at 10°C compared to 25°C.[36] The phospholipid composition of these yeasts was unchanged. The increase in unsaturated fatty acids in *Candida utilis* as growth temperatures were lowered from 30°C to 5°C is shown in Table 16–6; linolenic acid increased at the expense of oleic acid at the lower temperatures.

In a comparative study of four *Vibrio* spp. that grew over the range of −5 to 15°C and four *Pseudomonas* spp. that grew over the range of 0–25 or 27°C, significant changes were observed

Table 16–6 Effects of Incubation Temperature on the Fatty Acid Composition of Stationary Cultures of *Candida utilis*

Incubation Temperature (°C)	Cell Concentration (mg/mL)	Fatty Acid Composition*				
		16:0	16:1	18:1	18:2	18:3
30	2.0	18.9	4.6	39.1	34.3	2.1
20	2.0	20.3	11.4	31.6	27.7	6.1
10	2.0	27.4	20.6	20.7	17.6	10.7
5	1.7	19.2	15.9	18.2	16.3	27.3

*Values quoted are expressed as percentages of the total fatty acids. Fatty acids are designated *x* : *y*, where *x* is the number of carbon atoms and *y* is the number of double bonds per molecule.

Source: From McMurrough and Rose,[43] copyright © 1973, American Society for Microbiology.

in total phospholipids of vibrios as growth temperatures were lowered from 15 to −5°C but not among the pseudomonads over their range of growth.[6,7,27] A change from saturated to unsaturated lipids would not be expected to occur in pseudomonads as growth temperatures are lowered because the psychrotrophic strains contain between 59% and 72% unsaturated lipids, making them more versatile than many other organisms. In contrast to most other psychrotrophs, *Micrococcus cryophilus* undergoes chain shortening in response to low temperatures, which apparently decreased the melting point of its membrane lipids.[59]

The widespread occurrence of low-temperature-induced changes in fatty acid composition suggests that they are associated with physiological mechanisms of the cell. It is known that an increase in the degree of unsaturation of fatty acids in lipids leads to a decrease in the lipid melting point. It has been suggested that increased synthesis of unsaturated fatty acids at low temperatures has the function of maintaining the lipid in a liquid and mobile state, thereby allowing membrane activity to continue to function. This concept, referred to as the *lipid solidification* theory, was first proposed by Gaughran[20] and Allen.[3] It has been shown by Byrne and Chapman[10] that the melting point of fatty-acid side chains in lipids is more important than the entire lipid structure.

When *L. monocytogenes* was cultured in the growth range of 12–13°C, there were significant deviations in the percentages of both *i*15:O (*i* = *iso*) and *a*15:O (*a* = *anteiso*) fatty acids together with a suggested deviation in *a*17:O, which resulted in a significant change in the total branched-chain fatty acids.[54] The cytoplasmic membrane of *L. monocyogenes* cultured at low temperatures contains a large proportion of shorter-chain fatty acids where branching switches from *iso* to *anteiso* (see reference 38).

Psychrotrophs synthesize high levels of polysaccharides. Well-known examples of this effect include the production of ropy milk and ropy bread dough, both of which are favored by low temperatures. The production of extracellular dextrans by *Leuconostoc* and *Pediococcus* spp. are known to be favored at temperatures below the growth optima of these organisms. The greater production of dextran at lower temperatures is due apparently to the fact that dextransucrase is very rapidly inactivated at temperatures in excess of 30°C.[53] A temperature-sensitive dextransucrase synthesizing system has also been shown for a *Lactobacillus* sp.[12]

From a practical standpoint, increased polysaccharide synthesis at low temperatures manifests itself in the characteristic appearance of low-temperature spoiled meats. Slime formation is characteristic of the bacterial spoilage of frankfurters, fresh poultry, and ground beef. The coalescence of surface colonies leads to the sliminess of such meats, and no doubt contributes to the increased hydration capacity that accompanies low-temperature meat spoilage. This extra polymeric material undoubtedly plays a role in biofilm formation (see Chapter 22).

Pigment production is favored. This effect appears to be confined to those organisms that synthesize phenazine and carotenoid pigments. The best-documented example of this phenomenon involves pigment production by *Serratia marcescens*. The organism produces an abnormally heat-sensitive enzyme that catalyzes the coupling of a monopyrrole and bipyrrole precursor to give prodigiosin (the red pigment).[72] The increased production of pigments at suboptimum temperatures is also reported by others.[65,72] It is interesting that a very large number of marine psychrotrophs (and perhaps psychrophiles) are pigmented. This is true for bacteria as well as yeasts. On the other hand, none of the more commonly studied thermophiles are pigmented.

Some strains display differential substrate utilization. It has been reported that sugar fermentation at temperatures below 30°C gives rise to both acid and gas, whereas above 30°C, only acid is produced.[23] Similarly, others have found psychrotrophs that fermented glucose and other sugars with the formation of acid and gas at 20°C and lower, but produced only acid at higher temperatures.[67] The latter was ascribed to a temperature-sensitive formic hydrogenase system. These investigators studied a similar effect and attributed the difference to a temperature-sensitive hydrogenase synthesizing system of the cell. Beef spoilage bacteria have been shown to liquefy gelatin and utilize water-soluble beef proteins more at 5°C than at 30°C,[33] but whether this effect is due to temperature-sensitive enzymes is not clear.

THE EFFECT OF LOW TEMPERATURES ON MICROBIAL PHYSIOLOGIC MECHANISMS

Of the effects that low incubation temperatures have on the growth and activity of foodborne microorganisms, five have received the most attention and are outlined below.

Psychrotrophs have a slower metabolic rate. The precise reasons as to why metabolic rates are slowed at low temperatures are not fully understood. Psychrotrophic growth decreases more slowly than that of mesophilic with decreasing temperatures. The temperature coefficients (Q_{10}) for various substrates such as acetate and glucose have been shown by several investigators to be lower for growing psychrotrophs than for mesophiles. The end products of mesophiles and psychrotrophic metabolism of glucose were shown to be the same, with the differences largely disappearing when the cells were broken.[29] In other words, the temperature coefficients are about the same for psychrotrophs and mesophiles when cell-free extracts are employed.

As the temperature is decreased, the rate of protein synthesis is known to decrease, and this occurs in the absence of changes in the amount of cellular DNA. One reason may be the increase in intramolecular hydrogen bonding that occurs at low temperatures, leading to increased folding of enzymes with losses in catalytic activity.[37] On the other hand, the decrease in protein synthesis appears to be related to a decreased synthesis of individual enzymes at low growth temperatures. Although the precise mechanism of reduced protein synthesis is not well understood, it has been suggested that low temperatures affect the synthesis of a repressor protein[40] and that the repressor protein itself is thermolabile.[64] Several investigators have suggested that low temperatures may influence the fidelity of the translation of messenger RNA (mRNA) during protein synthesis. For example, in studies with *E. coli*, it was shown that a leucine-starved auxotroph of this mesophile incorporated radioactive leucine into protein at 0°C.[22] It was suggested that at this temperature, all essential steps in protein synthesis apparently go on and involve a wide variety of proteins. The rate of synthesis at 0°C was estimated to be about 350 times slower than at 37°C for this organism. It has been suggested that the cessation of RNA synthesis in general may be the controlling factor in determining low temperature growth,[26] and the lack of polysome formation in *E. coli* when shifted to a temperature below its growth minimum has been demonstrated. The formation of polysomes is thus sensitive to low temperatures (at least in some organisms), and protein synthesis would be adversely affected.

Whatever the specific mechanism of lowered metabolic activity of microorganisms as growth temperature is decreased, psychrotrophs growing at low temperature have been shown to possess good enzymatic activity, since motility, endospore formation, and endospore germination will occur at 0°C.[63] *P. fragi*, among other organisms, produces lipases within 2–4 days at −7°C, within 7 days at −18°C, and within 3 weeks at −29°C.[2] The minimum growth temperature may be determined by the structure of the enzymes and the cell membrane, as well as by enzyme synthesis.[63] The lack of production of enzymes at high temperatures by psychrotrophs, on the other hand, is due apparently to the inactive nature of enzyme-synthesizing reactions rather than to enzyme inactivation,[63] although the latter is known to occur (see below). With respect to individual groups of enzymes, yields of endocellular proteolytic enzymes are greater in *Pseudomonas fluorescens* grown at 10°C than at either 20°C or 35°C,[56] whereas other investigators have shown that *P. fragi* preferentially produces lipase at low temperatures, with none being produced at 30°C or higher.[51,52] *P. fluorescens* has been found to produce just as much lipase at 5°C as at 20°C, but only a slight amount was produced at 30°C.[1] On the other hand, a proteolytic enzyme system of *P. fluorescens* showed more activity on egg white and hemoglobin at 25°C than at 15 and 5°C.[28]

It has been suggested that there are preformed elements in microbial cells grown at any temperature that are selectively temperature sensitive.[35] Microorganisms may cease to grow at a certain low temperature because of excessive sensitivity in one or several control mechanisms, the effectors of which cannot be supplied in the growth medium.[30] According to the latter investigators, the interaction between effector molecules and the corresponding allosteric proteins may be expected to be a strong function of temperature.

Psychrotroph membranes transport solutes more efficiently. It has been shown in several studies that upon lowering the growth temperature of mesophiles within the psychrotrophic range, solute uptake is decreased. Studies by Baxter and Gibbons[4] indicate that the minimum growth temperature of mesophiles is determined by the temperature at which transport permeases are inactivated. Farrell and Rose[16] offered three basic mechanisms by which low temperature could affect solute uptake: (1) inactivation of individual permease proteins at low temperature as a result of low-temperature-induced conformational changes that have been shown to occur in some proteins; (2) changes in the molecular architecture of the cytoplasmic membrane that prevent permease action; and (3) a shortage of energy required for the active transport of solutes. Although the precise mechanisms of reduced uptake of solutes at low temperatures are not clear at this time, the second mechanism seems the most likely.[16]

From studies of four psychrophilic vibrios, maximum uptake of glucose and lactose occurred at 0°C and decreased when temperatures were raised to 15°C, whereas with four psychrotrophic pseudomonads, maximum uptake of these substrates occurred in the 15–20°C range and decreased as temperatures were reduced to 0°C.[27] The vibrios showed significant changes in total phospholipids at 0°C, whereas no meaningful changes occurred with pseudomonads as their growth temperature was lowered. In studies with *Listeria monocytogenes* at 10°C, metabolism at low temperatures was believed to be the result of a cold-resistant sugar transport system that provided high concentrations of intracellular substrates.[71] The latter investigators noted that a cold-resistant sugar transport system is the property most readily identified as a fitness trait for psychrotrophy and that it applies not only to *Listeria monocytogenes*, but also to *Erysipelothrix rhusiopathiae* and *Brochothrix thermosphacta*.[70] It has been suggested that the minimum growth temperature of an organism may be defined by the inhibition of substrate uptake.

As noted above, psychrotrophs tend to possess in their membrane lipids that enable the membrane to be more fluid. The greater mobility of the psychrotrophic membrane may be expected to facilitate membrane transport at low temperatures. In addition, the transport permeases of psychrotrophs are apparently more operative under these conditions than are those of other mesophiles. Whatever the specific mechanism of increased transport might be, it has been demonstrated that psychrotrophs are more efficient than other mesophiles in the uptake of solutes at low temperatures. Baxter and Gibbons[4] showed that a psychrotrophic *Candida* sp. incorporated glucosamine more rapidly than a mesophilic *Candida*. The psychrotroph transported glucosamine at 0°C, whereas scarcely any was transported by the mesophile at this temperature or even at 10°C.

Cold-shock proteins (CsPs) have been demonstrated to develop in at least some microorganisms when they are first cultured at 35–37°C and then shifted to grow at or near their minimum growth temperatures. CsPs are RNA-binding proteins that mediate transcription elongation and message stability. They are referred to as RNA-chaperones. They are typically 7 kDa in size, and are found in most bacteria. With *L. monocytogenes* shifted to 5°C from 37°C, 12 CsPs were detected.[5] The

12 CsPs had a molecular weight range of 14,400 to 48,000. In another study, 32 CsPs were formed under similar conditions, and 4 were designated cold acclimation proteins (Caps, see reference 38). Caps result from increased synthesis during balance growth at low temperatures.

The adaptation of *L. monocytogenes* to low temperature led to the production of CsPs, which occurred also after HHP treatment for 10 minutes at 200 MPa at 30°C. A temperature decrease from 37 to 10°C resulted in a 10-fold increase in CsP1 and a 3.5-fold increase in CsP3.[68] Increases in CsPs following HHP were not as great as for cold shock. However, the level of survival of cold-shock cells following HHP was 100-fold higher than 37°C-grown cells. A study employing 9 strains of *L. monocytogenes* found that cold shock leads to an increase in thermal sensitivity.[46] The D_{60} values were reduced by 13–37% over controls, with the greatest effect occurring in cold-shocked, stationary-phase cells compared to lag or log phase cells.[46]

From a study of gene expression in *L. monocyogenes* growing at 10°C, it was concluded that acclimation to growth at 10°C likely involves amino acid starvation, oxidative stress, aberrant protein synthesis, cell surface remodeling, alterations in degradative metabolism, and induction of global regulatory responses.[38] When the growth of *E. coli* was reduced from 37 to 5°C, 12 CsPs formed. Increases in CsPs and CaPs has been demonstrated in the following organisms following a downshift in growth temperature: a *Vibrio* sp., *Pseudomonas fluorescens*; *Lactococcus lactis*, *Bacillus subtilis*, and *Vibrio vulnificus* (see reference 5).

In a study of the cold-shock response by two strains of *E. coli* 0157:H7 compared to two non-pathogenic strains, growth at 10°C vs. 20°C followed by freezing at −18°C for up to 24 hours in four substrates led to an increase in the ability of 25–35% of the pathogenic strains to survive frozen storage in brain heart infusion agar and apple juice compared to only 5% for the nonpathogens.[24] Freezing in frozen yogurt or ground beef did not show the same effect. Whether cryotolerance is a characteristic of pathogenic *E. coli* strains is unclear. A study of cold shock in *E. coli* 0157:H7 in different substrates found no increased survival in beef and pork, but increased survival did occur in milk, whole egg, or sausage.[8] Growth of the organism in trypticase soy broth at pH 5.0 appeared to negate the protective effect of the cold-shock treatment. The cold-shock protection appeared to be associated with the appearance of a new protein.

Some psychrotrophs produce larger cells. Yeasts, molds, and bacteria have been found to produce larger cell sizes when growing under psychrotrophic conditions than under mesophilic conditions. With *Candida utilis*, the increased cell size was believed to be due to increases in RNA and the protein content of cells.[58] Low-temperature synthesis of additional RNA has been reported by others, but one group found no increase in the amount of RNA at 2°C when *Pseudomonas* strain 92 cells were grown at 2°C and 30°C under the same conditions.[19] The latter authors found no increase in cell size, protein content, or catalase activity. On the other hand, psychrotrophic organisms are generally regarded as having higher levels of both RNA and proteins.[27]

Flagella synthesis is more efficient. Examples of the more efficient production of flagella at low temperatures include *E. coli*, *Bacillus inconstans*, *Salmonella* Paratyphi B, and other organisms, including some psychrophiles.

Psychrotrophs are favorably affected by aeration. The effect of aeration on the generation time of *P. fluorescens* at temperatures from 4°C to 32°C, employing three different carbon sources, is presented in Table 16–7. The greatest effect of aeration (shaking) occurred at 4°C and 10°C, whereas at 32°C, aerated cultures produced a longer generation time.[55] The significance of this effect is not clear. In a study of facultatively anaerobic psychrotrophs under anaerobic conditions, the organisms were shown to grow more slowly, survive longer, die more rapidly at higher temperatures, and produce lower maximal cell yields under anaerobic conditions than under aerobic conditions.[66] It has been commonly observed that plate counts on many foods are higher with incubation at low temperatures than at temperatures of 30°C and above. The generally higher counts are due in part to the increased

Table 16–7 Effect of Growth Temperature, Carbon Source, and Aeration on Generation Times (Hours) of *Pseudomonas fluorescens*

Growth Medium*	Culture	Growth Temperature					
		4°C	*10°C*	*15°C*	*20°C*	*25°C*	*32°C*
Glucose	Stationary	8.20	3.52	2.02	1.47	0.97	1.19
	Aerated	5.54	2.61	2.00	1.46	0.93	1.51
Citrate	Stationary	8.20	3.46	2.00	1.43	1.01	1.24
	Aerated	6.68	2.95	2.02	1.26	0.98	1.45
Casamino acids	Stationary	7.55	3.06	1.78	1.36	1.12	0.95
	Aerated	4.17	2.57	1.56	1.12	0.87	1.10

*Basal salts + 0.02% yeast extract + the carbon source indicated.

Source: Taken from Olsen and Jezeski.[55]

solubility and consequently, the availability of O_2.[61] The latter investigators found that equally high cell yields can be obtained at both low and high incubation temperatures when O_2 is not limiting. The greater availability of O_2 in refrigerated foods undoubtedly exerts selectivity on the spoilage biota of such foods. The vast majority of psychrotrophic bacteria studied are aerobes or facultative anaerobes, and these are the types associated with the spoilage of foods stored at refrigerator temperatures. Relatively few anaerobic psychrotrophs have been isolated and studied. One of the first was *Clostridium putrefaciens*.[42]

Some psychrotrophs display an increased requirement for organic nutrients. In one study, the generation times for unidentified aquatic bacterial isolates in low-nutrient media were two to three times longer than in high-nutrient media.[69]

NATURE OF THE LOW HEAT RESISTANCE OF PSYCHROTROPHS/PSYCHROPHILES

It has been known for years that psychrotrophic microorganisms are generally unable to grow much above 30–35°C. Among the first to suggest reasons for this limitation of growth were Edwards and Rettger,[14] who concluded that the maximum growth temperatures of bacteria may bear a definite relationship to the minimum temperatures of destruction of respiratory enzymes. Their conclusion has been borne out by results from a large number of investigators. It has been shown that many respiratory enzymes are inactivated at the temperatures of maximal growth of various psychrotrophic types (Table 16–8). Thus, the thermal sensitivity of certain enzymes of psychrotrophs is at least one of the factors that limit the growth of these organisms to low temperatures.

When some psychrotrophs are subjected to temperatures above their growth maxima, cell death is accompanied by the leakage of various intracellular constituents. The leakage substances have been shown to consist of proteins, DNA, RNA, free amino acids, and lipid phosphorus. The last was thought to represent phosphorus of the cytoplasmic membrane. Although the specific reasons for the release of cell constituents are not fully understood, it would appear to involve rupture of the cell membrane. These events appear to follow those of enzyme inactivation.

Table 16–8   Some Heat-Labile Enzymes of Psychrotrophic Microorganisms

Enzyme	Organism	Temperature of Maximum Growth (°C)	Temperature of Enzyme Inactivation (°C)
Extracellular lipase*	*P. fragi*		30
α-Oxoglutarate-synthesizing enzymes and others	*Cryptococcus*	~28	30
Alcohol dehydrogenase	*Candida* sp.	<30	
Formic hydrogen lyase	Psychrophile 82	35	45
Hydrogenase	Psychrophile 82	35	>20
Malic dehydrogenase	Marine *Vibrio*	30	30
Pyruvate dehydrogenase	*Candida* sp.	~20	25
Isocitrate dehydrogenase	*Arthrobacter* sp.	~35	37
Fermentative enzymes	*Candida* sp. P16	~25	35
Reduced NAD oxidase	Psychrophile 82	35	46
Cytochrome *c* reductase	Psychrophile 82	35	46
Lactic and glycerol dehydrogenase	Psychrophile 82	35	46
Pyruvate clastic enzymes	Psychrophile 82	35	46
Protein and RNA synthesizing	*Micrococcus cryophilus*	25	30

*Enzyme-forming system inactivated.

Whatever the true mechanism of psychrotroph death at temperatures a few degrees above their growth maxima is, their destruction at these relatively low temperatures is characteristic of this group of organisms. This is especially true of those that have optimum growth temperatures at and below 20°C. Reports on psychrotrophs isolated and studied over the past four or so decades reveal that all are capable of growing at 0°C with growth optima at either 15°C or between 20°C and 25°C and growth maxima between 20°C and 35°C. Included among these organisms are Gram-negative rods, Gram-positive aerobic and anaerobic rods, spore formers and non-spore formers, Gram-positive cocci, vibrios, and yeasts. One of these, *Vibrio fisheri (marinus)*, was shown by Morita and Albright[48] to have an optimum growth temperature at 15°C and a generation time of 80.7 minutes at this temperature. In almost all cases, the growth maxima of these organisms were only 5–10 above the growth optima.

Somewhat surprisingly, the proteinases of many psychrotrophic bacteria found in raw milk are heat resistant. This is true of pseudomonads as well as spore formers. The typical raw milk psychrotrophic pseudomonad produces a heat stable metalloproteinase with molecular weight in the 40- to 50-kDa range, which has a D value at 70°C of 118 minutes or higher.[15] The spores of some psychrotrophic bacilli have D values at 90°C of 5–6 minutes.[44]

REFERENCES

1. Alford, J.A., and L.E. Elliott. 1960. Lipolytic activity of microorganisms at low and intermediate temperatures. I. Action of *Pseudomonas fluorescens* on lard. *Food Res.* 25:296–303.

2. Alford, J.A., and D.A. Pierce. 1961. Lipolytic activity of microorganisms at low and intermediate temperatures. III. Activity of microbial lipases at temperatures below 0°C. *J. Food Sci.* 26:518–524.

3. Allen, M.B. 1953. The thermophilic aerobic sporeforming bacteria. *Bacteriol. Rev.* 17:125–173.

4. Baxter, R.M., and N.E. Gibbons. 1962. Observations on the physiology of psychrophilism in a yeast. *Can J. Microbiol.* 8:511–517.

5. Bayles, D.O., B.A. Annous, and B.J. Wilkinson. 1996. Cold stress proteins induced in *Listeria monocytogenes* in response to temperature downshock and growth at low temperatures. *Appl. Environ. Microbiol.* 62:1116–1119.

6. Bhakoo, M., and R.A. Herbert. 1979. The effect of temperature on psychrophilic *Vibrio* spp. *Arch. Microbiol.* 121:121–127.

7. Bhakoo, M., and R.A. Herbert. 1980. Fatty acid and phospholipid composition of five psychrotrophic *Pseudomonas* spp. grown at different temperatures. *Arch. Microbiol.* 126:51–55.

8. Bollman, J., A. Ismond, and G. Blank. 2001. Survival of *Escherichia coli* 0157:H7 in frozen foods: impact of the cold shock response. *Int. J. Food Microbiol.* 64:127–138.

9. Bonde, G.J. 1981. Phenetic affiliation of psychrotrophic *Bacillus*. In *Psychrotrophic Microorganisms in Spoilage and Pathogenicity*, ed. T.A. Roberts, G. Hobbs, J.H.B. Christian, and N. Skovgaard, 39–54. New York: Academic Press.

10. Byrne, P., and D. Chapman. 1964. Liquid crystalline nature of phospholipids. *Nature* 202:987–988.

11. Desrosier, N.W. 1963. *The Technology of Food Preparation*. Westport, CT: AVI.

12. Dunican, L.K., and H.W. Seeley. 1963. Temperature-sensitive dextransucrase synthesis by a lactobacillus. *J. Bacteriol.* 86:1079–1083.

13. Eddy, B.P. 1960. The use and meaning of the term "psychrophilic." *J. Appl. Bacteriol.* 23:189–190.

14. Edwards, O.F., and L.F. Rettger. 1937. The relation of certain respiratory enzymes to the maximum growth temperatures of bacteria. *J. Bacteriol.* 34:489–515.

15. Fairbairn, D.J., and B.A. Law. 1986. Proteinases of psychrotrophic bacteria: Their production, properties, effects, and control. *J. Dairy Res.* 53:139–177.

16. Farrell, J., and A. Rose. 1967. Temperature effects on micro-organisms. *Annu. Rev. Microbiol.* 21:101–120.

17. Fennema, O., and W. Powrie. 1964. Fundamentals of low-temperature food preservation. *Adv. Food Res.* 13:219–347.

18. Fennema, O.R., W.D. Powrie, and E.H. Marth. 1973. *Low-Temperature Preservation of Foods and Living Matter*. New York: Marcel Dekker.

19. Frank, H.A., A Reid, L.M. Santo, N.A. Lum, and S.T. Sandler. 1972. Similarity in several properties of psychrophilic bacteria grown at low and moderate temperatures. *Appl. Microbiol.* 24:571–574.

20. Gaughran, E.R.I., 1947. The thermophilic microorganisms. *Bacteriol. Rev.* 11:189–225.

21. Georgala, D.L., and A. Hurst. 1963. The survival of food poisoning bacteria in frozen foods. *J. Appl. Bacteriol.* 26:346–358.

22. Goldstein, A., D.B. Goldstein, and L.I. Lowney. 1964. Protein synthesis at 0°C in *Escherichia coli*. *J. Mol. Biol.* 9:213–235.

23. Greene, V.W., and J.J. Jezeski. 1954. The influence of temperature on the development of several psychrophilic bacteria of dairy origin. *Appl. Microbiol.* 2:110–117.

24. Grzadkowska, D., and M.W. Griffiths. 2001. Cryotolerance of *Escherichia coli* 0157:H7 in laboratory media and food. *J. Food Sci.* 66:1169–1173.

25. Gunderson, M.F., and K.D. Rose. 1948. Survival of bacteria in a precooked fresh-frozen food. *Food Res.* 13:254–263.

26. Harder, W., and H. Veldkamp. 1968. Physiology of an obligately psychrophilic marine *Pseudomonas* species. *J. Appl. Bacteriol.* 31:12–23.

27. Herbert, R.A. 1981. A comparative study of the physiology of psychrotrophic and psychrophilic bacteria. In *Psychrotrophic Microorganisms in Spoilage and Pathogenicity*, ed. T.A. Roberts, G. Hobbs, J.H.B. Christian, and N. Skovgaard, 3–16. New York: Academic Press.

28. Hurley, W.C., F.A. Gardner, and C. Vanderzant. 1963. Some characteristics of a proteolytic enzyme system of *Pseudomonas fluorescens*. *J. Food Sci.* 28:47–54.

29. Ingraham, J.L., and G.F. Bailey. 1959. Comparative study of effect of temperature on metabolism of psychrophilic and mesophilic bacteria. *J. Bacteriol.* 77:609–613.

30. Ingraham, J.L., and O. Maaløe. 1967. Cold-sensitive mutants and the minimum temperature of growth of bacteria. In *Molecular Mechanisms of Temperature Adaptation*, ed. C.L. Prosser, 297–309. Pub. No. 84. Washington, DC: American Association for the Advancement of Science.

31. Ingraham, J.L., and J.L. Stokes. 1959. Psychrophilic bacteria. *Bacteriol. Rev.* 23:97–108.

32. Ingram, M. 1951. The effect of cold on microorganisms in relation to food. *Proc. Soc. Appl. Bacteriol.* 14:243.

33. Jay, J.M. 1967. Nature, characteristics, and proteolytic properties of beef spoilage bacteria at low and high temperatures. *Appl. Microbiol.* 15:943–944.

34. Jay, J.M. 1987. The tentative recognition of psychrotrophic Gram-negative bacteria in 48 h by their surface growth at 10°C. *Int. J. Food Microbiol.* 4:25–32.

35. Jezeski, J.J., and R.H. Olsen. 1962. The activity of enzymes at low temperatures. In *Proceedings, Low Temperature Microbiology Symposium—1961*, 139–155. Camden, NJ: Campbell Soup Co.

36. Kates, M., and R.M. Baxter. 1962. Lipid comparison of mesophilic and psychrotrophic yeasts (*Candida* species) as influenced by environmental temperature. *Can. J. Biochem. Physiol.* 40:1213–1227.

37. Kavanau, J.L. 1950. Enzyme kinetics and the rate of biological processes. *J. Gen. Physiol.* 34:193–209.

38. Liu, S., J.E. Graham, L. Bigelow, P.D. Morse, II, and B.J. Wilkinson. 2002. Identification of *Listeria monocytogenes* genes expressed in response to growth at low temperatures. *Appl. Environ. Microbiol.* 68: 1697–1705.

39. Luyet, B. 1962. Recent developments in cryobiology and their significance in the study of freezing and freeze-drying of bacteria. In *Proceedings, Low Temperature Microbiology Symposium—1961*, 63–87, Camden, NJ: Campbell Soup Co.

40. Marr, A.G., J.L. Ingraham, and C.L. Squires. 1964. Effect of the temperature of growth of *Escherichia coli* on the formation of β-galactosidase. *J. Bacteriol.* 87:356–362.

41. Mazur, P. 1966. Physical and chemical basis of injury in single-celled microorganisms subjected to freezing and thawing. In *Cryobiology*, ed. H.T. Merryman, Chap. 6. New York: Academic Press.

42. McBryde, C.N. 1911. *A Bacteriological Study of Ham Souring*. Bulletin No. 132. Beltsville, MD: U.S. Bureau of Animal Industry.

43. McMurrough, I., and A.H. Rose. 1973. Effects of temperature variation on the fatty acid composition of a psychrophilic *Candida* species. *J. Bacteriol.* 114:451–452.

44. Meer, R.R., J. Baker, F.W. Bodyfelt, and M.W. Griffiths. 1991. Psychrotrophic *Bacillus* spp. in fluid milk products: A review. *J. Food Protect.* 54:969–979.

45. Michener, H., and R. Elliott. 1964. Minimum growth temperatures for food-poisoning, fecal-indicator, and psychrophilic microorganisms. *Adv. Food Res.* 13:349–396.

46. Miller, A.J., D.O. Bayles, and S. Eblen. 2000. Cold shock induction of thermal sensitivity in *Listeria monocytogenes*. *Appl. Environ. Microbiol.* 66:4345–4350.

47. Morita, R.Y. 1975. Psychrophilic bacteria. *Bacteriol. Rev.* 39:144–167.

48. Morita, R.Y., and L.J. Albright. 1965. Cell yields of *Vibrio marinus*, an obligate psychrophile, at low temperatures. *Can. J. Microbiol.* 11:221–227.

49. Mossel, D.A.A., M. Jansma, and J. De Waart. 1981. Growth potential of 114 strains of epidemiologically most common salmonellae and arizonae between 3 and 17°C. In *Psychrotrophic Microorganisms in Spoilage and Pathogenicity*, ed. T.A. Roberts, G. Hobbs, J.H.B. Christian, and N. Skovgaard, 29–37. New York: Academic Press.

50. Mossel, D.A.A., and H. Zwart. 1960. The rapid tentative recognition of psychrotrophic types among Enterobacteriaceae isolated from foods. *J. Appl. Bacteriol.* 23:183–188.

51. Nashif, S.A., and F.E. Nelson. 1953. The lipase of *Pseudomonas fragi*. I. Characterization of the enzyme. *J. Dairy Sci.* 36:459–470.

52. Nashif, S.A., and F.E. Nelson. 1953. The lipase of *Pseudomonas fragi*. II. Factors affecting lipase production. *J. Dairy Sci.* 36:471–480.

53. Neely, W.B. 1960. Dextran: Structure and synthesis. *Adv. Carbohydr. Chem.* 15:341–369.

54. Nichols, D.S., K.A. Presser, J. Olley, T. Ross, and T.A. McMeekin. 2002. Variation of branched-chain fatty acids marks the normal physiological range for growth in *Listeria monocytogenes*. *Appl. Environ. Microbiol.* 68: 2809–2813.

55. Olsen, R.H., and J.J. Jezeski. 1963. Some effects of carbon source, aeration, and temperature on growth of a psychrophilic strain of *Pseudomonas fluorescens*. *J. Bacteriol.* 86:429–433.

56. Peterson, A.C., and M.F. Gunderson. 1960. Some characteristics of proteolytic enzymes from *Pseudomonas fluorescens*. *Appl. Microbiol.* 8:98–104.

57. Reuter, G. 1981. Psychrotrophic lactobacilli in meat products. In *Psychrotrophic Microorganisms in Spoilage and Pathogenicity*, ed. T.A. Roberts, G. Hobbs, J.H.B. Christian, and N. Skovgaard, 253–258. New York: Academic Press.

58. Rose, A.H. 1968. Physiology of microorganisms at low temperature. *J. Appl. Bacteriol.* 31:1–11.

59. Russell, N.J. 1971. Alteration in fatty acid chain length in *Micrococcus cryophilus* grown at different temperatures. *Biochim. Biophys. Acta* 231:254–256.

60. Scott, W.J. 1962. Available water and microbial growth. In *Proceedings, Low Temperature Microbiology Symposium—1962*, 89–105. Camden, NJ: Campbell Soup Co.

61. Sinclair, N.A., and J.L. Stokes. 1963. Role of oxygen in the high cell yields of psychrophiles and mesophiles at low temperatures. *J. Bacteriol.* 85:164–167.

62. Stead, D., and S.F. Park. 2000. Roles of Fe superoxide dismutase and catalase in resistance of *Campylobacter coli* to freeze-thaw stress. *Appl. Environ. Microbiol.* 66:3110–3112.

63. Stokes, J.L. 1967. Heat-sensitive enzymes and enzyme synthesis in psychrophilic microorganisms. In *Molecular Mechanisms of Temperature Adaptation*, ed. C.L. Prosser, 311–323. Pub. No. 84. Washington, DC: American Association for the Advancement of Science.

64. Udaka, S., and T. Horiuchi. 1965. Mutants of *Escherichia coli* having temperature sensitive regulatory mechanism in the formation of arginine biosynthetic enzymes. *Biochem. Biophys. Res. Commun.* 19:156–160.

65. Uffen, R.L., and E. Canale-Parola. 1966. Temperature-dependent pigment production by *Bacillus cereus* var. *alesi. Can. J. Microbiol.* 12:590–593.

66. Upadhyay, J., and J.L. Stokes. 1962. Anaerobic growth of psychrophilic bacteria. *J. Bacteriol.* 83:270–275.

67. Upadhyay, J., and J.L. Stokes. 1963. Temperature-sensitive formic hydrogenlyase in a psychrophilic bacterium. *J. Bacteriol.* 85:177–185.

68. Wemekamp-Kamphuis, H.H., A.K. Karatzas, J.A. Wouters, and T. Abee. 2002. Enhanced levels of cold shock proteins in *Listeria monocytogenes* LO28 upon exposure to low temperature and high hydrostatic pressure. *Appl. Environ. Microbiol.* 68: 456–463.

69. Wiebe, W.J., W.M. Sheldon, Jr., and L.R. Pomeroy. 1992. Bacterial growth in the cold: Evidence for an enhanced substrate requirement. *Appl. Environ. Microbiol.* 58:359–364.

70. Wilkins, P.O. 1973. Psychrotrophic Gram-positive bacteria: Temperature effects on growth and solute uptake. *Can J. Microbiol.* 19:909–915.

71. Wilkins, P.O., R. Bourgeois, and R.G. Murray. 1972. Psychrotrophic properties of *Listeria monocytogenes. Can. J. Microbiol.* 18:543–551.

72. Williams, R.P., M.E. Goldschmidt, and C.L. Gott. 1965. Inhibition by temperature of the terminal step in biosynthesis of prodigiosin. *Biochem. Biophys. Res. Commun.* 19:177–181.

CHAPTER 17

Food Protection with High Temperatures, and Characteristics of Thermophilic Microorganisms

The use of high temperatures to preserve food is based on their destructive effects on microorganisms. By high temperatures are meant any and all temperatures above ambient. With respect to food preservation, there are two temperature categories in common use: pasteurization and sterilization. Pasteurization by use of heat implies either the destruction of all disease-producing organisms (e.g., pasteurization of milk) or the destruction or reduction in the number of spoilage organisms in certain foods, as in the pasteurization of vinegar. The pasteurization of milk is achieved by heating at one of the following time/temperature combinations:

145°F (63°C) for 30 minutes (low temperature, long time [LTLT])
161°F (72°C) for 15 sec (primary high temperature, short time [HTST] method)
191°F (89°C) for 1.0 sec
194°F (90°C) for 0.5 sec
201°F (94°C) for 0.1 sec
212°F (100°C) for 0.01 sec

These treatments are equivalent and are sufficient to destroy the most heat resistant of the non-sporeforming pathogenic organisms—*Mycobacterium tuberculosis* and *Coxiella burnetti*. When six different strains of *M. avium* subsp. *paratuberculosis* were added to milk at levels from 40 to 100,000 cfu/ml followed by pasteurization by LTLT or HTST, no survivors were detected on suitable culture media incubated for 4 months.[26]

Milk pasteurization temperatures are sufficient to destroy, in addition, all yeasts, molds, Gram-negative bacteria, and many Gram positives. The two groups of organisms that survive milk pasteurization are placed into one of two groups: thermodurics and thermophiles. *Thermoduric* organisms are those that can survive exposure to relatively high temperatures but do not necessarily grow at these temperatures. The non-sporeforming organisms that survive milk pasteurization generally belong to the genera *Streptococcus* and *Lactobacillus*, and sometimes to other genera. *Thermophilic* organisms are those that not only survive relatively high temperatures but *require* high temperatures for their growth and metabolic activities. The genera *Bacillus, Clostridium, Alicyclobacillus, Geobacillus, and Thermoanaerobacter* contain the thermophiles of greatest importance in foods (see later section in this chapter). Pasteurization (to destroy spoilage biota) of beers in the brewing industry is carried out usually for 8–15 minutes at 60°C.

Sterilization means the destruction of all viable organisms as may be measured by an appropriate plating or enumerating technique. Canned foods are sometimes called "commercially sterile" to indicate that no viable organisms can be detected by the usual cultural methods employed or that the number of survivors is so low as to be of no significance under the conditions of canning and storage. Also, microorganisms may be present in canned foods that cannot grow in the product by reason of undesirable pH, oxidation–reduction potential (Eh), or temperature of storage.

The processing of milk and milk products can be achieved by the use of ultrahigh temperatures (UHT). Milk so produced is a product in its own right and is to be distinguished from pasteurized milk. The primary features of the UHT treatment include its continuous nature, its occurrence outside the package necessitating aseptic storage and aseptic handling of the product downstream from the sterilizer, and the very high temperatures (in the range 140–150°C) and the correspondingly short time (a few seconds) necessary to achieve commercial sterility.[24]

UHT-processed milks have higher consumer acceptability than the conventionally heated pasteurized products, and because they are commercially sterile, they may be stored at room temperatures for up to 8 weeks without flavor changes.

FACTORS AFFECTING HEAT RESISTANCE OF MICROORGANISMS

Equal numbers of bacteria placed in physiologic saline or nutrient broth at the same pH are not destroyed with the same ease by heat. Some 12 factors or parameters of microorganisms and their environment have been studied for their effects on heat destruction, and they are presented below.[22]

Water

The heat resistance of microbial cells increases with decreasing humidity, moisture, or water activity (a_w), and this is illustrated in Table 17–1 for spores of *Bacillus cereus*. For example, at a_w of 1.00 and pH 6.5, D_{95} was 2.386 minutes while at a_w of 0.86, D_{95} was 13.842 minutes.[18]

Dried microbial cells placed in test tubes and then heated in a water bath are considerably more heat resistant than moist cells of the same type. Because it is well established that protein denaturation occurs at a faster rate when heated in water than in air, it is suggested that protein denaturation is either the mechanism of death by heat or is closely associated with it (see a later section in this chapter). The precise manner in which water facilitates heat denaturation of proteins is not entirely clear, but it has been pointed out that the heating of wet proteins causes the formation of free–SH groups with a consequent increase in the water-binding capacity of proteins. The presence of water allows for thermal breaking of peptide bonds, a process that requires more energy in the absence of water and, consequently, confers a greater refractivity to heat.

Fat

In the presence of fats, there is a general increase in the heat resistance of some microorganisms (Table 17–2). This is sometimes referred to as fat protection and is presumed to increase heat resistance

Table 17–1 Influence of Temperature, a_w, and pH on *D* Values of *Bacillus cereus* Spores

		D (Minutes)		
°C	a_w	6.5	5.5	4.5
95	1.00	2.386	1.040	0.511
95	0.95	5.010	2.848	1.409
95	0.86	13.842	14.513	7.776
85	1.00	63.398	13.085	5.042
85	0.86	68.909	91.540	33.910

Source: Adapted with permission from S. Gaillard et al.[18] Model for combined effects of temperature, pH and water activity on thermal inactivation of *Bacillus cereus* spores, *J. Food Sci.* 63: 887–889, copyright © 1998, Institute of Food Technologists.

by directly affecting cell moisture. Sugiyama[56] demonstrated the heat-protective effect of long-chain fatty acids on *Clostridium botulinum*. It appears that the long-chain fatty acids are better protectors than short-chain acids.

Salts

The effect of salt on the heat resistance of microorganisms is variable and dependent on the kind of salt, concentration employed, and other factors. Some salts have a protective effect on microorganisms, and others tend to make cells more heat sensitive. It has been suggested that some salts may decrease water activity and thereby increase heat resistance by a mechanism similar to that of drying, whereas others may increase water activity (e.g., Ca^{2+} and Mg^{2+}) and, consequently, increase sensitivity to heat. It has been shown that supplementation of the growth medium of *Bacillus megaterium* spores with $CaCl_2$ yields spores with increased heat resistance, whereas the addition of L-glutamate, L-proline, or increased phosphate content decreases heat resistance.[31]

Table 17–2 The Effect of the Medium on the Thermal Death Point of *Escherichia coli*

Medium	Thermal Death Point (°C)
Cream	73
Whole milk	69
Skim milk	65
Whey	63
Bouillon (broth)	61

Note: Heating time: 10 minutes.

Source: From Carpenter.[12] Courtesy of W.B. Saunders Co., Philadelphia.

Table 17–3 *D* Values (60°C) of Starved *Listeria monocytogenes* Cells in Pork Slurry with NaCl and Sodium Pyrophosphate (SPP) Alone and in Combinations (Summarized from Lihono et al.[32])

%SPP	%NaCl	D Values (Standard Deviation)
0	0	1.18 (0.15)
0	3	1.56 (0.32)
0	6	3.50 (0.46)
0.5	0	1.46 (0.09)
0.5	3	1.80 (0.23)
0.5	6	3.07 (0.35)

The effect of NaCl and sodium pyrophosphate (SPP) on $D_{60°C}$ values of starved *L. monocytogenes* cells is presented in Table 17–3 where the value at 6% NaCl was significantly higher (3.50) than with no added NaCl (1.18; reference 32). In another study of starved *L. monocytogenes* cells, $D_{60°C}$ in mushroom was 1.6 compared to non-starved cells, which was 0.7.[35]

Carbohydrates

The presence of sugars in the suspending menstruum causes an increase in the heat resistance of microorganisms suspended therein. This effect is at least in part due to the decrease in water activity caused by high concentrations of sugars. There is great variation, however, among sugars and alcohols relative to their effect on heat resistance, as can be seen in Table 17–4 for *D* values of *Salmonella* Senftenberg 775W. At identical a_w values obtained by the use of glycerol and sucrose, wide differences in heat sensitivity occurred.[3,21] Corry[13] found that sucrose increased the heat resistance of *S.* Senftenberg more than any of four other carbohydrates tested. The following decreasing order was found for the five tested substances: sucrose > glucose > sorbitol > fructose > glycerol.

pH

Microorganisms are most resistant to heat at their optimum pH of growth, which is generally about 7.0. As the pH is lowered or raised from this optimum value, there is a consequent increase in heat sensitivity (Figure 17–1, Table 17–1). Advantage is taken of this fact in the heat processing of high-acid foods, where considerably less heat is applied to achieve sterilization compared to foods at or near neutrality. The heat pasteurization of egg white provides an example of an alkaline food product that is neutralized prior to heat treatment, a practice not done with other foods. The pH of egg white is about 9.0. When this product is subjected to pasteurization conditions of 60–62°C for 3.5–4 minutes, coagulation of proteins occurs along with a marked increase in viscosity. These changes affect the volume and texture of cakes made from such pasteurized egg white. Cunningham and Lineweaver[14] reported that egg white may be pasteurized the same manner as whole egg if the pH is reduced to about 7.0. This reduction of pH makes both microorganisms and egg white proteins more heat stable. The addition of salts of iron or aluminum increases the stability of the highly heat-labile egg protein conalbumin sufficiently to permit pasteurization at 60–62°C. Unlike their resistance to heat in other materials, bacteria are more resistant to heat in liquid whole egg at pH values of 5.4–5.6 than at values

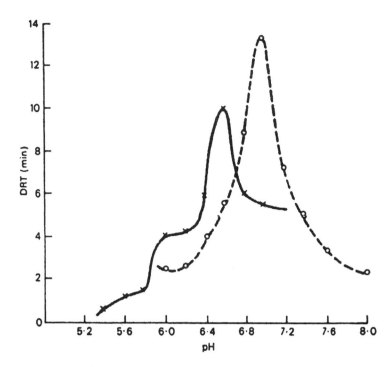

Figure 17–1   Effect of pH on the decimal reduction time (DRT) of *Enterococcus faecalis* (C and G) exposed to 60°C in citrate–phosphate buffer (crosses) and phosphate buffer (circles) solutions at various pH levels. *Source:* From White.[61]

of 8.0–8.5 (Table 17–4). This is true when the pH is lowered with an acid such as HCl. When organic acids such as acetic or lactic acid are used to lower the pH, a decrease in heat resistance occurs.

Proteins and Other Substances

Proteins in the heating menstrum have a protective effect on microorganisms. Consequently, high-protein-content foods must be heat processed to a greater degree than low-protein-content foods in order to achieve the same end results. For identical numbers of organisms, the presence of colloidal-sized particles in the heating menstruum also offers protection against heat. For example, under identical conditions of pH, numbers of organisms, and so on, it takes longer to sterilize pea purée than nutrient broth.

Numbers of Organisms

The larger the number of organisms, the higher is the degree of heat resistance (Table 17–5). It has been suggested that the mechanism of heat protection by large microbial populations is due to the production of protective substances excreted by the cells, and some investigators claim to have demonstrated the existence of such substances. Because proteins are known to offer some protection

Table 17–4   Reported D values of *Salmonella* Senftenberg 775 W

Temperature (°C)	D Values	Conditions
61	1.1 min	Liquid whole egg
61	1.19 min	Tryptose broth
60	9.5 min*	Liquid whole egg, pH ~ 5.5
60	9.0 min*	Liquid whole egg, pH ~ 6.6
60	4.6 min*	Liquid whole egg, pH ~ 7.4
60	0.36 min*	Liquid whole egg, pH ~ 8.5
65.6	34–35.3 sec	Milk
71.7	1.2 sec	Milk
70	360–480 min	Milk chocolate
55	4.8 min	TSB,[†] log phase, grown 35°C
55	12.5 min	TSB,[†] log phase, grown 44°C
55	14.6 min	TSB,[†] stationary, grown 35°C
55	42.0 min	TSB,[†] stationary, grown 44°C
57.2	13.5 min*	a_w 0.99 (4.9% glyc.), pH 6.9
57.2	31.5 min*	a_w 0.90 (33.9% glyc.), pH 6.9
57.2	14.5 min*	a_w 0.99 (15.4% sucro.), pH 6.9
57.2	62.0 min*	a_w 0.90 (58.6% sucro.), pH 6.9
60	0.2–6.5 min[‡]	HIB,[§] pH 7.4
60	2.5 min	a_w 0.90, HIB, glycerol
60	75.2 min	a_w 0.90, HIB, sucrose
65	0.29 min	0.1M phosphate buffer, pH 6.5
65	0.8 min	30% sucrose
65	43.0 min	70% sucrose
65	2.0 min	30% glucose
65	17.0	70% glucose
65	0.95 min	30% glycerol
65	0.70 min	70% glycerol
55	35 min	a_w 0.997, tryptone soya agar, pH 7.2

*Mean/average values.
[†]Trypticase soy broth.
[‡]Total of 76 cultures.
[§]Heart infusion broth.

against heat, many of the extracellular compounds in a culture would be expected to be protein in nature and, consequently, capable of affording some protection. Of perhaps equal importance in the higher heat resistance of large cell populations over smaller ones is the greater chance for the presence of organisms with differing degrees of natural heat resistance.

Age of Organisms

Bacterial cells tend to be most resistant to heat while in the stationary phase of growth (old cells) and less resistant during the logarithmic phase. This is true for *S*. Senftenberg (see Table 17–4), whose

Table 17–5 Effect of Number of Spores of *Clostridium botulinum* on Thermal Death Time at 100°C

Number of Spores	Thermal Death Time (minutes)
72,000,000,000	240
1,640,000,000	125
32,000,000	110
650,000	85
16,400	50
328	40

Source: From Carpenter.[12] Courtesy of W.B. Saunders Co., Philadelphia.

stationary phase cells may be several times more resistant than log phase cells.[42] Heat resistance has been reported to be high also at the beginning of the lag phase but decreases to a minimum as the cells enter the log phase. Old bacterial spores are reported to be more heat resistant than young spores. The mechanism of increased heat resistance of less active microbial cells is undoubtedly complex and not well understood.

Growth Temperature

The heat resistance of microorganisms tends to increase as the temperature of incubation increases, and this is especially true for sporeformers. Although the precise mechanism of this effect is unclear, it is conceivable that genetic selection favors the growth of the more heat-resistant strains at succeedingly high temperatures. *S.* Senftenberg grown at 44°C was found to be approximately three times more resistant than cultures grown at 35°C (Table 17–4).

Inhibitory Compounds

A decrease in heat resistance of most microorganisms occurs when heating takes place in the presence of heat-resistant antibiotics, SO_2, and other microbial inhibitors. The use of heat plus antibiotics and heat plus nitrite has been found to be more effective in controlling the spoilage of certain foods than either alone. The practical effect of adding inhibitors to foods prior to heat treatment is to reduce the amount of heat that would be necessary if used alone (see Chapter 13).

Time and Temperature

One would expect that the longer the time of heating, the greater the killing effect of heat. All too often, though, there are exceptions to this basic rule. A more dependable rule is that the higher the temperature, the greater the killing effect of heat. This is illustrated in Table 17–6 for bacterial spores. As temperature increases, time necessary to achieve the same effect decreases.

These rules assume that heating effects are immediate and not mechanically obstructed or hindered. Also important is the size of the heating vessel or container and its composition (glass, metal, plastic).

Table 17–6   Effect of Temperature on Thermal Death Times of Spores

Temperature	Clostridium botulinum (60 billion spores suspended in buffer at pH 7)	A thermophile (150,000 spores per ml of corn juice at pH 6.1)
100°C	260 minutes	1,140 minutes
105°C	120	
110°C	36	180
115°C	12	60
120°C	5	17

Source: From Carpenter.[12] Courtesy of W.B. Saunders Co., Philadelphia.

It takes longer to effect pasteurization or sterilization in large containers than in smaller ones. The same is true of containers with walls that do not conduct heat as readily as others.

Effect of Ultrasonics

The exposure of bacterial endospores to ultrasonic treatments just before or during heating results in a lowering of spore heat resistance (see the section Manothermosonication in Chapter 19).

RELATIVE HEAT RESISTANCE OF MICROORGANISMS

In general, the heat resistance of microorganisms is related to their optimum growth temperatures. Psychrophilic microorganisms are the most heat sensitive, followed by mesophiles and thermophiles. Sporeforming bacteria are more heat resistant than non-sporeformers, and thermophilic sporeformers are, in general, more heat resistant than mesophilic sporeformers. With respect to Gram reaction, Gram-positive bacteria tend to be more heat resistant than Gram negatives, with cocci, in general, being more resistant than non-sporeforming rods. Yeasts and molds tend to be fairly sensitive to heat, with yeast ascospores being only slightly more resistant than vegetative yeasts. The asexual spores of molds tend to be slightly more heat resistant than mold mycelia. Sclerotia are the most heat resistant of these types and sometimes survive and cause trouble in canned fruits. The relative heat resistance of some bacteria and fungi that cause spoilage of high-acid foods is indicated in Table 17–7. Of the genus *Alicyclobacillus*, *A. acidoterrestris* is one of the highly resistant species found in some fruit juice products.

Spore Resistance

The extreme heat resistance of bacterial endospores is of great concern in the thermal preservation of foods. In spite of intense study over several decades, the precise reason why bacterial spores are so heat resistant is still not entirely clear.

Spore heat resistance has been associated with protoplast dehydration, mineralization, and thermal adaptation. The compound dipicolinic acid, which is unique to bacterial spores, was once believed

Table 17–7 *D* Values of Some Organisms That Cause Spoilage of Acid and High-Acid Foods

Organisms	Substrate	°C	D (minutes)	z	Reference
Neosartorya fischeri	PO$_4$ buffer, pH 7.0	85	35.25	4.0	45
Neosartorya fischeri	PO$_4$ buffer, pH 7.0	87	11.1	4.0	45
Neosartorya fischeri	PO$_4$ buffer, pH 7.0	89	3.90	4.0	45
Neosartorya fischeri	Apple juice	87.8	1.4	5.6	48
Neosartorya fischeri	Blueberry fruit filling	91	<2.0	5.4–11*	10
Talaromyces flavus	Blueberry fruit filling	91	2.5–5.4	9.7–16.6*	10
Talaromyces flavus	Apple juice	90.6	2.2	5.2	48
Alicyclobacillus	Berry juice	91.1	3.8	–	37
Alicyclobacillus	Berry juice	95	1.0	–	37
Alicyclobacillus	Berry juice	87.8	11.0	–	37
Alicyclobacillus	Concord grape juice, 30	85.0	76.0	6.6	52
Alicyclobacillus	Concord grape juice, 30	90	18.0	6.6	52
Alicyclobacillus	Concord grape juice, 30	95	2.3	6.6	52

*Range for three different strains.

to be responsible for thermal resistance, especially as a calcium–dipicolinate complex. However, it has been found that heat resistance is independent of this complex, and just what role it plays in heat resistance is unclear. Small, acid-soluble proteins (SASP) of the α/β type are found in spores (they prevent depuration of spore DNA) and thus they contribute to heat resistance. Heat resistance appears to be associated with a contractile cortex that either reduces the water content of the protoplast or maintains it in a state of dehydration. That protoplast dehydration and diminution are major factors of spore thermal resistance has been substantiated,[8] but other factors are known to have an additive effect.[41]

The endospores of a given species grown at maximum temperature are more heat resistant than those grown at lower temperatures.[62] It appears that protoplast water content is lowered by this thermal adaptation, resulting in a more heat-resistant spore.[7] Heat resistance is affected extrinsically by changes in mineral content. Although all three factors noted contribute to spore thermal resistance, dehydration appears to be the most important.[7] For more information on bacterial spores relative to food microbiology, see Setlow and Johnson.[49]

The recognized genera of heterotrophic sporeforming bacteria are listed in Table 17–8. Some of these genera are not known to be important in foods but since they are found in the general environment, it is possible that they have been incorrectly identified. One of the more interesting is *Anaerobacter polyendosporus*, which produces up to 5–7 spores/cell.[51] The effect of some chemicals on spores is presented in Chapter 13.

THERMAL DESTRUCTION OF MICROORGANISMS

In order to better understand the thermal destruction of microorganisms relative to food protection and canning, it is necessary to understand certain basic concepts associated with this technology. Following are listed some of the more important concepts, but for a more extensive treatment of thermobacteriology, the monograph by Stumbo[54] should be consulted.

Table 17–8   Currently Recognized Genera of Heterotrophic Sporeforming Bacteria that are Known to Occur in Foods Along with Some Whose Foodborne Status is Unknown

Genus	Gram Rx	Morphology	In Foods?
Aerobes			
Alicyclobacillus	+ positive	R rod	+ yes
Amphibacillus	+	R	+
Aneurinibacillus	+	R	? unknown
Bacillus	+	R	+
Brevibacillus	+	R	+
Desulfotomaculum	+[a]	R	+
Geobacillus	+	R	+
Gracilibacillus	+	R	?
Halobacillus	+	R	+
Paenibacillus	+	R	+
Salibacillus	+	R	?
Serratia marcescens subsp. Sakuensis	−	R	?
Sporolactobacillus	+	R	+
Sporosarcina	+	Coccus	?
Virgibacillus	+	R	?
Thermoactinomycetes	+	R	?
Ureibacillus	+	R	?
Anaerobes			
Anaerobacter	+	R	?
Caloramater	+	R	?
Clostridium	+	R	+
Filifactor	+	R	?
Moorella	+	R	+
Oxobacter	+	R	?
Oxolophagus	+	R	?
Sporohalobacter	−	R	?
Syntrophospora	−	R	?
Thermoanaerobacter	+	R	+
Thermoanaerobacterium	+	R	+

[a]Often reported as Gram negative.

Thermal Death Time

Thermal death time (TDT) is the time necessary to kill a given number of organisms at a specified temperature. By this method, the temperature is kept constant and the time necessary to kill all cells is determined. Of less importance is the thermal death point, which is the temperature necessary to kill a given number of microorganisms in a fixed time, usually 10 minutes. Various means have been proposed for determining TDT: the tube, can, "tank," flask, thermoresistometer, unsealed tube, and capillary tube methods. The general procedure for determining TDT by these methods is to place a known number of cells or spores in a sufficient number of sealed containers in order to get the desired number of survivors for each test period. The organisms are then placed in an oil bath and heated

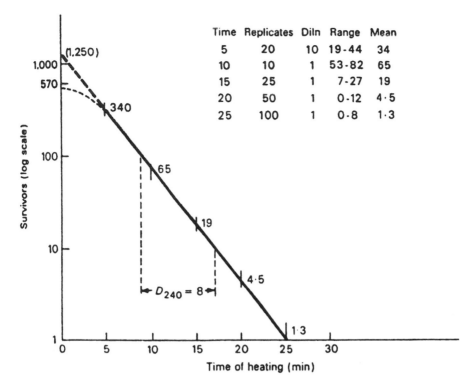

Time	Replicates	Diln	Range	Mean
5	20	10	19-44	34
10	10	1	53-82	65
15	25	1	7-27	19
20	50	1	0-12	4·5
25	100	1	0-8	1·3

Figure 17–2   Rate of destruction curve. Spores of strain F.S. 7 heated at 240°F in canned pea brine pH 6.2. *Source:* From Gillespy,[20] courtesy of Butterworths Publishers, London.

for the required time period. At the end of the heating period, containers are removed and cooled quickly in cold water. The organisms are then placed on a suitable growth medium, or the entire heated containers are incubated if the organisms are suspended in a suitable growth substrate. The suspensions or containers are incubated at a temperature suitable for growth of the specific organisms. Death is defined as the inability of the organisms to form visible colonies after extended incubations.

D Value

This is the decimal reduction time, or the time required to destroy 90% of the organisms. This value is numerically equal to the number of minutes required for the survivor curve to traverse one log cycle (Figure 17–2). Mathematically, it is equal to the reciprocal of the slope of the survivor curve and is a measure of the death rate of an organism. When D is determined at 250°F, it is often expressed as D_r. The effect of pH on the D value of *C. botulinum* in various foods is presented in Table 17–9, and D values for *S.* Senftenberg 775W under various conditions are presented in Table 17–4. D values of 0.20–2.20 minutes at 150°F have been reported for *S. aureus* strains, D 150°F (65.5°C) of 0.50–0.60 minute for *Coxiella burnetii*, and D 150°F of 0.20–0.30 minute for *Mycobacterium tuberculosis*.[54] For pH-elevating strains of *Bacillus licheniformis* spores in tomatoes, a $D_{95°C}$ of 5.1 minutes has been reported, whereas for *B. coagulans*, a $D_{95°C}$ of 13.7 minutes has been found.[39]

Table 17–9   Effect of pH on *D* Values for Spores of *C. botulinum* 62A Suspended in Three Food Products at 240°F (115°C)

pH	Spaghetti, Tomato Sauce, and Cheese	D Value (minutes) Macaroni Creole	Spanish Rice
4.0	0.128	0.127	0.117
4.2	0.143	0.148	0.124
4.4	0.163	0.170	0.149
4.6	0.223	0.223	0.210
4.8	0.226	0.261	0.256
5.0	0.260	0.306	0.266
6.0	0.491	0.535	0.469
7.0	0.515	0.568	0.550

Source: From Xezones and Hutchings,[65] copyright © 1965 by Institute of Food Technologists.

D values of some yeasts and molds that cause fruit spoilage are listed in Table 17–10 along with $D_{60°C}$ values.[50] It may be noted that *Saccharomyces cerevisiae* was the most heat resistant of the seven listed.

$D_{70°C}$ *values of Listeria innocua* and a mixture of six *Salmonella* serotypes determined in six meat and poultry products are noted in Table 17–11.[40] Overall, the heat resistance of the salmonellae was only slightly higher than for *L. innocua*. Thermal *D* values of salmonellae in various products have been reviewed.[16] The highest *D* values for salmonellae occurred in liquid egg white and yolks.

Using a pilot-scale pasteurized and five strains of *Mycobacterium avium* subsp. *paratuberculosis*, the mean $D_{72°C}$ was determined to be <2.03.[46] From this finding, a 15-sec exposure of this organism at 72°C is sufficient to effect $> 10^7$ cell kill.

z Value

The *z* value refers to the degrees Fahrenheit required for the thermal destruction curve to traverse one log cycle. Mathematically, this value is equal to the reciprocal of the slope of the TDT curve (Figure 17–3). Whereas *D* reflects the resistance of an organism to a specific temperature, *z* provides

Table 17–10   *D* Values of Seven Fruit Spoilage Fungi Determined in 0.1 M Citric Acid Buffer at 60°C and pH 4.0 (Extracted from Shearer et al.[50])

Penicillium citrinum	0.009
Torulaspora delbrueckii	0.018
Rhodotorula mucilaginosa	0.158
Zygosaccharomyces rouxii	0.008
Penicillium roquefortii	0.290
Aspergillus niger	0.449
Saccharomyces cerevisiae 0.28	

Table 17–11 $D_{70°C}$ Values of a Six-Serotype* Mixture of *Salmonella enterica* Serotypes and *Listeria innocua* Determined in Five Food Products (Extracted from Murphy et al.[40])

Food Specimen	Salmonella	L. innocua
Chicken patties	0.32	0.21
Chicken tenders	0.32	0.29
Frankfurters	0.39	0.36
Beef patties	0.25	0.29
Beef-turkey patties	0.37	0.18

*S. California, Heidelberg, Mission, Montevideo, Senftenberg, and Typhimurium.

information on the relative resistance of an organism to different destructive temperatures; it allows for the calculation of equivalent thermal processes at different temperatures. If, for example, 3.5 minutes at 140°F is considered to be an adequate process and $z = 8.0$, either 0.35 minute at 148°F or 35 minutes at 132°F would be considered equivalent processes.

D and z values of the beer-spoilage bacteria, *Pectinatus* spp., are noted in Table 17–12 where *P. frisingensis* had the highest $D_{60°C}$ and z values of the three cultures tested.[59] These investigators noted how the thermal properties of the tested organisms differed from strain to strain, and from one heating menstruum to another. The mean $D_{60°C}$ values for an eight-strain cocktail of salmonellae were found to be 1.30 and 5.48 in beef (containing 12.5% fat), and 5.70 in chicken (containing 7% fat), all determined by linear regression methods.[25]

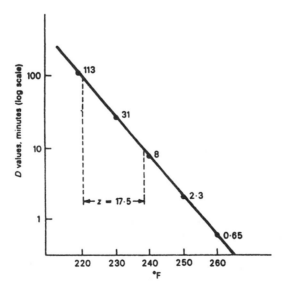

Figure 17–3   Thermal death time curve. Spores of strain F.S. 7 heated in canned pea brine pH 6.2. *Source:* From Gillespy,[20] courtesy of Butterworths Publishers, London.

Table 17–12 D and z Values of Three *Pectinatus* spp. Determined at 60°C in pH 5.2 Fresh Wort (Extracted from Watier et al.[59])

Organisms	D_{60} Values	$Z(°C)$
Pectinatus cerevisiiphilus	0.12	3.53
P. frisingensis	1.69	8.49
Pectinatus sp.	1.17	6.13

F Value

This value is the equivalent time, in minutes, at 250°F (121.1°C) of all heat considered, with respect to its capacity to destroy spores or vegetative cells of a particular organism. The integrated lethal value of heat received by all points in a container during processing is designated F_s or F_0. This represents a measure of the capacity of a heat process to reduce the number of spores or vegetative cells of a given organism per container. When we assume instant heating and cooling throughout the container of spores, vegetative cells, or food, F_0 may be derived as follows:

$$F_0 = D_r(\log a - \log b),$$

where a is the number of cells in the initial population and b is the number of cells in the final population.

Thermal Death Time Curve

For the purpose of illustrating a thermal destruction curve and D value, data are employed from Gillespy[20] on the killing of flat sour spores at 240°F (115°C) in canned pea brine at pH 6.2. Counts were determined at intervals of 5 minutes with the mean viable numbers indicated as follows:

Time (minutes)	Mean Viable Count
5	340.0
10	65.0
15	19.0
20	4.5
25	1.3

The time of heating in minutes is plotted on semi-log paper along the linear axis, and the number of survivors is plotted along the log scale to produce the TDT curve presented in Figure 17–2. The curve is essentially linear, indicating that the destruction of bacteria by heat is logarithmic and obeys a first-order reaction. Although difficulty is encountered at times at either end of the TDT curve, process calculations in the canning industry are based on a logarithmic order of death. From the data presented in Figure 17–2, the D value is calculated to be 8 minutes, or $D_{240} = 8.0$.

D values may be used to reflect the relative resistance of spores or vegetative cells to heat. The most heat-resistant strains of *C. botulinum* types A and B spores have a D_r value of 0.21, whereas the most heat-resistant thermophilic spores have D_r values of around 4.0–5.0. Putrefactive anaerobe (PA) 3679 was found by Stumbo et al.[55] to have a D_r value of 2.47 in cream-style corn, whereas flat-sour (FS) spore strain 617 was found to have a D_r of 0.84 in whole milk.

The approximate heat resistance of spores of thermophilic and mesophilic spoilage organisms may be compared by use of D_r values.

Geobacillus stearothermophilus:	4.0–5.0
Thermoanaerobacterium thermosaccharolyticum:	3.0–4.0
Clostridium nigrificans:	2.0–3.0
C. botulinum (types A and B):	0.10–0.2
C. sporogenes (including PA 3679):	0.10–1.5
B. coagulans:	0.01–0.07

The effect of pH and suspending menstrum on D values of *C. botulinum* spores is presented in Table 17–9. As noted above, microorganisms are more resistant at and around neutrality and show different degrees of heat resistance in different foods.

In order to determine the z value, D values are plotted on the log scale, and degrees Fahrenheit are plotted along the linear axis. From the data presented in Figure 17–3, the z value is 17.5. Values of z for *C. botulinum* range from 14.7 to 16.3, whereas for PA 3679, the range 6.6–20.5 has been reported. Some spores have been reported to have z values as high as 22. Peroxidase has been reported to have a z value of 47; and 50 has been reported for riboflavin, and 56 for thiamine.

12-D Concept

The 12-D concept refers to the process lethality requirement long in effect in the canning industry and implies that the minimum heat process should reduce the probability of survival of the most resistant *C. botulinum* spores to 10^{-12}. Because *C. botulinum* spores do not germinate and produce toxin below pH 4.6, this concept is observed only for foods above this pH value. An example from Stumbo[54] illustrates this concept from the standpoint of canning technology. If it is assumed that each container of food contains only one spore of *C. botulinum*, F_0 may be calculated by use of the general survivor curve equation with the other assumptions noted above in mind:

$$F_0 = D_r(\log a - \log b),$$
$$F_0 = 0.21(\log 1 - \log 10^{-12}),$$
$$F_0 = 0.21 \times 12 = 2.52$$

Processing for 2.52 minutes at 250°F, then, should reduce the *C. botulinum* spores to one spore in 1 of 1 billion (10^{12}) containers. When it is considered that some flat-sour spores have D_r values of about 4.0 and some canned foods receive F_0 treatments of 6.0–8.0, the potential number of *C. botulinum* spores is reduced to nondetectable levels.

SOME CHARACTERISTICS OF THERMOPHILES

On the basis of growth temperatures, thermophiles may be characterized as organisms with a minimum of around 45°C, an optimum between 50°C and 60°C, and a maximum of 70°C or above. By this definition, thermophilic species/strains are found among the cyanobacteria, archaebacteria, actinomycetes,[58] the anaerobic photosynthetic bacteria, thiobacilli, algae, fungi, bacilli, clostridia,

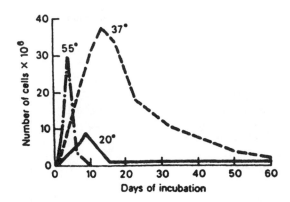

Figure 17–4   Growth curves of a bacterial strain incubated at 20°C, 37°C, and 55°C. *Source:* Reprinted with permission from Tanner and Wallace.[57]

lactic acid bacteria, and other groups. Those of greatest importance in foods belong to the genera *Bacillus*, *Clostridium*, *Alicyclobacillus*, *Geobacillus*, and *Thermoanaerobacterium*.

In thermophilic growth, the lag phase is short and sometimes difficult to measure. Spores germinate and grow rapidly. The logarithmic phase of growth is of short duration. Some thermophiles have been reported to have generation times as short as 10 minutes when growing at high temperatures. The rate of death or "die off" is rapid. Loss of viability or "autosterilization" below the thermophilic growth range is characteristic of organisms of this type. The growth curves of a bacterium at 55°C, 37°C, and 20°C are compared in Figure 17–4.

Why some organisms require temperatures of growth that are destructive to others is of concern not only from the standpoint of food preservation but also from that of the overall biology of thermophilism. Some of the known features of thermophiles are summarized below.

Enzymes

The enzymes of thermophiles can be divided into three groups:

1. Some enzymes are stable at the temperature of production but require slightly higher temperatures for inactivation—for example, malic dehydrogenase, adenosinetriphosphatase (ATPase), inorganic pyrophosphatase, aldolase, and certain peptidases.
2. Some enzymes are inactivated at the temperature of production in the absence of specific substrates—for example, asparagine deamidase, catalase, pyruvic acid oxidase, isocitrate lyase, and certain membrane-bound enzymes.
3. Some enzymes and proteins are highly heat resistant—for example, α-amylase, some proteases, glyceraldehyde-3-phosphate dehydrogenase, certain amino-acid-activating enzymes, flagellar proteins, esterases, and thermolysin.

In general, the enzymes of thermophiles produced under thermophilic growth conditions are more heat resistant than those of mesophiles (Table 17–13). Of particular note is α-amylase produced by a strain of *G. stearothermophilus*, which retained activity after being heated at 70°C for 24 hours. In one

Table 17–13 Comparison of Thermostability and Other Properties of Enzymes from Mesophilic and Thermophilic Bacteria

Species	Enzyme	(%)	Heat Stability*	Half-Cystine (mole/mole of Protein)	Molecular Weight	Metal Required for Stability
B. subtilis	Subtilisin BPN'	45	(50°C, 30 minutes)	0	28,000	Yes
B. subtilis	Neutral protease	50	(60°C, 15 minutes)	0	44,700	Yes
P. aeruginosa	Alkaline protease	80	(60°C, 10 minutes)	0	48,400	Yes
P. aeruginosa	Elastase	86	(70°C, 10 minutes)	4.6	39,500	Yes
		10	(75°C, 10 minutes)			
Group A streptococci	Streptococcal protease	0	(70°C, 30 minutes)	1	32,000	
C. histolyticum	Collagenase	1.5	(50°C, 20 minutes)	0	90,000	
S. griseus	Pronase	60	(60°C, 10 minutes)			Yes
B. thermoproteolyticus	Thermolysin	95	(60°C, 120 minutes)	0	42,700	Yes
		50	(80°C, 60 minutes)			
B. subtilis	α-Amylase	55	(65°C, 20 minutes)	0	50,000	Yes
G. stearothermophilus	α-Amylase	100	(70°C, 24 hours)	4	15,500	

*Activity remaining after heat treatment shown in parentheses.

Source: Matsubara,[34] copyright © 1967 by The American Association for the Advancement of Science.

study, the optimum temperature for the activity of amylase from *G. stearothermophilus* was found to be 82°C with a pH optimum of 6.9.[53] The enzyme required Ca^{2+} for thermostability. The heat stability of cytoplasmic proteins isolated from four thermophiles was greater than that from four mesophiles.[28]

Several possibilities exist as to why the enzymes of thermophiles are thermostable. Among these is the existence of higher levels of hydrophobic amino acids than exist in similar enzymes from mesophiles. A more hydrophobic protein would be more heat resistant. Regarding amino acids, it has been shown that lysine in place of glutamine decreased the thermostability of an enzyme, whereas replacements with other amino acids enhanced thermostability.[30] Another factor has to do with the binding of metal ions such as Mg^{2+}. The structural integrity of the membrane of *G. stearothermophilus* protoplasts was shown to be affected by divalent cations.[63]

Overall, the proteins of thermophiles are similar in molecular weight, amino acid composition, allosteric effectors, subunit composition, and primary sequences to their mesophilic counterparts. Extremely thermophilic and obligately thermophilic organisms synthesize macromolecules that have sufficient intrinsic molecular stability to withstand thermal stress.[1]

Ribosomes

In general, the thermal stability of ribosomes corresponds to the maximal growth temperature of a microorganism (Table 17–14). Heat-resistant ribosomes have been reported but not DNA. In a study of the ribosomes of *G. stearothermophilus*, no unusual chemical features of their proteins could be found that could explain their thermostability,[2] and in another study, no significant differences in either the size or the arrangement of surface filaments of *G. stearothermophilus* and *Escherichia coli* ribosomes could be found.[4] The base composition of ribosomal RNA (rRNA) has been shown to affect thermal stability. In a study of 19 organisms, the G–C content of rRNA molecules increased and the A–U content decreased with increasing maximal growth temperatures.[43] The increased G–C content makes for a more stable structure through more extensive hydrogen bonding. On the other hand, the thermal stability of soluble RNA from thermophiles and mesophiles appears to be similar.

Flagella

The flagella of thermophiles are more heat stable than those of mesophiles, with the former remaining intact at temperatures as high as 70°C, whereas those of the latter disintegrate at 50°C.[27,29] The thermophilic flagella are more resistant to urea and acetamide than those of mesophiles, suggesting that more effective hydrogen bonding occurs in thermophilic flagella.

OTHER CHARACTERISTICS OF THERMOPHILIC MICROORGANISMS

Nutrient Requirements

Thermophiles generally have a higher nutrient requirement than mesophiles when growing at thermophilic temperatures. Although this aspect of thermophilism has not received much study, changes in nutrient requirements when incubation temperature is raised may be due to a general lack of efficiency on the part of the metabolic complex. Certain enzyme systems might well be affected by the increased temperature of incubation, as well as the overall process of enzyme synthesis.

Table 17–14 Maximal Growth Temperatures and Ribosome Melting

Organism and Strain Number	Maximum Growth Temperature (°C)	Ribosome T_m (°C)
1. *V. marinus* (15381)	18	69
2. 7E-3	20	69
3. 1–1	28	74
4. *V. marinus* (15382)	30	71
5. 2–1	35	70
6. *D. desulfuricans* (*cholinicus*)*	40	73
7. *D. vulgaris* (8303)*	40	73
8. *E. coli* (B)	45	72
9. *E. coli* (Q13)	45	72
10. *S. itersonii* (SI–1)[†]	45	73
11. *B. megaterium* (Paris)	45	75
12. *B. subtilis* (SB-19)	50	74
13. *B. coagulans* (43P)	60	74
14. *D. nigrificans* (8351)[‡]	60	75
15. Thermophile 194	73	78
16. *G. stearothermophilus* (T-107)	73	78
17. *G. stearothermophilus* (1503R)	73	79
18. Thermophile (Tecce)	73	79
19. *G. stearothermophilus*	73	79

*Desulfovibrio.
[†]Spirillum.
[‡]Desulfotomaculum.

Source: From Pace and Campbell.[43]

Oxygen Tension

Thermophilic growth is affected by oxygen tension. As the temperature of incubation is increased, the growth rate of microorganisms increases, thereby increasing the oxygen demand on the culture medium while reducing the solubility of oxygen. This is thought by some investigators to be one of the most important limiting factors of thermophilic growth in culture media. Downey[15] has shown that thermophilic growth is optimal at or near the oxygen concentration normally available in the mesophilic range of temperatures—143 to 240 μM. Although it is conceivable that thermophiles are capable of high-temperature growth due to their ability to consume and conserve oxygen at high temperatures, a capacity that mesophiles and psychrophiles lack, further data in support of this notion are wanting.

Cellular Lipids

The state of cellular lipids affects thermophilic growth. Because an increase in degree of unsaturation of cellular lipids is associated with psychrotrophic growth, it is reasonable to assume that a reverse effect occurs in the case of thermophilic growth. This idea finds support in the investigations of many authors. Gaughran[19] found that mesophiles growing above their maximum range showed decreases in lipid content and more lipid saturation. According to this investigator, cells cannot grow at temperatures below the solidification point of their lipids. Marr and Ingraham[33] showed a progressive increase in saturated fatty acids and a corresponding decrease in unsaturated fatty acids in *E. coli* as the temperature of growth increased. The general decrease in the proportion of unsaturated fatty acids as growth temperature increase has been found to occur in a large variety of animals and plants. Saturated fatty acids form stronger hydrophobic bonds than do unsaturated. Among the saturated fatty acids are branched-chain acids. The preferential synthesis of branched heptadecanoic acid and the total elimination of unsaturated fatty acids by two thermophilic *Bacillus* spp. have been demonstrated.[60]

Mesophilic bacteria display changes in their membrane lipids when grown significantly above or below their normal growth range. When $D_{57°C}$ values of four heat-adapted strains of *E. coli* (including *E. coli* 0157:H7 and a *rpoS* mutant) were determined, the values were up to 3.9 minutes longer than controls for all strains.[64] Palmitic acid (16:0) and *cis*-vaccenic acid (18:1ϖ7c) increased in the membrane of two strains. Total intracellular verotoxin decreased in the heat-adapted strain but extracellular toxin increased in the heat-adapted cells, apparently the result of increased membrane fluidity.[64]

Cellular Membranes

The nature of cellular membranes affects thermophilic growth. Brock[11] reported that the molecular mechanism of thermophilism is more likely to be related to the function and stability of cellular membranes than to the properties of specific macromolecules. This investigator pointed out that there is no evidence that organisms are killed by heat because of the inactivation of proteins or other macromolecules, a view that is widely held. According to Brock, an analysis of thermal death curves of various microorganisms shows them to be a first-order processes compatible with an effect of heat on some large structure such as the cell membrane, as a single hole in the membrane could result in leakage of cell constituents and subsequent death. Brock has also pointed out that thermal killing due to the inactivation of heat-sensitive enzymes, or heat-sensitive ribosomes, of which there are many copies in the cell, should not result in simple first-order kinetics. The leakage of ultraviolet light-absorbing and other material from cells undergoing "cold shock" would tend to implicate the membrane in high-temperature death. Because most animals die when body temperatures reach between 40°C and 45°C and most psychrophilic bacteria are killed at about this temperature range, the suggestion that lethal injury is due to the melting of lipid constituents of the cell or cell membrane is not only plausible, it has been supported by the findings of various investigators. The unit cell membrane consists of layers of lipid surrounded by layers of protein and depends on the lipid layers for its biological functions. The disruption of this structure would be expected to cause cell damage and perhaps death. In view of the changes in cellular lipid saturation noted above, cellular membrane integrity appears to be critical to growth and survival at thermophilic temperatures.

Effect of Temperature

Brock[11] called attention to the fact that thermophiles apparently do not grow as fast at their optimum temperatures as one would predict or is commonly believed. Arrhenius plots of thermophile growth compared to *E. coli* over a range of temperatures indicated that, overall, the mesophilic types were more efficient. Brock noted that thermophile enzymes are inherently less efficient than mesophiles because of thermal stability; that is, the thermophiles have had to discard growth efficiency in order to survive at all.

Genetics

A significant discovery toward an understanding of the genetic bases of thermophilism was made by McDonald and Matney.[36] These investigators effected the transformation of thermophilism in *B. subtilis* by growing cells of a strain that could not grow above 50°C in the presence of DNA extracted from one that could grow at 55°C. The more heat-sensitive strain was transformed at a frequency of 10^{-4}. These authors noted that only 10–20% of the transformants retained the high-level streptomycin resistance of the recipient, which indicated that the genetic loci for streptomycin resistance and that for growth at 55°C were closely linked.

Although much has been learned about the basic mechanisms of thermophilism in microorganisms, the precise mechanisms underlying this high-temperature phenomenon remain a mystery. The facultative thermophiles such as some *B. coagulans* strains present a picture as puzzling as the obligate thermophiles. The facultative thermophiles display both mesophilic and thermophilic types of metabolism. In their studies of these types from the genus *Bacillus*, which grew well at both 37°C and 55°C, Bausum and Matney[6] reported that the organisms appear to shift from mesophilism to thermophilism between 44°C and 52°C.

CANNED FOOD SPOILAGE

Although the objective in the thermal canning of foods is the destruction of microorganisms, these products nevertheless undergo microbial spoilage under certain conditions. The main reasons for this are underprocessing, inadequate cooling, contamination of the can resulting from leakage through seams, and preprocess spoilage. Because some canned foods receive low-heat treatments, it is to be expected that a rather large number of different types of microorganisms may be found upon examining such foods.

As a guide to the type of spoilage that canned foods undergo, the following classification of canned foods based on acidity is helpful.

Low Acid (pH > 4.6)

This category includes meat and marine products, milk, some vegetables (corn, lima beans), meat and vegetable mixtures, and so on. These foods are spoiled by the thermophilic flat-sour group (*Geobacillus stearothermophilus, B. coagulans*), sulfide spoilers (*Clostridium nigrificans, C. bifermentans*), and/or gaseous spoilers (*Thermoanaerobacterium thermosaccharolyticum*). Mesophilic spoilers include putrefactive anaerobes (especially PA 3679 types). Spoilage and toxin production by proteolytic *C. botulinum* strains may occur if they are present. Medium-acid foods are those with a pH range of 5.3–4.6, whereas low-acid foods are those with pH ≥ 5.4.

Acid (pH 3.7–4.0 to 4.6)

In this category are fruits such as tomatoes, pears, and figs. Thermophilic spoilers include *B. coagulans* types. Mesophiles include *P. polymyxa, P. macerans (B. betanigrificans), C. pasteurianum, C. butyricum, Thermoanaerobacterium thermosaccharolyticum*, lactobacilli, and others.

High Acid (pH < 4.0–3.7)

This category includes fruits and fruit and vegetable products—grapefruit, rhubarb, sauerkraut, pickles, and so forth. These foods are generally spoiled by non-sporeforming mesophiles—yeasts, molds, *Alicyclobacillus* spp., and/or lactic acid bacteria. *Alicyclobacillus* spp. can grow in and cause spoilage of apple and tomato juice and white grape juice.[52] The fungus *Byssochlamys* can grow at pH as low as 2.0, and *Neosartorya fischeri* can grow as low as pH 3.0.[9]

Canned food spoilage organisms may be further characterized as follows:

- Mesophilic organisms
  - Putrefactive anaerobes
  - Butyric anaerobes
  - Aciduric flat sours
  - Lactobacilli
  - Yeasts
  - Molds
- Thermophilic organisms
  - Thermophilic anaerobes producing sulfide
  - Flat-four spores
  - Thermophilic anaerobes not producing sulfide

The canned food spoilage manifestations of these organisms are presented in Table 17–15.

With respect to the spoilage of high-acid and other canned foods by yeasts, molds, and bacteria, several of these have been repeatedly associated with certain foods. The yeasts *Torula lactis-condensi* and *T. globosa* cause blowing or gaseous spoilage of sweetened condensed milk, which is not heat processed. The mold *Aspergillus repens* is associated with the formation of "buttons" on the surface of sweetened condensed milk. *Lactobacillus brevis* (*L. lycopersici*) causes a vigorous fermentation in tomato catsup, Worcestershire sauce, and similar products. *Leuconostoc mesenteroides* has been found to cause gaseous spoilage of canned pineapples and ropiness in peaches. The mold *Byssochlamys fulva* causes spoilage of bottled and canned fruits, and its actions cause disintegration of fruits as a result of pectin breakdown.[5] *Torula stellata* has been found to cause the spoilage of canned bitter lemon, and to grow at a pH of 2.5.[44]

Frozen concentrated orange juice sometimes undergoes spoilage by yeasts and bacteria. Hays and Reister[23] investigated samples of this product spoiled by bacteria. The orange juice was characterized as having a vinegary to buttermilk off-odor with an accompanying off-flavor. From the spoiled product were isolated *L. plantarum* var. *mobilis*, *L. brevis*, *Leuconostoc mesenteroides*, and *Leuconostoc dextranicum*. The spoilage characteristics could be reproduced by inoculating the above isolates into fresh orange juice.

Minimum growth temperatures of spoilage thermophiles are of some importance in diagnosing the cause of spoiled canned foods. *B. coagulans* (*B. thermoacidurans*) has been reported to grow only slowly at 25°C but grows well between 30°C and 55°C. *G. stearothermophilus* does not grow at 37°C, its optimum temperature being around 65°C with smooth variants showing a shorter generation time at this temperature than rough variants.[17] *T. thermosaccharolyticum* does not grow at 30°C but has been reported to grow at 37°C. For reviews on the spoilage of acid and low-acid food products, see references 9, 38, and 52.

Also of importance in diagnosing the cause of canned food spoilage is the appearance of the unopened can or container. The ends of a can of food are normally flat or slightly concave. When

Table 17–15 Spoilage Manifestations in Acid and Low-Acid Canned Foods

Type of Organism	Appearance and Manifestations of Can	Condition of Product
Acid products		
1. *B. thermoacidurans* (flat sour: tomato juice)	Can flat; little change in vacuum	Slight pH change; off-odor and flavor
2. Butyric anaerobes (tomatoes and tomato juice)	Can swells; may burst	Fermented; butyric odor
3. Non-sporeformers (mostly lactics)	Can swells, usually bursts, but swelling may be arrested	Acid odor
Low-acid products		
1. Flat sour	Can flat; possible loss of vacuum on storage	Appearance not usually altered; pH markedly lowered—sour; may have slightly abnormal odor; sometimes cloudy liquor
2. Thermophilic anaerobe	Can swells; may burst	Fermented, sour, cheesy, or butyric odor
3. Sulfide spoilage	Can flat; H_2S gas absorbed by product	Usually blackened; "rotten egg" odor
4. Putrefactive anaerobe	Can swells; may burst	May be partially digested; pH slightly above normal; typical putrid odor
5. Aerobic sporeformers (odd types)	Can flat; usually no swelling, except in cured meats when NO_3 and sugar are present	Coagulated evaporated milk, black beets

Source: From Schmitt.[47]

microorganisms grow and produce gases, the can goes through a series of changes that are visible from the outside. The can is designated a *flipper* when one end is made convex by striking or heating the can. A *springer* is a can with both ends bulged when one or both remain concave if pushed in or when one end is pushed in and the other pops out. A *soft swell* refers to a can with both ends bulged that may be dented by pressing with the fingers. A *hard swell* has both ends bulged, so that neither end can be dented by hand. These events tend to develop successively and become of value in predicting the type of spoilage that might be in effect. Flippers and springers may be incubated under wraps at a temperature appropriate to the pH and type of food in order to allow for further growth of any organisms that might be present. These effects on cans do not always represent microbial spoilage. Soft swells often represent microbial spoilage, as do hard swells. In high-acid foods, however, hard swells are often *hydrogen swell*, which result from the release of hydrogen gas by the action of food acids on the iron of the can. The other two most common gases in cans of spoiled foods are CO_2 and H_2S, both of which are the result of the metabolic activities of microorganisms. Hydrogen sulfide may be noted by its characteristic odor, whereas CO_2 and hydrogen may be determined by the following test. Construct an apparatus of glass or plastic tubing attached to a hollow punch fitted with a large rubber stopper. Into a test tube filled with dilute KOH, insert the free end of this apparatus and invert it in a beaker filled with dilute KOH. When an opening is made in one end of the can with the hollow

Table 17–16 Some Features of Canned Food Spoilage Resulting from Understerilization and Seam Leakage

Feature	Understerilization	Leakage
Can	Flat or swelled; seams generally normal	Swelled; may show defects
Product appearance	Sloppy or fermented	Frothy fermentation; viscous
Odor	Normal, sour, or putrid but generally consistent	Sour, fecal, generally varying from can to can
PH	Usually fairly constant	Wide variation
Microscopic and cultural	Pure cultures, sporeformers; growth at 98°F and/or 113°F; may be characteristic on special media, e.g., acid agar for tomato juice	Mixed cultures, generally rods and cocci; growth only at usual temperatures
History	Spoilage usually confined to certain portions of pack. In acid products, diagnosis may be less clearly defined. Similar organisms may be involved in understerilization and leakage	Spoilage scattered

Source: From Schmitt.[47]

punch, the gases will displace the dilute KOH inside the tube. Before removing the open end from the beaker, close the tube by placing the thumb over the end. To test for CO_2, shake the tube and look for a vacuum as evidenced by suction against the finger. To test for hydrogen, repeat the test and apply a match near the top of the tube and then quickly remove the thumb. A "pop" indicates the presence of hydrogen. Both gases may be found in some cans of spoiled foods.

"Leakage-type" spoilage of canned foods is characterized by a biota of non-sporeforming organisms that would not survive the heat treatment normally given to heat-processed foods. These organisms enter cans at the start of cooling through faulty seams, which generally result from can abuse. The organisms that cause leakage-type spoilage can be found either on the cans or in the cooling water. This problem is minimized if the cannery cooling water contains <100 bacteria/ml. This type of spoilage may be further differentiated from that caused by understerilization (see Table 17–16).

REFERENCES

1. Amelunxen, R.E., and A.L. Murdock. 1978. Microbial life at high temperatures: Mechanisms and molecular aspects. In *Microbial Life in Extreme Environments*, ed. D.J. Kushner, 217–278. New York: Academic Press.

2. Ansley, S.B., L.L. Campbell, and P.S. Sypherd. 1969. Isolation and amino acid composition of ribosomal proteins from *Bacillus stearothermophilus. J. Bacteriol.* 98:568–572.

3. Baird-Parker, A.C., M. Boothroyd, and E. Jones. 1970. The effect of water activity on the heat resistance of heat sensitive and heat resistant strains of salmonellae. *J. Appl. Bacteriol.* 33:515–522.

4. Bassel, A., and L.L. Campbell. 1969. Surface structure of *Bacillus stearothermophilus* ribosomes. *J. Bacteriol.* 98:811–815.

5. Baumgartner, J.G., and A.C. Hersom. 1957. *Canned Foods.* Princeton, NJ: D. Van Nostrand.

6. Bausum, H.T., and T.S. Matney, 1965. Boundary between bacterial mesophilism and thermophilism. *J. Bacteriol.* 90:50–53.

7. Beaman, T.C., and P. Gerhardt. 1986. Heat resistance of bacterial spores correlated with protoplast dehydration, mineralization, and thermal adaptation. *Appl. Environ. Microbiol.* 52:1242–1246.

8. Beaman, T.C., J.T. Greenamyre, T.R. Corner, H. S. Pankratz, and P. Gerhardt. 1982. Bacterial spore heat resistance correlated with water content, wet density, and protoplast/sporoplast volume ratio. *J. Bacteriol.* 150:870–877.

9. Beuchat, L.R. 1998. Spoilage of acid products by heat-resistant molds. *Dairy Food Environ. Sanit.* 18:588–593.

10. Beuchat, L.R. 1986. Extraordinary heat resistance of *Talaromyces flavus* and *Neosartorya fischeri* ascospores in fruit products. *J. Food Sci.* 52:1506–1510.

11. Brock, T.D. 1967. Life at high temperatures. *Science* 158:1012–1019.

12. Carpenter, P.L. 1967. *Microbiology*, 2nd ed. Philadelphia: W.B. Saunders.

13. Corry, J.E.L. 1974. The effect of sugars and polyols on the heat resistance of salmonellae. *J. Appl. Bacteriol.* 37:31–43.

14. Cunningham, F.E., and H. Lineweaver, 1965. Stabilization of egg-white proteins to pasteurizing temperatures above 60°C. *Food Technol.* 19:1442–1447.

15. Downey, R.J. 1966. Nitrate reductase and respiratory adaptation in *Bacillus stearothermophilus*. *J. Bacteriol.* 91:634–641.

16. Doyle, M.E., and A.S. Mazzotta. 2000. Review of studies on the thermal resistance of salmonellae. *J. Food Protect.* 63:779–795.

17. Fields, M.L. 1970. The flat sour bacteria. *Adv. Food Res.* 18:163–217.

18. Gaillard, S., I. Leguerinel, and P. Mafart. 1998. Model for combined effects of temperature, pH and water activity on thermal inactivation of *Bacillus cereus* spores. *J. Food Sci.* 63:887–889.

19. Gaughran, E.R.L. 1947. The saturation of bacterial lipids as a function of temperature. *J. Bacteriol.* 53:506.

20. Gillespy, T.G. 1962. The principles of heat sterilization. *Recent Adv. Food Sci.* 2:93–105.

21. Goepfert, J.M., I.K. Iskander, and C.H. Amundson. 1970. Relation of the heat resistance of salmonellae to the water activity of the environment. *Appl. Microbiol.* 19:429–433.

22. Hansen, N.H., and H. Riemann. 1963. Factors affecting the heat resistance of nonsporing organisms. *J. Appl. Bacteriol.* 26:314–333.

23. Hays, G.L., and D.W. Riester. 1952. The control of "off-odor" spoilage in frozen concentrated orange juice. *Food Technol.* 6:386–389.

24. Jelen, P. 1982. Experience with direct and indirect UHT processing of milk—A Canadian viewpoint. *J. Food Protect.* 45:878–883.

25. Juneja, V.K., B.S. Eblen, and G.M. Ransom. 2001. Thermal inactivation of *Salmonella* spp. in chicken broth, beef, pork, turkey, and chicken: Determination of D- and z-values. *J. Food Sci.* 66:146–152.

26. Keswani, J., and J.F. Frank. 1998. Thermal inactivation of *Mycobacterium paratuberculosis* in milk. *J. Food Protect.* 61:974–978.

27. Koffler, H. 1957. Protoplasmic differences between mesophiles and thermophiles. *Bacteriol. Rev.* 21:227–240.

28. Koffler, H., and G.O. Gale. 1957. The relative thermostability of cytoplasmic proteins from thermophilic bacteria. *Arch. Biochem. Biophys.* 67:249–251.

29. Koffler, H., G.E. Mallett, and J. Adye. 1957. Molecular basis of biological stability to high temperatures. *Proc. Natl. Acad. Sci. USA* 43:464–477.

30. Koizumi, J.I., M. Zhang, T. Imanaka, and S. Aiba. 1990. Does single-amino-acid replacement work in favor of or against improvement of the thermostability of immobilized enzyme? *Appl. Environ. Microbiol.* 56:3612–3614.

31. Levinson, H.S., and M.T. Hyatt. 1964. Effect of sporulation medium on heat resistance, chemical composition, and germination of *Bacillus megaterium* spores. *J. Bacteriol.* 87:876–886.

32. Lihono, M.A., A.F. Mendonca, J.S. Dickson, and P.M. Dixon. 2003. A predictive model to determine the effects of temperature, sodium pyrophosphate, and sodium chloride on thermal inactivation of starved *Listeria monocytogenes* in pork slurry. *J. Food Protect.* 66:1216–1221.

33. Marr, A.G., and J.L. Ingraham. 1962. Effect of temperature on the composition of fatty acids in *Escherichia coli*. *J. Bacteriol.* 84:1260–1267.

34. Matsubara, H. 1967. Some properties of thermolysin. In *Molecular Mechanisms of Temperature Adaptation*, ed. C.L. Prosser, Pub. No. 84, 283–294. Washington, DC: American Association for the Advancement of Science.

35. Mazzotta, A.S. 2001. Heat resistance of *Listeria monocytogenes* in vegetables: Evaluation of blanching processes. *J. Food Protect.* 64:385–387.

36. McDonald, W.C., and T.S. Matney. 1963. Genetic transfer of the ability to grow at 55°C in *Bacillus subtilis*. *J. Bacteriol.* 85:218–220.

37. McIntyre, S., J.Y. Ikawa, N. Parkinson, J. Hagland, and J. Lee. 1995. Characteristics of an acidophilic *Bacillus* strain isolated from shelf-stable juices. *J. Food Protect.* 58:319–321.

38. Morton, R.D. 1998. Spoilage of acid products by butyric acid anaerobes—A review. *Dairy Food Environ. Sanit.* 18:580–584.

39. Montville, T.J., and G.M. Sapers. 1981. Thermal resistance of spores from pH elevating strains of *Bacillus licheniformis*. *J. Food Sci.* 46:1710–1712.

40. Murphy, R.Y., L.K. Duncan, E.R. Johnson, M.D. Davis, and J.N. Smith. 2002. Thermal inactivation D- and z-values of *Salmonella* serotypes and *Listeria innocua* in chicken patties, chicken tenders, franks, beef patties, and blended beef and turkey patties. *J. Food Protect.* 65:53–60.

41. Nakashio, S., and P. Gerhardt. 1985. Protoplast dehydration correlated with heat resistance of bacterial spores. *J. Bacteriol.* 162:571–578.

42. Ng, H., H.G. Bayne, and J.A. Garibaldi. 1969. Heat resistance of *Salmonella*: The uniqueness of *Salmonella* Senftenberg 775W. *Appl. Microbiol.* 17:78–82.

43. Pace, B., and L.L. Campbell. 1967. Correlation of maximal growth temperature and ribosome heat stability. *Proc. Natl. Acad. Sci. USA* 57:1110–1116.

44. Perigo, J.A., B.L. Gimbert, and T.E. Bashford. 1964. The effect of carbonation, benzoic acid, and pH on the growth rate of a soft drink spoilage yeast as determined by a turbidostatic continuous culture apparatus. *J. Appl. Bacteriol.* 27:315–332.

45. Rajashekhara, E., E.R. Suresh, and S. Ethiraj. 1996. Influence of different heating media on thermal resistance of *Neosartorya fischeri* isolated from papaya fruit. *J. Appl. Bacteriol.* 81:337–340.

46. Pearce, L.E., H.T. Truong, R.A. Crawford, G.F. Yates, S. Cavaignac, and G.W. de Lisle. 2001. Effect of turbulent-flow pasteurization on survival of *Mycobacterium avium* subsp. *paratuberculosis* added to raw milk. *Appl. Environ. Microbiol.* 67:3964–3969.

47. Schmitt, H.P. 1966. Commercial sterility in canned foods, its meaning and determination. *Assoc. Food Drug Off. U.S., Q. Bull.* 30:141–151.

48. Scott, V.N., and D.T. Bernard. 1987. Heat resistance of *Talaromyces flavus* and *Neosartorya fischeri* isolated from commercial fruit juices. *J. Food Protect.* 50:18–20.

49. Setlow, P., and E.A. Johnson. 2001. Spores and their significance. In *Food Microbiology—Fundamentals and Frontiers*, 2nd ed., ed. M.P. Doyle, L.R. Beuchat, and T.J. Montville, 33–70. Washington, DC: ASM Press.

50. Shearer, A.E.H., A.S. Mazzotta, R. Chuyate, and D.E. Gombas. 2002. Heat resistance of juice spoilage microorganisms. *J. Food Protect.* 65:1271–1275.

51. Siunov, A.V., D.V. Nikitin, N.E. Suzina, V.V. Dmitriev, N.P. Kuzmin, and V.I. Duda. 1999. Phylogenetic status of *Anaerobacter polyendosporus*, an anaerobic, polysporogenic bacterium. *Int. J. Syst. Bacteriol.* 49:1119–1124.

52. Splittstoesser, D.F., C.Y. Lee, and J.J. Churey. 1998. Control of *Alicyclobacillus* in the juice industry. *Dairy Food Environ. Sanit.* 18:585–587.

53. Srivastava, R.A.K., and J.N. Baruah. 1986. Culture conditions for production of thermostable amylase by *Bacillus stearothermophilus*. *Appl. Environ. Microbiol.* 52:179–184.

54. Stumbo, C.R. 1973. *Thermobacteriology in Food Processing*, 2nd ed. New York: Academic Press.

55. Stumbo, C.R., J.R. Murphy, and J. Cochran. 1950. Nature of thermal death time curve for P.A. 3679 and *Clostridium botulinum*. *Food Technol.* 4:321–326.

56. Sugiyama, H. 1951. Studies on factors affecting the heat resistance of spores of *Clostridium botulinum*. *J. Bacteriol.* 62:81–96.

57. Tanner, F.W., and G.I. Wallace. 1925. Relation of temperature to the growth of thermophilic bacteria. *J. Bacteriol.* 10:421–437.

58. Tendler, M.D., and P.R. Burkholder. 1961. Studies on the thermophilic *Actinomycetes*. I. Methods of cultivation. *Appl. Microbiol.* 9:394–399.

59. Watier, D., I. Leguerinel, J.P. Hornez, I. Chowdhury, and H.C. Dubourguier. 1995. Heat resistance of *Pectinatus* sp., a beer spoilage anaerobic bacterium. *J. Appl. Bacteriol.* 78:164–168.

60. Weerkamp, A., and W. Heinen. 1972. Effect of temperature on the fatty acid composition of the extreme thermophiles, *Bacillus caldolyticus* and *Bacillus caldotenax*. *J. Bacteriol.* 109:443–446.

61. White, H.R. 1963. The effect of variation in pH on the heat resistance of cultures of *Streptococcus faecalis*. *J. Appl. Bacteriol.* 26:91–99.

62. Williams, O.B., and W.J. Robertson. 1954. Studies on heat resistance. VI. Effect of temperature of incubation at which formed on heat resistance of aerobic thermophilic spores. *J. Bacteriol.* 67:377–378.

63. Wisdom, C., and N.E. Welker. 1973. Membranes of *Bacillus stearothermophilus*. Factors affecting protoplast stability and thermostability of alkaline phosphatase and reduced nicotinamide adenine dinucleotide oxidase. *J. Bacteriol.* 114:1336–1345.

64. Yuk, H.-G., and D.L. Marshall. 2003. Heat adaptation alters *Escherichia coli* 0157:H7 membrane lipid composition and verotoxin production. *Appl. Environ. Microbiol.* 69:5115–5119.

65. Xezones, H., and I.J. Hutchings. 1965. Thermal resistance of *Clostridium botulinum* (62A) spores as affected by fundamental food constituents. *Food Technol.* 19:1003–1005.

CHAPTER 18

Protection of Foods by Drying

The preservation of foods by drying is based on the fact that microorganisms and enzymes need water in order to be active. In preserving foods by this method, one seeks to lower the moisture content to a point where the activities of food-spoilage and food-poisoning microorganisms are inhibited. Dried, desiccated, or low-moisture (LM) foods are those that generally do not contain more than 25% moisture and have a water activity (a_w) between 0.00 and 0.60. These are the traditional dried foods. Freeze-dried foods are also in this category. Another category of shelf-stable foods are those that contain between 15% and 50% moisture and an a_w between 0.60 and 0.85. These are the intermediate-moisture (IM) foods. Some of the microbiological aspects of IM and LM foods are dealt with in this chapter.

PREPARATION AND DRYING OF LOW-MOISTURE FOODS

The earliest uses of food desiccation consisted of exposing fresh foods to sunlight until drying had been achieved. Through this method of drying, which is referred to as sun drying, certain foods may be successfully preserved if the temperature and relative humidity (RH) allow. Fruits such as grapes, prunes, figs, and apricots may be dried by this method, which requires a large amount of space for large quantities of the product. The drying methods of greatest commercial importance consist of spray, drum, evaporation, and freeze-drying.

Preparatory to drying, foods are handled in much the same manner as for freezing, with a few exceptions. In the drying of fruits such as prunes, alkali dipping is employed by immersing the fruits into hot lye solutions of between 0.1% and 1.5%. This is especially true when sun drying is employed. Light-colored fruits and certain vegetables are treated with SO_2 so that levels of between 1,000 and 3,000 ppm may be absorbed. The latter treatment helps to maintain color, conserve certain vitamins, prevent storage changes, and reduce the microbial load. After drying, fruits are usually heat pasteurized at 150–185°F (65.6–85°C) for 30–70 minutes.

Similar to the freezing preparation of vegetable foods, blanching or scalding is a vital step prior to dehydration. This may be achieved by immersion from 1 to 8 minutes, depending on the product. The primary function of this step is to destroy enzymes that may become active and bring about undesirable changes in the finished product. Leafy vegetables generally require less time than peas, beans, or carrots. For drying, temperatures of 140–145°F (60–62.8°C) have been found to be safe for many vegetables. The moisture content of vegetables should be reduced below 4% in order to have satisfactory storage life and quality. Many vegetables may be made more stable if given a treatment with SO_2 or a sulfite. The drying of vegetables is usually achieved by use of tunnel-, belt-, or cabinet-type driers.

Meat is usually cooked before being dehydrated. The final moisture content after drying should be approximately 4% for beef and pork.

Milk is dried as either whole milk or nonfat skim milk. The dehydration may be accomplished by either the drum or spray method. The removal of about 60% water from whole milk results in the production of evaporated milk, which has about 11.5% lactose in solution. Sweetened condensed milk is produced by the addition of sucrose or glucose before evaporation, so that the total average content of all sugar is about 54%, or over 64% in solution. The stability of sweetened condensed milk is due in part to the fact that the sugars tie up some of the water and make it unavailable for microbial growth.

Eggs may be dried as whole egg powder, yolks, or egg white. Reducing the glucose content prior to drying increases dehydration stability. Spray drying is the method most commonly employed.

In *freeze drying* (lyophilization, cryophilization), actual freezing is preceded by the blanching of vegetables and the precooking of meats. The rate at which a food material freezes or thaws is influenced by the following factors:[12]

1. the temperature differential between the product and the cooling or heating medium;
2. the means of transferring heat energy to, from, and within the product (conduction, convection, radiation);
3. the type, size, and shape of the package;
4. the size, shape, and thermal properties of the product.

Rapid freezing has been shown to produce products that are more acceptable than slow freezing. Rapid freezing allows for the formation of small ice crystals and, consequently, causes less mechanical damage to the food structure. Upon thawing, fast-frozen foods take up more water and, in general, display characteristics more like the fresh product than the slow-frozen foods. After freezing, the water in the form of ice is removed by sublimation. This process is achieved by various means of heating plus vacuum. The water content of protein foods can be placed into two groups: freezable and unfreezable. Unfreezable (bound) water has been defined as that which remains unfrozen below $-30°C$. The removal of freezable water takes place during the first phases of drying, and this phase of drying may account for the removal of anywhere from 40% to 95% of the total moisture. The last water to be removed is generally bound water, some of which may be removed throughout the drying process. Unless heat treatment is given prior to freeze drying, freeze-dried foods retain their enzymes. In studies on freeze-dried meats, it has been shown that 40–80% of the enzyme activity is not destroyed and may be retained after 16 months of storage at $-20°C$.[24] The final product moisture level in freeze-dried foods may be about 2–8% or have an a_w of 0.10–0.25.[36]

Freeze drying is generally preferred to high-temperature vacuum drying. Among the disadvantages of the latter compared to the former are the following:[17]

1. pronounced shrinkage of solids;
2. migration of dissolved constituents to the surface when drying solids;
3. extensive denaturation of proteins;
4. case hardening: the formation of a relatively hard, impervious layer at the surface of a solid, caused by one or more of the first three changes, that slows the rates of both dehydration and reconstitution;
5. formation of hard, impervious solids when drying liquid solution;
6. undesirable chemical reactions in heat-sensitive materials;
7. excessive loss of desirable volatile constituents;
8. difficulty of rehydration as a result of one or more of the other changes.

EFFECT OF DRYING ON MICROORGANISMS

Although some microorganisms are destroyed in the process of drying, this process is not lethal per se to microorganisms, and, indeed, many types may be recovered from dried foods, especially if poor-quality foods are used for drying and if proper practices are not followed in the drying steps.

Bacteria require relatively high levels of moisture for their growth, with yeasts requiring less and molds still less. Because most bacteria require a_w values above 0.90 for growth, they play no role in the spoilage of dried foods. With respect to the stability of dried foods, Scott[29] has related a_w levels to the probability of spoilage in the following manner. At a_w values of between 0.80 and 0.85, spoilage occurs readily by a variety of fungi in 1–2 weeks. At a_w values of 0.75, spoilage is delayed, with fewer types of organisms in those products that spoil. At an a_w of 0.70, spoilage is greatly delayed and may not occur during prolonged holding. At an a_w of 0.65, very few organisms are known to grow, and spoilage is most unlikely to occur for even up to 2 years. Some investigators have suggested that dried foods to be held for several years should be processed so that the final a_w is between 0.65 and 0.75, with 0.70 suggested by most.

At a_w levels of about 0.90, the organisms most likely to grow are yeasts and molds. This value is near the minimum for most normal yeasts. Even though spoilage is all but prevented at an a_w less than 0.65, some molds are known to grow very slowly at a_w 0.60–0.62. Osmophilic yeasts such as *Zygosaccharomyces rouxii* strains have been reported to grow at an a_w of 0.65 under certain conditions. The most troublesome group of microorganisms in dried foods is the mold, with the *Aspergillus glaucus* group being the most notorious at low a_w values. The minimum a_w values reported for the germination and growth of molds and yeasts are presented in Table 18–1. Pitt and Christian[26] found the predominant spoilage molds of dried and high-moisture prunes to be members of the *A. glaucus* group and *Xeromyces bisporus*. Aleuriospores of *X. bisporus* were able to germinate in 120 days at an a_w of 0.605. Generally, higher moisture levels were required for both asexual and sexual sporulation.

As a guide to the storage stability of dried foods, the "alarm water" content has been suggested. The alarm water content is the water content that should not be exceeded if mold growth is to be avoided.

Table 18–1   Minimum a_w Reported for the Germination and Growth of Food Spoilage Yeasts and Molds

Organism	Minimum a_w
Candida utilis	0.94
Botrytis cinerea	0.93
Rhizopus stolonifer (*nigricans*)	0.93
Mucor spinosus	0.93
Candida scottii	0.92
Trichosporon pullulans	0.91
Candida zeylanoides	0.90
Saccharomycopsis vernalis	0.89
Alternaria citri	0.84
Aspergillus glaucus	0.70
Aspergillus echinulatus	0.64
Zygosaccharomyces rouxii	0.62

Note: See Tables 3–5 for other organisms.

Table 18–2   "Alarm Water" Content for Miscellaneous
Foods

Foods	% Water
Whole milk powder	~ 8
Dehydrated whole eggs	10–11
Wheat flour	13–15
Rice	13–15
Milk powder (separated)	15
Fat-free hydrated meat	15
Pulses	15
Dehydrated vegetables	14–20
Starch	18
Dehydrated fruit	18–25

Note: RH = 70%; temperature = 20°C.

Source: From Mossel and Ingram.[25]

Although these values may be used to advantage, they should be followed with caution because a rise of only 1% may be disastrous in some instances.[29] The alarm water content for some miscellaneous foods is presented in Table 18–2. In freeze-dried foods, the rule of thumb has been to reduce the moisture level to 2%. Burke and Decareau[7] pointed out that this low level is probably too severe for some foods that might keep well at higher levels of moisture, without the extra expense of removing the last low levels of water.

Although drying destroys some microorganisms, bacterial endospores survive, as do yeasts, molds, and many Gram-negative and Gram-positive bacteria. In their study of bacteria from chicken meat after freeze drying and rehydration at room temperature, May and Kelly[23] were able to recover about 32% of the original biota. These workers showed that *Staphylococcus aureus* added prior to freeze-drying could survive under certain conditions. Some or all foodborne parasites, such as *Trichinella spiralis*, have been reported to survive the drying process.[11] The goal is to produce dried foods with a total count of not more than 100,000/g. It is generally agreed that the coliform count of dried foods should be zero or nearly so, and no food-poisoning organisms should be allowed with the possible exception of low numbers of *Clostridium perfringens*. With the exception of those that may be destroyed by blanching or precooking, relatively fewer organisms are destroyed during the freeze-drying process. More are destroyed during freezing than during dehydration. During freezing, between 5% and 10% of water remains "bound" to other constituents of the medium. This water is removed by drying. Death or injury from drying may result from denaturation in the still-frozen, undried portions, due to concentration resulting from freezing, the act of removing the "bound" water, and/or recrystallization of salts or hydrates formed from eutectic solutions.[24] When death occurs during dehydration, the rate is highest during the early stages of drying. Young cultures have been reported to be more sensitive to drying than old cultures.[13]

The freeze-drying method is one of the best known ways of preserving microorganisms. Once the process has been completed, the cells may remain viable indefinitely. Upon examining the viability of 277 cultures of bacteria, yeasts, and molds that had been lyophilized for 21 years, Davis[10] found that only three failed to survive.

STORAGE STABILITY OF DRIED FOODS

In the absence of fungal growth, desiccated foods are subject to certain chemical changes that may result in the food's becoming undesirable upon holding. In dried foods that contain fats and oxygen, oxidative rancidity is a common form of chemical spoilage. Foods that contain reducing sugars undergo a color change known as *Maillard* reaction or nonenzymic browning. This process is brought about when the carbonyl groups of reducing sugars react with amino groups of proteins and amino acids, followed by a series of other more complicated reactions. Maillard-type browning is quite undesirable in fruits and vegetables not only because of the unnatural color, but also because of the bitter taste imparted to susceptible foods. Freeze-dried foods also undergo browning if the moisture content is about 2%, thus the moisture content should be held below 2%.

With regard to a_w, the maximal browning reaction rates in fruits and vegetable products occur in the 0.65–0.75 range, whereas for nonfat dry milk browning, it seems to occur most readily at about 0.70[36] Other chemical changes that take place in dried foods include a loss of vitamin C in vegetables, general discolorations, structural changes leading to the inability of the dried product to rehydrate fully, and toughness in the rehydrated, cooked product.

Conditions that favor one or more of the above changes in dried foods generally tend to favor all, so preventive measures against one are also effective against others to varying degrees. At least four methods of minimizing chemical changes in dried foods have been offered:

1. Keep the moisture content as low as possible. Gooding[14] has pointed out that lowering the moisture content of cabbage from 5% to 3% doubles its storage life at 37°C.
2. Reduce the level of reducing sugars as low as possible. These compounds are directly involved in nonenzymic browning, and their reduction has been shown to increase storage stability.
3. When blanching, use water in which the level of leached soluble solids is kept low. Gooding[14] has shown that the serial blanching of vegetables in the same water increases the chances of browning. The explanation given is that the various extracted solutes (presumably reducing sugars and amino acids) are impregnated on the surface of the treated products at relatively high levels.
4. Use sulfur dioxide. The treatment of vegetables prior to dehydration with this gas protects vitamin C and retards the browning reaction. The precise mechanism of this gas in retarding the browning reaction is not well understood, but it apparently does not block reducing groups of hexoses. It has been suggested that it may act as a free-radical acceptor.

One of the most important considerations in preventing fungal spoilage of dried foods is the RH of the storage environment. If improperly packed and stored under conditions of high RH, dried foods will pick up moisture from the atmosphere until some degree of equilibrium has been established. Because the first part of the dried product to gain moisture is the surface, spoilage is inevitable; surface growth tends to be characteristic of molds due to their oxygen requirements.

INTERMEDIATE-MOISTURE FOODS

Intermediate-moisture foods (IMF) are characterized by a moisture content of around 15–50% and an a_w between 0.60 and 0.85. These foods are shelf-stable at ambient temperatures for varying periods of time. Although impetus was given to this class of foods during the early 1960s with the development and marketing of intermediate-moisture dog food, foods for human consumption that meet the basic

Table 18–3   Traditional Intermediate Moisture Foods

Food Products	a_w Range
Dried fruits	0.60–0.75
Cake and pastry	0.60–0.90
Frozen foods	0.60–0.90
Sugars, syrups	0.60–0.75
Some candies	0.60–0.65
Commercial pastry fillings	0.65–0.71
Cereals (some)	0.65–0.75
Fruit cake	0.73–0.83
Honey	0.75
Fruit juice concentrates	0.79–0.84
Jams	0.80–0.91
Sweetened condensed milk	0.83
Fermented sausages (some)	0.83–0.87
Maple syrup	0.90
Ripened cheeses (some)	0.96
Liverwurst	0.96

criteria of this class have been produced for many years now. These are referred to as traditional IMFs to distinguish them from the newer IMFs. In Table 18–3 are listed some traditional IMFs along with their a_w values. All of these foods have lowered a_w values, which are achieved by withdrawal of water by desorption, adsorption, and/or the addition of permissible additives such as salts and sugars. The developed IMFs are characterized not only by a_w values of 0.60–0.85 but also by the use of additives such as glycerol, glycols, sorbitol, sucrose, and so forth, as humectants, and by their content of fungistats such as sorbate and benzoate.

Preparation of IMF

Because *S. aureus* is the only bacterium of public health importance that can grow at a_w values near 0.86, an IMF can be prepared by formulating the product so that its moisture content is between 15% and 50%, adjusting the a_w to a value below 0.86 by use of humectants, and adding an antifungal agent to inhibit the rather large number of yeasts and molds that are known to be capable of growth at a_w values above 0.70. Additional storage stability is achieved by reducing the pH. Although this is essentially all that one needs to produce an IMF, the actual process and the achievement of storage stability of the product are considerably more complicated.

The determination of the a_w of a food system is discussed in Chapter 3. One can also use Raoult's law of mole fractions where the number of moles of water in a solution is divided by the total number of moles in the solution.[3]

$$a_w = \frac{\text{Moles of H}_2\text{O}}{\text{Moles of H}_2\text{O} + \text{Moles of solute}}$$

For example, a liter of water contains 55.5 moles. Assuming that the water is pure,

$$a_w = \frac{55.5}{55.2 + 0} = 1.00$$

If, however, 1 mole of sucrose is added,

$$a_w = \frac{55.5}{55.5 + 1} = 0.98$$

This equation can be rearranged to solve the number of moles of solute required to give a specified a_w value. Although the foregoing is not incorrect, it is highly oversimplified, as food systems are complex by virtue of their content of ingredients that interact with water and with each other in ways that are difficult to predict. Sucrose, for example, decreases a_w more than expected, so that calculations based on Raoult's law may be meaningless.[4]

In preparing IMF, water may be removed either by adsorption or desorption. By adsorption, food is first dried (often freeze dried) and then subjected to controlled rehumidification until the desired composition is achieved. By desorption, the food is placed in a solution of higher osmotic pressure so that at equilibrium, the desired a_w is reached.[28] Although identical a_w values may be achieved by these two methods, IMF produced by adsorption is more inhibitory to microorganisms than that produced by desorption (see below). When sorption isotherms of food materials are determined, adsorption isotherms sometimes reveal that less water is held than for desorption isotherms at the same a_w. The sorption isotherm of a food material is a plot of the amount of water adsorbed as a function of the relative humidity or activity of the vapor space surrounding the material. It is the amount of water that is held after equilibrium has been reached at a constant temperature.[21] Sorption isotherms may be either adsorption or desorption, and when the former procedure results in the holding of more water than the latter, the difference is ascribed to a hysteresis effect. This, as well as other physical properties associated with the preparation of IMF, has been discussed by Labuza,[21] Sloan et al.,[33] and others, and will not be dealt with further here. The sorption properties of an IMF recipe, the interaction of each ingredient with water and with other ingredients, and the order of mixing of ingredients, all add to the complications of the overall IMF preparation procedures, and both direct and indirect effects on the microbiology of these products may result.

The following general techniques are employed to change the water activity in producing an IMF[20]:

1. *Moist infusion.* Solid food pieces are soaked and/or cooked in an appropriate solution to give the final product the desired water level (desorption).
2. *Dry infusion.* Solid food pieces are first dehydrated and then infused by soaking in a solution containing the desired osmotic agents (adsorption).
3. *Component blending.* All IMF components are weighed, blended, cooked, and extruded or otherwise combined to give the finished product the desired a_w.
4. *Osmotic drying.* Foods are dehydrated by immersion in liquids with a water activity lower than that of the food. When salts and sugars are used, two simultaneous countercurrent flows develop: solute diffuses from solution into food, and water diffuses out of food into solution.

The foods in Table 18–4 were prepared by moist infusion for military use. The 1-cm-thick slices equilibrated following cooking at 95–100°C in water and holding overnight in a refrigerator. Equilibration is possible without cooking over prolonged periods under refrigeration.[6] IMF deep-fried

Table 18–4 Preparation of Representative Intermediate Moisture Foods by Equilibration

Initial Material	% H₂O	Processing	Equilibrated Product		Ratio: Initial Weight / Solution Weight	% Components of Solution					
			% H_2O	a_w		Glycerol	Water	NaCl	Sucrose	Potassium Sorbate	Sodium Benzoate
Tuna, canned water pack pieces, 1 cm thick	60.0	Cold soak	38.8	0.81	0.59	53.6	38.6	7.1	–	0.7	–
Carrots, diced 0.9 cm, cooked	88.2	Cook 95–98°C, refrig.	51.5	0.81	0.48	59.2	34.7	5.5	–	0.6	–
Macaroni, elbow, cooked, drained	63.0	Cook 95–98°C, refrig.	46.1	0.83	0.43	42.7	48.8	8.0	–	0.5	–
Pork loin, raw, 1 cm thick	70.0	Cook 95–98°C, refrig.	42.5	0.81	0.73	45.6	43.2	10.5	–	0.7	–
Pineapple, canned, chunks	73.0	Cold soak	43.0	0.85	0.46	55.0	21.5	–	23.0	0.5	–
Celery, 0.6 cm cross cut, blanch	94.7	Cold soak	39.6	0.83	0.52	68.4	25.2	5.9	–	0.5	–
Beef, ribeye, 1 cm thick	70.8	Cook 95–98°C, refrig.	–	0.86	2.35	87.9	–	10.1	–	–	2.0

Source: From Brockmann,[6] copyright © 1970 by Institute of Food Technologists.

450

Protection of Foods by Drying   451

Table 18–5 Typical Composition of Soft Moist or Intermediate-Moisture Dog Food

Ingredient	%
Meat byproducts	32.0
Soy flakes	33.0
Sugar	22.0
Skimmed milk, dry	2.5
Calcium and phosphorus	3.3
Propylene glycol	2.0
Sorbitol	2.0
Animal fat	1.0
Emulsifier	1.0
Salt	0.6
Potassium sorbate	0.3
Minerals, vitamins, and color	0.3

Source: From Kaplow,[19] copyright © 1970 by Institute of Food Technologists.

catfish, with raw samples of about 2 grams each, has been prepared by the moist infusion method.[9] Pet foods are more often prepared by component blending. The general composition of one such product is given in Table 18–5. The general way in which a product of this type is made is as follows. The meat and meat products are ground and mixed with liquid ingredients. The resulting slurry is cooked or heat treated and later mixed with the dry ingredient mix (salts, sugars, dry solids, and so on). Once the latter is mixed into the slurry, an additional cook or heat process may be applied prior to extrusion and packaging. The extruded material may be shaped in the form of patties or packaged in loose form. The composition of a model IMF product called Hennican is given in Table 18–6. According to Acott and Labuza,[1] this is an adaptation of pemmican, an Indian trail and winter storage food made of buffalo

Table 18–6 Composition of Hennican

Components	Amount (wt. basis, %)
Raisins	30
Water	23
Peanuts	15
Chicken (freeze dried)	15
Nonfat dry milk	11
Peanut butter	4
Honey	2

Note: Moisture content = 41 g water/100 g solids; $a_w = 0.85$.

Source: From Acott and Labuza,[1] copyright © 1975 by Institute of Food Technologists.

meat and berries. Hennican is the name given to the chicken-based IMF. Both moisture content and a_w of this system can be altered by adjustment of ingredient mix.

The humectants commonly used in pet food formulations are propylene glycol, polyhydric alcohols (sorbitol, for example), polyethylene glycols, glycerol, sugars (sucrose, fructose, glucose, and corn syrup), and salts (NaCl, KCl, and so on). The commonly used mycostats are propylene glycol, potassium sorbate, sodium benzoate, and others. The pH of these products may be as low as 5.4 and as high as 7.0.

Microbial Aspects of IMF

The general a_w range of IMF products make it unlikely that Gram-negative bacteria will proliferate. This is true also for most Gram-positive bacteria with the exception of cocci, some spore formers, and lactobacilli. In addition to the inhibitory effect of lowered a_w, antimicrobial activity results from an interaction of pH, oxidation–reduction (Eh), added preservatives (including some of the humectants), the competitive microflora, generally low storage temperatures, and the pasteurization or other heat processes applied during processing.

The fate of *S. aureus* S-6 in IM pork cubes with glycerol at 25°C is illustrated in Figure 18–1. In this desorption IM pork at an a_w of 0.88, the numbers remained stationary for about 15 days and then

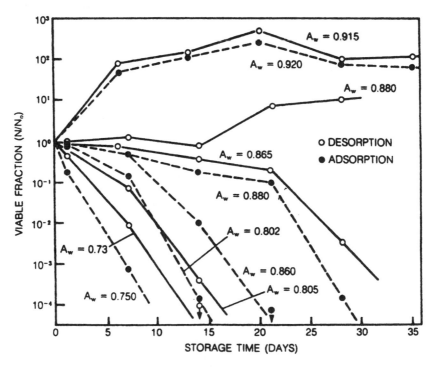

Figure 18–1 Viability of *Staphylococcus aureus* in IMF systems: pork cubes and glycerol at 25°C. *Source*: From Plitman et al.,[27] copyright © 1973 by Institute of Food Technologists.

increased slightly, whereas in the adsorption IM system at the same a_w, the cells died off slowly during the first three weeks and thereafter, more rapidly. At all a_w values below 0.88, the organisms died off, with the death rate considerably higher at 0.73 than at higher values.[27] Findings similar to these have been reported by Haas et al.,[15] who found that an inoculum of 10^5 staphylococci in a meat–sugar system at an a_w of 0.80 decreased to 3×10^3 after 6 days and to 3×10^2 after 1 month. Although growth of *S. aureus* has been reported to occur at an a_w of 0.83, enterotoxin is not produced below an a_w of 0.86.[34] It appears that enterotoxin A is produced at lower values of a_w than enterotoxin B.[35]

Using the model IM Hennican at pH 5.6 and a_w 0.91, Boylan et al.[5] showed that the effectiveness of the IM system against *S. aureus* F265 was a function of both pH and a_w. Adsorption systems are more destructive to microorganisms than desorption systems. Labuza et al.[22] found that the reported minimum a_w s apply in IMF systems where desorption systems are involved, but that growth minima are much higher if the food is prepared by an adsorption method. *S. aureus* was inhibited at a_w 0.9 in adsorption, whereas values between 0.75 and 0.84 were required for desorption systems. A similar effect was noted for molds, yeasts, and pseudomonads.

With regard to the effect of IMF systems on the heat destruction of bacteria, heat resistance increases as a_w is lowered and the degree of resistance is dependent on the compounds employed to control a_w (see Table 17–3). In a study of the death rate of salmonellae and staphylococci in the IM range of about 0.8 at pasteurization temperatures (50–65°C), it has been found that cell death occurs under first-order kinetics.[18] These investigators confirmed the findings of many others that the heat destruction of vegetative cells is at a minimum in the IM range, especially when a solid menstrum is employed. Some D values for the thermal destruction of *Salmonella* Senftenberg 775W at various a_w values are given in Table 17–3.

With respect to molds in IMF systems, these products would be made quite stable if a_w were reduced to around 0.70, but a dry-type product would then result. A large number of molds are capable of growth in the 0.80 range, and the shelf life of IM pet foods is generally limited by the growth of these organisms. The interaction of various IM parameters on the inhibition of molds was shown by Acott et al.[2] In their evaluation of seven chemical inhibitors used alone and in combination to inhibit *Aspergillus niger* and *A. glaucus* inocula, propylene glycol was the only approved agent that was effective alone. None of the agents tested could inhibit alone at a_w 0.88, but in combination, the product was made shelf stable. All inhibitors were found to be more effective at pH 5.4 and a_w 0.85 than at pH 6.3. Growth of the two fungi occurred in 2 weeks in the a_w 0.85 formulation without inhibitors, but did not occur until 25 weeks when potassium sorbate and calcium propionate were added (Table 18–7). Growth of *Staphylococcus epidermidis* was inhibited by both fungistats, with inhibition being greater at a_w 0.85 than at 0.88. This is probably an example of the combined effects of pH, a_w, and other growth parameters on the growth inhibition of microorganisms in IMF systems.

Storage Stability of IMF

The undesirable chemical changes that occur in dried foods occur also in IMF. Lipid oxidation and Maillard browning are at their optima in the general IMF ranges of a_w and percentage moisture. However, there are indications that the maximum rate for Maillard browning occurs in the $0.4a_w$–$0.5a_w$ range, especially when glycerol is used as the humectant.[36]

The storage of IMFs under the proper conditions of humidity is imperative in preventing moldiness and for overall shelf stability. The measurement of equilibrium relative humidity (ERH) is of

Table 18–7   Time for Growth of Microbes in Inoculated Dog Food with Inhibitors, pH 5.4

	Storage Conditions	
Inhibitor	$a_w = 0.85$; 9-Month Storage	$a_w = 0.88$; 6-Month Storage
No inhibitor added	A. niger—2 weeks	A. niger—1 week
	A. glaucus—1 week	A. glaucus—1 week
	S. epidermidis—2 weeks	S. epidermidis— 1/2 week
Potassium sorbate (0.3%)	No mold	A. niger—5 weeks
	S. epidermidis—25 weeks	S. epidermidis—$3\frac{1}{2}$ weeks
Calcium propionate (0.3%)	A. niger—25 weeks	A. glaucus—2 weeks
	A. glaucus—25 weeks	S. epidermidis—$1\frac{1}{2}$ weeks
	S. epidermidis—$3\frac{1}{2}$ weeks	

Note: Mold—first visible sign; bacteria—2 log cycle increase.

Source: From Acott et al.,[2] copyright © 1976 by Institute of Food Technologists.

importance in this regard. ERH is an expression of the desorbable water present in a food product and is defined by the following equation:

$$\text{ERH} = (P_{\text{equ}}/P_{\text{sat}}), \quad T, P = 1\,\text{atm}$$

where P_{equ} is partial pressure of water vapor in equilibrium with the sample in air at 1 atm total pressure and temperature T, P_{sat} is the saturation partial vapor pressure of water in air at a total pressure of 1 atm and temperature T.[16] A food in moist air exchanges water until the equilibrium partial pressure at that temperature is equal to the partial pressure of water in the moist air, so that the ERH value is a direct measure of whether moisture will be sorbed or desorbed. In the case of foods packaged or wrapped in moisture-impermeable materials, the relative humidity of the food-enclosed atmosphere is determined by the ERH of the product, which, in turn, is controlled by the nature of the dissolved solids present, the ratio of solids to moisture, and the like.[30] Both traditional and newer IMF products have longer shelf stability under conditions of lower ERH.

In addition to the direct effect of packaging on ERH, gas-impermeable packaging affects the Eh of packaged products with consequent inhibitory effects on the growth of aerobic microorganisms.

IMF and Glass Transition

The use of a_w values for IMF has been questioned by some investigators who suggest that "water dynamics" may be a better predictor of microbial activity in such systems. "Water dynamics" refers to the amorphous matrix of food components that are sensitive to changes in moisture content and temperature. The matrix may exist either as a very viscous "glassy", or as a more liquid-like "rubbery", amorphous, structure. Thus, the glassy state refers to the increased viscosity of an aqueous amorphous and rubbery system that results in inhibited or slowed flow. In the glassy state, crystallization of constituents is limited. The glass-rubbery transition occurs at the characteristic temperature, T_g, and it decreases with increasing moisture. T_g has been proposed as a parameter that is a better predictor of shelf-stability in IMF systems than a_w.[31,32]

The validity and utility of the above concept for IMF systems remains to be established on a wider scale. It has been examined by one group and has not been found to be a better alternative than a_w for predicting microbial activity in foods.[8]

REFERENCES

1. Acott, K.M., and T.P. Labuza. 1975. Inhibition of *Aspergillus niger* in an intermediate moisture food system. *J. Food Sci.* 40:137–139.

2. Acott, K.M., A.E. Sloan, and T.P. Labuza. 1976. Evaluation of antimicrobial agents in a microbial challenge study for an intermediate moisture dog food. *J. Food Sci.* 41:541–546.

3. Bone, D. 1973. Water activity in intermediate moisture foods. *Food Technol.* 27(4):71–76.

4. Bone, D.P. 1969. Water activity—Its chemistry and applications. *Food Prod. Dev.* 3(5):81–94.

5. Boylan, S.L., K.A. Acott, and T.P. Labuza. 1976. *Staphylococcus aureus* challenge study in an intermediate moisture food. *J. Food Sci.* 41:918–921.

6. Brockmann, M.C. 1970. Development of intermediate moisture foods for military use. *Food Technol.* 24:896–900.

7. Burke, R.F., and R.V. Decareau. 1964. Recent advances in the freeze-drying of food products. *Adv. Food Res.* 13:1–88.

8. Chirife, J., and M.D.P. Buera. 1994. Water activity, glass transition and microbial stability in concentrated/semimoist food systems. *J. Food Sci.* 59:921–927.

9. Collins, J.L., and A.K. Yu. 1975. Stability and acceptance of intermediate moisture, deep-fried catfish. *J. Food Sci.* 40:858–863.

10. Davis, R.J. 1963. Viability and behavior of lyophilized cultures after storage for twenty-one years. *J. Bacteriol.* 85:486–487.

11. Desrosier, N.W. 1963. *The Technology of Food Preservation*. New York: Van Nostrand Reinhold.

12. Fennema, O., and W.D. Powrie. 1964. Fundamentals of low-temperature food preservation. *Adv. Food Res.* 13:219–347.

13. Fry, R.M., and R.I.N. Greaves. 1951. The survival of bacteria during and after drying. *J. Hyg.* 49:220–246.

14. Gooding, E.G.B. 1962. The storage behaviour of dehydrated foods. In *Recent Advances in Food Science*, ed. J. Hawthorn and J.M. Leitch, Vol. 2, 22–38. London: Butterworths.

15. Haas, G.J., D. Bennett, E.B. Herman, and D. Collette. 1975. Microbial stability of intermediate moisture foods. *Food Prod. Dev.* 9(4):86–94.

16. Hardman, T.M. 1976. Measurement of water activity. Critical appraisal of methods. In *Intermediate Moisture Foods*, ed. R. Davies, G.G. Birch, and K.J. Parker, 75–88. London: Applied Science.

17. Harper, J.C., and A.L. Tappel. 1957. Freeze-drying of food products. *Adv. Food Res.* 7:171–234.

18. Hsieh, F.-H., K. Acott, and T.P. Labuza. 1976. Death kinetics of pathogens in a pasta product. *J. Food Sci.* 41:516–519.

19. Kaplow, M. 1970. Commercial development of intermediate moisture foods. *Food Technol.* 24:889–893.

20. Karel, M. 1976. Technology and application of new intermediate moisture foods. In *Intermediate Moisture Foods*, ed. R. Davies, G.G. Birch, and K.J. Parker, 4–31. London: Applied Science.

21. Labuza, T.P. 1968. Sorption phenomena in foods. *Food Technol.* 22:263–272.

22. Labuza, T.P., S. Cassil, and A.J. Sinskey. 1972. Stability of intermediate moisture foods 2. Microbiology. *J. Food Sci.* 37:160–162.

23. May, K.N., and L.E. Kelly. 1965. Fate of bacteria in chicken meat during freeze-dehydration, rehydration, and storage. *Appl. Microbiol.* 13:340–344.

24. Meryman, H.T. 1966. Freeze-drying. In *Cryobiology*, ed. H.T. Meryman, Chap. 13. New York: Academic Press.

25. Mossel, D.A.A., and M. Ingram. 1955. The physiology of the microbial spoilage of foods. *J. Appl. Bacteriol.* 18:232–268.

26. Pitt, J.I., and J.H.B. Christian. 1968. Water relations of xerophilic fungi isolated from prunes. *Appl. Microbiol.* 16:1853–1858.

27. Plitman, M., Y. Park, R. Gomez, and A.J. Sinskey. 1973. Viability of *Staphylococcus aureus* in intermediate moisture meats. *J. Food Sci.* 38:1004–1008.

28. Robson, J.N. 1976. Some introductory thoughts on intermediate moisture foods. In *Intermediate Moisture Foods*, ed. R. Davies, G.G. Birch, and K.J. Parker, 32–42. London: Applied Science.

29. Scott, W.J. 1957. Water relations of food spoilage microorganisms. *Adv. Food Res.* 1:83–127.

30. Seiler, D.A.L. 1976. The stability of intermediate moisture foods with respect to mould growth. In *Intermediate Moisture Foods*, ed. R. Davies, G.G. Birch, and K.J. Parker, 166–181. London: Applied Science.

31. Slade, L., and H. Levine. 1991. Beyond water activity: Recent advances based on an alternative approach to the assessment of food quality and safety. *CRC Crit. Rev. Food Sci. Nutr.* 30:115–360.

32. Slade, L., and H. Levine. 1987. Structural stability of intermediate moisture foods—A new understanding. In *Food Structure—Its Creation and Evaluation*, ed. J.R. Mitchell and J.M.V. Blanshard, 115–147. London: Butterworths.

33. Sloan, A.E., P.T. Waletzko, and T.P. Labuza. 1976. Effect of order-of-mixing on a_w-lowering ability of food humectants. *J. Food Sci.* 41:536–540.

34. Tatini, S.R. 1973. Influence of food environments on growth of *Staphylococcus aureus* and production of various enterotoxins. *J. Milk Food Technol.* 36:559–563.

35. Troller, J.A. 1972. Effect of water activity on enterotoxin A production and growth of *Staphylococcus aureus*. *Appl. Microbiol.* 24:440–443.

36. Troller, J.A., and J.H.B. Christian. 1978. *Water Activity and Food*. New York: Academic Press.

CHAPTER 19

Other Food Protection Methods

The methods presented in the previous six chapters are well established and in wide use throughout the world. Presented in this chapter are some methods that are less widely used but which show promise of being of greater importance in the future.

HIGH HYDROSTATIC PRESSURES (HHP, HPP)

The use of high pressure processing (HPP) or Pascalization to reduce or destroy microorganisms in foods dates back to 1884.[11] In 1899, Hite successfully used hydrostatic pressures to improve the quality of milk,[18] and in 1914 he demonstrated the susceptibility of fruitborne organisms to hydrostatic pressures.[19] Thus, the utility of this process to control microorganisms and preserve foods has a long history but only relatively recently has it received detailed study. The current interest is due apparently to consumer demands for minimally processed foods, and to the lower cost and greater availability of processing equipment. HHP treatments may be applied at room temperature, and with the exception of some vegetables, shape, color, and nutrients of most foods are not affected. At least 10 HPP-treated foods, including fruit purées, jams, fruit juices, and rich cakes, have been commercially available in Japan since the early 1990s.[8]

To carry out HPP, high hydrostatic pressures (HHP) are used, and one needs a suitable mechanical chamber (steel cylinder) and pressure pumps to generate pressures of several hundred megaPascal (MPa) (1 MPa = 10 atm; 100 MPa = 1 kbar). Come-up and come-down times for pressure are important, and rates of 2–3 MPa/sec are not uncommon. After the food is placed in suitable containers and sealed, the food packs are placed in the cylinder containing a low-compressibility liquid such as water. Pressure is generated with a pump, and it may be applied continuously (static) or in an oscillatory manner. For the latter, 2–4 pressure cycles may be applied with varying holding periods for each cycle. In a study on the inactivation of *Zygosaccharomyces bailii*, continuous and oscillatory treatments were compared and the latter was found to be the more effective.[38] With an initial inoculum of about 1.6×10^6 colony-forming units (cfu)/ml, oscillatory treatments with holding times totaling 20 minutes at 276 MPa reduced the numbers to <10 cfu/ml (Figure 19–1). The cells were suspended in Sabouraud's glucose 2% broth with sucrose added to adjust water activity (a_w) to 0.98. In an earlier study with spores of *Geobacillus stearothermophilus*, 10^6 spores/ml were destroyed after six 5-minute cycles (60 minutes total) at 600 MPa and 70°C while with static and 800 MPa at 60°C for 60 minutes, spores were reduced to 10^2/ml.[15] By either method, the action is instant and uniform throughout a container regardless of size, and HPP is equally effective for both liquid and solid foods.

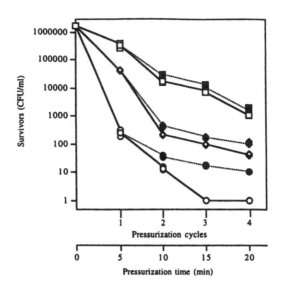

Figure 19–1 *Zygosaccharomyces bailii* survivor counts after cycles of pressurization (5 minutes each) at 207 (□), 241 (◇), or 276 MPa (○) or continuous pressurization at 207 (■), 241 (◆), or 276 MPa (●). *Source:* Reprinted with permission from E. Palou (University of Washington, Pullman, Washington) et al.,[39] oscillatory high hydrostatic pressure inactivation of *Zygosaccharomyces bailii, J. Food Protect.* 61:1214, copyright © 1998, held by the International Association of Milk, Food and Environmental Sanitarians, Inc.

For typical antimicrobial actions, pressures in the range of 200–1,000 MPa are needed depending on other parameters. For more information on the application of HHP to food preservation, see Cheftel.[8]

Some Principles and Effects of HHP on Foods and Organisms

Among the known effects of HHP, the following are some that are of interest in food preservation.

1. Hydrostatic pressures are nonthermal, and covalent bonds are not broken, so that flavor is unaffected. They are effective at ambient and refrigerator temperatures, and hydrogen bonds appear to be strengthened.
2. Between 400 and 600 MPa, proteins are readily denatured.
3. Up to 450 MPa will inactivate vegetative cells with sensitivity decreasing in the following order: eucaryotic cells, Gram-negative bacteria, fungi, Gram-positive bacteria, and bacterial endospores. Cells in the stationary phase tend to be more resistant than those in the logarithmic phase.[16]
4. Microorganisms in dehydrated foods such as spices are highly resistant to HHP (baroresistant). In general, baroresistance increases as a_w is lowered.
5. In general, baroresistance tends to parallel thermal resistance but this is not consistent for all organisms.
6. Between 450 and 800 MPa are needed to destroy spore formers under the most optimal conditions. Some spores require >1,000 MPa.

7. Cell morphology is altered, and ribosomes are destroyed.

8. Changes occur in the lipid–protein complex of cell membranes, and increased membrane fluidity is one consequence. The leakage of nucleic acids from cells exposed to 200–400 MPa has been demonstrated.

9. Adenosinetriphosphatase (ATPase) is inactivated, leading to shortage of cellular ATP, but oxidative enzymes of fruits are baroresistant.

10. Although HHP is generally ineffective against the walls of bacteria, there is synergism between HHP treatments and bacteriocins with both Gram-positive and Gram-negative bacteria, and with heat, low pH, CO_2, and lysozyme. Thus, HHP can be used as a hurdle in multiplex systems.

11. Since HHP inflicts cell injury, injured cells have been shown to resuscitate in the food product and grow over time, and this phenomenon should be anticipated.[31]

12. Bacterial endospores display high resistance, but when inactivated, it appears to be the result of induced germination with subsequent destruction of the vegetative cells.

13. *G. stearothermophilus* spores may be reduced by use of a rapid decompression method that involves the use of 200 MPa at 75°C for 60 minutes.[15]

14. From a study of the effect of HHP on spore germination, an 100 MPa induced spore germination in *B. subtilis* by activating germinant receptors, and 550 MPa opened the channels for release of dipicolinic acid leading to the later steps in spore germination.[36]

Effects of HHP on Specific Foodborne Organisms

D_{MPa} value determinations from HHP treatments are often difficult to calculate because of the "tailing" of survivor curves that has been shown by a number of investigators. From the inactivation curves of *Listeria monocytogenes* in Figure 19–2, the tailing effect can be seen between 20 and 30

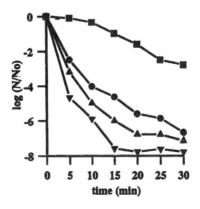

Figure 19–2 Pressure inactivation of *L. monocytogenes* 2433 in 10 mM phosphate-buffered saline (pH 7.0) at 20°C for 300 MPa (■), 350 MPa (●), 375 MPa (▲), and 400 MPa (▼). *No* = initial number; *N* = number of survivors. Each point is a mean of three values. *Source*: Reprinted with permission from M.F. Patterson,[42] (Queen's University of Belfast, N. Ireland) et al., sensitivity of vegetative pathogens to high hydrostatic pressure treatment in phosphate-buffered saline and foods, *J. Food Protect.* 58:525, copyright © 1995, held by the International Association of Milk, Food and Environmental Sanitarians, Inc.

minutes after the MPa treatments at 375 and 400.[31] D_{MPa} values for two salmonellae were reported by Metrick et al.[31] at 340 MPa and 23°C in buffer and chicken. For *Salmonella* serotype Typhimurium in buffer and chicken, D_{MPa} values were 7.40 and 7.63, respectively, while for *Salmonella* serotype Senftenberg the respective values were 4.20 and 7.13. These investigators were able to calculate D values in spite of the "tailing" of survivor curves. In a more recent study, the D value for *Salmonella* serotype Typhimurium in fresh pork loin samples at 25°C and 414 MPa was 1.48.[3] In the same study, the $D_{MPa=414}$ for *L. monocytogenes* was 2.17. In another study, the $D_{MPa=350}$ was 8.52 minutes for strain Scott A of *L. monocytogenes* on pork chops.[33] The latter investigators found this organism to be more resistant than the indigenous biota of pork chops.

In a study of *S. cerevisiae* in orange juice, Zook et al.[64] found the following D values: 10.81 minutes with 300 MPa; 0.97 with 400 MPa; and 0.18 minute with 500 MPa. The z value was around 117 MPa, and results were similar for apple juice.[64]

The effect of a_w and potassium sorbate on the inactivation of *Z. bailii* at 21°C and pH 3.5 in a laboratory system has been studied,[38] and the times required to inactivate (detection limit <10 cfu/ml) under the test conditions are indicated below:

$$a_w = 0.98 + \text{potassium sorbate} \quad \geq 345\,\text{MPa} =< 2\,\text{minutes}$$
$$a_w = 0.98 \,(\text{no sorbate}) \quad 517\,\text{MPa} => 4\,\text{minutes}$$
$$a_w = 0.95 + \text{potassium sorbate} \quad \geq 517\,\text{MPa} = 4\,\text{minutes}$$
$$a_w = 0.95 \,(\text{no sorbate}) \quad \geq 517\,\text{MPa} = 10\,\text{minutes}$$

These findings demonstrate the antagonistic effect of low a_w on HHP and the potentiating effect of potassium sorbate. These investigators concluded from this study that approximately 10^5 cells of *Z. bailii* could be inactivated at 689 MPa regardless of a_w, 1,000 ppm potassium sorbate, or duration of treatment.[38] In another study, 304 MPa for 10 minutes at 25°C in citrate buffer at pH 3.0 produced the total inactivation of 10^8 cfu/ml of *Z. bailii*.[40] No effect was observed in the same menstruum when 152 MPa was applied for 30 minutes.

The effect of pH on HHP inactivation of *E. coli* 0157:H7 was assessed by using orange juice inoculated with 10^8 cfu/ml and adjusting the pH of samples from 3.4 to 5.0.[26] The conditions that allowed for a 6-log reduction of the organism were as follows: 550 MPa for 5 minutes in orange juice with pH 3.4, 3.6, 3.9, or 4.5 but not pH 5.0. A similar reduction was achieved at pH 5.0 by combining HHP treatment with mild heat (30°C).[26]

The combined effect of nisin and HHP on *Listeria innocua* and *E. coli* in liquid whole egg (pH 8.0) was studied, and with nisin at 5 mg/l and 450 MPa at 20°C for 10 minutes, a 5-log reduction of *E. coli* and a 6-log reduction of *L. innocua* were obtained.[44] In an earlier study, nisin and the bacteriocin Pediocin AcH were shown to increase the lethality of HHP treatments.[23] These and a number of other food additives increase the effectiveness of HHP by lowering the baroresistance of bacteria.[45] When mechanically recovered poultry meat containing 100 ppm nisin and 1% glucono-delta-lactone was exposed to about 350 MPa, shelf life was extended during the 36-day storage at refrigerator temperatures.[63]

The relative sensitivity of six foodborne pathogens in buffer, milk, and poultry meat was investigated by Patterson et al.,[42] who found *Yersinia enterocolitica* to be the most sensitive with $>10^5$ reduction in numbers occurring in pH 7.0 buffer with MPa as indicated below for 15 minutes. To achieve a similar reduction for the other five under similar conditions, the following HHPs were required:

275   for   *Y. enterocolitica*
350   for   *Salmonella* serotype Typhimurium
375   for   *L. monocytogenes*
450   for   *Salmonella* serotype Enteritidis
700   for   *Staphylococcus aureus*
700   for   *Escherichia coli* 0157:H7

The effectiveness of HHP against *V. parahaemolyticus* was demonstrated by Styles et al.[58] who found that 170 MPa at 23°C for 10 minutes would inactivate about 10^6 cells/ml in clam juice. On the other hand, 340 MPa in 80 minutes at 23°C was required to inactivate about 10^6 cfu/ml of *L. monocytogenes* in ultrahigh temperature (UHT) milk.

When whole (3.5% fat) and skim (0.3% fat) milk were subjected to 400 MPa at 25°C for 30 minutes, shelf life was extended to 45 days at refrigerator temperatures while the untreated milk had a shelf life of 15 days.[13] However, since plasmin was not inactivated, casein hydrolysis occurred, leading to flavor defects upon prolonged storage. In another study, combining HHP and mild heating was shown to be very effective in destroying *E. coli* 0157:H7 and *S. aureus*.[41] In UHT whole milk or poultry meat, each treated with 400 MPa at 50°C for 15 minutes, *E. coli* 0157:H7 showed a 6-log reduction in poultry and a 5-log reduction in the milk compared to <1-log reduction for a treatment of 400 MPa at 20°C. Interestingly, *S. aureus* was inactivated more efficiently in milk than in the poultry meat.

The effect of HPP on *Vibrio* species was studied by Berlin et al.[5] who found that 250 MPa for 15 minutes or 300 MPa for 5 minutes at 25°C reduced a mixture of five species to nondetectable levels. The species and strains included were *V. parahaemolyticus*, *V. vulnificus*, *V. cholerae* 01 and non-01, *V. hollisae*, and *V. mimicus*. Using oysters, *V. parahaemolyticus* at a level up to 8.1×10^7 cfu/g or *V. vulnificus* at up to 2.5×10^7 cfu/g were reduced to <10 cfu/g with 200 MPa for 10 minutes at 25°C.

The microbiota of vegetables was essentially unaffected by 100 and 200 MPa treatments at 20°C for 10 minutes or 10°C for 20 minutes, but at 300 MPa significant reductions occurred.[4] *Saccharomyces cerevisiae* was effectively reduced by 300 MPa at 10°C for 20 minutes while 350 MPa was needed to reduce most Gram-negative bacteria and molds. The Gram positives were not completely reduced with 400 MPa. However, these investigators noted some undesirable changes in vegetables with pressures >300 MPa. For example, the skin of tomatoes loosened and peeled away, and lettuce, while remaining firm, underwent browning. In another study, spinach leaves lost nutrients after treatment at 400 MPa for 30 minutes at 5°C, while cauliflower was more acceptable under the same conditions.[47] Thus, some fresh vegetables appear to be unsuitable for HHP preservation treatments.

CO_2 under high pressure is considerably more antimicrobial than under atmospheric conditions. In one study, CO_2 at 6.18 MPa for 2 hours reduced about 10^9 cfu/ml of *L. monocytogenes* to undetectable levels in distilled water or broth, while N_2 under the same condition did not.[60] At 13.7 MPa, CO_2 was effective against *L. monocytogenes* and *Salmonella* serotype Typhimurium in chicken meat, egg yolk, shrimp, and orange juice. In another study, the concentration of dissolved CO_2 was increased by using microbubbles of pressured CO_2. At 6 MPa, 35°C, and a residence time of 15 minutes, the following results were achieved.[54]: *Lactobacillus brevis* was completed inhibited at $\geq 11\gamma$ ($\gamma =$ Kuenen's gas absorption coefficient); *E. coli* and *Saccharomyces cerevisiae* were inhibited at ≥ 17; *Torulopsis versatilis* required ≥ 21; and *Z. rouxii* could be sterilized at 10 MPa and 26. The greater resistance of *Z. rouxii* compared to *T. versatilis* and *S. cerevisiae* is shown in Figure 19–3.

Encapsulated viruses such as cytomegalovirus and herpes simplex type 1 have been found to be inactivated at 300 MPa and 25°C for 10 minutes. The pressure appears to damage the viral capsule and prevent the binding of viral particles to host cells. On the other hand, sindbis virus resisted 700 MPa.[8]

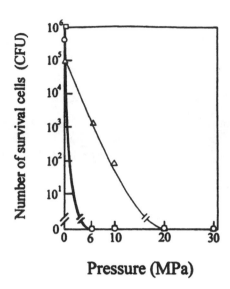

Pressure (MPa)

Figure 19–3   Effect of pressure on inactivation of *S. cerevisiae* (O), *T. versatilis* (□), and *Z. rouxii* (Δ). The treatment was carried out at CO_2 flow rate of 2.0 kg/hour and 35°C. *Source*: reprinted with permission from M. Shimoda et al.,[54] antimicrobial effects of pressured carbon dioxide in a continuous flow system, *J. Food Sci.*, 63:712, copyright © 1998, Institute of Food Technologists.

A $\log_{10}$ 3.5–5.0 cfu/ml reduction (from initial of 10^6) of spores of *B. subtilis* in milk, *B. coagulans* in tomato juice, and *Alicyclobacillus* sp. in both tomato and apple juice was effected by a combined treatment of ≤1.0% sucrose laurate and 392 MPa for 10 minutes at 45°C.[53] The $D_{500\,MPa}$ values of spores of a strain of *Bacillus anthracis* that was deficient in the production of toxin components were found to be 4 minutes at 75°C and 160 minutes at 20°C.[9] At 20°C, the D under atmospheric pressure (0.1 MPa) and 75°C was 348 minutes (5.8 hours).

D and *z* values of *Saccharomyces cerevisiae* ascospores were determined in a fruit juice buffer at pH 3.5–5.0 containing ca. 10^6 ascospores/ml. The $D_{500\,MPa}$ was 8 sec with *z* values between 115 and 121 MPa; and the $D_{300\,MPa}$ was 10.8 minutes.[64] Ascospores of *Talaromyces macrosporus* were reduced by <2 logs by exposure to 700 MPa for 60 minutes at 20°C.[52] Two strains of *C. botulinum* type A spores in a crab meat blend were reduced by logs 2.7 and 3.2 at 827 MPa for 15 minutes.[51]

Rotaviruses were shown to be extremely sensitive to HHP with 200 MPa for 2 minutes at 25°C effecting ca. $\log_{10}$ 8 inactivation as assessed by a tissue culture assay.[24] A small fraction of viruses remained resistant up to 800 MPa for 10 minutes. The viruses were resistant to PEF at 20–29 kV/cm².

Vibrio parahaemolyticus in broth and live oysters was reduced to nondetectable levels by 345 MPa for 30 sec in broth, and 90 sec in oysters.[7] A treatment with 345 MPa at 50°C for 5 minutes effected >8-log reduction of the following organisms: *L. monocytogenes* (two strains); *E. coli* 0157:H7 (two strains); and one strain each of *S.* Enteritidis, *S.* Typhimurium, and *Staphylococcus aureus*. With the addition of either citric or lactic acid at pH 4.5, 1.3–3.9 additional log cycle reductions were achieved.[1] The APC of shelled oysters treated at 400 MPa in two 5-minute periods at 7°C was reduced by ca. 5 logs, and they remained stable for 41 days at 2°C.[28]

An APC reduction of ca. 4 logs was achieved with 500 MPa for 5 or 15 minutes at 65°C in vacuum-packaged sausages that were stored at 2 or 8°C.[62] The psychrotrophs and enteric bacteria

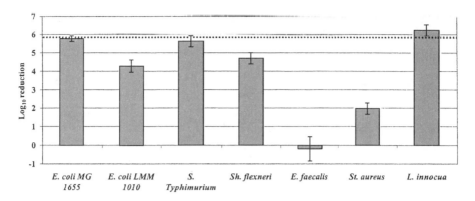

Figure 19–4 Inactivation of bacteria inoculated on garden cress seeds by HHP (300 MPa, 20°C, 15 minutes). The dotted line represents the limit of detection,[61] copyright © 2003, International Association of Food Protection.

were destroyed by the heat treatment, and no *L. monocytogenes* or *S. aureus* were found in treated samples.

Plastic bags of finfish (ca. 100 g/bag) containing 13–118 *Anisakis simplex* larvae were exposed to different levels of pressure, and the treatments that effected a 100% kill of the larvae were as follows: 30–60 sec with 414 MPa; 90–180 sec with 276 MPa; and 180 sec at 207 MPa.[10] A significant increase in whiteness of flesh occurred with each of these treatments. In another study, all *A. simplex* larvae were killed at 200 MPa for 10 minutes at 0–15°C, but at 140 MPa, about an hour was needed.[32]

In regard to seeds for sprouts, garden cress, sesame, radish, and mustard seeds were inoculated and subjected to HHP under various conditions. With garden cress seeds inoculated with seven bacterial species, a treatment of 300 MPa for 15 minutes at 30°C effected a >6-log cycle reduction of *S.* Typhimurium, *E. coli* 1655, and *L. innocua*: a >4-log reduction of *E. coli* 1010 and *Shigella flexneri*; and a 2-log reduction of *S. aureus*.[61] *Enterococcus faecalis* was essentially unaffected (see Figure 19–4).

Overall, the effectiveness of HHP treatment of certain foods to control microorganisms is well documented. The most logical application of this methodology appears to be for shelf-life extension of high-acid and semipreserved foods. In combination with mild heat and ionophores such as nisin, HHP can be used to destroy vegetative cell pathogens. Considerably more research is needed before HHP-treated foods can be equated to thermally processed foods relative to safety and shelf life. For a review of the effects of HHP on vegetative cells, see Smelt et al.,[56] and for spores, see Heinz and Knorr.[17]

PULSED ELECTRIC FIELDS

This physical method consists of the application of short pulses (microseconds) of high electric fields to foods placed between two electrodes. It is a nonthermal process similar in this regard to HPP described above. The lethal effect is essentially a function of pulse intensity, pulse width, and pulse repetition rate. Pulsed electric field (PEF) generation requires a pulsed power supply and a treatment chamber.

The use of electric currents to destroy microorganisms was studied in the 1920s, but those early studies consisted of applying continuous current to liquid foods, which resulted in heat buildup and free radical formation. The use of PEF dates back to the mid-1960s. The pulses used may be of the square-wave or the exponentially decaying types, and the former are more lethal than the latter. In one study, a 99% decrease in *E. coli* numbers was produced by a square wave after 100 microseconds at 7°C compared to 93% by the exponentially decaying method.[46]

Among the general properties and features of PEF as applied to foods are the following:

1. Gram-negative bacterial cells are more sensitive than Gram positives or yeasts.
2. Vegetative cells are more sensitive than spores.
3. Microbial cells are more sensitive in the log phase of growth than in the stationary phase.
4. Cell death by PEF appears to be due to disruption of cell membrane function and by electroporation (production of pores in membranes by the electric current). It has been suggested that bacterial inactivation by PEF may be an "all or nothing" event since sublethal injury could not be detected.[55]
5. Overall, the antimicrobial effects of PEF are functions of electric field strength, treatment time, and treatment temperature, with cells being more sensitive when treated at higher temperatures. The temperature effect on *L. monocytogenes* is shown in Figure 19–5.

A typical PEF application consists of the following components listed in order:

- Pulse intensity (from 10 to 90 kV/cm is common)
- Pulse number (number of pulses varies widely from 10 to at least 70)

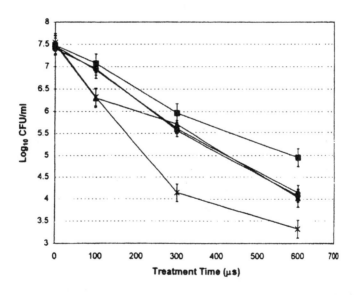

Figure 19–5 PEF inactivation of *Listeria monocytogenes* in whole milk at 10°C (●), 25°C (■), 30°C (◆), 43°C (▲), and 50°C (×). Treatment conditions: field strength, 30 kV/cm; flow rate, 7 ml/sec; pulse duration time, 1.5 microseconds; and frequency, 1,700 Hz. *Source*: Reprinted with permission from L.D. Reina (North Carolina State University, Raleigh, North Carolina) et al.[50] inactivation of *Listeria monocytogenes* in milk by pulsed electric field, *J. Food Protect*. 61: 1205, © 1998 copyright held by the International Association of Milk, Food and Environmental Sanitarians, Inc.

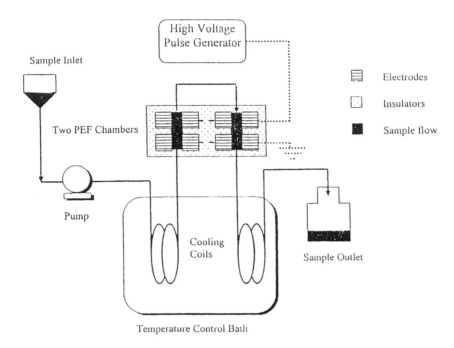

Figure 19–6   Configuration of a PEF system.[50] See text for details, copyright © 1998 International Association for Food Protection, used with permission.

- Pulse duration (in microseconds, 2μ is common)
- Flow rate (time in minutes/hour for given volume to pass).
- Treatment parameters (the main ones are temperature and pH; others could be a_w, presence of additives, and the like).

The configuration of a PEF system is illustrated in Figure 19–6, and some examples of PEF applications to food protection are summarized below. For a review of PEF processing of foods, see reference 22.

With *L. monocytogenes*, cells were more sensitive to PEF when grown at 4°C than at 37°C; more resistant at low a_w; more sensitive under acidic conditions; and more resistant in stationary phase than log phase of growth.[2] When orange juice was subjected to 30 kV/cm and 50 kVcm at 50°C, up to a 5-log cycle reduction of *Leuconostoc mesenteroides*,*E. coli*, and *Listeria innocua* was achieved.[30] At 50 kV/cm and 50°C, *S. cerevisiae* ascospores were reduced by a maximum of 2.5 logs. In another study employing orange juice, a reduction of the APC of >6 log cycles occurred in fresh orange juice with a treatment of 80 kV/cm, 20 pulses, pH 3.5, temperature of 44°C, and 100 IU/ml nisin.[20] The treated juice had a 40°C-shelf life of at least 28 days.

A $\log_{10}$ 5.9 reduction of *S*. Typhimurium in orange juice was demonstrated with 90 kV/cm and 50 pulses at 55°C.[25] Nisin and lysozyme acted synergistically, and combined with PEF, an additional 1.37 log-cycle reduction of the pathogen was noted. The synergy between nisin and lysozyme supports the view that the plasma membrane is the PEF target. A 5.35-$\log_{10}$ cfu/ml reduction of *E. coli* 0157:H7 added to apple cider was achieved with 80 kV/cm and 30 pulses at 42°C.[21] With 90 kV/cm and ten pulses at 42°C, a 5.91-log reduction occurred but when cinnamon powder (2%) or nisin (2.5%) was added, a 6- to 8-log reduction occurred.[21]

Using raw skim milk, APC was reduced by 7 logs with 80 kV/cm, 50 pulses at 52°C, and both nisin (38 IU/ml) and lysozyme (1,638 IU/ml) added.[57] Against vegetative cells of *Bacillus cereus*, a treatment with 16.7 kV/cm, 50 pulses each at 2 μsec plus 0.06 ppm nisin, an increased reduction of 1.8 log units was observed over what PEF and nisin alone achieved.[43] A 5-log reduction of *E. coli* 0157:H7 was obtained in a simulated milk medium by 5 kV/cm + 1,200 IU nisin/ml at a_w of 0.95.[59] It was found that NaCl and nisin reduced the effectiveness of PEF.

In a study that compared PEF with HHP and heat for controlling ascospores of *Z. bailii* in fruit juices, two pulses of 32–36.5 kV/cm reduced vegetative cells 4.5–5 and ascospores 3.5–4 log cycles, and this compared with a nearly 5-log reduction of vegetative cells by HHP but only a 0.5–1 log reduction of ascospores with a 5-minute treatment at 300 MPa.[49] Overall, two pulses of 32–36.5 kV/cm reduced vegetative cells or ascospores 3.5–5 log cycles for each of the five juices tested. The ascospores were 5–8 times more heat resistant than the vegetative cells.

Regarding *E. coli*, when 10^6 cfu/ml were added to pea soup and treated with two 16-pulse steps at 35 kV/cm for a total of 2 sec, cells could not be detected by plate count.[48] In an earlier study, bacteriophages of *Lactococcus cremoris* were found to be more sensitive to electric shock than four species of bacteria, including spores of *Bacillus subtilis*.[14] More information on the PEF treatment of foods can be found in reference 48.

ASEPTIC PACKAGING

In traditional canning methods, nonsterile food is placed in nonsterile containers, followed by container closure and sterilization. In aseptic packaging, sterile food under aseptic conditions is placed in sterile containers, and the packages are sealed under aseptic conditions as well. Although the methodology of aseptic packaging was patented in the early 1960s, the technology was little used until 1981, when the Food and Drug Administration approved the use of hydrogen peroxide for the sterilization of flexible multilayered packaging materials used in aseptic processing systems.

In general, any food that can be pumped through a heat exchanger can be aseptically packaged. The widest application has been to liquids such as fruit juices, and a wide variety of single-serve products of this type has resulted. The technology for foods that contain particulates has been more difficult to develop, with microbiological considerations only one of the many problems to overcome. In determining the sterilization process for foods pumped through heat exchangers, the fastest-moving components (those with the minimum holding time) are used, and where liquids and particulates are mixed, the latter will be the slower moving. Heat-penetration rates are not similar for liquids and solids, making it more difficult to establish minimum process requirements that will effectively destroy both organisms and food enzymes.

Some of the advantages of aseptic packaging are as follows:

1. Products such as fruit juices are more flavorful and lack the metallic taste of those processed in metal containers.
2. Flexible multilayered cartons can be used instead of glass or metal containers.
3. The time a product is subjected to high temperatures is minimized when ultrahigh temperatures are used.
4. The technology allows the use of membrane filtration of certain liquids.
5. Various container headspace gases such as nitrogen may be used.

Among the disadvantages are that packages may not be equivalent to glass or metal containers in preventing the permeation of oxygen, and the output is lower than that for solid containers.

A wide variety of aseptic packaging techniques now exists, with more under development. Sterilization of packages is achieved in various ways, one of which involves the continuous feeding of rolls of packaging material into a machine where hot hydrogen peroxide is used to effect sterilization, followed by the forming, filling with food, and sealing of the containers. Sterility of the filling operation may be maintained by a positive pressure of air or gas such as nitrogen. Aseptically packaged fruit juices are shelf stable at ambient temperatures for 6–12 months or longer.

The spoilage of aseptically packaged foods differs from foods in metal containers. Whereas hydrogen swells occur in high-acid foods in the latter containers, aseptic packaging materials are nonmetallic. Seam leakage may be expected to be absent in aseptically packaged foods, but the permeation of oxygen by the nonmetal and nonglass containers may allow for other types of spoilage in low-acid foods.

MANOTHERMOSONICATION (THERMOULTRASONICATION)

When bacterial spores are simultaneously exposed to ultrasonic waves and heat, there is a reduction in spore resistance. The effect is greatest when the two treatments are simultaneous, although some reduction in resistance occurs when ultrasonication is carried out just before heating. This phenomenon has been studied by workers in Spain and designated manothermosonication (MTS) or thermoultrasonication.[34] In addition to spores, MTS has been shown to be effective in reducing the thermal resistance of the enzymes peroxidase, lipooxygenase, and polyphenol oxidase.[27] Manothermosonication employs a lethal temperature while *manosonication* employs a sublethal temperature. In essence, these methods employ ultrasonic waves under pressure.

Manosonication (117 μm and 200 kPa) at 60°C was tested on *S.* Senftenberg, and a 3-log cycle reduction was achieved compared to a heat treatment at 60°C, which effected only a 0.5-log cycle reduction.[29]

In an early study of the effect of MTS on heat resistance using quarter-strength Ringer's solution, the D at 110°C of a *B. cereus* strain was reduced from 11.5 to ~1.5 minutes, and that of a *B. licheniformis* strain at 99°C from 5.5 to 3 minutes.[6] In another study using whole milk and two strains of *B. subtilis*, D values at 100°C were reduced from 2.59 to 1.60 for one strain, and from 11.30 to 1.82 minutes in another.[12,35] Comparable z values were 9.12–9.37 and 6.72–6.31, respectively. Ultrasonication was carried out at 20 kHz and 150 W. The z value results seem to confirm the minimal effect that MTS has on z values.[34]

As to the possible mechanism by which the heat resistance of bacterial spores is reduced by ultrasonic treatments, a study using *G. stearothermophilus* found that the ultrasonic treatment effected the release of calcium, dipicolinic acid, fatty acids, and other low-molecular-weight components.[37] The effect of this on spores was believed to lead to a modified hydration state and, thus, lowered heat resistance. This would not explain the effect of MTS on enzymes.

REFERENCES

1. Alpas, H., N. Kalchayanaand, F. Bozoglu, and B. Ray. 2000. Interactions of high hydrostatic pressure, pressurization temperature and pH on death and injury of pressure-resistant and pressure-sensitive strains of foodborne pahogens. *Int. J. Food Microbiol.* 60:33–42.

2. Alvarez, I., R. Pagán, J. Raso, and S. Condón. 2002. Environmental factors influencing the inactivation of *Listeria mono-cytogenes* by pulsed electric fields. *Lett. Appl. Microbiol.* 35:489–493.

3. Ananth, V., J.S. Dickson, D.G. Olson, and E.A. Murano. 1998. Shelf life extension, safety, and quality of fresh pork loin treated with high hydrostatic pressure. *J. Food Protect.* 61:1649–1656.

4. Arroyo, G., P.D. Sanz, and G. Préstamo. 1997. Effect of high pressure on the reduction of microbial populations in vegetables. *J. Appl. Microbiol.* 82:735–742.

5. Berlin, D.L., D.S. Herson, D.T. Hicks, and D.G. Hoover. 1999. Response of pathogenic *Vibrio* species to high hydrostatic pressure. *Appl. Environ. Microbiol.* 65:2776–2780.

6. Burgos, J., J.A. Ordóñez, and F. Sala. 1972. Effect of ultrasonic waves on the heat resistance of *Bacillus cereus* and *Bacillus licheniformis* spores. *Appl. Microbiol.* 24:497–498.

7. Calik, H., M.T. Morrisey, P.W. Reno, and H. An. 2002. Effect of high-pressure processing on *Vibrio parahaemolyticus* stains in pure culture and Pacific oysters. *J. Food Sci.* 67:1506–1510.

8. Cheftel, J.C. 1995. Review: high-pressure, microbial inactivation and food preservation. *Food Sci. Technol. Int.* 1:75–90.

9. Cléry-Barraud, C., A. Gaubert, P. Masson, and D. Vidal. 2004. Combined effects of high hydrostatic pressure and temperature for inactivation of *Bacillus anthracis* spores. *Appl. Environ. Microbiol.* 70:635–637.

10. Dong, F.M., A.R. Cook, and R.P. Herwig. 2003. High hydrostatic pressure treatment of finfish to inactivate *Anisakis simplex*. *J. Food Protect.* 66:1924–1926.

11. Earnshaw, R.G., J. Appleyard, and R.M. Hurst. 1995. Understanding physical inactivation processes: Combined preservation opportunities using heat, ultrasound and pressure. *Int. J. Food Microbiol.* 28:197–219.

12. Garcia, M.L., J. Burgos, B. Sanz, and J.A. Ordoñez. 1989. Effect of heat and ultrasonic waves on the survival of two strains of *Bacillus subtilis*. *J. Appl. Bacteriol.* 67:619–628.

13. García-Risco, M.R., E. Cortés, A.V. Carrascosa, and R. López-Fandiño. 1998. Microbiological and chemical changes in high-pressure-treated milk during refrigerated storage. *J. Food Protect.* 61:735–737.

14. Gilliland, S.E., and M.L. Speck. 1967. Inactivation of microorganisms by electrohydraulic shock. *Appl. Microbiol.* 15:1031–1037.

15. Hayakawa, I., T. Kanno, K. Yoshiyama, and Y. Fujio. 1994. Oscillatory compared with continuous high pressure sterilization on *Bacillus stearothermophilus* spores. *J. Food Sci.* 59:164–167.

16. Hayakawa, I., S. Furukawa, A. Midzunaga, H. Horiuchi, T. Nakashima, Y. Fujio, Y. Yano, T. Ishikura, and K. Sasaki. 1998. Mechanism of inactivation of heat-tolerant spores of *Bacillus stearothermophilus* IFO 12550 by rapid decompression. *J. Food Sci.* 63:371–374.

17. Heinz, V., and D. Knorr. 2001. Effect of high pressure on spores. In *Ultra High Pressure Treatments of Foods*, ed. M.E.G. Hendrickx and D. Knorr, 77–113. New York: Kluwer Academic Publishers.

18. Hite, B.H. 1899. The effect of pressure in the preservation of milk. *W.V. Agric. Exp. Sta. Bull.* 58:15–35.

19. Hite, B.H., N.J. Giddings, and C.E. Weakley. 1914. The effect of pressure on certain microorganisms encountered in the preservation of fruits and vegetables. *W.V. Agric. Exp. Sta. Bull.* 146:3–67.

20. Hodgins, A.M., G.S. Mittal, and M.W. Griffiths. 2002. Pasteurization of fresh orange juice using low-energy pulsed electrical field. *J. Food Sci.* 67:2294–2299,

21. Iu, J., G.S. Mittal, and M.W. Griffiths. 2001. Reduction in levels of *Escherichia coli* 0157:H7 in apple cider by pulsed electric fields. *J. Food Protect.* 64:964–969.

22. Jeyamkondan, S., D.S. Jayas, and R.A. Holley. 1999. Pulsed electric field processing of foods: A review. *J. Food Protect.* 62:1088–1096.

23. Kalchayanand, N., A. Sikes, C.P. Dunne, and B. Ray. 1994. Hydrostatic pressure and electroporation have increased bactericidal efficiency in combination with bacteriocins. *Appl. Environ. Microbiol.* 60:4174–4177.

24. Khadre, M.A., and A.E. Yousef. 2002. Susceptibility of human rotavirus to ozone, high pressure, and pulsed electric field. *J. Food Protect.* 65:1441–1446.

25. Liang, Z., G.S. Mittal, and M.W. Griffiths. 2002. Inactivation of *Salmonella* Typhimurium in orange juice containing antimicrobial agents by pulsed electric field. *J. Food Protect.* 65:1081–1087.

26. Linton, M., J.M.J. McClements, and M.F. Patterson. 1999. Inactivation of *Escherichia coli* 0157:H7 in orange juice using a combination of high pressure and mild heat. *J. Food Protect.* 62:277–279.

27. López, P., F.J. Sala, J.L. de la Fuente, S. Condón, J. Raso, and J. Borgos. 1994. Inactivation of peroxidase, lipoxygenase, and polyphenol oxidase by manothermosonication. *J. Agric. Food Chem.* 42:252–256.

28. López-Caballero, M.E., M. Pérez-Mateos, P. Montero, and A.J. Borderías. 2000. Oyster preservation by high pressure treatment. *J. Food Protect.* 63:196–201.

29. Mañas, P., R. Pagán, J. Raso, F.J. Sala, and S. Condón. 2000. Inactivation of *Salmonella* Enteritidis, *Salmonella* Typhimurium, and *Salmonella* Senftenberg by ultrasonic waves under pressure. *J. Food Protect.* 63:451–456.

30. McDonald, C.J., S.W. Lloyd, M.A. Vitale, K. Petersson, and F. Innings. 2000. Effects of pulsed electric fields on microorganisms in orange juice using electric field strengths of 30 and 50 kV/cm. *J. Food Sci.* 65:984–989.

31. Metrick, C., D.G. Hoover, and D.F. Farkas. 1989. Effects of high hydrostatic pressure on heat resistant and heat-sensitive strains of *Salmonella*. *J. Food Sci.* 54:1547–1549, 1564.

32. Molina-García, A.D., and P.D. Sanz. 2002. *Anisakis simplex* larva killed by high-hydrostatic-pressure processing. *J. Food Protect.* 65:383–388.

33. Mussa, D.M., H.S. Ramaswamy, and J.P. Smith. 1999. High-pressure destruction kinetics of *Listeria monocytogenes* on pork. *J. Food Protect.* 62:40–45.

34. Ordoñez, J.A., M.A. Aguilera, M.L. Garcia, and B. Sanz. 1987. Effect of combined ultrasonic and heat treatment (thermoultrasonication) on the survival of a strain of *Staphylococcus aureus*. *J. Dairy Res.* 54:61–67.

35. Ordoñez, J.A., and J. Burgos. 1976. Effect of ultrasonic waves on the heat resistance of *Bacillus* spores. *Appl. Environ. Microbiol.* 32:183–184.

36. Paidhungat, M., B. Setlow, W.B. Daniels, D. Hoover, E. Papafragkou, and P. Setlow. 2002. Mechanisms of induction of germination of *Bacillus subtilis* spores by high pressure. *Appl. Environ. Microbiol.* 68:3172–3175.

37. Palacios, P., J. Burgos, L. Hoz, B. Sanz, and J.A. Ordóñez. 1991. Study of substances released by ultrasonic treatment from *Bacillus stearothermophilus* spores. *J. Appl. Bacteriol.* 71:445–451.

38. Palou, E., A. López-Malo, G.V. Barbosa-Cánovas, J. Welti-Chanes, and B.G. Swanson. 1997. High hydrostatic pressure as a hurdle for *Zygosaccharomyces bailii* inactivation. *J. Food Sci.* 62:855–857.

39. Palou, E., A. López-Malo, G.V. Barbosa-Cánovas, J. Welti-Chanes, and B.G. Swanson. 1998. Oscillatory high hydrostatic pressure inactivation of *Zygosaccharomyces bailii*. *J. Food Protect.* 61:1213–1215.

40. Pandya, Y., F.F. Jewett, Jr., and D.G. Hoover. 1995. Concurrent effects of high hydrostatic pressure, acidity and heat on the destruction and injury of yeasts. *J. Food Protect.* 58:301–304.

41. Patterson, J.F., and D.J. Kilpatrick. 1998. The combined effect of high hydrostatic pressure and mild heat on inactivation of pathogens in milk and poultry. *J. Food Protect.* 61:432–436.

42. Patterson, M.F., M. Quinn, R. Simpson, and A. Gilmour. 1995. Sensitivity of vegetative pathogens to high hydrostatic pressure treatment in phosphate-buffered saline and foods. *J. Food Protect.* 58:524–529.

43. Pol, I.E., H.C. Mastwijk, P.V. Bartels, and E.J. Smid. 2000. Pulsed-electric field treatment enhances the bactericidal action of nisin against *Bacillus cereus*. *Appl. Environ. Microbiol.* 66:428–430.

44. Ponce, E., R. Pla, E. Sendra, B. Guamis, M. Mor-Mur. 1998. Combined effect of nisin and high hydrostatic pressure on destruction of *Listeria innocua* and *Escherichia coli* in liquid whole egg. *Int. J. Food Microbiol.* 43:15–19.

45. Popper, L., and D. Knorr. 1990. Applications of high-pressure homogenization for food preservation. *Food Technol.* 44(4):84–89.

46. Pothakamury, U.R., U. Vega, Q. Zhang, G.V. Barbosa-Cánovas, and B.G. Swanson. 1996. Effect of growth stage and processing temperature on the inactivation of *E. coli* by pulsed electric fields. *J. Food Protect.* 59:1167–1171.

47. Préstamo, G., and G. Arroyo. 1998. High hydrostatic pressure effects on vegetable structure. *J. Food Sci.* 63:878–881.

48. Qin, B.-L., U.R. Pothakamury, H. Vega, O. Martín, G.V. Barbosa-Cánovas, and B.G. Swanson. 1995. Food pasteurization using high-intensity pulsed electric fields. *Food Technol.* 49(12):55–60.

49. Raso, J., M.L. Calderón, M. Góngora, G.V. Barbosa-Cánovas, and B.G. Swanson. 1998. Inactivation of *Zygosaccharomyces bailii* in fruit juices by heat, high hydrostatic pressure and pulsed electric fields. *J. Food Sci.* 63:1042–1044.

50. Reina, L.D., Z.T. Jin, Q.H. Zhang, and A.E. Yousef. 1998. Inactivation of *Listeria monocytogenes* in milk by pulsed electric field. *J. Food Protect.* 61:1203–1206.

51. Reddy, N.R., H.M. Solomon, R.C. Tetzloff, and E.J. Rhodehamel. 2003. Inactivation of *Clostridium botulinum* type A spores by high-pressure processing at elevated temperatures. *J. Food Protect.* 66:1402–1407.

52. Reyns, K.M.F.A., E.A. Veraverbeke, and C.W. Michiels. 2003. Activation and inactivation of *Talaromyces macrosporus* ascospores by high hydrostatic pressure. *J. Food Protect.* 66:1035–1042.

53. Shearer, A.E.H., C.P. Dunne, A. Sikes, and D.G. Hoover. 2000. Bacterial spore inhibition and inactivation in foods by pressure, chemical preservatives, and mild heat. *J. Food Protect.* 63:1503–1510.

54. Shimoda, M., Y. Yamamoto, J. Cocunubo-Castellanos, H. Tonoike, T. Kawano, H. Ishikawa, and Y. Osajima. 1998. Antimicrobial effects of pressured carbon dioxide in a continuous flow system. *J. Food Sci.* 63:709–712.

55. Simpson, R.K., R. Whittington, R.G. Earnshaw, and N.J. Russell. 1999. Pulsed high electric field causes all "all or nothing" membrane damage in *Listeria monocytogenes* and *Salmonella* Typhimurium, but membrane H^+-ATPase is not a primary target. *Int. J. Food Microbiol.* 48:1–10.

56. Smelt, J.P., J.C. Hellemons, and M. Patterson. 2001. Effects of high pressure on vegetative microorganisms. In *Ultra High Pressure Treatments of Foods*, ed. M.E.G. Hendrickx and D. Knorr, 55–76. New York: Kluwer Academic Publishers.

57. Smith, K., G.S. Mittal, and M.W. Griffiths. 2002. Pasteurization of milk using pulsed electrical field and antimicrobials. *J. Food Sci.* 67:2304–2308.

58. Styles, M.F., D.G. Hoover, and D.F. Farkas. 1991. Response of *Listeria monocytogenes* and *Vibrio parahaemolyticus* to high hydrostatic pressure. *J. Food Sci.* 56:1404–1407.

59. Terebiznik, M., R. Jagus, P. Cerrutti, M.S. de Huergo, and A.M.R. Pilosof. 2002. Inactivation of *Escherichia coli* by a combination of nisin, pulsed electric fields, and water activity reduction by sodium chloride. *J. Food Protect.* 65:1253–1258.

60. Wei, C.I., M.O. Balaban, S.Y. Fernando, and A.J. Peplow. 1991. Bacterial effect of high pressure CO_2 treatment on foods spiked with *Listeria* or *Salmonella*. *J. Food Protect.* 54:189–193.

61. Wuytack, E.Y., A.M.J. Diels, K. Meersseman, and C.W. Michiels, 2003. Decontamination of seeds for seed sprout production by high hydrostatic pressure. *J. Food Protect.* 66:918–923.

62. Yuste, J., R. Pla, M. Capellas, E. Ponce, and M. Mor-Mur. 2000. High-pressure processing applied to cooked sausages: Bacterial populations during chilled storage. *J. Food Protect.* 63:1093–1099.

63. Yuste, J., M. Mor-Mur, M. Capellas, B. Guamis, and R. Pla. 1998. Microbiological quality of mechanically recovered poultry meat treated with high hydrostatic pressure and nisin. *Food Microbiol.* 15:407–414.

64. Zook, C.D., M.E. Parish, R.J. Braddock, and M.O. Balaban. 1999. High pressure inactivation kinetics of *Saccharomyces cerevisiae* ascospores in orange and apple juices. *J. Food Sci.* 64:533–535.

PART VI

Indicators of Food Safety and Quality, Principles of Quality Control, and Microbiological Criteria

The use of microorganisms and/or their products as quality indicators is presented in Chapter 20, along with the use of coliforms and enterococci as safety indicators. The principles of the hazard analysis critical control point (HACCP) system and Food Safety Objective (FSO) are presented in Chapter 21 as methods to control pathogens in foods. This chapter also contains an introduction to sampling plans and examples of microbiological criteria. The whole area of food quality control has been given much attention, and some of the varying approaches and views can be obtained from the following references:

Blackburn, C., and P. McClure, eds. 2002. *Foodborne Pathogens—Hazards, Risk Analysis and Control.* Boca Raton, FL: CRC Press. Assesses techniques used to manage foodborne hazards in the food industry.

ICMSF. 2002. *Microorganisms in Foods—Microbiological Testing in Food Safety Management.* A most authoritative source on sampling, testing, and process control methods.

Novak, J.S., G.M. Sapers, and V.K. Juneja, eds. 2003. *Microbial Safety of Minimally Processed Foods.* Boca Raton, FL: CRC Press. General coverage of pathogens relative to minimally processed foods along with strategies for their control is covered along with the application of the HACCP system.

Stevenson, K.E., and D.T. Bernard, eds. 1995. *HACCP—Establishing Hazard Analysis Critical Control Point Programs. A Workshop Manual.* Washington, DC: Food Processors Institute. Provides step-by-step procedures for setting up and monitoring HACCP programs.

CHAPTER 20

Indicators of Food Microbial Quality and Safety

Indicator organisms may be employed to reflect the microbiological quality of foods relative to product shelf life or their safety from foodborne pathogens. In general, indicators are most often used to assess food safety/sanitation, and most of this chapter treats them in this context. However, quality indicators may be used, and some general aspects of this usage are outlined in the following section.

SOME INDICATORS OF PRODUCT QUALITY

Microbial product quality or shelf-life indicators are organisms and/or their metabolic products whose presence in given foods at certain levels may be used to assess existing quality or, better, to predict product shelf life. When used in this way, the indicator organisms should meet the following criteria:

1. They should be present and detectable in all foods whose quality (or lack thereof) is to be assessed.
2. Their growth and numbers should have a direct negative correlation with product quality.
3. They should be easily detected and enumerated and be clearly distinguishable from other organisms.
4. They should be enumerable in a short period of time, ideally within a working day.
5. Their growth should not be affected adversely by other components of the food microbiota.

In general, the most reliable indicators of product quality tend to be product specific; some examples of food products and possible quality indicators are listed in Table 20–1. The products noted have restricted biota, and spoilage is typically the result of the growth of a single organism. When a single organism is the cause of spoilage, its numbers can be monitored by selective culturing or by a method such as impedance with the use of an appropriate selective medium. The overall microbial quality of the products noted in Table 20–1 is a function of the number of organisms noted, and shelf life can be increased by their control. In effect, microbial quality indicators are spoilage organisms whose increasing numbers result in loss of product quality.

Table 20–1 Some Organisms that are Negatively Correlated with Product Quality

Organisms	Products
Acetobacter spp.	Fresh cider
Bacillus spp.	Bread dough
Byssochlamys spp.	Canned fruits
Clostridium spp.	Hard cheeses
Flat-sour spores	Canned vegetables
Geotrichum spp.	Fruit cannery sanitation
Lactic acid bacteria	Beers, wines
Lactococcus lactis	Raw milk (refrigerated)
Leuconostoc mesenteroides	Sugar (during refinery)
Pectinatus cerevisiiphilus	Beers
"*Pseudomonas putrefaciens*"	Butter
Yeasts	Fruit juice concentrates
Zygosaccharomyces bailii	Mayonnaise, salad dressing

Metabolic products may be used to assess and predict microbial quality in some products; and some examples are listed in Table 20–2. The diamines (cadaverine and putrescine), histamine, and polyamines have been found to be of value for several products (discussed further in Chapter 4). Diacetyl was found to be the best negative predictor of quality in frozen orange juice concentrates, where it imparts a buttermilk aroma at levels of 0.8 ppm or above.[62] A 30-minute method for its detection was developed by Murdock.[61] Ethanol has been suggested as a quality index for canned salmon, where 25–74 ppm were associated with "offness," and levels higher than 75 ppm indicated spoilage.[35] Ethanol was found to be the most predictive of several alcohols in fish extracts stored at 5°C, where 227 of 241 fish-spoilage isolates produced this alcohol.[3] Lactic acid was the most frequently found organic acid in spoiled canned vegetables, and a rapid (2-hour) silica-gel plate method was developed for its detection.[1] The production of trimethylamine (TMA) from trimethylamine-N-oxide by fish spoilers has been used by a large number of investigators as a quality or spoilage index. Various procedures have been employed to measure total volatile substances as indicators of

Table 20–2 Some Microbial Metabolic Products That Negatively Correlate with Food Quality

Metabolites	Applicable Food Product
Cadaverine and putrescine	Vacuum-packaged beef
Diacetyl	Frozen juice concentrate
Ethanol	Apple juice, fishery products
Histamine	Canned tuna
Lactic acid	Canned vegetables
Trimethylamine (TMA)	Fish
Total volatile bases (TVB), total volatile nitrogen (TVN)	Seafoods
Volatile fatty acids	Butter, cream

fish quality, including total volatile bases (TVB)—ammonia, dimethylamine, and TMA—and total volatile nitrogen (TVN), which includes TVB and other nitrogen compounds that are released by steam distillation of fish products. See Chapter 5.

Total viable count methods have been used to assess product quality. They are of greater value as indicators of the existing state of given products than as predictors of shelf life because the portion of the count represented by the ultimate spoilers is difficult to ascertain.

Overall, microbial quality indicator organisms can be used for food products that have a biota limited by processing parameters and conditions where an undesirable state is associated consistently with a given level of specified organisms. Where product quality is significantly affected by the presence and quantity of certain metabolic products, they may be used as quality indicators. Total viable counts are generally not reliable in this regard, but they are better than direct microscopic counts.

INDICATORS OF FOOD SAFETY

Microbial indicators are employed more often to assess food safety and sanitation than quality. Ideally, a food safety indicator should meet certain important criteria. It should

1. be easily and rapidly detectable
2. be easily distinguishable from other members of the food biota
3. have a history of constant association with the pathogen whose presence it is to indicate
4. always be present when the pathogen of concern is present
5. be an organism whose numbers ideally should correlate with those of the pathogen of concern (Figure 20–1)

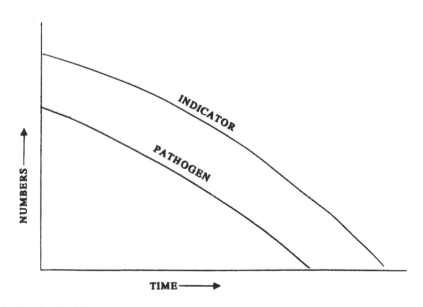

Figure 20–1 Idealized relationship between an indicator organism and the relevant pathogen(s). The indicator should exist in higher numbers than the pathogen during the existence of the latter.

6. possess growth requirements and a growth rate equaling or exceeding that of the pathogen
7. have a die-off rate that at least parallels that of the pathogen and ideally persists slightly longer than the pathogen of concern (Figure 20–1)
8. be absent from foods that are free of the pathogen except perhaps at certain minimum numbers

These criteria apply to most, if not all, foods that may be vehicles of foodborne pathogens, regardless of their source to the foods. In the historical use of safety indicators, however, the pathogens of concern were assumed to be of intestinal origin, resulting from either direct or indirect fecal contamination. Thus, such sanitary indicators were used historically to detect fecal contamination of waters and thereby the possible presence of intestinal pathogens. The first fecal indicator was *Escherichia coli*. When the concept of fecal indicators was applied to food safety, some additional criteria were stressed, and those suggested by Buttiaux and Mossel in 1961[11] are still valid:

1. Ideally, the bacteria selected should demonstrate specificity, occurring only in intestinal environments.
2. They should occur in very high numbers in feces so as to be encountered in high dilutions.
3. They should possess a high resistance to the extraenteral environment, the pollution of which is to be assessed.
4. They should permit relatively easy and fully reliable detection even when present in very low numbers.

Following the practice of employing *E. coli* as an indicator of fecal pollution of waters, other organisms were suggested for the same purpose, and most are covered below.

Coliforms

While attempting to isolate the etiologic agent of cholera in 1885, Escherich[24] isolated and studied the organism that is now *E. coli*. It was originally named *Bacterium coli commune* because it was present in the stools of each patient he examined. Schardinger[77] was the first to suggest the use of this organism as an index of fecal pollution because it could be isolated and identified more readily than individual waterborne pathogens. A test for this organism as a measure of drinking water potability was suggested in 1895 by T. Smith.[82] This marked the beginning of the use of coliforms as indicators of pathogens in water, a practice that has been extended to foods.

Species/Strains

In a practical sense, coliforms are Gram-negative asporogeneous rods that ferment lactose within 48 hours and produce dark colonies with a metallic sheen on Endo-type agar.[4] By and large, coliforms are represented by four or five genera of the family Enterobacteriaceae: *Citrobacter*, *Enterobacter*, *Escherichia*, and *Klebsiella*. Since the relatively new genus *Raoultella* was formerly a part of *Klebsiella*, it could represent the fifth coliform genus. Occasional strains of *Arizona hinshawii* and *Hafnia alvei* ferment lactose but generally not within 48 hours, and some *Pantoea agglomerans* strains are lactose positive within 48 hours.

Since *E. coli* is more indicative of fecal pollution than the other genera and species noted (especially *E. aerogenes*), it is often desirable to determine its incidence in a coliform population. The IMViC formula is the classic method used, where I = indole production, M = methyl red reaction,

V = Voges–Proskauer reaction (production of acetoin), and C = citrate utilization. By this method, the two organisms noted have the following formulas:

	I	M	V	C
E. coli	+	+	−	−
E. aerogenes	−	−	+	+

The IMViC reaction + + − − designates *E. coli* type I; *E. coli* type II strains are − + − −. The MR reaction is the most consistent for *E. coli*. *Citrobacter* spp. have been referred to as intermediate coliforms, and delayed lactose fermentation by some strains is known. All are MR+ and VP−. Most are citrate +, whereas indole production varies. *Klebsiella* isolates are highly variable with respect to IMViC reactions, although *K. pneumoniae* is generally MR−, VP+, and C+, but variations are known to occur in the MR and I reactions. Fluorogenic substrate methods for differentiating between *E. coli* and other coliforms are discussed in Chapter 11.

Fecal coliforms are defined by the production of acid and gas in EC broth between 44°C and 46°C, usually 44.5°C or 45.5°C. (EC broth, for *E. coli*, was developed in 1942 by Perry and Hajna.)[68] A test for fecal coliforms is essentially a test for *E. coli* type 1, although some *Citrobacter* and *Klebsiella* strains fit the definition. Notable exceptions are the EHEC strains that do not grow at 44.5°C in the standard EC medium formulation but will grow when the bile salts content in the medium is reduced from 0.15% to 0.112%.[86] A schematic for the detection and differentiation of coliforms, fecal coliforms, and *E. coli* is presented in Figure 20–2.

Five species of the genus *Escherichia* other than *E. coli* are listed in Table 20–3. *E. hermannii* is the only species that produces a yellow pigment, and it shares this feature with *Enterobacter sakazakii*,

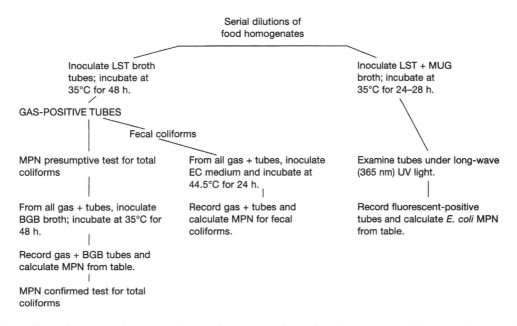

Figure 20–2 Summary of most probable number methods for total coliforms, fecal coliforms, and *Escherichia coli*. *Source:* Reprinted from reference 39, p. 542, by courtesy of Marcel Dekker. Jay, J.M. (2001). Indicator organisms in foods. In *Foodborne Disease Handbook*, Vol. 1, eds. Y.H. Hui, J.R. Gorham, K.D. Murrell, and D.O. Cliver, 537–546. Marcel Dekker, Inc., New York.

Table 20–3   Some Biochemical Reactions of Six *Escherichia* Species (Extracted from references 6, 7, 37).

Species	Lactose	Sorbitol	Indole	Methylred	Voges Prosk.	Yel.	Decarboxylase
E. albertii	−	−	−	+	−	−	+
E. blattae	−	−	−	+	−	−	+
E. fergusonii	−	−	+	+	−	−	+
E. vulneris	−/+[a]	−	−	+	−	−/+	+
E. hermannii	−/+	−	+	+	−	+	−
E. coli	+	+	+	+	−	−	+

[a]Most are negative.

which is discussed in Chapter 31. Since five of them do not produce gas from lactose, they are not included among coliforms. However, *E. albertii*, *E. vulneris*, and *E. hermannii* were originally recovered from human specimens. *E. albertii* was recovered as a diarrheagenic strain from the stools of Bangladeshi children;[37] *E. vulneris* was recovered from human wounds, sputum, and lung specimens;[7] and *E. hermannii* was recovered from human clinical specimens.[6] As noted above, the genus *Klebsiella* has been delimited by the transfer of several species to the new genus *Raoultella*.[21] Because the new genus produces gas from lactose, it properly belongs with the coliforms. The hallmark of this new genus is the capacity to grow at 10°C in contrast to the re-defined *Klebsiella* genus.[21]

Edwardsiella tarda is associated with the gastrointestinal tract of humans, and it is an oportunistic pathogen of humans. It is more common in the gut of cold-blooded animals, and is pathogenic to eels and other fish and rarely found in the feces of healthy humans.

Growth

Like most other nonpathogenic Gram-negative bacteria, coliforms grow well on a large number of media and in many foods. They have been reported to grow at temperatures as low as −2°C and as high as 50°C. In foods, growth is poor or very slow at 5°C, although several investigators have reported the growth of coliforms at 3–6°C. Coliforms have been reported to grow over a pH range of 4.4–9.0. *E. coli* can be grown in a minimal medium containing only an organic carbon source such as glucose and a source of nitrogen such as $(NH_4)_2SO_4$ and other minerals. Coliforms grow well on nutrient agar and produce visible colonies within 12–16 hours at 37°C. They can be expected to grow in a large number of foods under the proper conditions.

Coliforms are capable of growth in the presence of bile salts, which inhibit the growth of Gram-positive bacteria. Advantage is taken of this fact in their selective isolation from various sources. Unlike most other bacteria, they have the capacity to ferment lactose with the production of gas, and this characteristic alone is sufficient to make presumptive determinations. The general ease with which coliforms can be cultivated and differentiated makes them nearly ideal as indicators, except that their identification may be complicated by the presence of atypical strains. The aberrant lactose fermenters, however, appear to be of questionable sanitary significance.[29]

One of the attractive properties of *E. coli* as a fecal indicator for water is its period of survival. It generally dies off about the same time as the more common intestinal bacterial pathogens, although some reports indicate that some bacterial pathogens are more resistant in water. It is not, however, as resistant as intestinal viruses. Buttiaux and Mossel[11] concluded that various pathogens may persist after *E. coli* is destroyed in foods that are frozen, refrigerated, or irradiated. Similarly, pathogens may persist in treated waters after *E. coli* destruction. Only in acid food does *E. coli* have a particular value as an indicator organism due to its relative resistance to a low pH.[11]

Detection and Enumeration

A large number of methods have been developed for the detection and enumeration of *E. coli* and coliforms, and some are discussed in Chapters 10 and 11. One of the standard references listed in Table 10–1 should be consulted for an appropriate method to use under specified conditions.

Distribution

The primary habitat of *E. coli* is the intestinal tract of most warm-blooded animals, although sometimes it is absent from the gut of hogs. The primary habitat of *E. aerogenes* is vegetation and, occasionally, the intestinal tract. It is not difficult to demonstrate the presence of coliforms in air and dust, on the hands, and in and on many foods. The issue is not simply the presence of coliforms but their relative numbers. For example, most market vegetables harbor small numbers of lactose-fermenting, Gram-negative rods of the coliform type, but if these products have been harvested and handled properly, the numbers tend to be quite low and of no real significance from the standpoint of public health.

Coliform Criteria and Standards

Although the presence of large numbers of coliforms and *E. coli* in foods is highly undesirable, it would be virtually impossible to eliminate all from fresh and frozen foods. The basic questions regarding numbers are as follows:

1. Under proper conditions of harvesting, handling, storage, and transport of foods by use of a hazard analysis critical control point (HACCP) system, what is the lowest possible and feasible number of coliforms to maintain?
2. At what quantitative level do coliforms or *E. coli* indicate that a product is unsafe?

In the case of water and dairy products, there is a long history of safety related to allowable coliform numbers. Some coliform and *E. coli* criteria and standards for water, dairy products, and other foods covered by some regulatory agencies are as follows:

1. not over 10/ml for Grade A pasteurized milk and milk products, including cultured products
2. not over 10/ml for certified raw milk and not over 1 for certified pasteurized milk
3. not over 10/ml for precooked and partially cooked frozen foods
4. not over 100/ml for crabmeat
5. not over 100/ml for custard-filled items

Low numbers of coliforms are permitted in sensitive foods at numbers ranging from 1 to not over 100/g or 100 ml. These criteria reflect both feasibility and safety parameters.

Some products for which coliform criteria have been recommended by the International Commission on the Microbiological Specifications for Foods (ICMSF)[38] are listed in Table 20–4. The

Table 20–4   Suggested Coliform/*E. coli* Criteria

	Indicator/Products	Class Plan	n	c	m	M
1. Coliforms	Dried milk	3	5	1	10	10^2
2. Coliforms	Pasteurized liquid, frozen, and dried egg products	3	5	2	10	10^3
3. Coliforms	Infants, children, and certain dietetic foods; coated or filled, dried shelf-stable biscuits	3	5	2	10	10^2
4. Coliforms	Dried and instant products requiring reconstitution	3	5	1	10	10^2
5. Coliforms	Dried products requiring heating to boiling before consumption	3	5	3	10	10^2
6. Coliforms	Cooked ready-to-eat crabmeat	3	5	2	500	5,000
7. Coliforms	Cooked ready-to-eat shrimp	3	5	2	100	10^3
8. *E. coli*	Fresh, frozen, cold-smoked fish; frozen raw crustaceans	3	5	3	11	500
9. *E. coli*	Precooked breaded fish; frozen cooked crustaceans	3	5	2	11	500
10. *E. coli*	Cooked, chilled, frozen crabmeat	3	5	1	11	500
11. *E. coli*	Frozen vegetables/fruits, pH > 4.5; dried vegetables	3	5	2	10^2	10^3
12. *E. coli*	Fresh/frozen bivalve mollusks	2	5	0	16	–
13. *E. coli*	Bottled water	2	5	0	0	–

Note: Items 6 and 7 are recommendations of the National Advisory Committee on the Microbiological Criteria for Foods, USDA/FDA, January 1990, and the criteria noted are for process integrity. All other items are from ICMSF.[38]

values noted are not meant to be used apart from the total suggested criteria for these products. They are presented here only to show the acceptable and unacceptable ranges of coliforms or *E. coli* for the products noted. Implicit in the recommendations for the first four products is that one or two of five subsamples drawn from a lot may contain up to 10^3 coliforms and yet be safe for human consumption.

Some Limitations for Food Safety Use

Although the coliform index has been applied to foods for many years, there are limitations to the use of these indicators for certain foods. As a means of assessing the adequacy of pasteurization, a committee of the American Public Health Association in 1920 recommended the use of coliform[53] and this method was well established in the dairy industry around 1930.[67] Coliform tests for dairy products are not intended to indicate fecal contamination but do reflect overall dairy farm and plant sanitation.[73] For frozen blanched vegetables, coliform counts are of no sanitary significance because some, especially *Enterobacter* types, have common associations with vegetation.[83] However, the presence of *E. coli* may be viewed as an indication of a processing problem. For poultry products, coliforms are not good sanitary indicators because salmonellae may exist in a flock prior to slaughter, and thus positive fecal coliform tests may be unrelated to postslaughter contamination.[88] The standard coliform test is not suitable for meats because of the widespread occurrence of psychrotrophic enterics and *Aeromonas* spp. in meat environments, but fecal coliform tests are of value.[63]

Coliform tests are widely used in shellfish sanitation, but they are not always good predictors of sanitary quality. The U.S. National Shellfish Sanitation Program was begun in 1925, and the presence of coliforms was used to assess the sanitation of shellfish-growing waters. Generally, shellfish from waters that meet the coliform criteria ("open waters") have a good history of sanitary quality, but some human pathogens may still exist in these shellfish. In oysters, there is no correlation between fecal coliforms and *Vibrio cholerae*[16,41] or between *E. coli* and either *Vibrio parahaemolyticus* or *Yersinia enterocolitica*.[52] Coliforms are of no value in predicting scombroid poisoning,[52] nor do they always predict the presence of enteric viruses (see the subsection Coliphages, below). For sanitizing surfaces in meat packing plants, *K. pneumoniae* (and possibly *Raoultella* spp.) might be a better choice than generic *E. coli*.[85]

In spite of the limitations noted, coliforms are of proven value as safety indicators in at least some foods. They are best employed as a component of a safety program such as the HACCP system described in Chapter 21.

Enterococci

Around 30 species of the genus *Enterococcus* are recognized; and 22 are summarized in Table 20–5. Prior to 1984, the "fecal streptococci" consisted of two species and three subspecies, and they, along with *S. bovis* and *S. equinus*, were placed together because each contained Lancefield group D antigens. The latter two species are retained in the genus *Streptococcus*.

Historical Background

Escherich was the first to describe the organism that is now *E. faecalis*, which he named *Micrococcus ovalis* in 1886. *E. faecium* was recognized first in 1899 and further characterized in 1919 by Orla-Jensen.[64] Because of their existence in feces, these classic enterococci were suggested as indicators of water quality around 1900. Ostrolenk et al.[65] and Burton[9] were the first to compare the classic enterococci to coliforms as indicators of safety. Pertinent features of the classic enterococci that led to their use as pollution indicators for water are the following:

1. They generally do not multiply in water, especially if the organic matter content is low.
2. They are generally less numerous in human feces than *E. coli*, with ratios of fecal coliforms to enterococci of 4.0 or higher being indicative of contamination by human waste. Thus, the classic enterococcal tests presumably reflect more closely the numbers of intestinal pathogens than fecal coliforms.
3. The enterococci die off at a slower rate than coliforms in waters and thus would normally outlive the pathogens whose presence they are used to indicate.

The simultaneous use of enterococci and coliforms was advocated in the 1950s by Buttiaux,[10] as in his opinion the presence of both suggested the occurrence of fecal contamination. In his review of the literature, Buttiaux noted that 100% of human and pig feces samples contained enterococci, whereas only 86–89% contained coliforms.[10]

Classification and Growth Requirements

Although the classic enterococci never achieved the status of coliforms as sanitation indicators for water or foods, their current classification in an expanded genus could, on one hand, make them more

Table 20–5 Characteristics of Some *Enterococcus* spp.

Property	E. faecalis	E. faecium	E. avium	E. casseliflavus	E. durans	E. malodoratus	E. gallinarum	E. hirae	E. mundtii	E. raffinosus	E. solitarius	E. pseudoavium	E. cecorum	E. saccharolyticus	E. columbae	E. dispar	E. flavescens	E. seriolicida	E. sulfureus	E. fallax	E. asini
Growth at/in																					
10°C	+	+	+	+	+	+	+	+	+	(+)	+	+	−	+		+	(+)	+	+		+/−
45°C	+	+	+	+	+	−	+	+	+	+	+	+	+	+		−	(+)	+	−		+/−
pH 9.6	+	+	+	+	+/−	+	+	+	+	+	+	−	(+)	(+)		+		+	+		
6.5% NaCl	+	+	+	+	+/−	+	+	+	+	+	+		+	+		+		+	+		−
40% bile	+	+	+	+	+	+	+	+	+				+		+			+	+		+
0.1% methylene blue	+	−	−	+	+	−	−	−	+				−			−		+			
0.04% K-tellurite	+	−	−	+	−		+								+				+		
0.01% Tetrazolium	+		+	+												+					
Resist 60°C/30 min	+	+	+*	+	+/−			+	+	+	+					+	−	+			
Serologic group D	+	+	+	−/+	+	−	+	+	+	−	−	−	−	−	−	−	−	+	−		−
Motility	−	−	+/−	+	−	−	+	−	−	−	−	−	−	−	−	−	−	−	−		−
Pigmented	−	−	−	yel	−	+	−	−	yel	−	−	−	−	−	−	−	yel	−	yel		+
Esculin hydrolysis	+/−	+	+	+	+	+	+	+	+	+	+	+	+	+	+	+	+	−	−		+
Hippurate hydrolysis	+/−	+	v	+/−	v	−	−	−	+	−	+	+	+	+	+	v	+	−	−		−
Arginine hydrolase	+	+	−	+	+	+	+	+	+	−	+	−	−	−	−	+	+	+	−		
Produces H₂S	−	−	+	−	+	−	−	−	−	−	−	−	−	−	−	−	−	−	−		
Acid from																					
Glycerol	+	+	+	−	−	v	−/+	v	v	+	−	−	−	−	−	+	−	+	−		−
Mannitol	+	+	+	+	−/+	+	+	−	+	+	+	−	−	−	+	−	+	−	−		−
Sucrose	+	v	+	+	−	+	+	+	+	+	+	+	+	+	+	+	+	+	+		−
Salicin	+	+	+/−	+	+	−	+	+	+	+	+	+	+	+	v	+	+	−	+		+
Lactose	+	+	+	+	−	−	+	+	+	+	−	+	+	+	+	+	+	−	−		+
Arabinose	−	+	+	+	−	−	+	−	−	+	−	−	−	−	+	−	+	−	+		−
Raffinose	−	+/−	−	−	−	+	+	+	+	+	−	−	+	−	+	−	+	−	+		−

Note: + = Positive; − = negative; +/−; v = variable reactions.

*Also group Q.

482

attractive as indicators or, on the other hand, make them less attractive and meaningful (Table 20–5). *E. faecalis* is found most frequently in the feces of a variety of mammals and *E. faecium* largely in hogs and wild boars;[56,81] the natural distribution of some other members of the new genus is less well understood. Prior to 1984, enterococci and "fecal strep" were essentially synonymous and consisted principally of only *E. faecalis*, *E. faecium*, and *E. durans*. Currently, a test for enterococci is of less significance as fecal, sanitary, or quality indicators than the classic species. An inspection of the features in Table 20–5 reveals that *E. cecorum* does not grow at 10°C or in 6.5% NaCl. Although *E. pseudoavium* grows at 10°C, it does not grow in the presence of 6.5% NaCl.[15] With the exception of *E. cecorum*, apparently all grow at 10°C, and some strains of *E. faecalis* and *E. faecium* have been reported to grow between 0°C and 6°C. Most of the enterococci grow at 45°C and some, at least *E. faecalis* and *E. faecium*, grow at 50°C. The phylogenetic relationship of enterococci, other lactic acid bacteria, *Listeria*, and *Brochothrix* is presented in Chapter 25 (Figure 25–1).

At least 13 species grow at a pH of 9.6 and in 40% bile, whereas at least three do not grow in 6.5% NaCl. *E. cecorum*, *E. columbae*, *E. dispar*, and *E. saccharolyticus* do not react with serologic group D antisera. In addition to reacting with group D antisera, *E. avium* alone reacts with group Q.[13] The murein type possessed by *E. faecalis* is Lys-Ala$_{2-3}$, whereas the other species contain the Lys-D-Asp murein. The mol% G + C content of DNA of the enterococci is 37–45. Regarding biochemical characteristics, esculin is hydrolyzed by all species. Four species produce a yellow pigment (*E. casseliflavus*, *E. flavescens*, *E. mundtii*, and *E. sulfureus*); two produce H_2S (*E. casseliflavus* and *E. malodoratus*); and all known strains of *E. gallinarum*[14] and *E. flavescens* are motile.

As is typical of other Gram-positive bacteria, enterococci are more fastidious in their nutritional requirements than Gram negatives but differ from most other Gram positives in having requirements for more growth factors, especially B vitamins and certain amino acids. The requirement for specific amino acid allows some strains to be used in microbiological assays for these compounds. They grow over a much wider range of pH than all other foodborne bacteria (see Chapter 3). Although they are aerobes, they do not produce catalase (except a pseudocatalase by some strains when grown in the presence of O_2), and they are microaerophiles that grow well under conditions of low oxidation–reduction potential (Eh).

Distribution

Although the two classic enterococcal species (*E. faecalis* and *E. faecium*) are known to be primarily of fecal origin, the new ones await further study of natural occurrence, especially regarding fecal occurrence. *E. hirae* and *E. durans* have been found more often in poultry and cattle than in six other animals, whereas *E. gallinarum* has been found only in poultry.[20] *E. durans* and *E. faecium* tend to be associated with the intestinal tract of swine more than does *E. faecalis*. The last appears to be more specific for the human intestinal tract than are other species. *E. cecorum* was isolated from chicken cecae, *E. columbae* from pigeon intestines, and *E. saccharolyticum* from cows. *E. avium* is found in mammalian and chicken feces; *E. casseliflavus* in silage, soils, and on plants; *E. mundtii* on cows, hands of milkers, soils, and plants; *E. hirae* in chicken and pig intestines; *E. dispar* in certain human specimens; and *E. gallinarium* in the intestines of fowls.

It is well established that the classic enterococci exist on plants and insects and in soils. The yellow-pigmented species are especially associated with plants, and *E. cecorum* appears to be closely associated with chicken cecae. In general, enterococci on insects and plants may be from animal fecal matter. Such enterococci may be regarded as temporary residents and are disseminated among vegetation by insects and wind, reaching the soil by these means, by rain, and by gravity.[58] Although *E. faecalis* is often regarded as being of fecal origin, some strains appear to be common on vegetation and thus have no sanitary significance when found in foods. Mundt[59] studied *E. faecalis* from humans, plants, and other sources and found that the nonfecal indicators could be distinguished from the more fecal types by their reaction in litmus milk and their fermentation reactions in melizitose and melibiose broths. In another study of 2,334 isolates of *E. faecalis* from dried and frozen foods, a high percentage of strains bore a close similarity to the vegetation-resident types and, therefore, were not of any sanitary significance.[57] When used as indicators of sanitary quality in foods, it is necessary to ascertain whether *E. faecalis* isolates are of the vegetation type or whether they represent those of human origin. Enterococci may also be found in dust. They are rather widely distributed, especially in such places as slaughterhouses and curing rooms, where pork products are handled.

With respect to the use of the classic enterococci as indicators of water pollution, some investigators who have studied their persistence in water have found that they die off at a faster rate than coliforms, whereas others found the opposite. Leininger and McCleskey[48] noted that enterococci do not multiply in water as coliforms sometimes do. Their more exacting growth requirements may be taken to indicate a less competitive role in water environments. In sewage, coliforms and the classic enterococci were found to exist in high numbers, but approximately 13 times more coliforms than enterococci were found.[50]

In a study conducted when the genus *Enterococcus* consisted of only eight species, Devriese et al.[20] studied 264 isolates of enterococci obtained from farm animal intestines. Strains were selected solely on the basis of their growth in 40% bile and 6.5% NaCl. Of the 264 isolates, 255 conformed to one of the 8 species, with *E. faecalis, E. faecium*, and *E. hirae*, representing 37.6%, 29.8%, and 23% of the isolates, respectively. Other species found were *E. durans* (5.1%), *E. gallinarum* (1.6%), *E. avium* (1.2%), *E. mundtii* (1.2%), and *E. casseliflavus* (<1%). These 255 isolates were obtained from eight animal species, with poultry, cattle, and pigs yielding the largest number of isolates.

In a later study of enterococci in foods of animal origin,[19] a total of 161 strains were isolated from the following foods in Belgium: meats,[84] cheeses,[54] fish and shellfish,[42] and cheese–meat combinations.[30] *E. faecium* accounted for 58.4% (94 of 161) and *E. faecalis* for 26.1% (42 of 161) of the isolates with 9.3% (15 of 161) being represented by *E. hirae/E. durans*. None of the last 10 or so species to be named was identified in either of the above studies.

Relationship to Sanitary Quality of Foods

In this section, the enterococci discussed are those that were defined prior to 1984. A large number of investigators found the classic enterococci to be better than coliforms as indicators of food sanitary quality, especially for frozen foods. In one study, enterococcal numbers were more closely related to aerobic plate counts (APC) than to coliform counts, whereas coliforms were more closely related to enterococci than to APC.[30] Enterococci have been found in greater numbers than coliforms in frozen foods (Table 20–6). In a study of 376 samples of commercially frozen vegetables, Burton[9] found that coliforms were more efficient indicators of sanitation than enterococci prior to freezing, whereas enterococci were superior indicators after freezing and storage. In samples stored at $-20°C$ for 1–3 months, 81% of enterococci and 75% of coliforms survived. After 1 year, 89% of enterococci survived but only 60% of coliforms. In another study, enterococci remained relatively constant for 400 days when stored at freezing temperatures. Enterococci were recovered from 57% of 14 samples of dried foods, whereas 87% of 13 different frozen vegetables yielded these organisms, many of which were of the vegetation-resident types.[57] The relative longevity of coliforms and enterococci in frozen fish sticks is presented in Table 20–7.

Overall, the elevation of the once "fecal strep" to the status of a genus and the expansion of the genus to include some species that appear to have no natural association with fecal matter raise

Table 20–6 Enterococci and Coliform Most Probable Number (MPN) Counts in Frozen Precooked Fish Sticks

Number	Enterococci MPN Count/100 g	Coliforms MPN Count/100 g
1	86,000	6
2	18,600	19
3	86,000	0
4	46,000	300
5	48,000	150
6	46,000	28
7	46,000	150
8	18,600	7
9	8,600	0
10	4,600	186
11	4,600	186
12	48,000	1,280
13	8,600	46
14	4,600	480
15	48,000	240
16	10,750	1,075
17	10,750	17,000
18	60,000	23,250
19	10,750	2,275
Average	32,339	2,457

Source: From Raj et al.[71]

questions about the utility of this group as sanitary indicators. During the 1960s and 1970s, enterococcal tolerances were suggested for a variety of foods, but they have been disregarded in this context in recent years. Interest in the enterococci as food safety indicators has clearly waned, probably because of the simultaneous interest in faster and more efficient ways to detect and enumerate *E. coli*. As indicators, the enterococci and coliforms are compared in Table 20–8.

Bifidobacteria

Around the year 1900, in the course of his research on the stools of infants Tissier[89] noted an organism that occurred with great frequency and named it *Bacillus bifidus*; it was later named *Lactobacillius bifidus* and is currently known as *Bifidobacterium bifidum*. The common occurrence of the bifidobacteria in stools led Mossel[55] to suggest the use of these Gram-positive anaerobic bacteria as indicators of fecal pollution, especially of waters. Interestingly, some bifidobacteria are employed in the production of fermented milks, yogurt, and other food products, and they are believed to provide some health benefits.

The genus *Bifidobacterium* consists of at least 25 species of catalase-negative, nonmotile rods whose minimum and maximum growth temperature ranges are 25 to 28°C and 43 to 45°C, respectively. They grow best in the pH range 5 to 8 and produce lactic and acetic acids as the major end products of their carbohydrate metabolism.

Table 20–7   Effect of −6°F Storage on the Longevity of Coliforms and Enterococci in Precooked Frozen Fish Sticks

Days in Storage at −6°F	Most Probable Number*	
	Coliform	Enterococci
0	5,600,000	15,000,000
7	6,000,000	20,000,000
14	1,400,000	13,000,000
20	760,000	11,300,000
35	440,000	11,200,000
49	600,000	20,000,000
63	88,000	11,000,000
77	395,000	15,000,000
91	125,000	41,000,000
119	50,000	5,400,000
133	136,000	7,400,000
179	130,000	5,600,000
207	55,000	3,500,000
242	14,000	4,000,000
273	21,000	4,000,000
289	42,000	3,200,000
347	20,000	2,300,000
410	8,000	1,600,000
446	260	2,300,000
481	66	5,000,000

*Average of four determinations.

Source: From Kereluk and Gunderson.[46]

Distribution

The bifidobacteria have been found in human feces at higher levels per gram (10^8–10^9) than *E. coli* (10^6–10^7), and this makes them more attractive as indicators of human fecal pollution. By using the bifidobacteria, it is possible to determine their origin among the following three sources: human feces, animal feces, or environmental conditions. The method for distinguishing between human and animal strains was devised by Gavini et al.[26] and it divides bifidobacteria into seven groups with those of human origin belonging to groups I, III, and VII. From a limited study of 50 samples of ground meats, 39 contained both *E. coli* and bifidobacteria.[5] Of the latter, only two were of human origin while the others were from animals. *B. adolescentis* and *B. longum* are most often isolated in the highest numbers—about 10^6/100 ml of raw sewage.[74] They have been suggested as indicators of recent fecal contamination in tropical freshwaters as they die off faster than either coliforms or enterococci.[60]

Overall, the close association of bifidobacteria with feces, their usual absence where fecal matter does not occur, their lack of growth in water, and the specific association of some only with human feces make these bacteria attractive as pollution indicators. On the other hand, because they are strict anaerobes, they tend to grow slowly and require several days for results. As they are more likely to

Table 20–8 Coliforms and Enterococci as Indicators of Food Sanitary Quality

Characteristic	Coliforms	Enterococci
Morphology	Rods	Cocci
Gram reaction	Negative	Positive
Incidence in intestinal tract	10^7-10^9/g feces	10^5-10^8/g feces
Incidence in fecal matter of various animal species	Absent from some	Present in most
Specificity to intestinal tract	Generally specific	Generally less specific
Occurrence outside of intestinal tract	Common in low numbers.	Common in higher numbers.
Ease of isolation and identification	Relatively easy	More difficult
Response to adverse environmental conditions	Less resistant	More resistant
Response to freezing	Less resistant	More resistant
Relative survival in frozen foods	Generally low	High
Relative survival in dried foods	Low	High
Incidence in fresh vegetables	Low	Generally high
Incidence in fresh meats	Generally low	Generally low
Incidence in cured meats	Low or absent	Generally high
Relationship to foodborne intestinal pathogens	Generally high	Lower
Relationship to nonintestinal foodborne pathogens	Low	Low

grow in meat and seafood products than in vegetables (because of the higher natural Eh of the latter), it is possible that they could serve as indicators for meats and seafoods.

Coliphages/Enteroviruses

Research during the 1920s revealed that bacteriophages occur in waters in association with their host bacteria, and this led Pasricha and DeMonte[66] to suggest that phages specific for several intestinal pathogens could be measured as indirect indicators of their host bacterial species. A coliphage assay procedure for water samples that contain five or more phages/100 ml and that can be completed in 4–6 hours is described in *Standard Methods for the Examination of Water and Wastewater.*[4] Thus, the utility of the coliphage assay for waters using *E. coli* strain C has been established. Of concern in the detection of coliphages is the capacity of the host strains used to allow plaque development by all viable phages. Although the American Public Health Association procedure recommends the use of *E. coli* strain C, other hosts may be used simultaneously to increase the plaque counts. There is no way of enumerating all *E. coli* phages or all phages of any other specific bacterium, suggesting the use of mixed indicators for best results.[70]

Because coliphage assay by use of *E. coli* hosts may reflect heterogeneous phages with different survival characteristics, the detection of male-specific phages is one method that leads to a more homogeneous phage population. Male-specific phages are single-stranded, homogeneous, and similar in structure and size to enteroviruses.[33] Although their standard hosts are F$^+$ or Hfr strains of *E. coli* K-12, host cells can be constructed by plasmid insertions in *Salmonella* Typhimurium. The latter cells contain F-pili, which serve as receptors for male-specific coliphages and are employed essentially in the same way as *E. coli* hosts.

Utility for Water

The prediction of fecal coliforms in water by the enumeration of their phages has been shown to be feasible by some investigators[34,42,47,92] and not feasible by others. In a study of coliphages and fecal and total coliforms in natural waters from 10 cities, a linear relationship was found between the two groups.[91]

Because bacteria and viruses possess different properties relative to their persistence in the environment, coliphages have attracted interest from those interested in indicators of enteroviruses, especially in water. The inability of the coliform index to predict correctly the presence of enteric viruses in waters has been reported by a number of investigators. The survival of coliphages in water has been shown to parallel that of human enteric viruses.[80] In a study of approved waters for oyster harvesting along the Gulf Coast, neither *E. coli* nor coliform levels were predictive of the presence of enteroviruses in oysters.[25] In recreational waters considered to be acceptable and safe by coliform standards, enteroviruses were detected 43% of the time and 35% of the time in waters acceptable for shellfish harvesting.[27] In a study of open waters along the North Carolina coast, enteric viruses were isolated from 3 of 13, 100-g clam samples from open beds, and 6 of 15 were positive from closed beds.[91] A well-documented outbreak of human hepatitis A was traced to the consumption of oysters taken from open waters.[69]

In a study comparing coliphages, total coliforms, fecal coliforms, enterococci, and standard plate counts on water from different treatment processes, coliphages correlated better with enteroviruses than either of the other groups noted.[84] When secondary sewage effluents were tested for male-specific coliphages, up to 8,200 plaque-forming units (pfu) were found,[33] but how assays for male-specific coliphages compare to the more traditional assay methods is unclear. Because some coliphages have been reported to have their natural habitat in environmental waters, their numbers may not correlate directly with fecal pollution.[78] Male-specific coliphages are more indicative of fecal pollution of waters than total coliphages because they do not form F-pili at temperatures less than 30°C, and thus cannot infect their F+ host cells.[79] In a later study of 1,081 samples consisting of feces from humans and 11 animal species in addition to human-associated wastewaters, male-specific phages appeared to be the indicator of choice for assessing the potential presence of human enteric viruses in estuarine and marine environments impacted by wastewater sources.[12] Although the 11 animal species contained male-specific phages, the numbers were generally low.

Similar to coliphages, *Bacteroides fragilis* phages are found in waters that contain high levels of human fecal waste. While their numbers tend to be lower than coliphages, they are more specific to human feces. In one study, they were not found in significant numbers in slaughterhouse wastewaters or in waters that contained fecal matter from wildlife only.[87] These investigators showed that *B. fragilis* phages multiplied only under anaerobic conditions.

The double-stranded DNA (dsDNA) phages of *Vibrio vulnificus* along with their host cells have been found in a variety of oyster tissues and fluids, and the phages were more abundant and diverse in molluscan shellfish than in other habitats, suggesting their possible use as indicators of *V. vulnificus* presence.[17] In another study, it was found that the organism was lowest in oysters during the month of January–March and highest during the summer and fall months when numbers per gram were 10^3–10^4 and the *V. vulnificus* phages were also present in highest numbers during that period.[18]

Regarding human enteric viruses, not only can at least some survive better in water than coliforms, but viruses also tend to be more resistant to destruction by chlorine. Whereas chlorine destroyed 99.999% of fecal coliforms, total coliforms, and fecal streptococci in primary sewage effluents, only 85–99% of viruses present were destroyed in one study.[53]

When four phages were compared by heating, drying, and ClO_2 exposure, hepatitis A virus was found to be more resistant than poliovirus I and small round RNA (MS2) and DNA (ΦX174) coliphages.[51] Bovine enteroviruses cause asymptomatic infections in cattle, and in one study they were found in the feces of 76% of 139 cattle, 38% of 50 white-tailed deer, and in one of three Canada geese.[49] In this study, all animals noted shared the same pasture and stream waters, and the bovine enteroviruses were found in oysters from river water down-stream of the farm.[49] Teschoviruses specifically infect pigs (PTV) and by use of a RT-PCR technique, 92 pigs of PTV RNA/ml of sample could be detected.[40] This group of viruses allows the determination of whether water pollution is of pig origin.

Utility for Foods

The utility of employing coliphage assays for coliforms in foods was first reported by Kennedy et al. in 1984.[43] They employed a 16- to 18-hour incubation at 35°C and recovered coliphages from all 18 fresh chicken and pork sausage samples. The highest numbers of coliphages were found on fresh chicken and ranged from 3.3 to 4.4 $\log_{10}$ pfu/100 g. High coliphage levels in general reflected products that contained high fecal coliform counts.[43] In a later study involving 120 samples of 12 products, coliphages at levels of 10 or more pfu/100 g were found in 56% of the samples and 11 of the products.[44] The highest numbers were recorded from fresh meats by the 16- to 18-hour incubation procedure, and chickens yielded the highest counts (2.66 to 4.04 $\log_{10}$ pfu/100 g). In general, coliphages correlated better with *E. coli* and fecal coliforms than total coliforms. The recovery of coliphages from foods was not affected by pH in the range 6.0–9.0.[45] Results could be achieved in 4–6 hours, but these investigators preferred incubations of 16–18 hours. On the other hand, male-specific coliphages employing *S.* Typhimurium hosts did not correlate with total coliforms, fecal coliforms, or aerobic plate counts in 472 samples of clams from the Chesapeake Bay.[13] The low numbers found may have been due to the general absence of sewage contamination in the clam waters.

The possible use of male specific (F+) coliphages as fecal indicators for fresh carrots was investigated, and of 25 carrots tested, 25% were positive for F+ phages, 8% for *E. coli*, but only 4% for *Salmonella*.[22] When ground beef and poultry meat were examined for the recovery efficiency of 3 different phage groups, 100% of 5 F+ coliphages; 69% of a somatic coliphage; and 65% of 3 salmonellae phages were recovered.[36] When 8 market foods were examined, F+ phages were found in 63% and somatic phages in 88%. The sensitivity of the phage assay was 38 pfu/100 g of beef or chicken.

Overall, the findings from water and sewage and the limited studies with foods suggest that coliphage assays may be suitable either as an alternative to *E. coli* or coliform determinations or as direct indicators for enteroviruses. Because results can be obtained in 4 to 6 hours and because coliphages appear to correlate better with enteroviruses than coliforms, further research seems indicated. Host cell systems need to be developed that will yield plaques from all coliphages without allowing plaque development by phages that normally parasitize other closely related enteric bacteria.

THE POSSIBLE OVERUSE OF FECAL INDICATOR ORGANISMS

The successful use of the coliform/fecal coliform index to assess the potability of drinking water led to its widespread use for the microbial safety of foods, and not only has this use been extended to a wide variety of food products but also to food handling surfaces and utensils. It is well established that coliforms as well as fecal coliforms may exist in high numbers on certain foods and in food-processing environments as well as in waters and yet not be related to safety. In a 2-year study of coliforms, fecal coliforms, and enterococci on market lettuce and fennel in Italy, it may be noted from Table 20–9 that

Table 20–9 Mean $\log_{10}$ Numbers/100 g of Coliforms, Fecal Coliforms, and Enterococci on Lettuce and Fennel

Product	APC	Coliforms	Fecal Coliforms	Enterococci
Lettuce	7.82	4.77	3.79	3.34
Fennel	6.37	4.89	3.89	3.49

Source: Reprinted with permission from G.L. Ercolani,[23] Bacteriological Quality Assessment of Fresh Marketed Lettuce and Fennel, *Applied Environmental Microbiology*, Vol. 31, pp. 847–852, copyright © 1976, American Society for Microbiology.

these two ready-to-use vegetables contained between 3 and 4 $\log_{10}$/100 g of each group.[23] Overall, about 10% of the total coliforms were of the fecal type on both products, and all coliform genera were present.

In a study of bird droppings around Boston Harbor that compared fecal coliforms, enterococci, and F-specific coliphages, the cfu/g of coliforms found in bird droppings were as follows: goose, 10^1–10^5; pigeon, 10^5–10^9; and herring gull, 10^3–10^8.[75] Further, up to 10^6 somatic coliphages, 10^8 enterococci, and 10^2 F-specific coliphages per gram of feces were found. The intestinal carriage of numbers of this magnitude by these healthy avians can lead to the unjustified rejection of water or seafood products from such environments.

The relative numbers of coliphages, fecal coliforms, and enterococci in the feces of seven animal species are compared to those of humans in Table 20–10, and it can be seen that overall all contained relatively high numbers of each group with the following exception.[32] Neither F-specific nor somatic phages were found in human feces in appreciable numbers, and fecal coliforms were equally high in human and pig feces.

Table 20–10 Arithmetic Mean of the Number of Bacteriophages (pfu/g) and Indicator Bacteria (cfu/g) in Human and Animal Feces*

Source	F-Specific RNA Phages	Somatic Coliphages	Thermotolerant Coliforms	Faecal Streptococci	Spores of s.r. Clostridia[†]
Pig	2.8×10^3	3.4×10^6	3.0×10^6	7.3×10^5	6.4×10^2
Broiler chicken	$>1.2 \times 10^6$	1.1×10^7	1.9×10^8	5.6×10^6	$<10^2$
Dog	<10	4.1×10^4	9.0×10^7	8.2×10^6	1.6×10^6
Cow	<10	4.0×10^5	5.6×10^5	1.1×10^5	9.8×10^2
Horse	<10	2.2×10^4	1.8×10^5	1.3×10^4	$<10^2$
Sheep	1.9×10^3	3.1×10^6	1.2×10^7	1.3×10^5	$<10^2$
Calf	5.8×10^4	2.2×10^7	3.2×10^7	1.1×10^6	8.0×10^3
Human	$<10^1$	6.1×10^4	1.9×10^8	3.7×10^5	$>1.8 \times 10^3$

*Results are the averages of ten mixed samples from three individuals each.
[†]s.r.: Sulphite reducing.

Source: Reprinted with permission from Bacteriophages and Indicator Bacteria in Human and Animal Feces, *Journal of Applied Bacteriology*, A. Havelaar et al.,[32] Vol. 60, p. 259, copyright © 1986, Blackwell Science Ltd.

In a study of fecal material from 39 diamond-back terrapins from brackish waters along the east-southeast U.S. coast, 51% of cloacal swabs were positive for fecal coliforms, and 80% of 10 fecal specimens were positive.[31]

Like many others, the above studies make it clear that coliforms and fecal coliforms exist in a number of places and in many raw food materials where their presence has little or no relationship to food safety. The excessive and unwise use of these indicators may, on the one hand, lead to the rejection of safe products; on the other hand, it may lead to the acceptance of unsafe products because the incorrect indicators were used.

PREDICTIVE MICROBIOLOGY/MICROBIAL MODELING

The presence/absence of indicator organisms as noted above is used to predict food safety. If a safety indicator is absent, the product is regarded as being safe relative to the hazard for which the indicator is used. On the other hand, a product can have extremely low numbers of a safety indicator and yet not pose a hazard. The latter is true for many foodborne pathogens such as enterotoxigenic staphylococci. When low numbers of an indicator or pathogen are present, it is important to know how either will behave in a food product over time. This future behavior calls into question the multiple parameters that affect the growth and activity of microorganisms in foods, and if predictions are to be made about the fate of low numbers of pathogens in a given product, how the pathogens and these parameters interact needs to be handled.

Microbial modeling or predictive microbiology is a rapidly emerging subdiscipline that entails the use of mathematical models/equations to predict the growth and/or activity of a microorganism in a food product over time. The predictive or modeling aspect is not new, for it is embodied in heat-process calculations in the canning of low-acid foods that are described in Chapter 17. What is new is the interest in broadening the use of this concept to a wider range of food poisoning and food spoilage organisms by the use of more sophisticated mathematical/computer models that can handle multiple growth parameters. For more information see references 2, 28, 54, 90, 93, 94.

As noted in Chapter 2 under temperature effects, predicting the growth of an organism for a single parameter is not too difficult. Difficulty arises when multiple parameters are involved, as relatively few studies have been conducted to determine their interplay on organisms. One such study is that of Buchanan and Phillips[8] who studied the interaction of five parameters on the growth of *Listeria monocytogenes*: pH, temperature, nitrite, NaCl, and gaseous atmosphere. After fitting data from 709(!) growth curves using a nonlinear regression analysis in conjunction with the Gompertz function, the investigators concluded that the model could provide reasonable "first estimates" on the behavior of *L. monocytogenes*.

The effective application of predictive microbiology requires the selection of appropriate models to reflect the effect of growth parameters. Among the many models that have been proposed and tested are two kinetic models—the nonlinear Arrhenius and Bělehradék types. The former is applied with the dependent variable expressed as ln rate, whereas with the latter square-root model, the dependent variable is expressed as $\sqrt{}$ rate. The further development of these models and their utility have been outlined and discussed by Ratkowsky et al.[72] Computer software packages for predictive microbiology are available from private and commercial sources.

One of the simplest applications of predictive microbiology is the use of Monte Carlo Simulation. This method entails probability distributions based on collected data to predict shelf life/safety relative to changes in environmental parameters such as pH, a_w, etc. When Monte Carlo was used to predict the shelf life of pasteurized milk, the data employed consisted of initial numbers of spoilage organisms in milk, their generation times, and milk storage temperature. With these and other relevant data subjected to an appropriate statistical package, it was found that 2.1°C decreases in milk storage temperature significantly increased simulated shelf life and even more so with one-half $\log_{10}$ decreases in microbial load.[76] If enough parameters are known about a given foodborne pathogen, Monte Carlo Simulation may be used to predict growth of the pathogen and disease outcome.

REFERENCES

1. Ackland, M.R., E.R. Trewhella, J. Reeder, and F.G. Bean. 1981. The detection of microbial spoilage in canned foods using thin-layer chromatography. *J. Appl. Bacteriol.* 51:277–281.

2. Adams, M.R., and M.O. Moss. 2000. *Food Microbiology*, 2nd ed. New York: Springer.

3. Ahmed, A., and J.R. Matches. 1983. Alcohol production by fish spoilage bacteria. *J. Food Protect.* 46:1055–1069.

4. American Public Health Association. 1985. *Standard Methods for the Examination of Water and Wastewater*, 16th ed. Washington, DC: APHA.

5. Beerens, H. 1998. Bifidobacteria as indicators of faecal contamination in meat and meat products: Detection, determination of origin and comparison with *Escherichia coli*. *Int. J. Food Microbiol.* 40:203–207.

6. Brenner, D.J., B.R. Davis, A.G. Steigerwalt, G.R. Fanning, C.F. Riddle, A.C. McWhorter, S.D. Allen, J.J. Farmer, III, and Y. Saitoh. 1982a. Atypical biogroups of *Escherichia coli* found in clinical specimens and description of *Escherichia hermannii* sp. nov. *J. Clin. Microbiol.* 15:703–713.

7. Brenner, D.J., A.C. McWhorter, J.K.L. Knutson, and A.G. Steigerwalt. 1982b. *Escherichia vulneris*: A new species of *Enterobacteriaceae* associated with human wounds. *J. Clin. Microbiol.* 15:1133–1140.

8. Buchanan, R.L., and J.G. Phillips. 1990. Response surface model for predicting the effects of temperature, pH, sodium chloride, sodium nitrite concentrations and atmosphere on the growth of *Listeria monocytogenes*. *J. Food Protect.* 53:370–376, 381.

9. Burton, M.C. 1949. Comparison of coliform and enterococcus organisms as indices of pollution in frozen foods. *Food Res.* 14:434–448.

10. Buttiaux, R. 1959. The value of the association *Escherichiae*–Group D streptococci in the diagnosis of contamination in foods. *J. Appl. Bacteriol.* 22:153–158.

11. Buttiaux, R., and D.A.A. Mossel. 1961. The significance of various organisms of faecal origin in foods and drinking water. *J. Appl. Bacteriol.* 24:353–364.

12. Calci, K.R., W. Burkhardt, III, W.D. Watkins, and S.R. Rippey. 1998. Occurrence of male-specific bacteriophage in feral and domestic animal wastes, human feces, and human-associated wastewaters. *Appl. Environ. Microbiol.* 64:5027–5029.

13. Chai, T.-J., T.-J. Han, R.R. Cockey, and P.C. Henry. 1990. Microbiological studies of Chesapeake Bay soft-shell clams (*Myarenaria*). *J. Food Protect.* 53:1052–1057.

14. Collins, M.D., D. Jones, J.A.E. Farrow, R. Kilpper-Balz, and K.H. Schleifer. 1984. *Enterococcus avium* nom. rev., comb. nov.; *E. casseliflavus* nom. ref., comb. nov.; *E. durans* nom. rev., comb. nov.; *E. gallinarum* comb. nov.; and *E. malodoratus* sp. nov. *Int. J. System. Bacteriol.* 34:220–223.

15. Collins, MD., R.R. Facklam, J.A.E. Farrow, and R. Williamson. 1989. *Enterococcus raffinosus* sp. nov.; *Enterococcus solitarus* sp. nov. and *Enterococcus pseudoavium* sp. nov. *FEMS Microbiol. Lett.* 57:283–288.

16. Colwell, R.R., R.J. Seidler, J. Kaper, S.W. Joseph, S. Garves, H. Lockman, D. Maneval, H. Bradford, N. Roberts, E. Remmers, I. Huq, and A. Huq. 1981. Occurrence of *Vibrio cholerae* serotype 01 in Maryland and Louisiana estuaries. *Appl. Environ. Microbiol.* 41:555–558.

17. DePaola, A., S. McLeroy, and G. McManus. 1997. Distribution of *Vibrio vulnificus* phage in oyster tissues and other estuarine habitats. *Appl. Environ. Microbiol.* 63:2464–2467.

18. DePaola, A., M.L. Motes, A.M. Chan, and C.A. Suttle. 1998. Phages infecting *Vibrio vulnificus* are abundant and diverse in oysters (*Crassostrea virginica*) collected from the Gulf of Mexico. *Appl. Environ. Microbiol.* 64:346–351.

19. Devriese, L.A., B. Pot, L. Van Damme, K. Kersters, and F. Haesebrouck. 1995. Identification of *Enterococcus* species isolated from foods of animal origin. *Int. J. Food Microbiol.* 26:187–197.

20. Devriese, L.A., A. van de Kerckhove, R. Kilpper-Balz, and K.H. Schleifer 1987. Characterization and identification of *Enterococcus* species isolated from the intestines of animals. *Int. J. System. Bacteriol.* 37:257–259.

21. Drancourt, M., C. Bollet, A. Carta, and P. Rousselier. 2001. Phylogenetic analyses of *Klebsiella* species delineate *Klebsiella* and *Raoultella* gen nov., with description of *Raoultella ornithinolytica* comb. nov., *Raoultella terrigena* comb. nov., and *Raoultella planticola* comb. nov. *Int. J. Syst. Evol. Microbiol.* 51:925–932.

22. Endley, S., L. Lu, E. Vega, M.E. Hume, and S.D. Pillai. 2003. Male-specific coliphages as an additional fecal contamination indicator for screening fresh carrots. *J. Food Protect.* 66:88–93.

23. Ercolani, G.L. 1976. Bacteriological quality assessment of fresh marketed lettuce and fennel. *Appl. Environ. Microbiol.* 31:847–852.

24. Escherich, T. 1885. Die Darmbacterien des Neugeborenen und Sauglings. *Fortschr. Med.* 3:515–522, 547–554.

25. Fugate, K.J., D.O. Cliver, and M.T. Hatch. 1975. Enteroviruses and potential bacterial indicators in Gulf coast oysters. *J. Milk Food Technol.* 38:100–104.

26. Gavini, F., A.M. Pourcher, C. Neut, D. Monget, C. Romand, C. Oger, and D. Izard. 1991. Phenotypic differentiation of bifidobacteria of human and animal origins. *Int. J. Syst. Bacteriol.* 41:548–557.

27. Gerba, C.P., S.M. Goyal, R.L. LaBelle, I. Cech, and G.F. Bodgan. 1979. Failure of indicator bacteria to reflect the occurrence of enteroviruses in marine waters. *Am. J. Public Health* 69:1116–1119.

28. Gibson, A.M., N. Bratchell, and T.A. Roberts. 1988. Predicting microbial growth: Growth responses of salmonellae in a laboratory medium as affected by pH, sodium chloride, and storage temperature. *Int. J. Food Microbiol.* 6:155–178.

29. Griffin, A.M., and C.A. Stuart. 1940. An ecological study of the coliform bacteria. *J. Bacteriol.* 40:83–100.

30. Hartman, P.A. 1960. *Enterococcus*: Coliform ratios in frozen chicken pies. *Appl. Microbiol.* 8:114–116.

31. Harwood, V.J., J. Butler, D. Parrish, and V. Wagner. 1999. Isolation of fecal coliform bacteria from the diamond-back terrapin (*Malaclemys terrapin centrata*). *Appl. Environ. Microbiol.* 65:865–867.

32. Havelaar, A.H., K. Furuse, and W.M. Hogeboom. 1986. Bacteriophages and indicator bacteria in human and animal faeces. *J. Appl. Bacteriol.* 60:255–262.

33. Havelaar, A.H., and W.M. Hogeboom. 1984. A method for the enumeration of male-specific bacteriophages in sewage. *J. Appl. Bacteriol.* 56:439–447.

34. Hilton, M.C., and G. Stotzky. 1973. Use of coliphages as indicators of water pollution. *Can. J. Microbiol.* 19:747–751.

35. Hollingworth, T.A., Jr., and H.R. Throm. 1982. Correlation of ethanol concentration with sensory classification of decomposition in canned salmon. *J. Food Sci.* 47:1315–1317.

36. Hsu, F.-C., Y.-S.C. Shieh, and M.D. Sobsey. 2002. Enteric bacteriophages as potential fecal indicators in ground beef and poultry meat. *J. Food Protect.* 65:93–99.

37. Huys, G., M. Cnockaert, J.M. Janda, and J. Swings. 2003. *Escherichia albertii* sp. nov., a diarrhoeagenic species isolated from stool specimens of Bangladeshi children. *Int. J. Syst. Evol. Microbiol.* 53:807–810.

38. International Commission on the Microbiological Specifications for Foods. 1986. *Microorganisms in Foods. 2. Sampling for Microbiological Analysis: Principles and Specific Application*, 2nd ed. Toronto: University of Toronto Press.

39. Jay, J.M. 2001. Indicator organisms in foods. In *Foodborne Disease Handbook*, 2nd ed. ed. Y.H. Hui, M.D. Pierson, and J.R. Gorham, Vol. 1, 645–653. New York: Marcel Dekker.

40. Jiménez-Clavero, M.A., C. Fernández, J.A. Ortiz, J. Pro, G. Carbonell, J.V. Tarazona, N. Roblas, and V. Ley. 2003. Teschoviruses as indicators of porcine fecal contamination of surface water. *Appl. Environ. Microbiol.* 69:6311–6315.

41. Kaper, J., H. Lockman, R.R. Colwell, and S.W. Joseph 1979. Ecology, serology, and enterotoxin production by *Vibrio cholerae* in Chesapeake Bay. *Appl. Environ. Microbiol.* 37:91–103.

42. Kenard, R.P., and R.S. Valentine. 1974. Rapid determination of the presence of enteric bacteria in water. *Appl. Microbiol.* 27:484–487.

43. Kennedy, J.E., Jr., J.L. Oblinger, and G. Bitton. 1984. Recovery of coliphages from chicken, pork sausage and delicatessen meats. *J. Food Protect.* 47:623–626.

44. Kennedy, J.E., Jr., C.I. Wei, and J.L. Oblinger. 1986. Methodology for enumeration of coliphages in foods. *Appl. Environ. Microbiol.* 51:956–962.

45. Kennedy, J.E., Jr., C.I. Wei, and J.L. Oblinger. 1986. Distribution of coliphages in various foods. *J. Food Protect.* 49:944–951.

46. Kereluk, K., and M.F. Gunderson. 1959. Studies on the bacteriological quality of frozen meats. IV. Longevity studies on the coliform bacteria and enterococci at low temperatures. *Appl. Microbiol.* 7:327–328.

47. Kott, Y., N. Roze, S. Sperber, and N. Betzer. 1974. Bacteriophages as viral pollution indicators. *Water Res.* 8:165–171.

48. Leininger, H.V., and C.S. McCleskey. 1953. Bacterial indicators of pollution in surface waters. *Appl. Microbiol.* 1:119–124.

49. Ley, V., J. Higgins, and R. Fayer. 2002. Bovine enteroviruses as indicators of fecal contamination. *Appl. Environ. Microbiol.* 68:3455–3461.

50. Litsky, W., M.J. Rosenbaum, and R.L. France. 1953. A comparison of the most probable numbers of coliform bacteria and enterococci in raw sewage. *Appl. Microbiol.* 1:247–250.

51. Mariam, T.W., and D.O. Cliver. 2000. Small round coliphages as surrogates for human viruses in process assessment. *Dairy Food. Environ. Sanit.* 20:684–689.

52. Matches, J.R., and C. Abeyta. 1983. Indicator organisms in fish and shellfish. *Food Technol.* 37(6):114–117.

53. McCrady, M.H., and E.M. Langevin. 1932. The coliaerogenes determination in pasteurization control. *J. Dairy Sci.* 15:321–329.

54. McMeekin, T.A., J.N. Olley, T. Ross, and D.A. Ratkowsky. 1993. *Predictive Microbiology: Theory and Application.* New York: John Wiley & Sons.

55. Mossel, D.A.A. 1958. The suitability of bifidobacteria as part of a more extended bacterial association, indicating faecal contamination of foods. In *Proceedings of the 7th International Congress on Microbiology, Abstracts of Papers*, 440–441. Uppsala: Almquist & Wikesells.

56. Mundt, J.O. 1982. The ecology of the streptococci. *Microbiol. Ecol.* 8:355–369.

57. Mundt, J.O. 1976. Streptococci in dried and frozen foods. *J. Milk Food Technol.* 39:413–416.

58. Mundt, J.O. 1961. Occurrence of enterococci: Bud, blossom, and soil studies. *Appl. Microbiol.* 9:541–544.

59. Mundt, J.O. 1973. Litmus milk reaction as a distinguishing feature between *Streptococcus faecalis* of human and nonhuman origins. *J. Milk Food Technol.* 36:364–367.

60. Munoa, F.J., and R. Pares. 1988. Selective medium for isolation and enumeration of *Bifidobacterium* spp. *Appl. Environ. Microbiol.* 54:1715–1718.

61. Murdock, D.I. 1968. Diacetyl test as a quality control tool in processing frozen concentrated orange juice. *Food Technol.* 22:90–94.

62. Murdock, D.I. 1967. Methods employed by the citrus concentrate industry for detecting diacetyl and acetyl-methylcarbinol. *Food Technol.* 21:643–672.

63. Newton, K.G. 1979. Value of coliform tests for assessing meat quality. *J. Appl. Bacteriol.* 47:303–307.

64. Orla-Jensen, S.H. 1919. The lactic acid bacteria. *Mem. Acad. Royal Soc. Denmark Ser. 8.* 5:81–197.

65. Ostrolenk, M., N. Kramer, and R.C. Cleverdon. 1947. Comparative studies of enterococci and *Escherichia coli* as indices of pollution. *J. Bacteriol.* 53:197–203.

66. Pasricha, C.L., and A.J.H. DeMonte. 1941. Bacteriophages as an index of water contamination. *Indian Med. Gaz.* 76:492–493.

67. Peabody, F.R. 1963. Microbial indexes of food quality: The coliform group. In *Microbiological Quality of Foods*, ed. L.W. Slanetz, C.O. Chichester, A.R. Gaufin, and Z.J. Ordal. 113–118. New York: Academic Press.

68. Perry, C.A., and A.A. Hajna. 1944. Further evaluation of EC medium for the isolation of coliform bacteria and *Escherichia coli*. *Am. J. Public Health* 34:735–738.

69. Portnoy, B.L., P.A. Mackowiak, C.T. Caraway, J.A. Walker, T.W. McKinley, and C.A. Klein, Jr. 1975. Oyster-associated hepatitis: Failure of shellfish certification programs to prevent outbreaks. *JAMA.* 233:1065–1068.

70. Primrose, S.B., N.D. Seeley, K.B. Logan, and J.W. Nicolson. 1982. Methods for studying aquatic bacteriophage ecology. *Appl. Environ. Microbiol.* 43:694–701.

71. Raj, H., W.J. Wiebe, and J. Liston. 1961. Detection and enumeration of fecal indicator organisms in frozen sea foods. *Appl. Microbiol.* 9:433–438.

72. Ratkowsky, D.A., T. Ross, T.A. McMeekin, and J. Olley 1991. Comparison of Arrhenius-type and Belehrádek-type models for prediction of bacterial growth in foods. *J. Appl. Bacteriol.* 71:452–459.

73. Reinbold, G.W. 1983. Indicator organisms in dairy products. *Food Technol.* 37(6):111–113.

74. Resnick, I.G., and M.A. Levin. 1981. Assessment of bifidobacteria as indicators of human fecal pollution. *Appl. Environ. Microbiol.* 42:433–438.

75. Ricca, D.M., and J.J. Cooney. 1998. Coliphages and indicator bacteria in birds around Boston Harbor. *J. Ind. Microbiol. Biotechnol.* 21:28–30.

76. Schaffner, D.W., J. McEntire, S. Duffy, R. Montville, and S. Smith. 2003. Monte Carlo Simulation of the shelf life of pasteurized milk as affected by temperature and initial concentration of spoilage organisms. *Food Protect. Trends* 23:1014–1021.

77. Schardinger, F. 1892. Uber das Vorkommen Gährung erregender Spaltpilze im Trinkwasser und ihre Bedeutung für die hygienische Beurtheilung desselben. *Wien, Klin. Wachr.* 5:403–405, 421–423.

78. Seeley, N.D., and S.B. Primrose. 1982. The isolation of bacteriophages from the environment *J. Appl. Bacteriol.* 53:1–17.

79. Seeley, N.D., and S.B. Primrose. 1980. The effect of temperature on the ecology of aquatic bacteriophages. *J. Gen. Virol.* 46:87–95.

80. Simkova, A., and J. Cervenka. 1981. Coliphages as ecological indicators of enteroviruses in various water systems. *Bull. WHO* 59:611–618.

81. Slanetz, L.W., and C.H. Bartley. 1964. Detection and sanitary significance of fecal streptococci in water. *Am. J. Public Health* 54:609–614.

82. Smith, T. 1895. Notes on *Bacillus coli commune* and related forms, together with some suggestions concerning the bacteriological examination of drinking water. *Am. J. Med. Sci.* 110:283–302.

83. Splittstoesser, D.F. 1983. Indicator organisms on frozen vegetables. *Food Technol.* 37(6):105–106.

84. Stetler, R.E. 1984. Coliphages as indicators of enteroviruses. *Appl. Environ. Microbiol.* 48:668–670.

85. Stiles, M.E. and L.-K. Ng. 1981. *Enterobacteriaceae* associated with meats and meat handling. *Appl. Environ. Microbiol.* 41:867–872.

86. Szabo, R.A., E.C.D. Todd, and A. Jean. 1986. Method to isolate *Escherichia coli* 0157:H7 from food. *J. Food Protect.* 49:768–772.

87. Tartera, C., F. Lucena, and J. Jofre. 1989. Human origin of *Bacteroides fragilis* bacteriophages present in the environment. *Appl. Environ. Microbiol.* 55:2696–2701.

88. Tompkin, R.B. 1983. Indicator organisms in meat and poultry products. *Food Technol.* 37(6):107–110.

89. Tissier, H. 1908. Recherches sur la flore intestinale normale des enfants agés d'un an à cinq ans. *Ann. Inst. Pasteur* 22:189–208.

90. Van Impe, J.F., B.M. Nicolai, M. Schellekens, T. Martens, and J.D. Baerdemaeker. 1995. Predictive microbiology in a dynamic environment: A system theory approach. *Int. J. Food Microbiol.* 25:227–249.

91. Wait, D.A., C.R. Hackney, R.J. Carrick, G. Lovelace, and M.D. Sobsey 1983. Enteric bacterial and viral pathogens and indicator bacteria in hard shell clams. *J. Food Protect.* 46:493–496.

92. Wentsel, R.S., P.E. O'Neill, and J.F. Kitchens. 1982. Evaluation of coliphage detection as a rapid indicator of water quality. *Appl. Environ. Microbiol.* 43:430–434.

93. Whiting, R.C., and R.L. Buchanan. 1994. Microbial modeling. *Food Technol.* 48(6):113–120.

94. Zwietering, M.H., T. Wijtzes, J.C. De Wit, and K. Van't Riet. 1992. A decision support system for prediction of the microbial spoilage in foods. *J. Food Protect.* 55:973–979.

CHAPTER 21

The HACCP and FSO Systems for Food Safety

Among the desirable qualities that should be associated with foods is freedom from infectious organisms. Although it may not be possible to achieve a zero tolerance for all such organisms under good manufacturing practices (GMP), the production of foods with the lowest possible numbers is the desirable goal. With fewer processors producing more products that lead to foods being held longer and shipped farther before they reach the consumers, new approaches are needed to ensure safe products. Classic approaches to microbiological quality control have relied heavily on microbiological determinations of both raw materials and end products, but the time required for results is too long for many products. The development and use of certain rapid methods have been of value, but these alone have not obviated the need for newer approaches to ensuring safe foods. The hazard analysis critical control point (HACCP) system is presented in this chapter as the method of choice for ensuring the safety of foods from farm to table. The outline of a newer concept, Food Safety Objective (FSO), is presented. When deemed necessary, microbiological criteria may be established for some ingredients and foods, and these in connection with sampling plans are presented as components of the HACCP system.

HAZARD ANALYSIS CRITICAL CONTROL POINT (HACCP) SYSTEM

The concept and early history of the HACCP system are presented in the previous edition of this text. The presentation that follows is not intended to be used alone to establish an HACCP program in either a food production plant or food service establishment. For these purposes, a more detailed HACCP reference should be consulted.[4,7,9,13,14,24] More general information and background can be found in references 5, 18, 22, 23.

The objective of this section is to provide a general overview of what HACCP is, and examples of how one might go about setting up an HACCP system.

HACCP is a system that should lead to the production of microbiologically safe foods by analyzing for the hazards of raw materials—those that may appear throughout processing and those that may occur from consumer abuse. It is a proactive, systematic approach to controlling foodborne hazards. Although some classic approaches to food safety rely heavily on end product testing, the HACCP system places emphasis on the quality of all ingredients and all process steps on the premise that safe products will result if these are properly controlled. The system is thus designed to control organisms

497

Table 21–1   Leading Factors Contributing to Outbreaks of Foodborne Illness in the United States

Factors	1961–1982
Improper cooling	44%
Lapse of 12 or more hours between preparation/eating	23%
Contaminated by handlers	18%
Raw ingredient added without subsequent heating/cooking	16%
Inadequate cooking/canning/heating	16%

Note: $N = 1918$.

Source: From Bryan.[1,2]

at the point of production and preparation. The five leading factors that contributed to foodborne illness in the United States for the years 1961–1982 are noted in Table 21–1, and it may be noted that events associated with the handling and preparation of foods were significant.[2] Mishandling of foods in food service establishments in Canada in 1984 was the reason for illness in about 39% of foodborne incidents.[26] Proper implementation of HACCP in food service establishments and the home will lead to a decrease in foodborne illness.

A subcommittee of the U.S. National Research Council, National Academy of Sciences, made the following recommendation in 1985.[16] Because the application of the HACCP system provides for the most specific and critical approach to the control of microbiological hazards presented by foods, use of this system should be required of industry. Accordingly, this subcommittee believes that government agencies responsible for control of microbiological hazards in foods should promulgate appropriate regulations that would require industry to utilize the HACCP system in their food protection programs. Before an HACCP program is developed, there are some prerequisite programs that should be in place.

Prerequisite Programs

Prerequisite programs include a wide range of activities and events that may have an impact on an HACCP system for a specific food product even though they are not parts of the HACCP system per se. Some examples of prerequisite programs are noted in reference 14, and they are explained in more detail in reference 21.

Briefly stated, prerequisite programs include concerns and aspects of the entire food environment *before* the HACCP system is initiated. They include the suitability of facilities, control of suppliers, safety and maintenance of production equipment, cleaning and sanitation of equipment and facilities, personal hygiene of employees, control of chemicals, pest control, and the like. These prerequisites include good manufacturing practices,[12] and they should be brought up to acceptable standards before the HACCP system is initiated.

Definitions

The following terms and concepts are valuable in the development and execution of an HACCP system and are taken from International Commission on Microbiological Specifications for Foods (ICMSF)[8] and/or National Advisory Committee on the Microbiological Criteria for Foods (NACMCF):[14]

- *Control point*: Any point in a specific food system where loss of control does not lead to an unacceptable health risk.
- *Critical control point (CCP):* Any point or procedure in a food system where control can be exercised and a hazard can be minimized or prevented.
- *Critical limit*: One or more prescribed tolerances that must be met to ensure that a CCP effectively controls a microbiological health hazard.
- *CCP decision tree*: A sequence of questions to assist in determining whether a control point is a CCP.
- *Corrective action*: Procedures followed when a deviation occurs.
- *Deviation*: Failure to meet a required critical limit for a CCP.
- *HACCP plan*: The written document that delineates the formal procedures to be followed in accordance with these general principles.
- *Hazard*: Any biological, chemical, or physical property that may cause an unacceptable consumer health risk (unacceptable contamination, toxin levels, growth, and/or survival of undesirable organisms).
- *Monitoring*: A planned sequence of observations or measurements of critical limits designed to produce an accurate record and intended to ensure that the critical limit maintains product safety.
- *Risk category*: One of six categories prioritizing risk based on food hazards.
- *Validation*: That element of verification focused on collecting and evaluating scientific and technical information to determine whether the HACCP plan, when properly implemented, will effectively control the hazards.
- *Verification*: Methods, procedures, and tests used to determine whether the HACCP system is in compliance with the HACCP plan.

HACCP Principles

Although interpreted variously, the ICMSF and NACMCF view HACCP as a natural and systematic approach to food safety and as consisting of the following seven principles:

1. Assess the hazards and risks associated with the growing, harvesting, raw materials, ingredients, processing, manufacturing, distribution, marketing, preparation, and consumption of the food in question.
2. Determine the CCP(s) required to control the identified hazards.
3. Establish the critical limits that must be met at each identified CCP.
4. Establish procedures to monitor the CCP(s).
5. Establish corrective actions to be taken when there is a deviation identified by monitoring a given CCP.
6. Establish procedures for verification that the HACCP system is working correctly.
7. Establish effective record-keeping systems that document the HACCP plan.

Each of these principles is discussed in more detail below.

Principle 1: Assess Hazards and Risks

Hazards and risks may be assessed for individual food ingredients from a flow diagram or by ranking the finished food product by assigning to it a hazard rating from A through F. A plus sign (+) is assigned when a hazard exists. Six hazard categories have been defined, representing an expansion of the three proposed by the National Research Council (NRC)[17] for salmonellae control. However, this system of ranking and hazard category assignment was not popular in the late 1990s and it may be ignored (see reference 14 for alternative). It is presented here for historical purposes:

A. This is a special class of foods that consist of nonsterile products designated and intended for consumption by individuals at risk, including infants, the aged, infirm, and immunoincompetent.
B. The product contains "sensitive" ingredients relative to microbiological hazards (e.g., milk, fresh meats).
C. There is no controlled processing step (such as heat pasteurization) that effectively destroys harmful microorganisms.
D. The product is subject to recontamination after processing but before packaging (e.g., pasteurized in bulk and then packaged separately).
E. Substantial potential for abusive handling exists in distribution and/or by consumers that could render the product harmful when consumed (e.g., products to be refrigerated are held above refrigerator temperatures).
F. There is no terminal heat process after packaging or when cooked in the home.

Next, the formulated product should be assigned to one of six hazard categories, expanded from four suggested by the NRC.[16]

VI. A special category that applies to nonsterile products designated and intended for individuals in hazard category A
V. Food products subject to all five general hazard characteristics (B, C, D, E, and F)
IV. Food products subject to any four general hazard characteristics
III. Products subject to any three of the general hazard characteristics
II. Products subject to any two general hazard characteristics
I. Products subject to any one of the general hazard characteristics
0. Products subject to no hazards

Principle 2: Determine CCP(s)

The ICMSF[9] recognized two types of CCPs: CCP1, to ensure control of a hazard, and CCP2, to minimize a hazard. Typical of CCPs are the following:

1. Heat process steps where time–temperature relations must be maintained to destroy given pathogens;
2. Freezing and time to freezing before pathogens can multiply;
3. The maintenance of pH of a food product at a level that prevents growth of pathogens;
4. Employee hygiene.

A decision tree such as the one in Figure 21–1 is often used to identify CCPs.

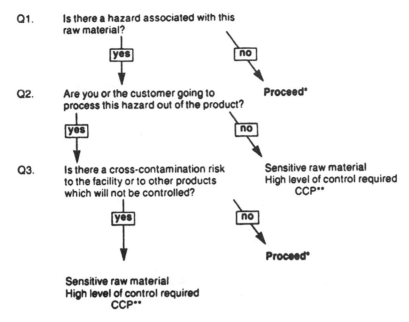

Q1. Is there a hazard associated with this raw material?

        yes   no

                   Proceed*

Q2. Are you or the customer going to process this hazard out of the product?

        yes   no

Q3. Is there a cross-contamination risk to the facility or to other products which will not be controlled?

                   Sensitive raw material
                   High level of control required
                   CCP**

        yes   no

                   Proceed*

Sensitive raw material
High level of control required
CCP**

* Proceed to your next raw material

** Following the hazard analysis, you are likely to find that this raw material must be managed as a CCP

Figure 21–1 Raw material control decision tree. *Source:* From Mortimore and Wallace,[13] copyright © 1994 by Kluwer Academic Publishers.

Principle 3: Establish Critical Limits

A critical limit is one or more prescribed tolerances that must be met to ensure that a CCP effectively controls a microbiological hazard. This could mean keeping refrigeration temperatures within a certain specific and narrow range, or making sure that a certain minimum destructive temperature is achieved and maintained long enough to effect pathogen destruction. Examples of the latter include adherence to the temperatures noted in Table 21–2 for the control of the respective organisms.

Principle 4: Establish Procedures to Monitor CCPs

The monitoring of a CCP involves the scheduled testing or observation of a CCP and its limits; monitoring results must be documented. If, for example, the temperature for a certain process step should not exceed 40°C, a chart recorder may be installed. Microbial analyses are not used to monitor since too much time is required to obtain results. Physical and chemical parameters such as time, pH, temperature, and water activity (a_w) can be quickly determined and the results obtained immediately.

Principle 5: Establish Corrective Actions

Establish corrective actions to be taken when deviations occur in CCP monitoring. The actions taken must eliminate the hazard that was created by any deviation from the plan. If a product is involved

Table 21–2 USDA Cooking and Cooling Parameters for Perishable Uncured Meat and Poultry Products

---

Cooking parameters

USDA/FSIS has established minimal internal temperatures required for cooking perishable uncured meat and poultry products. These temperature requirements are referenced in Title 9 of the CFRs (CFR 301–390) or in policies disseminated through the FSIS Policy Book or Notices.

Cooking requirements*

Cooked beef and roast beef (9 CFR 318.17)	130–145°F (54.4–62.7°C)
(121 min at 130°F to instantaneous at 145°F)	
Baked meatloaf (9 CFR 317.8)	160°F (71.1°C)
Baked pork cut (9 CFR 317.8)	170°F (76.7°C)
Pork (to destroy *trichinae*) (9 CFR 318.10)	120°–144°F (48.9°–62.2°C)
(21 hours at 120°F to instantaneous at 144°F)	
Cooked poultry rolls and other uncured poultry products	160°F (71.1°C)
(9 CFR 381.150)	
Cooked duck, salted (FSIS Policy Book)	155°F (68.3°C)
Jellied chicken loaf (FSIS Policy Book)	160°F (71.1°C)
Partially cooked, comminuted products (FSIS Notice 92-85)	$\geq$151°F for 1 minute
	$\geq$148°F for 2 minutes
	$\geq$146°F for 3 minutes
	$\geq$145°F for 4 minutes
	$\geq$144°F for 5 minutes

Cooling parameters

Similarly, parameters for cooling and storing refrigerated products, including temperatures and times, are reflected in agency regulations (9 CFR) and policies.

Cooling requirements

Guidelines for refrigerated storage temperature and internal temperature control point	40°F (4.4°C)
Recommended refrigerated storage temperature for periods exceeding 1 week (FSIS Directive 7110.3)	35°F (1.7°C)

Cooling procedures require that the product's internal temperature not remain between 130°F (54.4°C) and 80°F (26.7°C) for more than 1.5 hours or between 80°F (26.7°C) and 40°F (4.4°C) for more than 5 hours (FSIS Directive 7110.3).

Cooling procedures for products consisting of intact muscle (e.g., roast beef) require that chilling be initiated within 90 minutes of the cooking cycle. Product shall be chilled from 120°F (48°C) to 55°F (12.7°C) in not more than 6 hours. Chilling shall continue and the product shall not be packed for shipment until it has reached 40°F (4.4°C).

Roast beef for export to the United Kingdom must be chilled to 68°F (20°C) or less within 5 hours after leaving the cooker and to 46°F (7°C) or less within the following 3 hours.

*Some temperature requirements are based on appearance and labeling characteristics rather than safety.

Note: USDA = United States Department of Agriculture; CFR = Code of Federal Regulations; FSIS = Food Safety and Inspection Service.

---

that may be unsafe as a result of the deviation, it must be removed. Although the actions taken may vary widely, in general they must be shown to bring the CCP under control.

Principle 6: Establish Procedures for Verification

Establish procedures for verification that the HACCP system is working correctly. Verification consists of methods, procedures, and tests used to determine that the system is in compliance with the plan. Verification confirms that all hazards were identified in the HACCP plan when it was developed, and verification measures may include compliance with a set of established microbiological criteria when established. Verification activities include the establishment of verification inspection schedules, including review of the HACCP plan, CCP records, deviations, random sample collection and analysis, and written records of verification inspections. Verification inspection reports should include the designation of persons responsible for administering and updating the HACCP plan, direct monitoring of CCP data while in operation, certification that monitoring equipment is properly calibrated, and deviation procedures employed.

Principle 7: Establish Effective Record-Keeping Systems

Establish effective record-keeping systems to document the HACCP plan. The HACCP plan must be on file at the food establishment and must be made available to official inspectors upon request. Forms for recording and documenting the system may be developed, or standard forms may be used with necessary modifications. Typically, these may be forms that are completed on a regular basis and filed away. The forms should provide documentation for all ingredients, processing steps, packaging, storage, and distribution.

Flow Diagrams

The development of an HACCP plan for a food establishment begins with the construction of a flow diagram for the entire process. The diagram should begin with the acquisition of raw materials and include all steps through packaging and subsequent distribution. A flow diagram for the production of frozen, cooked beef patties is illustrated in Figure 21–2. To begin the HACCP process, the three questions in Figure 21–1 should be raised. When this is done, the answer to all three is yes, as outlined below:

Q1. Is there a hazard? Raw ground beef patties are known to be vehicles for *Escherichia coli* 0157:H7, *Toxoplasma gondii*, and salmonellae.
Q2. Will the hazard be processed out? This will be achieved in step 5 (cooking).
Q3. Is there a risk of cross-contamination? This can occur in steps 7, 8, and 10.

Application of HACCP Principles

This section deals with an application of the seven HACCP principles to the manufacture of frozen, cooked beef patties as outlined in Figure 21–2, and the steps referred to are those on the flow diagram.

Principle 1—Hazards and Risks

Raw meat is a sensitive ingredient and the cooked product is subject to recontamination after processing and during distribution.

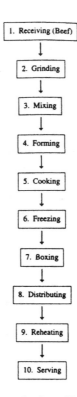

Figure 21–2   Example of a flow diagram for the production of frozen, cooked beef patties. *Source:* Reprinted with permission from The International Committee on Microbiological Specifications for Foods of the International Union of Microbiological Societies (ICMSF), *Journal of Food Protection*, Vol. 61, p. 1255, © 1998. Copyright held by the International Association of Milk, Food and Environmental Sanitarians, Inc.

Principle 2—CCPs

An important concern about step 1 is the overall condition of the beef carcasses or cuts. The comments below are based on the assumption that the beef has been produced and handled under GMP. Step 5 is the indisputable CCP1, since it can eliminate the hazards. CCP2s may be assigned to steps 6 and 8, and possibly to step 7.

Principle 3—Critical Limits

Temperature is the critical parameter from steps 1 to 9, and it consists of proper refrigeration temperature in steps 1 to 4; proper cooking temperature in step 5; freezing in steps 6 to 8; and heating in step 9. The overall objective is to keep the fresh beef at or below 40°F (4.4°C) at all times, cook patties to 160°F (71.1°C), freeze to −20°F or 0°C, and store at the same temperature.

Principle 4—Monitoring HACCP

Use chart recorders for steps 2 to 4, use thermometers for steps 5 and 6, and temperature recorders for step 8.

Principle 5—Corrective Actions

These refer to deviations from critical limits identified during monitoring of CCPs. Specific corrective actions to be taken should be clearly spelled out. For example, if the target temperature in step 5 is not reached, will the batch be discarded, reprocessed, or assigned to another use?

Principle 6—Verification

Overall, this is an assessment of how effective the HACCP system is performing. Typically, some microbial analyses are in order—for example, were all relevant pathogens destroyed in step 5? Have the products in retail stores been contaminated after being cooked?

Principle 7—Record-Keeping

This should be done by product lot number in such a way that records are available to verify the events in steps 2 to 4. Where room temperatures are involved, chart recorder tracings should be kept.

A flow diagram for the production of roast beef is presented in Figure 21–3. Cooking is the most important CCP for this product (CCP1), followed by chilling and prevention of recontamination after cooking. The cooking temperature should reach 145°F (62.3°C) or otherwise be sufficient to effect a 4-log cycle reduction of *Listeria monocytogenes*. This will not destroy *Clostridium perfringens*

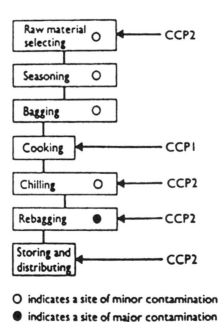

O indicates a site of minor contamination
● indicates a site of major contamination
CCP1 effective CCP. CCP2 not absolute

Figure 21–3   Flow diagram for production of roast beef. *Source:* From ICMSF,[9] copyright © 1988 by Blackwell Scientific Publications. Used with permission.

spores, and their germination and growth must be controlled by proper chilling and storage. Cooking and cooling parameters for perishable uncured meats are presented in Table 21–2.

Some Limitations of HACCP

Although it is the best system yet devised for controlling microbial hazards in foods from the farm to the table, the uniform application of HACCP in the food manufacturing and service industries will not be without some debate. Among the lingering questions and concerns raised by Tompkin[27] are the following:

1. HACCP requires the education of nonprofessional food handlers, especially in the food service industry and in homes; whether this will be achieved remains to be seen. The failure of these individuals to get a proper understanding of HACCP could lead to its failure.
2. To be effective, this concept must be accepted not only by food processors but also by food inspectors and the public. Its ineffective application at any level can be detrimental to its overall success for a product.
3. It is anticipated that experts will differ as to whether a given step is a CCP and how best to monitor such steps. This has the potential of eroding the confidence of others in HACCP.
4. The adoption of HACCP by industry has the potential of giving false assurance to consumers that a product is safe, and, therefore, there is no need to exercise the usual precautions between the purchase and consumption of a product. Consumers need to be informed that most outbreaks of foodborne illness are caused by errors in food handling in homes and food service establishments and that no matter what steps a processor takes, HACCP principles must be observed after foods are purchased for consumption.

Food Safety Objective (FSO)

An FSO is a statement of the frequency or maximum concentration of a microbiological hazard in a food considered acceptable for consumer protection.[28] It has been endorsed by the ICMSF, and the steps that need to be taken for its development are outlined in Figure 21–4. Among examples of specific FSOs are the following:[28] (1) Staphylococcal enterotoxin in cheese must not exceed 1 μg/100 g; (2) Aflatoxin in peanuts should not exceed 15 μg/kg; (3) *Listeria monocytogenes* in ready-to-eat foods should not exceed 100/g at the time of consumption; and (4) salmonellae on raw poultry meat should be <10%. For more on the establishment of FSOs, see reference 28.

In regards to an FSO for *S. aureus* in cream-filled baked goods, a recent survey revealed that 37.3% of 1438 food handlers in four countries carried this organism.[25] From eleven staphylococcal outbreaks in eight countries, the level of *S. aureus* found in the vehicle bakery products ranged from >10^6 to 10^9 cfu/g; and most produced staphylococcal enterotoxin A (SEA) in nine of eleven outbreaks where the enterotoxin type was identified. One outbreak was caused by SED. About 40% of 536 food handlers carried enterotoxigenic strains of *S. aureus*.[25]

MICROBIOLOGICAL CRITERIA

The concept that microbial limits be assigned to at least some foods to designate their safety or overall quality was suggested as early as 1903 by Marxer, who suggested an aerobic plate count (APC) limit of

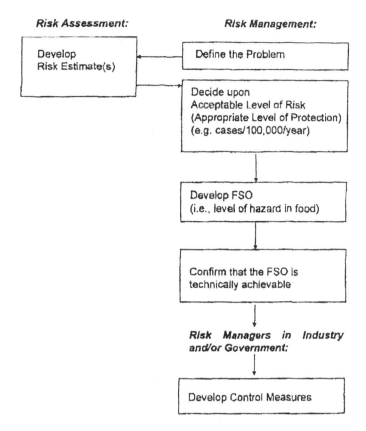

Figure 21–4   Steps leading to the development of a food safety objective and related control measures,[28] copyright © 1998, Elsevier Publishing. Used with permission.

10^6 for hamburger meat. Similarly, APC and indicator organism limits were suggested for many other products through the 1920s and 1930s, with pasteurized milk being notable among those for which limits were widely accepted. The early history of microbiological limits for foods has been reviewed.[6] In an effort to eliminate confusion and to agree upon an international language, the Codex Alimentarius Commission[3] has established definitions. The ICMSF has endorsed the Codex definitions with some modifications. The Codex definitions are summarized below, with ICMSF modifications noted.

Definitions

Microbiological criteria fall into two main categories: mandatory and advisory. A *mandatory criterion* is a microbiological standard that normally should contain limits only for pathogens of public health significance, but limits for nonpathogens may be set. The ICMSF[8] regards a *standard* as being part of a law or regulation that is enforceable by the regulatory agency having jurisdiction. An *advisory criterion* is either a microbiological end product specification intended to increase assurance that hygienic significance has been met (it may include spoilage organisms), or a microbiological guideline that is applied in a food establishment at a point during or after processing to monitor hygiene (it, too, may include nonpathogens). Before recommending a criterion, the ICMSF[8] notes that each product must be in international trade, must have associated with it good epidemiological evidence that it has been implicated in foodborne disease, and have associated with it good evidence that a criterion will reduce the potential hazard(s) in Principle 2.

The Codex definition of a microbiological criterion consists of five components: (1) a statement of the organisms of concern and/or their toxins, (2) the analytical methods for their detection and quantitation, (3) a sampling plan, including when and where samples are to be taken, (4) microbiological limits considered appropriate to the food, and (5) the number of sample units that should conform to these limits. These five components are embodied in a sampling plan.

Sampling Plans

A sampling plan is a statement of the criteria of acceptance applied to a lot based on appropriate examinations of a required number of sample units by specified methods. It consists of a sampling procedure and decision criteria and may be a two-class or a three-class plan.

A two-class plan consists of the following specifications: n, c, m; a three-class plan requires n, c, m, and M, where

$n =$ the number of sample units (packages, beef patties, and so forth) from a lot that must be examined to satisfy a given sampling plan.

$c =$ the maximum acceptable number, or the maximum allowable number of sample units that may exceed the microbiological criterion m. When this number is exceeded, the lot is rejected.

$m =$ the maximum number or level of relevant bacteria per gram; values above this level are either marginally acceptable or unacceptable. It is used to separate acceptable from unacceptable foods in a two-class plan, or, in a three-class plan, to separate good quality from marginally acceptable quality foods. The level of the organism in question that is acceptable and attainable in the food product is m. In the presence/absence situations for two-class plans, it is common to assign $m = 0$. For three-class plans, m is usually some nonzero value.

$M =$ a quantity that is used to separate marginally acceptable quality from unacceptable quality foods. It is used only in three-class plans. Values at or above M in any sample are unacceptable relative to health hazard, sanitary indicators, or spoilage potential.

A two-class plan is the simpler of the two and in its simplest form may be used to accept or reject a larger batch (lot) of food in a presence/absence decision by a plan such as $n = 5$, $c = 0$, where $n = 5$ means that five individual units of the lot will be examined microbiologically for, say, the presence of salmonellae, and $c = 0$ means that all five units must be free of the organisms by the method of examination in order for the lot to be acceptable. If any unit is positive for salmonellae, the entire lot is rejected. If it is desired that two of the five samples may contain coliforms, in a presence/absence test, for example, the sampling plan would be $n = 5$, $c = 2$. By this plan, if three or more of the five-unit samples contained coliforms, the entire lot would be rejected. Although the presence/absence situations generally obtained for salmonellae, an allowable upper limit for indicator organisms such as coliforms is more often the case. If it is desired to allow up to 100 coliforms/g in two of the five units, the sampling plan would be $n = 5$, $c = 2$, $m = 10^2$. After the five units have been examined for coliforms, the lot is acceptable if no more than two of the five contain as many as 10^2 coliforms/g but is rejected if three or more of the five contain 10^2 coliforms/g. This particular sampling plan may be made more stringent by increasing n (e.g., $n = 10$, $c = 2$, $m = 10^2$) or by reducing c (e.g., $n = 5$, $c = 1$, $m = 10^2$). On the other hand, it can be made more lenient for a given size n by increasing c.

Whereas a two-class plan may be used to designate acceptable/unacceptable foods, a three-class plan is required to designate acceptable/marginally acceptable/unacceptable foods. To illustrate a typical three-class plan, assume that for a given food product, the standard plate count (SPC) shall not exceed $10^6/g$ (M) or be higher than $10^5/g$ from three or more of five units examined. The specifications are thus $n = 5$, $c = 2$, $m = 10^5$, $M = 10^6$. If any of the five units exceeds $10^6/g$, the entire lot is rejected (unacceptable). If not more than c sample units give results above m, the lot is acceptable. Unlike two-class plans, the three-class plan distinguishes values between m and M (marginally acceptable).

With either two- or three-class attribute plans, the numbers n and c may be employed to find the probability of acceptance (P_a) of lots of foods by reference to appropriate tables.[8] The decision to employ a two-class or three-class plan may be determined by whether presence/absence tests are desirable, in which case a two-class plan is required, or whether count or concentration tests are desired, in which case a three-class plan is preferred. The latter offers the advantages of being less affected by nonrandom variations between sample units and of being able to measure the frequency of values in the m to M range. The ICMSF report and recommendations[8] should be consulted for further details on the background, uses, and interpretations of sampling plans. Further information may also be obtained from Kilsby.[10]

Microbiological Criteria and Food Safety

The application of criteria to products in the absence of an HACCP program is much less likely to be successful than when the two are combined. Thus, microbiological criteria are best applied as part of a comprehensive program. When criteria are not applied as components of a systematic approach to food safety or quality, the results are known to be less than satisfactory, as found by Miskimin et al.[11] and Solberg et al.[20] These investigators studied over 1,000 foods consisting of 853 ready-to-eat and 180 raw products. They applied arbitrary criteria for APC, coliforms, and *E. coli* and tested the efficacy of the criteria to assess safety of the foods with respect to *Staphylococcus aureus*, *C. perfringens*, and salmonellae. An APC criterion of less than $10^6/g$ for raw foods resulted in 47% of the samples being accepted even though one or more of the three pathogens were present, whereas 5% were rejected from which pathogens were not isolated, for a total of 52% wrong decisions. An APC of less than $10^5/g$ for ready-to-eat foods resulted in only 5% being accepted that contained pathogens, whereas 10% that did not yield pathogens were rejected. In a somewhat similar manner, a coliform criterion of less than $10^2/g$ resulted in a total of 34% wrong decisions for raw and 15% for ready-to-eat foods. The lowest percentage of wrong decisions for ready-to-eat foods (13%) occurred with an *E. coli* criterion of less than 3/g, whereas 30% of the decisions were wrong when the same criterion was applied to raw foods. Although the three pathogens were found in both types of foods, no foodborne outbreaks were reported over the 4-year period of the study, during which time more than 16 million meals were consumed.[19]

The above findings represent some initial data from the Rutgers Foodservice Program. After a 17-year experience with modifications in surveillance tests, food audits, laundry evaluations, and more than 30 million meals served, this HACCP-type system has been very effective.[19] The microbial guidelines employed by this program for raw and ready-to-eat foods are presented in Table 21–3. Of over 1,600 food samples examined over the period 1983–1989, only 1.24% contained pathogens, with protein salads most often contaminated (4.3%). Among the foods that failed microbial surveillance were raw vegetables (they had excessive coliform numbers).[19] The Rutgers Foodservice

Table 21–3 ICMSF Sampling Plans and Recommended Microbiological Limits

Products	Tests	Case	Class Plan	N	c	m	M	Comments
Precooked breaded fish	APC	2	3	5	2	5×10^5	10^7	
	E. coli	5	3	5	2	11	500	Products likely to be mishandled
	S. aureus	8	3	5	1	10^3	10^4	
Raw chicken (fresh or frozen), during processing	APC	1	3	5	3	5×10^5	10^7	In-plant processing
Frozen vegetables and fruit, pH 4.5	E. coli	5	3	5	2	10^2	10^3	m value is an estimate
Comminuted raw meat (frozen) and chilled carcass meat	APC	1	3	5	3	10^6	10^7	In-plant control
Cereals	Molds	5	3	5	2	10^2–10^4	10^5	m values are estimates
Frozen entrées containing rice or corn flour as a main ingredient	S. aureus	8	3	5	1	10^3	10^4	m value is estimated
Noncarbonated natural mineral and bottled noncarbonated waters	Coliforms	5	2	5	0	0	—	Not for use in infant formula or use by highly susceptibles
Roast beef	Salmonella	12	2	20	0	0	—	
Frozen raw crustaceans	S. aureus	7	3	5	2	10^3	10^4	
	V. parahaemolyticus	8	3	5	1	10^2	10^3	
	Salmonella*	10	2	5	0	0	—	
	APC†	2	3	5	2	5×10^5	10^7	
	E. coli†	5	3	5	2	11	500	
	S. aureus†	8	2	5	0	10^3	—	

Note: Except where noted for in-plant use, they are intended primarily for foods in international trade and are cited here primarily to illustrate the assignment of products to case and limits on a variety of organisms. The ICMSF[8] should be consulted for methods of analysis and more details in general.

*Normal plans and limits.
†Additional tests where appropriate.

510

HACCP-based system is an example of how microbial criteria can be integrated to provide for safe foods; in the 17-year program, no foodborne illness occurred.[19]

Microbiological Criteria for Various Products

Prior to the development of the HACCP and sampling plan concepts, microbiological criteria (generally referred to as standards at the time) were applied to a variety of products.

Presented below are foods and food ingredients that are covered under microbiological standards of various organizations (in the United States) along with federal, state, and city standards in effect (after W.C. Frazier, *Food Microbiology*, 1968, courtesy of McGraw-Hill Publishing Company).

1. Standards for Starch and Sugar (National Canners Association)
   A. *Total thermophilic spore count*: Of the five samples from a lot of sugar or starch none shall contain more than 150 spores per 10 g, and the average for all samples shall not exceed 125 spores per 10 g.
   B. *Flat-sour spores*: Of the five samples, none shall contain more than 75 spores/10 g, and the average for all samples shall not exceed 50 spores per 10 g.
   C. *Thermophilic anaerobe spores*: Not more than three (60%) of the five samples shall contain these spores, and in any one sample, not more than four (65%) of the six tubes shall be positive.
   D. *Sulfide spoilage spores*: Not more than two (40%) of the five samples shall contain these spores, and in any one sample, there shall be no more than five colonies per 10 g (equivalent to two colonies in the six tubes).
2. Standard for "Bottlers" Granulated Sugar, Effective July 1, 1953 (American Bottlers of Carbonated Beverages)
   A. *Mesophilic bacteria*: Not more than 200 per 10 g.
   B. *Yeasts*: Not more than 10 per 10 g.
   C. *Molds*: Not more than 10 per 10 g.
3. Standard for "Bottlers" Liquid Sugar, Effective in 1959 (American Bottlers of Carbonated Beverages). All figures based on dry-sugar equivalent (D.S.E.)
   A. *Mesophilic bacteria* (a) Last 20 samples average 100 organisms or less per 10 g of D.S.E.; (b) 95% of last 20 counts show 200 or less per 10 g; (c) 1 of 20 samples may run over 200; other counts as in (a) or (b).
   B. *Yeasts*: (a) Last 20 samples average 10 organisms or less per 10 g of D.S.E.; (b) 95% of last 20 counts show 18 or less per 10 g; (c) 1 of 20 samples may run over 18; other counts as in (a) and (b).
   C. *Molds*: Standards like those for yeasts.
4. Standards for Dairy Products
   A. From 1965 recommendations of the U.S. Public Health Service.
      a. *Grade A raw milk for pasteurization*: Not to exceed 100,000 bacteria per milliliter prior to commingling with other producer milk; and not exceeding 300,000 per milliliter as commingled milk prior to pasteurization.
      b. *Grade A pasteurized milk and milk products* (except cultured products): Not over 20,000 bacteria per milliliter, and not over 10 coliforms per milliliter.
      c. *Grade A pasteurized cultured products*: Not over 10 coliforms per milliliter. *Note*: Enforcement procedures for (a), (b), and (c) require a three-out-of-five compliance by samples.

Whenever two of four successive samples do not meet the standard, a fifth sample is tested; and if this exceeds any standard, the permit from the health authority may be suspended. It may be reinstated after compliance by four successive samples has been demonstrated.

B. *Certified milk* (American Association of Medical Milk Commissions, Inc.)

    a. Certified milk (raw): Bacterial plate count not exceeding 10,000 colonies per milliliter; coliform colony count not exceeding 10 per milliliter.

    b. Certified milk (pasteurized): Bacterial plate count not exceeding 10,000 colonies per milliliter before pasteurization and 500 per milliliter in route samples. Milk not exceeding 10 coliforms per milliliter before pasteurization and 1 coliform per milliliter in route samples.

C. Milk for manufacturing and processing (USDA, 1955)

    a. Class 1: Direct microscopic clump count (DMC) not over 200,000 per milliliter.

    b. Class 2: DMC not over 3 million per milliliter.

    c. Milk for Grade A dry milk products: must comply with requirements for Grade A raw milk for pasteurization (see above).

D. *Dry milk*

    a. Grade A dry milk products: at no time a standard plate count over 30,000 per gram, or coliform count over 90 per gram (U.S. Public Health Service).

    b. Standards of Agricultural Marketing Service (USDA):

        (1) Instant nonfat: U.S. Extra Grade, a standard plate count not over 35,000 per gram, and coliform count not over 90 per gram.

        (2) Nonfat (roller or spray): U.S. Extra Grade, a standard plate count not over 50,000 per gram; U.S. Standard Grade, not over 100,000 per gram

        (3) Nonfat (roller or spray): Direct microscopic clump count not over 200 million per gram; and must meet the requirements of U.S. Standard Guide. U.S. Extra Grade, such as used for school lunches, has an upper limit of 75 million per gram.

    c. Dried milk (International Dairy Federation proposed microbiological specifications, 1982).

        Mesophilic count: $n = 5, c = 2, m = 5 \times 10^4, M = 2 \times 10^5$

        Coliforms: $n = 5, c = 1, m = 10, M = 100$

        Salmonella: $n = 15, c = 0, m = 0$.

E. Frozen desserts

    States and cities that have bacterial standards usually specify a maximal count of 50,000 to 100,000 per milliliter or gram. The U.S. Public Health Ordinance and Code sets the limit at 50,000 and recommends bacteriological standards for cream and milk used as ingredients. Few localities have coliform standards.

5. Standard for Tomato Juice and Tomato Products—Mold-count Tolerances (Food and Drug Administration)

The percentage of positive fields tolerated is 2% for tomato juice and 40% for other comminuted tomato products, such as catsup, purée, paste, and so forth. A microscopic field is considered positive when an aggregate length of not more than three mold filaments present exceeds one-sixth of the diameter of the field (Howard mold count method). This method has also been applied to raw and frozen fruits of various kinds, especially berries.

Other Criteria/Guidelines

1. Sampling plans and microbiological limits for nine products as recommended by ICMSF[8] are presented in Table 21–3 (for an explanation of plan stringency or case, see Table 21–4). The

Table 21–4   Plan Stringency in Relation to Degree of Health Hazard and Conditions of Use

Type of Hazard	Conditions in Which Food is Expected to be Handled and Consumed After Sampling		
	Reduce Degree of Hazard	*Cause No Change in Hazard*	*May Increase Hazard*
No direct health hazard			
Utility (e.g., general contamination, reduced shelf life, and spoilage)	Case 1	Case 2	Case 3
Health hazard			
Low, indirect (indicator)	Case 4	Case 5	Case 6
Moderate, direct, limited spread	Case 7	Case 8	Case 9
Moderate, direct, potentially extensive spread	Case 10	Case 11	Case 12
Severe, direct	Case 13	Case 14	Case 15

Source: ICMSF,[8] copyright © 1986 by University of Toronto Press, used with permission.

examples presented were selected to reflect different plan stringencies (for two- and three-class plans) and limits for a variety of organisms.

2. Suggested guidelines for further processed deboned poultry products studied in Canada. See Table 21–5.
3. Canadian criteria for cottage cheese and ice cream:[16]
   Coliforms: $n = 5, c = 1, m = 10, M = 10^3$ (for cottage cheese and ice cream)
   Aerobic plate count: $n = 5, c = 2, m = 10^5, M = 10^6$ (for ice cream only).
4. Recommended criteria for cooked ready-to-eat shrimp: [15]
   S. aureus: $n = 5, c = 2, m = 50, M = 50$
   Coliforms: $n = 5, c = 2, m = 10^2, M = 10^3$.
5. Recommended criteria for cooked ready-to-eat crabmeat:[15]
   S. aureus: $n = 5, c = 2, m = 102, M = 10$
   Coliforms: $n = 5, c = 2, m = 500, M = 5,000$.

Table 21–5   Suggested Guidelines for Further Processed Deboned Poultry Products

Tests/Conditions	N	c	m	M
APC (heat before serving)	5	3	10^4	10^5
APC (cook before serving)	5	3	10^6	10^7
APC (bring to boil before serving)	5	3	10^5	10^6
S. aureus	5	1	10^2	10^4
E. coli	5	2	10	10^2

Note: No salmonellae, yersinae, or campylobacters allowed.

Source: From Warburton et al.[29]

Both products in criteria 4 and 5 should be free of salmonellae and *L. monocytogenes*. The coliform criteria are recommended for process integrity.

REFERENCES

1. Bryan, F.L. 1990. Application of HACCP to ready-to-eat chilled foods. *Food Technol.* 44(7):70–77.

2. Bryan, F.L. 1988. Risks of practices, procedures and processes that lead to outbreaks of foodborne diseases. *J. Food Protect.* 51:663–673.

3. Codex Alimentarius Commission, 14th Session. 1981. *Report of the 17th Session of the Codex Committee on Food Hygiene.* Alinorm 81/13. Rome: Food and Agriculture Organization.

4. Corlett, D.A., Jr. 1998. *HACCP User's Manual.* Gaithersburg, MD: Aspen Publishers, Inc.

5. Dean, K.H. 1990. HACCP and food safety in Canada. *Food Technol.* 44(5):172–178.

6. Elliott, H.P., and H.D. Michener. 1961. Microbiological standards and handling codes for chilled and frozen foods: A review. *Appl. Microbiol.* 9:452–468.

7. Forsythe, S.J., and P.R. Hayes. 1998. *Food Hygiene, Microbiology and HACCP.* Gaithersburg, MD: Aspen Publishers, Inc.

8. ICMSF. 1986. *Microorganisms in Foods 2. Sampling for Microbiological Analysis: Principles and Specific Applications,* 2nd ed. Toronto: University of Toronto Press.

9. ICMSF. 1988. *Microorganisms in Foods 4. Application of the Hazard Analysis Critical Control Point (HACCP) System to Ensure Microbiological Safety and Quality.* London: Blackwell Scientific Publications.

10. Kilsby, D.C. 1982. Sampling schemes and limits. In *Meat Microbiology,* ed. M.H. Brown, 387–421. London: Applied Science Publishers.

11. Miskimin, D.K., K.A. Berkowitz, M. Solberg, W.E. Riha, Jr., W.C. Franke, R.L. Buchanan, and V. O'Leary. 1976. Relationships between indicator organisms and specific pathogens in potentially hazardous foods. *J. Food Sci.* 41:1001–1006.

12. Moberg, L. 1989. Good manufacturing practices for refrigerated foods. *J. Food Protect.* 52:363–367.

13. Mortimore, S.E., and C.A. Wallace. 1994. *HACCP: A Practical Approach.* New York: Kluwer Academic Publishers.

14. NACMCF. 1998. Hazard analysis and critical control point principles and application guidelines. *J. Food Protect.* 61:1246–1259.

15. National Advisory Committee on Microbiological Criteria for Foods. 1990. *Recommendations of the Seafood Working Group for Cooked Ready-To-Eat Shrimp and Cooked Ready-To-Eat Crabmeat.* Washington, DC: U.S. Department of Agriculture.

16. National Research Council (U.S.A.). 1985. *An Evaluation of the Role of Microbiological Criteria for Foods and Food Ingredients.* Washington, DC: National Academy Press.

17. National Research Council (U.S.A.). 1969. *An Evaluation of the Salmonella Problem.* Washington, DC: National Academy of Sciences.

18. Simonsen, B., F.L. Bryan, J.H.B. Christian, T.A. Roberts, R.B. Tompkin, and J.H. Silliker. 1987. Prevention and control of food-borne salmonellosis through application of Hazard Analysis Critical Control Point (HACCP). *Int. J. Food Microbiol.* 4:227–247.

19. Solberg, M., J.J. Buckalew, C.M. Chen, D.W. Schaffner, K. O'Neill, J. McDowell, L.S. Post, and M. Boderck. 1990. Microbiological safety assurance system for foodservice facilities. *Food Technol.* 44(12):68–73.

20. Solberg, M., D.K. Miskimin, B.A. Martin, G. Page, S. Goldner, and M. Libfeld. 1977. Indicator organisms, foodborne pathogens and food safety. *Assoc. Food Drug. Off. Quart. Bull.* 41(1):9–21.

21. Sperber, W.H., K.E. Stevenson, D.T. Bernard, K.E. Deibel, L.J. Moberg, L.R. Hontz, and V.N. Scott. 1998. The role of prerequisite programs in managing a HACCP system. *Dairy Food Environ. Sanit.* 18:418–423.

22. Sperber, W.H. 1991. The modern HACCP system. *Food Technol.* 45(6):116–120.

23. Stevenson, K.E. 1990. Implementing HACCP in the food industry. *Food Technol.* 44(5):179–180.

24. Stevenson, K.E., and D.T. Bernard, eds. 1995. *HACCP—Establishing Hazard Analysis Critical Control Point Programs: A Workshop Manual,* 2nd ed. Washington, DC: Food Processors Institute.

25. Stewart, C.M., M.B. Cole, and D.W. Schaffner. 2003. Managing the risk of staphylococcal food poisoning from cream-filled baked goods to meet a food safety objective. *J. Food Protect.* 66:1310–1325.

26. Todd, E.C.D. 1989. Foodborne and waterborne disease in Canada 1984: Annual summary. *J. Food Protect.* 52:503–511.

27. Tompkin, R.B. 1990. The use of HACCP in the production of meat and poultry products. *J. Food Protect.* 53:795–803.

28. Van Schothorst, M. 1998. Principles for the establishment of microbiological food safety objectives and related control measures. *Food Control.* 9:379–384.

29. Warburton, DW., K.F. Weiss, G. Lachapelle, and D. Dragon. 1988. The microbiological quality of further processed deboned poultry products sold in Canada. *Can. Inst. Food Sci. Technol. J.* 21:84–89.

PART VII

Foodborne Diseases

Most of the human diseases that are contracted from foods are covered in Chapters 23–31. Chapter 22 is intended to provide an overview of foodborne pathogens and to focus on the specific factors that set foodborne pathogens apart from nonpathogens. Much of what is known about each of the syndromes covered in Chapters 22–31 goes beyond the scope of this text; readers are referred to the following references for more extensive information.

de Leon, S.Y., S.L. Meacham, and V.S. Claudio. 2003. *Global Handbook on Food and Water Safety* (for the education of food industry management, food handlers, and consumers). Springfield, IL: Chas. C. Thomas. An easy-to-read coverage of a number of foodborne hazards with particular attention to practices in a number of countries.

Baker, H.F. 2001. *Molecular Pathology of the Prions.* Totowa, NJ: Humana Press. A 279-page broad coverage of these proteins.

Cary, J.W., J.E. Linz, and D. Bhatnagar, ed. 2000. *Microbial Foodborne Diseases: Mechanisms of Pathogenesis and Toxin Synthesis.* Lancaster, PA: Technomic Publishing. Presents and discusses cellular and molecular mechanisms by which certain foodborne organisms cause disease. A well referenced 547-page monograph

Duffy, G., P. Garvey, and D.A. McDowell, ed. 2001. *Verocytoxigenic E. coli.* Trumbull, CT: Food & Nutrition Press. This 450+ page work is devoted entirely to the titled foodborne pathogen.

Hui, Y.H., M.D. Pierson, and J.R. Gorham, ed. 2001. Foodborne Disease Handbook. *Diseases Caused by Bacteria*, Vol. 1, 2nd ed. New York: Marcel Dekker. Provides excellent coverage of foodborne diseases of bacterial origin, including the brucellae. Also, foodborne disease surveillance, foodborne disease investigations, and indicator organisms are covered.

Kaper, J.B., and A.D. O'Brien, ed. 1998. *Escherichia coli* 0157:H7*and Other Shiga Toxin-Producing E. coli Strains.* Washington, DC: ASM Press. Thorough coverage of all aspects of the titled organisms by researchers active in the respective areas.

Labbe, R.G., and S. Garcia. 2001. *Guide to Foodborne Pathogens.* New York: Wiley. Covers a large number of foodborne pathogens.

Miliotis, M.D., and J.W. Bier, ed. 2003. *International Handbook of Foodborne Pathogens.* New York: Marcel Dekker. Extensive coverage of foodborne pathogens from a global standpoint.

Orlandi, P.A., D.-M.T. Chu, J.W. Bier, and G.J. Jackson. 2002. Parasites and the food supply. *Food Technol.* 56(4):72–81. An easy-to-read Status Summary report of the Institute of Food Technologists relative to the titled organisms.

Ryser, E.T., and E. Marth. 2001. *Listeria, Listeriosis, and Food Safety,* 2nd ed. New York: Marcel Dekker. Very thorough coverage of the listeriae, including some laboratory methodology.

Sinha, K.K., and D. Bhatnagar, ed. 1998. Mycotoxins in Agriculture and Food Safety. New York: Marcel Dekker. Rather extensive coverage of a large number of the titled compounds with information on production, mode of action, and control.

Yousef, A.E., and V.K. Juneja, ed. 2003. *Microbial Stress Adaptation and Food Safety.* Boca Raton, FL: CRC Press. Presents and discusses the effect of various stresses on foodborne pathogens.

CHAPTER 22

Introduction to Foodborne Pathogens

INTRODUCTION

Although a number of different infectious diseases may be contracted from foods under certain circumstances, there are those that are contracted exclusively or predominantly from the consumption of food products. Two examples of the former are hemorrhagic colitis and listeriosis and of the latter are botulism and staphylococcal food poisoning. Anthrax and brucellosis are two diseases that have in decades past been contracted from eating diseased animals, but, with the prevalence of these diseases being so low, they are rarely if ever contracted via the foodborne route. The recognized foodborne pathogens include multicellular animal parasites, protozoa, fungi, bacteria, viruses, and prions (Exhibit 22–1). An overview of these organisms is presented in this chapter relative to their general habitats, their entry into foods, and general mechanisms of pathogenesis, and how they differ from closely related nonpathogenic species/strains. More details on each can be found in the respective chapters that follow.

Foodborne Illness Cases in the United States

The Centers for Disease Control and Prevention (CDCP) is the federal agency that gathers, analyzes, maintains, and reports statistics on all human diseases in addition to its other vast activities related to human health. In spite of the research and surveillance activities of this agency, there is no exact count of the number of foodborne illnesses for any year, and there are many reasons for this. First and foremost is the fact that not all foodborne illnesses are reported to health authorities at any level. Another important factor is the lack of a reporting requirement for foodborne illnesses caused by *B. cereus, C. perfringens, S. aureus,* and some other agents; and this leads to an undercount of these syndromes in any given year. For example, for *B. cereus* cases, the CDCP estimates the number of cases to be 38 times the number of reported cases (see reference 72). Although botulism cases are reported, the CDCP assumes that the actual number for a given year is two times the reported number, and that 100% of reported cases are foodborne. On the other hand, *L. monocytogenes* is reported and is covered by sentinel site surveillance (see below) but the total cases are estimated to be two times the reported cases. Foodborne shigellosis is estimated to be 20 times the reported cases but only 20% of this number is estimated to be foodborne (see reference 72).

519

Exhibit 22–1   Groups of Foodborne Pathogens

Flatworms	**Bacteria**
Flukes	Gram positive
Fasciola	*Staphylococcus*
Fasciolopsis	*Bacillus cereus*
Paragonimus	*B. anthracis*
Clonorchis	*Clostridium botulinum, C. argentinensis*
Tapeworms	*C. perfringens*
Diphyllobothrium	*Listeria monocytogenes*
Taenia	*Mycobacterium avium subsp.*
Roundworms	*paratuberculosis*
Trichinella	Gram negative
Ascaris	*Salmonella*
Anisakis	*Shigella*
Pseudoterranova	*Escherichia*
Toxocara	*Yersinia*
Protozoa	*Vibrio*
Giardia	*Campylobacter*
Entamoeba	*Aeromonas (?)*
Toxoplasma	*Brucella*
Sarcocystis	*Plesiomonas (?)*
Cryptosporium	**Viruses**
Cyclospora	Hepatitis A
Fungi—mycotoxin producers	Noroviruses (Norwalk, etc.)
Aflatoxins	Rotaviruses
Fumonisins	**Prions**
Alternaria toxins	Creutzfeldt-Jakob disease (new variant form)
Ochratoxins	**Toxigenic phytoplanktons**
	Paralytic shellfish poison
	Domoic acid
	Pfiesteria piscicida (?)
	Ciguatoxin

Based on all of the CDCP and surveillance activities, the annual number of foodborne illnesses in the United States is estimated to be 76,000,000 with 5,000 deaths, and the genesis of these numbers is indicated in Figure 22–1. The leading causes of gastroenteritis are noroviruses (see Chapter 31), accounting for ca. 67% of cases and 7% of deaths, followed by salmonellosis (26%) and campylobacteriosis (17%). About 75% of foodborne deaths are caused by the following pathogens: *L. monocytogenes, Salmonella,* and *Toxoplasma.*[72] Unknown agents account for ca. 81% of foodborne illnesses and 64% of deaths.[72]

CDCP Surveillance

The FoodNet data gathering system in the United States collects data on nine foodborne pathogens from sites in nine states (California, Colorado, Connecticut, Georgia, Maryland, Minnesota, New

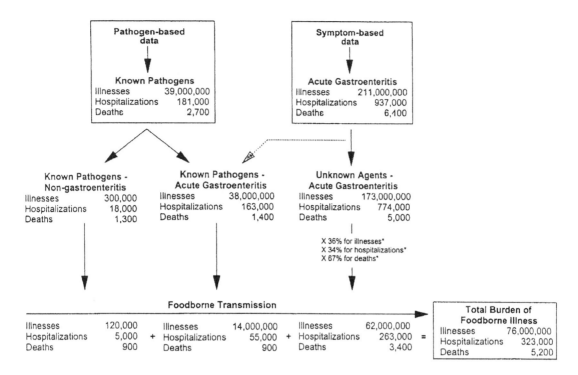

* Percentages derived from observed frequency of foodborne
transmission of acute gastroenteritis caused by known pathogens

Figure 22–1   Estimated frequency of foodborne illness in the United States.[72]

York, Oregon, and Tennessee). The system was begun in 1996 covering five sites with a population of 14.2 million persons, and it was increased in 2000 to cover nine sites, which represent 37.4 million (13% of the U.S. population).

The nine diseases covered and the number of laboratory-confirmed cases for 2002 are as follows:[19]

Salmonellosis (6,028)   Yersiniosis (166)
Campylobacteriosis (5,006)   Vibriosis (103)
Shigellosis (3,875)   Listeriosis (101)
Escherichia coli 0157 (647)   Cyclospora (43)
Cryptosporidium (541)

Among salmonellae, the three most common serotypes were *S.* Typhimurium (19%), Enteritidis (15%), and Newport (14%). The most common non-0157:H7 *E. coli* serotypes in 2002 were 026 and 0111.[19]

The FoodNet data are an important component of the estimates of foodborne illness throughout the United States. It should be noted that not all laboratory-confirmed isolates can be substantiated as being of food origin and, thus, the estimated numbers for some foodborne syndromes are probably too high. This is especially true for *Campylobacter* infections.

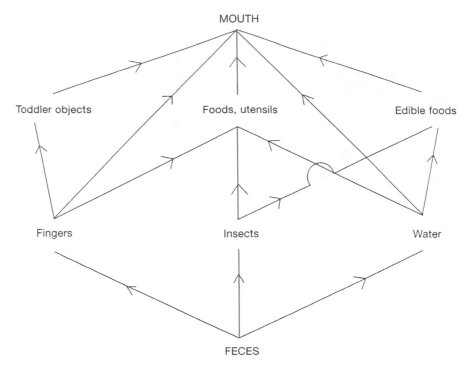

Figure 22–2   Fecal–oral routes of transmission of foodborne intestinal pathogens. The direction is from bottom to top.

The Fecal–Oral Transmission of Foodborne Pathogens

It is rather obvious that either a foodborne pathogen or its preformed toxic products must be ingested in order to initiate a foodborne disease. The common vehicle foods are noted in the chapters that follow. Except for botulinal toxins, the mycotoxins, and the phytoplankton toxins, just about all of the foodborne agents noted above may be contracted via the fecal–oral route, which is illustrated in Figure 22–2. Pathogens may be transmitted from contaminated feces via the fingers of unsanitary food handlers, by flying or crawling insects, or from water. While this route is not as common for syndromes such as staphylococcal food poisoning, it is the primary route of infection for the foodborne viruses and enteropathogenic protozoa and bacteria.

HOST INVASION

"Universal" Requirements

There are several hurdles that an intestinal pathogen must overcome in order to cause illness.

1. It must survive passage through the extremely acidic environment of the stomach. Some pathogens are aided in this process by the protective effect of food, and some survive acidity by the use of their adaptive acid tolerance mechanisms (see Acid Tolerance below).

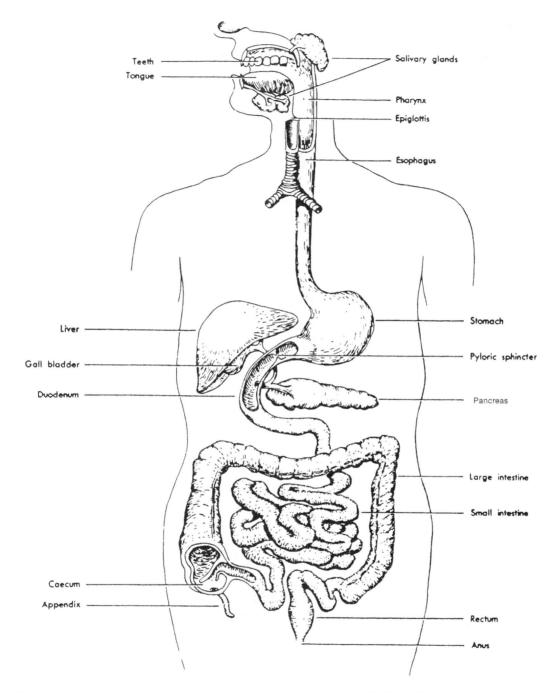

Figure 22–3   Diagram of the human digestive system. Courtesy of John W. Kimball, © 1965, Andover, Massachusetts.

2. It needs to attach to or colonize the intestinal walls in such way that it can increase in numbers. The mucus layer that covers the intestinal mucosa is regarded as being the first line of defense encountered by enteric pathogens.[24] But, in the case of *Listeria monocytogenes*, it has been suggested that it overcomes the mucus barrier by removing mucus through the aid of listeriolysin O (LLO).[24] With a pathogen such as *C. perfringens*, it appears that it does not need to attach to intestinal tissues.

3. It must possess the capacity to defend itself against host defense mechanisms such as gut-associated lymphoid tissue.

4. It must be able to compete with the large heterogeneous microbiota of the gut. This is the gist of competitive exclusion in that the harmless biota, once attached to all available sites on the intestinal walls, will exclude pathogens (see Chapter 26). Also, the gastrointestinal tract is a low-O_2 environment where the predominant organisms are anaerobes, but it has been observed that growth of *S*. Typhimurium in such environments actually induces its ability to enter mammalian cells.[60]

5. Once attached, the organisms need to be able to either elaborate toxic products (e.g., *Vibrio cholerae* non-01) or cross the epithelial wall and enter phagocytic or somatic cells (e.g., *L. monocytogenes*).

The inability of most microorganisms to meet the above requirements in all probability is the reason why they have not been demonstrated to be foodborne pathogens. The attachment sites and mechanisms are important virulence factors for foodborne pathogens, and this aspect is discussed further below.

Attachment Sites

A diagram of the human digestive system is presented in Figure 22–3, and a list of pathogens that can adhere to or enter at each site is presented in Exhibit 22–2. *Helicobacter* is listed since it is apparently the only bacterium that colonizes stomach walls. The obligate anaerobe, *Sarcina ventriculi*, is known to grow in the human stomach but it is not a foodborne pathogen. Whether *H. pylori* is a foodborne pathogen is yet to be proven. The pH of the stomach during food intake is in the 3.0–5.0 range but it may be as low as 1.5 during fasting.

QUORUM SENSING

This is one of the demonstrated means by which cell-to-cell communication occurs between bacteria. It allows for the expression of certain physiologic and phenotypic functions based on population density, and it is more fully described and illustrated below. It is presented here because of its demonstrated occurrence in the virulence of some bacteria, and its apparent involvement in other infections.

The prototype system for quorum sensing is LuxI-LuxR, first described in *Vibrio fisheri* around 1970 (*lux* = luminescence gene). How quorum sensing works is depicted in Figure 22–4.[35] On the left side of Figure 22–4 is a *V. fischeri* cell that is secreting autoinducer (AI) molecules made by LuxI autoinducer synthase. With low cell numbers, the AI continues to be produced with no evidence of cell-to-cell communication in effect. After a certain high level (a quorum) is attained, AI re-enters the producing and closely related cells. As the AI reenters, it binds to the LuxR protein, which is a transcriptional activator, on the right side of Figure 22–4, and gene expression is activated. Some of the phenotypic expressions that have been demonstrated to occur in various organisms are listed in Table 22–1. If the cell on the left side of Figure 22–4 is assumed to be *V. fischeri*, it is non-luminescent

Exhibit 22–2 Sites of Pathogenesis of Foodborne and Related Organisms in Humans

Skeletal muscles
Trichinella spiralis
Stomach
Helicobacter pylori
Liver
Clonorchis—liver flukes
Listeria monocytogenes
Hepatitis A and E
Small intestine
Astroviruses
Bacillus cereus
Campylobacter jejuni (distal ileum; ileum is
lowest part of small intestine)
Clostridium perfringens
Cryptosporidium parvum
Cyclospora cayetanensis
Escherichia coli—EPEC and ETEC strains
Giardia lamblia
Hepatitis A (also the liver)
Listeria monocytogenes

Rotaviruses
Salmonellae (nontyphoid)—terminal ileum
S. Typhi (distal small intestine)
Shigellae (terminal ileum and jejunum when
watery diarrhea is produced)
Toxoplasma gondii
Tapeworms
Vibrio cholerae
V. parahaemolyticus
Yersiniae
Large intestine/colon
Campylobacter (small intestine/colon)
Escherichia coli (enterohemorrhagic [EHEC]
and enteropathogenic [EPEC] strains,
especially the ascending and transverse colon)
Entamoeba histolytica
Plesiomonas shigelloides (apparently)
Salmonella Enteritidis
Shigella, especially *S. dysenteriae*

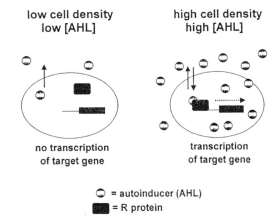

Figure 22–4 Quorum sensing in Gram-negative organisms involves two regulatory components: the transcriptional activator protein (R protein) and the AI molecule produced by the autoinducer synthase. Accumulation of AI occurs in a cell-density-dependent manner until a threshold level is reached. At this time, the AI binds to and activates the R protein, which in turn induces gene expression. The R protein consists of two domains: the N terminus of the protein that interacts with AI and the C terminus that is involved in DNA binding. Typically, Gram-negative AI molecules are *N*-acyl-HSLs; however, other types of signal molecules do exist,[35] copyright © 2000, American Society for Microbiology. Used with permission.

Table 22–1 Some of the Phenotypic Responses Demonstrated to Occur in Some Gram-Negative Bacteria as a Consequence of Quorum Sensing (Taken from the Literature)

Organisms	Demonstrated Responses
Vibrio fischeri	Bioluminescence
Escherichia coli strains	Stx toxin production
Escherichia coli LuxS mutant	Decreased swimming speed
Escherichia coli	Formation of att/eff
Escherichia coli	SOS response
EHEC and EPEC strains of *E. coli*	Type III secretion system
Serratia liquefaciens	Swarming motility
Serratia marcescens	Prodigiosin production; carbapenum synthesis
Pantoea stewartii	Increased polysaccharide synthesis
Pectobacterium carotovorum	Plant wall degrading enzymes produced
Pectobacterium chrysanthemi	Pectate lyases produced
Burkholderia cepacia	Proteases, siderophores produced
Aeromonas hydrophila	Exoprotease produced
Pseudomonas aeruginosa et al.	Normal biofilm structure

but the one on the right is bioluminescent as a result of acquiring a "quorum" of AI that binds to LuxR as noted above.

Autoinducer-2 (AI-2) is an alternate quorum signal compound in *V. harveyi* where it regulates bioluminescence in conjunction with AI-1. The AI-2 system has been demonstrated in a number of Gram-negative pathogens.

The minimum number of cells needed to produce a "quorum" is rarely reported but in one study of psychrotrophic *Enterobacteriaceae* of food origin, at least 10^6 cfu/g were found to be necessary to elicit a positive response to the biosensors that were used.[51]

The best known and most widely studied AI substances for Gram-negative bacteria are N-acetylhomoserine lactones (AHLs). These compounds are composed of homoserine (with a lactone ring) + an acyl side chain that varies from 3 to > 10 carbons; and the structures of two are in Figure 22–5. Not all Gram-negative bacteria employ the LuxI-LuxR. For example, *V. harveyi*, *E. coli*, and *S*. Typhimurium employ a related but different autoinducer production system.[94] In addition to the AHLs, some Gram-negative bacteria produce cyclic(cyclo) dipeptides that are involved in quorum sensing either alone or in combination with AHLs.[34,54] Structures of two of the cyclo-dipeptides identified by Degrassi et al.[34] are presented in Figure 22–5.

Among foodborne bacteria, an increase in Stx toxins has been demonstrated, and the Stx genes were induced by quorum sensing.[93] Although it is similar to the LuxI system (which produces AI-1), LuxS produces AI-2 and it has been demonstrated in *E. coli* 0157:H7, *S*. Typhimurium, and *Campylobacter jejuni* during their growth in milk and chicken broth.[23] As noted above, a number of Gram-negative psychrotrophs have been shown to produce AHLs in naturally contaminated foods when cell numbers reached 10^5–10^7 cfu/g.[51] Quorum sensing is important in biofilm formation, and this is discussed further below. For a review of quorum sensing in Gram-negative bacteria, see references 35 and 47.

Although most studied in Gram-negative bacteria, quorum sensing occurs among Gram-positive bacteria.[38] The AIs for Gram positives are peptides and peptide pheromones, and nisin is perhaps the best known. As described by Kleerebezem et al.,[62] cell density regulation in these systems appears to follow a common theme in which the signal molecule is a post-translationally processed peptide that

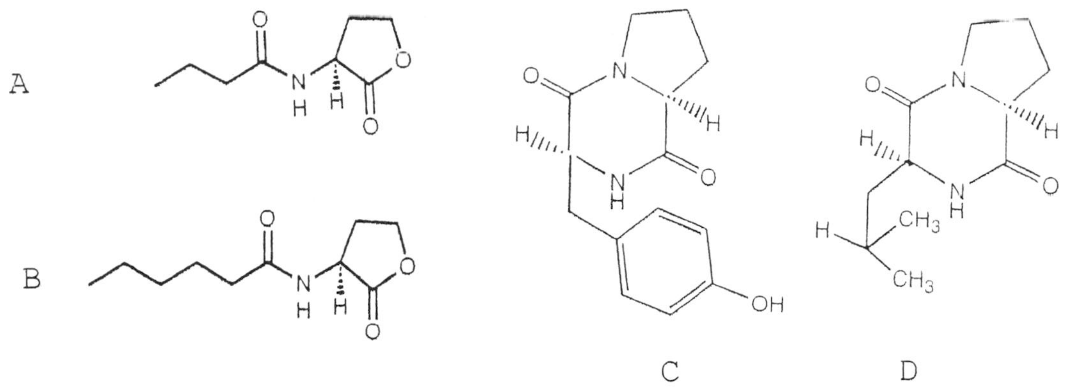

Figure 22–5 Structures of four quorum sensing autoinducers. A = *N*-butanoyl-L-homoserine lactone; B = *N*-hexanoyl-L-homoserine lactose; C = cyclo(L-Tyr-L-Pro); D = cyclo(L-Leu-L-Pro). Structures C and D are from reference 34.

is secreted by a dedicated ATP-binding-cassette exporter. The secreted peptide pheromone functions as the input signal for a specific sensor component of a two-component signal-transduction system. Interestingly, some Gram-positive bacteria such as *Bacillus* spp. produce lactonases that specifically degrade AHLs produced by Gram-negative bacteria.[36]

Among other phenotypic and physiologic activities demonstrated among Gram-positive bacteria is a virulence response in *Staphylococcus aureus*,[59] and the production of antimicrobial peptides other than nisin. Nisin has been shown to induce its own synthesis.[62] Quorum sensing has been demonstrated in *S. aureus* and *S. epidermidis*, and when the two were co-cultured, *S. epidermidis* appeared to be favored leading to the suggestion that this could be the reason why this species is more predominant on the skin where autoinducing pheromones are more likely to be effective than when inside the body.[78] In *S. aureus*, an octapeptide pheromone effects virulence by activating the expression of the *agr* locus.[59] The extent to which quorum sensing occurs in vivo is problematic because of the general lack of opportunities for AI substances to reach a quorum (see Biofilms section below).

BIOFILMS

The importance of biofilms in food safety warrants a better understanding of their biology, structure, and function. They are covered in this chapter relative to virulence properties of certain pathogens.

A biofilm consists of the growth of bacteria, fungi, and/or protozoa alone or in combination bound together by an extracellular matrix that is attached to a solid or firm surface. Common examples include the slimy surfaces on rocks or logs in bodies of running water, dental plaques, and the slime layer on refrigerator-spoiled fresh meats, fish, and poultry. They form on surfaces in large part because nutrients are found in higher concentrations than in the open liquid (planktonic) area. In laboratory studies, surface adherence is best in rich media.[11] Attachment is facilitated by the microbial excretion of an exopolysaccharide matrix sometimes referred to as a glycocalyx. Microcolonies form within this microenvironment in a manner that leads to microbial communities that allow water channels to form between and around the microcolonies. The latter has been likened to a primitive circulatory system where nutrients are brought in and toxic by-products are carried out. Microbial cells in liquids that are not in a biofilm are in a planktonic (free-floating) state.

From the standpoint of food safety and spoilage, biofilms are important because of their accumulation on foods, food utensils, and surfaces; and because of the difficulty of their removal. While under natural conditions, biofilms tend to be composed of mixed cultures, pure culture systems are often used in laboratory studies. Some of the solid surfaces employed to study foodborne bacteria include floor sealant, glass slides, nylon, polycarbonate, polypropylene, rubber, stainless steel, and Teflon. Glass and stainless steel are widely used. From some of the many studies that have been reported in food environments, the following summaries can be made:

1. Although biofilm formation by single cultures in rich media (e.g., tryptic soy broth) may be evident after 24 hours when appropriate growth temperatures are used, 3–4 days or more are necessary for maximum development. On glass slides in a culture medium for 3 days at 24°C, *L. monocytogenes* grew to about 6–7 $\log_{10}/cm^2$[21].
2. Not all strains of the same species are equally capable of initiating biofilm formation,[74] and surface attachment and biofilm development are different processes.[63]
3. Microorganisms in biofilms may exhibit different physiologic reactions than planktonic forms, and the biofilm may contain cells in the viable but nonculturable state.[18,22]
4. Microorganisms in biofilms are considerably more resistant to removal by commonly used cleaning and sanitizing agents, and cleaners and sanitizers used in combination appear to be more effective in removing biofilm growth.[1,77]
5. The attachment of a given pathogen to surfaces may be aided by the formation of a mixed-culture biofilm,[17,68,88] and an example is presented below.

In a biofilm with *L. monocytogenes* and a *Flavobacterium* sp. growing together on stainless steel, the attachment of *L. monocytogenes* was increased and persisted longer in mixed culture than when it grew alone.[15] In addition, sublethally injured *L. monocytogenes* cells increased significantly in mixed culture.

Shewanella putrefaciens readily formed biofilms on inert food processing surfaces, and when nutrients were supplied, multilayered structures were formed.[7] Three strains of *L. monocytogenes*, each from a human outbreak, produced unique "honeycomb" structured biofilms on stainless steel coupons, and strain Scott A was the most conspicuous.[71] A laboratory strain did not form a biofilm. In another study, the *L. monocytogenes* strains that produced the most extra polymeric substance (EPS) produced a three-dimensional biofilm structure in contrast to control strains.[14] Biofilm formation could not be correlated with serotypes. Employing a strain of *Pseudomonas aeruginosa*, extracellular DNA was found to be essential for biofilm formation using a flow-chamber system.[103] Although the source of DNA was not determined, the application of DNase I dissolved the biofilm, suggesting that DNA was an integral part of the biofilm structure.

As noted above, cells in the viable but nonculturable (VBNC) state can form biofilms, and the VBNC cells of *Enterococcus faecalis* adhered to Caco-2 and Girardi heart cells, but at a reduced capacity compared to controls.[82]

The inhibition of biofilm formation by *Bacillus subtilis* has been demonstrated using furanone ([5Z]-4-brome-5-[bromomethyhlene]-3-butyl-2[5H]-furanone), and it inhibited both growth and swarming motility of *B. subtilis*.[83] It was originally isolated from a marine alga. For more on biofilms, see references 26, 45, and 109.

Apparent Role of Quorum Sensing

The first published demonstration of the possible role of quorum sensing in biofilms was that of McLean et al.[73] who recovered AHLs from aquatic biofilms growing on submerged stones in the San Marcos River in Texas. The direct involvement of AHLs was demonstrated with *Pseudomonas aeruginosa* where a mutant that could not synthesize AHL produced atypical biofilms (contained no water channels) that were sensitive to sodium dodecyl sulfate in contrast to wild-type strains.[30]

Biofilm formation on indwelling medical devices and by organisms such as *P. aeruginosa* and *Burkholderia cepacia* that wreak havoc on cystic fibrosis patients is well documented (as is biofilm formation by *L. monocytogenes* in food processing environments). The relationship between biofilm formation, quorum sensing, virulence or pathogenicity of foodborne pathogens is unclear, but it is not inconceivable that relationships exist.

SIGMA (δ) FACTORS

Sigma is one of the four subunits of RNA polymerase, and its role is in the recognition of the promoter (where RNA polymerase binds to DNA and transcription begins). Sigma is involved only in the initial RNA polymerase-DNA complex. After a small portion of mRNA has formed, δ dissociates. δ^A (or δ^{70}, the number refers to the molecular size in kilodaltons) recognizes the majority of genes that encode essential cell functions, and its closest homologue is δ^S. Among the known sigma factors are the following: δ^{28} is involved in flagellar synthesis in *Salmonella*, and also in the type III secretion system. δ^{32} (RpoH) is involved in heat shock proteins (HSPs), some of which are molecular chaperones or proteases that eliminate those that cannot be repaired. δ^{54} (RpoN) regulates *harp* (hypersensitive response and pathogenicity) genes in at least some *Pseudomonas syringae* pathovars. δ^B endows *L. monocytogenes* with resistance to lethal acidic conditions. It is more abundant at 25°C than at 42°C in *E. coli*. It is involved in stress responses in *Bacillus subtilis*. δ^S is present in the γ-subclass of the *Proteobacteria* including the vibrios. It aids *V. vulnificus* in its resistance to adverse environmental conditions. It is discussed further below under Alternative Sigma Factors. A change in the cell's environment that is stressful (starvation, low pH, increased osmotic pressure, etc.) leads to the induction of alternative sigma factors that aid the cell in coping with its unfavorable state (see reference 5).

The general responses that bacteria make upon exposure to acidic conditions consist of the following:[27] (1) proton pumps come into play where a proton motive force (PMF) can facilitate the extrusion of protons from the cytoplasm, which results in a drop in intracellular pH; (2) repair of macromolecules such as DNA, and proteins such as RecA; (3) changes in cell membrane components (e.g., fatty acids); (4) regulation of gene expression by alternative sigma factors; (5) cell density and biofilm formation (which protects cells from certain adverse environmental influences); and (6) alteration of metabolic pathways. Biofilms are covered in the previous section of this chapter, and alternative sigma factors are discussed further below.

Alternative Sigma Factors

The alternative sigma factor, δ^{38} (sigma-38, δ^S) is encoded by the *rpoS* gene, and it regulates at least 30 proteins, and it along with δ^B is discussed in this section. Exposure to acid stress leads to the synthesis of proteins that protect the bacterium. With log-phase cells at or below pH 4.5, at least

43 proteins are induced. When stationary-phase cells are shifted to or below pH 4.5, they synthesize 15 proteins that are distinct from log-phase cells.

Acid Tolerance Response

With respect to the acid tolerance response (ATR) of *L. monocytogenes*, the pH minimum for growth of two strains was 3.5 and 4.0 in a chemically defined medium using HCl.[81] pH values below these were lethal unless the strains were previously exposed to pH of 4.8 and 3.5. With mutants of *L. monocytogenes* that showed increased ATR, they demonstrated increased lethality for mice compared to wild type strains[76], suggesting that acid conditions could be selective for strains with increased virulence. The mutants were recovered after exposure to pH 3.5 for up to 2 hours at 37°C.[76] Similarly, acid-adapted *Yersinia enterocolitica* cells grown at pH 7.5, then shifted to pH 5.0, were significantly more enteropathogenic than controls when tested using a suckling mouse model.[106]

δ^B has been identified in *L. monocytogenes*, *B. subtilis*, and *S. aureus*; and its function has been compared to those of RpoS/δ^S in Gram-negative bacteria. In *B. subtilis*, δ^B influences the regulon of 100 genes in response to environmental and energy stresses. *B. subtilis* mutants are more sensitive to heat, ethanol, acid, freezing, drying, etc.[27] One δ^B reduces virulence in *L. monocytogenes*. Interestingly, a hydrostatic pressure resistant strain of *L. monocytogenes* (survived 400 MPa for 20 minutes) displayed increased resistance to heat, acid, and hydrogen peroxide.[61] It has been found that the prior adaptation of *L. monocytogenes* that leads to acid resistance may depend upon a number of other growth parameters.[65]

The δ^B protein is necessary for full resistance of *L. monocytogenes* to lethal acid exposure, and it along with δ^S is associated with general stress responses in both Gram-positive and Gram-negative bacteria.[44] After acid adaptation (pH 3.0, HCl), *L. monocytogenes* strain Scott A, 11 proteins showed induced expression while 12 were repressed.[31] On the other hand, acid adaptation did not protect this species on beef treated with 2% lactic or acetic acid.[56] *L. monocytogenes* on fresh meats may be made more acid sensitive by the Gram-negative bacteria of the fresh-meat biota.[86] As noted above, δ^B plays a role in acid resistance of stationary phase cells, oxidative and osmotic stress resistance, and in low-temperature growth of *L. monocytogenes*. Acid adaptation of *L. monocytogenes* provides protection against HHP and freezing.[102]

Acid-adapted strains (pH 5.0–3.25, brain heart infusion, BHI, broth) of *Shigella flexneri* and *S. sonnei* survived up to 14 days in tomato and apple juice stored at 7°C.[6] The minimum pH for growth in acidified BHI was 4.75 and 4.5 for *S. flexneri* and *S. sonnei*, respectively. In one study, acid-adapted (pH 2.5, tryptic soy broth), *E. coli* 0157:H7 remained on beef carcasses after a 2% acetic acid wash more so than nonadapted cells.[9] In yet another study of acid tolerance in *E. coli* 0157:H7, acid tolerance decreased after exposure to non-acid washes.[85]

Acid resistance (percentage of cells that survive exposure to pH 2.5 for 2 hours) is well studied in shigellae. Gorden and Small[50] found that among the cultures they examined, 9 of 12 shigellae were acid resistant; 11 of 15 generic *E. coli* (including strain K-12) showed the same level of acid resistance; 3 of 8 enteroinvasive (EIEC) strains were resistant but none of 2 enteropathogenic (EPEC) strains or 12 salmonellae were acid resistant. As to why so few shigellae cells are needed to cause disease, these investigators hypothesized that after these organisms leave the colon, they enter the stationary phase outside the host. Upon ingestion by another host, they are already acid resistant, and low numbers can survive through the acidity of the stomach.[50] In another study, Stx-producing strains of *E. coli* that could not survive at pH 2.5 were made acid resistant by the introduction of the *rpoS* gene on a plasmid.[100] When Stx-producing strains of *E. coli* were grown in broth at pH 4.6–4.7, they became 1.1- to 2.0-fold more resistant to radiation than control strains.[16] It has been suggested that this response may lower the number of cells needed to initiate infection.[97] For instance, it has been

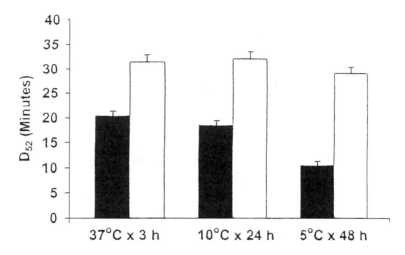

Figure 22–6 Starvation-induced thermal tolerance (D_{52}-value [minutes]) of *E. coli* 0157:H7. The starvation times, 3 hours at 37°C, 24 hours at 10°C, and 2 days at 5°C, were determined on the basis of the maximum expression of the uspA and *grpE* genes; (<), control; (), starvation. Values are significantly different for all treatment temperatures ($P < 0.01$),[108] © 2003, International Association for Food Protection.

noted that while the human infectious dose of *Salmonella* is approximately 10^5 when administered under defined conditions, disease can actually be caused by 50–100 cells when consumed as part of a contaminated food.[101] Others have noted that salmonellosis may be caused by as few as 10 cells. A mouse-virulent strain of *S.* Typhimurium has been shown to be much more acid resistant than avirulent strains.[105]

Employing macrophages and Int407 cell cultures, an acid-adapted strain (DT104) of *S.* Typhimurium was found to be no more invasive than non-DT104 strains.[46] Acid adaptation did result in increased resistance to a low-pH environment for all strains tested in apple cider, acetic acid, and a synthetic gastric fluid, all at pH 2 and 3.[46]

Although the ATR response (along with certain heat shock proteins) is generally induced by exposure of cells to acidic pH values, exposure to starvation has the same effect, at least in *E. coli.* When a strain of *E. coli* 0157:H7 was stressed at 37, 10, and 5°C, there was a significant increase in heat resistance at each temperature compared to controls and this effect is shown in Figure 22–6.[108] In the latter study, thermal tolerance was correlated with the proteins UspA and GrpE with the latter at 5°C being positively linked but negatively related to UspA.

When a *S.* Enteritidis PT (phage type) four strain was exposed in broth at pH between 3.0 and 6.0 and then challenged at pH 2.5–2.9, an increase in acid resistance occurred in ca. 5 minutes.[55] In another study employing laboratory-prepared mayonnaise, cells remained viable for 4 weeks at 4°C when they were first exposed to pH 5.8 followed by pH 4.5.[69] The pH of the mayonnaise remained at 4.2–4.5 during the period of storage. In the case of *E. coli* 0157:H7, heat-adapted cells showed an increase in membrane fluidity, which may have increased Stx toxin secretion.[107]

In a study of the relative resistance of *Vibrio vulnificus* and its phages, both were found to be sensitive to a pH < 3.0 but the phages were more acid resistant than their host cells.[64] Cold or cold-acid stress of *E. coli* 0157:H7 had no effect on the production of virulence factors but growth in acidic media (pH 5.5) enhanced the expression of *eaeA* and *hlyA* genes.[40] Cold stress was achieved by exposure to 4°C.

Overall, the full significance of the acid tolerance response and other sigma factors in foodborne bacterial infections is not fully understood. However, enough information has been presented for some pathogens to suggest the need for more research in this area.

PATHOGENESIS

When one looks at the many different types of causative organisms, it should not be surprising that there are many mechanisms that lead to the initiation and course of foodborne illness. The flat- and roundworms are contracted by ingesting infected meat or fish, and upon entry into the gastrointestinal (GI) tract, different paths are taken by these organisms, including passage to the liver, to skeletal muscles, or simply remaining in the GI tract. The foodborne protozoa remain in the gut with the notable exception of *Toxoplasma gondii*, which can cross the placental barrier and inflict severe damage to a fetus. The phytoplankton toxins and mycotoxins are ingested preformed, and these chemical compounds have affinities for specific tissue or cell targets (e.g., aflatoxins for DNA). The pathogenic mechanisms of the foodborne bacteria are more involved, and they are discussed below. Additional information on each group noted can be found in the respective chapters that follow. A glossary of relevant terminology is presented in Exhibit 3.

Gram-Positive Bacteria

In general, Gram-positive pathogens produce exocellular substances that typically account for most, if not all, of the virulence factors for this group and this is typified by *Staphylococcus aureus*. Virulent strains are known to produce a number of exotoxic factors that are absent in avirulent strains. In the case of the gastroenteritis syndrome presented in Chapter 23, the enterotoxins constitute the only agents of importance. Regardless of the number and type of other extracellular products that may be elaborated, enterotoxin-negative strains do not cause the gastroenteritis syndrome. What is known about the mode of action of staphylococcal enterotoxins is presented in Chapter 23.

Like the gastroenteritis-causing strains of the staphylococci, foodborne diseases caused by *Clostridium botulinum*, *C. perfringens*, and *Bacillus cereus* are also due to exotoxins. The only toxin of importance in botulism is a potent neurotoxin, which is elaborated by cells growing in susceptible foods. The *C. perfringens* enterotoxin (CPE) is a spore-associated protein that is produced during sporulation of bacterial cells in the GI tract. The emetic toxin of *B. cereus* is an exotoxin, but the toxic components that cause the diarrheal syndrome are not as well understood. More on what is known about the mode of action of these toxins is presented in Chapter 24.

It was widely assumed for many decades that a bacterium causing foodborne gastroenteritis "must" produce an enterotoxin in the manner of the staphylococci, and reasons for this view can be ascribed to this bacterium as being the prototype of foodborne disease organisms. Also, with the exception of neurotoxigenic strains of *C. botulinum*, *S. aureus* was the first foodborne pathogen whose mode of pathogenicity was established. First studied by Denys in 1894 and next by Barber in 1914, who produced in himself the symptoms of foodborne disease, it was proven conclusively by Dack et al. in 1930[29] when they showed that all symptoms of the disease could be produced by feeding a culture filtrate of *S. aureus* (to volunteer graduate students at that time!). This neat and crisp prototype has been sought for all foodborne pathogens and, in a sense, led to what in retrospect appears to have been undue efforts to find an enterotoxin in all foodborne pathogens, including the Gram negatives.

Listeria monocytogenes

Although Gram positive, this bacterium is significantly different from those noted above. The most notable difference is that it is an intracellular pathogen. More specifically, it resides in the cytosol where it replicates; and it apparently uses host-derived lipoic acid for its replication.[79] In order to enter epithelial cells, the invasin *internalin* interacts with *E-cadherin* on human host cells (mouse and rat E-cadherins are not receptors for internalin). The *L. monocytogenes* cells that lack internalin are noninvasive[67] Although virulent strains produce the exocellular thiol-activated, pore-forming substance *listeriolysin O* (LLO), it does not per se cause the foodborne gastroenteritis syndrome. LLO is a hemolysin involved in the invasion of the gut epithelium, and it contributes to the cell-to-cell spread of the organism. Unlike the other syndromes above caused by Gram-positive bacteria (with the exception of *Clostridium perfringens*), the ingestion of viable cells is necessary for listeric infection to occur.

The PEST (P, pro.; E. glu.; S, ser.; T, thr.) sequence of LLO is essential for virulence and the intracellular state of *L. monocytogenes*. This sequence induces host macrophages to degrade LLO once it escapes from lysosomes. Mutants lacking the PEST sequence entered host cell cytosol and killed the host cells.[33] LLO lacking PEST accumulated in the cytosol, suggesting that it targets LLO for degradation.[33] In effect, PEST prevents LLO from prematurely destroying host cells.

Virulent strains of *L. monocytogenes* can breach the mucous barrier as noted above, and they then enter epithelial cells, but just how is unclear. Also, gastrointestinal symptoms are seen in only around one-third of human cases.[49] The body's early defense against these organisms consists of resident macrophages, especially the Kupffer cells of the liver.[91] They gain entry into these cells by being internalized in M cells nondestructively.[58] This is followed by the induction of host T-cell-mediated immunity, which is further described in Chapter 25. Polymorphonuclear neutrophils (PMNs) lyse *Listeria*-infected parenchymal cells and thus expose the bacteria to professional phagocytes. The PMNs contain superoxide anions, proteolytic enzymes, and other factors. When PMNs interact with *L. monocytogenes*, they show increases in *cytokines* such as interleukin-Iβ, interleukin-6 (IL-6), and tumor necrosis factor (TNF).[91] Once listeriae are phagocytized, the cells escape by lysing the vacuolar membrane with the aid of LLO, move about the cytosol by actin filaments, and then spread to neighboring cells where the process is repeated. Although virulent strains contain other substances that may contribute to virulence, what sets this species apart from the nonpathogenic *Listeria* is the capacity to adhere to and breach the mucosal/epithelial barrier, and to spread from cell to cell with the aid of LLO. It may be postulated that these virulence factors in listeriae were acquired independently, probably from other Gram-positive bacteria that produce thiol-activated toxins.

Gram-Negative Bacteria

The pathogenesis and virulence properties of this group are considerably different and far more complex than for Gram-positive bacteria. Great effort was devoted to finding enterotoxins for most, and while these pursuits were successful for some, the significance of the enterotoxins in foodborne pathogenesis seems questionable. The findings on some of these organisms during the past two decades or so are summarized below.

Salmonellae

It is estimated that *Salmonella* and *Escherichia* arose from a common ancestor about 120–160 million years ago.[39] All *S. enterica* serovars carry pathogenicity islands 1 and 2 (SPI-1, SPI-2), which were acquired via horizontal transfer either by plasmids or phages.[8] In *S*. Typhimurium, at least 60 genes are required for virulence,[52] and the two SPIs are known to contain at least 42 of these genes. By comparison of the 16S and 23S rRNA sequence data, the salmonellae have been shown to be closely related to *E. coli* and shigellae, with the monophasic salmonellae serovars being adapted to mammals and the diphasic to reptiles.[21] Regarding their evolution, up to 35 kilobases (kb) of the DNA that encompasses the SPI region of *Salmonella* at centisome 63 may have been acquired as a block from another microorganism as they evolved toward becoming pathogens.[48] This is supported by the observation that *S. enterica* and *E. coli* contain a high incidence of mutator phenotypes that lead to increased mutation rates and enhanced recombination among the diverse species.[70]

A 29-kDa polypeptide enterotoxin has been demonstrated in *S*. Typhimurium that has the following features: it cross-reacts with cholera toxin, it activates adenylate cyclase, its preferred host cell receptor is ganglioside GM_1, and it is positive in the ileal loop test.[75] This suggests that the toxins could play a role in causing the diarrheal part of the salmonellae syndrome, but their role in intracellular invasion and the subsequent pathogenesis is unclear. Production of other cytotoxic proteins has also been reported in nontyphoid salmonellae.[28]

Virulent strains of *S. enterica* initiate infection in nonphagocytic cells by attaching to the intestinal mucosa with the aid of fimbrial adhesins encoded by a gene on SPI-1.[96] This is followed by the penetration of the intestinal mucosa, mainly at the lymphoid follicles of Peyer's patches. Their initial site of infection is the ileum of the small intestine. Once inside, they invade the M cells of Peyer's patches[60] From the vesicles of these cells, they enter the lysosome. Virulent strains of *S. enterica* secrete into the cytoplasm a protein (SpiC) that prevents the fusion of vesicles with lysosomes. *S*. Typhimurium contains fimbriae that selectively adhere to M cells, and although they can enter any intestinal epithelial cell type, M cells are preferred. Their entry into nonphagocytic cells is aided by a type III (also known as "contact") secretion system. As Galán[48] has noted, this entry mechanism involves a rather intimate interaction between the bacterium and host cells that results in "cross-talk." As a consequence, cytoskeletal rearrangements, membrane ruffling, and bacterial uptake by macropinocytosis take place. The migration of neutrophils across the epithelial cells occurs and cytokines (e.g., interleukin-8) are produced. Once inside these cells, they remain inside membrane-bound vacuoles during their entire intracellular stage.[84] Following multiplication, the cells ultimately burst and the pathogen is spread. The entry of salmonellae into macrophages is accompanied also by membrane ruffling and macropinocytosis.[84] Once inside, they are found inside the membrane-bound phagosomes, which become enlarged. *Salmonella* Typhimurium induces apoptosis in macrophages. For a review of *Salmonellae* Enteritis, see reference 99.

The nontyphoid salmonellae serovars differ in their degree of human pathogenicity with *S*. Pullorum and *S*. Gallinarium being among the least pathogenic, and *S*. Choleraesuis, *S*. Dublin, and *S*. Enteritidis being the most pathogenic. *Salmonella* serovar Choleraesuis is isolated from blood more often than from stools of victims, and it, along with *Salmonella* serovar Dublin, is associated with higher mortality than other serovars.[87] In the case of *S*. Choleraesuis, intestinal involvement and excretion are rare but septicemia is common. In one study of 19 cases of salmonellosis caused by this serovar, all victims had septicemia.[3] Just what sets these serovars apart from the more commonly occurring *S*. Typhimurium relative to the locus of enterocyte effacement (LEE) and secretion system is unclear.

Escherichia coli

The disease-causing strains of this organism are placed in 5–6 virulence or pathogenicity groups and they are discussed in Chapter 27. The two groups discussed here are enteropathogenic (EPEC) and enterohemorrhagic (EHEC).

As noted above, molecular genetic data suggest that the genera *Escherichia* and *Salmonella* arose from a common ancestor, and thus it should not be surprising that virulence genes were exchanged between them via horizontal transfer. The pathogenicity island on the chromosome of EHEC and EPEC includes LEE, which contains the *eae* gene that encodes the intimin protein that is essential for attachment–effacement (A/E).[12] The *eae* gene and LEE apparently were transferred horizontally within EHEC.[12] EPEC strains contain the EPEC-secreted protein (*esp*B) that makes them similar to EHEC. It appears that EHEC strains evolved from EPEC via acquisition of phage-encoded Shiga toxins.[80] Evidence has been presented showing how EHEC evolved sequentially from an EPEC O55:H7 ancestor by first acquiring the *Stx2* gene and then by diverting into two branches.[43] The strains in one branch are β-glucuronidase and sorbitol negative (the O157:H7 clone) and in the other are nonmotile but sorbitol and glucuronidase positive (the O157:H7 clone). These investigators came to this conclusion by, among other methods, subjecting EPEC and EHEC strains to multilocus enzyme electrophoresis (see Chapter 11). They postulated that the *Stx2* gene was acquired early and has been evolving in the O157:H7 genome for a longer time than other virulence factors. Acid resistance was also acquired early on, but whether it preceded *Stx2* is unclear. In addition to the *Stx* genes, adhesins also appear to have been acquired via horizontal transfer.[104]

EHEC strains require intimin for colonization but it alone is not sufficient to cause A/E. The possible use of intimin-based vaccines to protect cattle against EHEC infections has been suggested.[32] The pathogenicity of EHEC is due to the possession of Stx toxins, endotoxins, and host-derived cytokines such as tumor necrosis factor alpha (TNF-a) and interleukin-1β. Stx1 and Stx2 toxins inhibit protein synthesis in endothelial cells, and their receptor is globotriasylceramide (Gb3). Human renal tissue contains large amounts of Gb3 and thus it is highly sensitive to the Stx toxins.[57] Stx2 toxin has been found to be more toxic than Stx1 to human intestinal microvascular endothelial cells, and this finding may be relevant to the preponderance of Stx2-producing EHEC in hemorrhagic colitis infection.[57]

EPEC strains require the plasmid-borne type IV bundle-forming pili (bfp) for adherence and auto-agglutination. Mutants that lacked bfp caused less severe diarrhea and were about 200-fold less virulent in human volunteers.[10] The A/E lesion and "pedestals" of densely clustered cytoskeletal protein (including actin) are regarded as the hallmarks of EPEC infection.[41] The A/E lesion begins as a nonintimate attachment of the bacterium, followed by the injection of type III proteins, which effect cytoskeletal changes and effacement of microvilli. Intimin is required for the latter event.[37] Although they are not proven foodborne pathogens, some strains of *Citrobacter freundii* and *Hafnia alvei* produce A/E, especially in certain animals.[89]

Yersiniae

Y. enterocolitica (and some other yersiniae) possess a chromosomal determinant that is involved in iron uptake, which is mediated by the siderophore yersiniabactin, and it is regarded as a pathogenicity island (PI). This PI is also found in EAggEC strains but rarely in EPEC, EIEC, and ETEC; it is absent from the EHEC, salmonellae, and shigellae strains tested.[90] It was probably acquired horizontally between *Y. pestis* and some strains of *E. coli*.[90]

The most significant pathogenic mechanism of *Y. enterocolitica* is contained in the yersiniae outer protein (Yop) virulon (see Exhibit 22–3), which is also possessed in *Y. pestis* and *Y. pseudotuberculosis*. This virulon allows yersiniae to survive and multiply in host lymphoid tissue, and it consists of four components as noted in Exhibit 22–3. Yop is encoded by a 70-kb plasmid, pYV, and it possesses high-pathogenicity island 1 that is necessary for virulence expression, and it determines Ca^{2+} dependency.[4,25]

Exhibit 22–3 Glossary of Terms Relevant to Some Foodborne Pathogens

Adherence factor (EAF) plasmid—A 70-kDa unit in EPEC that contains genes for bundleforming pili

Apoptosis—Programmed cell death

Attachment-effacement (Att-eff, A/E)—The intimate (tight) adherence of bacteria to epithelial cells that leads to effacement of intestinal microvilli and changes in host cell cytoskeleton. Found in EPEC and EHEC. The genes for A/E are located on about the 34-kb region of chromosomal DNA of the LEE

Biovar, biotype—Subspecies that is physiologically different

Bundle-forming pili—Located on surface of pathogen, encoded by the *bfp* gene cluster that is located on the EAF plasmid

Diarrhea (*dia*, Gr.; *Rhein*, to flow through)—watery discharge, mainly from the small intestine; the "runs"

Dysentery (Gr., *dys*, bad + *entera*, bowels)—Frequent but smaller volume stools than diarrhea that contain blood and/or pus from mucosal damage; the "squirts"

Genomovar—Phenotypically similar but genotypically distinct groups of strains

Integrins—Host cell receptors; transmembrane proteins on the surfaces of many eukaryotic cells, especially the M cells of Peyers patches

Intimin—A 94-kDa outer membrane protein adhesin encoded by the chromosomal *eae* gene that is required for host cytoskeletal proteins beneath adhering bacteria; needed for colonization

Lamina propria—Connective tissue under the mucosal epithelium of the gut

Locus of enterocyte effacement (LEE)—Example of a pathogenicity island in EHEC and EPEC strains of *E. coli* that contain the genes for A/E and EPEC-secreted protein B. The entire gene sequence of LEE from one *E. coli* 0157:H7 consisted of 43,359 bp, and it included a prophage[34]

M Cells (microfold or membranous)—Part of Peyers patches; have only a small amount of mucous coating. They present antigens to immunocompetent cells of the lamina propria

Pathogen—Organism with a demonstrated capacity to cause disease

Pathogenicity island (PI)—Specific regions of bacterial chromosomal DNA that include a number of virulence genes, e.g., pathogenicity island 2 (SPI-2) of salmonellae

Pathovar—A biovar that has different host ranges

Pedestals—Structure 10 μm or so upward that forms beneath attached bacteria following destruction of brush border microvilli. They consist of densely clustered cytoskeletal proteins, including actin. Their formation is initiated by translocated intimin receptor (Tir)

Peyers patches or gland—Large subepithelial, oval patches of closely aggregated lymphoid follicles or nodules in the walls of the gut, especially abundant in the ileum. Its M cells are used by pathogens such as *Y. enterocolitica, C. jejuni*, shigellae, and *S. Typhimurium* as their primary portal of host entry (see M cells above)

Phagovar (phagotype)—Different phage or lysotype

RpoS (Stationary-phase sigma factor)—Regulates, among other things, acid resistance and starvation responses in some pathogens

Virulence—Relative degree of pathogenicity

Secretion system[a]

Type III—Proteins end up in periplasm. To get them outside the cell, special apparatus is needed, e.g., Yops. The secreted effector proteins induce uptake of bacteria by host cells

Type II—Proteins secreted to periplasm and directly across outer membrane

Type I—Proteins secreted directly to the environment by two cytoplasmic and one outer membrane protein

Serotype, serovar—Subdivision of a species based on antigenic differences

Tir (translocated intimin receptor) protein—Protein that is translocated from bacterium to host cell where it serves as receptor for intimin. Active in pedestal formation

Yops (yersiniae outer proteins)—*Yersinia* virulon that is encoded by a 70-kb plasmid, pYV. It is composed of four elements: (1) type III secretion system that is devoted to the secretion of Yop proteins, (2) a system that delivers bacterial proteins into host cells (YopB and YopD), (3) a control element (YopN), and (4) a set of effector Yop proteins

[a]Type IV secretion system is best known in *Agrobacterium tumefaciens*; not demonstrated to occur in foodborne human pathogens.

Yops are synthesized at 37°C and translocated into mammalian cells upon contact. Gram-positive bacteria can secrete proteins directly out of their cell since there is no outer membrane. In the type I secretory system of Gram-negative bacteria, bacterial proteins are secreted directly from the cytoplasm to the environment by two cytoplasmic and one outer membrane proteins. However, in a type III secretion system, the bacterial proteins need a specialized apparatus to exit the producing cells, and Yops is an example of such apparatus.

The yersiniae secretion apparatus is normally kept closed at the outer membrane by YopN, which acts as a cork. YopN can be removed (system uncorked) by removing Ca^{2+}, at which time Yops are secreted from the cytoplasm to the outside. YopP is responsible for the suppression of TNF-α release by infected macrophages.[13] When Yops contact a eucaryotic cell, a microinjection device is formed that allows Yops to pass via the type III secretion system and directly into the eucaryotic cell.[41] This process has been described by Silhavy[92] as death of macrophages by lethal injection. Falkow[41] has stated that "shigellae cause the macrophage to commit suicide." The type III secretion system in *S*. Typhimurium has been described as a supramolecular structure that spans the inner and outer membranes.[66] Type III systems are also possessed by the plant pathogenic strains of *Erwinia*, *Xanthomonas*, *Pseudomonas*, and *Ralstonia*.[2] As noted in Exhibit 22–3, a type IV system is possessed by the plant pathogen, *Agrobacterium tumefaciens*.

Shigellae

The M cells of Peyer's patches in the terminal ileum are invaded by shigellae as well as some salmonellae, some EPEC, and some viruses.[41] Shigellae invade macrophages of the colonic and rectal M cells and the macrophages die by apoptosis. The result is an acute inflammatory response with dysentery. This is especially true for invasive strains of *S. flexneri*.[110] This type of damage leads to the loss of blood and mucus in the intestinal lumen. Since colonic absorption of water is inhibited, the result is the passage of scanty (squirts) dysenteric stools. When shigellosis is accompanied by watery diarrhea, it is due to the transient multiplication of the organisms as they pass through the jejunum. Of the shigellae species, *S. sonnei* causes diarrhea most often. As to minimum infectious dose, as few as 10 cells caused disease in 10% of volunteers, and when using 500 cells, 50% became infected.[39]

The Shiga toxin of *S. dysenteriae* type 1 binds to galabiose and begins the inhibition of mammalian protein synthesis. Although hemolytic uremic syndrome (HUS) is most often associated with EHEC strains of *E. coli*, it may also be caused by *S. dysenteriae*.

Vibrios

In contrast to the Gram-negative bacteria discussed above, vibrios are not members of the family Enterobacteriaceae, and those associated with foodborne illness are also noninvasive. In the case of *V. parahaemolyticus*, its pathogenesis is associated with the production of a 46-kDa homodimer—thermostable direct hemolysin (TDH). The latter appears to be responsible for the following events: hemolysis, pore-forming capacity, cytotoxic effects, lethality in small animals, and enterotoxigenicity as assessed by its activity in ileal loops. For more details on TDH, see Chapter 28.

The 01 strains of *V. cholerae* colonize the epithelium of the small intestine with preference to M cells, and this leads to profuse diarrhea. The two primary virulence factors of this organism are (1) toxin-coregulated pili (TCP) that are required for intestinal colonization, and (2) cholera toxin (CT) that is an enterotoxin.[98] The CT genes (*ctx*AB) are part of a larger genetic element, CTX, which constitutes the genome of a filamentous bacteriophage designated CTXø.[42,98] The latter can be propagated in recipient *V. cholerae* strains in which it either integrates chromosomally to form stable lysogens or is maintained extrachromosomally.[42] The latter investigators showed that CTXø isolated from ten clinical or environmental strains of *V. cholerae* infected CT-negative strains. However, they noted that phage induction may not occur inside the human intestines. This pathogenicity locus appears to be an example of horizontal gene transfer that can lead to the emergence of new pathogenic strains, and the CTXø element is related to coliphage M13.[98] Among the unusual features of *V. cholerae* is its possession of two circular chromosomes.[95] The large one contains >2.96 million bases, and it contains the housekeeping genes and some that are involved in virulence. The small chromosome contains >1.07 million bases and many genes of unknown function. The cholera toxin B (CTB) subunit binds to GM_1 ganglioside cell surface receptors (it is closely related to subunit B of the heat-labile (LT) toxin of *E. coli*).

The role of bacteriophages in the transmission of virulence genes is illustrated by the CTX genetic element noted above. Among foodborne pathogens, genes for the following toxins are known to be carried by phages: Staphylococcal enterotoxin A, Stx1 and Stx2 of EHEC strains of *E. coli*, and botulinal toxins. It has been noted that while virulence-associated genes may be on plasmids in one organism, they may be on the chromosome in others, suggesting that genes may integrate following transmission.[20] A similar pattern for phage-mediated genes seems plausible.

SUMMARY

Much new information has been obtained during the past two decades on the specific mechanisms used by foodborne pathogens to cause human disease, and this is especially true of the Gram-negative bacteria. Beyond their role in intestinal fluid accumulation (diarrhea), not much more has been learned about the enterotoxins that are produced by Gram-negative bacteria. Their role in host cell invasion and subsequent pathogenesis seems minimal.

The concept of pathogenicity islands (PIs) in salmonellae, yersiniae, and EPEC and EHEC strains of *Escherichia coli* is a significant development. They are located on extrachromosomal DNA as well as parts of phage genomes, and they are not found in nonpathogenic bacteria (for an extensive review and discussion, see Hacker and Kaper[53]).

Molecular genetic studies have shed more light on the importance of plasmid and bacteriophage transfer of virulence genes between some of the Enterobacteriaceae, and within the genus *Vibrio*. The

Table 22–2 Examples of Some Gram-Negative Bacteria That Possess at Least One Virulence Factor Often Associated with Established Foodborne Pathogens

Organisms	Virulence Factor Demonstrated
Aeromonas caviae	Enterotoxin
A. hydrophila	Cytotoxic enterotoxin
Bacteroides fragilis	Loop-positive enterotoxin
Citrobacter freundii	Heat-stable enterotoxin; A/E lesions
Enterobacter cloacae	Heat-stable enterotoxin
Hafnia alvei	Produce A/E lesions
Klebsiella pneumoniae	Heat-stable enterotoxin
Plesiomonas shigelloides	Heat-stable enterotoxin

Table 22–3 The Most Recently Recognized Primary Foodborne
Pathogens

Pathogen/Syndrome	First Recognized
Infant botulism	1976
Yersinia enterocolitica	1976
Cyclospora cayetanensis	1977
Norwalk and related viruses	1978
Vibrio cholerae non-01	1979
Listeria monocytogenes	1981
Enterohemorrhagic *E. coli*	1982
New variant-Creutzfeldt-Jakob disease (vCJD)	1996

finding that nontyphoid salmonellae and Stx-producing *E. coli* strains exhibit high levels of mutability suggests that the emergence of new enteropathogenic variants may be expected among these groups.

The first requirement that an intestinal invasive pathogen must meet is that of intestinal adhesion. Recent findings have confirmed the importance of mobile genetic elements in the transfer of this property between avirulent and virulent strains. The degree to which avirulent strains of pathogenic species or phylogenetically related species can acquire, maintain, and express adherence/adhesive genes may be a crucial factor in the possible emergence of new enteropathogens.

In Table 22–2 are listed eight Gram-negative bacteria that possess at least one property or factor that is often associated with foodborne pathogens. It may be assumed that they are not primary foodborne pathogens due to a lack of other virulence properties such as the capacity to adhere to and enter epithelial cells. *Aeromonas hydrophila* and *Plesiomonas shigelloides* have been on the "watch list" of food microbiologists for at least two decades but neither has been demonstrated to cause foodborne gastroenteritis in the absence of another enteropathogen.

The slowness of the process of a nonpathogen becoming a pathogen may be inferred from Table 22–3, which lists the last eight recognized foodborne disease pathogens. Most of these may be presumed to have existed long before they were demonstrated to be a foodborne pathogen. The clear exceptions are the enterohemorrhagic colitis strains of *E. coli*. These strains were first recorded in 1975, and as is noted above, molecular genetic studies indicate that they evolved from *E. coli* O55:H7 apparently by the bacteriophage transfer of virulence genes. nvCJD may be the very newest foodborne disease. While the time frame for the emergence of foodborne pathogens may be slow and indefinite, once demonstrated they seem to persist forever. No foodborne pathogen ever recognized has been eliminated.

REFERENCES

1. Arizcun, C., C. Vasseur, and J.C. Labadie. 1998. Effect of several decontamination procedures on *Listeria monocytogenes* growing in biofilms. *J. Food Protect.* 61:731–734.

2. Alfano, J.R., and A. Collmer. 1997. The type III (Hrp) secretion pathway of plant pathogenic bacteria: Trafficking harpins, Avr proteins, and death. *J. Bacteriol.* 179:5655–5662.

3. Allison, M.J., H.P. Dalton, M.R. Escobar, and C.J. Martin. 1969. *Salmonella choleraesuis* infections in man: A report of 19 cases and a critical literature review. *South. Med. J.* 62:593–596.

4. Anderson, D.M., and O. Schneewind. 1997. A mRNA signal for the type III secretion of Yop proteins by *Yersinia enterocolitica*. *Science* 278:1140–1143.

5. Archer, D.L. 1996. Preservation microbiology and safety: Evidence that stress enhances virulence and triggers adaptive mutations. *Trends Food Sci. Technol.* 7:91–95.

6. Bagamboula, C.F., M. Uyttendaele, and J. Debevere. 2002. Acid tolerance of *Shigella sonnei* and *Shigella flexneri. J. Appl. Bacteriol.* 93:479–486.

7. Bagge, D., M. Hjelm, C. Johansen, I. Huber, and L. Gram. 2001. *Shewanella putrefaciens* adhesion and biofilm formation on food processing surfaces. *Appl. Environ. Microbiol.* 67:2319–2325.

8. Baumler, A.J., R.M. Tsolis, T.A. Ficht, and L.G. Adams. 1998. Evolution of host adaptation in *Salmonella enterica. Infect Immun.* 66:4579–4587.

9. Berry, E.D., and C.N. Cutter. 2000. Effects of acid adaptation of *Escherichia coli* 0157:H7 on efficacy of acetic acid spray washes to decontaminate beef carcass tissue. *Appl. Environ. Microbiol.* 66:1493–1498.

10. Bieber, D., S.W. Ramer, and C.-Y. Wu. 1998. IV pili, transient bacterial aggregates, and virulence of enteropathogenic *Escherichia coli. Science* 280:2114–2118.

11. Blackman, I.C., and J.F. Frank. 1996. Growth of *Listeria monocytogenes* as a biofilm on various food-processing surfaces. *J. Food Protect.* 59:827–831.

12. Boerlin, P., S. Chen, and J.K. Colbourne. 1998. Evolution of enterohemorrhagic *Escherichia coli* hemolysin plasmids and the locus for enterocyte effacement in Shiga toxin-producing *E. coli. Infect. Immun.* 66:2553–2561.

13. Boland, A., and G.R. Cornelis. 1998. Role of YopP in suppression of tumor necrosis factor alpha release by macrophages during *Yersinia* infection. *Infect. Immun.* 66:1878–1884.

14. Borucki, M.K., J.D. Peppin, D. White, F. Loge, and D.R. Call. 2003. Variation in biofilm formation among strains of *Listeria monocytogenes. Appl. Environ. Microbiol.* 69:7336–7342.

15. Bremer, P.J., I. Mond, and C.M. Osborne. 2001. Survival of *Listeria monocytogenes* attached to stainless steel surfaces in the presence or absence of *Flavobacterium* spp. *J. Food Protect.* 64:1369–1376.

16. Buchanan, R.L., S.G. Edelson, and G. Boyd. 1999. Effects of pH and acid resistance on the radiation resistance of enterohemorrhagic *Escherichia coli. J. Food Protect.* 62:219–228.

17. Buswell, C.M., Y.M. Herlihy, L.M. Lawrence, J.T. McGuiggan, P.D. Marsh, C.W. Keevil, and S.A. Leach 1998. Extended survival and persistence of *Campylobacter* spp. in water and aquatic biofilms and their detection by immunofluorescent-antibody and -rRNA staining. *Appl. Environ. Microbiol.* 64:733–741.

18. Carpenter, B., and O. Cerf. 1993. Biofilms and their consequences, with particular reference to hygiene in the food industry. *J. Appl. Bacteriol.* 75:499–511.

19. Centers for Disease Control and Prevention. 2003. Preliminary FoodNet data on the incidence of foodborne illnesses—Selected sites, United States, 2002. *Morb. Mortal. Wkly. Rep.* 52:340–343.

20. Cheetham, B.F., and M.E. Katz. 1995. A role for bacteriophages in the evolution and transfer of bacterial virulence determinants. *Mol. Microbiol.* 18:201–208.

21. Christensen, H., S. Nordentoft, and J.E. Olsen. 1998. Phylogenetic relationships of *Salmonella* based on rRNA sequences. *Int. J. Syst. Bacteriol.* 48:605–610.

22. Chumkhunthod, P., H. Schraft, and M.W. Griffiths. 1998. Rapid monitoring method to assess efficacy of sanitizers against *Pseudomonas putida* biofilms. *J. Food Protect.* 61:1043–1046.

23. Cloak, O.M., B.T. Solow, C.E. Briggs, C.-Y. Chen, and P.M. Fratamico. 2002. Quorum sensing and production of autoinducer-2 in *Campylobacter* spp., *Escherichia coli* 0157:H7, and *Salmonella enterica* serovar Typhimurium in foods. *Appl. Environ. Microbiol.* 68:4666–4671.

24. Coconnier, M.-H, E. Dlissi, and N. Robard. 1998. *Listeria monocytogenes* stimulates mucus exocytosis in cultured human polarized mucosecreting intestinal cells through action of listeriolysin O. *Infect. Immun.* 66:3673–3681.

25. Cornelis, G.R., and H. Wolf-Watz. 1997. The *Yersinia* Yop virulon: A bacterial system for subverting eukaryotic cells. *Mol. Microbiol.* 23:861–867.

26. Costerton, J.W. 1994. Biofilms, the customized microniche. *J. Bacteriol.* 176:2137–2142.

27. Cotter, P.D., and C. Hill. 2003. Surviving the acid test: Responses of Gram-positive bacteria to low pH. *Microbiol. Mol. Biol. Rev.* 67:429–453.

28. D'Aoust, J.Y. 1997. *Salmonella* species. In *Food Microbiology—Fundamentals and Frontiers*, ed. M.P. Doyle, L.R. Beuchat, and T.J. Montville, 129–158. Washington, DC: ASM Press.

29. Dack, G.M., W.E. Cary, O. Woolpert, and H. Wiggers. 1930. An outbreak of food poisoning proved to be due to a yellow hemolytic staphylococcus. *J. Prev. Med.* 4:167–175.

30. Davies, D.G., M.R. Parsek, J.P. Pearson, B.H. Iglewski, J.W. Costerton, and E.P. Greenberg. 1998. The involvement of cell-to-cell signals in the development of a bacterial biofilm. *Science* 280:295–298.

31. Davis, M.J., P.J. Coote, and C.P. O'Byrne. 1996. Acid tolerance in *Listeria monocytogenes*: The adaptive acid tolerance response (ATR) and growth-phase-dependent acid resistance. *Microbiology* 142:2975–2982.

32. Dean-Nystrom, E.A., B.T. Bosworth, H.W. Moon, and A.D. O'Brien. 1998. *Escherichia coli* O157:H7 requires intimin for enteropathogenicity in calves. *Infect. Immun.* 66:4560–4563.

33. Decatur, A.L., and D.A. Portnoy. 2000. A PEST-like sequence in listeriolysin O essential for *Listeria monocytogenes* pathogenicity. *Science* 290:992–995.

34. Degrassi, G., A. Anguilar, M. Bosco, S. Zahariev, S. Pongor, and V. Venturi. 2002. Plant growth-promoting *Pseudomonas putida* WCS358 produces and secretes four cyclic dipeptides: Cross-talk with quorum sensing bacterial sensors. *Curr. Microbiol.* 45:250–254.

35. De Kievit, T.R., and B.H. Iglewski. 2000. Bacterial quorum sensing in pathogenic relationships. *Infect. Immun.* 68:4839–4849.

36. Dong, Y.-H., A.R. Gusti, Q. Zhang, J.-L. Xu, and L.-H. Zhang. 2002. Identification of quorum-quenching *N*-acyl-homoserine lactonases from *Bacillus* species. *Appl. Environ. Microbiol.* 68:1754–1759.

37. Donnenberg, M.S., J.B. Kaper, and B.B. Finlay. 1997. Interactions between enteropathogenic *Escherichia coli* and host epithelial cells. *Trends Microbiol.* 5:109–114.

38. Dunny, G.M., and B.A.B. Leonard. 1997. Cell–cell communication in Gram-positive bacteria. *Ann. Rev. microbiol.* 51:527–564.

39. DuPont, H.L., M.M. Levine, and R.B. Hornick. 1989. Inoculum size in shigellosis and implications for expected mode of transmission. *J. Infect. Dis.* 159:1126–1128.

40. Elhanafi, D., B. Leenanon, W. Bang, and M.A. Drake. 2004. Impact of cold and cold-acid stress on poststress tolerance and virulence factor expression of *Escherichia coli* O157:H7. *J. Food Protect.* 67:19–26.

41. Falkow, S. 1996. The evolution of pathogenicity in *Escherichia, Shigella,* and *Salmonella.* In *Escherichia coli and Salmonella—Cellular and Molecular Biology,* 2nd ed., ed. F.C. Neidhardt, 2723–2729. Washington, DC: ASM Press.

42. Faruque, S.M., Asadulghani, A.R.M. Abdul Alim, M.J. Albert, K.M.N. Islam, and J.J. Mekalanos. 1998. Induction of the lysogenic phage encoding cholera toxin in naturally occurring strains of toxigenic *Vibrio cholerae* O1 and O139. *Infect. Immun.* 66:3752–3757

43. Feng, P.K., A. Lampel, and H. Karch. 1998. Genotypic and phenotypic changes in the emergence of *Escherichia coli* O157:H7. *J. Infect. Dis.* 177:1750–1753.

44. Ferreira, A., D. Sue, C.P. O'Byrne, and K.J. Boor. 2003. Role of *Listeria monocytogenes* δ^B in survival of lethal acidic conditions and in the acquired acid tolerance response. *Appl. Environ. Microbiol.* 69:2692–2698.

45. Frank, J.F., and R.A. Koffi. 1990. Surface-adherent growth of *Listeria monocytogenes* is associated with increased resistance to surfactant sanitizers and heat. *J. Food Protect.* 53:550–554.

46. Fratamico, P.M. 2003. Tolerance to stress and ability of acid-adapted and non-acid adapted *Salmonella enterica* serovar Typhimurium DT104 to invade and survive in mammalian cells in vitro. *J. Food Protect.* 66:1115–1125.

47. Fuqua, W.C., S.C. Winans, and E.P. Greenberg. 1994. Quorum sensing in bacteria: The Lux-R-LuxI family of cell density-responsive transcriptional regulators. *J. Bacteriol.* 176:269–275.

48. Galán, J.E. 1996. Molecular genetic bases of *Salmonella* entry into host cells. *Mol. Microbiol.* 20:263–271.

49. Gellin, B.G., and C.V. Broome. 1989. Listeriosis. *JAMA* 261:1313–1320.

50. Gorden, J., and P.L.C. Small. 1993. Acid resistance in enteric bacteria. *Infect. Immun.* 61:364–367.

51. Gram, L., A.B. Christensen, L. Ravn, S. Molin, and M. Givskov. 1999. Production of acylated homoserine lactones by psychrotrophic members of the *Enterobacteriaceae* isolated from foods. *Appl. Environ. Microbiol.* 65:3458–3463.

52. Groisman, E.A., and H. Ochman. 1997. How *Salmonella* became a pathogen. *Trends Microbiol.* 9:343–349.

53. Hacker, J., and J.B. Kaper. 2000. Pathogenicity islands and the evolution of microbes. *Ann. Rev. Microbiol.* 54:641–679.

54. Holden, M.T.G., S.R. Chhabra, R. de Nys, P. Stead, N.J. Bainton, P.J. Hill, M. Manefield, N. Kumar, M. Labatte, D. England, S. Rice, M. Givskov, G.P.C. Salmond, G.S.A.B. Stewart, B.W. Bycroft, S. Kjelleberg, and P. Williams. 1999. Quorum-sensing cross talk: Isolation and chemical characterization of cyclic dipeptides from *Pseudomonas aeruginosa* and other Gram-negative bacteria. *Mol. Microbiol.* 33:1254–1266.

55. Humphrey, T.J., N.P. Richardson, K.M. Statton, and R.J. Rowbury. 1993. Acid habituation in *Salmonella* Enteritidis PT4: Impact of inhibition of protein synthesis. *Lett. Appl. Microbiol.* 16:228–230.

56. Ikeda, J.S., J. Samelis, P.A. Kendall, G.C. Smith, and J.N. Sofos. 2003. Acid adaptation does not promote survival or growth of *Listeria monocytogenes* on fresh beef following acid and nonacid decontamination treatments. *J. Food Protect.* 66:985–992.

57. Jacewicz, M.S., D.W.K. Acheson, D.G. Binion, G.A. West, L.L.Lincicome, C. Fiocchi, and G.T. Keusch. 1999. Responses of human intestinal microvascular endothelial cells to Shiga toxins 1 and 2 and pathogenesis of hemorrhagic colitis. *Infect. Immun.* 67:1439–1444.

58. Jensen, V.B., J.T. Harty, and B.D. Jones. 1998. Interactions of the invasive pathogens. *Salmonella* Typhimurium,*Listeria monocytogenes*, and *Shigella flexneri* with M cells and murine Peyer's patches. *Infect. Immun.* 66:3758–3766.

59. Ji, G.Y., R.C. Beavis, and R.P. Novick. 1995. Cell density control of staphylococcal virulence mediated by an octapeptide pheromone. *Proc. Natl. Acad. Sci. USA* 92:12055–12059.

60. Jones, B.D., and S. Falkow. 1994. Identification and characterization of a *Salmonella* Typhimurium oxygen-regulated gene required for bacterial internalization. *Infect. Immun.* 62:3745–3752.

61. Karatzas, K.A.G., and M.H.J. Bennikk. 2002. Characterization of a *Listeria monocytogenes* Scott A isolate with high tolerance towards high hydrostatic pressure. *Appl. Environ. Microbiol.* 68:3183–3189.

62. Kleerebezem, M., L.E.N. Quadri, O.P. Kulpers, and W.M. de Vos. 1997. Quorum sensing by peptide pheromones and two-component signal-transduction systems in Gram-positive bacteria. *Mol. Microbiol.* 24:895–904.

63. Kim, K.Y., and J.F. Frank. 1995. Effect of nutrients on biofilm formation by *Listeria monocytogenes* on stainless steel. *J. Food Protect.* 58:24–28.

64. Koo, J., A. DePaola, and D.L. Marshall. 2000. Impact of acid on survival of *Vibrio vulnificus* and *Vibrio vulnificus* phage. *J. Food Protect.* 63:1049–1052.

65. Koutsoumanis, K.P., P.A. Kendall, and J.N. Sofos. 2003. Effect of food processing-related stresses on acid tolerance of *Listeria monocytogenes. Appl. Environ. Microbiol.* 69:7514–7516.

66. Kubori, T., Y. Matsushima, D. Nakamura, J. Uralil, M. Lara-Tejero, A. Sukhan, J.E. Galán, and S.-I. Aizawa. 1998. Supramolecular structure of the *Salmonella* Typhimurium type III protein secretion system. *Science* 280:602–605.

67. Lecuit, M., S. Vandormael-Pournin, J. Lefort, M. Huerre, P. Gounon, C. Dupuy, C. Babinet, and P. Cossart. 2001. A transgenic model for listeriosis: Role of internalin in crossing the intestinal barrier. *Science* 292:1722–1725.

68. LeClerc, J.E., B. Li, W.L. Payne, and T.A. Cebula. 1996. High mutation frequencies among *Escherichia coli* and *Salmonella* pathogens. *Science* 274:1208–1211.

69. Leuschner, R.G.K., and M.P. Boughtflower. 2001. Standardized laboratory-scale preparation of mayonnaise containing low levels of *Salmonella enterica* serovar Enteritidis. *J. Food Protect.* 64:623–629.

70. LeClerc, J.E., B. Li, W.L. Payne, and T.A. Cebula 1996. High mutation frequencies among *Escherichia coli* and *Salmonella* pathogens. *Science* 274:1208–1211.

71. Marsh, E.J., H. Luo, and H. Wang. 2003. Characteristics of biofilm development by *Listeria monocytogenes* strains. *FEMS Microbiol. Lett.* 228:203–210.

72. Mead, P.S., L. Slutsker, V. Dietz, l. F. McCaig, J.S. Bresee, C. Shapiro, P.M. Griffin, and R.V. Tauxe. 1999. Food-related illness and death in the United States. *Emerg. Infect. Dis.* 5:607–625.

73. McLean, R.J.C., M. Whiteley, D.J. Stickler, and W.C. Fuqua. 1997. Evidence of autoinducer activity in naturally occurring biofilms. *FEMS Microbiol. Lett.* 154:259–263.

74. Michiels, C.W., M. Schellekens, C.C.F. Soontjens, and K.J.A. Hauben. 1997. Molecular and metabolis typing of resident and transient fluorescent pseudomonad flora from a meat mincer. *J. Food Protect.* 60:1515–1519.

75. O'Brien, A.D., and R.K. Holmes. 1996. Protein toxins of *Escherichia coli* and *Salmonella*. In *Escherichia coli and Salmonella—Cellular and Molecular Biology*, 2nd ed., ed. F.C. Neidhardt, 2788–2802. Washington, DC: ASM Press.

76. O'Driscoll, B., C.G.M. Gahan, and C. Hill. 1996. Adaptive acid tolerance response in *Listeria monocytogenes*: Isolation of an acid-tolerant mutant which demonstrates increased virulence. *Appl. Environ. Microbiol.* 62:1693–1698.

77. Oh, D.-H, and D.L. Marshall. 1996. Monolaurin and acetic acid inactivation of *Listeria monocytogenes* attached to stainless steel. *J. Food Protect.* 59:249–252.

78. Otto, M., H. Echner, W. Voelter, and F. Götz. 2001. Pheromone cross-inhibition between *Staphylococcus aureus* and *Staphylococcus epidermidis. Infect. Immun.* 69:1957–1960.

79. O'Riordan, M., M.A. Moors, and D.A. Portnoy. 2003. *Listeria* intracellular growth and virulence require host-derived lipoic acid. *Science* 302:462–464.

80. Perna, N.T., G.F. Mayhew, G. Pósfai, S. Elliott, M.S. Donnenberg, J.B. Kaper, and F.R. Blattner. 1998. Molecular evolution of a pathogenicity island from enterohemorrhagic *Escherichia coli* O157:H7. *Infect. Immun.* 66:3810–3817.

81. Phan-Thanh, L., F. Mahouin, and S. Aligé. 2000. Acid responses of *Listeria monocytogenes*. *Int. J. Food Microbiol.* 55:121–126.

82. Pruzzo, C., R. Tarsi, M. del Mar Lleò, C. Signoretto, M. Zampini, R.R. Colwell, and P. Canepari. 2002. In vitro adhesion to human cells by viable but nonculturable *Enterococcus faecalis*. *Curr. Microbiol.* 45:105–110.

83. Ren, D., J.j. Sims, and T.K. Wood. 2002. Inhibition of biofilm formation and swarming of *Bacillus subtilis* by (5*Z*)-4-brome-5-(bromomethylene)-3-butyl-2-(5*H*)-furanone. *Lett. Appl. Microbiol.* 34:293–299.

84. Richter-Dahlfors, A.A., and B.B. Finlay. 1997. *Salmonella* interactions with host cells. In *Host Response to Intracellular Pathogens*, ed. S.H.E. Kaufmann, 251–270. Austin, TX: R.G. Landes Co.

85. Samelis, J., J.N. Sofos, J.S. Ikedak, P.A. Kendall, and G.C. Smith. 2002. Exposure to non-acid fresh meat decontamination washing fluids sensitizes *Escherichia coli* O157:H7 to organic acids. *Lett. Appl. Microbiol.* 34:7–12.

86. Samelis, J., J.N. Sofos, P.A. Kendall, and G.C. Smith. 2001. Influence of the natural microbial flora on the acid tolerance response of *Listeria monocytogenes* in a model system of fresh meat decontamination fluids. *Appl. Environ. Microbiol.* 67:2410–2420.

87. Saphra, I., and M. Wassermann. 1954. *Salmonella choleraesuis*: A clinical and epidemiological evaluation of 329 infections identified 1940 and 1954 in the New York *Salmonella* Center. *Am. J. Med. Sci.* 228:525–533.

88. Sasahara, K.C., and E.A. Zottola. 1993. Biofilm formation by *Listeria monocytogenes* utilizes a primary colonizing microorganism in flowing systems. *J. Food Protect.* 56:1022–1028.

89. Schauer, D.B., and S. Falkow. 1993. Attaching and effacing locus of a *Citrobacter freundii* biotype that causes transmissible murine colonic hyperplasia. *Infect. Immun.* 61:2486–2492.

90. Schubert, S., A. Rakin, H. Karch, E. Carriel, and J. Heesemann. 1998. Prevalence of the "high-pathogenicity island" of *Yersinia* species among *Escherichia coli* strains that are pathogenic to humans. *Infect. Immun.* 66:480–485.

91. Sibelius, U., E.-C. Schulz, F. Rose, K. Hattar, T. Jacobs, S. Weiss, T. Chakraborty, W. Seeger and F. Grimminger. 1999. Role of *Listeria monocytogenes* exotoxins listeriolysin and phosphatidylinositol-specific phospholipase C in activation of human neutrophils. *Infect. Immun.* 67:1125–1130.

92. Silhavy, T.J. 1997. Death by lethal injection. *Science* 278:1085–1086.

93. Sperandio, V., A.G. Torres, J.A. Girón, and J.B. Kaper. 2001. Quorum sensing is a global regulatory mechanism in enterohemorrhagic *Escherichia coli* O157:H7. *J. Bacteriol.* 183:5187–5197.

94. Surette, M.G., M.B. Miller, and B.L. Bassler. 1999. Quorum sensing in *Escherichia coli*, *Salmonella* Typhimurium, and *Vibrio harveyi*: A new family of genes responsible for autoinducer production. *Proc. Natl. Acad. Sci. USA* 96:1639–1644.

95. Trucksis, M., J. Michalski, Y.K. Deng, and J.B. Kaper. 1998. The *Vibrio cholerae* genome contains two unique circular chromosomes. *Proc. Natl. Acad. Sci.USA* 95:14459–14464.

96. van der Velden, A.W.M., A.J. Bäumler, R.M. Tsolis, and F. Heffron. 1998. Multiple fimbrial adhesins are required for full virulence of *Salmonella* Typhimurium in mice. *Infect. Immun.* 66:2803–2808.

97. Venturi, V. 2003. Control of *rpoS* transcription in *Escherichia coli* and *Pseudomonas*: Why so different? *Mol. Microbiol.* 49:1–9.

98. Waldor, M.K., and J.J. Mekalanos. 1996. Lysogenic conversion by a filamentous phage encoding cholera toxin. *Science* 272:1910–1914.

99. Wallis, T.S., and E.E. Galyov. 2000. Molecular basis of *Salmonella* induced enteritis. *Mol. Microbiol.* 36:997–1005.

100. Waterman, S.R., and P.L.C. Small. 1996. Characterization of the acid resistance phenotype and *rpoS* alleles of Shiga-like toxin-producing *Escherichia coli*. *Infect. Immun.* 64:2808–2811.

101. Waterman, S.R., and P.L.C. Small. 1998. Acid-sensitive enteric pathogens are protected from killing under extremely acidic conditions of pH 2.5 when they are inoculated onto certain solid food sources. *Appl. Environ. Microbiol.* 64:3882–3886.

102. Wemekamp-Kamphuis, H.H., J.A. Wouters, P.P.L.A. de Leeuw, T. Hain, T. Chakraborty, and T. Abee. 2004. Identification of sigma factor δ^B–controlled genes and their impact on acid stress, high hydrostatic pressure, and freeze survival in *Listeria monocytogenes* EGD-e. *Appl. Environ. Microbiol.* 70:3457–3466.

103. Whitechurch, C.B., T. Tolker-Nielsen, P.C. Ragas, and J.S. Mattick. 2002. Extracellular DNA required for bacterial biofilm formation. *Science* 295:1487.

104. Whittam, T.S. 1996. Genetic variation and evolutionary processes in natural populations of *Escherichia coli*. In *Escherichia coli and Salmonella—Cellular and Molecular Biology*, 2nd ed., ed. F.C. Neidhardt, 2708–2720. Washington, DC: ASM Press.

105. Wilmes-Riesenberg, M.R., B. Bearson, J.W. Foster, and R. Curtiss, III. 1996. Role of the acid tolerance response in virulence of *Salmonella* Typhimurium. *Infect. Immun.* 64:1085–1092.

106. Wong, H.-C., P.-Y. Peng, J.-M. Han, C.-Y. Chang, and S.-L. Lan. 1998. Effect of mild acid treatment on the survival, enteropathogenicity, and protein production in *Vibrio parahaemolyticus*. *Infect. Immun.* 66:3066–3071.

107. Yuk, H.-G., and D.L. Marshall. 2003. Heat adaptation alters *Escherichia coli* O157:H7 membrane lipid composition and verotoxin production. *Appl. Environ. Microbiol.* 69:5115–5119.

108. Zhang, V., and M.W. Griffiths. 2003. Induced expression of the heat shock protein genes *uspA* and *grpE* during starvation at low temperatures and their influence on thermal resistance of *Escherichia coli* O157:H7. *J. Food Protect.* 66:2045–2050.

109. Zottola, E.A. 1994. Microbial attachment and biofilm formation: A new problem for the food industry? *Food Technol.* 48(7):107–114.

110. Zychlinsky, A., M.C. Prevost, and P.J. Sansonetti. 1992. *Shigella flexneri* induces apoptosis in infected macrophages. *Nature* 358:167–169.

CHAPTER 23

Staphylococcal Gastroenteritis

The staphylococcal food-poisoning or food-intoxication syndrome was first studied in 1894 by J. Denys and later in 1914 by M.A. Barber, who produced in himself the signs and symptoms of the disease by consuming milk that had been contaminated with a culture of *Staphylococcus aureus*. The capacity of some strains of *S. aureus* to produce food poisoning was proved conclusively in 1930 by Dack et al.[28] who showed that the symptoms could be produced by feeding culture filtrates of *S. aureus*. Although some authors refer to food-associated illness of this type as food intoxication rather than food poisoning, the designation *gastroenteritis* obviates the need to indicate whether the illness is an intoxication or an infection.

Staphylococcal gastroenteritis is caused by the ingestion of food that contains one or more *enterotoxins*, which are produced only by some staphylococcal species and strains. Although enterotoxin production is believed generally to be associated with *S. aureus* strains that produce coagulase and thermonuclease (TNase), many species of *Staphylococcus* that produce neither coagulase nor TNase are known to produce enterotoxins.

An extensive literature exists on staphylococci and the food-poisoning syndrome, much of which goes beyond the scope of this chapter.

SPECIES OF CONCERN IN FOODS

The genus *Staphylococcus* includes over 30 species, and those of real and potential interest in foods are listed in Table 23–1. Of the 18 species and subspecies noted in the table, only 6 are coagulase positive, and they generally produce thermostable nuclease (TNase). Ten of the coagulase-negative species have been shown to produce enterotoxins, and they do not produce nuclease, or those that do, produce a thermolabile form. The coagulase-negative enterotoxigenic strains are not consistent in their production of hemolysins or their fermentation of mannitol. The long-standing practice of examining foods for coagulase-positive staphylococci as the strains of importance has undoubtedly led to underestimations of the prevalence of enterotoxin producers.

The relationship between TNase and coagulase production in staphylococci is discussed in Chapter 11. It is common to assume that TNase and coagulase-positive strains are the only staphylococci that warrant further investigations when found in foods, but the existence of both TNase and coagulase-negative enterotoxin-producing strains has been known for some time.

Among coagulase-positive species, *S. intermedius* is well known as an enterotoxin producer. This species is found in the nasal passages and on the skin of carnivores and horses, but rarely in humans.

Table 23–1   Staphylococcal Species and Subspecies Known to Produce Coagulase, Nuclease, and/or Enterotoxins

Organisms	Coagulase	Nuclease	Enterotoxin	Hemolysis	Mannitol	G + C of DNA
S. aureus subsp.						
anaerobius	+	TS	−	+	−	31.7
aureus	+	TS	+	+	+	32–36
S. intermedius	+	TS	+	+	(+)	32–36
S. hyicus	(+)	TS	+	−	−	33–34
S. delphini	+	−		+	+	39
S. schleiferi subsp.						
coagulans	+	TS		+	(+)	35–37
schleiferi	−	TS		+	−	37
S. caprae	−	TL	+	(+)	−	36.1
S. chromogens	−	−w	+	−	v	33–34
S. cohnii	−	−	+	−	v	36–38
S. epidermidis	−	−	+	v	−	30–37
S. haemolyticus	−	TL	+	+	v	34–36
S. lentus	−		+	−	+	30–36
S. saprophyticus	−	−	+	−	+	31–36
S. sciuri	−		+	−	+	30–36
S. simulans	−	V		v	+	34–38
S. warneri	−	TL	+	− w	+	34–35
S. xylosus	−	−	+	+	v	30–36

Note: + = positive; − = negative; −w = negative to weakly positive; (+) = weak reaction; v = variable; TS = thermostable; TL = thermolabile.

They are well known as pathogens in dogs. From pyrodermatitis in dogs in Brazil, 73 staphylococci were recovered, of which 52 were *S. intermedius*.[47] Of the 52, all were coagulase positive in rabbit plasma but negative in human plasma, and 13 (25%) were enterotoxigenic. Four produced staphylococcal enterotoxin D (SED), five produced SEE, and one each produced SEB, SEC, SED/E, and SEA/C. All 13 were TNase positive, and 3 produced the toxic shock syndrome toxin (TSST). A large number of *S. hyicus* strains are coagulase positive, and it appears that some produce enterotoxins. In one study, *S. hyicus* strains elicited positive enterotoxin responses in cynomologus monkeys, but the enterotoxin was not one of the known types—SEA through SEE.[3,49] In a study of goat isolates, two of six coagulase-positive *S. hyicus* produced SEC.[111] Enterotoxin production by *S. delphini*, *S. simulans*, and *S. schleiferi* subsp. *coagulans* has not been reported.

At least 10 of the coagulase-negative staphylococcal species listed in Table 23–1 produce enterotoxins. *S. cohnii*, *S. epidermidis*, *S. haemolyticus*, and *S. xylosus* were recovered from sheep milk along with *S. aureus*.[9] The one isolate of *S. cohnii* produced SEC; three isolates of *S. epidermidis* produced SEC and SEB/C/D (two strains); five isolates of *S. haemolyticus* produced SEA, SED, SEB/C/D, and SEC/D (two strains); whereas the four isolates of *S. xylosus* all produced SED.[9] These investigators noted that mannitol fermentation was best to distinguish between enterotoxin-positive and enterotoxin-negative strains. In another study, 1 of 20 coagulase-negative food isolates was found to be an enterotoxigenic strain of *S. haemolyticus* that produced both SEC and SED.[35] In a study of staphylococcal isolates from healthy goats, 74.3% of 70 coagulase positives produced enterotoxins and 22% of 272 coagulase negatives were enterotoxin positive.[111]). SEC was the most frequently found enterotoxin among the goat isolates. Seven species of the goat isolates produced more than one enterotoxin (*S. caprae, S. epidermidis, S. haemolyticus, S. saprophyticus, S. sciuri, S. warneri,* and *S. xylosus*) and two species produced only one—SEC by *S. chromogens* and SEE by *S. lentus.*[111] From cooked ready-to-eat crabmeat, the following species were identified among 100 staphylococcal suspect isolates: *S. lentus*—31; *S. hominis*—21; *S. epidermidis*—10; *S. kloosi*—8; *S. capitis*—5; and 3 each of *S. aureus,S. saprophyticus,* and *S. sciuri.*[32] Fewer than three of five other species were found. From Spanish dry-cured hams, an SEC-producing *S. epidermidis* was isolated.[69]

Among other foodborne staphylococcal species are *S. condimenti,S. piscifermentans,* and *S. fleurettii,* all of which are coagulase, TNase, and enterotoxin negative. *S. condimenti* was isolated from soy sauce mash,[87] *S. piscifermentans* from fermented fish in Thailand,[105] and *S. fleurettii* from goat's milk cheeses.[114] The former *S. caseolyticus* has been transferred to the genus *Macrococcus* as *M. caseolyticus*[64]

HABITAT AND DISTRIBUTION

The staphylococcal species are host-adapted with about one-half of the known species inhabiting humans solely (e.g., *S. cohnii* subsp. *cohnii*) or humans and other animals (e.g., *S. aureus*). The largest numbers tend to be found near openings to the body surface such as the anterior nares, axillae, and the inguinal and perineal areas where in moist habitats, numbers per square centimeter may reach 10^3–10^6, and in dry habitats, 10–10^3.[63] The two most important sources to foods are nasal carriers and individuals whose hands and arms are inflicted with boils and carbuncles, who are permitted to handle foods.

Most domesticated animals harbor *S. aureus.* Staphylococcal mastitis is not unknown among dairy herds, and if milk from infected cows is consumed or used for cheese making, the chances of contracting food intoxication are excellent. There is little doubt that many strains of this organism that cause bovine mastitis are of human origin. However, some are designated as "animal strains." In one study, staphylococcal strains isolated from parts of raw pork products were essentially all of the animal strain type. However, during the manufacture of pickled pork products, these animal strains were gradually replaced by human strains during the production process, to a point where none of the original animal strains could be detected in finished products.[95]

With regard to some of the non-*S. aureus* species, *S. cohnii* is found on the skin of humans and occasionally in urinary tract and wound infections. Human skin is the habitat of both *S. epidermidis* and *S. haemolyticus,* and the latter is associated with human infections. *S. hyicus* is found on the skin of pigs, where it sometimes causes lesions, and it has been found in milk and on poultry. The skin of lower primates and other mammals is the habitat of *S. xylosus,* and the skin of humans and other primates is the habitat of *S. simulans. S. schleiferi* subsp. *schleiferi* was found in clinical specimens from human patients with decreased resistance to infection,[37] and subsp. *coagulans* was isolated from ear infections in dogs. *S. aureus* subsp. *anaerobius* causes disease in sheep, and *S. delphini* was recovered from dolphins.[113] *S. sciuri* is found on the skin of rodents and *S. lentus* and *S. caprae* are associated with goats, especially goat milk.

Although many of the coagulase-negative species noted adapt primarily to nonhuman hosts, their entry into human foods is not precluded. Once in susceptible foods, their growth may be expected to lead to the production of enterotoxins. All of these species grow in the presence of 10% NaCl.

Since *S. aureus* has been studied most as a cause of staphylococcal foodborne gastroenteritis, most of the information that follows is about this species.

INCIDENCE IN FOODS

In general, staphylococci may be expected to exist, at least in low numbers, in any or all food products that are of animal origin or in those that are handled directly by humans, unless heat-processing steps are applied to effect their destruction. They have been found in a large number of commercial foods by many investigators (see Chapters 4, 5, and 9; Tables 4–3, 4–14, 5–6, and 9–1).

NUTRITIONAL REQUIREMENTS FOR GROWTH

Staphylococci are typical of other Gram-positive bacteria in having a requirement for certain organic compounds in their nutrition. Amino acids are required as nitrogen sources, and thiamine and nicotinic acid are required among the B vitamins. When grown anaerobically, they appear to require uracil. In one minimal medium for aerobic growth and enterotoxin production, monosodium glutamate serves as C, N, and energy sources. This medium contains only three amino acids (arginine, cystine, and phenylalanine) and four vitamins (pantothenate, biotin, niacin, and thiamine), in addition to inorganic salts[73] Arginine appears to be essential for enterotoxin B production.[116]

TEMPERATURE GROWTH RANGE

Although it is a mesophile, some strains of S. aureus can grow as low as 6.7°C.[5] The latter investigators found three food-poisoning strains that grew in custard at 114°F (45.6°C) but decreased at 116–120°F (46.7–48.9°C), with time of incubation. They grew in chicken à la king at 112°F (44.4°C) but failed to grow in ham salad at the same temperature. In general, growth occurs over the range 7–47.8°C, and enterotoxins are produced between 10°C and 46°C, with the optimum between 40°C and 45°C.[98] These minimum and maximum temperatures of growth and toxin production assume optimal conditions relative to the other parameters, and the ways in which they interact to raise minimum growth or lower maximum growth temperatures are noted below.

EFFECT OF SALTS AND OTHER CHEMICALS

Although S. aureus grows well in culture media without NaCl, it can grow well in 7–10% concentrations, and some strains can grow in 20%. The maximum concentrations that permit growth depend on other parameters such as temperature, pH, water activity (a_w), and oxidation–reduction potential (Eh) (see below).

S. aureus has a high degree of tolerance to compounds such as tellurite, mercuric chloride, neomycin, polymyxin, and sodium azide, all of which have been used as selective agents in culture media. S. aureus can be differentiated from other staphylococcal species by its greater resistance to acriflavine. In the case of borate, S. aureus is sensitive, whereas S. epidermidis is resistant.[59] With novobiocin, S. saprophyticus is resistant, whereas S. aureus and S. epidermidis are not. The capacity to tolerate high levels of NaCl and certain other compounds is shared by Micrococcus and Kocuria, which are widely distributed in nature and occur in foods generally in greater numbers than staphylococci, thus making the recovery of the latter more difficult. The effect of other chemicals on S. aureus is presented in Chapter 13.

EFFECT OF pH, WATER ACTIVITY, AND OTHER PARAMETERS

Regarding pH, *S. aureus* can grow over the range 4.0–9.8, but its optimum is in the range 6–7. As is the case with the other growth parameters, the precise minimum growth pH is dependent on the degree to which all other parameters are at optimal levels. In homemade mayonnaise, enterotoxins were produced when the initial pH was as low as 5.15 and when the final growth pH was not below 4.7.[44] SEB was produced at a level of 158 ng/100 g with an inoculum of approximately 10^5/g. In general, SEA production is less sensitive to pH than SEB. The buffering of a culture medium at pH 7.0 leads to more SEB than when the medium is unbuffered or buffered in the acid range.[71] A similar result was noted at a controlled pH of 6.5 rather than 7.0.[58]

With respect to a_w, the staphylococci are unique in being able to grow at values lower than any other nonhalophilic bacteria. Growth has been demonstrated as low as 0.83 under otherwise ideal conditions, although 0.86 is the generally recognized minimum a_w.

NaCl and pH

Using a protein hydrolysate medium incubated at 37°C for 8 days, growth and enterotoxin C production occurred over the pH range 4.00–9.83 with no NaCl. With 4% NaCl, the pH range was restricted to 4.4–9.43 (Table 23–2). Toxin was produced at 10% NaCl with a pH of 5.45 or higher, but none was produced at 12% NaCl.[39]

It has been shown that *S. aureus* growth is inhibited in broth at a pH of 4.8 and 5% NaCl. Growth and enterotoxin B production by strain S-6 occurred in 10% NaCl at pH 6.9 but not with 4% at pH 5.1.[41] The general effect of increasing NaCl concentration is to raise the minimum pH of growth. At a pH of 7.0 and 37°C, enterotoxin B was inhibited by 6% or more NaCl (see Figure 23–1).

pH, a_w, and Temperature

No growth of a mixture of *S. aureus* strains occurred in brain heart infusion (BHI) broth containing NaCl and sucrose as humectants either at pH 4.3, a_w of 0.85, or 8°C. No growth occurred with a combination of pH <5.5, 12°C, and a_w of 0.90 or 0.93; and no growth occurred at pH <4.9, 12°C and a_w of 0.96.[79]

Table 23–2 The Effect of pH and NaCl on the Production of Enterotoxin C by an Inoculum of 10^8 Cells/ml of *S. aureus* 137 in a Protein Hydrolysate Medium Incubated at 37°C for 8 Days

	pH Range				
	4.00–9.83	4.4–9.43	4.50–8.55	5.45–7.30	4.50–8.50
NaCl content (%)	0	4	8	10*	12
Enterotoxin production	+	+	+	+	−

*Enterotoxin was detected also with an inoculum of 3.6×10^6 at pH 6.38–7.30.

Source: From Genigeorgis et al.,[39] copyright © 1971 by American Society for Microbiology.

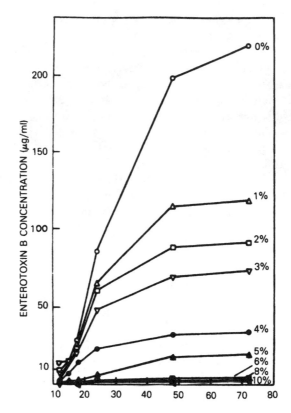

Figure 23–1   Staphylococcal enterotoxin B production in different NaCl concentrations in 4% NZ-Amine NAK medium at pH 7.0 and 37°C. *Source*: From Pereira et al.,[86] copyright © 1982 by International Association for Food Protection.

NaNO₂, Eh, pH, and Temperature of Growth

S. aureus strain S-6 grew and produced enterotoxin B in cured ham under anaerobic conditions with a brine content up to 9.2% but not below a pH of 5.30 and 30°C, or below a pH of 5.58 at 10°C. Under aerobic conditions, enterotoxin production occurred sooner than under anaerobic conditions. As the concentration of HNO₂ increased, enterotoxin production decreased.[40]

STAPHYLOCOCCAL ENTEROTOXINS: TYPES AND INCIDENCE

Thirteen staphylococcal enterotoxins (SEs) were identified through 2001, and they are listed in Table 23–4 along with some of their reported biological and chemical properties. The gene for SEG was reported first in 1992, but the toxin was not reported until 1998 along with SEI.[76] SEG was reported in 1995,[104] SEH in 1995,[104] and SEI in 1998.[76] SEJ was reported in 1998,[117] and SEK in 2001.[81] The latter authors reported evidence for the existence of SEL but specific data were not presented. Very little if any information exists on the last six SEs relative to their incidence/prevalence in foods. In

Table 23–3 Incidence of Staphylococcal Enterotoxins Alone and in Combination

| Source | No. of Cultures | Percentage Enterotoxic | Enterotoxins | | | | | Reference |
			A	B	C	D	E	
Human specimens	582	–	54.5	28.1	8.4	41.0	–	21
Raw milk	236	10	1.8	0.8	1.2	6.8	–	21
Frozen foods	260	30	3.4	3.0	7.4	10.4	–	21
Food-poisoning outbreaks	80	96.2	77.8	10.0	7.4	37.5	–	21
Foods	200	62.5	47.5	3.5	12.0	18.5	6.5	85
Poultry	139	25.2	1.4	0	0.7	23.7	0	46
Humans	293	39	7.8	17.7	7.2	6.8	0.7	83
Poultry	55	62	60.0	1.8	3.6	0	0	42
Spanish dry-cured hams	135	85.9	54.3	2.6	10.3	–	–	69
Various in Belgium and Zaire	285	16.2	6.5	4.5	2.7	0.5	–	56
Raw milk in Trinidad	230	40.4*	7.5	9.7	34.4	8.6	–	1

*Includes combinations.

regard to SEK, 14 of 36 clinical isolates of *S. aureus* were positive.[81] Reviews on the SEs have been presented by Dinges et al.[31] and Balaban and Rasooly.[8]

SEC$_3$ is chemically and serologically related to but not identical to SEC$_1$ and SEC$_2$.[90] Antibodies to each of the SECs cross-react with each other, although they differ slightly from each other antigenically. SEC$_3$ shares 98% nucleotide sequence with SEC$_1$,[25] and SEC$_1$ and SEB share 68% amino acid homology. There is cross-reaction between SEA and SEE, and some antibodies against SEB cross-react with the SECs.[15] What was believed to be SEF in the early 1980s turned out to be TSST. Some enterotoxin-producing strains also produce TSST, and some of the symptoms of Toxic Shock Syndrome appear to be caused by SEA, B, and C$_1$.[16] The genes for SEA, B, C$_1$, and E are reported by some authors to be chromosomal, and SED as being plasmid-borne.[54] More recently, however, SEB and SEK have been reported to be borne on a *S. aureus* pathogenicity island (see reference 81).

The relative incidence of five enterotoxins is presented in Table 23–3. In general, SEA is recovered from food-poisoning outbreaks more often than any of the others, with SED being second most frequent. The fewest number of outbreaks are associated with SEE. The incidence of SEA among 3,109 and SED among 1,055 strains from different sources, and by a large number of investigators, was 23% and 14%, respectively.[101] For SEB, SEC, and SEE, 11%, 10%, and 3%, respectively, were found among 3,367, 1,581, and 1,072 strains.

The relative incidence of specific enterotoxins among strains recovered from various sources varies widely. Whereas from human specimens in the United States, over 50% of isolates secrete SEA alone or in combination,[22] from human isolates in Sri Lanka SEA producers constituted only 7.8%.[83] Unlike other reports, the latter study found more SEB producers than any other types. Wide variations are found among *S. aureus* strains isolated from foods. Whereas in one study, Harvey et al.[46] found SED to be associated more with poultry isolates than human strains, in another study the investigators found no SED producers among 55 poultry isolates.[42] In yet another study, two of three atypical *S.*

aureus isolates that produced a slow, weak positive or negative coagulase reaction, and were negative for the anaerobic fermentation of mannitol, produced SED.[33] The isolates were from poultry. From Nigerian ready-to-eat foods, about 39% of 248 isolates were enterotoxigenic, with 44% of these producing SED.[2] Among 449 coagulase-positive *S. aureus* isolates from a variety of Nigerian foods, 57%, 15%, 6%, and 5% were SEA, SEB, SED, and SEC, respectively.[99] From sheep milk, SEA and SED constituted 35% each of 124 strains, including four coagulase-negative strains.[9] Of 48 isolates from dairy and 134 from meat products, 46% and 49%, respectively, were enterotoxigenic,[84] and of 80 strains from food-poisoning outbreaks, 96% produced SEA.[22] SEC was produced by 67.9% of 342 isolates of both coagulase-positive and coagulase-negative species from healthy goats.[111] SEA, SEB, and SEC were detected in the milk of 17 of 133 healthy goats.[111]

In a study of *S. aureus* strains that produce SEH, 10 of 21 that induced emesis in monkeys but which were negative for SEA through F produced SEH at levels from 13 to 230 ng/ml.[103] When another set of 20 strains that were known to produce at least one SE was examined for SEH, one SEC strain produced 142 ng/ml of SEH, and two SED strains produced 52 and 164 ng/ml, respectively.[103] An ELISA method was developed for SEH and its minimum detection level was about 2.5 ng/ml.

Regarding the percentage of strains that are enterotoxigenic, widely different percentages have been found depending on the source of isolates. Only 10% of 236 raw milk isolates were enterotoxigenic[22] whereas 62.5% of 200 food isolates were positive.[85] In one study, 33% of 36 food isolates were enterotoxigenic.[97]

An outbreak of 50 cases of staphylococcal food poisoning occurred in Brazil, and the vehicle food was Minas cheese.[20] The etiologic agent was a thermostable-nuclease (TNase) positive strain of *S. aureus*; and SEA, SEB, and SEC were found. Another outbreak with 328 victims was traced to raw milk, and the etiologic agent was a TNase-negative staphylococcal species other than *S. aureus*.[20] The milk contained SEC and SED, and $>2 \times 10^8$ cfu/ml of the etiologic agent. Bovine mastitis and food handlers appeared to be the sources of the causative organisms.

In a study of SEs from mastitic cows in Italy in 1999, the 2,343 composited quarter samples yielded 160 isolates that contained *S. aureus*.[23] Of the 160 isolates, 22 produced SEs with 7.5% being SED; 4.4% SEC; and 1.9% SEC and SEA. From 504 cafeteria foods examined in Spain, 19 (3.8%) yielded SE-positive isolates with 10 of the SEs being SEC, 4 SED, 3 SEB, and 2 SEA.[100]

Attempts to associate enterotoxigenicity with other biochemical properties of staphylococci such as gelatinase, phosphatase, lysozyme, lecithinase, lipase, and DNase production or the fermentation of various carbohydrates have been unsuccessful. Enterotoxigenic strains appear to be about the same as other strains in these respects. Attempts to relate enterotoxigenesis with specific bacteriophage types have been unsuccessful also. Most enterotoxigenic strains belong to phage group III, but all phage groups are known to contain toxigenic strains. Of 54 strains from clinical specimens that produced SEA, 5.5%, 1.9%, and 27.8% belonged, respectively, to phage groups I, II, and III, with 20.4% being untypeable.[21] Among poultry isolates, 49% were phage nontypeable.[21] In a study of 452 strains from meat plant workers, veterinary students, and meat plants, along with meat isolates and isolates from meat animals, 29.6% were nontypeable and 22.5% of the typeables belonged to group III.[57] Of 230 raw milk isolates in Trinidad, 50.2% were typeable with 23.6% and 9.8% belonging to phage groups I and III, respectively.[1]

Chemical and Physical Properties

Some of the properties of SEs that have been studied are summarized in Table 23–4. All are simple proteins which, upon hydrolysis, yield 18 amino acids, with aspartic, glutamic, lysine, and tyrosine

Table 23–4 Some Information on the Known Staphylococcal Enterotoxins (SEs)

SEs	Emetic Dose	Molecular Weight (Da)	Iso. Pt	Year Identified	Toxin Gene Locus*
A	5	27,100	6.8	1960	Chromo.
B	5	28,366	8.6	1959	SaPI
C_1	5	34,100	8.6	1967	–
C_2	5–10	34,000	7.0	1984	–
C_3	<10[a]	26,900	8.15	1984	–
D	20	27,300	7.4	1979	Plasmid
E	10–20	29,600	7.0	1971	Chromo.
G	–	27,043	–	1992	–
H	<30	27,300	5.7	1995	–
I	Weak	34,928	–	1998	–
J	–	–	–	1998	–
K	–	26,000	7.0–7.5	2001	SaPI
L	–	–	–	2001	–

*Chromo. = chromosomal; SaPI = *S. aureus* pathogenicity island.
[a]Per os; 0.05 μg/kg by the intravenous (IV) route.[90]

being the most abundant. The amino acid sequence of SEB was determined first.[50] Its N-terminal is glutamic acid, and lysine is the C-terminal amino acid. SEA, SEB, and SEE are composed of 239–296 amino acid residues. SEC$_3$ contains 236 amino acid residues, and the N-terminal is serine, whereas the N-terminal of SEC$_1$ is glutamic acid.[90] In their activate states, the enterotoxins are resistant to proteolytic enzymes such as trypsin, chymotrypsin, rennin, and papain, but sensitive to pepsin at a pH of about 2.[13] TSST-1 is more susceptible to pepsin than the SEs. Although the various enterotoxins differ in certain physiochemical properties, each has about the same potency. Although biological activity and serological reactivity are generally associated, it has been shown that serologically negative enterotoxin may be biologically active (see below). Based upon amino acids, SEA, SED, SEE, and SEI fall into one group while SEB, the SECs, and SEG fall into another.[76] The SECs are separated on the basis of minor epitopes.

The enterotoxins are quite heat resistant. The biological activity of SEB was retained after heating for 16 hours at 60°C and pH 7.3.[93] Heating of one preparation of SEC for 30 minutes at 60°C resulted in no change in serological reactions.[17] The heating of SEA at 80°C for 3 minutes or at 100°C for 1 minute caused it to lose its capacity to react serologically.[13] In phosphate-buffered saline, SEC has been found to be more heat resistant than SEA or SEB. The relative thermal resistance of these three enterotoxins was SEC > SEB > SEA.[109] The thermal inactivation of SEA based on cat emetic response was shown by Denny et al.[30] to be 11 minutes at 250°F (F_{250}^{48} = 11 minutes). When monkeys were used, thermal inactivation was F_{250}^{46} = 8 minutes. These enterotoxin preparations consisted of a 13.5-fold concentration of casamino acid culture filtrate employing strains 196-E. Using double gel-diffusion assay, Read and Bradshaw[89] found the heat inactivation of 99+% pure SEB in veronal buffer to be F_{250}^{58} = 16.4 minutes. The end point for enterotoxin inactivation by gel diffusion was identical to that by intravenous injection of cats. The slope of the thermal inactivation curve for SEA in beef bouillon at a pH of 6.2 was found to be around 27.8°C (50°F) using three different toxin concentrations (5, 17, and 60 μg/ml).[29] Some D values for the thermal destruction of SEB are presented in

Table 23–5 D Values for the Heat Destruction of Staphylococcal Enterotoxin B and Staphylococcal Heat-Stable Nuclease (Taken from the Literature)

Conditions	$D(C)$
Veronal buffer	$D_{110} = 29.7^*$
Veronal buffer	$D_{110} = 23.5^\dagger$
Veronal buffer	$D_{121} = 11.4^*$
Veronal buffer	$D_{121} = 9.9^\dagger$
Veronal buffer, pH 7.4	$D_{110} = 18$
Beef broth, pH 7.4	$D_{110} = 60$
Staph, nuclease	$D_{130} = 16.5$

*Crude toxin.
$\dagger$99+% purified.

Table 23–5. Crude toxin preparations have been found to be more resistant than purified toxins.[88] It may be noted from Table 23–5 that staphylococcal thermonuclease displays heat resistance similar to that of SEB (see Chapter 11 for more information on this enzyme). In one study, SEB was found to be more heat sensitive at 80°C than at 100°C or 110°C.[92] The thermal destruction was more pronounced at 80°C than at either 60°C or 100°C when heating was carried out in the presence of meat proteins. SEA and SED in canned infant formula were immunologically non-reactive after thermal processing but were biologically active when injected in kittens.[11]

S. aureus cells are considerably more sensitive to heat than the enterotoxins, as may be noted from D values presented in Table 23–6 from various heating menstra. The cells are quite sensitive in Ringer's solution at pH 7.2 ($D_{140°F} = 0.11$) and much more resistant in milk at pH 6.9 ($D_{140°F} = 10.0$). In frankfurters, heating to 71.1°C was found to be destructive to several strains of *S. aureus*,[84] and microwave heating for 2 minutes was destructive to over 2 million cells/g.[115]

The maximum growth temperature and heat resistance of *S. aureus* strain MF 31 were shown to be affected when the cells were grown in heart infusion broth containing soy sauce and monosodium glutamate (MSG). Without these ingredients in the broth, maximum growth temperature was 44°C, but with them, the maximum was above 46°C.[53] The most interesting effect of MSG was on $D_{60°C}$ values determined in Tris buffer at pH 7.2. With cells grown at 37°C, the mean $D_{60°C}$ value in buffer was 2.0 minutes, but when 5% MSG and 5% NaCl were added to the buffer, $D_{60°C}$ was 15.5 minutes. Employing cells grown at 46°C, the respective $D_{60°C}$ values were 7.75 and 53.0 minutes in buffer and buffer–MSG–NaCl. It is well known that heat resistance increases with increasing growth temperature, but changes of this magnitude in vegetative cells are unusual.

Production

In general, enterotoxin production tends to be favored by the optimum growth conditions of pH, temperature, Eh, and so on. It is well established that staphylococci can grow under conditions that do not favor enterotoxin production.

With respect to a_w, enterotoxin production (except for SEA) occurs over a slightly narrower range than growth. In precooked bacon incubated aerobically at 37°C, *S. aureus* A100 grew rapidly at an a_w

Table 23–6 *D* and *z* Values for the Thermal Destruction of *S. aureus* 196E in Various Heating Menstra at 140°F

Products	D(F)	Z	Reference
Chicken à la king	5.37	10.5	6
Custard	7.82	10.5	6
Green pea soup	6.7–6.9	8.1	107
Skim milk	3.1–3.4	9.2	107
0.5% NaCl	2.2–2.5	10.3	107
Beef bouillon	2.2–2.6	10.5	107
Skim milk alone	5.34	–	61
Raw skim milk + 10% sugar	4.11	–	61
Raw skim milk + 25% sugar	6.71	–	61
Raw skim milk + 45% sugar	15.08	–	61
Raw skim milk + 6% fat	4.27	–	61
Raw skim milk + 10% fat	4.20	–	61
Tris buffer, pH 7.2	2.0	–	53
Tris buffer, pH 7.2, 5.8% NaCl or 5% MSG	7.0	–	53
Tris buffer, pH 7.2 + 5.8% NaCl + 5% MSG	15.5	–	53

as low as 0.84 and produced SEA.[66] The production of the individual enterotoxins is more inherent to the toxin than to the strain that produces them.[96] SEA but not SEB has been shown to be produced by L-phase cells.[26] SEB, C, and D require a functional *agr* gene for maximal production. In pork, SEA production occurred at a_w of 0.86 but not at 0.83, and in beef at 0.88 but not at 0.86.[106] SEA can be produced under conditions of a_w that do not favor SEB.[110] SED has been produced at an a_w of 0.86 in 6 days at 37°C in BHI.[36] In general, SEB production is sensitive to a_w, whereas SEC is sensitive to both a_w and temperature. Regarding NaCl and pH, enterotoxin production has been recorded at pH 4.0 in the absence of NaCl (see Table 23–2). The effect of NaCl on SEB synthesis by strain S-6 at pH 7.0 at 37°C is presented in Figure 23–1.

With respect to growth temperature, SEB production in ham at 10°C has been recorded,[40] as well as small amounts of SEA, SEB, SEC, and SED in cooked ground beef, ham, and bologna at 10°C. Production has been observed at 46°C, but the optimum temperature for SEB and SEC was 40°C in a protein hydrolysate medium[112] and for SEE, 40°C at pH 6.0.[108] The growth of *S. aureus* on cooked beef at 45.5°C for 24 hours has been demonstrated, but at 46.6°C, the initial inoculum decreased by 2 log cycles over the same period.[18] The optimum for SEB in a culture medium at pH 7.0 was 39.4°C.[86] Thus, the optimum temperature for enterotoxin production is in the 40–45°C range.

Staphylococcal enterotoxins have been reported to appear in cultures as early as 4–6 hours (Figure 23–2) and to increase proportionately through the stationary phase[67] and into the transitional phase (Figure 23–3). Enterotoxin production has been shown to occur during all phases of growth,[27] although earlier studies revealed that with strain S-6, 95% of SEB was released during the latter part of the log phase of growth. Chloramphenicol inhibited the appearance of enterotoxin, suggesting that the presence of toxin was dependent on de novo protein synthesis.[95] In ice cream pies, 3.9 ng/g of SEA was produced in 18 hours at 25°C and 4.8 ng/g in 14 hours at 30°C.[48] In the same study, TNase was detectable before SEA, with 72 ng/g being found after 12 hours at 37°C. With a 3% pancreatic digest of casein as substrate and incubation at 37°C, SEC_1 and SEC_2 were produced during the

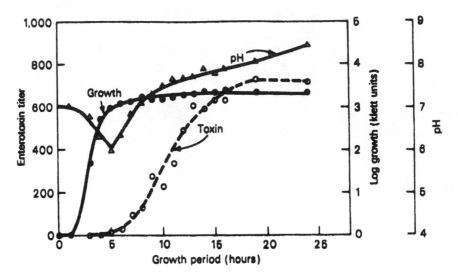

Figure 23–2   Enterotoxin B production, growth, and pH changes in *Staphylococcus aureus* at 37°C. *Source*: From McLean et al.[70] copyright © 1968 by American Society for Microbiology.

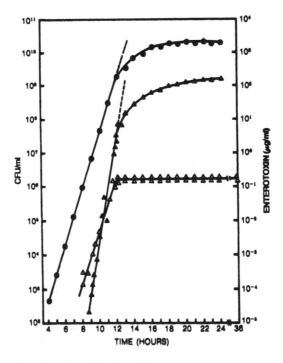

Figure 23–3   Rates of growth and enterotoxins A and B synthesis by *Staphylococcus aureus* S-6. Symbols:●, CFU/ml; △, enterotoxin A; ▲, enterotoxin B. *Source*: From Czop and Bergdoll,[27] copyright © 1974 by American Society for Microbiology.

exponential growth phase and at the beginning of the stationary phase.[82] SEC_1 was detected after 10 hours (2 ng/ml) at an *S. aureus* population of 8.3×10^7 cfu/ml, while TNase was detected after 5 hours with a cell count of 1.3×10^4 cfu/ml. SEC_2 and TNase first appeared after 7 hours with a cell population of 10^7 cfu/ml.[82] With both enterotoxins, TNase production ceased before enterotoxin production.

In regard to quantities of enterotoxins produced, levels of 375 and 60 μg/ml or more of SEB and SEC, respectively, have been recorded.[91] In a protein hydrolysate medium, up to 500 μg/ml\l [Author: Please check the unit "μg/ml/l" here]of SEB may be produced.[14] Employing a sac culture assay method, 289 ng/ml of SEA were produced by *S. haemolyticus*, 213 ng/ml of SEC by *S. aureus*, and 779 ng/ml of SED, also by *S. aureus*.[9] Chitin has been shown to enhance SEA production. With 0.5% crude chitin in BIH broth, SEA production increased by around 52%.[4] SEA thermostability was also increased, but cell growth was apparently not affected. The production of SEH was favored by aeration and controlled pH with about 275 ng/ml having been produced in a fermentor at pH 7.0 with aeration at 300 ml/minute.[102] Around 5 μg/ml of SEG and SEI have been reported.[76]

SEB production in unbuffered media has been found to be repressed by excess glucose in the medium.[74] Streptomycin, actinomycin D, acriflavine, Tween 80, and other compounds have been found to inhibit SEB synthesis in broth.[38] SEB production is inhibited by 2-deoxyglucose, and the inhibition is not restored by glucose, indicating that this toxin, at least, is not under catabolite control.[55] While actinomycin D has been shown to inhibit SEB synthesis in strain S-6, the inhibition occurred about 1 hour after cellular synthesis ceased. The latter was immediately and completely inhibited. A possible conclusion from this finding is that the messenger RNA (mRNA) responsible for enterotoxin synthesis is more stable than that for cellular synthesis.[62]

The lowest number of cells of *S. aureus* required to produce the minimum level of enterotoxin considered necessary to cause the gastroenteritis syndrome in humans (1 ng/g) appears to differ for substrates and for the particular enterotoxin. Detectable SEA has been found with as few as $\sim 10^4$ cfu/g.[48] In milk, SEA and SED were detected with counts of 10^7 but not below this level.[77] Employing a strain of *S. aureus* that produces SEA, SEB, and SED, SEB and SED were detected when the count reached 6×10^6/ml and the enterotoxin level was 1 ng/ml, while SEA at a level of 4 ng/ml was detected with a count of 3×10^7 cfu/ml.[78] In imitation cheese with pH of 5.56–5.90 and a_w of 0.94–0.97, enterotoxins were first detected at the following counts: SEA at 4×10^6/g; SEC at 1×10^8; SED at 3×10^6; SEE at 5×10^6; and SEC and SEE at 3×10^6/g.[12] In precooked bacon, SEA was produced by strain A100 with cells $> 10^6$/g.[95] In meat products and vanilla custard, SEA was produced with $\geq \log_{10} 7.2$ cells/g, but in certain vegetable products SEA production was delayed and detected only when numbers of cells were $\geq \log_{10} 8.9$/g.[80] In the latter study, no SE could be detected in spinach and french beans after 72 hours at 22°C when cell numbers were $\log_{10} 6.7$–8.7/g. All staphylococcal enterotoxins are resistant to pepsin (see exception in previous section).

Mode of Action

All staphylococcal enterotoxins, along with the toxic shock syndrome toxin (TSST), are bacterial *superantigens* (pyrogenic toxin superantigens—PTSags) relative to in vivo antigen recognition in contrast to conventional antigens. With the latter, a CD4 T cell facilitates contact between T cell antigen receptors and major histocompatibility complex (MHC) class II molecules. Staphylococcal superantigens bind directly to T cell receptor β chains without processing. Once bound to MHC class II molecules, SEs stimulate helper T cells to produce cytokines such as the interleukins (IL), gamma-interferon, and tumor necrosis factor. Superantigens are thus proteins that activate many different T cell clones. Among the cytokines, an overabundance of IL-2 is produced,[60] and it appears to be responsible for many or most of the symptoms of staphylococcal gastroenteritis (see below). The activity of superantigens can be demonstrated in the laboratory by exposing murine splenocytes to SEs. A positive response consists of T cell proliferation with concomitant production of IL-2 and gamma-interferon. The administration of IL-2 produces many of the symptoms caused by the enterotoxin. Studies with SEC_1 concluded that SEC_1 binds to an alpha helix of MHC class II such that the interaction between antigen-presenting cells and T cells is stabilized, leading to cytokine production and subsequent lymphocyte proliferation.[51] The region of SEs responsible for emetic activity is unclear although this activity has been separated from superantigenicity (see reference 81). SEl-SEL appear to be only weakly emetic if at all.

Regarding the pathogenesis of enterotoxins in humans, many or most of the symptoms are caused by IL-2,[60] including vomiting and diarrhea, and these symptoms can be produced by intravenous (IV) injections.

The C-terminus of the staphylococcal enterotoxin molecules is critical to several functions. In one study using SEB, the deletion of only nine amino acids from this region led to complete loss of T cell-stimulating activity.[72] The C-terminus is believed to be critical to the three-dimensional conformation of the SEB molecule.[72]

Emetic and T cell proliferation activities can be disassociated. When SEA was altered by deletion of three C-terminal residues, T-cell proliferation activity was retained, but the emetic activity was lost.[52] Using mutant copies of SEA and SEB, it has been shown that the MHC class II binding property alone is not sufficient for emesis in monkeys.[45]

THE GASTROENTERITIS SYNDROME

The symptoms of staphylococcal food poisoning usually develop within 4 hours of the ingestion of contaminated food, although a range of 1–6 hours has been reported. The symptoms—nausea, vomiting, abdominal cramps (which are usually quite severe), diarrhea, sweating, headache, prostration, and sometimes a fall in body temperature—generally lasting from 24 to 48 hours, and the mortality rate is very low or nil. The usual treatment for healthy persons consists of bed rest and maintenance of fluid balance. Upon cessation of symptoms, the victim possesses no demonstrable immunity to recurring attacks, although animals become resistant to enterotoxin after repeated oral doses.[14] Because the symptoms are referable to the ingestion of preformed enterotoxin, it is conceivable that stool cultures might be negative for the organisms, although this is rare. Proof of staphylococcal food poisoning is established by recovering enterotoxigenic staphylococci from leftover food and from the stool cultures of victims. Attempts should be made to extract enterotoxin from suspect foods, especially when the number of recoverable viable cells is low.

The minimum quantity of enterotoxin needed to cause illness in humans is about 20 ng (see the outbreak reported in the section below). This value is derived from an outbreak of staphylococcal gastroenteritis traced to 2% chocolate milk. From 12 cartons of milk, SEA was found at levels from 94 to 184 ng per carton, with a mean of 144 ng.[34] The attack rate was associated with the quantity of milk consumed and somewhat with age; those aged 5–9 years were more sensitive than those aged 10–19 years. Earlier findings indicated a dose of 20–35 μg of pure SEB for adults.[88] From 16 incidents of staphylococcal gastroenteritis, SE levels of less than 0.01–0.25 μg/g of food were found.[43]

Table 23–7 Staphylococcal Foodborne Gastroenteritis Outbreaks and Cases in the United States, 1973–1987

Years	Outbreaks	Cases	Percentage of All Cases
1973–1987	367	17,248	14.0
1983	14	1,257	15.9
1984	11	1,153	14.1
1985	14	421	1.8
1986	7	250	4.3
1987	1	100	1.0

Source: Data from Bean and Griffin.[10]

INCIDENCE AND VEHICLE FOODS

The incidence/prevalence of staphylococci in meats and seafoods are presented in Chapter 4 (Table 4–3) and Chapter 5 (Table 5–6). These organisms may be expected to occur in a wide variety of foods not given heat treatments for their destruction.

With regard to vehicle foods for staphylococcal enteritis, a large number has been incriminated in outbreaks, usually products made by hand and improperly refrigerated after being prepared. Outbreaks and cases of foodborne gastroenteritis reported to the Centers for Disease Control for the years 1973–1987 totaled 367 and 17,248, respectively (Table 23–7). From a high of around 16% in 1983, this syndrome accounted for only 1.0% of cases in 1987. The reported cases constitute only a small part of the actual number, however; estimates place the number of cases of staphylococcal foodborne gastroenteritis at between 1 million and 2 million per year in the United States. The six leading vehicle foods for 1973–1987 are listed in Table 23–7, with pork and pork products accounting for more outbreaks than the other five combined.

One of the largest outbreaks ever recorded occurred in June–July 2000 in the Kansai District in Japan.[7] There were 13,420 victims and the primary vehicle food was powdered skim milk from a single source. An SE-producing strain of *S. aureus* was the etiologic agent. Symptoms appeared in 83.4% of interviewed victims within 6 hours with 3–4 h being the peak period. Vomiting was reported by 73.3% and diarrhea by 75.9% of victims. A low-fat milk product contained ≤ 0.38 μg/ml of SEA, and powdered skim milk contained ca. 3.7 ng/g.[7] It was estimated that the average amount of SEA consumed/person was 20–100 ng.

For the years 1981–1995 in Korea, 64 outbreaks of staphylococcal food poisoning were recorded with 2,430 cases representing 16.5% of all foodborne outbreaks during this period.[65] During the same period in Japan, 9.9% of all foodborne cases and 15.9% of outbreaks were staphylococcal.[65] For the years 1980–1999 in Japan, there were 2,525 outbreaks of staphylococcal food poisoning and 59,964 cases with 3 deaths (see reference 94) (see also Table 23–8). The leading vehicle foods were rice, rice balls, and bean curds; and SEA along with the combination of SEA and SEB were the most common enterotoxins. Some other staphylococcal foodborne outbreaks are noted in Chapter 20.

As noted in Chapter 22, the problem is one of reporting where all too often the small outbreaks that occur in homes are not reported to public health officials. A large percentage of the reported cases of all types are those that result from banquets, generally involving large numbers of persons. An unusual

Table 23–8 Leading Food Sources for Staphylococcal
Gastroenteritis Outbreaks in the United States, 1973–1987

Food Sources	Number of Outbreaks
Pork	96
Bakery products	26
Beef	22
Turkey	20
Chicken	14
Eggs	9

Source: From Bean and Griffin.[10]

outbreak was caused by SEA and SED and traced to wild mushrooms in vinegar.[68] The food contained
10 ng SEA and 1 ng SED per gram.

ECOLOGY OF *S. AUREUS* GROWTH

In general, the staphylococci do not compete well with the normal biota of most foods, and this is
especially true for those that contain large numbers of lactic acid bacteria where conditions permit
the growth of the latter organisms (see Chapter 13). A large number of investigators have shown the
inability of *S. aureus* to compete in both fresh and frozen foods. At temperatures that favor staphylococcal growth, the normal food saprophytic biota offers protection against staphylococcal growth
through antagonism, competition for nutrients, and modification of the environment to conditions less
favorable to *S. aureus*. Bacteria known to be antagonistic to *S. aureus* growth include *Acinetobacter,Aeromonas,Bacillus,Pseudomonas,S. epidermidis*, the Enterobacteriaceae, the Lactobacillaceae,
enterococci, and others.[75] SEA has been shown to be resistant to a variety of environmental stresses,
but growth of several lactic acid bacteria did lead to its reduction and to a suggestion that toxin
reduction might have resulted from specific enzymes or other metabolites of the lactic acid bacteria.[24]

PREVENTION OF STAPHYLOCOCCAL AND OTHER FOOD-POISONING SYNDROMES

When susceptible foods are produced with low numbers of staphylococci, they will remain free
of enterotoxins and other food-poisoning hazards if kept either at or *below* 40°F (4.4°C) or *above*
140°F (60°C) until consumed. For the years 1961–1972, over 700 foodborne-disease outbreaks were
investigated by Bryan[19] relative to the factors that contributed to the outbreaks, and of the 16 factors
identified, the 5 most frequently involved were the following:

1. inadequate refrigeration;
2. preparing foods far in advance of planned service;
3. infected persons' practicing poor personal hygiene;
4. inadequate cooking or heat processing;
5. holding food in warming devices at bacterial growth temperatures.

Table 23–9 Leading Factors that Led to the Outbreaks of Staphylococcal Foodborne Gastroenteritis in the United States, 1973–1987

Causes	Number of Outbreaks
Improper holding temperatures	98
Poor personal hygiene	71
Contaminated equipment	43
Inadequate cooking	22
Food from unsafe source	12
Others	24

Source: From Bean and Griffin.[10]

Inadequate refrigeration alone comprised 25.5% of the contributing factors. The five listed contributed to 68% of outbreaks. For the period 1973–1987, the five leading identified causes are listed in Table 23–9; notice that the leading factors for 1961–1972 continued to be among the leading factors for the later years. Susceptible foods should not be held within the staphylococcal growth range for more than 3–4 hours.

REFERENCES

1. Adesiyun, A.A., L. Webb, and S. Rahaman. 1995. Microbiological quality of raw cow's milk at collection centers in Trinidad. *J. Food Protect.* 58:139–146.

2. Adesiyun, A.A. 1984. Enterotoxigenicity of *Staphylococcus aureus* strains isolated from Nigerian ready-to-eat foods. *J. Food Protect.* 47:438–440.

3. Adesiyun, A.A., S.R. Tatini, and D.G. Hoover. 1984. Production of enterotoxin(s) by *Staphylococcus hyicus. Vet. Microbiol.* 9:487–495.

4. Anderson, J.E., R.B. Beelman, and S. Doores. 1997. Enhanced production and thermal stability of staphylococcal enterotoxin A in the presence of chitin. *J. Food Protect.* 60:1351–1357.

5. Angelotti, R., M.J. Foter, and K.H. Lewis. 1961. Time–temperature effects on salmonellae and staphylococci in foods. *Am. J. Public Health* 51:76–88.

6. Angelotti, R., M.J. Foter, and K.H. Lewis. 1960. Time–temperature effects on salmonellae and staphylococci in foods. II. Behavior at warm holding temperatures. *Thermal-death-time studies.* Cincinnati, OH: Public Health Service, U.S. Department of Health, Education and Welfare.

7. Asao, T., Y. Kumeda, T. Kawai, T. Shibata, H. Oda, K. Haruki, N. Nakazawa, and S. Kozaki. 2003. An extensive outbreak of staphylococcal food poisoning due to low-fat milk in Japan: Estimation of enterotoxin A in the incriminated milk and powdered skim milk. *Epidemiol. Infect.* 130:33–40.

8. Balaban, M., and A. Rasooly. 2000. Staphylococcal enterotoxins. *Int. J. Food Microbiol.* 61:1–10.

9. Bautista, L., P. Gaya, M. Medina, and M. Nunez. 1988. A quantitative study of enterotoxin production by sheep milk staphylococci. *Appl. Environ. Microbiol.* 54:566–569.

10. Bean, N.H., and P.M. Griffin. 1990. Foodborne disease outbreaks in the United States, 1973–1987: Pathogens, vehicles, and trends. *J. Food Protect.* 53:804–817.

11. Bennett, R.W., and M.R. Berry, Jr. 1987. Serological reactivity and in vivo toxicity of *Staphylococcus aureus* enterotoxins A and D in selected canned foods. *J. Food Sci.* 52:416–418.

12. Bennett, R.W., and W.T. Amos. 1983. *Staphylococcus aureus* growth and toxin production in imitation cheeses. *J. Food Sci.* 48:1670–1673.

13. Bergdoll, M.S. 1967. The staphylococcal enterotoxins. In *Biochemistry of Some Foodborne Microbial Toxins*, ed. R.I. Mateles and G.N. Wogan, 1–25. Cambridge, MA: MIT Press.

14. Bergdoll, M.S. 1972. The enterotoxins. In *The Staphylococci*, ed. J.O. Cohen, 301–331. New York: Wiley-Interscience.

15. Bergdoll, M.S. 1990. Staphylococcal food poisoning. In *Foodborne Diseases*, ed. D.O. Cliver, 85–106. New York: Academic Press.

16. Betley, M.J., D.W. Borst, and L.B. Regassa. 1992. Staphylococcal enterotoxins, toxic shock syndrome toxin and streptococcal pyrogenic exotoxins: A comparative study of their molecular biology. *Chem. Immunol.* 55:1–35.

17. Borja, C.R., and M.S. Bergdoll. 1967. Purification and partial characterization of enterotoxin C produced by *Staphylococcus aureus* strain 137. *J. Biochem.* 6:1457–1473.

18. Brown, D.F., and R.M. Twedt. 1972. Assessment of the sanitary effectiveness of holding temperatures on beef cooked at low temperature. *Appl. Microbiol.* 24:599–603.

19. Bryan, F.L. 1974. Microbiological food hazards today—based on epidemiological information. *Food Technol.* 28(9):52–59.

20. Carmo, L.S., R.S. Dias, V.R. Linardi, M.J. de Sena, D.A. Santos, M.E. de Faria, E.C. Pena, M. Jett, and L.G. Heneine. 2002. Food poisoning due to enterotoxigenic strains of *Staphylococcus* present in Minas cheese and raw milk in Brazil. *Food Microbiol.* 19:9–14

21. Casman, E.P. 1965. Staphylococcal enterotoxin. In *The Staphylococci: Ecologic Perspectives*. Ann. N.Y. Acad. Sci. 28, 128: 124–133.

22. Casman, E.P., R.W. Bennett, A.E. Dorsey, and J.A. Issa. 1967. Identification of a fourth staphylococcal enterotoxin, enterotoxin D. *J. Bacteriol.* 94:1875–1882.

23. Cenci-Goga, B.T., M. Karama, P.V. Rossitto, R.A. Morgante, and J.S. Culler. 2003. Enterotoxin production by *Staphylococcus aureus* isolated from mastitic cows. *J. Food Protect.* 66:1693–1696.

24. Chordash, R.A., and N.N. Potter. 1976. Stability of staphylococcal enterotoxin A to selected conditions encountered in foods. *J. Food Sci.* 41:906–909.

25. Couch, J.L., and M.J. Betley. 1989. Nucleotide sequence of the type C_3 staphylococcal enterotoxin gene suggests that intergenic recombination causes antigenic variation. *J. Bacteriol.* 171:4507–4510.

26. Czop, J.K., and M.S. Bergdoll. 1970. Synthesis of enterotoxins by L-forms of *Staphylococcus aureus*. *Infect. Immun.* 1:169–173.

27. Czop, J.K., and M.S. Bergdoll. 1974. Staphylococcal enterotoxin synthesis during the exponential, transitional, and stationary growth phases. *Infect. Immun.* 9:229–235.

28. Dack, G.M., W.E. Cary, O. Woolpert, and H. Wiggers. 1930. An outbreak of food poisoning proved to be due to a yellow hemolytic staphylococcus. *J. Prev. Med.* 4:167–175.

29. Denny, C.B., J.Y. Humber, and C.W. Bohrer. 1971. Effect of toxin concentration on the heat inactivation of staphylococcal enterotoxin A in beef bouillon and in phosphate buffer. *Appl. Microbiol.* 21:1064–1066.

30. Denny, C.B., P.L. Tan, and C.W. Bohrer. 1966. Heat inactivation of staphylococcal enterotoxin. *J. Food Sci.* 31:762–767.

31. Dinges, M.M., P.M. Orwin, and P.M. Schlievert. 2000. Exotoxins of *Staphylococcus aureus*. *Clin. Microbiol. Rev.* 13:16–34.

32. Ellender, R.D., L. Huang, S.L. Sharp, and R.P. Tettleton. 1995. Isolation, enumeration, and identification of Gram-positive cocci from frozen crabmeat. *J. Food Protect.* 58:853–857.

33. Evans, J.B., G.A. Ananaba, C.A. Pate, and M.S. Bergdoll. 1983. Enterotoxin production by atypical *Staphylococcus aureus* from poultry. *J. Appl. Bacteriol.* 54:257–261.

34. Evenson, M.L., M.W. Hinds, R.S. Bernstein, and M.S. Bergdoll. 1988. Estimation of human dose of staphylococcal enterotoxin A from a large outbreak of staphylococcal food poisoning involving chocolate milk. *Int. J. Food Microbiol.* 7:311–316.

35. Ewald, S. 1987. Enterotoxin production by *Staphylococcus aureus* strains isolated from Danish foods. *Int. J. Food Microbiol.* 4:207–214.

36. Ewald, S., and S. Notermans. 1988. Effect of water activity on growth and enterotoxin D production of *Staphylococcus aureus*. *Int. J. Food Microbiol.* 6:25–30.

37. Freney, J., Y. Brun, M. Bes, H. Meugnier, F. Grimont, P.A.D. Grimont, C. Nervi, and J. Fleurette. 1988. *Staphylococcus lugdunensis* sp. nov. and *Staphylococcus schleiferi* sp. nov., two species from human clinical specimens. *Int. J. System. Bacteriol.* 38:168–172.

38. Friedman, M.E. 1966. Inhibition of staphylococcal enterotoxin B formation in broth cultures. *J. Bacteriol.* 92:277–278.

39. Genigeorgis, C., M.S. Foda, A. Mantis, and W.W. Sadler. 1971. Effect of sodium chloride and pH on enterotoxin C production. *Appl. Microbiol.* 21:862–866.

40. Genigeorgis, C., H. Riemann, and W.W. Sadler. 1969. Production of enterotoxin B in cured meats. *J. Food Sci.* 34:62–68.

41. Genigeorgis, C., and W.W. Sadler. 1966. Effect of sodium chloride and pH on enterotoxin B production. *J. Bacteriol.* 92:1383–1387.

42. Gibbs, P.A., J.T. Patterson, and J. Harvey. 1978. Biochemical characteristics and enterotoxigenicity of *Staphylococcus aureus* strains isolated from poultry. *J. Appl. Bacteriol.* 44:57–74.

43. Gilbert, R.J., and A.A. Wieneke. 1973. Staphylococcal food poisoning with special reference to the detection of enterotoxin in food. In *The Microbiological Safety of Food*, ed. B.C. Hobbs and J.H.B. Christian, 273–285. New York: Academic Press.

44. Gomez-Lucia, E., J. Goyache, J.L. Blanco, J.F.F. Garayzabal, J.A. Orden, and G. Suarez. 1987. Growth of *Staphylococcus aureus* and enterotoxin production in homemade mayonnaise prepared with different pH values. *J. Food Protect.* 50:872–875.

45. Harris, T.O., D. Grossman, J.W. Kappler, P. Marrack, R.R. Rich, and M.J. Betley. 1993. Lack of complete correlation between emetic and T-cell stimulatory activities of staphylococcal enterotoxins. *Infect. Immun.* 61:3175–3183.

46. Harvey, J., J.T. Patterson, and P.A. Gibbs. 1982. Enterotoxigenicity of *Staphylococcus aureus* strains isolated from poultry: Raw poultry carcasses as a potential food-poisoning hazard. *J. Appl. Bacteriol.* 52:251–258.

47. Hirooka, E.Y., E.E. Muller, J.C. Freitas, E. Vicente, Y. Yashimoto, and M.S. Bergdoll. 1988. Enterotoxigenicity of *Staphylococcus intermedius* of canine origin. *Int. J. Food Microbiol.* 7:185–191.

48. Hirooka, E.Y., S.P.C. DeSalzberg, and M.S. Bergdoll. 1987. Production of staphylococcal enterotoxin A and thermonuclease in cream pies. *J. Food Protect.* 50:952–955.

49. Hoover, D.G., S.R. Tatini, and J.B. Maltais. 1983. Characterization of staphylococci. *Appl. Environ. Microbiol.* 46:649–660.

50. Huang, I.-Y., and M.S. Bergdoll. 1970. The primary structure of staphylococcal enterotoxin B. III. The cyanogen bromide peptides of reduced and aminoethylated enterotoxin B and the complete amino acid sequence. *J. Biol. Chem.* 245:3518–3525.

51. Hoffmann, M.L., L.M. Jablonski, K.K. Crum, S.P. Hackett, Y.-I. Chi, C.V. Stauffacher, D.L. Stevens, and G.A. Bohach. 1994. Predictions of T-cell receptor and major histocompatibility complex-binding sites on staphylococcal enterotoxin C1. *Infect. Immun.* 62:3396–3407.

52. Hufnagle, W.O., M.T. Tremaine, and M.J. Betley. 1991. The carboxyl-terminal region of staphylococcal enterotoxin type A is required for a fully active molecule. *Infect. Immun.* 59:2126–2134.

53. Hurst, A., and A. Hughes. 1983. The protective effect of some food ingredients on *Staphylococcus aureus* MF 31. *J. Appl. Bacteriol.* 55:81–88.

54. Iandolo, J.J. 1989. Genetic analysis of extracellular toxins of *Staphylococcus aureus*. *Ann. Rev. Microbiol.* 43:375–402.

55. Iandolo, J.J., and W.M. Shafer. 1977. Regulation of staphylococcal enterotoxin B. *Infect. Immun.* 16:610–616.

56. Isigidi, B.K., A.M. Mathieu, L.A. Devriese, C. Godard, and J. van Hoof. 1992. Enterotoxin production in different *Staphylococcus aureus* biotypes isolated from food and meat plants. *J. Appl. Bacteriol.* 72:16–20.

57. Isigidi, B.K., L.A. Devriese, C. Godard, and J. van Hoof. 1990. Characteristics of *Staphylococcus aureus* associated with meat products and meat workers. *Lett. Appl. Microbiol.* 11:145–147.

58. Jarvis, A.W., R.C. Lawrence, and G.G. Pritchard. 1973. Production of staphylococcal enterotoxins A, B, and C under conditions of controlled pH and aeration. *Infect. Immun.* 7:847–854.

59. Jay, J.M. 1970. Effect of borate on the growth of coagulase-positive and coagulase-negative staphylococci. *Infect. Immun.* 1:78–79.

60. Johnson, H.W., J.K. Russell, and C.H. Pontzer. 1992. Super-antigens in human disease. *Sci. Am.* 266(4):92–101.

61. Kadan, R.S., W.H. Martin, and R. Mickelsen. 1963. Effects of ingredients used in condensed and frozen dairy products on thermal resistance of potentially pathogenic staphylococci. *Appl. Microbiol.* 11:45–49.

62. Katsuno, S., and M. Kondo. 1973. Regulation of staphylococcal enterotoxin B synthesis and its relation to other extracellular proteins. *Japan. J. Med. Sci. Biol.* 26:26–29.

63. Kloos, W.E., and T.L. Bannerman. 1994. Update on clinical significance of coagulase-negative staphylococci. *Clin. Microbiol. Rev.* 7:117–140.

64. Kloos, W.E., D.N. Ballard, C.G. George, J.A. Webster, R.J. Hubner, W. Ludwig, K.H. Schleifer, F. Fiedler, and K. Schubert. 1998. Delimiting the genus *Staphylococcus* through description of *Macrococcus caseolyticus* gen. nov., comb. nov. and *Macrococcus equipercicus* sp. nov., *Macrococcus bovicus* sp. nov. and *Macrococcus carouselicus* sp. nov. *Int. J. Syst. Bacteriol.* 48:859–877.

65. Lee, W.-C., M.-J. Lee, J.-S. Kim, and S.-Y. Park. 2001. Foodborne illness outbreaks in Korea and Japan studied retrospectively. *J. Food Protect.* 64:899–902.

66. Lee, R.Y., G.J. Silverman, and D.T. Munsey. 1981. Growth and enterotoxin A production by *Staphylococcus aureus* in precooked bacon in the intermediate moisture range. *J. Food Sci.* 46:1687–1692.

67. Lilly, H.D., R.A. McLean, and J.A. Alford. 1967. Effects of curing salts and temperature on production of staphylococcal enterotoxin. *Bacteriol. Proc.* 12.

68. Lindroth, S., E. Strandberg, A. Pessa, and M.J. Pellinen. 1983. A study of the growth potential of *Staphylococcus aureus* in *Boletus edulis*, a wild edible mushroom, prompted by a food poisoning outbreak. *J. Food Sci.* 48:282–283.

69. Marin, M.E., M.C. de la Rosa, and I. Cornejo. 1992. Enterotoxigenicity of *Staphylococcus* strains isolated from Spanish dry-cured hams. *Appl. Environ. Microbiol.* 58:1067–1069.

70. McLean, R.A., H.D. Lilly, and J.A. Alford. 1968. Effects of meat-curing salts and temperature on production of staphylococcal enterotoxin B. *J. Bacteriol.* 95:1207–1211.

71. Metzger, J.F., A.D. Johnson, W.S. Collins, II, and V. McGann. 1973. *Staphylococcus aureus* enterotoxin B release (excretion) under controlled conditions of fermentation. *Appl. Microbiol.* 25:770–773.

72. Metzroth, B., T. Marx, M. Linnig, and B. Fleischer. 1993. Concomitant loss of conformation and superantigenic activity of staphylococcal enterotoxin B deletion mutant proteins. *Infect. Immun.* 61:2445–2452.

73. Miller, R.D., and D.Y.C. Fung. 1973. Amino acid requirements for the production of enterotoxin B by *Staphylococcus aureus* S-6 in a chemically defined medium. *Appl. Microbiol.* 25:800–806.

74. Morse, S.A., R.A. Mah, and W.J. Dobrogosz. 1969. Regulation of staphylococcal enterotoxin B. *J. Bacteriol.* 98:4–9.

75. Mossel, D.A.A. 1975. Occurrence, prevention, and monitoring of microbial quality loss of foods and dairy products. *CRC Crit. Rev. Environ. Control.* 5:1–140.

76. Munson, S.H., M.T. Tremaine, M.J. Betley, and B.A. Welch. 1998. Identification and characterization of staphylococcal enterotoxin types G and I from *Staphylococcus aureus. Infect. Immun.* 66:3337–3348.

77. Noleto, A.L., and M.S. Bergdoll. 1980. Staphylococcal enterotoxin production in the presence of nonenterotoxigenic staphylococci. *Appl. Environ. Microbiol.* 39:1167–1171.

78. Noleto, A.L., and M.S. Bergdoll. 1982. Production of enterotoxin by a *Staphylococcus aureus* strain that produces three identifiable enterotoxins. *J. Food Protect.* 45:1096–1097.

79. Notermans, S., and C.J. Heuvelman. 1983. Combined effect of water activity, pH and suboptimal temperature on growth and enterotoxin production of *Staphylococcus aureus. J. Food Sci.* 48:1832–1835, 1840.

80. Notermans, S., and R.L.M. van Otterdijk. 1985. Production of enterotoxin A by *Staphylococcus aureus* in food. *Int. J. Food Microbiol.* 2:145–149.

81. Orwin, P.M., D.Y.M. Leung, H.L. Donahue, R.P. Novick, and P.M. Schlievert. 2001. Biochemical and biological properties of staphylococcal enterotoxin K. *Infect. Immun.* 69:360–366.

82. Otero, A., M.L. Garcia, M.C. Garcia, B. Moreno, and M.S. Bergdoll. 1990. Production of staphylococcal enterotoxins C_1 and C_2 and thermonuclease throughout the growth cycle. *Appl. Environ. Microbiol.* 56:555–559.

83. Palasuntheram, C., and M.S. Beauchamp. 1982. Enterotoxigenic staphylococci in Sri Lanka. *J. Appl. Bacteriol.* 52:39–41.

84. Palumbo, S.A., J.L. Smith, and J.C. Kissinger. 1977. Destruction of *Staphylococcus aureus* during frankfurter processing. *Appl. Environ. Microbiol.* 34:740–744.

85. Payne, D.N., and J.M. Wood. 1974. The incidence of enterotoxin production in strains of *Staphylococcus aureus* isolated from foods. *J. Appl. Bacteriol.* 37:319–325.

86. Pereira, J.L., S.P. Salzberg, and M.S. Bergdoll. 1982. Effect of temperature, pH and sodium chloride concentrations on production of staphylococcal enterotoxins A and B. *J. Food Protect.* 45:1306–1309.

87. Probst, A.J., C. Hertel, L. Richter, L. Wassill, W. Ludwig, and W.P. Hammes. 1998. *Staphylococcus condimenti* sp. nov., from soy sauce mash, and *Staphylococcus carnosus* (Schleifer and Fischer 1982) subsp. *utilis* subsp. nov. *Int. J. Syst. Bacteriol.* 48:651–658.

88. Raj, H.D., and M.S. Bergdoll. 1969. Effect of enterotoxin B on human volunteers. *J. Bacteriol.* 98:833–834.

89. Read, R.B., and J.G. Bradshaw. 1966. Thermal inactivation of staphylococcal enterotoxin B in veronal buffer. *Appl. Microbiol.* 14:130–132.

90. Reiser, R.F., R.N. Robbins, A.L. Noleto, G.P. Khoe, and M.S. Bergdoll. 1984. Identification, purification, and some physiochemical properties of staphylococcal enterotoxin C₃. *Infect. Immun.* 45:625–630.

91. Reiser, R.F., and K.F. Weiss. 1969. Production of staphylococcal enterotoxins A, B, and C in various media. *Appl. Microbiol.* 18:1041–1043.

92. Satterlee, L.D., and A.A. Kraft. 1969. Effect of meat and isolated meat proteins on the thermal inactivation of staphylococcal enterotoxin B. *Appl. Microbiol.* 17:906–909.

93. Schantz, E.J., W.G. Roessler, J. Wagman, L. Spero, D.A. Dunnery, and M.S. Bergdoll. 1965. Purification of staphylococcal enterotoxin B. *J. Biochem.* 4:1011–1016.

94. Shimizu, A., M. Fugita, H. Igarashi, M. Takagi, N. Nagase, A. Sasaki, and J. Kawano. 2000. Characterization of *Staphylococcus aureus* coagulase type VII isolates from staphylococcal food poisoning outbreaks (1980–1995) in Tokyo, Japan, by pulsed-field gel electrophoresis. *J. Clin. Microbiol.* 38:3746–3749.

95. Siems, H., D. Husch, H.-J. Sinell, and F. Untermann. 1971. Vorkommen und Eigenschaften von Staphylokokken in verschiedenen Produktionsstufen bei der Fleischverarbeitung. *Fleischwirts.* 51:1529–1533.

96. Silverman, G.J., D.T. Munsey, C. Lee, and E. Ebert. 1983. Interrelationship between water activity, temperature and 5.5 percent oxygen on growth and enterotoxin A secretion by *Staphylococcus aureus* in precooked bacon. *J. Food Sci.* 48:1783–1786, 1795.

97. Simkovicova, M., and R.J. Gilbert. 1971. Serological detection of enterotoxin from food-poisoning strains of *Staphylococcus aureus*. *J. Med. Microbiol.* 4:19–30.

98. Smith, J.L., R.L. Buchanan, and S.A. Palumbo. 1983. Effect of food environment on staphylococcal enterotoxin synthesis: A review. *J. Food Protect.* 46:545–555.

99. Sokari, T.G., and S.O. Anozie. 1990. Occurrence of enterotoxin producing strains of *Staphylococcus aureus* in meat and related samples from traditional markets in Nigeria. *J. Food Protect.* 53:1069–1070.

100. Soriano, J.M., G. Font, H. Rico, J.C. Miltó, and J. Mañes. 2002. Incidence of enterotoxigenic staphylococci and their toxins in foods. *J. Food Protect.* 65:857–860.

101. Sperber, W.H. 1977. The identification of staphylococci in clinical and food microbiology laboratories. *CRC Crit. Rev. Clin. Lab. Sci.* 7:121–184.

102. Su, Y.-C., and A.C.L. Wong. 1998. Production of staphylococcal enterotoxin H under controlled pH and aeration. *Int. J. Food Microbiol.* 39:87–91.

103. Su, Y.-C., and A.C.L. Wong. 1996. Detection of staphylococcal enterotoxin H by an enzyme-linked immunosorbent assay. *J. Food Protect.* 59:327–330.

104. Su, Y.-C., and A.C.L. Wong. 1995. Identification and purification of a new staphylococcal enterotoxin, H. *Appl. Environ. Microbiol.* 61:1438–1443.

105. Tanasupawat, S., Y. Hoshimoto, T. Ezaki, M. Kozaki, and K. Komagata. 1992. *Staphylococcus piscifermentans* sp. nov., from fermented fish in Thailand. *Int. J. Syst. Bacteriol.* 42:577–581.

106. Tatini, S.R. 1973. Influence of food environments on growth of *Staphylococcus aureus* and production of various enterotoxins. *J. Milk Food Technol.* 36:559–563.

107. Thomas, C.T., J.C. White, and K. Longree. 1966. Thermal resistance of salmonellae and staphylococci in foods. *Appl. Microbiol.* 14:815–820.

108. Thota, F.H., S.R. Tatini, and R.W. Bennett. 1973. Effects of temperature, pH and NaCl on production of staphylococcal enterotoxins E and F. *Bacteriol. Proc.* 1.

109. Tibana, A., K. Rayman, M. Akhtar, and R. Szabo. 1987. Thermal stability of staphylococcal enterotoxins A, B and C in a buffered system. *J. Food Protect.* 50:239–242.

110. Troller, J.A. 1972. Effect of water activity on enterotoxin A production and growth of *Staphylococcus aureus*. *Appl. Microbiol.* 24:440–443.

111. Valle, J., E. Gomez-Lucia, S. Piriz, J. Goyache, J.A. Orden, and S. Vadillo. 1990. Enterotoxin production by staphylococci isolated from healthy goats. *Appl. Environ. Microbiol.* 56:1323–1326.

112. Vandenbosch, L.L., D.Y.C. Fung, and M. Widomski. 1973. Optimum temperature for enterotoxin production by *Staphylococcus aureus* S-6 and 137 in liquid medium. *Appl. Microbiol.* 25:498–500.

113. Veraldo, P.E., R. Kilpper-Balz, F. Biavasco, G. Satta, and K.H. Schleifer. 1988. *Staphylococcus delphini* sp. nov., a coagulase-positive species isolated from dolphins. *Int. J. System. Bacteriol.* 38:436–439.

114. Vernozy-Rozand, C., C. Mazuy, H. Meugnier, M. Bes, Y. Lasne, f. Fiedler, J. Etienne, and J. Freney. 2000. *Staphylococcus flurettii* sp. nov., isolated from goat's milk cheeses. *Int. J. Syst. Evol. Microbiol.* 50:1521–1527.

115. Woodburn, M., M. Bennion, and G.E. Vail. 1962. Destruction of salmonellae and staphylococci in precooked poultry products by heat treatment before freezing. *Food Technol.* 16:98–100.

116. Wu, C.-H., and M.S. Bergdoll. 1971. Stimulation of enterotoxin B production. *Infect. Immun.* 3:784–792.

117. Zhang, S., J.J. Iandolo, and G.C. Stewart. 1998. The enterotoxin D plasmid of *Staphylococcus aureus* encodes a second enterotoxin determinant (sej). *FEMS Microbiol. Lett.* 168:227–233.

CHAPTER 24

Food Poisoning Caused by Gram-Positive Sporeforming Bacteria

At least three Gram-positive sporeforming rods are known to cause bacterial food poisoning: *Clostridium perfringens* (*welchii*), *C. botulinum*, and *Bacillus cereus*. The incidence of food poisoning caused by each of these organisms is related to certain specific foods, as is food poisoning in general.

CLOSTRIDIUM PERFRINGENS FOOD POISONING

The causative organism of this syndrome is a Gram-positive, anaerobic, spore-forming rod, widely distributed in nature. Based on their ability to produce certain exotoxins, five types are recognized: types A, B, C, D, and E. The food-poisoning strains belong to type A, as do the classic gas gangrene strains, but unlike the latter, the food-poisoning strains are generally heat resistant and produce only traces of alpha toxin. Some type C strains produce enterotoxin and may cause a food-poisoning syndrome. The classic food-poisoning strains differ from type C strains in not producing beta toxin. The latter, which have been recovered from enteritis necroticans, are compared to type A heat-sensitive and heat-resistant strains in Table 24–1. Type A heat-resistant strains produce theta toxin, which is perfringolysin O (PLO), a thiol-activated hemolysin similar to listeriolysin O (LLO), produced by *Listeria monocytogenes* (discussed in Chapter 25). Like LLO, PLO has a molecular weight of 60 kDa, and it has been sequenced and cloned.

Although *C. perfringens* has been associated with gastroenteritis since 1895, the first clearcut demonstration of its etiological status in food poisoning was made by McClung,[79] who investigated four outbreaks in which chicken was incriminated. The first detailed report of the characteristics of this food-poisoning syndrome was that of Hobbs et al.[50] in Great Britain. Although the British workers were more aware of this organism as a cause of food poisoning during the 1940s and 1950s, few incidents were recorded in the United States prior to 1960. It is clear now that *C. perfringens* food poisoning is widespread in the United States and many other countries.

Table 24–1   Toxins of *Clostridium welchii* (*perfringens*) Types A and C

Clostridium welchii	Toxins										
	α	β	γ	δ	ε	ϕ	ι	κ	λ	μ	ν
Heat-sensitive type A	+++	−	−	−	−	++	−	++	−	+ or −	+
Heat-resistant type A	± or tr	−	−	−	−	−	−	+ or −	−	+++ or −	−
Heat-resistant type C	+	+	+	−	−	−	−	−	−	−	+

Source: From B.C. Hobbs. 1962. *Bacterial Food Poisoning.* London: Royal Society of Health.

Distribution of *C. perfringens*

The food-poisoning strains of *C. perfringens* exist in soils, water, foods, dust, spices, and the intestinal tract of humans and other animals. Various investigators have reported the incidence of the heat-resistant, nonhemolytic strains to range from 2% to 6% in the general population. Between 20% and 30% of healthy hospital personnel and their families have been found to carry these organisms in their feces, and the carrier rate of victims after 2 weeks may be 50% or as high as 88%.[26] The heat-sensitive types are common to the intestinal tract of all humans. *C. perfringens* gets into meats directly from slaughter animals or by the subsequent contamination of slaughtered meat from containers, handlers, or dust. Because it is a spore former, it can withstand the adverse environmental conditions of drying, heating, and certain toxic compounds.

Characteristics of the Organism

Food poisoning as well as most other strains of *C. perfringens* grow well on a variety of media if incubated under anaerobic conditions or if provided with sufficient reducing capacity. Strains of *C. perfringens* isolated from horse muscle grew without increased lag phase at an oxidation–reduction potential (Eh) of -45 or lower, whereas more positive Eh values had the effect of increasing the lag phase.[12] Although it is not difficult to obtain growth of these organisms on various media, sporulation occurs with difficulty and requires the use of special media, such as those described by Duncan and Strong[27] or the employment of special techniques such as dialysis sacs.

C. perfringens is mesophilic, with an optimum between 37°C and 45°C. The lowest temperature for growth is around 20°C, and the highest is around 50°C. Optimum growth in thioglycollate medium for six strains was found to occur between 30°C and 40°C, and the optimum for sporulation in Ellner's medium was 37–40°C.[98] Growth at 45°C under otherwise optimal conditions leads to generation times as short as 7 minutes. Regarding pH, many strains grow over the range 5.5–8.0 but generally not below 5.0 or above 8.5. The lowest reported water activity (a_w) values for growth and germination of spores lie between 0.97 and 0.95 with sucrose or NaCl, or about 0.93 with glycerol employing a fluid thioglycollate base.[57] Spore production appears to require higher a_w values than the above minima. Although growth of type A was demonstrated at pH 5.5 by Labbe and Duncan,[65] no sporulation or toxin production occurred. A pH of 8.5 appears to be the highest for growth. *C. perfringens* is not as strict an anaerobe as are some other clostridia. Its growth at an initial Eh of +320 mV has been observed.[92] At least 13 amino acids are required for growth, along with biotin, pantothenate, pyridoxal, adenine, and other related compounds. It is heterofermentative, and a large number of carbohydrates are attacked. Growth is inhibited by around 5% NaCl.

The endospores of food-poisoning strains differ in their resistance to heat, with some being typical of other mesophilic spore formers and some being highly resistant. A $D_{100°C}$ value of 0.31 for *C. perfringens* (ATCC 3624) and a value of 17.6 for strain NCTC 8238 have been reported.[123] For eight strains that produced reactions in rabbits, $D_{100°C}$ values ranged from 0.70 to 38.37; strains that did not produce rabbit reactions were more heat sensitive.[115] Differences in heat sensitivity among *C. perfringens* strains is associated with the carriage of the *cpe* gene. $D_{100°C}$ values for 13 meat, poultry, and fish isolates where the *cpe* gene is chromosomal ranged from a low of 43 to a high of 170, while a non-outbreak strain that carried *cpe* on plasmids was 3.[124] These authors noted that outbreak strains typically possess a $D_{100°C}$ >40 while non-outbreak strains are typically <2.

In view of the practice of cooking roasts in water baths for long times at low temperatures (LTLT), the heat destruction of vegetative cells of *C. perfringens* has been studied by several groups. For strain ATCC 13124 in autoclaved ground beef, $D_{56.8°C}$ was 48.3 minutes, essentially similar to the $D_{56.8°C}$ or $D_{47.9°C}$ for phospholipase C.[32] Employing strain NCTC 8798, D values for cells were found to increase with increasing growth temperatures in autoclaved ground beef. For cells grown at 37°C, $D_{59°C}$ was 3.1 minutes; cells grown at 45°C had $D_{59°C}$ of 7.2; and cells grown at 49°C had $D_{59°C}$ of 10.6 minutes.[99] Although the wide differences in heat resistance between the two strains noted may in part be due to strain differences, the effect of fat in the heating menstruum may also have played a role. With beef roasts cooked in plastic bags in a water bath at 60–61°C, holding the product to an internal temperature of 60°C for at least 12 minutes eliminated salmonellae and reduced the *C. perfringens* population by about 3 log cycles. To effect a 12-log reduction of numbers for roasts weighing 1.5 kg, holding at 60°C for 2.3 hours or longer was necessary.[104] The thermal destruction of *C. perfringens* enterotoxin in buffer and gravy at 61°C required 25.4 and 23.8 minutes, respectively.[15]

Regarding the wide variations in heat resistance recorded for *C. perfringens*, similar variations have not been recorded for *C. botulinum*, especially types A and B. The latter organisms are less common in the human intestinal tract than *C. perfringens* strains. An organism inhabiting environments as diverse as these may be expected to show wide variations among its strains. Another factor that is important in heat resistance of bacterial spores is that of the chemical environment. Alderton and Snell[4] pointed out that spore heat resistance is largely an inducible property, chemically reversible between a sensitive and resistant state. Using this hypothesis, it has been shown that spores can be made more heat resistant by treating them in calcium acetate solutions—for example, 0.1 or 0.5 M at pH 8.5 for 140 hours at 50°C. The heat resistance of endospores may be increased 5- to 10-fold by this method.[3] On the other hand, heat resistance may be decreased by holding spores in 0.1 N HCl at 25°C for 16 hours or as a result of the exposure of endospores to the natural acid conditions of some foods. It is not inconceivable that the high variability of heat resistance of *C. perfringens* spores may be a more or less direct result of immediate environmental history.

The freezing survival of *C. perfringens* in chicken gravy was studied by Strong and Canada,[113] who found that only around 4% of cells survived when frozen to t17.7°C for 180 days. Dried spores, on the other hand, displayed a survival rate of about 40% after 90 days but only about 11% after 180 days.

For epidemiological studies, serotyping has been employed, but because of the many serovars, there appears to be no consistent relationship between outbreaks and given serovars. The bacteriocin typing of type A has been achieved, and of 90 strains involved in food outbreaks, all were typeable by a set of eight bacteriocins and 85.6% consisted of bacteriocin types 1–6.[101]

The Enterotoxin

The causative factor of *C. perfringens* food poisoning is an enterotoxin. It is unusual in that it is a spore-specific protein; its production occurs together with that of sporulation. All known food-poisoning cases by this organism are caused by type A strains. An unrelated disease, necrotic enteritis, is caused by beta toxin produced by type C strains and is rarely reported outside New Guinea. Although necrotic enteritis due to type C has been associated with a mortality rate of 35–40%, food poisoning due to type A strains has been fatal only in elderly or otherwise debilitated persons. Some type C strains have been shown to produce enterotoxin, but its role in disease is unclear.

The enterotoxin of type A strains was demonstrated by Duncan and Strong.[28] The purified entero-toxin has a molecular weight of 35,000 daltons and an isoelectric point of 4.3.[47] It is heat sensitive (biological activity destroyed at 60°C for 10 minutes) and pronase sensitive but resistant to trypsin, chymotrypsin, and papain.[112] L-forms of *C. perfringens* produce the toxin, and in one study they were shown to produce as much enterotoxin as classic forms.[77]

The enterotoxin is synthesized by sporulating cells in association with late stages of sporulation. The peak for toxin production is just before lysis of the cell's sporangium, and the enterotoxin is released along with spores. Conditions that favor sporulation also favor enterotoxin production, and this was demonstrated with raffinose,[67] caffeine, and theobromine.[66] The latter two compounds increased enterotoxin from undetectable levels to 450 μg/ml of cell extract protein. It has been shown to be similar to spore structural proteins covalently associated with the spore coat. Cells sporulate freely in the intestinal tract and in a wide variety of foods. In culture media, the enterotoxin is normally produced only when endospore formation is permitted (Figure 24–1), but vegetative cells are known to produce enterotoxin at low levels.[41,42] As noted above, a single gene, *cpe*, is responsible for the enterotoxin, and in Type A food poisoning strains it is chromosomal, while in nonfood poisoning strains it is on a plasmid (see references 73 and 124).

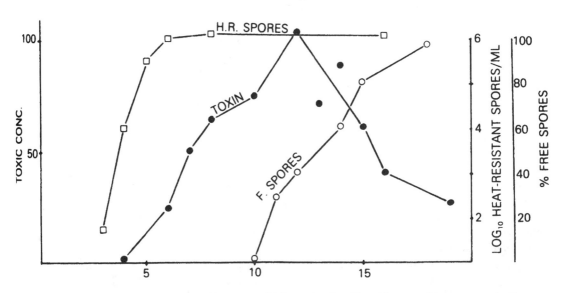

Figure 24–1 Kinetics of sporulation and enterotoxin formation by *Clostridium perfringens* type A. *Source*: Redrawn from Labbe,[64] copyright © 1980 by Institute of Food Technologists.

The enterotoxin may appear in a growth and sporulation medium about 3 hours after inoculation with vegetative cells,[25] and from 1 to 100 μg/ml of enterotoxin production has been shown for three strains of *C. perfringens* in Duncan–Strong (DS) medium after 24–36 hours.[33] It has been suggested that preformed enterotoxin may exist in some foods and, in infrequent cases, contribute to the early onset of symptoms. Purified enterotoxin has been shown to contain up to 3,500 mouse LD/mg of N.

The enterotoxin may be detected in the feces of victims. From one case, 13–16 μg/g of feces were found, and from another victim with a milder case, 3–4 μg/g were detected.[102]

Mode of Action

The *C. perfringens* enterotoxin (CPE) is not a superantigen[63] as are the staphylococcal enterotoxins (see subsection Mode of Action in Chapter 23). The CPE protein contains 319 amino acids, and it binds to either claudine and/or a 50 kDa eukaryotic membrane receptor that leads to the formation of a 90-kDa CPE-containing complex in host cell membranes. A larger complex, >160-kDa, is formed by the addition of host cell membrane proteins that ultimately leads to the induction of permeability changes in cell membranes and death of host cells (see reference 60).

Vehicle Foods and Symptoms

Symptoms appear between 6 and 24 hours, especially between 8 and 12 hours, after the ingestion of contaminated foods. The symptoms are characterized by acute abdominal pain and diarrhea; nausea, fever, and vomiting are rare. Except in the elderly or in debilitated persons, the illness is of short duration—a day or less. The fatality rate is quite low, and no immunity seems to occur, although circulating antibodies to the enterotoxin may be found in some persons with a history of the syndrome.

The true incidence of *C. perfringens* food poisoning is unknown. Because of the relative mildness of the disease, it is quite likely that only those outbreaks and cases that affect groups of people are ever reported and recorded. The confirmed outbreaks reported to the U.S. Centers for Disease Control for the years 1983–1987 are noted in Table 24–2, along with cases. The average number of cases was under 100 for each outbreak.

The foods involved in *C. perfringens* outbreaks are often meat dishes prepared one day and eaten the next. The heat preparation of such foods is presumably inadequate to destroy the heat-resistant

Table 24–2 Outbreaks, Cases, and Deaths from *C. perfringens* Foodborne Gastroenteritis in the United States, 1983–1987

Years	Outbreaks/Cases/Deaths
1983	5/353/0
1984	8/882/2
1985	6/1016/0
1986	3/202/0
1987	2/290/0

Source: From N.H. Bean, P.M. Griffin, J.S. Goulding, and C.D. Ivey. 1990. *J. Food Protect.* 53:711–728.

endospores, and when the food is cooled and rewarmed, the endospores germinate and grow. Meat dishes are most often the cause of this syndrome, although nonmeat dishes may be contaminated by meat gravy. The greater involvement of meat dishes may be due in part to the slower cooling rate of these foods and also to the higher incidence of food-poisoning strains in meats. Strong et al.[114] found the overall incidence of the organism to be about 6% in 510 American foods. The incidence for various foods was 2.7% for commercially prepared frozen foods; 3.8% for fruits and vegetables; 5% for spices; 1.8% for home-prepared foods; and 16.4% for raw meat, poultry, and fish. Hobbs et al.[50] found that 14–24% of veal, pork, and beef samples examined contained heat-resistant endospores, but all 17 samples of lamb were negative. In Japan, enterotoxigenic strains were recovered from food handlers (6% of 80), oysters (12% of 41), and water (10% of 20 samples).[100] More recently, only ca. 1.4% of *cpe*-positive isolates were detected in ca. 900 non-outbreak retail foods in the United States.[124]

An outbreak of food poisoning involving 375 persons where 140 became ill was found to be caused by both *C. perfringens* and *Salmonella* Typhimurium.[94] *C. perfringens* has been demonstrated to grow in a large number of foods. A study of retail, frozen precooked foods revealed that half were positive for vegetative cells, and 15% contained endospores.[121] The latter investigators inoculated meat products with the organism and stored them at t29°C for up to 42 days. Although spore survival was high, vegetative cells were virtually eliminated during the holding period. The survival of inoculated cells in raw ground beef was studied by Goepfert and Kim,[39] who found decreased numbers upon storage at temperatures between 1°C and 12.5°C. The raw beef contained a natural biota, and the finding suggests that *C. perfringens* is unable to compete under these conditions.

Prevention

The *C. perfringens* gastroenteritis syndrome may be prevented by proper attention to the leading causes of food poisoning of all types noted in previous chapters. Because this syndrome often occurs in institutional cafeterias, some special precautions should be taken. Upon investigating a *C. perfringens* food-poisoning outbreak in a school lunchroom in which 80% of students and teachers became ill, Bryan et al.[16] constructed a time–temperature chart in an effort to determine when, where, and how the turkey became the vehicle (Figure 24–2). It was concluded that meat and gravy, but not dressing, were responsible for the illness. As a means of preventing recurrences of such episodes, these investigators suggested nine points for the preparation of turkey and dressing:

1. Cook turkey until the internal breast temperature reaches at least 165°F (74°C), or preferably higher.
2. Thoroughly wash and sanitize all containers and equipment that previously had contact with raw turkey.
3. Wash hands and use disposable plastic gloves when deboning, deicing, or otherwise handling cooked turkey.
4. Separate turkey meat and stock before chilling.
5. Chill the turkey and stock as rapidly as possible after cooking.
6. Use shallow pans for storing stock and deboned turkey in refrigerators.
7. Bring stock to a rolling boil before making gravy or dressing.
8. Bake dressing until all portions reach 165°F or higher.
9. Just prior to serving, heat turkey pieces submerged in gravy until largest portions of meat reach 165°F.

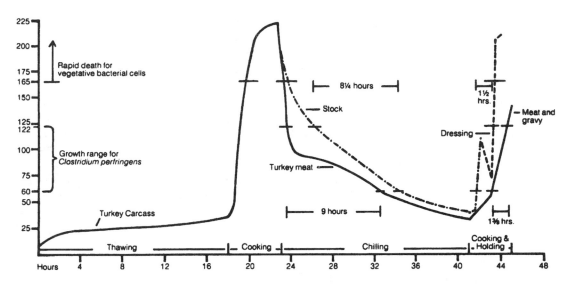

Figure 24–2   Illustration of possible time–temperature relationships during turkey preparation in a school lunch kitchen. *Source*: From Bryan et al.,[16] copyright © 1971 by International Association of Milk, Food, and Environmental Sanitarians.

BOTULISM

Unlike *C. perfringens* food poisoning, in which large numbers of viable cells must be ingested, the symptoms of botulism are caused by the ingestion of a highly toxic, soluble exotoxin produced by the organism while growing in foods.

Among the earliest references to what in all probability was human botulism was the order by Emperor Leo VI, one of the Macedonian-era rulers of the Byzantium, during the period AD 886–912, forbidding the eating of blood sausage because of its harmful health effects. An outbreak of "sausage poisoning" occurred in 1793 in Wildbad Württemberg, Germany, with 13 cases and 6 deaths. It was traced to blood sausage (pig gut filled with blood and other ingredients). The filled gut was tied, boiled briefly, smoked, and stored at room temperature. Between 1820 and 1822, Justinius Kerner studied 230 cases of "sausage poisoning" in Württemberg and noted that the product did not become toxic if air pockets were left in casings and that toxic sausage had always been boiled. In 1896, 24 music club members in Ellezelles ate raw salted ham; 23 became ill, and 3 died. E.P.M. Van Ermengen of the University of Ghent studied the outbreak. He found that the ham was neither cooked nor smoked, and the same organism was recovered from the ham and the spleen of a victim. Van Ermengen named the causative organism *Bacillus botulinus* (*botulus*, L "sausage"). This strain was later determined to be a type B.

Botulism is caused by certain strains of *C. botulinum*, a Gram-positive, anaerobic, sporeforming rod with oval to cylindrical, terminal to subterminal spores. On the basis of the serological specificity of their toxins, seven types are recognized: A, B, C, D, E, F, and G. Types A, B, E, F, and G cause disease in humans; type C causes botulism in fowls, cattle, mink, and other animals; and type D is associated with forage poisoning of cattle, especially in South Africa. The types are also differentiated on the basis of their proteolytic activity. Types A and G are proteolytic, as are some types B and F strains. Type E is nonproteolytic, as are some B and F strains (see Table 24–3). The proteolytic activity of type G is slower than that for type A, and its toxin requires trypsin potentiation. All strains that produce the Type G toxin are placed in the species *C. argentinense*.[116] It has been recovered from soils in Argentina, Switzerland, and the United States.

All toxin-producing strains have been placed into one of four groups—I, II, III, or IV. Group I contains the proteolytics, group II the nonproteolytics, and group IV serological type G. Group III consists of types C and D. Interestingly, two botulinal toxins have been detected in species other than *C. botulinum*. Type F is produced by *Clostridium baratii*,[45] and a neurotoxin that is antigenically similar but not identical to type E is elaborated by *Clostridium butyricum*.[80] In the case of the latter, the toxin gene was transferred from toxigenic *C. butyricum* to nontoxigenic *C. botulinum* type-E-like recipients by transduction of a defective bacteriophage that is made infective by a helper strain.[129] The type E toxin gene is chromosomal in both species. It can only be presumed that *C. baratii* acquired the toxin by a similar mechanism. The baratii toxin has a molecular weight of ~140 kDa.[36] Strains of *C. butyricum* that produce Type E toxin have been found to do so at a pH of 4.8 over a period of 43 and 44 days.[9] The lowest growth temperature was found to be 12°C, and at this temperature, Type E toxin was detectable after 15 days while at 25°C, toxin was detectable in 5 days in mascarpone cheese and pesto sauce.[9]

Distribution of *C. botulinum*

This organism is indigenous to soils and waters. In the United States, type A occurs more frequently in soils in the western states, and type B is found more frequently in the eastern states and in Europe. Soils and manure from various countries have been reported to contain 18% type A and 7% type B spores. Cultivated soil samples examined showed 7% to contain type A and 6% type B endospores. Type E spores tend to be confined more to waters, especially marine waters. In a study of mud samples from the harbor of Copenhagen, Pederson[93] found 84% to contain type E spores, whereas 26% of soil samples taken from a city park contained the organism. From a study of 684 environmental samples from Denmark, the Faroe Islands, Iceland, Greenland, and Bangladesh, 90% of aquatic samples from Denmark and 86% of marine samples from Greenland contained type E.[52] This strain was not found in Danish soils and woodlands, whereas type B was. Based on these results, Huss[52] suggested that type E is a truly aquatic organism that proliferates in dead aquatic animals and sediments and is disseminated by water currents and migrating fish. Type E spores have been known for some time to exist in waters off the shores of northern Japan. Prior to 1960, the existence of the organisms in the Great Lakes and the Gulf Coast waters was not known, but their presence in these waters as well as in the Gulf of Maine and the gulfs of Venezuela and Darien is well established. Ten percent of soil samples tested in Russia were found positive for *C. botulinum*, with type E strains being predominant. In a study of 333 samples from a Finnish trout farm, type E was found in 95% of those from 21 farms, in 68% of sediment samples, 15% of fish intestinal samples, and 5% of fish skin samples.[48] No types A, B, or F were found. According to these investigators, the Baltic Sea has the highest level of Type E contamination in the world.

As to the overall incidence of *C. botulinum* in soils, it has been suggested that the numbers per gram are probably less than 1. The nonproteolytic types are associated more with waters than soils, and it may be noted from Table 24–3 that the discovery of these types occurred between 1960 and 1969. The late recognition is probably a consequence of the low heat resistance of the nonproteolytics, which would be destroyed if specimens were given their usual heat treatment for spore recovery.

Table 24–3 Summary Comparison of *C. botulinum* Strains and Their Toxins

Property	Serologic Types						
	A	B	B	E	F	F	G
Year discovered	1904	1896	1960	1936	1960	1965	1969
Proteolytic (+), nonproteolytic (−)	+	+	−	−	+	−	+ (weak)
Group	I	I	II	II	I	II	IV
Primary habitat	Terrestrial	Terrestrial	Aquatic	Aquatic	Aquatic	Aquatic	Terrestrial
Minimum growth temp. (°C)	~10	~10	3.3	3.3	~10	3.3	~12
Maximum growth temp. (°C)	~50	~50	~45	~45	~50	~45	n.d.
Minimum pH for growth (see text)	4.7	4.7	4.7	4.8	4.8	4.8	4.8
Minimum a_w for growth	0.94	0.94	~0.97	~0.97	0.94?	~0.97	n.d.
Thermal D values for endospores (°C)	$D_{110}=2.72-2.89$	$D_{110}=1.34-1.37$	n.d.	$D_{80}=0.80$	$D_{110}=1.45-1.82$	$D_{82.2}=0.25-0.84$	$D_{110}=0.45-0.54$
Radiation D values of spores (kGy)	1.2-1.5	1.1-1.3	n.d.	1.2	1.1; 2.5	1.5	n.d.
Maximum NaCl for growth (%)	~10	~10	5-6	5-6	8-10	5-6	n.d.
Relative frequency of food outbreaks	High	High	n.d.	Highest for seafoods	1 outbreak	1 outbreak	None
H_2S production	+	+	−	−	+	−	++
Casein hydrolysis	+	+	−	−	+	−	+
Lipase production	+	+	+	+	+++	+	−
Glucose fermentation	+	+	+	+	+	+	−
Mannose fermentation	−	−	+	+	−	−	−
Propionic acid produced	+	+	n.d.	n.d.	+	n.d.	n.d.

Note: + = positive; ++ = strongly positive; − = negative; n.d. = no data.

The first type F strains were isolated by Moller and Scheibel[86] from a homemade liver paste incriminated in an outbreak of botulism, involving one death, on the Danish island Langeland. Since that time, Craig and Pilcher[20] have isolated type F spores from salmon caught in the Columbia River; Eklund and Poysky[29] found type F spores in marine sediments taken off the coasts of Oregon and California; Williams-Walls[127] isolated two proteolytic strains from crabs collected from the York River in Virginia; and Midura et al.[82] isolated the organism from venison jerky in California.

The type G strain was isolated first in 1969 from soil samples in Argentina,[37] and it was isolated later from five human corpses in Switzerland.[109] These deaths were not food associated. It has not been incriminated in food-poisoning outbreaks to date, and the reason might be due to the fact that this strain produces considerably less neurotoxin than type A. It has been shown that type G produced 40 LD_{50}/ml of toxin in media in which type A normally produces 10,000 to 1,000,000 LD_{50}/ml, but that under certain conditions the organism could be induced to produce up to 90,000 LD_{50}/ml of medium.[19]

For a review of the prevalence of *C. botulinum* spores in other environmental samples, see reference 10.

Growth of *C. botulinum* Strains

Some of the growth and other characteristics of the strains that cause botulism in humans are summarized in Table 24–3. The discussion that follows emphasizes the differences between the proteolytic and nonproteolytic strains irrespective of serological type. The proteolytic strains, unlike the nonproteolytics, digest casein and produce H_2S. The latter, on the other hand, ferment mannose, whereas the proteolytics do not. The proteolytics and nonproteolytics have been shown to form single groups relative to somatic antigens as evaluated by agglutination.[108] The absorption of antiserum by any one of a group removes antibodies from all three of that group.

The nutritional requirements of these organisms are complex, with amino acids, B vitamins, and minerals being required. Synthetic media have been devised that support growth and toxin production of most types. The proteolytic strains tend not to be favored in their growth by carbohydrates, whereas the nonproteolytics are. At the same time, the nonproteolytics tend to be more fermentative than the proteolytic types.

The proteolytics generally do not grow below 12.5°C, although a few reports exist in which growth was detected at 10°C. The upper range for types A and proteolytic B, and presumably for the other proteolytic types, is about 50°C. On the other hand, the nonproteolytic strains can grow as low as 3.3°C with the maximum about 5° below that for proteolytics. Minimum and maximum temperatures of growth of these organisms are dependent on the state of other growth parameters, and the minima and maxima noted may be presumed to be at totally optimal conditions relative to pH, a_w, and the like. In a study of the minimum temperature for growth and toxin production by nonproteolytic types B and F in broth and crabmeat, both grew and produced toxin at 4°C in broth, but in crabmeat, growth and toxin production occurred only at 26°C and not at 12°C or lower.[107] A type G strain grew and produced toxin in broth and crabmeat at 12°C but not at 8°C.[107]

The minimum pH that permits growth and toxin production of *C. botulinum* strains has been the subject of many studies. It is generally recognized that growth does not occur at or below pH 4.5, and it is this fact that determines the degree of heat treatment given to foods with pH values below this level (see Chapter 17). Because of the existence of botulinal toxins in some high-acid, home-canned foods, this area has been the subject of recent studies. In one study, no growth of types A and B occurred in tomato juice at a pH around 4.8, but when the product was inoculated with *Aspergillus gracilis*, toxin was produced at pH 4.2 in association with the mycelial mat.[90] In another study with the starting pH of tomato juice at 5.8, the pH on the underside of the mold mat increased to 7.0 after 9 days and to 7.8 after 19 days.[51] The tomato juice was inoculated with type A botulinal spores, a *Cladosporium* sp., and a *Penicillium* sp. The topmost 0.5 ml of product showed pH increases from 5.3 to 6.4 or 7.5 after 9 and 19 days, respectively. One type B strain was shown to produce gas in tomato juice at pH 5.24 after 30 days and at pH 5.37 after 6 days. In food systems consisting of whole shrimp, shrimp purée, tomato purée, and tomato and shrimp purée acidified to a pH of 4.2 and 4.6 with acetic or citric acid, none of three type E strains grew or produced toxin at 26°C after 8 weeks.[96] Growth and toxin production of a type E strain at pH 4.20 and 26°C in 8 weeks was demonstrated when citric acid but not acetic acid was used to control the pH of a culture medium.[122] In general, the pH minima are similar for proteolytic and nonproteolytic strains.

With the use of aqueous suspensions of soy proteins inoculated with four type A and two type B strains with incubation at 30°C, growth occurred at pH 4.2, 4.3, and 4.4.[103] The inoculum was 5×10^6 spores/ml, and 4 weeks were required for detectable toxin at pH 4.4 when pH was adjusted with either HCl or citric acid. When lactic or acetic acid was used, 12 and 14 weeks, respectively, were required for toxin at a pH of 4.4. Inocula of 10^3–10^4/g of botulinal spores represent considerably higher numbers of these organisms than would be found on foods naturally (see below). That growth may occur at a pH lower than 4.5 with large inocula does not render invalid the widely held view that this organism does not grow at or below pH 4.5 in raw foods with considerably smaller numbers of spores.

Regarding the interaction of pH, NaCl, and growth temperature, a study with Japanese noodle soup (*tsuyu*) revealed that with types A and B spores, no toxin developed at (1) pH <6.5, 4% NaCl, and 20°C; (2) pH <5.0, 1% NaCl, and 30°C; (3) pH <5.5, 3% NaCl, and 30°C; and (4) pH <6.0, 4% NaCl, and 30°C.[55]

The minimum a_w that permits growth and toxin production of types A and proteolytic B strains is 0.94, and this value is well established. The minimum for type E is around 0.97. Although all strains have not been studied equally, it is possible that the other nonproteolytic strains have a minimum similar to that of type E. The way in which a_w is achieved in culture media affects the minimum values obtained. When glycerol is used as humectant, a_w values tend to be a bit lower than when NaCl or glucose is used.[110] Salt at a level of about 10%, or 50% sucrose, will inhibit growth of types A and B, and 3–5% salt has been found to inhibit toxin production in smoked fish chubs.[18] Lower levels of salt are required when nitrites are present (see Chapter 13).

With respect to heat resistance, the proteolytic strains are much more resistant than nonproteolytics (Table 24–3). Although the values noted in the table suggest that type A is the most heat resistant, followed by proteolytic F and then proteolytic B, these data should be taken only as representative, as heating menstra, previous history of strains, and other factors are known to affect heat resistance (Chapter 17). Of those noted, all were determined in phosphate buffer. Among type E, the Alaska and Beluga strains appear to be more heat resistant than others, and in ground whitefish chubs, $D_{80°C}$ of 2.1 and 4.3 have been reported,[22] whereas in crabmeat, $D_{82.2°C}$ of 0.51 and 0.74 have been reported, respectively, for Alaska and Beluga.[76] With regard to smoked whitefish chubs, it was determined in one study that heating to an internal temperature of 180°C for 30 minutes produced a nontoxigenic product,[31] whereas in another study, 10% or 1.2% of 858 freshly smoked chubs given the same heat treatment were contaminated, mostly with type E strains.[91] (The heat destruction of bacterial endospores is dealt with further in Chapter 17.)

With regard to type G, the Argentine and Swiss strains both produce two kinds of spores: heat labile and heat resistant. The former, which are destroyed at around 80°C after 10 minutes, represent about

99% of the spores in a culture of the Swiss strains; in the Argentine strain, only about 1 in 10,000 endospores is heat resistant.[75] The $D_{230°F}$ (110°C) of two heat-resistant strains in phosphate buffer was 0.45–0.54 minute, whereas for two heat-labile strains. $D_{180°F}$ (82.2°C) was 1.8–5.9 minutes.[75] The more heat-resistant spores of type G have not yet been propagated.

Toxigenic strains of *C. butyricum* grew and produced toxin at pH 5.2 but did not grow at pH 5.0.[88] The heat-resistant strains (nontoxigenic) grew at pH 4.2 and were considerably more heat resistant than the nontoxic strains.

Unlike heat, radiation seems to affect the endospores of proteolytic and nonproteolytic strains similarly, with D values of 1.1–2.5 kGy having been reported (see Chapter 15). However, the D value of one nonproteolytic type F strain was found to be 1.5 kGy, which was similar to the D value for a type A strain, but a proteolytic type F strain produced a D of 1.16 kGy.[7]

Ecology of *C. botulinum* Growth

It has been shown that this organism cannot grow and produce its toxins in competition with large numbers of other microorganisms. Toxin-containing foods are generally devoid of other types of organisms because of heat treatments. In the presence of yeasts, however, *C. botulinum* has been reported to grow and produce toxin at a pH as low as 4.0. Whereas a synergistic effect between clostridia and lactic acid bacteria has been reported on the one hand, lactobacilli will antagonize growth and toxin production; indirect evidence for this is the absence of botulinal toxins in milk. Yeasts are presumed to produce growth factors needed by the clostridia to grow at low pH, whereas the lactic acid bacteria may aid growth by reducing the Eh or inhibit growth by "lactic antagonism" (see Chapter 13). In one study, type A was inhibited by soil isolates of *C. sporogenes*, *C. perfringens*, and *B. cereus*.[105] Some *C. perfringens* strains produced an inhibitor that was effective on 11 type A strains, on 7 type B proteolytic and 1 nonproteolytic strains, and on 5 type E and 7 type F strains.[58] It is possible for *C. botulinum* spores to germinate and grow in certain canned foods where the pH is less than 4.5 when *Bacillus coagulans* is present. In a study with tomato juice of pH 4.5 inoculated with *B. coagulans*, the pH increased after 6 days at 35°C to 5.07, and to 5.40 after 21 days, thus making it possible for *C. botulinum* to grow.[6] Kautter et al.[58] found that type E strains are inhibited by other nontoxic organisms whose biochemical properties and morphological characteristics were similar to type E. These organisms were shown to effect inhibition of type E strains by producing a bacteriocin-like substance designated "boticin E." In a more detailed study, proteolytic A, B, and F strains were found to be resistant to boticin E elaborated by a non-toxic type E, but toxic E cells were susceptible.[5] The boticin was found to be sporostatic for nonproteolytic types B, E, and F; and nonproteolytic type E.

A report on the ecology of type F showed that the absence of this strain in mud samples during certain times of the year was associated with the presence of *Bacillus licheniformis* in the samples during these periods, when the bacillus was apparently inhibiting type F strains.[125]

In a search for antimicrobial compounds against *C. botulinum* strains, 200 *Bacillus/Paenibacillius* isolates from chilled foods containing vegetables were screened against Type A, proteolytic Type B, and Type E strains using a well diffusion assay. Antibotulinal activity was displayed by 19 (9.5%) of the screened culture supernatants, and *P. polymyxa* cultures showed the highest activity.[38] The inhibitory factor appeared in culture supernatants from late-log/early stationary phases of growth after 7–10 days at 10°C; and after 2–3 days at 20°C under aerobic and anaerobic conditions. The *P. polymyxa* factor was a heat-resistant peptide, and it was inhibitory to other bacteria, suggesting its relationship to polymyxin.

Concerns for Sous Vide and Related Food Products

Special concerns for the growth and toxin production by *C. botulinum* strains are presented by *sous vide* processed foods. By this method, developed in France around 1980, raw food is placed in high-barrier bags and cooked under vacuum (*sous vide*, "under vacuum"). Most, if not all, vegetative cells are destroyed, but bacterial spores survive. Thus, the sous vide product is one that contains bacterial spores in an O_2-depressed environment with no microbial competitors. In low-acid foods such as meats, poultry, and seafoods, spores of *C. botulinum* can germinate, grow, and produce toxin. Holding temperature and time are the two parameters that must be carefully monitored to avoid toxic products.

Whereas the proteolytic strains do not grow in the refrigerator temperature range, the nonproteolytic strains can. A summary of published data on the incidence of botulinal spores in meat and poultry reveals that the numbers are extremely low—well below 1 spore/g (Table 24–4). Assuming a mean spore load of 1 spore/g and constant storage at 3–5°C, low-acid sous vide meat products should be safe for at least 21 days. With raw rockfish fillets inoculated at a level of 1 spore per sample with a mixture of 13 strains of types E, B, and F, no toxin could be detected in 21 days when stored at 4°C[54] or in red snapper homogenates after 21 days at 4°C.[71] In inoculated modified-atmosphere-packaged (MAP) pork stored at 5°C, no toxin could be detected in 44 days.[69] Whatever storage time is possible under constant low-temperature storage is shortened by temperature abuse. Products that have secondary barriers such as a_w <0.93 or pH <4.6 may be held safely for longer periods of time even with some temperature abuse. Because *Bacillus* spp. spores may be more abundant than botulinal and because some can grow at a pH <4.6, it is not inconceivable that these forms can germinate, grow, and elevate pH during temperature abuse. Botulinal toxin was detectable in anaerobically stored noodles with an initial pH <4.5 when the pH was increased by microbial growth.[53] Although it is widely assumed that fish contains more botulinal spores than land animals and consequently should be of more concern, a recent study of 1,074 test samples of commercial vacuum-packaged fresh fish that were held at 12°C

Table 24–4 Incidence of *Clostridium botulinum* Spores in Meat and Poultry Products over a 14-Year Period

Country	Product	Number Positive/ Number Tested	C. botulinum Type (No.)	No./g
United Kingdom	Bacon	36/397	A (23) B (13)	0.00217
United States	Cooked ham	5/100	A (5)	0.00166
	Smoked turkey	1/41	B	0.0081
	Other meats	0/231	—	—
	Frankfurters	1/10	B	0.0066
	Other meats	0/80	—	—
	Luncheon meats	1/73	B	0.0057
	"Sausages"	0/17	—	—
United States and Canada	Raw chicken	1/1078	C	0.0031
	Raw beef, pork	0/1279	—	—
Canada	Sliced meats	0/436	—	—

Source: From Tompkin.[120]

for 12 days failed to develop botulinal toxins.[70] Inoculated type E strains grew in controls, suggesting that either the samples contained no botulinal spores or they were overgrown by other members of the biota.

The development of mathematical models designed to predict the probability of growth and toxin production in sous vide and MAP foods has been undertaken by several groups of investigators. By the use of factorial designs, these models are designed to integrate the individual and combined effects of the parameters of temperature, a_w, pH, inoculum size, and storage time. Equations were developed in one series of studies that could predict the length of time to toxin production and the probability of toxigenesis by a single spore under defined conditions using cooked, vacuum-packed potatoes.[24] From the latter study, the response by mixtures of five each of types A and B spores was shown to be linear, whereas to a_w, the response was curvilinear. In another series of studies employing MAP-stored fresh fish inoculated with nonproteolytic strains, 74.6% of experimental variation in the final multiple linear regression model was accounted for by temperature of storage, with spore load accounting for only 7.4%.[11] The earliest time to toxicity at 20°C was 1 day, but at 4°C the time increased to 18 days. With type E spores, no growth was observed in chopped meat medium at 3°C in 170 days, but in vacuum-packaged herring inoculated with 10^4 type E spores per gram, toxigenesis was detected after 21 days at 3.3°C.[11] Among other models reported is one that includes sorbic acid up to 2,270 ppm in combination with some of the other parameters noted.[72]

Nature of the Botulinal Neurotoxins

The neurotoxins are formed within the organism and released upon autolysis. They are produced by cells growing under optimal conditions, although resting cells have been reported to form toxin as well. The botulinal neurotoxins (BoNT) are the most toxic substances known, with purified type A reported to contain about 30 million mouse LD_{50}/mg. The minimum lethal dose for mice has been reported also to be 0.4–2.5 ng/kg by intravenous or intraperitoneal injection, and a 50% human lethal dose of about 1 ng/kg of body weight has been reported. The first of these toxins to be purified was type A, which was achieved by Lamanna et al. and by Abrams et al., both in 1946. The purification of B, E, and F has been achieved.

The genes for BoNT A, B, E, and F are chromosomal, whereas type G is plasmidborne.[128] The type-G-producing strains cluster apart from the other botulinal strains consistent with their placement in group IV. The gene for BoNT type B has been sequenced and cloned.[126]

BoNT is produced as a single polypeptide chain that is posttranslationally nicked to form a di-chain consisting of a 100-kDa heavy chain and a 50-kDa light chain held together by a disulfide bond. It is composed of three domains: binding, translocation, and catalytic. The binding domain has been used as an immunogen that affords protection against challenge doses of BoNT.[17] After BoNT binds to nerve cell receptors, it is believed to be internalized into an endosome followed by proteolytic cleavage of synaptobrevins (protein components of synaptic vesicles) that block neurotransmitter release.[17]

Type A toxin has been reported to be more lethal than B or E. Type B has been reported to beassociated with a much lower case mortality than type A, and case recoveries from type B have occurred even when appreciable amounts of toxin could be demonstrated in the blood.

Symptoms of botulism can be produced by either parenteral or oral administration of the toxins. They may be absorbed into the blood stream through the respiratory mucous membranes, as well as through the walls of the stomach and intestines. The toxins are not completely inactivated by the proteolytic enzymes of the stomach, and, indeed, those produced by nonproteolytics may be activated. The high-molecular-weight complexes or the progenitor possess higher resistance to acid and pepsin.[117] While the derivative toxin was rapidly inactivated, the progenitor was shown to be resistant to rat intestinal juice in vitro. The progenitor was more stable in the stomachs of rats. It appears that the nontoxic component of the progenitor provides protection to the toxin activity. After botulinal toxins are absorbed into the blood stream, they enter the peripheral nervous system. *C. botulinum* toxins consist of a series of seven related toxins with the major one having a molecular weight of ca. 150,000. The toxin binds to presynaptic terminal membranes at nerve-muscle junctions where the release of acetylcholine is blocked. This leads to flaccid paralysis, which is the basis of the cosmetic uses of Botox. In contrast to the botulinum toxin, tetanus toxin blocks a nerve factor that allows muscle relaxation and leads to spastic paralysis. Unlike the staphylococcal enterotoxins and heat-stable toxins of other foodborne pathogens, the botulinal toxins are heat sensitive and may be destroyed by heating at 80°C (176°F) for 10 minutes, or boiling temperatures for a few minutes.

The Adult Botulism Syndrome: Incidence and Vehicle Foods

Symptoms of botulism may develop anywhere between 12 and 72 hours after the ingestion of toxin-containing foods. Even longer incubation periods are not unknown. Symptoms consist of nausea, vomiting, fatigue, dizziness, and headache; dryness of skin, mouth, and throat; constipation, lack of fever, paralysis of muscles, double vision, and, finally, respiratory failure and death. The duration of the illness is from 1 to 10 or more days, depending on host resistance and other factors. The mortality rate varies between 30% and 65%, with the rate being generally lower in European countries than in the United States. All symptoms are caused by the exotoxin, and treatment consists of administering specific antisera as early as possible. Although it is assumed that the tasting of toxin-containing foods allows for absorption from the oral cavity, Lamanna et al.[68] found that mice and monkeys are more susceptible to the toxins when administered by stomach tube than by exposure to the mouth. The botulinal toxins are neurotoxins and attach irreversibly to nerves. Early treatment by use of antisera brightens the prognosis.

Prior to 1963, most cases of botulism in the United States in which the vehicle foods were identified were traced to home-canned vegetables and were caused by types A and B toxins. In almost 70% of the 640 cases reported for the period 1899–1967, the vehicle food was not identified. Among the 640 cases, 17.8% were associated with vegetables, 4.1% fruits, 3.6% fish, 2.2% condiments, 1.4% meats and poultry, and 1.1% for all others. Reported foodborne cases in the United States for the years 1977–1997 reveal that the largest number of cases occurred in 1977 in a restaurant in Pontiac, Michigan, following consumption of a hot sauce prepared from home-canned jalapeño peppers. No deaths occurred, and type B toxin was identified. Total cases from all sources in the United States rarely exceed 50 per year, with the highest 10-year period being 1930–1939, when 384 cases were reported from noncommercial foods. Between 1899 and 1963, 1561 cases were reported from noncommercial foods, whereas 219 were reported from commercial foods between 1906 and 1963, with 24 in 1963 alone.

Of 404 verified cases of type E botulism through 1963, 304 or 75% occurred in Japan. No outbreak of botulism was recorded in Japan prior to 1951. For the period May 1951 through January 1960, 166 cases were recorded, with 58 deaths for a mortality rate of 35%. Most of these outbreaks were traced to a home-prepared food called *izushi*, a preserved food consisting of raw fish, vegetables, cooked rice, malted rice (*koji*), and a small amount of salt and vinegar. This preparation is packed tightly in a wooden tub equipped with a lid and held for 3 weeks or longer to permit lactic acid fermentation. During this time, the Eh potential is lowered, thus allowing for the growth of anaerobes.

Sixty-two outbreaks of botulism resulting from commercially canned foods were recorded for the period 1899–1973,[74] with 41 prior to 1930. Between 1941 and 1982, 7 outbreaks occurred in the United States involving commercially canned foods in metal containers, with 17 cases and 8 deaths.[89] Three of these outbreaks were caused by type A and the remainder by type E. In five of the outbreaks, can leakage or underprocessing occurred.[89] Canned mushrooms have been incriminated in several botulism outbreaks. A study in 1973 and 1974 turned up 30 cans of mushrooms containing botulinal toxin (29 were type B). An additional 11 cans contained viable spores of *C. botulinum* without preformed toxin.[74] The capacity of the commercial mushroom (*Agaricus bisporus*) to support the growth of inoculated spores of *C. botulinum* was studied by Sugiyama and Yang.[119] Following inoculation of various parts of mushrooms, they were sealed with plastic film and incubated. Toxin was detected as early as 3–4 days later, when products were incubated at 20°C. Although the plastic film used to wrap the inoculated mushrooms allowed for gas exchange, the respiration of the fresh mushrooms apparently consumed oxygen at a faster rate than it entered the film. No toxin was detected in products stored at refrigerator temperatures.

An unusual outbreak of 36 cases occurred in 1985 with victims in three countries: Canada, the Netherlands, and the United States. The vehicle food was chopped garlic in soybean oil that was packaged in glass bottles. Although labeled with instructions to refrigerate, unopened bottles were stored unrefrigerated for 8 months. The product was used to make garlic-buttered bread, which in turn was used to prepare beef dip sandwiches. Proteolytic type B spores were found, and toxin was produced within 2 weeks when proteolytic and nonproteolytic B strains were inoculated into bottles of chopped garlic and held at 25°C.[111] Type A toxin was produced in bottled chopped garlic in 20 days at 35°C when inoculated with 1 spore/g and by type B toxin in 20 days.[106] In the latter study, highly toxic bottles looked and smelled acceptable. An outbreak of type E involving 91 cases with 20 deaths occurred in Egypt in 1991, and the vehicle was an ungutted, salted raw fish product called *faseikh*.[84] In 1994, there were 30 victims in El Paso, Texas, after consuming potato-based dip and eggplant-based dip, both of which contained baked potato. The baked potatoes were wrapped in aluminum foil and held at room temperature for several days and, thus, they became toxigenic.[8] Although not common, this is not the first time that botulinal toxin was detected in baked potatoes.

One of the recorded outbreaks of botulism (five cases with one death) due to type F involved homemade liver paste. The U.S. outbreak occurred in 1966 from home-prepared venison jerky, with three clinical cases.[82]

The greatest hazards of botulism come from home-prepared and home-canned foods that are improperly handled or given insufficient heat treatments to destroy botulinal spores. Such foods are often consumed without heating. The best preventive measure is the heating of suspect foods to boiling temperatures for a few minutes, which is sufficient to destroy the neurotoxins.

Infant Botulism

First recognized as such in California in 1976, infant botulism has since been confirmed in most states in the United States and in many other countries. In the adult form of botulism, preformed toxins are ingested; in infant botulism, viable botulinal spores are ingested, and upon germination in the intestinal tract, toxin is synthesized. Although it is possible that in some adults under special conditions botulinal endospores may germinate and produce small quantities of toxin, the colonized intestinal tract does not favor spore germination. Infants over one year of age tend not to be affected by this syndrome because of the establishment of a more normal intestinal biota. The disease is mild in some infants; in others, it can be severe. High numbers of spores are found in the feces of infants during the acute phase of the disease, and as recovery progresses, the number of organisms abate.

This syndrome is diagnosed by demonstrating botulinal toxins in infant stools and by the use of the mouse lethality test. Because *Clostridium difficile* produces mouse-lethal toxins in the intestinal tract of infants, it is necessary to differentiate between these toxins and that of *C. botulinum.*[35]

Infants get viable spores from infant foods and possibly from their environment. Vehicle foods are those that do not undergo heat processing to destroy endospores; the two most common products are syrup and honey. Of 90 samples of honey examined, 9 contained viable spores. Six of these had been fed to babies who developed infant botulism.[83] Of the nine, seven were type B and two were type A. Of 910 infant foods from 10 product classes, only 2 classes were positive for spores: honey and corn syrup.[59] Of 100 honey samples, 2 contained type A, and 8 of 40 corn syrup samples yielded type B. In Canada, only 1 of 150 samples of honey contained viable botulinal spores (type A), 1 of 40 dry cereal samples (type B), whereas 43 syrup samples were negative.[56] Reported cases in the United States through 1997 revealed 62 cases for 1982, which occurred among infants aged 2–48 weeks, and involved type A and B toxins equally. The first two cases reported in Rome, Italy, were caused by type E toxin, which was produced by *Clostridium butyricum.*[21] The first reported infant case caused by type B in Japan occurred in 1995, and the toxin was found to possess a lower toxicity and possibly a lower binding capacity than the adult form.[62]

Animal models for the study of this syndrome consist of 8- to 11-day-old mice[118] and 7- to 13-day-old rats.[85] In the mouse model, botulinal toxin was found in the lumen of the large intestine, and it was not associated with the ileum. (The sensitivity of these animal models is noted in Chapter 12.)

BACILLUS CEREUS GASTROENTERITIS

Bacillus cereus is an aerobic, spore-forming rod normally present in soil, dust, and water. It has been associated with food poisoning in Europe since at least 1906. Among the first to report this syndrome with precision was Plazikowski. His findings were confirmed by several other European workers in the early 1950s. The first documented outbreak in the United States occurred in 1969, and the first in Great Britain occurred in 1971.

Low numbers of this bacterial species can be found in a number of food products, including fresh and processed foods. In a study of raw meats, meat products, and product additives, *B. cereus* was found in 6.6% of 534, 18.3% of 820, and 39.1% of 609 samples, respectively,[61] with levels of 10^2–10^4/g. It is unclear if any were enterotoxigenic. Enterotoxigenic strains were recovered from a variety of foods in another study, with 85% of 83 strains from raw milk being positive for the diarrheagenic toxin.[43] Other enterotoxin-producing species are listed below.

In addition to *B. cereus*, *B. mycoides* strains from milk have been shown to produce diarrheagenic enterotoxin in nine days at temperatures between 6°C and 21°C.[43] Varying numbers of isolates of the following species were found also to be enterotoxin producers: *P. circulans*, *B. lentus*, *B. thuringiensis*, *B. pumilus*, *P. polymyxa*, *B. carotarum*, and *B. pasterurii.*[43] *B. thuringiensis* has been isolated from foods, and it apparently produces a Vero-cell active toxin.[23]

This bacterium has a minimum growth temperature around 4–5°C, with a maximum around 48–50°C. Growth has been demonstrated over the pH range 4.9–9.3.[40] Its spores possess a resistance to heat typical of other mesophiles.

B. cereus Toxins

This bacterium produces a wide variety of extracellular toxins and enzymes, including lecithinase, proteases, β-lactamase, sphingomyelinase, cereolysin (mouse lethal toxin, hemolysin I), and

Table 24–5 *Bacillus* and
Paenibacillus Species that have
been Demonstrated to Produce
the HBL Enterotoxin Normally
Associated with *B. cereus* (taken
from 95, 97, and other references)

B. cereus	*B. mycoides*
B. amyloliquefaciens	*B. pasteurii*
P. circulans	*B. pseudomycoides*
B. coagulans	*B. subtilis*
B. lentimorbis	*B. thuringiensis*
B. lentus	*B. weihenstephanensis*
B. licheniformis	*P. polymyxa*
B. megaterium	

hemolysin BL. Cereolysin is a thiol-activated toxin analogous to perfringolysin O. It has a molecular weight of 55 kDa and apparently plays no role in the foodborne gastroenteritis syndromes.

The diarrheagenic syndrome appears to be produced by a tripartite complex composed of components B, L_1, and L_2 and designated hemolysin BL (HBL). Together, this complex exhibits hemolysis, cytolysis, dermonecrosis, vascular permeability, and enterotoxic activity. It accounts for about 50% of the retinal toxicity of *B. cereus* culture supernatants in endophthalmitis.[14] Although no single enterotoxin has been demonstrated, it appears that HBL is responsible for the diarrheagenic syndrome.[13] By the use of a commercial test kit that detects the L_2 component, it has been shown that it is produced during the logarithmic growth phase. About 10^7 cells/ml appear to be needed for demonstrable toxic activity, and production is favored over the pH range 6.0–8.5. Several strains have been shown to produce toxin between 6°C and 21°C.[43] A polymerase chain reaction (PCR) assay based on the *hbl*A gene (that encodes the B component) has been developed and found to be faster than the test kits for the diarrheal enterotoxin.

Although the diarrheal syndrome is generally associated with *B. cereus* and the HBL toxin complex, at least 14 other species of the genera *Bacillus* and *Paenibacillus* have been shown to produce this foodborne illness (Table 24–5). While it has been suggested that *B. cereus*, *B. thuringiensis*, and *B. anthracis* should be a single species based on a genetic analysis,[49] the latter species apparently does not produce the *B. cereus* toxin complex. *B. anthracis* is, however, a member of the *B. cereus* group, which includes *B. mycoides*, *B. pseudomycoides*, *B. weihenstephanensis*, and *B. thuringiensis*.[97]

For a review of the possible relationship of *B. anthracis* to food safety, see reference 30.

The *B. cereus* virulence factor is a hemolytic enterotoxin complex designated HBL as noted above, and it exhibits hemolysis, dermonerosis, and vascular permeability properties.

Diarrheal Syndrome

This syndrome is rather mild, with symptoms developing within 8–16 hours, more commonly within 12–13 hours, and lasting for 6–12 hours.[47] Symptoms consist of nausea (with vomiting being rare), cramplike abdominal pains, tenesmus, and watery stools. Fever is generally absent. The similarity between this syndrome and that of *C. perfringens* food poisoning has been noted.[34]

Vehicle foods consist of cereal dishes that contain corn and cornstarch, mashed potatoes, vegetables, minced meat, liver sausage, meat loaf, milk, cooked meat, Indonesian rice dishes, puddings, soups, and others.[34] Reported outbreaks between 1950 and 1978 have been summarized by Gilbert,[34] and when plate counts on leftover foods were recorded, they ranged from 10^5 to 9.5×10^5/g, with many in the 10^7–10^8/g range. The first well-studied outbreaks were those investigated by Hauge[46], which were traced to vanilla sauce; the counts ranged from 2.5×10^7 to 1×10^8/g. From meat loaf involved in a U.S. outbreak in 1969, 7×10^7/g were found.[81] Serovars found in diarrheal outbreaks include types 1, 6, 8, 9, 10, and 12. Serovars 1, 8, and 12 have been associated with this as well as with the emetic syndrome.[34]

Emetic Syndrome

This form of *B. cereus* food poisoning is more severe and acute than the diarrheal syndrome. The incubation period ranges from 1 to 6 hours, with 2 to 5 hours being most common.[87] Its similarity to the staphylococcal food-poisoning syndrome has been noted.[34] It is often associated with fried or boiled rice dishes. In addition to these, pasteurized cream, spaghetti, mashed potatoes, and vegetable sprouts have been incriminated.[34] Outbreaks have been reported from Great Britain, Canada, Australia, the Netherlands, Finland, Japan, and the United States. The first U.S. outbreak was reported in 1975, with mashed potatoes as the vehicle food.

The numbers of organisms necessary to cause this syndrome seem to be higher than those for the diarrheal syndrome, with numbers as high as 2×10^9/g having been found.[34] *B. cereus* serovars associated with the emetic syndrome include 1, 3, 4, 5, 8, 12, and 19.[34]

The emetic (vomiting type) toxin has been determined to be cereulide, an ionophoric, water-insoluble peptide that is closely related to the peptide antibiotic valinomycin.[1] It has a molecular weight of about 1.2 kDa. It induces the formation of vacuoles in HEp-2 cells (see Chapter 12), and neither this activity nor the emetic is lost after heating for 30 minutes at 121°C.[1,2] The house musk or shrew (*Suncus murinus*) has been found to be a suitable experimental animal for the emetic activity.[2]

Using 3 *B. cereus* strains and either tryptic soy or trypticase soy media, cereulide production commenced at the end of the log phase of growth and toxin production was independent of sporulation.[44] The strains produced 80 to 166 μg of cereulide/ml at 21°C in 1 to 3 days during the stationary growth phase when cell numbers reached 2×10^8 to 6×10^8 cfu/ml. Production at 40 and below 8°C was minimal.[44]

The emetic toxin strains grow over the range 15–50°C, with an optimum between 35°C and 40°C.[56] Whereas the emetic syndrome is most often associated with rice dishes, growth of the emetic toxin strains in rice is not favored in general over other *B. cereus* strains, although higher populations and more extensive germination have been noted in this product.[56]

REFERENCES

1. Agata, N., M. Ohta, M. Mori, and M. Isobe. 1995. A novel dodecadepsipeptide, cereulide, is an emetic toxin of *Bacillus cereus*. *FEMS Microbiol. Lett.* 129:17–20.

2. Agata, N., M. Mori, M. Ohta, S. Suwan, I. Ohtani, and M. Isobe. 1994. A novel dodecadepsipeptide, cereulide, isolated from *Bacillus cereus* causes vacuole formation in HEp-2 cells. *FEMS Microbiol. Lett.* 121:31–34.

3. Alderton, G., K.A. Ito, and J.K. Chen. 1976. Chemical manipulation of the heat resistance of *Clostridium botulinum* spores. *Appl. Environ. Microbiol.* 31:492–498.

4. Alderton, G., and N. Snell. 1969. Bacterial spores: Chemical sensitization to heat. *Science.* 163:1212–1213.

5. Anastasio, K.L., J.A. Soucheck, and H. Sugiyama. 1971. Boticinogeny and actions of the bacteriocin. *J. Bacteriol.* 107:143–149.

6. Anderson, R.E. 1984. Growth and corresponding elevation of tomato juice pH by *Bacillus coagulans*. *J. Food Sci.* 49:647, 649.

7. Anellis, A., and D. Berkowitz. 1977. Comparative dose-survival curves of representative *Clostridium botulinum* type F spores with type A and B spores. *Appl. Environ. Microbiol.* 34:600–601.

8. Angulo, F.J., J. Getz, J.P. Taylor, K.A. Hendricks, E.L. Hathaway, S.S. Barth, H.M. Solomon, A.E. Larson, E.A. Johnson, L.N. Nickey, and A.A. Reis. 1998. A large outbreak of botulism: The hazardous baked potato. *J. Infect. Dis.* 178:172–177.

9. Anniballi, F.L. Fenicia, G. Franciosa, and P. Aureli. 2002. Influence of pH and temperature on the growth of and toxin production by neurotoxigenic strains of *Clostridium butyricum* type E. *J. Food Protet.* 65:1267–1270.

10. Austin, J.W., and K.C. Dodds. 2001. *Clostridium botulinum*. In *Foodborne Disease Handbook*, Vol. 1, 2nd ed., ed. Y.H. Heu, M.D. Pierson, and J.R. Gorham, 107–138. New York: Marcel Dekker.

11. Baker, D.A., and C. Genigeorgis. 1990. Predicting the safe storage of fresh fish under modified atmospheres with respect to *Clostridium botulinum* toxigenesis by modeling length of the lag phase of growth. *J. Food Protect.* 53:131–140.

12. Barnes, E., and M. Ingram. 1956. The effect of redox potential on the growth of *Clostridium welchii* strains isolated from horse muscle. *J. Appl. Bacteriol.* 19:117–128.

13. Beecher, D.J., J.L. Schoeni, and A.C.L. Wong. 1995. Enterotoxic activity of hemolysin BL from *Bacillus cereus*. *Infect. Immun.* 63:4423–4428.

14. Beecher, D.J., J.S. Pulido, N.P. Barney, and A.C.L. Wong. 1995. Extracellular virulence factors in *Bacillus cereus* endophthalmitis: Methods and implication of involvement of hemolysin BL. *Infect. Immun.* 63:632–639.

15. Bradshaw, J.G., G.N. Stelma, V.I. Jones, J.T. Peeler, J.G. Wimsatt, J.J. Corwin, and R.M. Twedt. 1982. Thermal inactivation of *Clostridium perfringens* enterotoxin in buffer and in chicken gravy. *J. Food Sci.* 47:914–916.

16. Bryan, F.L., T.W. McKinely, and B. Mixon. 1971. Use of time-temperature evaluations in detecting the responsible vehicle and contributing factors of foodborne disease outbreaks. *J. Milk Food Technol.* 34:576–582.

17. Byrne, M.P., T.J. Smith, V.A. Montgomery, and L.A. Smith. 1998. Purification, potency, and efficacy of the botulinum neurotoxin type A binding domain from *Pichia pastoris* as a recombinant vaccine candidate. *Infect. Immun.* 66:4817–4822.

18. Christiansen, L.N., J. Deffner, E.M. Foster, and H. Sugiyama. 1968. Survival and outgrowth of *Clostridium botulinum* type E spores in smoked fish. *Appl. Microbiol.* 16:133–137.

19. Ciccarelli, A.S., D.N. Whaley, L.M. McCroskey, D.F. Gimenez, V.R. Dowell, Jr., and C.L. Hatheway. 1977. Cultural and physiological characteristics of *Clostridium botulinum* type G and the susceptibility of certain animals to its toxin. *Appl. Environ. Microbiol.* 34:843–848.

20. Craig, J., and K. Pilcher. 1966. *Clostridium botulinum* type F: Isolation from salmon from the Columbia River. *Science.* 153:311–312.

21. Creti, R., L. Fenicia, and P. Aureli. 1990. Occurrence of *Clostridium botulinum* in the soil of the vicinity of Rome. *Curr. Microbiol.* 20:317–321.

22. Crisley, F.D., J.T. Peeler, R. Angelotti, and H.E. Hall. 1968. Thermal resistance of spores of five strains of *Clostridium botulinum* type E in ground whitefish chubs. *J. Food Sci.* 33:411–416.

23. Damgaard, P.H., H.D. Larsen, B.M. Hansen, J. Bresciani, and K. Jorgensen. 1996. Enterotoxin-producing strains of *Bacillus thuringiensis* isolated from food. *Lett. Appl. Microbiol.* 23:146–150.

24. Dodds, K.L. 1989. Combined effect of water activity and pH on inhibition of toxin production by *Clostridium botulinum* in cooked, vacuum-packed potatoes. *Appl. Environ. Microbiol.* 55:656–660.

25. Duncan, C.L. 1973. Time of enterotoxin formation and release during sporulation of *Clostridium perfringens* type A. *J. Bacteriol.* 113:932–936.

26. Duncan, C.L. 1976. *Clostridium perfringens*. In *Food Microbiology: Public Health and Spoilage Aspects*, ed. M.P. deFigueiredo and D.F. Splittstoesser, 170–197. Westport, CT: AVI.

27. Duncan, C.L., and D.H. Strong. 1968. Improved medium for sporulation of *Clostridium perfringens*. *Appl. Microbiol.* 16:82–89.

28. Duncan, C.L., and D.H. Strong. 1969. Ileal loop fluid accumulation and production of diarrhea in rabbits by cell-free products of *Clostridium perfringens*. *J. Bacteriol.* 100:86–94.

29. Eklund, M., and F. Poysky. 1965. *Clostridium botulinum* type E from marine sediments. *Science* 149–306.

30. Erickson, M.C., and J.L. Kornacki. 2003. *Bacillus anthracis*: Knowledge in contamination of food. *J. Food Protect.* 66:691–699.

31. Fantasia, L.D., and A.P. Duran. 1969. Incidence of *Clostridium botulinum* type E in commercially and laboratory dressed white fish chubs. *Food Technol.* 23:793–794.

32. Foegeding, P.M., and F.F. Busta. 1980. *Clostridium perfringens* cells and phospholipase C activity at constant and linearly rising temperatures. *J. Food Sci.* 45:918–924.

33. Genigeorgis, C., G. Sakaguchi, and H. Riemann. 1973. Assay methods for *Clostridium perfringens* type A enterotoxin. *Appl. Microbiol.* 26:111–115.

34. Gilbert, R.J. 1979. *Bacillus cereus* gastroenteritis. In *Food-borne infections and intoxications*, ed. H. Riemann and F.L. Bryan, 495–518. New York: Academic Press.

35. Gilligan, P.H., L. Brown, and R.E. Berman. 1983. Differentiation of *Clostridium difficile* toxin from *Clostridium botulinum* toxin by the mouse lethality test. *Appl. Environ. Microbiol.* 45:347–349.

36. Giménez, J.A., M.A. Giménez, and B.R. DasGupta. 1992. Characterization of the neurotoxin isolated from a *Clostridium baratii* strain implicated in infant botulism. *Infect. Immun.* 60:518–522.

37. Gimenez, D.F., and A.S. Ciccarelli. 1970. Another type of *Clostridium botulinum*. *Zentral. Bakteriol. Orig. A* 215:221–224.

38. Girardin, H., C. Albagnac, C. Dargaignaratz, C. Nguyen-The, and F. Carlin. 2002. Antimicrobial activity of foodborne *Paenibacillus* and *Bacillus* spp. against *Clostridium botulinum*. *J. Food Protect.* 65:806–813.

39. Goepfert, J.M., and H.U. Kim. 1975. Behavior of selected foodborne pathogens in raw ground beef. *J. Milk Food Technol.* 38:449–452.

40. Goepfert, J.M., W.M. Spira, and H.U. Kim. 1972. *Bacillus cereus*: Food poisoning organism. A review. *J. Milk Food Technol.* 35:213–227.

41. Goldner, S.B., M. Solbert, S. Jones, and L.S. Post. 1986. Enterotoxin synthesis by nonsporulating cultures of *Clostridium botulinum*. *Appl. Environ. Microbiol.* 52:407–412.

42. Granum, P.E., W. Telle, Ø. Olsvik, and A. Stavn. 1984. Enterotoxin formation by *Clostridium perfringens* during sporulation and vegetative growth. *Int. J. Food Microbiol.* 1:43–49.

43. Griffiths, M.W. 1990. Toxin production by psychrotrophic *Bacillus* spp. present in milk. *J. Food Protect.* 53:790–792.

44. Häggblom, M.M., C. Apetroaie, M.A. Andersson, and M.S. Salkinoja-Salonen. 2002. Quantitative analysis of cereulide, the emetic toxin of *Bacillus cereus*, produced under various conditions. *Appl. Environ. Microbiol.* 68:2479–2483.

45. Hall, J.D., L.M. McCroskey, B.J. Pincomb, and C.L. Hatheway. 1985. Isolation of an organism resembling *Clostridium barati* which produces type F botulinal toxin from an infant with botulism. *J. Clin. Microbiol.* 21:654–655.

46. Hauge, S. 1955. Food poisoning caused by aerobic spore-forming bacilli. *J. Appl. Bacteriol.* 18:591–595.

47. Hauschild, A.H., and H. Hilsheimer. 1971. Purification and characteristics of the enterotoxin of *Clostridium perfringens* type A. *Can. J. Microbiol.* 17:1425–1433.

48. Hielm, S., J. Björkroth, E. Hyytiä, and H. Korkeala. 1998. Prevalence of *Clostridium botulinum* in Finnish trout farms: Pulsed-field gel electrophoresis typing reveals extensive genetic diversity among type E isolates. *Appl. Environ. Microbiol.* 64:4161–4167.

49. Helgasoon, E.O., A. Økstad, D.A. Caugant, H.A. Johansen, A. Fouet, M. Mock, I. Hegna, and A.-B. Kolsto. 2000. *Bacillus anthracis*, *Bacillus cereus*, and *Bacillus thuringiensis*—One species on the basis of genetic evidence. *Appl. Environ. Microbiol.* 66:2627–2630.

50. Hobbs, B., M. Smith, C. Oakley, G. Warrack, and J. Cruickshank. 1953. *Clostridium welchii* food poisoning. *J. Hyg.* 51:75–101.

51. Huhtanen, C.N., J. Naghski, C.S. Custer, and R.W. Russell. 1976. Growth and toxin production by *Clostridium botulinum* in moldy tomato juice. *Appl. Environ. Microbiol.* 32:711–715.

52. Huss, H.H. 1980. Distribution of *Clostridium botulinum*. *Appl. Environ. Microbiol.* 39:764–769.

53. Ikawa, J.Y. 1991. *Clostridium botulinum* growth and toxigenesis in shelf-stable noodles. *J. Food Sci.* 56:264–265.

54. Ikawa, J.Y., and C. Genigeorgis. 1987. Probability of growth and toxin production by nonproteolytic *Clostridium botulinum* in rockfish fillets stored under modified atmospheres. *Int. J. Food Microbiol.* 4:167–181.

55. Imai, H., K. Oshita, H. Hashimoto, and D. Fukushima. 1990. Factors inhibiting the growth and toxin formation of *Clostridium botulinum* types A and B in "tsuyu" (Japanese noodle soup). *J. Food Protect.* 53:1025–1032.

56. Johnson, K.M., C.L. Nelson, and F.F. Busta. 1983. Influence of temperature on germination and growth of spores of emetic and diarrheal strains of *Bacillus cereus* in a broth medium and in rice. *J. Food Sci.* 48:286–287.

57. Kang, C.K., M. Woodburn, A. Pagenkopf, and R. Cheney. 1969. Growth, sporulation, and germination of *Clostridium perfringens* in media of controlled water activity. *Appl. Microbiol.* 118:798–805.

58. Kautter, D.A., S.M. Harmon, R.K. Lynt, Jr., and T. Lilly, Jr. 1966. Antagonistic effect on *Clostridium botulinum* type E by organisms resembling it. *Appl. Microbiol.* 14:616–622.

59. Kautter, D.A., T. Lilly, Jr., H.M. Solomon, and R.K. Lynt. 1982. *Clostrium botulinum* spores in infant foods: A survey. *J. Food Protect.* 45:1028–1029.

60. Kokai-Kun, J.F., K. Benton, E.U. Wieckowski, and B.A. McClane. 1999. Identification of a *Clostridium perfringens* enterotoxin region required for large complex formation and cytotoxicity by random mutagenesis. *Infect. Immun.* 67:5634–5641.

61. Konuma, H., K. Shinagawa, M. Tokumaru, Y. Onoue, S. Konno, N. Fujino, T. Shigehisa, H. Kurate, Y. Kuwabara, and C.A.M. Lopes. 1988. Occurrence of *Bacillus cereus* in meat products, raw meat and meat product additives. *J. Food Protect.* 51:324–326.

62. Kozaki, S., Y. Kamata, T.-I. Nishiki, H. Kakinuma, H. Maruyama, H. Takahashi, T. Karasawa, K. Yamakawa, and S. Nakamura. 1998. Characterization of *Clostridium botulinum* type B neurotoxin associated with infant botulism in Japan. *Infect. Immun.* 66:4811–4816.

63. Krakauer, T., B. Fleischer, D.L. Stevens, B.A. McClane, and B.G. Stiles. 1997. *Clostridium perfringens* enterotoxin lacks superantigenic activity but induces an interleukin-6 response from human peripheral blood mononuclear cells. *Infect. Immun.* 65:3485–3488.

64. Labbe, R.G. 1980. Relationship between sporulation and enterotoxin production in *Clostridium perfringens* type A. *Food Technol.* 34(4):88–90.

65. Labbe, R.G., and C.L. Duncan. 1974. Sporulation and enterotoxin production by *Clostridium perfringens* type A under conditions of controlled pH and temperature. *Can. J. Microbiol.* 20:1493–1501.

66. Labbe, R.G., and L.L. Nolan. 1981. Stimulation of *Clostridium perfringens* enterotoxin formation by caffeine and theobromine. *Infect. Immun.* 34:50–54.

67. Labbe, R.G., and D.K. Rey. 1979. Raffinose increases sporulation and enterotoxin production by *Clostridium perfringens* type A. *Appl. Environ. Microbiol.* 37:1196–1200.

68. Lamanna, C., R.A. Hillowalla, and C.C. Alling. 1967. Buccal exposure to botulinal toxin. *J. Infect. Dis.* 117:327–331.

69. Lambert, A.D., J.P. Smith, and K.L. Dodds. 1991. Combined effect of modified atmosphere packaging and low-dose irradiation on toxin production by *Clostridium botulinum* in fresh pork. *J. Food Protect.* 54.94–101.

70. Lilly, T., Jr., and D.A. Kautter. 1990. Outgrowth of naturally occurring *Clostridium botulinum* in vacuum-packaged fresh fish. *J. Assoc. Off. Anal. Chem.* 73:211–212.

71. Lindroth, S., and C. Genigeorgis. 1986. Probability of growth and toxin production by nonproteolytic *Clostridium botulinum* in rock fish stored under modified atmospheres. *Int. J. Food Microbiol.* 3:167–181.

72. Lund, B.M., A.F. Graham, S.M. George, and D. Brown. 1990. The combined effect of incubation temperature, pH and sorbic acid on the probability of growth of nonproteolytic type B *Clostridium botulinum*. *J. Appl. Bacteriol.* 69:481–492.

73. Lukinmaa, S., E. Takkunen, and A. Siitonen. 2002. Molecular epidemiology of *Clostridium perfringens* related to foodborne outbreaks of disease in Finland from 1984 to 1999. *Appl. Environ. Microbiol.* 68:3744–3749.

74. Lynt, R.K., D.A. Kautter, and R.B. Read, Jr., 1975. Botulism in commercially canned foods. *J. Milk Food Technol.* 38:546–550.

75. Lynt, R.K., H.M. Solomon, and D.A. Kautter. 1984. Heat resistance of *Clostridium botulinum* type G in phosphate buffer. *J. Food Protect.* 47:463–466.

76. Lynt, R.K., H.M. Solomon, T. Lilly, Jr., and D.A.Kautter. 1977. Thermal death time of *Clostridium botulinum* type E in meat of the blue crab. *J. Food Sci.* 42:1022–1025, 1037.

77. Mahony, D.E. 1977. Stable L-forms of *Clostridium perfringens*: Growth, toxin production, and pathogenicity. *Infect. Immun.* 15:19–25.

78. Mäntynen, V., and K. Lindström. 1998. A rapid PCR-based test for enerotoxic *Bacillus cereus*. *Appl. Environ. Microbiol.* 64:1634–1639.

79. McClung, L. 1945. Human food poisoning due to growth of *Clostridium perfringens* (*C. welchii*) in freshly cooked chicken: Preliminary note. *J. Bacteriol.* 50:229–231.

80. McCroskey, L.M., C.L. Hatheway, L. Fenicia, B. Pasolini, and P. Aureli. 1986. Characterization of an organism that produces type E botulinal toxin but which resembles *Clostridium butyricum* from the feces of an infant with type E botulism. *J. Clin. Microbiol.* 23:201–202.

81. Midura, T., M. Gerber, R. Wood, and A.R. Leonard. 1970. Outbreak of food poisoning caused by *Bacillus cereus*. *Public Health Rep*. 85:45–47.

82. Midura, T.F., G.S. Nygaard, R.M. Wood, and H.L. Bodily. 1972. *Clostridium botulinum* type F: Isolation from venison jerky. *Appl. Microbiol*. 24:165–167.

83. Midura, T.F., S. Snowden, R.M. Wood, and S.S. Arnon. 1979. Isolation of *Clostridium botulinum* from honey. *J. Clin. Microbiol*. 9:282–283.

84. Mishu, B., A. Darweigh, J.T. Weber, C.L. Hathewahy, S. El-Sharkaway, and A. Corwin. 1991. A foodborne outbreak of type E botulism in Cairo, Egypt, April, 1991. *Am. J. Trop. Med. Hyg*. 45(3S):109.

85. Moberg, L.J., and H. Sugiyama. 1980. The rat as an animal model for infant botulism. *Infect. Immun*. 29:819–821.

86. Moller, V., and I. Scheibel. 1960. Preliminary report on the isolation of an apparently new type of *Cl. botulinum*. *Acta Pathol. Microbiol. Scand*. 48:80.

87. Mortimer, P.R., and G. McCann. 1974. Food-poisoning episodes associated with *Bacillus cereus* in fried rice. *Lancet* 1:1043–1045.

88. Morton, R.D., V.N. Scott, D.T. Bernard, and R.C. Wiley. 1990. Effect of heat and pH on toxigenic *Clostridium butyricum*. *J. Food Sci*. 55:1725–1727, 1739.

89. NFPA/CMI Task Force. 1984. Botulism risk from post-processing contamination of commercially canned foods in metal containers. *J. Food Protect*. 47:801–816.

90. Odlaug, T.E., and I.J. Pflug. 1979. *Clostridium botulinum* growth and toxin production in tomato juice containing *Aspergillus gracilis*. *Appl. Environ. Microbiol*. 37:496–504.

91. Pace, P.J., E.R. Krumbiegel, R. Angelotti, and H.J.Wieniewski. 1967. Demonstration and isolation of *Clostridium botulinum* types from whitefish chubs collected at fish smoking plants of the Milwaukee area. *Appl. Microbiol*. 15:877–884.

92. Pearson, C.B., and H.W. Walker. 1976. Effect of oxidation-reduction potential upon growth and sporulation of *Clostridium perfringens*. *J. Milk Food Technol*. 39:421–425.

93. Pederson, H.O. 1955. On type E botulism. *J. Appl. Bacteriol*. 18:619–629.

94. Peterson, D., H. Anderson, and H. Detels. 1966. Three outbreaks of foodborne disease with dual etiology. *Public Health Rep*. 81:899–904.

95. Phelps, R.J., and J.L. McKillip. 2002. Enterotoxin production in natural isolates of *Bacillus* outside the *Bacillus cereus* group. *Appl. Environ. Microbiol*. 68:3147–3151.

96. Post, L.S., T.L. Amoroso, and M. Solberg. 1985. Inhibition of *Clostridium botulinum* type E in model acidified food systems. *J. Food Sci*. 50:966–968.

97. Pruss, B.M., R. Dietrich, B. Nibler, E. Märtlbauer, and S. Scherer. 1999. The hemolytic enterotoxin HBL is broadly distributed among species of the *Bacillus cereus* group. *Appl. Environ. Microbiol*. 65:5436–5442.

98. Rey, C.R., H.W. Walker, and P.L. Rohrbaugh. 1975. The influence of temperature on growth, sporulation, and heat resistance of spores of six strains of *Clostridium perfringens*. *J. Milk Food Technol*. 38:461–465.

99. Roy, R.J., F.F. Busta, and D.R. Thompson. 1981. Thermal inactivation of *Clostridium perfringens* after growth at several constant and linearly rising temperatures. *J. Food Sci*. 46:1586–1591.

100. Saito, M. 1990. Production of enterotoxin by *Clostridium perfringens* derived from humans, animals, foods, and the natural environment in Japan. *J. Food Protect*. 53:115–118.

101. Satija, K.C., and K.G. Narayan. 1980. Passive bacteriocin typing of strains of *Clostridium perfringens* type A causing food poisoning for epidemiologic studies. *J. Infect. Dis*. 142:899–902.

102. Skjelkvâle, R., and T. Uemura. 1977. Detection of enterotoxin in faeces and anti-enterotoxin in serum after *Clostridium perfringens* food-poisoning. *J. Appl. Bacteriol*. 42:355-363.

103. Smelt, J.P.P.M., G.J.M. Raatjes, J.S. Crowther, and C.T. Verrips. 1982. Growth and toxin formation by *Clostridium botulinum* at low pH values. *J. Appl. Bacteriol*. 52:75–82.

104. Smith, A.M., D.A. Evans, and B.M. Buck. 1981. Growth and survival of *Clostridium perfringens* in rare beef prepared in a water bath. *J. Food Protect*. 44:9–14.

105. Smith, L.D.S. 1975. Inhibition of *Clostridium botulinum* by strains of *Clostridium perfringens* isolated from soil. *Appl. Microbiol*. 30:319–323.

106. Solomon, H.M., and D.A. Kautter. 1988. Outgrowth and toxin production by *Clostridium botulinum* in bottled chopped garlic. *J. Food Protect*. 51:862–865.

107. Solomon, H.M., D.A. Kautter, and R.K. Lynt. 1982. Effect of low temperatures on growth of nonproteolytic *Clostridium botulinum* types B and F and proteolytic type G in crabmeat and broth. *J. Food Protect.* 45:516–518.

108. Solomon, R.M., R.K. Lynt, Jr., D.A. Kautter, and T. Lilly, Jr. 1971. Antigenic relationships among the proteolytic and nonproteolytic strains of *Clostridium botulinum*. *Appl. Microbiol.* 21:295–299.

109. Sonnabend, O., W. Sonnabend, R. Heinzle, T. Sigrist, R. Dirnhofer, and U. Krech. 1981. Isolation of *Clostridium botulinum* type G and identification of type G botulinal toxin in humans: Report of five sudden unexpected deaths. *J. Infect. Dis.* 143:22–27.

110. Sperber, W.H. 1983. Influence of water activity on foodborne bacteria—A review. *J Food Protect.* 46:142–150.

111. St. Louis, M.E., S.H.S. Peck, D. Bowering, G.B. Morgan, J. Blatherwick, S. Banarjee, G.D.M. Kettyla, W.A. Black, M.E. Milling, A.H.W. Hauschild, R.V. Tauxe, and P.A. Blake. 1988. Botulism from chopped garlic: Delayed recognition of a major outbreak. *Ann. Intern. Med.* 108:363–368.

112. Stark, R.L., and C.L. Duncan. 1971. Biological characteristics of *Clostridium perfringens* type A enterotoxin. *Infect. Immun.* 4:89–96.

113. Strong, D.H., and J.C. Canada. 1964. Survival of *Clostridium perfringens* in frozen chicken gravy. *J. Food Sci.* 29:479–482.

114. Strong, D.H., J.C. Canada, and B. Griffiths. 1963. Incidence of *Clostridium perfringens* in American foods. *Appl. Microbiol.* 11:42–44.

115. Strong, D.H., C.L. Duncan, and G. Perna. 1971. *Clostridium perfringens* type A food poisoning. II. Response of the rabbit ileum as an indication of enteropathogenicity of strains of *Clostridium perfringens* in human beings. *Infect. Immun.* 3:171–178.

116. Suen, J.C., C.L. Hatheway, A.G. Steigerwalt, and D.J. Brenner. 1988. *Clostridium argentinense* sp. nov.: a genetically homogeneous group composed of all strains of *Clostridium botulinum* toxin type G and some nontoxigenic strains previously identified as *Closridium subterminale* or *Clostridium hastiforme*. *Int. J. Syst. Bacteriol.* 38:375–381.

117. Sugii, S., I. Ohishi, and G. Sakaguchi. 1977. Correlation between oral toxicity and in vitro stability of *Clostridium botulinum* types A and B toxins of different molecular sizes. *Infect. Immun.* 16:910–914.

118. Sugiyama, H., and D.C. Mills. 1978. Intraintestinal toxin in infant mice challenged intragastrically with *Clostridium botulinum* spores. *Infect. Immun.* 21:59–63.

119. Sugiyama, H., and K.H. Yang. 1975. Growth potential of *Clostridium botulinum* in fresh mushrooms packaged in semipermeable plastic film. *Appl. Microbiol.* 30:964–969.

120. Tompkin, R.B. 1980. Botulism from meat and poultry products—A historical perspective. *Food Technol.* 34(5):229–236, 257.

121. Trakulchang, S.P., and A.A. Kraft. 1977. Survival of *Clostridium perfringens* in refrigerated and frozen meat and poultry items. *J. Food Sci.* 42:518–521.

122. Tsang, N., L.S. Post, and M. Solberg. 1985. Growth and toxin production by *Clostridium botulinum* in model acidified systems. *J. Food Sci.* 50:961–965.

123. Weiss, K.F., and D.H. Strong. 1967. Some properties of heat-resistant and heat-sensitive strains of *Clostridium perfringens*. I. Heat resistance and toxigenicity. *J. Bacteriol.* 93:21–26.

124. Wen, Q., and B.A. McClane. 2004. Detection of enterotoxigenic *Clostridium perfringens* type A isolates in American retail foods. *Appl. Environ. Microbiol.* 70:2685–2691.

125. Wentz, M., H. Scott, and J. Vennes. 1967. *Clostridium botulinum* type F: Seasonal inhibition by *Bacillus licheniformis*. *Science* 155:89–90.

126. Whelan, S.M., M.J. Elmore, N.J. Dodsworth, J.K. Brehm, T. Atkinson, and N.P. Minton. 1992. Molecular cloning of the *Clostridium botulinum* structural gene encoding the type B neurotoxin and determination of its entire nucleotide sequence. *Appl. Environ. Microbiol.* 58:2345–2354.

127. Williams-Walls, N.J. 1968. *Clostridium botulinum* type F: Isolation from crabs. *Science* 162:375–376.

128. Zhou, Y., H. Sugiyama, H. Nakano, and E.A. Johnson. 1995. The genes for the *Clostridium botulinum* type G toxin complex are on a plasmid. *Infect. Immun.* 63:2087–2091.

129. Zhou, Y., H. Sugiyama, and E.A. Johnson. 1993. Transfer of neurotoxigenicity from *Clostridium butyricum* to a nontoxigenic *Clostridium botulinum* type E-like strain. *Appl. Environ. Microbiol.* 59:3825–3831.

CHAPTER 25

Foodborne Listeriosis

The suddenness with which *Listeria monocytogenes* emerged as the etiological agent of a foodborne disease is unparalleled. The acquired immunodeficiency syndrome (AIDS) and legionellosis are examples of two other human diseases that appeared suddenly, but unlike foodborne listeriosis, the etiological agents of these syndromes were previously unknown as human pathogens, and they proved to be difficult to culture. Not only is *L. monocytogenes* rather easy to culture, but also listeriosis was well documented as a disease of many animal species, and human cases were not unknown. For more information on the listeriae, see references 26 and 69.

TAXONOMY OF *LISTERIA*

The listeriae are Gram-positive, non-spore-forming, and non-acid-fast rods that were once classified as "*Listerella*." The generic name was changed in 1940 to *Listeria*. In many ways they are similar to the genus *Brochothrix*. Both genera are catalase positive and tend to be associated in nature, along with *Lactobacillus*. All three genera produce lactic acid from glucose and other fermentable sugars, but unlike *Listeria* and *Brochothrix*, the lactobacilli are catalase negative. At one time the listeriae were believed to be related to coryneform bacteria and, in fact, were placed in the family Corynebacteriaceae, but it is now clear that they are more closely related to *Bacillus*, *Lactobacillus*, and *Streptococcus*. From 16S ribosomal RNA (rRNA) sequence data, *Listeria* places closest to *Brochothrix*, and these two genera, together with *Staphylococcus* and *Kurthia*, occupy a position between the *Bacillus* group and the *Lactobacillus/Streptococcus* group within the *Clostridium–Lactobacillus–Bacillus* branch, where the mol% G + C of all members is less than 50.[65] Genetic transfers occur among *Listeria*, *Bacillus*, and *Streptococcus*, and immunological cross-reactions occur among *Listeria*, *Streptococcus*, *Staphylococcus*, and *Lactobacillus*. *Brochothrix* shares 338 common purine and pyrimidine bases with *Listeria*.[85] Although *Erysipelothrix* is in the *Mycoplasma* line, it shares at least 23 oligonucleotides in common with *Listeria* and *Brochothrix*.[85] *Listeria* spp. contain teichoic and lipoteichoic acids, as do the bacilli, staphylococci, streptococci, and lactobacilli, but unlike these groups, their colonies form a blue–green sheen when viewed by obliquely transmitted light.

Six species of *Listeria* are recognized, and they, with some differentiating characteristics, are listed in Table 25–1. The former *L. murrayi* has been merged with *L. grayi*.[107] It can be seen from Figure 25–1 that the two former species occupy a position away from the other five species. *L. ivanovii* is represented by two subspecies—*L. ivanovii* subsp. *ivanovii* and *L. ivanovii* subsp. *londoniensis*.[10] The

Table 25–1 Some Differentiating Characteristics of the Species of *Listeria*

Species	Xylose	Lactose	Galactose	Rhamnose	Mannitol	Hippurate Hydrolysis	CAMP test		Beta Hemolysis	Mol% G+C	Serovars
							S. aureus	R. equi			
L. monocytogenes	–	v	v	+	–	+	+	+	+	37–39	*
L. innocua	–	+	–	(+)	–	+	–	–	+	36–38	4ab, 6a, 6b
L. seeligeri	+		–	–	–		+	–	w	36	†
L. welshimeri	+		v	v	–		–	–	–	36	6a, 6b
L. ivanovii	+	+	v	–	–	+	–	+	++	37–38	5
L. grayi	–	+	+	–	+	–	–	–	–	41–42	5

Note: v = variable; w = weak; + = most strains positive.

*1/2a, b, c; 3a, b, c; 4a, ab, b, c, d, e; "7."

†Same as for *L. monocytogenes* and *L. innocua* but no 5 or "7."

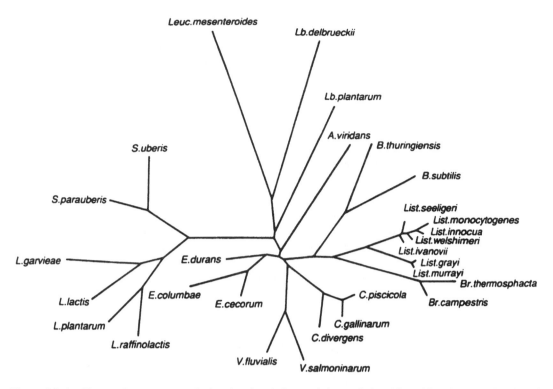

Figure 25–1 Unrooted tree or network showing the phylogenetic interrelationships of listeriae and other low-G + C-content Gram-positive taxa. The tree is based on a comparison of a continuous stretch of 1,340 nucleotides: the first and last bases in the sequence used to calculate K_{nuc} values correspond to positions 107 (G) and 1,433 (A), respectively, in the *E. coli* sequence. Abbreviations: *A.*, *Aerococcus*; *B.*, *Bacillus*; *Br.*, *Brochothrix*; *C.*, *Carnobacterium*; *E.*, *Enterococcus*; *L.*, *Lactococcus*; *Lb.*, *Lactobacillus*; *Leuc.*, *Leuconostoc*; *List.*, *Listeria*; *S.*, *Streptococcus*; *V.*, *Vagococcus*. *Source*: From M.D. Collins et al.,[20] copyright © 1991 by American Society for Microbiology, used with permission. *Note*: *L. grayi* and *L. murrayi* were subsequently shown to belong to the same species.

former can be distinguished from the latter by its ability to ferment ribose and its inability to ferment *N*-acetyl-β-D-mannosamine.[10]

Using polymerase chain reaction (PCR)-based DNA fingerprinting techniques to examine genetic relatedness between *L. innocua* and *L. welshimeri*, the two were found to share a high degree and that *L. grayi* is homogeneous and is clearly related to the other five species.[126] Poly(ribitolphosphate)-type teichoic acids are the prevalent accessory cell wall polymer in *Listeria* spp. The lipoteichoic acid of *L. grayi* is of the modified type, further separating it from the other species.[108] It appears that the teichoic acids are recognized by the bacteriophages as cell wall associated ligands.[81]

The CAMP (Christie–Atkins–Munch–Petersen) test is considered by many to be the definitive test for *L. monocytogenes*. An isolate that is CAMP positive with either *S. aureus* or *R. equi* must be considered a presumptive *L. monocytogenes* isolate, but not necessarily a virulent one.[91] The stimulation of hemolysis in the presence of *S. aureus* appears to be due to either a phosphatidylinositol-specific

Table 25–2   A Comparison of the Genera *Listeria* and *Erysipelothrix*

Genera	Motility	Catalase	H_2S Production	Major Diamino Acid	Mol% G + C
Listeria	+	+	−	Meso-DAP	36–38
Erysipelothrix	−	−	+	L-Lysine	36–40

or phosphatidyl-choline-specific phospholipase C from *L. monocytogenes*, and a sphingomyelinase from *S. aureus*.[91]

Members of the genus *Erysipelothrix* are often associated with *Listeria*, and some differences between the two genera are noted in Table 25–2. Unlike *Listeria*, *Erysipelothrix* is nonmotile, catalase negative, and H_2S positive and contains L-lysine as the major diamino acid in its murein. Like *L. monocytogenes*, *E. rhusiopathiae* causes disease in animals—in this case, swine erysipelas. The latter organism is also infectious for humans, in which it causes erysipeloid. Although *Listeria* spp. normally produce catalase, catalase-negative strains of *L. monocytogenes* have been isolated from foods.[58]

Serotypes

The six species of *Listeria* are characterized by the possession of antigens that give rise to 17 serovars. The primary pathogenic species, *L. monocytogenes*, is represented by 13 serovars, some of which are shared by *L. innocua* and *L. seeligeri*. Although *L. innocua* is represented by only 2 serovars (6a/6b), it is sometimes regarded as the nonpathogenic variant of *L. monocytogenes*. The greater antigenic heterogeneity of the outer envelope of the latter species may be related to the wide number of animal hosts in which it can proliferate.

The most commonly isolated of these serotypes are types 1/2 and 4. Prior to the 1960s, it appeared that type 1 existed predominantly in Europe and Africa and type 4 in North America, but this pattern appears to have changed. It has been noted that serotypes of listeriae in no way are related to host, disease process, or geographical origin, and this is generally confirmed by food isolations (see below), although serovars 1/2a and 4b do show some geographical differences.[113] In the United States and Canada, serovar 4b has accounted for 65–80% of all strains.

The 1998–1999 outbreak in the United States that was traced to wieners was caused by a rare strain of serovar 4b. Between January 1, 1966, and June 30, 1996, 60% of the 2,232 isolates from human cases in the United Kingdom were 4b with 17, 11, and 4% caused by 1/2a, 1/ab, and 1/2c, respectively.[94] In general, 4b strains are more often associated with outbreaks while 1/2 strains are associated with food products. The most frequently reported in Eastern Europe, West Africa, Central Germany, Finland, and Sweden is serovar 1/2a, whereas serovars 1/2a and 4b are more often reported in France and The Netherlands in about equal proportions.[113]

Subspecies Typing

In addition to serotyping, a variety of other methods has been applied to species and subspecies characterizations of *L. monocytogenes*, and they are summarized in Chapter 11. Among the methods are bacteriophage typing, multilocus enzyme electrophoresis (MEE) typing, restriction enzyme analysis

(REA), pulsed field gel electrophoresis (PFGE), restriction fragment length polymorphisms (RFLP), and ribotyping.

GROWTH

The nutritional requirements of listeriae are typical of those for many other Gram-positive bacteria. They grow well in many common media such as brain heart infusion, trypticase soy, and tryptose broths. Although most nutritional requirements have been described for *L. monocytogenes*, the other species are believed to be similar. At least four B vitamins are required—biotin, riboflavin, thiamine, and thioctic acid (α-lipoic acid; a growth factor for some bacteria and protozoa)—and the amino acids cysteine, glutamine, isoleucine, leucine, and valine are required. Glucose enhances growth of all species, and L(+)-lactic acid is produced. Although all species utilize glucose by the Embden–Meyerhof pathway, various other simple and complex carbohydrates are utilized by some. *Listeria* spp. resemble most enterococci in being able to hydrolyze esculin, and grow in the presence of 10 or 40% (w/v) bile, in about 10% NaCl, 0.025% thallous acetate, and 0.04% potassium tellurite, but unlike the enterococci, they do not grow in the presence of 0.02% sodium azide. The listeriae possess a bile-salt hydrolyase, which permits their growth in the gall bladder. Unlike most other Gram-positive bacteria, they grow on MacConkey agar. Although iron is important in its in vivo growth, *L. monocytogenes* apparently does not possess specific iron-binding compounds, and it obtains its needs through the reductive mobilization of free iron, which binds to surface receptors.

Effect of pH

Although the listeriae grow best in the pH range 6–8, the minimum pH that allows growth and survival has been the subject of a large number of studies. Most research has been conducted with *L. monocytogenes* strains, and whether the findings for this species are similar for other listerial species can only be assumed. In general, some species/strains will grow over the pH range of 4.1 to around 9.6 and over the temperature range of 1°C to around 45°C. (Details of these parameters follow.)

In general, the minimum growth pH of a bacterium is a function of temperature of incubation, general nutrient composition of growth substrate, water activity (a_w), and the presence and quantity of NaCl and other salts or inhibitors. Growth of *L. monocytogenes* in culture media has been observed at pH 4.4 in less than 7 days at 30°C,[45] at pH 4.5 in tryptose broth at 19°C, [12] and at pH 4.66 in 60 days at 30°C.[18] In the first study, growth at pH 4.4 occurred at 20°C in 14 days and at pH 5.23 at 4°C in 21 days.[45] In the second study, growth at pH 4.5 was enhanced by a restriction of oxygen. In the third study, growth of *L. monocytogenes* was observed at pH 4.66 in 60 days at 30°C, the minimum at 10°C was pH 4.83, whereas at 5°C, no growth occurred at pH 5.13. In yet another study, four strains of *L. monocytogenes* grew at pH 4.5 after 30 days in a culture medium incubated at 30°C,[102] but no growth occurred at pH 4.0 or lower. pH values of 3.8–4.0 were more destructive to one strain than pH 4.2–5.0 when held in orange serum at 30°C for 5 days (Figure 25–2).

When the pH of tryptic soy broth was adjusted with various acids, minimum pH for growth of four strains of *L. monocytogenes* was shown to be a function of the acid employed. At the same pH, the antimicrobial activity was acetic acid > lactic acid > citric acid > malic acid > HCl.[120] Growth occurred at a pH of 4.6 at 35°C between 1 and 3 days, and some strains grew at pH 4.4. The growth of two strains of this species in cabbage juice containing no added NaCl has been observed at pH 4.1 within

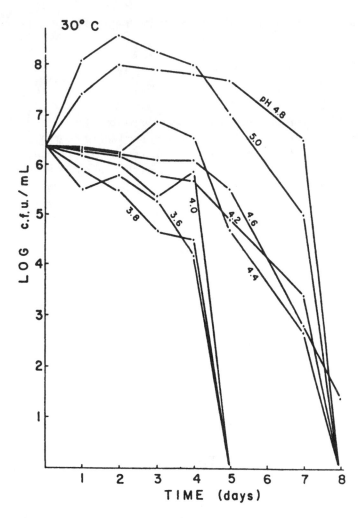

Figure 25–2　Change in cell populations of strains F5069/(4b) in pH-adjusted orange serum incubated at 30°C. Initial cell concentration was 2.2×10^6 colony-forming units (cfu)/ml. *Source*: From Parish and Higgins,[102] copyright © 1989 by International Association of Milk, Food and Environmental Sanitarians, used with permission.

8 days when incubated at 30°C, but death occurred at 30°C when the organism was inoculated into sterile cabbage juice adjusted to a pH <4.6 with lactic acid.[21] At a pH of 5.05 and incubation at 5°C, strain Scott A did not grow in cottage cheese with an inoculum of $\sim 10^3$ cfu/g.[104]

Combined Effect of pH and NaCl

The interaction of pH with NaCl and incubation temperature has been the subject of several studies.[19,21] The latter investigators used factorially designed experiments to determine the interaction of these parameters on the growth and survival of a human isolate (serovar 4b); some of their

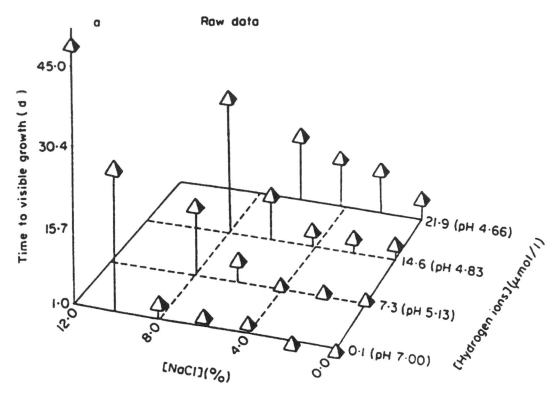

Figure 25–3 Effect of salt and hydrogen-ion concentration on the time to reach visible growth of *Listeria monocytogenes*. Three-dimensional scatterplot for the effect of salt (%, *x* axis) and hydrogen-ion concentration (μmol/l, *Z* axis) on the time to reach visible growth (days, *y* axis) representing at least a 100-fold increase in numbers of *Listeria monocytogenes* (a) at 30°C. The mean actual values are compared with predicted values determined from polynomial equations (1) and (2) (not shown). *Source*: From Cole et al.[19]

findings are illustrated in Figure 25–3. At pH 4.66, time to visible growth was 5 days at 30°C with no NaCl added, 8 days at 30°C with 4.0% NaCl, and 13 days at 30°C with 6.0% NaCl, all at the same pH.[19] Growth at 5°C occurred only at a pH of 7.0 in 9 days with no added NaCl, but 15 days were required for 4.0% NaCl and 28 days for 6.0% NaCl. The pH and NaCl effects were determined to be purely additive and not synergistic in any way.

Effect of Temperature

The mean minimum growth temperature on trypticase soy agar of 78 strains of *L. monocytogenes* was found to be 1.1 ± 0.3°C, with a range of 0.5–3.0°C.[67] Two strains grew at 0.5°C, and eight grew at or below 0.8°C in 10 days as determined with a plate-type continuous temperature gradient incubator. With 22 other strains (19 *L. innocua* and 1 each of *L. welshimeri*, *L. grayi*, and "*L. murrayi*"), minimum growth temperature ranged from 1.7° to 3.0°C, with a mean of 1.7° + 0.5°C.[67] That the *L. monocytogenes* strains had about a 0.6°C lower minimum temperature than the other species suggested to these investigators that the hemolysin may enhance growth and survival of *L. monocytogenes* in cold environments even though the growth of serovars 1/2a, 1/2b, and 4b was lower at around 3.0°C than those with OI antigens. The maximum growth temperature for listeriae is around 45°C.

Effect of a_w

Using brain heart infusion (BHI) broth, three humectants, and 30°C incubation, the minimum a_w that permitted growth of serotypes 1, 3a, and 4b of *L. monocytogenes* revealed the following: 0.90 with glycerol, 0.93 with sucrose, and 0.92 with NaCl.[34] In another study using trypticase soy broth base at a pH of 6.8 and 30°C incubation, the minimum a_w that permitted growth was 0.92 with sucrose as humectant.[103] In view of these findings, *L. monocytogenes* is second only to the staphylococci as a foodborne pathogen in being able to grow at a_w values <0.93.

DISTRIBUTION

The Environment

The listeriae are widely distributed in nature and can be found in decaying vegetation and in soils, animal feces, sewage, silage, and water. In general, listeriae may be expected to exist where the lactic acid bacteria, *Brochothrix*, and some coryneform bacteria occur. Their association with certain dairy products and silage is well known, as is the association with these products of some other lactic acid producers. In a study of gull feces, rooks, and silage in Scotland, gulls feeding at sewage works had a higher rate of carriage than those elsewhere, and fecal samples from rooks generally had low numbers of listeriae.[39] *L. monocytogenes* and *L. innocua* were most often found with only one sample containing *L. seeligeri*. In the same study, *L. monocytogenes* and *L. innocua* were found in 44% of moldy silage samples and in 22.2% of big bale silage. In Denmark, 15% of the 75 silage samples were positive for *L. monocytogenes*, as was 52% of the 75 fecal samples from cows.[118] The organism was found in silage with a pH above and below 4.5. *L. monocytogenes* was isolated from 8.4 to 44% of samples taken from grain fields, pastures, mud, animal feces, wildlife feeding grounds, and related sources.[129] Its survival in moist soils for 295 days and beyond has been demonstrated.[130] From California coastal waters, 62% of 37 samples of freshwater or low-salinity water and 17.4% of 46 sediment samples were positive for *L. monocytogenes*, but none could be recovered from 35 oyster samples.[18] Some of the ways in which *L. monocytogenes* is disseminated throughout the environment, along with the many sources of the organism to humans, are illustrated in Figure 25–4.

Foods and Humans

It is well established that any fresh food product of animal or plant origin may harbor varying numbers of *L. monocytogenes*. In general, the organism has been found in raw milk; soft cheeses; fresh and frozen meat, poultry, and seafood products; and on fruits and vegetable products. Its prevalence in milk and dairy products has received much attention because of early outbreaks. In bulk tank raw milk from 260 farms in Scotland examined over a year, 25 of the 160 had positive samples usually only once, but 7 were positive three or more times, usually with <1 cfu/ml with the single highest being 35 cfu/ml.[40] Of 5779 retail foods examined in The Netherlands in 1988, 3.0% were positive for this organism at ≥10/g. The lowest prevalence was in ice cream, in which only 0.2% of 649 samples were positive, and the highest was in fresh meat, with 7.5% of 416 samples being positive. Also, in The Netherlands, 4.6% of 929 samples of soft cheeses made from raw milk were positive for *L. monocytogenes* for a

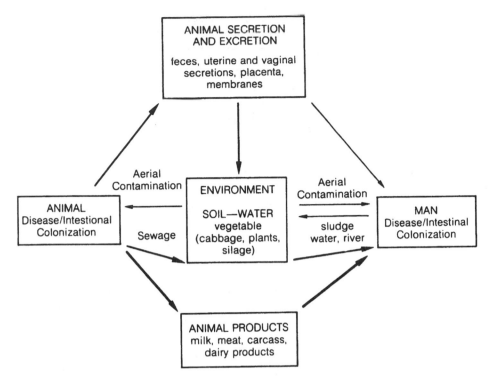

Figure 25–4 Ways in which *L. monocytogenes* is disseminated in the environment, animals, foods, and humans. *Source*: From Audurier and Martin.[3]

rate of 3.48%. In England and Wales, this organism was found in 4% of 56,959 ready-to-eat foods. Over a 39-month period in the United States, 7.1% of 1,727 raw beef samples collected throughout the country were positive for *L. monocytogenes*, and over a 21-month period, 19.3% of 3,700 raw broiler necks and backs were positive.[52] From the same survey, 2.8% of a variety of ready-to-eat meats from 4,105 processing plants throughout the United States were positive for the organism.

The most common serovar in meat products from six countries was 1/2.[63] Serotype 4 was isolated from meat products in five countries, and serotype 3 was recovered from products in only two countries. Serotype 1/2 was found more often than serotype 4 in raw milk[83,105] and cheese,[106] and serotype 1 has been found in seafoods[128,131] and vegetables.[56] Serovar 4b was associated with a cluster of human cases in Boston where raw vegetables appeared to be the source, whereas from potatoes and radishes, serovars 1/2a and 1/2 were the most frequently occurring.[56] The three most prevalent serovars isolated from foods, in decreasing order, are 1/2a, 1/2b, and 4b, whereas from human listeriosis, 4b, 1/2a, and 1/2b are the most prevalent.[22] The incidence/prevalence of *L. monocytogenes* in meats and poultry can be found in Chapter 4—Table 4–6.

Of serovars isolated from humans, 59% of 722 *L. monocytogenes* isolated in Britain were 4b, followed by 18, 14, and 4%, respectively, for 1/2a, 1/2b, and 1/2c.[93] From pathological specimens throughout the world, 98% of isolates belonged to the following scrovars: 1/2a, 1/2b, 1/2c, 3a, 3b, 3c, 4b, and 5.[113] Serovar 4b is by far the most often found in outbreak cases, and it appears that it possesses virulence properties far greater than others.

Table 25–3 High Numbers of *L. monocytogenes* per Gram or
Milliliter Reported for Various Food Products

Chocolate milk (USA, 1994)	$\sim 10^9$
Goat's milk soft cheese (England, 1989)	$>10^7$
Cheese outbreak (Switzerland, 1983–1987)	10^4–10^6
Temperature abused ricotta cheese	3.6×10^6
Smoked mussels (Tasmania, 1991)	$>10^6$
Chicken roll (USA, 1990)	1.9×10^5
Pâté (Great Britain, 1990)	10^3–10^6
Raw pork skins (USA, 1991)	4.3×10^4
Roast beef (USA, 1991)	3.6×10^4
Vacuum-packaged corned beef, 1992	3.3×10^4
Pâté (Australia, 1990), mean number	8.8×10^3
Cabbage (USA, 1991)	1.4×10^3

With respect to other *Listeria* spp. in foods, *L. innocua* is rather common in meats, milk, frozen seafoods, semisoft cheese, whole egg samples, and vegetables. In general, it is the most prevalent listerial species found in dairy products.[81] It was found in 8–16% of raw milk where its presence was reported, in 46% of 57 frozen seafoods,[128] and in 36% of 42 liquid whole egg samples.[76] In the last study, it was the most frequently found listerial species and occurred in all 15 positive samples. In one study, this species was found in 42% of beef and poultry examined, and overall it was found two times as often as *L. monocytogenes*.[118] It was found in 83% of mettwurst and 47% of pork tested in Germany,[112] and in 22% of fresh salads in Britain.[116]

L. welshimeri has been found in raw milk (from 0.3 to 3% of samples), meat roasts, vegetables, and turkey meat. In the latter, it was found in 16% of samples and thus was the most prevalent of listerial species. It was found in 24% of mettwurst and 30% of pork in Germany,[112] and in frozen ground beef and deli products in France.[100] The only other species of *Listeria* reported from foods are *L. grayi*, found in raw milk, beef and poultry; and *L. seeligeri*, found in raw milk, vegetables, cabbage, radishes, pork, and mettwurst. For more information on listeriae in meat and poultry products, see reference 62.

Prevalence

Because of the need for culture enrichments, numbers of *L. monocytogenes* per gram or milliliter in foods are often not reported. For bulk tank raw milk in the United States (specifically in California and Ohio), the number was estimated to be about 1 cell/ml or less.[84] Although the numbers of *L. monocytogenes* in foods tend to be so low that direct enumeration methods are without value, samples are sometimes found that contain numbers $>10^3$/g. Some of the highest numbers reported for food products are summarized in Table 25–3.

THERMAL PROPERTIES

Although *L. monocytogenes* cells were not isolated in the 1983 Massachusetts outbreak of human listeriosis in which pasteurized milk was incriminated, the adequacy of standard milk pasteurization protocols to destroy this organism was brought into question. Since 1985, a large number of studies have

Table 25–4 Summary of Some Findings on the Thermal Destruction of *L. monocytogenes*

Strains Tested/State	Number of Cells (ml)	Heating Menstrum	Heating Temperature (°C)	D Value (sec)	z Value (°C)	Reference
Scott A, free suspension	$\sim 10^5$	Sterile skim milk	71.7	1.7	6.5	9
	$\sim 10^5$	Sterile skim milk	71.7	2.0	6.5	109
	$\sim 10^5$	Sterile skim milk	71.7	0.9	6.3	11
Scott A, intracellular	$\sim 10^5$	Whole raw milk	71.7	1.9	6.0	15
Scott A, free suspension	$\sim 10^5$	Whole raw milk	71.7	1.6	6.1	15
F5069, intracellular	$\sim 10^6$	Sterile whole milk	71.7	5.0	8.0	14
F5069, free suspension	$\sim 10^6$	Sterile whole milk	71.7	3.1	7.3	14
Scott A, free suspension	$\sim 10^5$	Ice cream mix	79.4	2.6	7.0	9
	$\sim 10^8$	pH 7.2, phosphate buffer	70.0	9.0	—	8
	$\sim 10^8$	pH 5.9, meat slurry	70.0	13.8	—	8
	$\sim 10^7$	Liquid whole egg	72.0	36.0	7.1	41
Ten strains	$\sim 10^7$	Irradiated ground meats	62.0	61.0	4.92	36
Chicken/meat isolate	$\sim 10^5$	Beef	70.0		7.2	88
	$\sim 10^5$	Minced chicken	70.0		6.7	88

been reported on its thermal destruction in dairy products. *D* values have been determined on many strains of *L. monocytogenes* in whole and skim milk, cream, ice cream, and various meat products. As this organism is an intracellular pathogen, several studies were undertaken to determine its relative heat resistance inside and outside phagocytes. Overall, standard pasteurization protocols for milk are adequate for destroying *L. monocytogenes* at levels of 10^5–10^6/ml, whether freely suspended or in intracellular state. Some of the specific findings are presented below. For a more extensive review, see Doyle et al.[29]

Dairy Products

A summary of thermal *D* and *z* values for some *L. monocytogenes* strains is presented in Table 25–4. The *D* values indicate that the high-temperature, short-time (HTST) protocol for milk (71.7°C for 15 seconds) is adequate to reduce normally existing numbers of this organism below detectable levels. The vat or low-temperature, long-time (LTLT) pasteurization protocol (62.8°C for 30 minutes) is even more destructive (see Chapter 17). Employing the Scott A strain (serovar 4b from the Massachusetts outbreak), *D* values ranged from 0.9 to 2.0 seconds with *z* values of 6.0–6.5°C. The F5069 strain (serovar 4b) appeared to be a bit more heat resistant than Scott A from these results, although Scott A was the most heat resistant of three other strains evaluated, not including F5069.[11]

The thermal resistance of *L. monocytogenes* is not affected by its intracellular position. With Scott A freely suspended in whole raw milk at mean levels of 2.6×10^5 cfu/ml and heating at 71.7°C for 15 seconds, no survivors could be found after five heating trials.[82] In seven heating trials with Scott A engulfed in vitro by bovine phagocytes, no survivors could be detected with a mean number of 5×10^4 cfu/ml. Further, these investigators experimentally infected cows with Scott A and were still unable to find survivors following 11 pasteurization trials at 71.7°C for 15 seconds with numbers of Scott A that ranged from 1.4×10^3 to 9.5×10^3 cfu/ml I90I employing five strains of *L. monocytogenes* in whole milk, skim milk, and 11% nonfat milk solids, Donnelly and Briggs[27] found that composition did not affect heat destruction and that at 62.7°C, the D values were 60 seconds or less. The five strains employed included serotypes 1, 3, and 4. When milk that was naturally contaminated with a serotype 1 strain at around 10^4/ml was subjected to an HTST protocol at temperatures ranging from 60° to 78°C, no viable cells could be detected at processing temperatures of 69°C or above.[37] In their review of the early studies on the thermal resistance of *L. monocytogenes* in milk, Mackey and Bratchell[89] concluded that normal pasteurization procedures will inactivate this organism but that the margin of safety is greater for the vat protocol (LTLT) than the HTST protocol. Their mathematical model predicted a 39 D reduction for vat and a 5.2 D reduction for HTST.

Nondairy Products

For liquid whole egg and meat products, D values are generally higher than for milk, a fact not unpredicted, considering the effect of proteins and lipids on the thermal resistance of microorganisms (discussed in Chapter 17). For one strain of *L. monocytogenes* isolated from a chicken product, D values at 70°C were 6.6–8.4 seconds; they were essentially the same in beef and two poultry meats.[88] In one study, viable cells could be recovered by enrichments from eight of nine samples following heating in ground beef to 70°C.[8] In a study of blue crabmeat, strain Scott A at levels of about 10^7 had a D value of 2.61 minutes with a z of 8.4°C, indicating that the crabmeat pasteurization protocol of 30 minutes at 85°C was adequate to render the product safe from this organism.[55] Processing frankfurters to an internal temperature of 160°F (71.1°C) has been shown to effect at least a 3-log cycle reduction of strain Scott A.[133] The cooking of meat products to an internal temperature of 70°C for 2 minutes will destroy *L. monocytogenes*.[43,83,89]

In liquid whole egg (LWE) exposed to 60°C for 3.5 minutes, the calculated D value for strain Scott A was 2.1 minutes.[5] However, the same strain in LWE + 10% NaCl heated at 63°C for 3.5 minutes had a D of 13.7 minutes, whereas LWE + 10% sucrose gave a D of 1.9 minutes under the same conditions. The 10% NaCl lowered the a_w from 0.98 to 0.915, which could account in part for the higher D value. Higher D values were found for seven serovars incubated at 4°C for 5 days followed by 37°C incubation for 7 days.[119] In saline, D_{60} values were 0.72–3.1 and D_{62} were 0.30–1.3 minutes.

In the sausage-type meat employed by Farber,[36] the D value at 62°C was 61 seconds, but when cure ingredients were added, the D value increased to 7.1 minutes, indicating some heat-protective effects of the cure compounds, which consisted of nitrite, dextrose, lactose, corn syrup, and 3% (w/v) NaCl. An approximate doubling in D value in ground beef containing 30% fat, 3.5% NaCl, 200 ppm nitrite, and 300 ppm nitrate was found by Mackey et al.[88] who attributed the increased heat resistance to the 3.5% NaCl. The destruction of strain Scott A by microwave cooking was investigated by Lund et al.,[86] where more than 10^7 cells/g were placed in chicken stuffing and 10^6–10^7/g on chicken skin. By use of a home-type microwave unit, the adequacy of heating to an internal temperature of 70°C for 1 minute was shown to give a 6-log reduction in numbers. The thermal destruction of *L. monocytogenes* is similar to that of most other bacteria relative to pH of suspending menstrum where resistance is higher at pH values closer to 7.0 than values in the acid range. This was demonstrated in cabbage juice, where D values were higher at a pH of 5.6 than at 4.6.[7]

In a study of rainbow trout from retail markets in east Tennessee, 51% of the 74 samples were positive for *L. monocytogenes*.[31] The $\log_{10}$ means for aerobic plate counts (APC) and coliforms were 6.2 and 3.2, respectively, and the higher percentage of *L. monocytogenes* was associated with samples that had the highest APC and coliform numbers. See Chapters 4, 5 and 9 for numbers of listeriae in a variety of foods.

Effect of Sublethal Heating on Thermotolerance

It is unclear whether sublethal heating of *L. monocytogenes* cells renders them more resistant to subsequent thermal treatments. Some investigators have reported no effect,[11,13] and others have reported increased resistance.[35,38,79] In one study, the heat shocking of strain Scott A at 48°C for 20 minutes resulted in a 2.3-fold increase in D values at 55°C.[79] In another study employing Scott A in broth and ultrahigh temperature (UHT)-treated milk, an increase in heat resistance was observed following exposure to 48°C for 60 minutes and subsequent exposure to 60°C.[38] Finally, in a study employing 10 strains at a level of about 10^7/g in a sausage mix and heat shocking at 48°C for 30 or 60 minutes, no significant increase in thermotolerance was observed at 62° or 64°C, but those shocked for 120 minutes did show an average 2.4-fold increase in D values at 64°C.[35] In this study, the thermotolerance was maintained for at least 24 hours when the cells were stored at 4°C. If sublethal heating does lead to greater thermoresistance, it would not pose a problem for milk that contains fewer than 10 cells/ml assuming that a twofold to threefold increase in D value occurs.

VIRULENCE PROPERTIES

Of listerial species, *L. monocytogenes* is the pathogen of concern for humans. Although *L. ivanovii* can multiply in the mouse model, it does so to a much less degree than *L. monocytogenes*, and up to 10^6 cells caused no infection in the mouse.[59] *L. innocua*, *L. welshimeri*, and *L. seeligeri* are nonpathogens, although the last produces a hemolysin. The most significant virulence factor associated with *L. monocytogenes* is listeriolysin O (LLO).

Listeriolysin O and Ivanolysin O

In general, the pathogenic/virulent strains of *L. monocytogenes* produce beta-hemolysis on blood agar and acid from rhamnose but not from xylose. Strains whose hemolysis can be enhanced with either the prepurified exo-substance or by direct use of the culture are potentially pathogenic.[117] Regarding hemolysis, the evidence is overwhelming that all virulent strains of this species produce a specific substance that is responsible for beta-hemolysis on erythrocytes and the destruction of phagocytic cells that engulf them. The substance and perfringolysin O (PFO) have been shown to be highly homologous to streptolysin O (SLO) and pneumolysin O (PLO). It has been purified and shown to have a molecular weight of 60,000 Da and to consist of 504 amino acids.[44,95] It is produced mainly during the exponential growth phase, with maximum levels after 8–10 hours of growth.[43] Less LLO is synthesized at 26°C than at 37°C with high glucose, and synthesis was found to be best with 0.2% glucose at 37°C.[24] Sorbate at a level of 2% inhibited LLO synthesis at 35°C under aerobic or anaerobic conditions.[72] LLO has been detected in all strains of *L. monocytogenes*, including some that were nonhemolytic, but not in *L. welshimeri* or *L. grayi*. The gene that encodes its production is chromosomal and it is designated *hly*. Its role in virulence is discussed below.

L. ivanovii and *L. seeligeri* produce thiol-dependent exotoxins that are similar but not identical to LLO. Large quantities are produced by *L. ivanovii* but only small quantities by *L. seeligeri*.[43] The *L. ivanovii* thiol-activated cytolysin is ivanolysin O (ILO). Antiserum raised to the *L. ivanovii* product cross-reacts with that from *L. monocytogenes* and SLO.[73] ILO-deficient mutants have been shown to be avirulent in mice and chick embryos.[1]

Purified LLO has been shown to share in common with SLO and PLO the following properties: activated by SH-compounds such as cysteine, inhibited by low quantities of cholesterol, and common antigenic sites as evidenced by immunological cross-reactivity. Unlike SLO, LLO is active at a pH of 5.5 but not at pH 7.0, suggesting the possibility of its activity in macrophage phagosomes (phagolysosomes). Its LD_{50} for mice is about 0.8 μg, and it induces an inflammatory response when injected intradermally.[44] It appears that LLO and the other poreforming toxins evolved from a single progenitor gene.

Intracellular Invasion

When *L. monocytogenes* is contracted via the oral route, it apparently colonizes the intestinal tract by mechanisms that are poorly understood. From the intestinal tract, the organism invades tissues, including the placenta in pregnant women, and enters the blood stream, from which it reaches other susceptible body cells. As an intracellular pathogen, it must first enter susceptible cells, and then it must possess means of replicating within these cells. In the case of phagocytes, entry occurs in two steps: directly into phagosomes and from the phagosomes into the phagocyte's cytoplasm.

Entry or uptake into nonphagocytic cells is different. In nonphagocytic cell lines, uptake requires surface-bound proteins of the bacterium designated In1A and In1B.[78] The former has a molecular weight of 88 kDa and the latter 65 kDa. They are involved in aiding the entry of *L. monocytogenes* cells into host cells. The In1A protein, i.e., *internalin*, and its mammalian surface receptor is *E-cadherin*. It is required for entry into cultured epithelial cells, whereas In1B is required for invasion of cultured mouse hepatocytes.[32] Another invasion-associated protein of *Listeria* is p60, a 60-kDa protein encoded by the *iap* gene. It is secreted by all species of *Listeria*. Another surface protein, ActA (90 kDa), is required for actin polymerization and it allows for intracytoplasmic movement of cells.[61] Ami (ca. 90 kDa) is located on the surface of *L. monocytogenes* and it is a bacteriolysin. From a study of 150 food isolates, Ami was present in 149 while 283 of 300 human isolates contained this protein.[61] All of the positive strains contained LLO, InlB, and ActA.

L. monocytogenes survives inside macrophages by escaping from phagolysosomal membranes into the cytoplasm (cytosol), and this process is facilitated in part by LLO. Once inside the cytosol, the surface protein ActA (encoded by *actA*) aids in the formation of actin tails that propel the organism toward the cytoplasmic membrane. At the membrane, double membrane vacuoles form. With LLO and the two bacterial phospholipases, the phosphatidylinositol-specific phospholipase C (encoded by *plcA*) and the broad-range phospholipase C (encoded by *plcB*), the bacteria are freed and the process is repeated upon entry of bacteria into adjacent host cells. The latter occurs following the pushing out of the membrane to form a filopodium (a projection), which is absorbed by an adjacent cell and the invasion process is repeated. Thus, the spread of *L. monocytogenes* from cell to cell occurs without the bacterium having to leave the inner parts of host cells. For more information, see reference 97 and Chapter 22.

Monocytosis-Producing Activity

An interesting yet incompletely understood part of the *L. monocytogenes* cell is a lipid-containing component of the cell envelope that shares at least one property with the lipopolysaccharide (LPS) that is typical of Gram-negative bacteria. In Gram-negative bacteria, LPS is located in the outer membrane, but listeriae and other Gram-positive bacteria do not possess outer membranes. The *L. monocytogenes* substance is lipoteichoic acid (LTA). It was shown several decades ago that phenol–water extracts of *L. monocytogenes* cells induce the production of monocytes, and it was because of this monocytosis-producing activity (MPA) factor that the organism was given the species name *monocytogenes*. This

LTA fraction accounts for about 6% of the dry weight of cells and is associated with the plasma membrane. It has a molecular weight of about 1,000 Da, contains no amino acids or carbohydrates, and stimulates only mononuclear cells.[42] It possesses low tissue toxicity and is serologically inactive[121] but it kills macrophages in vitro.[42] It has been shown to share the following properties with LPS: it is pyrogenic and lethal in rabbits, produces a localized Schwartzman reaction, contains acylated hydroxy fatty acids, produces a positive reaction with the *Limulus* amoebocyte lysate (LAL) reagent, contains 2-keto-3-deoxyoctonic acid (KDO), and contains heptose. Regarding its LAL reactivity, 1 μg/ml was required to produce a positive reaction[115] whereas the same can be achieved with picogram quantities of LPS.

Sphingomyelinase

L. ivanovii is known to be infectious for sheep, in which it causes abortions, and to be a prolific producer of hemolysin on sheep erythrocytes. It has been shown to possess an LLO-like hemolysin (ILO), sphingomyelinase, and lecithinase.[73] Sphingomyelinase has a molecular weight of 27,000 Da.[127] Whereas the LLO-like agent is responsible for the inner complete zone of hemolysis on sheep erythrocytes, the halo of incomplete hemolysis that is enhanced by *Rhodococcus equi* appears to be caused by the two enzymes noted. In one study, a mutant defective in sphingomyelinase and another protein exhibited lower virulence than wild-type strains.[1]

ANIMAL MODELS AND INFECTIOUS DOSE

The first animal model employed to test the virulence of *L. monocytogenes* was the administration of a suspension of cells into the eye of a rabbit or guinea pig (Anton's test), where 10^6 cells produced conjunctivitis.[2] Chicken embryos have been studied by a large number of investigators. Inocula of 100 cells of *L. monocytogenes* into the allantoic sac of 10-day-old embryos led to death within 2–5 days, and the LD_{50} was less than 6×10^2 cells for virulent strains. *L. ivanovii* is also lethal by this method. Injections of 100–30,000 cells/egg into the chorioallantoic membrane of 10-day-old chick embryos resulted in death within 72 hours compared to about 5 days for mice.[123] Although Anton's test and chick embryos may be used to assess the relative virulence of strains of listeriae, the mouse is the model of choice for the additional information that it gives relative to cellular immunity.

Not only the mouse is the most widely used laboratory animal for virulence studies of listeriae, but also it is widely used in studies of T cell immunity in general. This model is employed by use of normal, baby, juvenile, or adult mice, as well as a variety of specially bred strains such as athymic (T cell-deficient) nude mice. Listerial cells have been administered intraperitoneally (IP), intravenously (IV), and intragastrically (IG). When normal adult mice are used, all smooth and hemolytic strains of *L. monocytogenes* at levels of 10^3–10^4/mouse multiply in the spleen.[59] With many strains, inocula of 10^5–10^6 are lethal to normal adult mice, although numbers as high as 7×10^9 have been found necessary to produce an LD_{50}. A low of 50 cells for 15-g mice has been reported (see below).

Whereas the IP route of injection is often used for mice, IG administration is employed to assess gastrointestinal behavior of listeriae. The administration of *L. monocytogenes* to 15-g mice by the IG route produced more rapid infection and more deaths in the first 3 days of the 6-day test than by IP.[105] By this method, the approximate 50% lethal dose (ALD_{50}) ranged from 50 to 4.4×10^5 cells for 15 food and clinical isolates of *L. monocytogenes*.[105] Six- to eight-week-old mice were given IP and oral challenges of a serovar 4b strain by one group to study their effect under normal and compromised states. Cells were suspended in 11% nonfat milk solids and administered to four groups of mice: normal, hydrocortisone treated, pregnant, and cimetidine treated. Minimum numbers of cells that caused a 50% infectious dose (ID_{50}) were 3.24–4.55 log cfu for normal mice, 1.91–2.74 for the cortisone-treated, and 2.48 for pregnant mice.[50] The ID_{50} for those administered cimetidine was similar to the normals. These investigators found no significant difference between IP and IG administration relative to ID_{50}. Employing neonatal mice (within 24 hours of birth), the LD_{50} by IP injection of *L. monocytogenes* was 6.3 × 10/cfu, but for 6- to 8-week-old mice, the LD_{50} was 3.2 × 10^6 by the same route of administration.[17] The neonatal mice were protected against a lethal dose of *L. monocytogenes* when γ-interferon was injected (see below). With 15- to 20-g Swiss mice treated with carrageenan, LD_{50} was found to range from about 6 to 3,100 cfu.[25]

When nude mice are challenged with virulent strains of *L. monocytogenes*, chronic infections follow, and for baby mice and macrophage-depleted adult mice, virulent strains are lethal. With the adult mouse model, rough strains of *L. monocytogenes* multiplied only weakly, and a weak immunity was induced; baby mice were killed, but nude mice survived.[59] In nonfatal infections by virulent strains, the organisms multiply in the spleen, and protection against reinfections results regardless of the serovar used for subsequent challenge.[59]

Overall, studies with the mouse model confirm the greater susceptibility to *L. monocytogenes* of animals with impaired immune systems than normal animals, as is the case with humans. The correspondence of minimal infectious doses for normal adult mice to humans is more difficult. It has been suggested that levels of *L. monocytogenes* less than 10^2/cfu appear to be inconsequential to healthy hosts.[50] From the nine cheeses reported by Gilbert and Pini[47] that contained 10^4–10^5/g of *L. monocytogenes*, no known human illness resulted.

INCIDENCE AND NATURE OF THE LISTERIOSIS SYNDROMES

Incidence

Although *L. monocytogenes* may have been described first in 1911 by Hülphers,[67] its unambiguous description was made in 1923 by Murray et al.[98] Since that time it has been shown to be a pathogen in over 50 mammals, including humans, in addition to fowls, ticks, fish, and crustaceans. The first human case of listeriosis was reported in 1929, and the disease has since been shown to occur sporadically throughout the world. *L. monocytogenes* is the etiological agent of about 98% of human and 85% of animal cases.[92] At least three human cases have been caused by *L. ivanovii* and one by *L. seeligeri*. There were around 60 human cases in the United Kingdom in 1981 but around 140 in 1985, along with a similar increase in animal cases.[93] Between 1986 and 1988, human listeriosis increased in England and Wales by 150%, along with a 100% increase in human salmonellosis. The overall mortality rate for 558 human cases in the United Kingdom was 46%, with 51 and 44%, respectively, for perinatal and adult cases.[93] For the period 1983–1987, 775 cases were reported in Britain, with 219 (28%) deaths, not including abortions. When the 44 abortions are added to the deaths, the fatality rate is 34%.[53] Prior to 1974, 15 documented cases were seen yearly in western France, but in 1975 and 1976 there were 115 and 54, respectively.[16] All but 3 of 145 strains that were serotyped were serotype 4. There were 687 cases in France in 1987.[22] In a 9-year period prior to early 1984, Lausanne, Switzerland, experienced a mean of 3 cases of human listeriosis per year, but in a 15-month period in 1983–1984, 25 cases were seen.[90] Thirty-eight of 40 strains examined were serovar 4b, and 92% had the same phage type.

By and large, foodborne outbreaks of human listeriosis seemed to have waned over the past several years with a few exceptions, as may be noted from Table 25–5. In the early to mid-1990s, the estimated number of cases/million persons in several countries were as follows:

Australia (1992)	2
Canada	2–4
Denmark	4–5
United Kingdom	2–3
United States	~4

The estimated number of cases in the United States for 1993 was 1092 with 248 deaths. Not all cases are of direct food origin, as other sources have been documented.

From a risk assessment study, a person on average is exposed 3.8 times via food to 5.0 $\log_{10}$ organisms and 0.8 times to $>10^6$ $\log_{10}$ organisms/year with about five to seven cases of listeriosis per year.[101] After considering other factors such as mouse infective doses, the investigators concluded that listeriosis is a rare disease in humans despite frequent exposure to the causative organism.

Source of Pathogens

With the incidence of human foodborne listeriosis being so low and sporadic, the source of the causative strains of *L. monocytogenes* is of great interest. Although the outbreaks traced to dairy products may be presumed to result from the shedding of virulent strains into milk, this is not always confirmed. In a study of 1,123 raw milk samples from the 27 farms that supplied milk to the incriminated cheese plant in California in 1985, Donnelly et al.[28] were unable to recover the responsible 4b serovar. A serotype 1 was isolated from 16 string samples from one control farm. In a review of the human cases through most of 1986, Hird[57] concluded that whereas the evidence was not conclusive in all cases, it nevertheless supported zoonotic transmission to some degree (zoonosis: disease transmissible under natural conditions from vertebrate animals to humans). Hird believes the healthy animal carrier is an important source of the organism, along with clinical listeriosis in livestock, but the relative degree to which each contributes to foodborne cases is uncertain.

L. monocytogenes was found to be shed in milk from the left forequarter of a mastitic cow, but milk from the other quarters was uninfected.[48] About 10% of healthy cattle tested in The Netherlands were positive for *L. monocytogenes*, and about 5% of human fecal samples from slaughterhouse workers in Denmark contained the organism.[68] The carriage rate for healthy humans seemed to be about the same regardless of their work position within food processing plants.[68] Over an 18-month period in the United Kingdom, 32 of 5,000 (0.6%) fecal samples were positive.[75] Cross-infection with *L. monocytogenes* from congenitally infected newborn infants to apparently healthy neonates in hospitals has been shown to occur.[94] Thus, although the organism is known to be fairly common in environmental specimens, it also exists in healthy humans at rates from less than 1% to around 15%. The relative importance of environmental, animal, and human sources to foodborne episodes awaits further study.

Among ready-to-eat (RTE) foods, meat and poultry products are the leading vehicles for human listeriosis. In its annual survey of RTE meat and poultry products, the Food Safety and Inspection Service (FSIS) of the U.S. Department of Agriculture found the following percent of these products to contain *L. monocytogenes* during the 9 years noted:

1995—3.02	2000—1.45
1996—2.91	2001 1.32
1997—2.25	2002—1.03
1998—2.54	2003—0.75 (January–September)
1999—1.91	

Table 25–5   Some of the Suspected and Proven Foodborne Listeriosis Outbreaks and Cases

Year	Source	Cases/Deaths	Location
1953	Raw milk	2/1	Germany
1959	Fresh meat/poultry*	4/2	Sweden
1960–1961	Various/unknown	81/?	Germany
1966	Milk/products	279/109	Germany
1979	Vegetables/milk?[†]	23/3	Boston
1980	Shellfish	22/6	New Zealand
1981	Cole slaw	41/18	Canada
1983	Pasteurized milk[†]	49/14	Boston
1983–1987	Vacherin Mont D'Or	122/34	Switzerland
1985	Mexican-style cheese	142/48	California
1986–1987	Vegetables?[†]	36/16	Philadelphia
1987–1989	Pâté	366/63	United Kingdom
1987	Soft cheese	1	United Kingdom
1988	Goats' milk cheese	1	United Kingdom
1988	Cooked-chld-chick.	1	United Kingdom
1988	Cooked-chld-chick.	2	United Kingdom
1988	Turkey franks	1	Oklahoma
1989	Pork sausage	1	Italy
1988	Alfalfa tablets	1	Canada
1989	Salted mushrooms	1	Finland
1989	Shrimp	9/1	United States (Conn.)
1989	Pork sausage	1	Italy
1990	Raw milk	1	Vermont
1990	Pork sausage	1	Italy
1990	Pâté	11/6	Australia
1991	Smoked mussels	3/0	Australia
1992	Smoked mussels	4/2	New Zealand
1992	Goat meat (from Calif.)	1	Canada
1992	Pork tongue in jelly	279/85	France
1993	Pork rillettes	39/0	France
1994	Chocolate milk	52/0	USA
1994	Pickled olives	1	Italy
1995	Brie cheese	17/0	France
1998–1999	Wieners	ca. 101/ca. 21	United States
1999–2000	Pork tongue in jelly	26/7	France
2000–2001	Homemade Mexican-style cheese	12/0	United States
2002	Deli turkey meat	46/7	10 USA states

*Suspected.
[†]Epidemiologically linked; organisms not found.

In an extensive review of this organism in the general processing environment, Tompkin[124] noted that foodborne listeriosis accounted for only ca. 0.02% of all foodborne illnesses in the United States. On the other hand, this disease accounted for ca. 28% of all deaths from foodborne illnesses. Foods that have been involved in human cases typically contain >1,000 cfu/g or ml; and outbreak strains typically have become established in the processing environment resulting in contamination of multiple lots of foods.[124]

In spite of the high fatality rate associated with foodborne listeriosis, overall the disease is relatively rare. Not all *L. monocytogenes* isolates are equally pathogenic, and just what makes a foodborne isolate infectious or noninfectious is poorly understood as are the reasons for the predominance in foods of serotypes 1/2a, 1/2b, and 4b.[69]

Syndromes

Listeriosis in humans is not characterized by a unique set of symptoms because the course of the disease depends on the state of the host. Non-pregnant healthy individuals who are not immunosuppressed are highly resistant to infection by *L. monocytogenes*, and there is little evidence that such individuals ever contract clinical listeriosis. However, the following conditions are known to predispose to adult listeriosis and to be significant in mortality rate: neoplasm, AIDS, alcoholism, diabetes (type 1, in particular), cardiovascular disease, renal transplant, and corticosteroid therapy. When susceptible adults contract the disease, meningitis and sepsis are the most commonly recognized symptoms. Of 641 human cases, 73% of victims had meningitis, meningoencephalitis, or encephalitis. Cervical and generalized lymphadenopathy are associated with the adult syndrome, and thus the disease may resemble infectious mononucleosis. Cerebrospinal fluid initially contains granulocytes, but in later states, monocytes predominate. Pregnant females who contract the disease (and their fetuses are often congenitally infected) may not present any symptoms, but when they do, they are typically mild and influenzalike. Abortion, premature birth, or stillbirth is often the consequence of listeriosis in pregnant females. When a newborn is infected at the time of delivery, listeriosis symptoms typically are those of meningitis, and they typically begin 1–4 weeks after birth, although a 4-day incubation has been recorded. The usual incubation time in adults ranges from 1 to several weeks. Among the 20 case patients studied from the cluster of cases in the Boston episode, 18 had bacteremia, 8 developed meningitis, and 13 complained of vomiting, abdominal pain, and diarrhea 72 hours before onset of symptoms.

The control of *L. monocytogenes* in the body is affected by T lymphocytes and activated macrophages, and thus any condition that adversely affects these cells will exacerbate the course of listeriosis. The most effective drugs for treatment are coumermycin, rifampicin, and ampicillin, with the last plus an aminoglycoside antibiotic being the best combination.[33] Even with that regimen, antimicrobial therapy for listeriosis is not entirely satisfactory because ill patients and compromised hosts are more difficult than competent hosts.

RESISTANCE TO LISTERIOSIS

Resistance or immunity to intracellular pathogens such as viruses, animal parasites, and *L. monocytogenes* is mediated by T cells, lymphocytes that arise from bone marrow and undergo maturation in the thymus (hence, T for thymus derived). Unlike B cells, which give rise to humoral immunity (circulating antibodies), activated T cells react directly against foreign cells. Once a pathogen is inside a host cell, it cannot be reached by circulating antibody, but the presence of the pathogen is signaled by structural changes in the parasitized cell, and T cells are involved in the destruction of this invaded host cell, which is no longer recognized as "self."

Macrophages are important to the actions of T cells, and their need for the destruction of *L. monocytogenes* and certain other intracellular pathogens was shown by Mackeness.[87] First, macrophages bind and "present" *L. monocytogenes* cells to T cells in such a way that they are recognized as being foreign. When T cells react with the organism, they increase in size and form clones specific for the same organism or antigen. These T cells, said to be activated, secrete interleukin-1 (IL-1). As the activated T cells multiply, they differentiate to form various subsets.

The most important subsets of T cells for resistance to listeriosis are helper or CD4 (L3T4[+]) and cytolytic (killer) or CD8 (Lyt2[+]).[71] The CD4 T cells react with the foreign antigen, after which they produce lymphokines (cytokines): IL-1, IL-2, IL-6, immune or γ-interferon, and others. γ-Interferon, whose production is aided also by tumor necrosis factor,[22] induces the production of IL-2 receptor expression on monocytes. Also, IL-2 may enhance the activation of lymphokine-activated killer cells that can lyse infected macrophages. As little as 0.6 μg per mouse of exogenously administered IL-2 has been shown to strengthen mouse resistance to L. monocytogenes.[54] γ-Interferon activates macrophages and CD8 T cells, and the latter react with L. monocytogenes-infected host macrophages—and cause their lysis. Both CD4 and CD8 T cells are stimulated by L. monocytogenes; they activate macrophages via their production of γ-interferon and contribute to resistance to listeriosis.[70] CD8 also secretes γ-interferon when exogenous IL-2 is provided, and both CD4 and CD8 can confer some passive immunity to recipient mice.[70]

Some of the events that occur in murine hosts following infection with L. monocytogenes are presumed to be the same events that occur in humans. When macrophages engulf L. monocytogenes, a factor-increasing monocytopoiesis (FIM) is secreted by the macrophages at the infection site. FIM is transported to the bone marrow, where it stimulates the production of more macrophages. Only viable L. monocytogenes cells can induce the T cell response and immunity to histeriosis. Because LLO is the virulence factor of L. monocytogenes that elicits the T cell response, this heat-labile protein is destroyed when cells are heat killed. The CD8 T cell subset appears to be the major T cell component responsible for antilisterial immunity, for it acts by eliciting lymphokine production by macrophages; passive immunity to L. monocytogenes can be achieved by transfer of the CD8 T cell subset.[4] The T cell response does not occur even when mice are injected with both killed cells and recombinant IL-1a.[60] Neither avirulent nor killed cells induce IL-1 in vitro; viable L. monocytogenes cells do,[96] indicating a critical role for IL-1 and γ-interferon in the initiation of the in vivo response, for it has been shown that simultaneously administered IL-1a and γ-interferon increased resistance to L. monocytogenes in mice better than either alone.[74] The combination was not synergistic, only additive. It appears that the primary role of γ-interferon is to elicit lymphokine production rather than acting directly, and it is known to increase the production of IL-1.[17] γ-Interferon is detectable in the blood stream and spleen of mice only during the first 4 days after infection.[99] Infection of mice by L. monocytogenes leads to an increase in IL-6, which is produced by nonlymphocyte cells.[80] Mice that are deficient in IL-6 have increased susceptibility to listeriosis.[23] IL-6 appears to act by stimulating the production of neutrophils.[23]

This synopsis of murine resistance to L. monocytogenes reveals some of the multifunctional roles of the lymphokines in T cell immunity (due to different E-cadherins needed for InlA receptors); and the apparent critical importance of LLO as the primary virulence factor of this organism. What makes immunocompromised hosts more susceptible to listeriosis is the dampening effect that immunosuppressive agents have on the T cell system. Possible therapy is suggested by the specific roles that some of the lymphokines play, but whether the effects are similar in humans is unclear.

PERSISTENCE OF *L. MONOCYTOGENES* IN FOODS

Because it can grow over the temperature range of about 1–45°C and the pH range of 4.1 to around 9.6, L. monocytogenes may be expected to survive in foods for long periods of time, and this has been confirmed. Strains Scott A and V7, inoculated at levels of 10^4–10^5/g, survived in cottage cheese for up to 28 days when held at 3°C.[111] When these two strains, with two others, were inoculated into a camembert cheese formulated with levels of 10^4–10^5, growth occurred during the first 18 days of ripening, and some strains attained levels of 10^6–10^7 after 65 days of ripening.[110] With an inoculum of 5×10^2/g and storage at 4°C, *L. monocytogenes* survived in cold-pack cheese for a mean of 130 days in the presence of 0.30% sorbic acid.[109] On the other hand, in manufactured and stored nonfat dry milk, a 1–1.5 log reduction in numbers of *L. monocytogenes* occurred during spray drying, and more than a 4-log decrease in colony-forming units occurred within 16 weeks when held at 25°C.[30]

In ground beef, an inoculum of 10^5–10^6 remained unchanged through 14 days at 4°C,[64] and inocula of *L. monocytogenes* of 10^3 or 10^5 remained unchanged in ground beef and liver for over 30 days, although the standard plate counts (SPC) increased threefold to sixfold during this time.[114] When added to a Finnish sausage mix that included 120 ppm $NaNO_2$ and 3% NaCl, the initial numbers of *L. monocytogenes* decreased only by about 1 log during a 21-day fermentation period.[66] When five strains of *L. monocytogenes* were added to eight processed meats that were stored at 4.4°C for up to 12 weeks, the organisms survived on all products and increased in numbers by 3 to 4 logs in most.[49] Best growth occurred in chicken and turkey products, due in part to the higher initial pH of these products. In vacuum-packaged beef in a film with barrier properties of 25–30 ml/m²/24 hours/101 kPa, an inoculated strain of *L. monocytogenes* increased about 4 log cycles on the fatty tissue of strip loins in 16 days and by about 3 log cycles in 20 days on lean meat when held between 5° and 5.5°C.[49,51] In meat, cheese, and egg ravioli stored at 5°C, a 3×10^5-cfu/g inoculum of strain Scott A survived 14 days.[6] Lettuce and lettuce juice supported the growth of *L. monocytogenes* held at 5°C for 14 days in one study, and the organism was recovered from two uninoculated lettuce samples.[122]

The studies noted are typical of others and show that the overall resistance of *L. monocytogenes* in foods is consistent with its persistence in many nonfood environmental specimens.

REGULATORY STATUS OF *L. MONOCYTOGENES* IN FOODS

Some countries have established legal limits on the numbers of organisms that are permissible in foods, especially ready-to-eat products, whereas others have suggested guidelines or criteria that do not have legal standing.

The United States government has the most rigid policy whereby *L. monocytogenes* has been designated as an "adulterant." This means that any ready-to-eat food that contains this organism can be considered adulterated and, thus, be subject to recall and/or seizure. The U.S. requirement is the absence of the organism in 50-g samples. Zero tolerance generally means the absence of the organism in 25-g samples, which is equivalent to $n = 5$, $c = 0$ in a sampling plan (see Chapter 20 for an explanation). The following summaries were believed to be in effect late in 1995, but it should be understood that some may have changed.

The European Community (EC) directive on milk and milk-based products specifies zero tolerance for soft cheeses, and absence of the organism in 1 g of other products.

Great Britain's provisional guidelines for some ready-to-eat foods establishes four quality groups based on numbers of *L. monocytogenes*. Not detected in 25 g is satisfactory; >10^2/25 g is fairly satisfactory; 10^2–10^3 is unsatisfactory; and numbers >10^3 make the product unacceptable.[46]

The proposed Canadian compliance criteria of 1993 placed ready-to-eat foods into three categories relative to *L. monocytogenes* actions. Category 1 includes products linked to outbreaks, category 2 includes those that have a self-life >10 days, and category 3 includes those that either support growth with a self-life ≤10 days or those that do not support growth. Among the latter are those that fit one or more of the following: pH 5.0–5.5 and a_w <0.95; pH <5.0 regardless of a_w; $a_w \geq 0.92$ regardless of pH; and frozen foods.[77] Recalls of category 3 foods require numbers >10^2/g.

The view in Germany is that zero tolerance is not only unrealistic but unnecessary, and foods are placed in four risk levels somewhat along the lines of Canada. Products that contain $>10^4$/g are subject to an automatic recall.

Australia requires the absence of *L. monocytogenes* in 5 × 25-g samples for many cheeses. France requires no *L. monocytogenes* in 25-g samples of foods for at-risk individuals. The French position seems to be that it is unrealistic to expect no *L. monocytogenes* in raw foods. It has been noted that the presence of this organism in food processing environments is inevitable, especially those for finished products; and that while the risks of final product contamination can be reduced, they cannot be eliminated.[125]

The International Commission on Microbiological Specification for Foods (ICMSF) has concluded that if this organism does not exceed 100/g of food at point of consumption, the food is considered acceptable for individuals who are not at risk (see FSO in Chapter 20). The ICMSF endorses the use of HACCP (see Chapter 21), and places *L. monocytogenes* in plan stringency cases nos. 10, 11, and 12 (see Table 21–4). The two-class sampling plan for case 10 is $n = 5$, $c = 0$; for cases 11, $n = 10$, $c = 0$; and for case 12, $n = 20$, $c = 0$. For the years 1994–1998, *L. monocytogenes* was responsible for 61% of the 1,328 food product recalls in the United States followed by 11% for salmonellae.[132] However, it is not correct to assume that all recalled foods would have caused listeriosis cases had they been sold. As noted above, not all strains of this organisms cause human illness; and RTE meat and poultry products are not handled the same way by different consumers.

REFERENCES

1. Ade, N., S. Steinmeyer, M.J. Loessner, H. Hof, and J. Kreft (Technical University of Munich, Weihenstephan, Freising, Germany). 1991. Personal communication.

2. Anton, W. 1934. Kritisch-experimenteller Beitrag zur Biologie des *Bacterium monocytogenes*. Mit basonderer Berueck-sichtigung seiner Beziehung zur infektiosen Mononucleose des Menschen. *Zbl. Bakteriol. Abt. I. Orig.* 131:89–103.

3. Audurier, A., and C. Martin. 1989. Phage typing of *Listeria monocytogenes*. *Int. J. Food Microbiol.* 8:251–257.

4. Baldridge, J.R., R.A. Barry, and D.J. Hinrichs. 1990. Expression of systemic protection and delayed-type hypersensitivity to *Listeria monocytogenes* is mediated by different T-cell subsets. *Infect. Immun.* 58:654–658.

5. Bartlett, F.M., and A.E. Hawke. 1995. Heat resistance of *Listeria monocytogenes* Scott A and HAL 957E1 in various liquid egg products. *J. Food Protect.* 58:1211–1214.

6. Beuchat, L.R., and R.E. Brackett. 1989. Observations on survival and thermal inactivation of *Listeria monocytogenes* in ravioli. *Lett. Appl. Microbiol.* 8:173–175.

7. Beuchat, L.R., R.E. Brackett, D.Y.-Y. Hao, and D.L. Conner. 1986. Growth and thermal inactivation of *Listeria monocytogenes* in cabbage and cabbage juice. *Can. J. Microbiol.* 32:791–795.

8. Boyle, D.L., J.N. Sofos, and G.R. Schmidt. 1990. Thermal destruction of *Listeria monocytogenes* in a meat slurry and in ground beef. *J. Food Sci.* 55:327–329.

9. Bradshaw, J.G., J.T. Peeler, J.J. Corwin, J.M. Hunt, and R.M. Twedt. 1987. Thermal resistance of *Listeria monocytogenes* in dairy products. *J. Food Protect.* 50:543–544.

10. Boerlin, P., J. Rocourt, F. Grimont, C. Jacquet, and J.-C. Piffaretti. 1992. *Listeria ivanovii* subsp. *londoniensis* subsp. nov. *Int. J. Syst. Bacteriol.* 42:69–73.

11. Bradshaw, J.G., J.T. Peeler, J.J. Corwin, J.M. Hunt, J.T. Tierney, E.P. Larkin, and R.M. Twedt. 1985. Thermal resistance of *Listeria monocytogenes* in milk. *J. Food Protect.* 48:743–755.

12. Buchanan, R.L., and L.A. Klawitter. 1990. Effects of temperature and oxygen on the growth of *Listeria monocytogenes* at pH 4.5. *J. Food Sci.* 55:1754–1756.

13. Bunning, V.K., R.G. Crawford, J.T. Tierney, and J.T. Peeler. 1990. Thermotolerance of *Listeria monocytogenes* and *Salmonella* Typhimurium after sublethal heat shock. *Appl. Environ. Microbiol.* 56:3216–3219.

14. Bunning, V.K., C.W. Donnelly, J.T. Peeler, E.H. Briggs, J.G. Bradshaw, R.G. Crawford, C.M. Beliveau, and J.T. Tierney. 1988. Thermal inactivation of *Listeria monocytogenes* within bovine milk phagocytes. *Appl. Environ. Microbiol.* 54:364–370.

15. Bunning, V.K., R.G. Crawford, J.G. Bradshaw, J.T. Peeler, J.T. Tierney,, and R.M. Twedt. 1986. Thermal resistance of intracellular *Listeria monocytogenes* cells suspended in raw bovine milk. *Appl. Environ. Microbiol.* 52:1398–1402.

16. Carbonnelle, B., J. Cottin, F. Parvery, G. Chambreuil, S. Kouyoumdjian, M. LeLirzin, G. Cordier, and F. Vincent. 1978. Epidemie de listeriose dans l'Ouest de la France (1975–1976). *Rev. Epidemiol. Santé Pub.* 26:451–467.

17. Chen, Y., A. Nakane, and T. Minagawa. 1989. Recombinant murine gamma interferon induces enhanced resistance to *Listeria monocytogenes* infection in neonatal mice. *Infect. Immun.* 57:2345–2349.

18. Colburn, K.G., C.A. Kaysner, C. Abeyta, Jr., and M.M. Wekell. 1990. *Listeria* species in a California coast estuarine environment. *Appl. Environ. Microbiol.* 56:2007–2011.

19. Cole, M.B., M.V. Jones, and C. Holyoak. 1990. The effect of pH, salt concentration and temperature on the survival and growth of *Listeria monocytogenes*. *J. Appl. Bacteriol.* 69:63–72.

20. Collins, M.D., S. Wallbanks, D.J. Lane, J. Shah, R. Nietupski, J. Smida, M. Dorsch, and E. Stackebrandt. 1991. Phylogenetic analysis of the genus *Listeria* based on reverse transcription sequencing of 16S rRNA. *Int. Syst. Bacteriol.* 41:240–246.

21. Conner, D.E., R.E. Brackett, and L.R. Beuchat. 1986. Effect of temperature, sodium chloride, and pH on growth of *Listeria monocytogenes* in cabbage juice. *Appl. Environ. Microbiol.* 52:59–63.

22. Cossart, P., and J. Mengaud. 1989. *Listeria monocytogenes*. A model system for the molecular study of intracellular parasitism. *Mol. Biol. Med.* 6:463–474.

23. Dalrymple, S.A., L.A. Lucian, R. Slattery, T. McNeil, D.M. Aud, S. Fuchino, F. Lee, and R. Murray. 1995. Interleukin-6-deficient mice are highly susceptible to *Listeria monocytogenes* infection: Correlation with inefficient neutrophilia. *Infect. Immun.* 63:2262–2268.

24. Datta, A.R., and M.H. Kothary. 1993. Effects of glucose, growth temperature, and pH on listeriolysin O production in *Listeria monocytogenes*. *Appl. Environ. Microbiol.* 59:3495–3497.

25. Del Corral, F., R.L. Buchanan, M.M. Bencivengo, and P.H. Cooke. 1990. Quantitative comparison of selected virulence associated characteristics in food and clinical isolates of *Listeria*. *J. Food Protect.* 53:1003–1009.

26. Donnelly, C.W. 2001. *Listeria monocytogenes*. In *Foodborne Disease Handbook*, ed. Y.H. Hui, M.D. Pierson, and J.R. Gorham, 2nd ed., 213–245. New York: Marcel Dekker.

27. Donnelly, C.W., and E.H. Briggs. 1986. Psychrotrophic growth and thermal inactivation of *Listeria monocytogenes* as a function of milk composition. *J. Food Protect.* 49:994–998.

28. Donnelly, C.W., E.H. Briggs, and G.J. Baigent. 1986. Analysis of raw milk for the epidemic serotype of *Listeria monocytogenes* linked to an outbreak of listeriosis in California. *J. Food Protect.* 49:846–847 (Abstract).

29. Doyle, M.E., A.S. Mazzotta, T. Wang, D.W. Wiseman, and V.N. Scott. 2001. Heat resistance of *Listeria monocytogenes*. *J. Food Protect.* 64:410–429.

30. Doyle, M.O., L.M. Meske, and E.H. Marth. 1985. Survival of *Listeria monocytogenes* during the manufacture and storage of nonfat dry milk. *J. Food Protect.* 48:740–742.

31. Draughon, F.A., B.A. Anthony, and M.E. Denton. 1999. *Listeria* species in fresh rainbow trout purchased from retail markets. *Dairy Food. Environ. Sanit.* 19:90–94.

32. Drevets, D.A., R.T. Sawyer, T.A. Potter, and P.A. Campbell. 1995. *Listeria monocytogenes* infects human endothelial cells by two distinct mechanisms. *Infect. Immun.* 63:4268–4276.

33. Espaze, E.P., and A.E. Reynaud. 1988. Antibiotic susceptibilities of *Listeria*: In vitro studies. *Infection* 16(Suppl. 2):160–164.

34. Farber, J.M., F. Coates, and E. Daley. 1992. Minimum water activity requirements for the growth of *Listeria monocytogenes*. *Lett. Appl. Microbiol.* 15:103–105.

35. Farber, J.M., and B.E. Brown. 1990. Effect of prior heat shock on heat resistance of *Listeria monocytogenes* in meat. *Appl. Environ. Microbiol.* 56:1584–1587.

36. Farber, J.M. 1989. Thermal resistance of *Listeria monocytogenes*. *Int. J. Food Microbiol.* 8:285–291.

37. Farber, J.M., G.W. Sanders, J.I. Speirs, J.-Y. D'Aoust, D.B. Emmons, and R. McKellar. 1988. Thermal resistance of *Listeria monocytogenes* in inoculated and naturally contaminated raw milk. *Int. J. Food Microbiol.* 7:277–286.

38. Fedio, W.M., and H. Jackson. 1989. Effect of tempering on the heat resistance of *Listeria monocytogenes*. *Lett. Appl. Microbiol.* 9:157–160.

39. Fenlon, D.R. 1985. Wild birds and silage as reservoirs of *Listeria* in the agricultural environment. *J. Appl. Bacteriol.* 59:537–543.

40. Fenlon, D.R., T. Stewart, and W. Donachie. 1995. The incidence, numbers and types of *Listeria monocytogenes* isolated from farm bulk tank milks. *Lett. Appl. Microbiol.* 20:57–60.

41. Foegeding, P.M., and N.W. Stanley. 1990. *Listeria monocytogenes* F5069 thermal death times in liquid whole egg. *J. Food Protect.* 53:6–8.

42. Galsworthy, S.B., and D. Fewster. 1988. Comparison of responsiveness to the monocytosis-producing activity of *Listeria monocytogenes* in mice genetically susceptible or resistant to listeriosis. *Infection* 16(Suppl. 2):118–122.

43. Geoffroy, C., J.-L. Gaillard, J.E. Alouf, and P. Berche. 1989. Production of thiol-dependent haemolysins by *Listeria monocytogenes*. *J. Gen. Microbiol.* 135:481–487.

44. Geoffroy, C., J.-L. Gaillard, J.E. Alouf, and P. Berche. 1987. Purification, characterization, and toxicity of the sulfhydryl-activated hemolysin listeriolysin O from *Listeria monocytogenes*. *Infect. Immun.* 55:1641–1646.

45. George, S.M., B.M. Lung, and T.F. Brocklehurst. 1988. The effect of pH and temperature on initiation of growth of *Listeria monocytogenes*. *Lett. Appl. Microbiol.* 6:153–156.

46. Gilbert, R.J. 1992. Provisional microbiological guidelines for some ready-to-eat foods sampled at point of sale: Notes for PHLS Food Examiners. *Public Health Serv. Lab. Q.* 9:98–99.

47. Gilbert, R.J., and P.N. Pini. 1988. Listeriosis and foodborne transmission. *Lancet* 1:472–473.

48. Gitter, M., R. Bradley, and P.H. Blampied. 1980. *Listeria monocytogenes* infection in bovine mastitis. *Vet. Rec.* 107:390–393.

49. Glass, K.A., and M.P. Doyle. 1990. Fate of *Listeria monocytogenes* in processed meat products during refrigerated storage. *Appl. Environ. Microbiol.* 55:1565–1569.

50. Golnazarian, C.A., C.W. Donnelly, S.J. Pintauro, and D.B. Howard. 1989. Comparison of infectious dose of *Listeria monocytogenes* F5817 as determined for normal versus compromised C57B1/6J mice. *J. Food Protect.* 52:696–701.

51. Grau, F.H., and P.B. Vanderline. 1990. Growth of *Listeria monocytogenes* on vacuum-packaged beef. *J. Food Protect.* 53:739–741.

52. Green, S.S. 1990. *Listeria monocytogenes* in meat and poultry products. Interim Rept. to Nat'l Adv. Comm. Microbiol. Spec. Foods. FSIS/USDA, Nov. 27.

53. Groves, R.D., and H.J. Welshimer. 1977. Separation of pathogenic from apathogenic *Listeria monocytogenes* by three in vitro reactions. *J. Clin. Microbiol.* 5:559–563.

54. Haak-Frendscho, M., K.M. Young, and C.J. Czuprynski. 1989. Treatment of mice with human recombinant interleukin-2 augments resistance to the facultative intracellular pathogen *Listeria monocytogenes*. *Infect. Immun.* 57:3014–3021.

55. Harrison, M.A., and Y.-W. Huang. 1990. Thermal death times for *Listeria monocytogenes* (Scott A) in crabmeat. *J. Food Protect.* 53:878–880.

56. Heisick, J.E., D.E. Wagner, M.L. Nierman, and J.T. Peeler. 1989. *Listeria* spp. found on fresh market produce. *Appl. Environ. Microbiol.* 55:1925–1927.

57. Hird, D.W. 1987. Review of evidence for zoonotic listeriosis. *J. Food Protect.* 50:429–433.

58. Hogen, C.J., E.R. Singleton, K.S. Kreuzer, E.M. Sloan, and J.N. Sofos.. 1998. Isolation of catalase-negative *Listeria monocytogenes* from foods. *Dairy Food Environ. Sanit.* 18:424–426.

59. Hof, H., and P. Hefner. 1988. Pathogenicity of *Listeria monocytogenes* in comparison to other *Listeria* species. *Infection* 16(Suppl. 2):141–144.

60. Igarashi, K.-I., M. Mitsuyama, K. Muramori, H. Tsukada, and K. Nomoto. 1990. Interleukin-1-induced promotion of T-cell differentiation in mice immunized with killed *Listeria monocytogenes*. *Infect. Immun.* 58:3973–3979.

61. Jacquet, C., E. Gouin, D. Jeannel, P. Cossart, and J. Rocourt. 2002. Expression of ActA, Ami, InlB, and listeriolysin O in *Listeria monocytogenes* of human and food origin. *Appl. Environ. Mirobiol.* 68:616–622.

62. Jay, J.M. 1996. Prevalence of *Listeria* spp. in meat and poultry products. *Food Control* 7:209–214.

63. Johnson, J.L., M.P. Doyle, and R.G. Cassens. 1990. *Listeria monocytogenes* and other *Listeria* spp. in meat and meat products. A review. *J. Food Protect.* 53:81–91.

64. Johnson, J.L., M.P. Doyle, and R.G. Cassens. 1988. Survival of *Listeria monocytogenes* in ground beef. *Int. J. Food Microbiol.* 6:243–247.

65. Jones, D. 1988. The place of *Listeria* among Gram-positive bacteria. *Infection* 16(Suppl. 2):85–88.

66. Junttila, J., J. Hirn, P. Hill, and E. Nurmi. 1989. Effect of different levels of nitrite and nitrate on the survival of *Listeria monocytogenes* during the manufacture of fermented sausage. *J. Food Protect.* 52:158–161.

67. Junttila, J.R., S.I. Niemela, and J. Hirn. 1988. Minimum growth temperatures of *Listeria monocytogenes* and nonhaemolytic listeria. *J. Appl. Bacteriol.* 65:321–327.

68. Kampelmacher, E.H., and L.M. Van Nooble Jansen. 1969. Isolation of *Listeria monocytogenes* from faeces of clinically healthy humans and animals. *Zentral. Bakt. Inf. Abt. Orig.* 211:353–359.

69. Kathariou, S. 2002. *Listeria monocytogenes* virulence and pathogenicity, a food safety perspective. *J. Food Protect.* 65:1811–1829.

70. Kaufmann, S.H.E. 1988. *Listeria monocytogenes* specific T-cell lines and clones. *Infection* 16(Suppl. 2):128–136.

71. Kaufmann, S.H.E., E. Hug, U. Vath, and I. Muller. 1985. Effective protection against *Listeria monocytogenes* and delayed-type hypersensitivity to listerial antigens depend on cooperation between specific L3T4+ and Lyt2+ T cells. *Infect. Immun.* 48:263–266.

72. Kouassi, Y., and L.A. Shelef. 1995. Listeriolysin O secretion by *Listeria monocytogenes* in broth containing salts of organic acids. *J. Food Protect.* 58:1314–1319.

73. Kreft, J., D. Funke, A. Haas, F. Lottspeich, and W. Goebel. 1989. Production, purification and characterization of hemolysins from *Listeria ivanovii* and *Listeria monocytogenes* Sv4b. *FEMS Microbiol. Lett.* 57:197–202.

74. Kurtz, R.S., K.M. Young, and C.J. Czuprynski. 1989. Separate and combined effects of recombinant interleukin-1a and gamma interferon on antibacterial resistance. *Infect. Immun.* 57:553–558.

75. Kwantes, W., and M. Isaac. 1971. Listeriosis. *Br. Med. J.* 4:296–297.

76. Leasor, S.B., and P.M. Foegeding. 1989. *Listeria* species in commercially broken raw liquid whole egg. *J. Food Protect.* 52:777–780.

77. Lammerding, A.M., and J.M. Farber. 1994. The status of *Listeria monocytogenes* in the Canadian food industry. *Dairy Food Environ. Sanit.* 14:146–150.

78. Lingnau, A., E. Domann, M. Hudel, M. bock, T. Nichterlein, J. Wehland, and T. Chakraborty. 1995. Expression of the *Listeria monocytogenes* EGD *inl*A and *inl*B genes, whose products mediate bacterial entry into tissue culture cell lines, by PrfA-dependent and -independent mechanisms. *Infect. Immun.* 63:3896–3903.

79. Linton, R.H., M.D. Pierson, and J.R. Bishop. 1990. Increase in heat resistance of *Listeria monocytogenes* Scott A by sublethal heat shock. *J. Food Protect.* 53:924–927.

80. Liu, Z., and C. Cheers. 1993. The cellular source of interleukin-6 during *Listeria* infection. *Infect. Immun.* 61:2626–2631.

81. Loessner, M.J., and M. Busse. 1990. Bacteriophage typing of *Listeria* species. *Appl. Environ. Microbiol.* 56:1912–1918.

82. Lovett, J., I.V. Wesley, M.J. Vandermaaten, J.G. Bradshaw, D.W. Francis, R.G. Crawford, C.W. Donnelly, and J.W. Messer. 1990. High-temperature short-time pasteurization inactivates *Listeria monocytogenes*. *J. Food Protect.* 53:734–738.

83. Lovett, J. 1988. Isolation and identification of *Listeria monocytogenes* in dairy products. *J. Assoc. Off. Anal. Chem.* 71:658–660.

84. Lovett, J., D.W. Francis, and J.M. Hunt. 1987. *Listeria monocytogenes* in raw milk: Detection, incidence, and pathogenicity. *J. Food Protect.* 50:188–192.

85. Ludwig, W., K.-H. Schleifer, and E. Stackebrandt. 1984. 16S rRNA analysis of *Listeria monocytogenes* and *Brochothrix thermosphacta*. *FEMS Microbiol. Lett.* 25:199–204.

86. Lund, B.M., M.R. Knox, and M.B. Cole. 1989. Destruction of *Listeria monocytogenes* during microwave cooking. *Lancet* 1:218.

87. Mackeness, G.B. 1971. Resistance to intracellular infection. *J. Infect. Dis.* 123:439–445.

88. Mackey, B.M., C. Pritchet, A. Norris, and G.C. Mead. 1990. Heat resistance of *Listeria*: Strain differences and effects of meat type and curing salts. *Lett. Appl. Microbiol.* 10:251–255.

89. Mackey, B.M., and N. Bratchell. 1989. The heat resistance of *Listeria monocytogenes*. *Lett. Appl. Microbiol.* 9:89–94.

90. Malinverni, R., J. Bille, Cl. Perret, F. Regli, F. Tanner, and M.P. Glauser. 1985. Listeriose epidemique. Observation de 25 cas en 15 mois au Centre hospitalier universitaire vaudois. *Schweiz. Med. Wschr.* 115:2–10.

91. McKellar, R.C. 1994. Use of the CAMP test for identification of *Listeria monocytogenes*. *Appl. Environ. Microbiol.* 60:4219–4225.

92. McLauchlin, J. 1997. The pathogenicity of *Listeria monocytogenes*: A public health perspective. *Rev. Med. Microbiol.* 8:1–14.

93. McLauchlin, J. 1987. *Listeria monocytogenes*, recent advances in the taxonomy and epidemiology of listeriosis in humans. *J. Appl. Bacteriol.* 63:1–11.

94. McLauchlin, J., A. Audurier, and A.G. Taylor. 1986. Aspects of the epidemiology of human *Listeria monocytogenes* infections in Britain 1967–1984. The use of serotyping and phage typing. *J. Med. Microbiol.* 22:367–377.

95. Mengaud, J., M.F. Vincente, J. Chenevert, J.M. Pereira, C. Geoffroy, B. Giequel-Sanzey, F. Baquero, J.-C. Perez-Diaz, and P. Cossart. 1988. Expression in *Escherichia coli* and sequence analysis of the listeriolysin O determinant of *Listeria monocytogenes*. *Infect. Immun.* 56:766–772.

96. Mitsuyama, M., K.-I. Igarashi, I. Kawamura, T. Ohmori, and K. Nomoto. 1990. Difference in the induction of macrophage interleukin-1 production between viable and killed cells of *Listeria monocytogenes*. *Infect. Immun.* 58:1254–1260.

97. Moors, M.A., B. Levitt, P. Youngman, and D.A. Portnoy. 1999. Expression of listeriolysin O and ActA by intracellular and extracellular *Listeria monocytogenes*. *Infect. Immun.* 67:131–139.

98. Murray, E.G.D., R.A. Webb, and M.B.R. Swann. 1926. A disease of rabbits characterized by large mononuclear leucocytosis caused by a hitherto undescribed bacillus *Bacterium monocytogenes* (n. sp.). *J. Pathol. Bacteriol.* 29:407–439.

99. Nakane, A., A. Numata, M. Asano, M. Kohanawa, Y. Chen, and T. Minagawa. 1990. Evidence that endogenous gamma interferon is produced early in *Listeria monocytogenes* infection. *Infect. Immun.* 58:2386–2388.

100. Nicolas, J.-A., and N. Vidaud. 1987. Contribution a l'étude des *Listeria* presentés dans les denrées d'origine animale destinées à la consommation humaine. *Rec. Med. Vet.* 163(3):283–285.

101. Notermans, S., J. Dufrenne, P. Teunis, and T. Chackraborty. 1998. Studies on the risk assessment of *Listeria monocytogenes*. *J. Food Protect.* 61:244–248.

102. Parish, M.E., and D.P. Higgins. 1989. Survival of *Listeria monocytogenes* in low pH model broth systems. *J. Food Protect.* 52:144–147.

103. Petran, R.L., and E.A. Zottola. 1989. A study of factors affecting growth and recovery of *Listeria monocytogenes* Scott A. *J. Food Sci.* 54:458–460.

104. Piccinin, D.M., and L.A. Shelef. 1995. Survival of *Listeria monocytogenes* in cottage cheese. *J. Food Protect.* 58:128–131.

105. Pine, L., G.B. Malcolm, and B.D. Plikaytis. 1990. *Listeria monocytogenes* intragastric and intraperitoneal approximate 50% lethal doses for mice are comparable, but death occurs earlier by intragastric feeding. *Infect. Immun.* 58:2940–2945.

106. Pini, P.N., and R.J. Gilbert. 1988. The occurrence in the U.K. of *Listeria* species in raw chickens and soft cheeses. *Int. J. Food Microbiol.* 6:317–326.

107. Rocourt, J., P. Boerlin, F. Grimont, C. Jaquet, and J.-C. Piffaretti. 1992. Assignment of *Listeria grayi* and *Listeria murrayi* to a single species, *Listeria grayi*, with a revised description of *Listeria grayi*. *Int. J. System. Bacteriol.* 42:171–174.

108. Ruhland, G.J., and F. Fiedler. 1987. Occurrence and biochemistry of lipoteichoic acids in the genus *Listeria*. *System. Appl. Microbiol.* 9:40–46.

109. Ryser, E.T., and E.H. Marth. 1988. Survival of *Listeria monocytogenes* in cold-pack cheese food during refrigerated storage. *J. Food Protect.* 51:615–621.

110. Ryser, E.T., and E.H. Marth. 1987. Fate of *Listeria monocytogenes* during the manufacture and ripening of Camembert cheese. *J. Food Protect.* 50:372–378.

111. Ryser, E.T., E.H. Marth, and M.P. Doyle. 1985. Survival of *Listeria monocytogenes* during manufacture and storage of cottage cheese. *J. Food Protect.* 48:746–750.

112. Schmidt, U., H.P.R. Seeliger, E. Glenn, B. Langer, and L. Leistner. 1988. Listerienfunde in rohen Fleischerzeugnissen. *Fleischwirtsch.* 68:1313–1316.

113. Seeliger, H.P.R., and K Höhne. 1979. Serotyping of *Listeria monocytogenes* and related species. *Meth. Microbiol.* 13:31–49.

114. Shelef, L.A. 1989. Survival of *Listeria monocytogenes* in ground beef or liver during storage at 4° and 25°C. *J. Food Protect.* 52:379–383.

115. Singh, S.P., B.L. Moore, and I.H. Siddique. 1981. Purification and further characterization of phenol extract from *Listeria monocytogenes*. *Am. J. Vet. Res.* 42:1266–1268.

116. Sizmur, K., and C.W. Walker. 1988. *Listeria* in prepackaged salads. *Lancet* 1:1167.

117. Skalka, B., J. Smola, and K. Elischerova. 1982. Routine test for in vitro differentiation of pathogenic and apathogenic *Listeria monocytogenes* strains. *J. Clin. Microbiol.* 15:503–507.

118. Skovgaard, N., and C.-A. Morgen. 1988. Detection of *Listeria* spp. in faeces from animals, in feeds, and in raw foods of animal origin. *Int. J. Food Microbiol.* 6:229–242.

119. Sörqvist, S. 1994. Heat resistance of different serovars of *Listeria monocytogenes*. *J. Appl. Bacteriol.* 76:383–388.

120. Sorrells, K.M., D.C. Enigl, and J.R. Hatfield. 1989. Effect of pH, acidulant, time, and temperature on the growth and survival of *Listeria monocytogenes*. *J. Food Protect.* 52:571–573.

121. Stanley, N.F. 1949. Studies on *Listeria monocytogenes*. I. Isolation of a monocytosis-producing agent (MPA). *Aust. J. Exp. Biol. Med.* 27:123–131.

122. Steinbruegge, E.G., R.B. Maxcy, and M.B. Liewen. 1988. Fate of *Listeria monocytogenes* on ready to serve lettuce. *J. Food Protect.* 51:596–599.

123. Terplan, G., and S. Steinmeyer. 1989. Investigations on the pathogenicity of *Listeria* spp. by experimental infection of the chick embryo. *Int. J. Food Microbiol.* 8:277–280.

124. Tompkin, R.B. 2002. Control of *Listeria monocytogenes* in the food-processing environment. *J. Food Protect.* 65:709–725.

125. Tompkin, R.B., L.N. Christiansen, A.B. Shapris, R.L.Baker, and J.M. Schroeder. 1992. Control of *Listeria monocytogenes* in processed meats. *Food Aust.* 44:370–376.

126. Vaneechoutte, M., P. Boerlin, H.-V. Tichy, E. Bannerman, B. Jäger, and J. Bille. 1998. Comparison of PCR-based DNA fingerprinting techniques for the identification of *Listeria* species and their use for atypical *Listeria* isolates. *Int. J. Syst. Bacteriol.* 48:127–139.

127. Vazquez-Boland, J.-A., L. Dominguez, E.-F. Rodriguez-Ferri, and G. Suarez. 1989. Purification and characterization of two *Listeria ivanovii* cytolysins, a sphingomyelinase C and a thiol-activated toxin (ivanolysin O). *Infect. Immun.* 57:3928–3955.

128. Weagant, S.D., P.N. Sado, K.G. Colburn, J.D. Torkelson, F.A. Stanley, M.H. Krane, S.C. Shields, and C.F. Thayer. 1988. The incidence of *Listeria* species in frozen seafood products. *J. Food Protect.* 51:655–657.

129. Weis, J., and H.P.R. Seeliger. 1975. Incidence of *Listeria monocytogenes* in nature. *Appl. Microbiol.* 30:29–32.

130. Welshimer, H.J. 1960. Survival of *Listeria monocytogenes* in soil. *J. Bacteriol.* 80:316–320.

131. Wesley, I.V., R.D. Wesley, J. Heisick, F. Harrel, and D. Wagner. 1990. Restriction enzyme analysis in the epidemiology of *Listeria monocytogenes*. In *Symposium on Cellular and Molecular Modes of Action of Selected Microbial Toxins in Foods and Feeds*, ed. J.L. Richard, 225–238. New York: Plenum Publishing.

132. Wong, S., D. Street, S.I. Delgado, and K.C. Klontz. 2000. Recalls of foods and cosmetics due to microbial contamination reported to the U.S. Food and Drug Administration. *J. Food Protect.* 63:1113–1116.

133. Zaika, L.L., S.A. Palumbo, J.L. Smith, F. Del Corral, S. Bhaduri, C.O. Jones, and A.H. Kim. 1990. Destruction of *Listeria monocytogenes* during frankfurter processing. *J. Food Protect.* 53:18–21.

CHAPTER 26

Foodborne Gastroenteritis Caused by *Salmonella* and *Shigella*

Among the Gram-negative rods that cause foodborne gastroenteritis, the most important are the members of the genus *Salmonella*. This syndrome and that caused by *Shigella* spp. are discussed in this chapter. The general prevalence of these organisms in a variety of foods can be found in Chapters 4, 5, 6, 7, and 9.

SALMONELLOSIS

The salmonellae are small, Gram-negative, non-sporing rods that are indistinguishable from *E. coli* under the microscope or on ordinary nutrient media. They are widely distributed in nature, with humans and animals being their primary reservoirs. *Salmonella* food poisoning results from the ingestion of foods containing appropriate strains of this genus in significant numbers.

Some significant changes have been adopted for the taxonomy of *Salmonella*. Although food microbiologists, scientists, and epidemiologists treat the 2,400 or so *Salmonella* serovars as though each was a species, all salmonellae have been placed in two species, *S. enterica* and *S. bongori*, with the 2,000 or so serovars being divided into five subspecies or groups, most of which are classified under *S. enterica*, the type species.[33] The major groups correspond to the following subspecies: group II (*S. enterica* subsp. *salamae*); group IIIa (*S. enterica* subsp. *arizonae*); group IIIb (*S. enterica* subsp. *diarizonae*); group IV (*S. enterica* subsp. *houtenae*); and group VI (*S. enterica* subsp. *indica*). The former group V organisms have been elevated to species status as *S. bongori*.[46] These changes are based on DNA-DNA hybridization and multilocus enzyme electrophoretic characterizations of the salmonellae (see Chapter 11). Thus, the long-standing practice of treating salmonellae serovars as species is no longer valid. For example, S. Typhimurium should be *S. enterica* serovar Typhimurium, or *Salmonella* Typhimurium (note that "typhimurium" is capitalized but not italicized).

For epidemiological purposes, the salmonellae can be placed into three groups:

1. Those that infect humans only: These include *S.* Typhi, *S.* Paratyphi A, *S.* Paratyphi C. This group includes the agents of typhoid and the paratyphoid fevers, which are the most severe of all diseases caused by salmonellae. Typhoid fever has the longest incubation time, produces the

highest body temperature, and has the highest mortality rate. *S*. Typhi may be isolated from blood and sometimes the stool and urine of victims prior to enteric fever. The paratyphoid syndrome is milder than that of typhoid.

2. The host-adapted serovars (some of which are human pathogens and may be contracted from foods): Included are *S*. Gallinarum (poultry), *S*. Dublin (cattle), *S*. Abortus-equi (horses), *S*. Abortus-ovis (sheep), and *S*. Choleraesuis (swine).

3. Unadapted serovars (no host preference): These are pathogenic for humans and other animals, and they include most foodborne serovars. The foodborne salmonellosis syndrome is described in a later section.

Serotyping of *Salmonella*

The serotyping of Gram-negative bacteria is described in Chapter 11. When applied to the salmonellae, species and serovars are placed in groups designated A, B, C, and so on, according to similarities in content of one or more O antigens. Thus, *S*. Hirschfeldii, *S*. Choleraesuis, *S*. Oranienburg, and *S*. Montevideo are placed in group C_1 because they all have O antigens 6 and 7 in common. *S*. Newport is placed in group C_2 due to its possession of O antigens K and 8 (Table 26–1). For further classification, the flagellar or H antigens are employed. These antigens are of two types: specific phase or phase 1, and group phase or phase 2. Phase 1 antigens are shared with only a few other species or varieties of *Salmonella*; phase 2 may be more widely distributed among several species. Any given culture of *Salmonella* may consist of organisms in only one phase or of organisms in both flagellar phases. The H antigens of phase 1 are designated with small letters, and those of phase 2 are designated by arabic numerals. Thus, the complete antigenic analysis of *S*. Choleraesuis is as follows: 6, 7, c, 1, 5, where 6 and 7 refer to O antigens, c to phase-1 flagellar antigens, and 1 and 5 to phase-2 flagellar antigens (Table 26–1). *Salmonella* subgroups of this type are referred to as serovars. With a relatively small number of O, phase 1, and phase 2 antigens, a large number of permutations are possible, allowing for the possibility of a large number of serovars.***

The naming of *Salmonella* is done by international agreement. Under this system, a serovar is named after the place where it was first isolated—*S*. London, *S*. Miami, *S*. Richmond, and so on. Prior to the adoption of this convention, species and subtypes were named in various ways—for example, *S*. Typhimurium as the cause of typhoid fever in mice.

S. Typhimurium definitive type 104 (DT104) is characterized by its resistance to five antimicrobials—amipicillin, chloramphenicol, streptomycin, sulfa drugs, and tetracyclines (the ACSSuT profile). It was first seen in the United Kingdom in 1984. In 1990 it represented about 7% of Typhimurium strains, about 28% in 1995, and 32% of human isolates in 1996. In addition to the antimicrobials noted, DT104 has acquired resistance to trimethoprim and the fluoroquinolones.

Distribution

The primary habitat of *Salmonella* spp. is the intestinal tract of animals such as birds, reptiles, farm animals, humans, and occasionally insects. Although their primary habitat is the intestinal tract, they may be found in other parts of the body from time to time. As intestinal forms, the organisms are excreted in feces from which they may be transmitted by insects and other living creatures to a large number of places. As intestinal forms, they may also be found in water, especially polluted water. When polluted water and foods that have been contaminated by insects or by other means are consumed by humans and other animals, these organisms are once again shed through fecal matter

Table 26–1   Antigenic Structure of Some Common Salmonellae

Group	Serovars (Serotypes)	O Antigens*	H Antigens Phase 1	Phase 2
A	S. Paratyphi A	1, 2, 12	a	(1, 5)
B	S. Schottmuelleri	1, 4, (5), 12	b	1, 2
	S. Typhimurium	1, 4, (5), 12	i	1, 2
C₁	S. Hirschfeldii	6, 7, (vi)	c	1, 5
	S. Choleraesuis	6, 7	(c)	1, 5
	S. Oranienburg	6, 7	m, t	–
	S. Montevideo	6, 7	g, m, s (p)	(1, 2, 7)
C₂	S. Newport	6, 8	e, h	1, 2
D	S. Typhi	9, 12, (Vi)	d	–
	S. Enteritidis	1, 9, 12	g, m	(1, 7)
	S. Gallinarum	1, 9, 12	–	–
E₁	S. Anatum	3, 10	e, h	1, 6

*The italicized antigens are associated with phage conversion. ( ) = May be absent.

with a continuation of the cycle. The augmentation of this cycle through the international shipment of animal products and feeds is in large part responsible for the worldwide distribution of salmonellosis and its consequent problems.

Although *Salmonella* spp. have been recovered repeatedly from a large number of different animals, their incidence in various parts of animals has been shown to vary. In a study of slaughterhouse pigs, Kampelmacher[30] found these organisms in the spleen, liver, bile, mesenteric and portal lymph nodes, diaphragm, and pillar, as well as in feces. A higher incidence was found in lymph nodes than in feces. The frequent occurrence of *Salmonella* spp. among susceptible animal populations is due in part to the contamination of *Salmonella*-free animals by animals within the population that are carriers of these organisms or are infected by them. A carrier is defined as a person or an animal that repeatedly sheds *Salmonella* spp., usually through feces, without showing any signs or symptoms of the disease.

Animal Feeds

The industrywide incidence rate of salmonellae in animal feeds in 1989 was about 49%. Among U.S. Department of Agriculture (USDA)-inspected packers and renderers, the rate was between 20% and 25%, and only 6% for pelleted animal foods.[23] In a study of breeder/multiplier and broiler houses, 60% of meat and bone meal contained salmonellae, and feed was considered to be the ultimate source of salmonellae to breeder/multiplier houses.[28] It has been noted that salmonellae contamination in U.S. broiler production changed little between 1969 and 1989.[28] Salmonellae contamination of rendered products is most likely due to recontamination. The primary serovars found in animal feeds are *S*. Senftenberg, *S*. Montevideo, and *S*. Cerro. *S*. Enteritidis has not been found in rendered products or finished feeds.

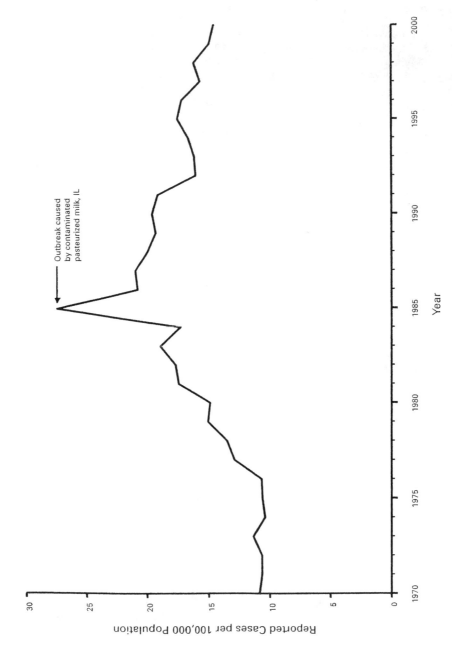

Figure 26–1   Salmonellosis cases (per 100,000 population) United States, 1970–2000. Centers for Disease Control and Prevention, 2002.

Food Products

Salmonellae have been found in commercially prepared and packaged foods with 17 of 247 products examined being positive.[1] Among the contaminated foods were cake mixes, cookie doughs, dinner rolls, and cornbread mixes. These organisms have been found in coconut meal, salad dressing, mayonnaise, milk, and many other foods. In a study of health foods, none of plant origin yielded salmonellae, but from two of three lots of beef liver powder from the same manufacturer were isolated *S*. Minnesota, *S*. Anatum, and *S*. Derby.[54] See Chapters 4–7 and 9.

Growth and Destruction of Salmonellae

These organisms are typical of other Gram-negative bacteria in that they are able to grow on a large number of culture media and produce visible colonies well within 24 hours at about 37°C. They are generally unable to ferment lactose, sucrose, or salicin, although glucose and certain other monosaccharides are fermented, with the production of gas. Although they normally utilize amino acids as N sources, in the case of *S*. Typhimurium, nitrate, nitrite, and NH_3 will serve as sole sources of nitrogen.[43] Although lactose fermentation is not usual for these organisms, some serovars can utilize this sugar.

The pH for optimum growth is around neutrality, with values above 9.0 and below 4.0 being bactericidal. A minimum growth pH of 4.05 has been recorded for some (with HCl and citric acids), but depending on the acid used to lower the pH, the minimum may be as high as 5.5.[15] The effect of acid used to lower the pH on minimum growth is presented in Table 26–2. Aeration was found to favor growth at the lower pH values. The parameters of pH, water activity (a_w), nutrient content, and temperature are all interrelated for salmonellae, as they are for most other bacteria.[56] For best growth, the salmonellae require a pH between 6.6 and 8.2. The lowest temperatures at which growth has been reported are 5.3°C for *S*. Heidelberg and 6.2°C for *S*. Typhimurium.[38] Temperatures of around 45°C have been reported by several investigators to be the upper limit for growth. Regarding available moisture, growth inhibition has been reported for a_w values below 0.94 in media with neutral pH, with higher a_w values being required as the pH is decreased toward growth minima.

Unlike the staphylococci, the salmonellae are unable to tolerate high salt concentrations. Brine above 9% is reported to be bactericidal. Nitrite is effective, with the effect being greatest at the lower pH values. This suggests that the inhibitory effect of this compound is referable to the undissociated HNO_2 molecule. The survival of *Salmonella* spp. in mayonnaise was studied by Lerche,[34] who found that they were destroyed in this product if the pH was below 4.0. It was found that several days may be required for destruction if the level of contamination is high, but within 24 hours for low numbers of cells. *S*. Thompson and *S*. Typhimurium were found to be more resistant to acid destruction than *S*. Senftenberg.

With respect to heat destruction, all salmonellae are readily destroyed at milk pasteurization temperatures. Thermal *D* values for the destruction of *S*. Senftenberg 775W under various conditions are given in Chapter 17. Shrimpton et al.[50] reported that *S*. Senftenberg 775W required 2.5 minutes for a 10^4–10^5 reduction in numbers at 54.4°C in liquid whole egg. This strain is the most heat resistant of all salmonellae serovars. This treatment of liquid whole egg has been shown to produce a *Salmonella*-free product and destroy egg α-amylase (see Chapter 17 for the heat pasteurization of egg white). It has been suggested[7] that the α-amylase test may be used as a means of determining the adequacy of heat pasteurization of liquid egg (compare with the pasteurization of milk and the enzyme phosphatase). In a study on the heat resistance of *S*. Senftenberg 775W, Ng et al.[42] found this strain to be more heat sensitive in the log phase than in the stationary phase of growth. These investigators also found that cells grown at 44°C were more heat resistant than those grown at either 15°C or 35°C.

With respect to the destruction of *Salmonella* in baked foods, Beloian and Schlosser[5] found that baked foods reaching a temperature of 160°F or higher in the slowest heating region can be considered *Salmonella* free. These authors employed *S.* Senftenberg 775W at a concentration of 7,000–10,000 cells/ml placed in reconstituted dried egg. With respect to the heat destruction of this strain in poultry, it is recommended that internal temperatures of at least 160°F be attained.[40] Although *S.* Senftenberg 775W has been reported to be 30 times more heat resistant than *S.* Typhimurium,[42] the latter organism has been found to be more resistant to dry heat than the former.[21] These investigators tested dry heat resistance in milk chocolate.

The destruction of *S.* Pullorum in turkeys was investigated by Rogers and Gunderson,[47] who found that it required 4 hours and 55 minutes to destroy an initial inoculum of 115 million in 10- to 11-lb turkeys with an internal temperature of 160°F, and for 18-lb turkeys with an initial inoculum of 320 million organisms, 6 hours and 20 minutes were required for destruction. The salmonellae are quite sensitive to ionizing radiation, with doses of 5–7.5 kGy being sufficient to eliminate them from most foods and feed. The decimal reduction dose has been reported to range from 0.4 to 0.7 kGy for *Salmonella* spp. in frozen eggs. The effect of various foods on the radiosensitivity of salmonellae is shown in a study by Ley et al.[35] These investigators found that for frozen whole egg, 5 kGy gave a 10^7 reduction in the numbers of *S.* Typhimurium, whereas 6.5 kGy was required to give a 10^5 reduction in frozen horsemeat, between 5 and 7.5 kGy for a 10^5–10^8 reduction in bone meal, and only 4.5 kGy to give a 10^3 reduction of *S.* Typhimurium in desiccated coconut. More extensive information on the effect of irradiation on salmonellae is presented in Chapter 15.

Table 26–2   Minimum pH at Which Salmonellae Would Initiate Growth under Optimum Laboratory Conditions

Acid	pH
Hydrochloric	4.05
Citric	4.05
Tartaric	4.10
Gluconic	4.20
Fumaric	4.30
Malic	4.30
Lactic	4.40
Succinic	4.60
Glutaric	4.70
Adipic	5.10
Pimelic	5.10
Acetic	5.40
Propionic	5.50

Note: Tryptone–yeast extract–glucose broth was inoculated with 10^4 cells per milliliter of *Salmonella* Anatum, *S.* Tennessee, or *S.* Senftenberg.

Source: From Chung and Goepfert,[15] copyright © 1970 by Institute of Food Technologists.

In dry foods, *S*. Montevideo was found to be more resistant than *S*. Heidelberg when inoculated into dry milk, cocoa powder, poultry feed, meat, and bone meal.[29] Survival was greater at a_w 0.43 and 0.52 than at a_w 0.75.

The *Salmonella* Food-Poisoning Syndrome

This syndrome is caused by the ingestion of foods that contain significant numbers of non-host-specific species or serotypes of the genus *Salmonella*. From the time of ingestion of food, symptoms usually develop in 12–14 hours, although shorter and longer times have been reported. The symptoms consist of nausea, vomiting, abdominal pain (not as severe as with staphylococcal food poisoning), headache, chills, and diarrhea. These symptoms are usually accompanied by prostration, muscular weakness, faintness, moderate fever, restlessness, and drowsiness. Symptoms usually persist for 2–3 days. The average mortality rate is 4.1%, varying from 5.8% during the first year of life, to 2% between the first and 50th year, and 15% in persons over 50 years. Among the different species of *Salmonella*, *S*. Choleraesuis has been reported to produce the highest mortality rate—21% (see Chapter 22).

Although these organisms generally disappear rapidly from the intestinal tract, up to 5% of patients may become carriers of the organisms upon recovery from this disease.

Numbers of cells on the order of 10^7–10^9/g are generally necessary for salmonellosis. That outbreaks may occur in which relatively low numbers of cells are found has been noted.[18] From three outbreaks, the numbers of cells found were as low as 100/100 g (*S*. Eastbourne in chocolate) to 15,000/g (*S*. Cubana in a carmine dye solution). In general, minimum numbers for gastroenteritis range between 10^5 and 10^6/g for *S*. Bareilly and *S*. Newport to 10^9–10^{10} for *S*. Pullorum.[8]

Salmonella Virulence Properties

Although an enterotoxin and a cytotoxin have been identified in pathogenic salmonellae, they seem to play only a minimal (if any) role in the gastroenteritis syndrome. The virulence mechanisms of the salmonellae continue to be unravelled, and summaries of what is known are presented in Chapter 22 along with other Gram-negative foodborne pathogens. A synopsis of the early history of salmonellae pathogenesis can be found in the previous edition of this text.

Incidence and Vehicle Foods

The precise incidence of salmonellae food poisoning in the United States is not known. However, the two largest recorded outbreaks of salmonellosis occurred under rather unusual circumstances. The largest occurred in 1994 and it involved more than 224,000 persons.[23] The vehicle food was ice cream produced from milk that was transported in tanker trucks that had previously hauled liquid eggs. The serovar was *S*. Enteritidis, and cases were seen in at least 41 U.S. states. The next largest outbreak occurred in 1985 and involved close to 200,000 persons.[48] (See Fig. 26–1 and Fig. 26–2.) The vehicle was 2% milk produced by a single dairy plant in Illinois, and *S*. Typhimurium was the etiological agent (see Figures 26–1 and 26–2). The third largest outbreak occurred in 1974 on the Navajo Indian Reservation, when 3,400 persons became ill.[26] The vehicle food was potato salad served to about 11,000 individuals at a barbecue. It was prepared and stored for up to 16 hours at improper holding temperatures prior to serving; the serovar isolated was *S*. Newport. It may be noted from Figure 26–2 that *S*. Typhimurium has been the single most

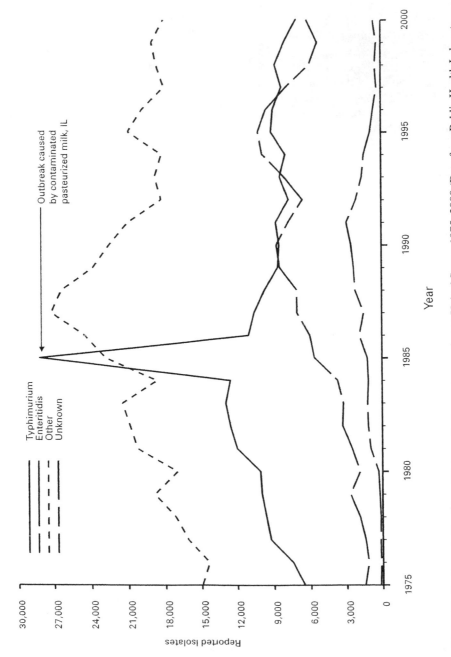

Figure 26–2 Reported isolates of *Salmonella* by serotype and year, United States, 1975–2000 (Data from Public Health Laboratory Information System, PHLIS, Centers for Disease Control and Prevention, 2002).

Table 26–3   Some Characteristics of All Outbreaks of *Salmonella* serotype Enteritidis Infection and Outbreaks in Health-Care Facilities, by Year—United States, 1985–1998 (Summarized from Reference 11)

Year		All outbreaks			Outbreaks in health-care facilities		
		Number of Outbreaks	Number ill	Deaths	Number of Outbreaks	Number ill	Deaths
1985		26	1,159	1	3	55	1
1986	47	1,444	6	6	96	5	
1987	58	2,616	15	8	489	14	
1988	48	1,201	11	8	227	9	
1989	81	2,518	15	19	505	13	
1990	85	2,656	3	12	265	3	
1991	74	2,461	5	8	118	4	
1992	63	2,348	4	2	42	2	
1993	66	2,215	6	6	66	4	
1994	51	5,492	0	2	32	0	
1995	56[a]	1,312	8	6	147	6	
1996	50	1,460	2	3	64	0	
1997	44	1,098	0	1	13	0	
1998	47	709	3	3	32	3	
Total		796	28,689	79(0.28%)	87	2,151	64(3%)

[a]Includes one outbreak associated with a Komodo dragon exhibit at a zoological park.

frequently isolated serotype since about 1975. The second most common is *S.* Enteritidis, and it is discussed further below.

For the years 1981–1995, 3,504 (23.8% of all) bacterial foodborne illnesses in Korea were caused by salmonella while in Japan for the same period, there were 101,395 (19.9%) cases.[32] Since the late 1970s, *S.* Enteritidis has been the cause of a series of outbreaks in the north-eastern part of the United States and in parts of Europe. The U.S. outbreaks for 1985–1998 are listed in Table 26–3. The cases and deaths from all outbreaks are compared to those among individuals in health-care facilities. Although it is the second most frequently isolated serotype in the United States, it is of significance because of its high association with eggs and its fatality rate. It may be noted from Table 26–3 that 796 outbreaks were recorded with 28,689 cases and 79 deaths for the period 1985–1998.[11] Raw or undercooked shell eggs accounted for 82% of the illness cases. The 79 deaths accounted for 0.27% of all cases but in health-care facilities, the death rate was 2.97%. In 1976, the number of cases in the United States per 100,000 population was 0.6 but by 1996 the rate was 3.6. The rate decreased to 2.2/100,000 in 1998.[11]

Because in the United States *S.* Enteritidis is so highly associated with the consumption of raw or undercooked eggs, the CDCP recommends the following[11]: (1) Raw or undercooked eggs should be avoided, especially by the young, elderly, and immunocompromised; (2) when eggs are not properly cooked, pasteurized egg products should be used; (3) eggs should be cooked at $\geq 145°F$ ($63°C$) for 15 seconds, or until both yolk and white are firm, and then should be eaten promptly; (4) casseroles and other dishes containing raw eggs should be cooked to $160°F$ ($71°C$); and (5) raw eggs should be stored at $\leq 45°F$ ($7.2°C$) at all times.

S. Enteritidis was the most common serovar in eggs in Spain[44] and the most predominant cause of foodborne salmonellosis in England and Wales in 1988, where it was found in both poultry meat and eggs.[24] Unlike U.S. outbreaks, those in Europe are caused by phage type 4 strains, which are more invasive for young chicks than phage types 7, 8, or 13a. [24]

Why the increased incidence of *S.* Enteritidis outbreaks are associated with eggs and poultry products is unclear. The organism has been found by some investigators inside the eggs and ovaries of laying hens,[44] but others have failed to find it in unbroken eggs. It was recovered from the ovaries of only 1 of 42 layer flocks located primarily in the southeast United States.[2] The *S.* Enteritidis outbreaks have occurred more in July and August than other months, suggesting growth of the organism in or on eggs and other poultry products. The strains in question are not heat resistant, and many of the outbreaks have occurred following the consumption of raw or undercooked eggs. In one study in which *S.* Enteritidis was inoculated into the yolk of eggs from normal hens, no growth occurred at 7°C in 94 days.[6] Growth in yolks at 37°C was faster from normal hens than in those from hens that were seropositive. The possible routes of *S.* Enteritidis to eggs are the following:[31]

1. Transovarial
2. Translocation from peritoneum to yolk sac or oviduct
3. Penetration of shell by organisms as eggs pass through the cloaca
4. Egg washing
5. Food handlers

Except for the association of *S.* Enteritidis with poultry and egg products, it is difficult to predict the association of most other salmonellae serovars with specific food products. The three outbreaks traced to fresh tomatoes in the United States during the years 1990, 1993, and 1997–1998 were caused by three different serotypes (see Table 26–4). *S.* Muenchen was the etiologic agent in an orange juice outbreak in 1999[13]; and it was also the cause of an alfalfa sprout outbreak during the same year.[45] A multi-state outbreak of *S.* Poona in the United States was traced to fresh cantaloupes from Mexico[10], and a raw-milk outbreak in 3 U. S. states in 2003 was caused by *S.* Tennessee[9].

Table 26–4　Synopses of Some Foodborne Outbreaks of Nontyphoidal Salmonellosis (From the Literature)

Products/Location/Year(s)	Number of Cases	Salmonella Serotype
Fresh tomatoes, 5 states USA, 1990	176	Javiana
Fresh tomatoes, USA, 1993	100	Montevideo
Fresh tomatoes, 8 states USA, 1998–1999	85	Baidon
Alfalfa sprouts, 7 states USA, 1999	157	Muenchen
Alfalfa sprouts, 4 states USA, 2001	31	Kottbus
Cantaloupes, 12 states USA, and Canada, 2002	47	Poona
Unpasteurized orange juice, 13 states USA, 2 Canadian Provinces, 1999	298	Muenchen
Raw/undercooked shell eggs, 4 states USA, 1997–1998	241	Enteritidis
Raw/undercooked ground beef, 5 states USA, 2002	47	Newport
Chocolate; Germany, Denmark, 8 other countries, 2001	>316	Oranienberg
Shandong peanuts; Australia, Canada, U.K., 2001	Ca. 102	Stanley

Synopses of the three leading serotypes recovered from six different sources are presented below.

1. Retail foods in Korea (1993–2001). Of the 1,334 samples tested, 2.2% were salmonellae positive. Most common serotypes: Enteritidis, Virginia, Haardt.[16]
2. Minced beef and pork in Germany (1996–1997). Of the 1,445 samples, 6.3% were salmonellae positive. Most common serotypes: Typhimurium, Derby, Typhimurium var. Copenhagen.[53]
3. Human food and beverages in Singapore (1998). Of the 2,617 samples, 1.4% were salmonellae positive. Most common serotypes: Typhimurium, Agona, and Dumfries and Enteritidis.[41]
4. Hen layer feed in Japan (1993–1998). Of 10,418 samples, 0.5% were salmonellae positive. Most common: Untypable, Eastbourne, Orion.[49]
5. Feedlot cattle feces in the United States (1996). Of the 4,977 samples, 5.5% were salmonellae positive. Most common: Anatum, Montevideo, Muenster.[20]
6. Beef cow feces in the United States (1998). Three most common serotypes: Oranienburg, Cerro, Anatum.[19]

Prevention and Control of Salmonellosis

The intestinal tract of humans and other animals is the primary reservoir of the etiological agents. Animal fecal matter is of greater importance than human, and animal hides may become contaminated from the fecal source. *Salmonella* spp. are maintained within an animal population by means of nonsymptomatic animal infections and in animal feeds. Both sources serve to keep slaughter animals reinfected in a cyclical manner, although animal feeds seem less important than once believed.

Secondary contamination is another of the important sources of salmonellae in human infections. Their presence in meats, eggs, and even air makes their presence in certain foods inevitable through the agency of handlers and direct contact of noncontaminated foods with contaminated foods.[25]

In view of the worldwide distribution of salmonellae, the ultimate control of foodborne salmonellosis will be achieved by freeing animals and humans of the organisms. This is obviously a difficult task but not impossible; only about 35 of the more than 2,400 serovars account for around 90% of human isolates and approximately 80% of nonhuman isolates.[37]

At the consumer level, the *Salmonella* carrier is thought to play a role, but just how important this role may be is not clear. Improper preparation and handling of foods in homes and food service establishments continue to be the primary factors in outbreaks.

With respect to the colonization of chickens by *S.* Enteritidis, one study used a phage type 8 strain administered orally 10^8 to adult laying hens.[31] Within two days, the organism was found throughout the body, including the ovary and oviduct. It was detected in some forming eggs, although its incidence was much lower in freshly laid eggs. Investigators concluded that forming eggs are subject to descending infection from colonized ovarian tissue, to ascending infections from colonized vaginal and cloacal tissues, and to lateral infections from colonized upper oviduct tissues.[31] The hatchery eggs are of critical importance because if they are contaminated, hatchlings may become infected at this early stage. Salmonellae rapidly penetrate freshly laid fertile eggs, become entrapped in the membrane, and may be ingested by an embryo as it emerges from the egg.

Competitive Exclusion to Reduce Salmonellae Carriage in Poultry

It is generally agreed that the primary source of salmonellae in poultry products is the gastrointestinal tract, including the ceca. If young chicks become colonized with salmonellae, the bacteria may be shed in feces, through which other birds become contaminated. Among the methods that may be employed to reduce or eliminate intestinal carriage is competitive exclusion (the Nurmi concept).

Under natural conditions where salmonellae exist when eggs hatch, young chicks develop a gastrointestinal tract flora that consists of these organisms and campylobacters, in addition to a variety of nonpathogens. Once the pathogens are established, they may remain and be shed in droppings during the entire lifetime of the bird. Competitive exclusion is a phenomenon whereby feces from salmonellae-free birds, or a mixed fecal culture of bacteria, are given to young chicks so that they will colonize the same intestinal sites that salmonellae employ and, thus, exclude the subsequent attachment of salmonellae or other enteropathogens. This concept was advanced in the 1970s and has been studied and found to be workable by a number of investigators relative to salmonellae exclusion.

The enteropathogen-free biota may be administered orally to newly hatched chicks through drinking water or by spray inoculation in the hatchery. Protection is established within a few hours and generally persists throughout the life of the fowl or as long as the biota remains undisturbed. Older birds can be treated by first administering antibacterial agents to eliminate enteropathogens, and they are then administered the competitive exclusion biota. Only viable cells are effective, and both aerobic and anaerobic components of the gut flora seem to be required. The crop and ceca appear to be the major adherence sites, with the ceca being higher in germ-free chickens. In one study, the protective flora remained attached to cecal walls after four successive washings.[52] Partial protection was achieved in 0.5–1.0 hour, but full protection required 6–8 hours after treatment of 1-day-old chicks.[50] In a review of the microbiology of competitive exclusion of *Salmonella* in poultry, the use of undefined cultures afford more protection than the use of defined cultures, especially under laboratory conditions.[51]

Field trials in several European countries have shown the success of the competitive exclusion treatment in preventing or reducing the entry of salmonellae in broilers and adult breeder birds.[39] In chicks pretreated with a cecal culture and later challenged with a *Salmonella* sp., the latter failed to multiply in the ceca over a 48-hour period, whereas in untreated control birds, more than 10^6/g of salmonellae were colonized in the ceca.[27]

The gist of competitive exclusion is that salmonellae and the native gut biota compete for the same adherence sites on gut walls. The precise nature of the bacterial adhesins is not entirely clear, although fimbriae, flagella, and pili have been suggested. In regard to the attachment of salmonellae to poultry skin, these bacterial cell structures were found not to be critical.[36] Extracellular polysaccharides of a glycocalyx nature may be involved, and if so, treatment of young chicks with this material may be as effective as the use of live cultures. Although the competitive exclusion treatment seems quite feasible for large hatcheries, its practicality for small producers seems less likely.

The sugar mannose is a receptor in the intestinal tract to which bacterial pathogens such as salmonellae bind. Since the yeast strain *Saccharomyces cerevisiae* var. *boulardii* contains mannose in its outer wall, some have suggested that the feeding of this yeast to susceptible poults should reduce the attachment of salmonellae. In essence, the yeast cell wall material would outcompete the gastrointestinal tract for the pathogens.

The possible use of probiotic cultures to exclude some Gram-negative pathogens from the gut biota has been investigated by several groups. When a 3-strain mixture of probiotic bacteria (competitive exclusion of *E. coli* strains) was used on weaned calves and challenged with *E. coli* serotypes 0111:NM, 026:H11, and 0157:H7, the probiotic-treated calves showed a significant reduction in the shedding of two of the three pathogens but not *E. coli* serotype 026:H11.[55] In another study, a mixed culture of *Lactobacillus crispatus* and *Clostridium lactatifermentans* cultured under cecal growth conditions inhibited the growth of *S*. Enteritidis.[57] When ca. 10^9 cells/chicken of a chicken isolate of *Enterococcus faecium* was orally administered to 30-hour-old broiler chicks followed by a challenge with 10^5 cells of *S.* Pullorum per chick, the chicks survived. However, chicks that were infected on the first day and then treated with the lactic culture died four days later.[2] These authors concluded that the *E. faecium* strain could prevent newly hatched chicks from *S.* Pullorum infections but it was not a good therapeutic agent.

SHIGELLOSIS

The genus *Shigella* belongs to the family Enterobacteriaceae, as do the salmonellae and escherichiae. Only four species are recognized: *S. dysenteriae,S. flexneri,S. boydii*, and *S. sonnei. S. dysenteriae* is a primary pathogen that causes classic bacillary dysentery; as few as 10 cfu are known to initiate infection in susceptible individuals. By applying data from two cruise ship outbreaks to a mathematical model, it was estimated that the outbreaks could have been due to ingestion of a mean of 344 *Shigella* cells per meal and 10.5 to 12 cells per glass of water.[17] Although this syndrome can be contracted from foods, it is not considered to be a food-poisoning organism in the same sense as the other three species, and it is not discussed further. Unlike the salmonellae and escherichiae, the shigellae have no known nonhuman animal reservoirs. Some of the many differences among the three genera are noted in Table 26–5. The shigellae are phylogenetically closer to the escherichiae than to the salmonellae.

The three species of concern as etiological agents of foodborne gastroenteritis are placed in separate serologic groups based on O antigens: *S. flexneri* in group B, *S. boydii* in group C, and *S. sonnei* in group D. They are nonmotile, oxidase negative, produce acid only from sugars, do not grow on citrate as sole carbon source, do not grow on KCN agar, and do not produce H_2S. In general, their growth on ordinary culture media is not as abundant as that of the escherichiae. Of shigellae isolated from humans in the United States in 1984, 64% were *S. sonnei*, 31% *S. flexneri*, 3.2% *S. boydii*, and 1.5% *S. dysenteriae*.[14]

The *Shigella* species of concern are typical of most other enteric bacteria in their growth requirements, with growth reported to occur at least as low as 10°C and as high as 48°C. In one study, growth of *S. flexneri* was not observed in brain heart infusion (BHI) broth at 10°C.[59] It appears that *S. sonnei* can grow at lower temperatures than the other three species. Growth at pH 5.0 has been recorded, with best growth occurring in the range of 6–8. With *S. flexneri*, no growth occurred at pH 5.5 at 19°C in BHI broth.[59] This species has been shown to be inhibited by nitrite as temperature and pH were decreased or as NaCl was increased.[58] It is unclear whether they can grow at a_w values below those for the salmonellae or escherichiae. Their resistance to heat appears to parallel that of *E. coli* strains.

Table 26–5 A Comparison of *Salmonella, Shigella,* and *Escherichia*

Genus	Glucose	Motility	H_2S	Indole	Citrate	Mol 1% G + C
Escherichia	AG	+*	−	+[†]	−	48–52
Salmonella	AG	+*	+	−	+	50–53
Shigella	A	−	−	−	−	49–53

*Usually.
[†]Type 1 strains.

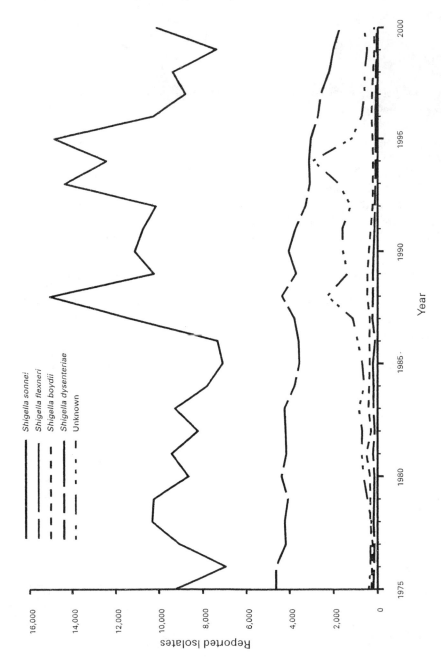

Figure 26-3 Reported isolates of *Shigella* by species and year, United States, 1975–2000. Centers for Disease Control and Prevention (2002).

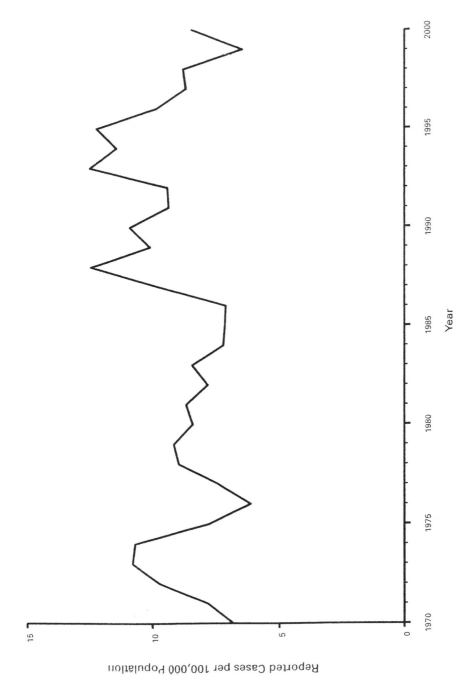

Figure 26–4 Reported cases of shigellosis per 1,000,000 population by year, United States, 1970–2000. Date from PHLIS, Centers for Disease Control and Prevention (2002).

Foodborne Cases

For the period 1973–1987, foodborne shigellosis accounted for 12% of reported food poisoning cases in the United States for which an etiological agent could be found, placing it third behind staphylococcal food poisoning (14%) and salmonellosis (45%).[4] Poor personal hygiene is a common factor in foodborne shigellosis, with shellfish, fruits and vegetables, chicken, and salads being prominent among vehicle foods. The prominence of these foods is due to the fecal–oral route of transmission. The shigellae are not as persistent in the environment as are salmonellae and escherichiae. The recorded species isolations of *Shigella* in the United States for the years 1975–2000 are noted in Figure 26–3. It should be noted that the isolations noted were from many sources including foods. *S. sonnei* is clearly the most frequent isolate followed by *S. flexneri*. The recorded shigellosis cases (per 100,000 population) in the United States for the years 1970–2000 can be seen from Figure 26–4, and they include foodborne as well as non-foodborne cases.

In 2000, an outbreak of shigellosis in three western states of the United States was caused by *S. sonnei*; the vehicle food was a 5-layer party dip, and there were 30 victims.[12] The dip layers consisted of bean/salsa/guacamole/nacho cheese/sour cream. An outbreak caused by *S. sonnei* occurred in Spain in 1995–1996 and it was traced to fresh pasteurized milk cheese;there were >200 victims, and an infected food handler appeared to be the source of the pathogen to susceptible individuals.

Virulence Properties

The virulence mechanisms of the shigellae are much more complex than previously thought, and they are discussed in Chapter 22 along with salmonellae and some *E. coli* strains.

REFERENCES

1. Adinarayanan, N., V.D. Foltz, and F. McKinley. 1965. Incidence of Salmonellae in prepared and packaged foods. *J. Infect. Dis.* 115:19–26.

2. Audisio, M.C., G. Oliver, and M.C. Apella. 2000. Protective effect of *Enterococcus faecium* J96, a potential probiotic strain, on chicks infected with *Salmonella* Pullorum. *J. Food Protect.* 63:1333–1337.

3. Barnhart, H.M., D.W. Dreesen, R. Bastien, and O.C. Pancorbo. 1991. Prevalence of *Salmonella* Enteritidis and other serovars in ovaries of layer hens at time of slaughter. *J. Food Protect.* 54:488–491.

4. Bean, N.H., and P.M. Griffin. 1990. Foodborne disease outbreaks in the United States, 1973–1987: Pathogens, vehicles, and trends. *J. Food Protect.* 53:804–817.

5. Beloian, A., and G.C. Schlosser. 1963. Adequacy of cooking procedures for the destruction of salmonellae. *Am. J. Public Health* 53:782–791.

6. Bradshaw, J.G., D.B. Shah, E. Forney, and J.M. Madden. 1990. Growth of *Salmonella* Enteritidis in yolk of shell eggs from normal and seropositive hens. *J. Food Protect.* 53:1033–1036.

7. Brooks, J. 1962. Alpha amylase in whole eggs and its sensitivity to pasteurization temperatures. *J. Hyg.* 60:145–151.

8. Bryan, F.L. 1977. Diseases transmitted by foods contaminated by wastewater. *J. Food Protect.* 40:45–56.

9. Centers for Disease Control and Prevention. 2003. Multistate outbreak of *Salmonella* serotype Typhimurium infections associated with drinking unpasteurized milk—Illinois, Indiana, Ohio, and Tennessee, 2002–2003. *Morb. Mort. Wkly. Rep.* 52:613–615.

10. Centers for Disease Control and Prevention. 2002. Multistate outbreaks of *Salmonella* serotype Poona infections associated with eating cantaloupe from Mexico—United States and Canada, 2000–2002. *Morb. Mort. Wkly. Rep.* 51:1044–1047.

11. Centers for Disease Control and Prevention. 2000a. Outbreak of *Salmonella* serotype Enteritidis infection associated with eating raw or undercooked shell eggs—United States, 1996–1998. *Morb. Mort. Wkly. Rep.* 49:73–79.

12. Centers for Disease Control and Prevention. 2000b. Outbreak of *Shigella sonnei* infections associated with eating a nationally distributed dip—California, Oregon, and Washington, January 2000. *Morb. Mort. Wkly. Rep.* 49:60–61.

13. Centers for Disease Control and Prevention. 1999. Outbreak of *Salmonella* serotype Muenchen infections associated with unpasteurized orange juice—United States and Canada, June 1999. *Morb. Mort. Wkly. Rep.* 48:582–585.

14. Centers for Disease Control. 1985. Shigellosis—United States, 1984. *Morb. Mort. Wkly. Rep.* 34:600.

15. Chung, K.C., and J.M. Goepfert. 1970. Growth of *Salmonella* at low pH. *J. Food Sci.* 35:326–328.

16. Chung, Y.H., S.Y. Kim, and Y.H. Chang. 2003. Prevalence and actibiotic susceptibility of *Salmonella* isolated from foods in Korea for 1993 to 2001. *J. Food Protect.* 66:1154–1157.

17. Crockett, C.S., C.N. Hass, A. Fazil. 1996. Prevalence of shigellosis in the U.S.: Consistency with dose-response information. *Int. J. Food Microbiol.* 30:87–99.

18. D'Aoust, J.Y., and H. Pivnick. 1976. Small infectious doses *of Salmonella. Lancet* 1:866.

19. Dargatz, D.A., P.J. Fedorka-Cray, S.R. Ladely, and K.E. Ferris. 2000. Survey of *Salmonella* serotypes shed in feces of beef cows and their antimicrobial susceptibility patterns. *J. Food Protect.* 63:1648–1653.

20. Fedorka-Cray, P.J., D.A. Dargatz, L.A. Thomas, and J.T. Gray. 1998. Survey of *Salmonella* serotypes in feedlot cattle. *J. Food Protect.* 61:525–530.

21. Goepfert, J.M., and R.A. Biggie. 1968. Heat resistance of *Salmonella* Typhimurium and *Salmonella* Senftenberg 775W in milk chocolate. *Appl. Microbiol.* 16:1939–1940.

22. Graber, G. 1991. Control of *Salmonella* in animal feeds. Division of Animal Feeds, Center for Veterinary Medicine, Food and Drug Administration. Report to the National Advisory Commission on Microbiological Criteria for Foods.

23. Hennessy, T.W., C.W. Hedberg, L. Slutsker. 1996. A national outbreak of *Salmonella* Enteritidis infections from ice cream. *N. Engl. J. Med.* 334:1281–1286.

24. Hinton, M., E.J. Threlfall, and B. Rowe. 1990. The invasive potential of *Salmonella* Enteritidis phage types for young chickens. *Lett. Appl. Microbiol.* 10:237–239.

25. Hobbs, B.C. 1961. Public health significance of *Salmonella* carriers in livestock and birds. *J. Appl. Bacteriol.* 24:340–352.

26. Horwitz, M.A., R.A. Pollard, M.H. Merson, and S.M. Martin. 1977. A large outbreak of foodborne salmonellosis on the Navajo Indian Reservation, epidemiology and secondary transmission. *Am. J. Public Health* 67:1071–1076.

27. Impey, C.S., and G.C. Mead. 1989. Fate of salmonellas in the alimentary tract of chicks pre-treated with a mature caecal microflora to increase colonization resistance. *J. Appl. Bacteriol.* 66:469–475.

28. Jones, F.T., R.C. Axtell, D.V. Rives, S.E. Schneideler, F.R. Tarver, Jr., R.L. Walker, and M.J. Wineland. 1991. A survey of *Salmonella* contamination in modern broiler production. *J. Food Protect.* 54:502–507.

29. Juven, B.J., N.A. Cox, J.S. Bailey, J.E. Thomson, O.W. Charles, and J.V. Shutze. 1984. Survival of *Salmonella* in dry food and feed. *J. Food Protect.* 47:445–448.

30. Kampelmacher, E.H. 1963. The role of salmonellae in foodborne diseases. In *Microbiological Quality of Foods*, ed. L.W. Slanetz et al., 84–101. New York: Academic Press.

31. Keller, L.H., C.E. Benson, K. Krotec, and R.J. Eckroade. 1995. *Salmonella* Enteritidis colonization of the reproductive tract and forming and freshly laid eggs of chickens. *Infect. Immun.* 63:2443–2449.

32. Lee, W.-C., M.-J. Lee, J.-S. Kim, and S.-Y. Park. 2001. Foodborne illness outbreaks in Korea and Japan studied restrospectively. *J. Food Protect.* 64:899–902.

33. Le Minor, L., and M.Y. Popoff. 1987. Designation of *Salmonella enterica* sp. nov., nom. rev., as the type and only species of the genus *Salmonella. Int. J. System. Bacteriol.* 37:465–468.

34. Lerche, M. 1961. Zur Lebenfahigkeit von *Salmonella* bakterien in Mayonnaise und Fleischsalat. *Wein. Tierarztl. Mschr.* 6:348–361.

35. Ley, F.J., B.M. Freeman, and B.C. Hobbs. 1963. The use of gamma radiation for the elimination of salmonellae from various foods. *J. Hyg.* 61:515–529.

36. Lillard, H.S. 1986. Role of fimbriae and flagella in the attachment of *Salmonella* Typhimurium to poultry skin. *J. Food Sci.* 51:54–56, 65.

37. Martin, W.J., and W.H. Ewing. 1969. Prevalence of serotypes of *Salmonella. Appl. Microbiol.* 17:111–117.

38. Matches, J.R., and J. Liston. 1968. Low temperature growth of *Salmonella. J. Food Sci.* 33:641–645.

39. Mead, G.C., and P.A. Barrow. 1990. *Salmonella* control in poultry by "competitive exclusion" or immunization. *Lett. Appl. Microbiol.* 10:221–227.

40. Milone, N.A., and J.A. Watson. 1970. Thermal inactivation of *Salmonella* Senftenberg 775W in poultry meat. *Health Lab. Sci.* 7:199–225.

41. Ng, D.L.K., B.B. Koh, L. Tay, and M. Yeo. 1999. The presence of *Salmonella* in local food and beverage items in Singapore. *Dairy Food Environ. Sanit.* 19:848–852.

42. Ng, H., H.G. Bayne, and J.A. Garibaldi. 1969. Heat resistance of *Salmonella*: The uniqueness of *Salmonella* Senftenberg 775W. *Appl. Microbiol.* 17:78–82.

43. Page, G.V., and M. Solberg. 1980. Nitrogen assimilation by *Salmonella* Typhimurium in a chemically defined minimal medium containing nitrate, nitrite, or ammonia. *J. Food Sci.* 45:75–76, 83.

44. Perales, I., and A. Audicana. 1988. *Salmonella* Enteritidis and eggs. *Lancet* 2:1133.

45. Proctor, M.E., M. Hamacher, M.L. Tortorello, J.R. Archer, and J.P. Davis. 2001. Multistate outbreak of *Salmonella* serovar Muenchen infections associated with alfalfa sprouts grown from seeds pretreated with calcium hypochlorite. *J. Clin. Microbiol.* 39:3461–3465.

46. Reeves, M.W., G.M. Evins, A.A. Heiba, B.D. Plikaytis, and J.J. Farmer. 1989. Clonal nature of *Salmonella* Typhi and its genetic relatedness to other salmonellae as shown by multilocus enzyme electrophoresis, and proposal of *Salmonella bongori* comb. nov. *J. Clin. Microbiol.* 27:311–320.

47. Rogers, R.E., and M.F. Gunderson. 1958. Roasting of frozen stuffed turkeys. I. Survival of *Salmonella* Pullorum in inoculated stuffing. *Food Res.* 23:87–95.

48. Ryan, C.A., M.K. Nickels, N.T. Hargrett-Bean, M.E. Potter, T. Endo, L. Mayer, C.W. Langkop, C. Gibson, R.C. McDonald, R.T. Kenney, N.D. Puhr, P.J. McDonnell, R.J. Martin, M.L. Cohen, and P.A. Blake. 1987. Massive outbreak of antimicrobial-resistant salmonellosis traced to pasteurized milk. *JAMA* 258:3269–3274.

49. Shirota, K., H. Katoh, T. Murase, T. Ito, and K. Otsuki. 2001. Monitoring of layer feed and eggs for *Salmonella* in eastern Japan between 1993 and 1998. *J. Food Protect.* 64:734–737.

50. Shrimpton, D.H., J.B. Monsey, B.C. Hobbs, and M.E. Smith. 1962. A laboratory determination of the destruction of alpha amylase and salmonellae in whole egg by heat pasteurization. *J. Hyg.* 60:153–162.

51. Stavric, S., and J.-Y. D'Aoust. 1993. Undefined and defined bacterial preparations for the competitive exclusion of *Salmonella* in poultry—A review. *J. Food Protect.* 56:173–180.

52. Stavric, S., T.M. Gleeson, B. Blanchfield, and H. Pivnick. 1987. Role of adhering microflora in competitive exclusion of *Salmonella* from young chicks. *J. Food Protect.* 50:928–932.

53. Stock, K., and A. Stolle. 2001. Incidence of *Salmonella* in minced meat produced in a European Union-approved cutting plant. *J. Food Protect.* 64:1435–1438

54. Thomason, B.M., W.B. Cherry, and D.J. Dodd. 1977. Salmonellae in health foods. *Appl. Environ. Microbiol.* 34:602–603.

55. Tkalcic, S., T. Zhao, B.G. Harmon, M.P. Doyle, C.A. Brown, and P. Zhao. 2003. Fecal shedding of enterohemorrhagic *Escherichia coli* in weaned calves following treatment with probiotic *Escherichia coli*. *J. Food Protect.* 66:1184–1189.

56. Troller, J.A. 1976. *Salmonella* and *Shigella*. In *Food Microbiology: Public Health and Spoilage Aspects*, ed. M.P. de-Figueiredo and D.F. Splittstoesser, 129–155. Westport, CT: AVI.

57. Van der Wielen, P.W.J.J., L.J.A. Lipman, F. van Knapen, and S. Biesterveld. 2002. Competitive exclusion of *Salmonella enterica* serovar Enteritidis by *Lactobacillus crispatus* and *Clostridium lactatifermentans* in a sequencing fed-batch culture. *Appl. Environ. Microbiol.* 68:555–559.

58. Zaika, L.L., A.H. Kim, and L. Ford. 1991. Effect of sodium nitrite on growth of *Shigella flexneri*. *J. Food Protect.* 54:424–428.

59. Zaika, L.L., L.S. Engel, A.H. Kim, and S.A. Palumbo. 1989. Effect of sodium chloride, pH and temperature on growth of *Shigella flexneri*. *J. Food Protect.* 52:356–359.

CHAPTER 27

Foodborne Gastroenteritis Caused by *Escherichia coli*

Escherichia coli was established as a foodborne pathogen in 1971 when imported cheeses turned up in 14 American states that were contaminated with an enteroinvasive strain that caused illness in nearly 400 individuals. Prior to 1971, at least five foodborne outbreaks were reported in other countries, with the earliest being from England in 1947. As a human pathogen, evidence suggests that it was recognized as a cause of infant diarrhea as early as the 1700s.[60] Since the meatborne outbreaks in the United States of 1982 and 1993, the status of this bacterium as a foodborne pathogen is unquestioned. *Escherichia coli* as an indicator of fecal contamination is discussed in Chapter 20, culture and isolation methods are covered in Chapter 10, and molecular and bioassay methods for its detection are covered in Chapters 11 and 12. For a more detailed history of *E. coli* 0157:H7, see reference 69.

SEROLOGICAL CLASSIFICATION

Pathogenic strains of *Escherichia* are serologically typed in the same way as other Enterobacteriaceae, and the procedure is described in Chapter 11. For *E. coli*, over 200 O serotypes have been recognized. Because the flagellar proteins are less heterogeneous than the carbohydrate side chains that make up the O groups, considerably fewer H antigenic types exist (around 30).

THE RECOGNIZED VIRULENCE GROUPS

Based on disease syndromes and characteristics, and also on their effect on certain cell cultures and serological groupings, five virulence groups of *E. coli* are recognized: enteroaggregative (EAggEC), enterohemorrhagic (EHEC), enteroinvasive (EIEC), enteropathogenic (EPEC), and enterotoxigenic (ETEC).

Enteroaggregative *E. coli* (EAggEC)

This group (also designated enteroadherent) is related to EPEC but the aggregative adherence displayed by these strains is unique. Strains exhibit a "stacked-brick-type" of adherence to HEp-2

Table 27–1 Some of the O Serotypes Found Among the Five
Virulence Groups

EAggEC	EHEC	EIEC	EPEC	ETEC
3	2	28ac	18ab	6
4	5	29	19ac	8
6	6	112a	55	15
7	4	124	86	20
17	22	135	111	25
44	26	136	114	27
51	38	143	119	63
68	45	144	125	78
73	46	147	126	80
75	82	152	127	85
77	84	164	128ab	101
78	88	167	142	115
85	91		158	128ac
111	103			139
127	113			141
142	104			147
162	111			148
	116			149
	118			153
	145			159
	153			167
	156			
	157			
	163			

Note: Some serotypes (e.g., 111) are listed under more than one virulence group.

cells, and carry a 60-MDa plasmid that is needed for the production of fimbriae that are responsible for the aggregative expression, and for a specific outer membrane protein (OMP). Antibodies raised against the OMP of a prototype strain prevented adherence to HEp-2 cells.[20] An EAggEC DNA probe has been constructed by using a 1.0-kilobase (kb) fragment from the 60-MDa plasmid of the prototype strain (03:H2), and it was found to be 99% specific for these strains.[4] Some EAggEC strains produce a heat-stable enterotoxin (ST), which has been designated EAST1.[75] The plasmid-borne gene for EAST1 is *astA*, which encodes a 38-amino-acid molecule in contrast to *estA*, which encodes the 72 amino acid enterotoxin STa (see reference 75). They produce an enterotoxin/cytotoxin that is about 108 kDa, and it is located on the large virulence plasmid. The distinguishing clinical feature of EAggEC strains is a persistent diarrhea that lasts >14 days, especially in children. These strains are not the primary cause of traveler's diarrhea.[15]

It is unclear whether members of this group are foodborne pathogens. Some of the serotypes in which EAggEC strains have been found are listed in Table 27–1. Two serotypes that were designated as prototype are O3:H2 and O4:H7, and one serotype (O44) contains both EAggEC and EPEC strains.[76]

Enterohemorrhagic *E. coli* (EHEC)

These strains are both similar and dissimilar to EPEC strains. They are similar to EPEC in their possession of the chromosomal gene *eaeA* (or one that is similar) and in the production of attachment–effacement lesions (see the subsection on EPEC). In contrast to EPEC, EHEC strains affect only the large intestine (in piglet models) and produce large quantities of Shiga-like toxins (SLT, Stx, see below). EHECs produce a 60-MDa plasmid that encodes fimbriae that mediate attachment to culture cells, and they do not invade HEp-2 or INT407 cell lines, although some strains have the ability to invade some human epithelial cell lines.[64] Some EHEC strains produce curli fimbriae that facilitate attachment of cells to surfaces.

The Toxins

Shigella dysenteriae produces a potent toxin that is referred to commonly as Shiga toxin (after K. Shiga who first isolated and studied the organism). The toxins of EHEC strains of *E. coli* have been referred to as Shiga-like toxins (verotoxin, verocytotoxin) and the two prototypes as SLT-I and SLT-II. However, new terminology has been applied, and what was once SLT-I is now Stx1 and the former SLT-II is Stx2.[9] The genes for Stx1 and Stx2 are encoded by temperate bacteriophages in some EHEC strains. Stx1 differs from Stx (Shiga-toxin) by three nucleotides and one amino acid, and is neutralized by antibodies to Stx. Stx1 and Stx2 are differentiated by a lack of cross-neutralization by homologous polyclonal antisera, and by a lack of DNA–DNA cross-hybridization of their genes under conditions of high stringency.[9] Both Stx2 and Stx2e (formerly SLT-IIv, VTe) are neutralized by antisera against Stx2 but not by anti-Stx toxin. Stx2e is a variant of Stx2 that is more toxic to Vero cells than HeLa cells, and like Stx its gene is chromosomal.[52,63] All Stxs are cytotoxic for Vero cells and lethal for mice, and produce positive rabbit ileal loop responses. All Stxs consist of a single enzymatically active A subunit and multiple B subunits. Stx-sensitive cells possess the toxin receptor, globotriaosylceramide (Gb$_3$), and sodium butyrate appears to play a role in sensitizing cells to Stxs.[53] Once toxins bind to Gb$_3$, internalization follows with transport to the trans-Golgi network. Once inside host cells, the A subunit binds to and releases an adenine residue from the 28S ribosomal RNA (rRNA) of the 60S ribosomal subunit and this inhibits protein synthesis. The B subunits form pentamers in association with a single A subunit and, thus, they are responsible for the binding of the toxin to the neutral glycolipid receptors. Although serotype O157:H7 is the prototype for this group, Stxs are produced by a number of serotypes, some of which are listed in g 27–1. For reasons that are not clear, Stx2 appears to be more significant in the etiology of hemorrhagic colitis (HC) and hemolytic uremic syndrome (HUS) than Stx1.[63]

Growth and Stx Toxin Production

The nutritional requirements of Stx-producing strains are not unlike those for most other *E. coli* strains (see Chapter 20). Reports on the effect of temperature on Stx production vary. In an early study, temperature was found to have no effect on Stx1 synthesis, whereas iron repressed synthesis.[91] In another study, Stx production occurred at all temperatures that supported growth (Figure 27–1), although less toxin was found when cells were grown at 21°C than at 37°C even though cell numbers were similar.[1] In a ground roasted beef slurry, strain O157:H7 was found to produce Stx at either 21° or 37°C within 24 hours.[1] In an earlier study employing an O157:H7 strain in milk and fresh ground beef, Stx1 was found to be produced at maximum levels at 37°C in both products but at

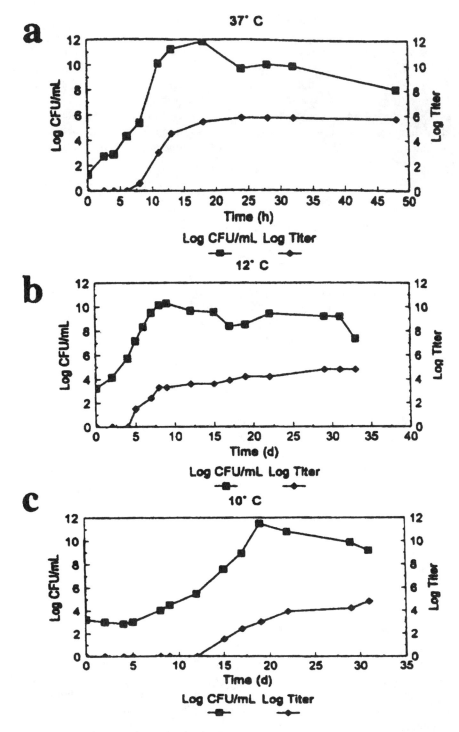

Figure 27–1 Growth and verotoxin production by *E. coli* A9124-1 at various temperatures: (a) = 37°C; (b) = 12°C; (c) = 10°C. *Source*: From Palumbo et al.,[68] copyright © 1995 by International Association of Milk, Food, and Environmental Sanitarians.

25° or 30°C only traces were found.[90] Highest levels were attained in fresh ground beef (452 ng/g), whereas in milk with agitation for 48 hours, the highest level found was 306 ng/ml. No toxin was detected in ground beef held at 8°C for up to 14 days. There is no evidence that preformed Stx plays any role in diseases caused by EHEC strains. The mean optimum growth temperature for 20 *E. coli* 0157:H7 strains was 40.2°C in Mueller–Hinton broth compared to a mean of 41.7°C for non-0157:H7 strains.[30]

With regard to minimum temperature for Stx production in brain heart infusion (BHI) broth, 4 of 16 strains grew at 8°C but not at 5°C, whereas 12 of these grew at 10°C but not at 8°C.[68] Three of 16 strains increased 1000-fold in numbers in 4–6 days at 10°C, and as noted above, Stx was produced at all temperatures that supported growth.[68] Concentrations of Stx1 were 63 and 85 ng/ml of slurry following incubations at 21° and 37°C, respectively.[1] Unlike most strains of *E. coli*, the O157:H7 strains do not grow in EC (*Escherichia coli*) medium at 44.5°C and their maximum in EC medium is around 42°C.[71]

Effect of Environmental and Physical Agents

Interest in the acid sensitivity of EHEC strains increased following an outbreak traced to fresh-pressed apple cider.[6] That product had a pH range of 3.7–3.9. In one study, EC O157:H7 survived for up to 56 days at pH ≥4.0 using tryptic soy broth and various acids for pH adjustment.[16] In another study, in which the pH of Luria broth was adjusted with HCl, no loss of viability of an EC O157:H7 strain was seen for at least 5 hours at pH 3.0–2.5 at 37°C.[5] In a more detailed study using apple ciders with pH values between 3.6 and 4.0 and EC O157:H7 inocula of 10^2–10^5, cells survived for 2–3 days at 25°C.[98] At 8°C, a 10^5/ml inoculum increased only about 1 log over 12 days and survived for 10–31 days at this temperature. Although potassium sorbate was only minimally effective, sodium sorbate shortened survival time at 8°C to 2–10 days, and to 1–2 days at 25°C.[98] Growth of EC O157:H7 was demonstrated in trypticase soy broth at pH 4.5 when HCl was used, but no growth occurred at this pH when lactic acid was used—the minimum was pH 4.6.[29]

In a study of EC O157:H7 survival in commercial mayonnaise, survival was noted for 35 days for products stored at 5° or 7°C, but cells could not be detected after 72 hours when stored at 25°C.[89] The mayonnaise had a pH of 3.65 and the inoculum was ~10^7 colony-forming units (cfu)/g. With inocula as high as log 6.23/g in commercial mayonnaise and storage at 5°, 20°, or 30°C, strain 0157:H7 did not grow and was approaching undetectable levels after 93 days at 5°C.[37] In another study, ≥6 log cfu of an EHEC strain was inoculated into five commercial real-mayonnaise-based and reduced-calorie and/or fat mayonnaise dressings and stored at 25°C.[24] The pH ranged from 3.21 to 3.94, and the products with pH <3.6 rapidly inactivated EHEC, producing a ≥7 log cfu decrease in ≤1–3 days. EHEC cells have been shown to have increased survival in acidic foods if they are first cultured in an acidic environment at around a pH of 5.0.[50] Two EHEC strains survived for 18 days at 4°C in four varieties of ground apples and the final pH of the four ranged from 3.91 to 5.11.[27] Fallen apples may be contaminated by EHEC strains in pastures, and also by contaminated fruit flies.[40] More information on the acid tolerance of *E. coli* strains is presented in Chapter 22.

With regard to salt tolerance of an EC O157:H7 strain, 4.5% NaCl in broth caused a threefold increase in doubling time, whereas at 6.5%, a 36-hour lag was noted with a generation time of 31.7 hours.[29] These investigators found that no growth occurred at ≥8.5% NaCl. In the same study, the EC O157:H7 survived sausage fermentation but did not grow when stored at 4°C for 2 months following inoculation at a level of 4.8×10^4.[29]

The thermal resistance of EHEC strains is not unlike that of most Gram-negative bacteria, and, in fact, these strains appear to be more heat sensitive than most salmonellae. A recent study found differences in thermal D values between different meat products, and the $D_{60^\circ C}$ values (minutes) and products are noted below:[2]

0.45–0.47	Beef
0.37–0.55	Pork sausage
0.38–0.55	Chicken
0.55–0.58	Turkey

The D values increased with increasing fat content, and this is a well-established phenomenon (see Chapter 17). These findings support a previous study where D and z values for high-fat and lean beef were as follows:[51]

30.5% fat	$D = 0.45$ minute	$z = 8.37^\circ F$
2.0% fat	$D = 0.30$ minute	$z = 8.30^\circ F$

In another study, an inoculum of $\sim 10^3/g$ of strain O157:H7 in low-fat ground beef was destroyed when cooked to internal temperatures of 66°, 68°, or 72°C.[25]

A recent study on the thermal properties of EC O157:H7 in apple juice revealed that a 4-D process could be achieved by heating at 60°C for ~ 1.6 minutes.[78] This is based on D values at 52°C obtained from 20 separate trials that ranged from 9.5 to 30 minutes with a mean of 18 minutes and a z of 4.8°C. Although EC O157:H7 cells became more heat sensitive in apple juice when L-malic acid was increased from 0.2 to 0.8%, or when pH was reduced from 4.4 to 3.6, benzoic acid at 1,000 ppm was the most effective additive in increasing heat sensitivity.[78]

The fate of EC O157:H7 along with *Listeria monocytogenes* and *Salmonella* Typhimurium in beef jerky was studied, and neither could be found in finished and dried products after 10 hours with inocula of $\sim 10^7/g$ or after storage for 8 weeks.[36]

Regarding radiation resistance of EHEC strains, there is no apparent basis for them to differ greatly from other enteric bacteria. Using chicken and an EC O157:H7 strain, the D value at 5°C was 0.27 kGy, whereas at -5°C, D was 0.42 kGy.[82] Employing a nonpathogenic strain of *E. coli*, Fielding et al.[26] found radiation D to be ~ 0.34 kGy in broth with a pH around 7.0, but the D was 0.24 kGy when cells were grown at pH 4.0 prior to irradiation. See Chapter 15 for more on irradiation of foods. In apple juice, nonacid adapted strains had a range of 0.12–0.21 kGy but when acid-adapted, the values increased to 0.22–0.31 kGy.[7]

In regard to survival in ovine and bovine manure, *E. coli* O157:H7 was found to survive in the former for 100 days at 4° or 10°C, and survival was unaffected by the possession of *stx* genes.[45]

Prevalence in Foods

Overall, the incidence and prevalence of EHEC strains in meat, milk, poultry, and seafood products are highly variable. Considerably more positives are found when DNA probes are used to detect for EHEC strains than when EC O157:H7 is tested alone. The first published study on the prevalence in meats of EHEC strains was that of Doyle and Schoeni,[22] who tested for EC O157:H7 and found this strain in 3.7% of 164 beef, 1.5% of 264 pork, 1.5% of 263 poultry, and 2.0% of 205 lamb samples. In Thailand, EC O157:H7 was recovered from 9% of retail beef, 8–28% of slaughterhouse beef, and 11–84% of cattle fecal specimens.[79] Although EC O157:H7 could not be recovered from sausage in the United Kingdom, a DNA probe gave positive results on 25% of 184 samples for other EHEC strains.[77] None were found in 112 samples from 71 chickens.[77] In a study of foods in the Seattle area following the 1993 outbreak, 17.3% of 294 foods were positive for colonies that contained Stx1 and/or

Stx2 strains.[74] Of the 51 positive colonies, 5 were Stx1, 34 were Stx2, and 12 were Stx1 and Stx2. The eight meat, poultry, and seafood products gave the following positive results: 63% of 8 veal, 48% of 21 lamb, 23% of 60 beef, 18% of 51 pork, 12% of 33 chicken, 10% of 62 fish, 7% of 15 turkey, and 4.5% of 44 shellfish.[74]

In their baseline studies of bacteria in or on beef and poultry carcasses and ground beef, the U.S. Department of Agriculture (USDA) findings are as follows: no *E. coli* O157:H7 was found in 563 samples of ground beef;[85] none on 1297 broiler carcasses;[86] and none on the carcasses of 2112 cows and bulls.[87] From steer and heifer carcasses, 4 of 2081 contained this organism at a maximum level of 0.93 most probable number (MPN)/cm^2.[88] Biotype 1 was found on 96% of these carcasses at numbers <10/cm^2.

When added to radish seeds and incubated at 18–25°C for 7 days, the organism was found in inner tissues and stomata of cotyledons as well as on the outer surfaces.[39] It could not be removed by immersion in 0.1% HgCl$_2$.

Immediately following the Pacific Northwest outbreak in 1993, the Food Safety and Inspection Service (FSIS) of the USDA undertook a multistate study of the prevalence and incidence of EC O157:H7 in both beef and dairy herds. The largest number per gram found was 15, and the average was around 4 cfu/g of fresh beef. Between 1994 and September 1998, the USDA in a nationwide survey found EC O157:H7 in 23 of 23,900 ground beef samples, about 1/1000 samples. The incidence and prevalence of *E. coli* O157:H7 and generic strains on meats and poultry are listed in Chapters 4, 5, 6, and 9.

In a risk assessment study of *E. coli* O157:H7 in hamburger, the three highest ranked predictive factors relative to the probability of illness by this organism from hamburgers were concentration of the organism in animal feces, host susceptibility, and carcass contamination.[11]

Prevalence in Dairy Cattle

Because more foodborne outbreaks of EHEC syndromes have been linked to beef than to any other single food source, it is widely believed that dairy herds are the primary reservoirs of these organisms. Whether this is true or not, dairy herds have been the subject of most studies. Overall, weaned calves have a higher prevalence of EHEC strains in their feces than either calves or adult cattle, and this is not surprising when one considers that the rumen biota of weaned calves is not as well established as for adult cattle. For example, of 1266 fecal samples from calves, heifers, and cows, only 18 (1.42%) were positive for EC O157:H7.[93] Only 1 of 662 cow samples was positive, whereas 5 of 210 calf and 12 of 394 heifer feces samples were positive.

Overall, these bacteria have been isolated from 0.3 to 2.2% of fecal samples collected from healthy calves or cattle in the United States, Canada, the United Kingdom, Germany, and Spain.[19] Of 23 raw milk samples examined from two farms, only 1 was positive for EC O157:H7. Using DNA probes, 28 different EHEC strains were found by these investigators with 8% from adult cows and 19% from heifers and calves.[93] Overall, EHEC strains were found on 80% of the farms examined. In another study of dairy herd fecal samples from 14 states in 1993, 31 of 965 (3.2%) were positive for EC O157:H7.[97] Of these 31, 16 were positive by direct plating with numbers of 10^3–10^5 cfu/g, whereas the other 15 were positive by enrichment only. Regarding toxin types, 19 of the 31 isolates produced Stx1 and Stx2, whereas 12 produced Stx2 only.[97]

In a study of experimentally infected calves and adult cattle, Cray and Moon[17] found that calves shed inoculated EC O157:H7 longer than adult cows, that the inoculated organisms were restricted to the gastrointestinal tract, and that most cattle infected with EC O157:H7 remain clinically normal. The infectious dose of in-vitro-grown EC O157:H7 for adult cattle was believed to be $\geq 10^7$ cfu.[17] In an earlier study in Germany, 10.8% of 1387 isolates from 259 healthy adult cattle hybridized with DNA

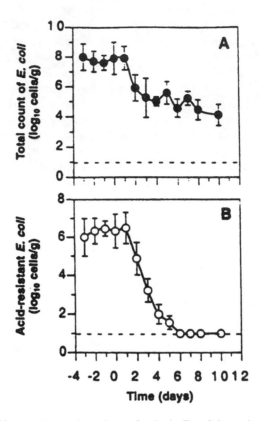

Figure 27–2   The effect of hay on the total numbers of colonic *E. coli* in cattle that had been consuming the 90% grain diet. (**A**) Cattle were switched from 90% grain to hay on day 0. (**B**) The numbers of *E. coli* were able to survive acid shock (pH 2.0, Luria broth, 1 hour). The bars indicate standard deviations of the mean (three animals, one replicate per animal, two independent experiments). The dotted lines show the detection limit of the enumerations. *Source*: Reprinted with permission from Diez-Gonzalez et al.,[21] copyright © 1998, American Association for the Advancement of Science.

probes for Stx1 and Stx2.[58] Hybridization to Stx1 alone occurred with 15.8% of the SLT positives, whereas 38.6% hybridized to Stx2 alone.

A number of studies have addressed the general microbial ecology of *E. coli* O157:H7 in the intestines and feces of bovines relative to animal rations. In one study, with cattle on high roughage for 4 days, there was a significantly lower number of *E. coli* O157:H7/g of feces but after 48 hours of fasting, they had significantly higher numbers.[41] The high-roughage ration consisted of 50% unprocessed alfalfa hay and 50% corn silage (monensin was absent). In another study, the presence of *E. coli* O157:H7 was significantly associated with corn silage feeding.[33] These investigators found *E. coli* O157:H7 to be higher in herds that were fed monensin and other additives and suggested that this ionophore could be a factor in the prevalence of this organism in cattle feces. In yet another study, cattle fed mostly grain had lower colonic pH and more acid-resistant *E. coli* than those fed only hay.[21] The grain-fed cattle had about 10^6-fold more acid-resistant *E. coli* than those fed hay, and the acid-resistant numbers decreased following a brief period of hay feeding (Figure 27–2). The association of acid resistance with virulence in some enteropathogens is discussed in Chapter 22.

From a study of *E. coli* 0157:H7 in fecal samples of commercial feedlot cattle in 2000 from 20 feedlot pens, 636 of 4790 samples (13%) were positive for this organism with the highest prevalence being in 10 pens supplied with chlorinated drinking water.[49] Sixty percent of the isolates belonged to a group of four PFGE (with *Xba*I) profiles that were present in each of eight pens throughout the sampling period. These investigators suggested that the farm environment is a potential reservoir of this organism. Of 9122 cattle fecal samples examined in the state of Washington, 1.01% were positive for *E. coli* 0157:H7,[72] while from 589 bovine fecal samples at the time of slaughter in Scotland, the animal-level prevalence was estimated to be 7.5%.[65] In the latter study, 44 infected cattle contained *E. coli* 0157:H7 at levels $>10^4$/g.

Human Disease Syndromes/Prevalence

The prototype strain for the syndromes below is EC 0157:H7. The H7 type was initially isolated in 1944 from a human diarrheal specimen, whereas the 0157 type was first isolated and named in 1972 from diarrheal swine feces.[66] However, the first 0157:H7 strain was recovered in 1975 from a patient with bloody diarrhea. Stx-producing strains of *E. coli* were identified in 1977 in the United States[61] and Canada[44]. Following its original isolation in 1975, the next recorded isolation of EC 0157:H7 was in 1978, when it was recovered from diarrheal stools in Canada.

HUS and HC are caused by Stx-producing strains of *E. coli*. It has been estimated that from 2 to 7% of infections by EC 0157:H7 will develop HUS.[31] HUS consists of hemolytic anemia, thrombocytopenia, and acute renal failure. Although not directly linked to EHEC strains until 1985, HUS was first described in 1955. In a German study on the duration of shedding of EC 0157:H7 in 53 children, the 28 children who had HC diarrhea shed the organism between 2 and 62 days (median of 13), whereas the 25 who developed HUS shed organisms for 5–124 days (median of 21).[42] HUS is associated more with strains that produce Stx2 alone than with those that produce Stx1, or Stx1 and Stx2[67] (see Chapter 22). Fifteen percent of 1275 individuals from which EHEC-positive cultures were recovered in the United Kingdom for the 3-year period 1989–1991 were reported as having developed HUS.[83]

Hemorrhagic colitis as a foodborne disease was first seen in 1982 in Oregon and Michigan, where in both instances victims had eaten sandwiches at a fast-food restaurant that contained undercooked ground beef.[73] Of 43 patients, all had bloody diarrhea and severe abdominal cramps, with 63% experiencing nausea, 49% vomiting, but only 7% fever. The mean incubation period was 3.8–3.9 days, and symptoms lasted for 3 to more than 7 days.[73] From other outbreaks, the time of onset of symptoms ranged between 3.1 and 8 days. The recovery of the etiological agent from stools requires examination of specimens within several days after symptoms. Stools tend to be negative for 7 or more days after onset of illness.[94] The bloody red stool is the telltale symptom for this syndrome and it reflects involvement of the etiological agent in the colon. Fever is rare, and the infectious dose is believed to be as low as 10 cfu.

Many of the reported outbreaks and cases of HC associated with food and water are summarized in Table 27–2. Although most were caused by *E. coli* 0157:H7, the 1993 outbreaks in New Hampshire and Rhode Island that were traced to raw salads were caused by serotype 06:NM (NM = nonmotile). The first established foodborne outbreak of an Stx-producer other than *E. coli* 0157:H7 in the United States was serotype 0104:H21, which was traced to contaminated pasteurized milk in 1994.[13] Serotype 0111:NM was traced to semidry fermented sausage in South Australia in 1995, and in 1992 the same serotype was the first ever Stx-producing strain associated with HUS in Italy when nine cases with one death occurred.[10]

Table 27-2 Some of the Reported Foodborne and Waterborne Outbreaks Caused by Stx-Producing Strains of *E. coli*

Year	Vehicle	Cases/Deaths	Location
1982	Hamburger meat	26/0	Oregon
1982	Hamburger meat	21/0	Michigan
1983	Hamburger meat	19/0	Alberta, Canada
1983	Hamburger meat	34/4	Nebraska
1984	Seafood Newberg	42/0	Maine
1985	Cold sandwiches/other*	73/19	Ontario, Canada
1985	Raw potatoes	24/0	United Kingdom
1986	Raw milk	46/0	Ontario, Canada
1986	Hamburger meat	37/2	Washington
1987	Frozen beef patties	15/2	Alberta, Canada
1987	Turkey rolls	26/0	United Kingdom
1987	Ground beef/other*	51/4	Utah
1988	Roast beef	61/0	Wisconsin
1988	Cooked frozen patties	32/0	Minnesota
1989	Water	243/4	Cabool, Missouri
1990	School lunch	10/0	Montana
1990	Roast beef	70/0	North Dakota
1991	Apple cider	23/0	Massachusetts
1992	Unknown[†]	9/1	Italy
1993	Hamburger meat	732/3	Washington, Idaho, California, Nevada
1993	Home-cooked burgers	10/0	California
1993	Garden salad[‡]	47/–	Rhode Island
1993	Tabouleh salad[‡]	121/0	New Hampshire
1994	Hamburgers	46/0	New Jersey
1994	Hamburgers (rare)	20/0	Virginia
1994	Dry-cured salami	23/0	Washington, California
1994	Contaminated pasteurized milk	17/0	Montana
1995	Semidry fermented sausage[†]	23/1	S. Australia
1995	Salad bar lettuce	>100/–	Montana
1996	Butcher shop foods	ca. 500	Scotland
1996	White radish sprouts	9492/3	Japan
1996	Apple cider (unpasteurized)	28	California, Colorado, Washington, British Columbia
1997	Alfalfa sprouts	108/0	Michigan, Virginia
1997	Ground beef	15	Colorado
1998	Drinking water	114	Wyoming
1998	Water park play pool	26/1	Georgia
1998	Fruit salads	47	Wisconsin
1998	Cake	20	California
1998	Coleslaw	33	Indiana
1998	Fresh cheese	55	Wisconsin
1998	Coleslaw	142	North Carolina
1999	Well water	775	New York state
1999	Salad bar salads	58	Texas
2001	Raw goat's milk	5	Canada
2002	Ground beef	28/5	Colorado + 6 others
2002[a]	Locally produced apple juice	64	Germany
2002	Cotton candy, ice cream	24	N. Wales

*Also person-to-person.
[†]EC O111:NM. The 23 victims had HUS.
[‡]EC O6:NM.EC O104:H21.
[a]Sorbitol-positive 0157:H-strains.

Table 27–3   Some Cases of EC O157:H7 Gastroenteritis from Drinking and Recreational Waters in the United States

Year	State	Vehicle	No. of Cases
1995	Minnesota	Spring water	33
1995	Illinois	Lake	12
1995	Minnesota	Lake	8
1995	Wisconsin	Lake	8
1996	Georgia	Pool	18
1996	Minnesota	Lake	6

Source: MMWR Morb Mort Wkly Rep 47, SS-5, 1998.

The numbers of cases of EC O157:H7 recorded by the U.S. Centers for Disease Control and Prevention (CDC) for the years 1994–1997 are as follows: 1420, 2139, 2741, and 2555, respectively, for 1994, 1995, 1996, and 1997.[12] In 1997, 1167 or 45.7% occurred in the months of July, August, and September. The gastroenteritis cases from drinking and recreational waters in the United States for 1995–1996 are summarized in Table 27–3. The outbreak that occurred in Scotland in 1996 is an example of what can happen when improper food handling and poor cleanup practices occur. There were around 500 cases (279 laboratory confirmed) that were traced to at least six foods, all from a single butcher shop.[3] All of the confirmed cases were caused by Stx2 toxin-producing strains.

In Japan, there were 29 outbreaks of EC O157:H7 human cases for the years 1991–1995. In 1996, there were 11,826 food-associated cases and 12 deaths caused by EC O157:H7.[54] The 1996 outbreak traced to white radish sprouts accounted for >9000 cases and three deaths. The existence of this organism in Northern Ireland is uncommon in cattle and human foods, and the case rates for Northern Ireland, England/Wales, and Scotland for 1997 were 1.8, 2.1, and 8.2/100,000.[96] In the United States, the rate for 1997 was 2.3, and 2.7 and 2.8 for 1996 and 1998, respectively.

Enteroinvasive *E. coli* (EIEC)

These strains generally do not produce enterotoxins as do ETECs, but they enter and multiply in colonic epithelial cells and then spread to adjacent cells in a manner similar to the shigellae.[14] Prior to the 1970s, some of these organisms were referred to as "paracolons." Like the shigellae, EIECs possess 140-MDa enteroinvasive plasmids (pINV) that are quite similar to those found in *Shigella flexneri* and are essential for their invasiveness (see Chapter 26). Plasmidless strains are not invasive. The classic EIEC strains are also Sereny positive. Members of this group have a predilection for the colon, and bloody or nonbloody but voluminous diarrhea is a consequence. Dysentery is rare, and the very young and very old are the most susceptible members of the population. The incubation period is between 2 and 48 hours with an average of 18 hours.[55] Some of the serotypes that include EIEC strains are listed in Table 27–1. At least one, O167, contains both EIEC and ETEC strains.[32]

Some of the early foodborne outbreaks are summarized in Table 27–4. The earliest recorded occurred in England in 1947 among school children, and salmon was the apparent food vehicle.[38] Although foods are a proven source for this syndrome, person-to-person transmission is known. EIEC strains have been isolated from persons with travelers' diarrhea, and they have been shown to be common in diarrheal stools from children.[81]

Table 27–4 Table 27–4 Synopses of the Earliest Known Foodborne Gastroenteritis Cases Caused by Pathogenic *E. coli* (Taken from the Literature)

Year	Location	Food/Source	No. of Victims/ No. of Risks	Toxin/Strain Type	Serotype
1947	England	Salmon (?)	47/300	EIEC	O124
1961	Rumania	Substitute coffee drink	10/50	EPEC	O86:B7; H34
1963	Japan	Ohagi	17/31	EIEC	O124
1966	Japan	Vegetables	244/435	EIEC	O124
1967	Japan	Sushi	835/1736	?	O11(?)
1971	United States*	Imported cheeses	387/?	EIEC	O124:B17
1980	Wisconsin	Food handler	500/>3000	ETEC	O6:H16
1981	Texas	Not identified	282/3000	ETEC(LT)†	O25:H+
1982	Oregon	Ground beef	26/?	EHEC	O157:H7

*In 14 states.
†LT = heat labile enterotoxin.

Enteropathogenic *E. coli* (EPEC)

These strains generally do not produce enterotoxins, although they can cause diarrhea. They exhibit localized adherence to tissue culture cells and autoagglutinate in tissue culture medium. They possess adherence factor plasmids that enable adherence to the intestinal mucosa. After colonizing the intestinal mucosa, attachment–effacement (att–eff, A/E) lesions are produced. The process starts upon initial contact and is believed to be aided by a plasmid-encoded bundle-forming pilus (see Chapter 22). EPEC-secreted proteins (Esps) block phagocytosis and lead to cytoskeletal rearrangement and tyrosine phosphorylation of Tir (see Chapter 22). When Tir binds with the outer membrane protein intimin, the attachment is intimate, resulting in destruction of brush border microvilli and formation of pedestals (see Chapter 22 for more on pathogenic mechanisms).

The A/E phenomenon appears to be the most important virulence factor of EPEC strains.[84] EPEC strains do not produce detectable quantities of Stxs. Some EPEC serotypes are listed in Table 27–1. First characterized in 1955, EPEC strains cause diarrhea in children generally under 1 year of age.

Enterotoxigenic *E. coli* (ETEC)

These strains attach to and colonize the small intestine by means of fimbrial colonization factor antigens (CFAs). There are four types of CFA—I, II, III, and IV—and they have been cloned and sequenced.[80] CFAs are plasmid encoded, generally on the same plasmid that encodes the heat-stable enterotoxin (see below), and they are not produced under 20°C. Once attached, they produce either one or two enterotoxins. Some of the ETEC serotypes are listed in Table 27–1. In a study of ETEC strains from 109 patients, the strains that produced both ST and LT were more restricted in O:K:H serotypes than those that produced only one of these toxins.[57] These toxins are further characterized below.

Unlike EPEC strains, which cause diarrhea primarily in the very young, ETEC strains cause diarrhea in both children and adults. These strains are among the leading causes of travelers' diarrhea. The ETEC disease syndromes are rarely accompanied by fever, and the diarrhea is sudden. It has been estimated that 10^8–10^{10} cfu are necessary for diarrhea by an ETEC strain in adult humans.[60]

The Enterotoxins

One of the *E. coli* enterotoxins is heat-labile (LT) and the other is heat-stable (STa or ST-1, and STb or ST-II). The LT toxin is destroyed at 60°C in about 30 minutes, whereas ST toxins can withstand 100°C for 15 minutes.

The LT toxin is a protein with a molecular weight of about 91 kDa,[18] and it possesses enzymatic activity similar to that of the cholera toxin (CT). Whereas CT is exported from the cytoplasm to the outside of producing cells, LT is deposited into the periplasm of producing cells. Further, antisera to CT neutralize LT and immunization with CT induces protection against both CT and LT challenges.

These enterotoxins are produced early in the growth phase of producing strains, with the maximum amount of ST produced after 7 hours of growth in one study in a Casamino acids yeast extract medium containing 0.2% glucose.[47] In a synthetic medium, ST appeared as early as 8 hours, but maximal production required 24 hours with aeration.[8] Although LT and ST appear to be produced under all conditions that allow cell growth, the release of LT from cells in enriched media was favored at a pH of 7.5–8.5.[59]

LT toxin is composed of two protomers: A, with a molecular weight of about 25.5 kDa, which when nicked with trypsin becomes an enzymatically active A_1 polypeptide chain of 21 kDa linked by a disulfide bond to an A_2-like chain; and B, which has a molecular weight of about 59 kDa and consists of five noncovalently linked individual polypeptide chains.[23] LTB is the binding subunit, whereas LTA stimulates the adenylate cyclase system. LTA and LTB have immunological properties similar to subunits A and B of the *Vibrio cholerae* toxin.[46] LTh and LTp designate human and porcine strains, respectively.

STa is methanol soluble and elicits a secretory response in infant mice. It is an 18–19 amino acid acidic peptide that contains three disulfide bonds and has a molecular weight of 1972 Da. It stimulates particulate intestinal guanylate cyclase. STa has been chemically synthesized.[43]

STb is methanol insoluble and is primarily of swine origin. It is the most prevalent toxin associated with diarrheagenic isolates of porcine origin, and it affects the small intestine and the ligated ileum of weaned piglets, and also the mouse intestinal loop when a protease inhibitor is added.[95] The STb gene (*estB*) has been sequenced and cloned.[48] The trypsin-sensitive STb toxin is synthesized as a 71-amino acid polypeptide that is later cleaved to yield the active 48-amino acid molecule with four cysteine residues that pass through the inner membrane to the periplasm. Although its mode of action is yet unclear, it has been shown to stimulate the synthesis of prostaglandin E_2.[35] Its receptor cell in mouse intestinal cells is a protein with a molecular weight of 25 kDa.[34]

Mode of Action of Enterotoxins. ETEC gastroenteritis is caused by the ingestion of 10^6–10^{10} viable cells per gram that must colonize the small intestines and produce enterotoxin(s). The colonizing factors are generally fimbriae or pili. The syndrome is characterized primarily by non-bloody diarrhea without inflammatory exudates in stools. The diarrhea is watery and similar to that caused by *V. cholerae*. Diarrhea results from enterotoxin activation of intestinal adenylate cyclase, which increases cyclic 3′,5′-adenosine monophosphate (cAMP).

With regard to LT, the B protomer mediates binding of the molecule to intestinal cells. LT binds to gangliosides, especially monosialogangliosides (GM_1).[23] CT also binds to GM_1 ganglioside, and CT and LT are known to share antigenic determinants among corresponding protomers, although they do not cross-react. Upon binding, the A polypeptide chain (of the A protomer) catalyzes ADP ribosylation of a G protein that activates adenylate cyclase and induces increases in intracellular cAMP.

Regarding ST, STa binds irreversibly to a specific high-affinity nonganglioside receptor and initiates a transmembrane signal to activate particulate guanylate cyclase, and triggers the production of intracellular cyclic guanosine monophosphate (cGMP). The increased levels of mucosal cGMP lead to loss of fluids and electrolytes. ST differs from CT in that only the particulate form of intestinal guanylate cyclase is stimulated by ST.[28] STa differs from LT in that the former stimulates guanylate cyclase, whereas the latter and CT activate adenylate cyclase. STb elevates luminal 5-hydroxytryptamine and prostaglandin E_2, both of which are mediators of intestinal secretions. STb does not activate guanylate cyclase, and genes controlling its production have been mapped[56] and subcloned from its plasmid and sequenced.[70]

The mechanisms of Shiga, Stx1, Stx2, Stx2e, and the castor bean protein, ricin, are the same. They are N-glycosidases that cleave a specific adenine residue from the 28S subunit of eukaryotic rRNA, leading to the inhibition of protein synthesis.[62,92]

Foodborne and Waterborne Outbreaks. Some of the earliest known foodborne outbreaks caused by ETEC and other strains are summarized in Table 27–4. Regarding the virulence groups, it may be noted that EIEC was first confirmed as the cause of a foodborne outbreak in 1947, and the first in the United States occurred in 1971. An EPEC strain was confirmed as the cause of a foodborne outbreak in 1961, an ETEC in 1980, and EHEC in 1982. The first well-documented outbreak of human disease by an ETEC was a waterborne outbreak that occurred in a national park in the state of Oregon in 1975. There were about 2200 victims who drank improperly chlorinated water. The causative ETEC strain was O6:H16.

PREVENTION

In general, the prevention/avoidance of foodborne illness by *E. coli* can be achieved by observing the factors noted in the last section of Chapter 23. However, because of the consequences to young children, special precautions need to be observed. The heat sensitivity of these organisms is such that cases should not occur when foods are properly cooked. In the case of ground beef, the recommendation is that it should be cooked to 160°F (71.1°C), or that the core temperature be brought to a minimum of 155°F (58.3°C) for at least 15 seconds and that the juices are clear (1993 recommendation of the U.S. Food and Drug Administration Food Code). Because of unevenness of hamburger patties, cooking at 155–160°F (58.3–71.1°C) provides a measure of safety. Once cooked, hamburgers as well as other meats should not be held between 40° and 140°F for more than 3–4 hours. Although the largest recorded foodborne outbreak was associated with ground beef, all raw meat, poultry, seafood; and some fruits and vegetables should be considered possible vehicles for hemorrhagic colitis.

TRAVELERS' DIARRHEA

E. coli is well established as one of the leading causes of acute watery diarrhea that often occurs among new arrivals in certain foreign countries. Among Peace Corps volunteers in rural Thailand, 57% of 35 developed the syndrome during their first 5 weeks in the country, and 50% showed evidence of infection by ETEC strains. In 1976, a shipboard outbreak of gastroenteritis was shown to be caused by serotype O25:K98:NM that produced only LT. Similar strains have been recovered from other victims of travelers' diarrhea in various countries along with EPEC and ST-producing strains.

Among other organisms associated with this syndrome are rotaviruses, noroviruses, *Entamoeba histolytica*, *Yersinia enterocolitica*, *Giardia lamblia*, *Campylobacter jejuni/coli*, *Shigella* spp. and possibly *Aeromonas hydrophila*, *Klebsiella pneumoniae*, and *Enterobacter cloacae*.

REFERENCES

1. Abdul-Raouf, U.M., L.R. Beuchat, T. Zhao, and M.S. Ammar. 1995. Growth and verotoxin I production by *Escherichia coli* O157:H7 in ground roasted beef. *Int. J. Food Microbiol.* 23:79–88.

2. Ahmed, N.M., D.E. Conner, and D.L. Huffman. 1995. Heat-resistance *of Escherichia coli* O157:H7 in meat and poultry as affected by product composition. *J. Food Sci.* 60:606–610.

3. Ahmed, S., and M. Donaghy. 1998. An outbreak of *Escherichia coli* O157:H7 in central Scotland. In *Escherichia coli O157:H7 and Other Shiga Toxin-Producing E. coli Strains*, ed. J.B. Kaper and A.D. O'Brien, 59–65. Washington, DC: ASM Press.

4. Baudry, B., S.J. Savarino, P. Vial, J.B. Kaper, and M.M. Levine. 1990. A sensitive and specific DNA probe to identify enteroaggregative *Escherichia coli*, a recently discovered diarrheal pathogen. *J. Infect. Dis.* 161:1249–1251.

5. Benjamin, M.M., and A.R. Datta. 1995. Acid tolerance of enterohemorrhagic *Escherichia coli*. *Appl. Environ. Microbiol.* 61:1669–1672.

6. Besser, R.E., S.M. Lett, J.T. Weber, M.P. Doyle, T.J. Barrett, J.G. Wells, and P.M. Griffin. 1993. An outbreak of diarrhea and hemolytic uremic syndrome from *Escherichia coli* O157:H7 in fresh-pressed apple cider. *JAMA* 269:2217–2220.

7. Buchanan, R.L., S.G. Edelson, K. Snipes, and G. Boyd. 1998. Inactivation of *Escherichia coli* O157:H7 in apple juice by irradiation. *Appl. Environ. Microbiol.* 64:4533–4535.

8. Burgess, M.N., R.J. Bywater, C.M. Cowley, N.A. Mullan, and P.M. Newsome. 1978. Biological evaluation of a methanol-soluble, heat-stable *Escherichia coli* enterotoxin in infant mice, pigs, rabbits, and calves. *Infect. Immun.* 21:526–531.

9. Calderwood, S.B., D.W.K. Acheson, G.T. Keusch, T.J. Barrett, P.M. Griffin, N.A. Strockbine, B. Swaminathan, J.B. Kaper, M.M. Levine, B.S. Kaplan, H. Karch, A.D. O'Brien, T.G. Obrig, Y. Takeda, P.I. Tarr, and I.K. Wachsmuth. 1996. Proposed new nomenclature for SLT (VT) family. *ASM News* 62:118–119.

10. Caprioli, A., I. Luzzu, F. Rosmini, C. Resti, A. Edefonti, F. Perfumo, C. Farina, A. Goglio, A. Gianviti, and G. Bizzoni. 1994. Communitywide outbreak of hemolytic-uremic syndrome associated with non-O157 verocytotoxin-producing *Escherichia coli*. *J. Infect. Dis.* 169:208–211.

11. Cassin, M.H., A.M. Lammerding, E.C.D. Todd, W. Ross, and R.S. McColl. 1998. Quantitative risk assessment for *Escherichia coli* O157:H7 in ground beef hamburger. *Int. J. Food Microbiol.* 41:21–44.

12. Centers for Disease Control and Prevention. 1998. Summary of notifiable diseases, United States, 1997. *Morb. Mort. Wkly. Rep.* 46(54).

13. Centers for Disease Control and Prevention. 1995. Outbreak of acute gastroenteritis attributable to *Escherichia coli* serotype O104:H21—Helena, Montana, 1994. *Morb. Mort. Wkly. Rep.* 44:501–503.

14. Cheasty, T., and B. Rowe. 1983. Antigenic relationships between the enteroinvasive *Escherichia coli* O antigens O28ac, O112ac, O124, O136, O143, O144, O152, and O164 and *Shigella* O antigens. *J. Clin. Microbiol.* 17:681–684.

15. Cohen, M.B., J.A. Hawkins, L.S. Weckbach, J.L. Staneck, M.M. Levine, and J.E. Heck. 1993. Colonization by enteroaggregative *Escherichia coli* in travelers with and without diarrhea. *J. Clin. Microbiol.* 31:351–353.

16. Conner, D.E., and J.S. Kotrola. 1995. Growth and survival of *Escherichia coli* O157:H7 under acidic conditions. *Appl. Environ. Microbiol.* 61:382–385.

17. Cray, W.C., Jr., and H.W. Moon. 1995. Experimental infection of calves and adult cattle with *Escherichia coli* O157:H7. *Appl. Environ. Microbiol.* 61:1586–1590.

18. Dallas, W.S., D.M. Gill, and S. Falkow. 1979. Cistrons encoding *Escherichia coli* heat-labile toxin. *J. Bacteriol.* 139:850–858.

19. Dean-Nystrom, E.A., B.T. Bosworth, W.C. Cray, Jr., and H.W. Moon. 1997. Pathogenicity of *Escherichia coli* O157:H7 in the intestines of neonatal calves. *Infect. Immun.* 65:1842–1848.

20. Debroy, C., J. Yealy, R.A. Wilson, M.K. Bhan, and R. Kumar. 1995. Antibodies raised against the outer membrane protein interrupt adherence of enteroaggregative *Escherichia coli*. *Infect. Immun.* 63:2873–2879.

21. Diez-Gonzalez, F., T.R. Callaway, H.G. Kizoulis, and J.B. Russell. 1998. Grain feeding and the dissemination of acid-resistant *Escherichia coli* from cattle. *Science* 281:1666–1668.

22. Doyle, M.P., and J.L. Schoeni. 1987. Isolation of *Escherichia coli* O157:H7 from retail fresh meats and poultry. *Appl. Environ. Microbiol.* 53:2394–2396.

23. Eidels, L., R.L. Proia, and D.A. Hart. 1983. Membrane receptors for bacterial toxins. *Microbiol. Rev.* 47:596–620.

24. Erickson, J.P., J.W. Stamer, M. Hayes, D.N. McKenna, and L.A. van Alstine. 1995. An assessment of *Escherichia coli* O157:H7 contamination risks in commercial mayonnaise from pasteurized eggs and environmental sources, and behavior in low-pH dressings. *J. Food Protect.* 58:1059–1064.

25. Fenton, L.L., L.W. Hand, T.G. Rehberger, F.K. Ray, and T.G. Harbolt. 1995. Fate of *Escherichia coli* O157:H7 in thermally processed low fat ground beef patties. *Proc. Inst. Food Technol.* 36.

26. Fielding, L.M., P.E. Cook, and A.S. Grandison. 1994. The effect of electron beam irradiation and modified pH on the survival and recovery of *Escherichia coli*. *J. Appl. Bacteriol.* 76:412–416.

27. Fisher, T.L., and D.A. Golden. 1998. Fate of *Escherichia coli* O157:H7 in ground apples used in cider production. *J. Food Protect.* 61:1372–1374.

28. Frantz, J.C., L. Jaso-Friedman, and D.C. Robertson. 1984. Binding of *Escherichia coli* heat-stable enterotoxin to rat intestinal cells and brush border membranes. *Infect. Immun.* 43:622–630.

29. Glass, K.A., J.M. Loeffelholz, J.P. Ford, and M.P. Doyle. 1992. Fate of *Escherichia coli* O157:H7 as affected by pH or sodium chloride and in fermented, dry sausage. *Appl. Environ. Microbiol.* 58:2513–2516.

30. Gonthier, A., V. Guérin-Faublée, B. Tilly, and M.-L. Delignette-Muller. 2001. Optimal growth temperature of 0157 and non-0157 *Escherichia coli* strains. *Lett. Appl. Microbiol.* 33:352–356.

31. Griffin, P.M., and R.V. Tauxe. 1991. The epidemiology of infections caused by *Escherichia coli* O157:H7, other enterohemorrhagic *E. coli*, and the associated hemolytic uremic syndrome. *Epidemiol. Rev.* 13:60–98.

32. Gross, R.J., L.V. Thomas, T. Cheasty, N.P. Day, B. Rowe, M.R.F. Toledo, and L.R. Trabulsi. 1983. Enterotoxigenic and enteroinvasive *Escherichia coli* strains belonging to a new O Group, O167. *J. Clin. Microbiol.* 17:521–523.

33. Herriott, D.E., D.D. Hancock, E.D. Ebel, L. V Carpenter, D.H. Rice, and T.E. Besser. 1998. Association of herd management factors with colonization of dairy cattle by Shiga toxin-positive *Escherichia coli* O157. *J. Food Protect.* 61:802–807.

34. Hitotsubashi, S., Y. Fujii, and K. Okamota. 1994. Binding protein for *Escherichia coli* heat-stable enterotoxin II in mouse intestinal membrane. *FEMS Microbiol. Lett.* 122:297–302.

35. Hitotsubashi, S., Y. Fujii, and H. Yamanaka. 1992. Some properties of purified *Escherichia coli* heat-stable enterotoxin II. *Infect. Immun.* 60:4468–4474.

36. Harrison, J., and M. Harrison. 1995. Fate of *Escherichia coli* O157:H7, *Listeria monocytogenes*, and *Salmonella* Typhimurium during preparation and storage of beef jerky. *Proc. Inst. Food Technol.* 30.

37. Hathcox, A.K., L.R. Beuchat, and M.P. Doyle. 1995. Death of enterohemorrhagic *Escherichia coli* O157:H7 in real mayonnaise and reduced-calorie mayonnaise dressing as influenced by initial population and storage temperature. *Appl. Environ. Microbiol.* 61:4172–4177.

38. Hobbs, B.C., M.E.M. Thomas, and J. Taylor. 1949. School outbreak of gastroenteritis associated with a pathogenic paracolon bacillus. *Lancet* 2:530–532.

39. Itoh, Y., Y. Sugita-Konishi, F. Kasuga, M. Iwaki, Y. Hara-Kudo, N. Saito, Y. Noguchi, H. Konuma, and S. Kumagai. 1998. Enterhemorrhagic *Escherichia coli* O157:H7 present in radish sprouts. *Appl. Environ. Microbiol.* 64:1532–1535.

40. Janisiewicz, W.J., W.S. Conway, M.W. Brown, G.M. Sapers, P. Fratamico, and R.L. Buchanan. 1999. Fate of *Escherichia coli* O157:H7 on fresh-cut apple tissue and its potential for transmission by fruit flies. *Appl. Environ. Microbiol.* 65:1–5.

41. Jordan, D., and S.A. McEwen. 1998. Effect of duration of fasting and a short-term high-roughage ration on the concentration of *Escherichia coli* biotype 1 in cattle feces. *J. Food Protect.* 61:531–534.

42. Karch, H., H. Rüssmann, H. Schmidt, A. Schwarzkopf, and J. Heeseman. 1995. Long-term shedding and clonal turnover of enterohemorrhagic *Escherichia coli* O157:H7 in diarrheal diseases. *J. Clin. Microbiol.* 33:1602–1605.

43. Klipstein, F.A., R.F. Engert, and R.A. Houghten. 1983. Properties of synthetically produced *Escherichia coli* heat-stable enterotoxin. *Infect. Immun.* 39:117–121.

44. Konowalchuk, J., J.I. Speirs, and S. Stavric. 1977. Vero response to a cytotoxin of *Escherichia coli*. *Infect. Immun.* 18:775–779.

45. Kudva, I.T., K. Blanch, and C.J. Hovde. 1998. Analysis of *Escherichia coli* O157:H7 survival in ovine or bovine manure and manure slurry. *Appl. Environ. Microbiol.* 64:3166–3174.

46. Kunkel, S.L., and D.C. Robertson. 1979. Purification and chemical characterization of the heat-labile enterotoxin produced by enterotoxigenic *Escherichia coli*. *Infect. Immun.* 25:586–596.

47. Lallier, R., S. Lariviere, and S. St-Pierre. 1980. *Escherichia coli* heat-stable enterotoxin: Rapid method of purification and some characteristics of the toxin. *Infect. Immun.* 28:469–474.

48. Lee, C.H., S.L. Moseley, H.W. Moon, S.C. Whipp, C.L. Gyles, and M. So. 1983. Characterization of the gene encoding heat-stable toxin II and preliminary molecular epidemiological studies of enterotoxigenic *Escherichia coli* heat-stable toxin II producers. *Infect. Immun.* 42:264–268.

49. LeJeune, J.T., T.E. Besser, D.H. Rice, J.L. Berg, R.P. Stilborn, and D.D. Hancock. 2004. Longitudinal study of fecal shedding of *Escherichia coli* O157:H7 in feedlot cattle: Predominance and persistence of specific clonal types despite massive cattle population turnover. *Appl. Environ. Microbiol.* 70:377–384.

50. Leyer, G.J., L.-L. Wang, and E.A. Johnson. 1995. Acid adaptation of *Escherichia coli* O157:H7 increases survival in acidic foods. *Appl. Environ. Microbiol.* 61:3752–3755.

51. Line, J.E., A.R. Fain, Jr., A.B. Moran, L.M. Martin, R.V. Lechowich, J.M. Carosella, and W.L. Brown. 1991. Lethality of heat to *Escherichia coli* O157:H7: *D*-value and *z*-value determinations in ground beef. *J. Food Protect.* 54:762–766.

52. Lior, H. 1994. *Escherichia coli* O157:H7 and verotoxigenic *Escherichia coli* (VTEC). *Dairy Food Environ. Sanit.* 14:378–382.

53. Louise, C.B., S.A. Kaye, B. Boyd, C.A. Lingwood, and T.G. Obrig. 1995. Shiga toxin-associated hemolytic uremic syndrome: Effect of sodium butyrate on sensitivity of human umbilical vein endothelial cells to Shiga toxin. *Infect. Immun.* 63:2766–2769.

54. Machino, H., K. Araki, S. Minami, T. Nakayama, Y. Ejima, K. Hiroe, H. Tanaka, N. Fujita, S. Usami, M. Yonekawa, K. Sadamoto, S. Takaya, and N. Sakai. 1998. Recent outbreaks of infections caused by *Escherichia coli* O157:H7 in Japan. In *Escherichia coli O157:H7 and Other Shiga Toxin-Producing E. coli Strains*, ed. J.B. Kaper and A.D. O'Brien, 73–81. Washington, DC: ASM Press.

55. Marier, R., J.G. Wells, R.C. Swanson, W. Callahan, and I.J. Mehlman. 1973. An outbreak of enteropathogenic *Escherichia coli* foodborne disease traced to imported French cheese. *Lancet* 2:1376–1378.

56. Mazaitis, A.J., R. Maas, and W.K. Maas. 1981. Structure of a naturally occurring plasmid with genes for enterotoxin production and drug resistance. *J. Bacteriol.* 145:97–105.

57. Merson, M.H., F. Orskov, I. Orskov, R.B. Sack, I. Huq, and F.T. Koster. 1979. Relationship between enterotoxin production and serotype in enterotoxigenic *Escherichia coli*. *Infect. Immun.* 23:325–329.

58. Montenegro, M.A., M. Bülte, T. Trumpf, S. Aleksic, G. Reuter, E. Bulling, and R. Helmuth. 1990. Detection and characterization of fecal verotoxin-producing *Escherichia coli* from healthy cattle. *J. Clin. Microbiol.* 28:1417–1421.

59. Mundell, D.H., C.R. Anselmo, and R.M. Wishnow. 1976. Factors influencing heat-labile *Escherichia coli* enterotoxin activity. *Infect. Immun.* 14:383–388.

60. Neill, M.A., P.I. Tarr, D.N. Taylor, and M. Wolf. 2001. *Escherichia coli*. In *Foodborne Disease Handbook: Diseases Caused by Bacteria*, ed. Y.H. Hui, M.D. Pierson, and J.R. Gorham, 2nd ed., 169–212. New York: Marcel Dekker.

61. O'Brien, A.D., M.R. Thompson, J.R. Cantey, and S.B. Formal. 1977. Production of *Shigella dysenteriae*-like toxins by pathogenic *Escherichia coli*. *Abstr., Amer. Soc. Microbiol.* 32.

62. O'Brien, A.D., and R.K. Holmes. 1987. Shiga and Shiga-like toxins. *Microbiol. Rev.* 51:206–220.

63. O'Brien, A.D., V.L. Tesh, A. Donohue-Rolfe, M.P. Jackson, S. Olsnes, K. Sandvic, A.A. Lindberg, and G.T. Keusch. 1992. Shiga toxin: Biochemistry, genetics, mode of action, and role in pathogenesis. *Curr. Top. Microbiol. Immunol.* 180:65–94.

64. Oelschlaeger, T.A., T.J. Barrett, and D.J. Kopecko. 1994. Some structures and processes of human epithelial cells involved in uptake of enterohemorrhagic *Escherichia coli* O157:H7 strains. *Infect. Immun.* 62:5142–5150.

65. Omisakin, F., M. MacRae, I.D. Ogden, and N.J.C. Strachan. 2003. Concentration and prevalence of *Escherichia coli* O157:H7 in cattle feces at slaughter. *Appl. Environ. Microbiol.* 69:2444–2447.

66. Ørskov, I., F. Ørskov, B. Jann, and K. Jann. 1977. Serology, chemistry, and genetics of O and K antigens of *Escherichia coli*. *Bacteriol. Rev.* 41:667–710.

67. Ostroff, S.M., P.I. Tarr, M.A. Neill, J.H. Lewis, N. Hargrett-Bean, and J.M. Kobayashi. 1989. Toxin genotypes and plasmid profiles as determinants of systemic sequelae in *Escherichia coli* O157:H7 infections. *J. Infect. Dis.* 160:994–998.

68. Palumbo, S.A., J.E. Call, F.J. Schultz, and A.C. Williams. 1995. Minimum and maximum temperatures for growth and verotoxin production by hemorrhagic strains of *Escherichia coli*. *J. Food Protect.* 58:352–356.

69. Park, S., R.W. Worobo, and R.A. Durst. 1999. *Escherichia coli* O157:H7 as an emerging foodborne pathogen: A literature review. *Crit. Rev. Fd. Sci. Nutr.* 39:481–502.

70. Picken, R.N., A.J. Mazaitis, W.K. Maas, M. Rey, and H. Heyneker. 1983. Nucleotide sequence of the gene for heat-stable enterotoxin II of *Escherichia coli*. *Infect. Immun.* 42:269–275.

71. Raghubeer, E.V., and J.R. Matches. 1990. Temperature range for growth of *Escherichia coli* serotype O157:H7 and selected coliforms in *E. coli* medium. *J. Clin. Microbiol.* 28:803–805.

72. Renter, D.G., J.M. Sargeant, R.D. Oberst, and M. Samadpour. 2003. Diversity, frequency, and persistence of *Escherichia coli* 0157 strains from range cattle environments. *Appl. Environ. Microbiol.* 69:542–547.

73. Riley, L.W., R.S. Remis, S.D. Helgerson, H.B. McGee, J.G. Wells, B.R. Davis, R.J. Hebert, E.S. Olcott, L.M. Johnson, N.T. Hrgrett, P.A. Blake, and M.L. Cohen. 1983. Hemorrhagic colitis associated with a rare *Escherichia coli* serotype. *N. Engl. J. Med.* 308:681–685.

74. Samadpour, M., J.E. Ongerth, J. Liston, N. Tran, D. Nguyen, T.S. Whittam, R.A. Wilson, and P.I. Tarr. 1994. Occurrence of Shiga-like toxin-producing *Escherichia coli* in retail fresh seafood, beef, lamb, pork, and poultry from grocery stores in Seattle, Washington. *Appl. Environ. Microbiol.* 60:1038–1040.

75. Savarino, S.J., A. Fasano, J. Watson, B.M. Martin, M.M. Levine, S. Guandalini, and P. Guerry. 1993. Enteroaggregative *Escherichia coli* heat-stable enterotoxin 1 represents another subfamily of *E. coli* heat-stable toxin. *Proc. Natl. Acad. Sci. USA* 90:3093–3097.

76. Smith, H.R., S.M. Scotland, G.A. Willshaw, B. Rowe, A. Cravioto, and C. Eslava. 1994. Isolates of *Escherichia coli* O44: 18 of diverse origin are enteroaggregative. *J. Infect. Dis.* 170:1610–1613.

77. Smith, H.B., T. Cheasty, D. Roberts, A. Thomas, and B. Rowe. 1991. Examination of retail chickens and sausages in Britain for vero cytotoxin-producing *Escherichia coli*. *Appl. Environ. Microbiol.* 57:2091–2093.

78. Splittstoesser, D.F., M.R. McClellan, and J.J. Churey. 1996. Heat resistance of *Escherichia coli* O157:H7 in apple juice. *J. Food Protect.* 59:226–229.

79. Suthienkul, O., J.E. Brown, J. Seriwatana, S. Tienthongdee, S. Sastravaha, and P. Escheverria. 1990. Shiga-like toxin-producing *Escherichia coli* in retail meats and cattle in Thailand. *Appl. Environ. Microbiol.* 56:1135–1139.

80. Taniguchi, T., Y. Fujino, K. Yamamoto, T. Miwatani, and T. Honda. 1995. Sequencing of the gene encoding the major pilin of pilus colonization factor antigen III (CFA/III) of human enterotoxigenic *Escherichia coli* and evidence that CFA/III is related to type IV pili. *Infect. Immun.* 63:724–728.

81. Taylor, D.N., P. Echeverria, O. Sethabutr, C. Pitarangsi, U. Leksomboon, N.R. Blacklow, B. Rowe, R. Gross, and J. Cross. 1988. Clinical and microbiologic features of *Shigella* and enteroinvasive *Escherichia coli* infections by DNA hybridization. *J. Clin. Microbiol.* 26:1362–1366.

82. Thayer, D.W., and G. Boyd. 1993. Elimination of *Escherichia coli* O157:H7 in meats by gamma irradiation. *Appl. Environ. Microbiol.* 59:1030–1034.

83. Thomas, A., H. Chart, T. Cheasty, H.R. Smith, J.A. Frost, and B. Rowe. 1993. Vero cytotoxin-producing *Escherichia coli*, particularly serogroup O157, associated with human infections in the United Kingdom: 1989–1991. *Epidemiol. Infect.* 110:591–600.

84. Tzioori, S., R. Gibson, and J. Montanaro. 1989. Nature and distribution of mucosal lesions associated with enteropathogenic and enterohemorrhagic *Escherichia coli* in piglets and the role of plasmid-mediated factors. *Infect. Immun.* 57:1142–1150.

85. U.S. Department of Agriculture. 1996. *Nationwide Federal Plant Raw Ground Beef Microbiological Survey*. Washington, DC: USDA.

86. U.S. Department of Agriculture. 1996. *Nationwide Broiler Chicken Microbiological Baseline Data Collection Program*. Washington, DC: USDA.

87. U.S. Department of Agriculture. 1996. *Nationwide Beef Microbiological Baseline Data Collection Program: Cows and Bulls*. Washington, DC: USDA.

88. U.S. Department of Agriculture. 1994. *Nationwide Beef Microbiological Baseline Data Collection Program: Steers and Heifers*. Washington, DC: USDA.

89. Weagant, S.D., J.L. Bryant, and D.H. Bark. 1994. Survival of *Escherichia coli* O157:H7 in mayonnaise and mayonnaise-based sauces at room and refrigerated temperatures. *J. Food Protect.* 57:629–631.

90. Weeratna, R.D., and M.P. Doyle. 1991. Detection and production of verotoxin 1 in *Escherichia coli* O157:H7 in food. *Appl. Environ. Microbiol.* 57:2951–2955.

91. Weinstein, D.L., R.K. Holmes, and A.D. O'Brien. 1988. Effects of iron and temperature on Shiga-like toxin I production by *Escherichia coli*. *Infect. Immun.* 56:106–111.

92. Weinstein, D.L., M.P. Jackson, L.P. Perera, R.K. Holmes, and A.D. O'Brien. 1989. In vivo formation of hybrid toxins comprising Shiga toxin and the Shiga-like toxins and role of the B subunit in localization and cytotoxic activity. *Infect. Immun.* 57:3743–3750.

93. Wells, J.G., L.D. Shipman, K.D. Greene, E.G. Sowers, J.H. Green, D.N. Cameron, F.P. Downes, M.L. Martin, P.M. Griffin, S.M. Ostroff, M.E. Potter, R.V. Tauxe, and I.K. Wachsmuth. 1991. Isolation of *Escherichia coli* serotype O157:H7 and other Shiga-like-toxin-producing *E. coli* from dairy cattle. *J. Clin. Microbiol.* 29:985–989.

94. Wells, J.G., B.R. Davis, K. Wachsmuth, L.W. Riley, R.S. Remis, R. Sokolow, and G.K. Morris. 1983. Laboratory investigation of hemorrhagic colitis outbreaks associated with a rare *Escherichia coli* serotype. *J. Clin. Microbiol.* 18:512–520.

95. Whipp, S.C. 1990. Assay for enterotoxigenic *Escherichia coli* heat-stable toxin b in rats and mice. *Infect. Immun.* 58:930–934.

96. Wilson, I.G., and J.C.N. Heaney. 1999. Surveillance for *Escherichia coli* and other pathogens in retail premises. *Dairy Food Environ. Sanit.* 19:170–179.

97. Zhao, T., M.P. Doyle, J. Shere, and L. Garber. 1995. Prevalence of enterohemorrhagic *Escherichia coli* O157:H7 in a survey of dairy herds. *Appl. Environ. Microbiol.* 61:1290–1293.

98. Zhao, T., M.P. Doyle, and R.E. Besser. 1993. Fate of enterohemorrhagic *Escherichia coli* O157:H7 in apple cider with and without preservatives. *Appl. Environ. Microbiol.* 59:2526–2530.

CHAPTER 28

Foodborne Gastroenteritis Caused by *Vibrio, Yersinia,* and *Campylobacter* Species

VIBRIOSIS (*Vibrio parahaemolyticus*)

Although most other known food-poisoning syndromes may be contracted from a variety of foods, *V. parahaemolyticus* gastroenteritis is contracted almost solely from seafood. When other foods are involved, they represent cross-contamination from seafood products. Another unique feature of this syndrome is the natural habitat of the etiological agent—the sea. In addition to its role in gastroenteritis, *V. parahaemolyticus* is known to cause extraintestinal infections in humans.

The genus *Vibrio* consists of at least 28 species, and 3 that are often associated with *V. parahaemolyticus* in aquatic environments and seafood are *V. vulnificus*, *V. alginolyticus*, and *V. cholerae*. Some of the distinguishing features of these species are noted in Table 28–1, and the syndromes caused by each are described below.

V. parahaemolyticus is common in oceanic and coastal waters. Its detection is related to water temperatures, with numbers of organisms being undetectable until the water temperature rises to around 19–20°C. A study of the Rhode River area of the Chesapeake Bay showed that the organisms survive in sediment during the winter and later are released into the water column, where they associate with the zooplankton from April to early June.[62] In ocean waters, they tend to be associated more with shellfish than with other forms.[79] They have been demonstrated to adsorb onto chitin particles and copepods, whereas organisms such as *Escherichia coli* and *Pseudomonas fluorescens* do not.[62] This species is generally not found in the open oceans, and it cannot tolerate the hydrostatic pressures of ocean depths.[111]

Growth Conditions

V. parahaemolyticus can grow in the presence of 1–8% NaCl, with best growth occurring in the 2–4% range.[110] It dies off in distilled water. It does not grow at 4°C, but growth between 5°C and 9°C has been demonstrated at a pH 7.2–7.3 and 3% NaCl, or at a pH of 7.6 and 7% NaCl (Table 28–2). Its growth at 9.5–10°C in food products has been demonstrated, although the minimum for growth in

Table 28–1   Differences Between *V. parahaemolyticus* and Three Other *Vibrio* spp.

Species	V. parahaemolyticus	V. alginolyticus	V. vulnificus	V. cholerae
Lateral flagella on solid media	+	+	−	−
Rod shape	S	S	C	d
Vogus–Proskauer reaction (VP)	−	+*	−	v
Growth in 10% NaCl	−	+	−	−
Growth in 6% NaCl	+	+	+	−
Swarming	−	+	−	−
Production of acetoin/diacetyl	−	+	−	+
Sucrose	−	+	−	+
Cellobiose	−	−	+	−
Utilization of putrescine	+	d	−	−
Color on thiosulfate–citrate–bile–sucrose (TCBS) agar	G	Y	G	Y

*24 Hours.
Note: S = straight; C = curved; G = green; Y = yellow; d = 11–90% of strains positive; v = variable; strain instability.

Source: From Krieg.[7]

open waters has been found to be 10°C.[62] The upper growth temperature is 44°C, with an optimum between 30°C and 35°C.[111] Growth has been observed over the pH range 4.8–11.0, with 7.6–8.6 being optimum. It may be noted from Table 28–2 that the minimum growth pH is related to temperature and NaCl content, with moderate growth of one strain observed at a pH of 4.8 when the temperature was 30°C and the NaCl content was 3%, but the minimum pH was 5.2 when the NaCl content was 7%.[8] Similar results were found for five other strains. Under optimal conditions, this organism has a generation time of 9–13 minutes (compared to about 20 minutes for *E. coli*). Optimum water activity (a_w) for growth corresponding to shortest generation time was found to be 0.992 (2.9% NaCl in

Table 28–2   Minimum pH of Growth of *V. parahaemolyticus* ATCC 107914 in TSB with 3% and 7% NaCl at Different Temperatures

| Temperature (°C) | Minimum pH at NaCl Concentration | |
	3%	7%
5	7.3	7.6
9	7.2	7.1
13	5.2	6.0
21	4.9	5.3
30	4.8	5.2

Source: From Beuchat.[8]

tryptic soy broth). After employing the latter medium at 29°C and various solutes to control a_w, minimum values were 0.937 (glycerol), 0.945 (KCl), 0.948 (NaCl), 0.957 (sucrose), 0.983 (glucose), and 0.986 with propylene glycol.[9] The organism is heat sensitive, with $D_{47°C}$ values ranging from 0.8 to 65.1 minutes having been reported.[10] With one strain, destruction of 500 cells/ml in shrimp homogenates was achieved at 60°C in 1 minute, but with 2×10^5 cells/ml, some survived 80°C for 15 minutes.[133] Cells are most heat resistant when grown at high temperatures in the presence of about 7% NaCl.

When the growth of *V. parahaemolyticus* was compared in estuarine water and a rich culture medium, differences were observed in cell envelope proteins and lipopolysaccharide and in alkaline phosphatase levels of K^+ and K^- strains.[99] Alkaline phosphatase was slightly higher in K^- strains grown in water. Changes in cell envelope composition may be associated with the capacity of *V. parahaemolyticus* to enter a viable yet nonculturable state in waters, making its recovery from water more difficult.[99]

Virulence Properties

The most widely used in vitro test of potential virulence for *V. parahaemolyticus* is the Kanagawa reaction, with most all virulent strains being positive (K^+) and most avirulent strains being negative (K^-). About 1% of sea isolates and about 100% of those from patients with gastroenteritis are K^+.[107] K^+ strains produce a thermostable direct hemolysin (TDH), K^- strains produce a heat-labile hemolysin, and some strains produce both. A thermostable-related hemolysin (TRH) has been shown to be an important virulence factor for at least some *V. parahaemolyticus* strains. Of 214 clinical strains tested, 52% produced TDH only, and 24% produced both TDH and TRH.[115] Of 71 environmental strains, 7% gave weak reactions to a TRH probe, but none reacted with a TDH probe. The Kanagawa reaction is determined generally by use of human red blood cells in Wagatsuma's agar medium. In addition to human red blood cells, those of the dog and rat are lysed, those of the rabbit and sheep give weak reactions, and those of the horse are not lysed.[111]

To determine the K reaction, the culture is surface plated, incubated at 37°C for 18–24 hours, and read for the presence of beta hemolysis. Of 2,720 *V. parahaemolyticus* isolates from diarrheal patients, 96% were K^+, whereas only 1% of 650 fish isolates were K^+. In general, isolates from water are K^-.

The TDH has a molecular weight of 42,000 daltons and is a cardiotrophic, cytotoxic protein that is lethal to mice[57] and induces a positive response in the rabbit ileal loop assay (see Chapter 12). Its mean mouse LD_{50} by intraperitoneal (i.p.) injection is 1.5 μg, and the rabbit ileal loop dose is 200 μg.[143] The hemolysin is under pH control and was found to be produced only when the pH is 5.5–5.6.[35] That K^+ hemolysin may aid cells in obtaining iron stems from the observation that lysed erythrocyte extracts enhanced the virulence of the organism for mice.[66] The membrane receptors of TDH are gangliosides G_{T1} and G_{D1a}, with the former binding hemolysin more firmly than the latter.[127] The resistance of horse erythrocytes to the hemolysin apparently is due to the absence of these gangliosides.[127]

A synthetic medium has been developed for the production of both the thermostable direct and the heat-labile hemolysins, and serine and glutamic acid were found to be indispensable.[65] The heat stability of TDH is such that it can remain in foods after its production. In Tris buffer at pH 7.0, $D_{120°C}$ and $D_{130°C}$ values of 34 and 13 minutes, respectively, were found for semipurified toxin, whereas in shrimp $D_{120°C}$ and $D_{130°C}$ values were 21.9 and 10.4 minutes, respectively.[18] Hemolysin was detected when cell counts reached 10^6/g, and its heat resistance was greater at pH 5.5–6.5 than at 7.0–8.0.[18]

The TDH gene (*tdh*) is chromosomal, and it has been cloned in *E. coli*. When the *tdh* gene was introduced into a K^- strain, it produced extracellular hemolysin.[94] The nucleotide sequence of the *tdh* gene has been determined,[94] and a specific *tdh* gene probe constructed, which consists of a 406-base pair.[93] Employing this probe, 141 *V. parahaemolyticus* strains were tested. All K$^+$ strains were gene-positive—86% of them were weak positives—and 16% of K$^-$ strains reacted with the probe. All gene-positive strains produced TDH as assessed by an enzyme-linked immunosorbent assay (ELISA). Of 129 other vibrios tested with the gene probe, including 19 named *Vibrio* spp., only *V. hollisae* was positive.[93] The transfer of R plasmids from *E. coli* to *V. parahaemolyticus* has been demonstrated.[48] Some clinical isolates that lack TDH contain the TDH-related hemolysin TRH (encoded by *trh*). Most isolates of *V. parahaemolyticus* from coastal waters of the United States contained both *tdh* and *trh* along with urease.[37]

At least 12 O antigens and 59 K antigens have been identified, but no correlations have been made between these and K$^+$ and K$^-$ strains, and the value of serotyping as an epidemiological aid has been minimal.

Because not all K$^+$ strains produce positive responses in the rabbit ileal loop assay and because some K$^-$ strains are associated with gastroenteritis and sometimes are the only strains isolated, the precise virulence mechanisms are unclear. In the U.S. Pacific Northwest, K$^-$ but urease-positive strains are associated with the syndrome.[69] Of 45 human fecal isolates in California and Mexico, 71% of 45 were urease positive, 91% were K$^+$, and the serovar was 04:K12.[1]

Adherence to epithelial cells is an important virulence property of Gram-negative bacteria, and it appears that *V. parahaemolyticus* produces cell-associated hemagglutinins that correlate with adherence to intestinal mucosa.[141] Pili (fimbriae) also play a role in intestinal tract colonization.[90]

Gastroenteritis Syndrome and Vehicle Foods

The identity of *V. parahaemolyticus* as a foodborne gastroenteritis agent was made first by Fujino in 1951.[45] Whereas the incidence of this illness is low in the United States and some European countries, it is high in Japan, accounting for 24% of bacterial food poisoning between 1965 and 1974.[13,110] The 1951 outbreak in Japan was traced to a boiled and semidried young sardines preparation, with 272 victims and 20 deaths.[110] The next two outbreaks occurred in Japan in 1956 and 1960.[110] In Korea and Japan, vibriosis accounted for 13.5% and 23.2%, respectively of all cases of foodborne illness during the years 1981–1995.[76] The respective percent of outbreaks were 17.4 and 32.3 in the two countries. The first outbreak in the United States occurred in 1971.[86] Steamed crabs and crab salad were the vehicle foods, and 425 of approximately 745 persons at risk became ill. The isolate from victims was K$^+$ and serotype 04:K11. Synopses of several outbreaks are presented below.

1. 1998. *V. parahaemolyticus* was the etiological agent among 23 persons living in Connecticut, New Jersey, and New York who ate raw oysters and clams harvested at Long Island Sound, NY.[23]
2. 1997. Raw oysters were the source of *Vibrio parahaemolyticus* in this outbreak in British Columbia, Washington, Oregon, and California, and there were 209 victims.[25]
3. 1981–1994. A study of raw oyster-associated infections in Florida between 1981 and 1994 included 237 (70%) with gastroenteritis and two deaths, and 102 (30%) who developed primary septicemia. Of the latter, 49% died and 80% of these were caused by *Vibrio vulnificus*.[55] Regarding species, 29% of the infections were caused by *V. parahaemolyticus*, 28% by *V. cholerae* non-01, 15% by *V. hollisae*, and 12% by *V. mimicus*.[55]

In regard to symptomatology, findings from a 1978 outbreak in Louisiana illustrate the typical features. The mean incubation period was 16.7 hours (range, 3–76 hours); symptoms lasted from 1 to 8 days, with a mean of about 4.6 days. Symptoms (along with percentage incidence of each) were diarrhea (95), cramps (92), weakness (90), nausea (72), chills (55), headache (48), and vomiting (12). Both sexes were equally affected, the age of victims ranging from 13 to 78 years. No illness occurred among 14 volunteers who ingested more than 10^9 cells, but illness did occur in one person from the accidental ingestion of approximately 10^7 K$^+$ cells.[110] In another study, 2×10^5 to 3×10^7 K$^+$ cells produced symptoms in volunteers, whereas 10^{10} cells of K$^-$ strains did not.[111,131] Some K$^-$ strains have been associated with outbreaks.[6,110]

Vehicle foods for outbreaks are seafoods such as oysters, shrimps, crabs, lobsters, clams, and related shellfish. Cross-contamination may lead to other foods as vehicles. The leading cause of food poisoning in Japan in 1996 and 1997 was salmonellosis with *V. parahaemolyticus* food poisoning being second. However, the latter became the leading cause in 1998 followed by salmonellosis. An outbreak in Japan in 1996 involved 691 cases traced to boiled crabs, and the serovar was 03:K6. This serovar has replaced 04:K8 since 1996. There were 209 and 23 cases in the United States in 1997 and 1998 respectively, all traced to raw oysters; and the serovar was 03:K6 (WHO Surveillance Newsletter No. 62, December 1999).

OTHER VIBRIOS

Vibrio cholerae

V. cholerae is best known as the cause of human cholera contracted from polluted water, and seven pandemics have been recorded. Prior to 1992, the strains that cause epidemic/pandemic cholera belonged to serovar O group 1, differentiated biochemically into two biotypes: classic and El Tor, and two serotypes, Inaba and Ogawa. Those strains of *V. cholerae* that do not agglutinate in O group I antiserum are referred to as non-01 or nonagglutinating vibrios (NAGs). The non-01 strains are considered to be autochthonous estuarine bacteria and they are widely distributed. Although generally nonpathogenic, non-01 strains are known to cause gastroenteritis, soft-tissue infections, and septicemia in humans.

The seven pandemics of cholera have been caused by *V. cholerae* 01. The seventh pandemic, caused by a 01 strain biotype El Tor, started in 1961 and waned after 1975. In 1992, a cholera epidemic occurred on the Indian subcontinent that was not caused by a 01 strain but by a new non-01 serotype, 0139. Because it was first isolated from the coastal areas of the Bay of Bengal, it was designated 0139 Bengal.[60] The 0139 serotype has been shown to be genetically similar to the seventh pandemic 01 El Tor biotype, and evidence has been presented to indicate that it evolved from seventh pandemic isolates.[64] Because 0139 lacks the 01 antigen gene cluster, this has been postulated by some as its evolutionary path from the El Tor biotype,[21] and by use of molecular fingerprinting methods, it appears that 0139 strains represent a clone that arose from an El Tor strain of the seventh pandemic.[104] Like 01, 0139 contains genes for the cholera toxin, but unlike 01, 0139 produces a capsule and its lipopolysaccharide (LPS) is reported to contain the sugar colitose.[104]

Among the earliest information linking non-01 *V. cholerae* to gastroenteritis in the United States are findings from 26 of 28 patients with acute diarrheal illness between 1972 and 1975. Although some had systemic infections, 50% of the 28 yielded noncholera vibrios from stools and no other pathogens.[58] In another retrospective study of non-01 *V. cholerae* cultures submitted to the Centers for Disease Control (CDC) in 1979, nine were from domestically acquired cases of gastroenteritis and each patient had eaten raw oysters within 72 hours of symptoms.[87] One of these isolates produced a heat-labile toxin, whereas none produced heat-stable toxins.

At least five documented gastroenteritis outbreaks of non-01 *V. cholerae* occurred prior to 1981. Those in former Czechoslovakia and Australia (1965 and 1973, respectively) were traced to potatoes and to egg and asparagus salads, and practically all victims experienced diarrhea. The third outbreak occurred in the Sudan, and well water was the source. Incubation periods from these three outbreaks ranged from 5 hours to 4 days. The fourth outbreak occurred in Florida in 1979 and involved 11 persons who ate raw oysters. Eight experienced diarrheal illness within 48 hours after eating oysters, and the other three developed symptoms 12, 15, and 30 hours after eating. The fifth outbreak, which occurred in 1980 mainly among U.S. soldiers in Venice, Italy, was traced to raw oysters. Of about 50 persons at risk, 24 developed gastroenteritis. The mean incubation period was 21.5 hours, with a range of 0.5 hours to 5 days; and the symptoms (and percentage complaining) were diarrhea (91.7), abdominal pain (50), cramps (45.8), nausea (41.7), vomiting (29.2), and dizziness (20.8). All victims recovered in 1–5 days, and non-01 strains were recovered from the stools of four.

In regard to *V. cholerae* 01, 6 outbreaks with 916 cases and 12 deaths were recorded by the CDC for the years 1973 through 1987. Prior to 1973, the last reported isolation of this organism in the United States was in 1911.[106] Of the six outbreaks, three were traced to shellfish and two to finfish. A single case of 01 infection occurred in Colorado in August 1988. The victim ate about 12 raw oysters that were harvested in Louisiana and within 36 hours had sudden onset of symptoms and passed 20 stools.[31] *V. cholerae* 01 El Tor serotype Inaba was recovered from stools. Three outbreaks are summarized below.

1. 1994. A woman in California developed cholera after consuming raw seaweed brought from the Philippines.[134] The causative strain was *V. cholerae* 01, serotype Ogawa; it was identified from stools.
2. 1994. Four persons in Indiana came down with cholera after consuming palm fruit brought from El Salvador 2 days earlier. *V. cholerae* 01, serotype Ogawa, biotype El Tor was the etiological agent.[28]
3. 1991. Four of six persons developed cholera in Maryland from the consumption of frozen fresh coconut milk imported from Thailand.[128] The strain responsible was *V. cholerae* 01, biotype El Tor.

With regard to distribution, non-01 strains of *V. cholerae* have been found in Asia and Mexico in stools of diarrheal patients along with enteropathogenic *E. coli*. *V. cholerae* non-01 was isolated from 385 persons with diarrhea in Mexico City in 1966–1967.[13] In July 1991, a serotype Inaba and biotype El Tor strain was recovered from an oyster-eating fish in Mobile Bay, Alabama.[32] This isolate was indistinguishable from the Latin American epidemic strain but differed from the endemic strains. Later in July and again in September, 1991, another isolate was made from an oyster. This strain continued to be present until August 1992 when oyster beds were opened. How the Latin American cholera outbreak strain got into this area is a matter of conjecture. In a study of relative retention rates, oysters accumulated higher concentrations of *V. cholerae* 01 than *E. coli* or *Salmonella* Tallahassee.

From Chesapeake Bay, 65 non-01 strains were isolated in one study.[63] Throughout the year, their numbers in waters were generally low, from 1 to 10 cells per liter. They were found only in areas where salinity ranged between 4% and 17%. Their presence was not correlated with fecal *E. coli*, whereas the presence of the latter did correlate with *Salmonella*.[63] Of those examined, 87% produced positive responses in Y-1 adrenal, rabbit ileal loop, and mouse lethality assays. Investigations conducted on waters along the Texas, Louisiana, and Florida coasts reveal that both 01 and non-01 *V. cholerae* are fairly common. Of 150 water samples collected along a Florida estuary, 57% were positive for *V. cholerae*.[38] Of 753 isolates examined, 20 were 01 and 733 were non-01 types. Of the 20 01 strains, 8 were Ogawa and 12 were Inaba serovars, and they were found primarily at a sewage treatment plant. The highest numbers of both 01 and non-01 strains occurred in August and November.[38] Neither the fecal coliform nor the total coliform index was an adequate indicator of the presence of *V. cholerae*, but the former was more useful than the latter. Along the Santa Cruz coast of, California, the highest numbers of non-01 strains occurred during the summer months and were associated with high coliform counts.[70] Both 01 and non-01 strains have been recovered from aquatic birds in Colorado,[96] and both types have been shown to be endemic in the Texas gulf as evidenced by antibody titers in human subjects.[59]

V. cholerae 01 El Tor synthesizes a 82-kDa preprotoxin and secretes it into culture media, where it is further processed into a 65-kDa active cytolysin.[139] Non-01 strains produce a cytotoxin and a hemolysin with a molecular weight of 60 kDa, which is immunologically related to the hemolysin of the El Tor strain. The OmpU outer membrane protein has been shown to be an adherence factor of *V. cholerae*, which may facilitate adherence to small intestines. Monoclonal antibodies raised to OmpU protected HeLa, HEp-2, Caco-2, and Henle 407 epithelial cells from invasion by viable organisms.[118]

From a patient with travelers' diarrhea a strain of 01 was isolated from which the *STa* (NAG-*STa*) gene was cloned.[95] The NAG-*STa* was chromosomal, and the toxin had a molecular weight of 8815. The NAG-*STa* shared 50% and 46% homology to *E. coli STh* and *STp*, respectively.[95] NAG-ST is methanol soluble, active in the infant mouse model, and similar to the ST of *Citrobacter freundii*.[126] Monoclonal antibodies to NAG-ST cross-react with the ST of *Yersinia enterocolitica*. Further, *V. mimicus* ST and *Y. enterocolitica* ST are neutralized by monoclonal antibodies to NAG-ST but not *E. coli* STh or STp.[126]

In a study of the survival of *V. cholerae* El Tor serotype Inaba in several foods, it was found that in meats with an inoculum of 2×10^3/g, cells remained viable for up to 90 days at $-5°C$ and for up to 300 days at $-25°C$.[36] The organism was not detected in milk after 34 days at $-5°C$ and 150 days at $-25°C$ with an inoculum of 2×10^4/ml. In milk at $7°C$, it survived 32 days on average but only 18–20 days in other foods. The virulence mechanisms of *V. cholerae* are discussed in Chapter 22.

Vibrio vulnificus

This organism is found in seawater and some seafoods. It is isolated more often from oysters and clams than from crustacean shellfish products. It has been isolated from seawater from the coast of Miami, Florida, to Cape Cod, Massachusetts, with most (84%) isolated from clams. Upon injection into mice, 82% of tested strains were lethal. *V. vulnificus*, along with other vibrios, have been recovered from mussels, clams, and oysters in Hong Kong at rates between 6% and 9%.[34] In a study of the estuarine waters of eastern North Carolina, *V. vulnificus* was isolated only when the water temperatures were between 15°C and 22°C.[103]

Following the warm summer of 1994 in Denmark during which 11 clinical cases of *V. vulnificus* were reported, a study was undertaken to determine the prevalence of the organism in Danish waters.[56] Upon testing suspect colonies with a DNA probe, from 0.8 to 19 colony-forming units (cfu) per liter were found in water between June and mid-September, and 0.04 to >11 cfu/g in sediment samples from July to mid-November. A strong correlation was found between the presence of *V. vulnificus* and water temperature. The organism was found in 7 of 17 mussels examined from 1 of 13 locations, and also from wild fish. Biotype 1 constituted 99.6% of 706 *V. vulnificus* isolates.[56] Like *V. alginolyticus* (see below), *V. vulnificus* causes soft-tissue infections and primary septicemia in humans, especially in the immunocompromised and those with cirrhosis. The fatality rate for those with septicemia is more than 50%, and more than 90% among those who become hypotensive.[139] These organisms are highly invasive, and they produce a cytotoxin with a molecular weight of about 56 kDa that is toxic to CHO cells and lytic to erythrocytes. However, the cytolysin appears not to be a critical virulence factor.[137]

Also, a hemolysin is produced with a molecular weight of about 36 kDa.[138] It also produces a zinc metalloprotease of the thermolysin family that induces a hemorrhagic reaction in skin by digesting type IV collagen, a key structure of the basement membrane.[85] The structural genes of *V. vulnificus* and the El Tor strain of *V. cholerae* 01 share areas of similarity, suggesting a common origin.[140] *V. vulnificus* induces fluid accumulation in the RITARD ligated rabbit loop (see Chapter 12), suggesting the presence of an enterotoxin.[119] *V. vulnificus* strains from the same oysters have been shown to display wide genomic diversity, suggesting that infections may be caused by mixed populations of cells or that only a few of the different strains are virulent.[20] *V. vulnificus* bacteriophages are discussed in Chapter 20 relative to their association with host cells and their possible use as indicators.

Infections are rather common in many countries, most occur between May and October, and most patients are men over 40 years of age. *V. vulnificus* is a significant pathogen in individuals with higher than normal levels of iron (e.g., in hepatitis and chronic cirrhosis) even though its virulence is not explained entirely by its capacity to sequester iron. For the period 1981–1992, 125 *V. vulnificus* cases were reported to the Florida Department of Health, and 25 persons (35%) died.[33] Raw oysters are the leading food source for this bacterium, and it is believed to be responsible for ∼95% of all seafood-associated deaths in the United States. In 1996, *V. vulnificus* was the cause of 16 cases and 3 deaths in Los Angeles associated with the consumption of raw oysters.[27] The oysters were traced back to Galveston Bay, Texas, and Eloi Bay, Louisiana. The addition of hot sauce to raw oysters has been shown to be ineffective in killing *V. vulnificus*[123] but diacetyl at a concentration of 0.05% decreased the numbers of *V. vulnificus* cells in oysters.[124]

Vibrio alginolyticus and V. hollisae

V. alginolyticus is a normal inhabitant of seawater and it has been found to cause soft-tissue and ear infections in humans. Human pathogenicity was first confirmed in 1973 but first suspected in 1969.[132] Wound infections occur on body extremities, with most patients being men with a history of exposure to seawater.

In coastal waters of the state of Washington, higher numbers of this organism were found in invertebrates and sediment samples than in open water, where the numbers were quite low.[5] Numbers found in oysters correlated with the temperature of overlying waters, with the highest numbers associated with warmer waters. In improperly stored oysters in Brazil, seven species of *Vibrio* were isolated at the rates shown[82]: *V. alginolyticus* (81%), *V. parahaemolyticus* (77%), *V. cholerae* non-01 (31%), *V. fluvialis* (27%), *V. furnissii* (19%), and *V. mimicus* and *V. vulnificus* (12% each).

First described in 1982, *V. hollisae* causes foodborne gastroenteritis, and for the period 1967–1990, 15 cases were recorded.[106] In contrast, only one case of human illness traced to shellfish consumption was reported for *V. alginolyticus* over the same period. *V. hollisae* produces an enterotoxin with a molecular weight of ∼33 kDa, and it is hemolytic on human and rabbit red blood cells.[74] Unlike *V. parahaemolyticus*, an isolate of *V. hollisae* from coastal fish produced a TDH-related hemolysin.[92] Using HeLa, Henle 407, and HCT-8 cell monolayers, *V. hollisae* has been shown to invade via microfilaments and microtubules.[84] The latter suggests that this organism may possess multiple modes of infection.

YERSINIOSIS (*Yersinia enterocolitica*)

In the genus *Yersinia*, which belongs to the family Enterobacteriaceae, 11 species and 5 biovars are recognized, including *Y. pestis*, the cause of plague. The species of primary interest in foods is

Table 28–3   Species of *Yersinia* Associated with *Y. enterocolitica* in the Environment and in Foods, and Minimum Biochemical Differences Between Them

Species	VP*	Sucrose	Rhamnose	Raffinose	Melibiose
Y. enterocolitica	+	+	−	−	−
Y. kristensenii	−	−	−	−	−
Y. fredcriksenii	+	+	+	−	−
Y. intermedia	+	+	+	+	+
Y. bercovieri	−	+	−	−	−
Y. mollaretti	−	+	−	−	−

VP = Voges–Proskauer reaction; + = positive reaction; − = negative reaction.

Y. enterocolitica. First isolated in New York State in 1933 by Coleman,[54] this Gram-negative rod is somewhat unique in that it is motile below 30°C but not at 37°C. It produces colonies of 1.0 mm or less on nutrient agar, is oxidase negative, ferments glucose with little or no gas, lacks phenylalanine deaminase, is urease positive, and is unique as a pathogen in being psychrotrophic. It is often present in the environment with at least three other of the yersiniae noted in Table 28–3.

Growth Requirements

Growth of *Y. enterocolitica* has been observed over the temperature range −2°C to 45°C, with an optimum between 22°C and 29°C. For biochemical reactions, 29°C appears to be the optimum. The upper limit for growth of some strains is 40°C, and not all grow below 4–5°C. Growth at 0–2°C in milk after 20 days has been observed. Growth at 0–1°C on pork and chicken has been observed,[77] and three strains were found to grow on raw beef held for 10 days at 0–1°C.[51] In milk at 4°C, *Y. enterocolitica* grew and attained up to 10^7 cells/ml in 7 days and competed well with the background biota.[2] The addition of NaCl to growth media raises the minimum growth temperature. In brain heart infusion (BHI) broth containing 7% NaCl, growth did not occur at 3°C or 25°C after 10 days. At a pH of 7.2, growth of one strain was observed at 3°C and very slight growth at pH 9.0 at the same temperature; no growth occurred at pH 4.6 and 9.6.[121] Although 7% NaCl was inhibitory at 3°C, growth occurred at 5% NaCl. With no salt, growth was observed at 3°C over the pH range 4.6–9.0.[121,125] Clinical strains were less affected by these parameters than were environmental isolates. With respect to minimum growth pH, the following values were found for six strains of *Y. enterocolitica* with the pH adjusted with HCl and incubated for 21 days: 4.42–4.80 at 4°C, 4.36–4.83 at 7°C, 4.26–4.50 at 10°C, and 4.18–4.36 at 20°C.[19] When organic acids were used to adjust pH, the order of their effectiveness was acetic > lactic > citric. On the other hand, the order of effectiveness of organic acids in tryptic soy broth was propionic ≥ lactic ≥ acetic > citric ≥ phosphoric ≥ HCl.[17]

A chemically defined growth medium has been devised, and it consists of four amino acids (L-methionine, L-glutamic acid, glycine, and L-histidine), inorganic salts, buffers, and potassium gluconate as carbon source.[3] *Y. enterocolitica* is destroyed in 1–3 minutes at 60°C.[50] It is rather resistant to freezing, with numbers decreasing only slightly in chicken after 90 days at 18°C.[77] The calculated $D_{62.8°C}$ for 21 strains in milk ranged from 0.7 to 17.8 sec, and none survived pasteurization.[42]

Distribution

Y. enterocolitica and the related species noted in Table 28–3 are widely distributed in the terrestrial environment and in lake, well, and stream waters, which are sources of the organisms to warm-blooded animals. It is more animal adapted and is found more often among human isolates than the other species in Table 28–3. Of 149 strains of human origin, 81%, 12%, 5.4%, and 2% were, respectively, *Y. enterocolitica, Yersinia intermedia, Yersinia frederiksenii,* and *Yersinia kristensenii*.[114] *Y. intermedia* and *Y. frederiksenii* are found mainly in fresh waters, fish, and foods, and only occasionally are isolated from humans. *Y. kristensenii* is found mainly in soils and other environmental samples as well as in foods but rarely isolated from humans.[7] Like *Y. enterocolitica,* this species produces a heat-stable enterotoxin. Many of the *Y. enterocolitica*-like isolates of Hanna et al.[52] were rhamnose positive and, consequently, are classified as *Y. intermedia* and/or *Y. frederiksenii,* and all grow at 4°C. Rhamnose-positive yersiniae are not known to cause infections in humans.

Animals from which *Y. enterocolitica* has been isolated include cats, birds, dogs, beavers, guinea pigs, rats, camels, horses, chickens, raccoons, chinchillas, deer, cattle, swine, lambs, fish, and oysters. It is widely believed that swine constitutes the single most common source of *Y. enterocolitica* to humans. Of 43 samples of pork obtained from a slaughterhouse and examined for *Y. enterocolitica, Y. intermedia, Y. kristensenii,* and *Y. frederiksenii,* 8 were positive and all four species were found.[53] Along with *Klebsiella pneumoniae, Y. enterocolitica* was recovered from crabs collected near Kodiak Island, Alaska, and was shown to be pathogenic.[4] In a study in the United States, 95 of 103 (92.2%) lots of market hogs carried at least one *Y. enterocolitica* isolate, and 98.7% of the pathogenic isolates were serotype 0:5 and 3.7% were 0:3.[46] In a study in Finland, 92% of 51 tongue and 25% of 255 ground meat samples contained *Y. enterocolitica*.[44] These investigators used a *yadA* gene-targeted polymerase chain reaction (PCR) along with a culture method and by the two methods >98% of the pork tongues were positive. Biotype 4 was the most common, as was serotype 0:3 (see below). A TaqMan assay was compared to two other methods for their respective recovery of *Y. enterocolitica* from fresh and frozen ground pork, and the sensitivity of TaqMan was 3 to 4 $\log_{10}$ cfu/g or ml.[138] Results could be obtained in 5 hours after an 18-hour enrichment. The thin agar layer Oxyrase (TALO) method could detect as few as 2 $\log_{10}$ cfu/g of freeze-injured cells. Using these two methods along with a standard selective medium, no *Y. enterocolitica* could be found in 100 ground pork samples.[138]

In regard to human carriage, an examination was made of 4,841 stool specimens from seven cities in as many U.S. states from November 1989 to January 1990, and the findings were as follows: 38%, 49%, 60%, and 98% contained, respectively, *Y. enterocolitica,* shigellae, *Campylobacter,* and salmonellae.[75] Of the *Y. enterocolitica* isolates, 92% were serotype 0:3.

Serovars and Biovars

The most commonly occurring *Y. enterocolitica* serovars (serotypes) in human infections are 0:3, 0:5,27, 0:8, and 0:9. Each of 49 isolates belonging to these serovars produced a positive HeLa cell response, whereas only 5 of 39 other serovars were positive.[88] Most pathogenic strains in the United States are 0:8 (biovars 2 and 3), and except for occasional isolations in Canada, it is rarely reported from other continents. In Canada, Africa, Europe, and Japan, serovar 0:3 (biovar 4) is the most common.[130] The second most common in Europe and Africa is 0:9, which has been reported also from Japan. Serovar 0:3 (biovar 4, phage type 9b) was practically the only type found in the province of Quebec, Canada, and it was predominant in Ontario.[130] The next most common were 0:5,27 and 0:6,30. From human infections in Canada, 0:3 represented 85% of 256 isolates, whereas for nonhuman sources,

Table 28–4 The Four Most Common Biovars of *Y. enterocolitica*

	Biovars			
Substrate/Product	1	2	3	4
Lipase (Tween 80)	+	–	–	–
Deoxyribonuclease	–	–	–	+
Indole	+	+	–	–
D-Xylose	+	+	+	–

0:5,27 represented 27% of 22 isolates.[130] Six isolates of 0:8 recovered from porcine tongues were lethal to adult mice,[130] and only 0:8 was found by Mors and Pai[88] to be Sereny positive. Employing HeLa cells, the following serovars were found to be infective: 0:1, 0:2, 0:3, 0:4, 0:5, 0:8, 0:9, and 0:21. Serovar 0:8 strains are not only virulent in humans, but they possess mouse lethality and invasiveness by the Sereny test. The four most common biovars of *Y. enterocolitica* are indicated in Table 28–4. It appears that only biovars 2, 3, and 4 carry the virulence plasmid.

Virulence Factors

Y. enterocolitica produces a heat-stable enterotoxin (ST) that survives 100°C for 20 minutes. It is not affected by proteases and lipases and has a molecular weight of 9,000–9,700 daltons, and biological activity is lost upon treatment with 2-mercaptoethanol.[97,98] When subjected to iso-electric focusing, two active fractions with iso-electric points (pIs) of 3.29 (ST-1) and 3.00 (ST-2) have been found.[97] Antiserum from guinea pigs immunized with the purified ST neutralized the activity of *Y. enterocolitica* ST and *E. coli* ST.[97] Like *E. coli* ST, it elicits positive responses in suckling mice and rabbit ileal loop assays and negative responses in the CHO and Y-1 adrenal cell assays (see Chapter 12). It is methanol soluble and stimulates guanylate cyclase and the cyclic adenosine monophosphate (cAMP) response in intestines but not adenylate cyclase.[98,107] It is produced only at or below 30°C,[100] and its production is favored in the pH range 7–8. Of 46 milk isolates, only 3 produced ST in milk at 25°C and none at 4°C. In a synthetic medium, enterotoxin production was favored by aeration but inhibited by high iron content.[3] At 25°C, more than 24 hours were required for ST production in a complex medium, and the gene that encodes its synthesis appears to be chromosomal.1996. *V. vulnificus* was the cause of 16 cases and 3 deaths in Los Angeles associated with the consumption of raw oysters.[27] The oysters were traced back to Galveston Bay, Texas, and Eloi Bay, Louisiana.

In a study of 232 human isolates, 94% produced enterotoxin, whereas only 32% of 44 from raw milk and 18% of 55 from other foods were enterotoxigenic.[101] Of the serovars 0:3, 0:8, 0:5,27, 0:6,30, and 0:9, 97% of 196 were enterotoxigenic. It has been found that most natural waters in the United States contain rhamnose-positive strains that are either serologically untypeable or react with multiple serovars.[54] In another study, 43 strains of *Y. enterocolitica* from children with gastroenteritis and 18 laboratory strains were examined for ST production, and all clinical and 7 laboratory strains produced ST as assessed by the infant mouse assay, and all were negative in the Y-1 adrenal cell assay.[100] Regarding the production of ST by species other than *Y. enterocolitica*, none of 21, 8, and 1 of *Y. intermedia, Y. frederiksenii,* and *Y. aldovae,* respectively, was positive in one study of species from raw milk, whereas 62.5% of *Y. enterocolitica* were ST positive.[135] On the other hand, about one-third of nonenterocolitica species, including *Y. intermedia* and *Y. kristensenii*, were positive for ST in two other studies.[129,136] *Y. bercovieri* produces a heatstable enterotoxin (YbST), and detectable levels are produced at 4°C after 144–168 hours.[122]

Although pathogenic strains of *Y. enterocolitica* produce ST, it appears that this agent is not critical to virulence. Some evidence for the lack of importance of ST was provided by Schiemann,[112] who demonstrated positive HeLa-cell and Sereny-test responses, with a 0:3 strain that did not produce enterotoxin. On the other hand, each of 49 isolates belonging to serovar 0:3 and the other four virulent serovars produced ST.[88] In addition to the diminished role of ST in *Y. enterocolitica* virulence, some other properties now seem less important.[22,81] The Yop virulon is the most significant virulence factor for yersiniae, and it along with some of the more recent findings on the pathogenesis of these organisms is discussed in Chapter 22.

Incidence of *Y. enterocolitica* in Foods

This organism has been isolated from cakes, vacuum-packaged meats, seafood, vegetables, milk, and other food products. It has been isolated also from beef, lamb, and pork.[77] Of all sources, swine appears to be the major source of strains pathogenic for humans. More specifically, pig tonsils have been found to be the primary sources of contamination of liver, heart, and kidneys.[43] The incidence and growth of *Y. enterocolitica* in milk are presented in Chapter 7, and for meats in Chapter 4.

Gastroenteritis Syndrome and Incidence

In addition to gastroenteritis, this organism has been associated with human pseudoappendicitis, mesenteric lymphadenitis, terminal ileitis, reactive arthritis, peritonitis, colon and neck abscesses, cholecystis, and erythema nodosum. It has been recovered from urine, blood, cerebrospinal fluid, and the eyes of infected individuals. It is, of course, recovered from the stools of gastroenteritis victims. Only the gastroenteritis syndrome is addressed below.

There is a seasonal incidence associated with this syndrome, with the fewest outbreaks occurring during the spring and the greatest number in October and November. The incidence is highest in the very young and the old. In an outbreak studied by Gutman et al.,[49] the symptoms (and percentage complaining of them) were fever (87), diarrhea (69), severe abdominal pain (62), vomiting (56), pharyngitis (31), and headache (18). The outbreak led to two appendectomies and two deaths.

Milk (raw, improperly pasteurized, or recontaminated) is a common vehicle food. The first documented outbreak in the United States occurred in 1976 in New York State, with serovar 0:8 as the responsible strain, and chocolate milk prepared by adding chocolate syrup to previously pasteurized milk was the vehicle food.[12] An outbreak of serotype 0:3 among 15 children occurred in Georgia in 1988–1989; the vehicle food was raw chitterlings.[30]

Symptoms of the gastroenteritis syndrome develop several days following ingestion of contaminated foods and are characterized by abdominal pain and diarrhea. Children appear to be more susceptible than adults, and the organisms may be present in stools for up to 40 days following illness.[4] A variety of systemic involvements may occur as a consequence of the gastroenteritis syndrome.

CAMPYLOBACTERIOSIS (*Campylobacter jejuni*)

The genus *Campylobacter* consists of at least species, and the one of primary importance in foods is *C. jejuni* subsp. *jejuni*. Unlike *C. jejuni* subsp. *doylei*, it is resistant to cephalothin, can grow at 42°C, and can reduce nitrates. Throughout this text, *C. jejuni* subsp. *jejuni* is referred to as *C. jejuni*.

The latter differs from *Campylobacter coli* in being able to hydrolyze hippurate. The campylobacters are more closely related to the genus *Arcobacter* than any other group. *C. jejuni* has the distinction of being the first foodborne pathogen whose genome was sequenced (it contains 1.64 million bases).

Prior to the 1970s, the campylobacters were known primarily to veterinary microbiologists as organisms that caused spontaneous abortions in cattle and sheep and as the cause of other animal pathologies. They were once classified as *Vibrio* spp.

C. jejuni is a slender, spirally curved rod that possesses a single polar flagellum at one or both ends of the cell. It is oxidase and catalase positive and will not grow in the presence of 3.5% NaCl or at 25°C. It is microaerophilic, requiring small amounts of oxygen (3–6%) for growth. Using an autobioluminescent strain of *C. jejuni*, its minimum, optimum, and maximum growth temperatures on solid media in a gradient-plate format were 30, 40, and 45°C.[68] At the optimum growth temperature of 40°C, this strain grew well at pH 5.5–8.0, and in the presence of up to 1.75% NaCl. Growth is actually inhibited in 21% oxygen. Carbon dioxide (about 10%) is required for good growth. When *C. jejuni* was inoculated into vacuum-packaged processed turkey meat, cell numbers decreased, but some remained viable for 28 days at 4°C.[105] Its metabolism is respiratory. In addition to *C. jejuni*, *C. coli*, *C. intestinalis*, and several other *Campylobacter* species are known to cause diarrhea in humans, but *C. jejuni* is by far the most important.

Because of their small cell size, they can be separated from most other Gram-negative bacteria by use of a 0.65-μm filter. *C. jejuni* is heat sensitive, with $D_{55°C}$ for a composite of equal numbers of five strains being 1.09 minutes in peptone and 2.25 minutes in ground, autoclaved chicken.[14] With internal heating of ground beef to 70°C, 10^7 cells/g could not be detected after about 10 minutes.[120] It appears to be sensitive to freezing, with about 10^5 cells per chicken carcass being greatly reduced or eliminated at -18°C, and for artificially contaminated hamburger meat, the numbers were reduced by 1 log cycle over a 7-day period.[47]

Distribution

Unlike *Y. enterocolitica* and *V. parahaemolyticus*, *C. jejuni* is not an environmental organism but rather is one that is associated with warm-blooded animals. A large percentage of all major meat animals have been shown to contain these organisms in their feces, with poultry being prominent. Its prevalence in fecal samples often ranges from around 30% to 100%. Reports on isolations by various investigators have been summarized by Blaser,[15] and the specimens and percentages positive for *C. jejuni* are as follows: chicken intestinal contents (39–83), swine feces (66–87), sheep feces (up to 73), swine intestinal contents (61), sheep carcasses (24), swine carcasses (22), eviscerated chicken (72–80), and eviscerated turkey (94).

In a 5-year longitudinal study on a small rearing farm in southern England in 1989–1994, 12,233 broilers were examined and 27% were positive for *C. jejuni*.[102] Of 251 shed flocks, 35.5% carried *C. jejuni*, but only 9.2% had the organism in successive flocks. Overall, there was a low level of transmission between flocks. A common source was suggested via vertical transmission rather than hatchery or transportation sources because of the lack of high diversity of types.[102] The ecology and prevalence of campylobacters in other fresh foods and the environment have been reviewed.[91]

Fecal specimens from humans with diarrhea yield *C. jejuni*, and it may be the single most common cause of acute bacterial diarrhea in humans. From 8,097 specimens submitted to eight hospital laboratories over a 15-month period in different parts of the United States, this organism was recovered from 4.6%, salmonellae from 2.3%, and shigellae from 1%.[16] The peak isolations for *C. jejuni* were in the age group 10–29 years. Peak isolations occur during the summer months, and it has been noted that 3–14% of diarrheal patients in developed countries yield stool specimens that contain *C. jejuni*.[15]

Peak isolations from individually caged hens occurred in October and late April–early May.[39] In the latter study, 8.1% of the hens were chronic excreters of the organism, whereas 33% were negative even though they were likely exposed. The most probable source of *C. jejuni* to a duck processing farm was found to be rat and mice droppings, with 86.7% of the former being positive for this species.[67] The consensus seems to be that this organism is not transmitted through the hatchery, but instead to broiler chicks by vermin as noted. The cecum is the principal site of colonization, and the organisms generally are not pathogenic to adult birds.

The numbers of *C. jejuni* on some poultry products range from log 2.00 to 4.26/g. Once this organism is established in a chicken house, most of the flock becomes infected over time. One study revealed that the organism appeared in all chicken inhabitants within a week once it was found among any of the inhabitants.[117] In addition to poultry, the other primary source of this organism is raw milk. Because the organism exists in cow feces, it is not surprising that it may be found in raw milk, and the degree of contamination would be expected to vary depending on milking procedures. In a survey of 108 samples from bulk tanks of raw milk in Wisconsin, only 1 was positive for *C. jejuni*, whereas the feces of 64% of the cows in a grade A herd were positive.[40] In the Netherlands, 22% of 904 cow fecal and 4.5% of 904 raw milk samples contained *C. jejuni*.[11]

The significance of cross-contamination as a source of this organism to humans is illustrated by the following outbreak: There were 14 cases of *Campylobacter* enteritis in Oklahoma traced to lettuce and lasagna.[26] The foods were prepared in a small area where raw chicken was cut, and in all probability, the vehicle foods became contaminated. See Chapters 4 and 9 for numbers of campylobacters in various foods.

Virulence Properties

At least some strains of *C. jejuni* produce a heat-labile enterotoxin (CJT) that shares some common properties with the enterotoxins of *V. cholerae* (CT) and *E. coli* (LT). CJT increases cAMP levels, induces changes in CHO cells, and induces fluid accumulation in rat ileal loops.[108] Maximal production of CJT in a special medium was achieved at 42°C in 24 hours, and the amount produced was enhanced by polymyxin.[72] The quantities produced by strains varied widely from none to about 50 ng/ml CJT protein. The amount of toxin was doubled as measured by Y-1 adrenal cell assay when cells were first exposed to lincomycin and then polymyxin.[83] CJT is neutralized by CT and *E. coli* LT antisera, indicating immunological homology with these two enterotoxins.[72] The *C. jejuni* LT appears to share the same cell receptors as CT and *E. coli* LT, and it contains a B subunit immunologically related to the B subunits of CT and LT of *E. coli*.[73] Also, a cytotoxin is produced that is active against Vero and HeLa cells. The enterotoxin and the cytotoxin induce fluid accumulation in rat jejunal loops but not in mice, pigs, or calves. Partially purified enterotoxin contained three fractions with molecular weights of 68, 54, and 43 kDa.[61] Of 202 strains of *C. jejuni* and *C. coli* recovered from humans with enteritis and from healthy laying hens, 34% and 22% of the *C. jejuni* and *C. coli* strains, respectively, produced enterotoxin as determined by CHO assay.[78]

From a comparison of poultry and human isolates of *Campylobacter jejuni*, invasiveness or Vero cell toxicity was higher for human than for poultry isolates.[89] Invasiveness was associated with biotypes 1 and 2 while cell toxicity (with CHO and INT-407 cells) was associated with biotypes 3 and 4.

C. jejuni enteritis appears to be caused in part by the invasive abilities of the organism. Evidence for this comes from the nature of the clinical symptoms, the rapid development of high agglutinin titers after infection, recovery of the organism from peripheral blood during the acute phase of the disease, and the finding that *C. jejuni* can penetrate HeLa cells.[80] However, *C. jejuni* is not invasive by either the Sereny or the Anton assay.

Caco-2 cells are invaded by an energy-dependent invasion mechanism, not by endocytosis.[109] Several sequelae are associated with campylobacteriosis, including Guillain–Barré syndrome (GBS: for a review, see reference 116). It is estimated that about one-third of patients with GBS develop symptoms 1–3 weeks after *C. jejuni* enteritis. In the Penner serotyping scheme, over 48 serotypes of *C. jejuni* are recognized, and serotype 19 is one that appears to be associated with GBS. This strain has an oligosaccharide structure that is identical to the terminal tetrasaccharide of host ganglioside GM_1. Because the gangliosides are surface components of nerve tissue, antibodies to the oligosaccharide structure of *C. jejuni* would exhibit antineural effects.[142]

Plasmids have been demonstrated in *C. jejuni* cells. Of 17 strains studied, 11 were found to carry plasmids ranging from 1.6 to 70 MDa, but their role and function in disease are unclear.

A serotyping scheme has been developed for *C. jejuni*. From chickens and humans, 82% and 98%, respectively, of isolates belonged to biovar 1.[113]

Overall, the specific modes of *Campylobacter* pathogenesis are still unclear. In one review, it is noted that motility and invasion play a role in pathogenesis, and that the roles of the toxins are far from clear.[71]

Enteritis Syndrome and Prevalence

From the first U.S. outbreak of *C. jejuni* traced to a water supply,[29] in which about 2,000 individuals contracted infections, the symptoms (and percentages of individuals affected) were as follows: abdominal pain or cramps (88), diarrhea (83), malaise (76), headache (54), and fever (52). Symptoms lasted from 1 to 4 days. In the more severe cases, bloody stools may occur, and the diarrhea may resemble ulcerative colitis, whereas the abdominal pain may mimic acute appendicitis.[15] The incubation period for enteritis is highly variable. It is usually 48–82 hours but may be as long as 7–10 days or more. Diarrhea may last 2–7 days, and the organisms may be shed for more than 2 months after symptoms subside.

Clinical laboratory isolates of *Campylobacter* in the United States for 1996–1998 exceeded salmonellae isolates (Figure 28–1). The isolates were from clinical laboratories in selected cities of five states This is the FoodNet Surveillance Network[24] described in Chapter 22. It should be noted that the numbers do not represent isolations from actual foodborne illness. It is assumed that the organisms noted were contracted from foods even though this connection is not demonstrated. In the case of *Campylobacter*, it is assumed that about 90% are of food origin. While this surveillance method might be a valid indicator of actual foodborne cases, it is unprecedented. It is unusual that such a fragile and environmentally sensitive organism should be the leading cause of foodborne illness. It is interesting to note that the largest recorded outbreak of *Campylobacter* enteritis, as noted above, was traced to the water supply of a Vermont town where about 2,000 persons were infected.[29]

PREVENTION

V. parahaemolyticus, *Y. enterocolitica*, and *C. jejuni* are all heat-sensitive bacteria that are destroyed by milk pasteurization temperatures. The avoidance of raw seafood products and care in preventing cross-contamination with contaminated raw materials will eliminate or drastically reduce the incidence of foodborne gastroenteritis caused by *V. parahaemolyticus* and *Y. enterocolitica*. To prevent wound

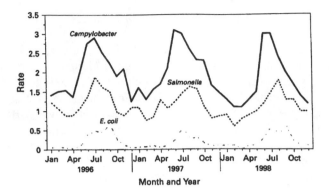

Figure 28–1   Rate (per 100,000 population) of laboratory-confirmed infections with selected pathogens detected by the Foodborne Diseases Active Surveillance Network (FoodNet), United States, 1966–1998. The results for 1998 are preliminary. 1999. *Morbidity and Mortality Weekly Report* 48:190 U. S. Center for Disease Control and Prevention.

infections by vibrios, individuals with body nicks or abrasions should avoid entering estaurine or seawaters. Yersinosis can be avoided or certainly minimized by not drinking water that has not been properly treated, and by avoiding raw or underprocessed milk. Campylobacteriosis can be avoided by not eating undercooked or unpasteurized foods of animal origin, especially milk and raw poultry.

REFERENCES

1. Abbott, S.L., C. Powers, C.A. Kaysner, Y. Takeda, M. Ishibashi, S.W. Joseph, and J.M. Janda. 1989. Emergence of a restricted bioserovar of *Vibrio parahaemolyticus* as the predominant cause of *Vibrio*-associated gastroenteritis on the West Coast of the United States and Mexico. *J. Clin. Microbiol.* 27:2891–2893.

2. Amin, M.K., and F.A. Draughon. 1987. Growth characteristics of *Yersinia enterocolitica* in pasteurized skim milk. *J. Food Protect.* 50:849–852.

3. Amirmozafari, N., and D.C. Robertson. 1993. Nutritional requirements for synthesis of heat-stable enterotoxin by *Yersinia enterocolitica*. *Appl. Environ. Microbiol.* 59:3314–3320.

4. Asakawa, Y., S. Akahane, N. Kagata, and M. Noguchi. 1973. Two community outbreaks of human infection with *Yersinia enterocolitica*. *J. Hyg.* 71:715–723.

5. Baross, J., and J. Liston. 1970. Occurrence of *Vibrio parahaemolyticus* and related hemolytic vibrios in marine environments of Washington state. *Appl. Microbiol.* 20:179–186.

6. Barrow, G.I., and D.C. Miller. 1976. *Vibrio parahaemolyticus* and seafoods. In *Microbiology in Agriculture, Fisheries and Food*, ed. F.A. Skinner and J.G. Carr, 181–195. New York: Academic Press.

7. Krieg, N.R. ed. 1984. *Bergey's Manual of Systematic Bacteriology*. Vol. 1. Baltimore: Williams & Wilkins.

8. Beuchat, L.R. 1973. Interacting effects of pH, temperature, and salt concentration on growth and survival of *Vibrio parahaemolyticus*. *Appl. Microbiol.* 25:844–846.

9. Beuchat, L.R. 1974. Combined effects of water activity, solute, and temperature on the growth of *Vibrio parahaemolyticus*. *Appl. Microbiol.* 27:1075–1080.

10. Beuchat, L.R., and R.E. Worthington. 1976. Relationships between heat resistance and phospholipid fatty acid composition of *Vibrio parahaemolyticus*. *Appl. Environ. Microbiol.* 31:389–394.

11. Beumer, R.R., J.J.M. Cruysen, and I.R.K. Birtantie. 1988. The occurrence of *Campylobacter jejuni* in raw cows' milk. *J. Appl. Bacteriol.* 65:93–96.

12. Black, R.E., R.J. Jackson, T. Tsai, M. Medvesky, M. Shayegani, J.C. Feeley, K.I.E. MacLeod, and A.M. Wakelee. 1978. Epidemic *Yersinia enterocolitica* infection due to contaminated chocolate milk. *N. Engl. J. Med.* 298:76–79.

13. Blake, P.A., R.E. Weaver, and D.G. Hollis. 1980. Diseases of humans (other than cholera) caused by vibrios. *Ann. Rev. Microbiol.* 34:341–367.

14. Blankenship, L.C., and S.E. Craven. 1982. *Campylobacter jejuni* survival in chicken meat as a function of temperature. *Appl. Environ. Microbiol.* 44:88–92.

15. Blaser, M.J. 1982. *Campylobacter jejuni* and food. *Food Technol.* 36(3):89–92.

16. Blaser, M.J., P. Checko, C. Bopp, A. Bruce, and J.M. Hughes 1982. *Campylobacter* enteritis associated with foodborne transmission. *Am J. Epidemiol.* 116:886–894.

17. Brackett, R.E. 1987. Effects of various acids on growth and survival of *Yersinia enterocolitica. J. Food Protect.* 50:598–601.

18. Bradshaw, J.G., D.B. Shah, A.J. Wehby, J.T. Peeler, and R.M. Twedt. 1984. Thermal inactivation of the Kanagawa hemolysin of *Vibrio parahaemolyticus* in buffer and shrimp. *J. Food Sci.* 49:183–187.

19. Brocklehurst, T.F., and B.M. Lund. 1990. The influence of pH, temperature and organic acids on the initiation of growth of *Yersinia enterocolitica. J. Appl. Bacteriol.* 69:390–397.

20. Buchrieser, C., V.V. Gangar, R.L. Murphree, M.L. Tamplin, and C.W. Kasper. 1995. Multiple *Vibrio vulnificus* strains in oysters as demonstrated by clamped homogeneous electric field gel electrophoresis. *Appl. Environ. Microbiol.* 61:1163–1168.

21. Calia, K.E., M. Murtagh, M.J. Ferraro, and S.B. Calderwood. 1994. Comparison of *Vibrio cholerae* 0139 with *V. cholerae* 01 classical and El Tor biotypes. *Infect. Immun.* 62:1504–1506.

22. Carter, P.B., R.J. Zahorchak, and R.R. Brubaker. 1980. Plague virulence antigens from *Yersinia enterocolitica. Infect. Immun.* 28:638–640.

23. Centers for Disease Control and Prevention. 1999. Outbreak of *Vibrio parahaemolyticus* infection associated with eating raw oysters and clams harvested from Long Island Sound—Connecticut, New Jersey, and New York, 1998. *Morb. Mortal. Wkly. Rep.* 48:48–51.

24. Centers for Disease Control and Prevention. 1998. Incidence of foodborne illnesses—FoodNet, 1997. *Morb. Mortal. Wkly. Rep.* 47:782–786.

25. Centers for Disease Control and Prevention. 1998. Outbreak of *Vibrio parahaemolyticus* infections associated with eating raw oysters—Pacific Northwest, 1997. *Morb. Mortal. Wkly. Rep.* 47:457–462.

26. Centers for Disease Control and Prevention. 1998. Outbreak of *Campylobacter* enteritis associated with cross-contamination of food—Oklahoma, 1996. *Morb. Mortal. Wkly. Rep.* 47:129–131.

27. Centers for Disease Control and Prevention. 1996. *Vibrio vulnificus* infections associated with eating raw oysters—Los Angeles, 1996. *Morb. Mortal. Wkly. Rep.* 45:621–624.

28. Centers for Disease Control and Prevention. 1995. Cholera associated with food transported from El Salvador—Indiana, 1994. *Morb. Mortal. Wkly. Rep.* 44:385–386.

29. Centers for Disease Control and Prevention. 1978. Waterborne *Campylobacter* gastroenteritis—Vermont. *Morb. Mortal. Wkly. Rep.* 27:207.

30. Centers for Disease Control and Prevention. 1990. *Yersinia enterocolitica* infections during the holidays in black families—Georgia. *Morb. Mortal. Wkly. Rep.* 39:819–821.

31. Centers for Disease Control and Prevention. 1989. Toxigenic *Vibrio cholerae* 01 infection acquired in Colorado. *Morb. Mortal. Wkly. Rep.* 38:19–20.

32. Centers for Disease Control and Prevention. 1993. Isolation of *Vibrio cholerae* 01 from oysters—Mobile Bay, 1991–1992. *Morb. Mortal. Wkly. Rep.* 42:91–93.

33. Centers for Disease Control and Prevention. 1993. *Vibrio vulnificus* infections associated with raw oyster consumption—Florida, 1981–1992. *Morb. Mortal. Wkly. Rep.* 42:405–407.

34. Chan, K.-Y., M.L. Woo, L.Y. Lam, and G.L. French. 1989. *Vibrio parahaemolyticus* and other halophilic vibrios associated with seafood in Hong Kong. *J. Appl. Bacteriol.* 66:57–64.

35. Cherwonogrodzky, J.W., and A.G. Clark. 1981. Effect of pH on the production of the Kanagawa hemolysin by *Vibrio parahaemolyticus. Infect. Immun.* 34:115–119.

36. Corrales, M.T., A.E. Bainotti, and A.C. Simonetta. 1994. Survival of *Vibrio cholerae* 01 in common food-stuffs during storage at different temperatures. *Lett. Appl. Microbiol.* 18:277–280.

37. DePaola, A., J. Ulaszek, C.A. Kaysner, B.J. Tenge, J.L. Nordstrom, J. Wells, N. Puhr, and S.M. Gendel. 2003. Molecular, serological, and virulence characteristics of *Vibrio parahaemolyticus* isolated from environmental, food, and clinical sources in North America and Asia. *Appl. Environ. Microbiol.* 69:3999–4005.

38. DePaola, A., M.W. Presnell, R.E. Becker, M.L. Motes, Jr., S.R. Zywno, J.F. Musselman, J. Taylor, and L. Williams. 1984. Distribution of *Vibrio cholerae* in the Apalachicola (Florida) Bay estuary. *J. Food Protect.* 47:549–553.

39. Doyle, M.P. 1984. Association of *Campylobacter jejuni* with laying hens and eggs. *Appl. Environ. Microbiol.* 47:533–536.

40. Doyle, M.P., and D.J. Roman. 1982. Prevalence and survival of *Campylobacter jejuni* in unpasteurized milk. *Appl. Environ. Microbiol.* 44:1154–1158.

41. Faghri, M.A., C.L. Pennington, L.B. Cronholm, and R.M. Atlas. 1984. Bacteria associated with crabs from cold waters, with emphasis on the occurrence of potential human pathogens. *Appl. Environ. Microbiol.* 47:1054–1061.

42. Francis, D.W., P.L. Spaulding, and J. Lovett. 1980. Enterotoxin production and thermal resistance of *Yersinia enterocolitica* in milk. *Appl. Environ. Microbiol.* 40:174–176.

43. Fredriksson-Ahomaa, M., T. Korte, and H. Korkeala. 2000. Contamination of carcasses, offals, and the environment with *yadA*-positive *Yersinia enterocolitica* in a pig slaughterhouse. *J. Food Protect.* 63:31–35.

44. Fredriksson-Ahomaa, M., S. Hielm, and H. Korkeala. 1999. High prevalence of *yadA*-positive *Yersinia enterocolitica* in pig tongues and minced meat at the retail level in Finland. *J. Food Protect.* 62:123–127.

45. Fujino, T., G. Sakaguchi, R. Sakazaki, and Y. Takeda. 1974. *International Symposium on Vibrio parahaemolyticus*. Tokyo: Saikon.

46. Funk, J.A., H.F. Troutt, R.E. Isaacson, and C.P. Fossler. 1998. Prevalence of pathogenic *Yersinia enterocolitica* in groups of swine at slaughter. *J. Food Protect.* 61:677–682.

47. Gill, C.O., and L.M. Harris. 1984. Hamburgers and broiler chickens as potential sources of human *Campylobacter* enteritis. *J. Food Protect.* 47:96–99.

48. Guerry, P., and R.R. Colwell. 1977. Isolation of cryptic plasmid deoxyribonucleic acid from Kanagawa-positive strains of *Vibrio parahaemolyticus*. *Infect. Immun.* 16:328–334.

49. Gutman, L.T., E.A. Ottesen, T.J. Quan, P.S. Noce, and S.L. Katz. 1973. An inter-familial outbreak of *Yersinia enterocolitica* enteritis. *N. Engl. J. Med.* 288:1372–1377.

50. Hanna, M.O., J.C. Stewart, Z.L. Carpenter, and C. Vanderzant. 1977. Heat resistance of *Yersinia enterocolitica* in skim milk. *J. Food Sci.* 42:1134, 1136.

51. Hanna, M.O., J.C. Stewart, D.L. Zink, Z.L. Carpenter, and C. Vanderzant. 1977. Development of *Yersinia enterocolitica* on raw and cooked beef and pork at different temperatures. *J. Food Sci.* 42:1180–1184.

52. Hanna, M.O., D.L. Zink, Z.L. Carpenter, and C. Vanderzant. 1976. *Yersinia enterocolitica*-like organisms from vacuum-packaged beef and lamb. *J Food Sci.* 41:1254–1256.

53. Harmon, M.C., B. Swaminathan, and J.C. Forrest. 1984. Isolation of *Yersinia enterocolitica* and related species from porcine samples obtained from an abattoir. *J. Appl. Bacteriol.* 56:421–427.

54. Highsmith, A.K., J.C. Feeley, and G.K. Morris. 1977. *Yersinia enterocolitica*: A review of the bacterium and recommended laboratory methodology. *Health Lab. Sci.* 14:253–260.

55. Hlady, W.G. 1997. *Vibrio* infections associated with raw oyster consumption in Florida, 1981–1994. *J. Food Protect.* 60:353–357.

56. Høi, L., J.L. Larsen, I. Dalsgaard, and A. Dalsgaard. 1998. Occurrence of *Vibrio vulnificus* biotypes in Danish marine environments. *Appl. Environ. Microbiol.* 64:7–13.

57. Honda, T., K. Goshima, Y. Takeda, Y. Sugino, and T. Miwatani. 1976. Demonstration of the cardiotoxicity of the thermostable direct hemolysin (lethal toxin) produced by *Vibrio parahaemolyticus*. *Infect. Immun.* 13:163–171.

58. Hughes, J.M., D.G. Hollis, E.J. Gangarosa, and R.E. Weaver. 1978. Noncholera vibrio infections in the United States: Clinical, epidemiologic and laboratory features. *Ann. Intern. Med.* 88:602–606.

59. Hunt, M.D., W.E. Woodard, B.H. Keswick, and H.L. Dupont 1988. Seroepidemiology of cholera in Gulf coastal Texas. *Appl. Environ. Microbiol.* 54:1673–1677.

60. Islam, M.S., M.K. Hasan, M.A. Miah, M. Yunus, K. Zaman, and M.J. Albert. 1994. Isolation of *Vibrio cholerae* 0139 synonym Bengal from the aquatic environment in Bangladesh: Implications for disease transmission. *Appl. Environ. Microbiol.* 60:1684–1686.

61. Kaikoku, T., M. Kawaguchi, K. Takama, and S. Suzuki. 1990. Partial purification and characterization of the enterotoxin produced by *Campylobacter jejuni*. *Infect. Immun.* 58:2414–2419.

62. Kancko, T., and R.R. Colwell. 1973. Ecology of *Vibrio parahaemolyticus* in Chesapeake Bay. *J. Bacteriol.* 113:24–32.

63. Kaper, J., H. Lockman, R.R. Colwell, and S.W. Joseph. 1979. Ecology, serology, and enterotoxin production of *Vibrio cholerae* in Chesapeake Bay. *Appl. Environ. Microbiol.* 37:91–103.

64. Karaolis, D.K.R., R. Lan, and P.R. Reeves. 1995. The sixth and seventh cholera pandemics are due to independent clones separately derived from environmental, nontoxigenic, non-01 *Vibrio cholerae. J. Bacteriol.* 177:3191–3198.

65. Karunsagar, I. 1981. Production of hemolysin by *Vibrio parahaemolyticus* in a chemically defined medium. *Appl. Environ. Microbiol.* 41:1274–1275.

66. Karunsagar, I., S.W. Joseph, R.M. Twedt, H. Hada, and R.R. Colwell. 1984. Enhancement of *Vibrio parahaemolyticus* virulence by lysed erythrocyte factor and iron. *Infect. Immun.* 46:141–144.

67. Kasrazedeh, M., and C. Genigeorgis. 1987. Origin and prevalence of *Campylobacter jejuni* in ducks and duck meat at the farm and processing plant level. *J. Food Protect.* 50:321–326.

68. Kelana, L.C., and M.W. Griffiths. 2003. Growth of autobioluminescent *Campylobacter jejuni* in response to various environmental conditions. *J. Food Protect.* 66:1190–1197.

69. Kelly, M.T., and E.M. Dan Stroh. 1989. Urease-positive, Kanagawa-negative *Vibrio parahaemolyticus* from patients and the environment in the Pacific Northwest. *J. Clin. Microbiol.* 27:2820–2822.

70. Kenyon, J.E., D.R. Piexoto, B. Austin, and D.C. Gilles. 1984. Seasonal variation in numbers of *Vibrio cholerae* (non-01) isolated from California coastal waters. *Appl. Environ. Microbiol.* 47:1243–1245.

71. Ketley, J.M. 1997. Pathogenesis of enteric infection by *Campylobacter. Microbiology* 143:5–21.

72. Klipstein, F.A., and R.F. Engert. 1984. Properties of crude *Campylobacter jejuni* heat-labile enterotoxin. *Infect. Immun.* 45:314–319.

73. Klipstein, F.A., and R.F. Engert. 1985. Immunological relationship of the B subunits of *Campylobacter jejuni* and *Escherichia coli* heat-labile enterotoxins. *Infect. Immun.* 48:629–633.

74. Kothary, M.H., and S.H. Richardson. 1987. Fluid accumulation in infant mice caused by *Vibrio hollisae* and its extracellular enterotoxin. *Infect. Immun.* 55:626–630.

75. Lee, L.A., J. Taylor, G.P. Carter, B. Quinn, J.J. Farmer, III, R.V. Tauxe, and the *Yersinia enterocolitica* Collaborative Study Group. 1991. *Yersinia enterocolitica* O:3: An emerging cause of pediatric gastroenteritis in the United States. The *Yersinia enterocolitica* Collaborative Study Group. *J. Infect. Dis.* 163:660–663.

76. Lee, W.-C., M.-J. Lee, J. -S. Kim, and S.-Y. Park. 2001. Foodborne illness outbreaks in Korea and Japan studied retrospectively. *J. Food Protect.* 64:899–902.

77. Leistner, L., H. Hechelmann, and R. Albert. 1975. Nachweis von *Yersinia enterocolitica* in Faeces and Fleisch von Schweinen, Hindern und Geflugel. *Fleischwirtschaft* 55:1599–1602.

78. Lindblom, G.-B., B. Kaijser, and E. Sjogren. 1989. Enterotoxin production and serogroups of *Campylobacter jejuni* and *Campylobacter coli* from patients with diarrhea and from healthy laying hens. *J. Clin. Microbiol.* 27:1272–1276.

79. Liston, J. 1973. *Vibrio parahaemolyticus.* In *Microbial Safety of Fishery Products*, ed. C.O. Chichester and H.D. Graham, 203–213. New York: Academic Press.

80. Manninen, K.I., J.F. Prescott, and I.R. Dohoo. 1982. Pathogenicity of *Campylobacter jejuni* isolates from animals and humans. *Infect. Immun.* 38:46–52.

81. Martinez, R.J. 1983. Plasmid-mediated and temperature-regulated surface properties of *Yersinia enterocolitica. Infect. Immun.* 41:921–930.

82. Matté, G.R., M.H. Matté, I.G. Rivera, and M.T. Martins. 1994. Distribution of potentially pathogenic vibrios in oysters from a tropical region. *J. Food Protect.* 57:870–873.

83. McCardell, B.A., J.M. Madden, and E.C. Lee. 1984. *Campylobacter jejuni* and *Campylobacter coli* production of a cytotonic toxin immunologically similar to cholera toxin. *J. Food Protect.* 47:943–949.

84. Miliotis, M.D., B.D. Tall, and R.T. Gray. 1995. Adherence to an invasion of tissue culture cells by *Vibrio hollisae. Infect. Immun.* 63:4959–4963.

85. Miyoshi, S.-I., H. Nakazawa, K. Kawata, K.-I. Tomochika, K. Tobe, and S. Shinoda. 1998. Characterization of the hemorrhagic reaction caused by *Vibrio vulnificus* metalloprotease, a member of the thermolysin family. *Infect. Immun.* 66:4851–4855.

86. Molenda, J.R., W.G. Johnson, M. Fishbein, B. Wentz, I.J. Mehlman, and T.A. Dadisman, Jr. 1972. *Vibrio parahaemolyticus* gastroenteritis in Maryland: Laboratory aspects. *Appl. Microbiol.* 24:444–448.

87. Morris, J.G., R. Wilson, B.R. Davis, I.K. Wachsmuth, C.F. Riddle, H.G. Wathen, R.A. Pollard, and P.A. Blake. 1981. Non-O group 1 *Vibrio cholerae* gastroenteritis in the United States. *Ann. Intern. Med.* 94:656–658.

88. Mors, V., and C.H. Pai. 1980. Pathogenic properties of *Yersinia enterocolitica. Infect. Immun.* 28:292–294.

89. Nadeau, E., S. Messier, and S. Quessy. 2003. Comparison of *Campylobacter* isolates from poultry and humans: Association between in vitro virulence properties, biotypes, and pulsed-field gel electrophoresis clusters. *Appl. Environ. Microbiol.* 69:6316–6320.

90. Nakasone, N., and M. Iwanaga. 1990. Pili of *Vibrio parahaemolyticus* strain as a possible colonization factor. *Infect. Immun.* 58:61–69.

91. National Advisory Committee on Microbiological Criteria for Foods. 1994. *Campylobacter jejuni/coli. J. Food Protect.* 57:1101–1121.

92. Nishibuchi, M., S. Doke, S. Toizumi, T. Umeda, M. Yoh, and T. Miwatani. 1988. Isolation from a coastal fish of *Vibrio hollisae* capable of producing a hemolysin similar to the thermostable direct hemolysin of *Vibrio parahaemolyticus. Appl. Environ. Microbiol.* 54:2144–2146.

93. Nishibuchi, M., M. Ishibashi, Y. Takeda, and J.B. Kaper. 1985. Detection of the thermostable direct hemolysin gene and related DNA sequences in *Vibrio parahaemolyticus* and other *Vibrio* species by the DNA colony hybridization test. *Infect. Immun.* 49:481–486.

94. Nishibuchi, M., and J.B. Kaper. 1985. Nucleotide sequence of the thermostable direct hemolysin gene of *Vibrio parahaemolyticus. J. Bacteriol.* 162:558–564.

95. Ogawa, A., J.-I. Kato, H. Watanabe, B.G. Nair, and T. Takeda. 1990. Cloning and nucleotide sequence of a heat-stable enterotoxin gene from *Vibrio cholerae* non-01 isolated from a patient with traveler's diarrhea. *Infect. Immun.* 58:3325–3329.

96. Ogg, J.E., R.A. Ryder, and H.L. Smith, Jr. 1989. Isolation of *Vibrio cholerae* from aquatic birds in Colorado and Utah. *Appl. Environ. Microbiol.* 55:95–99.

97. Okamoto, K., T. Inoue, H. Ichikawa, Y. Kawamoto, and A. Miyama. 1981. Partial purification and characterization of heat-stable enterotoxin produced by *Yersinia enterocolitica. Infect. Immun.* 31:554–559.

98. Okamoto, K., T. Inoue, K. Shimizu, S. Hara, and A. Miyama. 1982. Further purification and characterization of heat-stable enterotoxin produced by *Yersinia enterocolitica. Infect. Immun.* 35:958–964.

99. Pace, J., and T.-J. Chai. 1989. Comparison of *Vibrio parahaemolyticus* grown in estuarine water and rich medium. *Appl. Environ. Microbiol.* 55:1877–1887.

100. Pai, C.H., and V. Mors. 1978. Production of enterotoxin by *Yersinia enterocolitica. Infect. Immun.* 119:908–911.

101. Pai, C.H., V. Mors, and S. Toma. 1978. Prevalence of enterotoxigenicity in human and nonhuman isolates of *Yersinia enterocolitica. Infect. Immun.* 22:334–338.

102. Pearson, A.D., M.H. Greenwood, R.K.A. Feltham, T.D. Healing, J. Donaldson, D.M. Jones, and R.R. Colwell. 1996. Microbial ecology of *Campylobacter jejuni* in a United Kingdom chicken supply chain: Intermittent common source, vertical transmission, and amplication by flock propagation. *Appl. Environ. Microbiol.* 62:4614–4620.

103. Peffer, C.S., M.F. Hite, and J.D. Oliver. 2003. Ecology of *Vibrio vulnificus* in estaurine waters of eastern North Carolina. *Appl. Environ. Microbiol.* 69:3526–3531.

104. Popovic, T., P.I. Fields, O. Olsvik, J.G. Wells, G.M. Evins, D.N. Cameron, J.J. Farmer III, C.A. Bopp, K. Wachsmuth, R.B. sack, M.J. Albert, G.B. Nair, T. Shimada, and J.C. Feeley. 1995. Molecular subtyping of toxigenic *Vibrio cholerae* 0139 causing epidemic cholera in India and Bangladesh, 1992–1993. *J. Infect. Dis.* 171:122–127.

105. Reynolds, G.N., and F.A. Draughon. 1987. *Campylobacter jejuni* in vacuum packaged processed turkey. *J. Food Protect.* 50:300–304.

106. Rippey, S.R. 1994. Infectious diseases associated with molluscan shellfish consumption. *Clin. Microbiol. Rev.* 7:419–425.

107. Robins-Browne, R.M., C.S. Still, M.D. Miliotis, and H.J. Koornhof. 1979. Mechanism of action of *Yersinia enterocolitica* enterotoxin. *Infect. Immun.* 25:680–684.

108. Ruiz-Palacios, G.M., J. Torres, E. Escamilla, B.R. Ruiz-Palacios, and J. Tamayo. 1983. Cholera-like enterotoxin produced by *Campylobacter jejuni. Lancet* 2:250–253.

109. Russell, R.G., and D.C. Blake, Jr. 1994. Cell association and invasion of Caco-2 cells by *Campylobacter jejuni. Infect. Immun.* 62:3773–3779.

110. Sakazaki, R. 1979. *Vibrio* infections. In *Food-Borne Infections and Intoxications*, ed. H. Riemann and F.L. Bryan, 173–209. New York: Academic Press.

111. Sakazaki, R. 1983. *Vibrio parahaemolyticus* as a food-spoilage organism. In *Food Microbiology*, ed. A.H. Rose, 225–241. New York: Academic Press.

112. Schiemann, D.A. 1981. An enterotoxin-negative strain of *Yersinia enterocolitica* serotype 0:3 is capable of producing diarrhea in mice. *Infect. Immun.* 32:571–574.

113. Shanker, S., J.A. Rosenfield, G.R. Davey, and T.C. Sorrell. 1982. *Campylobacter jejuni*: Incidence in processed broilers and biotype distribution in human and broiler isolates. *Appl. Environ. Microbiol.* 43:1219–1220.

114. Shayegani, M., I. Deforge, D.M. McGlynn, and T. Root. 1981. Characteristics of *Yersinia enterocolitica* and related species isolated from human, animal and environmental sources. *J. Clin. Microbiol.* 14:304–312.

115. Shirai, H., H. Ito, T. Hirayama, Y. Nakamoto, N. Nakabayashi, K. Kumagai, Y. Takeda, and M. Nishibuchi. 1990. Molecular epidemiologic evidence for association of thermostable direct hemolysin (TDH) and TDH-related hemolysin of *Vibrio parahaemolyticus* with gastroenteritis. *Infect. Immun.* 58:3568–3573.

116. Smith, J.L. 2002. *Campylobacter jejuni* infection during pregnancy: Long-term consequences of associated bacteremia, Guillain–Barré syndrome, and reactive arthritis. *J. Food Protect.* 65:696–708.

117. Smitherman, R.E., C.A. Genigeorgis, and T.B. Farver. 1984. Preliminary observations on the occurrence of *Campylobacter jejuni* at four California chicken ranches. *J. Food Protect.* 47:293–298.

118. Sperandio, V., J.A. Girón, W.D. Silveira, and J.B. Kaper. 1995. The OmpU outer membrane protein, a potential adherence factor of *Vibrio cholerae*. *Infect. Immun.* 63:4433–4438.

119. Stelma, G.N., Jr., P.L. Spaulding, A.L. Reyes, and C.H. Johnson. 1988. Production of enterotoxin by *Vibrio vulnificus* isolates. *J. Food Protect.* 51:192–196.

120. Stern, N.J., and A.W. Kotula. 1982. Survival of *Campylobacter jejuni* inoculated into ground beef. *Appl. Environ. Microbiol.* 44:1150–1153.

121. Stern, N.J., M.D. Pierson, and A.W. Kotula. 1980. Effects of pH and sodium chloride on *Yersinia enterocolitica* growth at room and refrigeration temperatures. *J. Food Sci.* 45:64–67.

122. Sulakvelidze, A., A. Kreger, A. Joseph, R.M. Robins-Browne, A. Fasang, G. Wauters, N. Harnet, L. DeTolla, and J.G. Morris, Jr. 1999. Production of enterotoxin by *Yersinia bercovieri*, a recently identified *Yersinia enterocolitica*-like species. *Infect. Immun.* 67:968–971.

123. Sun, Y., and J.D. Oliver. 1995. Hot sauce: No elimination of *Vibrio vulnificus* in oysters. *J. Food Protect.* 58:441–442.

124. Sun, Y., and J.D. Oliver. 1994. Effects of GRAS compounds on natural *Vibrio vulnificus* populations in oysters. *J. Food Protect.* 57:921–923.

125. Swaminathan, B., M.C. Harmon, and I.J. Mehlman. 1982. *Yersinia enterocolitica*. *J. Appl. Bacteriol.* 52:151–183.

126. Takeda, T., G.B. Nair, K. Suzuki, and Y. Shimonishi. 1990. Production of a monoclonal antibody to *Vibrio cholerae* non-01 heat-stable enterotoxin (ST) which is cross-reactive with *Yersinia enterocolitica* ST. *Infect. Immun.* 58:2755–2759.

127. Takeda, Y. 1983. Thermostable direct hemolysin of *Vibrio parahaemolyticus*. *Pharm. Ther.* 19:123–146.

128. Taylor, J.L., J. Tuttle, T. Pramukul, K. O'Brien, T.J. Barrett, B. Jolbitado, Y.L. Lim, D. Vugia, J. Glenn Morris, Jr., R.V. Tauxe, and D.M. Dwyer. 1993. An outbreak of cholera in Maryland associated with imported commercial frozen fresh coconut milk. *J. Infect. Dis.* 167:1330–1335.

129. Tibana, A., M.B. Warnken, M.P. Nunes, L.D. Ricciardi, and A.L.S. Noleto. 1987. Occurrence of *Yersinia* species in raw and pasteurized milk in Rio de Janeiro, Brazil. *J. Food Protect.* 50:580–583.

130. Toma, S., and L. Lafleur. 1974. Survey on the incidence of *Yersinia enterocolitica* infection in Canada. *Appl. Microbiol.* 28:469–473.

131. Twedt, R.M., J.T. Peeler, and P.L. Spaulding. 1980. Effective ileal loop dose of Kanagawa-positive *Vibrio parahaemolyticus*. *Appl. Environ. Microbiol.* 40:1012–1016.

132. Twedt, R.M., P.L. Spaulding, and H.E. Hall. 1969. Morphological, cultural, biochemical, and serological comparison of Japanese strains of *Vibrio parahaemolyticus* with related cultures isolated in the United States. *J. Bacteriol.* 98:511–518.

133. Vanderzant, C., and R. Nickelson. 1972. Survival of *Vibrio parahaemolyticus* in shrimp tissue under various environmental conditions. *Appl. Microbiol.* 23:34–37.

134. Vugia, D.J., A.M. Shefer, J. Douglas, K.D. Greene, R.G. Bryant, and S.B. Werner. 1997. Cholera from raw seaweed transported from the Philippines to California. *J. Clin. Microbiol.* 35:284–285.

135. Walker, S.J., and A. Gilmour. 1990. Production of enterotoxin by *Yersinia* species isolated from milk. *J. Food Protect.* 53:751–754.

136. Warnken, M.B., M.P. Nunes, and A.L.S. Noleto. 1987. Incidence of *Yersinia* species in meat samples purchased in Rio de Janeiro, Brazil. *J. Food Protect.* 50:578–579, 583.

137. Wright, A.C., and J.G. Morris, Jr. 1991. The extracellular cytolysin of *Vibrio vulnificus*: Inactivation and relationship to virulence in mice. *Infect. Immun.* 59:192–197.

138. Wu, V.C.H., D.Y.C. Fung, and R.D. Oberst. 2004. Evaluation of a 5′-nuclease (TaqMan) assay with the thin agar layer oxyrase method for the detection of *Yhersinia enterocolitica* in ground pork samples. *J. Food Protect.* 67:271–277.

139. Yamamoto, K., Y. Ichinose, H. Shinagawa, K. Makino, A. Nakata, M. Iwanaga, T. Honda,and T. Miwatani. 1990. Two-step processing for activation of the cytolysin/hemolysin of *Vibrio cholerae* 01 biotype E1 Tor: Nucleotide sequence of the structural gene (*hly*A) and characterization of the processed products. *Infect. Immun.* 58:4106–4116.

140. Yamamoto, K., A.C. Wright, J.B. Kaper, and J.G. Morris, Jr. 1990. The cytolysin gene of *Vibrio vulnificus*: Sequence and relationships to the *Vibrio cholerae* E1 Tor hemolysin gene. *Infect. Immun.* 58:2706–2709.

141. Yamamoto, T., and T. Yokota. 1989. Adherence targets of *Vibrio parahaemolyticus* in human small intestines. *Infect. Immun.* 57:2410–2419.

142. Yuki, N., T. Taki, M. Takahashi, K. saito, T. Tai, T. Miyatake, and S. Hande. 1994. Penner's serotype 4 of *Campylobacter jejuni* has a lipopolysaccharide that bears a GM1 ganglioside epitope as well as one that bears a GD1a epitope. *Infect. Immun.* 62:2101–2103.

143. Zen-Yoji, H., Y. Kudoh, H. Igarashi, K. Ohta, and K. Fukai. 1975. Further studies on characterization and biological activities of an enteropathogenic toxin of *Vibrio parahaemolyticus*. *Toxicon* 13:134–135.

CHAPTER 29

Foodborne Animal Parasites

The animal parasites that can be contracted by eating certain foods belong to three distinct groups: protozoa, flatworms, and roundworms. Several of the more important members of each group of concern in human foods are examined in this chapter along with their classification.

In contrast to foodborne bacteria, animal parasites do not proliferate in foods, and their presence must be detected by direct means, as they cannot grow on culture media. Because all are larger in size than bacteria, their presence can be detected rather easily by the use of appropriate concentration and staining procedures. Because many are intracellular pathogens, resistance to these diseases is often by cellular phenomena similar to that for listeriosis (see Chapter 25). Finally, another significant way in which some animal parasites differ from bacteria is their requirement for more than one animal host in which to carry out their life cycles. The *definitive host* is the animal in which the adult parasite carries out its sexual cycle; the *intermediate host* is the animal where larval or juvenile forms develop. In some instances, there is only one definitive host (e.g., cryptosporidiosis); in others, more than one animal can serve as definitive host (e.g., diphyllobothriasis); and in still other cases, both larval and adult stages reside in the same host (e.g., trichinosis).

PROTOZOA

The protozoa belong to the kingdom Protista (Protoctista), which also comprises the algae and flagellate fungi. They are the smallest and most primitive of animal forms, and the five genera of concern in foods are classified as follows:

Kingdom Protista
  Phylum Sarcomastigophora
    Phylum Sarcomastigophora
      Class Zoomastigophorea
        Order Diplomonadida
          Family Hexamitidae
            Genus *Giardia*

Subphylum Sarcodina
  Superclass Rhizopoda
    Class Lobosea

Order Amoebida
  Family Endamoebidae
    Genus *Entamoeba*

Phylum Apicomplexa (= Sporozoa)
  Class Sporozoea
    Class Lobosea
      Order Eucoccidiida
        Family Sarcocystidae
          Genus *Toxoplasma*
          Genus *Sarcocystis*
        Family Cryptosporidiidae
          Genus *Cryptosporidium*
          Genus *Cyclospora*

Giardiasis

Giardia lamblia is a flagellate protozoan that exists in environmental waters at a higher level than *Entamoeba histolytica*. The protozoal cells (trophozoites) produce cysts, which are the primary forms in water and foods. The cysts are pear shaped, with a size range of 8–20 μm in length and 5–12 μm in width. The trophozoites have eight flagella that arise on the ventral surface near the paired nuclei and give rise to "falling-leaf" motility.

Upon ingestion, *Giardia* cysts excyst in the gastrointestinal tract with the aid of stomach acidity and proteases and give rise to clinical giardiasis in some individuals. Excystation of the trophozoites occurs somewhere in the upper small intestine, and this step is regarded as being equivalent to a virulence factor.[9] The trophozoites are not actively phagocytic, and they obtain their nutrients by absorption. Occasionally, bile ducts are invaded, leading to cholecystitis. Compared to some of the other intestinal protozoal parasites, *Giardia* trophozoites do not penetrate deeply in parenteral tissues.

Environmental Distribution

Water is the second most common source of giardiasis. The first recorded outbreak occurred at a ski resort in Aspen, Colorado, in 1965 with 123 cases.[22] Between 1965 and 1977, 23 waterborne outbreaks were recorded that affected over 7000 persons.[23] Between 1971 and 1985, 92 outbreaks were reported in the United States.[22] *Giardia* cysts are generally resistant to the levels of chlorine used in the water supply. Beavers and muskrats have been shown to be the major sources of this organism in bodies of water. In a study of 220 muskrat fecal specimens collected from natural waters in southwestern New Jersey, 70% contained *Giardia* cysts.[59] It is estimated that up to 15% of the U.S. population is infected with this organism.

Syndrome, Diagnosis, and Treatment

The incubation period for clinical giardiasis is 1–4 weeks, and cysts appear in stools after 3–4 weeks. Asymptomatic cyst passage is the most benign manifestation of *G. lamblia* infection in humans, but when clinical giardiasis occurs, symptoms may last from several months to a year or more. Up to 9.0 × 10^8 cysts are shed each day by patients, and they may survive as long as 3 months in sewage sludge.[3]

G. lamblia is generally noninvasive, and malabsorption often accompanies the symptomatic disease.[96] Growth of the organism is favored by the high bile content in the duodenum and upper jejunum.[79]

From an outbreak of giardiasis among 1400 Americans on the Madeira Island in 1976, the symptoms, along with the percentage incidence among victims, were as follows: abdominal cramps (75%), abdominal distention (72%), nausea (70%), and weight loss (40%). The median incubation period was 4 days, and *G. lamblia* was recovered from 47% of 58 ill patients. The consumption of tapwater and the eating of ice cream or raw vegetables were significantly associated with the illness.[66] The 29 victims of the 1979–1980 outbreak traced to home-canned salmon (see below) displayed the following symptoms: diarrhea (100%), fatigue (97%), abdominal cramps (83%), fever (21%), vomiting (17%), and weight loss (59%), among others.[77] From another study of 183 patients, the five leading symptoms (and percentage complaining) were diarrhea (92%), cramps (70%), nausea (58%), fever (28%), and vomiting (23%).[96] Weight loss of about 5 lb or so is a common feature of giardiasis, and it was associated with the 1985 outbreak traced to noodle salad.[78]

Giardiasis is a highly contagious disease. It has been documented in daycare centers where unsanitary conditions prevailed. The human infection rate ranges from 2.4 to 67.5%.[19] The minimum infectious dose of *G. lamblia* cysts for humans is 10 or less.[82]

Giardiasis is diagnosed by the demonstration of trophozoites in stool specimen by microscopic examinations using either wet mounts or stained specimens. *G. lamblia* can be grown in axenic culture, but this does not lend itself to rapid diagnosis. Effective enzyme-linked immunosorbent assay (ELISA) tests have been developed. Both circulating antibodies and T lymphocytes are elicited during infection by *G. lamblia*. Because no enterotoxin has been demonstrated, diarrhea is caused by other factors.[96]

The drug of choice for the treatment of giardiasis is quinacrine, an acridine derivative. Also effective are metronidazole and tinidazole.[96]

Incidence in Foods and Foodborne Cases

Giardia has been shown to occur in some vegetables, and it may be presumed that the organism occurs on foods that are washed with contaminated water or contaminated by unsanitary asymptomatic carriers. Of 64 heads of lettuce examined in Rome, Italy, in 1968, 48 contained *Giardia* cysts, and cysts were recovered from strawberries grown in Poland in 1981.[3]

As early as 1928, it was suggested that hospital food handlers were the likely source of protozoal infections of patients. Of 844 private patients in an urban center, 36% contracted giardiasis, and it was believed the infections were acquired by eating cyst-contaminated raw fruits and vegetables. These and some other early incidences of possible foodborne giardiasis have been discussed by Barnard and Jackson.[3] Following is a list of suspected and proved foodborne giardiasis:

1. Three of four members of a family who in 1960 ate Christmas pudding thought to have been contaminated by rodent feces became victims.[21] *Giardia*-like cysts were found.
2. In their surveillance of foodborne diseases in the United States for 1968–1969, Gangarosa and Donadio[39] recorded an outbreak of giardiasis with 19 cases for 1969 but provided no further details.
3. In December 1979, 29 of 60 school employees in a rural Minnesota community contracted the disease from home-canned salmon prepared by a worker after changing the diaper of an infant later shown to have an asymptomatic *Giardia* infection.[77] This was the first well-documented common-source outbreak.

4. In July 1985, 13 of 16 individuals at a picnic in Connecticut met the case definition of giardiasis, and the most likely vehicle food was a noodle salad.[78] Although most victims developed symptoms between 6 and 20 days after the picnic, the salad preparer became ill the day after the food was eaten by others. This was the second well-documented common-source outbreak traced to a food product.
5. In 1988, 21 of 108 members of a church youth group in Albuquerque, New Mexico, were victims. Taco ingredients were the most likely vehicles from dinners prepared by parents at a church.[18]

The U.S. Centers for Disease Control and Prevention (CDCP) recorded foodborne giardiases outbreaks in 1985 and 1986, with 1 outbreak and 13 cases in 1985 and 2 outbreaks and 28 cases in 1986.[5] The common occurrence of this organism suggests that it may be a more frequent cause of foodborne infection than is reported. The incubation period of 7 days plus could be a factor in the apparent underreporting. Another possible factor is the need to demonstrate the organism in stools and leftover foods by microscopic examination, a practice that is not routine in the microbiological examination of foods in foodborne gastroenteritis outbreaks.

Amebiasis

Amebiasis (amoebic dysentery), caused by *Entamoeba histolytica*, is often transmitted by the fecal–oral route, although transmission is known to occur by water, food handlers, and foods. According to Jackson[48] there is better documentation of food transmission of amebic dysentery than for the other intestinal protozoal diseases. The organism is unusual in being anaerobic, and the trophozoites (ameba stages) lack mitochondria. It is an aerotolerant anaerobe that requires glucose or galactose as its main respiratory substrate.[69] The trophozoites of *E. histolytica* range in size from 10 to 60 μm, whereas the cysts usually range between 10 and 20 μm. The trophozoites are motile; the cysts are not. It is often found with *Entamoeba coli*, with which it is associated in the intestine and stools. In warm stools from a case of active dysentery, *E. histolytica* is actively motile and usually contains red blood cells that the protozoan ingests by pseudopodia. Although generally outnumbered in stools by *Entamoeba coli*, the latter never ingests red blood cells. Although the trophozoites do not persist under environmental conditions, the encysted forms can survive as long as 3 months in sewage sludge.[3] A person with this disease may pass up to 4.5×10^7 cysts each day.[3]

The possible transmission of cysts to foods becomes a real possibility when poor personal restroom hygiene is practiced. The incidence of amebiasis varies widely, with a rate of 1.4% reported for Tacoma, Washington, to 36.4% in rural Tennessee.[19] It is estimated that 10% of the world's population is infected with *E. histolytica* and that up to 100 million cases of amebic colitis or liver abscesses occur each year.

In its trophozoite stage, the organism induces infection in the form of abscesses in intestinal mucosal cells and ulcers in the colon. Its adherence to host-cell glycoproteins is mediated by a galactose-specific lectin. It reproduces by binary fission in the large intestine. It encysts in the ileum, and cysts may occur free in the lumen. The organism produces an enterotoxic protein with molecular weight of 35,000–45,000 Da.[19]

Syndrome, Diagnosis, and Treatment

The incubation period for amebiasis is 2–4 weeks, and symptoms may persist for several months. Its onset is often insidious, with loose stools and generally no fever. Mucus and blood are characteristic of stools from patients. Later symptoms consist of pronounced abdominal pain, fever, severe diarrhea, vomiting, and lumbago, and somewhat resemble those of shigellosis. Weight loss is common, and all patients have heme-positive stools. According to Jackson,[48] fulminating amebiasis with ulceration of the colon and toxicity occur in 6–11% of cases, especially in women stressed by pregnancy or nursing. Masses of amebae and mucus may form in the colon, leading to intestinal obstruction. Amebiasis may last in some individuals for many years, in contrast to giardiasis, where disease symptoms rarely exceed 3 months.[3] Under some conditions, amebiasis may result from a synergistic relationship with certain intestinal bacteria.

Amebiasis is diagnosed by demonstrating trophozoites and cysts in stools or mucosal scrapings. Immunological methods such as indirect hemagglutination, indirect immunofluorescence, latex agglutination, and ELISA are useful. The sensitivity of these tests is high with extraintestinal amebiasis, and a titer of 1:64 by indirect hemagglutination is considered significant.

This syndrome can be treated with the amebicidal drugs metronidazole and chloroquine. Resistance is mediated by cell immunity. Lymphocytes from patients in the presence of *E. histolytica* antigens have been shown to produce γ-interferon, which activates macrophages that display amebicidal properties.[90]

Toxoplasmosis

This disease is caused by *Toxoplasma gondii*, a coccidian protozoan that is an obligate intracellular parasite. The generic name is based on the characteristic shape of the ameba stage of the protozoan (Gr. *toxo*, "arc"). It was first isolated in 1908 from an African rodent, the gondi—hence, its species name. In most instances, the ingestion of *T. gondii* oocysts causes no symptoms in humans, or the infection is self-limiting. In these cases, the organism encysts and becomes latent. However, when the immunocompetent state is abated, life-threatening toxoplasmosis results from the breaking out (recrudescence) of the latent infection.

Domestic and wild cats are the only definitive hosts for the intestinal or sexual phase of this organism, making them the primary sources of human toxoplasmosis. Normally, the disease is transmitted from cat to cat, but virtually all vertebrate animals are susceptible to the oocysts shed by cats. As few as 100 oocysts can produce clinical toxoplasmosis in humans, and the oocysts can survive over a year in warm, moist environments.[33] Pigs are the major animal food source to humans.

Symptoms, Diagnosis, and Treatment

In most individuals, toxoplasmosis is symptomless, but when symptoms occur, they consist of fever with rash, headache, muscle aches and pain, and swelling of the lymph nodes. The muscle pain, which is rather severe, may last up to a month or more. At times, some of the symptoms mimic infectious mononucleosis. The incubation period in adults is 6–10 days while in infants it is congenital.

The disease is initiated upon the ingestion of oocysts (if from cat feces), which pass to the intestine where digestive enzymes affect the release of the eight motile sporozoites. Oocysts are ovoid shaped, measure 10–12 μm in diameter, and possess a thick wall. Sporozoites are crescent shaped and measure about 3×7 μm. They cannot survive for long outside animal host tissues, nor can they survive the activities of the stomach. When freed in the intestines, these forms pass through intestinal walls and multiply rapidly in many other parts of the body, giving rise to clinical symptoms. The most rapidly multiplying forms are designated tachyzoites (Gr. *tachy*, "rapid"), and in immunocompetent individuals, they eventually give rise to clusters that are surrounded by a protective wall. This is a tissue cyst, and the protozoa inside are designated bradyzoites (Gr. *bradus*, "slow"). These cysts are

10–200 μm in diameter, and the bradyzoites are smaller in size than the more active tachyzoites. Bradyzoites may persist in the body for the lifetime of an individual, but if the cysts are mechanically broken or break down under immunosuppression, bradyzoites are freed and begin to multiply rapidly as tachyzoites and thus bring on another active infection. The development of a cyst wall around bradyzoites coincides with the development of permanent host immunity. The cysts are normally intracellular in host cells. *T. gondii* infections are asymptomatic in the vast majority of human cases (immunocompetents), but in congenital infections and in immunocompromised hosts, such as patients with acquired immunodeficiency syndrome (AIDS), the disease is much more severe. In pregnant mothers with newly contracted toxoplasmosis, the tachyzoites are reported to cross the placenta about 45% of the time.

Unlike certain other intestinal protozoal diseases, toxoplasmosis cannot be diagnosed by demonstrating oocysts in stools, as these forms occur only in cat feces. Various serological methods are widely used to diagnose acute infection. A fourfold rise in immunoglobulin G (IgG) antibody titer between acute and convalescent serum specimens is indicative of acute infection. A more rapid confirmation of acute infection can be made by the detection of immunoglobulin M (IgM) antibodies, which appear during the first week of infection and peak during the second to fourth weeks.[80] Among other diagnostic methods are the methylene blue dye test, indirect hemagglutination, indirect immunofluorescence, and immunoelectrophoresis. With the indirect hemagglutination test, antibody titers above 1:256 are generally indicative of active infection.

Although toxoplasma infection induces protective immunity, it is, in part, cell mediated. In many bacterial infections where phagocytes ingest the cells, their internal granules release enzymes that destroy the bacteria. During this process, aerobic respiration gives way to anaerobic glycolysis, which results in the formation of lactic acid and the consequent lowering of pH. The latter contributes to the destruction of the ingested bacteria along with the production of superoxide, which, at the acid pH, yields singlet oxygen ($^1O_2$). The latter is quite toxic. *T. gondii* tachyzoites are unusual in that once they are phagocytosed, the production of H_2O_2 is not triggered, and neither do the acid pH nor the singlet oxygen events occur. Also, they reside in vacuoles of phagocytes that do not fuse with preexisting secondary lysosomes. Thus, it appears that their mode of pathogenicity involves an alteration of phagocyte membranes in such a way that they fail to fuse with other endocytic or biosynthetic organelles, in addition to the other events noted.[55] T cells play a role in immunity to *T. gondii*, and this has been demonstrated by use of nude rats, where T cells from *T. gondii*-infected normal rats conferred to nude rats the ability to resist infection by a highly virulent strain of *T. gondii*.[29]

Antimicrobial therapy for toxoplasmosis consists of sulfonamides, pyrimethamine, pyrimethamine plus clindamycin, or fluconazole. Pyrimethamine is a folic acid antagonist that inhibits dihydrofolate reductase.

Distribution of *T. gondii*

Toxoplasmosis is regarded as a universal infection, with the incidence being higher in the tropics and lower in colder climes. It is estimated that 50% of Americans have circulating antibodies to *T. gondii* by the time of adulthood.[80] In a study of U.S. Army recruits, 13% were positive for toxoplasma antibodies.[36] In the United States, it is estimated that over 3000 babies are infected each year with *T. gondii* because their mothers acquire the infection during pregnancy.[33] Fetal infections occur in 17% of first-trimester and 65% of third-trimester cases, with the first-trimester cases being more severe.[80] Among 3000 pregnant women tested for *T. gondii* antibodies, 32.8% were positive.[58]

Table 29–1 Estimated Number of Clinically Significant Cases of Protozoal Infections in the United States, 1985

Infections	Cases
Amebiasis	12,000
Cryptosporidiosis	50
Giardiasis	120,000
Toxoplasmosis*	2,300,000

*Excluding congenital.

Source: From Bennett et al.[7]

Extensive surveys of *T. gondii* antibodies in meat animals have been reviewed by Fayer and Dubey[33] who reported that of more than 16,000 cattle surveyed, an average of 25% contained antibodies, and infectious cysts administered from cats persisted as long as 267 days, with most being found in the liver. In more than 9000 sheep, an average of 31% had antibodies, and oocysts administered persisted 173 days, with most protozoa found in the heart. Similarly for pigs, 29% had antibodies, and oocysts persisted for 171 days, with most in the brain and heart, whereas for goats, oocysts persisted in the animals for 441 days, with most being found in skeletal muscles. Because the meat animals noted are herbivores, Fayer and Dubey[33] concluded that the contamination of feed and water with oocysts from cat feces must be the ultimate source of infection, aided by the practice on some farms of keeping cats to kill mice.

Food-Associated Cases

The number of cases of toxoplasmosis that are contracted from foods is unknown, but the estimated number in the United States from all sources for 1985 has been put at 2.3 million (Table 29–1). This estimated number far exceeds the recorded cases for the total of all other protozoal diseases.

Fresh meats may contain toxoplasma oocysts. As early as 1954, undercooked meat was suspected to be the source of human toxoplasmosis.[51] In a study in 1960 of freshly slaughtered meats, 24% of 50 porcine, 9.3% of 86 ovine, but only 1 of 60 bovine samples contained oocysts.[53] *T. gondii* is more readily isolated from sheep than other meat animals.[51] The following cases have been proved or suspected:

1. In France in the early 1960s, 31% of 641 children in a tuberculosis hospital became seropositive for *T. gondii* after admission. When two additional meals per day of undercooked mutton were served, toxoplasmosis cases doubled. The investigators concluded that the custom of this hospital to feed undercooked meat was the cause of the high number of infections.[28] At an educational institution, 771 mothers were questioned about their preferences for meat. Of those who preferred well-done meat, 78% had toxoplasma antibody; of those who liked less well-done meats, 85% were antibody positive; and of those who ate meat rare or raw, 93% had toxoplasma antibodies.[28] The investigators were unable to make distinctions among beef, mutton, or horse meat. They further noted that 50% of children in France are infected with *T. gondii* before age 7 and believe this is due to the consumption of undercooked meats.

2. Eleven of 35 medical students in New York City in 1968 had an increase in toxoplasma antibodies following the consumption of hamburger cooked rare at the same snack bar, and 5 contracted clinical toxoplasmosis.[58]
3. In 1974, a 7-month-old infant who consumed unpasteurized goat's milk developed clinical toxoplasmosis. Although *T. gondii* could not be recovered from milk, some goats in the herd had antibody titers to *T. gondii* as high as 1:512, and the child had a titer of over 1:16,000.[83]
4. In 1978, 10 of 24 members of an extended family in northern California contracted toxoplasmosis after drinking raw milk from infected goats.[89]
5. In São Paulo, Brazil, 110 university students suffered acute toxoplasmosis after eating undercooked meat.[19]

Since most of the above cases have been traced to meats, the consumption of raw or undercooked meats carries the risk of this infection. Other documented meatborne outbreaks have been reviewed.[95]

Control

Toxoplasmosis in humans can be prevented by avoiding environmental contamination with cat feces (from cat litter boxes, for example) and by avoiding the consumption of meat and meat products that contain viable tissue cysts. The cysts of *T. gondii* can be destroyed by heating meats above 60°C or by irradiating at a level of 30 krad (0.3 kGy) or higher.[33] The organism may be destroyed by freezing, but because the results are variable, freezing should not be relied on to inactivate oocysts. See reference 8 for a detailed review of the life cycle of *Toxoplasma gondii*.

Sarcocystosis

Of more than 13 known species of the genus *Sarcocystis*, two are known to cause an extraintestinal disease in humans. One of these is obtained from cattle (*S. hominis*) and the other from pigs (*S. suihominis*). Humans are the definitive hosts for both species: the intermediate host for *S. hominis* is bovines, and pigs for *S. suihominis*.

When humans ingest a sarcocyst, bradyzoites are released and penetrate the lamina propria of the small intestine, where sexual reproduction occurs that leads to sporocysts. The latter pass out of the bowel in feces. When sporocysts are ingested by pigs or bovines, the sporozoites are released and spread throughout the body. They multiply asexually and lead to the formation of sarcocysts in skeletal and cardiac muscles. In this stage, they are sometimes referred to as Miescher's tubules. The bradyzoite-containing sarcocysts are visible to the unaided eye and may reach 1 cm in diameter.[19]

Several studies have been conducted to determine the relative infectivity of *Sarcocystis* spp. Of 20 human volunteers in five studies who ate raw beef infected with *S. hominis*, 12 became infected and shed oocysts, but only 1 had clinical illness.[34] Symptoms occurred within 3–6 hours and consisted of nausea, stomachache, and diarrhea. In 15 other volunteers who ate raw pork infected with *S. suihominis*, 14 became infected and shed oocysts, and 12 of these had clinical illness 6–48 hours after eating the pork.[34] Six who ate well-cooked pork did not contract the disease. In another study of another species of *Sarcocystis*, dogs did not become infected when fed beef cooked medium (60°C) or well done (71.1–74.4°C), but the beef was infective when fed raw or cooked rare (37.8–53.3°C). Dogs fed the same raw beef after storage for 1 week in a home freezer did not become infected.[35] In another study, two human volunteers passed sporocysts for 40 days after eating 500 g of raw ground beef diaphragm muscle infected with *Sarcosporidia*.[85]

Because bovine and porcine animals serve as intermediate hosts for these parasites, their potential as foodborne pathogens to humans is obvious.

Cryptosporidiosis

The protozoan *Cryptosporidium parvum* was first described in 1907 in asymptomatic mice, and for decades now it has been known to be a pathogen of at least 40 mammals and varying numbers of reptiles and birds. Although the first documented human case was not recorded until 1976, this disease has a worldwide prevalence of 1–4% among patients with diarrhea,[104] and it appears to be increasing. In England and Wales for the 5 years 1985–1989, the numbers of identified cases were 1874, 3694, 3359, 2838, and 7769, respectively.[2] This disease was the fourth most frequent cause of diarrhea during the period noted. It is estimated to cause infections in from 7 to 38% of AIDS patients in some hospitals.[104] The prevalence of *C. parvum* in diarrheal stools is similar to that of *Giardia lamblia*.[104] In humans, the disease is self-limiting in immunocompetent individuals, but it is a serious infection in the immunocompromised, such as AIDS patients. The protozoan is known to be present in at least some bodies of water (see below) and thus exists the potential for food transmission. The fecal–oral route of transmission is the most important, but indirect transmission by food and milk is known to occur.

C. parvum is an obligate intracellular coccidian parasite that carries out its life cycle in one host. Following ingestion of the thick-walled oocysts, they excyst in the small intestine and free sporozoites and penetrate the microvillous region of host enterocytes, where sexual reproduction leads to the development of zygotes. They invade host cells by disrupting their own membrane as well as that of the host. Host cell actin polymerization at the interface between the parasite and the host cell cytoplasm has been found to be necessary for infection.[30] About 80% of the zygotes form thick-walled oocysts that sporulate within host cells.[24] The environmentally resistant oocysts are shed in feces, and the infection is transmitted to other hosts when they are ingested.

The oocysts of *C. parvum* are spherical to ovoid and average 4.5–5.0 μm in size. Each sporulated oocyst contains four sporozoites. The oocysts are highly resistant in the natural environment and may remain viable for several months when kept cold and moist.[24] They have been reported to be destroyed by treatments with 50% or more ammonia and 10% or more formalin for 30 minutes.[24] The latter investigator has reported that temperatures above 60°C and below −20°C may kill *C. parvum* oocysts. The organism is destroyed by high-temperature, short time (HTST) milk pasteurization. Holding oocysts at 45°C for 5–20 minutes has been reported to destroy their infectivity.[1] In one study, infectivity was lost after 2 months when cysts were stored in distilled water or at 15–20°C within 2 weeks or at 37°C in 5 days.[92] In the latter study, cysts did not survive freezing even when stored in a variety of cryoprotectants. The survival of *C. pavum* oocysts of human and ovine origins in still natural mineral waters was assessed in a study in Scotland. When added to mineral waters and held at 20°C, a progressive decline in viability of both types was noted, but viability remained unaltered when stored for 12 weeks at 4°C.[73] Commonly used disinfectants are ineffective against the oocysts,[10] and this has been demonstrated for ozone and chlorine compounds. For a 90% or more inactivation of *C. parvum* oocysts, 1 ppm ozone required 5 minutes, 1.3 ppm chlorine dioxide required 60 minutes, and 80 ppm each of chlorine and monochloramine required about 90 minutes.[62] The oocysts were 14 times more resistant to ClO_2 than *Giardia* cysts, and these investigators suggested that disinfection alone should not be relied upon to inactivate *C. parvum* oocysts in water.

Human cryptosporidiosis may be acquired by at least one of five known transmission routes: zoonotic, person-to-person, water, nosocomial (hospital acquired), or food. Zoonotic transmission (from vertebrate animals to humans) is most likely where infected animals (such as calves) deposit fecal matter to which humans are exposed. The disease may be contracted by drinking untreated water. Oocysts at levels of 2–112 per liter were found in 11 samples of water from four rivers in Washington and California.[74] Although the minimum infectious dose for humans is not known, two of two primates became infected after the ingestion of 10 oocysts.[2] The organism has been shown to be an etiological agent of travelers' diarrhea.[102]

Symptoms, Diagnosis, and Treatment

The clinical course of cryptosporidiosis in humans depends on the immune state, with the most severe cases occurring in the immunocompromised. In immunocompetent individuals, the organism primarily parasitizes the intestinal epithelium and causes diarrhea. The disease is self-limiting, with an incubation period of 6–14 days, and symptoms typically last 9–23 days. In the immunocompromised, diarrhea is profuse and watery, with as many as 71 stools per day and up to 17 l per day reported.[32] Diarrhea is sometimes accompanied by mucus but rarely blood. Abdominal pain, nausea, vomiting and low-grade (<39°C) fever are less frequent than diarrhea, and symptoms may last for more than 30 days in the immunocompromised but generally less than 20 days (range: 4–21 days) in the immunocompetent. From the Milwaukee outbreak (see below), symptoms revealed by 285 of the victims were as follows: watery diarrhea, 93%; abdominal cramps, 84%; fever, 57%; and vomiting, 48%.[67] The median duration of illness was 9 days (range: 1–55), and the median maximal number of stools per day was 12 (range: 1–90). In an outbreak associated with a swimming pool in California in 1988, the following symptoms (and percentages affected) were given by the 44 of 60 victims: watery diarrhea (88%), abdominal cramps (86%), and fever (60%).[17] The organism was identified from stool cultures of some patients by a modified acid-fast stain. Oocysts generally persist beyond the diarrheal stage.

Diagnosis of cryptosporidiosis requires the identification of oocysts in stools of victims. Staining methods are used, including modified acid-fast procedures, negative staining, and sugar flotation. Another diagnostic method is a direct immunofluorescence test used for the detection of oocysts in feces.[101] The latter method employs a monoclonal antibody against an oocyst wall antigen.

Over 100 chemotherapeutic regimens have been tested and found to be ineffective,[25] although spiramycin, fluconazole, and amphotericin B show some promise. More recently, the aminoglycoside antibiotics paromomycin and geneticin were found to inhibit the growth of intracellular *C. parvum* in Caco-2 cells.[41]

Waterborne and Foodborne Outbreaks

The first demonstrated waterborne outbreak of cryptosporidiosis occurred in Braun Station, Texas, in 1984 following the consumption of artesian well water. There were actually two outbreaks— one in May and the other in July, with 79 victims.[26] A second outbreak with 13,000 victims occurred in Carrollton, Georgia, in 1987 and oocysts were found in the stools of 58 of 147 victims.[43] Three separate outbreaks occurred in the United Kingdom in 1988–1989. In one, there were 500 confirmed cases that resulted from the consumption of treated water, and as many as 5000 persons may have been affected.[97] In another outbreak, 62 cases were traced to contaminated swimming pool water. Early in 1990, there was an outbreak in Scotland.

Although foodborne cryptosporidiosis was suspected in the 1980s, only relatively recently have clear-cut outbreaks been documented, and some are summarized in Table 29–2. The 1993 apple cider outbreak occurred among at least 759 students and staff who attended a 1-day school fair.[70] The median incubation period was 6 days (range of 10 hours to 13 days). In the chicken salad outbreak, the food

Table 29–2   Summary of Foodborne Outbreaks of Cryptosporidiosis

Vehicle Foods	Year	Place	No. of Victims	Reference
Apple cider*	1993	Maine	ca. 150	70
Chicken salad	1995	Minnesota	15	14
Apple cider†	1996	New York	20	13
Raw green onions	1997	Washington	54	12

*Fresh pressed and unpasteurized.
†Unpasteurized.

preparer operated a daycare home and admitted to having changed diapers of toddlers 2 weeks before the outbreak.[14]

The single largest waterborne outbreak on record occurred in Milwaukee, Wisconsin, during the spring of 1993. It was estimated that 403,000 persons were infected.[67] The oocysts passed through one of the city's water treatment plants and the peak numbers were accompanied by increased turbidity in treated water. Infectious oocysts of *C. parvum* have been recovered from Chesapeake Bay oysters in an area that had low coliform numbers.[31]

The heating of cider for 10 of 20 seconds at either 70 or 71.7°C has been shown to affect a 4.9-log kill of oocysts.[27] The effectiveness of UV irradiation to destroy *Cryptosporidium parvum* oocysts in cider has been demonstrated. With up to 10^6 oocysts in cider, UV at 14.32 mJ inactivated all as assessed by injection in BALB/c mice.[42] For a review of *C. parvum*, *Cyclospora*, and *Giardia*, see reference 86.

Cyclosporiasis

The protozoan that causes this disease, *Cyclospora cayetanensis*, is a coccidian that is closely related to the cryptosporidia, and some human infections by the latter have been misdiagnosed as cyclosporiasis. Prior to the 1990s, this organism was thought to be an alga or a cyanobacterium by its microscopic appearance under ultraviolet (UV) light, and it was referred to as "cyanobacterium-like." The present classification was established by Ortega et al.[75,76] For an early review, see reference 98.

The *C. cayetanensis* oocysts measure approximately 8–10 μm in diameter, and they contain two sporocysts (about 4 μm wide and 6 μm long). Each sporocyst contains two crescent-shaped sporozoites about 1 μm wide and 9 μm long.[75] The oocysts are acid-fast and sensitive to drying, but resistant to chlorine. The oocysts sporulate between 5 and 13 days in culture,[76] and best at 22 or 30°C but not at either 4 or 37°C.[94] Person-to-person transmission of this disease is unlikely since the excreted oocysts must sporulate to become infectious. In a study of its prevalence in the stools of children below age 2.5 years in Peru, 6 and 18% were positive in two groups.[75] Unlike cryptosporidia, this protozoan is susceptible to Bactrim (trimethoprim-sulfamethoxazole).

The oocysts in stools can be identified by wetmount, phase microscopy, acid-fast staining, or epifluorescence microscopy. Confirmation or detection can be made by polymerase chain reaction (PCR). An example of the latter method was developed and could detect as few as 19 *C. cayetanensis* per PCR test or 10 of a highly similar protozoan, *Eimeria tenella*.[54]

C. cayetanensis is an intestinal pathogen that appears to parasitize epithelial cells (enterocytes) of the jejunum. The disease symptoms mimic those of cryptosporidiosis. Diarrhea is prolonged but self-limiting, lasting a mean of 43 ± 24 days,[93] and it is more severe in human immunodeficiency virus (HIV)-infected individuals.[98] The incubation period ranges between 2 and 11 days, with a mean of about 7 days. In the large outbreak in 1996, the leading symptoms and the percentages experiencing them were diarrhea (98.8%), loss of appetite (92.9%), fatigue (92.4%), and weight loss (90.7%).[43]

Prevalence and Outbreaks

It appears that the first documented human case of cyclosporiasis occurred in 1977 in Papua, New Guinea.[98] The first outbreak in the United States occurred in July 1990 in a physicians' hospital-dormitory in Chicago, and there were 21 case patients.[47] Tap water was the source, and it came from water-storage tanks on top of the building.[47] As an intestinal pathogen, one would expect to find this organism in feces-contaminated waters, and its oocysts have been detected in waste water and verified by PCR.[99] The disease was contracted by British soldiers and defenders in Nepal in 1994 from chlorinated water stored in tanks.[81] Cases have occurred in persons on cruise ships.

The largest clear-cut foodborne outbreak occurred in 1996 in 20 U.S. states and two Canadian Provinces. There were at least 1465 cases, and 66.8% were laboratory confirmed.[44] The vehicle food was raspberries imported from Guatemala. This same product was the vehicle in a later outbreak in Ontario, Canada, where at least 29 persons became ill.[15] The food item involved was a berry garnish that contained raspberries, blackberries, strawberries, and possibly blueberries; and 26% of the 108 who ate this garnish became ill. Raspberries from Guatemala were statistically associated with the organism. In addition to raspberries, mesclun lettuce has been suspected of being the source of cyclosporiasis.[16] An outbreak of cyclosporiasis occurred in Toronto in May 1999 among individuals who attended a wedding. There were around 79 victims, but the vehicle was unknown.

FLATWORMS

All flatworms belong to the animal phylum Platyhelminthes, and the genera discussed in this chapter belong to two classes:

Phylum Platyhelminthes
   Class Trematoda (flukes)
     Subclass Digenea
       Order Echinostomata
         Family Fasciolidae
           Genus *Fasciola*
           Genus *Fasciolopsis*
       Order Plagiorchiata
         Family Troglotrematidae
           Genus *Paragonimus*
       Order Opisthorchiata
         Family Opisthorchiidae
           Genus *Clonorchis*

Class Cestoidea
   Subclass Eucestoda (tapeworms)

Order Pseudophyllidea
 Family Diphyllobothriidae
  Genus *Diphyllobothrium*
Order Cyclophyllidea
 Family Taeniidae
  Genus *Taenia*

Fascioliasis

This syndrome (also known as parasitic biliary cirrhosis and liver rot) is caused by the digenetic trematode *Fasciola hepatica*. The disease among humans is cosmopolitan in distribution, and the organism exists where sheep and cattle are raised; they, along with humans, are its principal definitive hosts.

This parasite matures in the bile ducts, and the large operculate eggs (150 × 90 μm in size) enter the alimentary tract from bile ducts and eventually exit the host in feces. After a period of 4–15 days in water, the miricidium develops, enters a snail, and is transformed into a sporocyst. The sporocyst produces mother rediae, which later become daughter rediae and cercariae. When the cercariae escape from the snail, they become free swimming, attach to grasses and watercress, and encyst to form metacercariae. When ingested by a definitive host, the metacercariae excyst in the duodenum, pass through the intestinal wall, and enter the coelomic cavity. From the body cavity, they enter the liver, feed on its cells, and establish themselves in bile ducts, where they mature.[19,79]

Fascioliasis in cattle and sheep is a serious economic problem that results in the condemnation of livers. Human cases are known, especially in France, and they are contracted from raw or improperly cooked watercress that contains attached metacercariae. Human cases are rare in the United States and are limited to the South.[48] Pharyngeal fascioliasis (halzoun) in humans results from eating raw *Fasciola*-laded bovine liver where young flukes become attached to the buccal or pharyngeal membranes, resulting in pain, hoarseness, and coughing.[19]

Symptoms, Diagnosis, and Treatment

Symptoms develop in humans about 30 days after the infection; they consist of fever, general malaise, fatigue, loss of appetite and weight, and pain in the liver region of the body. The disease is accompanied typically by eosinophilia. Fascioliasis can be diagnosed by demonstrating eggs in stools or biliary or duodenal fluids. Effective treatment is achieved upon the administration of praziquantel.[79]

Fasciolopsiasis

Fasciolopsiasis is caused by *Fasciolopsis buski*, and the habitat of this organism is similar to that of *F. hepatica*. Humans serve as definitive host, several species of snails as first intermediate hosts, and water plants (watercress nuts) as second intermediate hosts. Unlike *F. hepatica*, this parasite occurs in the duodenum and jejunum of humans and pigs, and human infection rates as high as 40% are found in parts of Thailand, where certain uncooked aquatic plants are eaten.[19]

Human symptoms of fasciolopsiasis are related to the number of parasites, with no symptoms occurring when only a few parasites exist in the body. When symptoms occur, they develop within 1–2 months after the initial infection and consist of violent diarrhea, abdominal pain, loss of weight, and generalized weakness. Death may occur in extreme cases.[79] Symptoms appear to be due to the general toxic effect of metabolic products of the flukes.

Diagnosis is made by demonstrating eggs in stools. The eggs of *F. buski* are 130–140 μm × 80–85 μm in size. Both niclosamide and praziquantel are effective in treating this disease.[79]

Paragonimiasis

This parasitic disease (also known as parasitic hemoptysis) is caused by *Paragonimus* spp., especially *P. westermani*. It is found primarily in Asia but also in Africa and South and Central America. *P. kellicotti* is found in North and Central America. In contrast to the trematodes, *P. westermani* is a lung fluke.

The eggs of this parasite are expelled in sputum from definitive hosts (humans and other animals), and the miricida develop in 3 weeks in moist environments. A miricidium penetrates a snail (first intermediate host) and later gives rise to daughter rediae and cercariae about 78 days after entering the snail.[19] The cercariae enter a second intermediate host (crab or crayfish) and encyst. The crustacean host in parts of the Orient and the Philippines are various species of freshwater crabs, where they usually form metacercarial cysts in leg and tail muscle.[79] In *P. kellicotti*, cysts form in the heart region.[79] When the definitive host ingests the infected crustacean, the metacercariae hatch out of their shells, bore their way as young flukes through the walls of the duodenum, and then move to the lungs, where they become enclosed in connective tissue cysts.[79] The golden-brown eggs may appear in sputum 2–3 months later.

Symptoms, Diagnosis, and Treatment

Paragonimiasis is accompanied by severe chronic coughing and sharp chest pains. Sputum is often reddish-brown or bloody. Other non-specific symptoms may occur when parasites lose their way to the lungs.[79] Diagnosis is made by demonstrating the golden-brown eggs in sputum or stools. The eggs of *P. westermani* are 80–120 μm in length and 50–60 μm in width. Also, a complement fixation test titer of at least 1:16 is diagnostic, and ELISA tests are available. The disease can be treated with praziquantel.[79]

Clonorchiasis

The Class Trematoda of the flatworms consists of parasites commonly referred to as flukes that infect the liver, lungs, or blood of mammals. *Clonorchis (Opisthorchis) sinensis* is the Chinese liver fluke that causes oriental biliary cirrhosis. Flukes typically have three hosts: two intermediate, where the larval or juvenile stage develops, and the definitive or final host, where the sexually mature adult develops. *C. sinensis* is an endoparasite whose anterior sucker surrounds the mouth. It also has a midventral sucker. Along with cats, dogs, pigs, and other vertebrates, humans may serve as definitive hosts.

When deposited in water, the eggs of *C. sinensis* hatch into ciliated larvae (miracidia), which invade the first host, usually a snail. As a larva enters the snail, it rounds up as a sporocyst and reproduces asexually to form embryos. Each embryo develops into a redia that escapes from the sporocyst and begins to feed on host tissues. Embryos within the rediae develop into cercariae, which escape from the rediae through a birth pore. A cercaria is a miniature fluke with a tail. Cercariae leave the snail and swim through water in search of their next host—usually fish, clams, and the like. They bore into the new host, shed the tail, and become surrounded by a cyst. Within the cyst, further development leads to metacercariae, which develop further in the final host, usually a vertebrate, including humans. Upon ingestion of metacercariae-containing fish, the cyst wall dissolves in the intestine, and the young flukes emerge. They then migrate through the body to their final site, the bile ducts of the liver in the case of *C. sinensis* where, among other problems, they may cause cirrhosis (see below).

Liver flukes are common in China, Korea, Japan, and parts of Southeast Asia. It is estimated that more than 20 million persons in Asia are infested with this parasite.[45] In China, it is often associated with the consumption of a raw fish dish called *ide*. Over 80 species of fish are known to be capable of harboring *C. sinensis*.[45]

Symptoms, Diagnosis, Treatment, and Prevention

Symptoms may not occur if the infection is mild, but in severe cases, damage to the liver may occur. The liver damage may lead to cirrhosis and edema, and cancer of the liver is seen occasionally.[79]

Diagnosis is made by repeated microscopic examinations of feces and duodenal fluid for eggs. An ELISA test is useful, but cross-reactions with other trematodes may occur. Praziquantel is an effective chemotherapeutic agent.

The prevention of this syndrome is achieved by avoiding the deposition of human feces in fishing waters, but this seems unlikely in view of its wide distribution. The avoidance of raw fish products and the proper cooking of fish are more realistic alternatives. *C. sinensis* can be inactivated in fish by the same procedures as for roundworms and flatworms. According to Rodrick and Cheng,[84] all captured fish must be considered to be potential carriers of parasites. This applies to all flukes, flatworms and roundworms, and protozoa.

Diphyllobothriasis

This infection is contracted from the consumption of raw or undercooked fish, and the causative organism, *Diphyllobothrium latum*, is often referred to as the broad fish tapeworm. The definitive hosts for *D. latum* are humans and other fish-eating mammals; intermediate hosts are various freshwater fish and salmon, where plerocercoid (or metacestode) larvae are formed.

When humans consume fish flesh that contains plerocercoid larvae, the larvae attach to the ileal mucosa by two adhesive grooves (bothria) on each scolex and develop in 3–4 weeks into mature forms. As a worm matures, its strobila, made up of proglottids, increases in length to 10 m or nearly 20 m, and each worm may produce 3000–4000 proglottids that are wider than they are long (hence, broad fish tape) (see Figure 29–1). Over 1 million eggs may be released each day into stools of victims. Eggs are more often seen in stools than proglottids, and they are not infective for humans.

When human feces are deposited in waters, the eggs hatch and release six-hooked, free-swimming larvae or coracidia (also known as oncospheres). When these forms invade small crustaceans (copepods or microcrustaceans such as *Cyclops* or *Diaphtomus*), they metamorphose into a juvenile stage designated metacestode or procercoid larvae. When a fish ingests the crustacean, the larvae migrate into its muscles and develop into plerocercoid larvae. If this fish is eaten by a larger fish, the plerocercoid migrates, but it does not undergo further development. Humans are infected when they eat fish containing these forms.

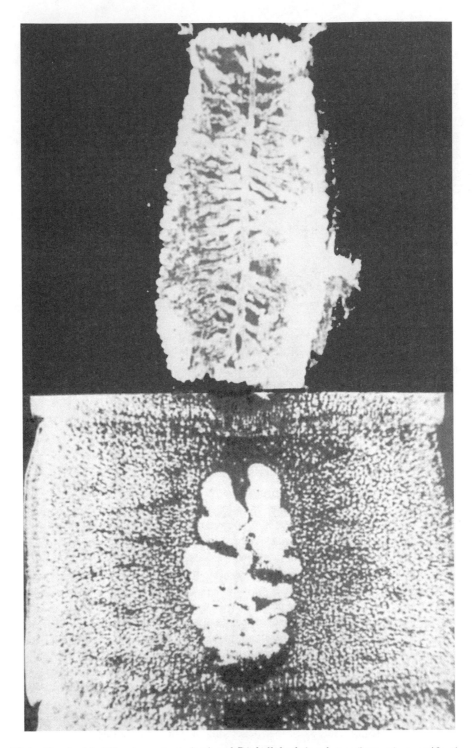

Figure 29–1 Proglottid of *Taenia saginata* (top) and *Diphyllobothrium latum* (bottom), magnification: 360×. *Source:* From S.H. Abadie, J.H. Miller, L.G. Warren, J.C. Swartzwelder, and M.R. Feldman, *Manual of Clinical Microbiology*, 2nd ed.; copyright © 1974 by American Society for Microbiology, used with permission.

Prevalence

Although the first human case was reported in 1906, it was the scattering of cases during the early 1980s that brought new attention to this disease in the United States and Canada. The cases in question resulted from the consumption of *sushi*, a raw fish product that has long been popular in parts of Asia but only relatively recently has become popular in the United States. The incidence of diphyllobothriasis is high in Scandinavia and the Baltic regions of Europe. It is estimated that 5 million cases occur in Europe, 4 million in Asia, and 100,000 in North America. However, only 1 of 275 asymptomatic natives of Labrador, Canada, examined in 1977 had a positive stool culture for this organism.[100]

Symptoms, Diagnosis, and Treatment

Although most cases of diphyllobothriasis are asymptomatic, victims may complain of epigastric pain, abdominal cramps, vomiting, loss of appetite, dizziness, and weight loss. Intestinal obstruction is not unknown. One of the consequences of this infestation is a vitamin B-12 deficiency, along with macrocytic anemia.

This disease is diagnosed by demonstrating eggs in stools. Treatment is the same as for taeniasis. The absence of overt symptoms does not mean the absence of the tapeworm in the intestines because the worms may persist for many years.

Prevention

Diphyllobothriasis can be prevented in humans by avoiding the consumption of raw or undercooked fish. Although the elimination of raw sewage from waters will undoubtedly help to reduce the incidence, it will not break the life-cycle chain of this organism, as humans are not the only definitive hosts. Cooking fish products to an internal temperature of 60°C for 1 minute or 65°C for 30 seconds will destroy the organism,[6] as will freezing fish to -20°C for at least 60 hours.[56,57]

Cysticercosis/Taeniasis

This syndrome in humans is caused by two species of flatworms: *Taenia saginata* (also *Taeniarhynchus saginatus*; beef tape) and *Taenia solium* (pork tape). They are unique among both flatworm and roundworm parasites in that humans are their definitive hosts; the adult and sexually mature stages develop in humans, whereas the larval or juvenile stage develops in herbivores. These helminths have no vascular, respiratory, or digestive systems nor do they possess a body cavity. They depend on the digestive activities of their human hosts for all of their nourishment. Their metabolism is primarily anaerobic.

The structure of a *T. saginata* proglottid is illustrated in Figure 29–1. The adult worm consists of a scolex (head) that is about 1 mm in size and lacks hooks but has four sucking discs. Behind the scolex is the generative neck, which segments to form the strobila composed of proglottids. The latter increase in length, with the oldest being the farthest away from the scolex. Each proglottid has a complete set of reproductive organs, and an adult worm may contain up to 2000 proglottids. These organisms may live up to 25 years and grow to a length of 4–6 m inside the intestinal tract. *T. saginata* sheds 8–9 proglottids daily, each containing 80,000 eggs. The eggs are not infective for humans.

When proglottids reach soil, they release their eggs, which are 30–40 μm in diameter, contain fully developed embryos, and may survive for months. When the eggs are ingested by herbivores, such as cattle, the embryos are released, penetrate the intestinal wall, and are carried to striated muscles of the tongue, heart, diaphragm, jaw, and hindquarters, where they are transformed into larval forms designated cysticerci. *Cysticercosis* is the term used to designate the existence of these parasites in the intermediate hosts. The cysticerci usually take 2 or 3 months to develop after eggs are ingested by a herbivore. When present in large numbers, the cysticerci impart a spotted appearance to the beef issue. Humans become infected upon the ingestion of meat that contains cysticerci.

The infection caused by the pork tape (*T. solium*) is highly similar to that described for the beef tape, but there are some significant differences. Although humans are also the definitive hosts, the larval stages develop in both swine and humans. In other words, humans can serve as intermediate (cysticercosis) and definitive (taeniasis) hosts, thus making autoinfections possible. For this reason, *T. solium* infections are potentially more dangerous than those of *T. saginata*. The infection caused by larval forms of *T. solium* is sometimes designated *Cysticercus* cellulosae. The *T. solium* scolex has hooks rather than sucking discs, and the strobila may reach 2–4 m and contain only about 1000 proglottids. Embryos of *T. solium* are carried to all tissues of the body, including the eyes and brain in contrast to *T. saginata*. Although *T. saginata* exists in both the United States and many other parts of the world, *T. solium* has been eliminated in the United States. However, it does exist in Latin America, Asia, Africa, and eastern Europe. The incidence of *T. saginata* in beef in the United States is below 1% as a result of federal and local meat inspections.

Symptoms, Diagnosis, and Treatment

Most cases of taeniasis are asymptomatic regardless of the *Taenia* species involved, but symptoms differ when humans serve as intermediate host. In these cysticercosis cases, the cysticerci develop in body tissues, including those of the central nervous system, and generally lead to eosinophilia.

Human taeniasis is diagnosed by demonstration of eggs or proglottids in stools and cysticercosis by tissue biopsies of calcified cysticerci or by immunological methods. Complement fixation, indirect hemagglutination, and immunofluorescence tests are valuable diagnostic aids.

A single-dose oral treatment with niclosamide, which acts directly on the parasites, is effective in ridding the body of adult worms. This drug apparently inhibits a phosphorylation reaction in the worm's mitochondria. Another effective chemotherapeutic agent is praziquantel. With cysticercosis, surgery may be indicated.

Prevention

The general approach in the prevention and elimination of diseases that require multiple hosts is to cut the cycle of transmission from one host to another. Because the eggs are shed in human feces, taeniasis can be eliminated by the proper disposal of sewage and human wastes, although *T. solium* infections in humans present a more complex problem. Cysticerci can be destroyed in beef and pork by cooking to a temperature of at least 60°C.[52] The freezing of meats to at least −10°C for 10–15 days or immersion in concentrated salt solutions for up to 3 weeks will inactivate these parasites. Freezing times and temperatures necessary to ensure the death of all cysticerci from infected calves were found by one group to be as follows: 360 hours at −5°C, 216 hours at −10°C, and 144 hours at −15, −20, −25, or −30°C.[46]

ROUNDWORMS

The disease-causing roundworms of primary importance in foods belong to two orders of the phylum Nematoda. The order Rhabditida includes *Turbatrix aceti* (the vinegar eel), which is not a human pathogen and is not discussed further.

Phylum Nematoda
　Class Adenophorea (= Aphasmidia)
　　Order Trichinellida
　　　Genus *Trichinella*

Class Secernentea (= Phasmidia)
　Order Rhabditida
　　Genus *Turbatrix*

Order Ascaridida
　Genus *Ascaris*
　　Subfamily Anisakinae
　　　Genus *Anisakis*
　　　Genus *Pseudoterranova*
　　　　(Phocanema)
　　　Genus *Toxocara*

Trichinosis

Trichinella spiralis is the etiological agent of trichinosis (trichinellosis), the roundworm disease that has been of greatest concern from the standpoint of food transmission, especially in the United States. The organism was first described in 1835 by J. Paget in London, and the first human case of trichinosis was seen in Germany in 1859.[61] Although most flatworm and roundworm diseases of humans are caused by parasites that require at least two different host animals, the trichinae are transmitted from host to host; no free-living stages exist. In other words, both larval and adult stages of *T. spiralis* are passed in the same host. It is contracted often from raw or improperly cooked pork or bear meats.

The adult forms of *T. spiralis* live in the duodenal and jejunal mucosas of mammals such as swine, canines, bears, marine mammals, and humans that have consumed trichinae-infested flesh. The adult females are 3–4 mm long, and adult males are about half this size. Although they may remain in the intestines for about a month, no symptoms are produced. The eggs hatch within female worms, and each female can produce around 1500. These larvae, each about 0.1 mm in length, burrow though the gut wall and pass throughout the body, ultimately lodging in certain muscles. Only those that enter skeletal muscles live and grow; the others are destroyed. The specific muscles affected include those of the eye, tongue, and diaphragm. When assaying for trichinae larvae in pork, the U.S. Department of Agriculture (USDA) employs diaphragm muscle or tongue tissues. In a recent study, the Crus muscle of the diaphragm was found to yield more larvae per gram than several others.[64] As the larvae burrow into muscles several weeks later, severe pain, fever, and other symptoms occur, which sometimes lead to death from heart failure (see below). The larvae grow to about 1 mm in muscles and then encyst by curling up and becoming enclosed in a calcified wall some 6–18 months later (Figure 29–2). The larvae develop no further until consumed by another animal (including humans), but they may remain viable for up to 10 years in a living host. When the encysted flesh is ingested by a second host, the encysted larvae are freed by the enzymatic activities in the stomach, and they mature in the lumen of the intestines.

Prevalence

About 75 species of animals can be infected by *T. spiralis*, but avians appear to be resistant.[69] During the 1930s and 1940s, about 16% of Americans were infected.[69] For the period 1966–1970, 4.7% of

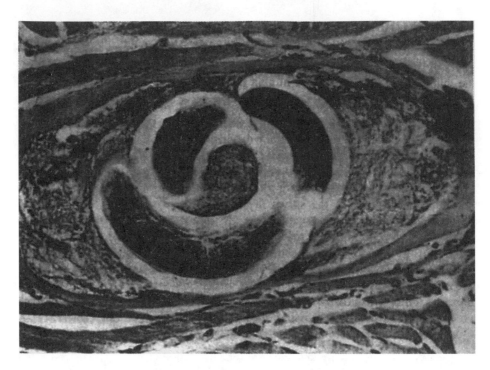

Figure 29–2 *Trichinella spiralis* in muscle, magnification: 350×. *Source:* From S.H. Abadie, J.H. Miller, L.G. Warren, J.C. Swartzwelder, and M.R. Feldman, *Manual of Clinical Microbiology*, 2d ed.; copyright © 1974 by American Society for Microbiology, used with permission.

pork contained trichinae in diaphragm muscles examined postmortem in the United States. For the 5-year period 1977–1981, 686 cases with 4 deaths were reported in the United States.[91] The CDC survey data for the years 1983–1987 show 33 outbreaks with 162 cases and 1 death, which represents a mean of around 32 cases per year for this 5-year period.[5] For the 15-year period 1973–1987, the CDC recorded 128 outbreaks and 843 cases for an average of 56 per year.[4] However, the actual number of cases in the United States in 1985 had been estimated to be 100,000.[7] Only three cases were recorded in Canada in 1982, with none in 1983 and 1984.[103] For the 3-year period 1987–1989, fewer than 50 annual cases were reported, but 120 were reported in 1990. Ninety of these occurred in Iowa among 250 immigrants from Southeast Asia who consumed raw pork sausage. An additional 15 cases occurred in Virginia; pork sausage was the vehicle food.

Pork was incriminated in 79% of the cases for the years 1975–1981, with bear meat in 14% and ground beef in 7%. Studies on pork in retail ground beef have revealed that from 3 to 38% of beef samples contained pork. The presence of pork in ground beef may be deliberate on the part of some stores, or it may result from using the same grinder for both products.

The reported cases in the United States (Fig. 29–3) for the 4-year period 1997–2001 are listed in Table 28–3.[88] Of the 33 cases, the average number per year was about 6, with 21 traced to bear meat. Although *T. spiralis* is best known as the cause of trichinosis, this disease can be caused also by *T. pseudospiralis* and *T. nativa*. In 1999, four cases by *T. pseudospiralis* were traced to undercooked, barbecued wild boar meat in France, and the incubation period was between 3 and 14 days. *T. nativa* causes trichinellosis in arctic and near-arctic regions, and the organism is freeze resistant.[37]

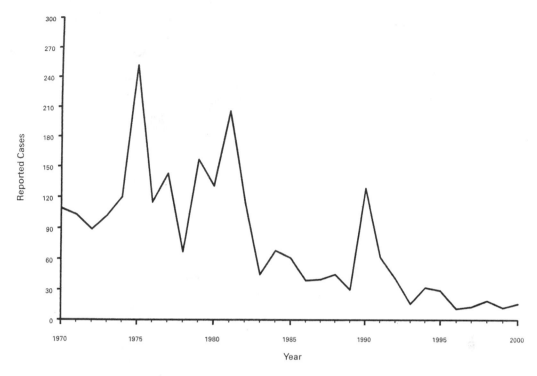

Figure 29–3   Reported cases of trichinellosis by year, United States, 1970–2000 (CDC, 02).

Symptoms and Treatment

One to 2 days after the ingestion of heavily encysted meat, trichinae penetrate the intestinal mucosa, producing nausea, abdominal pain, diarrhea, and sometimes vomiting. When only a few larvae are ingested, the incubation period may be as long as 30 days. The symptoms may persist for several days, or they may abate and be overlooked. The larvae begin to invade striated muscles about 7–9 days after

Table 29–3   Outbreaks of Trichinellosis Among Patients, by Year, State, Number of Cases, Month of Illness Onset, and Implicated Meat—United States, 1997–2001[88]

Year	State	No. of Cases	Month of Illness Onset	Implicated Meat
1997	Montana	5	December	Bear jerky
1998	Ohio	8	October–November	Bear roast, ground bear meat
1999	Illinois	2	March–May	Pork sausage, pork jerky
2000	Illinois	2	January	Pork sausage, smoked pork
2000	Alaska	4	August–September	Bear steak (fried)
2001	California	2	May	Home-raised pork
2001	California	6	May–June	Home-raised pork (raw)
2001	California	2	August	Bear
2001	California	2	November	Bear
Total		33		

the initial symptoms. Where 10 or fewer larvae are deposited per gram of muscle tissue, there are usually no symptoms. When 100 or more per gram are deposited, symptoms of clinical trichinosis usually develop, whereas for 1000 or more per gram of tissue, serious and acute consequences may occur. Muscle pain (myalgia) is the universal symptom of muscle involvement, and difficulty in breathing, chewing, and swallowing may occur.[69] About 6 weeks after the initial infection, encystment occurs, accompanied by tissue pain, swelling, and fever. Resistance to reinfection develops, and it appears to be T cell mediated. Thiabendazole and mebendazole have been shown to be effective drugs for this disease.

Diagnosis

Because the trichinae exist as coiled larvae in ovoid capsular cysts in skeletal muscles, biopsies are sometimes performed on the deltoid, biceps, or gastrocnemius muscles. A significant eosinophilia usually develops during the second week of the disease. Antibodies can be detected after the third week of infection; immunological methods that may be used include bentonite flocculation, cholesterol-lecithin flocculation, and latex agglutination. A bentonite titer of 1:5 is significant, but this test is not positive until at least 3 weeks after infection. This disease is positively diagnosed if a serologic test (e.g., ELISA) is positive for IgG and/or IgM antibodies to *Trichinella* in the serum of victims.

Prevention and Control

Trichinosis can be controlled by avoiding the feeding of infected meat scraps or wild game meats to swine, and by preventing the consumption of infested tissues by other animals. The feeding of uncooked garbage to swine helps to perpetuate this disease. Where only cooked garbage is fed to pigs, the incidence of trichinosis has been shown to fall sharply.

This disease can be prevented by the thorough cooking of meats such as pork or bear meat. In a study on the heat destruction of trichina larvae in pork roasts, all roasts cooked to an internal temperature of 140°F (60°C) or higher were subsequently found to be free of organisms.[11] Larvae were found in all roasts cooked at 130°F (54.4°C) and lower, and in some roasts cooked at 135°F. The USDA recommendation for pork products is that the product be checked with a thermometer after standing and if any part does not attain 76.7°C (170°F), the product should be cooked further.[105]

Freezing will destroy the encysted forms, but freezing times and temperatures depend on the thickness of the product and the specific strain of *T. spiralis* (Table 29–4). The lower the temperature of freezing, the more destructive it is to *T. spiralis*, as was demonstrated in the following study. Four selected temperatures were chosen for the freezing of infected ground pork that was stuffed into casings and packed into boxes. When frozen and stored at −17.8°C, the trichinae lost infectivity between 6 and 10 days; at −12.2°C, infectivity was lost between 11 and 15 days.[91] When frozen at −9.4°C, they remained infective up to 56 days, and for up to 71 days when frozen at −6.7°C.[108] Freezing in dry ice (−70°C) and liquid nitrogen (−193°C) destroys the larvae.[63] The destruction of trichina larvae by irradiation is discussed in Chapter 15.

The effect of curing and smoking on the viability of trichina in pork hams and shoulders was investigated by Gammon et al.[38] They employed the meat of hogs experimentally infested with *T. spiralis* as weanling pigs. After curing, the meat was hung for 30 days, followed by smoking for approximately 24 hours at 90–100°F, with subsequent aging. Live trichinae were found in both hams and shoulders 3 weeks after smoking, but none could be detected after 4 weeks. The effect of NaCl concentration, water activity (a_w), and fermentation method on viability of *T. spiralis* in Genoa salami

Table 29–4   Required Period of Freezing at Temperatures Indicated

Temperature (°C)	Group 1 (days)	Group 2 (days)
−15	20	30
−23	10	20
−29	6	12

Note: Group 1 = less than 15.24 cm in depth; group 2 = more than 15.24 cm in depth. From Sec. 18.10. Regulations Governing the Meat Inspection of the United States Department of Agriculture (9 CFR 18.10, 1960).

Source: From A.W. Kotula, K.D. Murrell, L. Acosta-Stein, L. Lamb, and L. Douglass. *J. Food Sci.* 48:765–768; copyright © 1983 by Institute of Food Technologists.

was evaluated by Childers et al.[20] Pork from experimentally infected pigs was used to prepare salami. The trichinae larvae were completely destroyed at day 30 and thereafter in salami made with 3.33% NaCl and given high-temperature (46.1°C) fermentation treatment, irrespective of product pH. No larvae were found in products made with 3.33% NaCl and given low-temperature fermentation after 30 days. In salami made with no salt, 25% of larvae were found at days 15–25, but none thereafter. A summary of the main control steps of prevention, detection, and inactivation are illustrated in Figure 29–4.

Microwave Cooking

The efficacy of microwave ovens in destroying *T. spiralis* larvae has been investigated by several groups. In a homemaker-oriented study in which most trichina-infected pork roasts were cooked in microwave ovens by time rather than product temperature, Zimmerman and Beach[107] found that of 51 products (48 roasts and 3 pork chops) cooked in 6 different ovens, 9 remained infective. Six of the 9 did not attain a midroast temperature of 76.7°C, whereas the other 3 exceeded this temperature at some point in the cooking cycle. The investigators noted that the experimentally infected pork used

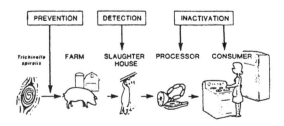

Figure 29–4   Various stages in the movement of pork from the farm to consumer at which control efforts may be applied. *Source:* From K.D. Murrell. 1985. *Food Technol.* 39(3):65–68, 110; copyright © Institute of Food Technologists.

in the study came from pigs infected with 250,000 *T. spiralis*, which produced around 1000 trichina per gram of tissue compared to about 1 trichina per gram in naturally infected pigs. Although the large number of trichinae per gram may have been a factor in their survival at the cooking procedures employed, the inherent unevenness of cooking in microwave ovens is of concern. In another study, whereas trichinae larvae were not inactivated at 77 or 82°C in microwave ovens, cooking to an internal temperature of 77°C in a conventional convection oven, flat grill, charbroiler, or deep fat did inactivate the larvae.[64] Further, infected larvae survived rapid cooking that involved thawing pork chops in an industrial microwave oven followed by cooking on a charbroiler to 71 or 77°C.[65]

The cooking of pork in microwave ovens is clearly a matter of concern relative to the destruction of trichinae larvae, and two factors may explain the greater efficiency of convection ovens over microwave ovens. First, microwave cooking is rapid, and herein may lie the problem. Oven heat has been shown to be more destructive to trichinae larvae in roasts when slow-cooked in conventional ovens at 200°F (93.3°C) than when fast-cooked at 350°F or 176.7°C.[11] Second, a convection oven is more uniformly heated than some microwave ovens. This is minimized if the product is rotated in the latter type ovens or if the oven is equipped with an automatic rotating device. Otherwise, uneven heating occurs, leading to undercooking of some parts of a roast while other parts may be overcooked. It has been shown that a set of criteria that leads to consistent doneness of pork products in microwave ovens will result in safe products.[106]

Anisakiasis

This roundworm infection is caused by two closely related genera and species: *Anisakis simplex* (the herringworm or whaleworm) and *Pseudoterranova decipiens* (formerly *Phocanema*; codworm or sealworm). Both of these organisms have several intermediate hosts and generally more than one definitive host. Humans are not final hosts for either, and human disease occurs as the result of humans being accidental interlopers in the life cycles of these worms.

The definitive hosts are marine mammals—whales in the case of *A. simplex* and gray (and other) seals in the case of *P. decipiens*. Feces of these animals contain thousands of eggs, which when they enter water undergo their first molt (stage L1 to stage L2). The free-swimming larvae that result are ingested by small crustaceans (copepods), and they, in turn, are ingested by larger crustaceans, which serve as intermediate hosts during the second molt (stage L2 to L3). A final host may ingest L3 larvae along with the crustacean intermediate, but more often, L3 is ingested by fish or squid, which may, in turn, be ingested by larger fish before reaching the final host. The last two molts (L3 and L4) lead to adults that mate, and these events take place in the final host. The infectious larva is L3, and it is usually found in tight, flat coils in or on fish viscera, and some larvae may occur in the belly flap muscles of fish. Because of its preference for whale hosts, *A. simplex* is found more often in fish from the northern Pacific.

In the case of *P. decipiens*, eggs in seal feces are ingested by copepods, and L2- or early-L3-stage larvae are ingested by the first intermediate host—fish. In fish, they penetrate the stomach wall, enter the body cavity, and many burrow into fish muscles. In fish, the L3 larvae grow to 25–50 mm in length and are red to brown in color. The final host, seals, tends to ingest the organisms principally from smelt and other small fish.

Human infections occur upon the ingestion of fish that contain L3- and L4-stage larvae. Thus, anisakids do not mature in humans. Disease symptoms arise from the activities of the juvenile worms. *A. simplex* larvae are more harmful than those of *P. decipiens* because they often penetrate the mucosal lining, whereas most *P. decipiens* larvae are passed in feces or are coughed up or vomited after irritating the mucosa. *Anisakis* is most often found in cases in Japan and the Netherlands; *Pseudoterranova* is more often seen in North America.

Symptoms, Diagnosis, and Treatment

Symptoms of human anisakiasis may develop within 4–6 hours after consumption of infested fish, and they consist of epigastric pain, nausea, and vomiting. In more severe cases, fever and bloody stools may occur within 7 days after ingesting infective fish. If the worms penetrate the mucosa, an eosinophilic granuloma may develop, or they may penetrate the gut wall and cause peritonitis. However, in the 23 North American cases through 1982, only 5 were caused by *Anisakis* sp., and only transient infection occurred.[60] Among the four cases reported by Kliks,[60] symptoms consisted of mild stomach pain and nausea from the time of ingestion up to 20 hours later, and worms were coughed up or found in the mouth up to 2 weeks after consumption of the infective raw salmon.

The diagnosis of this syndrome is made difficult by the absence of eggs or other parts of the worms in feces. Larvae in the intestinal tract can be viewed by endoscopy, and surgical resection of the affected tissue can be carried out. Complement fixation and indirect immunofluorescence tests are of some diagnostic value. Thiabendazole is an effective chemotherapeutic agent for treating this disease.

Prevalence and Distribution

A synopsis of the known cases of anisakiasis through 1989 is presented in Table 29–5. The first clear-cut case occurred in 1955 in the Netherlands, and between 1955 and 1965 over 149 cases were reported in that country. When freezing of herring to $-20°C$ for 24 hours was legislated, no cases were seen the following year.[49] Over 1000 cases were reported in Japan for the period 1964–1976.[72] In both instances, raw fish products such as sushi and sashimi were the vehicle foods, although lightly salted herring ("green herring") was a common vehicle in the Netherlands. Through 1976, six cases were recorded in the United States and one each in Canada, England, and Greenland.[68] This disease is associated with the raw fish dish ceviche in South America, where *D. pacificum* is the usual etiological agent.[84] The first documented case in North America occurred in Boston in 1973, and through 1990 fewer than 100 cases were reported in the United States by both etiological agents. Anisakiasis has been seen in Belgium, Britain, Chile, Denmark, France, Germany, Korea, and Taiwan. Over a 15-year period beginning in the early 1960s, some 1200 cases were seen in Japan. Raw mackerel is probably

Table 29–5 Summary of Some Cases of Anisakiasis

1955	First clear-cut case recorded in the Netherlands
1955–1965	About 149 cases reported in the Netherlands
1964–1976	Over 1,000 cases reported in Japan
1973	First documented case in North America (in Boston)
1980	Over 500 cases reported in Japan
1977–1981	About five cases recorded in California (two by *A. simplex* and three by *P. decipiens*)
1981–1989	Approximately 50 cases caused by *A. simplex* and 30 by *P. decipiens* seen in North America

Source: From Jackson,[49] Margolis,[68] and Myers.[72]

the most significant fish source in Asia,[84] although over 160 teleost species are believed to harbor these organisms.[45]

In regard to the prevalence of anisakid larvae in fish, two extensive investigations were conducted in the late 1970s. In one, 1010 fish belonging to 20 genera and 23 species were examined in the Washington, DC, area and of 703 that contained parasitic nematodes, 6547 nematodes were found, of which only 11 were *Anisakis* sp.[50] The mean content of nematode larvae/fish was 6.48, with an overall infection or 69.60%. A 2-year survey in 1974–1975 of fish and shell-fish from U.S. Pacific coastal waters off Washington, Oregon, and California included 2074 specimens.[65] *Anisakis* sp. was the most frequently found, and most often on fish viscera. In fish caught off the California coast, 41.6% contained anisakid larvae.[69] *Anisakis* was higher than *Phocanema* (*Pseudoterranova*), due largely to the number of whales, a situation that is reversed in the eastern Canadian waters of the Gulf of St. Lawrence.[65] No anisakid larvae were found in over 2000 shellfish examined. In yet another survey in Ann Arbor, Michigan, larval densities from 63 to 91/kg of salmon tissue were found, and they were in viable form.[87]

There is some controversy as to whether there is an increased incidence of anisakid larvae in commercial fish compared to decades ago. Clearly, these organisms are not new in fishing waters since their presence in fish was recognized as early as 1767. One view is that the disease in humans began to flourish when refrigeration was taken aboard fishing boats in the mid-1950s. Prior to this time, fish were caught and eviscerated, and the infective organisms were thus discarded. When fish are kept on ice for several days, some investigators believe the parasites migrate from the mesenteries to muscles following the death of fish.[45,71] On the other hand, the migration of parasites in dead fish has not been substantiated by all investigators.

Prevention

Anisakiasis can be prevented by avoiding the consumption of raw or undercooked fish. Sushi, ceviche, and sashimi should be consumed only when properly prepared from fish that has undergone inspection for absence of infective larvae. Infective forms can be destroyed by cooking fish to an internal temperature of 60°C for 1 minute or 65°C for 30 seconds.[6] Freezing at −20°C or below for at least 60 hours is reported to render the larvae uninfective.[56] Although some North American species survived after 52 hours at −20°C.[6] Brining for 4 weeks has been found to render larvae uninfective.[40]

REFERENCES

1. Anderson, B.C. 1985. Moist heat inactivation of *Cryptosporidium* sp. *Am. J. Public Health* 75:1433–1434.

2. Barer, M.R., and A.E. Wright. 1990. *Cryptosporidium* and water. *Lett. Appl. Microbiol.* 11:271–277.

3. Barnard, R.J., and G.J. Jackson. 1984. *Giardia lamblia*: The transfer of human infections by foods. In *Giardia and Giardiasis: Biology, Pathogenesis, and Epidemiology*, ed. S.L. Erlandsen and E.A. Meyer, 365–378. New York: Plenum.

4. Bean, N.H., and P.M. Griffin. 1990. Foodborne disease outbreaks in the United States, 1973–1987: Pathogens, vehicles, and trends. *J. Food Protect.* 53:804–817.

5. Bean, N.H., P.M. Griffin, J.S. Goulding, and C.B. Ivey. 1990. Foodborne disease outbreaks, 5-year summary, 1983–1987. *J. Food Protect.* 53:711–728.

6. Bier, J.W. 1976. Experimental anisakiasis: Cultivation and temperature tolerance determinations. *J. Milk Food Technol.* 39:132–137.

7. Bennett, J.V., S.D. Holmberg, M.F. Rogers, and S.L. Solomon. 1987. Infectious and parasitic diseases. In *The Burden of Unnecessary Illness*, ed. R.W. Amler and H.B. Dull, 102–114. New York: Oxford University Press.

8. Black, M.W., and J.C. Boothroyd. 2000. Lytic cycle of *Toxoplasma gondii*. *Microbiol. Mol. Biol. Rev.* 64:607–623.

9. Boucher, S.-E.M., and F.D. Gillin. 1990. Excystation of in vitro-derived *Giardia lamblia* cysts. *Infect. Immun.* 58:3516–3522.

10. Campbell, L., S. Tzipori, G. Hutchinson, and K.W. Angus. 1982. Effect of disinfectants on survival of cryptosporidium oocysts. *Vet. Rec.* 111:414–415.

11. Carlin, A.F., C. Mott, D. Cash, and W. Zimmerman. 1969. Destruction of trichina larvae in cooked pork roasts. *J. Food Sci.* 34:210–212.

12. Centers for Disease Control and Prevention. 1998. Foodborne outbreak of cryptosporidiosis—Spokane, Washington, 1997. *Morb. Mort. Wkly Rep.* 47:565–567.

13. Centers for Disease Control and Prevention. 1997. Outbreaks of *Escherichia coli* 0157:H7 infection and cryptosporidiosis associated with drinking unpasteurized apple cider—Connecticut and New York, October 1996. *Morb. Mort. Wkly. Rep.* 46:4–8.

14. Centers for Disease Control and Prevention. 1996. Foodborne outbreak of diarrheal illness associated with *Cryptosporidium parvum*—Minnesota, 1995. *Morb. Mort. Wkly. Rep.* 45:783–784.

15. Centers for Disease Control and Prevention. 1998. Outbreak of cyclosporiasis—Ontario, Canada, May 1998. *Morb. Mort. Wkly. Rep.* 47:806–809.

16. Centers for Disease Control and Prevention. 1997. Update: Outbreaks of cyclosporiasis—United States and Canada, 1997. *Morb. Mort. Wkly. Rep.* 46:521–523.

17. Centers for Disease Control and Prevention. 1990. Swimming-associated cryptosporidiosis—Los Angeles County. *Morb. Mort. Wkly. Rep.* 39:343–345.

18. Centers for Disease Control and Prevention. 1989. Common-source outbreak of giardiasis—New Mexico. *MMWR Morb. Mort. Wkly. Rep.* 38:405–407.

19. Cheng, T.C. 1986. *General Parasitology*, 2nd ed. New York: Academic Press.

20. Childers, A.B., R.N. Terrell, T.M. Craig, T.J. Kayfus, and G.C. Smith. 1982. Effect of sodium chloride concentration, water activity, fermentation method and drying time on the viability of *Trichinella spiralis* in Genoa salami. *J. Food Protect.* 45:816–819.

21. Conroy, D.A. 1960. A note on the occurrence of *Giardia* sp. in a Christmas pudding. *Rev. Iber. Parasitol.* 20:567–571.

22. Craun, G. 1988. Surface water supplies and health. *J. Am. Water Works Assoc.* 80:40–52.

23. Craun, G.F. 1979. Waterborne giardiasis in the United States: A review. *Am. J. Public Health* 69:817–819.

24. Current, W.L. 1988. The biology of *Cryptosporidium*. *ASM News* 54(11):605–611.

25. Current, W.L. 1987. *Cryptosporium:* Its biology and potential for environmental transmission. *CRC Crit. Rev. Environ. Cont.* 17:21–51.

26. D'Antonio, R.G., R.E. Winn, J.P. Taylor, T.L. Gustafson, W.L. Current, M.M. Rhodes, G.W. Gary, Jr., and R.A. Zajac. 1985. A waterborne outbreak of cryptosporidiosis in normal hosts. *Ann. Intern. Med.* 103:886–888.

27. Deng, M.Q., and D.O. Cliver. 2001. Inactivation of *Cryptosporidium parvum* oocysts in cider by flash pasteurization. *J. Food Protect.* 64:523–527.

28. Desmonts, G., J. Couvreur, F. Alison, J. Baudelot, J. Gerbeaux, and M. Lelong. 1965. Etude epidemiologique sur la toxoplasmose: De l'influence de la cuisson des viandes de boucherie sur la frequence de l'infection humaine. *Rev. Fr. Etudes Clin. Biol.* 10:952–958.

29. Duquesne, V., C. Auriault, F. Darcy, J.-P. Decavel, and A. Capron. 1990. Protection of nude rats against *Toxoplasma* infection by excreted-secreted antigen-specific helper T cells. *Infect. Immun.* 58:2120–2126.

30. Elliott, D.A., D.J. Coleman, M.A. Lane, R.C. May, L.A. Machesky, and D.P. Clark. 2001. *Cryptosporidium parvum* infection requires host cell actin polymerization. *Infect. Immun.* 69:5940–5942.

31. Fayer, R., T.K. Graczyk, E.J. Lewis, J.M. Trout, and C.A. Farley. 1998. Survival of infectious *Cryptosporidium parvum* oocysts in seawater and eastern oysters (*Crassostrea virginica*) in the Chesapeake Bay. *Appl. Environ. Microbiol.* 64:1070–1074.

32. Fayer, R., and B.L.P. Ungar. 1986. *Cryptosporidium* spp. and cryptosporidiosis. *Microbiol. Rev.* 50:458–483.

33. Fayer, R., and J.P. Dubey. 1985. Methods for controlling transmission of protozoan parasites from meat to man. *Food Technol.* 39(3):57–60.

34. Fayer, R. 1982. Other protozoa: *Eimeria, Isospora, Cystoisospora, Besnoitia, Hammondia, Frenkelia, Sarcocystis, Cryptosporidium, Encephalitozoon*, and *Nosema*. In *CRC Handbook Series in Zoonosis*, ed. J.H. Steele, 187–197. Boca Raton, FL: CRC Press.

35. Fayer, R. 1975. Effects of refrigeration, cooking and freezing on *Sarcocystis* in beef from retail food stores. *Proc. Helm. Soc. Wash.* 42:138–140.

36. Feldman, H.A., and L.T. Miller. 1956. Serological study of toxoplasmosis prevalence. *Am. J. Hyg.* 64:320–335.

37. Forbes, L.B., I. Measures, A. Gajadhar, and C. Kapel. 2003. Infectivity of *Trichinella nativa* in traditional northern (country) foods prepared with meat from experimentally infected seals. *J. Food Protect.* 66:1857–1863.

38. Gammon, D.L., J.D. Kemp, J.M. Edney, and W.Y. Varney. 1968. Salt, moisture and aging times effects on the viability of *Trichinella spiralis* in pork hams and shoulders. *J. Food Sci.* 33:417–419.

39. Gangarosa, E.J., and J.A. Donadio. 1970. Surveillance of foodborne disease in the United States. *J. Infect. Dis.* 62:354–358.

40. Grabda, J., and J.W. Bier. 1988. Cultivation as an estimate for infectivity of larval *Anisakis simplex* from processed herring. *J. Food Protect.* 51:734–736.

41. Griffiths, J.K., R. Balakrishnan, G. Widmer, and S. Tzipori. 1998. Paromomycin and geneticin inhibit intracellular *Cryptosporidium parvum* without trafficking through the host cell cytoplasm: Implications for drug delivery. *Infect. Immun.* 66:3874–3883.

42. Hanes, D.E., R.W. Worobo, P.A. Orlandi, D.H. Burr, M.D. Miliotis, M.G. Robl, J.W. Bier, M.J. Arrowood, J.J. Churey, and G.J. Jackson. 2002. Inactivation of *Cryptosporidium parvum* oocysts in fresh apple cider by UV irradiation. *Appl. Environ. Microbiol.* 68:4168–4172.

43. Hayes, E.B., T.D. Matte, T.R. O'Brien, T.W. McKinely, G.S. Logsdon, J.B. Rose, B.L.P. Ungar, D.M. Word, P.F. Pinsky, M.L. Cummings, M.A. Wilson, E.G. Long, E.S. Hurwitz, and D.J. Juranek. 1989. Large community outbreak of cryptosporidiosis due to contamination of a filtered public water supply. *N. Engl. J. Med.* 320:1372–1376.

44. Herwaldt, B.L., M.-L. Ackers, and the Cyclospora Working Group. 1997. An outbreak in 1996 of cyclosporiasis associated with imported raspberries. *N. Engl. J. Med.* 336:1548–1556.

45. Higashi, G.I. 1985. Foodborne parasites transmitted to man from fish and other aquatic foods. *Food Technol.* 39(3):69–74, 111.

46. Hilwig, R.W., J.D. Cramer, and K.S. Forsyth. 1978. Freezing times and temperatures required to kill cysticerci of *Taenia saginata* in beef. *Vet. Parasitol.* 4:215–219.

47. Huang, P., J.T. Weber, D.M. Sosin, P. M. Griffin, E.G. Long, J.J. Murphy, F. Kocka, C. Peters, and C. Kallick. 1995. The first reported outbreak of diarrheal illness associated with *Cyclospora* in the United States. *Ann. Intern. Med.* 123:409–414.

48. Jackson, G.J. 1990. Parasitic protozoa and worms relevant to the U.S. *Food Technol.* 44(5):106–112.

49. Jackson, G.J. 1975. The "new disease" status of human anisakiasis and North American cases: A review. *J. Milk Food Technol.* 38:769–773.

50. Jackson, G.J., J.W. Bier, W.L. Payne, T.A. Gerding, and W.G. Knollenberg. 1978. Nematodes in fresh market fish of the Washington, D.C. area. *J. Food Protect.* 41:613–620.

51. Jackson, M.H., and W.M. Hutchinson. 1989. The prevalence and source of *Toxoplasma* infection in the environment. *Adv. Parasitol.* 28:55–105.

52. Jacobs, L. 1962. Parasites in food. In *Chemical and Biological Hazards in Food*, ed. J.C. Ayres et al., 248–266. Ames: Iowa State University Press.

53. Jacobs, L., J.S. Remington, and M.L. Melton. 1960. A survey of meat samples from swine, cattle, and sheep for the presence of encysted *Toxoplasma. J. Parasitol.* 46:23–28.

54. Jinneman, K.C., J.H. Wetherington, W.E. Hill, A.M. Adams, J.M. Johnson, B.J. Tenge, N-L. Dang, R.J. Manger, and M.M. Wekell. 1998. Template preparation for PCR and RFLP of amplification products for the detection and identification of *Cyclospora* sp. and *Eimeria* spp. oocysts directly from raspberries. *J. Food Protect.* 61:1497–1503.

55. Joiner, K.A., S.A. Furhman, H.M. Miettinen, L.H. Kasper, and I. Mellman. 1990. *Toxoplasma gondii*: Fusion competence of parasitophorous vacuoles in Fc receptor-transfected fibroblasts. *Science* 249:641–646.

56. Karl, H. 1988. Vorkommen von Nematoden in Konsumfischen. *Verfah. Feststel. Abtoetung. Rundsch. Fleisch. Lebensmittelhyg.* 40:198–199.

57. Karl, H., and M. Leinemann. 1989. Ueber lebensfaehigkeit von Nematodenlarven (*Anisakis* sp.) in gefrosteten Heringen. *Arch. Lebensmittelhyg.* 40:14–16.

58. Kean, B.H., A.C. Kimball, and W.N. Christenson. 1969. An epidemic of acute toxoplasmosis. *J. Am. Med. Assoc.* 208:1002–1004.

59. Kirkpatrick, C.E., and C.E. Benson. 1987. Presence of *Giardia* spp. and absence of *Salmonella* spp. in New Jersey muskrats (*Ondatra zibethicus*). *Appl. Environ. Microbiol.* 53:1790–1792.

60. Kliks, M.M. 1983. Anisakiasis in the western United States: Four new case reports from California. *Am. J. Trop. Med. Hyg.* 32:526–532.

61. Kolata, G. 1985. Testing for trichinosis. *Science* 227:621, 624.

62. Korich, D.G., J.R. Mead, M.S. Madore, N.A. Sinclair, and C.R. Sterling. 1990. Effects of ozone, chlorine dioxide, chlorine, and monochloramine on *Cryptosporidium parvum* oocyst viability. *Appl. Environ. Microbiol.* 56:1423–1428.

63. Kotula, A.W., A.K. Sharar, E. Paroczay, H.R. Gamble, K.D. Murrell, and L. Douglass. 1990. Infectivity of *Trichinella spiralis* from frozen pork. *J. Food Protect.* 53:571–573.

64. Kotula, A.W., P.J. Rothenberg, J.R. Burge, and M.B. Solomon. 1988. Distribution of *Trichinella spiralis* in the diaphragm of experimentally infected swine. *J. Food Protect.* 51:691–695.

65. Kotula, A.W. 1983. Postslaughter control of *Trichinella spiralis*. *Food Technol.* 37(3):91–94.

66. Lopez, C.E., D.D. Juranek, S.P. Sinclair, and M.G. Schultz. 1978. Giardiasis in American travelers to Madeira Island, Portugal. *Am. J. Trop. Med. Hyg.* 27:1128–1132.

67. MacKenzie, W.R., N.J. Hoxie, M.E. Proctor, M.S. Gradus, K.A. Blair, D.E. Peterson, J.J. Kazmierczak, D.G. Addiss, K.R. Fox, J.B. Rose, and J.P. Davis. 1994. A massive outbreak in Milwaukee of cryptosporidium infection transmitted through the public water supply. *N. Engl. J. Med.* 331:161–167.

68. Margolis, L. 1977. Public health aspects of "codworm" infection: A review. *J. Fish. Res. Bd. Canada* 34:887–898.

69. Marquardt, W.C., and R.S. Demaree. 1985. *Parasitology*. New York: Macmillan.

70. Millard, P.S., K.F. Gensheimer, D.G. Addiss, D.M. Sosin, G.A. Beckett, A. Houck-Jakoski, and A. Hudson. 1994. An outbreak of cryptosporidiosis from freshpressed apple cider. *JAMA* 272:1592–1596.

71. Myers, B.J. 1979. Anisakine nematodes in fresh commercial fish from waters along the Washington, Oregon and California coasts. *J. Food Protect.* 42:380–384.

72. Myers, B.J. 1976. Research then and now on the anisakidae nematodes. *Trans. Am. Microscop. Soc.* 95:137–142.

73. Nichols, R.A.B., C.A. Paton, and H.V. Smith. 2004. Survival of *Cryptosporidium parvum* oocysts after prolonged exposure to still natural mineral waters. *J. Food Protect.* 67:517–523.

74. Ongerth, J.E., and H.H. Stibbs. 1987. Identification of *Cryptosporidium* oocysts in river water. *Appl. Environ. Microbiol.* 53:672–676.

75. Ortega, Y.R., R.H. Gilman, and C.R. Sterling. 1994. A new coccidian parasite (Apicomplexa: Eimeriidae) from humans. *J. Parasitol.* 80:625–629.

76. Ortega, Y.R., C.R. Sterling, R.H. Gilman, V.A. Cama, and F. Diaz. 1993. *Cyclospora* species—A new protozoan pathogen of humans. *N. Engl. J. Med.* 328:1308–1312.

77. Osterholm, M.T., J.C. Forfang, T.L. Ristinen, A.G. Daan, J.W. Washburn, J. R. Godes, R.A. Rude, and J.G. McCullough. 1981. An outbreak of foodborne giardiasis. *N. Engl. J. Med.* 304:24–28.

78. Petersen, L.R., M.L. Cartter, and J.L. Hadler. 1988. A foodborne outbreak of *Giardia lamblia*. *J. Infect. Dis.* 157:846–848.

79. Piekarski, G. 1989. *Medical Parasitology*. New York: Springer-Verlag.

80. Plorde, J.J. 1984. Sporozoan infections. In *Medical microbiology: An Introduction to Infectious Diseases*, ed. J.C. Sherris, K.J. Ryan, C.G. Ray, et al., 469–483. New York: Elsevier.

81. Rabold, J.G., C.W. Hoge, D.R. Shlim, C. Kefford, R. Rajah, and P. Echeverria. 1994. *Cyclospora* outbreak associated with chlorinated drinking water. *Lancet* 344:360 361.

82. Rendtorff, R.C. 1954. The experimental transmission of human intestinal protozoan parasites. II. *Giardia lamblia* cysts given in capsules. *Am. J. Hyg.* 59:209–220.

83. Riemann, H.P., M.E. Meyer, J.H. Theis, G. Kelso, and D.E. Behymer. 1975. Toxoplasmosis in an infant fed unpasteurized goat milk. *J. Pediatr.* 87:573–576.

84. Rodrick, G.E., and T.C. Cheng. 1989. Parasites: Occurrence and significance in marine animals. *Food Technol.* 43(11):98–102.

85. Rommel, M., and A.-O. Heydorn. 1972. Beiträge zum Lebenszyklus der Sarkosporidien. III. *Isospora hominis* (Railliet and Lucet, 1891) Wenyon, 1923, eine Dauerform der Sarkosporidien des Rindes und des Schweins. *Berl. Munchen. Tierarztl. Wochens.* 85:143–145.

86. Rose, J.B., and T.R. Slifko. 1999. *Giardia, Cryptosporidium*, and *Cyclospora* and their impact on foods: A review. *J. Food Protect.* 62:1059–1070.

87. Rosset, J.S., K.D. McClatchey, G.I. Higashi, and A.S. Knisely. 1982. *Anisakis* larval type I in fresh salmon. *Am. J. Clin. Pathol.* 78:54–57.

88. Roy, S.L., A.S. Lopez, and P.M. Schantz. 2003. Trichinellosis surveillance—United States, 1997–2001. *Morb. Mort. Wkly Rep.* 52(SS-6):1–7.

89. Sacks, J.J., R.R. Roberto, and N.F. Brooks. 1982. Toxoplasmosis infection associated with raw goat's milk. *J. Am. Med. Assoc.* 248:1728–1732.

90. Salata, R.A., A. Martinez-Palomo, L. Canales, H.W. Murray, N. Revino, and J.I. Ravdin. 1990. Suppression of T-lymphocyte responses to *Entamoeba histolytica* antigen by immune sera. *Infect. Immun.* 58:3941–3946.

91. Schantz, P.M. 1983. Trichinosis in the United States—1947–1981. *Food Technol.* 37(3):83–86.

92. Sherwood, D., K.W. Angus, D.R. Snodgrass, and S. Tzipori. 1982. Experimental cryptosporidiosis in laboratory mice. *Infect. Immun.* 38:471–475.

93. Shlim, D.R., M.T. Cohen, M. Eaton, R. Rajah, E.G. Long, and B.L.P. Unger. 1991. An alga-like organism associated with an outbreak of prolonged diarrhea among foreigners in Nepal. *Am. J. Trop. Med. Hyg.* 45:383–389.

94. Smith, H.V., C.A. Paton, M.M.A. Mtambo, and R.W.A. Girdwood. 1997. Sporulation of *Cyclospora* sp. oocysts. *Appl. Environ. Microbiol.* 63:1631–1632.

95. Smith, J.L. 1993. Documented outbreaks of toxoplasmosis: Transmission of *Toxoplasma gondii* to humans. *J. Food Protect.* 56:630–639.

96. Smith, P.D. 1989. *Giardia lamblia*. In *Parasitic Infections in the Compromised Host*, ed. P.D. Walzer and R.M. Genta, 343–384. New York: Marcel Dekker.

97. Smith, H.V., R.W.A. Girdwood, W.J. Patterson, R. Hardie, L.A. Green, C. Benton, W. Tulloch, J.C.M. Sharp, and G.J. Forbes. 1988. Waterborne outbreak of cryptosporidiosis. *Lancet* 2:1484.

98. Soave, R. 1996. *Cyclospora*: An overview. *Clin. Infect. Dis.* 23:429–437.

99. Sturbaum, G.D., Y.R. Ortega, R.H. Gilman, C.R. Sterling, L. Cabrera, and D.A. Klein. 1998. Detection of *Cyclospora cayetanensis* in wastewater. *Appl. Environ. Microbiol.* 64:2284–2286.

100. Sole, T.D., and N.A. Croll. 1980. Intestinal parasites in man in Labrador, Canada. *Am. J. Trop. Med. Hyg.* 29(3):364–368.

101. Sterling, C.R., and M.J. Arrowood. 1986. Detection of *Cryptosporidium* sp. infections using a direct immunofluorescent assay. *Pediatr. Infect. Dis.* 5:139–142.

102. Sterling, C.R., K. Seegar, and N.A. Sinclair. 1986. *Cryptosporidium* as a causative agent of traveler's diarrhea. *J. Infect. Dis.* 153:380–381.

103. Todd, E.C.D. 1989. Foodborne and waterborne disease in Canada—1983 annual summary. *J. Food Protect.* 52:436–442.

104. Tzipori, S. 1988. Cryptosporidiosis in perspective. *Adv. Parasitol.* 27:63–129.

105. U.S. Department of Agriculture. 1982. USDA advises cooking pork to 170 degrees Fahrenheit throughout. News release, USDA, Washington, DC.

106. Zimmermann, W.J. 1983. An approach to safe microwave cooking of pork roasts containing *Trichinella spiralis*. *J. Food Sci.* 48:1715–1718, 1722.

107. Zimmermann, W.J., and P.J. Beach. 1982. Efficacy of microwave cooking for devitalizing trichinae in pork roasts and chops. *J. Food Protect.* 45:405–409.

108. Zimmermann, W.J., D.G. Olson, A. Sandoval, and R.E. Rust. 1985. Efficacy of freezing in eliminating infectivity of *Trichinella spiralis* in boxed pork products. *J. Food Protect.* 48:196–199.

CHAPTER 30

Mycotoxins

A very large number of molds produce toxic substances designated mycotoxins. Some are mutagenic and carcinogenic, some display specific organ toxicity, and some are toxic by other mechanisms. Although the clear-cut toxicity of many mycotoxins for humans has not been demonstrated, the effect of these compounds on experimental animals and their effect in in vitro assay systems leaves little doubt about their real and potential toxicity for humans. At least 14 mycotoxins are known to be carcinogens, with the aflatoxins being the most potent.[84] It is generally accepted that about 93% of mutagenic compounds are carcinogens. With mycotoxins, microbial assay systems reveal an 85% level of correlation between carcinogenicity and mutagenesis.[84]

Mycotoxins are produced as secondary metabolites. The primary metabolites of fungi as well as for other organisms are those compounds that are essential for growth. Secondary metabolites are formed during the end of the exponential growth phase and have no apparent significance to the producing organism relative to growth or metabolism. In general, it appears that they are formed when large pools of primary metabolic precursors such as amino acids, acetate, pyruvate, and so on, accumulate. The synthesis of mycotoxins represents one way the fungus has of reducing the pool of metabolic precursors that it no longer requires for metabolism.

Some methods for the detection of mycotoxins in foods can be found in Chapter 11. For an extensive review of these substances in fruits, fruit juices, and dried fruits, see reference 27.

AFLATOXINS

Aflatoxins are clearly the most widely studied of all mycotoxins. Knowledge of their existence dates from 1960, when more than 100,000 turkey poults died in England after eating peanut meal imported from Africa and South America. From the poisonous feed were isolated *Aspergillus flavus*, and a toxin produced by this organism that was designated aflatoxin (*Aspergillus flavus* toxin—A-fla-toxin). Studies on the nature of the toxic substances revealed the following four components:

A. *flavus* produces AFB_1 and AFB_2; and A. *parasiticus* produces all four of the major aflatoxins (B_1, G_1, B_2, and G_2). AFB_1 is produced by all aflatoxin-positive strains, and it is the most potent of all. Other known aflatoxin producers include A. *nominus*,[47] A. *bombycis*, A. *pseudotamartii*, and A. *ochraceoroseus* among the aspergilli; and *Emericella venezuelensis*.

These compounds are highly substituted coumarins, and at least 18 closely related toxins are known. AFM_1 is a hydroxylated product of AFB_1, and it appears in milk, urine, and feces as a metabolic product.[30] AFL, $AFLH_1$, AFQ_1, and AFP_1 are all derived from AFB_1, AFB_2 is the 2,3-dihydro form of AFB_1, and AFG_2 is the 2,3-dihydro form of AFG_1. The toxicity of the six most potent aflatoxins decreases in the following order: $B_1 > M_1 > G_1 > B_2 > M_2 \neq G_2$.[4] When viewed under ultraviolet (UV) light, six of the toxins fluoresce as noted:

B_1 and B_2—blue
G_1—green
G_2—green–blue
M_1—blue–violet
M_2—violet

They are polyketide secondary metabolites whose carbon skeleton comes from acetate and malonate.

The proposed partial pathway for AFB_1 synthesis is as follows: acetate > norsolorinic acid > averantin > averufanin > averufin > versiconal hemiacetal acetate > versicolorin A > sterigmatocystin > O-methylsterigmatocystin > AFB_1. Versicolorin A is the first in the pathway to contain the essential C_2–C_3 double bond.

Requirements for Growth and Toxin Production

No aflatoxins were produced by 25 isolates of A. *flavus/parasiticus* on wort agar at 2, 7°, 41°, or 46°C within 8 days, and none was produced below 7.5°C or over 40°C even under otherwise favorable conditions.[75] In another study employing Sabouraud's agar, maximal growth of A. *flavus* and A. *parasiticus* occurred at 33°C when pH was 5.0 and water activity (a_w) was 0.99.[41] At 15°C, growth occurred at a_w 0.95 but not at 0.90, while at 27° and 33°C, slight growth was observed at an a_w of 0.85. The optimum temperature for toxin production has been found by many to be between 24° and 28°C. In one study, maximal growth of A. *parasiticus* was 35°C, but the highest level of toxin was produced at 25°C.[78]

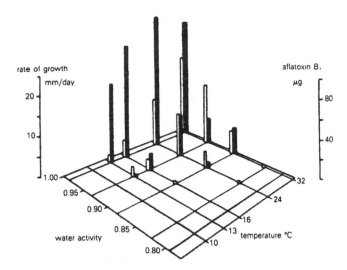

Figure 30–1 Growth and aflatoxin B_1 production on malt extract-glycerine agar at various water activity values and temperatures. White columns: rate of growth; black columns: average AFB_1 production. *Source*: From Northolt et al.,[63] copyright © 1976 by International Association of Milk, Food and Environmental Sanitarians.

The limiting moisture content for AFB_1 and AFB_2 on corn was 17.5% at a temperature of 24°C or higher, with up to 50 ng/g being produced.[96] No toxin was produced at 13°C. Overall, toxin production has been observed over the a_w range of 0.93–0.98, with limiting values variously reported as being 0.71–0.94.[58] In another study, no detectable quantities of AFB_1 were formed by *A. parasiticus* at a_w values of 0.83 at 10°C.[63] The optimum temperature at a_w 0.94 was 24°C (Figure 30–1). Growth without demonstrable toxin appeared possible at a_w 0.83 on malt agar containing sucrose. It has been observed by several investigators that rice supports the production of high levels of aflatoxins at favorable temperatures but none is produced at 5°C on either rice or cheddar cheese.[68]

Overall, the minimal and maximal parameters that control growth and toxin production by these eukaryotic organisms are not easy to define, in part because of their diverse habitats in nature and in part because of their eukaryotic status. It seems clear that growth can occur without toxin production.

AFG_1 is produced at lower growth temperatures than AFB_1, and while some investigators have found more AFB_1 than AFG_1 at around 30°C, others have found equal production. With regard to *A. flavus* and *A. parasiticus*, the former generally produces only AFB_1 and AFG_1.[22] Aeration favors aflatoxin production, and amounts of 2 mg/g can be produced on natural substrates such as rice, corn, soybeans, and the like.[22] Up to 200–300 mg/l can be produced in broth containing appropriate levels of Zn^{2+}. The release of AFB_1 by *A. flavus* appears to involve an energy-dependent transport system.

Production and Occurrence in Foods

With respect to production in foods, aflatoxin has been demonstrated on fresh beef, ham, and bacon inoculated with toxigenic cultures and stored at 15°, 20°, and 30°C,[9] and on country-cured hams during aging when temperatures approached 30°C, but not at temperatures below 15°C or relative humidity (RH) above 75%.[10] They have been found in a wide variety of foods, including milk, beer, cocoa, raisins, soybean meal, and so on (see below). In fermented sausage at 25°C, 160 and 426 ppm of AFG_1 were produced in 10 and 18 days, respectively, and 10 times more AFG_1 was found than B_1.[51] Aflatoxins have been produced in whole-rye and wholewheat breads, in tilsit cheese, and in apple juice at 22°C. They have been demonstrated in the upper layer of 3-month-old cheddar cheese held at room temperature[52] and on brick cheese at 12.8°C by *A. parasiticus* after 1 week but not for *A. flavus*.[78] AFB_1 was found in 3 of 63 commercial samples of peanut butter at levels less than 5 ppb.[69] From a 5-year survey of around 500 samples of Virginia corn and wheat, aflatoxins were detected in about 25% of corn samples for every crop year, with 18–61% of samples containing 20 ng/g or more and 5–29% containing more than 100 ng/g.[80] The average quantity detected over the 5-year period was 21–137 ng/g. Neither aflatoxins nor zearalenone and ochratoxin A were detected in any of the wheat samples. The 1988 drought led to an increase in the amount of aflatoxin produced in corn in some midwestern U.S. states that received less than 2 inches of rain in June and July. About 30% of samples contained more than 20 ppb compared to 2–3 ppb levels during normal rainfall.[83]

In a study of AFB_1 in foods and feedstuffs in Cuba for the years 1990–1996, 17% of 4529 samples were positive with 83% of sorghum and 40% of peanuts being the most contaminated.[32] The incidence in corn was 23 and 25% in wheat. In Botswana, 120 peanut samples were tested for aflatoxin and 78% were positive at levels from 12 to 329 μg/kg, and 49% contained >20 μg/kg.[59] Twenty-one percent of these samples contained cyclopiazonic acid at levels from 1 to 10 μg/kg.

Cyclopiazonic acid is produced by some *A. flavus* strains and it is thought to contribute to the toxicity of aflatoxin. In an examination of seven truckloads of corn, it was found in four at levels of 25–250 ng/g.[49] Also found in four of five loads was deoxynivalenol (DON, vomitoxin) at levels of 46–676 ng/g.

The effect of temperature cycling between 5° and 25°C on production in rice and cheese has been investigated. *A. parasiticus* produced more toxin under cycling temperatures than at 15°, 18°, or 25°C, while *A. flavus* produced less under these conditions.[68] On cheddar cheese, however, less aflatoxin was produced than at 25°C, and these investigators noted that cheese is not a good substrate for aflatoxin production if it is held much below the optimum temperature for toxin production.

Aflatoxin production has been demonstrated to occur on an endless number of food products in addition to those noted. Under optimal conditions of growth, some toxin can be detected within 24 hours—otherwise within 4–10 days.[18] On peanuts, Hesseltine[40] has made the following observations:

1. Growth and formation of aflatoxin occur mostly during the curing of peanuts after removal from soil.
2. In a toxic lot of peanuts, only comparatively few kernels contain toxin, and success in detecting the toxin depends on collecting a relatively large sample, such as 1 kg, for assay.
3. The toxin will vary greatly in amount even within a single kernel.
4. The two most important factors affecting aflatoxin formation are moisture and temperature.

The U.S. Food and Drug Administration (FDA) has established allowable action levels of aflatoxins in foods as follows: 20 ppb for food, feeds, Brazil nuts, peanuts, peanut products, and pistachio nuts; and 0.5 ppb for milk.[48] A committee of the Codex Alimentarius Commission has recommended the following maximum levels of mycotoxins in specific foods: 15 μg/kg of aflatoxins in peanuts for further processing; 0.05 μg/kg of aflatoxin M_1 in milk; 50 μg/kg of patulin in apple juice and apple juice ingredients in other beverages; and 5 μg/kg of ochratoxin A in cereals and cereal products.[61]

Relative Toxicity and Mode of Action

For the expression of mutagenicity, mammalian metabolizing systems are essential for aflatoxins, especially AFB_1. Also essential is their binding with nucleic acids, especially DNA. Although nuclear DNA is normally affected, AFB_1 has been shown to bind covalently to liver mitochondrial DNA, preferentially to nuclear DNA.[62] Cellular macromolecules other than nucleic acids are possible sites for aflatoxins. The site of the aflatoxin molecule responsible for mutagenicity is the C_2–C_3 double bond in the dihydrofurofuran moiety. Its reduction to the 2,3-dihydro (AFB_2) form reduces mutagenicity by 200- to 500-fold.[84] Following binding to DNA, point mutations are the predominant genetic lesions induced by aflatoxins, although frameshift mutations are known to occur. The mutagenesis of AFB_1 has been shown to be potentiated twofold by butylated hydroxyanisole (BHA) and butylated hydroxytoluene (BHT) and much less by propyl gallate employing the Ames assay, but whether potentiation occurs in animal systems is unclear.[77]

The LD_{50} of AFB_1 for rats by the oral route is 1.2 mg/kg, and 1.5–2.0 mg/kg for AFG_1.[12] The relative susceptibility of various animal species to aflatoxins is presented in Table 30–1. Young ducklings and young trout are among the most sensitive, followed by rats and other species. Most species of susceptible animals die within 3 days after administration of toxins and show gross liver damage, which, upon postmortem examination, reveals the aflatoxins to be hepatocarcinogens.[100] The toxicity is higher for young animals and males than for older animals and females, and the toxic effects are enhanced by low protein or cirrhogenic diets.

Circumstantial evidence suggests that aflatoxins are carcinogenic to humans. Among conditions believed to result from aflatoxins is the EFDV syndrome of Thailand, Reye's syndrome of Thailand and New Zealand,[12,13] and an acute hepatoma condition in a Ugandan child. In the last a fatal case of acute hepatic disease revealed histological changes in the liver identical to those observed in monkeys treated with aflatoxins, and an aflatoxin etiology was strongly suggested by the findings.[76]

Table 30–1 Comparative Lethality of Single Doses of Aflatoxin B_1

Animal	Age (or Weight)	Sex	Route	LD_{50} (mg/kg)
Duckling	1 day	M	PO	0.37
	1 day	M	PO	0.56
Rat	1 day	M–F	PO	1.0
	21 days	M	PO	5.5
	21 days	F	PO	7.4
	100 g	M	PO	7.2
	100 g	M	IP	6.0
	150 g	F	PO	17.9
Hamster	30 days	M	PO	10.2
Guinea pig	Adult	M	IP	ca. 1
Rabbit	Weanling	M–F	IP	ca. 0.5
Dog	Adult	M–F	IP	ca. 1
	Adult	M–F	PO	ca. 0.5
Trout	100 g	M–F	PO	ca. 0.5

Note: PO = oral; IP = intraperitoneal.

Source: Wogan.[100]

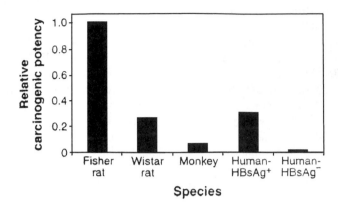

Figure 30–2 Relative carcinogenic potency in different species. Potency is expressed as cases per year per nanogram of aflatoxin B$_1$ per kilogram of body weight per day.[39] Copyright © by 1999 Amer. Assoc. Adv. of Science, used with permission.

Two researchers who worked with purified aflatoxin developed colon carcinoma.[24] On the other hand, it has been noted that no mycotoxin has been linked with a specific cancer in humans in the absence of chronic infection with hepatitis B virus.[87] The risk of liver cancer from aflatoxin consumption is ca. 30 times higher in individuals who have been exposed to hepatitis B (HbsAg$^+$) than in HbsAg$^-$ individuals.[39] The relative potency of carcinogenic activity in humans and some other animals is shown in Figure 30–2.

Degradation

AFB$_1$ and AFB$_2$ can be reduced in corn by bisulfite. When dried figs were spiked with 250 ppb of AFB$_1$ and subjected to several treatments, 1% sodium bisulfite affected a 28.2% reduction in 72 hours; 0.2% H$_2$O$_2$ (added 10 minutes before sodium bisulfite) affected a 65.5% reduction; heating at 45°–65°C for 1 hour affected a 68.4% reduction; and ultraviolet (UV) radiation affected a 45.7% reduction.[3] Aflatoxin-contaminated cottonseed treated with ammonia and fed to cows led to lower levels of AFB$_1$ and AFM$_1$ in milk than nontreated product.[42] When yellow dent corn naturally contaminated with 1600 ppm aflatoxin was treated with 3% NaOH at 100°C for 4 minutes, further processed, and fried, 99% of the aflatoxin was destroyed.[14]

As noted in the section below, *Flavobacterium aurantiacum* has been shown to remove AFB$_1$ from a variety of foods. In one study, this bacterium at 10^{10} cfu/ml and culture filtrates from older (72 hours) cultures were more efficient than younger (24 or 48 hours) cultures.[53] In a later study, Mg^{2+} and Ca^{2+} were found to enhance the destructive activity of this bacterium.[28] Although they do not degrade aflatoxins, bifidobacteria[64] and lactic acid bacteria have been shown to bind to AFB$_1$ and AFM$_1$ in culture and food substrates.[66] Both live and dead cells have been shown to bind these mycotoxins. A soil-borne myxobacterium (*Nannocystis exedens*) was tested against spores, hyphae, and sclerotia of the aflatoxin producing molds *Aspergillus flavus* and *A. parasiticus* and the results showed that both fungi were inhibited after 14 days at 28°C.[91] The myxobacterium actually lysed mold colony growth after 24 hours. It is not inconceivable that the active principal can be extracted and used in purified form against other fungi and possibly some bacteria. Findings from this study suggest that some of the myxomycetes (slime molds) may be effective in destroying mycotoxigenic fungi.

ALTERNARIA TOXINS

Several species of *Alternaria* (including *A. citri*, *A. alternata*, *A. solani*, and *A. tenuissima*) produce toxic substances that have been found in apples, tomatoes, blueberries, grains, and other foods.[85,86] The toxins produced include alternariol, alternariol monomethyl ether, altenuene, tenuazonic acid, and altertoxin-I.[85] On slices of apples, tomatoes, or crushed blueberries incubated for 21 days at 21°C, several *Alternaria* produced each of the toxins noted at levels up to 137 mg/100 g.[85] In another study, tenuazonic acid was the main toxin produced in tomatoes, with levels as high as 13.9 mg/100 g; on oranges and lemons, *A. citri* produced tenuazonic acid, alternariol, and alternariol monomethyl ether at a mean concentration of 1.15–2.66 mg/100 g.[86] The fruits were incubated at room temperature for 21–28 days.

In a study of 150 sunflower seed samples in Argentina, 85% contained alternariol (mean of 187 μg/kg), 47% contained alternariol monomethyl ether (mean of 194 μg/kg), and 65% contained tenuazonic acid (mean of 6692 μg/kg).[19] Following fermentation for 28 days by *A. alternata* and separation into oil and meal, no alternariol, 1.6–2.3% of tenuazonic, and 44–45% alternariol monomethyl ether were found in oil, but none of these toxins were in the meal.[19] An *A. alternata* strain produced stemphyltoxin III, which was mutagenic by the Ames assay.[23] More information on the alternaria toxins can be found in reference 17.

CITRININ

The citrinin mycotoxin is produced by *Penicillium citrinum*, *P. viridicatum*, and other fungi. It has been recovered from polished rice, moldy bread, country-cured hams, wheat, oats, rye, and other similar products. Under long-wave UV light, it fluoresces lemon yellow. It is a known carcinogen. Of seven strains of *P. viridicatum* recovered from country-cured hams, all produced citrinin in potato dextrose broth and on country-cured hams in 14 days at 20°–30°C but not at 10°C.[101] Growth was found to be poor at 10°C. Citrinin was identified from moldy foods examined in Germany, and it along with some other mycotoxins could be produced in a synthetic mediun.[50]

Citrinin

Although citrinin-producing organisms are found on cocoa and coffee beans, this mycotoxin as well as others are not found to the extent of growth. The apparent reason is the inhibition of citrinin in *P. citrinum* by caffeine. The inhibition of citrinin appears to be rather specific, since only a small decrease in growth of the organisms occurs.[6]

OCHRATOXINS

The ochratoxins consist of a group of at least seven structurally related secondary metabolites of which ochratoxin A (OA) is the best known and the most toxic. OB is dechlorinated OA and along with OC, it may not occur naturally. OA is produced by a large number of storage fungi, including *A. ochraceus*, *A. alliaceus*, *A. ostianus*, *A. mellus*, *A. niger* and *A. carbonarius*. The latter was found in Spain to be the main producer of OA in dried vine fruits (currants, raisins, and sultanas) with 97% of 91 isolates found to be OA producers.[1] Among penicillia that produce OA are *P. viridicatum*, *P. cyclopium*, *P. variable*, and others. OA is produced maximally at around 30°C and a_w 0.95.[5] The minimum a_w supporting OA production by *A. ochraceus* at 30°C in poultry feed is 0.85.[5] Its oral LD_{50} in rats is 20–22 mg/kg, and it is both hepatotoxic and nephrotoxic.

Ochratoxin A

These mycotoxins have been found in corn, dried beans, cocoa beans, soybeans, oats, barley, citrus fruits, Brazil nuts, moldy tobacco, country-cured hams, peanuts, coffee beans, and other similar products. Two strains of *A. ochraceus* isolated from country-cured hams produced OA and OB on rice, defatted peanut meal, and when inoculated into country-cured hams.[31] Two-thirds of the toxin penetrated to a distance of 0.5 cm after 21 days, with the other one-third located in the mycelial mat. Of six strains of *P. viridicatum* recovered from country-cured hams, none produced ochratoxins. From a study of four chemical inhibitors of both growth and OA production by two OA producers at pH 4.5, the results were potassium sorbate > sodium propionate > methyl paraben > sodium bisulfite; while at a pH of 5.5, the most effective two were methyl paraben and potassium sorbate.[94] Like most other mycotoxins, OA is heat stable. In one study, the highest rate of destruction achieved by cooking faba beans was 20%, and the investigators concluded that OA could not be destroyed by normal cooking procedures.[29] Under UV light, OA fluoresces greenish, while OB emits blue fluorescence. It induces abnormal mitosis in monkey kidney cells.

PATULIN

Patulin (clavicin, expansin) is produced by a large number of penicillia, including *P. claviforme*, *P. expansum*, *P. patulum*; by some aspergilli (*A. clavatus*, *A. terreus*, and others); and by *Byssochlamys nivea* and *B. fulva*.[22]

Its biological properties are similar to those of penicillic acid. Some patulin-producing fungi can produce the compound below 2°C.[4] This mycotoxin has been found in moldy bread, sausage, fruits (including bananas, pears, pineapples, grapes, and peaches), apple juice, cider, and other products. In apple juice, levels as high as 440 μg/l have been found, and in cider, levels up to 45 ppm. Along with citrinin and ochratoxin A, it was identified from moldy foods examined in Germany.[50]

Minimum a_w for growth of *P. expansum* and *P. patulum* has been reported to be 0.83 and 0.81, respectively. In potato dextrose broth incubated at 12°C, patulin was produced after 10 days by

P. patulum and *P. roquefortii*, with the former organism producing up to 1033 ppm.[8] Patulin was produced in apple juice also at 12°C by *B. nivea*, but the highest concentration was attained after 20 days at 21°C after a 9-day lag.[71] The next highest amount was produced at 30°C, with much less at 37°C. These investigators confirmed that patulin production is favored at temperatures below the growth optimum, as was previously found by Sommer et al.[82] The latter investigators used *P. expansum* and found production over the range 5°–20°C, with only small amounts produced at 30°C. In five commercial samples in Georgia, patulin levels from 244 to 3993 μg/l were found, with a mean of 1902 μg/l.[99] The overall incidence of patulin in apple juice has been reviewed.[38] Atmospheres of CO_2 and N_2 reduced production compared to that in air. To inhibit production, SO_2 was found more effective than potassium sorbate or sodium benzoate.[71]

Patulin

The LD_{50} for patulin in rats by the subcutaneous route is 15–25 mg/kg, and it induces subcutaneous sarcomas in some animals. Both patulin and penicillic acid bind to –SH and –NH_2 groups, forming covalently linked adducts that appear to abate their toxicities. Patulin causes chromosomal aberrations in animal and plant cells and is a carcinogen.

PENICILLIC ACID

This mycotoxin has biological properties similar to patulin. It is produced by a large number of fungi, including many penicillia (*P. puberulum*, for example) as well as members of the *A. ochraceus* group. One of the best producers is *P. cyclopium*. It has been found in corn, beans, and other field crops and has been produced experimentally on Swiss cheese. Its LD_{50} in mice by the subcutaneous route is 100–300 mg/kg, and it is a proven carcinogen.

Penicillic acid

Of 346 penicillia cultures isolated from salami, about 10% produced penicillic acid in liquid culture media, but 5 that were inoculated into sausage failed to produce toxin after 70 days.[20] In another study, some 183 molds were isolated from Swiss cheese; 87% were penicillia, 93% of which were able to grow at 5°C. Thirty-five percent of penicillia extracts were toxic to chick embryos, and from 5.5% of the toxic mixtures were recovered penicillic acid as well as patulin and aflatoxins.[7] Penicillic acid was produced at 5°C in 6 weeks by 4 of 33 fungal strains.

STERIGMATOCYSTIN

This mycotoxin is structurally and biologically related to the aflatoxins, and like the latter, it causes hepatocarcinogenic activity in animals. At least eight derivatives are known. Among the producing organisms are *Aspergillus versicolor*, *A. nidulans*, *A. rugulosus*, and others. The LD_{50} for rats by intraperitoneal injection is 60–65 mg/kg. Under UV light, the toxin fluoresces dark brick-red. Although not often found in natural products, they have been found in wheat, oats, Dutch cheese, and coffee beans. Although related to the aflatoxins, they are not as potent. They act by inhibiting DNA synthesis.

FUMONISINS

The fumonisins are produced by *Fusarium* spp. on corn and other grains, and certain diseases of humans and animals are associated with the consumption of grains and grain products that contain high levels of these molds.

The species demonstrated to produce fumonisins include *F. sacchari*, *F. subglutinans*, *F. thapsinum*, *F. globosum*, *F. anthophilum*, *F. dlamini*, *F. napiforme*, *F. nygami*, *F. moniliforme*, and *F. proliferatum*.[60] The latter species produces large quantities. *F. moniliforme* (formerly *F. verticilliodes*; *Gibberella fujikuroi*) was the first to be associated with the mycotoxin and it is the best studied of the three. The prevalence of *F. moniliforme* is significantly higher in corn from areas where a high rate of human esophageal cancer occurred than in low esophageal cancer rate areas.[54]

There are at least 15 fumonisins, with the best known being FB_1, FB_2, FB_3, FB_4, FA_1, FA_2, and FA_3. The major ones are FB_1–FB_3, and the others are considered to be minor and less well characterized. Of the three major toxins, FB_1 (also designated macrofusine) is produced in the largest quantities by producing strains. For example, among nine strains of *F. moniliforme*, the range of FB_1 produced on autoclaved corn was 960–2350 μg/g while for FB_2 the range was 120–320 μg/g.[72]

Fusarin C is produced by *F. moniliforme* but apparently is not involved in hepatocarcinogenic activity.[35] It is mutagenic in the Ames test but only after liver fraction activation.[98] In a culture medium, more was produced at pH <6.0 than above, and the highest yields were achieved between days 2 and 6 at around 28°C.[33] Corn isolates were shown to produce about 19–332 μg/g when grown on corn.[33]

Growth and Production

In regard to optimum growth temperature and pH, the maximum yield of FB_1 by a strain of *F. moniliforme* in a corn culture occurred in 13 weeks at 20°C with a yield of 17.9 g/kg dry weight.[2] The higher growth rate of the fungus occurred at 25°C, not at 20°C, and the stationary phase was reached in 4–6 weeks at either temperature.[2] In the same study, FB_1 production commenced after 2 weeks of active growth and decreased after 13 weeks. Overall, the optimum time and temperature for FB_1 production was 7 weeks at 25°C. Good growth by an *F. moniliforme* strain at 25 and 30°C over the pH range of 3–9.5 has been demonstrated.[97] Little growth occurred at 37°C over the same pH range. Culture media were used with acidic pH values adjusted with phosphoric acid.[97] Neither *F. moniliforme* nor *F. proliferatum* produces much toxin at a_w 0.925.[57] For one strain of the former grown on sterile corn for 6 weeks at 25°C, the ppm of FB_1 produced were 6.8, 14.4, 93.6, and 102.6 at the respective a_w values of 0.925, 0.944, 0.956, and 0.968.[57] The level of FB_1 was ca. 5 times higher in degermed maize kernals than in germ tissue. The pH of the degermed decreased from 6.4 to 4.7 after 10 days while that of colonized germ tissue increased to pH 8.5.[79] Acidic conditions were conducive to FB_1 production while alkaline conditions were repressive. When *F. moniliforme* and *F. proliferatum* were added to irradiated maize, wheat, and barley grains, FB_1 was produced only on maize, not on the other two.[55]

The preservative compounds benzoic acid, BHA, and carvacrol have been shown to inhibit or retard the mycelial growth of a number of *Fusarium* spp., with benzoic acid being the most effective followed by carvacrol and BHA.[93] The simultaneous effect on fumonisin production is unclear.

Prevalence in Corn and Feeds

It has been observed since the mid-1980s that leukoencephalomalacia (LEM) in horses, pulmonary edema (PE) of porcines, and esophageal cancer (EC) in humans occur in areas of the world where high levels of fumonisins are found in grain-based foods.[102] For example, the highest rate of human EC in southern Africa occurs in the Transkei where high levels of FB_1 and FB_2 are found in corn. Concomitant with the occurrence of the fumonisins is the presence of *Fusarium* spp., especially *F. moniliforme*. The incidence of fumonisin FB_1 in a high-risk county in China was about two times higher than in a low-risk area, although the differences were not statistically significant.[102] Trichothecenes (mainly deoxynivalenol) in addition to fumonisins were found in corn from the high-risk area. From corn and corn-based products from four states in America, FB_1 was found in 65% of 34 samples while FB_2 was found in only 29%.[37] The upper level for FB_1 was 2679 μg/kg, and 797 μg/kg for FB_2. *Fusarium* numbers ranged from 10^2 to $>10^5$.

The incidence and prevalence of FB_1 in corn and some corn products in six countries are summarized in Table 30–2. The highest levels found were in corn from an area in the Transkei, South Africa, where EC occurred at high rates. The range for these six samples was 3020–117,520 ng/g with a mean of 53,740 ng/g.[89] These levels exceeded those found in 12 samples of moldy corn from the same general area, where the mean was 23,900 ng/g.[70] Overall, FB_1 was found at lower levels in corn grits, while in corn meal levels tended to be higher (Table 30–2).

Feed samples from 11 U.S. states were examined for FB_1.[72] Of the 83 equine feeds that were associated with equine LEM, 75% contained >10 μg/g with a range of <1.0–126 μg/g. Of the 42 associated with porcine PE syndrome, 71% contained >10 μg/g with a range of <1.0–330 μg/g. On the other hand, all 51 samples of nonproblem feeds had <9 μg/g of FB_1, with 94% of the 51 being <6 μg/g.[72] Of 71 retail samples of corn-based and other grain products in Michigan, 11 contained fumonisins including 10 of 17 corn-based products.[65] The highest level of FB_1 found using ELISA was 15.6 μg/g in blue corn meal.[65] Based on literature reports of FB_1 in maize, it has been estimated that people in the Netherlands may be exposed to an intake of 1000 ng/day.[25]

When *Fusarium*-contaminated corn that was associated with outbreaks of mycotoxicosis in various animals in Brazil was examined for FB_1 and FB_2, 20 of 21 samples revealed FB_1 levels that ranged from 200 to 38,500 ng/g, and 18/21 had an FB_2 range of 100–12,000 ng/g.[88] Except for one isolate from this corn, all were acutely toxic to ducklings. In a 1996–1997 study of fumonisins in Spanish beers, 14 were positive at levels from 4.76 to 85.53 ng/ml.[95]

Physical/Chemical Properties of FB_1 and FB_2

The chemical structure of FB_1 and FB_2 is indicated below.[26] The two differ only by FB_1 having an –OH group in lieu of an H on carbon 10. These toxins differ from most others in this chapter in two ways: they do not possess cyclic or ring groups, and they are water soluble. On the other hand, they are heat-stable, as are many other mycotoxins. In one study, lyophilized culture materials containing FB_1 were boiled for 30 minutes and then oven dried at 60°C for 24 hours without loss of toxic activity.[2]

$$CH_3-[CH_2]_3-\underset{16}{\overset{21}{CH}}-\overset{15}{CH}-\overset{14}{CH}-CH_2-\underset{12}{CH}-CH_2-\overset{10}{\underset{\boxed{R}}{CH}}-[CH_2]_4-\overset{5}{\underset{OH}{CH}}-CH_2-\overset{3}{\underset{OH}{CH}}-\overset{2}{\underset{NH_2}{CH}}-\overset{1}{CH_3}$$

[1] R = OH

[2] R = H

fumonisins B_1 [1] and B_2 [2].

Table 30–2 Incidence and Prevalence of Fumonisin B_1 in Corn and Corn Products from Several Countries

Products	Country	Samples (Pos./Total)	Fumonisin Range(ng)	Mean (ng/g)	Reference
Corn grits	Switzerland	34/55	0–790	260	67
Corn grits	South Africa	10/18	0–190	125	90
Corn grits	U.S.A.	10/10	105–2545	601	90
Corn meal/muffin mix	U.S.A.	10/17	<200–15,600	–	65
Corn meal	Switzerland	2/7	0–110	85	67
Corn meal	South Africa	46/52	0–475	138	90
Corn meal	U.S.A.	15/16	0–2790	1048	90
Corn meal	Canada	1/2	0–50	50	90
Corn meal	Egypt	2/2	1780–2980	2380	90
Corn meal	Peru	1/2	0–660	660	90
Corn*	Charleston, SC	7/7	105–1915	635	90
Corn*	Transkei, S. Africa	6/6	3020–117,520	53,740	90
Corn (good)*	Transkei, S. Africa	12/12	50–7900	1600	70
Corn (good)[†]	Transkei, S. Africa	2/12	0–550	375	70
Corn (moldy)[†]	Transkei, S. Africa	12/12	3450–46900	23,900	70
Corn (moldy)[†]	Transkei, S. Africa	11/11	450–18,900	6520	70
White corn meal	U.S.A. (MD)		3500–7450		15
Yellow corn meal	U.S.A. (MD)		500–4750		15
Tortilla, white	U.S.A. (MD)		200–400		15
Yellow corn meal	U.S.A. (AZ)		450–650		15
Yellow corn meal	U.S.A. (NE)		500–2500		15
Tortilla, white	U.S.A. (AZ)		250–1450		15
Tortilla, white	U.S.A. (NE)		200–550		15
Maize + meal	Botswana	28/33	20–1270	247	81
				μg/kg	
Maize	Kenya	92/197	110–12,000	670	44

*From areas of respective countries where human esophageal cancer was high.
[†]From areas where esophageal cancer was low.

In another study, the thermal stability of these toxins at a level of 5 μg/g of FB$_1$ in processed corn products was assessed.[16] No significant loss was found upon baking at 204°C for 30 minutes. Almost complete loss occurred upon roasting corn meal samples at 218°C for 15 minutes. Significant but not total reduction was noted in cornbread at 232°C for 20 minutes. In regard to the thermal stability in canned foods, 5 μg/g were added to canned foods and then recanned. No significant loss occurred in creamed corn for infants and canned dog food, but significant reductions occurred in cream-style corn and whole-kernal corn, although it was not eliminated.[16] Overall, roasting was more effective than baking.

Pathology

In experimental animals, the liver is the primary target of FB$_1$. In a study using rats over a 26-month period, all animals that either died or were killed after 18 months had micro- and macronodular cirrhosis and large expansile nodules of cholangiofibrosis at the hilus of the liver.[34] (Cholangiofibrosis is considered to be a precursor lesion for cholangiocarcinoma in rats.) Of 15 rats that died or were killed between 18 and 26 months, 66% developed primary hepatocellular carcinoma. Some involvement of the kidneys occurred but only toward the end of this study. No esophageal lesions were noted in test animals, and no neoplastic changes were noted in the 25 controls.[34] The hepatocarcinogenic activity of FB$_1$ in rats was demonstrated by adding 50,000 ng/g in food rations over a 26-month period.[34] In an earlier study, FB$_1$ was shown to possess cancer-promoting activity by its capacity to elevate γ-glutamyltranspeptidase activity in rats.[2]

Leukoencephalomalacia (LEM) was reproduced in a horse by the intravenous (i.v.) injection of seven daily doses of FB$_1$ at a level of 0.125 mg/g live mass spread over 10 days (see reference 73). LEM was produced in two horses via the oral administration of FB$_1$ at a level of 1.25–4 mg/g body weight, and symptoms occurred in around 25 days (see reference 81). Pulmonary edema was produced in a pig after daily injections of 0.4 mg FB$_1$/g body weight for 4 days.[92] The prevalence of human esophageal cancer in the Transkei, South Africa, is statistically correlated with high levels of FB$_1$ and FB$_2$ in corn.[90]

SAMBUTOXIN

The sambutoxin mycotoxin was first reported in 1994,[45] and its structure is shown below. It is associated with dry-rotted potatoes and is produced primarily by strains of *Fusarium sambucinum* and *F. oxysporum*. Of 13 *Fusarium* species examined, about 90% of strains of the two species noted produced this toxin. From rotten potato samples in Korea, 9 of 21 contained 15.8–78.1 ng/g of sambutoxin with a mean of 49.2 ng/g.[46] Using wheat media, levels of 1.1–101 μg/g of sambutoxin were produced. The toxin was found in potatoes from parts of Iran that had a high incidence of esophageal cancer.[15]

Sambutoxin causes hemorrhage in the stomach and intestines of rats, and the animals refuse feed and lose weight.[45] Rats died within 4 days when their diets contained 0.1% sambutoxin. It is toxic to chick embryos with an LD_{50} of 29.6 μg/egg.[45]

ZEARALENONE

There are at least five naturally occurring zearalenones, and they are produced by *Fusarium* spp., mainly *F. graminearum* (formerly *F. roseum*, = *Gibberella zeae*) and *F. tricinctum*. Associated with corn, these organisms invade field corn at the silking stage, especially during heavy rainfall. If the moisture levels remain high enough following harvesting, the fungi grow and produce toxin. Other crops, such as wheat, oats, barley, and sesame, may be affected in addition to corn.

Zearalenone

The toxins fluoresce blue–green under long-wave UV and greenish under short-wave UV. They possess estrogenic properties and promote estrus in mice and hyperestrogenism in swine. Although they are nonmutagenic in the Ames assay, they produce a positive response in the *Bacillus subtilis* Rec assay.[84]

CONTROL OF PRODUCTION

A number of organisms, especially other fungi, have been shown to control the growth of toxigenic fungi and to inhibit toxin production (for reviews, see references 36 and 74). Among the early studies on the detoxification of aflatoxins was that of Ciegler et al.,[21] who showed that the bacterium *Flavobacterium aurantiacum* removed aflatoxins from solution. Actively growing yeasts have been shown to degrade patulin.[11] Among lactobacilli, *L. acidophilus* was found to be an efficient inhibitor of growth and toxin production by *A. flavus*.[43] Colonization of maize by *Fusarium* spp. has been shown to be clearly inhibited by *Aspergillus* and *Penicillium* spp. at 25°C, depending on a_w and the species tested.[56] Interactions that led to decreased colonization by *Fusarium* did not negatively affect fumonisin production.

Attempts to control the growth of *Botrytis cinerea* on apples have included testing *Burkholderia cepacia*, *Erwinia* sp., *Pichia guilliermondii*, *Cryptococcus* sp., *Acremonium breve*, and *Trichoderma pseudokoningii*, and all were found to be effective.[26] The most effective was an *Erwinia* sp., especially under ambient conditions.

REFERENCES

1. Abarca, M.L., F. Accensi, M.R. Bragulat, G. Castellá, and F.J. Cabañes. 2003. *Aspergillus carbonarius* as the main source of ochratoxin A contamination in dried vine fruits from the Spanish market. *J. Food Protect.* 66:504–506.

2. Alberts, J.F., W.C.A. Gelderblom, P.G. Thiel, W.F.O. Marasas, D.J. van Schalkwyk, and Y. Behrend. 1990. Effects of temperature and incubation period on production of fumonisin B₁ by *Fusarium moniliforme*. *Appl. Environ. Microbiol.* 56:1729–1733.

3. Altug, T., A.E. Yousef, and E.H. Marth. 1990. Degradation of aflatoxin B₁ in dried figs by sodium bisulfite with or without heat, ultraviolet energy or hydrogen peroxide. *J. Food Protect.* 53:581–582.

4. Ayres, J.C., J.O. Mundt, and W.E. Sandine. 1980. *Microbiology of Foods*, 658–683. San Francisco: Freeman.

5. Bacon, C.W., J.G. Sweeney, J.D. Hobbins, and D. Burdick. 1973. Production of penicillic acid and ochratoxin A on poultry feed by *Aspergillus ochraceus*: Temperature and moisture requirements. *Appl. Microbiol.* 26:155–160.

6. Buchanan, R.L., M.A. Harry, and M.A. Gealt. 1983. Caffeine inhibition of sterigmatocystin, citrinin, and patulin production. *J. Food Sci.* 48:1226–1228.

7. Bullerman, L.B. 1976. Examination of Swiss cheese for incidence of mycotoxin producing molds. *J. Food Sci.* 41:26–28.

8. Bullerman, L.B. 1984. Effects of potassium sorbate on growth and patulin production by *Penicillium patulum* and *Penicillium roquefortii*. *J. Food Protect.* 47:312–316.

9. Bullerman, L.B., P.A. Hartman, and J.C. Ayres. 1969. Aflatoxin production in meats. I. Stored meats. *Appl. Microbiol.* 18:714–717.

10. Bullerman, L.B., P.A. Hartman, and J.C. Ayres. 1969. Aflatoxin production in meats. II. Aged dry salamis and aged country cured hams. *Appl. Microbiol.* 18:718–722.

11. Burroughs, L.F. 1977. Stability of patulin to sulfur dioxide and to yeast fermentation. *J. Assoc. Off. Anal. Chem.* 60:100–103.

12. Busby, W.F., Jr., and G.N. Wogan. 1979. Food-borne mycotoxins and alimentary mycotoxicoses. In *Foodborne Infections and Intoxications*, ed. H. Riemann and F.L. Bryan, 519–610. New York: Academic Press.

13. Butler, W.H. 1974. Aflatoxin. In *Mycotoxins*, ed. I.F.H. Purchase, 1–28. New York: Elsevier.

14. Camou-Arriola, J.P., and R.L. Price. 1989. Destruction of aflatoxin and reduction of mutagenicity of naturally-contaminated corn during production of a corn snack. *J. Food Protect.* 52:814–817.

15. Castelo, M.M., S.S. Sumner, and L.B. Bullerman. 1998. Occurrence of fumonisins in corn-based food products. *J. Food Protect.* 61:704–707.

16. Castelo, M.M., S.S. Sumner, and L.B. Bullerman. 1998. Stability of fumonisins in thermally processed corn products. *J. Food Protect.* 61:1030–1033.

17. Chelkowski, J., and A. Visconti, eds. 1992. *Alternaria: Biology, Plant Diseases and Metabolites*. New York: Elsevier.

18. Christensen, C.M. 1971. Mycotoxins. *CRC Crit. Rev. Environ. Cont.* 2:57–80.

19. Chulze, S.N., A.M. Torres, A.M. Dalcero, M.G. Etcheverry, M.L. Ramirez, and M.C. Farnochi. 1995. *Alternaria* mycotoxins in sunflower seeds: Incidence and distribution of the toxins in oil and meal. *J. Food Protect.* 58:1133–1135.

20. Ciegler, A., H.-J. Mintzlaff, D. Weisleder, and L. Leistner. 1972. Potential production and detoxification of penicillic acid in mold-fermented sausage (salami). *Appl. Microbiol.* 24:114–119.

21. Ciegler, A., E.B. Lillehoj, R.E. Peterson, and L. Leistner. 1966. Microbial detoxification of aflatoxin. *Appl. Microbiol.* 14:934–939.

22. Davis, N.D., and U.L. Diener. 1987. Mycotoxins. In *Food and Beverage Mycology*, 2nd ed., ed. L.R. Beuchat, 517–570. New York: Kluwer Academic Publishers.

23. Davis, V.M., and M.E. Stack. 1991. Mutagenicity of stemphyltoxin III, a metabolite of *Alternaria alternata*. *Appl. Environ. Microbiol.* 57:180–182.

24. Deger, G.E. 1976. Aflatoxin—human colon carcinogenesis? *Ann. Intern. Med.* 85:204–205.

25. de Nijs, M., H.P. van Egmond, M. Nauta, F. Rombouts, and S.H.W. Notermans. 1998. Assessment of human exposure to fumonisin B₁. *J. Food Protect.* 61:879–884.

26. Dock, L.L., P.V. Nielsen, and J.D. Floros. 1998. Biological control of *Botrytis cinerea* growth on apples stored under modified atmospheres. *J. Food Protect.* 61:1661–1665.

27. Drush, S., and W. Ragab. 2003. Mycotoxins in fruits, fruit juices, and dried fruits. *J. Food Protect.* 66:1514–1527.

28. D'Souza, D.H., and R.E. Brackett. 2000. The influence of divalent cations and chelators on aflatoxin B₁ degradation by *Flavobacterium aurantiacum*. *J. Food Protect.* 63:102–105.

29. El-Banna, A.A., and P.M. Scott. 1984. Fate of mycotoxins during processing of foodstuffs. III. Ochratoxin A during cooking of faba beans (*Vicia faba*) and polished wheat. *J. Food Protect.* 47:189–192.

30. Enomoto, M., and M. Saito. 1972. Carcinogens produced by fungi. *Annu. Rev. Microbiol.* 26:279–312.

31. Escher, F.E., P.E. Koehler, and J.C. Ayres. 1973. Production of ochratoxins A and B on country cured ham. *Appl. Microbiol.* 26:27–30.

32. Escobar, A., and O.S. Regueiro. 2002. Determination of aflatoxin B_1 in food and feedstuffs in Cuba (1990 through 1996) using an immunoenzymatic reagent kit (Aflacen). *J. Food Protect.* 65:219–221.

33. Farber, J.M., and G.W. Sanders. 1986. Fusarin C production by North American isolates of *Fusarium moniliforme*. *Appl. Environ. Microbiol.* 51:381–384.

34. Gelderblom, W.C.A., N.P.J. Kriek, W.F.O. Marasas, and P.G. Thiel. 1991. Toxicity and carcinogenicity of the *Fusarium moniliforme* metabolite, fumonisin B_1, in rats. *Carcinogenesis* 12:1247–1251.

35. Gelderblom, W.C.A., K. Jaskiewicz, W.F.O. Marasas, P.G. Thiel, R.M. Horak, R. Vleggaar, and N.P.J. Kriek. 1988. Fumonisins—novel mycotoxins with cancer-promoting activity produced by *Fusarium moniliforme*. *Appl. Environ. Microbiol.* 54:1806–1811.

36. Gourama, H., and L.B. Bullerman. 1995. Antimycotic and antiaflatoxigenic effect of lactic acid bacteria: A review. *J. Food Protect.* 58:1275–1280.

37. Gutema, T., C. Munimbazi, and L.B. Bullerman. 2000. Occurrence of fumonisins and moniliformin in corn and corn-based food products of U.S. origin. *J. Food Protect.* 63:1732–1737.

38. Harrison, M.A. 1989. Presence and stability of patulin in apple products: A review. *J. Food Saf.* 9:147–153.

39. Henry, S.H., F.X. Bosch, T.C. Troxell, and P.M. Bolger. 1999. Reducing liver cancer—global control of aflatoxin. *Science* 286:2453–2454.

40. Hesseltine, C.W. 1967. Aflatoxins and other mycotoxins. *Health Lab. Sci.* 4:222–228.

41. Holmquist, G.U., H.W. Walker, and H.M. Stahr. 1983. Influence of temperature, pH, water activity and antifungal agents on growth of *Aspergillus flavus* and *A. parasiticus*. *J. Food Sci.* 48:778–782.

42. Jorgenssen, K.V., D.L. Park, S.M. Rua, Jr., and R.L. Price. 1990. Reduction of mutagenic potentials in milk: Effects of ammonia treatment on aflatoxin-contaminated cotton-seed. *J. Food Protect.* 53:777–778, 817.

43. Karunaratne, A., E. Wezenberg, and L.B. Bullerman. 1990. Inhibition of mold growth and aflatoxin production by *Lactobacillus* spp. *J. Food Protect.* 53:230–236.

44. Kedera, C.J., R.D. Plattner, and A.E. Desjardins. 1999. Incidence of *Fusarium* spp. and levels of fumonisin B_1 in maize in western Kenya. *Appl. Environ. Microbiol.* 65:41–44.

45. Kim, J.-C., and Y.-W. Lee. 1994. Sambutoxin, a new mycotoxin produced by toxic *Fusarium* isolates obtained from rotted potato tubers. *Appl. Environ. Microbiol.* 60:4380–4386.

46. Kim, J.-C., Y.-W. Lee, and S.-H. Yu. 1995. Sambutoxin-producing isolates of *Fusarium* species and occurrence of sambutoxin in rotten potato tubers. *Appl. Environ. Microbiol.* 61:3750–3751.

47. Kurtzman, C.P., B.W. Horn, and C.W. Hesseltine. 1987. *Aspergillus nominus*, a new aflatoxin-producing species related to *Aspergillus flavus* and *Aspergillus tamarii*. *Antonie van Leeuwenhoek* 53:147–158.

48. Labuza, T.P. 1983. Regulation of mycotoxins in food. *J. Food Protect.* 46:260–265.

49. Lee, Y.J., and W.M. Hagler, Jr. 1991. Aflatoxin and cyclopiazonic acid production by *Aspergillus flavus* isolated from contaminated maize. *J. Food Sci.* 56:871–872.

50. Leistner, L., and C. Eckardt. 1979. Vorkommen toxigener Penicillien bei Fleischerzeugnissen. *Fleischwirtsch.* 59:1892–1896.

51. Leistner, L., and F. Tauchmann. 1979. Aflatoxinbildung in Rohwurst durch verschiedene *Aspergillus flavus*-Stämme und einer *Aspergillus parasiticus*-Stamm. *Fleischwirtschaft* 50:965–966.

52. Lie, J.L., and E.H. Marth. 1967. Formation of aflatoxin in cheddar cheese by *Aspergillus flavus* and *Aspergillus parasiticus*. *J. Dairy Sci.* 50:1708–1710.

53. Line, J.E., R.E. Brackett, and R.E. Wilkinson. 1994. Evidence for degradation of aflatoxin B_1 by *Flavobacterium aurantiacum*. *J. Food Protect.* 57:788–791.

54. Marasas, W.F.O., K. Jaskiewicz, F.S. Venter, and D.J. van Schalkwyk. 1988. *Fusarium moniliforme* contamination of maize in oesophageal cancer areas in Transkei. *S. Afr. Med. J.* 74:110–114.

55. Marin, S., N. Magan, J. Serra, A.J. Ramos, R. Canela, and V. Sanchis. 1999. Fumonisin B_1 production and growth of *Fusarium moniliforma* and *Fusarium proliferatum* on maize, wheat, and barley grain. *J. Food Sci.* 64:921–924.

56. Marin, S., V. Sanchis, F. Rull, A.J. Ramos, and N. Magan. 1998. Colonization of maize grain by *Fusarium moniliforme* and *Fusarium proliferatum* in the presence of competing fungi and their impact on fumonisin production. *J. Food Protect.* 61:1489–1496.

57. Marin, S., V. Sanchis, I. Vines, R. Canela, and N. Magan. 1995. Effect of water activity and temperature on growth and fumonisin B_1 and B_2 production by *Fusarium proliferatum* and *F. moniliforme* on maize grain. *Lett. Appl. Microbiol.* 21:298–301.

58. Marth, E.H., and B.G. Calanog. 1976. Toxigenic fungi. In *Food Microbiology: Public Health and Spoilage Aspects*, ed. M.P. deFigueiredo and D.F. Splittstoesser, 210–256. New York: Kluwer Academic Publishers.

59. Mphande, F.A., B.A. Siame, and J.E. Taylor. 2004. Fungi aflatoxins, and cyclopiazonic acid associated with peanut retailing in Botswana. *J. Food Protect.* 67:96–102.

60. Nelson, P.E., R.D. Plattner, D.D. Shackelford, and A.E. Desjardins. 1992. Fumonisin B_1 production by *Fusarium* species other than *F. moniliforme* in section *Liseola* and by some related species. *Appl. Environ. Microbiol.* 58:984–989.

61. Newsome, R. 1999. Issues in international trade: Looking to the Codex Alimentarius Commission. *Food Technol.* 53(6):26.

62. Niranjan, B.G., N.K. Bhat, and N.G. Avadhani. 1982. Preferential attack of mitochondrial DNA by aflatoxin B_1 during hepatocarcinogenesis. *Science* 215:73–75.

63. Northolt, M.D., C.A.H. Verhulsdonk, P.S.S. Soentoro, and W.E. Paulsch. 1976. Effect of water activity and temperature on aflatoxin production by *Aspergillus parasiticus*. *J. Milk Food Technol.* 39:170–174.

64. Oatley, J.T., M.D. Rarick, G.E. Ji, and J.E. Linz. 2000. Binding of aflatoxin B_1 to bifidobacteria in vitro. *J. Food Protect.* 63:1133–1136.

65. Pestka, J.J., J.I. Azcona-Olivera, R.D. Plattner, F. Minervini, M.B. Doke, and A. Visconti. 1994. Comparative assessment of fumonisin in grain-based foods by ELISA, GC–MS, and HPLC. *J. Food Protect.* 57:169–172.

66. Pierides, M., H. El-Nezami, K. Peltonen, S. Salminen, and J. Ahokas. 2000. Ability of dairy strains of lactic acid bacteria to bind aflatoxin M_1 in a food model. *J. Food Protect.* 63:645–650.

67. Pittet, A., V. Parisod, and M. Schellenberg. 1992. Occurrence of fumonisins B_1 and B_2 in corn-based products from the Swiss market. *J. Agric. Food Chem.* 40:1352–1354.

68. Park, K.Y., and L.B. Bullerman. 1983. Effect of cycling temperatures on aflatoxin production by *Aspergillus parasiticus* and *Aspergillus flavus* in rice and cheddar cheese. *J. Food Sci.* 48:889–896.

69. Ram, B.P., P. Hart, R.J. Cole, and J.J. Pestka. 1986. Application of ELISA to retail survey of aflatoxin B_1 in peanut butter. *J. Food Protect.* 49:792–795.

70. Rheeder, J.P., W.F.O. Marasas, P.G. Thiel, E.W. Sydenham, and D.J. van Schalkwyk. 1992. *Fusarium moniliforme* and fumonisins in corn in relation to human esophageal cancer in Transkei. *Phytophathology* 82:353–357.

71. Roland, J.O., and L.R. Beuchat. 1984. Biomass and patulin production by *Byssochlamys nivea* in apple juice as affected by sorbate, benzoate, SO_2 and temperature. *J. Food Sci.* 49:402–406.

72. Ross, P.F., L.G. Rice, R.D. Plattner, G.D. Osweiler, T.M. Wilson, D.L. Owens, H.A. Nelson, and J.L. Richard. 1991. Concentration of fumonisin B_1 in feeds associated with animal health problems. *Mycopathology* 114:129–135.

73. Ross, P.F., P.E. Nelson, J.L. Richard, G.D. Osweiler, L.G. Rice, R.D. Plattner, and T.M. Wilson. 1990. Production of fumonisins by *Fusarium moniliforme* and *Fusarium proliferatum* isolates associated with equine leukoencephalomalacia and a pulmonary edema syndrome in swine. *Appl. Environ. Microbiol.* 56:3225–3226.

74. Schillinger, U., R. Geisen, and W.H. Holzapfel. 1996. Potential of antagonistic microorganisms and bacteriocins for the biological preservation of foods. *Trends Food Sci. Technol.* 7:158–164.

75. Schindler, A.F. 1977. Temperature limits for production of aflatoxin by twenty-five isolates of *Aspergillus flavus* and *Aspergillus parasiticus*. *J. Food. Protect.* 40:39–40.

76. Serck-Hanssen, A. 1970. Aflatoxin-induced fatal hepatitis? *Arch. Environ. Health* 20:729–731.

77. Shelef, L.A., and B. Chin. 1980. Effect of phenolic antioxidants on the mutagenicity of aflatoxin B_1. *Appl. Environ. Microbiol.* 40:1039–1043.

78. Shih, C.N., and E.H. Marth. 1974. Some cultural conditions that control biosynthesis of lipid and aflatoxin by *Aspergillus parasiticus*. *Appl. Microbiol.* 27:452–456.

79. Shim, W.-B., J.E. Flaherty, and C.P. Woloshuk. 2003. Comparison of fumonisin B_1 biosynthsis in maize germ and degermed kernels by *Fusarium verticillioides*. *J. Food Protect.* 66:2116–2122.

80. Shotwell, O.L., and C.W. Hesseltine. 1983. Five-year study of mycotoxins in Virginia wheat and dent corn. *J. Assoc. Off. Anal. Chem.* 66:1466–1469.

81. Siame, B.A., S.F. Mpuchane, B.A. Gashe, J. Allotey, and G. Teffera. 1998. Occurrence of aflatoxins, fumonisin B$_1$, and zearalenone in foods and feeds in Botswana. *J. Food Protect.* 61:1670–1673.

82. Sommer, N.F., J.R. Buchanan, and R.J. Fortlage. 1974. Production of patulin by *Penicillium expansum*. *Appl. Microbiol.* 28:589–593.

83. Stahr, H.M., R.L. Pfeiffer, P.J. Imerman, B. Bork, and C. Hurburgh. 1990. Aflatoxins—The 1988 outbreak. *Dairy Food Environ. Sanit.* 10:15–17.

84. Stark, A.A. 1980. Mutagenicity and carcinogenicity of mycotoxins: DNA binding as a possible mode of action. *Annu. Rev. Microbiol.* 34:235–262.

85. Stinson, E.E., D.D. Bills, S.F. Osman, J. Siciliano, M.J. Ceponis, and E.G. Heisler. 1980. Mycotoxin production by *Alternaria* species grown on apples, tomatoes, and blueberries. *J. Agric. Food Chem.* 28:960–963.

86. Stinson, E.E., S.F. Osman, E.G. Beisler, J. Siciano, and D.D. Bills. 1981. Mycotoxin production in whole tomatoes, apples, oranges, and lemons. *J. Agric. Food Chem.* 29:790–792.

87. Stoloff, L. 1987. Carcinogenicity of aflatoxins. *Science* 237:1283–1284 (letter to editor with two responses).

88. Sydenham, E.W., W.F.O. Marasas, G.S. Shephard, P.G. Thiel, and E.Y. Hirooka. 1992. Fumonisin concentrations in Brazalian feeds associated with field outbreaks of confirmed and suspected animal mycotoxicoses. *J. Agric. Food Chem.* 40:994–997.

89. Sydenham, E.W., G.S. Shepard, P.G. Thiel, W.F.O. Marasas, and S. Stockenström. 1991. Fumonisin contamination of commercial corn-based human foodstuffs. *J. Agric. Food Chem.* 39:2014–2018.

90. Sydenham, E.W., P.G. Thiel, W.F.O. Marasas, G.S. Shephard, D.J. van Schalkwyk, and K.R. Koch. 1990. Natural occurrence of some *Fusarium* mycotoxins in corn from low and high esophageal cancer prevalence areas of the Transkei, southern Africa. *J. Agric. Food Chem.* 38:1900–1903.

91. Taylor, W.J., and F.A. Draughon. 2001. *Nannocystis exedens*: A potential biocompetitive agent against *Aspergillus flavus* and *Aspergillus parasiticus*. *J. Food Protect.* 64:1030–1034.

92. Thiel, P.G., W.F.O. Marasas, E.W. Sydenham, G.S. Shephard, and C.A. Gelderblom. 1992. The implications of naturally occurring levels of fumonisins in corn for human and animal health. *Mycopathologia* 117:3–9.

93. Thompson, D.P. 1997. Effect of phenolic compounds on mycelial growth of *Fusarium* and *Penicillium* species. *J. Food Protect.* 60:1262–1264.

94. Tong, C.-H., and F.A. Draughon. 1985. Inhibition by antimicrobial food additives of ochratoxin A production by *Aspergillus sulphureus* and *Penicillium viridicatum*. *Appl. Environ. Microbiol.* 49:1407–1411.

95. Torres, M.R., V. Sanchis, and A.J. Ramos. 1998. Occurrence of fumonisins in Spanish beers analyzed by an enzyme-linked immunosorbent assay method. *Int. J. Food Microbiol.* 39:139–143.

96. Trenk, H.L., and P.A. Hartman. 1970. Effects of moisture content and temperature on aflatoxin production in corn. *Appl. Microbiol.* 19:781–784.

97. Wheeler, K.A., B.F. Hurdman, and J.I. Pitt. 1991. Influence of pH on the growth of some toxigenic species of *Aspergillus*, *Penicillium* and *Fusarium*. *Int. J. Food Microbiol.* 12:141–150.

98. Wiebe, L.A., and L.F. Bjeldanes. 1981. Fusarin C, a mutagen from *Fusarium moniliforme* grown on corn. *J. Food Sci.* 46:1424–1426.

99. Wheeler, J.L., M.A. Harrison, and P.E. Koehler. 1987. Presence and stability of patulin in pasteurized apple cider. *J. Food Sci.* 52:479–480.

100. Wogan, G.N. 1966. Chemical nature and biological effects of the aflatoxins. *Bacteriol. Rev.* 30:460–470.

101. Wu, M.T., J.C. Ayres, and P.E. Koehler. 1974. Production of citrinin by *Penicillium viridicatum* on country-cured ham. *Appl. Microbiol.* 27:427–428.

102. Yoshizawa, T., A. Yamashita, and Y. Luo. 1994. Fumonisin occurrence in corn from high- and low-risk areas for human esophageal cancer in China. *Appl. Environ. Microbiol.* 60:1626–1629.

CHAPTER 31

Viruses and Some Other Proven and Suspected Foodborne Biohazards

VIRUSES

Much less is known about the incidence of viruses in foods than about bacteria and fungi, for several reasons. First, being obligate parasites, viruses do not grow on culture media as do bacteria and fungi. The usual methods for their cultivation consist of tissue culture and chick embryo techniques. Second, because viruses do not replicate in foods, their numbers may be expected to be low relative to bacteria, and extraction and concentration methods are necessary for their recovery. Although much research has been devoted to this methodology, it is difficult to effect more than about a 50% recovery of virus particles from products such as ground beef. Third, laboratory virological techniques are not practiced in many food microbiology laboratories. Finally, not all viruses of potential interest to food microbiologists can be cultured by existing methods (the Norwalk virus is one example). However, the development of reverse transcription–polymerase chain reaction (RT–PCR) detection methodology has allowed the direct detection of some foodborne viruses in oyster and clam tissues.[5]

The efficacy of the RT–PCR technique to detect viruses in foods has been demonstrated by a number of researchers. In one study, four concentration and extraction methods were compared for the recovery of added astrovirus, hepatitis A, and poliovirus from mussels, and the glycine solution and borate buffer methods were found to be best.[106] Using RT–PCR, several combinations were effective in allowing detection of the three viruses from the analyzed samples. By using oysters spiked with 10^1 to 10^5 plaque-forming units (pfu) of poliovirus 1 or hepatitis A and concentration with polyethylene glycol, the combined concentration and purification scheme permitted the detection of poliovirus and hepatitis A, and 10^5 RT–PCR amplifiable units of the Norwalk agent.[53] In another study, dot-blot hybridization detection of amplicons from the RT–PCR allowed the detection of as few as 8 pfu of hepatitis A per gram of oyster meat.[32] The use of other concentration methods for hardshell clams spiked with poliovirus 1 or hepatitis A at 10^3 pfu allowed the recovery of 7 to 50% of poliovirus 1 and 0.3–8% of hepatitis A.[34] When clam meat was spiked with the Norwalk agent, detection at levels as low as 450 RT–PCR units/50 g of clam extract was achieved.[34]

Because it has been demonstrated that under unsanitary conditions any intestinal bacterial pathogen may be found in foods, the same may be presumed for intestinal viruses, even though they do not proliferate in foods. Cliver et al.[30] noted that virtually any food can serve as a vehicle for virus transmission, and they have stressed the importance of the anal–oral mode of transmission, especially for viral hepatitis of food origin. Just as nonintestinal bacteria of human origin are sometimes found in foods, the same may be true for viruses, but because of their tissue affinities, foods would serve as vehicles only for the intestinal or enteroviruses. These agents may be accumulated by some shellfish up to 900-fold.[41] Viral gastroenteritis is believed to be second only to the common cold in frequency.

Incidence in Foods and the Environment

A common food source of gastroenteritis-causing viruses is shellfish. Although crustaceans do not concentrate viruses, molluskan shellfish do because they are filter feeders. When poliovirus 1 was added to waters, blue crabs were contaminated, but they did not concentrate the virus.[47] Shucked oysters artificially contaminated with 10^4 pfu of a poliovirus retained viruses during refrigeration for 30–90 days with a survival rate of 10–13%.[33] It has been reported that the uptake of enteroviruses by oysters and clams is not likely when viruses in the water column are less than 0.01 pfu/ml.[66] The recovery method employed by the latter authors was capable of detecting 1.5–2.0 pfu per shellfish.

Although the coliform index is of proven value as an indicator of intestinal bacterial pathogens in waters, it appears to be inadequate for enteroviruses, which are more resistant to adverse environmental conditions than bacterial pathogens.[90] In a study of more than 150 samples of recreational waters from the upper Texas gulf, enteroviruses were detected 43% of the time when by coliform index the samples were judged acceptable, and 44% of the time when judged acceptable by fecal coliform standards.[42] In the same study, enteroviruses were found 35% of the time in waters that met acceptable standards for shellfish harvesting, and the investigators concluded that the coliform standard for waters does not reflect the presence of viruses. From a study of hard-shell clams off the coast of North Carolina, enteric viruses were found in those from open and closed beds.[113] (Closed waters are those not open to commercial shellfishing because of coliform counts.) From open beds, 3 of 13 100-g samples were positive for viruses, whereas all 13 were negative for salmonellae, shigellae, or yersiniae. From closed waters, 6 of 15 were positive for salmonellae, and all were negative for shigellae and yersiniae.[113] The latter investigators found no correlation between enteric viruses and total coliforms or fecal coliforms in shellfish waters, or total coliforms, fecal coliforms, "fecal streptococci," or aerobic plate counts (APC) in clams. Although enteric viruses may be found in shellfish from open waters, less than 1% of shellfish samples examined by the Food and Drug Administration (FDA) contained viruses.[67] (See Chapter 20 for further discussion of safety indicators and intestinal viruses.)

With respect to the capacity of certain viruses to persist in foods, it has been shown that enteroviruses persisted in ground beef up to 8 days at 23°C or 24°C and were not affected by the growth of spoilage bacteria.[48] In a study of 14 vegetable samples for the existence of naturally occurring viruses, none were found, but coxsackievirus B5 inoculated onto vegetables did survive at 4°C for 5 days.[62] In an earlier study, these investigators showed that coxsackievirus B5 had no loss of activity when added to lettuce and stored at 4°C under moist conditions for 16 days. Several enteric viruses failed to survive on the surfaces of fruits, and no naturally occurring viruses were found in nine fruits that were examined.[63] Echovirus 4 and poliovirus 1 were found in 1 each of 17 samples of raw oysters examined by Fugate et al.,[38] and poliovirus 3 was found in 1 of 24 samples of oysters. Of seven food-processing plants surveyed for human viruses, none was found in a vegetable-processing plant or in three that processed animal products.[64] The latter investigators examined 60 samples of market foods but were unable to detect viruses in any. They concluded that viruses in the U.S. food supply are very low.

Destruction in Foods

The survival of the hog cholera (HCV) and African swine fever viruses (ASFV) in processed meats was studied by McKercher et al.[72] From pigs infected with these viruses, partly cooked canned hams and dried pepperoni and salami sausages were prepared; whereas virus was not recovered from the partly cooked canned hams, they were recovered from hams after brining but not after heating. The ASFV retained viability in the two sausage products following the addition of curing ingredients and starters but were negative after 30 days. HCV also survived the addition of curing ingredients and starter and retained viability even after 22 days.

The effect of heating on destruction of the foot-and-mouth virus was evaluated by Blackwell et al.[8] When ground beef was contaminated with virus-infected lymph node tissue and processed to an internal temperature of 93.3°C, the virus was destroyed. However, in cattle lymph node tissue, the virus survived for 15 but not 30 minutes at 90°C. In another study, the thermal destruction of foot-and-mouth disease virus (FMDV) along with two others was achieved within 3 minutes at 67°C, or at 3 minutes at 60°C or 62°C in a pig slurry.[108] The boiling of crabs was found sufficient to inactivate 99.9% of poliovirus 1, and a rotavirus and an echovirus were destroyed within 8 minutes.[47] A poliovirus was found to survive stewing, frying, baking, and steaming of oysters.[33] In broiled hamburgers, enteric viruses could be recovered from 8 of 24 patties cooked rare (to 60°C internally) if the patties were cooled immediately to 23°C.[99] No viruses were detected if the patties were allowed to cool for 3 minutes at room temperature before testing.

Hepatitis A Virus

Prior to the 1990s, there were more documented outbreaks of hepatitis A traced to foods than any other viral infection. The virus belongs to the family Picornaviridae, as do the polio, echo, and coxsackie viruses, and all have single-stranded RNA (ss RNA) genomes. The incubation period for infectious hepatitis ranges from 15 to 45 days, and lifetime immunity usually occurs after an attack. The fecal–oral route is the mode of transmission, and raw or partially cooked shellfish from polluted waters is the most common vehicle food.

In the United States in 1973, 1974, and 1975, there were 5, 6, and 3 outbreaks, respectively, with 425, 282, and 173 cases. The 1975 outbreaks were traced to salad, sandwiches, and glazed doughnuts served in restaurants. The recorded outbreaks and cases in the United States for 1983–1987 are presented in Table 31–1. According to the Centers for Disease Control and Prevention (CDC), hepatitis A increased in the United States between 1983 and 1989 by 58%—from 9.2 to 14.5 per 100,000 persons.[25] Of the

Table 31–1 Outbreaks, Cases, and Deaths Associated with Viral Foodborne Gastroenteritis in the United States, 1983–1987

| Year | Outbreaks/Cases/Deaths | | |
	Hepatitis A	*Norwalk Agent*	*Other Viruses*
1983	10/530/1	1/20/0	–
1984	2/29/0	1/137/0	1/444/0
1985	5/118/0	4/179/0	1/114/0
1986	3/203/0	3/463/0	–
1987	9/187/0	1/365/0	–
Totals	29/1067/1	10/1164/0	2/558/0

Source: From N.H. Bean, P.M. Griffin, J.S. Goulding, and C.B. Ivey, 1990. *J. Food Protect.* 53:711–728.

cases in 1988, 7.3% were either foodborne or waterborne.[25] Among 88 elementary school students and their teachers in 1990 in the state of Georgia, 15 came down with hepatitis A. Among 641 residents and staff at an institution for the disabled in Montana, 13 contracted hepatitis A. Strawberry shortcake was the vehicle in both outbreaks. The frozen strawberries came from the same processing plant in California.[81]

The largest foodborne outbreak of hepatitis A ever recorded in the United States occurred in November 2003, and there were ca. 600 victims with 3 deaths. The vehicle food was imported green onions (scallions) served by a fast-food restaurant chain. Between 1992 and 2001, an estimated 230,000 cases of hepatitis A were reported to the CDCP. In 2001, an outbreak of >46 cases occurred in Massachusetts and it was associated with the consumption of sandwiches, which apparently were contaminated by a food handler.[16]

Noroviruses

The former Norwalk, Norwalk-like, and small-round-structured viruses (SRSV) have been placed in this group as the genus *Norovirus* of the human caliciviruses (HuCV). The noroviruses are in two genogroups, I and II. The former Norwalk group is the prototype for genogroup I, and the former Snow Mounain viruses are in genogroup II. All of these viruses are unenveloped and ssRNA with a diameter of 27–40 nm and their genome consists of 7,300 to 8,300 base pairs. The caliciviruses also include the genus *Sapovirus*. For more information, see references 40, 86, 96.

The Norwalk virus was first recognized in a school outbreak in Norwalk, Ohio in 1968, and water was suspected, but not proven, as the source. It is the most prevalent of the noroviruses in foods. The virus is more resistant to destruction by chlorine than other enteric viruses. In volunteers, 3.75 ppm chlorine in drinking water failed to inactivate the virus, whereas poliovirus type 1 and human and simian rotaviruses were inactivated.[58] Some Norwalk viruses remained infective at residual chlorine levels of 5–6 ppm. Hepatitis A viruses are not as resistant as Norwalk, but both are clearly more resistant to chlorine than the rotaviruses. The exposure of noroviruses to 0.37 mg/l of ozone at pH 7 and 5°C for up to 5 minutes in water effected a >3 $\log_{10}$ reduction after a 10 second exposure.[97]

Of 430 foodborne outbreaks in the United States in 1979, 4% displayed the pattern of Norwalk gastroenteritis.[57] This virus was thought to be the cause of more foodborne gastroenteritis than any single bacterium in the state of Minnesota in 1985.[65] Noroviruses are now the leading cause of gastroenteritis in the United States with an estimated 23 million cases per year.[17] From several cruise ship outbreaks in 2002, there were at least 1,786 victims. Up to 21 ship outbreaks occurred during the same general time period, and the virus appeared to come from unspecified environmental sources.

Of 1,412 outbreaks of foodborne intestinal diseases in England and Wales for the years 1992–1999, 82 (5.8%) were caused by noroviruses (all previously designated SRSVs). From the same survey, 12 of 60 (20%) were traced to fruits and vegetables.[96] Since they have yet to be cultured in the laboratory, the method of choice is RT–PCR (see Chapter 11).

During the fall of 2001, a rare waterborne outbreak of norovirus gastroenteritis occurred in the state of Wyoming, and there were around 84 victims. Ground water was contaminated with sewage, which was the source of the human HuCV. The etiologic agents were identified by use of RT–PCR, which demonstrated mainly genogroup I, subtype 3; but one stool sample yielded a genogroup II, subtype 6 strain.[86]

Among the earliest reported outbreaks that involved noroviruses is one that occurred in 1976 in England. Between December 21, 1976, and January 10, 1977, 33 outbreaks and 797 cases occurred, and cockles were incriminated.[4] The incubation period was 24–30 hours, and from 12 of 14 stool samples, small, round virus particles measuring 25–26 nm in diameter were demonstrated, but they were not found in cockles. Although the investigators believed that the agents were neither Norwalk nor Hawaii, these outbreaks are regarded by some as Norwalk virus outbreaks. The 1978 outbreak in Australia that involved at least 2,000 persons was well documented, and the vehicle food was oysters.[74] The virus was found in 39% of fecal specimens examined by electron microscopy, and antibody responses were demonstrated in 75% of paired sera tested. The incubation period ranged from 18 to 48 hours, with most cases occurring in 34–38 hours. Nausea was the first symptom, usually accompanied by vomiting, nonbloody diarrhea, and abdominal cramps, with symptoms lasting 2–3 days. Another outbreak in Australia was traced to bottled oysters and symptoms occurred in 24–48 hours.[36] The oysters had an APC of 2.2×10^4/g and a fecal coliform count of 500/100 g. The first documented food source outbreaks in the United States are those that occurred in New Jersey in 1979, where lettuce was the vehicle food, and the Florida outbreak in 1980 that was traced to raw oysters. In the latter, the agent was identified by a radioimmunoassay method.

Rotaviruses

The first demonstration of these viruses occurred in 1973 in Australia, and they were first propagated in the laboratory in 1981. Six groups have been identified, and three are known to be infectious for humans. Group A is the most commonly encountered among infants and young children throughout the world. Group B causes diarrhea in adults, and they have been seen only in China. Rotaviruses belong to the family Reoviridae; they are about 70 nm in diameter, are nonenveloped, and contain double-stranded RNA (dsRNA). The fecal–oral route is the primary mode of transmission.

Rotaviruses cause an estimated one-third of all hospitalizations for diarrhea in children below age 5, and the peak season for infection occurs during the winter months. Most susceptible are children between the ages of 6 months and 2 years, and it has been reported that virtually every child in the United States is infected by age 4.[24] Although most persons are immune by age 4, high inoculum or lowered states of immunity can lead to milder illness among older children and adults.[24] These viruses are known to be transmitted among children in day care centers and by water. A community waterborne outbreak occurred in Eagle-Vail, Colorado, in 1981, and 44% of 128 persons, most of them adults, became ill.[49] They are believed to be only infrequent causes of foodborne gastroenteritis.[30]

The incubation period for rotavirus gastroenteritis is 2 days. Vomiting occurs for 3 days accompanied by watery diarrhea for 3–8 days, and often abdominal pain and fever also occur.[24] They are known to be associated with travelers' diarrhea. It appears that these viruses induce diarrhea by activating the enteric nervous system (ENS) based on the inhibition of ENS functions in mice and in vitro by four drugs that inhibit ENS functions.[70]

For the 23-month period between January 1989 and November 1990, 48,035 stool specimens were examined in the United States, with 9,639 (20%) being positive for rotavirus.[22] The highest percentage of positive stools occurred in February (36%) and the lowest in October (6%). Between 1979 and 1985, an annual average of 500 children died from diarrheal illness in the United States, and 20% were caused by rotavirus infections.[22]

The host-cell-receptor protein for rotavirus also serves as the β-adrenergic receptor. Once inside cells, they are transported to lysosomes where uncoating occurs.

Rotaviral infections can be diagnosed by immunoelectron microscopy, RT–PCR, enzyme-linked immunosorbent assay (ELISA), and latex agglutination methods.

BACTERIA

Enterobacter sakazakii

This bacterium, once classified as a yellow-pigmented *Enterobacter cloaceae*, has been identified as the cause of neonatal necrotizing enterocolitis (NEC); neonatal meningitis; and sepsis, dating back to 1961. The usual vehicle food is powdered milk formulas. Although it is considered to be an opportunistic pathogen, some strains produce an enterotoxin, and they are lethal to suckling mice. *Citrobacter freundii* has been identified as the cause of neonatal infections and the vehicle foods were infant formulas (see reference 105).

In one study, 4 of 18 *E. sakazakii* strains produced enterotoxin, and at 10^8 cfu/mouse by the intraperitoneal route, all 18 isolates were lethal to suckling mice (16–18 days old) and 2 were lethal by the peroral route.[84] In addition to suckling mice, potential virulence is manifested in monolayers of CHO, Vero, and Y-1 adrenal cells. Infant mortality rates range from 40 to 60%. In an outbreak of 12 cases in 1998 in an intensive care unit in Belgium, 2 infants died and *E. sakazakii* was recovered from unused prepared formula and from unopened cans of a single batch.[110]

In a study of enteric bacteria in 141 powdered milk formulas from 35 countries, 25% contained *Pantoea agglomerans*, 21% *E. cloaceae*, and 14% *E. sakazakii*.[75] From a study of 120 dried infant formulas in Canada, 8 (6.7%) contained *E. sakazakii*.[80] The minimum growth temperature of tested isolates in the latter study was 5.5 to 8°C, and the maximum growth temperature ranged from 41 to 45°C with a mean of 42.5°C for 11 isolates. No isolates grew at 4°C.[79]

In regards to the thermal resistance of *E. sakazakii*, it appears to be higher than that of most Gram-negative bacteria. In a study of 10 strains (5 clinical and 5 food isolates) in reconstituted dried-infant formula, a mean $D_{60°C}$ of 2.5 minutes and $z - 5.82°C$ were found.[80] The mean $D_{60°C}$ for the 5 clinical isolates was 2.15, and 3.06 for the 5 food isolates. In another study, the $D_{58°C}$ of 12 strains in rehydrated infant formula ranged from 30.5 to 591.9 seconds (0.508 to 9.865 minutes).[35] By adding the most heat-resistant strain of *E. sakazakii* ($z = 5.6°C$) found in the last study to dry infant formula rehydrated at 70°C, > 4-log reduction was achieved.[35] According to these investigators, if it is assumed that the typical level of this organism in infant formula is 1 cfu/100 g of dry formula, a 4-D treatment should assure its absence after cooling for infant feeding.[35] The heat-resistant strain employed was the most resistant of 12 tested.

Stationary phase cells of *E. sakazakii* have been reported to be more resistant to osmotic and dry stress than *E. coli* and some other bacteria; and this increased resistance appeared to be associated with the accumulation of trehalose by stationary phase cells.[10] The latter investigators found $D_{58°C}$ values of 0.27–0.50 for 5 *E. sakazakii* strains compared to 0.40–0.50 for three salmonellae.

Histamine-Associated (Scombroid) Poisoning

Illness contracted from eating scombroid fish or fish products containing high levels of histamine is often referred to as scombroid poisoning. Among the scombroid fishes are tuna, mackerel, bonito, and others. In one report, histamine poisoning was associated with sailfish, a nonscombroid.[51] The histamine is produced by bacterial decarboxylation of the generally large quantities of histidine in the muscles of this group. Sufficient levels of histamine may be produced without the product being organoleptically unacceptable, with the result that scombroid poisoning may be contracted from both fresh and organoleptically spoiled fish. The history of this syndrome has been reviewed by Hudson and Brown,[50] who questioned the etiological role of histamine. This is discussed further below.

The bacteria most often associated with this syndrome are *Morganella* spp., especially *M. morganii*, of which all strains appear to produce histamine at levels >5,000 ppm. Among other bacteria shown to produce histidine decarboxylase are *Raoulella planticola* and *R. ornithinolytica*,[55] and *Hafnia alvei, Citrobacter freundii, Clostridium perfringens, Enterobacter aerogenes, Vibrio alginolyicus* and *Proteus* spp. A *Morganella morganii* isolate from temperature-abused albacore produced 5,253 ppm histamine in a tuna fish infusion medium at 25°C, and 2,769 ppm at 15°C.[59] Neither growth nor histamine production occurred at 4°C. P *phosphoreum* produces histamine at temperatures at and below 10°C.

From room-temperature spoiled skipjack tuna, 31% of bacterial isolates produced from 100 to 400 mg/dl of histamine in broth.[83] The strong histamine formers were *M. morganii,Proteus* spp., and a *Raoutella* sp., whereas weak formers included *H. alvei* and *Proteus* spp. Skipjack tuna spoiled in seawater at 38°C contained *C. perfringens* and *V. alginolyticus* among other histidine decarboxylase producers.[117] A strain of *M. morganii* isolated from anchovies was shown to produce 2,377 ± 350 ppm histamine in a culture medium at 37°C in 24 hours.[91] This strain also produced detectable levels of putrescine and cadaverine. From an outbreak of scombroid poisoning associated with tuna sashimi, *K. pneumoniae* was recovered and shown to produce 442 mg/dl of histamine in a tuna fish infusion broth.[102] This syndrome has been associated with foods other than scombroid fish, particularly cheeses, including Swiss cheese, which in one case contained 187 mg/dl of histamine; the symptoms associated with the outbreak occurred in 30 minutes to 1 hour after ingestion.[103]

The number of outbreaks reported to the CDC for the years 1972–1986 were 178 with 1,096 cases but no deaths.[26] The largest outbreaks of 51, 29, and 24 occurred in Hawaii, California, and New York, respectively. The three most common vehicle foods were mahi mahi (66 outbreaks), tuna (42 outbreaks), and bluefish (19 outbreaks). Although fresh fish normally contains 1 mg/dl of histamine, some may contain up to 20 mg/dl, a level that may lead to symptoms in some individuals. The FDA hazardous level for tuna is 50 mg/dl.[26] The cooking of toxic fish may not lead to safe products.

The histamine content of stored skipjack tuna can be estimated if incubation times and temperatures of storage are known. Frank et al.[37] found that 100 mg/dl formed in 46 hours at 70°F, in 23 hours at 90°F, and in 17 hours at 100°F. A nomograph was constructed over the temperature range of 70–100°F, underscoring the importance of low temperatures in preventing or delaying histamine formation. Vacuum packaging is less effective than low-temperature storage in controlling histamine production.[114] The culture medium of choice for detecting histamine-producing bacteria is that of Niven et al.[82]

Histamine production is favored by low pH, but it occurs more when products are stored above the refrigerator range. The lowest temperature for production of significant levels was found to be 30°C for *H. alvei, C. freundii, and E. coli;* and 15°C for two strains of *M. morganii*.[6]

The syndrome is contracted by eating fresh or processed fish of the type noted; symptoms occur within minutes and for up to 3 hours after ingestion of toxic food, with most cases occurring within 1 hour. Typical symptoms consist of a flushing of the face and neck accompanied by a feeling of intense heat and general discomfort, and diarrhea. Subsequent facial and neck rashes are common. The flush is followed by an intense, throbbing, headache tapering to a continuous dull ache. Other symptoms include dizziness, itching, faintness, burning of the mouth and throat, and the inability to swallow.[50] The minimum level of histamine thought necessary to cause symptoms is 100 mg/dl. Large numbers of *M. morganii* in fish of the type incriminated in this syndrome and a level of histamine more than 10 mg/dl is considered significant relative to product quality.

The first 50 incidents in Great Britain occurred between 1976 and 1979, with all but 19 occurring in 1979. Canned and smoked mackerel was the most common vehicle, with bonita, sprats, and pilchards involved in one outbreak each. The most common symptom among the 196 cases was diarrhea.[43]

Regarding etiology, Hudson and Brown[50] believe the evidence does not favor histamine per se as the agent responsible for the syndrome. They suggest a synergistic relationship involving histamine and other as yet unidentified agents such as other amines or factors that influence histamine absorption. This view is based on the inability of large oral doses of histamine or histamine-spiked fish to produce symptoms in volunteers. On the other hand, the suddenness of onset of symptoms is consistent with a histamine reaction, and the association of the syndrome with scombroid fish containing high numbers of histidine–decarboxylase-producing bacteria cannot be ignored. Although the precise etiology may yet be in question, bacteria do play a significant if not indispensable role.

Aeromonas

This genus consists of several species that are often found in gastrointestinal specimens. Among these are *A. caviae*, *A. eucrenophila*, *A. schubertii*, *A. sobria*, *A. veronii*, and *A. hydrophila*. An enterotoxin has been identified in *A. caviae*[76] and *A. hydrophila* (see below), and the other species noted are associated with diarrhea. As *A. hydrophila* has received the most study, the discussion that follows is based on this species. The aeromonads are basically aquatic forms that are often associated with diarrhea, but their precise role in the etiology of gastrointestinal syndromes is not clear.

A. hydrophila is an aquatic bacterium found more in salt waters than in fresh waters. It is a significant pathogen to fish, turtles, frogs, snails, and alligators and is also a human pathogen, especially in compromised hosts. It is a common member of the bacterial biota of pigs. Diarrhea, endocarditis, meningitis, soft-tissue infections, and bacteremia are caused by *A. hydrophila*.

Virulent strains of *A. hydrophila* produce a 52-kDa single polypeptide that possesses enterotoxic, cytotoxic, and hemolytic activities. This multifunctional molecule displays immunological cross-reactivity with the cholera toxin.[92] According to some investigators,[116] it resembles aerolysin while others contend that it is aerolysin.[78] Aerolysin is a pore- or channel-forming toxin that kills cells by forming discrete channels in their plasma membranes.[12] Ion channels are created by the oligomerization of toxin molecules. Cytotonic activity has been associated with an *A. hydrophila* toxin, which induced rounding and steroidogenesis in Y-1 adrenal cells. Also, positive responses in the rabbit ileal loop, suckling mouse, and CHO assays have been reported for a cytotonic toxin.[27]

A large number of studies have been conducted on *A. hydrophila* isolates from various sources. In one study, 66 of 96 (69%) isolates produced cytotoxins, whereas 32 (80%) of 40 isolates from diarrheal disease victims were toxigenic, with only 41% of nondiarrheal isolates being positive for cytotoxin production. Most enterotoxigenic strains are VP (Voges-Proskauer test) and hemolysin positive and arabinose negative[13] and produce positive responses in the suckling mouse, Y-1 adrenal cell, and rabbit ileal loop assays. In a study of 147 isolates from patients with diarrhea, 91% were enterotoxigenic, whereas only 70% of 94 environmental strains produced enterotoxin as assessed by the suckling mouse assay.[14] All but four of the clinical isolates produced hemolysis of rabbit red blood cells. Of 116 isolates from the Chesapeake Bay, 71% were toxic by the Y-1 adrenal cell assay, and toxicity correlated with lysine decarboxylase and VP reactions.[56] In yet another study, 48 of 51 cultures from humans, animals, water, and sewage, produced positive responses in rabbit ileal loop assays with 10^3 or more cells, and cell-free extracts from all were loop positive.[3]

Isolates from meat and meat products have been shown to possess biochemical markers that are generally associated with toxic strains of other species, with the mouse median lethal dose (LD_{50}) being log 8–9 colony-forming units (cfu) for most strains tested.[85] The latter investigators suggested the possibility that immunosuppressive states are important factors in food-associated infections by this organism, a suggestion that could explain the difficulty of establishing this organism as the sole etiological agent of foodborne gastroenteritis.

With regard to growth temperature and habitat, 7 of 13 strains displayed growth at 0–5°C, 4 of 13 at 10°C, and 1 at a minimum of 15°C.[93] The psychrotrophs had optimum growth between 15°C and 20°C. The maximum growth temperature for some strains was 40–45°C with optimum at 35°C.[46] Regarding distribution, the organism was found in all but 12 of 147 lotic and lentic habitats.[46] Four of those habitats that did not yield the organism were either hypersaline lakes or geothermal springs. Some waters contained up to 9,000/ml. An ecological study of *A. hydrophila* in the Chesapeake Bay revealed numbers ranging from $< 0.3/l$–5×10^3/ml in the water column, and about 4.6×10^2/g of sediment.[56] The presence of this organism correlated with total, aerobic, viable, and heterotrophic bacterial counts, and its presence was inversely related to dissolved O_2 and salinity, with the upper salt level being about 15%. Fewer were found during the winter than during the summer months. For reviews, see references 2, 52. Their occurrence in some ready-to-eat foods is presented in Chapter 9 (Table 9–3).

Plesiomonas

P. shigelloides is found in surface waters and soil and has been recovered from fish, shellfish, other aquatic animals, as well as from terrestrial meat animals. It differs from *A. hydrophila* in having $G + C$ content of DNA of 51%, versus 58–62% for *A. hydrophila*. It has been isolated by many investigators from patients with diarrhea and is associated with other general infections in humans. It produces a heat-stable enterotoxin, and serogroup 0:17 strains react with *Shigella* group D antisera.[1] In a study of 16 strains from humans with intestinal illness, *P. shigelloides* did not always bind Congo red, the strains were noninvasive in HEp-2 cells, and they did not produce Shiga-like toxin on Vero cells.[1] Although a low-level cytolysin was produced consistently, the mean LD_{50} for outbred Swiss mice was 3.5×10^8 cfu. Heat-stable enterotoxin was not produced by any of the 16 strains, and it was the conclusion of these investigators that this organism possesses a low pathogenic potential.[1]

P. shigelloides was recovered by Zajc-Satler et al.[118] from the stools of six diarrheal patients. It was believed to be the etiological agent, although salmonellae were recovered from two patients. Two outbreaks of acute diarrheal disease occurred in Osaka, Japan, in 1973 and 1974, and the only bacterial pathogen recovered from stools was *P. shigelloides*. In the 1973 outbreak, 978 of 2,141 persons became ill, with 88% complaining of diarrhea, 82% of abdominal pain, 22% of fever, and 13% of headaches.[107] The symptoms lasted 2 to 3 days. Of 124 stools examined, 21 yielded *P. shigelloides* 017:H2. The same serovar was recovered from tap water. In the 1974 outbreak, 24 of 35 persons became ill with symptoms similar to those noted. *P. shigelloides* serovar 024:H5 was recovered from three of eight stools "virtually in pure culture."[107] The organism was recovered from 39% of 342 water and mud samples, as well as from fish, shellfish, and newts.

A 15-year-old female contracted gastroenteritis, and 6 hours after she took one tablet of trimethoprim-sulfadiazine, *P. shigelloides* could be recovered from her blood.[87] The latter investigators noted that 10 of the previously known 12 cases of *P. shigelloides* bacteremia were in patients who were either immunocompromised or presented with other similar conditions. The 15-year-old had a temperature of 39°C and passed up to 10 watery stools daily. The isolated strain reacted with *S. dysenteriae* serotype 7 antiserum, placing it in O group 22 of *P. shigelloides*.[83]

Growth of *P. shigelloides* has been observed at 10°C,[93] and 59% of 59 fish from Zaire waters contained the organism.[111] In the latter study, the organism was found more in river fish than lake fish. It appeared not to produce an enterotoxin since only 4 of 29 isolates produced positive responses in rabbit ileal loops.[95] Foodborne cases have not been documented, but the organism has been incriminated in at least two outbreaks.[73]

Bacteroides fragilis

This obligately anaerobic, Gram-negative bacterium is of potential significance as a foodborne pathogen since it produces an ileal loop-positive enterotoxin and is often associated with human diarrhea, as are *A. hydrophila* and *P. shigelloides*. The enterotoxin was first demonstrated in 1984, and enterotoxic strains of *B. fragilis* were first associated with human diarrhea in 1987.

B. fragilis is estimated to constitute between 1% and 2% of the human intestinal biota. As a non-pore former, it is more sensitive to aerated environments than the clostridia and yet it has been recovered from municipal sewage. This species differs from most other *Bacteroides* in being catalase positive, and like most others it can grow in the presence of 20% bile.

The *B. fragilis* enterotoxin is produced as a single chain with a molecular weight of about 20,000 Da. It differs from the classic bacterial enterotoxins in belonging to a class of zinc-binding metalloprotease, designated metzincins. The enterotoxin has a wide range of protein substrates, and it undergoes autodigestion. The intestinal damage that it causes is believed to be due, at least in part, to its proteolytic action. It elicits a positive response in ileal loops of lambs and other animals.

Since the etiological agent is identified in only around 50% of foodborne outbreaks in the United States, it is clear that previously unrecognized agents need to be included. *B. fragilis* along with *Klebsiella pneumoniae*[60] and *Enterobacter cloacae*[61] may warrant more attention. The latter two organisms produce heat-stable enterotoxins that are similar to the heat-stable enterotoxin (ST) of *E. coli*, and their potential significance in foods has been noted.[109]

Erysipelothrix rhusiopathiae

This bacterium (E·ry·si·pe'·lo·thrix rhu·si·o·pa'·thi·ae) is phylogenetically closely related to *Listeria* (see Chapter 25), and like *L. monocytogenes*, it causes disease in animals and humans. It is the cause of erysipelas in swine and erysipeloid in humans. Because of these similarities, it seems to be a "logical" candidate for a foodborne pathogen although such cases do not seem to have been reported. In general, erysipeloid is a localized disease of the hands and arms of handlers of fresh meat and fish, but systemic involvements are not unknown. Erysipeloid in pigs is characterized by diamond-skin lesions.

The organism is a facultative anaerobe, catalase negative (in contrast to the listeriae), oxidase negative, and generally produces H_2S. At least 23 serovars are known. The only other species is *E. tonsillarum*, which was separated from *E. rhusiopathiae* based on its primary habitat of porcine tongues, and because of serovar differences.[100]

One of the first studies of the incidence of this organism in foods is that of Ternström and Molin,[104] who in 1982 undertook a study of foodborne pathogens in meats in Sweden. They examined 135 samples consisting of equal numbers of chicken, beef, and pork, and found *E. rhusiopathiae* in 36% and 13%, respectively, of pork and chicken, but none in beef. In one plant, 54% of pork loins were positive, and many of the isolates possessed mouse virulence. Of 112 retail pork samples examined in Japan, 34% contained this bacterium, and the 38 isolates represented 14 serovars.[98] In a study of meat samples from 93 wild boar and 36 deer in Japan, 44% of the wild boar and 50% of the deer samples contained *E. rhusiopathiae*, representing 13 serovars.[54] In a study of 750 chickens in Japan, *Erysipelothrix* spp. were recovered from 15.7% of skin samples, and from 59.2% of 179 feather samples.[76] *E. rhusiopathiae* represented 273 of 297 isolates and the remainder were *E. tonsillarum*. In another study of 153 chicken samples in Japan, 30% contained *Erysipelothrix* spp. with 65 of 67 being *E. rhusiopathiae*.[77]

Klebsiella pneumoniae

About 6 hours after consuming a fast-food chain hamburger, an individual complained of not feeling well. After hospital admission, this organism along with generic *E. coli* was isolated from leftover hamburger and from the patient's blood, and the two matched by cultural methods.[94] The strain of *K. pneumoniae* was LT+ and ST−. The coliform count in leftover hamburger was 3.0×10^6/g, and 1.9×10^5/g of bun.

Streptococcus iniae

There have been at least six human infections by this organism traced to a fish product. *S. iniae* was first recognized in 1972 as the cause of a disease in Amazon dolphins.[20] It was next recorded in Israel in 1986 as the cause of disease in tilapia and trout, and later seen in Taiwan and the United States.[89] The first human case was recorded in 1991 in Texas, and the second in 1994 in Ottawa.[20] Four human cases occurred in Ontario, Canada, in 1995–1996 and the organism was isolated from both fish and patients. The fish was tilapia that was imported from fish farms in the United States.

S. iniae appears to be a fish pathogen that causes disease in humans. In the Ontario cases, it appeared that the organism entered the body through hand lesions. It is beta-hemolytic on sheep blood. In humans, the organism produces fulminant soft tissue infections.[39]

PRION DISEASES

Prions are unique proteins in that they can convert other proteins into damaging ones by causing them to alter their shape. The normal cell prion protein (PrP) exists in the brain cell membrane where it carries out some vital functions and is then degraded by proteases. However, the pathogenic form is distorted and is resistant to proteases, and thus it accumulates in brain tissue and gives rise to disease (see below). It has been postulated that the distorted prion molecule, acting as a template, converts normal protein to a distorted form.[9] The normal protein (α-helical form) takes on a protease-resistant β-flat form (PrP^{Sc}, PrPres) when it becomes pathogenic. The pathogenic forms tend to aggregate into amyloid fibrils where they cause nerve cell degeneration, which leads to clinical signs of disease. Although the evidence of a prion etiology of these diseases seems strong, the possibility that a virus is the agent has been raised.[28]

These particles were named around 1982 by Stanley Prusiner, who was awarded the 1997 Nobel Prize in physiology for his pioneering work.[112] Prions cause the disease scrapie in sheep, goats, and hamsters; and kuru in humans. Another prion disease of humans is Creutzfeldt-Jakob disease (CJD). Bovine spongiform encephalopathy (BSE) is a prion disease of cattle and sheep referred to as "mad cow disease." All of these belong to a family of diseases called transmissible spongiform encephalopathies (TSEs).

Bovine spongiform encephalopathy (BSE)

BSE ("mad cow disease") was first recognized in Great Britain in 1984 and specifically diagnosed in cattle in 1986. Four years later, over 14,000 confirmed cases out of a population of 10 million cattle had been recognized in Great Britain. The epidemic seemed to peak around 1,000 new cases per week in 1993. By February 1998, a total of 172,324 cases were seen in cattle in the United Kingdom.[9] A total of 600 cases were recorded in 8 countries outside the U.K. with 256 (42.7%) in Switzerland.[9]) Since 1996, 4.5 million cows have been destroyed. Between 1986 and November 2003, 183,634 cattle were diagnosed with BSE in the U.K. with 4,469 in 22 other countries.[29] Around 84 human deaths were recorded in the U.K. in the year 2000. The first confirmed case in Japan occurred in September 2001, and about 9 cases were seen through 2003. The first case confirmed in North America was announced on May 20, 2003, and the animal was a cow in Alberta, Canada. The first confirmed case in the United States occurred in December 2003, in a Holstein downer cow in Moses Lake, Washington. The cow was 6.5 years old and, thus, might have consumed currently banned feed that contained animal matter. Around 35 million beef animals up to age 24 months are slaughtered in the United States annually, with an additional 6 million or so of older dairy cows.

Testing for prion proteins consists of an immunohistochemistry method (considered to be the gold standard), a conformation-dependent immunoassay (CDI) method (developed in 2003), a screening test developed by the Bio-Rad Corporation (TeSeE), another method developed by the IDEXX Laboratories (Herdchek[R]), and 5 to 6 other postmortem tests for central nervous system tissue of cattle. The two latter tests were approved early in 2004 by the U.S. Department of Agriculture for animal testing. Several of the available methods are ELISA based.

Creutzfeldt-Jakob Diseases (CJD, vCJD)

Since humans are susceptible to the prions that cause CJD, the early concern was whether humans could contract BSE from cattle. In March 1996, a new variant of CJD (nvCJD, vCJD) was reported in the United Kingdom in a small group of people, all of whom were much younger than most individuals with CJD. This prompted speculation that vCJD had been contracted from cattle. Normally, CJD appears in persons around age 60 or older, but in the U.K. vCJD was found to afflict individuals in the late teens to the early 40s. It has been concluded that vCJD is the human equivalent of BSE,[115] and the agents for BSE and vCJD appear to be the same based on studies using mice.[11]

Between February 1994 and October 1995, 10 persons in the United Kingdom were found to have the new variant form of CJD, and 8 died. Most were under age 30 (in the United States, most CJD victims are over age 55; see below). For the years 1995–1998, there were 39 cases of vCJD.[14] Through 2003, around 150 vCJD cases were seen in humans throughout Europe with about one-third in the United Kingdom.

For the 5-year period 1991–1995, 94 CJD deaths were recorded in the United States and 9 were below age 55.[21] None conformed to vCJD. Over 85% of CJD patients in the United States die within one year of onset. Between 1991 and 1995, the average annual death rate from CJD in the United States was 1.2/1 million population.[7] BSE is thought to have been contracted by cattle through specified bovine offal (that contained brain, spinal cord, and the like) from infected animals, a practice that was banned in 1989. Using a mouse assay, prions could not be detected in beef muscle and milk from infected cattle.[9] The incubation period for BSE is between 1 and 15 years.

Regarding the heat destruction of the prions of vCJD, studies are wanting. However, data on scrapie and CJD have been presented and summarized.[15] The latter investigator suggested that the brain tissue of a TSE-infected cow can be expected to contain about 10^{11} prions per gram. Assuming that the nerve tissue is ground with muscle tissue, about 10^8 prions per gram may be expected in ground beef, or 10^{10} prions in a 100-g portion. To effect a 12-D reduction, 22D is required ($10 + 12 = 22$). Thus, some of the times in minutes needed to achieve a 22D were calculated as follows: $D_{160^\circ C} = 1.0$; $D_{140^\circ C} = 11.0$; $D_{120^\circ C} = 110$.[15] It has been suggested that there is a need for new processing or packaging technologies such that high-temperature short-time treatments can be carried out in order to render products free of prions.[15] For more information, see references 9, 31, 101.

Chronic wasting disease (CWD)

This is a prion or TSE disease first detected in captive mule deer in 1967 in the state of Colorado. It has been diagnosed in wild deer and elk in the states of Wyoming, Colorado, Nebraska, and in the Saskatchewan Province of Canada. It has been seen on elk farms in other states. It appears to be transmitted through saliva or feces. It is estimated that 4–6% of mule deer and <1% of free-ranging elk are infected in the endemic areas. The primary symptoms in elk are emaciation and drooling. In December 2003, the U.S. Department of Agriculture put in place a herd certification program along with restrictions on the interstate movement of captive deer and elk. At least two tests for postmortem testing of central nervous system tissues have been developed and used.

TOXIGENIC PHYTOPLANKTONS

Paralytic Shellfish Poisoning

This syndrome is contracted by eating toxic mussels, clams, oysters, scallops, or cockles. These bivalves become toxic after feeding on certain dinoflagellates of which *Gonyaulax catenella* is representative of the United States Pacific Coast biota. Along the North Atlantic Coast of the United States and over to northern Europe, *G. tamarensis* is found, and its poison is more toxic than that of *G. catenella*. *G. acatenella* is found along the coast of British Columbia. Masses or blooms of these toxic dinoflagellates give rise to the red tide condition of seas. In 1996, about 150 manatees were killed during a red tide off the coast of Florida. Another dinoflagellate, *Karenia brevis* (= *Gymnodinium breve*), produces brevetoxin, which can cause respiratory distress and food poisoning in humans.[45] It has caused massive fish kills along the east coast of the United States, and has been implicated in the death of bottlenose dolphins and manatees (see reference 45).

The paralytic shellfish poison (PSP) is saxitoxin, and its structural formula is as follows:

Saxitoxin exerts its effect in humans through cardiovascular collapse and respiratory failure. It blocks the propagation of nerve impulses without depolarization, and there is no known antidote. It is heat stable, water soluble, and generally not destroyed by cooking. It can be destroyed by boiling for 3–4 hours at pH of 3.0. A *D* value at 250°F (121.1°C) of 71.4 minutes in soft-shell clams has been reported.[44]

Symptoms of PSP develop within 2 hours after ingestion of toxic mollusks, and they are characterized by paresthesia (tingling, numbness, or burning), which begins about the mouth, lips, and tongue and later spreads over the face, scalp, and neck; and to the fingertips and toes. The mortality rate is variously reported to range from 1 to 22%.

Between 1793 and 1958, some 792 cases were recorded, with 173 (22%) deaths.[71] In the 15-year period 1973–1987, 19 outbreaks (with a mean of 8 cases) were reported by state health departments to the CDC. In 1990, there were 19 cases from two outbreaks in the states of Massachusetts and Alaska alone. In the former, six fishermen became ill after eating boiled mussels that contained 4,280 μg/100 g saxitoxin.[23] The raw mussels contained 24,400 μg/100 g. The 13 cases in Alaska resulted in 1 death, and gastric contents from the victim who died contained 370 μg/100 g of PSP toxin, whereas a sample of the butterclam that was consumed contained 2,650 μg/100 g.[23] The maximum safe level of PSP toxin is 80 μg/100 g.[23]

Outbreaks of PSP seem to occur between the months of May and October on the United States West Coast and between August and October on the East Coast. Mollusks may become toxic in the absence of red tides. Detoxification of mollusks can be achieved by their transfer to clean water, and a month or more may be required.

Ciguatera Poisoning

This syndrome is contracted from the ingestion of any one of over 300 fish species (barracuda, grouper, sea bass, etc.) that feed on herbivorous or reef fishes, which in turn feed on phytoplankton, especially the dinoflagellates. The responsible dinoflagellate is *Gambierdiscus toxicus*, which produces ciguatoxin. This toxin is concentrated more in fish organs such as the liver than in muscle tissue.

Upon ingestion of toxic fish, symptoms occur within 3–6 hours (about the same as for staphylococcal food poisoning), and consist of nausea and paresthesia about the mouth, tongue, and throat. In general, the symptoms are quite similar to those for paralytic shellfish poisoning. Respiratory paralysis is the consequence in the absence of appropriate therapy. The disease has been associated with farm-raised salmon fish. For a review, see reference 68.

For the years 1983–1992, 129 outbreaks were reported to the CDC involving 508 persons with no deaths.[18] An outbreak in Texas in 1997 involved 17 crewmembers of a cargo ship, and the vehicle food was barracuda.[18]

Domoic Acid

This is an uncommon amino acid that antagonizes glutamic acid in the central nervous system. It is produced by a diatom, *Pseudonitzschia pungens*, and its structure is as indicated. (Diatoms are single-celled algae with walls of silicon.)

Domoic acid causes amnesic shellfish poisoning (ASP) following the consumption of mussels or scallops harvested from marine waters with a bloom of the diatom noted. The first recorded outbreak of human cases occurred in eastern Canada in 1988 following the consumption of mussels from Prince Edward Island,[88] and there were 107 victims and three deaths. Since this episode, domoic acid-producing diatoms have been found in other parts of the world. An ASP episode affected scallops in northwest Spain in 1996.[69] The largest quantity of domoic acid was found in the hepatopancreas— from 52% to 88% of the total.[69] During frozen storage, some domoic acid transferred to other parts of the scallops. It was found that canning of scallops did not destroy this toxic principal. According to Leira et al.,[69] the Canadian regulatory level is 20 μg/g of tissue for fresh bivalve mollusks.

Pfiesteria piscicida

This dinoflagellate was first recognized in the early 1990s as the cause of death of thousands of fish in tributaries of the Chesapeake Bay. It is an animal-like organism that produces potent toxins. One toxin stuns fish within a few seconds, and the animals die within a few minutes. It is heat stable. Another toxin causes the fish epidermis to slough off. The dinoflagellate reproduces sexually after a fish kill, and it can encyst.

The exact identity of the toxins is unclear, as are their effect on humans. Those who have been exposed have a history of memory loss, confusion, acute skin burning, and usually general symptoms such as headache, skin rash, muscle cramps, and the like.[19]

REFERENCES

1. Abbott, S.L., R.P. Kokka, and J.M. Janda. 1991. Laboratory investigations on the low pathogenic potential of *Plesiomonas shigelloides*. *J. Clin. Microbiol.* 29:148–153.

2. Albert, M.J., M. Ansaruzzaman, K.A. Talukder, A.K. Chopra, I. Kuhn, M. Rahman, A.S.G. Faruque, M.S. Islam, R.B. Sack, and R. Mollby. 2000. Prevalence of enterotoxin genes in *Aeromonas* spp. isolated from children with diarrhea, healthy controls, and the environment. *J. Clin. Microbiol.* 38:3785–3790.

3. Annapurna, E., and S.C. Sanyal. 1977. Enterotoxicity of *Aeromonas hydrophila*. *J. Med. Microbiol.* 10:317–323.

4. Appleton, H., and M.S. Pereira. 1977. A possible virus aetiology in outbreaks of food-poisoning from cockles. *Lancet* 1:780–781.

5. Atmar, R.L., F.H. Neill, J.L. Romalde, F. LeGuyader, C.M. Woodley T.G. Metcalf, and M.K. Estes. 1995. Detection of Norwalk virus and hepatitis A virus in shellfish tissues with the PCR. *Appl. Environ. Microbiol.* 61:3014–3018.

6. Behling, A.R., and S.L. Taylor. 1982. Bacterial histamine production as a function of temperature and time of incubation. *J. Food Sci.* 47:1311–1314, 1317.

7. Belay, E.D. 1999. Transmissible spongiform encephalopathies in humans. *Ann. Rev. Microbiol.* 53:283–314.

8. Blackwell, J.H., D. Rickansrud, P.D. McKercher, and J.W. McVicar. 1982. Effect of thermal processing on the survival of foot-and-mouth disease virus in ground meat. *J. Food Sci.* 47:388–392.

9. Blanchfield, J.R. 1998. Bovine spongiform encephalopathy (BSE)—A review. *Int. J. Food Sci. Technol.* 33:81–97.

10. Breeuwer, P., A. Lardeau, M. Peterz, and H.M. Joosten. 2003. Desiccation and heat tolerance of *Enterobacter sakazakii*. *J. Appl. Microbiol.* 95:967–973.

11. Bruce, M.E., R.G. Will, J.W. Ironside, I. McConnell, D. Drummond, A.Suttie, L. McCardle, A. Chree, J. Hope, C. Birkett, S. Cousens, H. Fraser, and C.J. Bostock. 1997. Transmissions to mice indicate that 'new variant' CJD is caused by the BSE agent. *Nature* 389:498–501.

12. Buckley, J.T., and S.P. Howard. 1999. The cytotoxic enterotoxin of *Aeromonas hydrophila* is aerolysin. *Infect. Immun.* 67:466–467.

13. Burke, V., J. Robinson, H.M. Atkinson, and M. Gracey. 1982. Biochemical characteristics of enterotoxigenic *Aeromonas* spp. *J. Clin. Microbiol.* 15:48–52.

14. Burke, V., J. Robinson, M. Cooper, J. Beamons, K. Partridge, D. Peterson, and M. Gracey. 1984. Biotyping and virulence factors in clinical and environmental isolates of *Aeromonas* species. *Appl. Environ. Microbiol.* 47:1146–1149.

15. Casolari, A. 1998. Heat resistance of prions and food processing. *Food Microbiol.* 15:59–63.

16. Centers for Disease Control and Prevention. 2003. Foodborne transmission of hepatitis A—Massachusetts, 2001. *Morb. Mort. Wkly. Rep.* 52:565–567.

17. Centers for Disease Control and Prevention. 2002. Outbreaks of gastroenteritis associated with noroviruses on cruise ships—United States. *Morb. Mort. Wkly. Rep.* 51:1112–1114.

18. Centers for Disease Control and Prevention. 1998. Ciguatera fish poisoning—Texas, 1997. *Morb. Mort. Wkly. Rep.* 47:692–694.

19. Centers for Disease Control and Prevention. 1997. Results of the public health response to *Pfiesteria* workshop—Atlanta, Georgia, September 29–30, 1997. *Morb. Mort. Wkly. Rep.* 46:951–952.

20. Centers for Disease Control and Prevention. 1996. Invasive infection with *Streptococcus iniae*—Ontario, 1995–1996. *Morb. Mort. Wkly. Rep.* 45:650–653.

21. Centers for Disease Control and Prevention. 1996. Surveillance for Crutzfeldt-Jakob disease—United States. *Morb. Mort. Wkly. Rep.* 45:665–668.

22. Centers for Disease Control and Prevention. 1991. Rotavirus surveillance—United States, 1989–1990. *Morb. Mort. Wkly. Rep.* 40:80–81, 87.

23. Centers for Disease Control and Prevention. 1991. Paralytic shellfish poisoning—Massachusetts and Alaska. 1990. *Morb. Mort. Wkly. Rep.* 40:157–161.

24. Centers for Disease Control and Prevention. 1990. Viral agents of gastroenteritis. Public health importance and outbreak management. *Morb. Mort. Wkly. Rep.* 39:1–23.

25. Centers for Disease Control and Prevention. 1990. Foodborne hepatitis A—Alaska, Florida, North Carolina, Washington. *Morb. Mort. Wkly. Rep.* 39:228–232.

26. Centers for Disease Control and Prevention. 1989. Scombroid fish poisoning—Illinois, South Carolina. *Morb. Mort. Wkly. Rep.* 38:140–142, 147.

27. Chakraborty, T., M.A. Montenegro, S.C. Sanyal, R. Helmuth, E. Bulling, and K.N. Timmis. 1984. Cloning of enterotoxin gene from *Aeromonas hydrophila* provides conclusive evidence of production of a cytotonic enterotoxin. *Infect. Immun.* 46:435–441.

28. Chesebro, B. 1998. BSE and prions: Uncertainties about the agent. *Science* 279:42–43.

29. Cliver, D.O. 2004. How now, mad cow? *Food Technol.* 58(1):100.

30. Cliver, D.O. (and the IFT Expert Panel on Food Safety and Nutrition). 1988. Virus transmission via foods. *Food Technol.* 42(10):241–248.

31. Collinge, J., and M.S. Palmer, eds. 1997. *Prion Diseases.* New York: Oxford University Press.

32. Cromeans, T.L., O.V. Nainan, and H.S. Margolis. 1997. Detection of hepatitis A virus RNA in oyster meat. *Appl. Environ. Microbiol.* 63:2460–2463.

33. DiGirolamo, R., J. Liston, and J.R. Matches. 1970. Survival of virus in chilled, frozen, and processed oysters. *Appl. Microbiol.* 20:58–63.

34. Dix, A.B., and L.-A. Jaykus. 1998. Virion concentration method for the detection of human enteric viruses in extracts of hard-shelled clams. *J. Food Protect.* 61:458–465.

35. Edelson-Mammel, S.G., and R.L. Buchanan. 2004. Thermal inactivation of *Enterobacter sakazakii* in rehydrated infant formula. *J. Food Protect.* 67:60–63.

36. Eyles, M.J., G.R. Davey, and E.J. Huntley. 1981. Demonstration of viral contamination of oysters responsible for an outbreak of viral gastroenteritis. *J. Food Protect.* 44:294–296.

37. Frank, H.A., D.H. Yoshinaga, and I.-P. Wu. 1983. Nomograph for estimating histamine formation in skipjack tuna at elevated temperatures. *Mar. Fish. Rev.* 45:40–44.

38. Fugate, K.J., D.O. Cliver, and M.T. Hatch. 1975. Enteroviruses and potential bacterial indicators in Gulf Coast oysters. *J. Milk. Food Technol.* 38:100–104.

39. Fuller, J.D., D.J. Bast, V. Nizet, D.E. Low, and J.C.S. de Azavedo. 2001. *Streptococcus iniae* virulence is associated with a distinct genetic profile. *Infect. Immun.* 69:1994–2000.

40. Gerba, C.P., and D. Kayed. 2003. Caliciviruses: A major cause of foodborne illness. *J. Food Protect.* 68:1136–1142.

41. Gerba, C.P., and S.M. Goyal. 1978. Detection and occurrence of enteric viruses in shellfish: A review. *J. Food Protect.* 41:743–754.

42. Gerba, C.P., S.M. Goyal, R.L. LaBelle, I. Cech, and G.F. Bodgous. 1979. Failure of indicator bacteria to reflect the occurrence of enteroviruses in marine waters. *Am. J. Public Health* 69:1116–1119.

43. Gilbert, R.J., G. Hobbs, G.K. Murray, J.G. Cruickshank, and S.E.J. Young. 1980. Scombrotoxic fish poisoning: Features of the first 50 incidents to be reported in Britain (1976–1979). *Br. Med. J.* 281:71–72.

44. Gill, T.A., J.W. Thompson, and S. Gould. 1985. Thermal resistance of paralytic shellfish poison in soft-shell clams. *J. Food Protect.* 48:659–662.

45. Gray, M., B. Wawrik, J. Paul, and E. Casper. 2003. Molecular detection and quantitation of the red tide dinoflagellate *Karenia brevis* in the marine environment. *Appl. Environ. Microbiol.* 69:5726–5730.

46. Hazen, T.C., C.B. Fliermans, R.P. Hirsch, and G.W. Esch 1978. Prevalence and distribution of *Aeromonas hydrophila* in the United States. *Appl. Environ. Microbiol.* 36:731–738.

47. Hejkal, T.W., and C.P. Gerba. 1981. Uptake and survival of enteric viruses in the blue crab, *Callinectes sapidus*. *Appl. Environ. Microbiol.* 41:207–211.

48. Herrmann, J.E., and D.O. Cliver. 1973. Enterovirus persistence in sausage and ground beef. *J. Milk Food Technol.* 36:426–428.

49. Hopkins, R.S., G.B. Gaspard, F.P. Williams, Jr., R.J. Karlin, G. Cukor, and N.R. Blacklow. 1984. A community waterborne gastroenteritis outbreak: Evidence for rotavirus as the agent. *Am. J. Public Health* 74:263–265.

50. Hudson, S.H., and W.D. Brown. 1978. Histamine (?) toxicity from fish products. *Adv. Food Res.* 24:113–154.

51. Hwang, D.-F., S.-H. Chang, C.-Y. Shiau, and C.-C. Cheng. 1995. Biogenic amines in the flesh of sailfish (*Istiophorus platypterus*) responsible for scombroid poisoning. *J. Food Sci.* 60:926–928.

52. Isonhood, J.H., and M. Drake. 2002. *Aeromonas* species in foods. *J. Food Protect.* 65:575–582.

53. Jaykus, L.-A., R. de Leon, and M.D. Sobsey. 1996. A virion concentration method for detection of human enteric viruses in oysters by PCR and oligoprobe hybridization. *Appl. Environ. Microbiol.* 62:2074–2080.

54. Kanai, Y., H. Hayashidani, K.-I. Kaneko, M. Ogawa, T. Takahashi, and M. Nakamura. 1997. Occurrence of zoonotic bacteria in retail game meat in Japan with special reference to *Erysipelothrix*. *J. Food Protect.* 60:328–331.

55. Kanki, M., T. Yoda, T. Tsukamoto, and T. Shibata. 2002. *Klebsiella pneumoniae* produces no histamine: *Raoultella planticola* and *Raoultella ornithinolytica* strains are histamine producers. *Appl. Environ. Microbiol.* 68:3462–3466.

56. Kaper, J.B., H. Lockman, R.R. Colwell, and S.W. Joseph. 1981. *Aeromonas hydrophila*: Ecology and toxigenicity on isolates from an estuary. *J. Appl. Bacteriol.* 50:359–377.

57. Kaplan, J.E., R. Feldman, D.S. Campbell, C. Lookabaugh, and G.W. Gary. 1982. The frequency of a Norwalk-like pattern of illness in outbreaks of acute gastroenteritis. *Am. J. Public Health* 72:1329–1332.

58. Keswick, B.H., T.K. Satterwhite, P.C. Johnson, H.L. DuPont, S.L. Secor, J.A. Bitsura, G.W. Gary, and J.C. Hoff. 1985. Inactivation of Norwalk virus in drinking water by chlorine. *Appl. Environ. Microbiol.* 50:261–264.

59. Kim, S.-H., B. Ben-Gigirey, J. Barros-Velázquez, R.J. Price, and H. An. 2000. Histamine and biogenic amine production by *Morganella morganii* isolated from temperature-abused albacore. *J. Food Protect.* 63:244–251.

60. Klipstein, F.A., and R.F. Engert. 1976. Purification and properties of *Klebsiella pneumoniae* heat-stable enterotoxin. *Infect. Immun.* 13:373–381.

61. Klipstein, F.A., and R.F. Engert. 1976. Partial purification and properties of *Enterobacter cloacae* heat-stable enterotoxin. *Infect. Immun.* 13:1307–1314.

62. Konowalchuk, J., and J.I. Speirs. 1975. Survival of enteric viruses on fresh vegetables. *J. Milk Food Technol.* 38:469–472.

63. Konowalchuk, J., and J.I. Speirs. 1975. Survival of enteric viruses on fresh fruit. *J. Milk Food Technol.* 38:598–600.

64. Kostenbader, K.D., Jr., and D.O. Cliver. 1977. Quest for viruses associated with our food supply. *J. Food Sci.* 42:1253–1257, 1268.

65. Kuritsky, J.N., M.T. Osterholm, J.A. Korlath, K.E. White, and J.E. Kaplan. 1985. A statewide assessment of the role of Norwalk virus in outbreaks of food-borne gastroenteritis. *J. Infect. Dis.* 151:568.

66. Landry, E.F., J.M. Vaughn, T.J. Vicale, and R. Mann. 1982. Inefficient accumulation of low levels of monodispersed and feces-associated poliovirus in oysters. *Appl. Environ. Microbiol.* 44:1362–1369.

67. Larkin, E.P. 1981. Food contaminants—Viruses. *J. Food Protect.* 44:320–325.

68. Lehane, L., and R.J. Lewis. 2000. Ciguatera: recent advances but the risk remains. *Int. J. Food Microbiol.* 61:91–125.

69. Leira, F.J., J.M. Vieites, L.M. Botana, and M.R. Vyeites. 1998. Domoic acid levels of naturally contaminated scallops as affected by canning. *J. Food Sci.* 63:1081–1083.

70. Lundgren, O., A. t. Peregrin, K. Persson, S. Kordasti, I. Uhnoo, and L. Svensson. 2000. Role of the enteric nervous system in the fluid and electrolyte secretion of rotavirus diarrhea. *Science* 287:491–495.

71. McFarren, E.F., M.L. Shafer, J.E. Campbell, K.H. Lewis, G.R. Davey, and R.H. Millsom. 1960. Public health significance of paralytic shellfish poison. *Adv. Food Res.* 10:135–179.

72. McKercher, P.D., W.R. Hess, and F. Hamdy. 1978. Residual viruses in pork products. *Appl. Environ. Microbiol.* 35:142–145.

73. Miller, M.L., and J.A. Koburger. 1985. *Plesiomonas shigelloides*: An opportunistic food and waterborne pathogen. *J. Food Protect.* 48:449–457.

74. Murphy, A.M., G.S. Grobmann, P.J. Christopher, W.A. Lopez, G.R. Davey, and R.H. Millsom. 1979. An Australia-wide outbreak of gastroenteritis from oysters caused by Norwalk virus. *Med. J. Austr.* 2:329–333.

75. Muytjens, H.L., H. Roelofs-Willemse, and G.H. Jaspar. 1988. Quality of powdered substitutes for breast milk with regard to members of the family Enterobacteriaceae. *J. Clin. Microbiol.* 26:743–746.

76. Nakazawa, H., H. Hayashidani, J. Higashi, K.-I. Kaneko, T. Takahashi, and M. Ogawa. 1998a. Occurrence of *Erysipelothrix* spp. in broiler chickens at an abattoir. *J. Food Protect.* 61:907–909.

77. Nakazawa, H., H. Hayashidani, J. Higashi, K.-I. Kaneko, T. Takahashi, and M. Ogawa. 1998b. Occurrence of *Erysipelothrix* spp. in chicken meat parts from a processing plant. *J. Food Protect.* 61:1207–1209.

78. Namdari, H., and E.J. Bottone. 1990. Cytotoxin and enterotoxin production as factors delineating enteropathogenicity of *Aeromonas caviae*. *J. Clin. Microbiol.* 28:1796–1798.

79. Nazarowec-White, M., and J.M. Farber. 1997a. Incidence, survival, and growth of *Enterobacter sakazakii* in infant formula. *J. Food Protect.* 60:226–230.

80. Nazarowec-White, M., and J.M. Farber. 1997b. Thermal resistance of *Enterobacter sakazakii* in reconstituted dried-infant formula. *Lett. Appl. Microbiol.* 24:9–13.

81. Niu, M. T., L. B. Polish, B. H. Robertson, B. K. Bhanna, B. A. Woodruff, C. N. Shapiro, M. A. Miller, J. D. Smith, J. K. Gedrose, M. J. Alter, and H. S. Margolis. 1992. Multistate outbreak of hepatitis A associated with frozen strawberries. *J. Infect. Dis.* 166:518–524.

82. Niven, C.F., Jr., M.B. Jeffrey, and D.A. Corlett, Jr. 1981. Differential plating medium for quantitative detection of histamine-producing bacteria. *Appl. Environ. Microbiol.* 41:321–322.

83. Omura, Y., R.J. Price, and H.S. Olcott. 1978. Histamine-forming bacteria isolated from spoiled shipjack tuna and jack mackerel. *J. Food Sci.* 43:1779–1781.

84. Pagotto, F.J., M. Nazarowec-White, S. Bidawid, and J.M. Farber. 2003. *Enterobacter sakazakii*: Infectivity and enterotoxin production in vitro and in vivo. *J. Food Protect.* 66:370–375.

85. Palumbo, S.A., M.M. Bencivengo, B. Del Corral, A.C. Williams, and R.L. Buchanan. 1989. Characterization of the *Aeromonas hydrophila* group isolated from retail foods of animal origin. *J. Clin. Microbiol.* 27:854–859.

86. Parshionikar, S.U., S. Willian-True, G.S. Fout, D.E. Robbins, S.A. Seys, J.D. Cassady, and R. Harris. 2003. Waterborne outbreak of gastroenteritis associated with a norovirus. *Appl. Environ. Microbiol.* 69:5263–5268.

87. Paul, R., A. Siitonen, and P. Karkkainen. 1990. *Plesiomonas shigelloides* bacteremia in a healthy girl with mild gastroenteritis. *J. Clin. Microbiol.* 28:1445–1446.

88. Perl, T. M., L. Bedard, T. Kosatsky, J. C. Hockin, E. C. Todd, and R. S. Remis. 1990. An outbreak of toxic encephalopathy caused by eating mussels contaminated with domoic acid. *N. Engl. J. Med.* 322:1775–1780.

89. Pier, G.B., S.H. Madin, and S. Al-Nakeeb. 1978. Isolation and characterization of a second isolate of *Streptococcus iniae*. *Int. J. System. Bacteriol.* 28:311–314.

90. Portnoy, B.L., P.A. Mackowiak, C.T. Caraway, J.A. Walker, T.W. McKinley, and C.A. Klein. 1975. Oyster-associated hepatitis: Failure of shellfish certification programs to prevent outbreaks. *JAMA* 233:1065–1068.

91. Rodriguez-Jerez, J.J., E.I. Lopez-Sabater, A.X. Roig-Sagues, and M.T. Mora-Ventura. 1994. Histamine, cadaverine and putrescine forming bacteria from ripened Spanish semipreserved anchovies. *J. Food Sci.* 59:998–1001.

92. Rose, J.M., C.W. Houston, D.H. Coppenhaver, J.D. Dixon, and A. Kurosky. 1989. Purification and chemical characterization of a cholera toxin-cross-reactive cytolytic enterotoxin produced by a human isolate of *Aeromonas hydrophila*. *Infect. Immun.* 57:1165–1169.

93. Rouf, M.A., and M.M. Rigney. 1971. Growth temperatures and temperature characteristics of *Aeromonas*. *Appl. Microbiol.* 22:503–506.

94. Sabota, J. M., W. L. Hoppes, J. R. Ziegler, Jr., H. DuPont, J. Mathewson, and G. W. Rutecki. 1998. A new variant of food poisoning: Enteroinvasive *Klebsiella pneumoniae* and *Escherichia coli* sepsis from a contaminated hamburger. *Am. J. Gastroenterol.* 93:118–119.

95. Sanyal, S.C., S.J. Singh, and P.C. Sen. 1975. Enteropathogenicity of *Aeromonas hydrophila* and *Plesiomonas shigelloides*. *J. Med. Microbiol.* 8:195–198.

96. Seymour, I.J., and H. Appleton. 2001. Foodborne viruses and fresh produce. *J. Appl. Microbiol.* 91:759–773.

97. Shin, G.-A., and M.D. Sobsey. 2003. Reduction of Norwalk virus, poliovirus 1, and bacteriophage MS2 by ozone disinfection of water. *Appl. Environ. Microbiol.* 69:3975–3978.

98. Shiono, H., H. Hayashidani, K.-I. Kaneko, M. Ogawa, and M. Muramatsu. 1990. Occurrence of *Erysipelothrix rhusiopathiae* in retail raw pork. *J. Food Protect.* 53:856–858.

99. Sullivan, R., R.M. Marnell, E.P. Larkin, and R.B. Read, Jr. 1975. Inactivation of poliovirus 1 and coxsackievirus B-2 in broiled hamburgers. *J. Milk Food Technol.* 38:473–475.

100. Takahashi, T., T. Fujisawa, Y. Tamura, S. Suzuki, M. Muramatsu, T. Sawata, Y. Benno, and T. Mitsuoka. 1992. DNA relatedness among *Erysipelothrix rhusiopathiae* strains representing all twenty-three serovars and *Erysipelothrix tonsillarum*. *Int. J. System. Bacteriol.* 42:469–473.

101. Taylor, D.M. 1998. Inactivation of the BSE agent. *J. Food Saf.* 18:265–274.

102. Taylor, S.L., L.S. Guthertz, M. Leatherwood, and E.R. Lieber. 1979. Histamine production by *Klebsiella pneumoniae* and an incident of scombroid fish poisoning. *Appl. Environ. Microbiol.* 37:274–278.

103. Taylor, S.L., T.J. Keefe, E.S. Windham, and J.F. Howell. 1982. Outbreak of histamine poisoning associated with consumption of Swiss cheese. *J. Food Protect.* 45:455–457.

104. Ternström, A., and G. Molin. 1987. Incidence of potential pathogens on raw pork, beef and chicken in Sweden, with special reference to *Erysipelothrix rhusiopathiae*. *J. Food Protect.* 50:141–146.

105. Thurm, V., and B. Gericke. 1994. Identification of infant food as a vehicle in a nosocomial outbreak of *Citrobacter freundii*: epidemiological subtyping by allozyme, whole-cell protein and antibiotic resistance. *J. Appl. Bacteriol.* 76:553–558.

106. Traore, O., C. Arnal, B. Mignotte, A. Maul, H. Laveran, S. Billaudel, and L. Schwartzbrod. 1998. Reverse transcriptase PCR detection of astrovirus, hepatitis A virus, and poliovirus in experimentally contaminated mussels: Comparison of several extraction and concentration methods. *Appl. Environ. Microbiol.* 64:3118–3122.

107. Tsukamoto, T., Y. Konoshita, T. Shimada, and R. Sakazaki. 1978. Two epidemics of diarrhoeal disease possibly caused by *Plesiomonas shigelloides*. *J. Hyg.* 80:275–280.

108. Turner, C., S.M. Williams, and T.R. Cumby. 2000. The inactivation of foot and mouth disease, Aujeszky's disease and classical swine fever viruses in pig slurry. *J. Appl. Microbiol.* 89:760–767.

109. Twedt, R.M., and B.K. Boutin. 1979. Potential public health significance of non-*Escherichia coli* coliforms in food. *J. Food Protect.* 42:161–163.

110. Van Acker, J., F. de Smet, G. Muyldermans, A. Bougatef, A. Naessens, and S. Lauwers. 2001. Outbreak of necrotizing enterocolitis associated with *Enterobacter sakazakii* in powdered milk formula. *J. Clin. Microbiol.* 39:293–297.

111. Van Damme, L.R., and J. Vandepitte. 1980. Frequent isolation of *Edwardsiella tarda* and *Plesiomonas shigelloides* from healthy Zairese freshwater fish: A possible source of sporadic diarrhea in the tropics. *Appl. Environ. Microbiol.* 39:475–479.

112. Vogel, G. 1997. Prusiner recognized for once-heretical prion theory. *Science* 278:214.

113. Wait, D.A., C.R. Hackney, R.J. Carrick, G. Lovelace, and M.D. Sobsey. 1983. Enteric bacterial and viral pathogens and indicator bacteria in hard shell clams. *J. Food Protect.* 46:493–496.

114. Wei, C.I., C.-M. Chen, J.A. Koburger, W.S. Otwell, and M.R. Marshall. 1990. Bacterial growth and histamine production on vacuum packaged tuna. *J. Food Sci.* 55:59–63.

115. Williams, N. 1997. New studies affirm BSE-human link. *Science* 278:31.

116. Xu, X.-J., M.R. Ferguson, V.L. Popov, C.W. Houston, J.W. Peterson, and A.K. Chopra. 1998. Role of cytotoxic enterotoxin in *Aeromonas*-mediated infections: Development of transposon and isogenic mutants. *Infect. Immun.* 66:3501–3509.

117. Yoshinaga, D.R., and H.A. Frank. 1982. Histamine-producing bacteria in decomposing shipjack tuna (*Katsuwonus pelamis*). *Appl. Environ. Microbiol.* 44:447–452.

118. Zajc-Satler, J., A.Z. Dragav, and M. Kumelj. 1972. Morphological and biochemical studies of 6 strains of *Plesiomonas shigelloides* isolated from clinical sources. *Zbt. Baktr. Hyg. Abt. Orig. A.* 219:514–521.

Grouping of Gram-Positive and Gram-Negative Bacterial Genera

Grouping of the genera of Gram-positive and Gram-negative bacteria is based on four phenotypic characters: Gram reaction (GP = positive; GN = negative), oxidase (+ or −), catalase (+ or −), and absence (n) or presence (p) of colony pigmentation. Groupings for most aerobic foodborne bacteria can be made within 24–48 hours after surface plating onto plate count agar with incubation at 30°C. Foodborne and environmental bacterial genera are not known for the following two groups: GP 3 (Gr + Ox + Cat − n) and GP 4 (Gr + Ox + Cat − n). Among Gram negatives, the genera in GN 3 (Gr − Ox + Cat − n) and GN 4 (Gr − Ox + Cat − p) are rarely if ever reported in foods.

Gram-Positive Groups

GP 1: Gr + Ox + Cat + n	*GP 2: Gr + Ox + Cat + p*
Alicyclobacillus	*Arthrobacter*
Aneurinibacillus	*Bacillus* (some)
Arthrobacter	*Brachybacterium*
Bacillus (some)	*Brevibacillus*
Brachybacterium	*Brevibacterium* (some)
Brevibacillus	*Corynebacterium*
Brochothrix	*Deinococcus*
Corynebacterium (some)	*Dermacoccus*
Dermacoccus	*Exiguobacterium*
Geobacillus	*Halobacillus*
Gracilibacillus	*Janibacter*
Janibacter	*Kocuria*
Macrococcus	*Luteococcus*
Micrococcus	*Macrococcus*
Nesterenkonia	*Micrococcus*
Paenibacillus	*Nesterenkonia*
Propioniflex	*Salinococcus*
Salibacillus	*Streptomyces* (most)
Sporosarcina	
Staphylococcus lentus,	

sciuri, vitulus
Stomatococcus
Streptomyces (some)
Terracoccus

GP 5: Gr + Ox − Cat + n	GP 6: Gr + Ox − Cat + p
Anaerobacter	*Bacillus* (some)
Bacillus (most)	Brachybacterium
Brevibacterium (most)	*Brevibacterium linens*
Brachybacterium	*Caseobacter*
Caseobacter	*Clavibacter*
Clavibacter	*Corynebacterium* (some)
Corynebacterium (some)	*Demetria*
Demetria	*Exiguobacterium*
Erysipelothrix	*Gordona*
Geobacillus (some)	*Janibacter*
Janibacter	*Kineococcus*
Jonesia	*Kocuria*
Kocuria	*Kytococcus*
Kurthia	*Microbacterium*
Kytococcus	*Planococcus*
Leucobacter	*Propionibacterium*
Listeria	*Rathayibacter*
Paenibacillus (some)	*Sanguibacter*
Propionibacterium	*Staphylococcus aureus*
Staphylococcus	
Terribacter	
Terracoccus	

GP 7: Gr + Ox − Cat − n	GP 8: Gr + Ox − Cat − p
Amphibacillus	*Clostridium* (some)
Bifidobacterium	*Lactobacillus* (some)
Clostridium	
Erysipelothrix	
Facklamia	
Helcococcus	
Lactic acid bacteria[a]	
Sporolactobacillus	
S. aureus subsp. *anaerobius*	

Gram-Negative Groups

GN 1: Gr − Ox + Cat + n	GN 2: Gr − Ox + Cat + p
Achromobacter	*Acidomonas*
Acidovorax	*Acidovorax*
Aeromonas	*Alteromonas*
Agrobacterium	*Aminobacter*

Alcaligenes
Alteromonas
Amaricoccus
Aminobacter
Arcobacter
Azomonas
Azotobacter
Bergeyella
Brevundimonas
Burkholderia
Campylobacter
Carnimonas
Comamonas
Delftia
Devosia
Enhydrobacter
Halomonas
Meniscus
Moraxella
Ochrobacter
Oligella
Pandoraea
Paracoccus
Pedobacter
Photobacterium
Plesiomona
Pseudoalteromonas
Pseudomonas
Psychrobacter
Ralstonia
Rhizomonas
Shewanella
Sphingomonas
Stenotrophomonas
Telluria
Vibrio
Xanthobacter

Azomonas
Azotobacter
Brevundimonas
Campylobacter (at least 2 spp.)
Chryseobacterium
Chromobacterim
Chryseomonas
Burkholderia cepacia
Duganella
Empedobacter
Flavobacterium
Hydrogenophaga
Hymenobacter actinosclerus
Janthinobacterium
Kingella
Methylobacterium
Myroides
Pandoraea (some)
Paracoccus
Pedobacter
Persicobacter
Pseudoalteromonas
Pseudoaminobacter
Rhizomonas
Sphingobacterium
Sphingomonas
Stenotrophomonas
Telluria chitinolytica
Variovorax
Vogesella
Xanthobacter

GN 3: *Gr − Ox + Cat − n*	GN 4: *Gr − Ox + Cat − p*
Campylobacter concisus	*Cytophaga*
Cardiobacterium	*Hydrogenophaga*
Eikenella	*Persicobacter*
Kingella	*Wolinella*
Suttonella	

GN 5: *Gr − Ox − Cat + n*	GN 6: *Gr − Ox − Cat + p*
Acetobacter	*Acinetobacter radioresistens*
Acidomonas	*Asaia*
Acinetobacter	*Azoarcus*
Asaia	*Chemohalobacter*

Burkholderia cepacia,   Citrobacter
B. cocovenenans   Deinobacter grandis
Campylobacter (some)   Erwinia
Enterobacteriaceae[b]   Flavimonas
Gluconobacter   Fraturia
Moraxella bovis, ovis   Pandoraea
Pandoraea   Pantoea
Pseudomonas (a few)   Pectobacterium
Raoultella   Pedobacter
Saccharobacter   Serratia
Stenotrophomonas (some)   Xanthomonas
Xylella   Xylophilus
Xanthomonas
Zymobacter
Zymomonas

GN 7: Gr − Ox − Cat − n	GN 8: Gr − Ox − Cat − p
Acidaminococcus	Prevotella nigrescens
Bacteroides	
Megasphera	
Pectinatus	
Streptobacillus	
Veillonella	

[a]All of the lactic acid genera listed in Chapter 7.

[b]Includes Enterobacter, Escherichia, Salmonella, Shigella, and the other enteric bacteria.

Index

CL

664.
001
579
JAY

5000257070